Groundwater and Wells

Second Edition

Fletcher G. Driscoll, Ph.D.
Principal Author and Editor

Published by Johnson Division, St. Paul, Minnesota 55112

Copyright © 1986 by Johnson Division, St. Paul, Minnesota 55112

Library of Congress Catalog Card Number 85-63577

ISBN 0-9616456-0-1

Second Printing 1987

Printed in the United States of America. All rights reserved. This book, or parts thereof, may not be reproduced without written permission from the publisher.

H. M. Smyth Company, Inc. printed this volume on Mead Corporation's sixty pound Moistrite stock. Typography was done by CTS Inc.; the text was set in 9 point Times Roman type face.

The information and recommendations contained in this book have been compiled from sources believed to be reliable and to represent the best opinion on the subject as of 1986. However, no warranty, guarantee, or representation, express or implied, is made by Johnson Division as to the correctness or sufficiency of this information or to the results to be obtained from the use thereof. It cannot be assumed that all necessary warnings, safety suggestions, and precautionary measures are contained in this book, or that any additional information or measures may not be required or desirable because of particular conditions or circumstances, or because of any applicable U.S.A. federal, state, or local law, or any applicable foreign law or any insurance requirements or codes. The warnings, safety suggestions, and precautionary measures contained herein do not supplement or modify any U.S.A. federal, state, or local law, or any applicable foreign law, or any insurance requirements or codes.

To Gerald F. Briggs

for his outstanding contributions to the success
of Johnson Division and to the worldwide
development of the groundwater industry

Preface

Groundwater is a priceless resource lying beneath most of the Earth's land surface. In the United States, groundwater supplies about 25 percent of the nation's domestic, agricultural, and industrial water. Fifty percent of the United States' population depends on groundwater for its drinking water supply. Reliance on groundwater has increased greatly over the past 35 years because of population shifts to areas where surface water is often not plentiful. Similarly, substantial increases in groundwater withdrawals have occurred recently in almost every country of the world.

The vulnerability of groundwater to overuse and water-quality degradation was not widely understood until recently. In the future, as much attention will be paid to water-quality preservation and resource conservation as to resource development. The principal objective of this book is to provide the necessary technical knowledge to locate, extract, treat, and protect groundwater so that current and future generations can depend on this resource to enhance their quality of life.

Johnson Division is the world's largest manufacturer of water well screens. Since 1904, its technical staff in the United States and abroad has designed tens of thousands of high-capacity wells for every conceivable application. From this collective experience has come the development of effective well design procedures for all types of aquifers. Through the years, countless practical suggestions have been received from water well contractors, well design engineers, consultants, and scientists. This book summarizes this knowledge so that groundwater resources can be developed economically in the widest range of hydrogeologic settings without damage to aquifers through either overpumping or contamination.

Significant advances have been made in almost all phases of groundwater technology in recent years. Enhanced understanding of the groundwater environment, new geophysical exploration techniques, advances in well drilling methods, improved drilling fluid additives and screen designs, the development of groundwater quality monitoring technology, and remarkably effective new water treatment methods are just some examples. In addition, the majority of groundwater scientists and technicians have only recently received their formal training. Thus the need for sound, practical information is at a historic high. In response to this need, this edition of *Groundwater and Wells* provides both new and experienced members of the groundwater industry with a single, comprehensive, up-to-date reference on all essential aspects of groundwater technology.

The approach of this book is pragmatic rather than theoretical. The material is presented so the interested reader, regardless of background, will gain a good understanding of the physics, chemistry, and hydraulics of groundwater and the various technologies used to develop this resource. In each subject area, a major effort is directed toward bringing out the practical elements; theoretical aspects are discussed wherever they serve to directly strengthen practical knowledge.

The book is written for four distinct reader groups — design engineers, water well contractors, government officials involved in well projects, and educators and students interested in the practical elements of hydrogeology. To be of maximum value to these four diverse reader groups, highly technical words from geology, hydrogeology, chemistry, physics, engineering, and hydraulics have been used sparingly. A certain number of technical words are necessary, but these words are generally defined where they are first used in the text. A large number of technical words are also defined in the glossary. To facilitate use of the book in countries using System International units, all basic equations are presented first in English units of measure and are followed immediately by the S.I. equivalent. Numerical values are given in both English and S.I. units. Problems, however, are worked out in English Engineering units only. References are listed at the end of the text. This reference list is not comprehensive but serves to indicate representative articles or textbooks that examine a particular subject in more detail.

The first edition of *Groundwater and Wells*, which has become a classic in its field with more than 60,000 copies sold worldwide, provided reliable technical information on many phases of water well technology. The Johnson Division staff is confident that this greatly expanded second edition will become an even more valuable resource for those involved in groundwater technology. If use of this book contributes to the public good, to the economic growth of any country or region, and to the advancement of the water-well industry, our efforts will have been rewarded.

JOHNSON DIVISION

Sam R. Goodman
General Manager and Vice President

Acknowledgments

Much of the value of this book is a direct reflection of the Johnson Division technical staff's competence, creativity, dedication, and willingness to share their knowledge. For this edition of *Groundwater and Wells,* the staff helped define the most practical and economical methods for designing, constructing, testing, and rehabilitating wells. Many of these practical methods were suggested initially by contractors eager to improve the design and construction standards for water wells. As a result, readers of this book are the beneficiaries of the wisdom derived from constructing countless wells in every type of geologic formation by all major drilling methods.

The value of this book has been enhanced further by the many groundwater and water well experts who have given generously of their time in reviewing individual chapters and providing advice on various topics. These include Henry A. Baski, Jon R. Carpenter, Walter R. Conley, Jeffrey J. Daniels, Arthur R. Hart, Richard L. Henkle, M. W. "Bo" Judd, W. Scott Keys, Gary V. Lanser, Ing. Jose Arreguin Mañon, Carl E. Mason, Michael P. McGrath, Norman E. Melhorn, Duane D. Nowlin, Hans Olaf Pfannkuch, Ing. Miguel Concha Pinochet, Thomas T. Renner, Eugenio Celedon S., Thomas J. Schutte, Donald I. Siegel, Frank B. Snelgrove, Thomas L. Stevens, William C. Walton, and Jerome W. Wojtkiewicz. John E. Voytek and the staff of the National Water Well Association were particularly helpful in our research efforts. Undoubtedly, other names should be added to this list because so many groundwater professionals have made significant, although less formal, contributions to this volume. Johnson Division is grateful to these named and unnamed individuals.

Several members of the Johnson Division staff have been particularly instrumental in providing editorial and production assistance. Tom Davis has added immeasurably to the overall accuracy of the text by exhaustive technical reviews of all chapters. With unending patience, Tim Stanley carefully drafted and keylined the illustrations. Bill Unumb and A. Phillips Beedon created the layout of the book and provided expert editorial assistance. Jan Smith provided assistance in photographing well screens and drilling fluid devices. As outside editor, P. T. Moyer's critical review of the text led to a great improvement in readability and overall style. Lyn Page managed the entire project with great skill and dedication.

Table of Contents

Chapter 1. Water: Mankind's Most Vital and Versatile Resource . . . 1

Chapter 2. Origin of Water 7
    Formation of the Ocean Floor and Continental Land Masses 10
    Origin of the Earth's Hydrosphere 14
    Evolution of the Earth's Atmosphere 16

Chapter 3. Formation of Aquifer Systems 19
    Weathering 19
    Erosion 20
    Alluvial Aquifers 21
    Sedimentary Rock Types 26
    Glacial Deposits 31
    Igneous and Metamorphic Rock Aquifers 42

Chapter 4. Weather Patterns and the Hydrologic Cycle 46
    Coriolis Effect 49
    How Precipitation Occurs 49
    Causes of Precipitation 51
    Effect of Precipitation on Groundwater 52
    Hydrologic Cycle 53
    Pathways of Water After it Falls to Earth 56

Chapter 5. Occurrence and Movement of Groundwater 59
    Types of Subsurface Water 59
    Energy Contained in Groundwater 62
    Aquifer Functions 66
    Flow Nets 79
    Groundwater Flow Velocities 81

Chapter 6. Groundwater Chemistry **86**
    Origin of the Chemical Constituents of Groundwater 88
    Units of Measure 89
    Important Properties of Water 90
    Groundwater Constituents 97
    Water Quality 107
    Methods to Present Water Quality Data 114
    Importance of Water Chemistry 116

Chapter 7. A Summary of Groundwater Resources
               of North America **118**
    Groundwater in the United States 119
    Canada 140
    Mexico 147

Chapter 8. Groundwater Exploration **150**
    Hydrogeologic Reports 151
    Maps 153
    Aerial Photographs 157
    Formation Sampling 160
    Geophysical Exploration Methods 168
        Surface Geophysical Methods 170
        Borehole Geophysical Methods 180
    Analysis of Aquifers Using Pumping Test Data 202

Chapter 9. Well Hydraulics **205**
    Definition of Terms 206
    Nature of Converging Flow 207
    Cone of Depression 211
    Equilibrium Well Equations 212
    Nonequilibrium Well Equation 218
    Modified Nonequilibrium Equation 219
    Hydrogeologic Conditions that Affect Time-Drawdown Graphs . . . 223
    Calculating Drawdown for Intermittent Pumping Situations 235
    Distance-Drawdown Graphs 236
    Well Interference 242
    Well Efficiency 244
    Radius of Influence 245
    Recharge and Boundary Conditions 246
    Combined Use of Semilog Graphs 247
    Effect of Partial Penetration 249
    Water-Level Recovery Data 252
    Theis Nonequilibrium Well Equation 260
    Other Methods of Aquifer Analysis 264

Chapter 10. Well Drilling Methods **268**
    Cable Tool Method 268

California Stovepipe Method 277
Direct Rotary Drilling 278
Drilling Fluids 286
Reverse Circulation Rotary Drilling 289
Air Drilling Systems 295
In-Verse Drilling 299
Dual-Wall Reverse Circulation Rotary Method 301
Drill-Through Casing Driver 304
Jet Drilling 307
Hydraulic-Percussion Method 309
Boring with Earth Augers 310
Driven Wells 313
Drilling Procedures When Boulders are Encountered 314
Fishing Tools 316
Grouting and Sealing Well Casing 317
Plumbness and Alignment 333

Chapter 11. Drilling Fluids 340

Types of Drilling Fluids 340
Functions of a Drilling Fluid 341
Properties of Water-Based Drilling Fluids 343
Treatment of Mix Water for Drilling Fluids 362
Mixing Additives into Water-Based Systems 364
Air Drilling 366
Drilling Fluid Additives 385
Guidelines for Solving Specific Drilling Fluid Problems 386
Typical Drilling Problem 392

Chapter 12. Well Screens and Methods of Sediment-Size
Analysis 395

Continuous-Slot Screen 396
Other Types of Well Screens 401
Sediment-Size Analysis 405

Chapter 13. Water Well Design 413

Casing Diameter 414
Casing Materials 418
Well Depth 431
Well Screen Length 432
Well Screen Slot Openings 434
Pressure-Relief Screens 447
Formation Stabilizer 447
Well Screen Diameter 449
Open Area 450
Entrance Velocity 450
Screen Transmitting Capacity 453
Selection of Material 454

Design of Domestic Wells 458
Design for Sanitary Protection 460
Special Well Designs 461

Chapter 14.   Installation and Removal of Well Screens **464**
Pull-Back Method 464
Open-Hole Methods for Screen Installation 472
Filter Packed Wells 476
Installation of Plastic Screens 483
Bail-Down Procedure 485
Wash-Down Method 487
Jetting Method 489
Installing Well Points 491
Removing Well Screens 492

Chapter 15.   Development of Water Wells **497**
Well Development 497
Factors that Affect Development 499
Well Development Methods 502
A Comparison of Three Development Methods 521
Use of Polyphosphates in Development 522
Development of Rock Wells 523
Allowable Sediment Concentration in Well Water 526
Aquifer Development Techniques 528

Chapter 16.   Collection and Analysis of Pumping Test Data **534**
Conducting a Pumping Test 535
Measuring Drawdown in Wells 547
Well Efficiency 554
Step-Drawdown Tests 555
Problems of Pumping Test Analysis 559

Chapter 17.   Water Well Pumps **580**
Variable Displacement Pumps 581
Positive Displacement Pumps 604
Pumps Used to Circulate Drilling Fluid 606
Air-Lift Pumping 608
Pump Selection 608
Water Storage 610

Chapter 18.   Water-Quality Protection for Wells and Nearby
   Groundwater Resources **612**
Choosing a Well Site 614
Predicting the Pollution Potential at a Drilling Site 616
Well Design 617

Disinfection Procedures Required to Maintain a Sanitary Well
    during Drilling 618
Disinfecting Wells and Piping 619
Sealing the Wellhead 624
Horizontal Suction Lines 625
Pitless Adaptors 627
Sealing Abandoned Wells 627

Chapter 19. Well and Pump Maintenance and Rehabilitation . . . 630
Major Causes of Deteriorating Well Performance 631
Well Failure Caused by Incrustation 633
Well Failure Caused by Iron Bacteria 646
Well Failure Caused by Physical Plugging of Screen and Surrounding
    Formation 655
Importance of Screen Design on Rehabilitation 657
Well Failure from Corrosion 658
Pump Maintenance 665

Chapter 20. Groundwater Law, Water Well Specifications, and
    Well Contract Problems 670
Groundwater Law 671
Water Well Specifications 675
Contract Problems 696

Chapter 21. Groundwater Monitoring Technology 702
Major Federal Legislation Pertaining to Groundwater Quality and
    Monitoring Procedures 703
Groundwater Contamination Sources 705
Effect of Aquifer Characteristics on the Spread of Groundwater
    Contamination 707
Delineating Contaminant Plumes 712
Monitoring Contaminant Movement (Transport) 714
Locating Monitoring Wells 715
Personnel Safety at Monitoring Sites 717
Design of Monitoring Wells 719
Sampling Monitoring Wells 726
The Task of Groundwater Protection 728
Aquifer Restoration 730

Chapter 22. Alternative Uses for Wells and Well Screens 734
Dewatering 734
Well-Point Systems Used for Water Supply 760
Infiltration Galleries 761
Collector Wells 768
Injection Wells 769
Pressure-Relief Wells 777

      Wells for Heat Pumps 783
      Surface-Water Withdrawal 789

Chapter 23. Water Treatment **796**
      Components of Water Treatment and Waste Treatment Systems . . . 798
      Treatment Technologies Appropriate for Meeting Drinking Water
        Regulations 799
      Point-of-Use Water Treatment Systems 824

Chapter 24. Wise Use of Groundwater **837**
      Estimating Groundwater Use, Recharge, and Volume in Storage . . . 838
      Impact of Droughts on Groundwater Supply and Use 843
      Managing Groundwater Supplies 846

References **861**

Glossary **885**

Appendices **893**

Index **1073**

Groundwater and Wells
Second Edition

CHAPTER 1
Water: Mankind's Most Vital and Versatile Resource

For years nearly everyone took water for granted. Few persons worried that water resources had finite limits, that they could be lost to contamination or outright removal, or that the pressures of a burgeoning population would create physical and chemical stresses on these resources never dreamed of only a generation ago. Since the 1960's, however, a keen awareness of the fragility of these resources has developed throughout the world. This is not to say that people living in water-short regions were not already aware that water supply was a major factor in their lives. For example, there has been a 400-ft (122-m) decline in the groundwater table in Phoenix, Arizona over the last 50 years. Elsewhere, the preservation of water quality can present major technological challenges. Repeated use of Colorado River water for irrigation has caused a marked increase in the salinity of the water as it travels from its headwaters in the Rocky Mountains to the Gulf of California. The increase is so large that the United States, in order to comply with treaty agreements with Mexico, must now reduce the salinity of the water before it enters Mexico.

Once a resource is endangered by some impact of man, society can expend large sums of money and time to determine what the problems are and to develop potential solutions. Therefore, in recent years many scientists have dedicated their professional lives to studying water. At the same time, other scientists have made discoveries that in some cases transcend all the discoveries made in the history of that science. Geologists, for example, have learned more about the major features of the Earth's surface since the early 1960's than in the previous 200-year history of the science. Laymen are now taking advantage of these studies to utilize natural resources more fully and wisely.

Water, one of the fundamental resources, is at once one of the most common substances and also one of the most unusual. Although its chemical formula is deceptively simple, the effect of water on almost everything in our environment is far more consequential than might be imagined.

Water is often called "the universal solvent" because of its extraordinary ability to dissolve a broad range of substances. In fact, it dissolves more substances in greater

Figure 1.1. Niagara Falls is one of nature's most powerful displays of the energy contained in water. The Canadian Horseshoe Falls is shown in the foreground and the smaller American Falls is in the center. *(Power Authority of the State of New York.)*

quantities than any other liquid. The salinity of the world's oceans is a direct result of water's ability to dissolve rock materials as water flows overland to the sea. Eventually, an element such as calcium becomes so abundant in sea water that it precipitates out, forming crystals of the mineral calcite. So much calcite is precipitated that thick layers of a sedimentary rock, called limestone, form on the ocean floor. This rock may later be uplifted by large-scale geologic forces to form part of the land areas of the world. If conditions are just right, the limestone may in time provide huge reservoirs for underground water resources.

Water has the highest heat of vaporization of any liquid. In other words, huge amounts of heat energy are required to evaporate even small quantities of water. The subsequent release of this energy through condensation during rainstorms provides an important energy source for driving weather systems. The energy responsible for generating tornados and hurricanes comes from the heat energy acquired by countless individual water molecules when they evaporate from a water surface (Figure 1.2).

Because of water's high heat capacity, the presence of oceans, lakes, and large rivers prevents extreme fluctuations in local temperatures. Coastal communities have much more uniform temperature regimes than do towns farther inland. Even within the human body, water is critical in maintaining uniform body temperatures. Without the large volume of water in our bodies (approximately 75 percent), we would warm up or cool down much more rapidly than we do.

With the steep rise in fuel costs during the 1970's, solar heating of homes and other buildings has become much more popular. Heat taken up from the sun's radiation is stored most effectively in water (Figure 1.3). Two physical characteristics of water are utilized in this process: its high heat capacity and its ability to conduct heat. Water can take up or release more heat than any other liquid, except for liquid ammonia, and thus water is ideally suited for heating buildings. In recent years, homeowners in all parts of the United States have shown interest in using water stored in the ground for space heating. The temperature of shallow groundwater reflects the local mean annual temperature, which may be 20° to 35°F (11° to 19°C) above typical midwinter temperatures, with the result that groundwater provides a large and relatively inexpensive heat source.

Water has other unusual physical and chemical characteristics that play a large

Figure 1.2. The energy driving a tornado comes mainly from the quick release of solar energy contained in water vapor. *(U.S. Department of Commerce, National Oceanic and Atmospheric Administration)*

but often unrecognized role in our daily lives. For example, part of any volume of water has a natural tendency to break down spontaneously into hydrogen (H^+) and hydroxyl (OH^-) ions*. This process is called dissociation. When an abundance of these charged ions is available, an electric current can be transmitted through the liquid. The dissociation process is enhanced when an acid is added to water. For instance, automobile batteries contain an electrolyte (acid added to water) which contains abundant H^+ and OH^- ions. The electrolyte permits batteries to store and give up electricity rapidly. The ability of water to conduct electrical charges can lead to severe corrosion of some metal objects placed in wet ground.

The Earth's atmosphere contains from 0.02 to 4 percent water by volume, depending on location. In addition to providing sources for precipitation, atmospheric water vapor affords two powerful safeguards for life on Earth. First, it intercepts some of the ultraviolet (short-wave) radiation from the sun. It is this type of solar radiation that produces skin cancer in humans. Second, much of the heat that the Earth receives from the sun is radiated back into space. Atmospheric water vapor, however, intercepts some of this potential heat loss and redirects part of it back to Earth, while part is retained in the atmosphere. These phenomena produce a warm atmospheric envelope around the Earth that prevents large daily temperature fluctuations similar to those found on the moon. The usual nighttime temperature on the moon is −280°F (−173°C), whereas daytime temperatures soar to 260°F (127°C) or more.

Another unusual feature of water related to its molecular structure is the great

---

*An ion is a charged atom or group of atoms.

Figure 1.3. The two solar panels on this home provide a substantial portion of total heating requirements, even though the mean annual temperature is 40°F (4.4°C). *(Sunwood Energy Products, Inc.)*

capacity of water molecules* to cling to one another. This characteristic, called hydrogen bonding, gives water the highest surface tension of any liquid. Therefore, it is relatively difficult to pass anything through a water surface. Because of this physical property, water vapor tends to form droplets when condensing in the atmosphere, rather than falling as fragmented masses of particles.

In hard rainstorms, droplets can loosen more than 100 tons of soil per acre (224 metric tons of soil per hectare) (Figure 1.4). With sufficient rainfall, some of these loosened particles are transported by overland flow to nearby streams and rivers. In time, the sediment is carried to the sea. Eventually, the coarsest and most stable of these particles are reworked by ocean waves into beach deposits. Over long periods, beach sands can become solidified into layers of sandstone, which are found over much of the Earth's surface. On a global scale, sandstone formations offer the best reservoir sites for groundwater storage. Thus, falling raindrops trigger a chain reaction that results in building massive reservoirs for groundwater storage.

Unlike most other liquids, water reaches its maximum density† at 39.2°F (4°C), which is well above its freezing point of 32°F (0°C). As a result, lakes in colder climates do not freeze from the bottom up, but instead are covered each winter with a relatively thin layer of ice. Biological activity would virtually cease in lakes if they were filled with ice each winter.

In the colder climates of the world, air temperatures pass back and forth through the freezing point many times each year. Water in shallow cracks and crevices of rocks found in these climates freezes and thaws as many as 70 times annually. The force exerted by the freezing water, as much as 30,000 psi (pounds per square inch) [207,000 kPa (kilo pascals)] is sufficient to crack even the most durable rock. This process, called frost wedging, can cause extensive destruction of large rock masses.

---

*A molecule is a unit of matter that is the smallest particle of an element or chemical combination of atoms (as a compound) capable of retaining chemical identity with the substance in mass.

†Density is a physical characteristic of a substance and is defined as mass per unit volume [in pounds per cubic ft (grams per cubic centimeter)]. Water has a density of 1 g/cm^3.

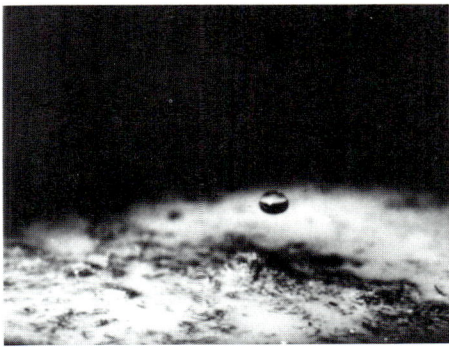 

Figure 1.4. A large raindrop lands on a wet soil surface. The impact loosens soil particles which can then be washed away to nearby streams. *(Official U. S. Navy photograph)*

In mountainous areas, the resulting angular fragments (talus) vary in size from house-size blocks to fine particles and can be seen at the base of mountain slopes (Figure 1.5). Eventually, these rock materials are reworked by the agents of erosion — running water, glacier ice, and wind — into various kinds of unconsolidated sediment. Some of these sediments serve as storage sites for groundwater.

Thus, water plays a major role in virtually every aspect of human life, as the examples discussed herein demonstrate. Regrettably, too few persons understand the physical and chemical properties of water well enough to effectively solve urgent and nearly universal problems relating to its cost, availability, distribution, and contamination. Water problems of any type stem largely from lack of knowledge and, therefore, from mismanagement of the natural system. These problems are intensified by the technological impact of man on that system.

The purpose of this book is to describe the groundwater system and the tools and

Figure 1.5. Large talus cones at the base of a mountain in the Canadian Rockies. *(Ray Atkeson)*

materials that can serve to utilize that system wisely. We will examine in detail the origin, physics, chemistry, storage sites, and flow characteristics of groundwater; methods of groundwater exploration and extraction; well drilling, development, and maintenance techniques; and the economic and legal considerations in groundwater use. The importance of groundwater cannot be overemphasized because it represents 97 percent of all the available freshwater supplies found on Earth.

CHAPTER 2
Origin of Water

Water is such a fundamental part of the Earth that anyone studying water must first understand how the Earth evolved and the changes that have taken place in and near the Earth's crust* over the last several billion years. In the past, geologists thought the Earth was more or less static; that is, its topographic features and internal structure remain relatively constant. Recent discoveries now demonstrate that the Earth is a dynamic planet. Subtle changes are occurring constantly in the arrangement of continents, the building and destruction of mountain chains, the creation and movement of the sea floor, and even the climatic conditions affecting the planet. For hundreds of years, the magnitude of these changes, and even their existence, escaped man's attention because they have occurred over geologic time, that is, over hundreds of thousands or tens of millions of years, rather than over a human lifetime.

Geologists now know much about the origin of continents and ocean basins, why mountains exist, why earthquakes occur only in certain regions of the Earth, why oceanic sediments overlie much of the continental areas, how the Earth's atmosphere evolved, and most important to us, why large volumes of water exist here on Earth and nowhere else in our solar system.

The explosion of geologic knowledge in the 1960's and early 1970's resulted from technological advances that began during World War II. These advances include the development of sophisticated computer technology, new techniques to determine the Earth's previous magnetic field as preserved in rocks, the ability to detect heat flow from deep within the Earth, the ability to determine the ages of rocks, and remarkable new ships capable of drilling into the deep-ocean floor. For example, computers can assimilate the huge quantities of geologic data being collected, analyze them, and provide answers almost instantaneously, thus saving years of tedius hand calculations. Aided by more powerful geophysical instruments, investigations of the magnetic characteristics of old rocks show that continental land masses have drifted considerable distances over the Earth's surface. In other studies, the detection of high heat flow zones led to the discovery that the crust was thinner where more heat escaped from the upper mantle. The mantle extends from beneath the Earth's crust to a depth of

---

*The rigid, outermost layer of the Earth generally ranging in thickness from 5 to 25 mi (8 to 40 km).

about 1,800 mi (2,900 km).

By the late 1960's and early 1970's, geochemists could determine the ages of rocks by comparing the ratios of certain radioactive elements in the rocks. These radioactive substances decay or change spontaneously to other substances at a constant rate. By measuring the relative quantities of the original and new materials, geochemists can approximate the ages of many rocks. For example, pure uranium changes to half uranium and half lead in 713 million years. Over another 713 million years, the same initial quantity of uranium will become three-quarters lead and only one-quarter uranium. Some of the rocks recently dated originated about 4 billion years ago, shortly after the formation of the Earth.

Until a few years ago, scientists knew little about the composition of the ocean floor because of its inaccessibility. Recent investigators, using new ships capable of remaining stationary and drilling in water depths of 23,000 ft (7,010 m) or more, have discovered that the ocean floor consists almost entirely of two rock types, basalt and gabbro. Continental land masses, however, contain numerous rock types. This discovery suggests a completely different mechanism for the origin of ocean basins as contrasted with continents. Therefore, it is best to examine the Earth from its very beginning to understand the origins of groundwater systems.

How did the Earth form, what was it like in the beginning, and what physical changes have occurred over the last several billion years? Geophysicists believe that the Earth formed from the remnants of an exploded star, a supernova. In time, individual pieces of rock fell together by gravitational attraction. Other bodies in our solar system formed in a similar manner. It is not known how long the Earth took to form, but it is known from the ages of rocks that the Earth is about 4.6 billion years old. This date of formation is substantiated by the ages of meteorites and the oldest rocks recovered from the moon's surface.

At its inception, the Earth most likely did not have an atmosphere, hydrosphere*, or the familiar crustal components seen today — that is, no distinct continents, ocean basins, or mountain chains existed. Probably the entire surface was a heterogeneous mass of rock debris. Thus, the structure of the Earth's atmosphere and surface has changed profoundly through time.

Ideas concerning the origin of the atmosphere and a plausible theory for the evolution of water on the Earth's surface developed within the framework of a theory called plate tectonics. The theory was first proposed in 1912 by Alfred Wegener, a German meteorologist. Initially, the theory was not taken seriously by many scientists because little data existed to justify Wegener's claims. Geologic information obtained with new scientific tools during the 1960's, however, convinced most scientists that Wegener had been correct. The evolution of the atmosphere and hydrosphere is best explained as a fundamental part of plate tectonics theory.

Plate theory proposes that the Earth's crust is broken up into a global mosaic of individual plates which vary greatly in size. In other words, the crust is not a continuous sheet consisting of various rock types, but instead comprises numerous irregularly shaped segments, called plates, which have a well-defined rock structure (Figure 2.1). Some of these plates underlie only the oceanic basins, whereas others make up large continental land masses as well as nearby ocean floors.

---

*The hydrosphere comprises all the liquid and solid water resources of the Earth.

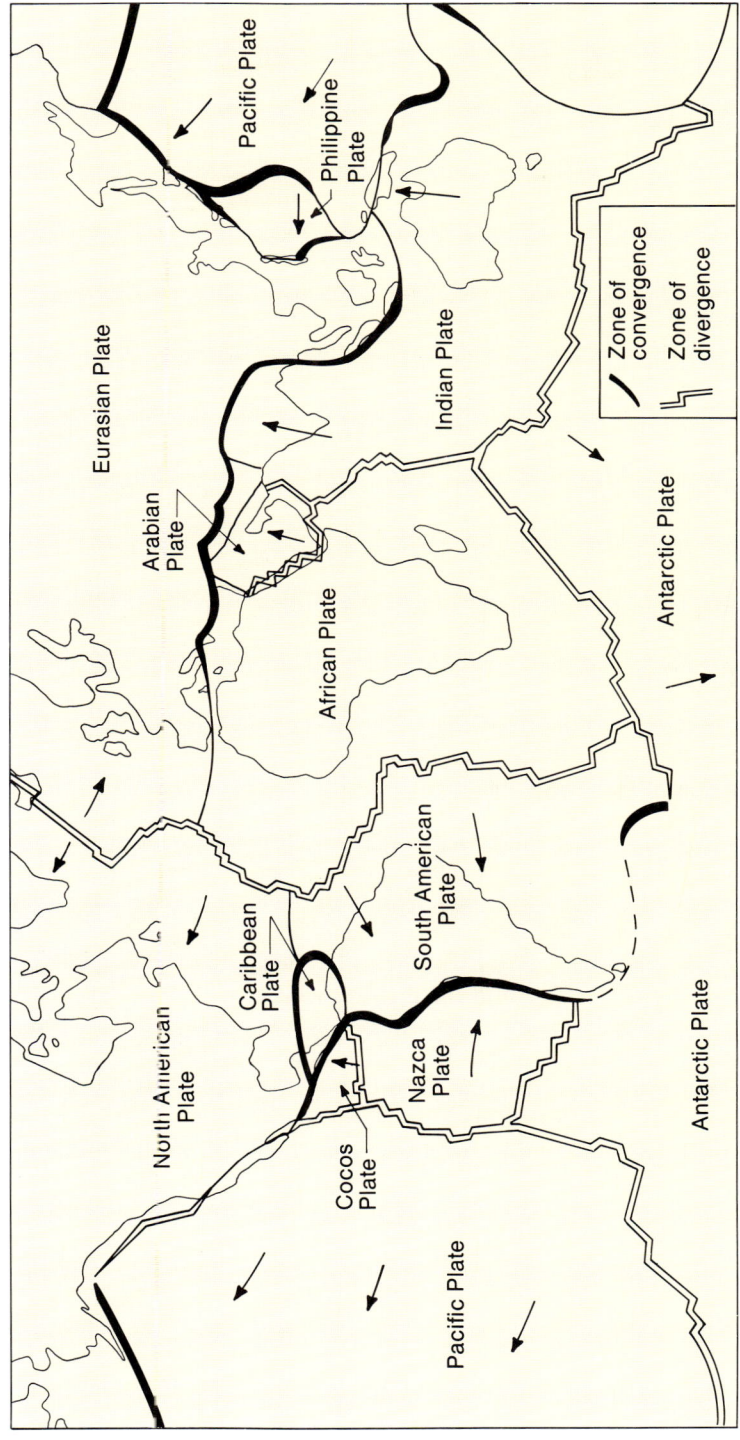

Figure 2.1. Major plates of the world. Mid-oceanic ridges from which the plates move apart are represented by double lines. Trenches and other subduction zones are marked by solid lines. Arrows indicate the assumed movement of the plates. *(From PUTNAM'S GEOLOGY, Fourth Edition, by Edwin E. Larson and Peter W. Birkeland, Copyright © 1964, 1971, 1978, and 1982 by Oxford University Press, Inc. Reprinted by permission.)*

The theory of plate tectonics suggests that the plates are in constant motion with respect to one another. Tectonics is the term used to denote the movement and deformation of these plates. Many geophysicists believe that giant convection cells operating in the upper mantle are responsible for these crustal movements. Each of the plates moves independently of its neighbors. Thus, some plates are colliding (convergent), some are pulling apart (divergent), and still others are merely slipping along one another (transverse). Earthquakes represent the most obvious and dramatic evidence of movement along plate boundaries. The most severe earthquakes usually occur along zones where plates are colliding (convergent). Recent quakes in China, Nicaragua, Iran, and Italy are examples of earthquakes produced by colliding plates.

In the mid-Atlantic and mid-Pacific ocean basins, areas between the divergent plates are being filled with basalt and gabbro, thus forming new sea floor. These spreading zones, called rift zones, were first detected from the presence of mountain chains on the sea floor and abnormal amounts of heat flowing from the Earth's interior. These two major rift zones, represented by double lines in Figure 2.1, have been plotted by studying the locations (epicenters) of numerous shallow earthquakes.

An example of two plates sliding along each other is found along the San Andreas Fault System in California. The westernmost part of California, between the San Francisco Peninsula and Baja California, is part of the Pacific Plate and is moving northwestward along the coast of North America at a rate of 2.5 in (64 mm) per year. The remainder of the United States, Canada, and Mexico are on the North American Plate. Because of the San Andreas Fault, people living in California can expect earthquakes from time to time.

FORMATION OF THE OCEAN FLOOR AND CONTINENTAL LAND MASSES

In a specific part of the upper mantle, the temperature and pressure are just right to allow partial melting of the mantle rock. This zone lies at a depth of 50 to 110 mi (80 to 177 km) and is called the low-velocity zone because some seismic waves move more slowly through this region. Radioactive elements in this zone continuously lose mass and, in the process, liberate heat. This heat is sufficient at the prevailing pressure to melt approximately one percent of the total rock mass in the low-velocity zone.

The elements that melt from the rocks of the upper mantle are usually those elements melting at the lowest temperature. Once melted, the rock material is lighter than the surrounding rock and therefore rises through crevices to rift zones where it spreads laterally on the Earth's surface. These rift zones extend at least 40,000 mi (64,400 km) over the Earth's surface, mainly in the major ocean basins. Generally, the zones are marked by suboceanic mountain ranges as high as 2.5 mi (4 km). New basaltic crust is being made in the rift zones at the rate of 2 to 4 in (51 to 102 mm) per year.

Oceanic crust is initially about 6 mi (9.7 km) thick and moves outward at right angles from the rift zones. In time, the new crustal material cools and thereby becomes denser, sinking deeper into the upper mantle; at the same time it thickens by solidification of mantle material from below (Figure 2.2). When the oceanic crust is new, no sediments exist on it. In time, clay particles, skeletal remains of tiny marine organisms, and precipitated elements such as calcium and manganese begin to cover

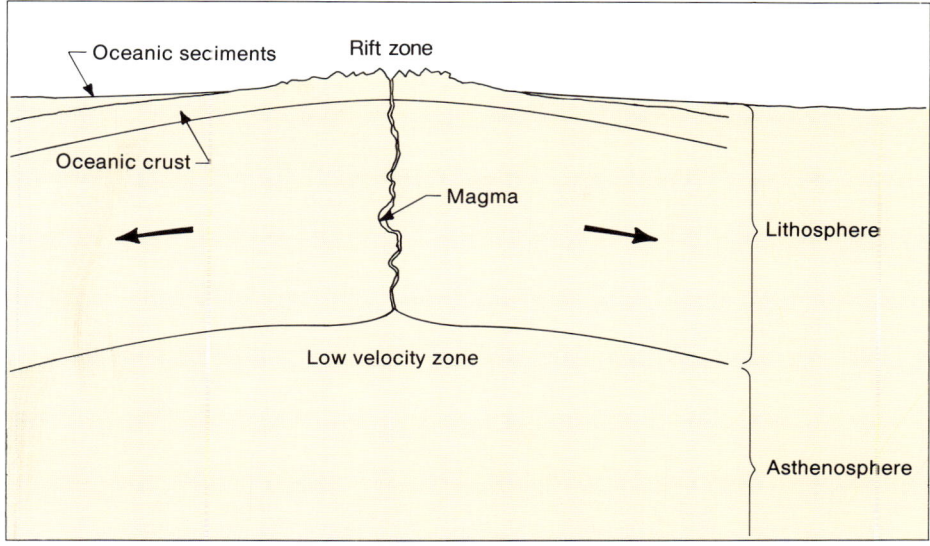

Figure 2.2. Cross section of the upper mantle and crust showing the relatively rigid lithospheric plate riding on the underlying rocks of the asthenosphere. These two rock layers are separated by the more deformable rocks of the low-velocity zone that permit the lithosphere to move over the asthenosphere.

the bottom. As the plates move along, the sediments may reach thicknesses of a mile or more.

Eventually, the newly created crust encounters other segments of the sea floor. Because the Earth cannot expand in surface area, some moving plates are forced to reenter the upper mantle, where the down-going edges apparently melt completely by the time they reach a depth of 435 mi (700 km); the inclined zones of contact are called Benioff zones. In this process, which scientists call subduction, heat is generated principally from the friction of the massive plates sliding along one another. Additional heat is supplied from below by radioactivity in the upper mantle. Thus, sufficient heat accumulates to cause partial melting along the Benioff zones, and minerals which melt at low temperatures are selectively melted from the descending oceanic plate. In most cases, the oceanic sediments resting on the plate are light enough so that they do not enter the mantle along with the plate. Instead, they accumulate in or near the trench zones (Figure 2.3). Some of the sediments, however, are caught on the descending plates and are then melted in the Benioff zones.

Magma (molten rock) that is produced in the subduction process consists primarily of quartz, micas, potassium feldspar, sodium feldspar, and ferromagnesian minerals. Some of this magma rises eventually to the surface through weak zones in the overlying plate and forms volcanic mountain ranges. Rocks formed from these magmas are typically granite and diorite when intrusive and rhyolite and andesite when extrusive. Intrusive igneous rocks form from magma injected beneath the Earth's surface, whereas extrusive rocks form from magma that flows out on the Earth's surface. This new crust, called continental crust, not only comprises rock types different from those in the ocean floor, but it is also much less dense. Therefore, it rests (floats) higher on the mantle. In time, the weight of the newly formed continental rocks will begin to depress the underlying oceanic plates.

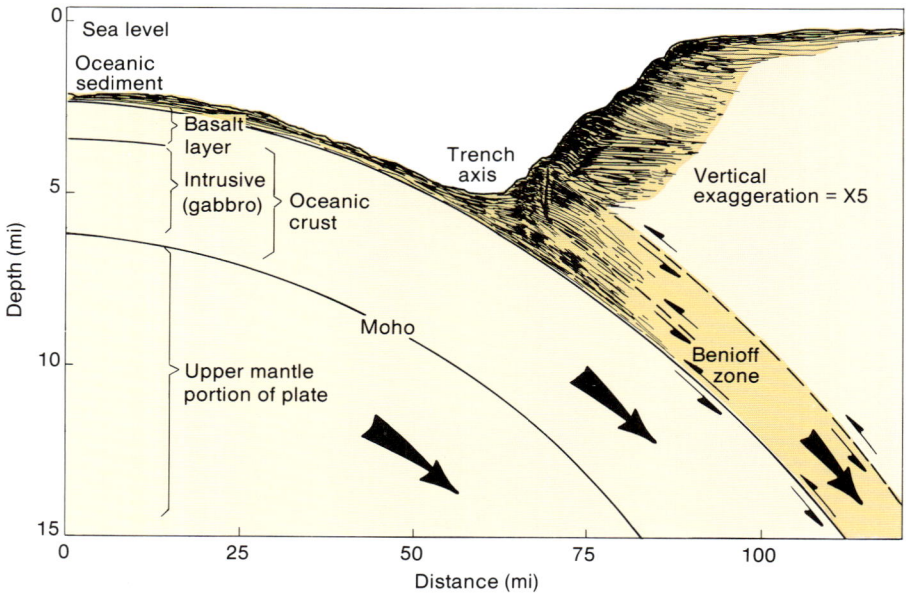

Figure 2.3. Cross section through down-going oceanic crust (lithosphere) at a convergent plate boundary marked by an ocean trench. Some of the oceanic sediments are folded and plastered to the inner trench wall, whereas others descend farther down the Benioff zone. *(Sawkins et al. Reprinted with permission of Macmillan Publishing Co., Inc. Copyright ©1974.)*

Examples of relatively new continental crust include most of the island arcs of the world. One of the youngest is the Aleutian chain, whereas the Japanese and Philippine Archipelagos are considerably older. Over time, repeated collisions with other plates continue to build the bulk of the continents. The crust in Minnesota, U.S.A., and southern Ontario, Canada, for example, comprises ancient and highly altered rocks that originated in an island arc at a plate margin. Similarly, the oldest rocks in other continents are found near the center of the land mass and in part have a volcanic origin.

In addition to the subduction process, rock mass is added to the continents in another major way. Oceanic sediments that have been deposited between continents are lighter than the underlying basalt and mantle rock (Figure 2.4). Thus, when two continents collide, these sediments have little tendency to subduct along with the oceanic plate. Caught between two land masses, the sediments are squeezed up into giant folds which form mountains consisting of former sea-floor sediments. The physical characteristics of most of these rocks make them ideal storage sites for groundwater.

A classic example of this squeezing process can be found along the eastern seaboard of North America and the northwestern coast of Africa. Four-hundred million years ago, the eastern flank of North America was attached to the northwestern part of Africa. The two continents subsequently split apart, and oceanic sediments were deposited over the new sea floor. For some reason the spreading stopped and the two continents then began to reapproach each other. The sediments caught between them were folded up into a great mountain chain. In North America, these folded sediments are called the Appalachian Mountains, and in North Africa they are called the Atlas Mountains. Eventually, the two continents separated again and have been drifting apart for the last 130 million years.

There are distinct differences in the densities of the rocks making up the continents, sea floor, and upper mantle. The average density is 2.7 g/cm^3 for continental rocks, 3.0 g/cm^3 for the basaltic sea floor, and approximately 3.3 g/cm^3 for the upper mantle. Because each of these rock units has a different density, they float in isostatic equilibrium relative to one another; Figure 2.5 shows this relationship. When material is

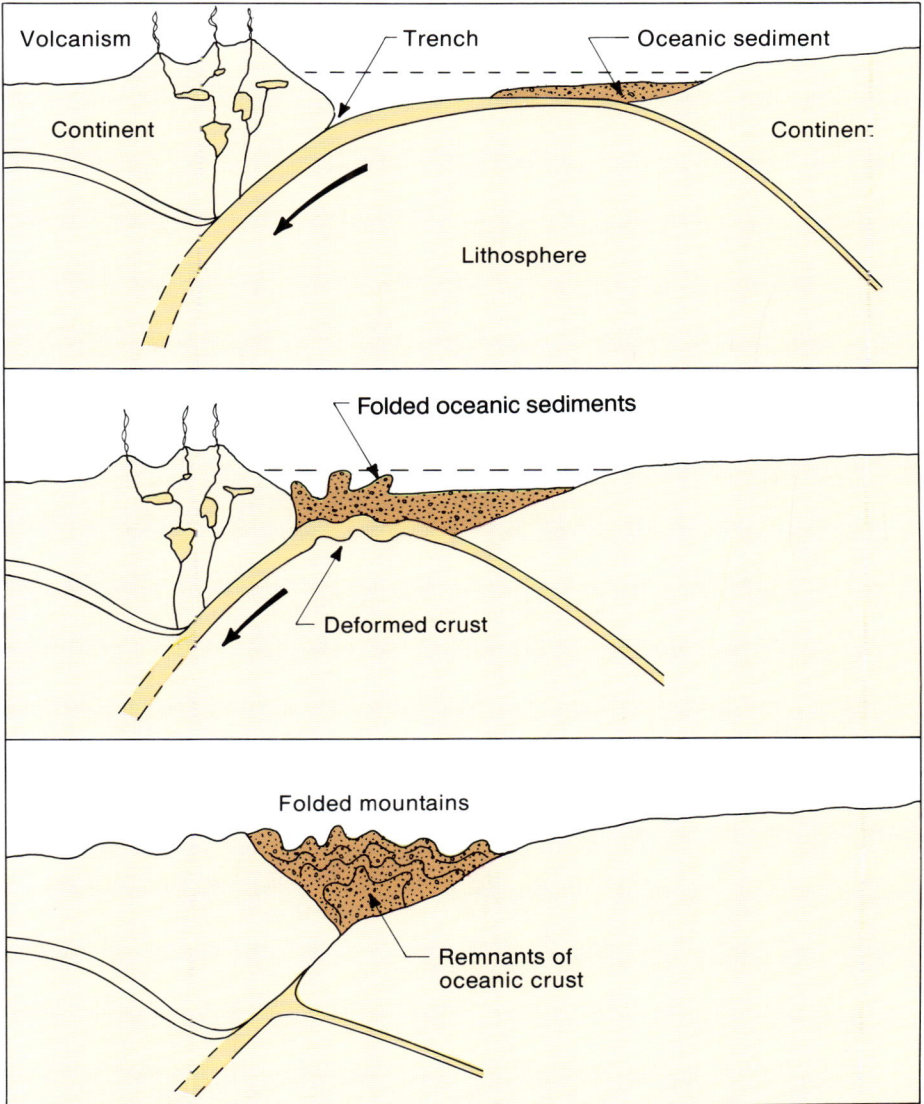

Figure 2.4. Sequence showing the collision of two continents and the resulting thickening of continental crust caused by the compression of oceanic sediments.

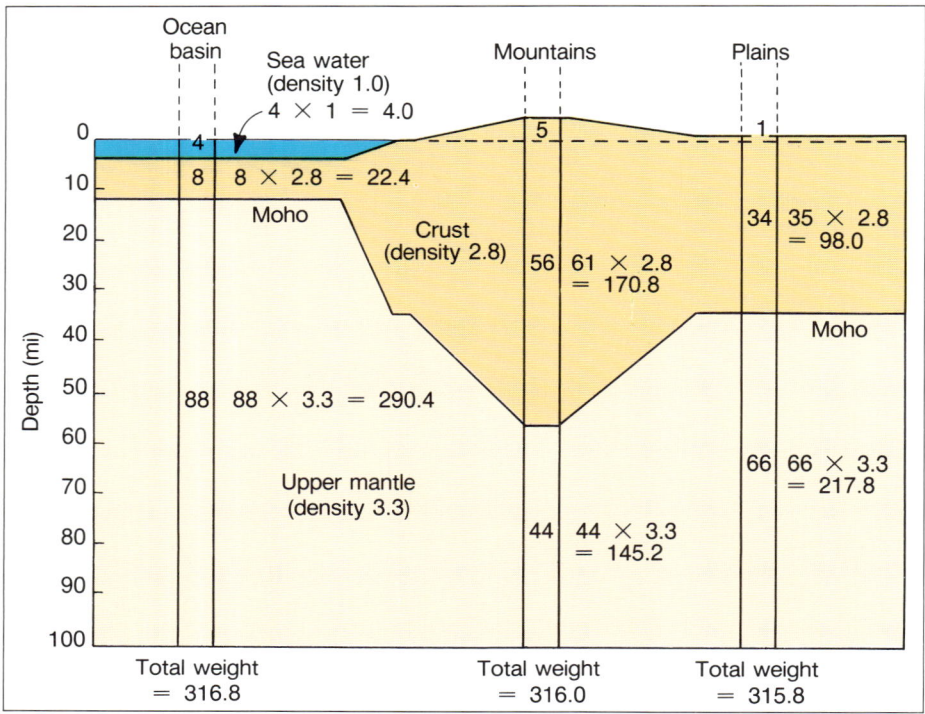

Figure 2.5. Diagram showing the isostatic adjustment taken by different crustal components. In this diagram, an average density is selected for both oceanic and continental crust that takes into account all the rock types present. Densities are given in g/cm³. *(Sawkins et al. Reprinted with permission of Macmillan Publishing Co., Inc. Copyright ©1974.)*

added to a continent by mountain building, the continental root penetrates farther into the mantle. When rock material is removed through erosion, the root cannot maintain its position and floats higher in the mantle, pushing the continental crust upward. The principle is the same as a block of ice floating in water, with the exception that the rock medium is much more viscous. The effects of isostatic equilibrium as it relates to the origin of groundwater reservoirs will be discussed in Chapter 3.

ORIGIN OF THE EARTH'S HYDROSPHERE

As suggested earlier, the Earth originated when fragments of rock material flying through space came together by gravitational attraction. It is unlikely that large volumes of free water existed on the Earth's surface at that time. If the Earth originally had little or no water, and water now covers three-fourths of the Earth's surface, what geologic processes produced the gigantic volumes of water available today?

When plate collisions began and magma formed in the subduction process, gases such as carbon dioxide (CO_2), nitrogen (N_2), hydrogen (H_2), hydrogen sulfide (H_2S), sulfur dioxide (SO_2), and carbon monoxide (CO) were produced in the accompanying volcanic eruptions. The principal gas released was water vapor, however, because hydrogen and oxygen exist in the chemical structure of many rock-forming minerals. When these rocks are melted, hydrogen and oxygen are released during volcanism and unite quickly in the atmosphere to form water vapor. Over geologic time, large amounts of oxygen were also contributed by photosynthesis; this oxygen can readily combine with hydrogen liberated in volcanic eruptions. The water formed from the melting of rocks is called juvenile water, that is, water never before on Earth in a combined form. It should be noted that some of the water vapor released in volcanic eruptions today is groundwater that was incorporated into the magma before the eruption (Figure 2.6).

Another source of water comes from the destruction of rocks at the Earth's surface by a process called weathering. Some rocks originate under high pressure and temperature at great depth in the Earth. Once exposed, these rocks are out of their original chemical and physical equilibrium, leading to gradual disintegration and the release of certain gases, including water vapor.

In addition to plate movements and rock weathering, what other processes control today's global water supplies? Water is lost from the Earth's surface in two ways: by atmospheric and subduction processes. Vaporized water in the atmosphere can, under certain conditions, escape the force of gravity and leave the Earth's atmosphere. This occurs when water molecules are broken up by ultraviolet light from the sun. The lighter hydrogen atoms can easily escape, leaving oxygen behind in the atmosphere. On a global scale, this loss is not great.

Major water loss occurs through sea-floor spreading, sedimentation, and plate sub-

Figure 2.6. Mount St. Helens, Washington, on May 18, 1980. Water vapor in the vicinity of active volcanoes condenses so rapidly that violent thunderstorms often occur near the summits. Occasionally, so much vapor condenses that the heat energy released is sufficient to cause tornadoes close to the mountains. *(David Frank, U.S. Geological Survey.)*

duction. Although the Earth's age is estimated at 4.6 billion years, the average age of the sea floor is about 65 million years, or only 1.5 percent the age of Earth. Thus, the sea floor is constantly being recycled through the upper mantle. As the newly formed plates move or spread from the oceanic rift zones, the upper layers of the plate (basalt) absorb water from the sea. This occurs quite slowly and the actual volume of water absorbed per square foot of basalt is not great, but the total surface of the sea floor is so large that huge quantities of water are held in the upper layers of the basalt. In addition, during the sedimentation process, water becomes trapped in the void spaces of newly formed sediment. As the sediment turns to rock, the water is entombed. In time, the basalt enters the subduction zone and some of the recently deposited sediments are carried into the upper mantle. Water trapped in the basalt and sediment is either released during the partial melting that occurs in subduction and expelled in volcanic activity, or carried down into the mantle where hydration of rock-forming minerals takes place.

In light of the ongoing geologic processes affecting water supply, is the Earth actually gaining or losing water today? Many scientists believe that the global water supply has remained stable for approximately the last 500 million years. Studies of the more stable continental areas have shown little real change in worldwide sea levels for that length of time. Of course, sea levels have fluctuated periodically as a result of glaciation, but those episodes were relatively brief and did not affect the long-term trends of water levels. Thus, for the last 500 million years, equilibrium has existed in the hydrosphere between the loss of water through atmospheric and subduction processes and the gain in water from volcanic and weathering processes.

EVOLUTION OF THE EARTH'S ATMOSPHERE

Great changes were occurring in the character of the atmosphere as the Earth's crust was evolving. Two evolutionary models have been proposed. Since the mid-1950's, most scientists have believed that the Earth's early atmosphere was devoid of oxygen, that is, anoxic (Cloud, 1968, 1973, 1976). Many forms of geologic evidence also suggest that the atmosphere was in a chemically reduced state and consisted almost entirely of nitrogen (N_2), methane (CH_4), water (H_2O), and possibly ammonia (NH_3). It is assumed that any hydrogen and helium present when the Earth formed would have been lost shortly thereafter because these gases are unusually light. The little oxygen that may have existed came from the breakdown of water vapor in the upper atmosphere. Because it quickly combined with ammonia and methane, no free oxygen was present.

About 1.9 billion years ago, oxygen began to be increasingly important as a part of the atmosphere. Sedimentary rocks in Africa contain evidence of primitive organic matter that is approximately 2.7 billions years old, which means that photosynthetic processes had started by that time. Plants use atmospheric carbon dioxide during photosynthesis and give off oxygen. As plants proliferated, photosynthetic activity contributed increasing amounts of oxygen. Moreover, oxygen was continually added to the atmosphere by the breakdown of atmospheric water vapor produced during plate activities.

Recent biological and interplanetary studies now indicate, however, that the early atmosphere may have contained significant amounts of oxygen and carbon dioxide

Figure 2.7. Weathering of marble headstones in a cemetery at Princeton, New Jersey. Carbon dioxide concentration in the atmosphere is partly influenced by weathering of carbonate minerals; marble consists of calcium carbonate. The stones are dated, left to right, 1796, 1828 and 1898. All were photographed in 1968. (S. Judson)

(Clemmey and Badham, 1982). These authors also point out several forms of geologic evidence that support this view. In addition, experimental work in biology during the late 1970's also suggests that oxygen may have been a major long-term constituent of the Earth's atmosphere (Henderson-Sellers et al., 1980).

Regardless of which theory will dominate in the future, there is little doubt that the Earth's atmosphere is constantly changing in response to emissions of volcanic gases during subduction processes, biological activity, and the weathering of rocks at the Earth's surface (Figure 2.7). The physical and chemical destruction of rocks produce small amounts of water, carbon, chlorine, sulfur, and nitrogen. Thus, both the creation and destruction of rocks lead to the liberation of gases. As suggested above, oxygen concentration in the atmosphere depends on the amount of water vapor present and its rate of breakdown, and the contribution of oxygen produced by plants.

The composition of today's atmosphere is given in Table 2.1. The only component that appears to be changing rapidly (in terms of man's time on Earth) is the carbon dioxide content. Since the start of the Industrial Revolution, huge amounts of fossil fuels have been burned to provide energy for industrial expansion and better living standards. In the burning of coal, oil, wood, and other fossil fuels, carbon dioxide is released while oxygen is consumed. So much carbon dioxide has been released in the

Table 2.1. Composition of the Earth's Atmosphere

Gas	Volume (%)
Nitrogen (N_2)	78.1
Oxygen (O_2)	20.9
Argon (A)	0.934
Water (H_2O)	up to 1.0 (variable)
Carbon dioxide (CO_2)	0.031
Neon (Ne)	0.0018
Helium (He)	0.00052

20th century that some scientists are worried that it is trapping more of the Earth's heat in the atmosphere. Rather than escaping into space, some of this extra heat is radiated back to Earth by the carbon dioxide and may cause a worldwide increase in temperature, leading to partial melting of the continental ice sheets on Greenland and Antarctica and massive flooding of coastal communities. Another adverse impact caused by burning large amounts of fossil fuels is the acidification of rainfall over large areas of industrialized countries. Biological activity in lakes has either ceased or is limited seriously where the bedrock does not naturally buffer (neutralize) this rainfall; for example, in Canada, New England, northern Minnesota, Florida, southern Norway, and Germany. In time, this phenomenon, called acid rain, may increase significantly the acidity of near-surface groundwater.

With the acceptance of the theory of plate tectonics, it is now possible to explain the origin of water, the distribution of rocks on the Earth's surface, and the formation of the Earth's atmosphere. By utilizing this information, the groundwater specialist is able to manage groundwater resources more wisely. Important aspects of the hydrologic cycle and the role of the atmosphere in precipitation are discussed in Chapter 4.

CHAPTER 3
Formation of Aquifer Systems

All igneous and metamorphic rocks exposed at or near the Earth's surface are in an unstable chemical and physical condition, and over geologic time these rocks break down into finer and finer components. Destruction of rocks and the redistribution and deposition of the rock particles play a significant role in producing three of the four major types of aquifer* systems — alluvial, sedimentary, glacial, and igneous/metamorphic. These particles are entrained and redistributed by the three agents of erosion — wind, running water, and glacier ice. Running water is the most effective of these agents because it operates continuously over most land areas of the Earth. Furthermore, it acts both as the primary agent in the creation of alluvial aquifers and as a major force in building or altering other types of aquifers. Before any major removal of a rock mass can take place, however, it must be broken down into particles that can be carried by these agents of erosion. These processes are called rock weathering.

WEATHERING

Weathering is the in-situ physical disintegration and chemical decomposition of rocks in response to the environmental conditions found at or near the Earth's surface. In general, the higher the temperature at which a rock formed, the more unstable it will be at the Earth's surface. Granite, which crystallizes from magma at approximately 1,110°F (500°C), is much more resistant to weathering than basalt, which forms at 2,190°F (1,200°C). Thus, the weathering rate depends on the temperature at which the rock formed, as well as climatic conditions, the availability of water, its chemistry, and rate of movement, and the chemistry of minerals making up the rocks. Because water has such a chemical affinity for almost all substances, and because of its unusual physical properties, rock weathering proceeds rapidly in the presence of water.

Weathering is not only the first step in the production of alluvial and sedimentary aquifer systems, it is also the process by which soils are produced. Although soil is a vital product of weathering, it is more important in the present context to focus on

---
*An aquifer is a water-bearing reservoir capable of yielding enough water to satisfy a particular demand.

the impact of weathering on the creation of groundwater reservoirs.

Most rocks decompose or disintegrate by a combination of both physical and chemical processes. Physical weathering is disintegration of rocks in place without associated major chemical change. During physical weathering, rocks are broken down into numerous small particles having a much larger combined surface area than the original rock. Heating and cooling, ice-crystal growth, and wetting and drying are ways in which rocks undergo reduction in size (comminution). In cold regions, physical weathering is the dominant process; in the rest of the world, chemical weathering is far more important. Once the physical size of the particles is reduced, decomposition is then accelerated by chemical attack on the larger surface area.

Chemical weathering of rocks is most effective in warm, humid regions where it is assisted by organisms. Few common rock-forming minerals can resist chemical decomposition; quartz and muscovite mica are two notable exceptions. In general, most silicate minerals break down into both relatively insoluble residues (such as clay minerals) and soluble substances which are carried away in solution. Clay minerals are produced through decomposition of feldspars. Limestone, made up of calcium carbonate, can be removed locally in solution if carbon dioxide (CO_2) is present in percolating water. Impurities contained in limestone, in the form of either clay or quartz particles, may remain after the calcium carbonate is completely removed. In general, all residual products of weathering are more stable in the presence of air and water than were the rocks from which they came.

Chemical weathering also contributes to the disintegration of rocks already attacked by physical forces. For example, chemical weathering dissolves and washes away some minerals, creating porosity (open space), and thereby reduces the resistance of the rock to crumbling. Some other minerals, when altered through weathering, increase in volume. Swelling causes the alteration products to break their bonds with adjacent material. Still other minerals are simply weakened by weathering, causing them to be less resistant to physical attack. The principal chemical changes that occur during weathering are brought about by oxidation, hydration, hydrolysis, chelating, and solution. All of these weathering processes are discussed in Chapter 6.

In general, the products of chemical weathering are clays rich in aluminum and iron; quartz or silica, either as individual sand grains (SiO_2) or in solution (Si); and solutes (a substance dissolved in another substance), mostly as sodium, calcium, potassium, and magnesium. Virtually all of the potassium is retained in the weathered soil zone, but most of the sodium, calcium, and magnesium may be carried away.

EROSION

The erosion cycle begins with weathering, when rocks are broken down into particles small enough to be carried by running water. Rainwater forms sheetwash, which carries away the smallest particles; sheetwash is water flowing sheet-like across a land surface and generally occurs only during heavy rains. Sheetwash becomes channelized within short distances into rills or rivulets, occasionally into gullies, then into small streams, and finally into rivers.

The eroding capacity of overland flow depends on the rate of precipitation and the length and steepness of the slope holding the weathered products. Less sediment will be removed if water can infiltrate the ground, the surface is highly resistant as in the case of bedrock, or the vegetation cover is extensive. Thus, the amount of material carried by any river or stream depends on many factors. In general, weathering and erosion of recently uplifted land in a moderately dry climate produces the maximum sediment yields from drainage basins. The most extensive alluvial aquifers are found adjacent to rivers flowing from mountainous regions that have moderately dry climates. Sediment accumulations in the floodplains of rivers draining these regions have coarse textures and may have thicknesses of 300 ft (91.5 m) or more in certain areas.

Streams and rivers transport material in three ways: as bedload, as suspended load, and in solution. Bedload comprises the coarser particles moved along the channel bottom by sliding, rolling, or saltation (jumping along the bed, propelled by impacts from other particles). In rapidly flowing streams, bedload may reach 50 percent or more of the total load in transport, whereas in times of low discharge there may be little or no movement of material along the bed. Typically, only about 10 percent of the total load in a stream is carried as bedload. Because of their coarse texture, bedload deposits often form the most productive layers in alluvial aquifers.

Suspended load is weathered material that a river transports in suspension. A river can always move a fine suspended load, even when bedload movement is nonexistent. The size of the suspended sediment depends primarily on the gradient* of the river and generally ranges from clay-sized particles to coarse sand.

Dissolved material in rivers is carried in solution, usually in the form of ions. River waters are particularly high in dissolved solids in areas of low relief, such as in the Atlantic and Gulf Coast states. Typical solution products in rivers include calcium, magnesium, iron, chlorine, sulfate, and sodium ions. These materials enter the sea where they may precipitate out when oversaturation is reached. In many cases, these precipitated materials can form thick sequences of rock that, once lifted above sea level, can become major aquifers.

It is difficult to imagine the amount of sediment carried by rivers. Figure 3.1 shows the Mono Reservoir, in California, filled to capacity, not with water but with sediment; the reservoir was filled completely only two years after construction.

ALLUVIAL AQUIFERS

Rivers and streams build groundwater reservoirs consisting of alluvial deposits. Each year about 30,000 mi³ (125,000 km³) of water falls as snow and rain on the land areas of the Earth. About 30 percent of this water returns more or less directly to the world's oceans by rivers and streams. As this water returns to the sea, it erodes the landscape, deposits sediment such as sand and gravel along river courses, and carries the remaining products of weathering to the sea.

Rivers cause significant changes in the landscape because they constantly work to reach equilibrium conditions. For example, all rivers tend to (1) extend their drainage systems both upstream and occasionally downstream as base level (sea level) changes, (2) downcut or aggrade (build up) their channels, and (3) widen their valley walls. These changes occur because streams are extremely sensitive to changes in sediment load, discharge, gradient, and velocity. Hydrologists suggest that streams continually make adjustments in their channel shapes and gradients to accommodate changes in discharge and sediment load. A river is "graded" when these changes cease to occur;

---

*Gradient refers to the steepness of a slope, that is, the loss in vertical height of a streambed as it flows over a specified horizontal distance.

Figure 3.1. The Mono Reservoir (California), built in 1935 following a severe fire which burned off much of the vegetation in the watershed area, was completely filled with silt and sand in only two years. This serves as a vivid reminder that rivers represent systems of moving water and sediment, not just water alone. *(USDA Forest Service)*

in other words, equilibrium exists among the independent factors influencing the river (discharge and sediment load) and the dependent variables (gradient, channel shape, and channel roughness).

Mackin (1948) defines a graded stream as "one in which, over a period of years, slope is delicately adjusted to provide, with available discharge and with prevailing channel characteristics, just the velocity required for transportation of the load supplied from the drainage basin." In reality, only short stretches of a river can be truly graded at any one time; usually downcutting occurs in the headwaters, while deposition occurs in the middle and lower reaches of a river. Downcutting, transportation, and deposition are occurring simultaneously at various places in every river. In terms of geologic time, erosion (downcutting and transportation) becomes the dominant process as land surfaces are reduced to near sea level. In terms of human time, however, all the sediment picked up by a river does not reach the sea but is redeposited along the way as a result of relatively short-term changes in gradient, discharge, or sediment load. Landforms such as floodplains, alluvial fans, and deltas form when rivers deposit rather than erode.

The development of a typical river valley is shown in Figure 3.2. In Figure 3.2A, the river runs through an uneroded land area. It is rapidly downcutting because the gradient is quite steep. After thousands of years, the relief has been reduced because the river has carried away much of the sediment brought to it from the surrounding upland (Figure 3.2B). The valley has become enlarged and surface relief is reduced considerably in Figure 3.2C. By this time the gradient of the river has lessened, suggesting that the suspended load and bedload are not only reduced in volume but are also smaller in particle size. Near the end of the erosion cycle, some of the sediment cannot be carried and is deposited as alluvium on a floodplain in the middle reaches of the river. Floodplains form in the valleys excavated by rivers (Figure 3.2D). During peak discharges, the river goes over its banks, the velocity of the water decreases, and suspended sediments are deposited. As the river meanders back and forth, sediments accumulate into an actual plain. Wells drilled in this plain will almost always be successful because the sediments are of similar size (well sorted) and the river provides continuous recharge to the sediments. In Figure 3.2E, virtually all the land has been reduced to sea level and the river merely meanders back and forth across the floodplain, although occupying only a small part of the wide floodplain at any time. The sediment carried to the sea will be deposited in a delta. Deltas form whenever rivers

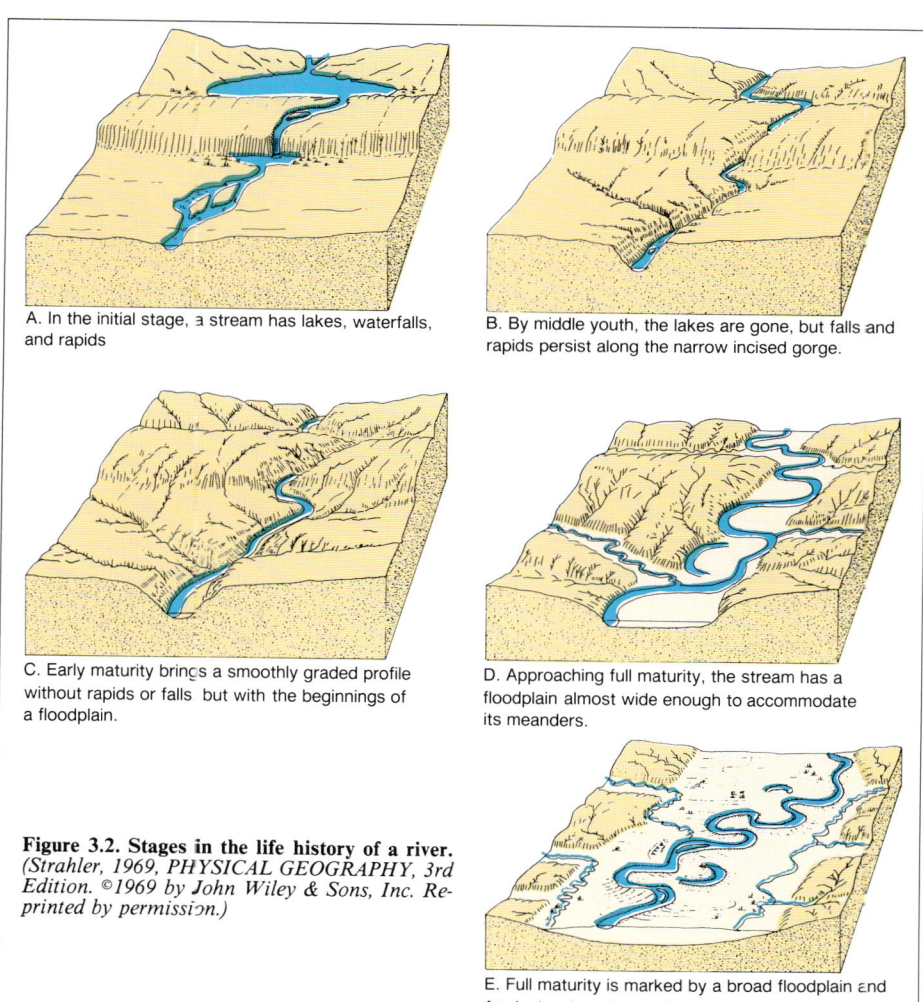

A. In the initial stage, a stream has lakes, waterfalls, and rapids

B. By middle youth, the lakes are gone, but falls and rapids persist along the narrow incised gorge.

C. Early maturity brings a smoothly graded profile without rapids or falls but with the beginnings of a floodplain.

D. Approaching full maturity, the stream has a floodplain almost wide enough to accommodate its meanders.

E. Full maturity is marked by a broad floodplain and freely developed meanders.

Figure 3.2. Stages in the life history of a river. *(Strahler, 1969, PHYSICAL GEOGRAPHY, 3rd Edition. ©1969 by John Wiley & Sons, Inc. Reprinted by permission.)*

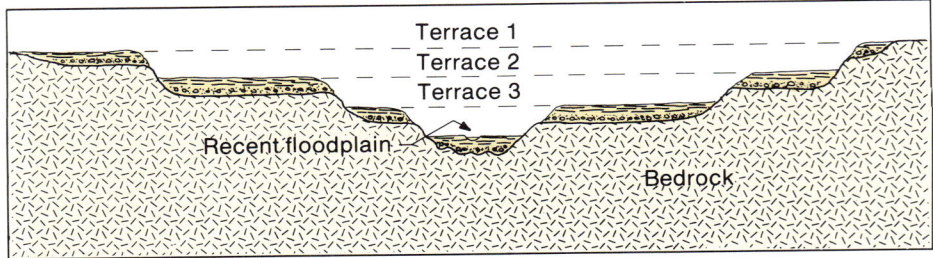

Figure 3.3. Cross section of a valley where the dashed lines represent the height of the old floodplains. The river valley has undergone three periods of renewed downcutting, leaving three distinct terrace deposits at successively lower elevations.

reach large bodies of standing water or even when fast-moving streams carrying a large suspended load or bedload enter slower moving rivers.

If the land surface is uplifted as the river approaches erosional maturity or base level falls because of continental glaciation (Figure 3.2D), the river cuts down into the floodplain deposits and leaves terraces along the valley walls. Terraced valleys are erosional remnants created by periodic downcutting. Once saturated, the sediments comprising the terrace generally become aquifers (Figure 3.3).

Rivers build one other landform, alluvial fans, that can be important as an aquifer (Figure 3.4). Alluvial fans are constructed at the base of mountain fronts in relatively dry climates when huge quantities of sediment are carried to the dry valley floors by ephemeral rivers* draining nearby mountains. Physical weathering produces large amounts of loose rock material that can be entrained easily into the swift mountain rivers that flow during the infrequent heavy rainstorms. Because the valley floors are nearly horizontal, the sediment load is dumped abruptly as the rivers enter the valleys. Deposition results from the sudden decrease in river velocity. Over time, huge aprons of sediment are constructed against the mountain fronts. As multiple fans form, they often coalesce into continuous aprons of sediment, or bajadas, set against the mountain fronts.

Alluvial fans are important to those interested in groundwater because, along with other valley sediments, fans often represent the only extensive unconsolidated deposits in mountainous regions capable of yielding high volumes of groundwater. Water from ephemeral rivers quickly infiltrates into the fans and thereafter flows into the valley-floor sediments.

Hydraulic Characteristics of Alluvial Aquifers

The average grain size of river-deposited sediments can vary considerably; floodplain deposits may consist of extremely fine silt, whereas coarse gravel or sand may be more typical of alluvial fan deposits. Floodplain deposits are usually fine grained, well rounded, and generally well sorted. Therefore, the porosity is excellent but the hydraulic conductivity† varies considerably, depending on the average grain size. If the gradient of a river steepens or the discharge increases, the sediment will become coarser and thus the hydraulic conductivity will be higher.

---

*A stream that flows briefly only in direct response to precipitation in the immediate locality and whose channel is at all times above the water table.

†Hydraulic conductivity is a measure of the capacity of a porous medium to transmit water (see Chapter 5).

Floodplain deposits are quite uniform except at point bars (inside the meander bends) and at the bottom of the rivers. When a river meanders, coarse sediment accumulates near or on the point bars. As the meanders continue to migrate both laterally and downstream, finer grained (floodplain) material covers the coarse sediment at the point bars. Similarly, bedload is buried during the meandering process. A cross section of a typical meandering-river alluvial sequence is shown in Figure 3.5. It is clear from this figure that coarse lenses of sediment can exist in otherwise finely textured floodplain sediments. The thickness of the coarse-grained layers is limited, as is their areal continuity. However, wells that intersect one or more of these coarse layers have higher yields than those in the finer floodplain deposits.

Alluvial fan deposits are highly irregular in grain size and degree of grain roundness; they are built by braided streams that continually deposit sediment as they flow over the fan surface. The constantly changing paths of the braided streams produce deposits that have an irregular areal distribution. Furthermore, the competencies* of these streams change so often, depending upon rainfall intensity, that grain sizes of the bedload vary considerably. The typical particles in an alluvial fan stream may not be as well rounded as those in a regular stream because the weathered material may not be transported as far before deposition. Fine and coarse sediments are intermixed, and thus the hydraulic conductivity and porosity of fan sediments may not be as good as in river alluvium. It is particularly important, therefore, to drill test holes before designing and constructing wells in alluvial fans. Because of their great thicknesses, these deposits can yield high volumes of water to wells.

In deltaic deposits, which are essentially bedload accumulations, porosity is good because the sediments are generally well sorted and well rounded. If a tributary stream

Figure 3.4. Alluvial fans formed by rivers draining the Panamint Mountains, Death Valley, California, are local sources of groundwater. The youngest gravels, the light colored areas on the fan, are important areas of recharge. *(Malde, U.S. Geological Survey)*

---

*The ability of a stream to transport sediment of a certain size is called its competency.

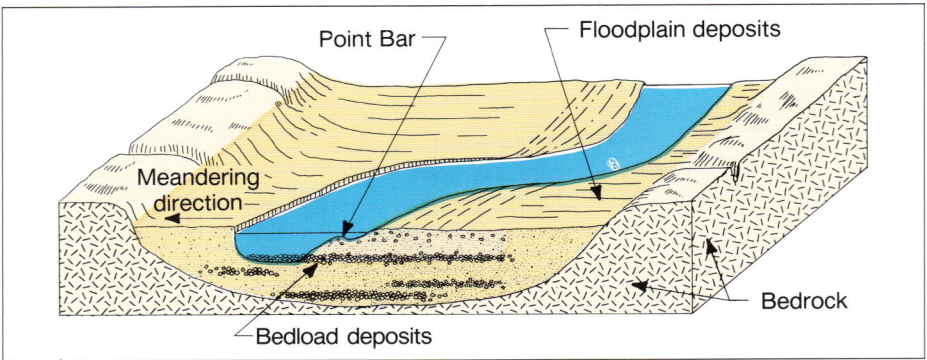

Figure 3.5. Point bar and bedload deposition creates local areas of relatively high hydraulic conductivity in typical finer grained floodplain deposits.

is considerably steeper than the river it enters, the bedload from the stream may be coarse and the hydraulic conductivity of the stream's delta, quite high. On the other hand, deltaic deposits may be so fine, as in the case of the Mississippi River, that the hydraulic conductivity is not adequate for high-yield wells.

The attitude or positioning of the sedimentary beds in deltas may reduce horizontal permeabilities*, as indicated in Figure 3.6. Wells completed in the foreset beds will yield less water than wells finished in either the bottomset or topset beds, assuming equal drawdown† potential. In the case of foreset beds, water cannot flow toward a well along the easiest path but must cross sediment boundaries, as shown in Figure 3.6. Hydraulic conductivity across these sediment boundaries is considerably lower than along individual sediment planes.

SEDIMENTARY ROCK TYPES

When rock material is eroded from the land, the continental masses are inhibited from rising instantaneously at a particular site because in most cases the continental rocks are strongly attached to other regional rock systems or nearby oceanic plates. Other factors such as the relatively viscous nature of upper mantle material, also contribute to inhibiting the rebound of continental landmasses.

In some cases, the rocks have such a strong hold on adjacent segments of a continent that no vertical movement occurs at all until a landmass is reduced to sea level. In time, however, the buoyant effect exerts itself and the continent rebounds gradually. When continents rebound, various sediments deposited on the formerly submerged continental slopes are uplifted. Once flushed of sea water, they form some of the largest and most prolific groundwater reservoirs.

Transgressive Seas

When rebound of a continental land mass is inhibited and erosion of the land persists for a number of years, the average elevation of the surface is reduced hundreds or thousands of feet. Eventually, the sea begins to encroach on the lowered surface.

---

*Permeability is the ease with which water will move through a porous medium.

†Drawdown is the difference between the static water level and the pumping water level. It is a measure of the force required to drive water into a well.

In this case, the sea encroaches because the land is continually being lowered by erosion, and not because the ocean has significantly more water. Land sediments that are deposited on the ocean floor do displace sea water, however, and produce an actual rise in sea level relative to the land surface.

As the sea rises and transgresses over the land, beach sand is deposited along the advancing shore. Nearly all the sand existing on a beach was brought to the shore zone by rivers. In inland areas, rocks are being continually broken down by chemical and physical means. Running water picks up the loose particles and transports them to the sea, where wave turbulence in the beach zone tends to separate the fine material from the coarse. The finer material (finer than sand) is kept continually in suspension by wave action until it eventually moves offshore to be deposited in quiet water. The coarsest material stays in the beach zone where it is picked up and redeposited time and time again by waves. Mechanical abrasion is so great that the sand particles, which are mostly quartz, become well rounded. The sizes of all particles tend to be the same because wave power at any one beach is more or less uniform through time, even though wave power does vary with wind velocity. Thus, beach sand is characteristically well rounded and uniform in size.

In time, the beach zone migrates farther inland and the water deepens at the former beach zone. Waves continue to affect the ocean bottom, but their power is reduced. Newly deposited sediment in the former beach zone is now considerably smaller in size than the original beach material; but it is still uniform in size because wave energy remains constant, although reduced in magnitude. Wave turbulence can only reach a depth equal to one half the actual wave length. Therefore, as the water continues to deepen, wave turbulence can no longer reach the bottom. Clay-sized particles [less than 0.0002 in (0.005 mm) in size] that were carried offshore begin to settle on top of the submerged beach sand. The upward change in particle size from beach sand, to finer offshore sand, to clay is quite gradual; one particle size grades into another almost imperceptibly.

A third type of sedimentary deposit occurs in warm, shallow seas when the shells and secretions from marine organisms form organic reefs of calcium carbonate. Great thicknesses of coral can grow if the water deepens slowly. If the water deepens sud-

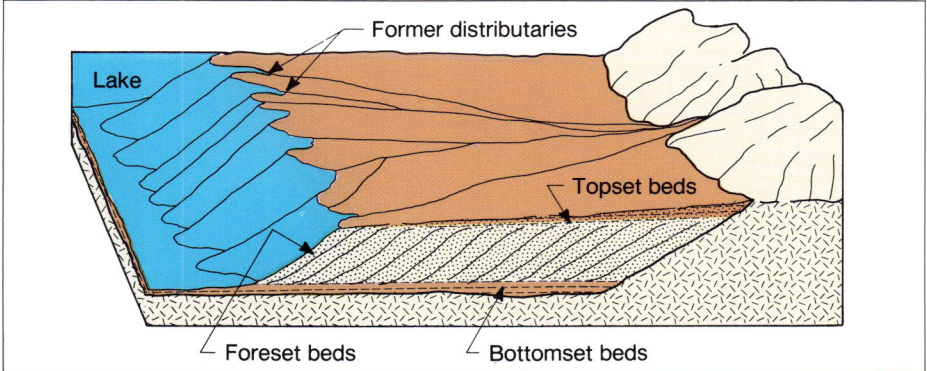

Figure 3.6. Structure of a simple delta, shown in vertical section. The bottomset beds are deposited almost horizontally in front of the advancing foreset beds. Later, topset beds form on top of the foreset beds and represent a continuation of the landward alluvial plain. The topset beds truncate or cover the upper edges of the inclined foreset beds.

denly, the coral will die because of restricted light. Sedimentation on top of the reefs is limited generally to clay particles if near shore, or it may include the skeletal remains of tiny marine organisms if the water becomes deep enough. The latter, called pelagic oozes, are either calcium or silica-rich, depending on the type of organisms from which they are derived. A related sediment type is also extremely important in the eventual formation of groundwater reservoirs. When sea water becomes completely saturated with calcium introduced in solution by rivers, some of the calcium precipitates out as discrete crystals of calcium carbonate (calcite). These tiny crystals accumulate on the bottom in thicknesses of 20 to 30 ft (6.1 to 9.1 m) or more. Thus, a calcium-rich deposit can form on the sea floor by inorganic processes (chemical precipitation) as well as organic processes (coral reef formation).

At a particular site, sand, silt, clay, coral reefs, or calcite crystals may be deposited (Figure 3.7). In reality, the sea transgresses at an irregular rate. Therefore, deposits of any particular sediment may vary radically in thickness from those at another site in the same general area. Moreover, elevation adjustments caused by isostatic rebound may modify the particle size or even the type of sediment found within a localized area at the same depth.

Modifications in land surface elevation result in oscillations of the shoreline across considerable horizontal distances. For example, clay lenses (layers) are often found in sand deposits. These lenses indicate that the sea had temporarily deepened at that site. As suggested above, the volume of water in the sea did not change significantly but the land elevation did. Typically, most sediments do reveal oscillations in sea level, as demonstrated by vertical gradations in particle sizes or actual changes in sediment type. Fortunately, the sea did rise steadily in many cases, producing highly uniform sediment structures of great thickness. These formations, particularly sand deposits, now form some of the greatest aquifer systems.

Regressive Seas

Isostatic rebound causes the seas to withdraw eventually from former land surfaces.

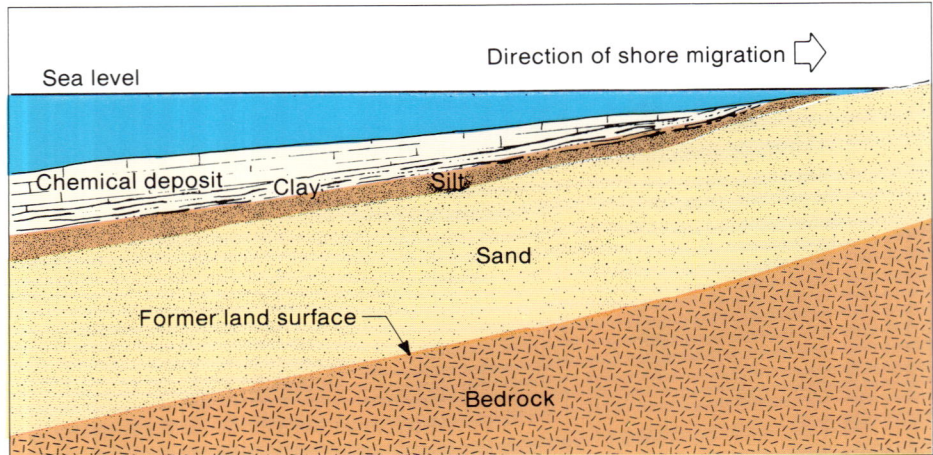

Figure 3.7. Typical cross section of sediments laid down by a transgressive sea on a recently inundated coastal plain. Note that a gradual upward fining of the sediment occurs at any site as the water deepens. The slope of the former land surface is exaggerated for clarity.

In retreat, the regressive seas deposit the same succession of sediments as found in transgressive seas, except the vertical order is reversed. The actual thickness of the sediments depends on the rate of retreat and the amount of material carried by rivers to the shore. Oscillations in water depth do occur during retreat stages; therefore, sediment types may change frequently as the sea encroaches again and then retreats.

Figure 3.8 shows the sequence of deposition in a typical transgressive-regressive situation. This is a classic sedimentary arrangement; that is, all sediment types are represented but no reversals of sediment type occur from sea-level fluctuations, and no sediments have been lost through erosion. Because land surfaces can rise and fall many times within even short intervals of geologic time, the thicknesses of sediments over the basement rocks (igneous and metamorphic rocks of continents) can be great. In Minnesota, sediments accumulated to depths of only 100 to 1,000 ft (30.5 to 305m) over much of the state, whereas sediment thicknesses in Texas are up to 26,000 ft (7,930 m) or more.

Once deposited, sediment accumulations are not always preserved through subsequent periods of uplift. For instance, when a beach deposit is raised above sea level and is exposed to erosion, all traces of the former beach may be removed within a short period of geologic time. Similarly, clay deposits can also be eroded easily. As a result, erosional breaks, or gaps, often occur in sedimentary sequences. Periods of erosion can be recognized in rock exposures by abrupt changes from one sediment type to another and by the non-parallel attitude of two adjacent rock formations.

Hydraulic Properties of Sand

Sediments laid down by transgressive-regressive seas are modified through time. Physical changes occur from compaction caused by the increased depth of overburden, chemical precipitation or solution (removal) effects within the void spaces of the sediment, and heat from magmatic intrusions. Some of these changes increase the usefulness of the sediment as a storage medium for groundwater. Other modifications reduce or even eliminate the storage capabilities of the sediment.

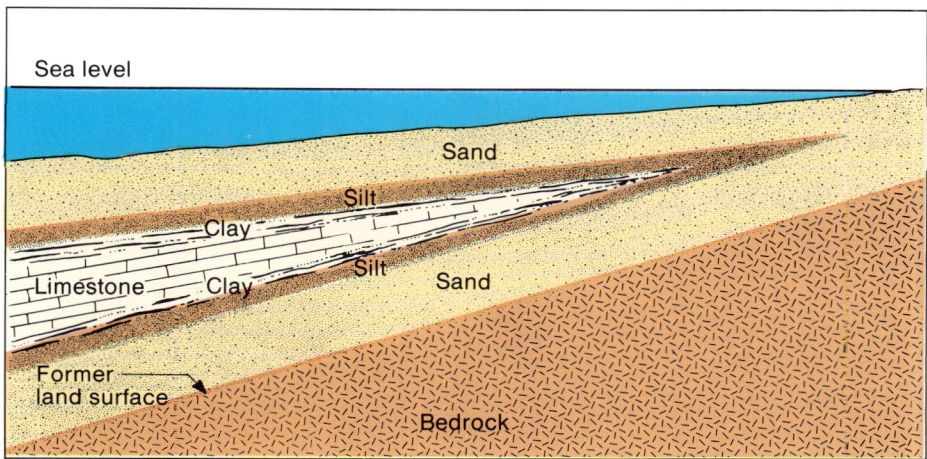

Figure 3.8. Complete transgressive-regressive sedimentary sequence. In many parts of central and southern United States, this sequence occurs repeatedly in the sedimentary structure of the crust. Occasional periods of erosion may remove one or more of the layers in the transgressive-regressive sequence.

Former beach deposits are the most valuable storage sites because the volume of void space is greatest; that is, beach sediment ordinarily can store more water per rock volume than any other type of sedimentary rock. Void space in a new beach sand may be as high as 25 to 40 percent. In time, this space may be reduced by settling and rearrangement of the grains, chemical precipitation, or heat, thereby producing a rock formation called sandstone.

In settling, the finer particles in a sediment tend to fill in the void spaces between the uniformly sized grains. The space that these small grains occupy reduces the open space or porosity of the sediment and thereby decreases its storage capacity.

Chemical precipitation also decreases the void space. Solutions percolating through sand often carry significant amounts of silica and calcium dissolved from overlying sediments. Precipitation of these elements cements the sand grains together (Figure 3.9). If the cementation is carried too far, the void space can be substantially reduced.

Sufficient heat associated with magmatic intrusions can partially melt the grains of sand, causing them to compress under the pressure of overlying rocks and fill in all or much of the void space. In addition, hydrothermal solutions associated with magmas contain high concentrations of many chemicals. When these solutions reach sand layers, the void space may be reduced by precipitation of various minerals. The physical and chemical changes in sandstone resulting from heat and high pressure produce a rock called quartzite which has virtually no void space.

Sandstone formations are the most important reservoir rock for storing large volumes of groundwater. Some individual sandstone layers are extensive; for example, the St. Peter Sandstone, a major aquifer in central United States, covers more than 290,000 mi^2 (751,000 km^2) and averages 80 to 160 ft (24.4 to 48.8 m) in thickness.

Hydraulic Properties of Clay

When originally laid down, clay has high porosity. For example, clays now being deposited on the Mississippi River Delta have 90 percent porosity. In time, compaction usually reduces the pore space considerably. Although the volume of void space is relatively high in clays, the actual size of the voids is extremely small. Water is strongly attracted to the large surface area of the clay particles and is less controlled by the groundwater gradient. Thus, water does not move easily through clay sediment.

Because water cannot move easily through clays, chemical precipitation is not important in the cementation of clay particles. Heat and pressure are much more likely

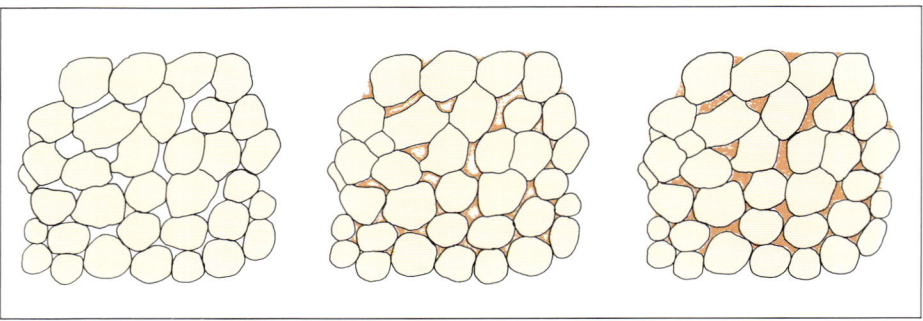

Figure 3.9. Grains of a former beach sand cemented together over time by precipitated mineral matter from percolating groundwater.

to be the primary factors controlling the consolidation of clay. When sufficient and prolonged pressures are applied to clay deposits, the clay changes to shale, which is weakly consolidated and breaks easily along depositional planes. When exposed to water during drilling, some shales can swell to much larger volumes. In shale-rich areas, drilling operations can be severely handicapped or even stopped by swelling shale. If greater heat and pressure are applied, the shale changes to slate, a hard, dense rock with virtually no storage space for groundwater.

Hydraulic Properties of Calcium-Rich Deposits

Initially, inorganic calcium-rich deposits are quite massive and no large amount of postdepositional chemical or physical action is required to change them to rock. Therefore, little void space exists for fluid storage. This is not always true of coral reefs (organic calcium deposits), where initial void spaces occur frequently but are spaced irregularly. Time and other factors, however, do make calcium-rich deposits more rigid and, in some cases, more dense. This type of sedimentary rock, known as limestone, is inflexible and can easily crack when supported unevenly. Therefore, almost all calcium-rich sedimentary deposits show extensive joint or fault systems because of uneven isostatic adjustment and local stresses produced by solution effects and erosion.

Under certain conditions, some of the calcium in limestone can be chemically replaced by magnesium from sea water, either during deposition or after the rock strata formed. Limestone with high concentrations of magnesium is called dolomite. It is not known with any certainty how this process happens, but the addition of small amounts of magnesium causes a pronounced toughening of the rock.

Once exposed at or near the land surface, limestone and dolomite can undergo remarkable change. Rainwater, in its descent to earth, commonly absorbs carbon dioxide from the air and thereby forms a weak acid, called carbonic acid. When rainwater enters cracks and crevices in carbonate rocks, the acidic water may dissolve small volumes of rock. Over time, the removal of limestone along the cracks can be extensive, and voids can be created (Figure 3.10). If these voids are close to the land surface, the roofs of the resulting caverns may cave in and thus produce ponds or sink holes. Lakes in this type of terrain, which is call karst topography, are intimately connected to the groundwater system. Water usually moves rapidly in certain directions through aquifers in karstic terrain and extreme care must be taken to prevent contamination of the groundwater.

Limestone does not initially offer much of a reservoir for storage, but through secondary solution many deposits of limestone and dolomite can become large-capacity reservoirs for groundwater storage. Other limestones are not cavernous and the only water in them exists in cracks and crevices.

GLACIAL DEPOSITS

Glacial aquifers are the second most important category of aquifer systems. Although not as extensive as aquifers found in sedimentary rocks, glacial aquifers occur throughout much of the highly populated regions of northern United States, Canada, and northern Europe. In many of these areas, the glacial aquifers are the only ones present, because the underlying igneous and metamorphic bedrock is practically devoid of water; or, if water is in the bedrock, it is of poor quality.

Physical scientists have known for some time that worldwide temperatures periodically turn downward for relatively short intervals of geologic time. When this occurs, great continental ice sheets, 1 to 3 mi (1.6 to 4.8 km) thick, develop over large areas of the Earth. Today, only 10 percent of the Earth's surface is covered by glacier ice, whereas only 18,000 to 20,000 years ago, glaciers covered about 30 percent. Changes brought about by a cooler climate during Pleistocene* time had far-reaching effects: sea levels rose and fell as much as 460 ft (140 m); belts of precipitation shifted; plants and animals altered their life patterns; the altitude of mountain snow lines in the middle latitudes fell about 2,500 ft (762 m); much land outside the glaciers was subjected to repeated freezing and thawing; and most important of all, glaciers removed massive amounts of rock and soil and deposited them elsewhere.

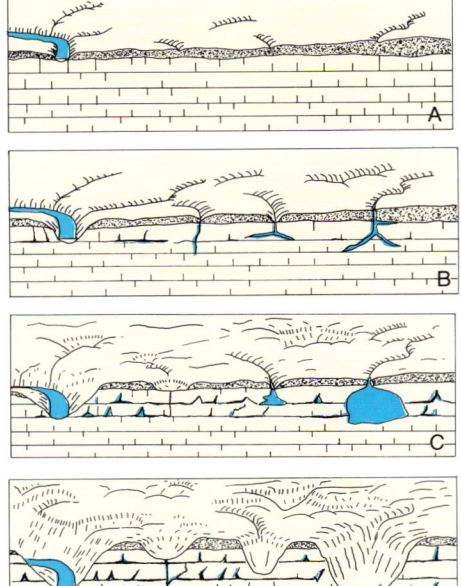

Figure 3.10. Over time, some of the limestone is taken into solution by percolating rain water causing an enlargement of the crevices, joints and fractures. As a cavern system develops, surface stream flow is diverted to underground flow. *(Strahler, 1969, PHYSICAL GEOGRAPHY, 3rd Edition. ©1969 by John Wiley & Sons, Inc. Reprinted by permission.)*

Debate over the mechanisms that cause worldwide climatic change has continued for at least 100 years among glacial geologists. Perhaps the most plausible single explanation for worldwide temperature oscillations is the astronomical theory proposed by M. Milankovitch, a Yugoslavian geophysicist. Milankovitch (1969) suggests that changes in the Earth's orientation relative to the sun and changes in distance from the sun cause periodic fluctuations in the distribution of solar insolation (radiation) received on Earth, although the total amount of insolation received annually remains much the same.

Other factors must be considered when reasons for glaciation are discussed. To achieve large-scale glaciation, land areas must exist at high latitudes and sufficiently high altitudes to offer sites for massive accumulations of snow. Periods of low temperatures most effectively bring about glaciation when plate movements and mountain building combine to locate land masses of sufficient size and high elevation in high-latitude positions. Currently, these conditions exist in higher latitudes of the Northern Hemisphere, and have existed for the past several million years.

Pleistocene glaciers have been active over much of the world at various times during the past 3 million years, especially in the Northern Hemisphere (Figure 3.11). The latest ice advances began about 80,000 years ago, with the ice withdrawing about 8,000

---

*Pleistocene is the name given to the most recent glacial period which began about 3 million years ago and lasted until 10,000 years ago. See Appendix 3.A for a complete geologic time scale.

to 10,000 years ago from most areas in the northern United States, southern Canada, and Scandinavia. The most recent Pleistocene glaciers were not as large as some of the earlier glaciers; therefore, they did not completely cover the landscape, but tended to follow topographic lows in a manner similar to modern valley glaciers. These later glaciers left sediments called glacial drift deposited irregularly in both areal extent and depth. Understanding the processes responsible for the distribution of glacial drift is important in locating and constructing wells for maximum yield.

During Pleistocene time in North America, ice accumulated in the high-plateau areas of Baffin Island, Keewatin, and Labrador-Ungava in Canada. Ice depths at the centers of these major accumulation areas commonly reached 2 mi (3.2 km) or more. In Western Europe, major ice centers developed in the mountains of Scandinavia and Switzerland.

Temperature conditions at the bases of these continental ice sheets, near their centers, were probably well below freezing. Under these conditions, basal ice is frozen to the ground and ice flow above the base occurs by differential slippage within individual ice crystals. As the ice flows outward from accumulation areas (generally toward the south in the Northern Hemisphere), heat is added to the ice mass by the friction between ice grains, from geothermal heat emanating from the upper mantle, and from the heat released when surface meltwater enters the glaciers and refreezes.

When the base of an ice sheet approaches the melting temperature, pressure melting begins to take place at the bottom. Melting occurs at the base, at a temperature less than 32°F (0°C), because of the weight of overlying ice and the pressure exerted on various parts of the basal ice by rocks frozen in the ground. When the advancing ice encounters a rock, melting takes place on the up-glacier side of the rock (Figure 3.12). This water flows to the down-glacier side of the rock, where refreezing occurs because

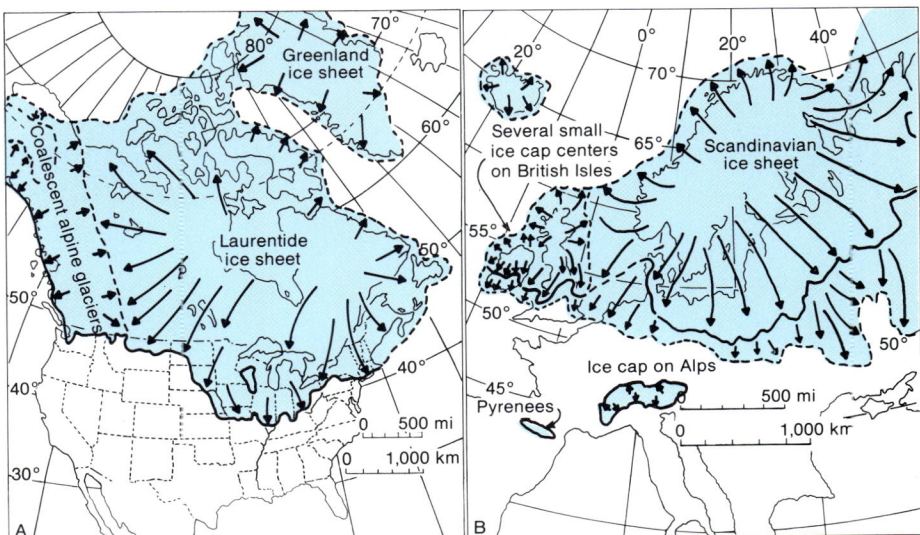

Figure 3.11. A. Pleistocene ice sheets of North America reached as far south as the present Ohio and Missouri Rivers. B. The Scandinavian ice sheet dominated northern Europe during the Pleistocene glaciations. Solid line shows limits of ice in the last glacial stage; dotted line on land shows maximum advance at any time. *(After R. F. Flint, 1971; Strahler, 1975, PHYSICAL GEOGRAPHY, 4th Edition. ©1975 by John Wiley & Sons, Inc. Reprinted by permission.)*

less pressure exists there and the temperature is below 32°F. The process of pressure melting is called regelation; it enables ice to move over and around impediments. Thus, glacier ice flows in two ways: by intracrystalline sliding and regelation at the glacier base when the temperature at the base is close to 32°F.

Regelation also plays a vital role in the entrainment of soil and rock debris at the base of an ice sheet. Regelation releases water that commonly runs into crevices in the ground, and if the surface is frozen the water refreezes almost immediately. As freezing occurs, the ice expands in the crevices and forces angular blocks of soil and rock upward into the overlying ice. The soil and rock become frozen to the base of the glacier and are carried toward the glacier terminus; this process is called plucking or quarrying (Figure 3.12). Quarrying is the principal method by which continental glaciers entrain rock debris.

Ice generally flows at rather uniform rates on flat topography; however, when the topography is uneven, the ice will flow faster in the depressions. Entrainment of debris is greatest in these depressions and the ice can deepen valleys to hundreds of feet below sea level. The deep depressions forming the Great Lakes of North America and the fiords of Norway are extreme examples of glacial quarrying in preexisting lowlands.

As the ice flows toward the terminus, glaciers continue to warm and eventually the ice near the margins reaches either the pressure-melting temperature or 32°F. Bottom melting releases debris which is then deposited on the underlying ground surface as the ice flows over it. This material, called lodgement till, rarely reaches significant thicknesses in any single advance of the ice. Bottom melting and sediment deposition generally occurred when the Pleistocene glaciers were approximately 30 to 90 mi (48 to 145 km) away from their terminuses.

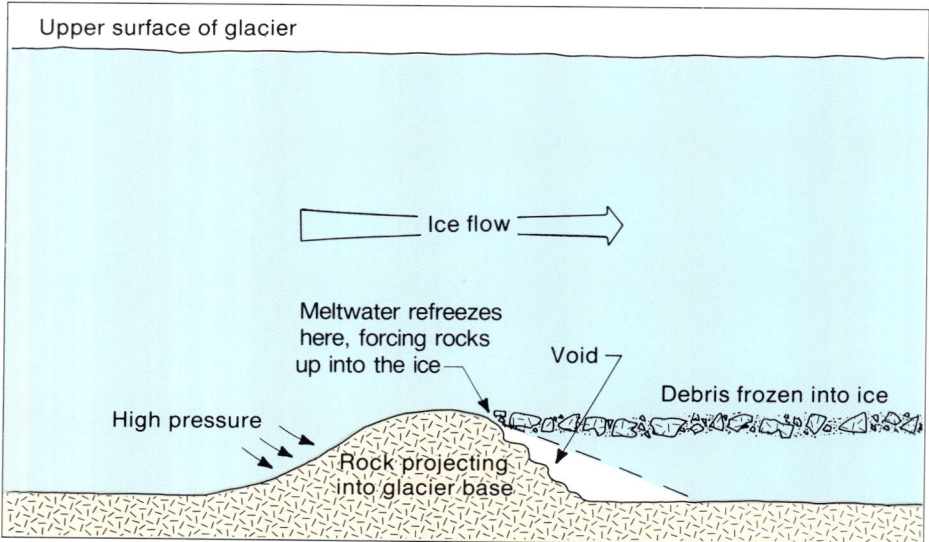

Figure 3.12. If ice is near its pressure-melting point, regelation or pressure melting can occur on the up-glacier (upstream) sides of rocks that project into the base of the ice sheet. Water runs around the projections and refreezes to the glacier sole. Usually, rock fragments become frozen to the base during the refreezing process; thus a quarrying or plucking operation exists on the down-ice side of projections. Ice depth is not to scale.

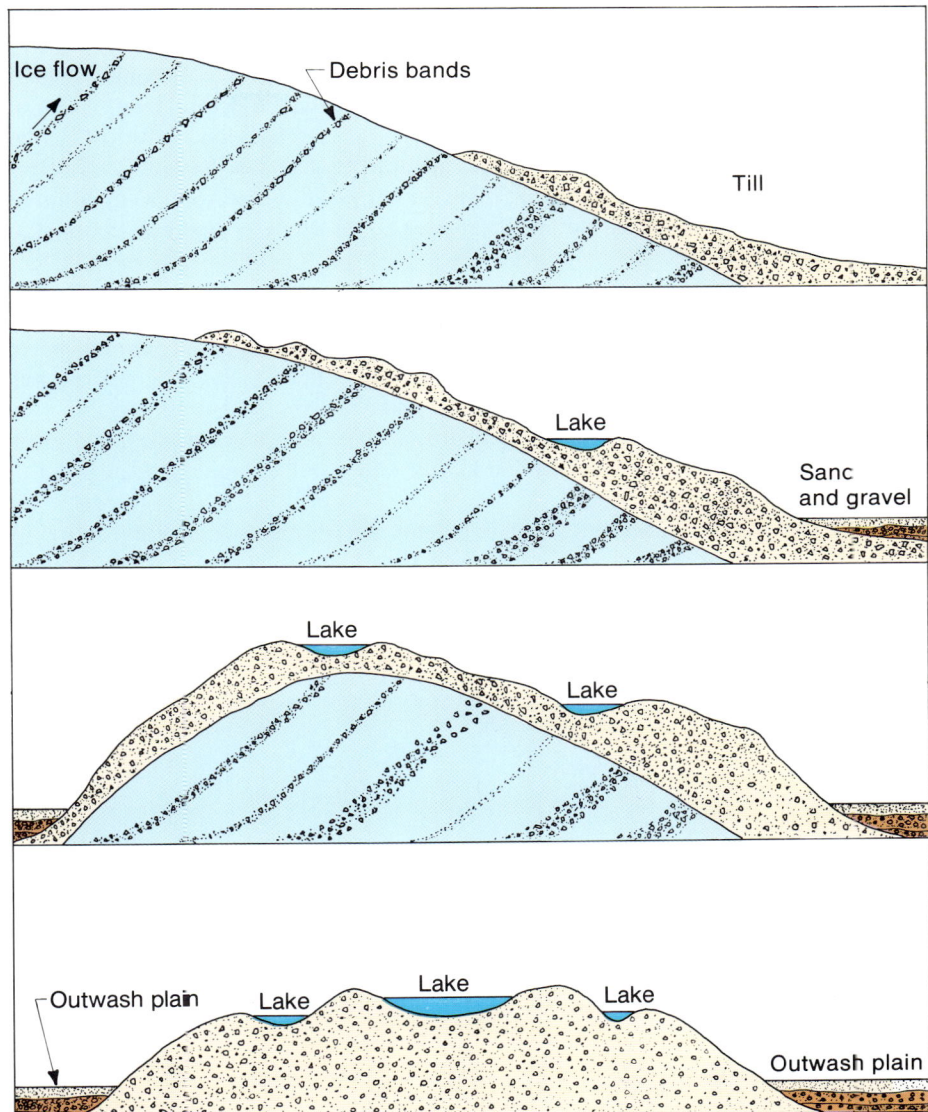

Figure 3.13. Stages in the formation of an end moraine. When the rates of ice advance and melting reach equilibrium resulting in a stillstand of the ice front, the moving ice continues to carry debris to the front. A ridge of till is deposited along the front, some of which is carried away by meltwater, then sorted and redeposited as stratified outwash. When melting exceeds accumulation, the ice front melts back, exposing ground moraine previously deposited.

Moraine Formation

Rock debris carried well up in a glacier and not deposited as lodgement till is eventually transported to the glacier terminus. This heterogeneous material is carried in discrete debris bands above the base of the ice. When the very end of an ice sheet becomes stagnant and temperatures are sufficiently warm to melt the ice at approximately the same rate as it flows into the area, the active ice flowing toward the front is forced upward at angles of 45 to 90 degrees before melting (Figure 3.13). An ice-cored moraine forms when debris melting out of the ice (ablation till) begins to accumulate near the terminus (Figure 3.14). More ice eventually becomes stagnant because of the overlying debris, and the active ice is forced to retreat. Debris continues to melt out of the stagnant ice by undermelting and is thereby added to the till blanket. In time, the ice core melts completely, leaving only the glacial sediment on the former land surface. The width of a terminal moraine varies from 20 ft (6.1 m) to 20 mi (32 km) or more, depending on the length of time equilibrium existed between melting and the influx of ice to the moraines.

Depositional Features of Moraines

Unlike sediments deposited by other agents of erosion (wind and running water), glacier ice can entrain and deposit all sizes of sediment in a single land form; huge rocks may be mixed with clay or fine sand. Small lenses of sand and gravel occur frequently in moraines when the newly deposited ablation till is reworked by running water. These lenses are usually limited in size and occur irregularly throughout the moraine.

Thick layers of clay without large fragments are also found in many moraines. Because clay is by definition extremely well sorted, the presence of thick clay beds in a typically heterogeneous till matrix should be explained. During the downmelting of an ice-cored moraine, many topographic low spots develop. These are filled with water from time to time by superglacial streams (streams that flow on the ice surface)

Figure 3.14. Ablation till melts out at the surface of the ice near the glacier terminus. The till consists of all size materials from huge boulders to clay particles. Some of the till will be reworked by meltwater streams into sinuous sand lenses.

or meltwater running down from the surrounding slopes into the depressions. Lakes may exist for many years in these depressions if cracks do not occur in the ice underlying the lake bottoms. During the melting season, superglacial streams continually carry sediments to the lakes. The finer or clay-sized material is deposited in the offshore areas, and in time a thick clay bottom may form. Although sloughing or partial erosion of the clay layers may occur later in the melting processes affecting the moraine, the very massiveness of the clays suggests that some of them will survive intact and be interspersed with till or sand and gravel lenses in the moraine. Even though a drilling contractor may pass through 50 ft (15.2 m) or more of clay, additional sand or gravel deposits are likely to be found beneath the clay layer.

Outwash Deposits

During the time a glacier remains in contact with its moraine, meltwater rushing off the clean ice courses through the moraine and picks up sediment. Because the meltwater is free of sediment before reaching the moraine, the transporting capability of these waters is high. Furthermore, the gradient of meltwater rivers is quite steep, usually much steeper than in rivers on land adjacent to the terminus, thus giving them additional capacity to remove sediment from the moraines.

When climatic conditions are favorable for rapid melting, meltwaters entrain gravel-sized debris. Most of this gravel is carried as bedload, that is, the individual stones are transported along the bottoms of rivers and streams. Once the rivers reach the nonglaciated areas outside the terminus where the gradient is much lower, the rivers can no longer transport all the material picked up in the moraine and thus begin to deposit some of it in their channels. Deposition continues until the rivers have the capacity to carry all of the remaining load. Clay-sized materials are usually carried far downstream by the meltwater rivers and deposited away from the outwash aprons. In general, the average particle size in each sediment layer tends to diminish downstream from the moraine.

If temperature conditions remain favorable for melting for several days or weeks, a thick layer of gravel may be deposited along the river bottoms. Climatic conditions near ice fronts are generally quite variable, with cool and cloudy conditions often prevailing. When cooler conditions prevail, the meltwater discharge is drastically reduced, which decreases the average particle size the rivers can transport as bedload. Therefore, the next layer of sediment atop the gravel will be finer grained, for example, medium-grained sand.

In time, stratified sand and gravel deposits called outwash build up in front of the moraine (Figure 3.15). Usually, the sediment aprons extend some 6 to 12 mi (9.7 to 19 km) outward from the terminus. If ice remains in contact with a moraine for hundreds of years, outwash deposits laid down by many rivers may form an extensive plain down-glacier from the moraine. Eventually, a long-term warming trend develops and the relatively clean ice up-glacier from the terminal moraine melts back rapidly from the stagnant, ice-cored moraine, thereby decreasing the rate of deposition on the outwash plain.

After the active ice melts back, moraines stand high over the surrounding landscape because of their massive ice cores. In Minnesota, for example, ice-cored moraines probably stood 650 to 1,000 ft (198 to 305 m) above the surrounding landscape at the end of Pleistocene time. Several thousand years may be required to completely

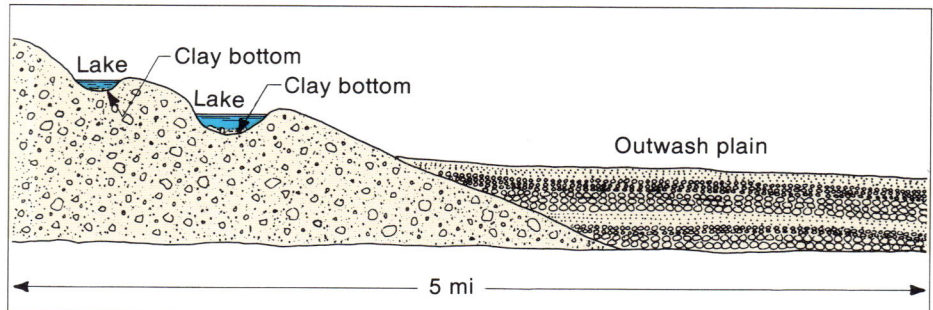

Figure 3.15. Cross section of an outwash plain formed down-ice from a moraine. The extensiveness of the moraine and size of the outwash plain suggest that active ice maintained contact with the moraine for hundreds of years.

melt the ice core because the debris cover on moraines prevents rapid melting. During this time, meltwater streams continue to flow off the moraines and less extensive, but nevertheless significant, secondary outwash plains capable of yielding high discharges to wells are built on the up-glacier sides of the moraines.

Outwash deposits and till sequences are also laid down between moraines by advancing or retreating ice fronts. During a long-term withdrawal, a glacier is periodically interrupted in its retreat by sudden surges of forward movement. Once the advancing ice reaches its new limit, another stillstand of the ice front may occur, and another moraine is created. Several repetitions of advance and retreat may occur before the ice withdraws at the conclusion of the glacial episode (Figure 3.16a). Thus, multiple moraines occur commonly in the terminal areas of many glaciers. Figure 3.16b shows a longitudinal cross section of a land surface covered by recent intermorainal glacial deposits.

Recognition of Moraines and Outwash Plains

Moraines deposited during the last 10,000 to 35,000 years are relatively easy to identify from topographic maps or by judicious use of a highway map showing lakes. Moraines are generally characterized by numerous abrupt changes in surface elevations over short horizontal distances. The characteristic hummocky or knolly topography develops by uneven deposition of the superglacial debris during melting of the ice core. If the climate is wet, moraines generally have numerous lakes and swamps which form in the abundant undrained depressions.

Outwash plains characteristically show only small differences in relief and are marked by shallow, well-rounded lakes. They are recognized easily in the field and on topographic maps by their essentially flat topography. When using a highway map, it must be assumed that the glaciers came from a northerly direction (in the Northern Hemisphere); therefore, the major outwash plains associated with particular moraines must be south of the moraines. Exceptions to this rule do occur, however, especially in glaciated regions near mountains.

Hydraulic Properties of Glacial Deposits

Two types of glacial sediments are generally recognized: till and outwash. Till has been subdivided into two types: lodgement till and ablation till. Lodgement till, often called hardpan by drillers, consists of glacial sediment deposited on the ground beneath the ice when the glacier temperature approaches 32°F (0°C). Ordinarily, individual layers of lodgement till are highly compressed, poorly drained, clay-rich sheets from 3 to 30 ft (0.9 to 9.1 m) thick; multiple ice advances may increase the total thickness to 300 ft (91.5 m) or more. Extensive till plains are found in Minnesota and Illinois, U.S.A., southern Ontario, Canada, and in large areas of northern Europe.

Ablation till consists of material released through surface melting at the glacier

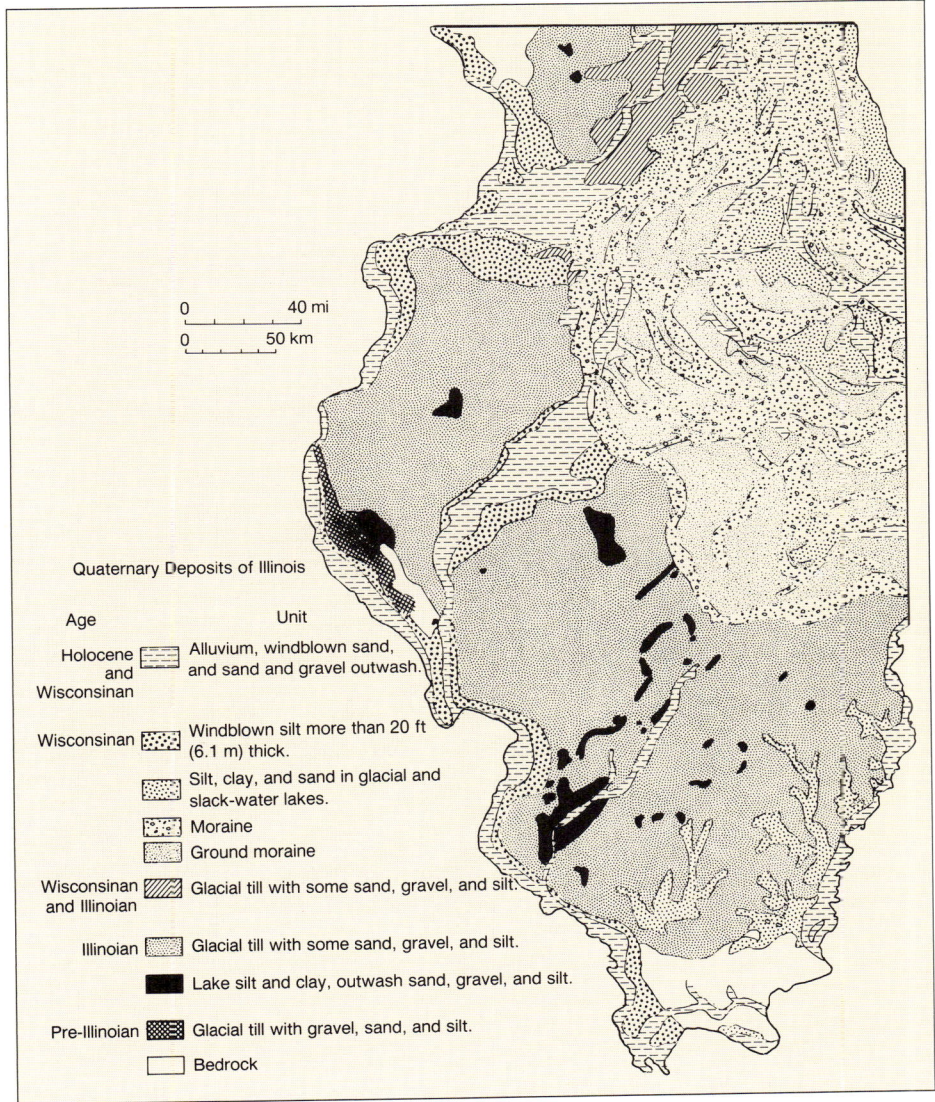

Figure 3.16a. Terminal and recessional moraines in Illinois. In areas where multiple advances of an ice sheet occurred over relatively flat land, tills are commonly separated by discontinuous sheets of outwash, lake bottom clays, or wind-deposited loess. In a typical advance, meltwater streams deposit an outwash apron in front of the advancing ice front. As the ice overrides the outwash, till is deposited. Eventually a recessional outwash is laid down on top of the till as the active ice begins to retreat. *(Lineback, 1981)*

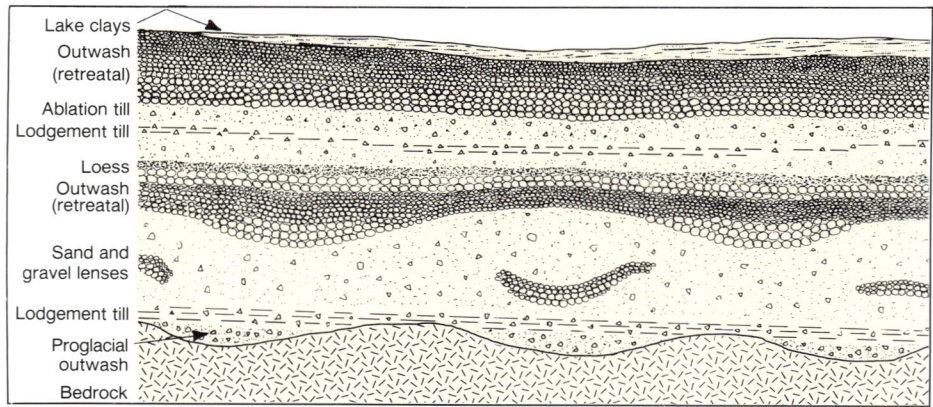

Figure 3.16b. Geologic cross section through an area separating two moraines.

terminus (Figure 3.14). Ablation till is loosely consolidated, clay poor, and contains all sizes of mostly angular to semirounded material. This type of till deposit forms huge curvilinear moraine complexes that are found in many northern areas of the United States and Europe.

Till, especially clay-rich till, has little pore space because abundant small particles fill in the voids between the larger grains. Thus, there is little storage area for significant volumes of groundwater in till. Water in substantial volumes can move easily only through the most sandy (ablation) tills or those that have developed joints. Well yields in excess of 15 gpm (81.8 m^3/day) are uncommon in till deposits.

Hydraulic conductivity in till deposits may be greater locally where small outwash lenses are present. Although these outwash bodies are usually discontinuous and therefore difficult to locate, useful volumes of water may exist in these lenses. In water well drilling, three difficulties arise: (1) the sand and gravel layers may be too thin to screen* effectively, (2) the lateral extent of the sand and gravel may be so limited that even relatively low yields cannot be sustained, and (3) recharge to the sand and gravel may be limited by the imperviousness of the enclosing till layers.

Outwash consists of well-stratified and well-sorted silts, sands, and gravels. Outwash sediments are more or less uniform in grain size and contain particles that are usually well rounded, loosely packed, and relatively uncemented; thus, porosities in outwash deposits are unusually high. Unlike wells in small outwash bodies in till deposits, wells in outwash plains can yield significant quantities of water because of their greater size. Yields will vary depending on the depth and areal extent of the sediments.

Two constructional features of outwash plains are significant. First, the thickness of outwash can vary as much as the relief in the former landscape. In some glaciated regions of North America, old preglacial valleys of the Mississippi, Minnesota, and Ohio Rivers are filled with 250 to 300 ft (76.2 to 91.5 m) of primarily outwash material, whereas the adjacent upland surfaces are covered with only 15 to 45 ft (4.6 to 13.7 m) of sediments. These preglacial valleys, called buried valleys, offer excellent sites for high-yield wells. Care should be used in extrapolating potential yields from one site to another in areas marked by buried valleys because of the variability in their physical extent over small distances. Methods to locate these buried valleys are dis-

---

*To install a filtering device that permits water to enter the well freely while excluding aquifer material.

cussed in Chapter 8.

Second, because of the manner in which meltwater streams deposit sediment, most outwash deposits that are not laid down in preexisting valleys have little uniformity in either vertical or horizontal directions. Clean, coarse sediments taken from a single exploratory well in an outwash plain do not guarantee high yields, because sediments may pinch out only a few feet away from the well bore. Sediments are much more uniform in a buried valley, and samples taken from a single well there are much more representative of true subsurface conditions.

Miscellaneous Glacial Deposits

Two other sediment types related to glacial activity are also potential reservoirs for groundwater. The first is loess, a nonstratified and unconsolidated sediment consisting mostly of silt-sized particles of quartz and feldspar. These particles are picked up by wind from outwash plains and deposited downwind as loess at varying distances from the ice fronts. Loess deposits occur extensively because of the abnormally high wind velocities associated with ice fronts, and the ready supply of silt-sized material continuously deposited by meltwater on the outwash plains. Outwash that is carried great distances downstream in rivers draining ice fronts also provides a source for loess. In the United States, some of the most extensive loess deposits are near the Mississippi, Missouri, and Illinois Rivers because these rivers carried vast amounts of sediment away from ice fronts during Pleistocene time. In Eastern Europe, thick loess blankets cover most of the countryside. These deposits are related to the outwash plains constructed in Pleistocene time between the Scandinavian and Alpine ice sheets. Dry climates in Eastern Europe also have favored the deposition of loess.

Loess found near its source areas may be 150 ft (45.7 m) or more thick, but the deposits thin rapidly in the downwind direction. Some loess blankets, however, may extend for hundreds of miles downwind from the source areas. Although the porosity of loess is quite high, the total surface area of the pores is so large that the hydraulic conductivity is limited. Permeability may be enhanced locally by former root holes which stand open because of the cohesive characteristics of loess. Yields from loess may be sufficient for domestic wells.

Valley-train deposits, consisting of coarse sand and gravel, are the second type of sediment indirectly associated with glacial activity. As indicated above, rivers draining the ice fronts carried huge quantities of outwash sand and gravel. At first, the local rivers were unable to carry all the sediment for any great distance, even during the high discharges of the summer season and despite the steeper gradients brought about by the reduction in sea level. In time, the river gradients near the ice fronts were steepened sufficiently by depositional processes for the rivers to carry the available load. Eventually, coarse sands and gravels were transported hundreds of miles downstream via major river systems; some of these sands were transported all the way to the Gulf of Mexico. Today, these valley-train deposits, although localized to a great extent in buried valleys and along existing rivers, represent significant sources for groundwater.

Yields from wells in outwash plain or valley-train deposits can be extraordinarily high, often 1,000 to 2,000 gpm (5,450 to 10,900 m³/day) or more. On the other hand, yields from till deposits and loess may be sufficient only for domestic wells. Every well drilled in glacial sediments has a unique stratigraphic record. Rarely will two neighboring wells penetrate exactly the same sediment at exactly the same depth.

IGNEOUS AND METAMORPHIC ROCK AQUIFERS

Plate tectonic theory suggests that the central or oldest part of every continent, called the shield, has a core of igneous and metamorphic rocks. About 20 percent of the Earth's land area consists of these rocks (Davis and DeWiest, 1966). In northern latitudes, these bedrock areas are covered by scattered and mostly thin glacial drift. Elsewhere, sinuous bands of alluvium occupy major river valleys. Locating adequate water supplies in these bedrock regions is extremely difficult because potentially good aquifers (glacial and alluvial aquifers) are widely scattered and the physical makeup of igneous and metamorphic rocks is generally unfavorable for storage or transmission of economically useful volumes of water. Nevertheless, much of the world's population lives on land consisting of these rock types, and whatever water supplies are available must be utilized.

The major intrusive igneous rock types are granite, diorite, and gabbro. As originally formed, these rocks do not have the necessary hydraulic characteristics required for adequate water supplies. They have a solid structure which precludes both significant water storage and transmission. Postemplacement structural and metamorphic changes, however, can produce significant alterations in these rock masses, thereby enhancing their potential usefulness for water supplies. The principal metamorphic rocks of the continental cores are slate, schist, and gneiss.

Extrusive igneous rocks, on the other hand, commonly possess physical features that can provide reasonable-to-large volumes of water to wells. Common extrusive rock types include basalt, andesite, rhyolite, and loosely consolidated volcanic deposits. Unfortunately, extrusive rocks constitute only small portions of the igneous and metamorphic rock areas.

Important changes affecting intrusive igneous and metamorphic rocks include jointing, fracturing, weathering, and solution. Most intrusive igneous rocks, formed during the subduction process, cooled slowly at great depths and under high pressures. These rock bodies, called plutons, crack or fracture during cooling. Over hundreds of millions of years, the overlying rocks are slowly removed by running water, glacier ice, and wind. As overburden is stripped away, the plutons tend to expand upward and additional cracks or fractures form that are more or less parallel to the erosional surface. These are called sheet-type joint systems.

Removal of many miles of overburden also causes continental rocks to float higher on the mantle. Therefore, adjustments take place within the pluton that produce vertical fractures or faults. Unlike joints, along which little or no movement occurs, large vertical displacements can take place along fractures that produce significant widening of the cracks. Rubble zones consisting of broken rocks may develop in fracture zones where they are wide enough and where vertical movement has been significant.

As the position of a pluton changes relative to the surface, other important physical and chemical changes take place. Because plutons are formed under high pressures and temperatures, they are not in chemical and physical equilibrium with conditions at the Earth's surface and thus begin to break down. Given enough time, weathering may alter the plutonic rocks to depths of 300 ft (91.5 m) or more. The massive structure of the original rock is destroyed in the weathering process as less resistant minerals are altered or removed in solution. In the weathered zone, the altered and somewhat more resistant minerals remain to form weak and crumbly disaggregated masses that have hydraulic characteristics drastically different from those of the original rock. Porosity, for example, may increase from virtually nothing in the original rock to 10 to 35 percent in the weathered zone.

The actual depth of the weathered zone depends on the length of time the rock has been exposed to surface or near-surface conditions and its original mineral composition. For example, granite is more resistant to weathering than is basalt because granite forms at lower temperatures and thus is less out of chemical and physical equilibrium when it becomes exposed at the Earth's surface. Drillers should remember that relatively recent glacial events have covered weathered igneous and metamorphic rocks in many areas with a mantle of glacial till and outwash. In western and central Minnesota, for example, recent glacial deposits [30 to 450 ft thick (9.1 to 137 m)] have covered older rocks that are weathered to depths of 450 ft. or more. Usually, however, the weathered zone is not more than 300 ft (91.5 m) thick.

In general, any rock underlying recent glacial drift has a weathered zone*. This zone is thickest in the low areas of the preglacial landscape and thinnest on the high areas. In this zone, drillers can expect to encounter marl (calcium-rich clay), iron-rich crusts representing old soil zones, and weakened quartzite, sandstone, and marble with numerous small-to-large cavities that may be arranged along old depositional surfaces in the original rock. In almost every case, a mushy clay zone of varying thickness exists between the glacial deposits and the more competent underlying weathered bedrock.

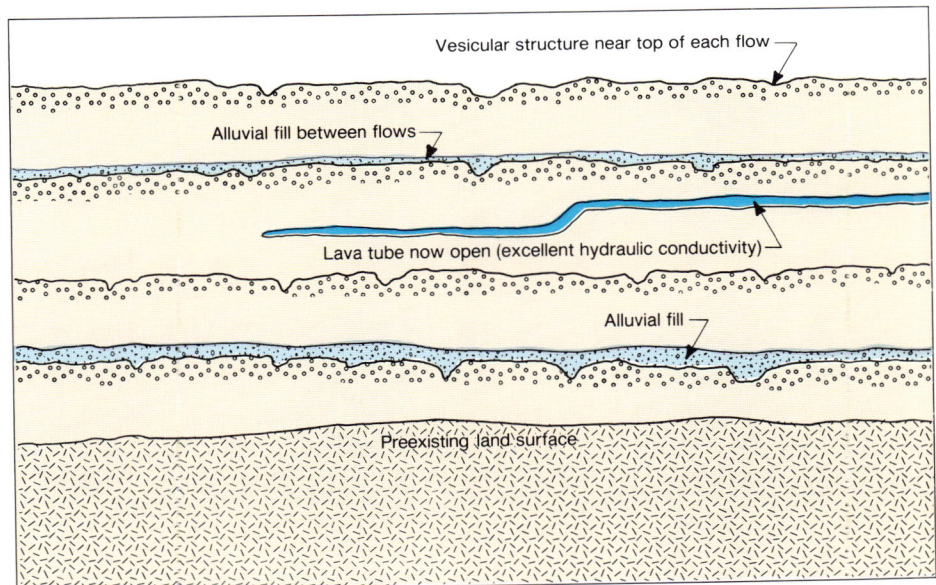

Figure 3.17. Basaltic lava flows possessing a vesicular structure can be highly permeable. Any alluvial deposits laid down between flows can also store and transmit significant volumes of water.

---

*In some areas the glacier ice may have scoured the rock so deeply that the weathered zone has been completely removed. These zones are usually 30 to 125 mi (48 to 201 km) up-glacier from the terminus, however.

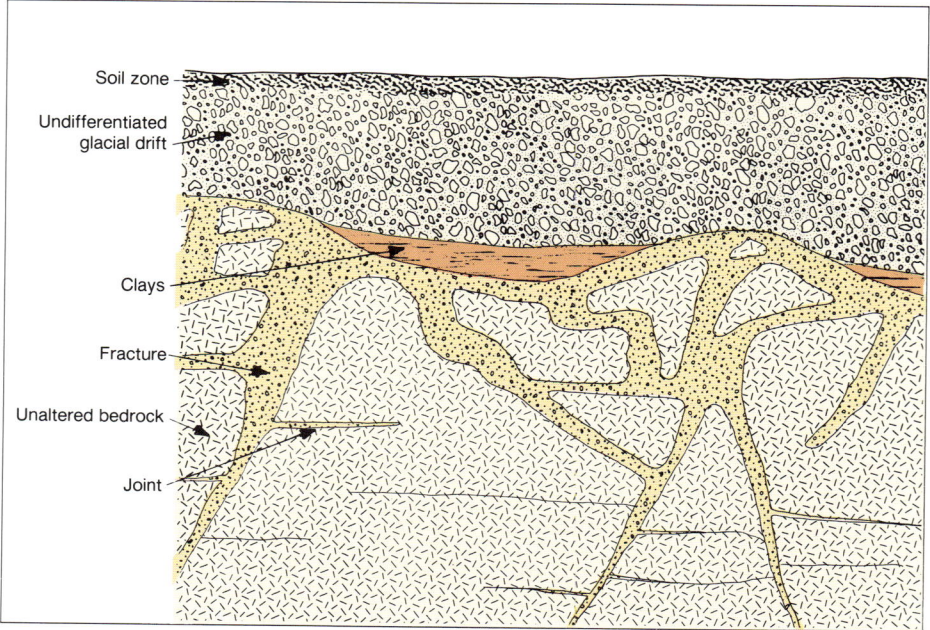

Figure 3.18. Stratigraphic relationship of deeply weathered igneous or metamorphic rock overlain by glacial drift. Clays are weathered from bedrock and collect on the surface as residual products.

Massive basalt flows (extrusive rocks) originate in the interior of some continents as a result of rifting events or so-called hot spots in the upper mantle. Basalt may have high porosity and hydraulic conductivity, depending on the way individual flows cooled or the length of time between flows. The openings in a sequence of basalt flows occur in several ways: as cracks formed during cooling where the hardened crust of a basalt flow has collapsed, as vesicles (gas holes) near the top of each flow, in empty lava tubes, and in alluvial sediments laid down between flows. Vesicles near the top of individual lava flows are a major cause of high porosity and hydraulic conductivity. These voids are produced when a crust forms on top of a cooling flow and traps rising gas bubbles. Each flow is marked by a horizontal zone of holes beneath the upper 1 to 3 ft (0.3 to 0.9 m) of crust. These voids provide excellent porosity and, if interconnected, extraordinarily high hydraulic conductivity (Figure 3.17). If substantial time intervals exist between lava flows, weathering, erosion, and deposition are likely to produce alluvial deposits between flows. Although these layers may be relatively thin, they can enhance the capacity of a basalt to store and transmit water. The presence of lava tubes can also increase the storage and transmission capabilities of basalt.

Figure 3.18 shows a typical igneous or metamorphic rock mass overlain by glacial deposits. Over long intervals of geologic time, the hydraulic capability of the rock improves so that it can store and transmit groundwater. Significant volumes of water can be found in fractures and joint systems and in the weathered zone. As stated above, the porosity in the weathered zone may reach 35 percent. However, hydraulic conductivity may vary drastically in rocks with fractures and joint systems, depending on the openness of these separations. For example, the yield may be quite high when

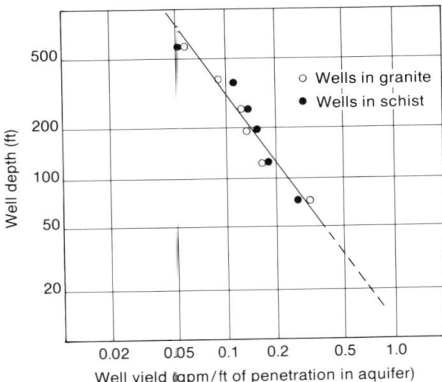

Figure 3.19. Well yields from crystalline rocks of the eastern United States. Wells reaching depths greater than 700 to 1,000 ft (213 to 305 m) in crystalline igneous and metamorphic rocks are not generally practical. *(Davis and Turk, 1964)*

a well intercepts a large fracture; other wells intercepting only tight joints that are poorly connected to fractures may be virtually dry. Methods of increasing yields from these wells using blasting techniques are discussed in Chapter 15. The driller should be aware, however, that as the depth of a well increases, fracture and joint systems become tighter, and it is therefore less likely that adequate water sources will be encountered. It is probably uneconomical to drill more than 700 to 1,000 ft (213 to 305 m) in any igneous or metamorphic rock terrain (Figure 3.19). Either blasting or hydrofracturing the hole or drilling another well at a new site should be considered.

Hydraulic Properties of Igneous and Metamorphic Rock Aquifers

Yields from bedrock wells generally vary from 1 to 50 gpm (5.5 to 273 m^3/day). The yield may be so low in some cases that the water standing in a well before pumping represents the total volume that can be pumped at any time. Nevertheless, this volume may be sufficient for most domestic purposes. Unusually large yields may occur in carbonate rocks where solution by rainwater has produced secondary porosity, but these yields are exceptional and are found rarely in the shield areas of the continents.

Wells in basalt have a much greater yield potential than do those in intrusive rocks. Although basalts exist on many continents, the ability of these rocks to serve as aquifers varies significantly. For example, incomplete rifting 1 billion years ago in the central United States produced a massive belt of basalt extending from the Lake Superior basin to Kansas. Because these rocks lie well beneath the surface over much of their length and contain relatively few broken or highly vesicular zones and almost no alluvial deposits, this basalt belt is an exceptionally poor reservoir for groundwater. On the other hand, the basalt of the Snake River Group in Idaho, Washington, and Oregon is one of the most permeable aquifers known; transmissivities* as high as 15 million gpd/ft (186,000 m^2/day) have been determined (McGuinness, 1963).

Andesite also tends to form flows, whereas rhyolitic material commonly explodes from volcanic cones and produces pyroclastic deposits such as tuff, pumice, ash, scoria, and breccia rather than distinct rock units. Pyroclastic deposits provide reservoirs for groundwater, although the hydraulic conductivity may be limited except where joints or fractures have developed. Porosity of these rocks is generally high, depending on fragment size, sorting, and subsequent degree of cementation.

In summary, the success rate for wells in igneous and metamorphic terrains is much lower than in the geologic settings discussed earlier in this chapter. Even when elaborate exploratory methods are employed, the yields from wells may be disappointingly small, and dry holes are common. Sophisticated aquifer stimulation methods, described in Chapter 15, may increase the yield significantly in some cases.

---

*See Chapter 5 for a complete discussion of transmissivity.

CHAPTER 4
Weather Patterns and the Hydrologic Cycle

The presence of water in underground reservoirs depends not only on the creation of the storage facilities, but also on nature's ability to keep them supplied. The reasons for local abundance or scarcity of water should be important to anyone involved in the groundwater industry. This chapter describes briefly the reasons for uneven distribution of precipitation over the Earth's surface and the processes involved in the movement of water from place to place.

The sun's energy is concentrated in equatorial zones between lat 35° N and lat 35° S. In this region, the sun's rays can strike the Earth most directly. Therefore, a net heat gain occurs here, whereas in polar regions there is a net heat loss. Although the Earth's orientation relative to the sun changes throughout the year, the Earth's long-term thermal environment remains relatively constant. In other words, incoming solar radiation equals the heat radiated by Earth into space.

Because more solar radiation is received near the equator, some of the surplus heat is transferred toward the poles by fluid movements of the oceans and atmosphere. Ocean currents carry away about 20 percent of the surplus heat as sensible heat* and the atmosphere transports the rest as latent heat. Thus, solar radiation provides the energy to move water, and water itself serves as a storage medium for thermal energy. Together they determine the three basic parameters of climate: air temperature, air pressure, and precipitation.

The actual volume of water in air is small, ranging generally from about 0.02 percent in desert regions to about 4 percent in humid areas. Of the total atmospheric pressure of 14.7 psi [(1,013 millibars (mb)]† at sea level, water vapor contributes about 0.15 psi (10 mb), whereas nitrogen contributes about 11.03 psi (760 mb) and oxygen about 3.48 psi (240 mb). Although the moisture content of air is seemingly insignificant, it

---

*Sensible heat is heat that can be felt.

†In the International System of Units, atmospheric pressure at sea level equals 101,300 pascals. This pressure is equivalent to one bar (1,000 mb) or 14.7 psi. One pascal equals pressure in mb times 100. By convention, millibars are used instead of pascals when referring to atmospheric pressures. Meteorologists generally use mb to quantify atmospheric pressure.

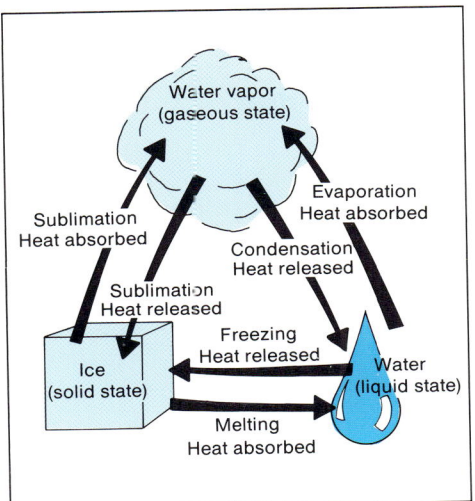

Figure 4.1. Three physical states (phases) of water.

is vital in influencing atmospheric conditions.

Water vapor affects weather and climate in three ways:

1. It is the source of all condensing water (rain, snow, hail, and frost). Water vapor is the only gas in the atmosphere that condenses under normal atmospheric temperatures.

2. It is the most important gas in the atmosphere for absorbing both shortwave solar radiation and long-wave terrestrial radiation (radiation by Earth). Thus, water vapor is critical in regulating air temperatures near the Earth's surface.

3. It is a source of latent or stored energy. This is the energy that drives the movement of air masses over the Earth. Latent heat is the heat energy gained and stored by water molecules as they change state (ice to water, water to vapor). Tremendous amounts of heat must be supplied to water molecules to induce them to change state. For example, 80 calories (335 joules)* of heat are required to change one gram of ice at 32°F (0°C) to water at a temperature of 32°F. An additional 100 calories (419 joules) of heat are required to raise the water temperature to near its boiling point of 212°F (100°C). But to actually vaporize one gram of water, 540 calories (2,260 joules) of heat are required. This is why steam is so much more dangerous than boiling water — it contains 5½ times as much heat! (Figure 4.1)

In changing from ice to water to vapor, a gram of water requires about 720 calories (3,010 joules). The heat-storage capacity of water is the highest of any liquid, with the exception of liquid ammonia. Water stored in the air as vapor contains huge amounts of energy that are ready to be released when environmental conditions change.

Air laden with moisture is lighter than dry air. At first this may not seem correct, but consider that a water molecule (H_2O) has an atomic weight of 18 (hydrogen = 1, oxygen = 16). A nitrogen molecule (N_2), the principal atmospheric constituent, has an atomic weight of 28. Oxygen (O_2) the other principal constituent, has an atomic weight of 32. Because these molecules are about the same size, a water molecule substituting for either a nitrogen or oxygen molecule will reduce the weight of a unit volume of air, thereby decreasing the atmospheric pressure. Thus, air laden with moisture is lighter than dry air.

Winds blowing over the Earth's surface reflect pressure differences created by the great movements of air associated with heat redistribution from the equatorial zones. What happens to air masses near the equator? A casual examination of the globe shows that about three-fourths of the area near the equator is covered by water; high insolation (incoming solar radiation) in this zone produces high evaporation rates. Continual heating causes the moist air to expand, become less dense, and eventually

---

*1 calorie (4.187 joules) is the heat required to raise 1 gram of water 1°C at about 15°C.

rise. As air rises, it encounters lower pressures; expansion of the air mass occurs, resulting in lower temperatures. At some point, the air becomes oversaturated with water (warm air holds more moisture than does cold air) and precipitation occurs. Because of the high insolation rates, land areas near the equator receive from 60 in (1,520 mm) to over 80 in (2,030 mm) of rain each year.

Equatorial air continues to rise and ultimately flows both north and south at high altitudes. As moisture is shed and temperatures drop, the air becomes denser. Near lat 30° N and lat 30° S, it reaches sufficient density to displace the air beneath it and gradually settles to Earth. The air mass heats up as it descends, and the little moisture it retains is securely held as vapor. Therefore, precipitation in regions near lat 30° N and lat 30° S is severely limited. The great deserts of the world are in these zones and are particularly widespread in the Northern Hemisphere, where most of the world's land areas exist (Figure 4.2).

Air masses descending near lat 30° N are deflected near the Earth's surface in two directions — toward the equator and toward the North Pole (Figure 4.3). As the air masses move in both directions over the Earth's surface, they evaporate moisture from surface waters and gain moisture from plant transpiration. The air mass moving toward the equator becomes less dense and begins to rise, with most of the ascension taking place in the equatorial zones. This great circulation cell is driven by pressure differences that arise from variations in moisture content. The air mass moving toward the North Pole also continually evaporates moisture until it approaches lat 65° N, where the air mass is so light that it begins to rise. Precipitation in this zone is potentially high, although other geographic factors may reduce it, especially in parts

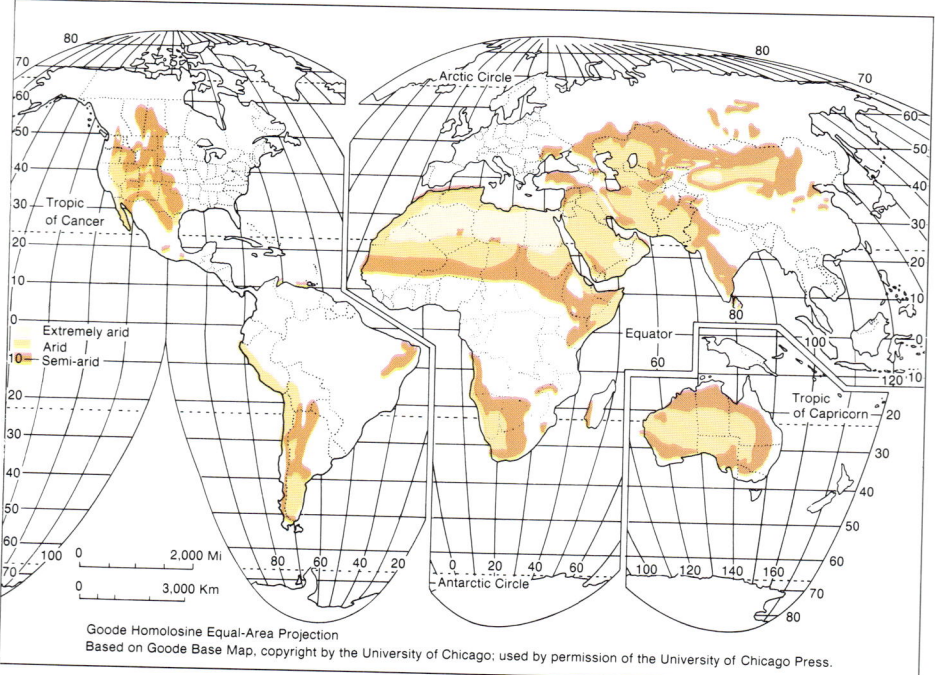

Figure 4.2. The great deserts of the world are in regions near lat 30° N and lat 30° S. *(Snead, 1980)*

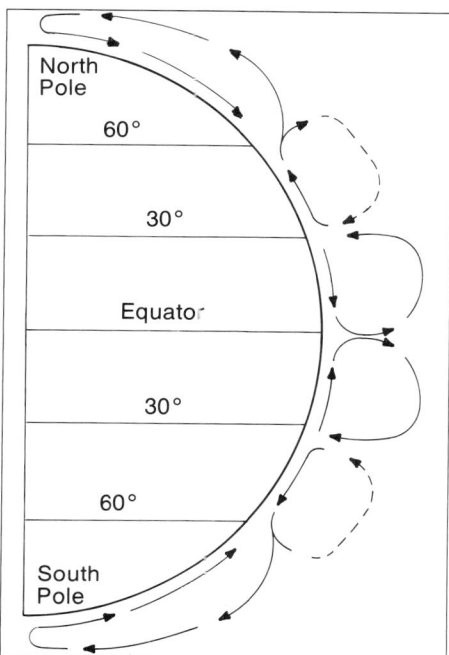

Figure 4.3. Generalized model of atmospheric circulation in which solar energy received near the equator is distributed toward the poles. The movement of these air masses determines zones of precipitation.

of North America and Eurasia. Part of the rising air mass returns to lat 30° N and the remainder continues on to the pole.

CORIOLIS EFFECT

Precipitation patterns determined by the movement of large air masses are affected greatly by the Coriolis effect. The Coriolis effect is the tendency for air masses, water, or other materials in motion and not directly attached to the Earth's surface to be deflected because of the Earth's rotation. The Coriolis effect is the resolution of Earth's rotational force and the pressure differences between air masses. If the Earth did not rotate, winds and ocean currents would follow the directions created by heat in air masses and heat and salinity differences in ocean currents. Deflections of large air masses occur, however, in response to the angular momentum imparted by the Earth's rotation, although the Earth itself has little direct effect on the air masses. Figure 4.4 shows the general path taken by large air masses as they move across the Earth's surface. Seasonally, the air masses migrate somewhat north and south of the locations shown in Figure 4.4. Wind deflections caused by the Coriolis effect are an important factor in determining precipitation patterns on land masses and ultimately the supply of groundwater.

HOW PRECIPITATION OCCURS

Precipitation can occur in many forms, including sleet, snow, dew, hoarfrost, fogdrip and rime, as well as rain. Snow and rain are dominant and, therefore, the processes that produce these forms of precipitation are the most important. Moisture-laden air must be lifted and cooled by some mechanism to induce any significant amount of precipitation. Hence, before discussing precipitation processes, the temperature regime of the lower part of the Earth's atmosphere must be examined because it plays a vital role in triggering these lifting and cooling mechanisms.

There are several concentric atmospheric layers surrounding the Earth (Figure 4.5). The lowest layer, called the troposphere, influences man's activities because all weather phenomena affecting the Earth's surface occur in this layer. The troposphere varies in thickness from about 4 mi (6.4 km) near the poles to about 12 mi (19.3 km) near the equator.

Air cools with increasing height throughout the troposphere. The rate at which it cools is called the lapse rate and equals about 3.6°F per 1,000 ft (6.5°C per km). For example, if the temperature is 68°F (20°C) at the Earth's surface near lat 45° N, the air temperature at the top of the troposphere [8 mi (12.9 km)] would be about -83°F (-63.9°C).

A temperature inversion occurs near the top of the troposphere, after which temperatures increase for about 20 mi (32.2 km). Thus, rising air masses reach an effective ceiling near the top of the troposphere, confining all weather phenomena to this zone. Recognizing this fact, commercial airliners fly somewhat above the top of the troposphere to avoid weather problems.

Another temperature mechanism affecting precipitation arises from adiabatic temperature changes. Because air is a relatively poor heat conductor, cooling in rising air masses is independent of the ambient temperature and pressure conditions surrounding the air masses. Adiabatic temperature changes are produced by changes in pressure and volume that occur within an air mass as it rises or falls; there is no heat transfer between the air parcel and the surrounding air. For example, a decrease in pressure (which happens when air masses rise) causes an increase in volume and a decrease in temperature.

There are two adiabatic cooling rates: dry adiabatic and wet adiabatic. The dry adiabatic lapse rate (when no condensation occurs) is 5.4°F per 1,000 ft (9.8°C per km), which is well above the usual tropospheric lapse rate. When condensation occurs in a rising air mass, the latent heat of vaporization is released, resulting in a wet adiabatic lapse rate. The wet adiabatic rate averages about 3.3°F per 1,000 ft (6°C per km), somewhat lower than the tropospheric lapse rate. The reason the wet adiabatic rate is much lower than the dry adiabatic rate is because the latent heat contained in the water vapor is released when the vapor condenses as rain, adding heat, and thereby lowering the rate of cooling to 3.3°F per 1,000 ft. Recall that each gram of water must receive over 600 calories (2,510 joules) of heat to evaporate. This amount of latent heat is retained in the gram of water vapor until it condenses as rain. Thus, the cloud containing the water vapor will continue to cool but at a rate less than the tropospheric lapse rate. Because the cloud cools more slowly than the surrounding air, it will continue to rise until condensation ceases or the surrounding air mass becomes warmer. At this point, the cloud becomes denser than the surrounding air and it begins to descend.

All three lapse rates play a part in producing precipitation in rising air masses (Figure 4.6). In all cases, the air is uplifted and cooled, causing water vapor to condense

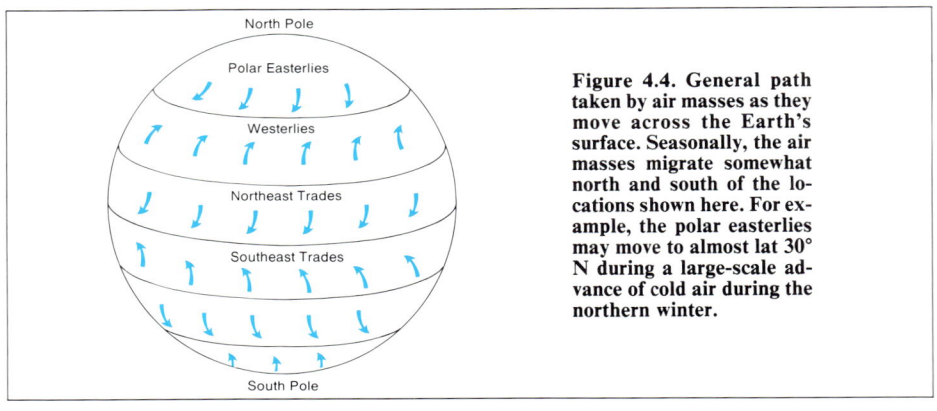

Figure 4.4. General path taken by air masses as they move across the Earth's surface. Seasonally, the air masses migrate somewhat north and south of the locations shown here. For example, the polar easterlies may move to almost lat 30° N during a large-scale advance of cold air during the northern winter.

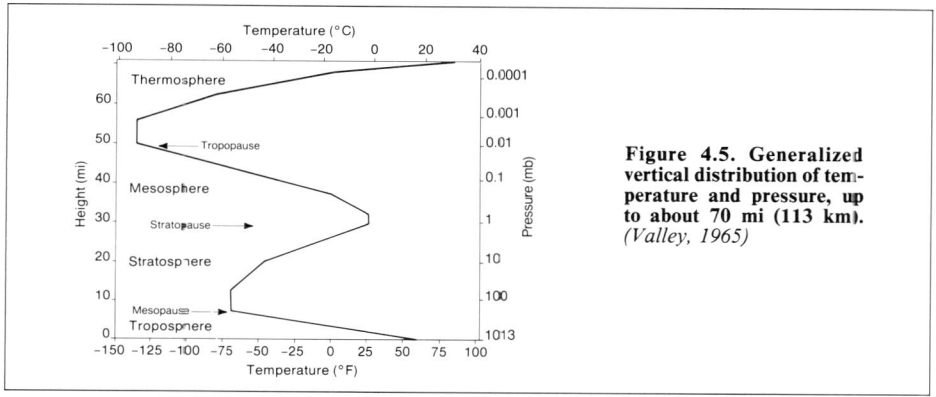

Figure 4.5. Generalized vertical distribution of temperature and pressure, up to about 70 mi (113 km). (Valley, 1965)

on salt or clay particles (dust) in the air. The amount and areal distribution of rainfall depends on the uplift mechanism involved. On a global basis, rainfall patterns generally can be predicted on the basis of physical principles governing air masses and tropospheric temperature conditions.

CAUSES OF PRECIPITATION

Convection is one of the three common mechanisms that cause precipitation. Heating of the ground surface causes the overlying air to become warmer and less dense than the surrounding air. The warmed air parcel begins to rise and at some point condensation takes place. As long as condensation is occurring, the temperature in the air parcel is governed by the wet adiabatic lapse rate, and the temperature within the parcel can remain somewhat above that of the surrounding air. So much heat is released during rapid condensation that air parcels can move upward at speeds of 100 ft (30.5 m) per second or more. The cloud formations associated with convection cells are the towering cumulus and cumulonimbus types. Often these have flat tops, showing that they are near the upper boundary of the troposphere and the beginning of the temperature inversion. Winds in the inversion zone tend to elongate the tops of the clouds. Rainfall from convection cells is spotty, often falling in bands parallel to weather fronts. Rainstorms of this type tend to last from 30 to 60 minutes and are occasionally violent.

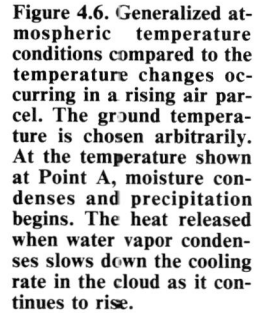

Figure 4.6. Generalized atmospheric temperature conditions compared to the temperature changes occurring in a rising air parcel. The ground temperature is chosen arbitrarily. At the temperature shown at Point A, moisture condenses and precipitation begins. The heat released when water vapor condenses slows down the cooling rate in the cloud as it continues to rise.

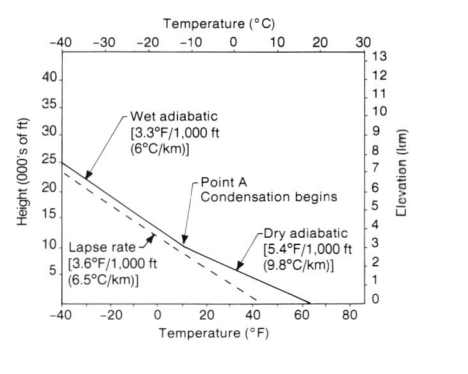

A second mechanism causing precipitation is cyclonic wedging. When high- and low-pressure air masses collide, the warm, moist (lighter) air of the low-pressure system is forced up over the dry, cooler air (Figure 4.7). As a consequence, precipitation occurs along a broad front between the two systems; rain falls over extensive areas and lasts for 6 to 12 hours. This type of rainfall is particularly effective in recharging groundwater aquifers. In the Northern Hemisphere, most storms of this type occur in the zone from lat 40° N to lat 65° N, and generally move from west to east.

Precipitation can also be caused by orographic effects. In many parts of the world, high mountains lie along coastlines and in other places where they intercept moist air. If the predominant wind direction is toward these mountains, air masses saturated with water vapor are constantly being forced to rise over the mountains. When this happens, the inevitable cooling causes condensation (Figure 4.8).

Rainfall on the windward slopes of mountains not only tends to be heavier but also occurs more frequently. Also, convection and cyclonic types of precipitation are more efficient in mountainous regions where moist air is available. Desert conditions may exist on the leeward side of the mountains, if they are high enough, because as the air descends it warms up and can hold more moisture. Not until the air again absorbs sufficient moisture will precipitation occur. A good example of this situation is in the western part of North America where great arid and semiarid regions lie east of mountains extending from Mexico to Alaska.

At any given time and place, the atmosphere can contain only about 1 in (25.4 mm) of water. Heavier precipitation in any area is caused by the movement of air masses containing moisture into that area during storms. A record rainfall of 75 in (1,910 mm) occurred on Reunion Island, near Madagascar, during a 24-hour period in March 1952.

EFFECT OF PRECIPITATION ON GROUNDWATER

Precipitation patterns are the key in determining whether groundwater resources

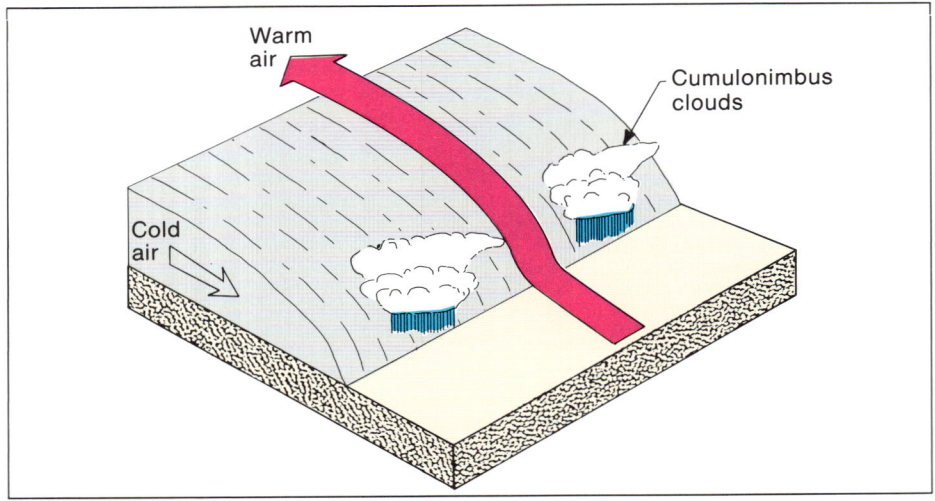

Figure 4.7. Precipitation occurs along a front created by colliding cold, dry air and warm, moist air.

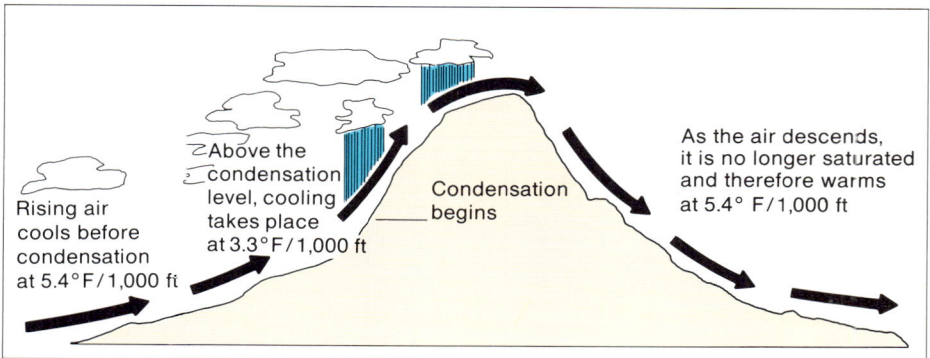

Figure 4.8. Precipitation caused by orographic effects. As air masses rise on the windward slopes of mountains, the air is cooled and precipitation occurs. Dry conditions prevail on the leeward side of the mountains because the air is warmed as it descends, thereby increasing its ability to hold moisture.

can offer dependable long-term supplies of water. Although potential aquifers are found in many parts of the world, recharge may not be sufficient to keep pace with demand. For example, critical groundwater situations exist in the rapidly growing cities of Phoenix and Tucson, Arizona. Precipitation in this region is extremely light because of orographic effects and geographic position; also, heavy demands have been placed on groundwater because little surface water is present. With inadequate recharge, the groundwater table in southern Arizona has fallen approximately 400 ft (122 m) in the last 50 years. On the other hand, in parts of the Sahara Desert of northern Africa, large groundwater resources are still available. Should these resources be extracted, however, there is little opportunity for recharge because dry air now dominates the region between lat 20° N and lat 30° N.

With few exceptions, long-term utilization of groundwater supplies must be governed by local and regional recharge rates. Water shortages exist occasionally in areas where abundant rainfall is normally found. During these periods of temporary shortage, overly stringent regulations may be enacted to limit long-term use of groundwater. In terms of the usual groundwater abundance, these regulations are not justified. For example, a groundwater table may fall 7 to 10 ft (2.1 to 3 m) or more during periods of drought. Such situations should not be worrisome, however, because immediately following the drought years recharge to the underground is substantial and recovery of the water table is rapid. A rapid recovery of the groundwater aquifers in California occurred following the severe three-year drought of 1975 to 1977. Within a year, most groundwater tables had recovered to former levels. On the other hand, some regulations may be required in areas where groundwater withdrawals consistently exceed recharge to the aquifers. Any regulations should be consistent with regional long-term precipitation trends.

In areas of limited precipitation and recharge, care should be taken in determining proper withdrawal rates. Every water-well contractor and well-design engineer should become familiar with methods that accurately predict long-term well yields. Some of these methods are discussed in Chapters 9 and 16.

HYDROLOGIC CYCLE

Terrestrial moisture is in constant motion, and all near-surface water participates in what is called the hydrologic cycle (Figure 4.9). The term "cycle" suggests that water comes from a single source and ultimately returns to that source. This is true, for once water has been released from rocks through plate processes it exists in greatest quantity within the oceans. Oceans thus serve as the primary reservoir for water participating in the hydrologic cycle. Water in the hydrologic cycle not only undergoes changes in its geographical location, but it also continually changes state. Water can exist as solid, liquid, or gas (vapor). During changes in physical state, huge amounts of energy are absorbed or released, and this energy drives all weather systems.

During evaporation, water molecules near a water surface acquire sufficient energy from solar radiation to vaporize. This occurs because added heat gives the molecules enough kinetic energy to overcome the surface tension of the liquid and change to a gaseous state. The rate of evaporation depends upon the difference between the volume of water vapor in the overlying air mass and the volume of vapor in the thin layer of air lying just above the water body and the rate of incoming solar radiation. Winds are also important because they continually cause unsaturated air to flow over the water surface, thereby increasing the evaporation rate.

Two other processes contribute to total evaporation: transpiration and sublimation. In transpiration, moisture given off by plants is returned to the atmosphere. In most parts of the world, transpiration and ordinary evaporation cannot be differentiated; thus, the loss of water from a land surface is called evapotranspiration. Sublimation occurs when ice and snow pass directly into the gaseous state without first becoming a liquid.

Figure 4.10 shows how much of the total precipitation supply undergoes movement

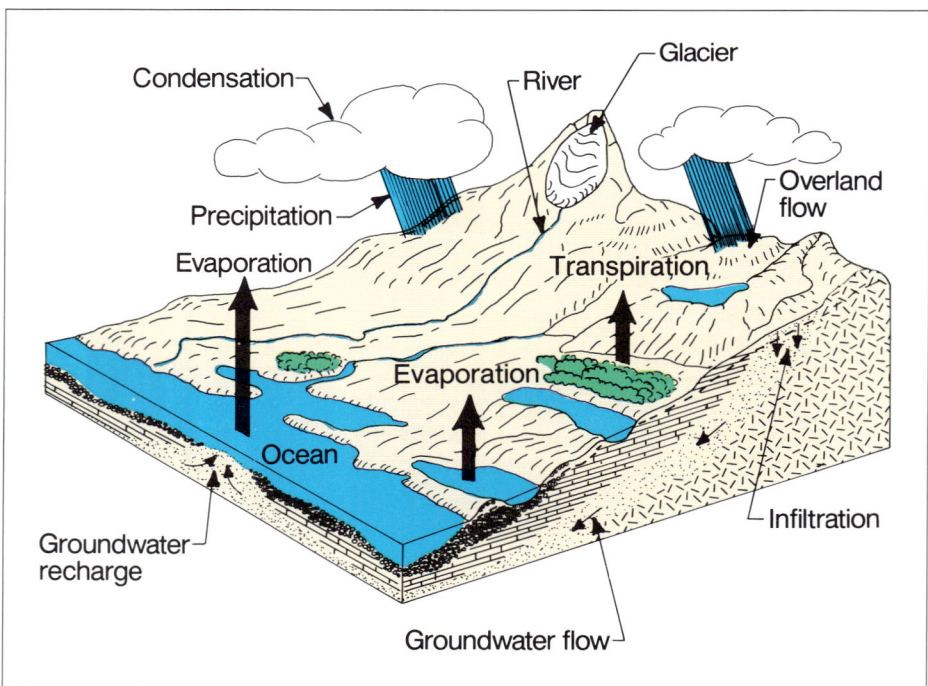

Figure 4.9. Hydrologic cycle.

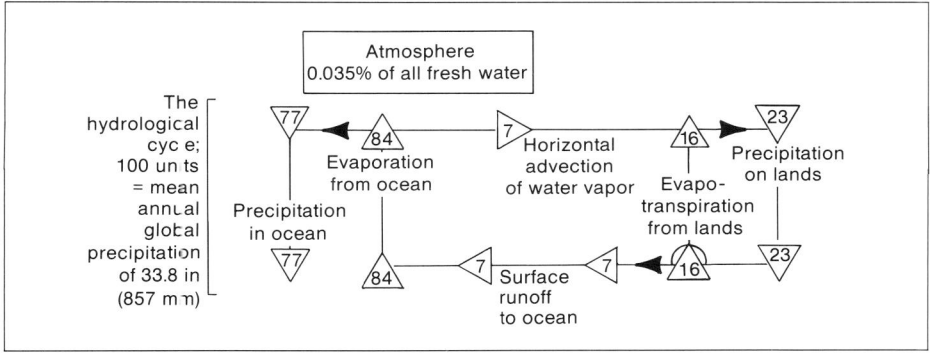

Figure 4.10. The hydrologic cycle and water storage of the globe. Exchanges in the cycle are referred to 100 units, which equal the mean annual global precipitation of 33.8 in (859 mm). *(More, 1967)*

in each part of the hydrologic cycle in a single year. The major reservoirs for the freshwater components of the hydrologic cycle are given in Table 4.1. Saltwater lake volumes are included for completeness.

Approximately 80,400 mi³ (335,000 km³) of water is evaporated annually from the ocean basins. Another 15,600 mi³ (65,000 km³) is evaporated from land areas, including lakes, rivers, and glaciers. To balance evaporation losses, precipitation must equal about 96,000 mi³ (400,000 km³) each year. Only about 24,000 mi³ (100,000 km³) falls on land. Some of this water is reevaporated but most of it returns rather quickly to the seas via rivers.

Evapotranspiration from land surfaces equals about 19 percent of the water evaporated from the world's oceans. Undoubtedly some of the precipitation falling on land undergoes evapotranspiration several times before it returns to the sea. In fact,

Table 4.1. Distribution of Water in the Conterminous United States

	Area (mi²)	Volume (mi³)	Annual circulation (million acre-ft per year)	Detention period (yr)
Frozen water:				
Glaciers	200	16	1.3	40
Liquid water:				
Fresh-water lakes*	61,000	4,500	150	100
Salt-water lakes	2,600	14	4.6	10
Average in stream channels		12	1,500	0.03
Groundwater:				
Shallow	3,000,000	15,000	250	200
Deep	3,000,000	15,000	5	10,000
Soil moisture				
(3-ft root zone)	3,000,000	150	2,500	0.2
Gaseous water:				
Atmosphere	3,000,000	45	5,000	0.03

*United States part of Great Lakes only.

(U.S. Geological Survey)

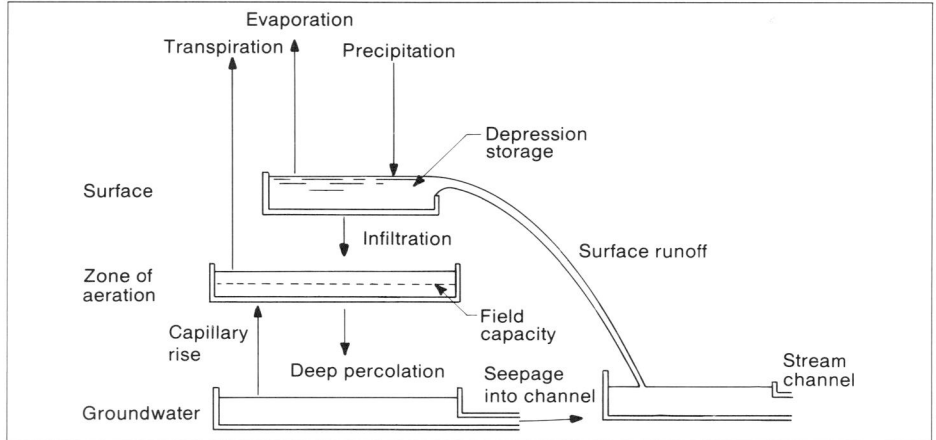

Figure 4.11. Watershed model showing how water moves in the near-surface environment during precipitation. *(D. R. Dowdy and T. O'Donnel, ASCE Journal of Hydraulics, Vol 91, Hy 4, 1965.)*

water may leave and reenter the atmosphere many times in one year. About 30 percent of the land precipitation does not become involved in the evapotranspiration cycle, but returns more or less directly to the sea.

The energy imparted to water as it moves through the hydrologic cycle is enormous. Recall that water absorbs energy through evaporative processes and releases it through condensation. A 1-in rainfall over 1 mi^2 produces about 72,000 tons of water*; the heat released in condensing this amount of water is roughly equivalent to that produced by burning 65,000 tons (59,000 metric tons) of high-grade coal. Everyone who works with water should recognize that enormous amounts of potential energy are present in atmospheric water.

PATHWAYS OF WATER AFTER IT FALLS TO EARTH

When precipitation falls to Earth as snow, rain, or hail, some part of it is intercepted by trees, plants, and buildings. This water does not reach the ground during brief or low-intensity storms but is rapidly evaporated. Thus, light rainfall is reevaporated into the atmosphere within a short time. This portion of the total rainfall is known as the interception loss.

During heavier precipitation, water does reach the ground and can follow several pathways (Figure 4.11). Almost immediately, some of it evaporates from the soil surface and returns to the atmosphere; another part enters the ground. If precipitation eventually exceeds the infiltration and evaporation rates, water will begin to collect on the surface. Water temporarily stored on the surface in low areas is called depression storage. If rainfall continues, overland flow commences as water in low-lying places begins to run together. Water soon collects into rills or small channels which flow into gullies leading to streams. The overland flow that enters streams is called surface runoff.

Water entering the ground may remain temporarily in the soil zone, it may flow laterally above the groundwater table until it reaches a stream or other low-lying body of water, or it may continue to infiltrate downward until it reaches the groundwater

---

*A 25-mm rainfall over 1 km^2 produces about 25,200 metric tons of water.

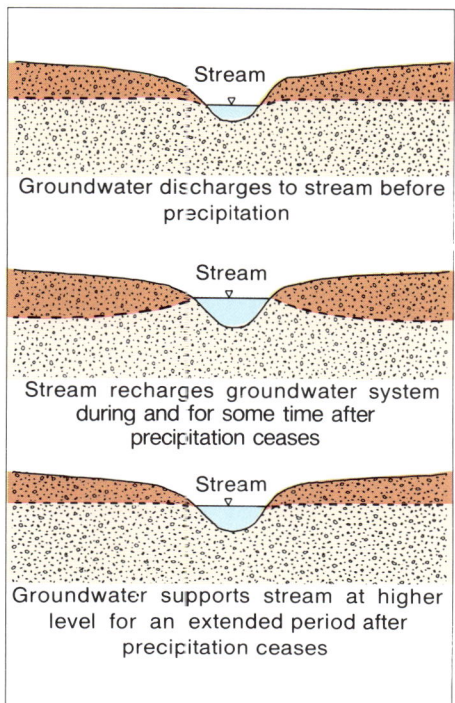

Figure 4.12. Between rainstorms, most streams are supported by the local groundwater table. During storms, overland flow raises the stream level above the groundwater table. In time, recharge from the stream raises the water table until a new water-level relationship is established with the stream. With no further precipitation, both the stream level and the groundwater table will gradually fall with the groundwater partially or completely supporting the stream.

table. Water remaining in the soil zone may be used by plants or evaporate directly. The maximum volume of water a soil zone can hold is called its field capacity; some fine-grained soils can hold more than 12 in (305 mm) of water in the soil zone, although the exact amount of water is difficult to determine.

Overland flow does not result from storms of short duration or low intensity. In more severe storms, however, overland flow will occur after some time. Interception, evaporation, and depression storage can account for all of the moisture falling during the early stages of a rainfall. But as interception becomes ineffective, the depressions are filled, and the infiltration capacity of the soil is exceeded, overland flow commences and the local streams begin to rise.

Even though overland flow carries water away from local areas, the water may not totally bypass the groundwater system. Ordinarily, most perennial streams and lakes are supported by the groundwater table. Many sediments near the shores of lakes and the banks of streams are highly permeable, and water can flow easily into the groundwater system once a stream or lake has temporarily risen above the groundwater table. Figure 4.12 demonstrates the hydraulic relationship between a stream and the groundwater system before, during, and after a storm of moderate intensity and duration. Inflow to a stream during a storm may consist not only of overland flow but also direct channel precipitation and interflow. Interflow is the water that moves toward a stream above the groundwater table, but underneath the soilwater zone. The stream recharges the groundwater during and some time after the storm.

The situation shown in Figure 4.12 suggests that a dynamic relationship exists between the groundwater system and streams. For much of the year the groundwater table may support the level of streams, but during periods of heavy precipitation or the spring melt in cold climates, streams may provide large volumes of surface water to recharge the underground. During dry seasons, some of this water will support the streams.

It might seem that lakes are quite influential in recharging groundwater levels because lakes exist throughout the year at relatively constant elevations. But studies have shown that the bottom sediments of most lakes are nearly impervious to water movement; clay sediments and organic muds produced in lakes effectively seal the

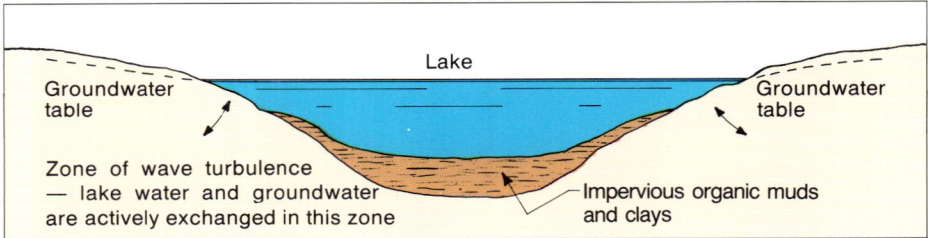

Figure 4.13. During periods of drought, the groundwater table and lake level fall. However, if the lake falls more than a few feet, the bottom muds will prevent lake water from entering the local groundwater system, even though the groundwater table continues to fall.

lake bottoms. Therefore, lake water actively interchanges with the groundwater system only in areas close to shore, or in other areas where sediments are removed by wave action, and at springs which may occur randomly in the lake bottoms. Except for springs, the water depths in these recharge-discharge zones are usually only a few feet deep. Thus, a lake may not support the local groundwater system during severe drought (Figure 4.13). Exceptions to this situation exist in newly formed lakes, karst regions, and relatively large, shallow lakes produced by glacial activity. In shallow glacial lakes, wave turbulence may prevent extensive organic mater from forming and also keep clay particles in suspension.

River stage and groundwater levels, on the other hand, are generally related closely and any study of a groundwater system usually includes a thorough analysis of the local stream hydrology. In fact, the relationship is so close that river levels in a community may be the governing factor in determining when groundwater withdrawals are reaching the critical stage. Because rivers are of many uses to man, they should not be lowered excessively through overuse of groundwater. Rivers offer means to transport goods; provide important recreational outlets; offer water resources for irrigation, industrial, and domestic uses; provide means to dispose of some of our wastes; and are also extremely important in an aesthetic sense. Therefore, reducing river levels to the point where some of these functions cannot be met may cause serious problems in community life.

In many areas, valuable surface water is lost each spring during periods of high runoff because the local groundwater table is relatively high and river recharge to the underground is only minimal. In each yearly cycle, more water could be retained in an area if the groundwater table were drawn down during the late summer, fall, and winter. Enhanced recharge during the following spring, from deep percolation and river recharge, would make up for withdrawals of the preceding year. Over the long term, the water balance in the region could be maintained, but greater volumes of water would be in use each year.

All facets of the hydrologic cycle are both fascinating and complicated. Each segment of the cycle is being studied by thousands of scientists throughout the world. This book must concentrate on only one aspect of the cycle, but sound engineering decisions involving groundwater require more than a cursory knowledge of other aspects of the hydrologic cycle. This is illustrated by the lake-stream-groundwater system discussed above. Geologic knowledge of aquifers and a thorough understanding of the physical and chemical impact of other segments of the hydrologic cycle on groundwater are mandatory for intelligent groundwater resource utilization.

CHAPTER 5
Occurrence and Movement of Groundwater

In earlier chapters several elements of physical science related to groundwater occurrence were examined. These include the way in which large-scale crustal events form potential groundwater reservoirs, how water originates on Earth, and the processes involved in precipitation and the reasons for its uneven distribution over the globe. Once water is in the ground, how does it behave and what physical constraints are imposed by the confining rock medium? Of equal importance is the chemical relationship of groundwater to its storage site and the effect each has on the other. Successful groundwater utilization depends on our ability to predict how water behaves in the ground, both physically and chemically. Groundwater hydrology deals largely with the unseen, and this is why the potential factors affecting groundwater occurrence and movement have been discussed in detail. In this chapter, important physical characteristics of groundwater are described quantitatively. Chemical characteristics of groundwater are reviewed in Chapter 6.

TYPES OF SUBSURFACE WATER

Regolith is a geologic term used to describe the loose and discontinuous blanket of decayed rock debris overlying solid bedrock. The term "soil" is sometimes used for this unconsolidated material, but soil is only the very uppermost part of the regolith where chemical and physical weathering are the most active. Herein, regolith includes the soil layer and the underlying loose material. In temperate regions, the regolith may be only 6 ft (1.8 m) deep, whereas in humid regions it may be 230 ft (70 m) or more deep. Regolith can act as a storage medium for water or it can transmit water vertically and horizontally to bedrock storage sites.

Water is introduced to the regolith by precipitation and streamflow; once in the regolith, water exists in several different environments that are illustrated in Figure 5.1. This classification, proposed by Davis and DeWiest (1966), suggests that water occurs generally in two types of environments in the regolith — a zone of vadose water and a zone of phreatic water. In the vadose zone, three separate types of water exist: soil water, intermediate vadose water, and capillary water.

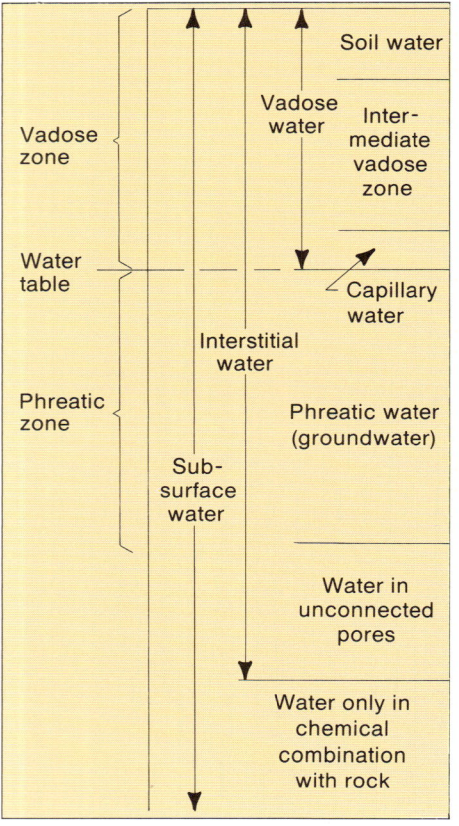

Figure 5.1. Classification of subsurface water. *(Davis & DeWiest, 1966)*

Soil water is particularly important to agriculture because it provides the water for plant growth. Water is lost from this zone by transpiration, evaporation, and percolation when oversaturation occurs. Soils undergo wide variations in moisture content, from complete saturation to total lack of moisture. The depth of the soil-water zone varies from 3 to 30 ft (0.9 to 9.1 m).

Water is held in soils by molecular attraction and capillarity acting against the force of gravity. Molecular attraction tends to hold water in a thin film on the surface of each soil particle. Capillarity holds water in the smallest spaces between soil particles. When the waterholding capacity of the capillary forces is exceeded, water begins to percolate downward under the force of gravity.

The region immediately below the soil-water zone is called the intermediate zone. Although most water in this zone is moving downward, some of it is retained, but no in-situ use for it exists and it cannot be recovered by man. In humid regions, this zone may be quite thin or even absent. Probably little water passes completely through the intermediate zone in dry regions and the little water that does reach the groundwater table comes from percolation through stream beds.

A capillary fringe lies at the bottom of the intermediate zone, where groundwater is drawn upward by capillary forces. The thickness of the capillary fringe is a function of the average grain size of the material in this zone. Capillarity is not effective in coarse sediments, but water may migrate upward 10 ft (3 m) or more in fine, well-sorted sediments. Fine sediments are often completely saturated within the capillary fringe zone, and the physical forces acting on the fluid are the same as those operating below the water table.

In desert regions, the deeply penetrating roots of some plants reach into the capillary fringe and even below. They do not use soil water, which is, of course, severely limited. These plants, known as phreatophytes, grow along streams where the capillary fringe is not deep and thus indicate near-surface sources of groundwater. Most phreatophytes have little value and their use of water in water-short areas is of major concern. Other phreatophytes, such as alfalfa, are of great value to man.

The groundwater table lies at the very bottom of the capillary zone. If a series of observation wells is drilled through the capillary fringe zone, the level of the water standing in each well marks the elevation of the groundwater table. Water below the water table is generally called groundwater, although strictly speaking, all water in the ground could be called groundwater. Some people refer to the region beneath the groundwater table as the zone of saturation. Unfortunately, this is not correct because the capillary fringe directly above can also be completely saturated if the sediment is fine enough. To avoid these ambiguous terms, Davis and DeWiest (1966) suggest the term "phreatic water," which is defined as water that enters freely into wells under both confined and unconfined conditions. In this text, we shall continue to use the term "groundwater," because generally it does refer to phreatic water. Groundwater, then, is underground water that can be removed by wells. All other water in the ground is termed "subsurface water" and is not available for man's use directly.

The bottom of the groundwater zone is nearly impossible to delineate because it grades almost imperceptibly into a region where openings in the rocks are more and more isolated. Water in these openings may not flow toward a well because individual pores are not connected. In areas of igneous rocks, the bottom of the groundwater zone, such as it is, may be as little as 500 to 900 ft (152 to 274 m) deep, whereas in sedimentary rocks it may be nearly 52,000 ft (15,900 m) deep. Below the zone of unconnected pores, water is chemically combined in the rock-forming minerals; only melting of the rocks will release the water. This phenomenon is described more fully in Chapter 2.

The groundwater zone may be imagined as a huge natural reservoir or system of reservoirs in rocks whose capacity is the total volume of pores or openings that are filled with water. Groundwater may be found in one continuous body or in several distinct rock or sediment layers at any one location. Thickness of the groundwater zone is governed by local geology, availability of pores or openings in the rock formation, recharge, and movement of water from areas of recharge toward points or areas of discharge.

It is nearly impossible to adequately summarize all types of geologic environments in which water can exist, but the list below presents some typical types of openings found in rocks.
 1. Intergrain pores in unconsolidated sand and gravel
 2. Intergrain pores in sandstone
 3. Intergrain pores in shale
 4. Systematic joints in metamorphic and igneous rocks
 5. Cooling fractures in basalt
 6. Solution cavities in limestone
 7. Gas-bubble holes and lava tubes in basalt
 8. Systematic joints in limestone
 9. Openings in fault zones

Unfortunately, rock masses are rarely homogeneous and adjacent rock types may vary significantly in their ability to hold water. Nevertheless, intelligent groundwater assessment or use requires an understanding of how water exists in each type of rock or sediment medium.

Aquifers

An aquifer is a saturated bed, formation, or group of formations which yields water in sufficient quantity to be economically useful. Water-bearing formations and groundwater reservoirs are synonyms for the word aquifer. To be an aquifer, a geologic formation must contain pores or open spaces (both of these are often called interstices) that are filled with water. These interstices must be large enough to transmit water toward wells at a useful rate.

Both the size of pores and the total number of pores in a formation can vary remarkably, depending on the types of material and the geologic and chemical history. Individual pores in a fine-grained sediment such as clay are extremely small, but the combined volume of the pores can be unusually large. For example, the total pore volume of a recently deposited clay may be as great as 95 percent. Subsequent compaction of clay reduces the pore space considerably. Although clay has a large water-holding capacity, water cannot move readily through the tiny open spaces. This means that a clay formation under normal conditions will not yield water to wells, and therefore it is not an aquifer even though it may be water-saturated. However, clays that are squeezed may yield economic amounts of water. Although this is an unusual circumstance, some wells in Houston, Texas are supplied in part from clay formations that have been squeezed as a result of pressure reduction in underlying formations.

Ordinarily a clay or shale formation is nearly impermeable and is called an aquiclude, or a formation through which virtually no water moves. Formations which do yield some water, but usually not enough to meet even modest demands, are called aquitards. In reality, almost all formations will yield some water, and therefore are classified as either aquifers or aquitards. In water-poor areas, a formation producing small quantities of water may be called an aquifer, whereas the same formation in a water-rich area would be an aquitard.

Water can exist in aquifers under two completely different physical conditions. The most common condition is when the water table is exposed to the atmosphere through openings in the overlying regolith. This type of aquifer is referred to as an unconfined or water-table aquifer; unconfined is the preferred term.

Groundwater may also occur under confined conditions. Confined groundwater is isolated from the atmosphere at the point of discharge by impermeable geologic formations, and the confined aquifer is generally subject to pressures higher than atmospheric pressure. Unconfined conditions exist, however, in recharge areas for confined aquifers. Figure 5.2 illustrates unconfined and confined groundwater conditions.

ENERGY CONTAINED IN GROUNDWATER

To understand how these two groundwater systems function, it is necessary to examine the forms of energy contained in groundwater. The total energy in any water mass consists of three components: pressure, velocity, and elevation head (energy derived from the elevation of the water body). The sum of these energy potentials, H, are expressed by the Bernoulli equation:

$$H = \frac{p}{\gamma} + \frac{V^2}{2g} + z \qquad (5.1)$$

where p is pressure, γ is the specific weight of water, V is the velocity of flow, g is the acceleration of gravity, and z is the elevation above a certain datum. The pressure head p/γ, is the energy contained in a water mass that can be attributed to the forces confining the water. A measure of this force is the movement or expansion of the water when the force is removed. The velocity head, $V^2/2g$ is the energy component

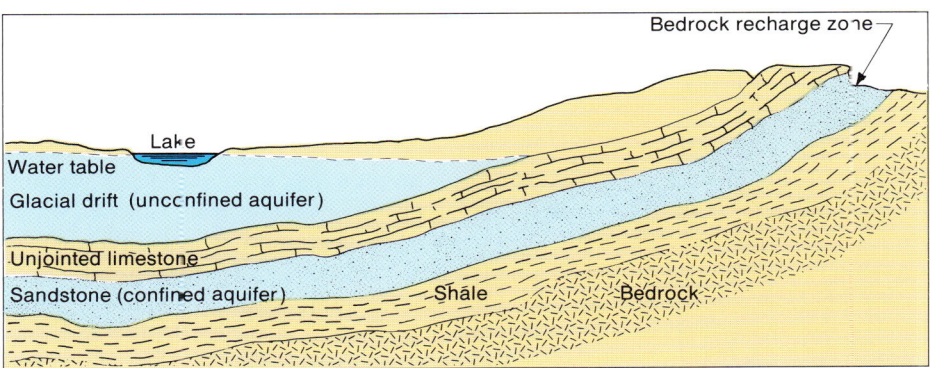

Figure 5.2. Groundwater exists in the underground in two major environments: unconfined and confined.

resulting from the movement of the water. The elevational head, z, is a latent form of energy; that is, if the water is allowed to fall, elevational energy can be converted to velocity or pressure energy.

As water moves from one site to another, it will lose part of its energy. This energy loss can be expressed by the equation:

$$\frac{p_1}{\gamma} + \frac{V_1^2}{2g} + z_1 = \frac{p_2}{\gamma} + \frac{V_2^2}{2g} + z_2 + h_L \qquad (5.2)$$

where h_L is the loss of head energy caused by movement of the water from one point to another point. The physical environment of the water determines what forms of energy the water possesses. In groundwater situations, the flow velocities are so low that velocity energy can be neglected in computing the energy contained in a groundwater system. Therefore, the velocity terms in the Bernoulli equation can be eliminated and the equation becomes:

$$\frac{p_1}{\gamma} + z_1 = \frac{p_2}{\gamma} + z_2 + h_L \qquad (5.3)$$

Unconfined Aquifers

Water at any depth in an unconfined aquifer is under the pressure exerted by the overlying water and atmospheric pressure. Because the pressure exerted by the atmosphere is more or less constant at a site, it is usually not taken into consideration when calculating the energy available to drive groundwater to a stream, spring, or well. Thus, the depth of the water column, z, represents the only form of energy actually available that can be converted to pressure energy to drive water through the aquifer materials into a well. The amount of energy required to move the water a certain distance through the groundwater system can be determined by the equation:

$$H = z_1 - z_2 = h_L \qquad (5.4)$$

and is equal to the difference in elevational heads between any two points in the groundwater system.

When a pump is turned on in a well, a significant elevational difference is created between the water surface in the well and the surrounding aquifer. This difference in elevational head forces the water in the aquifer to flow toward the well. The maximum energy head available is the difference in elevation between the natural water level in the aquifer and the pump intake. For a specific well, the rate of flow toward the well will increase as the elevational difference is increased.

In some geologic settings, a local zone of saturation may exist at some level above the regional water table. This situation can occur where an impervious stratum within the vadose zone intercepts downward-percolating water and causes some of it to accumulate above the stratum. The upper surface of the groundwater in this case is called a perched water table. Perched water tables commonly occur where clay lenses exist, as they do, for example, in thick sequences of clay-rich glacial drift (Figure 5.3).

Under unconfined conditions the water table is free to rise and fall. During periods of drought, the water table may drop 3 to 7 ft (0.9 to 2.1 m) or more, as outflow to springs, streams, and wells reduces the volume of water in storage. When precipitation begins again, aquifer recharge is generally rapid. If heavy precipitation persists for many months or years, the groundwater table may rise well above its established mean level. High groundwater levels generally provide unusually high discharges at nearby springs and streams, and cause lake levels to rise.

Confined Aquifers

When a well is drilled through an overlying impervious layer into a confined aquifer, water rises in the well to some level above the top of the aquifer. The water level in the well represents the confining pressure at the top of the aquifer. Confined pressure is defined as the vertical distance between the water level in the well and the top of the aquifer. This is equivalent to the hydrostatic head, expressed in feet (meters) of water.

The elevation to which water rises in a well that taps a confined aquifer is called its potentiometric level. Under confined conditions, the potentiometric surface is an imaginary surface representing the confined pressure (hydrostatic head) throughout all or part of a confined aquifer. This imaginary surface is analogous to an actual water surface, that is, the water table as it would be in an unconfined aquifer. Figure 5.4 shows a typical potentiometric surface map for an area affected by overpumping.

The hydrostatic pressure within a confined aquifer is occasionally sufficient to cause the water to rise in a well high enough so it flows out on the land surface; a flowing

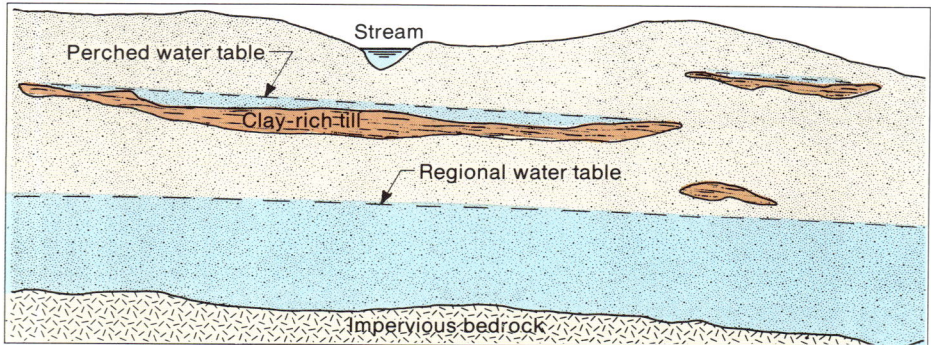

Figure 5.3. Perched water table supported by stringers of clay-rich till.

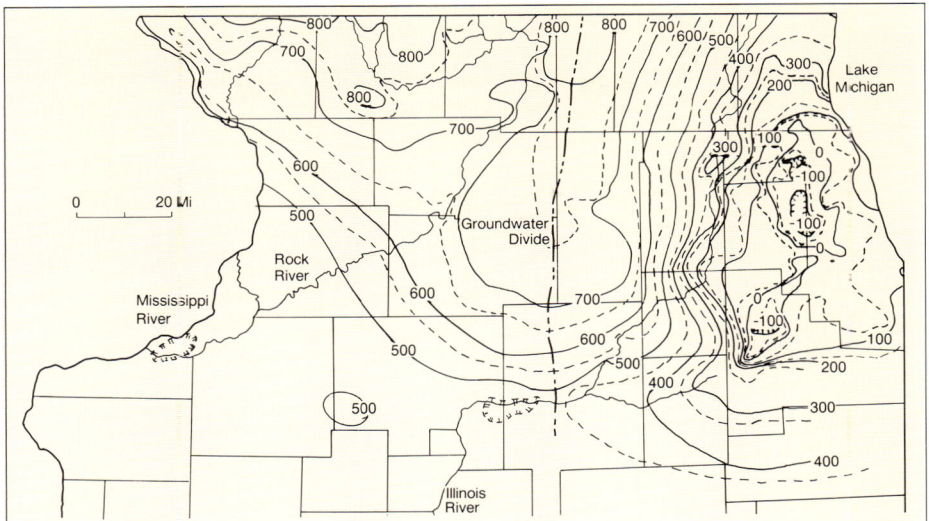

Figure 5.4. Elevation of the potentiometric surface (in feet above mean sea level) of the Cambrian-Ordovician aquifer in the Chicago area of northeastern Illinois, October, 1980. *(Sasman et al., 1982)*

artesian well results. The static water level in this case is above ground and can be measured in the well if the casing (or pipe) extends high enough above the ground surface to prevent flow. The hydrostatic head can also be determined by capping the well near ground level and then measuring the shut-in head with a pressure gauge. For every 1 psi (6.9 kPa) of pressure measured, the water will rise approximately 2.31 ft (0.7 m) above the gauge.

Just as in an unconfined aquifer, the potentiometric surface in a confined aquifer is free to rise and fall in response to the volume of water in the aquifer. When a well penetrating a confined aquifer is pumped, internal aquifer pressure is reduced and the overlying sediments compact the aquifer. But the pores in the rock remain saturated as long as the pumping water level remains above the aquifer. Should the potentiometric surface of a confined aquifer fall below the top of the aquifer itself during pumping, it then becomes a partially unconfined aquifer.

Removal of water from the pores of an unconfined aquifer is easy to visualize, because water is simply drained from the openings in the sediment. Because the unconfined water is at atmospheric pressure, the volume of water drained from the aquifer is essentially the same as that existing in the pores although some water will cling to the particles making up the aquifer and cannot be drained. On the other hand, the physical aspect of removing water from a confined aquifer is considerably more complex. Meinzer (1942) observed that more water could be removed from a well in a confined aquifer than was calculated to be flowing toward the well. Meinzer and other investigators concluded that the major sources of water from a well drilled in a confined aquifer comprise (1) water moving through the aquifer toward the well, (2) water forced from the aquifer by compaction caused by the weight of overlying sediments, (3) water expansion resulting from reduced pressures in the aquifer, and (4) water forced from surrounding aquicludes by compaction.

The hydraulic behavior of confined and unconfined aquifers is dissimilar, and, furthermore, the mathematical analysis of confined aquifers is complicated by certain geologic environments. Confined aquifers occur because the overlying and basal beds are relatively impermeable to water flow; this impermeability varies according to the geologic structure. For example, a water-bearing formation may be overlain by an aquitard which permits water to move slowly upward out of the aquifer or vertically downward into the aquifer, depending on the hydrostatic head in the aquifer. Where the aquitard is under the aquifer, water may be lost to or gained from the rocks below. Thus, confining conditions are sometimes less than perfect, making an accurate mathematical analysis of yields from wells in these aquifers much more difficult. Confined aquifers that lose or receive water from the surrounding formations are called leaky confined aquifers, and special mathematical methods have been devised to analyze them. In Chapter 9, an empirical analysis of the effects of leaky confined aquifer conditions on pumping test data is presented.

AQUIFER FUNCTIONS

An aquifer performs two important functions — a storage function and a conduit function. The interstices of a water-bearing formation act as storage sites and are part of a network of conduits. Groundwater is constantly moving through these conduits under the local hydraulic gradient. Rates of movement vary from feet per year to feet per day. Thus, water contained in any aquifer is in temporary storage, and if not used, will be discharged to springs, lakes, streams, or oceans.

Our previous examination of groundwater environments indicates that openings in aquifers comprise three general classes:

1. Openings between individual particles in sandstone and sand and gravel formations.
2. Crevices, joints, faults, and gas holes in igneous and metamorphic rocks.
3. Solution channels, caverns, and vugs (openings) in limestone and dolomite.

The shape of the openings in the rock or sediment, their size, volume, and interconnection all play a vital part in the hydraulic characteristics of an aquifer.

Storage

Two important properties of an aquifer that are related to the storage function are porosity and specific yield. The porosity of a water-bearing formation is determined by that part of its volume consisting of openings or pores. Porosity is an index of how much groundwater can be stored in a saturated medium and is usually expressed as a percentage of the bulk volume of the material.

$$Porosity = \eta = \frac{Volume\ of\ pore\ space}{Volume\ of\ bulk\ solid} \cdot 100 \qquad (5.5)$$

For example, if 1 ft^3 (1 m^3) of sand contains 0.3 ft^3 (0.3 m^3) of open space or pores, its porosity is 30 percent. Typical porosities for some common materials are presented in Table 5.1.

Although the volume of water contained in a particular segment of aquifer is of interest, of more concern is how much water can be actually released from storage per unit area of aquifer, per unit change in head. Whereas porosity represents the

Table 5.1 Porosities for Common Consolidated and Unconsolidated Materials

Unconsolidated Sediments	η (%)	Consolidated Rocks	η (%)
Clay	45–55	Sandstone	5–30
Silt	35–50	Limestone/dolomite (original &	
Sand	25–40	secondary porosity	1–20
Gravel	25–40	Shale	0–10
Sand & gravel mixes	10–35	Fractured crystalline rock	0–10
Glacial till	10–25	Vesicular basalt	10–50
		Dense, solid rock	<1

volume of water an aquifer can hold, it does not indicate how much water the aquifer will yield.

When water is drained from a saturated material under the force of gravity, the material releases only part of the total volume stored in its pores. The quantity of water that a unit volume of unconfined aquifer gives up by gravity is called its specific yield (Figure 5.5). Specific yields for certain rocks and sediment types are presented in Table 5.2. Some water is retained in the pores by molecular attraction and capillarity. The amount of water that a unit volume of aquifer retains after gravity drainage is called its specific retention. The smaller the average grain size, the greater is the percent of retention; the coarser the sediment, the greater will be the specific yield when compared to the porosity. The surface area for different-size sand grains is shown in Table 5.3. Note the large increase in surface area for the finest sediment. As the surface area increases, a larger percentage of the water in the pores is held by surface tension or other adhesive forces. Therefore, finer sediments have lower specific yields compared to coarser sediments, even if they both have the same porosity.

Specific yield plus specific retention equals the porosity of an aquifer. Both specific yield and specific retention are expressed as decimal fractions or percentages. Specific yields of unconfined aquifers (equivalent to their storage coefficients*) range from 0.01 to 0.30. Specific yields cannot be determined for confined aquifers because the aquifer materials are not dewatered during pumping.

Storage coefficients are much lower in confined aquifers because they are not drained during pumping, and any water released from storage is obtained primarily by compression of the aquifer and expansion of the water when pumped. During

Table 5.2. Representative Specific Yield Ranges for Selected Earth Materials

Sediment	Specific Yield, %
Clay	1–10
Sand	10–30
Gravel	15–30
Sand and Gravel	15–25
Sandstone	5–15
Shale	0.5– 5
Limestone	0.5– 5

(Walton, 1970)

---

*The coefficient of storage is fully defined in Chapter 9. Briefly, it is the volume of water taken into or released from storage per unit change in head per unit area.

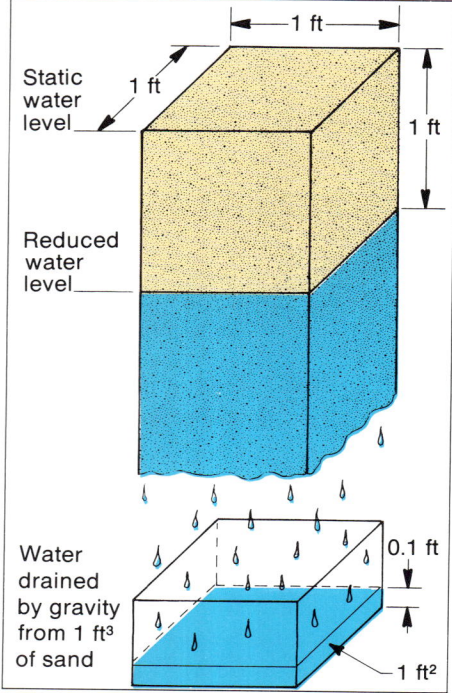

Figure 5.5. Specific yield of sand can be visualized from this diagram. Its value here is 0.1 ft³ per ft³ of aquifer material.

pumping, the pressure is reduced in the aquifer, but the aquifer is not dewatered. Typical storage coefficients for confined aquifers range from 10^{-5} to 10^{-3}. Figure 5.6 demonstrates how equal declines in the water table for unconfined aquifers and the potentiometric surface for confined aquifers produce significantly different volumes of water.

An example may serve to illustrate some of the points just made. If 1 ft³ (0.03 m³) of saturated sand under unconfined conditions has a porosity of 25 percent and a specific yield of 10 percent, its storage coefficient is also 10 percent. Specific retention is 15 percent. This means that 40 percent of the water residing in the aquifer is available for man's use. If this sediment extends over 20 mi² (51.8 km²) with an average thickness of 40 ft (12.2 m), the total volume of the upper 5 ft (1.5 m) of the aquifer is 2.8×10^9 ft³ (7.9×10^7 m³) and contains 7×10^8 ft³ (2×10^7 m³) of water. If the upper 5 ft of the aquifer were drained by pumping, the total yield would be about 2.8×10^8 ft³ (7.9×10^6 m³). This quantity could supply five wells pumping 790 gpm (4,310 m³/day), 12 hours each day, for 736 days, or slightly more than 2 years. This pumping rate could be sustained by the groundwater stored in the upper 5 ft of the aquifer in the total absence of any recharge to the aquifer during the 2-year period.

This simple example shows how the storage function of an aquifer permits the use of groundwater at a constant rate even though recharge to the aquifer may be intermittent or at irregular intervals. Because of their enormous capacity, groundwater reservoirs serve much more effectively during periods of droughts than do surface-water reservoirs.

Storage Capabilities of Aquifers

Recharge to groundwater aquifers from precipitation occurs unpredictably in time

Table 5.3. Total Surface Area of Grains in Samples Composed of Uniform Spheres

Diameter of Sphere (mm)	Number of Particles in 1-mm cube*	Approximate Surface Area (mm²)	
		One Particle	All Particles
1.0	1	3.1	3.1
0.062	4.1×10^3	1.2×10^{-2}	49
0.004	1.7×10^7	5.0×10^{-5}	850

*Cubic packing assumed.
(After Baver, 1956)

and space and amount of precipitation. In the long term, precipitation follows well-established averages, but in the short term, deviations from the usual rate of precipitation may be extreme. In the north-central part of the United States, rainfall can vary from as little as 16 inches (406 mm) in one year to 36 inches (914 mm) the next. This is more than a 100-percent deviation.

Unlike recharge, discharge from near-surface aquifers is more or less constant and generally occurs at an accelerated rate during times of precipitation surplus. Because precipitation is generally the only source of recharge, aquifer levels fall during periods of drought. If water levels fall below the lowest level at which the aquifer naturally discharges to streams, no more water can be lost except through pumping.

Groundwater, although a renewable resource, can be temporarily depleted when aquifers are overpumped. Good management practices demand adequate information on how much water is in storage and how this volume varies with time. Data on groundwater storage are obtained by periodic measurements of the depth to water or the height of the potentiometric or water table surface from some reference point.

Several methods are used to present water-level data. The most common are well hydrographs, water-level profiles, water-level contour maps, and water-level change

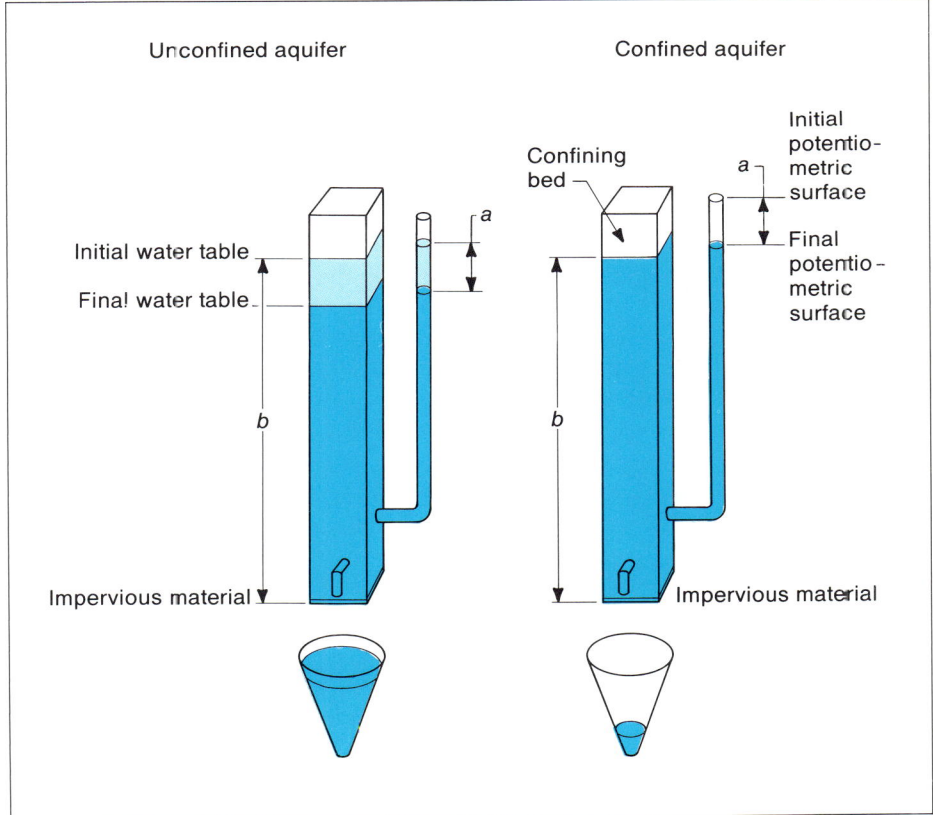

Figure 5.6. Unit prisms of unconfined and confined aquifers illustrating differences in storage coefficients. For equal declines in head, the yield from an unconfined aquifer is much greater than that from a confined aquifer. *(After Heath and Trainer, 1968)*

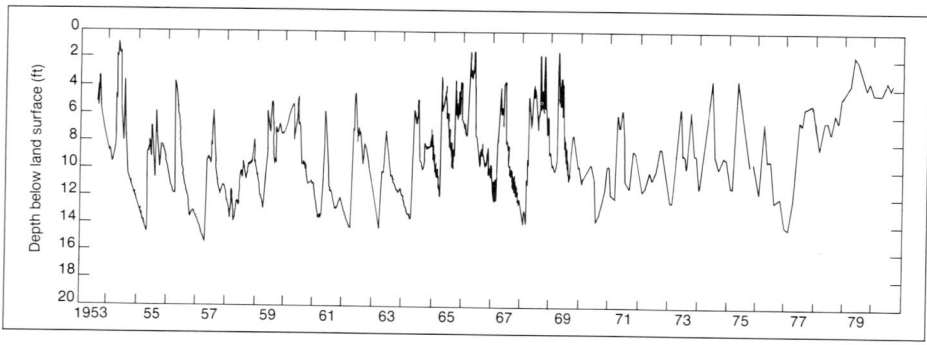

Figure 5.7. Hydrograph showing long-term trends of water level for a well in St. Louis County, Minnesota. Fewer readings were taken after mid-1969. *(U.S. Geological Survey, 1981)*

maps. Figure 5.7 shows an example of the hydrograph method where the level of the groundwater table varies significantly from year to year, but is relatively stable over greater lengths of time.

When precipitation is inadequate to compensate for discharges from an aquifer, the water level in an aquifer will fall over time. The rate of decline, with no pumping, is governed by the storage coefficient, hydraulic conductivity of the aquifer, and the hydraulic gradient. Heavy pumping could hasten the decline considerably. During times of drought, the water-level decline closely resembles the curve shown in Figure 5.8. Longer periods of falling water levels are interrupted by brief intervals of recharge. Note that the recovery of water levels is not complete in any of the precipitation episodes.

The water-level curve shown in Figure 5.9 is typical where recharge and discharge are out of balance through time. These data are from a well in the Phoenix, Arizona area. Water use in this region began to accelerate in the middle 1950's as the population rose and farming operations expanded. As energy costs began to escalate in the early 1970's, more people began to move to the southwestern United States which increased demands for groundwater and caused an accelerated decline in water levels. To provide water for the burgeoning population, municipal water suppliers have had to augment their supplies by purchasing water from nearby irrigation farmers. Groundwater in the Phoenix area is being "mined" and major hydrogeologic studies are being conducted to find additional sources of water for the city.

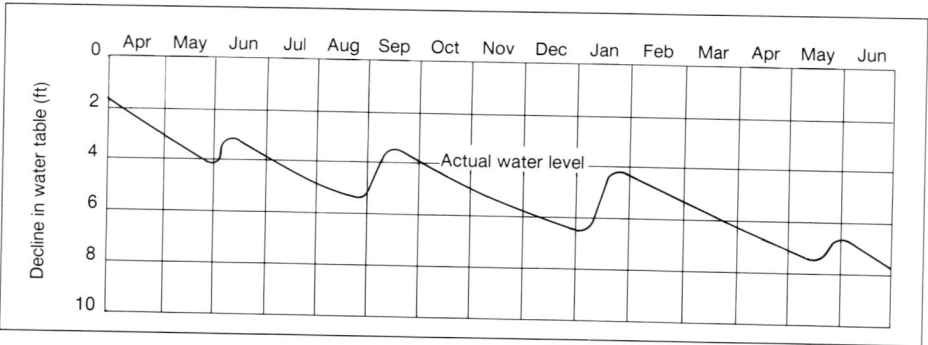

Figure 5.8. A 15-month, drought-induced decline in water level, interrupted by periods of recharge.

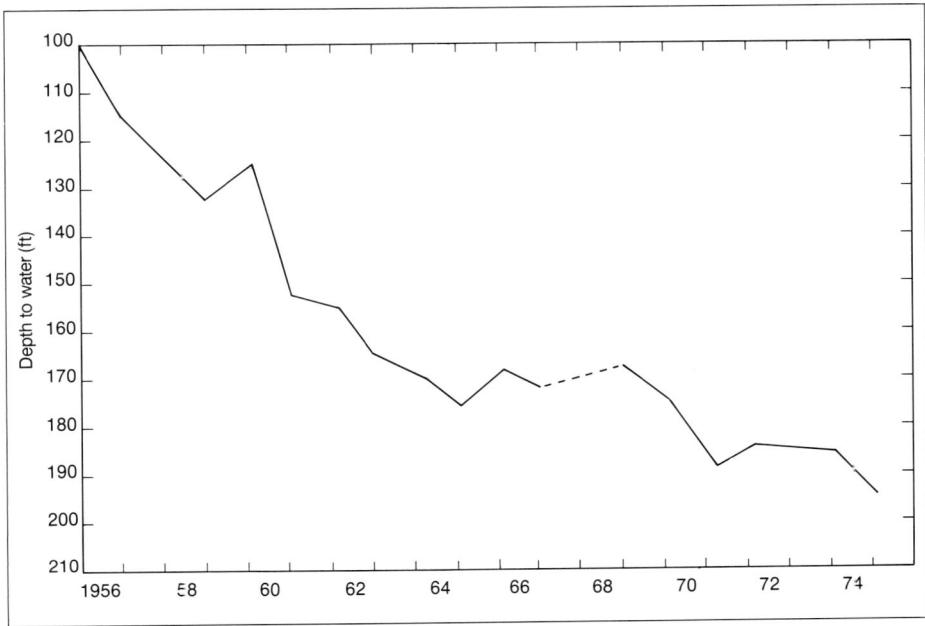

Figure 5.9. Hydrograph from a well in Phoenix, Arizona where the recharge and discharge are out of balance through time. *(Laney et al., 1978)*

Hydraulic Conductivity

The property of a water-bearing formation that relates to its pipeline or conduit function is called hydraulic conductivity (K) and is defined as the capacity of a porous medium to transmit water. In the past, this characteristic has been called the coefficient of permeability. Because permeability refers to the ease with which any fluid moves through a formation, the term is probably more aptly applied to the so-called intrinsic permeability; the intrinsic permeability, k, is a function of the medium alone and can be expressed as:

$$k = K \frac{\mu}{\rho g} \tag{5.6}$$

where K is the hydraulic conductivity, μ is dynamic viscosity, ρ is fluid density, and g is acceleration of gravity. The notion of intrinsic permeability is used in the petroleum industry because fluids of different phases and density [gas (vapor), oil, and water] are analyzed for their rate of movement through the rock medium.

Hydraulic conductivity is governed by the size and shape of the pores, the effectiveness of the interconnection between pores, and the physical properties of the fluid. If the interconnecting tubes are small, as shown in Figure 5.10a, the volume of water passing from pore to pore is restricted and the resulting hydraulic conductivity is quite low. In a reasonably coarse sediment, on the other hand, the connecting tubes are large relative to the pores and the hydraulic conductivity will be high (Figure 5.10b). The fluid properties of water vary with temperature, and the hydraulic conductivity is defined on the basis of fluid viscosity and density at a particular tem-

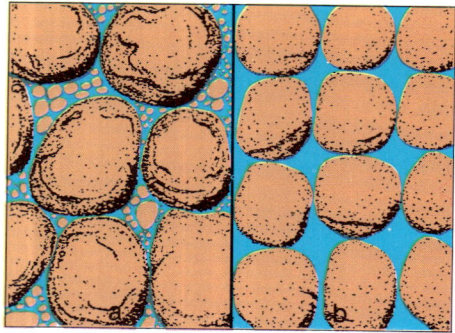

Figure 5.10. Permeability is affected by size of interconnections between pores in storage sites.

perature. Groundwater is usually considered incompressible when calculating the hydraulic conductivity.

Movement of groundwater takes place whenever a difference in head exists between two points. Thus, water can move from locations of high head to locations of low head. In addition, water in confined aquifers may move from low-pressure areas to high-pressure areas. Although this statement may seem contradictory, the diagram in Figure 5.11 illustrates how this occurs. At point *a*, the water has high potential energy with low pressure energy; during its movement toward point *b*, the potential energy (head) is transformed into pressure energy. At point *b*, the potential energy is lower, but the pressure energy is higher than at point *a*.

Two basic types of flow occur in groundwater, with one more prevalent than the other. Ordinarily, water moves so slowly through the ground that laminar flow takes place. In this condition, the water particles tend to flow in ribbonlike patterns through the pore openings, although water moving in the center of the pores does move faster than water closer to the walls (Figure 5.12a). There is no intermixing of individual layers. Occasionally, another type of flow occurs near wells and other points where relatively large volumes of water must converge through constricted openings. This type is generally found in streams and rivers and is called turbulent flow. In turbulent flow, individual water particles intermix completely and follow irregular paths through the pores (Figure 5.12b). Experiments have shown that velocity determines which

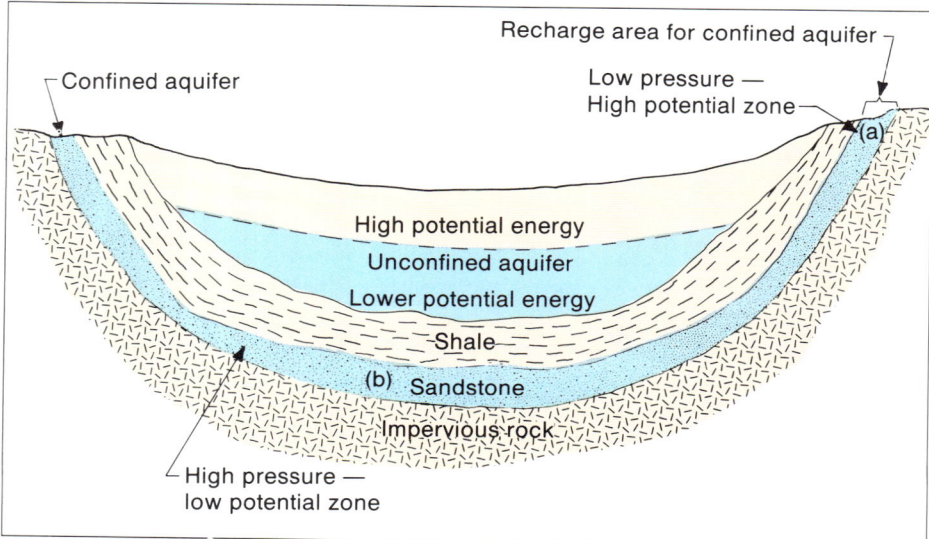

Figure 5.11. Groundwater may follow unusual flow paths because it can move from areas of low pressure toward areas of high pressure. This pressure energy is gained at the expense of available potential (head) energy.

one of these flow regimes is present. Laminar flow is dominant at very low velocities, but at some point (at a Reynolds Number* of approximately 10), turbulent flow begins as velocity increases.

Henri Darcy (1856), a French engineer, recognized that the flow of water through the ground is analogous to pipeflow. He conducted a series of experiments on a vertical pipe filled with sand. Darcy

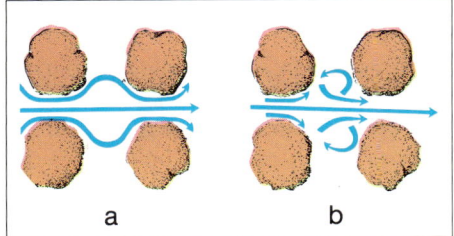

Figure 5.12. Schematic representation of (a) laminar and (b) turbulent flow in a granular deposit.

learned that the rate of flow through a column of saturated sand is proportional to the difference in hydraulic head at the ends of the column and inversely proportional to the length of the column. His published work contains what is now known as Darcy's law, which is the basic equation describing the flow of groundwater:

$$V_{Df} = \frac{K(h_1 - h_2)}{L} \quad (5.7)$$

where V_{Df} is the Darcy flux velocity (specific discharge), $(h_1 - h_2)$ is the difference in hydraulic head, L is the distance along the flowpath between the points where h_1 and h_2 are measured, and K is the hydraulic conductivity (Figure 5.13). V_{Df} is called the Darcy flux velocity because it assumes that the discharge occurs throughout the entire material cross section in spite of the fact that solids make up a large part of the cross-sectional area. The equation for the actual flow velocity through the pores is presented later in this chapter. The law provides a way to quantify the energy (head) required to move water through an aquifer. Energy is lost because of friction between the moving water and the confining walls of the pores. Equation 5.7 simply states that energy loss is proportional to the velocity of flow under laminar conditions, or, the faster the flow, the higher the energy loss. Energy losses caused by the onset of turbulent flow near a well are discussed in Chapter 16.

The hydraulic gradient, I, is by definition the difference in hydraulic head, $(h_1 - h_2)$, divided by the distance, L, along the flowpath. If:

$$I = \frac{h_1 - h_2}{L} \quad (5.8)$$

then:

$$V_{Df} = KI \quad (5.9)$$

From pipeflow, we know that the discharge, Q, is equal to the flow velocity times the cross-sectional area, A, or:

$$Q = VA \quad (5.10)$$

Substituting the equivalent value of V_{Df} from Equation 5.9, we have:

$$Q = KIA \quad (5.11)$$

---

*A dimensionless number that relates diameter of a void and the velocity, density, and viscosity of a moving fluid to determine the type of fluid flow, that is, whether laminar or turbulent.

This form of Darcy's equation is more interesting because Q (discharge) of the aquifer is generally the most important variable*. Discharge is the quantity of flow per unit time, such as gallons per minute (gpm) or cubic meters per day (m³/day).

In these equations, K is called the hydraulic conductivity of the porous medium. Its value depends on the size and arrangement of the particles in an unconsolidated formation, on the size and character of the crevices, fractures, and solution openings in a consolidated formation, and the viscosity of the fluid as determined by the temperature. Hydraulic conductivity may change with any variation in these physical parameters.

The hydraulic conductivity indicates the quantity of water that will flow through a unit cross-sectional area of a porous medium per unit time under a hy-

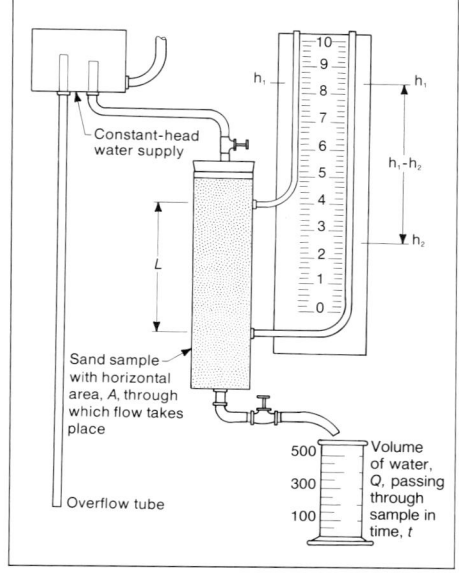

Figure 5.13. An apparatus used to determine the hydraulic conductivity of sediments. With this constant-head permeameter, hydraulic conductivity is found by measuring the area of the sample, rate of flow, and hydraulic gradient.

draulic gradient of 1 (100 percent) at a specified temperature. A hydraulic gradient of 1 means that the head falls 1 ft (1 m) for every 1 ft (1 m) of flow travel. For convenience, K is expressed as the flow in gallons per day through a cross section of one square foot of water-bearing medium under a hydraulic gradient of 1 at a temperature of 60°F (cubic meters per day through a cross section of one square meter under a hydraulic gradient of 1 at a temperature of 15.6°C). That is:

$$1\ K\ Unit = \frac{1\ gal\ of\ water\ at\ 60°\ F/day}{1\ ft^2\ (-1\ ft/ft)} \qquad 1\ K\ Unit = \frac{1\ m^3\ of\ water\ at\ 15.5°\ C/day}{1\ m^2\ (-1\ m/m)}$$

The range for K is generally from 10 to 5,000 gpd/ft² in the English engineering system (0.4 to 204 m/day in the International System of Units) for natural aquifer material (Figure 5.14).

From our preliminary examination of groundwater flow, three important points have been established:

1. Flow from one point to another in an aquifer is always related to a difference in head between the two points.

2. The difference in head reflects the frictional resistance that develops in the pores of an aquifer during flow.

3. The hydraulic gradient is the head loss in feet per foot or meters per meter of travel through the medium.

Darcy's experiments show that flow in a saturated sand varies directly with the

---

*Velocity is much more important in contaminant-transport studies.

hydraulic gradient. If the hydraulic gradient (head loss per unit length of travel) is doubled, the rate of flow in a given sand is also doubled. Conversely, doubling of the flow rate requires doubling of the hydraulic gradient. These ratios apply only to laminar flow, however. If turbulent flow is present, the flow rate does not change in direct proportion with the hydraulic gradient; doubling of the hydraulic gradient may increase the flow rate by only 1.5 times. The information in this paragraph is vital to understanding water-well hydraulics, which is presented in Chapter 9.

The slope of the water table or potentiometric surface is the hydraulic gradient under which groundwater movement takes place. The total flow through any vertical section of an aquifer can be calculated if we know the thickness of the aquifer, its width, its average hydraulic conductivity, and the hydraulic gradient. The flow, q, through each foot of aquifer width is:

$$q = KbI \tag{5.12}$$

where K is the hydraulic conductivity averaged over the height of the aquifer, b is the aquifer thickness in feet, and I is the hydraulic gradient.

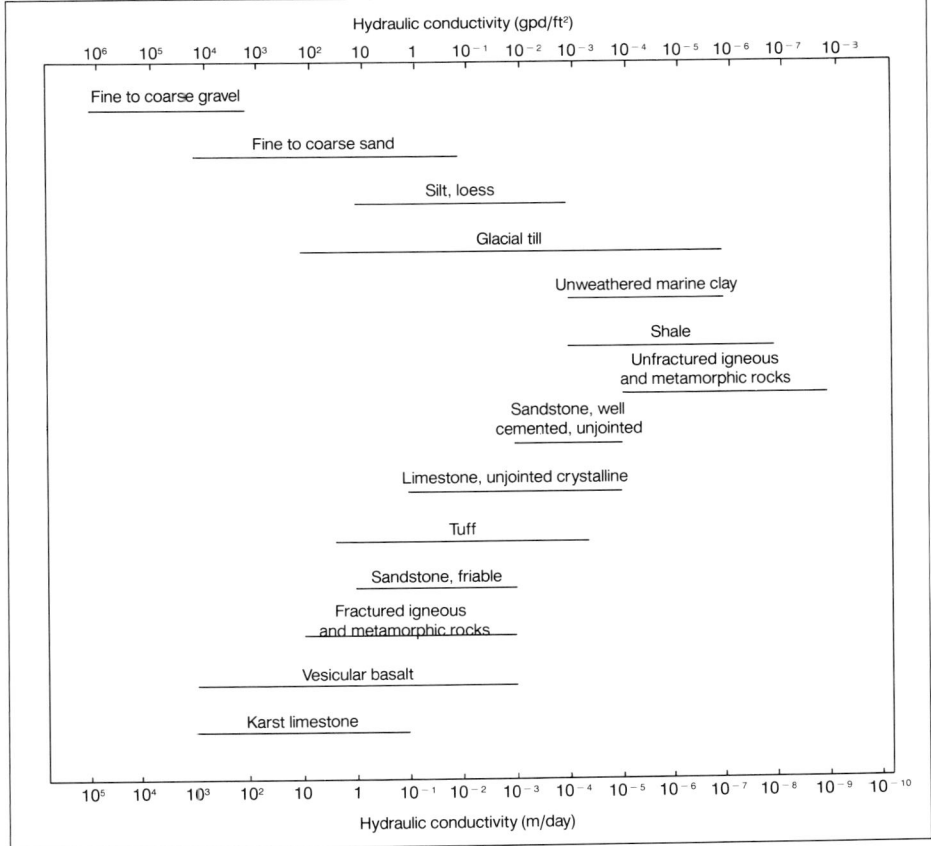

Figure 5.14. Typical K values for consolidated and unconsolidated aquifers. *(After Davis, 1969; Dunn and Leopold, 1978; Freeze and Cherry, 1979).*

Transmissivity

Theis (1935) pointed out the convenience of using the product of Kb as a single term to represent the transmission capability of the entire thickness of an aquifer. He introduced the term "coefficient of transmissibility," defining it as the rate of flow in gallons per minute through the vertical section of an aquifer one foot wide and extending the full saturated height of an aquifer under a hydraulic gradient of 1. The term "transmissibility" is still in use although it has been largely replaced by the word "transmissivity;" the latter term will be used in this text. In the International System of Units, transmissivity units are cubic meters per day per meter (m^3/day/m) or square meters per day (m^2/day). The temperature assumed in this definition is the ambient temperature of the groundwater in the aquifer. When the coefficient of transmissivity, T, is introduced in the Darcy equation, flow through any vertical section of an aquifer is expressed as:

$$Q = TIw \qquad (5.13)$$

where w is the width of the vertical section through which the flow occurs; the other terms have their standard definitions.

Three general methods are used to determine transmissivity: using data collected during pumping tests, analyzing the hydraulic properties of aquifer material, and calculations based on laboratory tests. The first method is based on observation of the decline in groundwater levels during a pumping test. Several equations are available that describe flow near wells and give the transmissivity value if the declines are known. These equations are discussed in Chapter 9. The second method involves determining hydraulic conductivities of the materials in an aquifer. Estimating is done from grain-size analysis of cuttings or samples that are taken from a well or other exploratory hole drilled into the aquifer. The transmissivity for an individual layer is determined by multiplying the strata thickness by the K value characterizing that layer. The individual T values for all strata are then added to determine the transmissivity for the entire aquifer. The third method involves testing field samples from an aquifer in an apparatus similar to that shown in Figure 5.13. When measured quantities of water are run through this device (called a permeameter), the head loss is determined and the hydraulic conductivity can be calculated. As in method two, the hydraulic conductivity for all layers are summed to yield an overall transmissivity value.

Of these three methods, the coefficients established during pumping tests are the most accurate and should reflect the true hydraulic conditions within an aquifer. Because of their importance, methods for collecting and analyzing pumping test data are presented in Chapters 9 and 16. Transmissivity values obtained by the two methods discussed below are less valid but can yield valuable information, depending on the ability of the individual investigator. Transmissivity values calculated on the basis of the latter two methods should never be used without judgment based on considerable experience. An explanation of the methods used to calculate transmissivity is presented later in this chapter.

Grain-Size Analysis

Hydraulic conductivity varies not only with porosity but also with the size, distribution, and continuity of pores. Formations made up entirely of coarse, unconsolidated sand or gravel produce high yields because the pores or voids are large. In finer grained sediments, the small pores offer more resistance to flow; therefore, more water will flow through coarser material under a given hydraulic gradient. The hydraulic conductivity is greater for the coarser material even though the porosity may be the same for both sediments.

It is difficult to discuss the effect of particle size on hydraulic conductivity without also discussing the grading of the sample, because the degree of uniformity of a mixture of fine to coarse material has an important bearing on transmitting capacity. When sediment consists of similar-size particles, the material is said to be uniformly graded. The graded condition suggests that the strength of the geologic forces that caused the deposition of the material were more or less constant during the time the sediment was deposited. Typical examples are waves on an ocean beach and a stream flowing at constant discharge for some length of time. Uniformly graded material has a higher porosity than does less uniform or nongraded mixtures of coarse and fine sediment. In the nongraded mixture, porosity is reduced because the finer materials occupy openings between the coarser fragments. The result is a more compact arrangement which reduces the volume of the void space. Even fine, well-sorted material has a higher porosity than do mixtures of larger nongraded particles.

Many attempts have been made to calculate hydraulic conductivity on the basis of grain size and degree of sorting alone. Grain-size distribution is determined by passing the disaggregated sample through a series of sieves. The sieves are generally mounted in a device that vibrates both vertically and horizontally for a certain length of time, usually five to ten minutes per sample. The cumulative weight of the particles caught on each sieve is then plotted as a percent of the total sample weight against grain size in inches (millimeters); degree of sorting and average grain size are clearly indicated. Figure 5.15 shows three sieve analyses of natural materials: beach sand, a mixture of glacial sands and gravels, and a typical deposit from a slow moving river. For a thorough treatment of the sieving process, see Chapter 12.

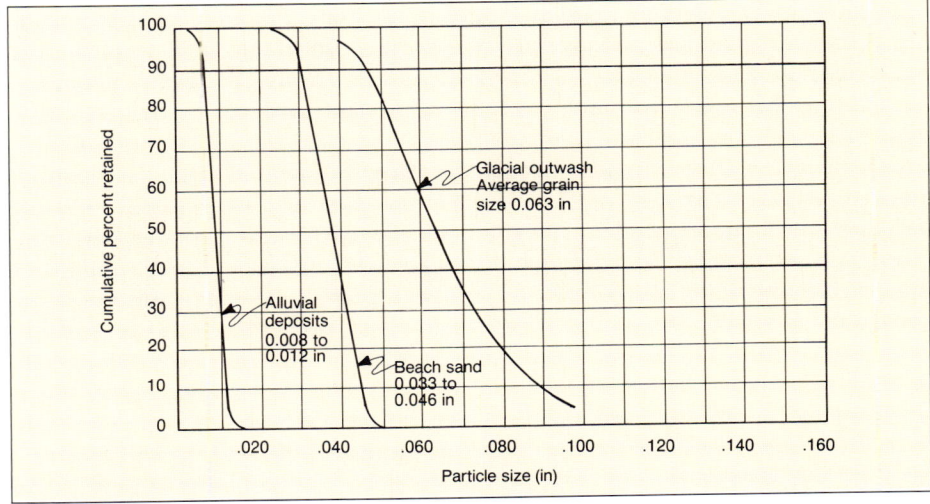

Figure 5.15. Sieve analysis for three naturally graded materials.

Table 5.4. Comparative Data for Filter Sands

	Fine Sand 0.008" to 0.012" (0.2 mm to 0.3 mm)	Coarse Sand 0.033" to 0.046" (0.84 mm to 1.17 mm)	Fine Gravel 0.046" to 0.093" (1.17 mm to 2.36 mm)
Effective size in inches (mm)	0.008 (0.2)	0.034 (0.86)	0.048 (1.22)
Uniformity coefficient	1.2	1.2	1.4
Hydraulic conductivity, in gpd/ft^2 (m/day)	540 (22)	9,600 (391)	13,000 (529)
Porosity, in percent	37	37	35

Table 5.4 presents an interesting comparison of several sands. Hydraulic conductivity measurements were made by laboratory flow tests. Note that the effective size* of the coarse sand is a little more than four times that of the fine sand, whereas its hydraulic conductivity is about 18 times greater. Hazen (1893), the investigator in this experiment, indicated that hydraulic conductivity varies in proportion to the square of the effective grain size. The values for the first two sands listed in Table 5.4 conform approximately to this ratio, since $4^2 = 16$ and the hydraulic conductivity ratio is about 18. Hazen's rule does not hold true for the third sediment. A ratio of the squares of the effective sizes of the fine gravel and the coarse sand is about 2, but the ratio of the hydraulic conductivity is only 1.4. The larger uniformity coefficient of the fine gravel apparently reduces the impact of the greater effective size of the larger gravel and its influence on hydraulic conductivity. Thus, hydraulic conductivity generally increases with an increase in effective size, but other sediment characteristics tend to produce varying relationships between hydraulic conductivity and grain size. In Chapter 13, a graph is presented that can serve as an approximate guide in estimating hydraulic conductivity based on grain-size distribution.

Permeameters

Modern permeameters are based on the apparatus used by Darcy in his experiments. The permeameter enables investigators to measure accurately the rate at which water percolates through various types of earth materials. There are two types of permeameters, falling head and constant head. In a constant-head device, the hydraulic gradient is kept constant while the discharge is recorded (Figure 5.13). In a falling-head permeameter, the hydraulic gradient decreases in time; therefore, the discharge will also decrease in time (Figure 5.16). Several tests are usually run with each sample under varying hydraulic gradients to obtain an average hydraulic conductivity.

Use of a permeameter to determine hydraulic conductivity is usually confined to large engineering firms or academic institutions because the equipment is not easily transportable, may be highly sophisticated, and must be operated with great care to produce good results. Some small permeameters are easy to use, however, and provide an initial estimate of the hydraulic conductivity of the material being tested. Field permeameters are based on the falling head principle and yield data on reasonably permeable materials within a matter of minutes, but their accuracy is limited.

---

*Effective particle size is defined as the 90 percent retained size, that is, 10 percent of the sediment is finer and 90 percent is coarser. Effective size and uniformity coefficient are discussed in Chapter 12.

Several problems can occur in attempting to determine a proper hydraulic conductivity with a permeameter. For example, trapped air within the sample can reduce the flow rate. In sophisticated permeameters, air may be driven out by passing carbon dioxide gas through the sample. Deaerated water is then allowed to enter the testing chamber. Any air that remains in the system is readily absorbed by the deaerated water, which also absorbs the carbon dioxide gas previously introduced. A combination of these two procedures assures that no air will remain in the sample chamber to impede the flow.

Packing of the grains presents a more difficult problem. When a sample is collected, the arrangement of individual grains is quite dense because component particles have settled together over tens of thousands or even tens of millions of years. Recent alluvial or late Pleistocene glaciofluvial sands and gravels, however, are not usually well packed and generally possess high porosities. Every sample becomes at least partly disaggregated during collection, transportation to the laboratory, and placement within the permeameter. The problem for the laboratory technician is clear: the original packing density must be reestablished if the measurements are to truly represent the percolation rates found in nature. Probably the best way to reestablish the original packing of a sample is by mechanical vibration; electrical vibrators are extremely useful for this purpose. Hand jarring or tamping also tends to reduce the bulk of the sample. In addition, it is possible to introduce water at a very slow rate into the sample chamber; as the water rises by capillarity, small particles are pulled downward into the voids.

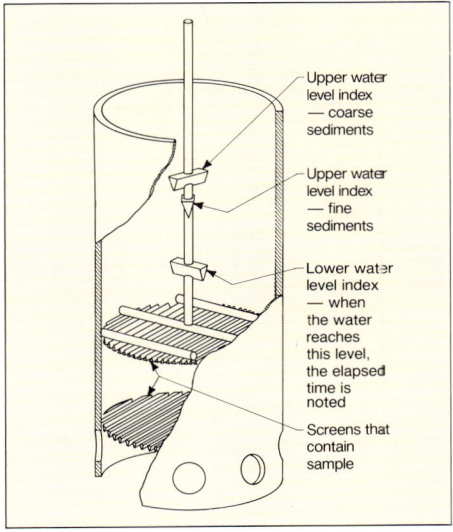

Figure 5.16. A field permeameter is based on the falling-head principle in which a certain volume of water is allowed to flow through a predetermined thickness of aquifer material. Once the time for the flow is known, an estimate of hydraulic conductivity can be made using Darcy's equation.

Results obtained from properly built and operated permeameters may be quite accurate, provided that a sample has been returned to its field condition. Of course, a relatively undisturbed sample is best for testing.

FLOW NETS

Flow in an unconfined groundwater environment is controlled by impermeable geologic boundaries and the water table, which also acts as a boundary. Because no flow can cross these boundaries, the water must flow more or less parallel to them in what are called flow lines. Flow occurs because the potential energy head drives the water from areas of higher head to areas of lower head (that is, for example, from beneath a hill to a stream channel) (Figure 5.17a). Flow lines are perpendicular to lines of equal water-table elevations (potential head elevations) (Figure 5.17b).

The dotted lines in Figure 5.17b represent the potential head at those places on the groundwater table; these lines are called equipotential lines. Along each dotted line,

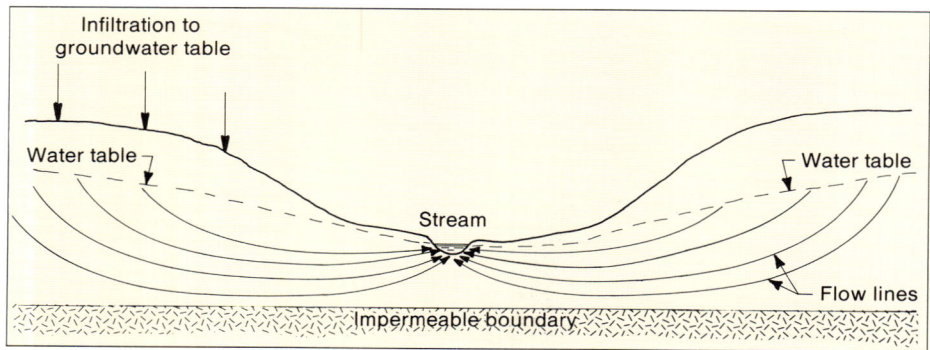

Figure 5.17a. Cross section through a stream valley showing flow lines in the groundwater system.

the potential energy head is the same. Flow lines can be easily drawn where the elevations of the water table are known because groundwater always flows perpendicular to these potentials in a down-gradient direction. This is strictly true only for isotropic aquifers where flow rates can be the same in all directions. The set of intersecting flow lines and equipotential lines is called a flow net. Figure 5.18 shows the arrangement of flow lines and equipotential lines around a pumping well in a homogeneous aquifer; the fault crossing the aquifer serves as an impermeable boundary.

Flow patterns and equipotential lines demonstrate how adjacent wells can have different water levels, depending on the hydraulic head at their intake points (Figure 5.19). This condition exists in many well fields but may be obscured because the elevation differences are minor or natural inhomogeneities of the aquifers cause distortion of the flow lines.

Knowing the direction of groundwater movement has become increasingly important because of the danger of contaminating groundwater supplies. Wells may become unsafe when sewage or other contaminants enter the ground at a higher head (gradient)

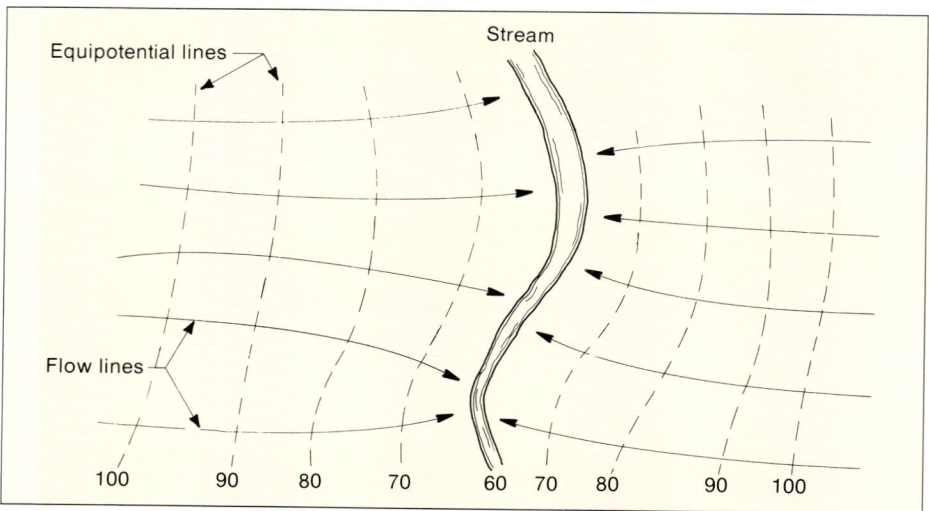

Figure 5.17b. Looking down on the stream valley from above with the water table exposed. The dotted lines represent points of equal groundwater elevation.

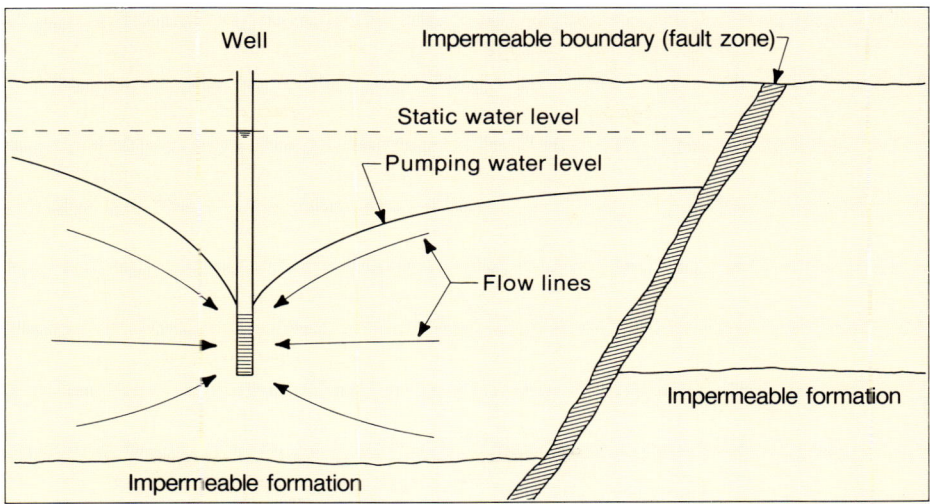

Figure 5.18a. Cross section near a pumping well showing the flow lines followed by water moving toward the well.

than exists in nearby shallow wells. The flow lines, or lines of water movement, can be determined by using water-elevation data from a minimum of three wells (Figure 5.20). The heavy solid line shown in Figure 5.20 shows the flowpath taken by water in the area. If sufficient information is available, flow lines may be drawn which indicate the general direction of groundwater flow in a specific area. Areas of recharge and discharge can also be defined by the use of a flow-net diagram. Flow lines diverge in areas of recharge and converge in areas of discharge.

GROUNDWATER FLOW VELOCITIES

Ordinarily, the rate of groundwater movement is of negligible interest to water well

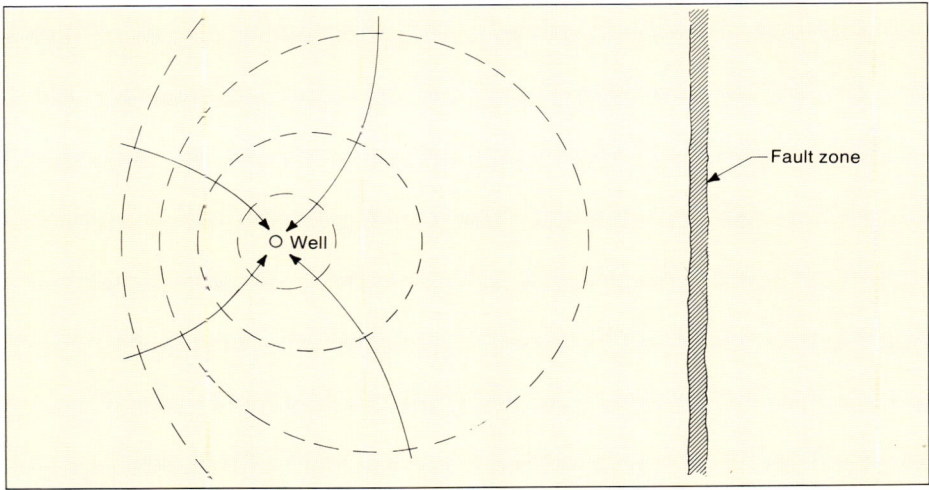

Figure 5.18b. Map of the potentiometric surface during long-term pumping shows that the impermeable boundary causes greater drawdown in the direction of the fault.

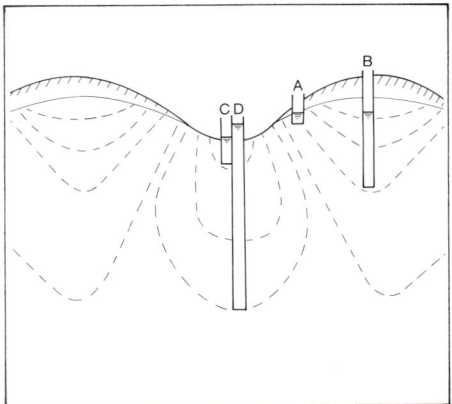

Figure 5.19. Water level in a well rises to the elevation of the hydraulic head represented by the potential at the intake end of the well. The water level in wells A and B is the same because both wells terminate on the same equipotential line. *(From APPLIED HYDROGEOLOGY, by C. W. Fetter, Jr., © 1980 Merrill Publishing Company, Columbus, Ohio)*

contractors or others concerned primarily with yields from wells. Yields are directly affected by flow velocities, of course, but velocity in an aquifer is difficult to study. Determination of the groundwater flow rate, however, has become more important as underground contamination problems multiply. The ability to predict the rate at which a plume of contaminated water may move downgradient from a source of pollution has become a vital water management objective.

The average groundwater flow rate is easily determined by combining Darcy's equation with the standard continuity equation of hydraulics. The continuity equation simply states that whatever goes into a system must come out. In Darcy's equation, the hydraulic conductivity has the dimensional units of velocity. When the conductivity is expressed as cubic feet (meters) per day passing through one square foot (meter) of aquifer, this is the same as expressing the flow rate in terms of feet (meters) per day under the defined hydraulic gradient of one. Multiplying this flow rate by the actual hydraulic gradient in the aquifer gives the velocity of groundwater movement in that aquifer.

Thus, using Darcy's equation:

$$Q = \frac{KA(h_1 - h_2)}{L} \qquad (5.14)$$

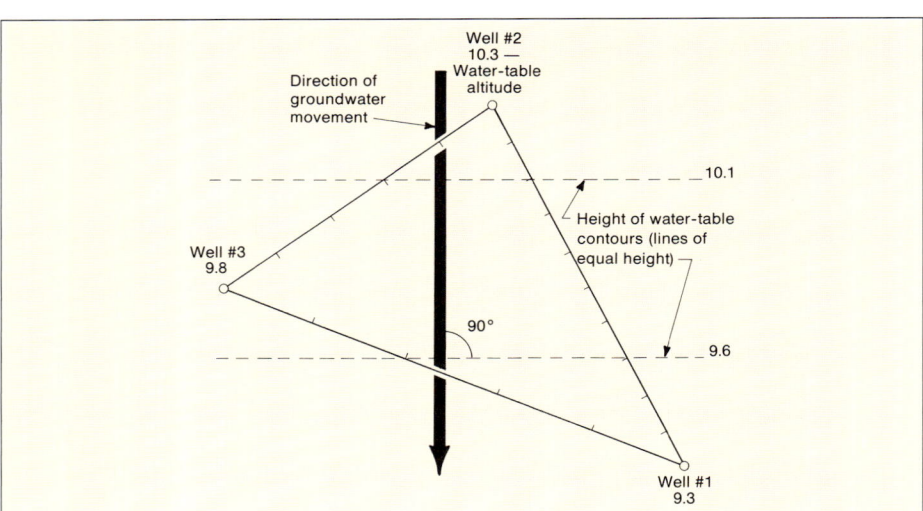

Figure 5.20. The direction of groundwater flow can be calculated by connecting a triangle linking the known potentiometric levels of three wells.

and the continuity equation:

$$Q_{in} = V_1 A_1 = V_2 A_2 = Q_{out} \tag{5.15}$$

we can make the substitution:

$$VA = \frac{KA(h_1 - h_2)}{L} \tag{5.16}$$

and by eliminating terms:

$$V_{Df} = \frac{K(h_1 - h_2)}{L} \tag{5.17}$$

Equation 5.17 is not yet complete, for actual groundwater flow velocities will be higher than the equation indicates because flow occurs only through the actual pore space, and not through the entire cross section of the porous medium. The actual velocity of groundwater flow, V_a, can now be calculated by knowing only the hydraulic conductivity, the hydraulic gradient, and the average porosity:

$$V_a = \frac{\frac{K(h_1 - h_2)}{L}}{\eta} \tag{5.18a}$$

where η equals the porosity of the porous medium.

In the English system of measurement, the hydraulic conductivity, K, is in gallons per day per square foot, and this volume must be converted to cubic feet to calculate a velocity in feet per day. To obtain the velocity of the water in feet per day through the actual openings in a formation, divide Equation 5.17 by 7.5 (the number of gallons in a cubic foot) and by the porosity of the aquifer. Equation 5.18a becomes:

$$V_a = \frac{\frac{K(h_1 - h_2)}{L}}{7.5 \cdot \eta} \tag{5.18b}$$

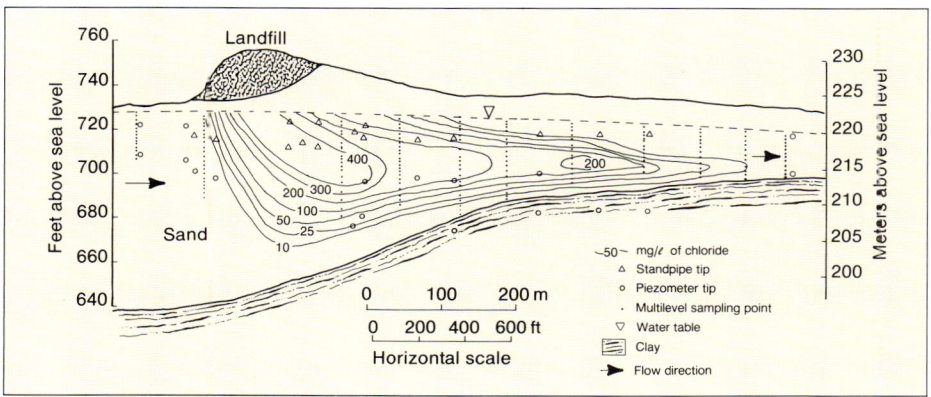

Figure 5.21. Plume of chloride leachate migrating from a sanitary landfill into a homogeneous, sandy aquifer; contaminated zone is represented by contours of chloride concentration in groundwater. *(R. Allen Freeze and John A. Cherry, GROUNDWATER, ©1979, page 439. Reprinted by permission of Prentice-Hall, Inc., Englewood Cliffs, NJ.)*

The reader should keep in mind that the velocity calculated in Equation 5.18b is only approximate. In certain parts of an aquifer, the interconnections between pores may be almost straight line in a downgradient direction, whereas elsewhere the water is forced to travel great distances to move rather short distances downgradient. Tor-

Table 5.5. Common Types of Tracers and Their Special Uses

Tracer	Lowest Detectable Concentrations	Advantages	Disadvantages
Dyes: Uranine (Sodium fluorescein) Rhodamine B Sulforhodamine G. Extra	0.01–0.04 $\mu g/\ell$	Easy to use, safe. Concentrations can be measured in the field.	Some dyes affected by pH, temperature, or are absorbed by clays and organic soils.
Strong Electrolytes Sodium Chloride Potassium Chloride Ammonium Chloride Lithium Chloride	1–0.001 mg/ℓ	Can be measured in field or laboratory by electrical conductivity or electrical resistivity. Smaller concentrations measured by atomic absorption spectroscopy.	Must use large amounts of salts if ordinary analytical methods are used.
Radioactive Isotopes Tritiated Water Iodine ion Many others	0.001 mg/ℓ or lower	Can be used in such small quantities that no effect on physical and chemical properties of water occurs. Concentrations easily measured by sophisticated equipment. Cost of tritiated water very low.	May be radiation danger (not in case of tritiated water). Requires expensive detection equipment.
Detergents Alkylbenzol- sulfonates	0.05 mg/ℓ	Easy to use, safe.	May be confused with sewage-related detergents. Material disperses aggregated soils thereby changing permeabilities.

tuosity is the special term applied to the length of the sinuous path followed by a fluid particle. Tortuosity is defined as the length of flow path divided by the overall length of sample. The arrival time of any contaminant depends partly on the tortuosity of its flowpath through the aquifer.

Measuring Groundwater Flow Velocities

Groundwater flow velocity can be measured by placing a tracer such as dye or salt in one well and noting the time of its arrival in a second well downgradient from the first. Tracers are used also to determine groundwater flow patterns, the age of groundwater, geologic and geophysical origin of groundwater, volume of water in an aquifer, and physical and chemical characteristics of an aquifer such as porosity, hydraulic conductivity, and dispersivity.

The ability of a tracer to indicate dispersion is particularly important because knowledge of the dilution rate of any contaminant is extremely important in assessing the severity of pollution problems. Except in karstic regions, dispersion of a contaminant usually results in a plume of contaminated water downgradient from the source (Figure 5.21). The width, depth, and spreading velocity of this plume can be determined by tracers.

The type of groundwater investigation followed will usually dictate which tracer should be used. But generally any tracer must be detectable in extremely low concentrations and must not react chemically or physically with the groundwater or the aquifer material. Table 5.5 lists some common types of tracers and their special uses.

Tracer analysis has improved greatly in recent years because new testing devices are capable of detecting extremely small concentrations of any chemical added to groundwater. Extensive tracer tests of an aquifer can yield valuable information concerning flow rates and other physical characteristics. But understanding the physical complexities of even small aquifers may be only partly achieved by tracer methods alone. A more thorough discussion of contaminants appears in Chapter 21.

CHAPTER 6
Groundwater Chemistry

Many well design engineers and water well contractors may not be familiar with water chemistry because their work focuses on the design and construction of wells that will yield adequate supplies of water. In many cases, however, the chemical nature of the groundwater, not the volume of water obtained, may be the major factor determining a well's suitability. Furthermore, the success of many wells depends on the drilling contractors' knowledge of chemical changes that occur in or near a pumping well and how their activities can cause serious chemical changes near the well, resulting in low yields.

In recent years, society has become aware of groundwater and the state of its chemical quality. New methods of measuring contaminants in extremely small concentrations (parts per billion or even parts per trillion), and the concurrent discovery of widespread underground contamination problems, have led health departments and other environmental agencies to become much more active in monitoring groundwater quality. Now, more than ever, the drilling contractor faces greater challenges in complying with new laws that address groundwater quality and in dealing with an ever-increasing number of industrial and domestic contamination problems. These challenges and problems can be handled satisfactorily only through greater understanding of the chemical reactions taking place in the ground.

Some of the physical properties of water were discussed in Chapter 1. A few of these properties are important in many phases of groundwater hydrology. One of the most unusual characteristics of water is its ability to dissolve a greater range of substances than any other liquid. The slow percolation of water through the ground results in prolonged contact of water with minerals in the soil and bedrock. Many of these minerals are dissolved slowly as groundwater passes over them, and in time a quasi-chemical equilibrium can be reached between the groundwater and the minerals in the soil and rock. The water is then saturated with dissolved solids derived from these minerals. This ability of water to dissolve minerals determines the chemical nature of groundwater.

What gives the water molecule its unusual dissolving power? Atoms and molecules are either positively or negatively charged before combining to form minerals or compounds. The charged atoms and molecules are called ions. Positively charged ions are called cations, and negatively charged ions are called anions. In the water molecule (H_2O), the two hydrogen ions (or atoms) have positive charges and the oxygen ion has a double negative charge; thus, the molecule has a balanced electrical charge. However, the arrangement of the atoms is not symmetrical, and the ends of the molecule have residual positive and negative charges (Figure 6.1). This kind of molecule is called dipolar, that is, it consists of two poles with opposite charges. On a larger scale, the Earth also is dipolar; the North Pole is positively charged and the South Pole is negatively charged.

Every rock-forming mineral consists of positively and negatively charged ions that are attracted to one another because of the plus and minus electrical charges. Ions can be removed from a mineral only if an electrochemical charge can be applied to break the electrical attractions between the mineral ions. Molecules of most solvents are balanced electrically and thus do not have the ability to attract positive and negative ions. Water is an exception. When water is in contact with a mineral surface, the positively and negatively charged ends of the water molecules attach themselves to the ions at the mineral surface. The water molecules tend to neutralize the attracting forces binding the mineral ions and, in many cases, are strong enough to remove the ions from the mineral surface. As long as unsaturated water remains on the mineral surfaces, the dissolving process will go on until chemical equilibrium is reached between the water and the mineral. This usually occurs when groundwater movement is slow and minerals are relatively soluble.

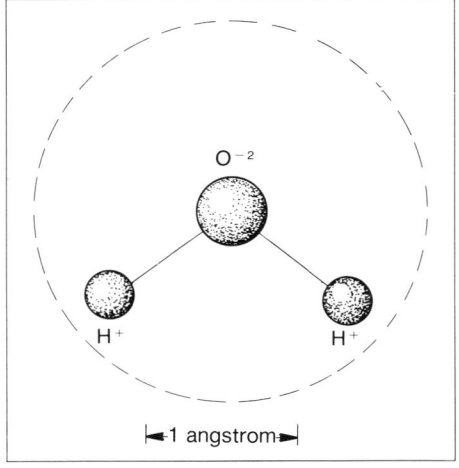

Figure 6.1. Dipolar structure of the water molecule. The hemisphere with the two hydrogens carries a net positive charge; the opposite hemisphere, with oxygen, carries a net negative charge.

Groundwater is not pure water because it usually contains dissolved mineral ions. The type and concentration of these dissolved minerals can affect the usefulness of groundwater for various purposes. If certain mineral constituents are present in excessive amounts, some type of treatment may be necessary to either change or remove the dissolved mineral before the water can be used for the intended purpose.

Most groundwater contains no suspended particles and practically no bacteria or organic matter. It is usually clear and odorless. These characteristics contrast with those of surface waters, which generally contain suspended matter and considerable quantities of bacteria. Therefore, groundwater is normally more hygienic, although it ordinarily contains more dissolved solids than does surface water. Some of these dissolved minerals are essential for good health; others, if too abundant, may cause taste or odor problems.

Drilling contractors in many states are required to take water samples from wells they have drilled. These samples are then submitted to a state agency or local laboratory for analysis. A typical chemical analysis includes hardness, specific electrical conductance, hydrogen ion activity (pH), free carbon dioxide, and total dissolved solids. Each of these properties is useful in evaluating the chemical character of groundwater. Sampling techniques, however, can affect the accuracy of the chemical analysis. In this chapter, correct sampling methods are described for more accurate and complete analyses.

ORIGIN OF THE CHEMICAL CONSTITUENTS OF GROUNDWATER

Table 6.1 shows the distribution of elements in the Earth's crust. About 95 percent of the crust by volume is composed of a mineral group called silicates. The basic building block for all silicates is the complex silica ion, SiO_4^{-4}*. Numerous metallic ions (cations) such as calcium, potassium, iron, magnesium, and sodium can bond with the silica ion because it is not balanced electrically. Although originally derived from igneous rocks, most of the calcium, magnesium, and bicarbonate ions found in groundwater come from the dissolution of carbonate rocks. Silicon and oxygen constitute approximately three-fourths of the mass of the Earth's crust (Figure 6.2). Even though the oxides of silicon, aluminum, and iron are the most common substances in the crust, they are only slightly soluble in water. However, the cations (sodium, calcium, magnesium, and potassium) that combine with these oxides are quite soluble and are present in most freshwater supplies. Table 6.2 lists typical cations and anions found in groundwater.

An increase in water temperature usually increases the solubility of most minerals, although there are important exceptions [for example, carbon dioxide (CO_2)]. Gen-

Table 6.1. Distribution of Elements in the Earth's Crust (expressed as oxides)

Constituent	Average Composition of Crust (wt %)
Silica (SiO_2)	63.5
Alumina (Al_2O_3)	15.9
Calcium (CaO)	4.9
Sodium (Na_2O)	3.3
Potassium (K_2O)	3.3
Iron (FeO)	3.3
Iron (Fe_2O_3)	2.9
Magnesium (MgO)	2.9
Carbon (CO_2)	—
Hydrogen (H_2O)	—

(Garrels and MacKenzie, 1971)

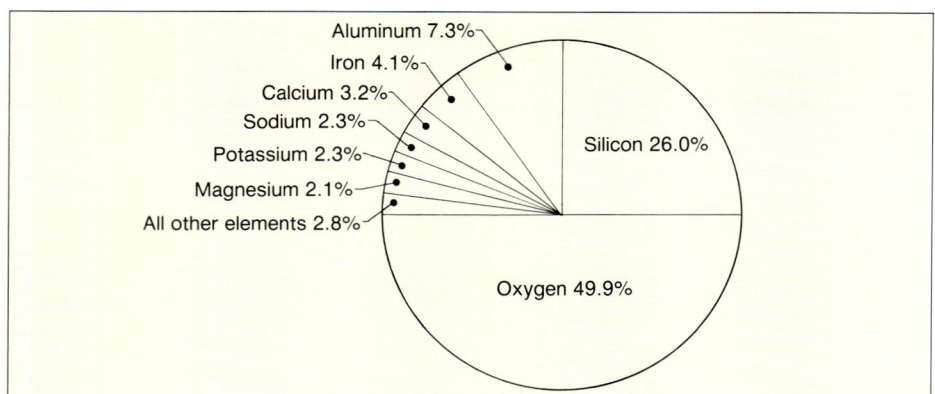

Figure 6.2. Distribution of elements in the Earth's crust. *(From THE NALCO WATER HANDBOOK, Frank N. Kemmer, Editor-in-chief. Copyright ©1979 by McGraw-Hill, Inc. Used with the permission of the McGraw-Hill Book Company.)*

---

*The charge of -4 for the complex ion SiO_4^{-4} is called its valence and indicates the capacity of this ion to react, unite, or interact with something else. Each electron that can be exchanged or shared is indicated in the chemical formula. In this case, 4 electrons are available to unite with cations.

Table 6.2. Major Cations and Anions Found in Groundwater

Cations	Anions
Calcium (Ca^{+2})	Bicarbonates (HCO_3^-)
Magnesium (Mg^{+2})	Sulfates (SO_4^{-2})
Sodium (Na^+)	Chlorides (Cl^-)
Potassium (K^+)	

(Gorham, 1961)

erally, near-surface groundwater has a nearly constant temperature, which ordinarily is close to the mean annual air temperature. Geothermal heat will increase groundwater temperatures about 1°F (0.6°C) per 100 ft (30.5 m) of depth beginning at approximately 100 ft. Therefore, most groundwater found at greater depths is more mineralized.

UNITS OF MEASURE

The most common measurement practice in the United States has been to report dissolved minerals in parts per million (ppm) by weight, that is, one part by weight of dissolved mineral contained in one million parts by weight of water. The concentration of minerals in water is also referred to as total dissolved solids (TDS). The dissolved minerals are classified as inorganic salts, thus the term "salinity" is another way to describe mineral concentration. Under the SI System, mass concentration of dissolved solids in a liquid is defined as kilograms per cubic meter (kg/m^3). The same ratio of mass to volume is maintained in the milligrams per liter (mg/ℓ) designation. Therefore, the mg/ℓ unit is used because it is both accurate and numerically equal to the ppm units for high-quality fresh water. For most practical purposes, water with less than about 10,000 mg/ℓ total dissolved solids (TDS), and at temperatures below 212°F (100°C), can be considered to have a density sufficiently close to 1 kg/ℓ so that 1 mg/ℓ equals 1 ppm (Freeze and Cherry, 1979). When water has a higher salinity or temperature, the equivalence between 1 mg/ℓ and 1 ppm no longer holds and density corrections must be made.

In countries where the English system of units is still used, hardness, or the dissolved mineral concentration, is often expressed in grains per gallon. Water treatment specialists generally use this unit of measure. One grain per United States gallon equals 17.12 mg/ℓ. To convert mg/ℓ to grains per United States gallon, multiply by 0.058.

Sometimes chemists use cation and anion concentrations to verify the validity of a chemical analysis of a water sample. By converting the mg/ℓ concentrations to equivalents per million (epm), also called milligram equivalents per liter (meq), the electrochemical composition of the ions in the water can be described. To determine the epm, an ion concentration in mg/ℓ is divided by the equivalent weight of a particular ion. Equivalent weight equals molecular weight divided by the valence. Because all solutions are electrochemically neutral, the sum of the equivalents per million of the positively charged ions must be equal to the sum of the epm of the negatively charged ions. In Figure 6.3, the equivalents per million are calculated for (a) 63 mg/ℓ magnesium and (b) 5 mg/ℓ nitrate. In (c), the concentration of ions in a water sample is checked on the basis of a mg/ℓ analysis. The cation and anion totals are not equal, suggesting that either some cations were missed in the analysis or measurement errors

a. To convert 63 mg/ℓ Mg^{+2} to epm:

   Atomic weight Mg = 24.32
   Valence = 2
   Equivalent weight = $\frac{24.32}{2}$ = 12.16

   63 mg/ℓ Mg^{+2} = $\frac{63}{12.16}$ = 5.19 epm

b. To convert 5 mg/ℓ NO_3^- to epm:

   Atomic weight N = 14.0
   Atomic weight O = 16.0
   Molecular weight NO_3^- = 62.0
   Valence = 1
   Equivalent weight = $\frac{62.0}{1}$ = 62

   5 mg/ℓ NO_3^- = $\frac{5}{62}$ = 0.08 epm

c. To check the following analysis, tabulate the epm of anions and cations and then add each column.

Ion	mg/ℓ	epm cations	epm anions
Ca^{+2}	42	2.10	
Mg^{+2}	27	2.22	
HCO_3^-	196		3.21
SO_4^{-2}	15		0.31
Cl^-	72		2.03
NO_3^-	5		0.08
Total		4.32	5.63

Figure 6.3. Calculating the equivalent parts per million (epm) of a water sample.

occurred in the identified substances. Common chemical conversion factors for substances found in groundwater are presented in Appendix 6.A.

Some important constituents of water are not electrically charged and therefore must be measured in different ways depending on the substance. These substances include some organic compounds, colloids*, and silica. Many organic comounds are soluble in water, but identification of specific carbonaceous materials may be difficult. At least ten tests can be performed to determine the non-specific organic content of water. Waters having high concentrations of organic materials are often stained a deep brown or reddish-brown color. Many heavy metals may be present in waters as colloidal material. Iron and manganese, for example, can be in either colloidal or ionized form, depending on the oxygen content of the water. If the water is deoxygenated, both iron and manganese will be in soluble form and can be measured as cations. Because most groundwater contains little or no oxygen, iron and manganese are usually in solution. Silica in the form of sodium silicate forms colloids which may be removed by filtration or coagulation techniques. The volumetric distribution of colloids in water is non-uniform and therefore the chemical forming the colloid is not measured as a concentration. Various major instruments or methods used to measure ions in solution, solids (colloids or incrustants), specific and total organics, and organisms are shown in Figure 6.4.

IMPORTANT PROPERTIES OF WATER

Hardness

The term "hardness" may be one of the oldest words in use to describe a property of water. Ever since the use of soap became prevalent, people noticed that, depending on the water source, different amounts of soap were needed to produce suds. Water requiring more soap was called hard water; that is, suds were hard to produce. Rainwater, on the other hand, required little soap to produce suds and was called soft

---

*Colloids are chemical substances that remain suspended in a liquid.

water. Hardness might be considered the soap-consuming property of water, because no suds can be produced in hard water until the minerals causing the hardness have been combined with the soap. The minerals that are removed by soap remain as an insoluble scum — the familiar bathtub ring.

Hardness in water is caused primarily by calcium and magnesium cations, although some heavy metals such as iron and manganese also consume soap. The principal chemical sources for the calcium and magnesium ions are the compounds of calcium and magnesium bicarbonate, carbonate, and sulfates. Almost all of the carbonate and bicarbonate ions in groundwater originate in soils from respiring organisms and decaying vegetation and from the dissolution of carbonate rocks such as dolomite and limestone.

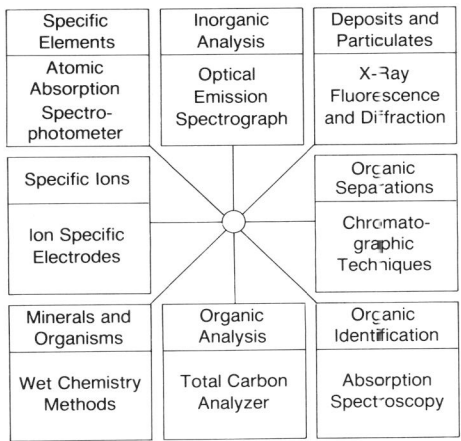

Figure 6.4. A display of the various devices and techniques commonly used for water analysis.

Hardness of water may be divided into two types, carbonate and noncarbonate. Carbonate hardness includes that portion of the calcium and magnesium that combines with bicarbonate and the small amount of carbonate present. This was once called temporary hardness because it can be removed by boiling, which precipitates calcium and magnesium carbonate and sulfate minerals. Hardness is usually expressed in terms of calcium carbonate.

Noncarbonate hardness is the difference between total hardness and carbonate hardness. It is caused by those amounts of calcium and magnesium that combine normally with the sulfate, chloride, and nitrate ions, plus the slight hardness contributed by minor constituents such as iron. Noncarbonate hardness cannot be removed by boiling.

Water that has a hardness of less than 50 mg/ℓ is considered soft. A hardness of 50 to 150 mg/ℓ is not objectionable for most purposes, but the amount of soap needed to reduce the calcium and magnesium increases with the mineral content. Laundries and other industries using large quantities of soap generally find it cost effective to reduce hardness to about 50 mg/ℓ. Water having 100 to 150 mg/ℓ hardness will deposit considerable scale in steam boilers. Hardness of more than 150 mg/ℓ is quite noticeable because of scale buildup or staining. At concentrations of greater than 200 mg/ℓ, water is commonly softened for household uses. When municipal water supplies are softened, the hardness is reduced to about 85 mg/ℓ. Further softening of municipal water is not considered economical.

Numerical values for hardness have meaning only in a relative sense. An individual living in New England, where waters normally contain only small amounts of dissolved solids, might consider water with 100 mg/ℓ hardness as very hard. In contrast, water of 100 mg/ℓ hardness in Iowa, Minnesota, or Nebraska might be considered soft by local residents.

Some of the calcium and magnesium contributes to incrustation that may develop when hard water undergoes changes in temperature and pressure. This type of incrustation results from a change in the solution balance which produces the insoluble carbonates of those metals and, in some instances, insoluble sulfates. The problem of scale deposits in tea kettles is familiar to everyone and causes little concern. But scale deposits in large evaporators such as cooling towers and, under certain conditions, in water wells, can lead to complete operational failure. A brief explanation for the scale deposits caused by changes in pressure and temperature is presented below.

Carbon dioxide that is loosely bound into the bicarbonate ion can be partly driven off as gas by heating the water. In this process, part of the bicarbonate changes to carbonate which then reacts with calcium and magnesium ions, resulting in a precipitate of insoluble calcium and magnesium carbonate scales. Calcium carbonate scale forms first be-

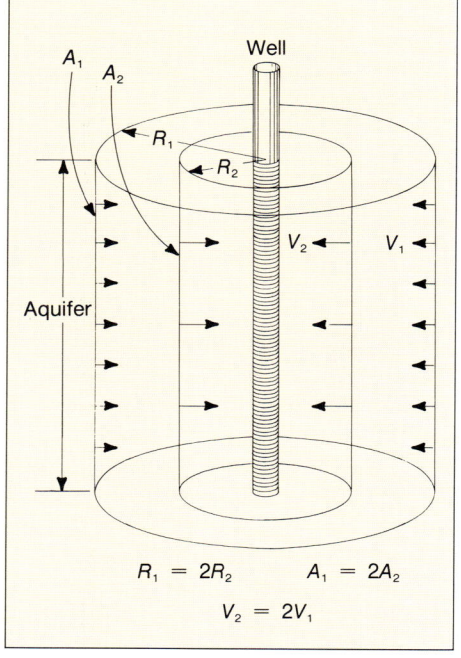

Figure 6.5. As water moves toward a well, its velocity varies with the distance from the well.

cause it is more insoluble than magnesium carbonate. In the absence of carbon dioxide, water can carry only about 14 mg/ℓ of calcium carbonate in solution. Under the same conditions, the solubility of magnesium carbonate is more than five times greater, or about 70 mg/ℓ.

Reduced water pressures also result in scale deposits; they are particularly troublesome in wells drilled in limestone and dolomite. As shown in Figure 6.5, the velocity of water must increase as it moves from an area that is away from a pumping well (cylinder A_1) to an area nearer the well (cylinder A_2). Because the same volume of water must pass through both imaginary cylinders, the velocity of the water is doubled as the distance to the well is halved. As shown in the discussion of the Bernoulli equation (Chapter 5), any increase in the velocity of groundwater flow results in a corresponding decrease in pressure. Therefore, the pressure decreases near the well as the water velocity increases. The decrease in pressure is especially significant within 3 to 6 ft (0.9 to 1.8 m) of the well. Reduced pressure in this zone leads to the release of carbon dioxide gas, causing some of the bicarbonate to react with magnesium and calcium to form carbonate scale. Depending on the hydraulic characteristics of the well and surrounding formations, scale deposition can cause serious losses in well efficiency or actual loss of the well in periods as short as one to three years.

Specific Electrical Conductance

The electrical conductance of a substance is its ability to conduct an electrical current. Current flows in ionized or mineralized water because the ions are electrically charged and move toward a current source that will neutralize them. For example,

Table 6.3. Conductivity of Two Salt Solutions

Concentration of Dissolved Salt in mg/ℓ	Specific Conductance at 77°F (25°C), Micromhos	
	NaCl	Ca(HCO$_3$)$_2$
50	93	62
100	187	125
200	370	250
400	750	500

when sodium chloride (NaCl) dissolves in water, the sodium (Na$^+$) ions become positively charged and the chlorine (Cl$^-$) ions become negatively charged. If current is introduced into the liquid by two electrodes, the sodium ions will move toward the negatively charged electrode, whereas the chlorine ions move toward the positively charged electrode. Anions (negative charge) move toward the positive electrode (anode), whereas cations (positive charge) are attracted to the negative electrode (cathode).

Specific electrical conductance is defined as the conductance of a cubic centimeter of any substance compared with the conductance of the same volume of water. Chemically pure water has a very low electrical conductance, indicating that it is a good insulator. Only a minute amount of dissolved mineral matter, however, will increase the conductance of the water.

The units of measurement for conductance are the inverse of ohms, the unit commonly used to express resistance. The conductivity units are written as mhos (pronounced "mose"). Values of specific conductance for groundwaters are reported in millionths of mhos (micromhos) [microsiemens (μS) in the SI system]. In solutions as dilute as most groundwaters, the specific conductance varies almost directly with the amount of dissolved minerals. However, the specific conductance for solutions of different minerals is not the same. For example, 100 mg/ℓ of sodium chloride, NaCl, gives water a specific conductance that is higher than for 100 mg/ℓ of calcium bicarbonate, Ca(HCO$_3$)$_2$. Table 6.3 presents conductance values for solutions of these two mineral salts. To be comparable, conductance mesurements must be performed at the same temperature.

Figure 6.6 shows the strong correlation between total dissolved solids and specific conductance. As a result, the concentration of dissolved salts can be estimated on the basis of electrical conductivity measurements. For most groundwaters, the specific conductance multiplied by a factor of 0.55 to 0.75 gives a reasonable estimate of the dissolved solids. For the values shown on Figure 6.6, the specific conductance multipled by a

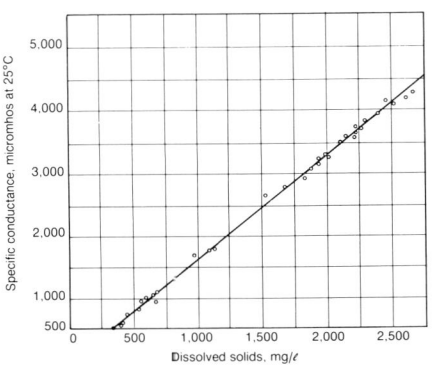

Figure 6.6. Dissolved solids concentration compared with specific conductance shows a constant ratio between conductance and mineral content of water, taken from Gila River at Bylas, Arizona, in a 12-month period. *(U.S. Geological Survey)*

factor of 0.55 to 0.63 gives the dissolved-solid concentration. The multiplication factor for saline waters is usually higher than 0.75, and for acidic water it may be much lower. Estimation of total dissolved solids from conductivity measurements is convenient because conductivity can be determined quickly in the field by rather simple equipment and procedures.

Water that has relatively high specific conductance can corrode iron and steel, even though other properties of the water may not indicate a corrosion problem. Because specific conductance reflects activity of the electrically charged ions, it follows that the higher the conductivity the greater will be the potential for electrochemical action. The importance of this statement as it relates to the long-term success of wells becomes clear in the chapters on well design (Chapter 13) and maintenance (Chapter 19).

Hydrogen Ion Concentration (pH)

Although water molecules are quite stable chemically, they still tend to break down or dissociate into their component parts, H^+ (hydrogen) ions and OH^- (hydroxyl) ions:

$$H_2O = HOH = H^+ + OH^-$$

Water is said to be either acidic or alkaline (basic), depending on the relative concentration of hydrogen ions. Hydrogen ions in water cause it to act as an acid. The capability of water to neutralize acid, that is, reduce the number of hydrogen ions in solution, is called alkalinity. The acidity-alkalinity characteristics of water are basic to an understanding of water chemistry.

The hydrogen ion concentration of water is expressed as pH. The pH is equal to the logarithm of the inverse of the hydrogen ion concentration, or

$$pH = \text{Log} \frac{1}{H^+}$$

This particular equation is used because the actual number of ions is very small. The pH range is from 0 to 14, with a pH value of 7 at 77°F (25°C) indicating a neutral solution in which H^+ and OH^- ions have the same concentration. A pH less than 7 indicates an acid solution, whereas a pH greater than 7 indicates an alkaline solution. Temperature plays a roll in determining the pH at which neutrality occurs. For example, at 32°F (0°C) the concentrations of positive and negative ions are equal at pH 7.53, whereas at 122°F (50°C) neutrality occurs at pH 6.65 (Freeze and Cherry, 1979). The pH of some common liquids is presented in Table 6.4.

Table 6.4. pH of Some Common Liquids

Lemon juice	2.2-2.4	Milk	6.3-6.6
Vinegar	3.0	Saliva, human	6.5-7.5
Tomato juice	3.5	Blood, human	7.3-7.5
Wine	2.8-3.8	Urine, human	4.8-8.4
Beer	4-5	Seawater	8.3
Cheese	4.8-6.4	Drinking water	6.5-8.0

Acid conditions generally lead to corrosive attack on metallic objects placed in groundwater. Corrosion becomes a problem when the pH is below 7, but corrosion rates for steel are minimal at a pH above 11. It should be noted, however, that the specific ions determining pH may also affect the corrosion rate. In some cases, a very low pH may cause little corrosion if the pH is produced by a particular group of ions. In other cases, corrosion may exist at pH values of about 11. In addition, some corrosion products form a protective film that may retard further corrosion.

There is some confusion regarding the term "alkalinity," because the presence of alkalinity does not mean that the pH of the water must be greater than the neutral value of 7. Groundwater with a pH value of less than 7 may still contain salts that will neutralize acids and, therefore, may have some measurable alkalinity. Carbonate and bicarbonate ions contribute to alkalinity, whereas chloride, sulfate, and nitrate ions do not. Hydroxyl (OH^-) ions affect alkalinity, but very few groundwaters contain enough hydroxyl ions to be of concern. Hydroxide (hydroxyl ions) may be found in treated water and in water that has been in contact with concrete.

The pH of most groundwaters results from the balance between dissolved carbon dioxide gas derived from the atmosphere and biological activity, and the dissolved carbonates and bicarbonates derived from carbonate rocks. This balance occurs naturally in the existing geologic and biologic environment at or near the Earth's surface. For example, water high in carbon dioxide will progressively dissolve any available carbonate rocks until the pH is generally between 7 and 9. An increase in soil alkalinity often produces greater biological activity, causing more carbon dioxide to be released to the atmosphere from respiration and decay. In time, percolating rainwater enriched in carbon dioxide reduces the pH of the groundwater to about 7.

According to Hem (1970), most groundwaters in the United States have pH values that range from about 6 to 8.5. Alkaline conditions generally occur in limestone and dolomite terrains and in regions recently glaciated by ice containing these rock types. Those waters with pH values lower than about 4.5 probably contain free-mineral acids from mine water, volcanic gases, or contamination by certain industrial wastes. Acids in mine waters, particularly in coal areas, are almost always sulfuric and come from the oxidation of iron pyrite or other metallic minerals. Fortunately, free-mineral acids are not common in groundwater in most areas.

The pH of water can be determined in numerous ways. For scientific purposes, pH is always measured with a pH meter equipped with an appropriate electrode. These units are generally accurate to 0.01 to 0.05 pH. Another common method uses acid-base indicators that undergo color change over a rather narrow pH range. The most widely used are methyl orange, phenolphthalein, and bromthymol blue. A universal indicator (pH paper), made by combining several acid-base indicators, may be used to determine the pH (within one unit) of any water. Indicators on the pH paper produce a series of colors ranging from deep red in strongly acidic solutions to deep blue in strongly basic solutions.

Alkalinity can be determined by adding a standard acid solution to water and noting the amount required to cause two different chemical indicators to turn color. Phenolphthalein and methyl orange are the two indicators used in alkalinity determinations. Figure 6.7 shows the general relationship of carbon dioxide, alkalinity, and pH values.

Determination of pH and dissolved minerals is important in the management of water wells, because these factors affect well screens. Samples taken from a pumping well, however, may not represent the true chemical character of water in the aquifer because depending on pressure changes and, to a lesser extent, temperature changes,

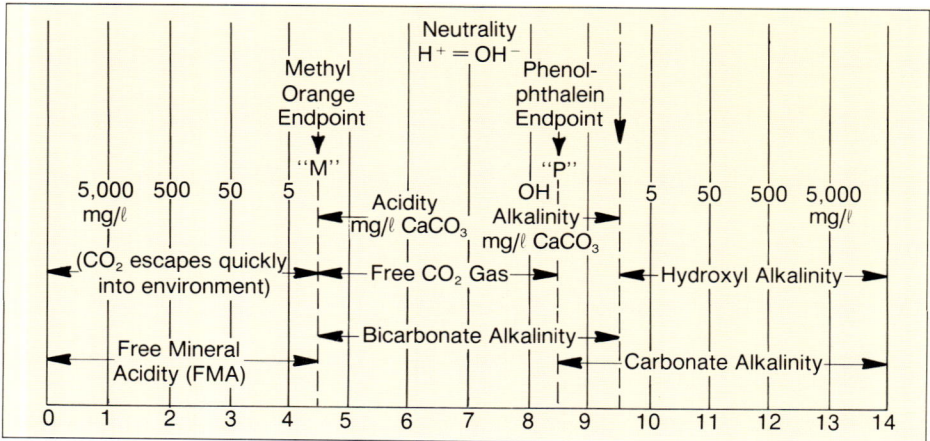

Figure 6.7. Acidity and various types of alkalinity and their pH ranges. *(From THE NALCO WATER HANDBOOK, Frank N. Kemmer, Editor-in-chief. Copyright ©1979 by McGraw-Hill, Inc. Used with the permission of McGraw-Hill Book Company.)*

some of the gas may escape during pumping before the sample has been collected. Carbon dioxide gas is released when pressure is reduced in the vicinity of the well. The greater the pressure drop near a well, the more likely it is that the true chemical character of the water has been altered. Immediate sealing of the sample bottle will retain the carbon dioxide that was in the water when the sample was taken.

Because the chemistry of groundwater is sensitive to environmental changes, several chemical properties should be determined while still in the field. Portable pH meters, for example, are available for field use, and several chemical companies make water-testing kits that require little knowledge of complicated water-testing procedures yet give excellent results. Reagents are premixed in many of the kits and the tester merely follows a simple step-by-step procedure.

Changes in Ionic Content of Groundwater through Time

Groundwater chemistry changes as water flows along in the underground environment, increasing in dissolved solids and most major ions (Chebotarev, 1955). The longer water remains underground, and the farther it travels, the more it resembles sea water; the principal anions change predictably. In this regard, Domenico (1972) has noted that groundwater chemistry changes with depth in large sedimentary basins:

1. Highest zone. Water is low in dissolved solids and high in bicarbonate (HCO_3^-) as active water movement effectively leaches* the rocks.

2. Middle zone. Water moves more slowly and gains in dissolved solids; the sulfate ion (SO_4^-) becomes dominant.

3. Deep zone. So little water moves through this zone that mineral leaching is extremely active, producing a high content of dissolved solids and a relative increase in the chloride ion (Cl^-).

Although this zonal analysis is somewhat simplified, it does indicate the general chemical changes that occur as water travels through the ground, particularly in sedimentary rocks. Further discussion of groundwater chemistry and man-induced con-

---

*Leaching is the removal of mineral salts by solution.

tamination is given in Chapter 21.

GROUNDWATER CONSTITUENTS

The total concentration of dissolved minerals in a water is a general indication of its suitability for any particular use. Total dissolved solids may be determined from the weight of dry residue remaining after a sample of water has evaporated. It may also be calculated by adding the concentrations of all ions in the water. This computed concentration, however, may be somewhat lower than the residue from evaporation. The difference may be 10 to 20 mg/ℓ for water containing 100 to 500 mg/ℓ of dissolved solids; for highly mineralized water, the difference may be even greater.

Water that contains abundant mineral matter is not suitable for certain uses. Categories of general water use include domestic and municipal water supplies, irrigation water, and industrial water used in manufacturing, mining, and power generation. Water that contains less than 500 mg/ℓ of dissolved solids is generally satisfactory for domestic use and many industrial purposes. Water with more than 1,000 mg/ℓ of dissolved solids usually contains minerals that give it a distinctive taste or make it unsuitable in other ways. Water high in dissolved solids should be viewed as potentially corrosive to well screens and other parts of well structures, regardless of other chemical characteristics of the water. Table 6.5 presents a groundwater classification system based on total dissolved solids.

The important mineral constituents in water are listed in Table 6.6. Fortunately, the concentrations of many of these substances are small and put few limitations on water

Table 6.5 Simple Groundwater Classification Based on Total Dissolved Solids

Category	Total Dissolved Solids (mg/ℓ)
Fresh water	0 - 1,000
Brackish water	1,000 - 10,000
Saline water	10,000 - 100,000
Brine water	More than 100,000

(Freeze and Cherry, 1979)

Table 6.6. Classification of Dissolved Inorganic Constituents in Groundwater

Major constituents (greater than 5 mg/ℓ)			
Bicarbonate	Silicon	Copper	Rubidium
Calcium	Sodium	Gallium	Ruthenium
Chloride	Sulfate	Germanium	Scandium
Magnesium		Gold	Selenium
Minor constituents (0.01-10.0 mg/ℓ)		Indium	Silver
Boron	Nitrate	Iodide	Thallium
Carbonate	Potassium	Lanthanum	Thorium
Fluoride	Strontium	Lead	Tin
Iron		Lithium	Titanium
Trace constituents (less than 0.1 mg/ℓ)		Manganese	Tungsten
Aluminum	Bromide	Molybdenum	Uranium
Antimony	Cadmium	Nickel	Vanadium
Arsenic	Cerium	Niobium	Ytterbium
Barium	Cesium	Phosphate	Yttrium
Beryllium	Chromium	Platinum	Zinc
Bismuth	Cobalt	Radium	Zirconium

(Davis and De Wiest, 1966)

use. On the other hand, even very small concentrations of iron and manganese can be troublesome. The particular substances discussed below are found naturally in groundwater in concentrations high enough to cause problems in operating wells.

Iron (Fe^{+2}, Fe^{+3})

Most water supplies contain some iron, because iron is common in many igneous rocks and is found in trace amounts in practically all sediments and sedimentary rocks. The iron content of water is important because small amounts seriously affect water's usefulness for some domestic and industrial purposes. Concentrations of 1 to 5 mg/ℓ in groundwater are common, although the concentration may fall to 0.1 mg/ℓ in water that is aerated during pumping or treatment. The standards of the U.S. Environmental Protection Agency recommend that the iron content of drinking water should not be greater than 0.3 mg/ℓ because iron in water stains plumbing fixtures, stains clothes during laundering, incrusts well screens, and clogs pipes. Some industrial plant processes, however, cannot tolerate more than 0.1 mg/ℓ of iron.

Water may dissolve iron upon contact with metal well casings, pump parts, and piping. Thus, large amounts of iron in any distribution system may not be derived from earth materials. The more corrosive the water (low pH or high oxygen content), the more metal it will dissolve.

Water standing in a well that has been idle may have a higher iron content than the natural waters in the aquifer. In sampling water, therefore, the pump should be operated long enough to remove all the standing water from the well. When the water is completely clear, the sample should be collected as close to the pump discharge as possible and before the water has come into contact with air.

The specific form iron takes in water depends on the amount of oxygen in the water and the pH. In natural groundwater systems where oxygen concentrations are low or absent and the pH is from 6.5 to 7.5, the iron occurs primarily as dissolved ferrous ions (Fe^{+2}). In most areas where iron problems exist, the iron concentrations range from 2 to 10 mg/ℓ. Ferrous ion concentrations may reach 50 mg/ℓ in water that has a pH of 7 and no oxygen (reduced conditions). Ferrous ions are unstable when in contact with oxygen. In the presence of air, they change to ferric ions (Fe^{+3}) and precipitate as ferric oxide or oxyhydroxides. Ferric iron is almost completely insoluble in alkaline or weakly acidic waters. When water with a pH of 7 to 8.5 is aerated, almost all of the iron becomes insoluble.

Most water problems that result from high iron content are associated with the sudden change from ferrous (dissolved) to ferric (semisolid) iron. Ferric oxides and oxyhydroxides come out of solution and coat surrounding surfaces. These coatings are precipitated from solution during aeration and also occur as rust on metal surfaces exposed to the atmosphere. Groundwater containing several mg/ℓ of iron may be completely clear and colorless when first pumped. Upon standing for a time in an open container, contact with the oxygen in air is sufficient to affect the dissolved iron. The water begins to cloud up slightly and later a deposit of rustcolored material appears at the bottom of the container.

Other precipitates are found within the well structure. For example, ferrous ions combine with carbonate ions to form iron bicarbonate. Although not as common as ferric oxides and oxyhydroxides, iron bicarbonates often contribute to the plugging of water well intake screens.

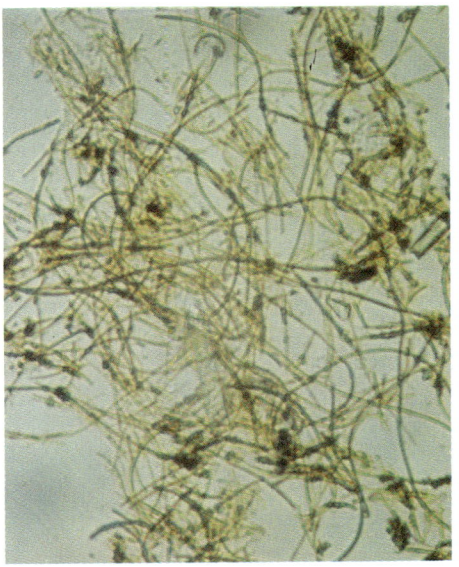

Figure 6.8. *Crenothrix* iron bacteria.

Iron-bearing waters also favor the growth of iron bacteria, such as *Crenothrix, Gallionella,* and *Leptothrix*. These growths form so abundantly in water mains, recirculating systems, and occasionally in wells that they have a marked clogging effect. In fact, their growth rate in wells may be so rapid that the water supply may be nearly shut off within months after a well is first put into operation. One of the most common bacteria, *Crenothrix,* are individually surrounded by a filamentous sheath (Figure 6.8). Iron bacteria can change certain forms of dissolved iron to insoluble ferric iron which is deposited in the sheaths of the organisms and in the voids of sand. The sheaths, developed as the bacteria multiply, create a gel-like slime that can seriously clog water-bearing formations and the openings in well screens. Unfortunately, these large gelatinous masses occasionally break loose and further clog the system. Iron bacteria thrive best in the dark in water containing little or no oxygen but with considerable carbon dioxide along with dissolved iron.

The original source of iron bacteria is not known. In some cases, the bacteria are already in the groundwater before the wells are drilled, and simply multiply as the localized food source is vastly increased through pumping. Alternatively, bacteria may be introduced into wells by drilling operations. For example, abundant iron bacteria are commonly present in some surface waters, especially swamps and marshes. If this water is used to mix drilling fluid without prior chlorination, bacteria will be injected into the ground. Dirty drill rods or bits can also carry bacteria into the well bores. Excessive growth may occur when sufficient iron is present to support these bacteria.

When the iron content of water is naturally high, treatment for removal is the ultimate solution. In Chapter 23, the Vyredox™ system for iron removal is described. In cases where the iron and carbon dioxide content of groundwaters are moderately low and there is no treatment, problems may be minimized by taking care to avoid aeration. Iron that is already oxidized can be removed by slow and rapid sand filters, a process that is described in Chapter 23 (Fair et al., 1971).

Manganese (Mn^{+2})

Manganese resembles iron in its chemical behavior and occurrence in groundwater but is less abundant than iron (Kemmer, 1977). Therefore, it is found less commonly in water and the concentration is generally much lower than iron, although in deep wells manganese may reach concentrations as high as 2 to 3 mg/ℓ. Drainage or waste water from mines and metallurgical operations commonly contain abundant manganese. Manganese is objectionable in water in the same general way as iron. It occurs

Figure 6.9. A drive point completely incrusted with manganese carbonate.

as soluble manganous bicarbonate which changes to insoluble manganese hydroxide when it reacts with atmospheric oxygen. Stains caused by manganese are more objectionable and harder to remove than those from iron. Therefore, drinking water regulations limit manganese concentrations to 0.05 mg/ℓ to avoid manganese staining.

Manganese bicarbonate precipitates out of solution as a black, sooty deposit when carbon dioxide is liberated from the water near a well. Manganese bicarbonates can virtually cement a poorly designed well point (screen) in the ground, making removal and replacement difficult. Figure 6.9 shows a well point and some of the surrounding formation cemented together by manganese carbonate. Slime-forming bacteria similar to iron bacteria also may cause oxidation of manganese compounds, forming an insoluble residue.

Both iron and manganese can be kept in solution by adding a small amount of sodium hexametaphosphate to the water. The polyphosphate stabilizes the iron and manganese compounds and delays their precipitation; the delay time varies with the amount of polyphosphate added. Stabilizing chemicals must be added before the water is exposed to air.

Silica (SiO_2)

Silicon (Si) is the second most abundant element in the Earth's crust; only oxygen is more abundant. Silicon in combination with oxygen, as the oxide SiO_2, is called silica. The common mineral quartz, in its many varieties, is essentially pure silicon dioxide. Silicon and oxygen combine readily with many other major elements such as potassium, magnesium, sodium, iron, calcium, and aluminum to create a broad range of rock-forming minerals.

Silica is not readily dissolved by water. Nevertheless, warm groundwaters sometimes contain as much as 100 mg/ℓ silica, and silica concentrations of 20 mg/ℓ are common. The origin of silica in groundwater is attributed to the chemical weathering of silicate minerals. Temperature, the rate of water movement through the rock, and natural acids such as carbonic acid affect weathering and the degree to which silica dissolves in the water. The principal dissolved silicon species in groundwater at typical pH values (6 to 9) is $Si(OH)_4$ (Freeze and Cherry, 1979).

Silica does not contribute to the hardness of water. It is, however, an important constituent of the incrusting material or scale formed by many groundwaters. As deposited, the scale is commonly calcium or magnesium silicate. Silicate scale cannot be dissolved by acids or other chemicals that are used for chemical treatment of wells. Therefore, silica-rich water that is used in boilers must be treated beforehand by absorption or ion-exchange techniques.

Table 6.7. Major Dissolved Chemical Species in Seawater

Constituent	mg/ℓ
Chloride (Cl^-)	19,000
Sodium (Na^+)	10,500
Sulfate (SO_4^{-2})	2,650
Magnesium (Mg^{+2})	1,300
Calcium (Ca^{+2})	400
Potassium (K^+)	380
Bicarbonate (HCO_3^-)	140
Bromide (Br^-)	65
Carbonate (CO_3^{-2})	18
Strontium (Sr^{+2})	8
Silica (SiO_2)	6
Boron (B)	4.5
Fluoride (F^-)	1.3
Aluminum (Al^{+3})	0.001

(Garrels et al., 1975)

Sodium (Na^+)

Sodium belongs to a metals group called the alkali metals, which includes the important element potassium. Even though the alkali metals have similar chemical traits, sodium is the only one to be found in significant quantities in natural waters.

Nearly all sodium salts are highly soluble in water and, once leached from rocks or sediments, they remain in solution. Sodium derived from rock weathering and carried to the sea is by far the most abundant metallic ion in seawaters, averaging about 10,000 mg/ℓ (Table 6.7). On the other hand, groundwater in limestone formations may have only a few mg/ℓ sodium out of several hundred mg/ℓ total dissolved solids. The usual concentration of sodium ions in groundwater is 10 to 100 mg/ℓ (Kemmer, 1977).

The solubility of sodium compounds is so high that they do not precipitate to form scale and thus do not plug wells. Also, sodium does not contribute to the hardness of water. Groundwater containing considerable quantities of sodium carbonate or sodium bicarbonate is alkaline, however, and may have pH values of 9 or more. This is important because some drilling fluid additives may be adversely affected when used in water with a high pH.

Chloride (Cl^-)

Chloride occurs as the predominant negatively charged ion in seawater; the average concentration is about 19,000 mg/ℓ (Table 6.7). Although very little chloride exists in crystalline rocks, the weathering of these rocks does contribute minute amounts to rivers and streams flowing to the sea. Data from White et al. (1963) indicate that in basaltic and gabbroic terrains the chloride content in groundwater averages about 12 to 13 mg/ℓ, whereas in granite the concentration is 6 mg/ℓ. Chloride in groundwater found elsewhere averages close to 6 mg/ℓ (Tardy, 1971).

Chloride ions eventually become trapped in the water-saturated pores of sedimentary rocks forming on the seafloor adjacent to continents. In addition, some chlorine-bearing evaporites may form in shallow marine basins. An evaporite is a sediment produced from sea water in tidal areas under arid conditions as a result of extensive evaporation over millions of years. Tectonic activity may subsequently raise the crustal elements supporting these sediments, leading to subaerial weathering and the slow release of the chlorides back into the environment where streams and rivers carry them once again to the sea. The residence time* for chlorides in sedimentary rocks is about 218 million years, whereas the residence time in oceans is approximately

---

*The total mass of a substance in a reservoir at a given time divided by the rate of inflow or outflow.

100 million years (Garrels et al., 1975). The slowness with which chlorides cycle from rocks to oceans and back into rocks is clear. If the Earth's oceans have existed for almost three billion years, a typical chloride ion has cycled only ten times.

Rainwater in coastal regions generally contains more chlorides than it does farther inland. Mairs (1967) has shown that the chloride concentration of lakes in Maine decreases from about 10 mg/ℓ near the coast to less than 1 mg/ℓ at 150 mi (241 km) inland. Other data suggest that atmospheric chlorides contribute about 0.4 mg/ℓ to the total chloride content of inland surface waters. The average chloride content of North American rivers is 8 mg/ℓ; therefore, much of the chloride must have come from the weathering of sedimentary and crystalline rocks. Locally, however, some chlorides may result from industrial activity. Abundant chlorides in groundwater may also indicate seepage from certain types of sewage facilities. Human wastes are generally high in chloride content, and once these wastes are deposited in sewage lagoons chlorides often move into the groundwater system. Because they are not absorbed by soil, chlorides can travel great distances.

When wells near the seacoast are continuously pumped at high yields, some salt water may move into the freshwater aquifer. Long-term records from observation wells indicate whether chloride contamination is occurring.

Water that contains less than 150 mg/ℓ chloride is satisfactory for most purposes. A chloride content of more than 250 mg/ℓ is generally objectionable for a municipal water supply, and water containing more than 350 mg/ℓ is objectionable for most irrigation and industrial uses. Water containing as much as 500 mg/ℓ chloride frequently has a disagreeable taste. Animals, however, can drink water that has much more chloride; for example, cattle can safely drink water with chloride concentrations as high as 3,000 to 4,000 mg/ℓ.

Fluoride (F^-)

Fluoride, derived from fluorite (the principal fluoride mineral in igneous rocks) and the minerals apatite and mica, is generally present in only low concentrations in groundwater. Volcanic or fumarolic gases may contain fluoride, and in some areas these may be the source of fluoride in groundwater.

It is important to know the amount of fluoride in water used by children. Excessive fluoride causes mottling of tooth enamel; furthermore, teeth may become brittle because fluoride affects tooth density. These defects are more evident in children who drink too much fluoridated water while their permanent teeth are still forming.

Although too much fluoride can harm teeth, minute amounts are quite beneficial. Public health laws in many parts of the United States require the addition of fluoride to public water supplies. The usual recommended fluoride content is 1.4 to 2.4 mg/ℓ, depending on the local temperature regime. In this range, fluoride helps to prevent tooth decay while avoiding the adverse effects of higher concentrations. Recent medical research has determined that regular ingestion of small amounts of sodium fluoride by pregnant women provides remarkable protection against tooth decay in babies long after birth.

Standards for water quality established by the U.S. Environmental Protection Agency (National Interim Drinking Water Regulations of 1975) set an upper limit of permissible fluoride concentration of 2.4 mg/ℓ for areas where the annual maximum

Table 6.8. Fluoride Limits Based on Annual Average Maximum Daily Air Temperature

Degrees F	Degrees C	Maximum Fluoride Limit (mg/ℓ)
53.7 and below	12 and below	2.4
53.8 to 58.3	12.1 to 14.6	2.2
58.4 to 63.8	14.7 to 17.6	2.0
63.9 to 70.6	17.7 to 21.4	1.8
70.7 to 79.2	21.5 to 26.2	1.6
79.3 to 90.5	26.3 to 32.5	1.4

(U.S. Environmental Protection Agency, 1975a)

daily air temperature averages between 50 and 53.7°F (10 to 12°C). Because water consumption is higher in warmer regions, limits of fluoride addition range downward, with a maximum permissible limit of 1.4 mg/ℓ where maximum daily temperatures average between 79.3 and 90.5°F (26.3 to 32.5°C). Table 6.8 gives temperatures and the corresponding recommended maximum fluoride limits.

Nitrate (NO_3^-)

Unlike most other elements in groundwater, nitrate is not derived primarily from the minerals in rocks that make up the groundwater reservoir. Instead, nitrate enters groundwater from another part of the nitrogen cycle in the Earth's hydrosphere and biosphere. Several nitrogen compounds are found in groundwater: nitrate (NO_3^-), nitrite (NO_2^-), and ammonia (NH_3). In water analyses, these species are reported as either the complex ion or as equivalent molecular nitrogen (N); 1 mg/ℓ nitrogen equals 4.5 mg/ℓ nitrate.

Nitrogen enters the ground from several sources. Certain plants, such as alfalfa and other legumes, fix atmospheric nitrogen and transfer it to the soil, where it is used by plants. Some of the surplus nitrogen, however, is removed in solution by downward-percolating soil water. Other sources of soil nitrogen are decomposing plant debris, animal waste, and nitrate fertilizers. Additional nitrogen may enter the ground from sewage discharges on land or from sewage lagoons. Also, many industrial waste chemicals contain high concentrations of nitrogen. Natural nitrate concentrations in groundwater range from 0.1 to 10 mg/ℓ (Davis and DeWiest, 1966). Concentrations may reach 600 mg/ℓ or more when enrichment from nitrate fertilizers or barnyard runoff has taken place.

High nitrate concentrations in well water are cause for concern. They originate from either direct discharge of contaminated surface water into a well or natural infiltration by contaminated surface water. Excessive nitrogen in well waters may indicate leakage from cesspools, barnyards, or sewage lagoons, and from the over-use of agricultural chemicals in sandy soils. Unfortunately, localized contamination of groundwater from these sources is common in many agricultural areas. Moreover, a high nitrate content can be considered an indicator, and thus a warning, that an aquifer should be tested for harmful (pathogenic) bacteria which may accompany contamination from these sources. Nitrate contamination is particularly troublesome in karstic regions where water movement through solution openings is rapid and allows for little attenuation (dilution) of the contaminant.

Nitrate in concentrations greater than 45 mg/ℓ is undesirable in domestic water supplies because of the potential toxic effect on young infants, although adults and older children are not affected. Methemoglobinemia is a disease caused by nitrates which convert to nitrites in the intestines, resulting in an overabundance of methemoglobin molecules. Signs of the disease include listlessness and a bluish tinge to the skin. Cattle are also extremely susceptible to the disease. Loss of milk production and aborted calves are two signs of nitrate poisoning.

The safe nitrate limit for domestic water is set at 45 mg/ℓ by the U.S. Environmental Protection Agency. This is equivalent to 10 mg/ℓ of elemental nitrogen (N). Water containing as much as 20 mg/ℓ nitrogen or 90 mg/ℓ nitrate is considered harmful to infants. It should be noted that nitrate cannot be removed from water by boiling, but must be treated by demineralization or distillation. Because nitrate in groundwater originates most often from sewage waste, its presence is taken as evidence of contamination. However, sewage contamination also adds chloride to groundwater. In fact, chloride travels especially well in the underground and is usually the first indication of contamination. Therefore, high nitrate together with high chloride is a more positive indication of sewage seepage or barnyard pollution than is nitrate alone.

Sulfate (SO_4^{-2})

Sulfate in groundwater is derived principally from the evaporite minerals gypsum (hydrous calcium sulfate, $CaSO_4 \cdot 2H_2O$) and anhydrite (calcium sulfate, $CaSO_4$); it may also come from the oxidation of pyrite, which is an iron sulfide mineral. Groundwater in igneous or metamorphic rocks generally contains less than 100 mg/ℓ sulfate (Davis and DeWiest, 1966)

Groundwater may have much higher sulfate content near evaporite deposits in sedimentary rocks. These deposits may include halite (sodium chloride, NaCl) and other chloride salts. In addition, groundwater may contain other minerals from evaporite deposits. For example, magnesium sulfate (Epsom salt, $MgSO_4 \cdot 7H_2O$) and sodium sulfate (Glauber's salt, $Na_2SO_4 \cdot 10H_2O$) impart a bitter taste to water if present in sufficient quantities. For people not accustomed to drinking high-sulfate water, these salts may act as a laxative.

Dissolved Gases

Dissolved gases are usually not determined in the routine analysis of water. Their presence in substantial amounts, however, critically affects the use of water for certain purposes. The most common dissolved gases include oxygen, hydrogen sulfide, carbon dioxide, nitrogen, sulphur dioxide, and ammonia. Of this group, the first three have the greatest significance for groundwater and its development. The solubility of gas decreases with temperature and increases with pressure. Dissolved gases are involved in corrosion of well casings and well screens and also the deposition of incrusting materials because of pressure changes near wells. Dissolved gases typically occur in concentrations of 1 to 100 mg/ℓ.

Dissolved Oxygen

The solubility of air in water at 32°F (0°C) at atmospheric pressure is about 29 mg/ℓ; about 10 mg/ℓ of this represents the oxygen portion. Solubility of oxygen in water decreases with higher temperatures and becomes almost zero at the boiling point. On the other hand, more oxygen can be held in water at greater pressures. Thus, water in a pneumatic tank may contain far more than 10 mg/ℓ of oxygen.

The oxygen content of groundwater at depths greater than 100 to 150 ft (30.5 to 45.7 m) is generally considered to be quite low. Most of the dissolved air presumably is used up in the oxidation of organic materials as the water percolates downward through the vadose zone.

Pure water, that is, highly demineralized water containing no dissolved gases, is not corrosive to metals. If dissolved oxygen is present, however, a corrosion potential exists. Water with dissolved oxygen corrodes metals more rapidly when the pH is low. However, a water with some dissolved oxygen and relatively high electrical conductance (caused by total dissolved solids) will be corrosive even though the pH may be 8 or more. Susceptible metals include iron, steel, galvanized iron, and brass. The rate of corrosion tends to increase directly with temperature, but the amount of oxygen in solution decreases with higher temperatures. Therefore, the corrosion rate may not vary significantly unless the heated water is under pressure.

Dissolved oxygen may cause water to attack galvanized iron and some kinds of brass as rapidly as black iron. The zinc of the galvanizing is oxidized and washed away much more rapidly than if the water were free of oxygen. Dissolved oxygen also removes zinc from brass alloys, leaving the metal porous and weakened. Iron oxide scale accumulates on the inner surface of iron pipes when iron put into solution by low pH water combines with dissolved oxygen to form the insoluble oxide. An oxide film then deposits on the corroded surfaces. The scale occupies a larger volume than did the original metal and gradually the pipe fills with scale, reducing capacity. One study shows that water mains lose carrying capacity at the rate of 1 to 2 percent per year over a period of 20 to 30 years.

Hydrogen Sulfide (H_2S)

Groundwater that contains dissolved hydrogen sulfide gas is easily recognized by its rotten-egg odor. As little as 0.5 mg/ℓ of hydrogen sulfide in cold water is noticeable, and the odor from 1 mg/ℓ is definitely offensive. Water with small amounts of hydrogen sulfide forms a weak acid and is usually corrosive.

Sulfate-reducing bacteria are found in some groundwaters. Conditions favorable for growth include absence of oxygen and fairly high sulfate content. These bacteria gain energy from the oxidation of organic compounds and, in the process, take oxygen from the sulfate ions. Reduction of the sulfate ions (removal of oxygen) produces hydrogen sulfide gas, which can be absorbed easily by the water. In an iron pipe, the water attacks the metal to form iron sulfide, which is deposited as insoluble iron scale in the pipe.

Carbon Dioxide (CO_2)

Carbon dioxide is dissolved by rainwater as it falls through the atmosphere, but a much larger amount is dissolved by water flowing through soil where plants are growing. Plant roots and decaying vegetation contribute carbon dioxide to the voids between soil particles above the groundwater table.

The presence of carbon dioxide in groundwater is especially significant where calcium and bicarbonate ions are in solution. Under ambient pressure conditions, the amount of carbon dioxide in solution remains constant. However, the pressure near a well is usually reduced by pumping and the carbon dioxide then comes out of solution as bubbles of gas. When chemical equilibrium is disturbed, calcium carbonate may precipitate until the solution is again in equilibrium. These chemical adjustments take place each time a well is pumped in areas where carbon dioxide occurs in waters high in calcium carbonate.

To minimize deposition of calcium carbonate when a well is pumped, head losses (pressure reduction) must be kept as low as possible. Low entrance velocities through well-screen openings help greatly. Minimal entrance velocity is obtained by using well screens that provide maximum inlet area.

Table 6.9. Free Carbon Dioxide Contained in Water

Bicarbonate Alkalinity, mg/ℓ as $CaCO_3$	Free Carbon Dioxide		
	at pH = 7.0 (mg/ℓ)	at pH = 7.5 (mg/ℓ)	at pH = 8.0 (mg/ℓ)
100	22	6	2
200	43	12	4
300	63	17	6
400	82	22	7

The interrelationship of pH, bicarbonate, and carbon dioxide, discussed earlier in this chapter, provides a way of using bicarbonate alkalinity and pH to calculate the free carbon dioxide content of water samples. Table 6.9 gives the amount of free carbon dioxide contained in water for several combinations of alkalinity and pH. The figures in the table are subject to slight correction for temperature and total dissolved mineral content. However, they are accurate enough for most purposes.

Radionuclides

Radioactive decay of certain unstable elements produces radiation called alpha (α), beta (β), and gamma (γ) radiation. The human body is highly susceptible to damage from alpha and gamma radiation; accumulated radiation exposure can cause leukemia, birth defects, mental retardation, and tumors. Use of atomic energy can increase the average exposure rate in certain localities (Eddy and Wilbur, 1980). With traditional energy sources dwindling, the anticipated worldwide expansion of nuclear facilities will lead to increased opportunities for radioactive contamination of both surface water and groundwater. However, recent efforts by the power industry to reduce or eliminate radioactive waste discharges have been significant.

Because water has such a long residence time in the ground, the potential exists for radioactive contamination. One of the common isotopes of radium*, Ra^{226}, is the most toxic of all inorganic substances (Davis and DeWiest, 1966), yet it is found naturally in the ground. For example, 15 community water systems in Texas are unable to meet proposed radium concentration limits. Nationwide there may be as many as 500 community water systems that have radium concentrations exceeding

---

*Radium is one of the decay products of uranium (U^{238}).

Table 6.10. Radionuclides

Radionuclide	Maximum Contaminant Level
Radium-226 and Radium-228	Combined limit of 5 pCi/ℓ*
Gross Alpha Particle (including radium-226 but excluding radon and uranium)	No more than 4 millirem/year
Tritium	20,000 pCi/ℓ
Strontium-90	8 pCi/ℓ

*pCi/ℓ = picocurie per liter — that quantity of radioactive material producing 2.22 nuclear transformations per minute.
(U.S. Environmental Protection Agency, 1975a)

the present recommended limit. Fortunately, naturally occurring radioactive contamination of groundwater is not a widespread problem, but in certain areas radioactive elements have adversely affected groundwater.

Table 6.10 lists common radioactive elements under study and those that have been assigned limits. Regulators have set the radiation limits so low that the biological effects are almost impossible to detect.

Minor Constituents

Natural waters may contain other minerals not previously discussed, but usually these occur in concentrations of less than 1 mg/ℓ. Determinations of minor constituents ordinarily are made only in more comprehensive analyses unless their presence is indicated by some special circumstance, such as (1) the geologic environment of the groundwater, (2) a noticeable change in the character of the water, (3) an accidental spillage of waste which may infiltrate the groundwater reservoir, or (4) a noticeable effect on agricultural crops. In certain areas, one or more of the minor constituents may be present in large amounts and may even be a major constituent for a particular water. Table 6.6 lists the most common minor constituents.

One minor constituent, boron, is often determined in arid and semiarid regions where the water is used for irrigation. Although boron in trace amounts is essential to plant growth, it is quite detrimental to many plants when present in excess of 1 mg/ℓ in irrigation water.

WATER QUALITY

The primary purpose of a water analysis is to determine the suitability of water for a proposed use. The three main classes of use are domestic (household), agricultural, and industrial. A supply intended for municipal use may include all three classes and accordingly require a standard of quality that is generally higher than that needed for any one class. On the other hand, water for use in a particular industry may require a quality that is substantially higher than the one considered acceptable for a municipal supply.

Domestic Use

In March 1975 the U.S. Environmental Protection Agency proposed the National Interim Primary Drinking Water Regulations under provisions of the Public Health

Service Act as amended by the Safe Drinking Water Act. Based in part on Public Health Service regulations developed in 1946 and 1962 and later modified, the interim

Table 6.11. National Primary Drinking Water Regulations

Constituent	Maximum Contaminant Level*
Inorganic Chemicals	
Arsenic (As)	0.05
Barium (Ba)	1.0
Cadmium (Cd)	0.01
Chromium (Cr)	0.05
Fluoride (F)	1.4 - 2.4†
Lead (Pb)	0.05
Mercury (Hg)	0.002
Nitrate (as N)	10.0
Selenium (Se)	0.01
Silver (Ag)	0.05
Organic Chemicals	
Chlorinated hydrocarbons	
Endrin	0.0002
Lindane	0.004
Methoxychlor	0.1
Toxaphane	0.005
Chlorophenoxys	
2, 4-D	0.1
2, 4, 5-TP Silvex	0.01
Total trihalomethanes	0.1
Microbiological Standards	
Membrane Filter Technique	Number of *coliform* bacteria shall not exceed 1 per 100 mℓ as the arithmetic mean of all samples‡
Fermentation Tube Method (10mℓ standard portion)	*Coliform* bacteria shall not be present in more than 10 percent of the tubes in any one month‡
Turbidity	1 turbidity unit (TU) as determined by a monthly average §

*Maximum contaminant levels are set according to health criteria.
†Limit depends on average air temperature of the region. See Table 6.8 for further details.
‡These maximum contaminant levels may be modified depending on the system size and thus, the number of samples taken.
§These levels are subject to further modifications.

(U.S. Environmental Protection Agency, 1975a)

regulations became final in June 1977, but are continually under review. These federal regulations specify maximum contaminant levels (MCL's) for drinking water supplies and apply to all public water systems. At the recommended maximum contaminant levels, no adverse health effects are known to exist.

A public water system is defined as "a system for the provision to the public of piped water for human consumption, if such system has at least 15 service connections or regularly serves at least 25 individuals." The regulations cover both privately and publicly owned systems that serve on either a regular or periodic basis. Campgrounds, lodges, and trailer parks operate under the regulation when the water system serves an average of 25 people for at least 60 days each year. Approximately 40,000 community water systems have been built in the United States; 58 percent of these are publicly owned and provide 88 percent of the total drinking water supply in the United States (U.S. EPA, 1975b). More than 177 million persons are served by these systems.

Two types of public water systems exist; one serves residents and the other serves transients or intermittent users. These two systems are distinguished because the potential effects of contaminants in drinking water vary according to whether the water is ingested on a regular basis or only intermittently. Therefore, different contaminant standards may apply to these two types of systems.

The maximum contaminant levels permitted under the 1975 primary regulations assumed that an individual consumes 2 qt (1.9 ℓ) of water per day. Although consumption in certain regions may be higher, the safety factors built into the maximum contaminant levels provide good protection to all consumers regardless of the amount of water ingested. The current primary standards for various inorganic substances are presented in Table 6.11. The secondary contaminant levels established by the Safe Drinking Water Act are set for both aesthetic and health reasons. Secondary regulations applying to drinking water were proposed in March 1977 and became final in August 1981. Contaminants covered by these regulations are those which may adversely affect the aesthetic qualities of drinking water such as taste, odor, color, and appearance and which thereby may deter public acceptance of drinking water provided by public water systems. Secondary maximum contaminant levels are established for chloride, color, copper, corrosivity, foaming agents, iron, manganese, odor, pH, sulfates, total dissolved solids, and zinc. At considerably higher concentrations, these contaminants may also be associated with adverse health implications. These secondary levels represent reasonable goals for drinking water quality, but are not federally enforceable. Instead, states are encouraged to implement these standards. The secondary contaminant levels are shown in Table 6.12.

Water-well contractors have a special interest in the chemical characteristics of water, because the usefulness of a water supply frequently depends upon the mineral content. The contractor should therefore understand the significance of the recommended concentrations specified in the drinking water standards. Occasionally, more than one water-bearing formation is encountered in a well and the composition of the respective waters may vary drastically. Under these conditions it may be desirable to exclude the poor-quality water from the principal supply.

Bacteriological quality of a water supply is determined by analyzing for coliform bacteria. The coliform group of organisms is used as an indicator of dangerous contaminant levels because some of these bacteria are excreted in large numbers from the human intestinal tract. Study of these organisms is advantageous because they

Table 6.12. National Secondary Drinking Water Regulations

Constituent	Recommended level*
Chloride (Cl)	250 mg/ℓ
Color	15 color units
Copper (Cu)	1.0 mg/ℓ
Corrosivity	Noncorrosive
Foaming agents	0.5 mg/ℓ
Iron (Fe)	0.3 mg/ℓ
Manganese (Mn)	0.05 mg/ℓ
Odor	3 threshold odor number
pH	6.5-8.5
Sulfate (SO$_4$)	250 mg/ℓ
Total dissolved solids (TDS)	500 mg/ℓ
Zinc (Zn)	5.0 mg/ℓ

*Recommended levels for these constituents are mainly to provide acceptable aesthetic and taste characteristics.

(U.S. Environmental Protection Agency, 1977)

are nonpathogenic, do not multiply outside the human body, and are easily identified and counted. Current standards specify that drinking water must not contain more than 1 coliform colony in 100 mℓ of water after undergoing a standardized laboratory procedure.

The water well contractor has little control over the preexisting bacteriological quality of a water supply, but is obligated to take certain precautions to prevent surface-water contamination from entering the well during drilling. The contractor must also construct the well in such a way that the physical structure will protect the original quality of the water in the aquifer. This may mean sealing off one or more poor-quality water zones in a well.

Tests to determine the presence of bacteria are usually done after a well has been completed and the installation sterilized. Some states provide the sampling flasks to be used by the drilling contractor for water samples. Water quality tests are usually conducted by public health authorities or other qualified laboratories.

Industrial Use

Quality requirements for industrial waters vary widely according to potential use. For example, salt and brackish waters are commonly used as cooling water, particularly when they are used only once (not recycled) and can be disposed of without polluting the environment. Disposal of these waters is a major problem; deep-well injection may provide the best, or often the only, disposal method available.

Industrial process waters must be of much higher quality than cooling waters. Municipal supplies are generally good enough to satisfy the quality requirements of most process waters, with the exception of those waters used in boilers. About 60 percent of the water used by industry must be treated to meet quality standards (Fair et al., 1971). Sanitary requirements for waters used in processing milk, canned goods, meats, and beverages exceed even those for drinking water.

In many cases, groundwater may be desirable for particular uses because of its low,

Table 6.13. Quality Tolerances for Industrial Process Waters*

Industry	Turbidity	Color	Hardness as mg/ℓ of CaCO₃	Alkalinity as mg/ℓ of CaCO₃	Fe + Mn, mg/ℓ	Total Solids, mg/ℓ	Other
Food products							
Baked goods	10	10	†	...	0.2	...	a
Beer	10	...	...	75-150	0.1	500-1000	a, b
Canned goods	10	...	25-75	...	0.2	...	a
Confectionery	...	...	...	...	0.2	100	a
Ice	5	5	...	30-50	0.2	300	a, c
Laundering	...	...	50	...	0.2	...	...
Manufactured products							
Leather	20	10-100	50-135	135	0.4	...	...
Paper	5	5	50	...	0.1	200	d
Paper pulp	15-50	10-20	100-180	...	0.1-1.0	200-300	e
Plastics, clear	2	2	...	...	0.02	200	...
Textiles, dyeing	5	5-20	20	...	0.25	...	f
Textiles, general	5	20	20	...	0.5	...	...

*Stated values are general averages only; there is much local variance.
†Some hardness is desirable.
a Must conform to standards for potable water.
b NaCl no more than 275 mg/ℓ.
c SiO₂ no more than 10 mg/ℓ; Ca and Mg bicarbonates are troublesome; sulfates and chlorides of Na, Ca, and Mg each no more than 300 mg/ℓ.
d No slime formation.
e Noncorrosive.
f Constant composition; residual alumina no more than 0.5 mg/ℓ.
(American Society for Testing Materials, 1960; Fair et al., 1971).

relatively constant temperature. In other cases, groundwater may be suitable because of its natural hardness, because distilleries, bakeries, and breweries prefer hard water. On the other hand, even small amounts of iron, manganese, or calcium can cause great harm in paper-making processes. Table 6.13 lists some typical quality tolerances for industrial process water.

Agricultural Use

Water quality, soil types, and cropping practices all play a role in successful irrigation. Good-quality water permits maximum yields consistent with proper soil and water management. Study of soil types determines the infiltration rates that can be expected for certain soils, thereby providing some guides to the amount of leaching of mineral salts that can be anticipated; leaching is essential to reduction of salinity in topsoils. Equally important, the tolerance for certain elements must be determined before specific crops can be selected. Table 6.14 lists the substances and qualities that are important when evaluating a particular water for irrigation purposes. Table 6.15 gives some recommended concentrations for trace elements in irrigation waters.

Water quality problems in irrigation include salinity and toxicity. Excessive salinity occurs when there is an accumulation of salts in topsoils. These salts can affect crop production because crop roots, especially in the upper root zone, have great difficulty extracting enough water and nutrients from saline solutions. Soil permeability can be reduced significantly by the buildup of salts in the soil zone. Consequently, crop production is limited because sufficient water cannot reach the root zone. Toxicity is also a problem in maintaining good yields. Some waters contain high enough concentrations of certain elements to retard or even eliminate the growth of some plants. Boron, chlorides, and sodium are common toxic substances.

Sodium has far-reaching effects on soils. Most sodium in natural water originates with the release of soluble products during the weathering of plagioclase feldspars (Davis and DeWiest, 1966). In addition, minor amounts of sodium may come from the mineral halite (table salt).

Of particular consequence is the ratio of sodium to calcium and magnesium. When sodium-rich water is applied to soil, some of the sodium is taken up by clay; the clay gives up calcium and magnesium in exchange. This reaction, called base exchange,

Table 6.14. Laboratory Determinations Needed to Evaluate Water Used for Irrigation

Acidity-Alkalinity	Iron[2]
Adjusted Sodium Adsorption Ratio	Lithium[2]
Ammonium-Nitrogen[1,2]	Magnesium
Bicarbonate	Nitrate-Nitrogen[1]
Boron	Phosphate Phosphorous[2]
Calcium	Potassium[2]
Carbonate	Sodium
Chloride	Sulphate
Electrical Conductivity	

[1]Nitrate-nitrogen (NO_3-N) is nitrogen in the form of nitrate (NO_3) and ammonium-nitrogen (NH_4-N) is nitrogen in the form of ammonia (NH_4), reported as nitrogen (N) in mg/ℓ
[2]Special situations only.

(Ayers and Wescot, 1976)

Table 6.15. Recommended Maximum Concentrations of Trace Elements in Irrigation Waters[1]

Element (Symbol)	For Waters Used Continuously on Soils (mg/ℓ)	For Use Up to 20 Years on Fine-Textured Soils of pH 6.0 to 8.5 (mg/ℓ)
Aluminum (Al)	5.0	20.0
Arsenic (As)	0.1	2.0
Beryllium (Be)	0.1	0.5
Boron (B)	[2]	2.0
Cadmium (Cd)	0.01	0.05
Chromium (Cr)	0.1	1.0
Cobalt (Co)	0.05	5.0
Copper (Cu)	0.2	5.0
Fluoride (F)	1.0	15.0
Iron (Fe)	5.0	20.0
Lead (Pb)	5.0	10.0
Lithium (Li)[3]	2.5	2.5
Manganese (Mn)	0.2	10.0
Molybdenum (Mo)	0.01	0.05[4]
Nickel (Ni)	0.2	2.0
Selenium (Se)	0.02	0.02
Vanadium (V)	0.1	1.0
Zinc (Zn)	2.0	10.0

[1] These levels normally do not adversely affect plants and soils. No data are available for mercury (Hg), silver (Ag), tin (Sn), titanium (Ti), or tungsten (W).
[2] No problem when less than 0.75 mg/ℓ; increasing problem when between 0.75 and 2.0 mg/ℓ; severe problem when greater than 2.0 mg/ℓ.
[4] For only acid fine-textured soils and acid soils with relatively high iron oxide content.

(National Academy of Sciences and National Academy of Engineering, 1972)

alters the physical characteristics of soil and can even lead to growth retardation. Clay that takes up sodium becomes sticky and slick when wet and has low permeability. When dry, the clay shrinks into hard clods that are difficult to cultivate. Even worse, high concentrations of sodium salts can produce alkali soils in which little or no vegetation can grow. On the other hand, when the same clay carries excess calcium or magnesium ions, it tills easily and has good permeability.

If an irrigation water contains calcium and magnesium ions sufficient to equal or exceed the sodium ions, enough calcium and magnesium is retained on clay particles to maintain good tilth and permeability. These waters serve well for irrigation, even though the total mineral content may be quite high.

The importance of sodium led to the adoption of a method to measure the effect of sodium ions. In 1954 the United States Salinity Laboratory proposed that the sodium effect be calculated by the sodium adsorption ratio (SAR method). The SAR is calculated from the following equation:

$$SAR = \frac{Na}{\sqrt{\dfrac{Ca + Mg}{2}}}$$

where sodium, calcium, and magnesium are in milliequivalents per liter from the water analysis. Development of excess sodium in soils will result from irrigation water that has a high SAR value (18 or above). Values below 10 indicate little danger of a sodium problem. See Appendix 6.B for a complete analysis of this method. Appendix 6.B also presents a convenient soils classification system that has been developed on the basis of the sodium adsorption ratio and the conductivity of the irrigation water.

Under ordinary conditions, plants take up little of the dissolved minerals from irrigation water. Most minerals carried by the water remain in the soil or remain dissolved in the unused portion of the water. Repeated irrigation may result in the accumulation of too much mineral salts and thus destroy the productivity of the irrigated soil. Therefore, some means of leaching salts from the soil must be found.

Under favorable water quality conditions, irrigation can be carried on successfully if management practices include (1) irrigating frequently to maintain an adequate soil-water supply for the crop, (2) planting crops that can tolerate an existing or potential salinity problem, (3) routinely using extra water to satisfy leaching requirements, (4) changing the irrigation method to one that gives better salt control, and (5) changing cultural practices. An analysis of the initial water quality will suggest effective management practices, but, as a normal part of the irrigation management system, water quality must be monitored annually to detect changes that could affect yields.

Proper management practices can, in some cases, permit use of poor-quality irrigation water. For example, the Pecos River irrigation project at Carlsbad, New Mexico, has salt-tolerant crops growing in sandy soils with water that has an electrical conductivity of 3,210 micromhos (3,210 microsiemens).

Chemical analyses of irrigation water do not indicate whether the water is free of harmful bacteria and hence suitable for domestic use. Most groundwaters, when sufficiently low in mineral content, are potable unless contaminated by man's activities. Nevertheless, their hygienic quality should be checked periodically if they are to be used without treatment to insure freedom from harmful bacteria.

METHODS TO PRESENT WATER QUALITY DATA

As a better understanding of the interaction between groundwater and aquifer materials has come about, the techniques to display water quality data have become more complex and instructive. Five techniques are commonly used to portray

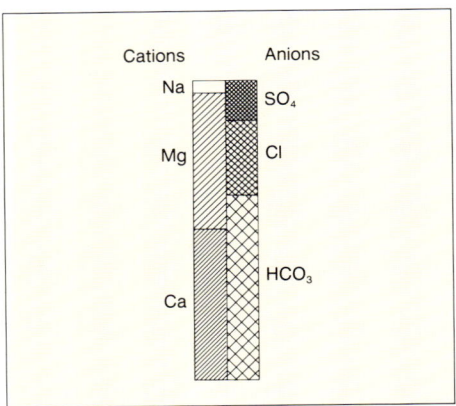

Figure 6.10. Analysis of a water sample using the graphical method developed by the U.S. Geological Survey.

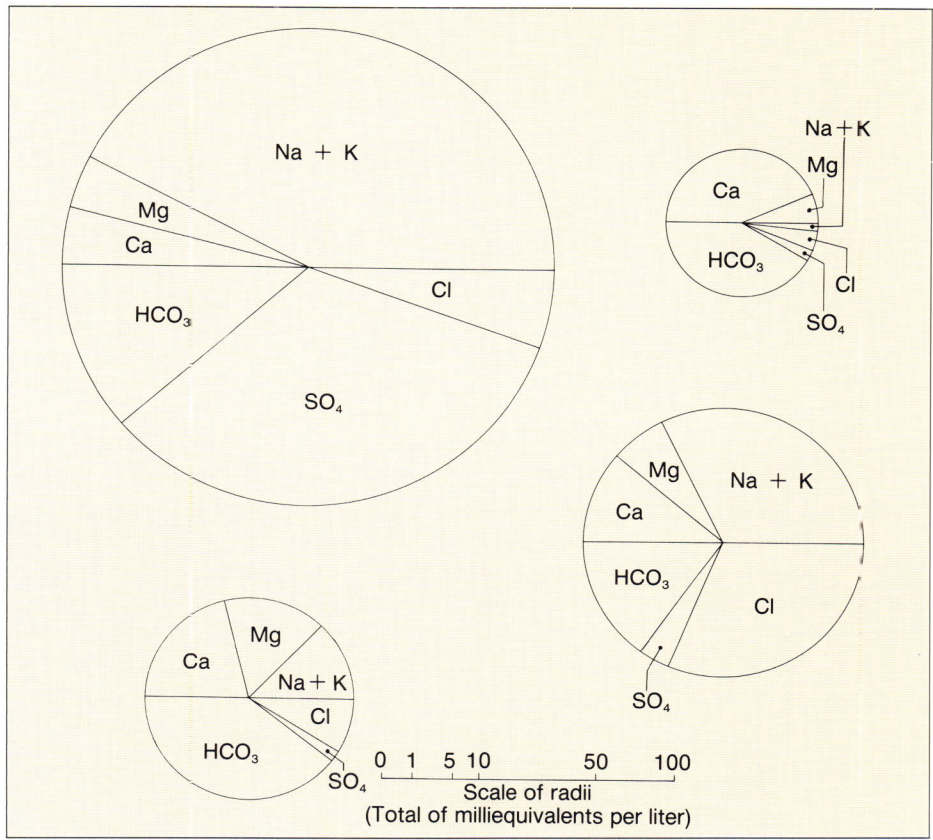

Figure 6.11. Four analyses represented by circular diagrams subdivided on the basis of percentage of total milliequivalents per liter. *(Hem, 1970)*

the chemical analyses of natural waters; four of these techniques are graphical. The simplest technique is to tabulate the data into tables indicating the specific ions present and their relative or absolute concentration. Collins (1923) presented the first graphical method in which the concentration of individual ions, both cations and anions, were indicated by color or patterns on a bar graph. The areal extent of a particular color (ion) indicated the milliequivalents of that ion compared to the others present in the water sample (Figure 6.10). If, as is usual, the amount of cations and anions were not equivalent, the height of the respective bars differed. This method is still used by the U.S. Geological Survey.

"Pie" diagrams show both the individual ions present in a water sample and the total milliequivalents per liter (Figure 6.11). The scale for the radius of the circle, and thus the area, represents the total ionic concentration, whereas subdivisions of the circle represent the proportions for individual ions.

Currently trilinear diagrams are widely used to depict chemical data. This method, proposed independently by Piper (1944) and earlier by Hill (1940), shows the relative concentrations of the major cations (Ca^{+2}, Mg^{+2}, and K^+) and anions (CO_3^-, HCO_3^-, Cl^-, and SO_4^-) (Figure 6.12). Cations are plotted on the left triangle and anions

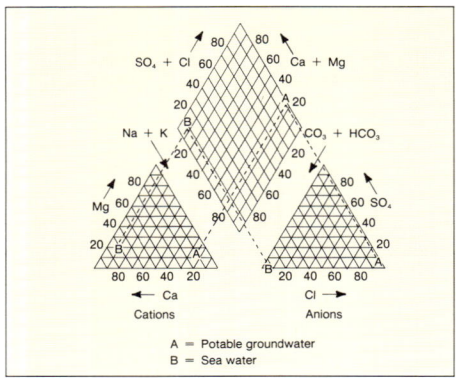

Figure 6.12. Chemical analyses of water represented as percentages of total equivalents per liter on the trilinear diagram developed by Hill (1940) and Piper (1944).

on the right triangle. In order to use this method, Na and K and CO_2 and HCO_3 are combined. The numbers along the sides of the triangle indicate the percentage of the specific ion in the sample. Thus, the relative concentrations of three ions or ion groups can be shown in each triangular diagram. If only one cation is present, the concentration point will fall on an apex of the triangle (100 percent). If two ions are present, the concentration point will fall on one side of the triangle. If all three ions are present, the point will fall within the triangle.

The diamond-shaped diagram above the cation and anion triangles can be used to present both anion and cation groups as a percentage of the sample. This point is found by extending the cation point parallel with the magnesium leg into the upper diagram. Similarly, another extension is made from the point in the anion triangle parallel with the sulfate leg into the upper diamond. This point indicates the relative composition of the water sample in regard to the cation-anion pairs that correspond to the four sides of the diamond-shaped area. Ideally, the relative percentages of both cations and anions should be 50 percent. However, many chemical analyses cannot be conducted with the required accuracy and, inevitably, one of the ion groups will appear to be larger than the other on a percentage basis.

One other graphical procedure has been developed to depict water chemistry. Stiff (1951) devised a simple pattern analysis that can be used to trace similar formation waters over large areas. The scale is used as a guide to plot the ion concentrations for a specific water sample. When the points are connected, the resulting pattern provides a pictorial representation of the water sample. Two typical samples are plotted in Figure 6.13. Once a pattern for a certain groundwater has been established, it is a simple matter to compare patterns as the water moves toward different hydrogeologic environments.

IMPORTANCE OF WATER CHEMISTRY

This chapter describes briefly some of the most important constituents of water. In most instances, the drilling contractor or the engineer in charge can become familiar with the chemical characteristics of the water through laboratory testing or by using one of the convenient field testing kits. They can then take action to avoid potential well failure brought about by chemical disequilibrium conditions during pumping or by the use of inappro-

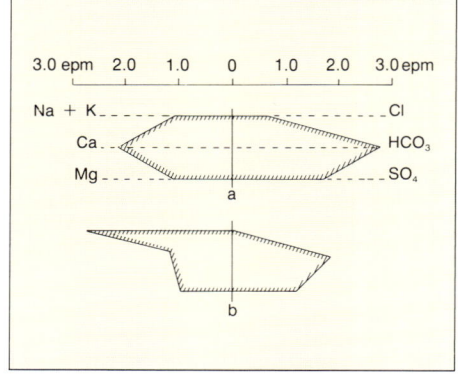

Figure 6.13. Pattern analysis developed by Stiff (1951) that can be used to trace similar formation waters over large areas.

priate well screens, casing, or pumps.

The validity of a chemical analysis reflects the care with which the water sample was obtained in the field and the accuracy of the analysis in the laboratory. Occasionally, aging of the sample and analytical errors in the laboratory may lead to erroneous data. Summers (1972) suggests that a complete analysis of every sample should be made to verify that the total weight of the cations approximates the total weight of the anions and, in turn, these weights equal the weight of the evaporated residue (total dissolved solids). Doubts about a particular analysis can be minimized if the sample is analyzed by more than one laboratory, if multiple samples are tested by one laboratory, or if repeated analysis of a single sample is done over time by one laboratory. Quality control and quality assurance of chemical analyses are enhanced if the laboratory understands thoroughly the objectives of the sampling program and the samples are delivered to the laboratory at the agreed-upon time.

CHAPTER 7
A Summary of Groundwater Resources of North America

Even though groundwater exists in economic quantities in virtually all countries of the world, it is not evenly distributed with respect to either quantity or quality. Two conditions must be met to have abundant groundwater supplies: local geologic conditions must be suitable for storage and transmission of large volumes of water, and climatic conditions must be favorable enough to keep the aquifers recharged. Furthermore, the value of any groundwater resource is affected by the presence or absence of local surface water supplies and the prevailing climate in the area. For example, nearly all of Mexico is classified as arid or semiarid, suggesting that surface water supplies are generally not available. Consequently, the presence of groundwater aquifers is enormously important for a prosperous economy. The opposite condition exists in England, where consistent and heavy precipitation makes surface water supplies perpetually available. Therefore, little groundwater development has taken place there.

In North America, rising standards of living have increased water demands. In response to these demands, the development of groundwater resources in Canada, Mexico, and the United States has greatly accelerated since the mid-1960's. Federal, state, and provincial expenditures for groundwater studies have soared during these years, reflecting the concern for adequate water supplies. It is estimated that approximately 900,000 wells are drilled annually in the United States *(Water Well Journal,* 1981), one and a half times the number of wells drilled in 1966. Water from these wells is needed to accommodate the great expansion in irrigation, mining, and industrial processes, as well as the demand brought about by increased population.

Groundwater development has also expanded significantly in Mexico and Canada. Population growth (especially high in Mexico), adoption of high-technology irrigation systems, greatly accelerated industrial expansion, and energy resource development have heightened demand for groundwater supplies. Improvement of general living conditions in Mexico, where two-thirds of the country has insufficient precipitation, will depend to a large extent on obtaining new groundwater resources. In Canada, energy and mineral resource development in Alberta, British Columbia, and Saskatchewan, and the expansion of irrigated agriculture in southwest Manitoba, have accelerated groundwater withdrawals.

It is not possible within this chapter to describe in detail the groundwater resources of North America — only an overview will be presented. The reader is urged to contact state, provincial, or federal water authorities who may be able to supply comprehensive studies for local regions. Several major sources of groundwater information are cited in this chapter.

GROUNDWATER IN THE UNITED STATES

The United States is reasonably well endowed with water when compared to many other countries. Much of the country is bordered by oceans which provide abundant atmospheric moisture, and the country is far enough north to be largely outside the desert zone lying near lat. 30° N. Except for large areas of the Far West, mountain ranges in the United States do not cause drier-than-normal conditions because of orographic effects.

On the average the conterminous United States receives about 30 in (762 mm) of precipitation as rain and snow. About 22 in (559 mm) of this water is directly evapotranspirated into the atmosphere, 2 in (51 mm) is withdrawn and used, and the remainder returns more or less directly to the sea. Twenty-five percent of the water withdrawn is used consumptively, that is, it is evaporated and therefore lost for further use in the local area.

Unfortunately, the distribution of precipitation is not uniform, especially over wide areas of western United States (Figure 7.1). Except for coastal areas and mountainous

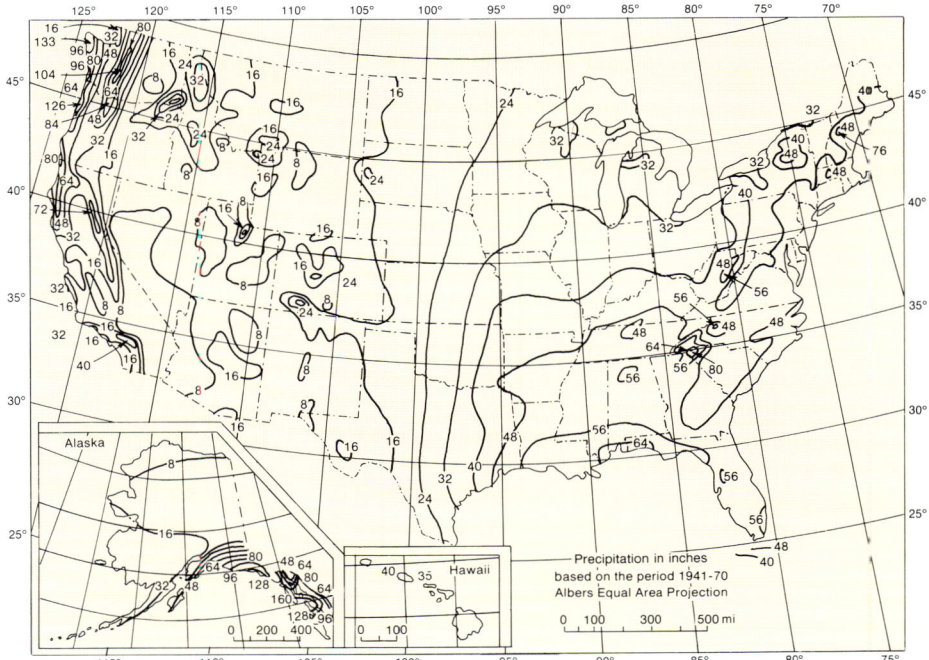

Figure 7.1. Average annual precipitation in the United States. *(National Climatic Data Center, 1983)*

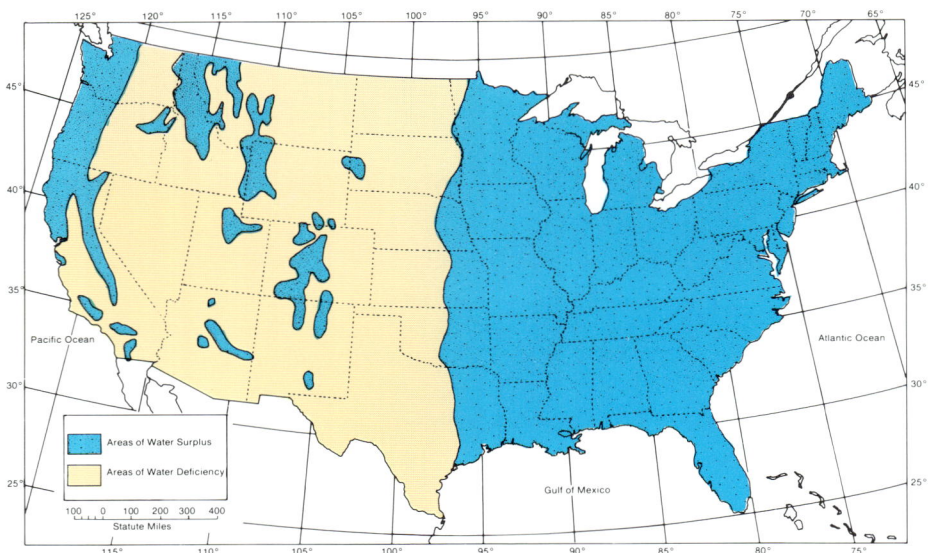

Figure 7.2. Areas of water surplus and deficiency. *(Miller et al., 1963)*

regions, most of the western half of the United States is water-deficient (Figure 7.2).

Meinzer (1923) divided the 48 conterminous states into 21 groundwater provinces. Thomas (1952) consolidated and rearranged Meinzer's 21 provinces into 10 groundwater regions, with Alaska and Hawaii considered separately. Until recently, Thomas's interpretation of the groundwater provinces had been widely recognized as the best broad classification of the distribution of groundwater.

In 1982, the U.S. Geological Survey further refined the groundwater regions proposed earlier by Thomas (Heath, 1982). First, a new region was designated that comprises the important alluvial valleys wherever they are found in the United States (Figure 7.3). Heretofore, these valleys had been considered a part of the region in which they occurred. The valleys shown in Figure 7.3 are occupied by perennial rivers that offer recharge to floodplain sediments. No abandoned watercourses or buried valleys are shown. A second change involved combining the Superior Uplands with the northeastern part of the United States. In both these areas, the glacial deposits are underlain by fractured crystalline rock. A third change is the separation of Florida and adjacent areas from the Atlantic and Gulf Coastal Plain. This distinction is based on the presence of the calcareous Floridan aquifer which does not occur over most of the Atlantic and Gulf Coastal Plain.

Five criteria are used to differentiate one groundwater system from another (Heath, 1982): (1) the aquifers and confining beds that make up the groundwater system; (2) the types of primary or secondary pore space, solution cavities, or fractures; (3) the composition of the dominant aquifer material, that is, whether it is soluble or insoluble or consists of both types of material; (4) the storage and transmission properties of the dominant aquifer, that is, storage coefficient and transmissivity; and (5) the recharge and discharge conditions of the dominant aquifer or the entire groundwater system.

Figure 7.4 shows the location of aquifer types in the conterminous United States.

Based on these rock types and the criteria listed above, the U.S. Geological Survey has proposed the 13 groundwater regions presented in Figure 7.5. Note that Hawaii and Alaska are considered as separate hydrogeologic regions, although Alaska will certainly be subdivided once sufficient information on groundwater resources is available.

Table 7.1 lists the 14 groundwater regions and the general geologic settings of aquifers in these regions. The hydraulic characteristics of the dominant aquifers also are shown, although the ranges in values must necessarily be large to accommodate all the physical conditions existing in aquifers within a specific region. Brief descriptions of individual regions are presented in this chapter. This chapter also includes a discussion of groundwater resources in Canada and Mexico.

The text of this chapter is based primarily on the detailed analyses by McGuinness (1963), who has summarized the work of Meinzer and Thomas. Many of the passages in this chapter are paraphrased from McGuinness, although they are not attributed to him each time they appear in the text. The interested reader is urged to consult Meinzer (1923), Thomas (1952), McGuinness (1963), and Heath (1982, 1984) for thorough analyses of groundwater resources in the United States.

Western Mountain Ranges

The Western Mountain Ranges, which include the northern Coast Ranges, Cascades, Sierra Nevadas, isolated ranges of the Basin and Range province, and Rocky Mountains, are the principal sources of water for the West. Most of the precipitation falls on these mountains and supplies streams which recharge aquifers throughout the West. The mountains consist mainly of impermeable rocks, although some water

Figure 7.3. The Alluvial Valleys groundwater region. The streams shown are perennial and provide recharge to wells constructed in the floodplain. *(Heath, 1982)*

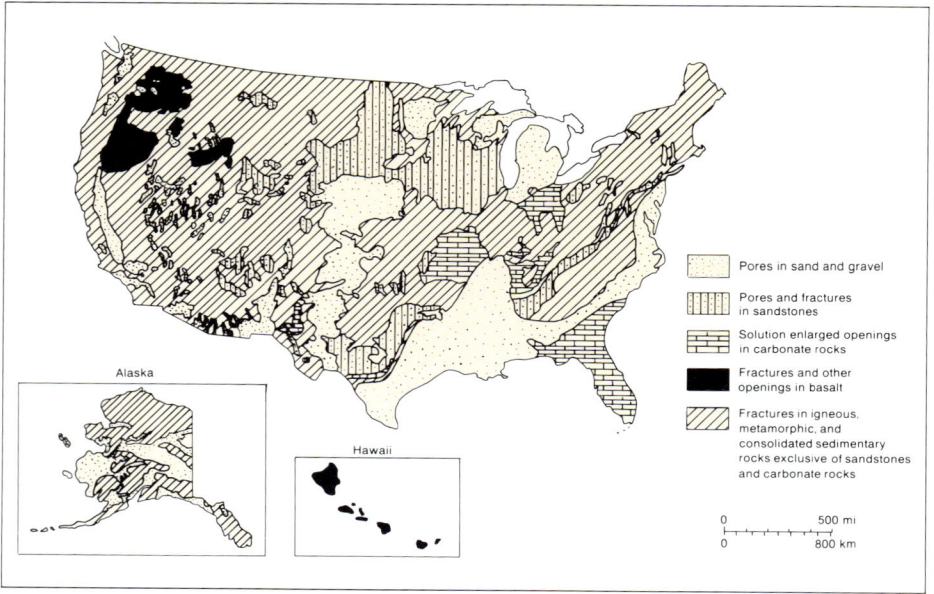

Figure 7.4. Areal extent of various types of aquifers in the United States. *(Heath, 1982)*

is temporarily stored or transmitted in the weathered surficial rocks. Locally, the more permeable rocks may take in water and carry it for some miles underground.

The Western Mountain Ranges have a limited population because of their remoteness, ruggedness, and general unsuitability for agriculture. The population requires only a small fraction of the available water. But, as Thomas (1952) points out, the ranges are surrounded by a great area having little precipitation or other water resources. Water reaches these areas from the mountains by way of the Sacramento, San Joaquin, Snake, Columbia, Missouri, Platte, Arkansas, Rio Grande, and Colorado Rivers, and by aquifers in the valley fill underlying lowlands adjacent to the mountains.

To some extent, faults or joints in the hard rocks may take in water and transmit it to the valley fill directly rather than by stream recharge; and in a few places relatively long, deep faults may carry water from one alluvial basin to another at a lower elevation. Most faults tend to act as barriers to groundwater movement rather than as conduits, except where they are in soluble rocks such as limestone and can be enlarged by solution.

Most aquifers in this region are bodies of alluvium in intermontane valleys, similar to but smaller than those in the alluvial basins of surrounding regions. Aquifers also consist of some consolidated rocks, mainly sandstone and limestone; the largest occurs in the Bighorn Basin of Wyoming. The most productive aquifers are glacial outwash deposits. A good example is found in the Spokane Valley-Rathdrum Prairie area of Washington and Idaho where outwash marks a former spillway for glacial meltwaters from the vicinity of Lake Pend Oreille. The wide, deep channel is filled with gravel approximately 300 ft (91.5 m) thick and is quite permeable in places. Individual large-diameter municipal wells at Spokane yield as much as 39,600 gpm (216,000 m³/day) with moderate drawdown; one well had a specific capacity of 10,600 gpm/ft of drawdown (190,000 m³/day/m of drawdown), one of the highest, if not the highest, on record.

Total groundwater storage capacity, however, is so small in most of this region that it has only a limited effect on streamflow. Thus, much of the runoff occurs during the spring snowmelt and in summer thunderstorms, although winter rainfall is the chief contributor to water supply along the Pacific coast. Because most water is used in the summer, huge artificial storage structures are necessary, and a large fraction of the existing reservoir space in the nation is in the Western Mountain region.

It is not easy to obtain groundwater by drilling in the mountainous areas of the Western Mountain Ranges. Fortunately, small springs are common and these, combined with wells in the valleys and small surface reservoirs, are sufficient to meet domestic needs in most places. On the whole, then, the Western Mountain Ranges constitute a region that readily meets the water needs of its own population and has a large surplus for export.

Alluvial Basins

The Alluvial Basins, consisting of vast low areas bounded by highlands, receive most of their water from adjacent mountain ranges. The Alluvial Basins include parts of the Basin and Range province of Nevada, Utah, and California, and an extension of this province easterly around the southern edge of the Colorado Plateau through southern Arizona and New Mexico, and westerly around the Sierra Nevada Mountains to include California's Central Valley and its southern, drier Coast Ranges and associated valleys.

The basins are filled with unconsolidated materials consisting of clay, silt, sand, gravel, and boulders that were eroded from the mountain slopes. The upper parts were deposited mainly during the last 3 to 4 million years, when glacial conditions

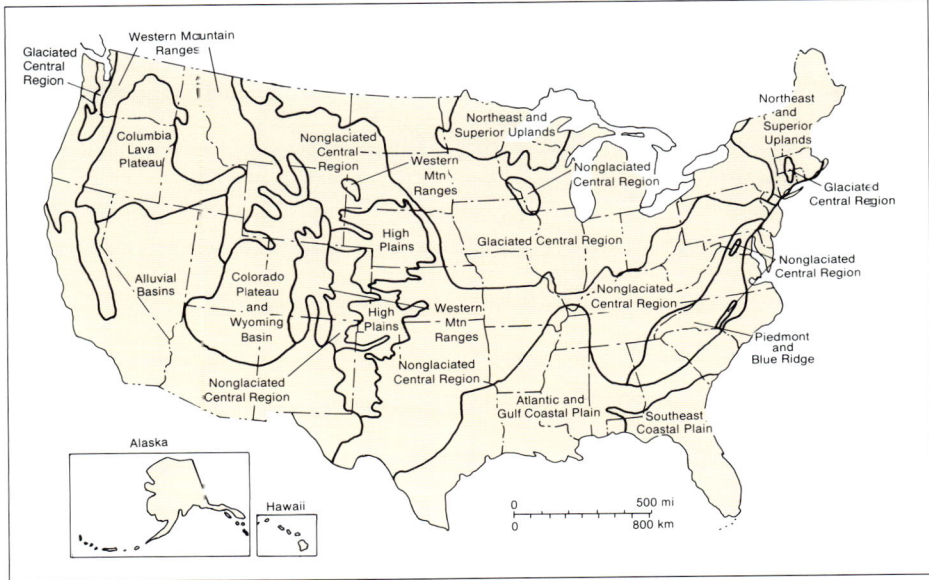

Figure 7.5. Groundwater regions of the United States, excluding the Alluvial Valleys region shown in Figure 7.3. *(Heath, 1982)*

occurred intermittently over much of North America. Much wetter conditions prevailed, and the debris washed down was relatively coarse and formed good aquifers. Huge lakes existed in many valleys during much of Pleistocene time, resulting in the formation of clay beds in the central parts of the valleys. A drier climate that began 10,000 years ago reduced streamflows and the debris washed out was finer, producing excellent raw materials for soils. The leveling action of the stream waters provided a surface ideally suited for large-scale cultivation. Thus, some of the world's most productive agricultural valleys were formed.

Precipitation in the Alluvial Basins region ranges from less than 3 in (76 mm) a year to more than 59 in (1,500 mm) with most parts receiving less than 15 in (381 mm). Unfortunately, precipitation in most of the region is poorly distributed throughout the year and is usually inadequate during the growing season. Water stored in mountain reservoirs is used during this time for irrigation.

The Alluvial Basins region greatly exceeds all others in the use of groundwater, particularly for irrigation. More than half of all the groundwater used in the nation is pumped from aquifers in this one region. Vast quantities of water in underground storage makes this possible. Valley fill is known to be hundreds of feet deep in many valleys, and in some places it exceeds 3,000 ft (915 m).

The huge groundwater basins that underlie the Central Valley, and other basins of California, have an estimated total capacity of 1.3 billion acre-ft (1.6 trillion m^3) with a sustainable yield estimated as high as 143 million acre-ft (176 billion m^3). Groundwater supplies 40 percent of the water used in the San Joaquin Valley during years with normal precipitation; in the drought year of 1977, however, groundwater provided about 80 percent of agriculture's needs. Approximately 9,000 new wells were drilled in this valley alone in 1977. More than 20,000 new wells were brought into production in California in 1977, thereby aggravating the substantial groundwater overdraft. In recent years, groundwater overdraft has approximated 2 million acre-ft (2.5 billion m^3) when precipitation remains normal; the groundwater overdraft in 1977, however, has been estimated at 4 to 10 million acre-ft (4.9 to 12 billion m^3) (*California Water Atlas,* 1979). Overdrafts in the future during drought years will probably be higher unless controls are imposed.

Groundwater overdrafts have become commonplace in many areas of the Alluvial Basins. Groundwater storage has been so great that in the early stages of development the resource was regarded as unlimited. Now, after 50 years of pumping, records show that the groundwater in storage has been permanently reduced in some basins. Average annual replenishment to these valley reservoirs is small; recharge is primarily seepage from streams that originate in the bordering mountain ranges.

Artificial recharge of groundwater has been practiced for many years, especially in southern California. The usual method involves infiltration of water released from reservoirs into natural stream channels or onto the adjacent parts of alluvial fans. Nevertheless, artificial recharge alone cannot make up for the groundwater pumped from the Alluvial Basins. Neither can salvage of water now "consumptively wasted" by phreatophytes (water-loving plants) having low economic value.

Because groundwater in many western basins is developed beyond its average rate of replenishment, previously stored groundwater is being mined. Thus, California officials have warned that "failure to control such overdrafts will increase energy requirements for pumping, decrease water availability, produce water of poorer qual-

Table 7.1 Common Ranges for the Hydraulic Characteristics of Groundwater Regions of the United States[1]

Region No.	Region	Geologic Situation	Transmissivity gpd/ft	Transmissivity m²/day	Hydraulic Conductivity gpd/ft²	Hydraulic Conductivity m/day	Recharge Rate in/yr	Recharge Rate mm/yr	Well Yield gpm	Well Yield m³/day
1	Western Mountain Ranges	Mountains with thin soils over fractured rocks, alternating with narrow alluvial and, in part, glaciated valleys	40–7,000	0.5–100	0.007–400	0.0003–15	0.1–2	3–50	10–100	50–500
2	Alluvial Basins	Thick alluvial (locally glacial) deposits in basins and valleys bordered by mountains	10,000–1,000,000	20–20,000	700–10,000	30–600	0.001–1	0.03–30	100–5,000	500–30,000
3	Columbia Lava Plateau	Thick lava sequence interbedded with unconsolidated deposits and overlain by thin soils	100,000–40,000,000	2,000–500,000	8,000–70,000	200–3,000	0.2–10	5–300	100–20,000	500–100,000
4	Colorado Plateau and Wyoming Basin	Thin soils over fractured sedimentary rocks	40–7,000	0.5–100	0.07–40	0.003–2	0.01–2	0.3–50	10–1,000	50–5,000
5	High Plains	Thick alluvial deposits over fractured sedimentary rocks	70,000–700,000	1,000–10,000	700–7,000	30–300	0.2–3	5–80	100–3,000	500–20,000
6	Nonglaciated Central Region	Thin regolith over fractured sedimentary rocks	20,000–700,000	300–10,000	70–7,000	3–300	0.2–20	5–500	100–5,000	500–30,000
7	Glaciated Central Region	Thick glacial deposits over fractured sedimentary rocks	7,000–100,000	100–2,000	40–7,000	2–300	0.2–10	5–300	50–500	300–3,000
8	Piedmont and Blue Ridge	Thick regolith over fractured crystalline and metamorphosed sedimentary rocks	700–10,000	9–200	0.02–20	0.001–1	1–10	30–300	50–500	300–3,000
9	Northeast and Superior Uplands	Thick glacial deposits over fractured crystalline rocks	4,000–40,000	50–500	40–700	2–30	1–10	30–300	20–200	100–1,000
10	Atlantic and Gulf-Coastal Plain	Complexly interbedded sands, silts, and clays	40,000–700,000	500–10,000	70–3,000	3–100	2–20	50–500	100–5,000	500–30,000
11	Southeast Coastal Plain	Thick layers of sand and clay over semiconsolidated carbonate rocks	70,000–7,000,000	1,000–100,000	700–70,000	30–3,000	1–20	30–500	1,000–20,000	5,000–100,000
12	Alluvial Valleys	Thick sand and gravel deposits beneath floodplains and terraces of streams	10,000–4,000,000	200–50,000	700–40,000	30–2,000	2–20	50–500	100–5,000	500–30,000
13	Hawaiian Islands	Lava flows segmented by dikes, interbedded with ash deposits, and partly overlain by alluvium	700,000–7,000,000	10,000–100,000	4,000–70,000	200–3,000	1–40	30–1,000	100–5,000	500–30,000
14	Alaska	Glacial and alluvial deposits in part perennially frozen and overlying crystalline, metamorphic, and sedimentary rocks	7,000–700,000	100–10,000	700–10,000	30–600	0.1–10	3–300	10–1,000	50–5,000

[1] All values have been rounded to one significant figure.
[2] An average thickness of about 16.4 ft (5 m) was used as the break point between thick and thin.

(Heath, 1982)

ity, encourage saltwater intrusion along the shores of saline bays and the ocean, and bring about significant and sometimes serious land subsidence" *(California Water Atlas,* 1979, p. 3).

Solution of the water problems of the Alluvial Basins rests partly on new technology. For example, economical methods of converting saline water into usable fresh water are at hand. By 1985, 50,700 gpm (276,000 m³/day) of Colorado River water were being treated by a reverse osmosis facility to reduce its salinity. This unit, located at Yuma, Arizona, is 20 times larger than any existing reverse osmosis facility. In addition, the United States may have to import water from the water-rich provinces of Canada, depending on future population trends. Much of this water, derived indirectly from Canada's extensive ice fields, flows directly to the sea and is currently of little use to man. And finally, numerous ways exist to extend the usefulness of the water already available to this area; they include using more efficient irrigation systems, recycling of water whenever possible, increasing natural recharge rates, reducing consumption in both industry and agriculture, and eliminating phreatophytes wherever possible.

Columbia Lava Plateau

The Columbia Lava Plateau is a large area formed principally by volcanic rocks, mainly lava flows, which are interbedded with or overlain by alluvium and lake sediments. It is a rather high and generally dry plateau, drained largely by the Columbia River and its tributaries. The most important tributary is the Snake River.

The plateau is bounded on the west, north, and east by the Cascade Range and the Rockies which, along with volcanic vents (openings) in the plateau itself, were sources of the lava flows. On the south it merges with the Great Basin of Nevada and western Utah, and overlaps the Alluvial Basins region on the southwest. A sizable plateau consisting of volcanic rocks, the structure is dominated by the block faulting typical of the Great Basin. The basins between mountain blocks are filled with alluvium.

The region gets part of its water from local precipitation, especially where elevations are highest [up to 10,000 ft (3,050 m)], but mainly it flows from the mountains on the west, north, and east. The climate is arid to humid, as the average annual precipitation ranges from less than 10 in (254 mm) to more than 30 in (762 mm).

The volcanic materials are less than 100 ft (30.5 m) thick in places near the borders of the region. More commonly, however, they are much thicker. One oil test well near Benton City, Washington, is reported to have penetrated 10,000 ft (3,050 m) of primarily basaltic lava without reaching the bottom of the deposit. These great thicknesses were produced by many individual lava flows, deposited one on top of another.

The rocks range widely in water-bearing properties. Zones between successive lava flows are generally the most permeable in the western part of the area; but farther east, in Idaho, pore space consists chiefly of fracture openings. In many places the solidified lava is so broken up that it is practically a rubble. Gas bubbles, lava tubes, and even small caverns add to the porosity and storage space for water. Permeable gravel and sand layers also occur between basalt layers.

The water supply of this region has been developed on a large scale for irrigation and power production and can be developed further for these and other uses, although, as Thomas (1952) points out, withdrawal for various needs will conflict to some extent. The deep dissection of the plateau by streams provides many good reservoir sites. In addition, large-scale uses of surface water supplies have developed on the Snake and Columbia Rivers. Some areas that are irrigated with surface water have become waterlogged, and these now present serious drainage problems. A dramatic example of increased groundwater flow resulting from irrigation is the Thousand Springs, and springs above and below them, in the stretch of the Snake River between Milner and King Hill, Idaho. The average inflow to the river in this stretch, derived mainly from spring flow, increased from about 1.7 million gpm (9.3 million m^3/day) in 1902, to 3.6 million gpm (19.6 million m^3/day) or more by 1942, largely as a result of irrigation north of the river (Thomas, 1952). The total flow in this region has been stable since the 1950's; today, the flow in the Thousand Springs area alone averages 2.9 million to 3.2 million gpm (15.8 million to 17.4 million m^3/day).

The more permeable parts of the groundwater reservoirs offer great promise as a medium for storing surface water during times of surplus. Artificial-recharge projects, intended to use the large storage capacity of rocks above the present water table, will need to be carefully designed to avoid waterlogging. One promising approach is to locate aquifers in which water can be dammed against structural barriers, such as folds and faults, and then recovered by wells drilled just upstream from the barriers.

This region still has large potential for water-resource development. Because of competition from irrigation, mining, energy development, and hydropower generation interests, careful analyses must be made to determine the hydrogeologic relationship between streams and aquifers.

Colorado Plateau and Wyoming Basin

The Colorado Plateau and Wyoming Basin is an extensive region comprising nearly flat-lying, interbedded sandstone and shale. In places, these beds are tilted, folded, or broken by faults into large blocks that form plateaus at different elevation levels. The region lies west of the Rocky Mountains but includes an area in Wyoming that is a structural basin largely enclosed by mountains, but is similar to the plateaus in other respects. The plateaus are deeply dissected by streams, principally the Colorado River and its tributaries and tributaries of the Missouri River in the northern and eastern parts of the Wyoming Basin. The Grand Canyon of the Colorado is the most spectacular "cut" in the region and, indeed, the continent. The plateaus are surmounted by a few mounts which mostly represent extinct volcanoes such as the San Francisco Peaks near Flagstaff, Arizona, and Mount Taylor between Gallup and Albuquerque, New Mexico.

The region is similar in some respects to the Columbia Lava Plateau. It is mostly high land, is considerably dissected, and is rather dry [precipitation ranges from less than 10 in (254 mm) to 30 in (762 mm), and averages about 15 in (381 mm)]. The resemblance ends there, however, because the sedimentary strata of the Colorado Plateau are much lower in average hydraulic conductivity than are the lavas of the Northwest. Also, the climate is warmer, and thus the precipitation is less effective in generating runoff and groundwater recharge. Most important, however, is that less surface water reaches the plateaus. The region is bounded by mountains along a smaller part of its perimeter. These mountains receive less precipitation than the Columbia Lava Plateau and streams coming into the region are cut below the plateau levels because the dissected edges of the plateaus stand higher than the adjacent land. A few of the higher plateaus receive relatively abundant precipitation, some of it as snow. There is less snow, however, and it melts more quickly than it does in the neighboring mountains. The aquifers have no great storage capacity and are not well situated for recharge, so the water runs off quickly and streamflow is not well sustained.

The area for most favorable recharge of the sandstone aquifers is the uptilted southern edge of the Colorado Plateau, the Mogollon Rim, where the land is high enough to receive abundant precipitation; some water also enters the aquifers exposed on the relatively gentle slopes to the north. But on the whole the Colorado Plateau-Wyoming Basin region is a water-poor area, with annual runoff of 0.5 in (13 mm) or less per year in most of the region. Many of the aquifers are not highly productive and are not well situated to receive recharge.

Most of the aquifers are sandstone, although limestone and alluvium contain water in a few places. Average productivity of the aquifers is low, and wells commonly yield 3 to 30 gpm (16.4 to 164 m^3/day). The most productive consolidated aquifer is the Coconino Sandstone which yields 250 to 500 gpm (1,360 to 2,730 m^3/day) to some wells. The sandstone outcrops on the north slope of the Mogollon Rim, dips northward beneath other rocks, and carries water downdip to the Little Colorado and Zuni Rivers, where much of the water comes up in springs or is tapped by both flowing and pumped wells. Farther north, circulation in the Coconino is restricted and the water is salty.

Water-bearing alluvial deposits are scattered and few are extensive. The largest lies at the foot of the mountains in the northern part of the region. Deposits in the Uinta Basin at the foot of the Uinta Range in northeastern Utah, and other deposits along the Wind River and Medicine Bow Mountains in Wyoming, are the principal alluvial aquifers and are potentially productive.

Because the region is sparsely populated, water requirements have been small. Recent power plant expansion and mining activities, however, have placed greater demand on available water supplies. Although the scarcity of productive aquifers limits future economic development, further utilization of groundwater resources, even though small in comparison with those of most other regions, is possible.

High Plains

The High Plains region comprises extensive semiarid to subhumid plains and plateaus east of the Rockies, extending from South Dakota to western Texas and eastern New Mexico. Precipitation varies from as little as 10 inches (254 mm) in west Texas to as much as 25 inches (635 mm) in eastern Nebraska.

The region is a remnant of a gigantic alluvial apron of late Tertiary age which extended eastward from the foot of the Rockies and mountains to the south in New Mexico to an unknown terminus probably some hundreds of miles east of its present edge. Alluvium was deposited on the eroded and gently undulating eastward-sloping body of interbedded sandstone, shale, and limestone of Paleozoic, Mesozoic, and Tertiary age. These north-south trending strata form a vast structural basin. Their sharply upturned edges lie along the foot of the Rockies and the Black Hills on the west, and gently upward-sloping eastern extensions outcrop in broad belts east of the High Plains.

The alluvium of this vast plain, which is essentially unconsolidated, covers the eroded older rocks to thicknesses of more than 500 ft (152 m). The bulk of it is uniform enough to be classified as a single stratigraphic unit, the Ogallala Formation. After the materials were deposited, the land was uplifted and erosion increased. The original depositional surface is almost intact in large areas, forming a flat and imperceptibly eastward-sloping tableland modified only by shallow depressions and sand dunes. Around the edges of the Plains, however, the alluvial apron has been stripped away by erosion and the Plains are now isolated from the Rockies. Elsewhere, the Canadian River has cut to bedrock across the High Plains in Texas and Oklahoma, and tributaries of the Red River to the south and the Smoky Hill River in western Kansas to the north have reduced the alluvial apron to narrow bands. The sediments of the apron have been reduced in thickness along all the other major streams, and have been stripped away entirely along stretches of some of them, especially along the Republican River and its tributaries in southwestern Nebraska and northwestern Kansas. This dissection is not completely detrimental to local groundwater resources, however, because the beds of the streams themselves contain water-bearing alluvium, derived largely from the adjacent material of the alluvial apron.

In Nebraska, between the Platte and Niobrara Rivers, an extensive deposit of windblown sand has been built up on the original depositional surface of the Plains, forming the well-known Sand Hills. These deposits are exceptionally permeable and increase the thickness of unconsolidated material available to serve as an aquifer.

The deposits of the Ogallala Formation are better sorted than the torrentially deposited alluvium of many of the Alluvial Basins and contain a high proportion of sand and fine gravel. However, abrupt variations from silt and clay to coarse sand and gravel occur within short distances, both vertically and horizontally. The underlying bedrock is generally of low permeability, and the little water it contains tends to be rather highly mineralized. In a few areas, however, the bedrock immediately below the alluvium yields substantial supplies of usable water. The Dakota Sandstone, a confined aquifer, is capable of yielding economic volumes of water, although it may be highly mineralized.

Groundwater in the High Plains is recharged by local precipitation. The sand and gravel aquifers formerly reached all the way to the Rockies and were fed by mountain streams, but they are now cut off from that source. Streams crossing the High Plains that have cut to bedrock beneath the Ogallala Formation, such as the Canadian River, no longer contribute water to the Ogallala. These streams probably receive small quantities of water from the Ogallala, but much of the water that does seep out at the base of the Ogallala along the valley walls is consumed by evapotranspiration. As pointed out, however, the larger streams have alluvium along all or part of their channels, and this alluvium receives recharge from the streams. Furthermore, some large streams have not cut all the way through the unconsolidated deposits, and the alluvium along them is in hydraulic continuity with the adjacent and underlying Ogallala Formation. This means that the streams may receive water from their own alluvium as well as from the Ogallala Formation in times of low water, and may contribute water to the unconsolidated deposits during floods or when wells near the streams are pumped heavily enough to lower the adjacent water table.

The High Plains region is water-poor because recharge lags far behind use. The land surface is flat and, because of the rather dry conditions, absorption by the soil and subsequent evaporation and transpiration are high; little surface runoff and groundwater recharge occur. Surface runoff varies from virtually zero to about 1 in (25 mm).

In the Sand Hills of Nebraska, however, infiltration is high and runoff ranges from 1 to 11 in (25 to 279 mm) per year.

Groundwater recharge from precipitation in the High Plains is generally slight except where surficial materials are unusually permeable. Recharge is impeded by tight subsurface layers, especially those of the lime-cemented rock known as caliche. Most importantly, however, it is reduced because insufficient precipitation falls to wet the soil beyond field capacity. In most of the High Plains, groundwater recharge occurs mainly in exceptionally wet years and meagerly or not at all in dry and normal years. Also, there is evidence that a considerable part of the recharge occurs from "sinks" in which storm runoff accumulates until it evaporates or infiltrates to the water table through solution cavities in the caliche.

The development of high-capacity wells in the High Plains, particularly in Texas, Nebraska, and New Mexico, has encouraged the expansion of irrigation. More than 93,500 irrigation wells are in operation in Texas, and in Lincoln and York Counties of Nebraska more than 850 irrigation wells existed in 1979. Because withdrawal exceeds recharge and the natural groundwater discharge to rivers continues unabated, it is apparent that virtually all the withdrawal in heavily pumped areas comes from storage, that is, the water is being mined from the aquifer. In the future, groundwater use in low-recharge areas will have to be reduced. Thus, the challenge is how to adjust to a decreasing water supply.

Reduction in groundwater storage in the High Plains has led to the adoption of numerous water conservation practices. These include making use of the surface water that gathers in topographic lows, either by applying it directly to crops or recharging it through pits or recharge wells for later use; distributing water in pipes instead of ditches; improving irrigation practices to prevent waste of water; reusing excess water that accumulates at the lower ends of fields; growing crops that need less water; and following land-treatment practices such as summer fallowing to conserve moisture derived from precipitation.

Water is repeatedly diverted from several major rivers in the High Plains and used for irrigation, and then returned to the river less the amount used consumptively on the fields. Numerous withdrawal cycles can cause a marked reduction in river discharge. The water also tends to become mineralized from this repeated use, in spite of dilution resulting from a climate that becomes wetter toward the east. At present, however, rivers furnish some relief in an otherwise rather unpromising groundwater situation.

Nonglaciated Central Region

The Nonglaciated Central region is a vast area in the interior of the United States. Except for the High Plains region just described, it includes the unglaciated part of the country east of the Rockies and Alluvial Basins, and west of the Appalachians and north of the Atlantic and Gulf Coastal Plain. It also includes an isolated area within the glaciated parts of Wisconsin, Minnesota, and Iowa, often referred to as the Driftless Area. On the southwest, it abuts the Alluvial Basins region.

This region is large and topographically complex and has a considerable range in climate. Geologically it is characterized mainly by plains and plateaus underlain by horizontal or gently dipping sedimentary rocks. The land surface is gently undulating in most of the region, but the area near the Rockies is considerably dissected and characterized by high plateaus and mesas.

The strata are bowed up sharply in the foothills of the Rockies and around the Black Hills of South Dakota. They are folded and faulted in the Wichita and Arbuckle Mountains of Oklahoma and the Ouachita Mountains of Oklahoma and Arkansas. Elsewhere the strata are bowed up more gently into structural domes which have been eroded and no longer stand much above the surrounding areas. In fact, in some places the upward arching of the strata has made the rocks more susceptible to erosion, and thus some of these areas are now lowlands rather than physiographic domes. Former structural domes are found in the Big Bend country in Texas, the Ozark Plateau in Missouri and Arkansas, the Nashville dome in Tennessee, and the Blue Grass region in Kentucky.

Substantial alluvial deposits are found only along the major rivers and streams; the most important exist along the upper and lower Missouri and its tributaries in Montana, Wyoming, Kansas, and Missouri, and the stretch of the Ohio River between Indiana and Kentucky. The Missouri and Ohio Rivers flow near the boundary between the Glaciated and Nonglaciated Central regions, and much of the alluvium was deposited during glacial episodes.

The bedrock aquifers in most of the region are limestone and sandstone of low-to-moderate productivity. Other areas of this region include some aquifers that are among the least productive in the United States because of low yield and salty water. These unproductive areas are found both in the dry western and southwestern parts of the region and in the much wetter eastern areas.

Few outstanding aquifers exist in this region, but the most productive include (1) the limestone in the Edwards Plateau in Texas, (2) limestone in the Ozark region from which issue some of the nation's largest springs, (3) limestone in northern Alabama and parts of Kentucky and Tennessee, (4) sandstone and glacial outwash in the Driftless Area in Wisconsin, and (5) the alluvium along some of the major rivers and streams. The Dakota Sandstone in the Dakotas and states to the south, and the Trinity Sands in northeast-central Texas, may also be important locally.

Limestone is the most productive type of rock in this region, although the sandstone of the Driftless Area is capable of yielding 1,000 gpm (5,450 m^3/day) or more to individual wells, and in places along the Ohio River, the alluvium can yield 3,000 gpm (16,400 m^3/day) which is readily replaced by infiltration of river water.

The region is characterized generally by consolidated-rock aquifers having low yield and saline water at shallow depth. Seawater trapped in the rocks has not been completely flushed, and salt beds can contaminate fresh water. In some other areas, however, the porous, fractured, or cavernous nature of the rocks and their geologic structure has allowed water to circulate to considerable depths and flush out whatever saline water was present in the zone of circulation. Examples include the Denver Basin, Balcones Fault in Texas, Dakota Sandstone, and Trinity Sands. In many areas of deep fresh-water circulation, however, there is some danger of contamination by saline water from adjacent, underlying, or even overlying aquifers as a result of the lowering of the fresh-water heads accompanying well development.

The Nonglaciated Central region has a generally subhumid to humid climate. The generous precipitation in the central and eastern parts of the region [30 to 50 in (762 to 1,270 mm)] is reflected in runoff rates that usually exceed 5 in (127 mm) per year. In the western part, runoff is 1 in (25 mm) or less because annual precipitation averages only 14 to 16 in (356 to 406 mm).

Glaciated Central Region

The Glaciated Central region includes the central area of the United States which at one time or another has been overrun by continental glaciers from Canada. This region extends from the Pacific Northwest to the Appalachians on the east, and from the Canadian border southward to the limits of glaciation. The present channel of the Missouri River from Great Falls, Montana, to St. Louis, Missouri, and that of the Ohio River from Pittsburgh, Pennsylvania, to its mouth at Cairo, Illinois, are close to the southern limits of glaciation. Within these boundaries, the only area not included is the Driftless Area of Wisconsin, Minnesota, and Iowa.

A westward extension of the Glaciated Central region occurs in the Puget Sound area. The Puget-Willamette Trough contains in its central and northern parts a thick sequence of glacial deposits. Some of these deposits are highly productive. At Tacoma, Washington, six production wells yield 50,000 gpm (273,000 m^3/day), with an average drawdown of 3.8 ft (1.2 m). These wells, located in coarse glacial gravels, are recharged by the Green River (Carr, 1976).

The region is similar to the Nonglaciated Central region in climate, physiography, and types of consolidated rocks, and differs fundamentally in only one important respect — a mantle of glacial drift has been deposited on the consolidated rocks. The glacial drift consists mostly of fine-grained rock debris, but like the valley fill of the West it contains enough beds of water-sorted permeable sand and gravel to constitute an important aquifer over large areas. Glacial drift reaches thicknesses of 300 to 1,000 ft (91.5 to 305 m) in parts of Michigan, Indiana, Ohio, Illinois, Iowa, Wisconsin, and Minnesota. The drift in these areas contains important groundwater aquifers that provide sufficient yields to wells, recharge other aquifers in the bedrock below, and sustain the base flow of streams.

The average climate of the region is cooler than that of the unglaciated region to the south. Thus, for a given amount of precipitation, the runoff tends to be greater in the north than in the south. The runoff in the Glaciated Central region exceeds 5 in (127 mm) east of a line running from northeastern Minnesota to northeastern Kansas. It is generally less than 1 inch (25 mm) in the Dakotas and Montana.

Water ordinarily is less difficult to obtain from a well in a glaciated region than in an unglaciated one. Larger average yields are relatively easy to obtain where substantial drift thicknesses contain at least a few beds of sand and gravel. These lenses may yield 5 to 65 gpm (27 to 354 m^3/day). Many areas in this region are covered by fine-grained glacial lake deposits, or consist of drift derived from shale, limestone, dolomite, or fine-grained sandstone bedrock which could not furnish large quantities of coarse-grained debris to the glaciers. In these areas the principal aquifers are along watercourses, where rapidly flowing meltwater streams were able to sort the glacial debris.

Irrigation has expanded greatly in the Dakotas, Minnesota, and Wisconsin since the early 1970's. Virtually all of this water comes from wells drilled in outwash deposits associated with terminal or lateral moraines. In some cases, deeply buried outwash is covered by one or more layers of clayey till laid down by subsequent ice advances. These wells average 200 to 300 ft (61 to 91.5 m) in depth and commonly yield 750 to 1,000 gpm (4,090 to 5,450 m^3/day). Thicknesses of the sand and gravel lenses range from only 6 ft (1.8 m) to more than 50 ft (15.2 m).

In the central and eastern parts of the region, full-scale irrigation is not needed for most crops; but farmers are becoming increasingly aware of the profit to be gained from applying 2 to 12 in (51 to 305 mm) of water at times when rainfall is inadequate, not only to save crops from failure but to obtain optimum yields. Thus, supplemental irrigation is a rapidly growing practice, both here and elsewhere in the Eastern states.

In areas where the glacial drift is not especially productive, the bedrock may be capable of furnishing good yields to wells. Typical examples are found in western Ohio, eastern Indiana, and central Missouri. In the subhumid to humid eastern part of the region, where limited groundwater resources exist, there is a substantial surface-water supply. In the western Dakotas and Montana, however, surface water is scarce except along major streams flowing from mountains.

Water problems are not serious in this region, but local shortages and poor quality occur frequently enough to cause concern in many communities. For example, the potentiometric surface in some Chicago-area aquifers has fallen more than 850 ft (183 m) since 1864. Future urban and industrial growth will strain existing water supplies in more communities, forcing the development of new water resources at increasingly greater costs.

Piedmont and Blue Ridge Region

The Piedmont and Blue Ridge region extends from Alabama on the southwest to Pennsylvania and New Jersey on the north. Together with the Northeast and Superior Uplands region to the north, the regions make up the area involved in compression and uplift which formed the Appalachian Mountains about 200 million years ago when the North American and African plates collided. This region is characterized by mountains and hilly uplands separated by broad valleys and bounded by central lowlands on the west and coastal plains on the east. The region includes four rather different physiographic segments: Piedmont, Blue Ridge, Valley and Ridge, and Appalachian Plateaus.

The Piedmont segment contains extensively weathered and eroded rocks which form an undulating low plateau that stands a little above the Coastal Plain. The Blue Ridge area is geologically similar but stands higher because it has been uplifted along faults and some of its rocks are also more resistant to erosion. The Valley and Ridge area is characterized by folded rocks, principally limestone. There are some large springs and productive wells in the area, but sites where high-yield wells can be constructed are difficult to find. The Appalachian Plateaus are underlain by limestone, sandstone, and shale. Plateaus are elevated and deep canyons have been cut. The limestone and sandstone are fairly productive, but locating a reliable aquifer is rather difficult and costly in most places.

The region receives generous precipitation, especially in the Blue Ridge province because of orographic effects. Precipitation exceeds 76 in (1,930 mm) a year in the Great Smoky Mountains at the south end of the Ridge, but tends to decrease northward from the Gulf of Mexico. Precipitation exceeds 40 in (1,020 mm) nearly everywhere in the Blue Ridge and Piedmont areas, but drops to between 30 and 40 inches (762 and 1,020 mm) in the northern part of the Valley and Ridge province.

Groundwater resources in this region are generally small to moderate and productivity may be somewhat erratic. The weathered and fractured crystalline rocks of the

Piedmont are one of the nation's most reliable aquifers for domestic supplies and are tapped by hundreds of thousands of rural and suburban wells, most of which are less than 150 ft (45.7 m) deep. Gneisses and schists are probably the best producers, storing and transmitting water only in the fractured and weathered zones. Yields in this area, however, rarely exceed 50 to 100 gpm (273 to 545 m^3/day).

Wells in all the crystalline rocks generally obtain the bulk of their water at the base of the weathered mantle and in the fractured rock immediately below. The thickness of the weathered mantle in undissected upland areas probably averages between 50 to 100 ft (15.2 to 30.5 m). Sizable fractures in the underlying rock generally extend no more than 30 to 50 ft (9.1 to 15.2 m) deeper, gradually becoming fewer and smaller with depth. Deeper holes occasionally encounter fractures at depths of 500 to 1,000 ft (152 to 305 m), but the deeper the hole, the less likely water will be found.

Except for the limestone areas, the consolidated rocks of this region are poor places to drill if large yields are desired. Larger and more reliable yields can be obtained from the limestone and fractured and jointed rock if sophisticated groundwater prospecting methods are used. The best prospects for large quantities of groundwater are sand and gravel beds in the floodplains of perennial streams.

Northeast and Superior Uplands

The Northeast and Superior Uplands region includes all of New England, eastern New York, northern New Jersey, northern Minnesota, northern Wisconsin, and the western part of the Upper Michigan Peninsula. In the Northeast, the rocks are similar to those of the unglaciated region to the south. The most productive bedrock aquifers are sandstones in parts of northern New Jersey, and sandstone and carbonate rocks in scattered areas elsewhere. Less than 10 percent of the region, however, is underlain by good bedrock aquifers.

Productive bedrock aquifers are found in small areas in Connecticut, Massachusetts, New Jersey, and New York, underlain by Triassic sandstone and shale; scattered areas along the Hudson River section of the Valley and Ridge province and at the southern edge of the Adirondack Mountains, underlain by limestone and sandstone of early Paleozoic age; and an east-west strip in central New York State, underlain by limestone of early Paleozoic age. Yields as high as 400 gpm (2,180 m^3/day) have been obtained from some wells in these rocks, but the average yield is about 50 gpm (273 m^3/day).

Except for these areas of relatively productive bedrock, the principal groundwater sources are in areas of unconsolidated deposits composed of glacial sand and gravel. These take two principal forms, outwash plains and channel fillings. A third, minor type is represented by "ice contact" or kame deposits which are irregular mounds of sand and gravel formed at the edges of melting ice sheets. Sand and gravel deposits making up the outwash plains absorb water freely from precipitation and yield it readily to wells. Outwash deposits in some places are only 30 ft (9.1 m) thick; for large supplies of water, these deposits must be tapped by closely spaced wells to take full advantage of available drawdown. Channel-filling deposits occur as (1) outwash deposits in valleys that are still occupied by streams; (2) deposits in valleys formed by glacial streams that were diverted elsewhere after the deposits were laid down (valleys now contain no sizable streams); and (3) buried-valley deposits.

Floodplain deposits are excellent sources of groundwater because when heavily pumped the deposits are replenished from streamflow. Thus, the maximum sustained yield of the deposits along a river channel is equivalent to the low flow of the stream, plus the volume obtained by pumping water from storage after the stream has dried up. The abandoned floodplains and buried-valley deposits yield just as much water initially as do the active watercourses, but because they are recharged mainly by precipitation, their perennial yield will be smaller.

Average annual precipitation varies from 40 to 50 inches (1,020 to 1,270 mm) in the highest mountains of Maine and New York and in the coastal lowlands. In the low-lying areas in the central part of the region, precipitation is 30 to 40 inches (762 to 1,020 mm). Annual runoff exceeds 30 inches in nearly the entire region, but few good storage sites exist.

In some cases, water demands in this region have exceeded or will exceed the capabilities of local supplies. This is especially true because the population is concentrated largely in the valleys, which constitute only a small part of the geographic area. Nevertheless, future water problems should be much less difficult to handle than in those regions with less abundant precipitation and runoff.

Precambrian bedrock in the Superior Uplands (northern Minnesota, northern Wisconsin, and part of the Upper Peninsula of Michigan) consists mainly of fractured and jointed crystalline and volcanic rock, although significant areas are underlain by sandstone. These rocks are covered discontinuously by various thicknesses of glacial drift. Deposits of outwash sand and gravel can yield relatively large volumes of groundwater to municipal and industrial wells, but more typically the wells in the drift yield only moderate water supplies. Water quality varies widely, depending on the geochemistry of the rocks contained in the drift. For example, in Minnesota, glacial deposits picked up in the Winnipeg Lowland contain larger amounts of calcium and magnesium, whereas drift deposited by the ice lobes surging out of the Superior Basin is dominated by rocks containing iron and manganese. Thus, water found in these distinctly different drift sheets will normally be affected by these cations.

Water in the bedrock occurs in joints and fractures in the crystalline rock and in the pores of the old Keweenawan or Cambrian sandstones. Yields are usually low in the crystalline rock, but can be increased significantly in granite by blasting (see Chapter 15). The sandstones yield larger volumes of water, but they underlie only a small part of the region and may be rather thin where they do occur.

Many of the major cities in this region are situated on Lake Superior or Lake Michigan. Duluth, Minnesota, for example, obtains its entire water supply from Lake Superior. Most communities, however, must obtain their water supplies from scattered lenses of outwash sands and gravels which generally rest on top of the bedrock.

Atlantic and Gulf Coastal Plain

The Atlantic and Gulf Coastal Plain comprises the coastal plains from Massachusetts southward along the Atlantic seaboard, excluding most of Florida and southern Georgia, and westward to Texas. The area includes the Mississippi Embayment as far north as the southern tip of Illinois. Nearly everywhere along its course, the landward edge of the Coastal Plain forms a boundary between elevated and dissected uplands and relatively low and undissected seaward-sloping plains. In places, the Coastal Plain is rather well dissected and has considerable relief at its landward edge, but it is generally lower than the adjacent inland area.

The Coastal Plain consists of a huge seaward-thickening wedge of unconsolidated sedimentary rocks of Cenozoic age and younger. The sediments are mainly clay, silt, sand, gravel, and marl, interbedded in alternating and intertonguing strata; some are locally consolidated. They represent the weathered rock debris washed into the sea from the eastern two-thirds of the country, beginning when the shoreline was essentially at the Plain's present boundary. As the sediments accumulated and the relative ocean level dropped as a result of isostatic rebound, the shoreline migrated seaward to its present position. As sediments accumulated offshore, the sea bottom sank gradually under their weight, but the water remained shallow. Therefore, the strata generally thicken seaward, and the slope of the bedrock surface and older Coastal Plain strata is steeper than the inclination of the younger strata; the youngest Tertiary strata and the Quaternary deposits dip only very gently seaward. The deposits vary in thickness from 300 ft (91.5 m) at places in the northeastern part of the region to 50,000 ft (15,200 m) or more at the mouth of the Mississippi River.

The region is subhumid to humid, with precipitation ranging from 20 to 30 inches (508 to 762 mm) in the western part to as much as 65 inches (1,650 mm) along the Gulf Coast. Runoff averages 15 inches (381 mm) or more over most of the region, but varies from as little as 1 inch (25 mm) in Texas to 30 inches (762 mm) in Mississippi. Huge discharges are also added from rivers that originate outside the region.

The Coastal Plain has some of the country's most extensive and productive aquifers. Nearly the entire plain has aquifers capable of yielding 50 gpm (273 m^3/day) to wells. The principal aquifers, which are beds of sand or sand and gravel, include glacial outwash deposits in New England and along the Ohio and Mississippi Rivers; broad sheets of marine sand and less extensive sand bodies deposited in deltas and estuaries; alluvium along present-day streams; and some large and abandoned watercourses in the broad valley of the Mississippi. Groundwater, however, is in meager supply in sizable areas of Texas and Louisiana, and in smaller areas of Mississippi and Alabama, where all sediments within feasible drilling depths are tight and consist of relatively impervious materials such as silt and clay.

It is likely that most permeable sediments of the Coastal Plain were saturated with saltwater at some time early in their history. Saltwater has been flushed from the aquifers at a rate determined by recharge in their upgradient portions and the ability of the water to escape or discharge downgradient. Permeable sediments exposed at the land surface have been flushed quickly, except for those in some low marshy tracts where saltwater is so readily available that it easily overcomes the diluting effect of precipitation. In large areas of the Coastal Plain, fresh water is usually found to depths of 1,100 to 1,300 ft (335 to 396 m), although some aquifers have been flushed to depths as great as 6,000 ft (1,830 m). Of course, the flushing process has proceeded unevenly and there is every conceivable combination of alternating fresh and salty strata.

Land subsidence results from groundwater withdrawal in some parts of the Coastal Plain. The strata are largely unconsolidated and capable of being compacted. The prevailing structure of seaward-dipping and alternating permeable and impermeable strata produces widespread confined conditions. When water is withdrawn from a confined aquifer, it comes not from the emptying of pore spaces but from the compaction of the aquifer, as well as from slight expansion of the water itself. Initially, each stratum is supported to some extent by the pressure of the confined water; when the support is reduced by a decline in the confined head, caused by flowing or pumped wells, the beds slump and the land subsides. The Coastal Plain is especially vulnerable to the effects of land subsidence, because only slight lowering of the land surface may cause extensive damage to harbor facilities, drainage structures, and buildings. In some areas, the cost of repairing the effects of land subsidence may be higher than the cost of securing water supplies from distant or less sensitive formations.

Groundwater in the Coastal Plain can support, and will need to meet, vastly increased demands in the future. Environmental considerations and rising heating costs in the north may combine to greatly increase the population, especially in the southern Coastal Plain area. To some extent the increased groundwater withdrawals will reduce the evapotranspirative discharge of groundwater and thus will salvage water that is now not retained in the region. Also, carefully designed groundwater developments will be able to use some of the water that now passes out beneath the coast, without seriously increasing the danger of saltwater encroachment. The increased use of groundwater, to the extent that it is consumptive, will be largely competitive with the use of surface water because pumping will result eventually in depletion of streamflow. Here, however, as in many other parts of the country, the competition will not be entirely destructive, because not all the streamflow depletion will occur in dry weather, and pumping of groundwater will create storage space that can be recharged in wet weather. Moreover, there may be some reduction in flood peaks on streams. The great areal extent and thickness of aquifers will provide opportunities for the integrated development of groundwater and surface-water resources of great potential importance.

Southeast Coastal Plain

Florida and the southern third of Georgia can be differentiated from the rest of the Atlantic and Gulf Coastal Plain because this entire area is dominated by the Floridan limestone aquifer. In southern Georgia, this aquifer furnishes most municipal and rural water supplies. Reaching thicknesses of 600 ft (183 m) near the coast, the aquifer yields good quality but hard water with a dissolved solids content of 600 mg/ℓ or less. Most of Florida is underlain by this aquifer, which rises above sea level in only scattered localities. The aquifer is recharged mainly in the lakes area of central Florida and the northeast-central part of the state. It is generally covered by younger deposits, commonly the Tampa Limestone and the Hawthorn Formation consisting of marl, clay, and some water-bearing sand and limestone beds. These and other overlying deposits cause the water in the Floridan aquifer to be under confined pressure.

Florida receives extremely high rainfall [53 in (1,350 mm) per year], and runoff averages about 14 in (356 mm) per year [40 billion gpd (151 million m^3/day)]. Retention of even a small part of this total discharge could increase Florida's water supply substantially.

Except for areas south of Lake Okeechobee and some areas near the St. Johns River in northeastern and east-central Florida, water quality in the Floridan aquifer is good. Saltwater intrusion is a problem along the coasts and canals where local overpumping is occurring. The Floridan aquifer also contains unflushed (saline) water in certain areas of the state. Apparently the aquifer is less permeable in these areas than where the water in the aquifer is fresher. Subsurface contamination is a severe problem in Florida because the water table is high and the openings in the aquifer are generally cavernous. Thus, pollutants can migrate underground rapidly with little attenuation of contaminant strength.

Alluvial Valleys

The principal geologic and hydraulic characteristics of alluvial aquifers are presented in Table 7.1. Because this type of aquifer occurs in widely varying geologic environments and was formed under diverse hydrologic regimes, it is nearly impossible to generalize as to areal extent, depth, hydrologic parameters, and water chemistry. For example, terrace deposits along the Ouachita River and the Red River (of the south) are generally less than 100 ft (30.5 m) thick and yield small to moderate water supplies. Elsewhere, the Quaternary alluvium of the Mississippi Alluvial Plain is as much as 200 ft (61 m) thick and can yield large quantities [up to 5,000 gpm (27,300 m^3/day)] of generally hard water to wells.

Water quality varies in some river valleys depending on the distance downstream. In the South Platte River alluvium, the groundwater in the main valley (in Colorado) tends to deteriorate in quality downriver. Re-use and evaporation of the water causes an increase in hardness which generally makes it unsuitable for domestic or municipal supplies. In general, most alluvial aquifers will have poor water quality if the evapotranspiration is high, that is, if the climate is dry.

In some areas of northeastern United States that are underlain by dense crystalline rock, the glaciofluvial deposits left by Pleistocene glaciers and more recent alluvial deposits are extremely important because they can yield higher volumes of water than the underlying bedrock. Yields of 1,000 gpm (5,450 m^3/day) or more are possible even though the stratified deposits are not deep. In the past, contamination of the rivers supplying these alluvial aquifers posed a significant water-quality problem for shallow wells in the alluvium. More thorough treatment of domestic and industrial wastes has reduced this threat considerably in recent years. Fortunately, the restoration of contaminated alluvial aquifers in the Northeast can occur rather quickly once the contaminating source is eliminated because of the high annual rainfall, small extent of most aquifers, and high rates of induced infiltration.

The presence of glacial outwash in modern river channels is also important in the Midwest where recent alluvium is fine grained and rather thin. Much of the Mississippi River Valley floor is covered by coarse glacial outwash. When wells constructed in the alluvial deposits are pumped heavily, high rates of induced infiltration may occur, depending on the distance between individual wells and the river.

The importance of alluvial aquifers cannot be overstressed because they often occur in areas where surface water supplies are absent, groundwater quality is poor in deeper aquifers, and climatic conditions are so dry that adequate recharge to deeper aquifers is prevented. Alluvial aquifers provide dependable sources of water for industrial, municipal, and irrigation wells.

Alaska

Alaska is nearly one-fifth the size of the other 48 continental states. Because of its size and isolation from the rest of the states, Alaska merits a separate description. So little of the state is developed, however, that information on groundwater resources is limited.

Groundwater conditions in Alaska are highly variable. The principal aquifers are bodies of water-sorted sand and gravel incorporated within the glacial drift, glacial outwash, and other alluvial deposits. The most productive aquifers are those in the vast Central Plateau, whereas productive alluvial deposits are scarce in the southeastern coastal area. Much of the groundwater, especially in the glacial deposits, is poor in quality; it is commonly hard, and much of it contains high iron concentrations. Organic matter also occurs in many glacial outwash deposits and causes problems of color, taste, and odor.

Alaska has a varied climate which ranges from extremely wet and rather mild near the southern coasts, to semiarid and extremely cold in the interior and far north. Potential evapotranspiration is generally less than precipitation, so a surplus of water remains to run off as streamflow or glacier ice.

Groundwater availability in Alaska is complicated by at least two major conditions. The most serious is the widespread distribution of permafrost (permanently frozen ground) which prevents easy access to or precludes the presence of groundwater. In many areas the frozen ground begins only 3 to 15 ft (0.9 to 4.6 m) below the surface and extends to great depths. Groundwater may occur above the permafrost, in thawed zones within the permafrost, or below the permafrost. Near large lakes or in areas adjacent to streams, permafrost may be absent or at some depth. In the Arctic Slope region, however, permafrost is virtually continuous. The second major problem is the scarcity of good aquifers in southeastern Alaska.

Alaska, like many other states in early pioneer times, has huge quantities of many natural resources, including water, and potentially bright economic prospects. While Alaska's total water resources are large, the cost to use these resources may be high.

Hawaii

The islands of Hawaii are built from basaltic volcanic rocks that have melted at a hot spot in the upper mantle and then risen through the oceanic floor. In most areas the basaltic lava is rather permeable. Openings consist of cavities between adjacent lava beds, shrinkage cracks, lava tubes, vesicles (gas bubble holes), and cavities left by the burning of trees overwhelmed by the lava. Central parts of the individual flows are tight, but numerous outcrops of permeable zones provide so many entrances for rain and surface water that the basalt as a whole has good permeability and porosity.

Fresh groundwater occurs throughout the island in the form of a huge lens floating on seawater. This is called basal groundwater. High-level groundwater occurs in some areas at elevations considerably above the main water table. This happens where igneous rock dikes of low permeability block normal vertical or horizontal groundwater movement. Some of these dikes act as underground dams within the porous rock, backing up groundwater in local areas. When the local water table is high enough to overflow the subsurface dike walls, water discharges naturally to porous zones at lower levels.

Rainfall varies greatly from island to island, ranging from 6 to 450 in (152 to 11,400 mm) per year. Much of the rainfall is orographic in origin. A large part of the precipitation runs off in spite of the heavy vegetation. Some of this water percolates down to the water table and thus maintains the basal water table at about 30 to 60 ft (9.1 to 18.3 m).

Hawaii, Maui, Oahu, and Kauai, the largest islands in the Hawaiian chain, have large resources of both surface water and groundwater, and they make extensive use of them for irrigation, municipal, and industrial water supplies. Most of the water used in these islands for domestic purposes comes from the ground. The island of Molokai has much smaller, though still substantial, water supplies; the smallest islands, Lanai, Niihau, and Kahoolawe, have relatively little groundwater.

CANADA

Canada* is predominantly a humid country where surface water is extremely abundant. Nevertheless, groundwater contributes about 10 percent of the water supplied by municipal water-supply systems serving communities with a population of more

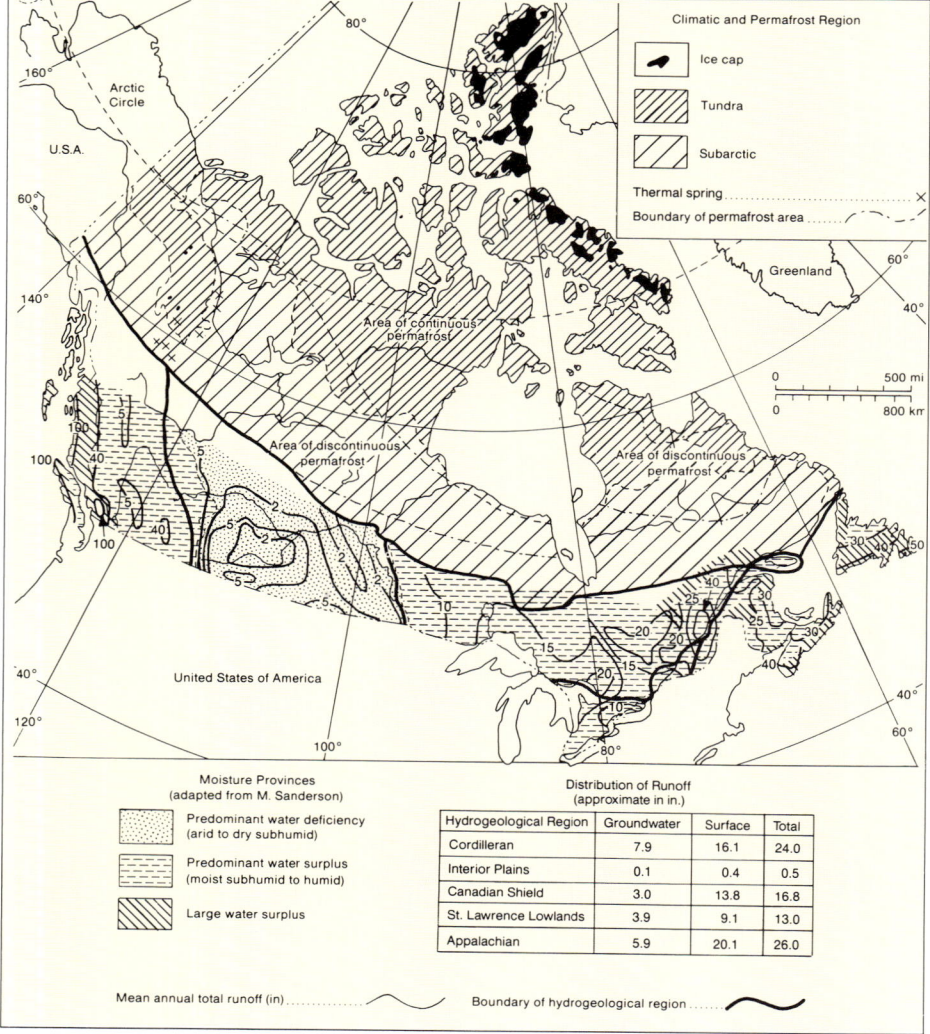

Figure 7.6. **Precipitation patterns throughout Canada.** *(United Nations, 1976)*

---

*The information for the remainder of this chapter has been paraphrased from a United Nations Report (1976) on groundwater resources in the Western Hemisphere. Federal or provincial water resource agencies should be contacted for more specific information (Appendix 7.A).

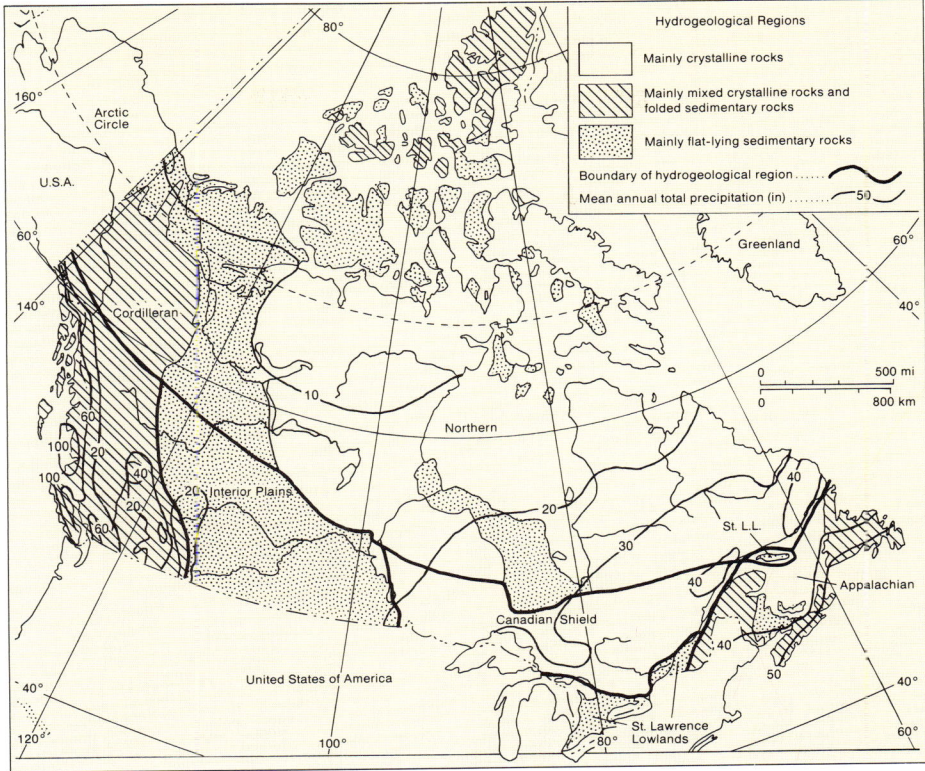

Figure 7.7. Six hydrogeologic regions of Canada. *(United Nations, 1976)*

than 1,000 and a greater proportion of the water used by individual homes. Large quantities of industrial and irrigation water are also obtained from groundwater.

Precipitation patterns vary greatly throughout Canada, from the large water surpluses of the West Coast to the water-deficient areas of southwestern Manitoba and Saskatchewan (Figure 7.6). Crystalline rocks form much of central and eastern Canada and represent the shield area. Flat-lying sedimentary rocks appear in large areas of Ontario, Saskatchewan, and Alberta. The mountains of the West are made up of crystalline and folded sedimentary rocks. Differences in precipitation patterns, geology, climate, and topography serve to divide Canada into six regions of differing hydrogeological conditions (Figure 7.7). Geology is the most important factor in defining these regions because it determines the amount of water that can be stored and transmitted. Precipitation and climate control the amount of water available for recharging groundwater flow systems and topography influences the length and depth of the systems.

Appalachian Region

The Appalachian region of Canada is characterized by crystalline rocks, folded to flat-lying sedimentary rocks, and comparatively thin surficial materials. It is an upland, sloping gently to the southeast, that is dissected by valleys and broken by broad lowland areas developed on belts of relatively weak rocks. Relief in the Gaspe region exceeds 2,000 ft (610 m), but elsewhere it is rarely more than 1,000 ft (305 m) and generally is about 325 ft (99 m). The climate is humid continental. Annual precipitation ranges from 30 to 47 in (762 to 1,190 mm), with the maximum along the Atlantic coast.

Yields from wells in the deformed rock area are seldom greater than 5 gpm (27.3 m^3/day), although the Mississippian limestones, dolomites, and conglomerates typically yield 25 gpm (136 m^3/day). Younger Late Carboniferous and Permian sandstones can produce 25 gpm. Some saline water is present in all these formations, partly as a result of overpumping and partly through contact with salt deposits.

Groundwater in the Appalachian region is generally low in total dissolved solids and bicarbonate-type waters because of the comparatively short groundwater flow systems coupled with the high rainfall and the preponderance of crystalline rocks and outwash deposits. The three main causes of poor water quality in this region are: (1) high iron, manganese, and sulphate content derived from pyrite in some of the Pre-Carboniferous rocks, especially in Nova Scotia; (2) high calcium sulphate and sodium chloride content derived from evaporites, especially in the Windsor Group in Nova Scotia; and (3) induced salt-water intrusion in the coastal areas where overpumping occurs.

St. Lawrence Lowlands Region

The St. Lawrence Lowlands, an area of about 42,500 mi^2 (110,000 km^2), is underlain by undeformed Paleozoic rocks and covered with a generally thick layer of surficial materials. Relief is low, rarely exceeding 200 ft (61 m). The climate is humid continental, with annual precipitation of 30 to 47 in (762 to 1,190 mm). The Lowlands region is divided into three parts, separated by Precambrian rocks. The western part (most of southern Ontario) is separated from the central part by the Frontenac axis. The central part (Paleozoic rocks of the Ottawa area) is separated from rocks of the St. Lawrence River area by the Beauharnois axis at the confluence of the Ottawa and St. Lawrence Rivers. The eastern part is separated from the central part by about 360 mi (579 km) of the St. Lawrence River, and consists of Anticosti Island, the Mingan Islands, and a narrow strip of the north shore of the St. Lawrence River.

The porosity and permeability of bedrock in the St. Lawrence Lowlands region are largely produced by fractures of various types. The surficial deposits consist of glacial and nonglacial sediments deposited during the last glacial stage of the Pleistocene Epoch, excluding relatively small areas of recent sediments that are generally associated with modern streams and rivers. Yields from the bedrock vary from less than 15 gpm (81.8 m^3/day) to 600 gpm (3,270 m^3/day), depending on the extent of joint and fracture development. Wells in glacial sands and gravels may yield from 15 to 500 gpm (81.8 to 2,730 m^3/day).

Throughout most of southern Ontario, the groundwater consists of calcium and magnesium bicarbonate waters that reflect the composition of the limestone and dolomite aquifers. Similarly, a zone extending from Caledonia northwestward to Southampton is characterized by water high in calcium sulphate caused by the underlying gypsiferous Salina Formation. Areas underlain by black shale yield water containing sodium bicarbonate and hydrogen sulphide. Sodium chloride is common in bedrock waters and is generally the dominant component in waters from deep bedrock aquifers.

Canadian Shield Region

The Canadian Shield region is underlain almost entirely by mixed crystalline rocks, with only small areas of sedimentary rocks. Surficial materials are distributed irregularly; thin deposits exist in the upland areas and thick deposits are confined to the valleys. The topography is extremely rugged, though relief is rarely more than 325 to 650 ft (99 to 198 m) except for a few areas where it exceeds 1,000 ft (305 m). The climate is humid continental, and total annual precipitation varies from 22 inches (559 mm) in the west near the Interior Plains to 43 inches (1,090 mm) in the east near the Gulf of St. Lawrence. In the western part, about one-third of the annual precipitation falls as snow; in the eastern part the proportion is nearly one-half. This precipitation is ample to fulfill theoretically the recharge needs of all groundwater basins; therefore, except for local areas of high pumpage, these basins can be considered always to be at their seasonal maximum.

Because of the abundance of good surface water, groundwater is not widely used. Only 20 percent of the total population, including rural, uses groundwater. Approximately 110,000 gpm (600,000 m^3/day) of water (not including industrial supplies) are used by municipalities, and of this an estimated 8,000 gpm (43,600 m^3/day) is groundwater.

Bedrock over much of the Canadian Shield is covered by ground moraine consisting basically of clay till, and groundwater supplies are not likely to be sufficient for more than domestic use. The best sources of groundwater are the coarser grained surficial materials deposited by meltwaters from the glaciers. These materials are commonly found along the existing valleys in the bedrock.

Both the crystalline rocks and the surficial materials derived from them are composed of nearly insoluble minerals; therefore, groundwater in the Canadian Shield is generally similar to the acidic surface waters. Although both the organic content and the stream stage fluctuate widely, the chemical content remains nearly constant throughout the year. This indicates that groundwater discharged to streams is similar in composition to the surface water, and strongly suggests that groundwater flow systems are comparatively short and shallow. Poor-quality water may be encountered near some mineral deposits. In most base-metal mines, the ore body contains sulphides and the groundwater is strongly acidic and corrosive.

Interior Plains Region

The Interior Plains region is south of the southern limit of discontinuous permafrost between the Rocky Mountains and the Canadian Shield. Except for a narrow folded belt along the western edge, the entire area is underlain by nearly horizontal strata of Paleozoic, Mesozoic, and Tertiary age. It is almost entirely covered by a thick layer of surficial deposits, and bedrock outcrops are rare except along deeply incised river valleys. Elevation varies from about 4,000 ft (1,220 m) in the west to 500 ft (152 m) along the eastern edge. Most of the surface is rolling, with relief of 100 to 200 ft (30.5 to 61 m), except for the flat glacial lake basins. The climate is humid continental, except for the area between Moose Jaw, Saskatchewan, and Lethbridge, Alberta, and as far north as lat. 52°N, where it is semiarid. Total annual precipitation varies from 10 to 20 in (254 to 508 mm), with the eastern and western parts of the region receiving the greater amount.

Surficial deposits fall into three categories: till, lacustrine sediments (mainly silt and clay), and outwash deposits. Even though shallow wells yield less than 5 gpm (27.3 m³/day) and cannot be relied upon for sustained yields, the till is an important aquifer because approximately 60 percent of all farm water supplies in the Prairie provinces is drawn from these sediments. The most reliable yields are from stratified sand and gravel deposits within the till. Potential yields from these sand and gravel deposits depend upon their areal extent. Such intertill aquifers near Regina, Saskatchewan have a combined safe yield of nearly 14,700 gpm (8,000 m³/day).

Outwash deposits are found along the shores of former glacial lakes and as glaciofluvial sands at the mouth of meltwater channels. They consist chiefly of well-sorted sand and fine gravel, and are commonly less than 15 ft (4.6 m) thick, although some are more extensive and may be 50 to 100 ft (15.2 to 30.5 m) thick. Their hydraulic conductivity is from 12 to 150 gpd/ft² (0.5 to 6.1 m/day), regardless of depth. Though limited in areal extent, their potential is the greatest of the various aquifers within the Prairie provinces. Industrial wells on the Carberry Sand Plains of Manitoba produce more than 750 gpm (4,090 m³/day) without apparent depletion during the few months of the year they are in use.

Bedrock aquifers of the Interior Plains are sufficient to meet domestic water requirements, but sandstone aquifers that can yield 50 gpm (273 m³/day) or more are limited to the Paskapoo Formation, parts of the Edmonton Formation, the Birchlake and Ribstone Creek Members of the Belly River Formation, and perhaps some parts of the Eastend and Ravenscrag Formations. Currently, only the Paskapoo Formation and, locally, the Edmonton Formation are tapped for municipal use, and the production of 90 gpm (491 m³/day) appears to be about the ultimate safe yield.

In western Alberta, where annual precipitation is more than 18 in (457 mm), groundwater in surficial deposits contains less than 800 mg/ℓ of total solids, predominantly calcium and magnesium bicarbonates. In east-central Alberta and central Saskatchewan, where rainfall is less than 16 in (406 mm), the total dissolved solids increase to 2,500 mg/ℓ. The predominant constituents are sodium, calcium, and magnesium sulphates. In southwestern Manitoba, the average total dissolved solids diminishes to 2,000 mg/ℓ, and in south-central Manitoba, to 1,100 mg/ℓ. In the latter area, rainfall is slightly greater and calcium and magnesium sulphates are the predominant dissolved salts. In central and eastern Manitoba, rainfall is considerably greater and the amount of total dissolved solids falls to 900 mg/ℓ with calcium and magnesium bicarbonate predominating. The chemistry of shallow groundwater on the prairies generally maintains this consistent relationship because of the uniformity of the surficial deposits. However, this simple relationship is modified where the groundwater infiltrates into the chemically less uniform bedrock.

Cordilleran Region

The Cordilleran region is largely underlain by crystalline rocks and steeply folded sedimentary rocks, with some flat-lying sedimentary and volcanic rocks in the central plateau and coastal areas. Surficial deposits are confined largely to the valleys and lowland areas. The relief is thousands of feet in the more mountainous belts but only about 325 ft (99 m) in the central plateau, except where valleys have been cut below this general surface. Total annual precipitation varies from less than 2 in (51 mm) to more than 100 in (2,540 mm), and extremes can occur as close as 30 mi (48 km) apart. The total annual precipitation for the region is estimated to be 264 trillion gal (1,000 billion m³), whereas the runoff amounts to 211 trillion gal (800 billion m³). About 53 trillion gal (200 billion m³) are evaporated or represent change in storage for glaciers and groundwater. This region has vast water resources that could be tapped by Canada for use in water-short areas in Canada or for export to the United States.

The Cordilleran hydrogeological region includes three major physiographic subdivisions: the western, interior, and eastern systems. The western system includes the St. Elias mountains and coastal ranges, with elevations of 13,100 ft (3,990 m) to more than 19,700 ft (6,010 m); the Queen Charlotte and Vancouver Island ranges; and the Cascade Mountains, with relief ranging from 3,300 to 23,000 ft (1,010 to 7,010 m). The higher mountains nourish Alpine glaciers and ice fields. Within this system, the Georgia Basin is bordered by the Naniamo Lowland on the west and the Georgia and Fraser Lowlands on the east. These comprise the humid coastal belt. The eastern system includes the Rocky Mountains and the adjacent foothills belt, and contains peaks of more than 9,800 ft (2,990 m) that also support alpine glaciers. The interior system comprises several major and minor mountain ranges and plateaus, including the Columbia Mountains and the extensive interior plateau.

Air masses that originate in the Pacific and Arctic regions ascend over the coastal mountains of the western system and in so doing are cooled and loose their moisture as heavy precipitation along the coastal belt. Precipitation averaging 60 to 100 in (1,520 to 2,540 mm) has been recorded at Henderson Lake, at the head of a valley off Barkley Sound, British Columbia. After crossing the mountain barrier, drier air, sometimes heated, descends on the lee side and produces an interior dry belt where precipitation varies from 12 to 16 in (305 to 406 mm) per year and, in the deep valleys, to less than 8 in (203 mm). The eastward-moving air masses then rise over the Columbia and Rocky Mountains and produce a second rain belt with precipitation exceeding 39 inches (991 mm) in places, and again there is a drier belt on the lee side of the mountains, and in the foothills and adjacent parts of the Interior Plains.

The Pleistocene geology is complex, with at least three major glaciations and a fourth minor glaciation in the valleys. The principal water-bearing materials include outwash sand and gravel; fluvial sand and gravel between stony clays; fluvial sand and gravel confined beneath ground moraine or till; and interconnected lenses of coarser material included in the silt, clay, and fine sand that fills some valleys to depths of more than 1,000 ft (305 m). Artesian wells with flows of 15 gpm (81.8 m³/day) exist in confined aquifers in the valleys.

In general, the groundwater is low in mineral content and hardness. The harder waters are usually higher in iron and manganese. Groundwater that is unsatisfactory for most uses occurs in certain geological environments, especially where (1) leaching produces a concentration of salts, (2) reducing conditions will increase the iron content, and (3) overpumping in coastal areas may induce sea-water intrusion.

Northern Region

The Northern region encompasses all of Canada north of the southern limit of discontinuous permafrost. Precipitation is low throughout this region, being less than 10 inches (254 mm) in the north and less than 20 inches (508 mm) in the south. The region is underlain by a wide variety of crystalline rocks and folded or flat-lying sedimentary rocks. Surficial materials cover much of the area and reach their greatest thicknesses in the valley bottoms along the Mackenzie River, and in the Yukon Territory. In the mountain belts west of the Mackenzie River and on Ellesmere Island, relief is several thousand feet; but elsewhere it is only about 300 ft (91.5 m) and seldom exceeds 1,000 ft (305 m).

Permafrost occurs everywhere beneath the ground surface in the continuous permafrost zone and is generally more than 300 ft (91.5 m) thick. This zone grades southward into the discontinuous zone, where permafrost co-exists with areas of unfrozen ground. Permafrost also occurs at higher altitudes in southern Labrador-Ungava and in the Cordilleran region for some distance south of the limits shown in Figure 7.7.

The occurrence of groundwater in permafrost areas differs from its occurrence in warmer climates. The main effect of permafrost on the hydrology of an area is that it restricts the movement of groundwater. In regions of low relief, this results in numerous lakes and swamps. Conversely, in areas of high relief, the runoff is rapid and complete because of a lack of infiltration.

Groundwater in permafrost areas may occur as suprapermafrost water, intrapermafrost water, or subpermafrost water. Suprapermafrost water is present nearly everywhere in permafrost regions during the summer thawing season. If thawing extends deep enough, it can create an appreciable reservoir of groundwater perched upon the underlying frozen material. Intrapermafrost water exists within the thawed zones of frozen ground. It commonly occurs in alluvium near rivers or in abandoned river channels and in glaciofluvial material covering the wide river valleys, such as the Pelly River Valley in the Yukon. Subpermafrost water occurs beneath large areas of permafrost. Examples of this situation are the flowing wells that occur along the Alaska Highway between Haines Junction and the Alaska boundary. The highway follows the Shakwak Trench, a great valley containing extensive deposits of coarse-grained, glaciofluvial sand and gravel. The water in these flowing wells is reported to be coming from beneath permafrost.

Canadian communities in the area of continuous permafrost do not obtain a permanent water supply from groundwater, but some do in the area of discontinuous permafrost. Some of these latter communities are next to large streams or lakes and obtain their water from wells in unfrozen material close to the surface water. Both Dawson and Hay River have water supplies of this type; Fort Smith and Whitehorse obtain water from such well fields in addition to supplies from rivers. These communities use groundwater to warm the river water during cold periods to prevent freezing in pipes. Many small communities along the Alaska Highway obtain groundwater from permeable materials near surface water.

Mineral content of the groundwater varies greatly in the Northern region. Water in the active zone generally has low mineral content, whereas water in the rocks beneath the permafrost may be highly mineralized.

Conclusions

Although much of Canada has abundant surface water supplies, groundwater development is occurring rapidly in most of the populated regions of the country, especially in Saskatchewan, Alberta, and British Columbia. The provincial and federal governments have taken an active interest in groundwater development and have produced a wide variety of informative publications on all phases of groundwater.

MEXICO

Nearly one-third of Mexico is classified as arid and another one-third is classified as semiarid. In the desert areas of the northwest, the average annual rainfall is less than 4 in (102 mm); during some years, there is no precipitation. The only humid areas are found close to the border between Mexico and Guatemala.

Approximately 80 percent of the average annual precipitation [28 in (711 mm)] falls between May and September. Most of the crops, however, are raised between February and April. Thus, irrigation is virtually indespensible for the 21 percent of the country that receives less than 12 in (305 mm) of precipitation, and necessary for the 40 percent that receives 12 to 28 in (305 to 711 mm) annually. About 4 trillion gallons (15 billion m^3) of groundwater are extracted each year, much of it used for irrigation. Yet, 7.9 trillion gal (30 billion m^3) would be needed to fully develop all suitable agricultural land.

It has been estimated that 75 percent of the 370 billion gal (1.4 billion m^3) of rain falling on Mexico each year is lost by evapotranspiration. Practically no runoff occurs from 25 percent of the land, whereas 47 percent of the annual runoff flows off 12 percent of the land. The resulting scarcity and uneven distribution of surface water supplies increase the value of local groundwater resources.

The principal groundwater units in Mexico are closely related to the eight physiographic units (Figure 7.8). A brief hydrogeologic evaluation of each province is presented below (United Nations, 1976).

Central Highlands

The Central Highlands comprises about 30 percent of the area of Mexico. The

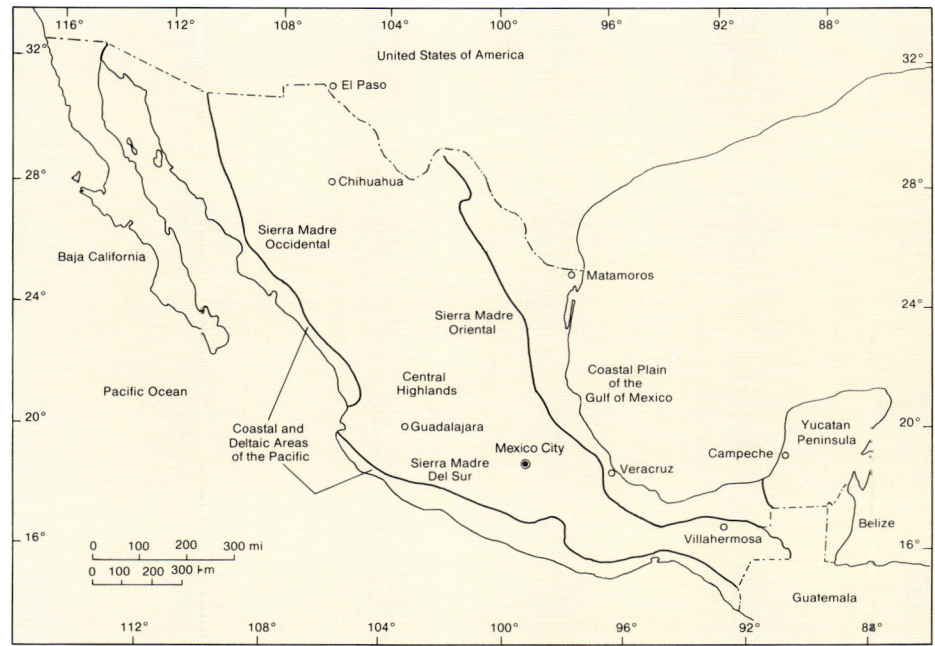

Figure 7.8. Groundwater regions of Mexico. *(United Nations, 1976)*

northern part is arid and includes basins and valleys of interior drainage at elevations of 1,300 to 6,500 ft (396 to 1,980 m). Aquifers are found in alluvium, volcanics, and limestones. Exploited alluvium and volcanic aquifers cover 2,700 mi^2 (6,990 km^2). About 770 billion gal (2.9 billion m^3) of water are extracted from about 26,000 wells used for irrigating these areas. Municipal water supplies are obtained only from groundwater sources. Most of the waters originate in Quaternary alluvium, but some pyroclastics and Tertiary lava flows also are tapped.

Cretaceous limestones of the Mexican geosyncline are 300 to 6,500 ft (91.5 to 1,980 m) thick and outcrop widely in the eastern mountains (Sierra Madre Oriental). Cretaceous limestones are also present beneath Upper Cretaceous and Cenozoic marine sediments in the coastal plains of the Gulf of Mexico. Some of the major oil fields of Mexico are developed in the folded structures within these formations.

Sierra Madre Oriental

In the Sierra Madre Oriental, groundwater is found in Cretaceous limestones in the narrow intermontane valleys. Groundwater is extracted here for the water supplies of such important cities as Monterrey, Saltillo, and Monclova; it is also exploited in the Carboniferous zone in the spurs of Sierra Madre Oriental, where wells have been drilled to depths of 650 to 6,500 ft (198 to 1,980 m). About 40 producing wells yield approximately 26 billion gal (100 million m^3) of water per year, which is equivalent to 50,300 gpm (274,000 m^3/day). Drains and tunnels that reach the limestone aquifers underground yield an additional 35,000 gpm (191,000 m^3/day).

The permeable Cretaceous limestones are of major hydrogeological significance, because they are easily replenished by rainfall and runoff. These aquifers contain fresh to moderately saline waters under confined pressure. Individual artesian wells may yield from 1,600 to 4,800 gpm (8,720 to 26,200 m^3/day).

In topographically closed valleys and basins, such as Tula and Jaumave near Ciudad Victoria, wells drilled in limestones yield a total of 79 billion gal (300 million m^3) per year. On the eastern slopes of the Sierra Madre Oriental, near the coastal plain of the Gulf of Mexico, the limestone aquifers sustain many springs which have a total discharge of 32 billion gal (120 million m^3) per year.

Sierra Madre Occidental

The Sierra Madre Occidental, about 150 mi (241 km) wide, is underlain mainly by Miocene volcanic rocks (rhyolites) and also plutonic rocks. These are not considered to be significant as aquifers. Most producing wells of this area are in pyroclastic rocks and unconsolidated sediments.

Sierra Madre del Sur

In the Sierra Madre del Sur, the southern continuation of the Sierra Madre Occidental, metamorphic rocks act as an impervious lower barrier. Groundwater is forced to flow upward through the Lower Cretaceous limestone, the main aquifer which feeds a series of large springs with a total yield of more than 264 billion gal (1 billion m^3) per year. The aquifer, however, is not heavily exploited.

Coastal Plain of the Gulf of Mexico

Groundwater in the Coastal Plain of the Gulf of Mexico is utilized mainly in the region of the United States-Mexico border. About 66 billion gal (250 million m³) are extracted annually from aquifers consisting of Upper Cretaceous-Miocene marine sedimentary formations. In the vicinity of Veracruz, about 32 billion gal (120 million m³) are extracted from Miocene-Oligocene marine sedimentary formations.

Coastal and Deltaic Areas of the Pacific Ocean

In the Sonora desert, as much as 450 billion gal (1.7 billion m³) of water are pumped annually from sand and gravel aquifers. In the Sinaloa plains, about 92 billion gal (350 million m³) of groundwater are extracted annually, mainly from volcanic tuffs, conglomerates, and sands. These waters are utilized for municipal water supplies and irrigation. In areas close to the shore, sea-water intrusion has occurred.

Peninsula of Baja California

About 45 billion gal (170 million m³) of groundwater are extracted annually from the Baja California Peninsula. Alluvial fills and fractured volcanics constitute the main aquifers.

Yucatan Peninsula

The Yucatan Peninsula is a flat region without rivers, underlain by limestones on which extensive Karst development has taken place. About 66 billion gal (250 million m³) of groundwater are pumped annually from the limestone aquifers.

Conclusions

Groundwater development in Mexico is proceeding rapidly. Some aquifers have been overdrawn, and severe subsidence has occurred [as much as 10 to 25 ft (3 to 7.6 m) in an 80-year period in the Mexico Valley]. In other places, the aquifers have been drawn down beneficially; that is, surface water that would evaporate is induced to infiltrate, and groundwater is utilized that otherwise might flow to areas where it could not be used. Some sea-water intrusion problems are encountered along the coast. Artificial recharge methods to control sea-water intrusion are difficult to implement because surface water is not available. Therefore, limits on pumping must be observed or the wells should be abandoned and new ones constructed farther inland. With the rapid growth rate of the Mexican population (one of the highest in the world, at 3.5 percent annually), the improvement of living conditions will require an ongoing effort in water resources development, including groundwater. Appendix 7.B lists some important sources for information on Mexican groundwater resources.

CHAPTER 8
Groundwater Exploration

The demand for groundwater resources has accelerated rapidly since 1960, but new sources of high-quality water are increasingly difficult to find because the most accessible groundwater reservoirs have already been tapped and are being heavily utilized. Therefore, future withdrawals must come from deeper aquifers where water quality may be a problem. Utilization of various groundwater exploration techniques is becoming essential in locating new aquifers of high-quality water.

Four major categories of exploration techniques exist: (1) hydrogeologic maps and reports compiled by federal, state, and local agencies, (2) geophysical surveys conducted at the ground surface, (3) borehole sampling procedures, and (4) geophysical logging of the borehole. The expertise required to utilize these different methods varies widely. Some exploration methods, for example, can be used by drilling contractors without any special training. Other methods may be so complex that only highly trained groundwater specialists are able to utilize them.

In recent years, many contractors have moved their rigs hundreds or even thousands of miles from their home offices to exploit new business opportunities. To drill efficiently in these new areas, the contractor should have topographic and hydrogeologic maps which provide surface elevations, stratigraphic sections, and typical yields for aquifers. The tracing of fracture zones (lineaments) by aerial photography in limestone, dolomite, or crystalline rock terrains has increased the success of water well contractors seeking groundwater. Consultants and governmental water resource planners often use satellite photographs to see if groundwater supplies are likely to be found.

Examination of borehole cuttings has always been an excellent way to estimate the yield from an aquifer. Sampling of the aquifer and measurement of its response to various pumping rates are common practices used by virtually every contractor.

Geophysical instruments provide information on the physical and chemical character of the subsurface environment. Geophysical methods can be used by any contractor willing to learn the physical basis for a particular instrument and then spend the time to learn its operation. Many water well contractors now use several different types of highly sophisticated geophysical well-logging equipment to select the most productive zones of wells that must be screened.

All of the methods discussed in this chapter tend to complement one another; that is, using two methods may be more than twice as useful as using only one. Most water well contractors drilling small-diameter wells, however, use only one exploratory approach — their past experience in the area. Having drilled in the same locale for many years, they are able to predict with remarkable precision the depth to water, its rate of flow to a well, and its quality. Should these drillers move into a new territory, they can avoid costly errors if they use one or more exploration methods to familiarize themselves with the new aquifer. In some areas, such as southwestern United States, drillers almost always use several exploration techniques because the boreholes are deep and water quality varies significantly with depth.

In this chapter, the major techniques of groundwater exploration are discussed from the viewpoint of the professional consultant and the drilling contractor. Both the contractor and the consultant must know what techniques are available, how each method works, how to interpret the data, and what problems or limitations apply to each method.

Two limitations must be kept firmly in mind. First, it is important to recognize that geophysical data by themselves may provide little definitive information. Therefore, all data obtained by geophysical prospecting methods must be correlated or verified with actual borehole samples. Second, the data should be collected and interpreted by an individual who is experienced with both the methodology and the local hydrogeology. Unfortunately, an expert from one area cannot provide accurate data analysis in a new area without some experience, no matter how able the person may be in understanding the operation of the equipment. In addition, geophysicists who are familiar with oil-field exploration data and analysis may have difficulty analyzing groundwater data.

The primary steps in water-well exploration are to:
1. Locate the best sites for test holes.
2. Obtain representative samples of the formations penetrated.
3. Run geophysical logs in the completed borehole.
4. Determine the depth to the static water level in each permeable formation.
5. Obtain water samples from potential aquifers to determine water quality.

Each of these steps is discussed in this chapter.

HYDROGEOLOGIC REPORTS

The number of sources for information on groundwater, and the exploration techniques used to locate groundwater resources, have multiplied rapidly in the last few years. Several publications are available that address groundwater from a scientific point of view, that is, where it exists, its quality, and how it flows toward wells. The principal publications in the United States are listed below:

Bulletin, International Association of Scientific Hydrology
Ground Water (Journal)
Journal, American Water Works Association
Journal of Hydrology
Transactions, American Society of Civil Engineers
Water Resources Bulletin
Water Resources Research
Water Well Journal

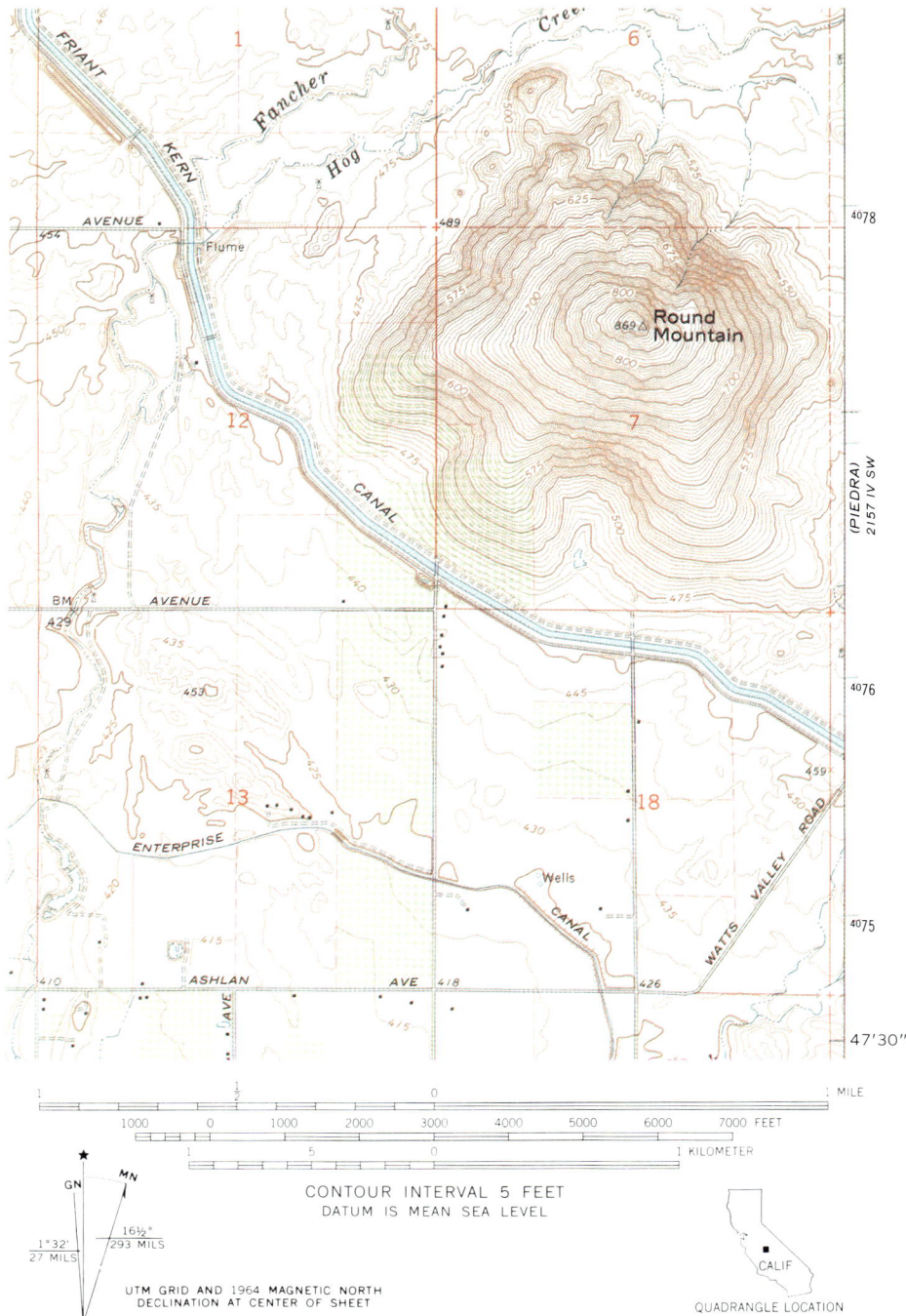

Figure 8.1. Topographic maps are useful to drilling contractors because they indicate the nature of the terrain, the presence of streams, springs, and lakes, and the location of highways, buildings, and railroads. *(U.S. Geological Survey)*

A more complete list of the publications relating to groundwater is provided in Appendix 8.A.

The water well contractor and the professional groundwater consultant can also obtain information from a wide range of federal, state (provincial), and local agencies. In the United States, the Water Resources Division of the U.S. Geological Survey (USGS) has conducted many studies of groundwater distribution, quantity, and quality. Reports based on these studies are available from the U.S. Government Printing Office, Washington, D.C. The Canadian government disseminates groundwater information through the Geological Survey of Canada and Environment Canada. In Mexico, the Secretaria de Desarrollo Urbano y Ecologia and the Secretaria de Agricultura y Recursos Hidraulicos are responsible for publishing many of the studies relating to groundwater. But these federal agencies, important as they are, provide only part of all the hydrogeologic information available today. States and local units of government have become extremely active in conducting research and issuing reports for selected geographic areas, especially those areas afflicted with water shortages or water-quality problems.

MAPS

The type of geologic materials and the topography of the Earth's surface in an area influence the location of groundwater. For example, groundwater will generally exist nearer the surface and in larger quantities in valleys rather than in upland areas. Topographic maps provide information on the size, shape, and distribution of features on the land surface, the location of lakes, swamps, springs, and streams, as well as important cultural information such as the location of buildings, railroads, and highways (Figure 8.1). Topographic maps constructed on a 7.5-minute (1 in:31,250 ft) or 15-minute (1 in:62,500 ft) base are the most useful to drilling contractors. Elevations on the map can be used to construct a profile section indicating high and low areas in the region of interest. Vegetation is also shown and may indicate where near-surface groundwater exists, especially in arid climates. The density (closeness) of streams and the pattern of surface drainage usually suggest where infiltration is occurring and also may reveal certain structural features such as faults, folds, or joint systems

Geologic maps indicate the nature of the consolidated or unconsolidated materials comprising the area being investigated. They show the rock type and the distribution of geologic structures (Lahee, 1952). Specific types of geologic maps are available that show the location of the formations (areal geologic map) and the structural contours of the subsurface geologic materials (structure contour map). The distribution of different rock types on a geologic map is shown by colors or patterns. Fault lines, contacts between rock types, and other breaks in the rock are indicated by lines on geologic maps. Many geologic maps have columnar sections that show the various formations, their thicknesses, and their stratigraphic relationships to other formations. Hydrogeologists then assess these stratigraphic columns on the basis of their potential to yield water to wells. All saturated formations are then classified as either aquifers or aquitards.

Hydrogeologic maps are similar to geologic maps in that the rock types are shown, but other valuable information is also illustrated (Figure 8.2). In glaciated terrains, for example, the location of buried bedrock valleys is indicated. These are particularly

Opc + Єj
Osp
+ Omdg
DODGE
OLMSTED
WINONA
MOWER
FILLMORE
ROCHESTER
Kenyon
Roscoe
Fairpoint
Pine Island
Oronoco
Potsdam
Plainview
Concord
Eden
Wasioja
Mantorville
Dodge Center
Kasson
Byron
Genoa
Douglas
Chester
Eyota
Dover
Utica
Saratoga
Weaver
Minneiska
Beaver
Elba
Altura
Rollingstone
Minnesota City
Goodview
Winona
Stockton
Lewiston
LOCK & DAM NO 5
Salem Corners
Golden
Marion
Simpson
Predmore
Vernon
Rock Dell
Hayfield
Oslo
Judge
Pleasant Grove
Stewartville
High Forest
Chatfield
Pilot Mound
Rushford
Money Creek
Sargeant
Waltham
Mayville
Brownsdale
Renova
Dexter
Washington
Racine
Hamilton
Fillmore
Fountain
Lanesboro
Peterson
Bratsberg
Grand Meadow
Spring Valley
Wykoff
Preston
Choice
AUSTIN
Nicolville
Elkton
Rose Creek
Ostrander
Cherry Grove
Greenleafton
Amherst
Adams
Taopi
Bristol
Canton
Mabel
Newhouse
Johnsburg
Le Roy
York
Granger
Prosper
900
1000
1100

10 0 10 Miles 20 30
10 0 10 20 30 40
Kilometers
Root River
Root River
Mill Creek
Trout River
Pine Creek
Rush Creek
Money Creek
Corey Creek
Q
Omdg
Osp
Opc + Єj
Єfig
Єec
Єsl
Єj

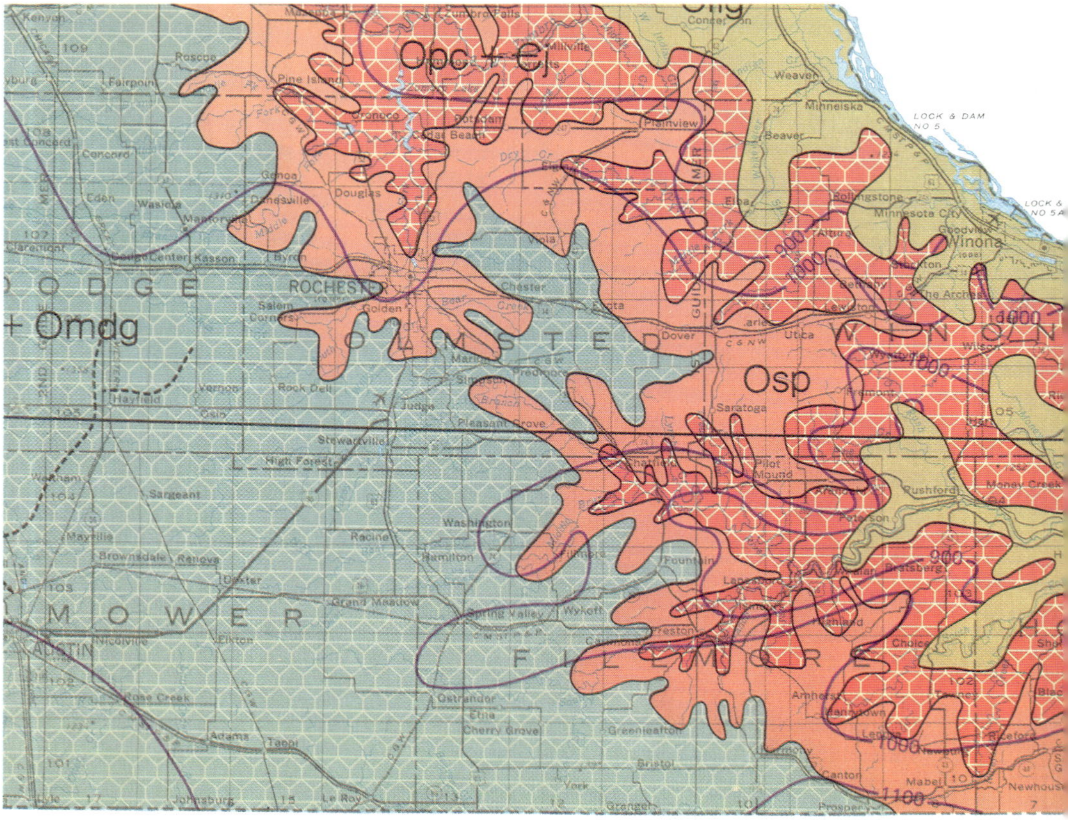

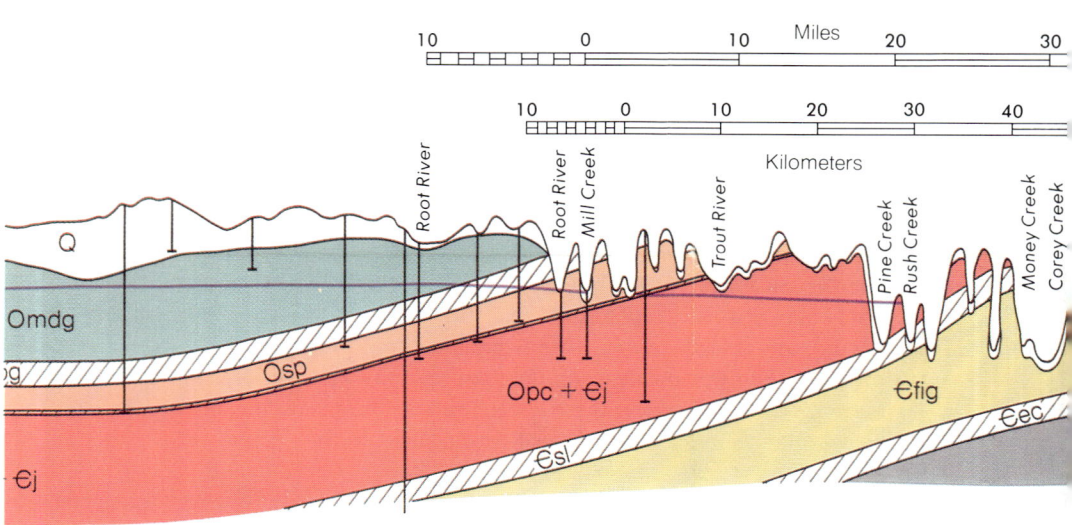

GROUNDWATER EXPLORATION

EXPLANATION

Undifferentiated Quaternary deposits (glacial drift, alluvium, etc.).

Cedar Valley—Maquoketa—Dubuque—Galena-Aquifer

Limestone, dolomite, dolomitic limestone, thin- to medium-bedded, vuggy, fractured; thin calcareous shale partings especially common in the Dubuque Formation (Middle Devonian-Late Ordovician age).

Decorah—Platteville—Glenwood Confining Bed

Decorah Shale, fissile to blocky; Platteville Formation, dolomite and dolomitic limestone, thin- to medium-bedded; Glenwood Formation, calcareous shale (Middle Ordovician age).

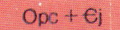

St. Peter Aquifer

Sandstone, quartzose, fine- to medium-grained, well-sorted and generally massive; thin beds of siltstone and shale near base (shown by diagonal pattern) form a confining bed (Middle Ordovician age).

Prairie du Chien—Jordan Aquifer

Dolomite, sandy, thin- to thick-bedded, vuggy, fractured, some shale partings. Sandstone, quartzose, fine- to coarse-grained, massive or thick- to thin-bedded, friable to well-cemented (Early Ordovician-Late Cambrian age).

St. Lawrence Confining Bed

Shale and siltstone, dolomitic; sandstone, quartzose, fine-grained, in part dolomitic and glauconitic, some beds of dolomite (Late Cambrian age).

Franconia—Ironton—Galesville Aquifer

Sandstone, fine-grained, and shale overlying sandstone, very fine- to coarse-grained, well- to poorly-sorted and silty; some interbedded shale (Late Cambrian age).

Eau Claire Confining Bed

Sandstone, quartzose, fine-grained, silty; some siltstone and shale (Late Cambrian age).

Mount Simon—Hinckley—Fond du Lac Aquifer

Sandstone, medium- to coarse-grained, well-sorted, friable to well-cemented; siltstone and shale in basal part of aquifer (Late Cambrian-Proterozoic age).

———— Approximate potentiometric surface of Prairie du Chien—Jordan Aquifer

⊥ Wells.

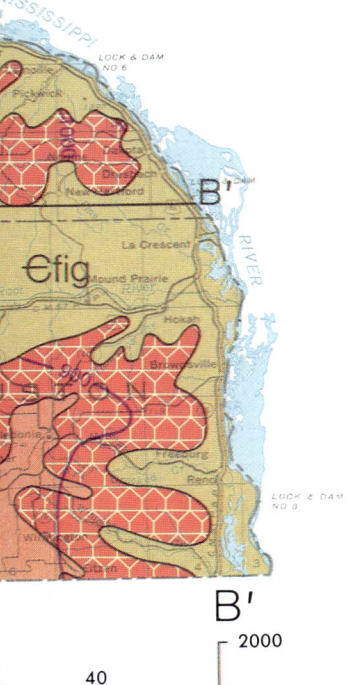

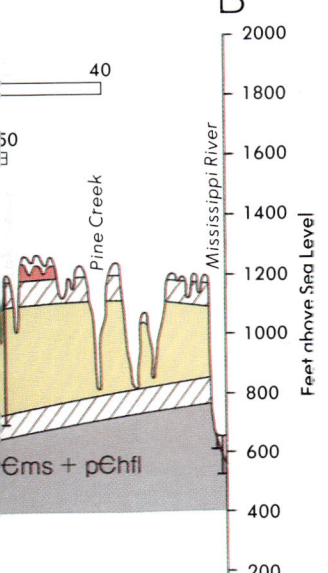

Figure 8.2. Hydrogeologic maps provide information on the structure of the Earth and the presence of groundwater. *(Minnesota Geological Survey)*

good sites for locating high-capacity wells, because thick sequences of sands and gravels are often present. In other cases, the hydrological characteristics of faults are presented. Usually the potentiometric surface is given for each aquifer. Cross sections are also provided and give an excellent in-depth view of the aquifers.

Geohydrochemical maps are specialized maps that indicate the chemical characteristics of groundwater. This type of map is useful in identifying high-quality water that can be used for drinking water or for other specific purposes. Scientists use these maps to explain how groundwater is modified either by natural processes or by the activities of man as it circulates in the underground environment. Eight general types of water have been defined as shown in Table 8.1; these are shown on geohydrochemical maps in the colors indicated in Figure 8.3. These maps are occasionally supplemented by cross sections or three-dimensional fence diagrams that suggest the spatial occurrence of the geohydrochemical features.

Other types of specialized geologic maps are useful to the water well contractor. One type of subsurface map, called a structure contour map, shows the elevations of the upper surface or base of a specific formation (Figure 8.4). Isopach maps show the thickness of the formation. These maps are particularly valuable in selecting the site and estimating the depth of wells in areas that are unfamiliar to the drilling contractor. Other maps show the anticipated yield from certain types of deposits. Recently some states in western United States have compiled maps showing the volume of water used in individual basins. Other specialized maps show one or more aquifer parameters such as transmissivity.

Table 8.1. Representative Colors of Eight Types of Water

Type of Water	Predominant Ions		Representative Color
	Anion	Cation	
Bicarbonate water	HCO_3^- (plus CO_3^{-2})		**Blue**
Calcium		Ca^{+2}	Light blue
Magnesium		Mg^{+2}	Violet blue
Sodium		Na^+ or ($Na^+ + K^+$)	Dark blue (Prussian blue)
Sulfate water	SO_4^{-2}		**Yellow and brown**
Calcium		Ca^{+2}	Yellow
Magnesium		Mg^{+2}	Light brown
Sodium		Na^+ or ($Na^+ + K^+$)	Dark brown
Chloride water	Cl^-		**Green**
Calcium		Ca^{+2}	Light green
Sodium		Na^+ or ($Na^+ + K^+$)	Dark green

(Table 4, from LEGENDS FOR GEOHYDROCHEMICAL MAPS, ©Unesco 1975. Reproduced by permission of Unesco.)

AERIAL PHOTOGRAPHS

Aerial photographs are of two major types — those taken from relatively near the Earth, and those taken from satellites which generally orbit the Earth at approximately 22 mi (35 km). Aerial photographs often reveal important hydrologic information that cannot be seen clearly at the ground surface. Faults, joint systems, old river courses, and the contact between moraines and outwash plains are some examples.

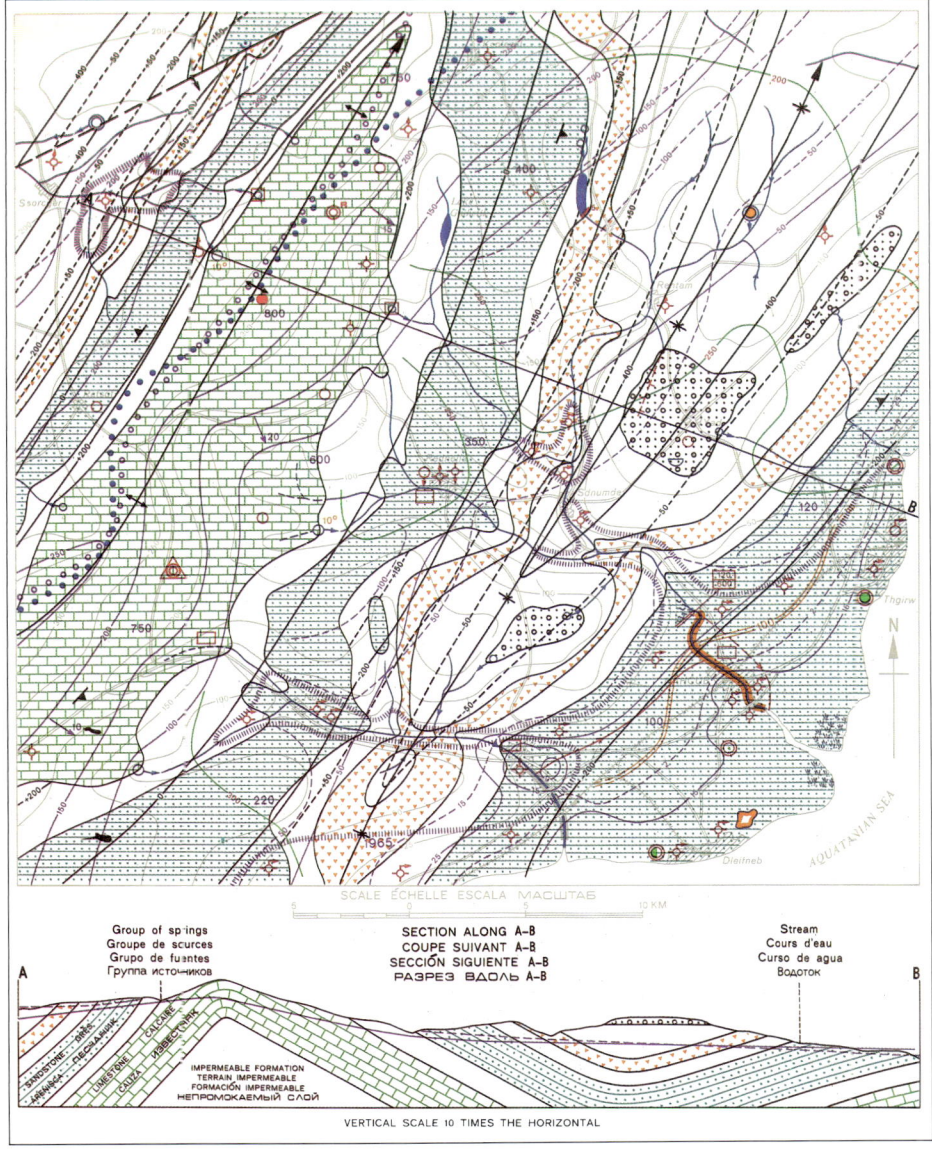

Figure 8.3. Geohydrochemical maps contain the maximum amount of information about groundwater resources and the geologic character of a specific region as it relates to these resources. A legend for this map is in the pocket. (©Unesco/IASH/IAH and the Institute of Geological Sciences, 1970)

Unlike topographic or geologic maps, however, aerial photographs contain major scale distortions; that is, the height of landscape features cannot be determined accurately without special equipment.

Aerial photography has been especially helpful in fracture-trace analysis. A fracture trace is defined as a natural linear feature less than 1 mi (1.6 km) long that can be identified by aerial photographs (Zall and Russell, 1979). Linear traces that are longer

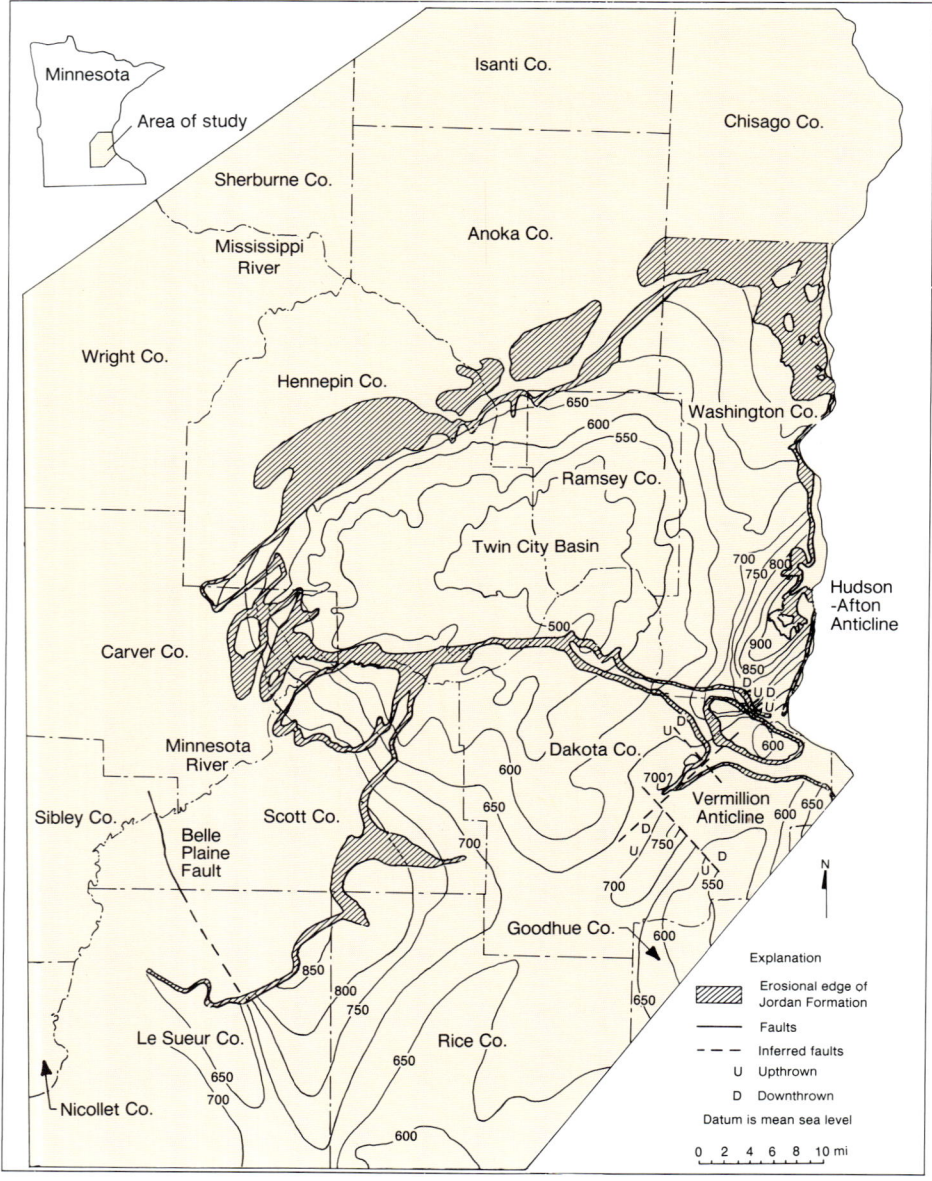

Figure 8.4. Structure contour map of top of the Jordan Sandstone, showing the configuration of the Twin City Basin, Minnesota. *(Mossler, 1972)*

than a mile are called lineaments. Crustal movement along faults in igneous and metamorphic rock terrains creates rubble zones in which relatively large volumes of water may be stored in these otherwise massive rocks. In the past, the success of wells constructed in massive limestone or dolomite aquifers was usually considered a matter of chance (Parizek and Drew, 1966). Virtually dry holes have been drilled near wells producing 2,000 to 3,000 gpm (10,900 to 16,400 m^3/day). Using aerial photographs and later satellite imagery, hydrogeologists now know that well yields in carbonate terrains depend on the presence of joints, fractures, and solution cavities and on how well these features are interconnected (Figure 8.5). The weathered width of fractures may range from 20 to 250 ft (6.1 to 76.2 m) or more. The trace is usually delineated by aligned sinkholes or other surface depressions. Thus, fracture traces represent zones of increased porosity and hydraulic conductivity in carbonate terrains, as well as in igneous and metamorphic rocks.

Under the National Aeronautic Space Administration's (NASA) Earth Resources

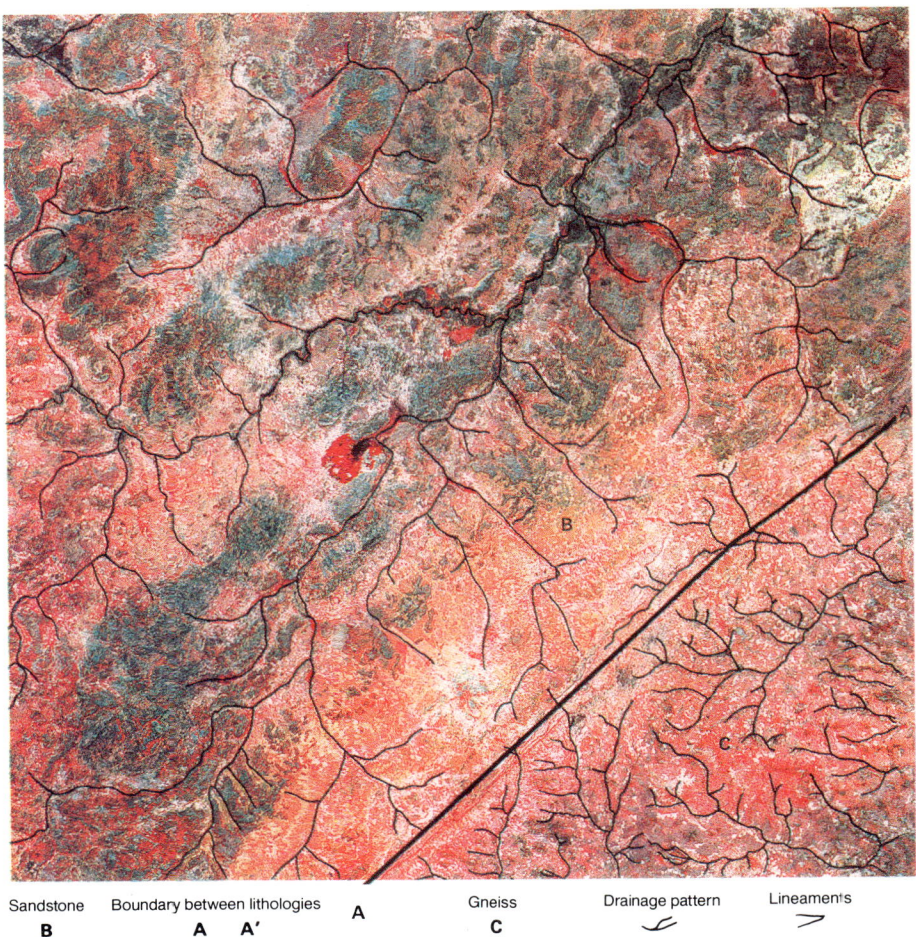

Figure 8.5. Computer-enhanced Landsat color composite of Bobo-Dioulasso Area, Upper Volta, West Africa. Acquired on March 31, 1976. Scale 1:500,000. *(Earth Satellite Corporation)*

Technology Satellite (ERTS) program, remote sensing of groundwater occurrence is a rapidly developing technique that can be used by groundwater consultants to investigate large areas. Each Landsat scene covers almost 13,500 mi^2 (35,000 km^2). A groundwater geologist can estimate groundwater supplies by studying the landforms, drainage patterns, and vegetation types and patterns (Cragwall, 1979). Surficial sand and gravel aquifers, for example, have been mapped using Landsat imagery by analyzing agricultural land use (Rahn and Moore, 1979). Certain row crops are grown on land that can be worked early in the spring. Floodplains, on the other hand, are normally wet in the spring and early tillage is impossible. Therefore, deep-rooted phreatophytes such as alfalfa are ideally suited to this type of land because they do not require early spring tillage and can tap either shallow or deep water tables. If alfalfa continues to grow vigorously during the dry parts of the summer, it is likely that there is a good groundwater supply. Schowengerdt et al. (1979) have shown that Landsat imagery may be used to survey large regions for lineaments to identify promising areas which can then be mapped at aerial-photography scales. In addition, recharge and discharge zones, as well as differences in the groundwater level, can be identified.

FORMATION SAMPLING

In every groundwater investigation, an analysis must be made of the geologic materials at a site and the capacity of these materials to yield water to wells. Reliable information on the geologic materials can be obtained by taking formation samples during the drilling operation, by interpreting the results of various geophysical tests performed after the borehole has been completed, and by test pumping the aquifer. It is best to combine two or more methods of investigation to verify the nature of the materials and their exact depth. Estimates of the potential well yield can be made with extremely simple test procedures that require little time, or with far more advanced methods that require several days. The relative importance of the well usually dictates at what confidence level geologic and hydrologic data must be obtained.

Most groundwater investigations usually involve drilling one or more test wells before construction of the production well is begun. Information gained from test wells is used for two general purposes: the work may be part of a study to evaluate the groundwater resources of a large area, or it may be preliminary to the design and construction of one or more wells at a particular site. In an area study, test holes are drilled to verify or supplement information obtained from other sources, such as geologic or hydrogeologic maps.

The surest way to learn the character of formations beneath the Earth's surface is to drill through them, obtain samples while drilling, and record the data. Lithologic logs record characteristic properties of the various strata encountered in the borehole. The actual logging procedure selected by the contractor will depend to some extent on the drilling method.

The most common type of lithologic log consists of the driller's description of the geologic character of each formation, the depth at which changes were observed, the thickness of the formation, and the depth to water. Ideally, the driller should collect representative samples at measured depths and at intervals that will show the complete lithologic character of the formations penetrated. For boreholes less than 200 ft (61 m) deep, samples should be taken at 5-ft (1.5-m) intervals and at every change in formation materials. Sampling every 5 ft is impractical for deep boreholes where formations may be hundreds of feet thick, and thus only major changes in formation materials are recorded. Furthermore, it may not be possible to correlate a particular sample with a specific depth in a deep rotary-drilled borehole because of contamination by materials eroded from the borehole walls and separation of materials coming up the long borehole. If erosion and separation are not a problem, however, the drilling contractor can estimate the depth for a particular sample by calculating the uphole velocity of the drilling fluid and cuttings.

The samples usually obtained by various drilling methods consist of cuttings produced by the action of the drill bit. While not entirely representative of the formations penetrated in some cases, they are commonly relied upon in groundwater exploration for identifying the best aquifer materials and determining the size of the screen openings.

Awareness of changes in drilling action is vital in compiling an accurate and informative log. Observations made by the driller should be included in the log because the drilling action and penetration rate indicate the character of the formation and especially the depth at which a formation change is encountered. Therefore, the person making the log must pay close attention to the noise made by the rig, the motion of the rig, and the changes in drilling fluid levels. When drilling by the rotary method, for example, the drilling action in clay or shale is smooth. An occasional chatter or temporary reduction in penetration rate may indicate scattered gravel in clay and glacial till, or concretions in shale. Continuous chatter usually indicates sand and gravel formations or sandstone. Smooth drilling with rapid penetration occurs in layers of fine sand. Drilling action is similar with auger rigs.

Experienced cable tool drillers can often estimate the character of the formation by the feel of the drill cable or by observing the water level in the well. The water level indicates whether the material just below the casing is impervious (clay), is taking water (dry sand), or is permitting water to flow into the hole (saturated sand). The extent to which sand heaves or moves up into the casing provides some idea of the looseness or compactness of the aquifer materials. Compact sand does not heave; loose sand may heave so much that it makes drilling difficult. Fluid pressure in the aquifer created by confining conditions may significantly enhance the looseness of the sand.

For rotary-drilled test holes, a drilling-time log supplies useful information about the formations encountered, because the character of the material largely determines the rate at which penetration proceeds. The drilling-time log is an accurate record of the time required to drill a certain distance. Clean sand formations usually drill very rapidly; muddy sand drills somewhat more slowly; loose sand drills more rapidly than cemented sand; and tight clay, shale, and hard rock drill more slowly than other materials. The drilling-time log is constructed as a curve or diagram showing penetration time for each length of drill rod. Each significant change indicates a difference in the material being drilled. The top, bottom, and thickness of each formation can be approximated from the diagram. Every driller notices whether the rate of penetration speeds up or slows down and then interprets this information in some general way based on experience; but the value of a systematic record of the drilling time for each interval is sometimes overlooked.

Factors other than formation character also affect the drilling rate. These include weight on the bit, sharpness of the bit, diameter of the hole, type of bit, velocity through the nozzles in the bit, and speed of rotation. For example, weight on the bit increases as the hole is deepened and additional drill pipe is added, thereby increasing the penetration rate. Interpretation of the time log is a relative matter, however, and this gradual increase in weight on the bit does not affect seriously the usefulness of the results. If hydraulic pulldown is used when penetrating hard formations, the force applied should be recorded and taken into consideration in the interpretation of the results. Careful time records show that none of the mechanical factors listed above, except for pulldown, influence drilling rate as much as the character of the formation being penetrated. Nevertheless, for best results these factors should be kept nearly constant so that drilling can proceed under reasonably uniform conditions.

Test-Drilling Methods and Sampling Procedures

Many methods are used to drill test wells, and several of the principal methods are discussed briefly below. A full explanation of each drilling method is given in Chapter 10. No one method is applicable for all situations, and often two different drilling methods are used in the same hole. For example, sampling by solid-stem augers may be done to the static water level, followed by direct rotary methods using a drilling fluid. The two most important criteria for successful test drilling are sample accuracy and drilling speed.

Formation samples may be taken by various methods depending on the type of drill rig used. Certain methods will provide more representative samples, that is, samples that truly reflect the grain-size distribution of the aquifer materials, whereas other methods are better for identifying the depths of formation contacts. For unconsolidated formations, representative samples are required so that the size of the screen slot openings can be chosen. If the borehole is in bedrock, the formation sample may suggest where water will enter the open borehole by indicating the presence of faults or the occurrence of a more permeable formation in an otherwise massive rock.

Direct Rotary Method

Direct rotary is the method used most often for test drilling and is particularly suitable for unconsolidated formations that do not include cobbles or boulders. This is usually the primary drilling method for test holes deeper than 1,000 ft (305 m). Most rotary-drilled test holes are 4 or 6 inches (102 or 152 mm) in diameter. A general disadvantage of the rotary method is that the depth to static water level usually cannot be measured unless casing is installed and most of the drilling fluid is removed.

Collection of representative samples when drilling by the direct rotary method presents several difficulties. Obtaining good samples depends largely on the skill and experience of the driller. The depth of the borehole will control the accuracy of the samples. For example, in the procedure described below, representative samples can be obtained to depths of 600 ft (183 m). Beyond this depth, separation of particles becomes a major problem, and the samples become increasingly less representative of the materials being drilled. For deep holes, other ways of checking the lithology and hydrology of the aquifer become more important. These methods are discussed later in this chapter.

Samples of sand, or sand and gravel, layers are washed to some extent by the drilling fluid as they are transported upward from the bottom of the hole. Separation of particles of different sizes can be minimized by proper control of the drilling fluid, but separation cannot be eliminated entirely. Fine and intermediate sizes of sand are carried upward by the drilling fluid ahead of the coarse particles. These separated fractions must be combined as the sample is collected at the surface.

A common sampling method involves two steps. First, the fluid is circulated without drilling until all cuttings have been removed from the borehole. The sample container is cleaned out at the same time. Next, the bit is allowed to penetrate the formation for a predetermined distance — for example, 2, 3, or 5 ft (0.6, 0.9, or 1.5 m). All the cuttings from this sample interval are then caught as circulation is continued without further drilling. Rotation of the drill pipe should be continued without allowing it to feed downward to maintain uniform flow in the annular space around the drill pipe. The drilling fluid lifts the cuttings from the sampling interval more readily under these conditions.

The cuttings that accumulate in the sample container should be placed in a tub, pail, or other container and allowed to settle. The excess drilling fluid should then be poured off carefully. The description of the cuttings may then be noted on the well log. After thoroughly mixing the material, a smaller representative sample of the cuttings should be taken, placed in a suitable container, and kept for shipment to the laboratory for grain-size-distribution and lithologic analysis.

If the rig action shows that the bit is entering a different type of material while drilling the predetermined sample interval, penetration should be stopped and a sample taken as described above. Proper notation of the new formation and depth is made on the log, and a new interval is then started from this point, repeating the operations cited above.

The rotation speed should not be too great in test drilling. Changes in the material being penetrated can be detected more easily with the bit rotating at slower speed but with the circulation kept at the suggested rate.

When drilling a 6-in (152-mm) test hole with 2⅞-in (73-mm) drill pipe, the vertical velocity of fluid for certain rates of circulation will be as follows:

Rate of Circulation	Vertical Velocity of Drilling Fluid
50 gpm (273 m³/day)	46 ft/min (14.0 m/min)
100 gpm (545 m³/day)	91 ft/min (27.7 m/min)
150 gpm (818 m³/day)	137 ft/min (41.8 m/min)
200 gpm (1,090 m³/day)	182 ft/min (55.5 m/min)

Uphole drilling fluid velocities of 125 to 200 ft/min (38 to 61 m/min) are desirable for representative sampling.

Choosing the type of bit for test drilling requires some experience in each locality. A drag bit or a cull fishtail bit is preferred by many drillers where alternate layers of clay and sand and gravel are encountered. A roller bit would be the next choice; a hammer bit is the poorest for test hole work, except in hard-rock formations. Care must be taken when drilling unconsolidated aquifers to avoid grinding of the cuttings so they do not arrive at the surface much smaller than they are in the formation. If the cuttings are diminished in size, it is nearly impossible to select the optimal size of screen opening.

The choice of drilling fluid will affect the quality of samples. If bentonitic clays are used for the drilling fluid additive, they tend to form part of the sample, although no clay may actually be present in the aquifer. This could lead to the selection of a screen with slot openings that are smaller than necessary. Bentonitic drilling fluids also have high gel strength; that is, the drilling fluid will "gel" and suspended particles will not drop out rapidly. This quality is important when particles must be suspended temporarily in the borehole when circulation is stopped, but this characteristic will keep some of the cuttings, especially the finer ones, from dropping out during the rather short time the fluid is being circulated in the mud pit. Thus, the collected samples may be missing an important percentage of fine material. If the size of the screen openings is based only on the coarser fraction, a sand-pumping well may result. Use of polymeric drilling fluid additives can increase the validity of the sample because drilling fluids made with these additives have little gel strength when at rest. Therefore, virtually all the fine material will drop out as the drilling fluid passes through the mud pit. The various types of drilling fluid additives are discussed in Chapter 11.

A problem sometimes arises in test drilling by the rotary method when material from overlying strata is eroded from the borehole wall by the ascending stream of drilling fluid and mixes with cuttings from the sample interval. This problem is likely to be more serious in the procedure described above, because circulation must be continued for a considerable time after each interval is drilled. Also, drilling of successive intervals is impractical for test holes of considerable depth because too much time is consumed in waiting for each sample.

Because of these difficulties, some drillers prefer to take samples while drilling more or less continuously, rather than take them by the sample-interval method. The materials caught at the surface while drilling continuously must be combined according to the judgment of the driller, and this obviously requires more skill and experience. The time required for the cuttings to reach the ground surface from the bottom of the borehole must be taken into account in determining the depth represented by the sample.

Use of the down-the-hole air hammer with rotary equipment provides a combined percussion-rotary method that penetrates rapidly in consolidated formations. Test holes are usually 6 inches (152 mm) in diameter when using this method. Occasionally, loosely consolidated alluvial sediments can be drilled out a few feet below the casing with an air hammer and will remain open as the casing is being driven. In most cases, however, conventional water-based drilling fluids must be used with a roller bit when drilling through unconsolidated overburden above bedrock. Exceptions to this occur when an air hammer is used to drive the casing after materials are blown out of the casing or when the rig is equipped with a casing driver.

If the rotary drilling machine is equipped with a casing driver, sampling by air can be accomplished in unconsolidated sediments such as bouldery tills and alluvial deposits. The depth to the water table can usually be detected in the casing driver method, although the depth to which the casing may be driven for test drilling operations may be limited in comparison with direct rotary methods using a water-based drilling fluid. Use of this method assures that the accuracy of the samples will be extremely good because the casing follows the bit down the borehole and no particles can fall at random from the borehole walls.

Side-Wall Coring

As suggested above, cuttings obtained at the surface may not be representative of the materials in deep formations. To verify the material present at a certain depth, side-wall cores are taken. Depending on the coring device used, the cores can either be continuous or be spaced evenly down the borehole. To obtain cores in formations that are not too hard, a special gun loaded with sample-taker bullets is lowered into the borehole. The bullets, which are hollow and usually about 1¾ in (44.5 mm) long by $^{13}/_{16}$ to 1 inch (20.6 to 25.4 mm) in diameter, are actually tiny core barrels. After being fired electrically, the bullets, which remain attached to the gun, are retrieved by the wireline used to lower the gun. The cores can be checked for hydraulic conductivity and porosity to estimate the yield.

Dual-Wall Method

The use of rotary drilling rigs equipped with dual-wall drill pipe is increasing in western and central areas of the United States. In this method, the drill pipe and bit are joined and advanced simultaneously. Either air or water can be used as the drilling fluid in this modification of the reverse-circulation technique.

Recovery of continuous samples when using the dual-wall rotary method is particularly easy if air is used as the drilling fluid. Because cuttings are immediately pulled into the drill pipe, contamination by overlying borehole materials is prevented. Therefore, the samples are representative of the actual formation, and for this reason test work using the dual-wall method is expanding. Samples from depths of 1,300 ft (396 m) or more can be collected. There is usually no grinding of cuttings, and the drilling fluid, if not air, can be clear water.

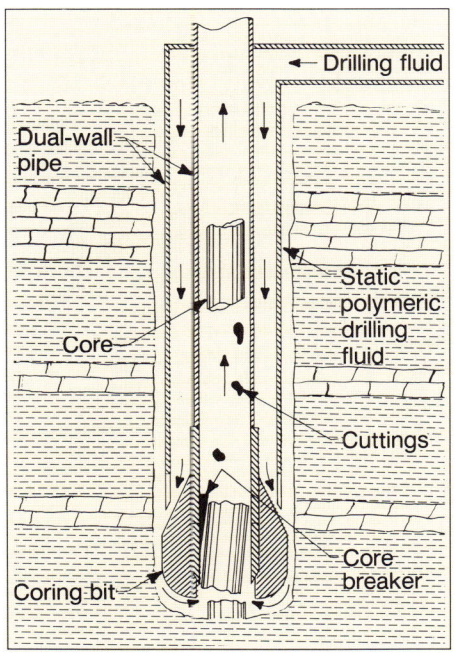

Figure 8.6. In this sampling procedure, a 4½-in (114-mm) OD drill pipe equipped with a diamond or carbide coring bit drills a 4⅞-in (124-mm) diameter hole while cutting a 2-in (51-mm) diameter core. *(Walker-Neer)*

Continuous cores can also be recovered by the dual-wall method. In this sampling procedure, a 4½-in (114-mm) OD drill pipe equipped with a diamond or carbide coring bit drills a 4⅞-in (124-mm) diameter hole while cutting a 2-in (51-mm) diameter core (Figure 8.6). The drilling fluid circulates downward between the two tubes and rises in the center tube, pushing the cores to the surface. A special automatic core breaker breaks off the cores in 5-in (127-mm) sections so they can be lifted by the drilling fluid. Cores have been recovered from depths to 2,000 ft (610 m).

Auger Drilling

Auger rigs have been used successfully in drilling test holes for foundation studies for bridges, roads, and other structures. The use of augers has increased for relatively shallow groundwater exploration purposes because the rig can be set up quickly, the rate of penetration is rapid, and, with most auger methods, reliable samples can be taken at any depth as drilling proceeds. The maximum depth of this sampling method may be somewhat limited, however, depending on the size of the auger rig and the diameter of the auger flights. For a small rig, the maximum depth is about 250 ft (76.2 m).

Samples can be recovered by four methods when using continuous-flight augers. Samples may be obtained from cuttings deposited at the top of the hole as the auger advances. This method is generally unsatisfactory because the samples are often mixed and originate at unknown depths. A better method is to pull the augers out of the hole at certain intervals and sample the material adhering to the auger bit or cutter head, thus providing a relatively unmixed sample from a known depth.

Figure 8.7. Undisturbed soil samples are obtained with a split spoon by driving the hollow sampler into the ground. After retrieval, the sampler is split open for visual observation.

A third method involves pulling the augers out at intervals and then driving a split-spoon or core-barrel sampler into the bottom of the hole (Figure 8.7). Undisturbed samples from known depths can be obtained in this way. The hollow-stem auger offers advantages in split-spoon sampling because the flights need not be removed while the sample is taken. The drill rods holding the bottom plug and cutter in place are withdrawn from the borehole in long sections, the sampler attached, the string reconnected, and the sample taken. The split spoon is usually driven out 1 to 2 ft (0.3 to 0.6 m), whereas the core barrel can be driven 5 to 20 ft (1.5 to 6.1 m), depending on its length. Once the sample is retrieved, the split spoon can be opened and the sample removed. In some of the early core barrels, the material had to be knocked out of the barrel, but now most barrels are constructed with an inner tube which contains the sample. The outer tube is first removed and the inner tube split to expose the sample, similar to a split spoon. Some inner tubes are made of transparent plastic so the sample can be inspected before the tube is cut open.

In the fourth method, called the wireline method, continuous cores can be taken as the augers advance (Figure 8.8). A thin-walled sample tube and special latching mechanism are placed within the lowest hollow-stem auger. The latching arrangement permits the tube to remain stationary while the auger rotates. When the sample tube is full, it is pulled to the surface by a wireline drum hoist and exchanged for an empty sampler. Drive samplers can also be driven out the bottom of the augers with the wireline method and a hammer arrangement. When sampling below the water table in loose sand formations, the water level within the auger must be kept at or above the groundwater level as the plug is pulled to prevent sand from rising up into the stem before the sampling tube is driven into the formation.

Cable Tool Drilling

The cable tool method is occasionally used for test drilling relatively shallow holes or when drilling in cavernous limestone, basalt, or hard tuff. Sampling unconsolidated materials by the cable tool method presents comparatively few difficulties. Much less water is needed for the drilling operation — an important factor where water must be hauled from some distance away. In addition, the depth from which the samples are obtained can be measured accurately. The principal drawback of test drilling by the cable tool method is the comparatively slow rate of penetration and the need for casing.

Collecting samples by the cable tool method involves driving casing a short distance and then using a bailer to clean out the plug of material. Compact plugs may have to be loosened and mixed by the drill bit before the material can be picked up by the bailer. The casing may be driven about 1 ft (0.3 m) in interbedded sand and clay, or several feet in thick sand, to isolate a sample.

Heaving sand interferes with sampling and logging when the cable tool method is used. There is no way to know what part of the sand formation is represented by the material inside the casing after the heave takes place. In addition, upward flow of the sand tends to separate fine fractions from coarse fractions. The usual practice is to discard material that moves up into the casing on a heave, and try to obtain a sample near the end of the casing. Some drillers add water to the casing to control heaving.

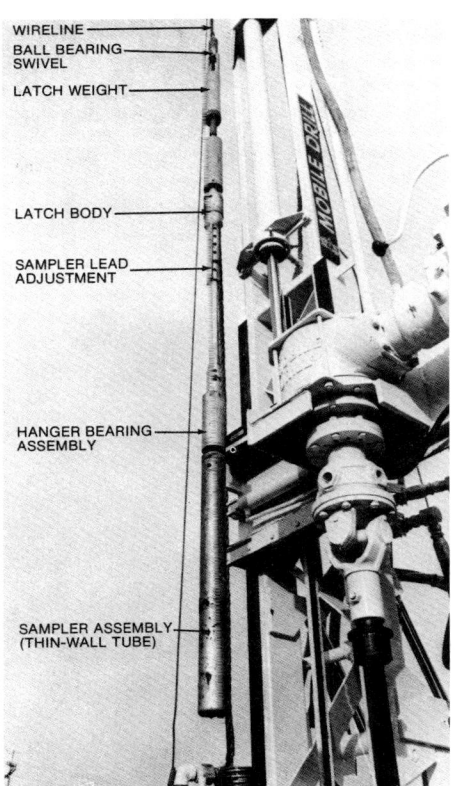

Figure 8.8. In wireline sampling, the sampler assembly is lowered and raised inside the hollow-stem auger string by means of a wireline drum hoist. Thus, the time required to obtain a sample is minimized. *(Mobile Drilling Company, Inc.)*

During any drilling and bailing procedure, the fines work toward the top of the material and the coarse particles settle to the bottom. More than one bailer load of material should be mixed together to provide a sample that is reasonably representative of the sampling interval. This is particularly important when sampling sand and gravel formations.

Several types of bailers can be used to remove the cuttings. A flat-valve bailer is worked down into a loose mass by a pumping action produced by lifting and dropping the bailer only a few inches. The driller often does this by pulling on the sand line. Raising and dropping the bailer more than a few inches is not effective. A sand pump with rod plunger is also useful for sampling work because the upward stroke of the plunger draws material up through the valve and into the bailer. The action produces some washing of the sample and this fact must be kept in mind. A dart-valve bailer is not as useful in sampling sand formations because it is effective only when enough clay is mixed with the sand to hold it in suspension in a mud slurry.

While not as widely used as it should be in cable tool test drilling, the method known as drive-core sampling provides the most accurate means of getting representative formation samples from unconsolidated strata. The method consists of driving a tube 2 to 4 ft (0.6 to 1.2 m) long into the material and then withdrawing it without jarring. To prevent loss of the core from the core barrel, the tube is overdriven — that is, it is driven a distance greater than its length in order to compact the material inside the tube. This practice permits recovery of the core in most cases, even when sampling clean sand or clean sand and gravel. The drive-core tube may be driven into a plug of material inside the casing after driving the casing a short distance, or it may be driven into the material below the bottom of the casing. The driller usually must determine the best procedure by trial in any given situation.

Long-stroke jars are used in driving the core barrel. The jars are connected directly to the upper end of the tube by a suitable substitute tool joint. A stem and rope socket are connected to the upper link of the jars and these two components provide the driving weight. Using the jars to drive downward is contrary to their primary purpose, but drillers who use the drive-core method report that unusual breakage does not occur and that good jars last long enough to justify their cost.

Handling Samples

Regardless of the sampling method, each sample [usually 8 to 16 oz (227 to 454 g)] should be completely and accurately identified. Excess water should be drained from the samples before sending them to the laboratory. If cloth mailing sacks are to be used, the samples should be dried over low heat. If waterproof containers are used, the samples need not be dried. Samples should not be washed. The depth from which the sample was taken, the thickness of material that it represents, and its sequence in the well log should be plainly written on the identification label. If the sample contains drilling fluid or natural clay, this fact should also be noted on the label.

GEOPHYSICAL EXPLORATION METHODS

Geophysical exploration methods are used either before or during well construction to obtain information on the character of formations and on the presence and chemical characteristics of groundwater. In addition, some methods are also extremely useful in determining the effectiveness of well construction. Certain methods are conducted at the surface, whereas others require a borehole. Regardless of the method used, however, the efficacy of each relies on the contrasting physical and physical-chemical properties of the water and the various layers of earth materials. The general physical differences between the unconsolidated and consolidated rock units making up the near-surface part of the Earth's crust were described in Chapters 2 and 3. It was shown that at most sites an unconsolidated sediment or series of sediments overlies one or more types of bedrock. Enough natural contrast usually exists between layers so that their general physical characteristics can be determined. But occasionally the ground may be so homogeneous that many of the best known geophysical exploration methods are of little value because no contrast exists. Yet the survey does indicate that the formation is uniform. Knowledge of how aquifer systems formed can be quite useful in the interpretation of geophysical data and, in turn, the use of various geophysical methods can minimize the amount of drilling needed to define the subsurface environment.

Surface and borehole geophysical techniques determine the density (porosity) and electrical, magnetic, nuclear, and acoustical properties of the geologic medium in which groundwater may exist. Surface methods include seismic refraction, seismic reflection, gravity, magnetic, resistivity, radar, and electromagnetic techniques. Borehole techniques include caliper, flow meter, temperature, resistivity, conductivity, spontaneous potential (SP), natural gamma, gamma-gamma, neutron, and acoustic (sonic). Each of these methods is particularly suitable in describing certain stratigraphic, lithologic, and other aquifer properties.

The signals received from any geophysical method are in response to a particular physical (geologic) property. Some geophysical methods measure a particular property directly, such as the gamma-ray emissions from a clay-rich sediment or the gravity difference between rocks. For others, an electrical signal or acoustic wave may be induced and the formation response measured.

In the past, most drilling contractors avoided doing their own logging work because they thought the equipment was too expensive or too complicated to use. Instead, they would call in a company specializing in one or more geophysical exploration methods. But as more drillers became involved in drilling deeper holes and their personnel achieved greater experience and education, they have added one or more geophysical techniques to the services they provide. The increasing number of drillers using geophysical methods is also attributable to the development of small, easy to operate devices that are relatively inexpensive to purchase and maintain.

The principal problem with most geophysical methods is that the interpretation of data requires training and experience. To be proficient, most operators must learn to "read" or interpret the data recorded by the geophysical instrument. Normally, new data must be correlated with cores or cuttings taken from a nearby borehole before the data can be interpreted. When going from one area to another, the operator must learn what type of data indicate a particular geologic material. Most of the time, however, the majority of drillers will operate in a rather small geographic area, and therefore can develop quickly a high level of expertise in interpreting results from a particular geophysical exploration method.

One other problem is important to recognize: signals or data recorded by a geophysical instrument consist of more than the signals from the aquifer under investigation. The signals are often affected by operator- and instrument-induced "noise" (any random and persistent disturbance that obscures or reduces the clarity or quality of a signal). In some cases, the noise is so great that the signal related to the geology cannot be determined. Usually, however, the operator learns quickly how to minimize the background noise in relation to the actual geological signal. Certain electronic techniques, such as signal enhancement, are extremely useful in emphasizing the true signal and minimizing the erratic signals caused by the operator or instrument. Again, experience plays an important role in helping the operator recognize various types of noise in the signal.

Geophysical prospecting methods can be divided into several major categories: mechanical, gravimetric, magnetic, electrical, nuclear, thermal, and acoustic. The method(s) selected will depend on the type of information needed, the nature of the aquifer materials and pore fluids, whether the test will be conducted at the surface or in the borehole, whether the borehole is cased or uncased, the professional competence of the contractor, and the amount of money available for geophysical prospecting.

None of the methods described are too difficult for a contractor who has the interest and the need to acquire this type of information. Descriptions of the methods in this chapter will focus first on the surface methods, then the borehole techniques. The reader must recognize, however, that some of these techniques, such as resistivity, may be useful both at the surface and in the borehole, although interpretation of the data is different.

A complete description of the theory, equipment used, field procedures, and data interpretation for each of these methods is not possible within the scope of this chapter. But an attempt is made herein to provide an overview of each method, its application, typical equipment requirements, type of data acquired, and how the drilling contractor can use this information. See Keys and MacCary (1971) for a complete discussion of borehole geophysical methods. Zohdy, Eaton, and Mabey (1974) describe the application of surface geophysics to groundwater investigations.

Surface Geophysical Methods

Surface geophysical methods provide specific information on the stratigraphy and structure of the local geologic environment as well as aquifer properties. Stratigraphic data may include information on the types and extent of surficial materials, and the nature and extent of underlying bedrock. Faults, fractures, folds, karstic terrain, and igneous rock intrusions (dikes) are common structural elements that can be located and defined by geophysical methods. In some instances, the rock type, thickness, dip, and structural character may be defined and assumptions may be made on the presence of groundwater, its rate and direction of movement, and its chemical quality. Electrical resistivity can be used to determine the slope of the groundwater table. Electrical resistivity and spontaneous potential provide information on the chemical quality of the water based on the concentration of ions present.

Sets of data collected by surface geophysical methods are called surveys. Thus, a gravity survey may delineate the structure of geologic materials at a site on the basis of their effect on the local gravity field. Usually a survey will combine specific types of data (magnetic, gravity, resistivity) for a site. Correlation of these data with a cuttings log from at least one borehole is always recommended.

Seismic Refraction/Reflection

Seismic refraction methods use seismic waves to determine the thickness and extent of aquifer materials where the formations are arranged so that the density increases with each successively lower layer. Seismic surveys are based on the velocity distribution of artificially generated seismic waves in the ground. Seismic waves can be produced by hammering on a metal plate, by dropping a heavy ball, or by using explosives. Energy from these sources is transmitted through the ground by elastic waves. The amount of energy is relatively low in comparison with the energy released by an earthquake, so the disturbance cannot be detected far from the point where the wave is initiated. The waves are called elastic because, as the waves pass a point in the rock, the particles are momentarily displaced or distorted, but immediately return to their original position or shape after the wave passes. Three types of waves can be created: compressional (P) waves, shear (S) waves, and surface waves. The arrival of a seismic wave is detected by geophones (seismometers) placed firmly in the ground. Compressional waves are the first to arrive at the geophones, and therefore are the most useful in seismic surveys. In general, the higher the density and elasticity of the rock unit, the faster the P wave will be transmitted. The velocity is much less and the energy is dissipated more quickly if the material is unconsolidated or poorly consolidated.

Three distinct paths are taken by compressional waves in the ground: direct (surface), refracted, and reflected. For a two-layered setting, these three wave paths are as shown in Figure 8.9. The exact arrival time of the seismic wave will depend on which path it takes and the density of the material. A single seismic impulse can be recorded as three separate arrivals at the geophone. In practice, however, only the first arrival can be readily recognized. At geophone-seismic source spacings generally used in refraction surveys, direct waves usually arrive at the geophone after refracted and reflected waves because surface materials are normally less dense than deeper materials. Because surface waves will not provide information about subsurface aquifer conditions, refracted and reflected waves are the most valuable for groundwater investigations. In most geologic settings, more than two distinct rock or sediment layers exist, thereby complicating the seismic analysis.

During a refraction survey, the investigator measures the time the seismic wave takes to reach one or more geophones placed at known distances from the seismic source. By plotting the distance-time relationship, the depth of several geologic units can be estimated at a particular site as long as each successively lower unit has a higher seismic velocity. The elapsed time and the distance traveled also provide information on the type of geologic material.

Each geologic formation has a characteristic seismic velocity that affects arrival time. Some representative seismic velocities are given in Table 8.2. In the field, local seismic velocities can also be estimated by measuring the travel time to and from a particular formation whose depth is already known from a drilling log. If a water-filled well is available, one or more geophones can be lowered into the well at suc-

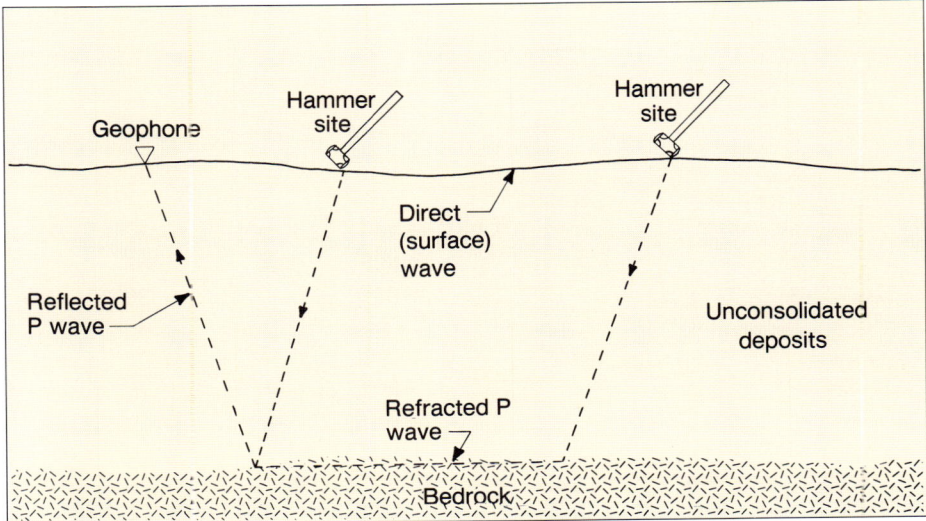

Figure 8.9. Waves from a seismic disturbance can travel as surface, reflected, and refracted waves. In water well exploration, analyses of refracted and reflected waves can determine the depth to bedrock at a potential drilling site.

cessively greater depths to obtain travel times from a seismic source placed at the surface (Mooney, 1981). Seismic velocities can also be obtained by conducting refraction surveys.

Seismic investigations are carried out from a point where borehole information is available or where the stratigraphy has been determined through other geophysical methods. To locate the water table with the refraction method, the water table should be well above the contact between the bedrock and the overlying alluvial deposits (Wallace, 1970). If seismic exploration methods are used in conjunction with other geophysical methods, exploratory drilling costs can be minimized.

In the seismic reflection method, the seismic wave produced by a hammer blow or other seismic source reflects off the bedrock and returns directly to the geophone, where the elapsed time is recorded. In order to maximize the reliability of the reflected wave energy, hammer stations are usually not more than 30 ft (9.1 m) from the geophone.

The most difficult problem with the reflection method is that the reflected wave is never the first to appear on the seismic record. Therefore, on an ordinary receiving device its arrival is almost impossible to recognize among the multitude of other wave arrivals. This problem can be overcome by using signal enhancement, which permits the operator to separate the primary reflected wave from other waves. In practice, the operator strikes a hammer plate one or more times at five to ten sites that are located within 30 ft (9.1 m) of the geophone. The seismic signals received from these sites are summed automatically by the seismograph. When the signals from several hammer sites are summed, the surface waves and other extraneous impulses which ordinarily obscure the primary reflected wave are canceled out, leaving the primary

Table 8.2. Approximate Range of Velocities of Compressional (P) Waves for Representative Materials Found in the Earth's Crust

Material	Velocity*	
	ft/sec	m/sec
Weathered surface material	1,000 - 2,000	305 - 610
Gravel, rubble, or sand (dry)	1,500 - 3,000	457 - 915
Sand (wet)	2,000 - 6,000	610 - 1,830
Clay	3,000 - 9,000	915 - 2,740
Water (depending on temperature and salt content)	4,700 - 5,500	1,430 - 1,680
Sea water	4,800 - 5,000	1,460 - 1,520
Sandstone	6,000 - 13,000	1,830 - 3,960
Shale	9,000 - 14,000	2,740 - 4,270
Chalk	6,000 - 13,000	1,830 - 3,960
Limestone	7,000 - 20,000	2,130 - 6,100
Salt	14,000 - 17,000	4,270 - 5,180
Granite	15,000 - 19,000	4,570 - 5,790
Metamorphic rocks	10,000 - 23,000	3,050 - 7,010
Ice	12,050	3,670

*The higher values in a given range are usually obtained at depth.
(Jakosky, 1950)

reflected wave prominently displayed on the cathode ray tube. This wave travels essentially the same distance even if the hammer sites are moved, as long as they are not too far from the geophone and the depth to bedrock is considerably greater than the spacing of the hammer sites.

Gravimetric Surveys

The Earth's gravitational attraction at a particular site is a function of the density of the surficial sediments and underlying rock units. Gravity meters can measure extremely small differences in the Earth's gravitational field caused by subsurface density variations. These density variations are usually produced by changes in rock type (that is, changes in porosity or grain density), degree of saturation, fault zones that may not be apparent at the surface, and varying thicknesses of unconsolidated sediments overlying bedrock. Gravity methods are probably most useful for locating buried valleys in glaciated areas.

Gravity methods are usually conducted at the surface for water well exploration, although borehole gravity investigations are common for engineering projects and for oil, gas, and mineral exploration. Gravity methods are useful for investigating large areas rapidly and inexpensively because the gravity meter is easily portable. The total elapsed time from the operator's arrival at the gravity station to the recording of the readings is usually less than 5 minutes. The number of readings that can be taken per day is limited only by the distance between stations, the speed at which the terrain can be traversed, and the accuracy required. The accuracy is dependent upon determining accurate elevations for the stations. Gravity anomalies associated with shallow geologic formations are less than 5 milligals*. A 1-ft (0.3-m) error in estimating the surface elevation of the station will result in a reading error of approximately 0.1 milligal (J. Daniels, personal communication). Because most high-quality altimeters are only accurate to plus or minus 2 ft (0.6 m), gravity stations should be surveyed. Once favorable areas are delineated gravimetrically, other geophysical methods or test drilling may be used to define the materials making up the potential aquifer.

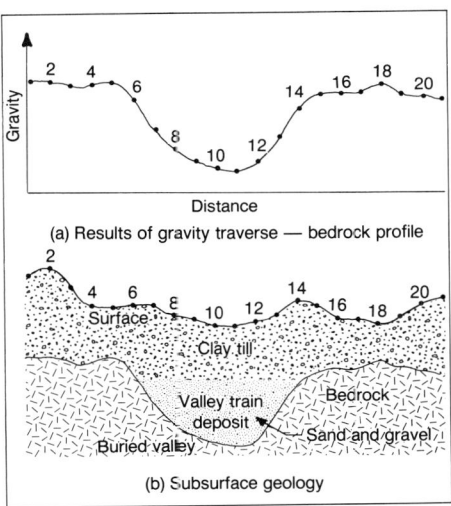

Figure 8.10. Gravity instruments are particularly useful in locating buried bedrock valleys lying beneath thick sequences of unconsolidated deposits. In glaciated terrains, these valleys often contain extensive sand and gravel deposits that will yield high volumes of groundwater. To determine the bedrock profile by the gravity method, readings are taken perpendicular to the valley. Because gravity is higher over high-density materials (bedrock), the lower density materials filling the valley will result in low gravity. The gravity readings are higher over the bedrock highs. Connecting the readings reveals the profile of the valley. (Actual depth to bedrock must be obtained from well logs, seismic refraction data, or density information.) *(Anderson, 1979)*

The gravity instrument is essentially a spring balance that is so sensitive it can

---

*A unit of acceleration used with gravity measurements; 10^{-3} gal = 10^{-5} m/sec².

measure the weight added by a single dust particle to the hood of a 4,000-lb (1,810-kg) automobile. Data from the gravity instrument can be used in several ways. Single traverses of buried valleys reveal the rough outline of the bedrock profile (Figure 8.10). The presence of faults, in which groundwater is likely to occur, can also be detected. It is possible to determine the depth to bedrock by using the traversing method. Three-dimensional maps can be constructed if the data are taken on a grid system (Figure 8.11). This procedure yields far more information on subsurface conditions than does a gravity profile survey, but data collection and map preparation are more time consuming. The actual readings from the gravity instrument must be corrected (Henry, 1973).

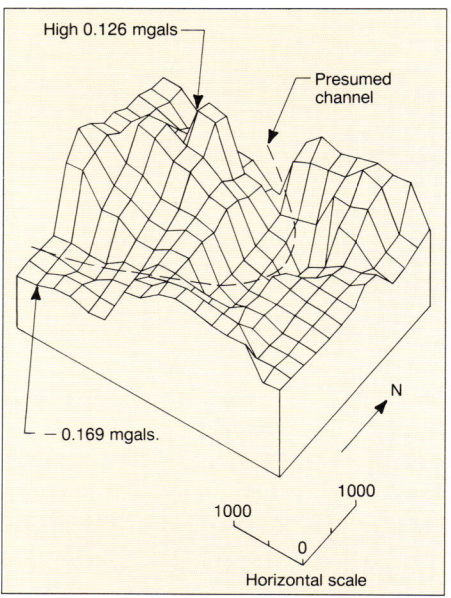

Figure 8.11. Gravity anomaly map for Hartford City, Indiana. This type of data presentation can be done by computer techniques. *(Carmichael, 1976)*

Electromagnetic Surveys

There are basically two ways in which electromagnetic methods are used in searching for groundwater and groundwater contamination; they are differentiated on the basis of operating frequency. At very high frequencies (above 100 MHz), electromagnetic waves propagate in the ground in a straight line to depths which vary from a few feet to a few tens of feet, depending on the electrical conductivity of the terrain; these microwave frequency instruments are called "ground penetrating radars." At low frequencies, the effective penetration can be much larger (several thousand feet), but in this case the electromagnetic waves diffuse slowly into the Earth, rather than traveling in a straight line. Thus, they sample a large volume of earth. Instruments operating at such low audio frequencies are known as "ground conductivity meters."

Use of ground-penetrating radar to obtain information about the subsurface environment is analogous to the seismic reflection technique because both methods record the time required for a wave to travel to an interface between two formations and then return (reflect) to the surface (Costello, 1980). However, significant differences exist between the methods. For example, the seismic interface is defined by the seismic wave velocity, which is controlled by the density of earth materials on either side of the interface. An electromagnetic interface, on the other hand, is defined at very high frequencies by changes in magnetic permeability, electrical conductivity, and dielectric constant; these three factors constitute an electromagnetic impedance. Electromagnetic waves travel through the ground at virtually the speed of light, unlike seismic waves which travel at the speed of sound, and so travel times must be measured with extreme care. Electromagnetic waves at such high frequencies provide excellent vertical and lateral resolution, but the effective penetration is limited from a few feet to 50 ft (15.2 m) depending on the frequency of the input signal used in the field investigation and the ground conductivity.

Although radar equipment is sophisticated, acquisition of data is quick and easy. Radar profile lines can be run (in grid patterns) at 2 to 3 mi (3.2 to 4.8 km) per hour for detailed surveys, and up to 10 mi (16.1 km) per hour for reconnaissance surveys (Figure 8.12). Profile lines can be run wherever the transducer can be towed or pushed either mechanically or manually. A radar transducer can emit a radar pulse, receive a pulse, or do both. Data are recorded by either a magnetic tape or a graphic recorder. The data generated take the form of a profile, with time in nanoseconds on one axis and distance on the other. Time is proportional to depth when the wave propagation (travel) velocity through the subsurface is known. This is usually accomplished by drilling to a known contact (reflector), measuring the depth, and then relating this depth to a measured velocity by a standard equation:

$$V_m = \frac{2D}{t} \qquad (8.1)$$

where V_m is the velocity of light at the site between the surface and the reflector, D is the measured depth to reflector interface, and t is the elapsed time between trans-

Figure 8.12. As an impulse radar system is towed across the ground, electromagnetic pulses are generated at the surface. Reflections from the surface and subsurface horizons are recorded on continuous-strip chart recorders. Because the velocity of the propagation waves into the earth is strongly dependent on moisture content, an estimate can be made on the degree of saturation. *(Geophysical Survey Systems, Inc.)*

mitted and reflected pulse.

Thus, for typical radar profiles, data processing entails converting travel times to depths once V_m is known for a sufficient number of points within the grid system. Data reduction is usually accomplished by computer because relatively complicated and voluminous calculations must be made.

Terrain conductivity is another method using electromagnetic waves. A variety of configurations is available, operating either in the frequency or time domain. Depths to 200 ft (61 m) can be mapped by using a two-person, frequency-domain configuration (Figure 8.13). Shallower depths [20 ft (6.1 m)] can be measured by one operator. In this method, a transmitter coil is energized with an alternating current at an audio frequency and placed on the ground. A receiver coil is located a short distance away. The magnetic field produced by the alternating current in the transmitter coil induces small electric currents in the earth, which in turn generate a secondary magnetic field which can be sensed (along with the primary field) by the receiver coil. The secondary magnetic field is a function of the intercoil spacing, the operating frequency, and the ground conductivity (McNeill, 1980a). The ratio of the secondary to the primary magnetic fields is linearly proportional to the terrain conductivity when the instruments have been designed appropriately. These instruments permit large areas to be surveyed rapidly with good resolution. They provide only moderate depth resolution, however, compared to the radar technique.

For water well and monitoring work, the terrain conductivity meter is useful in locating gravel, saline intrusions, cavities in carbonate rock, contaminated plumes of groundwater, and bedrock topography. Tracing of plumes in urban environments is particularly easy with a terrain conductivity instrument because no drilling or probe placement in the ground is required.

The time-domain method is similar to the ground penetrating radar described above except that instead of working at microwave (high) frequencies, the transmitters operate at a few tens of Hz. A square, single-turn transmitter loop, which can vary from 150 to 1,500 ft (45.7 to 457 m) on a side, is laid on the ground and a direct current of tens of amps is caused to flow through it. The current is abruptly terminated electronically, causing transient currents to diffuse into the ground to depths as great as 3,000 ft (915 m). By measuring the magnetic field associated with these currents, the instrument can detect different rock units to depths of several thousand feet with reasonable vertical resolution. The penetration depth is governed by the size of the loop; in general, the depth is 1.5 times the length of one side of the loop or 1.5 times the diameter of the loop if laid out in cir-

Figure 8.13. Hand-carried terrain conductivity devices use electromagnetic waves to measure the conductivity of earth materials. Direct contact with the ground is not required during data gathering. Thus, subsurface information can be obtained quickly in both highly urbanized and rural environments. *(Geonics Limited)*

cular form. Thus, the larger the configuration, the deeper will be the penetration. This method is called time domain because the decay with depth of a single frequency signal is measured during the extremely short time intervals between individual signal impulses.

The time-domain technique is appropriate when searching for groundwater at great depths and for detecting salt-water intrusions along coastal areas. Laying out the large transmitter loop requires some effort, thus survey speeds of 10 to 15 stations per day are typical.

Electrical Resistivity Method

The electrical resistivity method is a major geophysical tool used in groundwater exploration efforts. Resistivity, the inverse of electrical conductivity, is defined as the ratio of the voltage gradient to the current density over a small, thin, surface element of a medium. Stated more simply, it is the resistance of the geologic medium to current flow when a potential (voltage) difference is applied, or:

$$R = \frac{V}{I} \qquad (8.2)$$

where R is the resistance, V is the voltage, and I is the current. For a given material with a characteristic resistivity, the resistance is proportional to the length of material being measured and inversely proportional to its cross-sectional area:

$$R = \frac{\rho L}{A} \text{ or } \rho = \frac{RA}{L} \qquad (8.3)$$

where ρ is the characteristic resistivity of the geologic medium, A is the unit cross-sectional area, and L is its length. Units of resistivity are usually given in ohm-ft or ohm-meters.

In a resistivity survey, a direct current or low-frequency alternating current is sent through the ground between two metal stakes or electrodes. Because earth materials offer resistance to the passage of a current, some voltage loss will occur as the current flows from one electrode to another. The voltage loss (drop in potential) that occurs as the current moves through the ground is measured at other electrodes placed between the current electrodes.

The ability of a rock unit to conduct an electrical current depends primarily on three factors: the amount of open space between particles (porosity), the degree of interconnection between those open spaces, and the volume and conductivity of the water in the pores (Minning, 1973). The presence of water and its chemical character are the principal controls on the flow of the electric current because most rock particles offer high resistance to electrical flow. Thus, resistivity decreases as porosity, hydraulic conductivity, water content, and water salinity increase. Clay and shale have low resistivities, and dry sand and gravel have higher resistivities than do saturated sand and gravel. Typical resistivity values for common materials are presented in Table 8.3.

Table 8.3. Ranges of Resistivity Values for Various Earth Materials

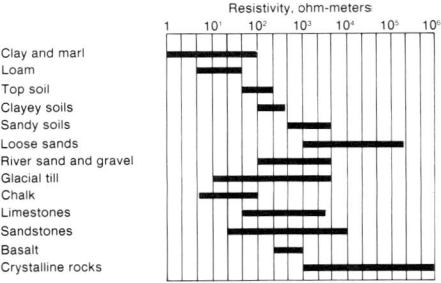

(After Culley et al., 1975; McNeill, 1980b)

Resistivity values are obtained by two different surface exploration methods. The first of these, called electrical sounding, involves vertical exploration. In this procedure, a series of stations is established and careful depth soundings are taken. These soundings are later transferred to a vertical cross-section chart. By evaluating the resistivity values, an understanding of subsurface materials can be developed. This exploration method is especially useful for estimating the depth to sand, gravel, bedrock, or water-bearing strata, or for estimating the thickness of selected formations. The electrical sounding method is accurate, however, only in the most simple geologic setting (horizontal layers parallel with the ground surface). Nevertheless, a qualitative analysis is still valuable because it provides a general indication of subsurface structure.

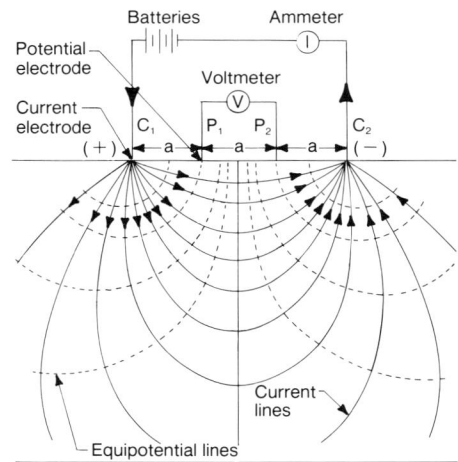

Figure 8.14. For surface application of the electric resistivity method, a Wenner electrode arrangement is common for groundwater exploration. Current is induced to flow between the two current electrodes and is measured at the potential electrodes. *(Minning, 1973)*

The second resistivity exploration technique is called electrical profiling. This method differs from electrical sounding in that material at only one depth is investigated. As in the case of sounding, numerous stations are selected. Resistivity measurements are then made — this time for the same depth — at each station. These values, once plotted, produce a numerical picture of the subsurface materials at the chosen depth across a horizontal plane. Electrical profiling is most often used in searching for ore bodies, faults or fault zones, for evaluating sand and gravel deposits, for delineating boundaries, and for finding dipping contacts of different earth materials.

Two general configurations of electrode arrays are used for electrical sounding and profiling to introduce current into the ground and record the drop in field values as it travels through the ground. The Wenner array is used primarily in North America, whereas the Schlumberger array is used mainly in Europe. In the Wenner configuration, four electrodes are arranged as indicated in Figure 8.14. The outer pair are the current electrodes, whereas the inner pair serve as the potential electrodes. Thus, if current is passed into the ground through two electrodes (C_1 and C_2), a potential difference can be measured in the other electrodes (P_1 and P_2). Note in Figure 8.14 that the path followed by the current is not a straight line, but fans out radially from one current electrode and converges at the other electrode, where it leaves the ground. Because the geologic materials offer resistance to the current flow, a drop in the potential (voltage) occurs. Apparent resistivity, ρ_a, is obtained from the ratio of the potential difference to the current, multiplied by the array (spacing) constant.

$$\rho_a = \frac{6.28\ aV}{I} \qquad (8.4)$$

where a is the spacing of the array, V is the voltage, and I is the current. The resistivity is called apparent because each resistivity value for a certain depth is really an average resistivity for all the materials above this depth. This apparent resistivity value results from the three-dimensional resistivity structure of the geologic materials and is a weighted average of all material encountered by the current flow. The potential difference is measured between the two inner electrodes. As resistivity values change, they indicate a change in subsurface conditions.

Care must be taken not to overvalue a resistivity reading for a particular depth. Each value represents an average resistivity for a 10- to 20-ft (3.1- to 6.1-m) vertical section of the formation. This value may indicate more than one combination of geologic materials and water, and experience must be called upon to make a correct interpretation.

The depth of electrical penetration is governed by the spacing of the electrodes — the larger the separation, the deeper the penetration. The electrode spacing can be progressively increased to determine the variation in resistivity with depth. If the resistivity changes with increased electrode spacing, it suggests that the formation changes as depth increases.

Surface resistivity exploration, either sounding or profiling, offers several distinct benefits. Large areas can be covered quickly at reasonable expense, the depth of the water table can be determined in most cases if within 150 ft (45.7 m) of the surface, and usually the particular types of formations can be identified. The method is not, however, without limitations. While these limitations are usually imposed by such unalterable considerations as terrain, subsurface composition, or other related factors, some measures can be taken to minimize their effects. In hilly or rough terrain, for example, measuring stations should be established, if possible, in creek beds. Areas with relatively flat surfaces having a high clay content will give more consistent resistivity readings. In areas where the surficial deposits are sand or gravel, useful results are more likely if the water table is within 10 to 15 ft (3.1 to 4.6 m) of the surface.

In general, surface resistivity methods will locate continuous and reasonably thick formations to depths of approximately 150 ft (45.7 m). Attempts to locate bedrock will usually be successful only if there is a clearly measureable difference in resistivity between the bedrock and overburden. In certain areas, such as those in which glacial drift is common, surface resistivity can be extremely helpful in locating buried valleys, outwash deposits, and sand and gravel lenses. Narrow, buried valleys of coarse sand and gravel might be the only source of groundwater in some areas. Thus, locating these deposits may be the difference between success and failure in obtaining reasonable groundwater yields in the area. Because surface resistivity is an important method of groundwater exploration, a case history of this method is discussed below.

A consultant was retained by a developer to locate a suitable groundwater supply for a proposed mobile home park. A test well drilled on the northwest portion of the property, to a depth of 250 ft, encountered about 50 to 60 ft of fine sand. The yield was about 100 gpm, much less than the developer required. The consultant recommended that a surface resistivity survey be conducted over the entire parcel to define the most promising area for another well.

A resistivity survey consisting of 44 stations was laid out on a grid, as shown in Figure 8.15. Earth resistivity soundings were taken with a Wenner array, using electrode spacings (a-spacings) of 10, 20, 40, 60, 80, 120, and 160 ft. These data enabled the consultant to construct an apparent resistivity/depth profile at each location. The apparent resistivity values were then corrected according to a standard technique. The corrected soundings were plotted to form profiles or cross sections of corrected resistivity (Figure 8.16). The profiles were contoured and then studied to determine whether any particular depth intervals displayed high values of resistivity, which would indicate the presence of saturated sand and gravel lenses. The resistivity data showed thick intervals of highly resistive materials. In such cases it is practical to average two or three depth values for plotting on the area map. Based upon this assessment, the consultant decided to average and then contour, as an areal map, the corrected resistivity values of the 40-, 60-, and 80-ft a-space readings (Figure 8.15). The cuttings from the test hole were helpful in establishing the meaning of the resistivity measurements for the area.

Interpretation of the resistivity values obtained from the survey indicate that aquifer conditions would be much more promising toward the eastern end of the property. Test drilling was recommended along the line of stations 8, 9, and 10. Ideally, drilling should have occurred along the line of stations 4, 5, and 6, but this area had already been developed with homes.

Subsequently a 12-in well was installed to a depth of 90 ft at the selected location. The boring encountered sand and gravel from a depth of about 5 ft to the bottom of the boring at 95 ft. During a 24-hour aquifer test, the well was pumped at 1,250 gpm; the specific capacity was 70 gpm/ft of drawdown. The well was rated at 1,570 gpm for 100 days of continuous pumping (no recharge), which far exceeded the short-term 100-gpm yield of the original test well a few hundred feet to the west.

Borehole Geophysical Methods

The second major method of investigating aquifers utilizes geophysical instruments

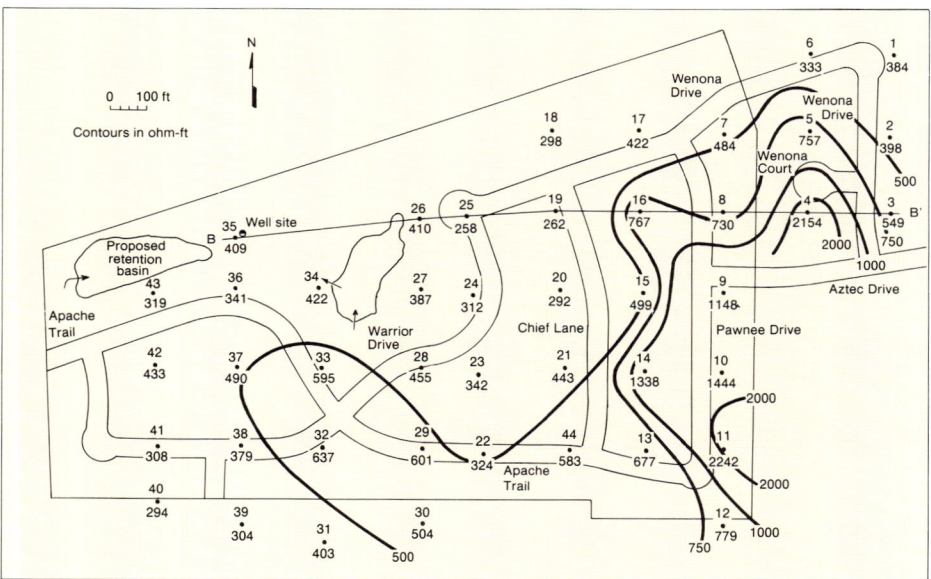

Figure 8.15. Resistivity layout for the proposed mobile home park. Values for the 40-, 60-, and 80-ft readings have been averaged. *(Keck Consulting Services, Inc.)*

in the borehole. Unlike surface exploration methods, borehole data collection and analysis are usually easier and can be done during the construction phase to facilitate screen emplacement, particularly in deep, high-capacity wells. Thus, many contractors who drill relatively deep wells now use one or more borehole methods. Once borehole geophysical data have been recorded on chart paper, they are referred to as a specific kind of borehole log. Several types of logging procedures are described below. They can be used to supplement data from a surface survey, or stand alone to guide the contractor or engineer in determining the thickness of formations, the zones of highest porosity, and water quality. They also assist in well design and verify the efficacy of well construction techniques. Because of the importance of well logging techniques and the rather limited discussion afforded here to each method, a list of general references is included in Appendix 8.B. The arrangement of a typical geophysical logging system is shown in Figure 8.17.

Borehole Resistivity Logs

Borehole resistivity logs, usually called electric logs when combined with the spontaneous potential curve, provide a useful tool for assuring good well design and construction. Many water well contractors have found that resistivity logging is easy to learn, the equipment costs are reasonable, and the information gained almost always increases the effectiveness of the well design. A good log gives a detailed picture of the character and thickness of the various strata at the well site and an indication of

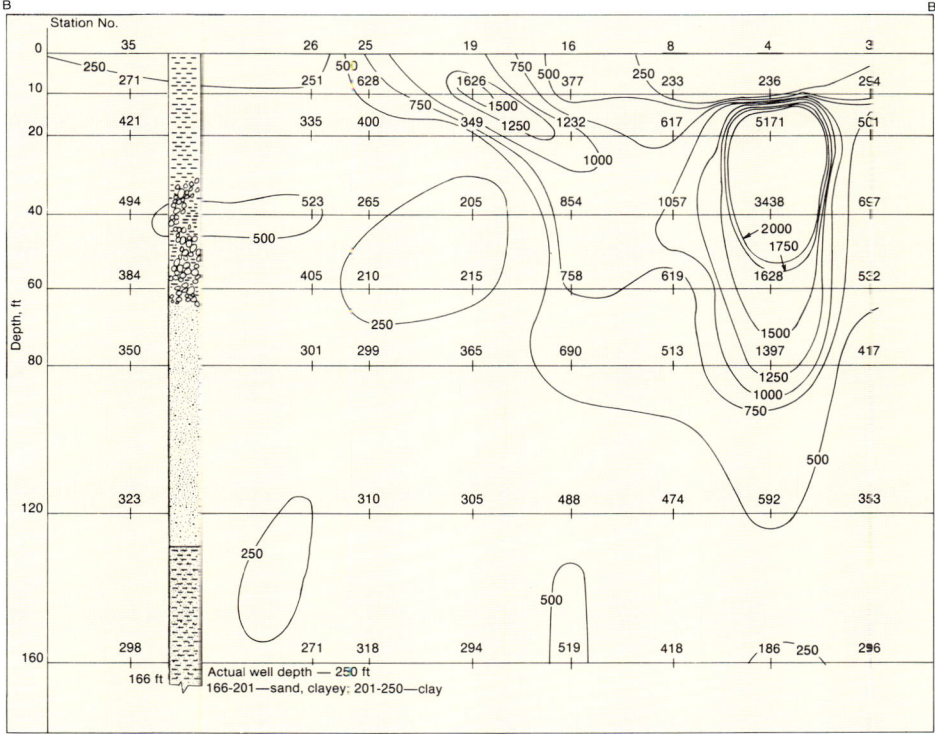

Figure 8.16. Profiles, or cross sections, of corrected resistivity. The readings indicate a thick sequence of saturated sand and gravel near the eastern border of the site. *(Keck Consulting Services, Inc.)*

the water quality by measuring the apparent resistivity of the materials surrounding the well bore. This permits the contractors to place the screen in the most desirable position with far more accuracy than merely relying on the cuttings log. Thus, electric logging (or other logging methods) should be an integral part of the well construction process for high-capacity wells because it provides a significant amount of detailed information quickly and at a reasonable cost. Electric logging offers several important advantages. These include locating the top and bottom of each distinct formation, determining relative water quality, and differentiating clean sand strata from silty sand and from sand strata with clay stringers.

To obtain an electric log, one or more electrodes are suspended on a conductor cable and lowered into a borehole filled with drilling fluid. An electric current is forced to flow from these electrodes to other electrodes that may be in the borehole or placed in the ground near the top of the well. The electric logging instrument then measures the current loss (resistance to flow) between two electrodes. Changes in electrical resistance of the entire circuit are recorded against depth to produce a graph or curve called an electric log, "E" log, or resistivity log.

Variations in resistivity are caused primarily by differences in the character of the subsurface strata and in the mineral content of the water contained in these formations. As in surface resistivity, all of the discrete layers measured between electrodes in the borehole affect the resistivity measurement. Thus, the reading is a weighted average representing individual resistivity values of the layers measured. The apparent resistivity of formation material lying outside of the two extreme electrodes is also measured.

A limiting factor in electric logging is that logging can be done only in boreholes that do not have casing and are filled with drilling fluid or water. Electric logging is common among contractors who use rotary drilling methods.

The fluid in the borehole will affect measured resistivities. Data recorded by the logger include the resistivities of the borehole fluid and filter cake as well as the formation. For example, if the drilling fluid is quite clayey or salty (low resistivity), the formation resistivity as detected by the logger will also be lowered. This "masking" influence of drilling fluids with clay additives can make it difficult to locate relatively thin, highly resistive zones which could be good aquifers. On the other hand, polymeric (highly resistive) drilling fluids do not have this masking effect, and the apparent resistivity as read on the logger will be closer to the actual value of the formation.

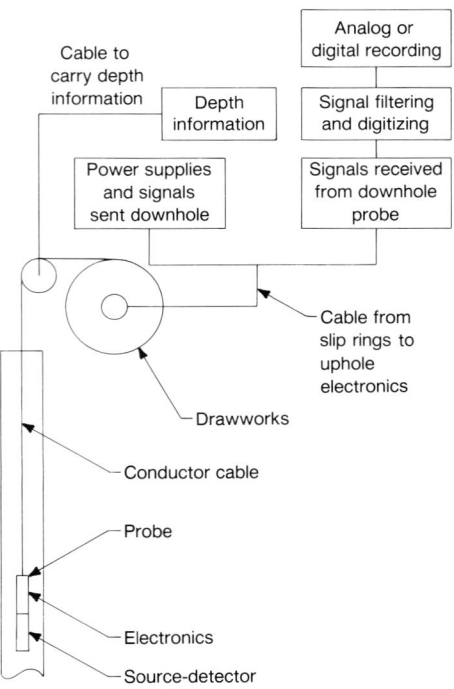

Figure 8.17. The basic components of a geophysical well-logging system. *(Daniels, 1984)*

Some electric loggers have a probe designed to measure the resistivity of the drilling fluid. The operator can then develop a feel for how much of the apparent resistivity as recorded by the instrument is actually a measure of the drilling fluid and not the formation. The data can then be adjusted to more truly reflect aquifer conditions.

The distribution of resistivity values around a borehole is shown in Figure 8.18. Because resistivity values change away from the borehole, several configurations of electrodes have been designed to provide specific information on materials at various distances from the well bore. The three common arrangements are shown in Figure 8.19.

The simplest electrical measurements can be taken with the single-point resistance method, in which one current electrode is suspended in the borehole and one electrode is at the surface. Current is directed down the borehole to the current electrode, whereupon it spreads out into the formation around the borehole. Part of the current returns to the surface electrode where the current drop is measured. Although the single-point method is no longer favored because the depth of investigation is limited and the resistance data are unduly affected by the drilling fluid, it does have several advantages. The single-point method has good vertical resolution, uses a single conductor cable, and is inexpensive to purchase and operate. The single-point method

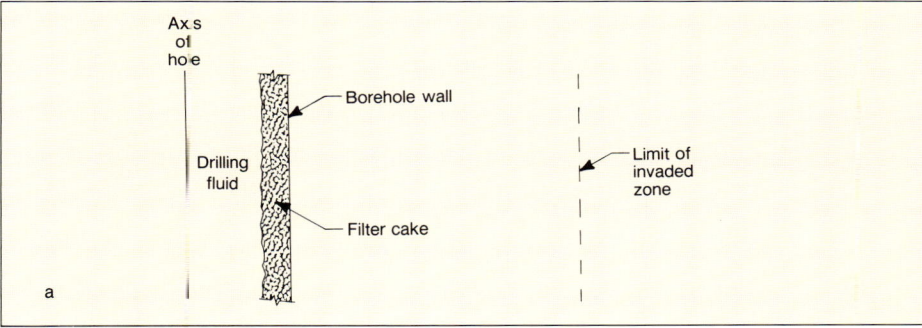

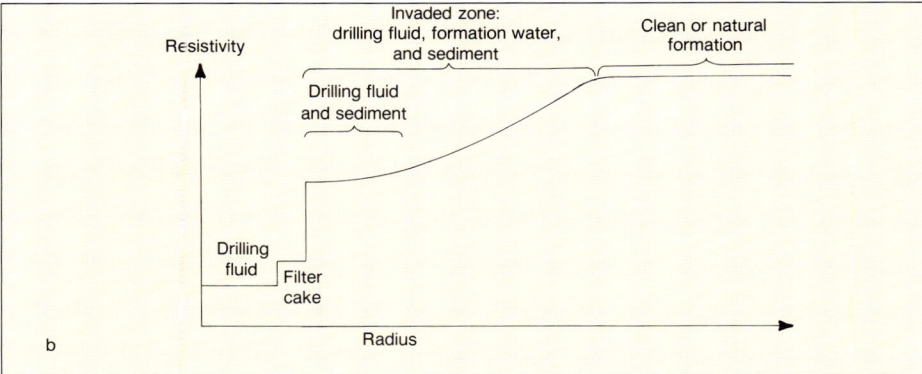

Figure 8.18. Variation of resistivity with radial distance from the axis of a borehole in sedimentary formations. (a) In the invaded zone closest to the borehole, the formation fluids are largely replaced by drilling fluid filtrate (drilling fluid solids form the filter cake); the rest of the invaded zone contains progressively less drilling fluid filtrate, until the resistivity stabilizes in the uninvaded formation. See Chapter 11 for factors influencing the depth of penetration of the drilling fluid. (b) Resistivity profile representing the borehole conditions presented in (a). *(After Beck, 1981)*

is a useful complement to other electric logs.

The two-electrode arrangement is called the normal log. When the separation of the electrodes is 16 in (406 mm) or less, the configuration is called a short normal log; if separated by 64 in (1,630 mm), it is called long normal. The spacing of the current and potential electrodes determines the depth of penetration into the formation for a given borehole diameter. The larger the spacing in relation to the borehole diameter, the deeper the penetration and the lower the bed resolution. The three-electrode arrangement consists of one current electrode and two potential electrodes in the borehole. The spacing between current and potential electrodes is 16 and 64 in. The distinguishing characteristic of this arrangement is that both potential electrodes are placed in the borehole.

An important factor influencing apparent resistivity is the diameter of the borehole in relation to the spacing of the downhole electrodes (Figure 8.20). In a large-diameter hole or with short spacing between the two electrodes, the resistivity will be heavily influenced by the drilling fluid in the borehole. This is simply because the "zone of influence" of the electrodes does not extend very far into the formation. But because the operator is reading a comparatively small part of the formation, this short-spacing of electrodes can be used to find tops and bottoms of very thin formations.

When the electrodes are spaced farther apart or when working in a smaller diameter borehole, greater penetration of the formation is achieved. Because of this, the resistivity indicated on the logger will more closely approach the true resistivity of the formation. This is true because the drilling fluid in the borehole represents a smaller

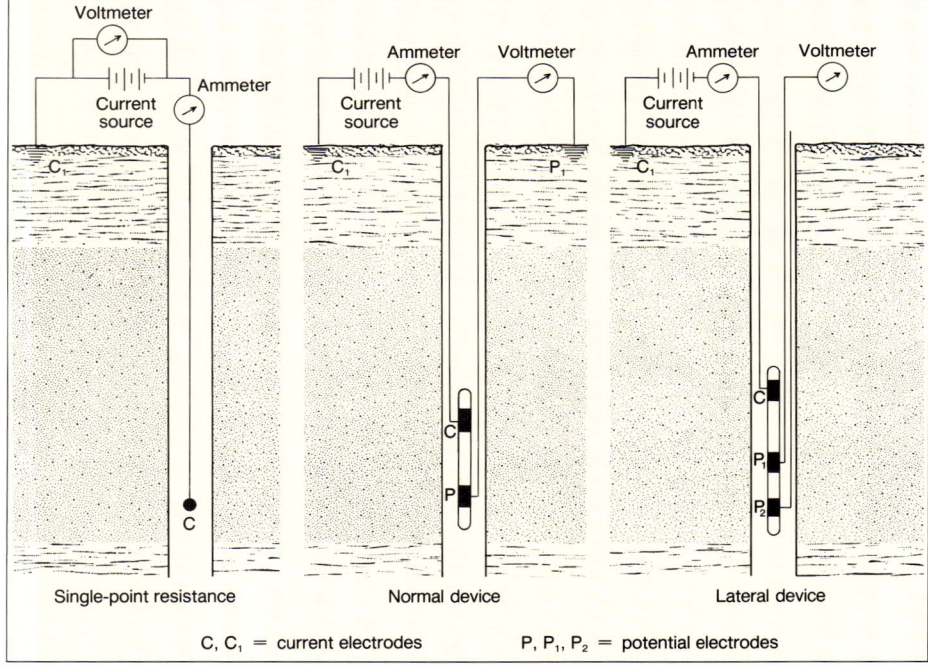

Figure 8.19. Electrode arrangements and circuits for three electric logging procedures. Each produces a resistivity curve that differs in some detail from the others; these differences assist in interpreting the logs.

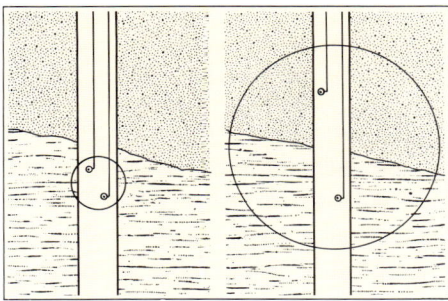

Figure 8.20. Apparent resistivity is affected by electrode spacing in relation to hole diameter. Short-space electrodes (left) read a smaller part of the formation, which makes possible more precise identification of formation interfaces. Wider electrode spacing (right) reads more of the formation, is less affected by borehole fluids, and delivers truer measurement of formation resistivity. Keys and MacCary (1971) note, however, that some reversals in resistivity values may occur for thin beds at larger electrode spacings.

percentage of the total volume being measured, so its influence is correspondingly reduced. If the formation thickness is greater than three times the electrode spacing, the effect of adjacent formations is negligible and the true resistivities of the aquifer can be determined (Beck, 1981). Electrode spacing of greater than 64 in (1,630 mm) is not recommended because thinner beds may exhibit less than true resistivity, or, at worst, may be shown as conductive beds if their thickness is equal to or less than the electrode spacing (Keys and MacCary, 1971).

Equipment for Measuring Resistivity in the Borehole

The usual electric logging instrument includes four principal components: (1) an electronic unit that feeds electric current to a down-the-hole electrode and measures the resistivity of the entire circuit, (2) a hoist or reel with conductor cable, (3) an electrode or probe from which current passes to the drilling fluid and formation surrounding the borehole, and (4) a recorder for automatically plotting values of resistivity against depth as a continuous curve.

The electronic unit includes control devices necessary for adjusting current flow to conditions in each well and for measuring changes in circuit resistance. The panel circuitry relays variations in resistivity that actuate a movable pen on the recorder. Some loggers are able to conduct two or more related logging functions simultaneously (Figure 8.21). For example, resistivity and spontaneous potential logs, or gamma-ray and single-point resistance logs can be run simultaneously. Similar logs, such as electric logs with different electrode spacings, can be overlapped to show similarities and differences (Figure 8.22).

The probe is lowered to the bottom of the borehole by a hoist or reel and then lifted slowly as the logging operation proceeds. The hoist or reel includes a device for indicating the depth of the probe at

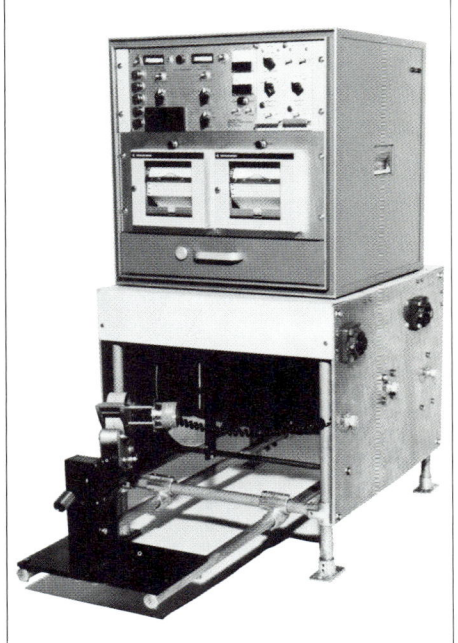

Figure 8.21. Some electronic loggers can perform two or more related functions simultaneously, thereby reducing the number of trips up the borehole. *(Keck Geophysical Instruments, Inc.)*

all times. Hand-operated hoists can be used to about 1,000 ft (305 m); at greater depths, a power-operated hoist is required. When the electric logger includes an automatic recorder, a cable-measuring device on the reel causes the recorder graph paper to move lengthwise at a rate corresponding to the movement of the probe.

The electrical probe consists of a cylindrical tool 1.5 to 2.5 inches (38.1 to 63.5 mm) in diameter. One or more electrodes in the form of short metal rings are mounted on the tool. Each electrode is connected to one of the wires of the conductor cable on which the tool is suspended.

Depending on the number of instrument modules, 11 or more functions can be logged by the instrument shown in Figure 8.21. Included are several resistivity func-

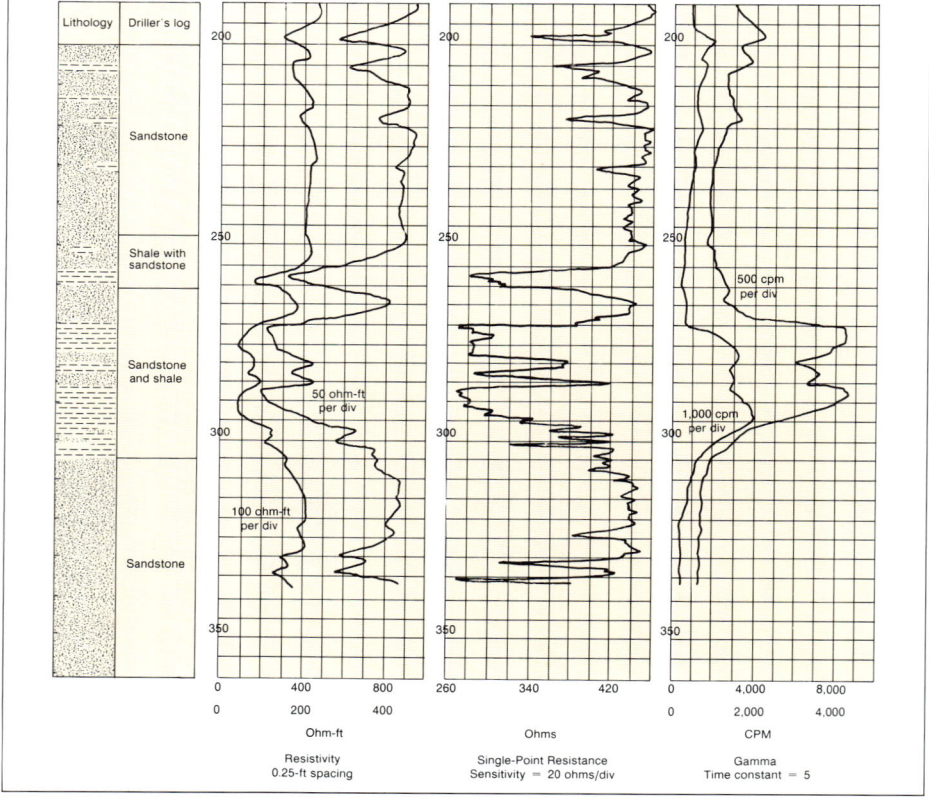

Figure 8.22. Three plots are shown here that depict different logging functions made in an open sandstone hole. The single-point resistance log shows a sharp deflection to the left, beginning at 254 ft (77.4 m) and ending at 260 ft (79.3 m). This indicates a low-resistance material (shale). The 0.25 normal resistivity also shows the beginning of low-resistance material at about 253 ft (77.1 m). In addition, the gamma-ray log shows a slightly higher gamma-ray count beginning at 254 ft, indicating an abundance of clay minerals contained in the shale. Notice the excellent correlation of the sandstone layer between 260 ft (79.3 m) and 270 ft (82.3 m) on the single-point and resistivity logs. This layer also shows up slightly as a small deflection to the left in the gamma-ray log. This indicates that the readings were influenced by actual changes in formation, and not simply by changes in borehole diameter (which can affect the single-point resistance log and the 0.25 normal resistivity log). Below 320 ft (97.6 m), single-point and resistivity logs show changes in material that are not shown in the gamma-ray log. In this case, the two resistance logs appear to be influenced by changes in the borehole diameter or the chemical quality of interstitial fluid, and not by actual changes in lithology. This theory could easily be checked by running a resistivity log with greater electrode spacing or a caliper log. *(Howe, 1979)*

tions utilizing different electrode spacings, spontaneous potential, gamma ray, temperature, and caliper. Each module plugs separately into the cabinet, which provides at least two advantages: the modules are readily available for service or repair, and later additions may be made to the unit without redesigning the cabinet.

Interpretation of Apparent Resistivity Values

The electric log, by itself, cannot be used to identify the material along the borehole because the measured resistivity is also a function of the resistivity of the borehole and interstitial fluids, the diameter of the borehole, and the distance between the electrodes on the borehole probe (Daniels, 1984). Samples recovered during the drilling operation are required for positive identification of specific geologic materials. Dry formations are poor electrical conductors and show very high resistivities. Saturation of a formation reduces its resistivity; the reduction in resistivity is partially controlled by the porosity. This occurs because water is an electrical conductor, so its presence in the interconnected pores reduces the overall formation resistivity. There are, however, general differences in the resistivity of various saturated formations which the electric logger can detect. Silt, clay, and shale, for example, have the lowest resistivity; silty or dirty aquifers have low to medium resistivity; sand and gravel with fresh water have moderate to high resistivity. The highest resistivity values are found in sandstone and limestone saturated with fresh water, and in dense igneous and metamorphic rocks such as granite and dry slate. The type of material, however, is only one factor influencing resistivity.

Water quality also affects resistivity. Water in the pores of clay is always highly mineralized because it dissolves minerals from the chemically active surfaces of the clay particles. As a result, wet clay formations show relatively low resistivity. In contrast, sand formations saturated with fresh water have relatively high resistivity values because the water picks up only small amounts of minerals from the sand particles. In general, the resistivity of a formation will vary inversely with the total dissolved solids (TDS) contained in the water. If all other conditions remain the same, resistivity decreases as TDS increases. For example, sand and gravel formations with salty water will register low resistivity values. The low resistivity of salty water almost completely overshadows the higher resistivity of the sand (Figure 8.23).

Correct interpretation of borehole resistivity data involves minimizing the effect of drilling fluid resistivity on the data. Ideally, the borehole diameter

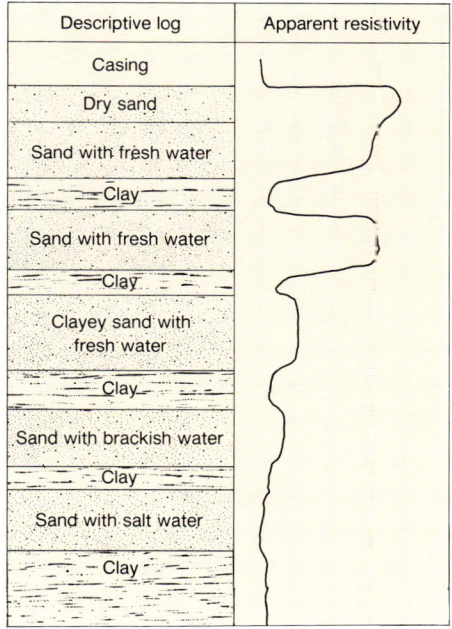

Figure 8.23. Electric log of a series of sand and clay beds. Brackish and salt water in the bottom sand formations causes lower electrical resistivities for these aquifers. As a result, the contrast between clay and sand disappears from the resistivity log.

should be kept as small as possible. In boreholes larger than 8 in (203 mm), the probes should be placed next to the borehole walls. The effect of the drilling fluid can be reduced if the electrode spacing is relatively large in comparison with the borehole diameter. It may be advisable to use all four electrodes on the downhole probe, rather than the normal array, when the resistivity of the borehole fluid is much less than the resistivity of the aquifer materials (Daniels, 1984).

Experience is an important factor in the interpretation of electric logs, but knowledge of the general trends described above is also important. When a clay layer underlies a sand layer saturated with fresh water, the electric log is deflected to the left as the probe moves upward past the clay. As the probe begins to measure the sand layer, the curve deflects to the right, indicating higher resistivity. The depth at which the electric log shows the change from lower resistivity to higher resistivity corresponds to the elevation of the bottom of the sand. Depths to top and bottom (thickness) of the layers can then be compared with the driller's log. In some cases, the resistivity change may be so small that the adjacent layers cannot be differentiated.

Spontaneous Potential Logs

Spontaneous potential logs are usually run in conjunction with resistivity logs. In fact, the SP curve was the first lithologic log done with electric logging techniques (Pirson, 1977). Spontaneous potentials are naturally occurring electrical potentials (voltages) that result from chemical and physical changes at the contacts between different types of subsurface geologic materials. For example, electrical potentials occur spontaneously at the contact surface between a clay layer and an underlying sand layer, or between a sand formation and an igneous rock formation. In a borehole, potentials also occur between the drilling fluid and the formation, and between the drilling fluid and the filter cake (Guyod, 1964). The SP log can also be affected by "streaming potentials" produced by fluid moving into or out of a permeable formation. These potentials become more pronounced when the pressure in the borehole greatly exceeds the pressure in the formation. The SP log measures these potentials.

To measure the spontaneous potential at various depths, an electrode is lowered into an uncased borehole filled with drilling fluid by means of an electric cable connected to one terminal of a millivolt meter and recorder. The other terminal of the instrument is connected to a ground terminal at the surface which is often placed in the mud pit. No external source of electric current is connected to this circuit. The downhole electrode is usually negative with respect to the surface electrode. Any current in the circuit which results from electrochemical action between the drilling fluid and the formation or formation water is conducted to the surface through the drilling fluid column. The millivolt meter connected between the two electrodes, therefore, measures the drop in potential in the drilling fluid column between the downhole electrode and the surface electrode.

As the downhole electrode is moved up and down in the borehole, the meter registers variations in spontaneous potentials of the different formations. A curve showing these potentials plotted against depth provides what is called the SP log. Although the SP curve may indicate the permeable zones, there is no definite relationship between the magnitude of the SP deflection and the permeability or porosity of the formation. Variations shown by the SP curve are interpreted along with variations in apparent resistivity shown by the resistivity curve. The two curves, taken together, constitute what is usually called the electric log. The SP log is plotted on the left-hand side of the curve sheet where it can be compared easily with the resistivity log on the right-hand side.

In oil-field work, the SP log usually exhibits a baseline, which is more or less vertical, with deflections or "peaks" to the left. Electric logs of oil wells show that the baseline frequently corresponds to impermeable beds such as clay or shale, whereas the peaks usually correspond to the positions of permeable strata. Many formations that are suitable for high-capacity water wells will have little shale or clay, and therefore a clay baseline will not occur on the log. If a shale or clay layer does exist, however, it will provide contrast and thereby increase the comparative value of the SP log. Without a baseline, it is much more difficult to interpret an SP curve.

On the idealized log shown in Figure 8.24, a vertical line drawn to connect the potentials corresponding to the various impermeable clay layers is the clay baseline. Deflections of the SP curve to the left of the baseline indicate permeable water-bearing sand formations. For sand formations containing fresh water, the SP variations correlate with the resistivity peaks of the apparent resistivity curve on the right. For sand formations containing salt water, however, only the SP log indicates the presence of a permeable layer because the resistivity of this formation is approximately that of the clay layers.

Most of the published information on interpretation of the SP curve is based on electric logs of oil wells. It is important to understand the differences between oil-well and water-well SP curves. Groundwater associated with oil is salt water. The electrical conductivity of this water is extremely high in comparison with the conductivity of water in the drilling fluid. Groundwater suitable for drinking water purposes has low dissolved solids, on the other hand, and therefore has a much lower conductivity than oil-field brine. Its electrical conductivity may be about the same as, or even less than, the conductivity of water in the drilling fluid. Thus, the electrochemical reaction between the formation water and the drilling fluid is quite different, depending on whether the formation water is considerably more salty than the drilling fluid (oil-field conditions), or whether the formation water is about the same salinity as the drilling fluid (water-well conditions).

When a permeable formation contains salt water and the drilling fluid is made with fresh water, the SP normally shows a relatively large deflection to the left in relation to the clay baseline. The SP deflection opposite the same formation containing fresh water, however, would be

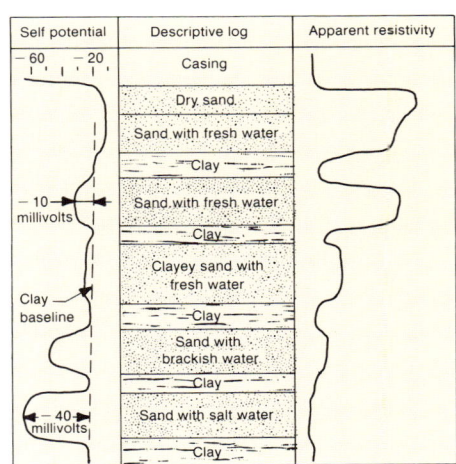

Figure 8.24. Idealized SP curve and resistivity curve showing electric-log responses corresponding to alternating sand and clay strata; sands are saturated with fresh water, brackish water, and salt water. Note relative deflections of SP curve opposite fresh-water and salt-water sands. In general, the resistivity and SP curves move in opposite directions if the drilling fluid is less saline, or fresher, than water in the formation. This is not the case for the upper sand layer, where the water in the formation has less total dissolved solids, and therefore is less active chemically, than the drilling fluid.

relatively small. This would result in the SP curve being essentially a straight line. Another way to describe the difference is to note that the formation with salt water shows a highly negative potential in relation to clay layers, whereas the formation with fresh water shows only a slightly negative potential. The deflections of the idealized SP curve in Figure 8.24 give an idea of the differences that can be expected among formations containing fresh water, brackish water, and salt water when the drilling fluid is made with fresh water.

The types of dissolved salts in the formation waters also differ in oil-well and water-well settings. The dominant salt in oil-field brine is sodium chloride (NaCl). In contrast, the important salts in many fresh groundwaters are calcium bicarbonate and magnesium bicarbonate, with only small amounts of sodium chloride. To interpret the SP log, the nature and concentration of the dissolved salts must be considered because both of these factors influence independently the SP curves for a given sequence of geologic strata.

Generally speaking, the following observations are helpful in interpreting SP logs for formations containing fresh water:

1. The SP curve is often difficult to interpret at shallow depths. On the other hand, SP deflections are more pronounced in moderately deep to deep wells because, as depth increases, the groundwater tends to become more highly mineralized.

2. The first step in interpretation is to establish the clay baseline (shale line) on the log. If no clay layers are present, the SP log adds little to the interpretation. Thus, for many wells the SP curve may be of little value because variations in the curve may be insignificant.

3. Note deflections to either the left (negative) side or the right (positive) side of the clay baseline. Formations having deflections to the left generally indicate groundwater having higher chemical activity than formations having deflections to the right. These deflections indicate the positions and thicknesses of aquifers containing fresh water. Deflections in the SP curve may be insignificant unless the strata are at least 4 ft (1.2 m) thick.

4. Conclusions drawn from the SP curve will generally correlate with data from the resistivity curve, although the curves will usually move in opposite directions.

5. The clay baseline may shift gradually or abruptly at increasing depths for no apparent reason.

6. SP curves should always be used in conjunction with resistivity or other logs because it may be particularly difficult to interpret the curves.

As already explained, the deflection opposite formations containing fresh water may be small when the drilling fluid water is about the same quality as the formation

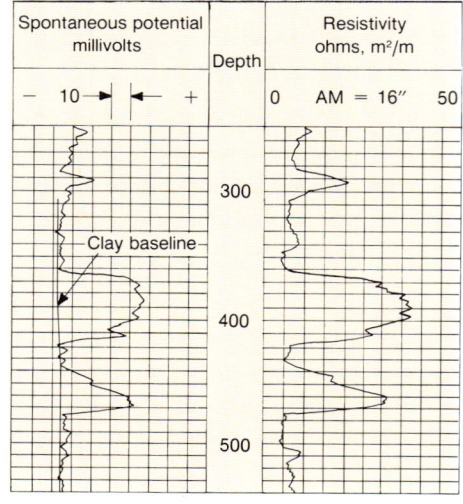

Figure 8.25. This SP curve of a water well in Texas was obtained after salt was added to drilling fluid. The salted drilling fluid reversed the polarity of SP opposite fresh-water sand formations. *(Schlumberger Well Surveying Corp.)*

water. In some instances, the deflection can be enhanced by adding salt to the drilling fluid to make it much more saline than the formation water. This reverses the polarity of the potential, causing a deflection to the right of the clay baseline when formations containing fresh water are encountered. Figure 8.25 is an example of an SP log where the drilling fluid was salted. The efficacy of this procedure varies with the types of formations in the borehole and the quality of the formation water. Better results may be obtained by using other logs, rather than changing the characteristics of the drilling fluid.

Gamma Logs

In gamma logging, measurements are made of naturally occurring radiation coming from the materials encountered in the borehole. The record of gamma radiation is used as a qualitative guide for stratigraphic correlation and permeability. In some areas, a direct relationship can be established between gamma radiation and permeability.

Gamma radiation is emitted from certain elements in geologic materials that are unstable and decay spontaneously into other more stable elements. Gamma rays are similar to X-rays in that they have a great ability to penetrate other materials, but gamma rays have a shorter wave length. A loss of mass from the atomic structure of the element results from gamma-ray emissions. Gamma emissions are one of the three kinds of radiation; the others are alpha and beta emissions.

Certain radioactive elements occur naturally in igneous and metamorphic rocks and as depositional particles in sedimentary rocks. Clays and shales contain high concentrations of radioactive isotopes, usually potassium. Mature sands and gravels, on the other hand, contain primarily silica, a stable substance, and therefore emit only very low levels of radiation. Limestone and dolomite also emit little radiation. Keys and MacCary (1971) note that for some immature sandstones or those highly cemented with calcium carbonate, much of the radiation may originate from uranium and thorium particles.

Although other types of energy emission are given off by natural radioactive minerals, only gamma rays are measured in well logging because only they can penetrate high-density materials such as casing and cement grout. Gamma-ray logging has a fundamental advantage over electrical logging: it can be done in either cased wells or in open boreholes containing air, water, or drilling fluid, whereas electrical logging can be done only in uncased boreholes filled with fluid. Therefore, gamma-ray equipment can log existing wells where the original logs have been lost or destroyed.

Detection of gamma-ray emissions involves two random processes. First, the rays are given off at random intervals by the radioactive minerals. This means that the number of energy pulses emitted per second or per minute varies within certain minimum and maximum values. Second, these irregularly spaced pulses collide randomly with the detecting element in the logging probe. Not all the pulses given off by the radioactive material are measured by the detector and the proportion that do strike the detector varies irregularly. Both this fact and the random nature of the gamma-ray emission must be considered in reading or interpreting gamma-ray logs.

The gamma-ray unit is a simple instrument to operate because no surface lines or stakes are required; all the necessary equipment is contained in the logger and probe. The probe consists of a scintillation-type receiver and counting circuit. Radiation intensity from a given geologic material is measured by the number of pulses detected by the instrument per unit of time. This intensity is expressed as the average number of counts per second or per minute*, and is usually recorded during a specified period. The period over which the pulses or counts are averaged is called the time constant.

Selection of the time constant will depend on the frequency of gamma emissions, the rate the probe is being raised or lowered in the borehole, and the resolution required. The time constant and the rate the probe is raised must be properly coordinated to obtain a good log; that is, a sufficient number of counts must be recorded over the time constant. For automated loggers in which the cable is raised and lowered mechanically, the cable speed is regulated so that enough counts can be recorded. With manually operated loggers, discrete measurements are made with the probe stationary for a specific period of time at each depth interval. With this type of logging system, measurements are recorded in the field and plotted manually in the office.

To illustrate the effect of the time constant, suppose 400 counts are recorded in 20 seconds with the probe positioned 200 ft (61 m) below ground. This would represent an average of 20 counts per second over the 20-second period, or the equivalent of 1,200 counts per minute. The actual number of counts per second will vary greatly from the average value.

The electronic circuit of a gamma-ray logger consists of a detector, a high-voltage power supply, a pulse amplifier, a voltage regulator, and an electronic timer. Most water well units are equipped with scintillation crystal detectors — thallium-activated sodium iodide — but these probes cannot be used at temperatures greater than 500°F (260°C), depending on the manufacturer. On the other hand, some older water well units are equipped with gas-filled Geiger-Mueller (G-M) tubes as detectors. The Geiger-Mueller tube is less sensitive than the scintillation probe, but is generally smaller in diameter.

The gamma rays detected by the probe originate from material within a short distance outside the borehole. It has been estimated that 90 percent of the gamma rays detected during logging originate within 6 to 12 in (152 to 305 mm) of the borehole wall. Thus, a relatively small, roughly spherical volume of material contributes most of the radiation that is picked up by the detector. The radius of this sphere is called the "radius of investigation" of the logger. The borehole is included within the radius of investigation. Thus, the size of the borehole and the position of the probe vis-a-vis the borehole wall have some effect on gamma-ray measurements.

Figure 8.26 shows the concept of the radius of investigation. A change in counts per second does not occur abruptly at the interface of the two beds. Instead, the shape of the log curve is somewhat rounded over short spans of depth as the probe passes from one formation to another. This must be taken into account when the log is used to determine the thickness of a particular formation. It must also be recognized that beds thinner than the radius of investigation may be obscured on the gamma-ray log.

The minerals normally found in sedimentary materials such as clay, limestone, and sandstone contain small amounts of radioactive potassium-40, and decay products

---

*Because the radioactivity of a particular substance can be measured in several ways, the American Petroleum Institute has established standard measurement units from tests conducted in pits constructed of certain materials (Belknap et al., 1959). Large service companies specializing in logging work calibrate their logs with the standard API units so that some correlation can be made from area to area.

of uranium and thorium. Potassium is an important constituent of clay, mica, feldspar, and shale. About 0.012 percent of the total potassium content in these materials is the radioactive isotope potassium-40, which emits gamma rays. Quartz sand contains no potassium or radioactive potassium-40. Quartz sand formations, therefore, emit gamma rays at extremely low levels. Normally, then, the gamma-ray log shows more counts per minute at depths corresponding to clay or shale layers, and fewer counts per minute at depths corresponding to sand or sandstone layers if the sand is mostly quartz.

A problem in log interpretation arises when sands include a high proportion of feldspar grains. Unlike quartz, feldspar does contain potassium and radioactive potassium-40. A feldspar-rich sand may emit gamma rays at an intensity approaching that of clay. The gamma-ray contrast between clay and this type of sand, therefore, is not as great as between clay and quartz. As a result, the log may not identify beds of feldspar-rich sands as clearly as might be desired. This is an important point to note in logging wells in glacial deposits or near igneous source rocks, where a significant number of sand grains consist of feldspar. Logging experience and knowledge of local geology are needed for making correct interpretations of logs in these areas.

Interpretation of gamma-ray logs is difficult where sandstone or other formations include volcanic rock fragments, such as rhyolites, that contain relatively high amounts of radioactive minerals. Ordinarily, however, uranium ores and other highly radioactive rocks occur over extremely small areas.

Another problem of interpretation is related to hole diameter. Where caving clays and shales are encountered and a washout occurs, the gamma-ray log will indicate lower levels of radioactivity opposite the enlarged sections of hole. Thus the log will appear to indicate a sand layer. Borehole samples, the driller's log, and a caliper log can be used to minimize this difficulty in interpretation.

Gamma-Gamma Logs

In gamma-gamma logging, an active source of gamma radiation (usually cesium-137 or cobalt-60) is lowered into the borehole along with a detector that is shielded so it counts only the back-scattered gamma rays. The source and detector are placed up to 15 in (381 mm) apart

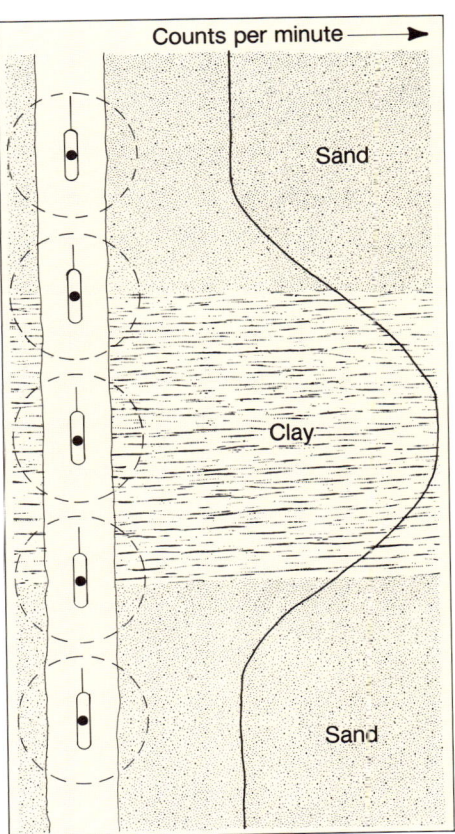

Figure 8.26. Successive positions of the gamma-ray probe show how the radius of investigation includes both clay and sand at top and bottom of the clay bed. Thus, the change in counts per minute occurs gradually as the probe moves past the contact of the two beds.

and are set against the borehole wall by mechanical arms (Figure 8.27). The gamma rays are directed into the formation surrounding the borehole. Because the amount of back-scattered radiation depends on the electron density of the formation, the recorded counts are approximately proportional to the bulk density. The gamma-gamma log is sometimes called the density log because this is the fundamental characteristic inferred from the log. In general, the higher the density, the lower the porosity will be. The gamma-gamma log can be used to calculate the porosity when the fluid and grain densities are known.

When using a gamma-gamma logger, a caliper log is first run in the hole to determine hole diameter because data from the gamma-gamma logger are strongly affected by drilling fluid density, filter cake thickness, and changes in hole diameter. The sidewall probe contains two detectors; one is mounted close to the radiation source and the other is mounted farther away. The short spacing is more sensitive to the filter cake and to small borehole irregularities. These readings can be used to correct data taken with the long spacing which more accurately measures the formation.

When interpreting a gamma-gamma log, the analyst must remember that the bulk density is affected not only by porosity, but also by the density of the materials — that is, where the rock matrix or grains are lighter, they may have less porosity than suggested by the data. However, if the bulk density of the rock is known (by gamma-gamma), the porosity can be calculated easily by using the equation:

$$\rho_b = \eta \rho_f + (1 - \eta) \rho_m \text{ or } \eta = \frac{\rho_m - \rho_b}{\rho_m - \rho_f} \qquad (8.5)$$

where ρ_b is the bulk density, η is the porosity, ρ_f is the fluid density, and ρ_m is the matrix (rock or grain) density. In order to make porosity estimates from this log, the grain density must be known.

Neutron Logs

Neutron logs are used primarily as an indicator of total porosity under saturated conditions. They are also valuable in measuring the amount of moisture in unsaturated zones. The neutron log is obtained by recording the number of neutrons impinging on a detector mounted some distance from a constant neutron source in the borehole. Before reaching the detector, many of the neutrons emitted from the source collide with various particles, lose energy, and eventually are captured. Most of the energy is lost in collisions with hydrogen ions. Because hydrogen is a principal component of water, the loss in energy indicates the amount of water present. If the energy

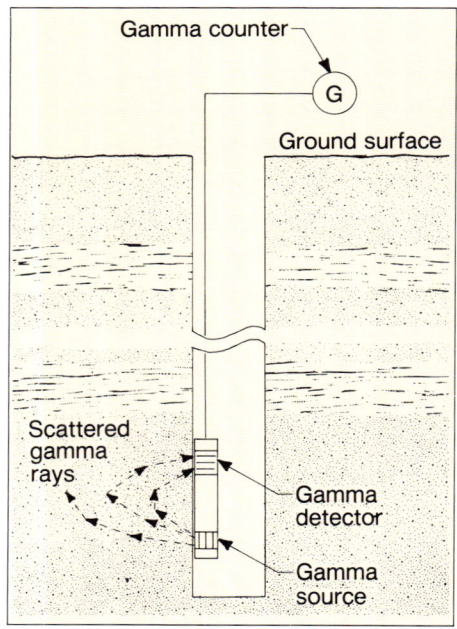

Figure 8.27. Basic elements of the borehole gamma-gamma measurement system. Special handling procedures are required for gamma-gamma investigations, and anyone handling radioactive materials should be trained and licensed. *(Costello, 1980)*

loss is large, the amount of hydrogen in the formation must be high, and therefore the porosity is large.

A common source of neutrons is americium-241/beryllium. The neutrons are emitted when alpha particles from americium-241 impinge on the beryllium. Plutonium/beryllium, and occasionally radium/beryllium, sources are also used. These radioactive sources are placed in sealed probes that require special handling to prevent exposure to radiation. Handling procedures and regulations for these and other radionuclides are found in several governmental publications. Anyone handling these materials should be trained and licensed.

The neutron source can be used in cased or open holes that can either be filled with drilling fluid or be dry (Figure 8.28). If the hole is dry, however, the results may be misleading if the hole has been drilled recently. Neutron loggers should be calibrated in American Petroleum Institute test pits and then standardized frequently in the field by using an unvarying high-hydrogen environment (Keys and MacCary, 1971). The depth of neutron penetration into a formation depends on the porosity, hole diameter, and spacing between the source and detector. For high-porosity materials, the depth of penetration may be 6 in (152 mm) or less, whereas for low-porosity materials it may be 2 ft (0.6 m).

Neutron logging theory and practice are somewhat complex, but this logging technique may be the most useful of all the borehole geophysical methods used in groundwater investigations (Keys and MacCary, 1971). Detailed explanations of this technique can be found in Keys and MacCary (1971), International Atomic Energy Agency (1971), and Beck (1981).

Acoustic (Sonic) Logs

Acoustic logs are useful in determining relative porosities of different formations and are widely used to verify how well the casing has been cemented to the formation. This latter log is called the cement bond log. The acoustic log is helpful in determining fracture patterns in the aquifer and thus is valuable in estimating where groundwater flow may be concentrated in semiconsolidated or consolidated rocks such as sandstone, conglomerates, and igneous rocks. Acoustic logs can also be used to locate the top of the static water level in deep holes and to detect perched water tables (Chagnon, 1981).

Acoustic logging measures the travel

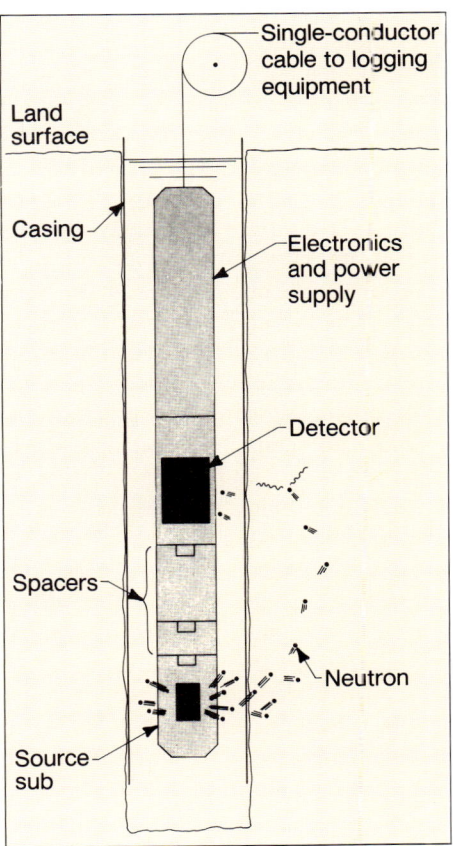

Figure 8.28. The equipment and operational principles of neutron logs. *(After Keys and MacCary, 1971)*

time and the attenuation of an acoustic signal created by an electromechanical source in the borehole. A transmitter centered in the borehole converts electrical energy to acoustic (sound) energy which travels through the formation to one or more receivers that in turn convert the acoustic wave back into an electrical impulse that can be measured at the surface (Figure 8.29). The average velocity of a sound wave passing through a formation is affected by the wave's velocity through the matrix material (for example, the sand grains or rock) and through the pore water; the rock or grain matrix conducts the acoustic wave more rapidly than does the fluid in the pore spaces.

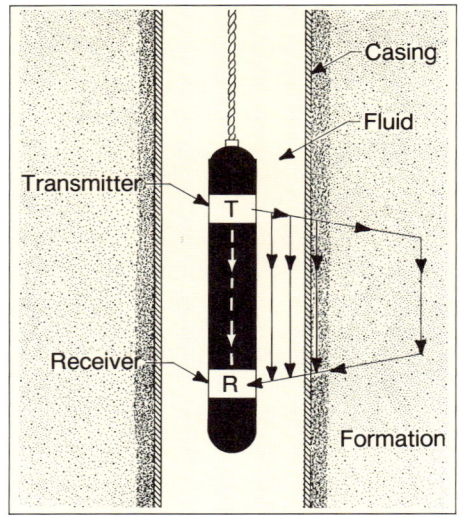

Figure 8.29. The operational characteristics of the acoustic instrument are similar to other borehole geophysical devices. *(Gearhart Industries, Inc., 1982)*

The matrix velocity is directly affected by lithology and can be determined from laboratory analysis of core samples (Costello, 1980). The measured travel times, then, will be somewhat slower than the laboratory velocities and will be a clear indicator of porosity. The presence of clays in the pore spaces of poorly sorted sedimentary formations may distort the readings, that is, reduce the speed of the acoustic waves. Therefore, the porosity may appear to be higher than it actually is. Typical travel times for some geologic materials are given in Table 8.2, page 172.

Analysis of sound waves is also important for verifying how well the casing has been cemented to a formation. In most cases, the sound waves passing through the casing arrive first, followed by the formation wave, and finally the fluid arrival (Figure 8.30). In practice, these arrivals tend to overlap each other. In a well-cemented borehole, most of the sound energy is carried by the cement and nearby formation materials because the casing is prevented from "ringing"; thus, the amplitude of the wave received is less and the travel time delayed, compared with waves traveling only through the casing. If the cement bond is poor, however, much more of the energy will be channeled through the steel casing (the wave is concentrated in the steel), with an obviously higher energy and earlier arrival time at the receiver. The compressive strength of the cement also affects the magnitude of the received signals. High compressive strengths lead to lower signal magnitudes. The actual acoustic signals (signatures) of an uncemented and well-cemented casing section are shown in Figure 8.31. The better the bond, the lower will be the amplitude of the sound wave for the casing signal, and the higher will be the amplitude for the formation signal. The method used by a computer to produce the graphic representation of the acoustic waves is shown in Figure 8.32.

Temperature Logging

Some investigators have shown that near-surface soil temperatures vary when the local geothermal gradient is disturbed by the presence of a shallow aquifer (Lovering and Goode, 1963). Thus, information on near-surface soil temperatures and on borehole temperature gradients provide a way to locate small or sinuous aquifers and to estimate the contribution of these separate zones (with water of different temperature) to total yield.

A temperature log is obtained by lowering a temperature sensor down the water-filled borehole at a slow but constant rate. The probe is constructed so that water flows by the sensor, which is mounted in a protective cage or tube (Keys and MacCary, 1971). Temperature logging is usually done by lowering the probe to assure that an undisturbed column can be logged and water can flow past the sensor more readily. The probe should be lowered at a constant rate so that the sensor can detect and transmit the temperature change for the depth recorded. The optimal speed for temperature logging is 20 ft (6.1 m) per minute. Slower logging speeds, however, will usually produce more detailed data (Figure 8.33). No temperature measurements should be taken immediately after drilling because the temperature equilibrium in the borehole has been disturbed.

Borehole temperature can be used to calculate the change in water temperature with depth (thermal gradient). The temperature at any point in a borehole will depend on several factors: (1) the distance from the surface (near-surface water temperatures may be affected by climate), (2) the background geothermal gradient at the site (this gradient can be much higher in areas of active volcanism), (3) the thermal conductivity of the rock, and (4) the vertical flow of fluids in the borehole caused by differences in head, water quality, or temperature. The geothermal gradient, or increase in temperature with depth, may be derived from a temperature log if the well fluid is in thermal equilibrium with the adjacent rocks, that is, if there is no vertical circulation

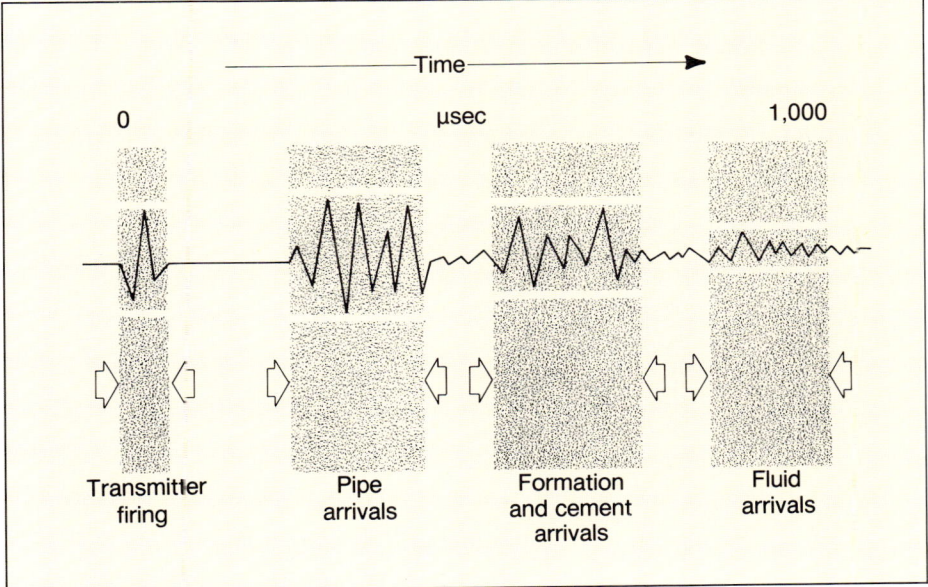

Figure 8.30. The relative arrival times for the sound waves are shown for the casing, formation materials, and formation fluids. Although separated in this diagram, the individual wave patterns will overlap one another in an actual seismic spectrum. Recognition of the first arrival times is possible because normally the first casing and first fluid arrival times can be predicted. *(Gearhart Industries, Inc., 1982)*

of fluids in the well bore. The temperature gradient is then largely determined by the thermal conductivity of the rocks. In most cases, however, the geothermal gradient in different rock units surrounding the borehole is altered because of vertical flow in the borehole; vertical flow is caused mainly by pressure differences, but also by temperature differences in discrete aquifers. If vertical flow is rapid, little temperature gradient exists in the borehole; if vertical flow is slight, the temperature gradient may be almost the same as that found in the formation. Ordinary geothermal temperature gradients range from 1 to 1.3°F (0.6 to 0.7°C) per 100 ft (30.5 m) of depth.

In general, the geothermal gradient is greater in formations with high hydraulic conductivity than in formations with low hydraulic conductivity. This relationship is usually governed by the rate of groundwater flow. Thus, interpretation of thermal data can suggest the relative hydraulic conductivity of the formation in the borehole. However, anyone who analyzes temperature data must remember that some distortion of the background geothermal gradient almost always occurs

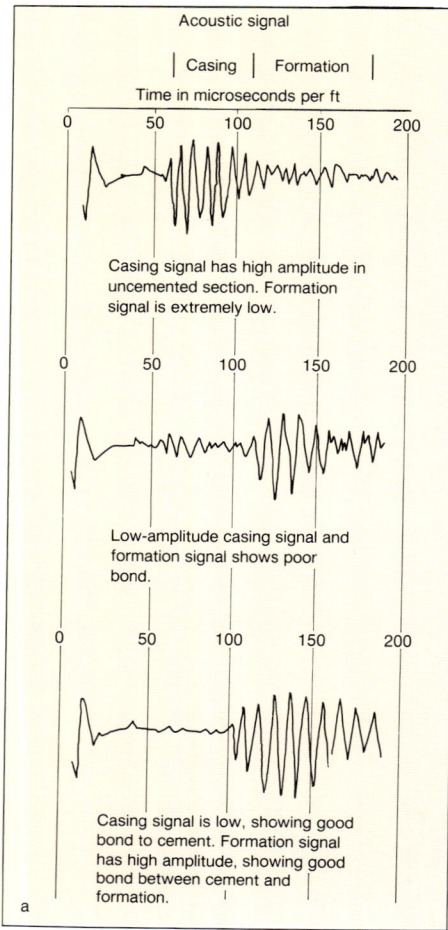

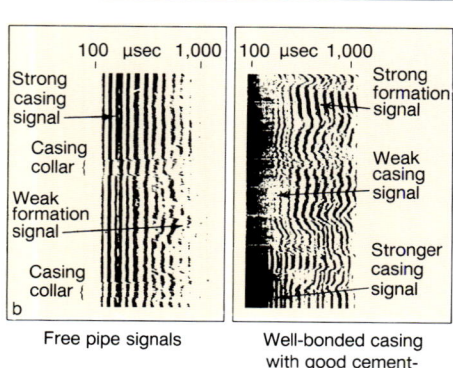

Figure 8.31. (a) The arrival time and amplitude of the acoustic wave vary according to how well the casing is cemented to the formation. (b) The camera signals also show where the cement forms a good bond to the formation.

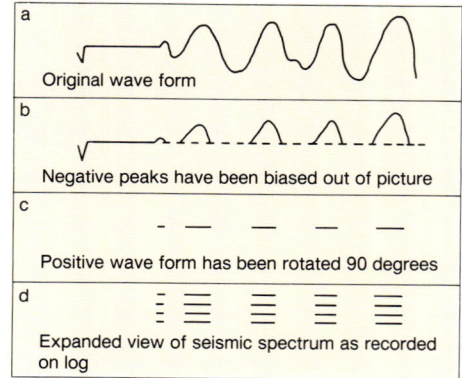

Figure 8.32. The seismic spectrum produced by the computer in the acoustic instrument transforms the typical wave form shown in (a) by clipping off the negative signal (b), rotating the positive signal 90 degrees (c), and then stacking the lines to represent signal strength with depth into the formation (d). *(Gearhart Industries, Inc., 1982)*

depending on the degree of vertical flow in the borehole. Disturbed geothermal gradients can provide information on permeable intervals. For example, flow through highly permeable zones can distort the geothermal gradient because the water in these zones has not equilibrated with temperature conditions in the earth.

Temperature logs are useful in detecting episodes of seasonal recharge (vertical movement of cooler or warmer water into the aquifer), because recharge upsets the usual temperature regime. In wells where only isolated aquifer zones having different temperatures are contributing most of the water, the actual temperature of the water pumped from the well is a good indicator of the zone contributing most of the water. Temperature logs are also valuable in identifying heat-pump recharge water, excess irrigation water, and industrial wastes, and are required for quantitative interpretation of resistivity logs.

Cement grout can be located by temperature logs as long as the borehole is filled with water above the top of the grout and the log is run within 24 hours of cementing. The more cement there is behind the casing, the higher will be the temperature anomaly. Typically, a neat portland-cement grout has a temperature of 160°F (71.1°C) after 4 to 8 hours, and 100°F (38°C) after 8 to 12 hours at depths less than 200 ft (61 m). Borehole temperatures at these depths are usually much less than 100°F (38°C). Recall

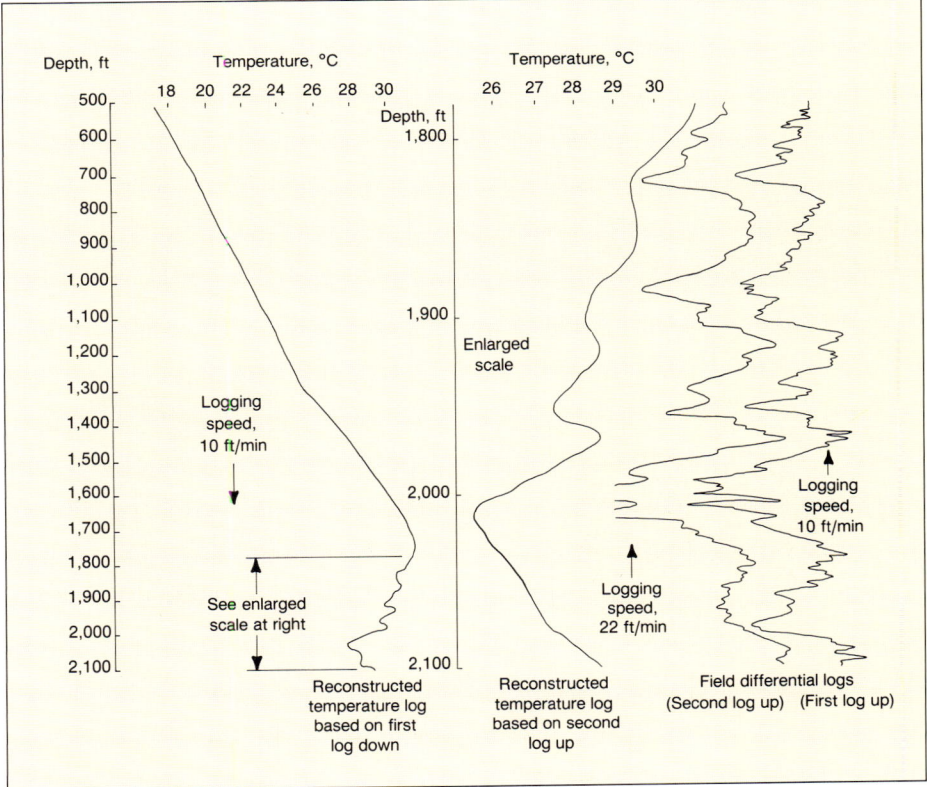

Figure 8.33. Field and reconstructed temperature logs (left). The interval from 1,780 to 2,100 ft (543 to 640 m) is enlarged (right) to show the detail and repeatability of differential temperature logs. *(After Keys and MacCary, 1971)*

that near-surface groundwater temperature is usually the local mean annual temperature.

Caliper Log

Although the caliper is the simplest logging device, it is one of the most useful, both by itself and when used in combination with other geophysical instruments. This mechanical logging device consists of a probe with 1 to 4 adjustable legs that can sense the diameter of the borehole. The caliper shown in Figure 8.34 can measure borehole diameters as small as 4 in (102 mm) or as large as 24 in (610 mm), to depths of 500 ft (152 m). Other models can measure boreholes to much greater depths. By knowing the exact diameter of the borehole, the contractor can determine the amount of borehole erosion that has taken place during drilling, the presence of swelling clays or resistant sandstone layers in an otherwise friable rock, fracture patterns in limestone or sandstone, the volume of filter pack or cement grout required for well completion, the positions of casing welds or joints, and areas where the casing has separated. Data from a hole caliper log is also extremely valuable in analyzing data for other types of logs where the readings are influenced by variations in hole diameter.

To operate the hole caliper, it is merely lowered into the borehole and the readings recorded at the surface as the caliper is withdrawn. Most calipers have a surface recorder that shows the diameter of the borehole as a function of depth. Some caliper cables are marked in feet or meters and can be lowered by hand or by a drum winch.

Caliper tools can be calibrated by placing the legs or feelers inside a cylinder or ring of known diameter. If any drift is anticipated, enough points should be plotted inside the casing (a known diameter) to establish the linearity of the system's response. One inch (25.4 mm) of chart width per inch of hole diameter provides enough sensitivity to locate fractures.

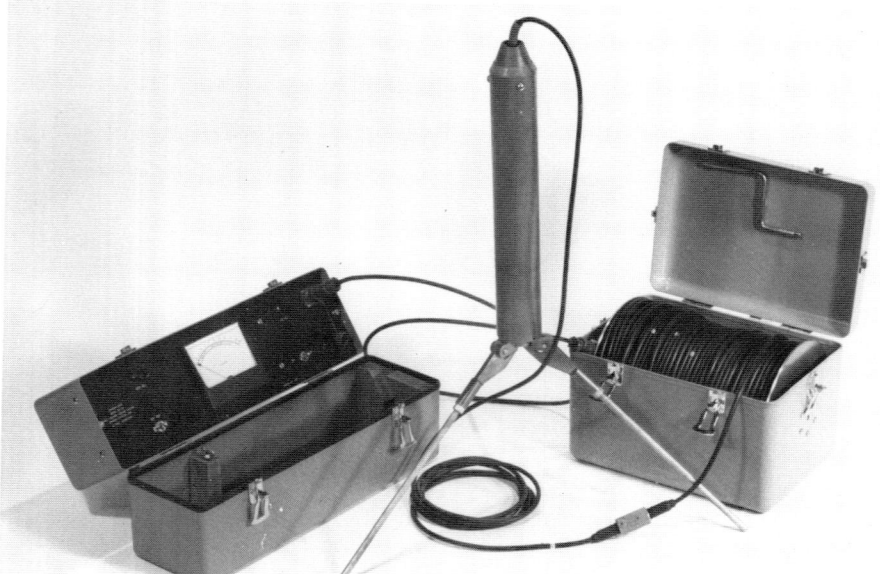

Figure 8.34. Calipers raised from the bottom of the borehole can provide information on the position of washouts, swelling clays, and casing separations. *(Keck Geophysical Instruments, Inc.)*

Fluid Movement Logs

Flow-meter logging is used to determine the production rate from any zone in a well. The flow of water into a well usually varies significantly over the intake area because strata having the highest hydraulic conductivity contribute a high percentage of the total flow. If, for economic reasons, only part of a deep well in semiconsolidated rock is to be screened, the flow-meter logs will indicate where these screens should be placed to intercept the high-flow areas. Flow-meter logging is also useful in locating lost-circulation zones or holes in the casing. Hole caliper logs are generally run in conjunction with a velocity log because the velocity at any point in the borehole is, in part, a function of the cross-sectional area of the borehole at that point.

Impeller-type velocity meters and thermal instruments are used to check both the horizontal and the vertical flow characteristics (Figure 8.35). One type of flow meter measures the movement of fluid by means of a low-inertia impeller that rotates on jewel-type bearings. Each time the shaft revolves, a small magnet on the shaft closes several magnetic switches. The action of these switches indicates the flow direction and the flow rate. Other flow meters measure flow by sensing the thermal changes in a small area caused by the movement of groundwater.

Logging of horizontal flow usually is done by recording continuous readings as the flow meter is lowered into the hole and as it is retrieved. The flow velocity at any point is equal to the average of the two readings recorded for that point. Vertical flow may also occur in the borehole when head differences exist between aquifers. These flow rates may be so low that stationary measurements are not possible because the flow is not sufficient to overcome the friction of the impeller. These rates may be obtained, however, by moving the flow meter up or down the well at a constant rate. The extra revolutions caused by the vertical flow are added or subtracted from the known revolution rate to obtain the actual flow rate in the borehole. Occasionally, where the cuttings suggest a highly permeable zone, the flow meter can be used at those depths to check on relative permeability under ambient head conditions.

Tracers such as salts, tritium, or dyes can also be used to check vertical flow velocities in boreholes. A tracer is a substance that affects the temperature, color, conductivity, or radioactive nature of the water depending on the tracer used, and is injected at a known point in the borehole and then measured at some distance away. The elapsed time between injection and detection provides a flow velocity.

Figure 8.35. Velocity meters can measure the horizontal or vertical velocity of groundwater flow in a borehole.

In summary, borehole geophysical techniques can provide valuable information about both the nature of the aquifer materials and the presence and quality of any water in a formation. A summary of common borehole logging techniques and the conditions affecting their validity is presented in Table 8.4. Newer techniques, such as the application of electromagnetic methods to borehole investigations, have not been addressed above because these methods generally have not been applied to water well exploration.

Conclusions

The principal methods of groundwater investigation used in the water well industry have been described. Most contractors use several of these methods, such as maps, groundwater reports, and various kinds of sampling procedures in their everyday operations. For deep, high-capacity wells, surface or borehole geophysical methods may be necessary for best results. When used and interpreted by an experienced operator, any geophysical method will yield valuable information that will almost always be worth the cost involved. Some of the theory underlying the operation of these methods may seem complicated, but the investigator does not have to have a thorough understanding of the theory to gain useful information. In fact, the greatest asset is experience gained by working in many kinds of geologic environments.

ANALYSIS OF AQUIFERS USING PUMPING TEST DATA

Use of hydraulic methods to verify the capacity of an aquifer to yield water constitutes the final step in the prospecting process. Although some idea of potential yields can be gained by analyzing the physical characteristics of the aquifer materials, actual field transmissivity and storage capabilities will depend on aquifer depth and continuity over a rather large area. Even the installation of many test holes may not

Table 8.4. Methods to Determine Geophysical Conditions in a Borehole

Log Type	Minimum Hole Diameter* in	mm	Borehole Fluid	Casing	Required Corrections
Resistivity and Resistance	2.25	57	Required[1]	Uncased	Drilling fluid resistivity, borehole diameter, and temperature log for quantitative uses.
Spontaneous Pontential	2.25	57	Required[1]	Uncased	Drilling fluid resistivity and borehole diameter for quantitative uses.
Natural Gamma	2.25	57	Not required but allowable	Any type or uncased	None for qualitative uses. Hole diameter, casing thickness, casing composition, casing size, and drilling fluid density for quantitative uses.
Gamma-Gamma	2.25	57	Not required but allowable	Any type or uncased	Same as natural gamma with addition of formation fluid and matrix density corrections.
Acoustic	2.25	57	Required[2]	Uncased	Hole diameter, formation fluid and matrix velocity corrections for quantitative uses.
Neutron	2.25	57	Not required but allowable	Any type or uncased	Same as natural gamma with addition of temperature, fluid salinity, and matrix composition corrections.
Caliper	2.00	51	Not required but allowable	Any type or uncased	None
Temperature	2.00	51	Required[2]	Any type or uncased	None
Fluid Movement	2.25	57	Required[2]	Uncased	Borehole diameter for velocity logging.

*Keys (personal communication) indicates that most of these probes can be obtained to log a 1½-in (38-mm) hole.
[1] Any type, but performance deteriorates in highly conductive or resistive fluids.
[2] Any type.
(Costello, 1980)

be as useful as running a single pumping test in which the well's yield and drawdown can be determined.

Aquifers are tested by several hydraulic methods. The four types of pumping tests most often done are: the bailer test, pump-in test, constant-rate pumping test, and step-drawdown pumping test. Data from the latter two tests are usually far more valuable than data from the bailer and pump-in tests. Because of the complexity of both the constant-rate and the step-drawdown tests, they are covered in detail in both Chapters 9 and 16.

For many years, cable tool drillers used an extremely simple method to determine the potential yield from their wells. Either before or after completing a well, depending on geologic conditions, they would bail the well and record the rate at which they were removing water. The drawdown was also recorded. After bailing at a more-or-less constant rate for some time, the drawdown would become somewhat stabilized. In this way, they could obtain a rough measure of the well's potential yield by determining its specific capacity, that is, the yield for a specified drawdown. In heaving formations, the well must be screened before this procedure is used. This method has been refined by applying a mathematical analysis to the hydraulic condition in the well during the withdrawal of water. This refined method, now known as a slug test, can be used to calculate hydraulic conductivity, transmissivity, and the storage coefficient. A description of the method is given in Chapter 16.

The pump-in test is the opposite of the bailer test; that is, the borehole is completed, casing and screen installed, and the borehole filled with water to a certain depth. The water level is constantly maintained so that the discharge into the formation and the water level in the borehole become equilibrated. Usually this method is applicable to rather shallow boreholes where a plentiful water supply is available. The hydraulic conductivity can then be calculated, but the value obtained relates primarily to the horizontal hydraulic conductivity (Bouwer, 1978). The pump-in test can also be called a slug test when a certain volume of water is added instantaneously to the borehole. This procedure is particularly appropriate for small-diameter piezometers or observation wells.

The constant-rate pumping test is used widely to obtain the specific capacity of a well and the transmissivity and storage values of the aquifer. Before a constant-rate test, the well is usually developed (see Chapter 15). One or more observation wells are installed at appropriate distances from the pumping well to record data during the test, because accurate drawdown data from the pumped well are normally difficult to obtain. During the pumping test, the well is pumped at a constant rate for either 24 or 72 hours, depending on the type of aquifer. During this time, periodic drawdown measurements are taken in the observation wells and pumped well. In most cases, the recovery of the water level should be recorded after the pump has been shut off at the conclusion of the test. These recovery data can be used to check the results from the actual pumping test. Drawdown data are plotted versus the time they were taken or the distance from the pumped well to obtain the transmissivity and storage coefficients. Once these factors are known, the well or aquifer performance under different pumping conditions can be predicted with good accuracy.

An alternate form of the constant-rate pumping test is the step-drawdown test. During this test, the pumping rate is increased in steps at regular intervals. For example, a well may be pumped at 100 gpm (545 m^3/day) for 2 hours, 200 gpm (1,090 m³/day) for the next 2 hours, 300 gpm (1,640 m³/day) for the next 2 hours, and so on for several more steps. As in the constant-rate test, drawdown data are taken in both the observation wells and the pumped well. The transmissivity and storage coefficients are calculated from data obtained during the first step, in this case after 2 hours of pumping at 100 gpm. Unfortunately, the validity of these values may be doubtful because they are based on data taken over such a short time. The real value of a step-drawdown test is that it shows the reduction in specific capacity with increasing yields.

Some types of pumping tests do not have a mathematical basis, but are nevertheless useful in estimating the specific capacity of a well. For wells drilled by the air rotary method, the driller can air-lift pump with the drill rods to estimate the yield. The volume of water air-lifted from the well is difficult to measure, however, so many drillers have developed an ability to visually estimate the volume. Drawdown is estimated from the length of the drill rods in the borehole. If the well has been drilled with a drilling fluid, the fluid should be removed before air-lift pumping is begun. Test pumping in unconsolidated formations is impossible unless the well screen and casing have been installed.

Ordinarily it would not be possible to test pump a well being drilled by the auger method. However, some auger flights are manufactured with an integral continuous-slot screen (Figure 8.36). Thus, the driller can test pump the borehole at any depth as drilling proceeds. Not only can the approximate specific capacity be determined, but also the water quality for any depth. The actual specific capacity of the finished well will be somewhat higher, however, because of development procedures.

Figure 8.36. Integral continuous-slot screens mounted in auger flights permit pumping tests to be conducted at selected depths. Water quality can also be determined as drilling proceeds. *(Keck Geophysical Instruments, Inc.)*

CHAPTER 9
Well Hydraulics

A well is a hydraulic structure which, when properly designed and constructed, permits the economic withdrawal of water from a water-bearing formation. Successful wells are designed and built by engineers and drilling contractors who:

1. Use materials that will provide an efficient well with a long service life.
2. Use techniques in drilling and well construction that take maximum advantage of the hydrogeologic conditions.
3. Apply the principles of hydraulics in a practical way to the analysis of wells and aquifer performance.

Individuals who design or construct wells must understand the fundamentals of well hydraulics. Some aspects of well hydraulics can be complicated, and few engineers have mastered all phases of the subject. In some cases, mathematicians have developed such complex solutions for specific well and aquifer conditions that practical application of the theory is nearly impossible in light of all the geologic and hydrologic uncertainties. Furthermore, certain geologic and aquifer environments are so complex that reliable analytical solutions for the flow patterns are almost unobtainable. Given the complexity of the underground environment, only the most fundamental hydraulic theories can be used successfully in everyday well design and construction. Remarkably, these basic methods regularly yield accurate results in most cases without laborious calculations. This chapter examines these fundamental and practical methods in detail.

The reader is cautioned, however, that extensive literature exists on hydraulics that is not treated in this chapter because of its complexity or narrow application. Good references particulary strong in hydraulic theory include works by Lohman (1979), McWhorter and Sunada (1977), Bennett (1976), Bear (1972, 1979), Deju (1971), and Walton (1970).

In Chapter 5, Darcy's law was introduced, the fundamental equations for groundwater movement were developed, and the storage and drainage characteristics of water in aquifers were examined. Application and extensions of these concepts can now be made to solve some typical problems of flow toward wells. Information in this chapter will enable practicing engineers and drilling contractors to make sound decisions on well design and construction for most situations.

DEFINITION OF TERMS

It is important to understand clearly the meaning of common terms related to pumping wells. Definitions are presented below, and several terms are defined diagramatically in Figure 9.1.

Static Water Level (SWL) — This is the level at which water stands in a well or unconfined aquifer when no water is being removed from the aquifer either by pumping or free flow. It is generally expressed as the distance from the ground surface (or from a measuring point near the ground surface) to the water level in the well. For example, when the static water level in a well is 15 ft (4.6 m), it means that water stands 15 ft below the measuring point when there is no pumping. For an artesian well which flows at the ground surface, the static water level is expressed as a height above the ground surface. When artesian flow is stopped or contained at the ground surface, the pressure developed is referred to as the shut-in head. If the well has a shut-in head of 3 psi (20.7 kPa) at the surface, it means that the confining pressure will cause the water to rise 7 ft (2.1 m) in a pipe extending above the ground surface.

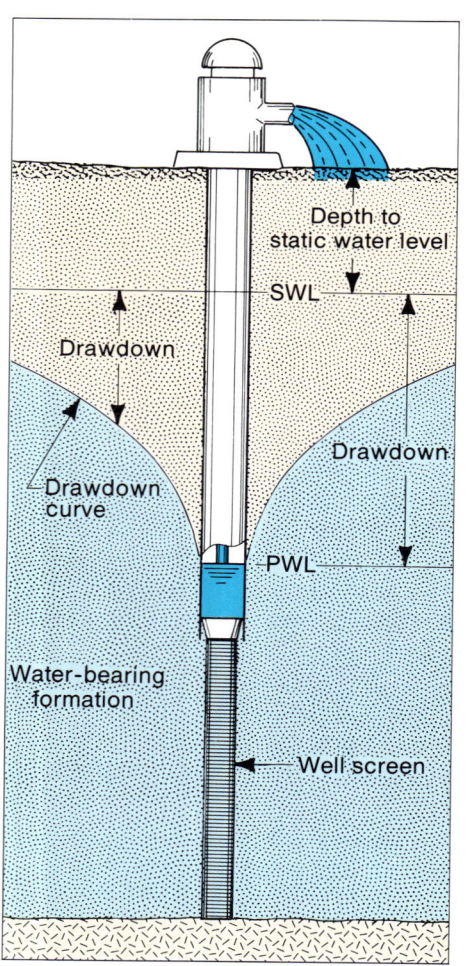

Figure 9.1. Terms relating to well performance.

Pumping Water Level (PWL) — This is the level at which water stands in a well when pumping is in progress. In the case of an artesian well, it is the aboveground level at which water is flowing from the well. The pumping water level is also called the dynamic water level as measured in the well.

Drawdown — Drawdown is the difference, measured in feet or meters, between the water table or potentiometric surface and the pumping water level. This difference represents the head of water (force) that causes water to flow through an aquifer toward a well at the rate that water is being withdrawn from the well. In the unconfined case, the head is represented graphically by the actual water level at a point along the drawdown curve. In confined conditions, the drawdown curve represents the pressure head at that point. To differentiate these two types of drawdown, all diagrams in this book show the water table for unconfined conditions as a solid line and the potentiometric surface for confined conditions as a dashed line.

Residual Drawdown — After pumping is stopped, the water level rises and approaches the static water level observed before pumping began. During water-level recovery, the distance between the water level and the initial static water level is called residual drawdown.

Well Yield — Yield is the volume of water per unit of time discharged from a well, either by pumping or free flow. It is measured commonly as a pumping rate in gallons per minute or cubic meters per day.

Specific Capacity — Specific capacity of a well is its yield per unit of drawdown, usually expressed as gallons of water per minute per foot (gpm/ft) of drawdown or cubic meters per day per meter (m^3/day/m) of drawdown, after a given time has elapsed, usually 24 hours. Dividing the yield of a well by the drawdown, when each is measured at the same time, gives the specific capacity. For instance, if the pumping rate is 1,000 gpm (5,450 m^3/day) and the drawdown is 30 ft (9.1 m), the specific capacity of the well is about 33.3 gpm per ft of drawdown (599 m^3/day/m of drawdown) at the time the measurements were taken. Specific capacity generally varies with duration of pumping — as pumping time increases, specific capacity decreases Also, specific capacity decreases as discharge increases in the same well. The reasons for decreasing specific capacity are discussed later in this chapter.

Static water level, pumping water level, drawdown, and residual drawdown apply similarly to a pumped well or other nearby wells and observation wells. For example, if the water level in an observation well located 80 ft (24.4 m) from a pumping well dropped 3 ft (0.9 m) as a result of the pumping, this lowering in the observation well is called its drawdown.

NATURE OF CONVERGING FLOW

The water level in the vicinity of a pumped well under unconfined conditions is lowered when pumping begins, with the greatest drawdown occurring in the well. As the pump removes water, an area of low pressure develops near the well bore. Because the water level is lower in a pumped well than at any place in the water-bearing formation surrounding it, water moves from the formation into the well to replace water being withdrawn by the pump. The pressure (force) that drives the water toward the well is called the head, which is the difference between the water level inside the well and the water level at any place outside the well. At some distance from the well a point is reached where the water level is essentially unaffected. This distance varies for different wells. It also varies for the same well, depending on both the pumping rate and the length of time the well is pumped.

In confined formations, the saturated thickness of the aquifer is generally not reduced during pumping. Hydrostatic pressure, however, is reduced in the aquifer, and the pressure drop is greatest at the well bore. The pressure drop is directly analogous to the dewatering effect in unconfined aquifers.

During pumping, water flows toward the well from every direction. As the water moves closer to the well, it moves through imaginary cylindrical sections that are successively smaller in area. Thus, as the water approaches the well, its velocity increases. In Figure 9.2, A_1 represents the area of a cylindrical surface 100 ft (30.5 m) from the center of the well, and A_2 represents the area of a similar surface 50 ft (15.2 m) from the well. Because A_1 is twice A_2 and the same quantity of water flows toward the pumped well through both cylinders, the velocity V_2 must be twice V_1.*

---

*The equation for the surface area of a cylinder is $A = 2\pi r h$, where $\pi = 3.14$, r is the radius of the cylinder, and h is its height.

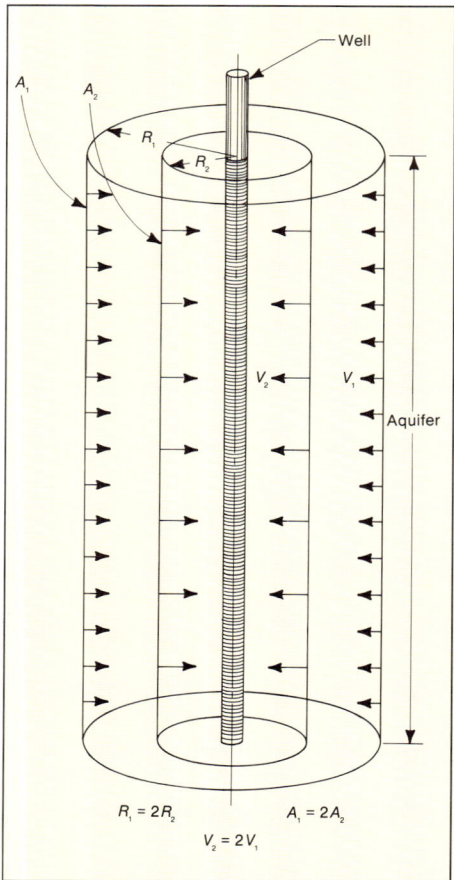

Figure 9.2. As flow converges toward a well, it passes through imaginary cylindrical surfaces that are successively smaller as the well is approached.

Darcy's law indicates that the velocity of flow through porous media varies directly with the hydraulic gradient. As the hydraulic gradient increases, velocity increases as flow converges toward a well. As a result, the lowered water surface develops a continually steeper slope toward the well. The form of this surface resembles a cone and is called the cone of depression. When pumped, all wells are surrounded by a cone of depression. Each cone differs in size and shape depending upon the pumping rate, pumping duration, aquifer characteristics, slope of the water table, and recharge within the cone of depression of the well.

Figure 9.3 shows two cones of depression around pumped wells that illustrate how transmissivity of an aquifer affects the shape of the cone. In a formation with low transmissivity, the cone is deep with steep sides and has a small radius. In a formation with high transmissivity, the cone is shallow with flat sides and has a large radius. The explanation for these different cone shapes is clear, for greater hydraulic head (feet of head) is required to move water through a less permeable formation than through a more permeable formation.

Figure 9.4 shows the levels at which water would be found in observation wells drilled at various distances from a pumped well. Only one side is shown; the other side is similar. This curve is called the drawdown curve and represents the lower limits of the cone of depression. In an unconfined aquifer, it represents the level to which the formation remains saturated. In a confined aquifer, it represents the hydrostatic pressure in the aquifer. Drawdown at any given point is the difference between the water level indicated by the curve and the static water level.

Head loss is a term used to describe the difference in head (pressure) that is required to cause flow from one point to another in an aquifer; it is a measure of the force required to overcome resistance to flow. The head losses from point to point along the pumping-water-level curve in Figure 9.4 are the differences in drawdown between these points.

Suppose, for example, that a well is pumped at a constant rate of 600 gpm (3,270 m³/day). At a distance of 28 ft (8.5 m) from the well, the drawdown is about 5 ft (1.5 m). This indicates that 5 ft of head are required to force 600 gpm of water through the formation from the outer limit of the cone of depression to within 28 ft of the well.

WELL HYDRAULICS

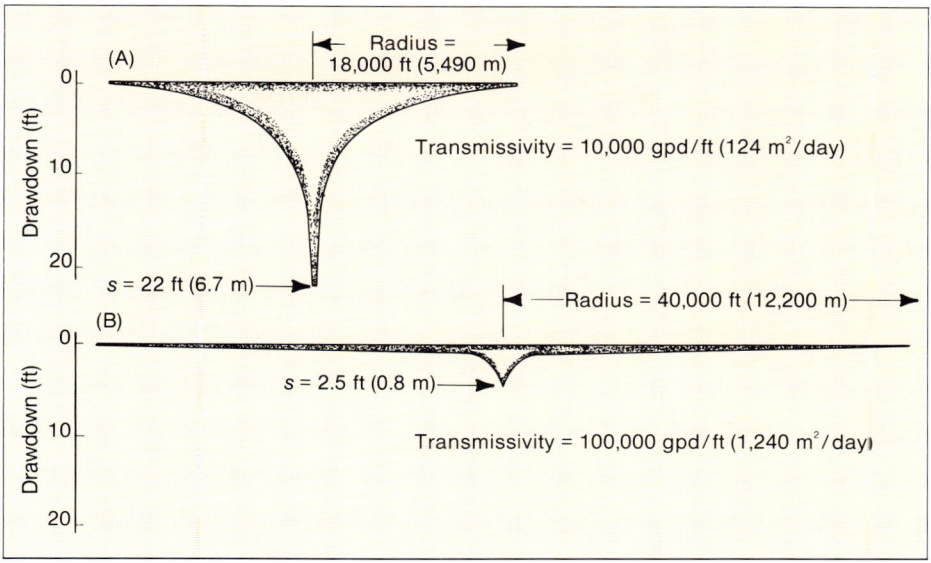

Figure 9.3. Effect of different coefficients of transmissivity on the shape, depth, and extent of the cone of depression. Pumping rate and other factors are constant.

Another 5 ft of head is required to move the same volume of water from 28 ft to within 14 ft (4.3 m) of the well. At this point, the drawdown is about 10 ft (3 m). The remainder of the total drawdown or head loss is used in pushing the water through the last 14 ft of the formation and through the well screen. The total drawdown of 20 ft (6.1 m) in the well is the head in feet required to move 600 gpm through the aquifer (within the cone of depression of the well) and into the well. This example shows that more head is expended for a given horizontal distance as the flow converges toward the well bore. The area through which the water moves decreases steadily while the velocity increases, resulting in increasing head loss along the flow path toward the well.

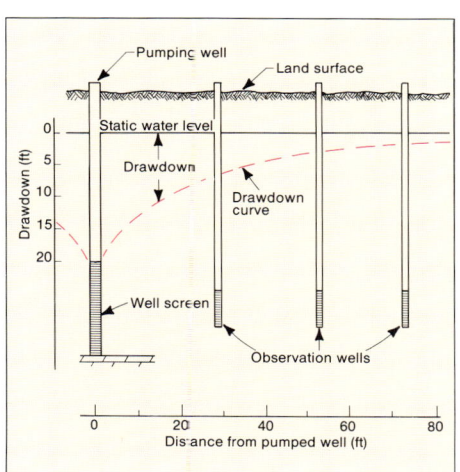

Figure 9.4. Trace of half a cone of depression showing variations in drawdown with distance from a pumped well.

Before proceeding to the equations describing groundwater flow, three important terms must be defined. Each of these terms describes a characteristic of the aquifer that can be determined by pumping.

Radius of influence, R, is the horizontal distance from the center of a well to the limit of the cone of depression. It is larger for cones of depression in confined aquifers than for those in unconfined aquifers. The reason for this difference will become clear later in this chapter.

Coefficient of storage, S, of an aquifer represents the volume of water released from storage, or taken into storage, per unit of aquifer storage area per unit change in head. In unconfined aquifers, S is the same as the specific yield of the aquifer. In confined aquifers, S is the result of compression of the aquifer and expansion of the confined water when the head (pressure) is reduced during pumping. The coefficient of storage is dimensionless. Values of S for unconfined aquifers range from 0.01 to 0.3; values for confined aquifers range from 10^{-5} to 10^{-3}. Figure 5.6 (page 69) demonstrates how different volumes of water are released from storage in unconfined and confined aquifers per unit head loss.

Coefficient of transmissivity, T, of an aquifer, as defined in Chapter 5, is the rate at which water flows through a vertical strip of the aquifer 1 ft or 1 m wide and extending through the full saturated thickness, under a hydraulic gradient of 1 (100 percent). Figure 9.5 illustrates the concepts of hydraulic conductivity and transmissivity. Values of T range from less than 1,000 to more than 1 million gpd/ft (12.4 to over 12,400 m²/day). If an aquifer has a transmissivity of less than 1,000 gpd/ft, it can supply only

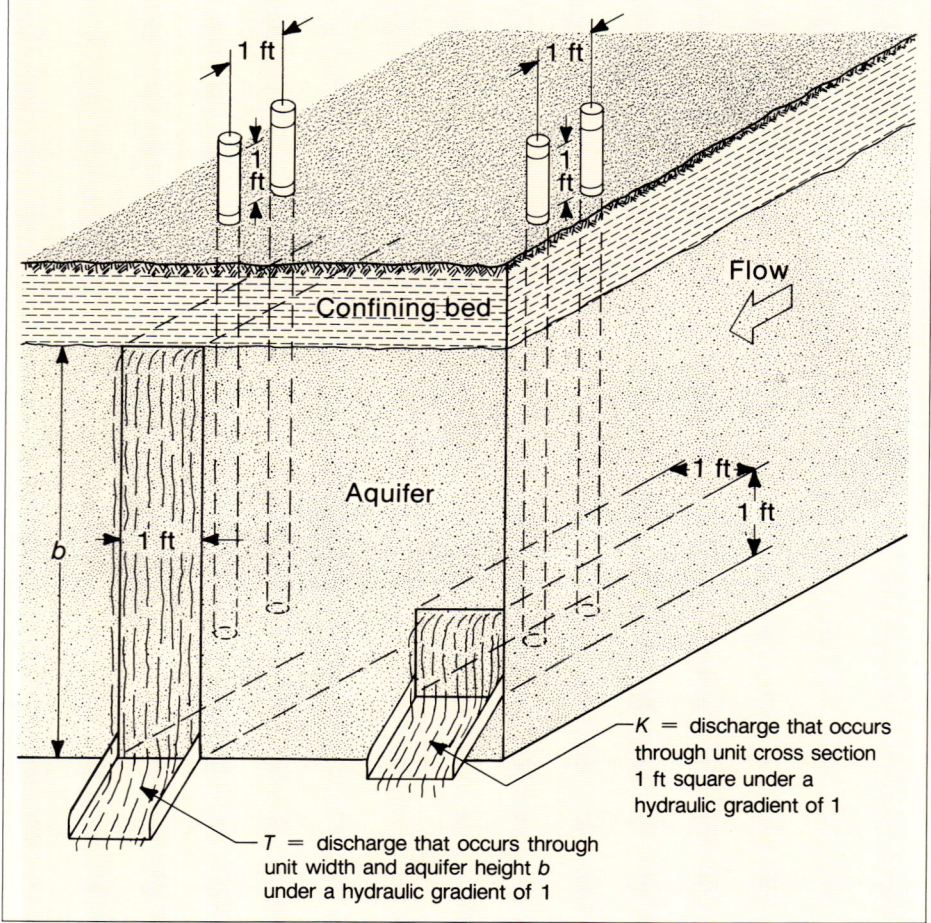

Figure 9.5. Illustration of the coefficients of hydraulic conductivity and transmissivity. Hydraulic conductivity multiplied by the aquifer thickness equals coefficient of transmissivity.

enough water for domestic wells or other low-yield uses. When the transmissivity is 10,000 gpd/ft (124 m²/day) or more, well yields can be adequate for industrial, municipal, or irrigation purposes.

The transmissivity and storage coefficients are especially important because they define the hydraulic characteristics of a water-bearing formation. The coefficient of transmissivity indicates how much water will move through the formation, and the coefficient of storage indicates how much can be removed by pumping or draining. If these two coefficients can be determined for a particular aquifer, predictions of great significance can usually be made. Some of these are:

1. Drawdown in the aquifer at various distances from a pumped well.
2. Drawdown in a well at any time after pumping starts.
3. How multiple wells in a small area will affect one another.
4. Efficiency of the intake portion of the well.
5. Drawdown in the aquifer at various pumping rates.

CONE OF DEPRESSION

When water is pumped from a well, the initial discharge is derived from casing storage and aquifer storage immediately surrounding the well (Figure 9.6). As pumping continues, more water must be derived from aquifer storage at greater distances from the well bore. This means that the cone of depression must expand. The radius of influence of the well increases as the cone expands. Drawdown at any point also increases as the cone deepens to provide the additional head required to move the water from greater distances. The cone expands and deepens more slowly with time, however, because an increasing volume of stored water is available with each additional foot of horizontal expansion.

Figure 9.7 illustrates how the cone of depression expands during equal intervals of time. Assume that after 10 hours of pumping the radius of the cone is 400 ft (122 m) and its depth is 6 ft (1.8 m) at the well bore. At the end of 20 hours, the cone's radius has expanded to 570 ft (174 m) and its depth has increased to 6.3 ft (1.9 m). In the second 10 hours, the cone

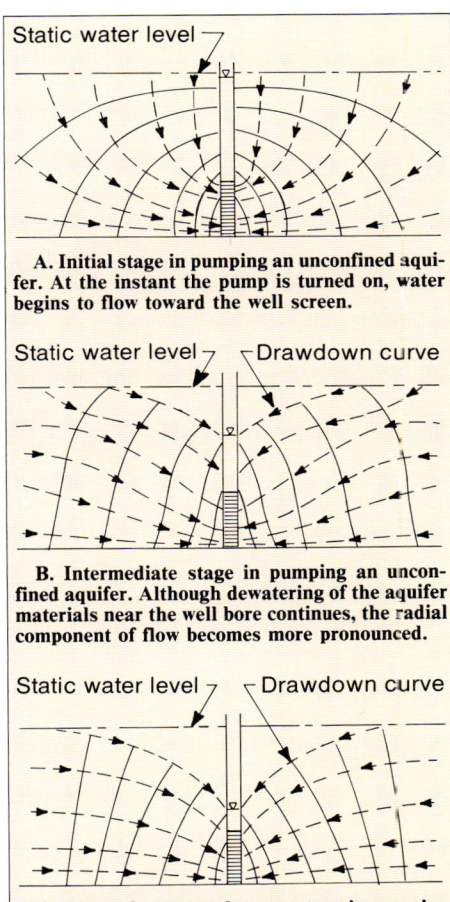

A. Initial stage in pumping an unconfined aquifer. At the instant the pump is turned on, water begins to flow toward the well screen.

B. Intermediate stage in pumping an unconfined aquifer. Although dewatering of the aquifer materials near the well bore continues, the radial component of flow becomes more pronounced.

C. Approximate steady state stage in pumping an unconfined aquifer. Profile of cone of depression is established. Nearly all water originates near the outer edge of the area of influence, and a stable, mainly radial flow pattern is established.

Figure 9.6. Development of flow distribution about a discharging well in an unconfined aquifer that is 33% screened. *(Water and Power Resources Service, 1981)*

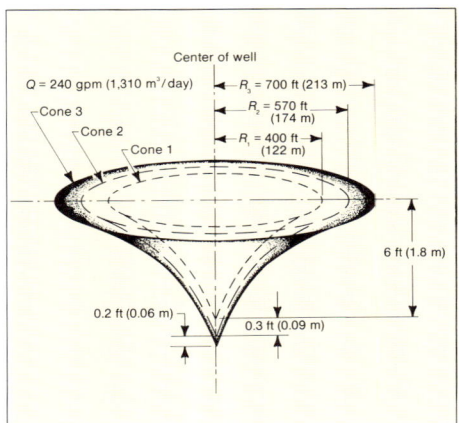

Figure 9.7. Changes in radius and depth of cone of depression after equal intervals of time, at constant pumping rate.

has only extended outward an additional 170 ft (51.8 m) and deepened by an additional 0.3 ft (0.09 m). An additional radial expansion of only 130 ft (39.6 m) and an increase in depth of only 0.2 ft (0.06 m) occurs in the next 10 hours. Calculations of the volume of each of the cones would show that cone 2 has twice the volume of cone 1, and cone 3 has three times the volume of cone 1. This occurs because, at a constant pumping rate, the same volume of water is discharged from the well during each 10-hour interval. Thus, the increase in volume of the cone of depression is constant over time if the well is being pumped at a constant rate and the aquifer is homogeneous.

It is evident from this example that after some hours deepening or expansion of the cone during short intervals of pumping is barely discernible. This often leads observers to conclude that the cone has stabilized and will not expand or deepen as pumping continues. The cone of depression will continue to enlarge, however, until one or more of the following conditions is met:

1. It intercepts enough of the flow in the aquifer to equal the pumping rate.
2. It intercepts a body of surface water from which enough additional water will enter the aquifer to equal the pumping rate when combined with all the flow toward the well.
3. Enough vertical recharge from precipitation occurs within the radius of influence to equal the pumping rate.
4. Sufficient leakage occurs through overlying or underlying formations to equal the pumping rate.

When the cone has stopped expanding because of one or more of the above conditions, equilibrium exists. There is no further drawdown with continued pumping. In some wells, equilibrium occurs within a few hours after pumping begins; in others, it never occurs even though the pumping period may be extended for years.

EQUILIBRIUM WELL EQUATIONS

More than a hundred years ago, engineers began work on adapting Darcy's basic flow equation to groundwater flow toward a pumping well. The objective was to derive simple mathematical expressions for describing the flow regime of water in the ground. Because direct observation of groundwater movement is impossible, mathematical analysis offers a convenient and reliable way to predict what happens to water in the ground.

Well discharge equations for equilibrium conditions were derived by various investigators (Slichter, 1899; Turneaure and Russell, 1901; Thiem, 1906). These equations relating well discharge to drawdown assumed two-dimensional radial flow toward a well (the vertical component of flow is ignored). There are two basic equations; one for unconfined conditions and the other for confined conditions. For both equa-

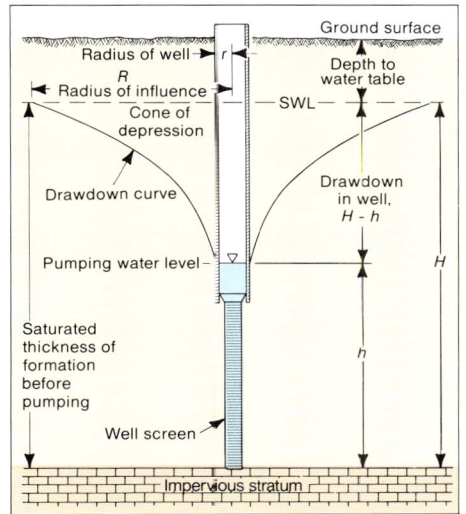

Figure 9.8. Well in an unconfined aquifer showing the meaning of the various terms used in the equilibrium equation.

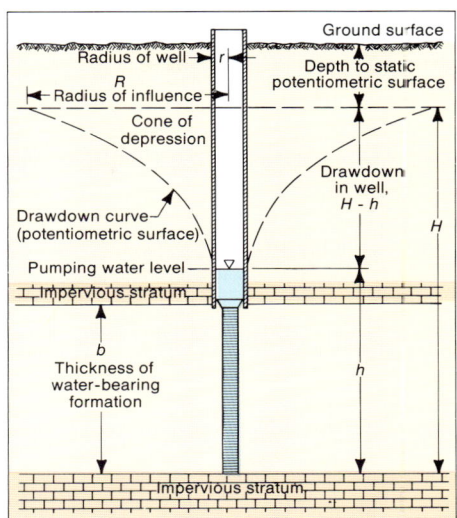

Figure 9.9. Well in a confined aquifer showing the meaning of various terms used in the equilibrium equation.

tions, all dynamic conditions in the well and ground are assumed to be in equilibrium; that is, the discharge is constant, the drawdown and radius of influence have stabilized, and water enters the well in equal volumes from all directions. Both assume horizontal flow everywhere in the aquifer with recharge occurring at the periphery of the cone of depression. Figure 9.8 shows a vertical section of a well constructed in an *unconfined* aquifer. The equation for the well yield of an unconfined aquifer is:

English Engineering Units*   International System of Units*

$$Q = \frac{K(H^2 - h^2)}{1,055 \log R/r} \qquad\qquad Q = \frac{1.366\ K(H^2 - h^2)}{\log R/r} \qquad (9.1)$$

where

Q = well yield or pumping rate, in gpm

K = hydraulic conductivity of the water-bearing formation, in gpd/ft^2

H = static head measured from bottom of aquifer, in ft

h = depth of water in the well while pumping, in ft

where

Q = well yield or pumping rate, in m^3/day

K = hydraulic conductivity of the water-bearing formation, in $m^3/day/m^2$ (m/day)

H = static head measured from bottom of aquifer, in m

h = depth of water in the well while pumping, in m

---

*The United States is slowly converting to the International System of Units from the older English system. For years the scientific community has used the metric system [now System International (SI)], but only recently have other business and technological entities adopted it. In the future, many engineers and drillers will have to present information in the metric system. Therefore, equations in both the SI and English systems of measure are given throughout this text; the SI notation is on the right of each page and the English system is on the left. The complete International System of Units as it applies to water resources is given in Appendix 9.A. A convenient unit conversion table is also provided. Symbols used in the text are given in Appendix 9.B.

R = radius of the cone of depression, in ft
r = radius of the well, in ft

R = radius of the cone of depression, in m
r = radius of the well, in m

Equation 9.1 is often called the equilibrium, or Thiem, equation.

Figure 9.9 is a vertical section of a well pumping from a *confined* aquifer. The equation for a well operating under confined conditions is:

$$Q = \frac{K b (H - h)}{528 \log R/r} \qquad\qquad Q = \frac{2.73 \, K b \, (H - h)}{\log R/r} \qquad (9.2)$$

where
b = thickness of aquifer, in ft
All other terms are as defined for Equation 9.1

where
b = thickness of aquifer, in m
All other terms are as defined for Equation 9.1

Derivations of the foregoing equations are based on the following simplifying assumptions:

1. The water-bearing materials have a uniform hydraulic conductivity within the radius of influence of the well.
2. The aquifer is not stratified.
3. For an unconfined aquifer, the saturated thickness is constant before pumping starts; for a confined aquifer, the aquifer thickness is constant.
4. The pumping well is 100-percent efficient, that is, the drawdown levels inside and just outside the well bore are at the same elevation (see Chapter 16). (Head losses in the vicinity of the well are minimal.)
5. The intake portion of the well penetrates the entire aquifer.
6. The water table or potentiometric surface has no slope.
7. Laminar flow exists throughout the aquifer and within the radius of influence of the well.
8. The cone of depression has reached equilibrium so that both drawdown and radius of influence of the well do not change with continued pumping at a given rate.

These assumptions appear to limit severely the use of the two equations. In reality, however, they do not. For example, uniform hydraulic conductivity is rarely found in a real aquifer, but the average hydraulic conductivity as determined from pumping tests has proved to be reliable for predicting well performance. In confined aquifers where the well is fully penetrating and open to the formation, the assumption of no stratification is not an important limitation.

Assumption of constant thickness is not a serious limitation because variation in aquifer thickness within the cone of depression in most situations is relatively small, especially in sedimentary rocks. Where changes in thickness do occur, as in glacial sediments, for example, they can be taken into account. The assumption that a well is 100-percent efficient can cause the calculated well yield to be seriously in error if the real well is inefficient because of improper design or construction. Factors contributing to inefficiency are discussed in Chapter 16.

The assumption that the water table or potentiometric surface is horizontal before pumping begins is not correct. The slope or hydraulic gradient, however, is usually almost flat and the effect on calculation of well yield is negligible in most cases. Slope of the water table or potentiometric surface does cause distortion of the cone of depression, making it more elliptical than circular.

Flow in all regions of an aquifer is considered to be laminar. Some investigators have theorized that turbulent flow near a well could result in relatively high head losses. Laboratory and field tests show, however, that some departure from laminar flow near a well causes only small additional head losses (Mogg, 1959).

Determining Aquifer Hydraulic Conductivity

Equations 9.1 and 9.2 can be modified to calculate hydraulic conductivity if Q, H, and R are determined from a pumping test, and b is known from the driller's log. For an unconfined aquifer, the equation for calculating K is:

$$K = \frac{1055 \; Q \log r_2/r_1}{(h_2^2 - h_1^2)} \qquad\qquad K = \frac{Q \log r_2/r_1}{1.366 \; (h_2^2 - h_1^2)} \qquad (9.3)$$

where

r_1 = distance to the nearest observation well, in ft

r_2 = distance to the farthest observation well, in ft

h_2 = saturated thickness, in ft, at the farthest observation well

h_1 = saturated thickness, in ft, at the nearest observation well

All other terms are as defined in Equation 9.1

where

r_1 = distance to the nearest observation well, in m

r_2 = distance to the farthest observation well, in m

h_2 = saturated thickness, in m, at the farthest observation well

h_1 = saturated thickness, in m, at the nearest observation well

All other terms are as defined in Equation 9.1

All the parameters on the right-hand side of Equation 9.3 can be determined from a pumping test. Two observation wells, located at distances r_1 and r_2 from the pumped well, are required to determine h_1 and h_2.

Figure 9.10 shows a sectional view of a pumping test layout in an unconfined formation for determining the hydraulic conductivity of the formation. All pertinent factors are easily measured in this kind of test, and the hydraulic conductivity of the aquifer can be determined accurately.

For confined conditions, the equation for determining the hydraulic conductivity from a test installation similar to Figure 9.10 is:

$$K = \frac{528 \; Q \log r_2/r_1}{b \; (h_2 - h_1)} \qquad\qquad K = \frac{Q \log r_2/r_1}{2.73 \; b \; (h_2 - h_1)} \qquad (9.4)$$

where

all terms except the following are the same as for Equation 9.3

b = thickness of the aquifer, in ft

h_2 = head, in ft, at the farthest observation well, measured from the bottom of the aquifer

h_1 = head, in ft, at the nearest obser-

where

all terms except the following are the same as for Equation 9.3

b = thickness of the aquifer, in m

h_2 = head, in m, at the farthest observation well, measured from the bottom of the aquifer

h_1 = head, in m, at the nearest observation well, measured from the bottom of the aquifer

vation well, measured from the bottom of the aquifer

In addition to providing accurate means for calculating the average hydraulic conductivity of an aquifer, the equilibrium equations are useful for studying the relationship of various factors to each other and to well yield. They show, for example, that if all other parameters are equal, well yield is directly proportional to hydraulic conductivity. A formation with twice the hydraulic conductivity of another should provide twice the yield. For a confined aquifer, Equation 9.2 indicates that the yield is directly proportional to the formation thickness when all other parameters are equal.

Relationship of Drawdown to Yield

Equation 9.2 for a well operating under confined conditions shows that yield is directly proportional to drawdown, $H - h$, as long as the drawdown does not exceed the distance from the static potentiometric surface to the top of the aquifer. If the drawdown exceeds this amount, b will then be reduced and the proportionality no longer holds true. Theoretically, this means that if the drawdown is doubled, the yield is doubled. Stated another way, the specific capacity of a well is constant at any pumping rate as long as the aquifer is not dewatered.

For a well in an unconfined aquifer, the part of the formation within the cone of depression is actually dewatered during the pumping. This affects the ratio of drawdown to yield. When the drawdown is doubled, the well yield is less than doubled because the saturated thickness is reduced. The specific capacity decreases with increased drawdown; in fact it decreases directly in proportion to the drawdown.

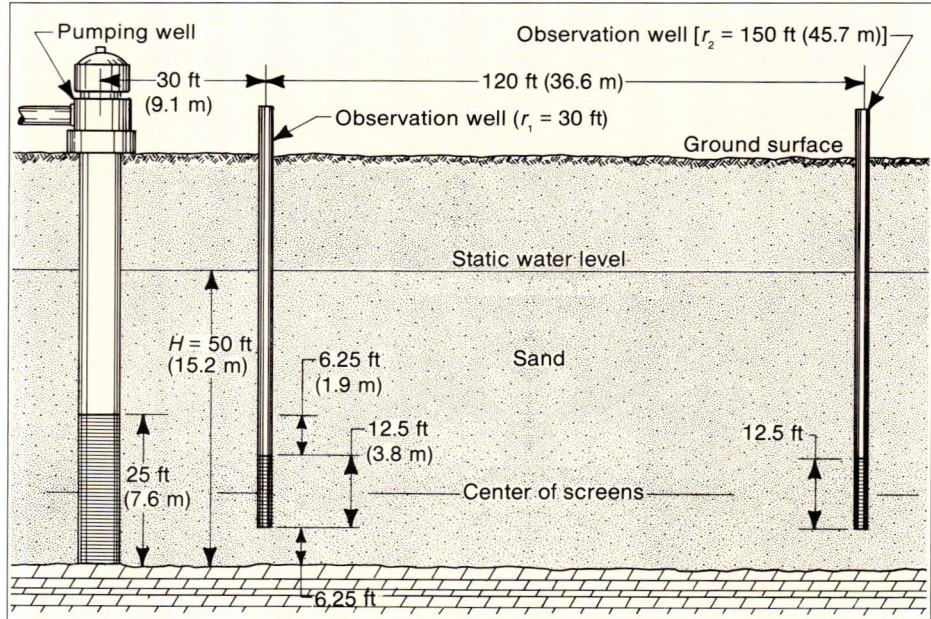

Figure 9.10. Typical arrangement of a pumped well and observation wells for obtaining field data required to calculate hydraulic conductivity from well-discharge equations. Observation wells can be placed farther away from a production well in confined conditions and still provide reliable data.

WELL HYDRAULICS

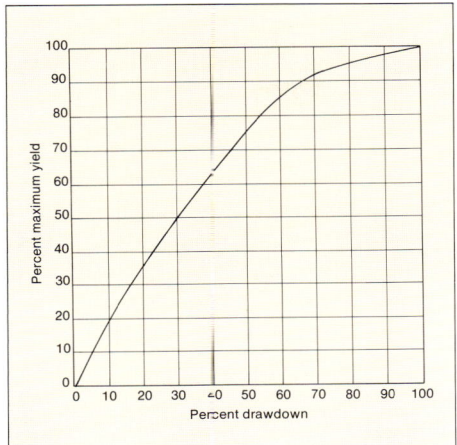

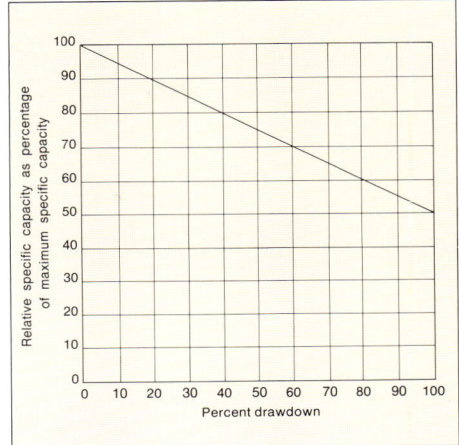

Figure 9.11. Comparison of yield with drawdown in an ideal unconfined aquifer that is fully penetrated and open to the well.

Figure 9.12. Relationship between specific capacity and drawdown in an unconfined aquifer that is fully penetrated and open to the formation.

Figure 9.11 shows the relationship between drawdown and yield for an unconfined aquifer. Maximum drawdown means lowering of the water level to the bottom of the well; 50-percent drawdown means lowering of the water level to a point halfway between the static water level and the bottom of the well. Maximum yield is the quantity a well will produce at maximum or 100-percent drawdown. For example, suppose that a well 40 ft (12.2 m) deep has a static water level of 5 ft (1.5 m) and the saturated thickness of the formation is 35 ft (10.7 m). During a test, the water was pumped at 16 gpm (87 m^3/day) and the pumping level stabilized at 15 ft (4.6 m) below the ground surface, or at a drawdown of 10 ft (3 m). How much will the yield be with 20 ft (6.1 m) of drawdown and the pumping level at 25 ft (7.6 m)?

In this case, 100-percent drawdown is 35 ft. The 10-ft drawdown during the test is thus 29 percent of the total possible drawdown. The curve in Figure 9.11 shows that at 29-percent drawdown, the yield is 50 percent of the obtainable maximum; thus 16 gpm is 50 percent of the maximum yield of the well. A drawdown of 20 ft is 57 percent of the total possible. The curve shows that this drawdown would give 82 percent of the maximum yield. If 16 gpm is 50 percent of the maximum, then 82 percent of the maximum would be 82/50 × 16 = 26 gpm (142 m^3/day). The well can be expected to yield 26 gpm at 20 ft of drawdown.

Figure 9.12 indicates how specific capacity varies with drawdown. Theoretically, maximum specific capacity corresponds to zero drawdown because there is no reduction in the saturated thickness; the minimum occurs when drawdown and yield are at the maximum. Note that the minimum specific capacity is 50 percent of the maximum. In the previous example, 85 percent of the maximum specific capacity would be obtained with 10 ft of drawdown and 71 percent with 20 ft of drawdown.

Figure 9.11 shows why it is uneconomical to operate a well with a drawdown greater than 67 percent of the maximum. At 67 percent of maximum drawdown, 90 percent of the maximum yield is obtained. To obtain the remaining 10 percent requires an additional 33-percent drawdown. Obviously, the extra pumping costs would be out of proportion to the increase in yield.

NONEQUILIBRIUM WELL EQUATION

Theis developed the nonequilibrium well equation in 1935. The Theis equation was the first to take into account the effect of pumping time on well yield. Its derivation was a major advance in groundwater hydraulics. By use of this equation, the drawdown can be predicted at any time after pumping begins. Transmissivity and average hydraulic conductivity can be determined during the early stages of a pumping test rather than after water levels in observation wells have virtually stabilized. Aquifer coefficients can be determined from the time-drawdown measurements in a single observation well rather than from two observation wells as required in Equations 9.3 and 9.4.

Derivation of the Theis equation is based on the following assumptions:

1. The water-bearing formation is uniform in character and the hydraulic conductivity is the same in all directions.
2. The formation is uniform in thickness and infinite in areal extent.
3. The formation receives no recharge from any source.
4. The pumped well penetrates, and receives water from, the full thickness of the water-bearing formation.
5. The water removed from storage is discharged instantaneously when the head is lowered.
6. The pumping well is 100-percent efficient.
7. All water removed from the well comes from aquifer storage.
8. Laminar flow exists throughout the well and aquifer.
9. The water table or potentiometric surface has no slope.

These assumptions are essentially the same as those for the equilibrium equation except that the water levels within the cone of depression need not have stabilized or reached equilibrium.

In its simplest form, the Theis equation is:

$$s = \frac{114.6 \, Q \, W(u)}{T} \qquad\qquad s = \frac{1}{4\pi} \frac{Q}{T} W(u) \qquad (9.5)$$

where

s = drawdown, in ft, at any point in the vicinity of a well discharging at a constant rate

Q = pumping rate, in gpm

T = coefficient of transmissivity of the aquifer, in gpd/ft

$W(u)$ = is read "well function of u" and represents an exponential integral

where

s = drawdown, in m, at any point in the vicinity of a well discharging at a constant rate

Q = pumping rate, in m³/day

T = coefficient of transmissivity of the aquifer, in m²/day

$W(u)$ = is read "well function of u" and represents an exponential integral

In the $W(u)$ function, u is equal to:

$$u = \frac{1.87 r^2 S}{Tt} \qquad\qquad\qquad u = \frac{r^2 S}{4Tt} \qquad (9.5a)$$

where

r = distance, in ft, from the center of a

where

r = distance, in m, from the center of a

pumped well to a point where the drawdown is measured
S = coefficient of storage (dimensionless)
T = coefficient of transmissivity, in gpd/ft
t = time since pumping started, in days

pumped well to a point where the drawdown is measured
S = coefficient of storage (dimensionless)
T = coefficient of transmissivity, in m²/day
t = time since pumping started, in days

The well function of u [$W(u)$] originated as a term to represent the heat distribution in a flat plate with a heating element at its center. Theis recognized that this same concept could be applied to the regular distribution of the groundwater head around a pumping well even though water flows toward the point source rather than away from it. The mathematical principles remain the same.

Analysis of pumping test data* using the Theis equation can yield transmissivity and storage coefficients for all nonequilibrium situations. In actual practice, however, the Theis method is often avoided because it requires curve-matching interpretation and is somewhat laborious. In fact, the work of applying the Theis method can be avoided in most cases. For example, if the pumping test is sufficiently long or the distance from the well to where the drawdown is measured is sufficiently small, the $W(u)$ function can be replaced by a simpler mathematical function which makes the analysis easier. The Theis method is developed at the end of this chapter, but at this point the simplified version is examined because it serves well in most cases.

MODIFIED NONEQUILIBRIUM EQUATION

In working with the Theis equation, Cooper and Jacob (1946) point out that when u is sufficiently small, the nonequilibrium equation can be modified to the following form without significant error:

$$s = \frac{264Q}{T} \log \frac{0.3\, Tt}{r^2 S} \qquad\qquad s = \frac{0.183Q}{T} \log \frac{2.25\, Tt}{r^2 S} \qquad (9.6)$$

where the symbols represent the same terms as in Equation 9.5 and 9.5a.

For values of u less than about 0.05, Equation 9.6 gives essentially the same results as Equation 9.5. The value of u becomes smaller as t increases and r decreases. Thus, Equation 9.6 is valid when t is sufficiently large and r is sufficiently small. Equation 9.6 is similar in form to the Theis equation except that the exponential integral function, $W(u)$, has been replaced by a logarithmic term which is easier to work with in practical applications of well hydraulics.

For a particular situation where the pumping rate is held constant, Q, T, and S are all constants. Equation 9.6 shows, therefore, that the drawdown, s, varies with $\log t/r^2$ when u is less than 0.05. From this relationship, two important relationships can be stated:

1. For a particular aquifer at any specific point (where r is constant), the terms s and t are the only variables in Equation 9.6. Thus, s varies as $\log C_1 t$, where C_1 represents all the constant terms in the equation.

2. For a particular formation and at a given value of t, the terms s and r are the

---

*The performance of newly completed wells is often checked by pumping tests. During the test, the drawdown in the pumping well and observation wells is measured at a constant discharge rate. When properly conducted, these tests yield information on transmissivity and storage capability. See Chapter 16 for a detailed analysis of pumping test procedures.

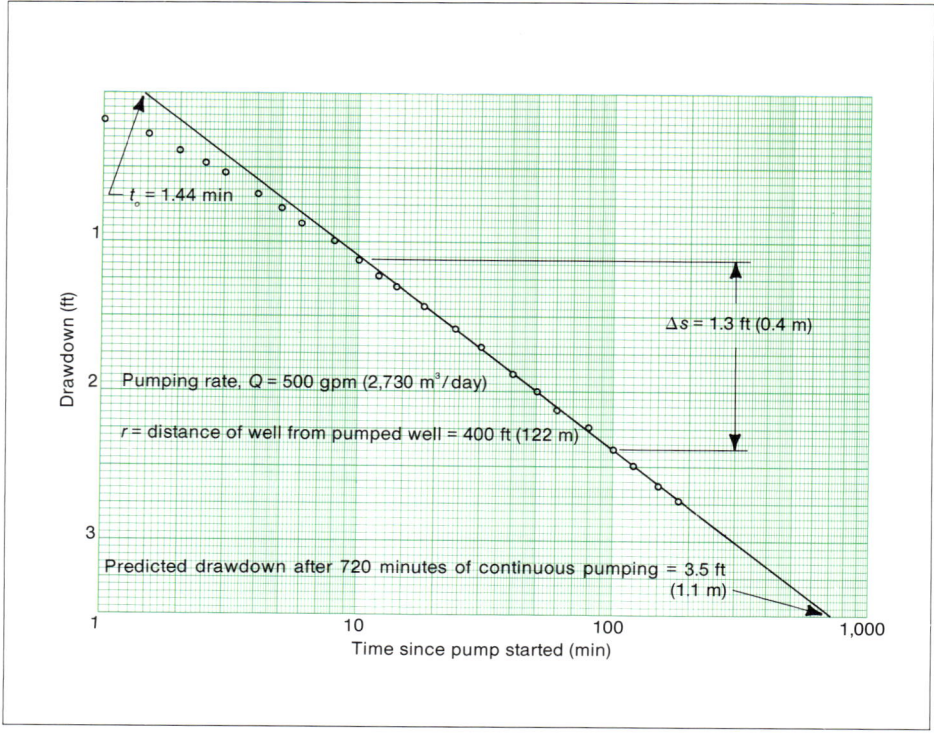

Figure 9.13. When data from Table 9.1 are plotted on semilogarithmic graph paper, most of the plotted points fall on a straight line. The reason for determining Δs and t_0 are explained in the text.

only variables in Equation 9.6. In this case, s varies as log C_2/r^2, where C_2 represents all the constant terms in the equation, including the specific value of t.

By using these simplified relationships based on Equation 9.6, it is possible to derive information on the hydraulic characteristics of the aquifer by plotting drawdown and time data taken during a pumping test. The data are plotted on semilogarithmic paper* as shown in Figure 9.13. Applying the first of the relationships developed above, time, t, is plotted horizontally on the logarithmic scale; drawdown, s, is plotted vertically on the arithmetic scale. Figure 9.13 shows the data from Table 9.1 plotted as a semilog diagram, where most of the points fall on a straight line.

All the points except those representing measurements made during the first 10 minutes of pumping fit the line. During the first 10 minutes, the value of u is larger than 0.05 and so the modified nonequilibrium equation is not applicable within that phase of the test.

Transmissivity

The coefficient of transmissivity is calculated from the pumping rate and the slope of the time-drawdown graph by using the following relationship developed from Equation 9.6:

---

*Semilogarithmic graph paper is constructed so that one scale is arithmetic and the other is based on the logarithm of the number being plotted. Thus, a straight-line relationship can be shown to exist between two variables whose relationship is actually changing in time.

$$T = \frac{264\ Q}{\Delta s} \qquad\qquad T = \frac{2.3}{4\pi}\frac{Q}{\Delta s} = \frac{0.183\ Q}{\Delta s} \qquad (9.7)$$

where

T = coefficient of transmissivity, in gpd/ft

Q = pumping rate, in gpm

Δs = (read "delta s") slope of the time-drawdown graph expressed as the change in drawdown between any two times on the log scale whose ratio is 10 (one log cycle)

where

T = coefficient of transmissivity, in m²/day

Q = pumping rate, in m³/day

Δs = (read "delta s") slope of the time-drawdown graph expressed as the change in drawdown between any two times on the log scale whose ratio is 10 (one log cycle)

In the example, Δs is 1.3 ft (0.4 m), which is the change in drawdown between 10 minutes and 100 minutes after the start of the pumping test, and Q equals 500 gpm (2,730 m³/day); so:

$$T = \frac{264 \cdot 500}{1.3} = 102{,}000\ \text{gpd/ft} \qquad\qquad T = \frac{0.183 \cdot 2{,}730}{0.4} = 1{,}250\ \text{m}^2/\text{day}$$

Table 9.1. Drawdown Measurements in an Observation Well 400 ft (122 m) from Pumped Well

Time since pump started, in min	Drawdown, s ft	m	Time since pump started, in min	Drawdown, s ft	m
1	0.16	0.05	24	1.58	0.48
1.5	0.27	0.08	30	1.70	0.52
2	0.38	0.12	40	1.88	0.57
2.5	0.46	0.14	50	2.00	0.61
3	0.53	0.16	60	2.11	0.64
4	0.67	0.20	80	2.24	0.68
5	0.77	0.23	100	2.38	0.73
6	0.87	0.27	120	2.49	0.76
8	0.99	0.30	150	2.62	0.80
10	1.12	0.34	180	2.72	0.83
12	1.21	0.37	210	2.81	0.86
14	1.30	0.40	240	2.88	0.88
18	1.43	0.44			

Coefficient of Storage

The coefficient of storage is also readily calculated from the time-drawdown graph by using the zero-drawdown intercept of the straight line as one of the terms in the equation. The following equation is derived from Equation 9.6:

$$S = \frac{0.3\ Tt_0}{r^2} \qquad\qquad S = \frac{2.25\ Tt_0}{r^2} \qquad (9.8)$$

where

S = storage coefficient

where

S = storage coefficient

T = coefficient of transmissivity, in gpd/ft

t_0 = intercept of the straight line at zero drawdown, in days

r = distance, in ft, from the pumped well to the observation well where the drawdown measurements were made

T = coefficient of transmissivity, in m²/day

t_0 = intercept of the straight line at zero drawdown, in days

r = distance, in m, from the pumped well to the observation well where the drawdown measurements were made

In our example, t_0 = 1.44 minutes or 0.001 day, T = 102,000 gpd/ft (1,270 m²/day), and r = 400 ft (122 m). Therefore:

$$S = \frac{0.3 \cdot 102{,}000 \cdot 0.001}{(400)^2}$$

$$= 1.9 \times 10^{-4}$$

$$S = \frac{2.25 \cdot 1{,}250 \cdot 0.001}{(122)^2}$$

$$= 1.9 \times 10^{-4}$$

The data used in this example are from an observation well 400 ft from a production well which was pumped at a rate of 500 gpm (2,730 m³/day) for 240 minutes. If measurements had been taken in another observation well still farther from the pumped well, and if the data were plotted on Figure 9.13, the points would fall along a straight line parallel to but above the one shown. If measurements in an observation well closer to the pumped well, or in the pumped well itself, were plotted on Figure 9.13, they would fall on a straight line parallel to but below the one shown*. Values of T and S calculated from all three sets of data would be the same, however.

Predicting Drawdown from the Time-Drawdown Graph

In addition to its use for calculating the aquifer constants, the time-drawdown diagram provides a graphical means of predicting future drawdown. The straight line in Figure 9.13 may be extended to the right to indicate the drawdown that would occur in the observation well [400 ft (122 m) from the pumped well] after any period of continuous pumping at 500 gpm (2,730 m³/day).

In our example, the predicted drawdown after 12 hours (720 minutes) of continuous pumping is the point where the extended line crosses the vertical line representing 720 minutes. The drawdown graph shows that when t = 720 minutes, s = 3.5 ft (1.1 m).

Should the drawdown after 120 hours (7,200 minutes) be desired, the value of Δs (1.3) can be added to the drawdown at 12 hours (720 minutes). Note that 120 hours is ten times 12, or one log cycle of time beyond 12 hours. By definition, Δs represents the increase in drawdown over one log cycle. Thus, after 5 days of continuous pumping at 500 gpm, the drawdown (s) 400 ft from the pumped well would be 3.5 + 1.3 = 4.8 ft (1.5 m).

The usefulness of this simple technique is readily seen. Once the slope of the time-drawdown graph has been established from a short-term pumping test, the straight line can be extrapolated to find the expected drawdown after longer periods of pumping at the same rate. Of course, at some point in the pumping test the hydraulic conditions in the aquifer may change, causing deflection of the straight line. Reasons for these changes are discussed later in this chapter.

---

*Storage coefficients calculated from drawdown data from the pumped well are generally not reliable.

It is good practice to pump a well in a confined aquifer for 24 hours to get data for the time-drawdown curve. A well in an unconfined aquifer should be pumped for 3 days. The 240-minute pumping period in our example is too short for most situations. A longer test provides data that define more reliably the slope and position of the line of best fit for the plotted points.

For a particular well, Equation 9.6 shows that when t is constant, the drawdown, s, is directly proportional to the pumping rate. Figure 9.13 shows that the drawdown in the observation well was 2.5 ft (0.8 m) after 120 minutes when the pumped well was discharging 500 gpm. If the pumping rate were 1,000 gpm (5,450 m^3/day), the drawdown in the observation well after 120 minutes would be twice as much, or 5 ft (1.5 m). Likewise, at $Q = 2,000$ gpm (10,900 m^3/day), s would be 10 ft (3 m).

For a given aquifer, Equation 9.7 shows that the slope, Δs, is directly proportional to the pumping rate. The test shows Δs to be 1.3 ft (0.4 m) at $Q = 500$ gpm. If Q were 1,000 gpm, Δs would be twice as great or 2.6 ft (0.8 m).

Good use can be made of these two relationships. Applying them to test data from one or more observation wells provides a means of predicting accurately the drawdown in the observation wells at rates of pumping different from the test rate. Use of these relationships also provides a way to make similar calculations for the performance of a pumped well if the well is efficient and the saturated thickness of the aquifer is not reduced significantly at potentially higher pumping rates.

HYDROGEOLOGIC CONDITIONS THAT AFFECT TIME-DRAWDOWN GRAPHS

This discussion of the Theis equation and the Jacob modification of it has followed the basic assumptions made earlier in this chapter in developing the equations for groundwater flow. At this point, it is helpful to review some of these assumptions to see the effects of some departures from them upon the time-drawdown relationship. Real aquifers do not conform fully to assumed geologic or hydrologic conditions. Thus, limits for the use of the Jacob equation must be set for those cases in which the differences are significant.

Recharge

The assumption that an aquifer receives no recharge during the pumping period is one of the six fundamental conditions upon which the nonequilibrium formulas are based. Therefore, all water discharged from a well is assumed to be taken from storage within the aquifer. This situation must occur because, as pumping continues, drawdown increases and the cone of depression expands. This fundamental concept makes it possible to calculate the transmissivity from the time-drawdown data, employing Equation 9.7. Assuming no recharge during pumping also permits extension of the time-drawdown curve at its initial slope to predict drawdowns at future times.

It is known, however, that most formations receive recharge. Hydrographs from long-term observation wells monitored by the U.S. Geological Survey, various state agencies, and similar data-gathering agencies in other parts of the world show that most water-bearing formations receive continual or intermittent recharge.

When recharge is intermittent because of seasonal effects, the aquifer may perform without recharge for periods of 1 to 3 months or even longer. Ambroggi (1978) has shown that even though an aquifer has been drawn down continuously for 10 years or more without recharge, water levels can recover completely within 1 year if the area receives normal precipitation. Thus, the time-drawdown curves presented here represent well performance during periods of no recharge. If a pumping test is performed carefully while abundant recharge is reaching the aquifer, the time-drawdown graph from the pumping test will reflect the recharge.

Figure 9.14 shows a time-drawdown graph for a pumped well operating under conditions of no recharge. The well was pumped at a constant rate of 350 gpm (1,910 m³/day) and the drawdown measurements were obtained at various intervals during 360 minutes of pumping. The points plotted on semilogarithmic paper define a straight line with a slope, or Δs value, of 9.3 ft (2.8 m). As seen earlier, the future drawdown in this well for any period of continuous pumping at 350 gpm can be quickly estimated by extending the straight line. The drawdown corresponding to 5,000 minutes of continuous pumping is 73.3 ft (22.3 m) in this case.

This method determines easily the anticipated pumping level and the pump setting required to provide adequate submergence for the pump bowls. Any desired safety factor can be applied to the calculated pumping level to offset changes in well performance resulting from incrustation or from interference effects of other wells that might later be constructed nearby.

Continuous pumping means operating 24 hours a day with no opportunity for recovery of water level. A well that is pumped for only part of each day will not show the same cumulative drawdown after 7, 30, or 90 days as would be predicted by a time-

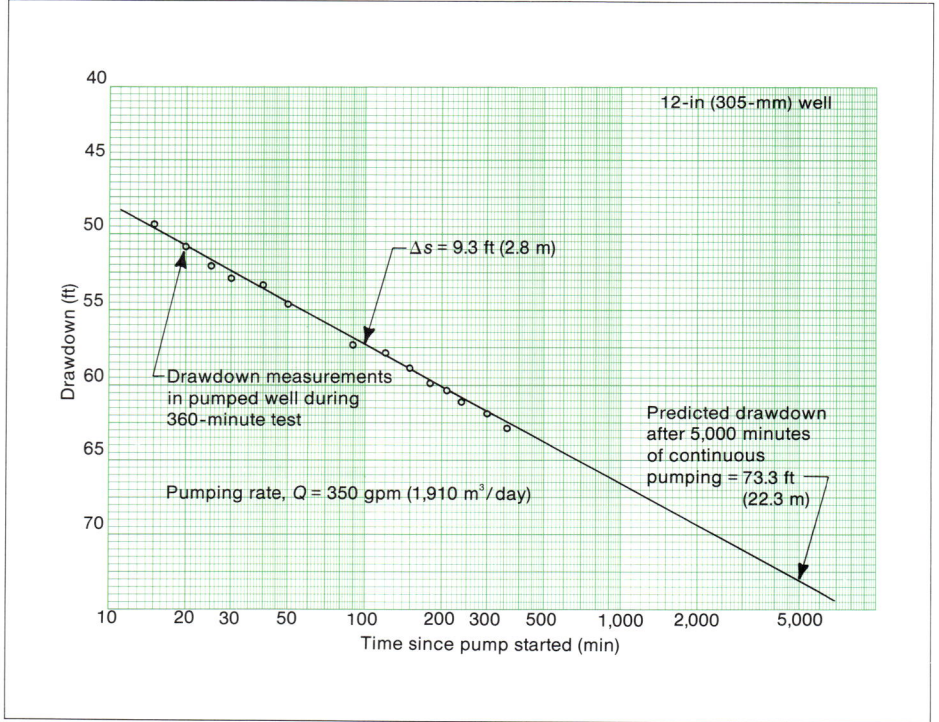

Figure 9.14. Time-drawdown graph for a pumped well (no recharge to aquifer) can be extended to predict drawdown for a period of continuous pumping longer than the test itself.

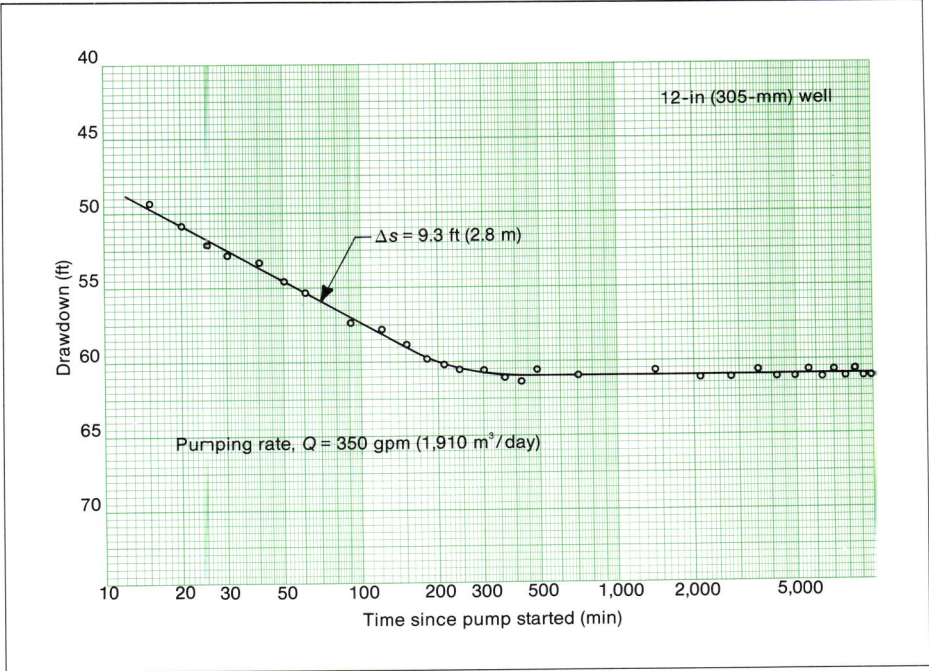

Figure 9.15. When recharge to the aquifer occurs within the zone of influence of the well, the slope of the time-drawdown curve becomes flatter. The horizontal leg indicates that recharge equals well discharge after 240 minutes of pumping.

drawdown graph such as Figure 9.14. Obviously, a well that is pumped on a cycle of 12 hours on and 12 hours off benefits from the recovery of water level during the 12-hour idle period.

If insufficient recharge takes place while the pump is off, the water level will not fully recover to the original static level. Each time pumping is resumed, drawdown starts from a new level which is slightly below the level at the start of the previous pumping period. A method used to estimate the cumulative drawdown under intermittent pumping conditions is discussed later in this chapter.

Drawdown will stabilize when recharge within the zone of influence of the pumping well equals the rate of discharge of the well. No further lowering of the water levels will occur as pumping continues at a constant rate. The time-drawdown curve then becomes horizontal, as in Figure 9.15.

The first part of the curve in Figure 9.15 shows that the cone of depression was enlarging during the first 240 minutes of pumping. After 240 minutes, the cone of depression, or area of influence of the well, encountered a source of recharge. In the second part of the curve, the rate of recharge within the area of influence was sufficient to equal the rate of pumping, resulting in stabilized water levels throughout the area of influence. Recharge generally occurs over a period of time; it does not occur instantaneously. Recharge may be from a lake or river located in only a part of the cone of depression. After the recharge boundary is encountered, the drawdown will increase slowly in areas away from the recharge source until equilibrium is established.

It sometimes happens that the recharge rate within the cone of depression is lower

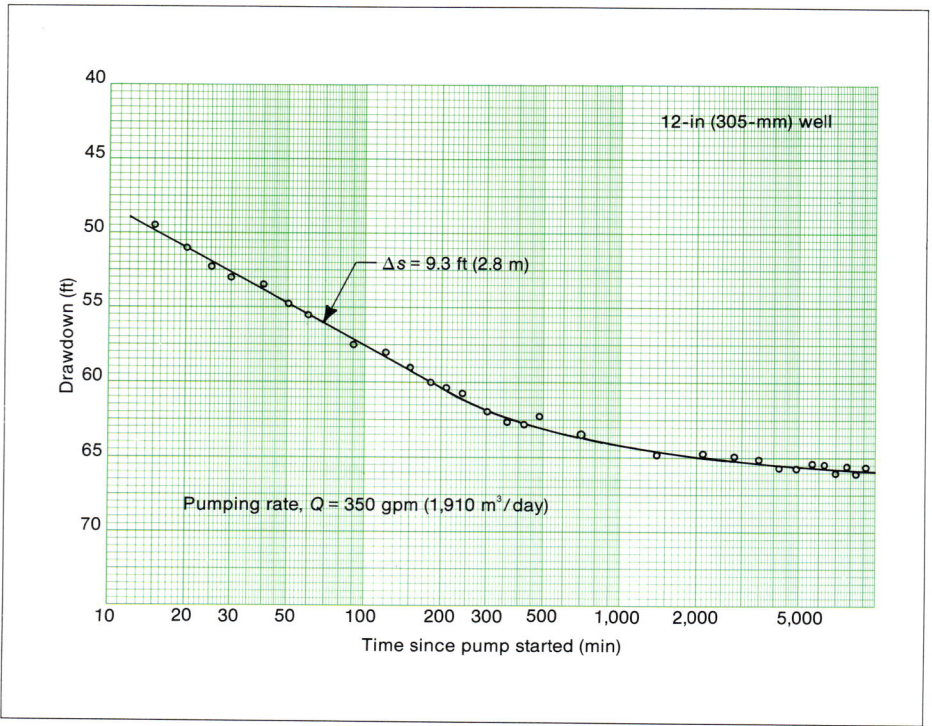

Figure 9.16. Recharge rate is somewhat less than the pump discharge; thus, the second part of the drawdown curve does not become horizontal.

than the pumping rate of the well. While this changes the slope of the time-drawdown curve, the second part may not become horizontal. Thus, the slope becomes flatter than the initial slope and indicates that the cone of depression is enlarging more slowly than during the first part of the pumping period (Figure 9.16). Future drawdown in a well may be predicted by extending the straight line of the second leg of the curve and reading the drawdown indicated for any future time.

When the slope of the drawdown curve changes after a period of continuous pumping, only graphical methods can be used to predict future drawdown. If the slope does not change, the future drawdown can be computed from the Theis equation or found by the graphical method. If the slope changes, however, the Theis equation cannot be applied to any time after the change from the initial slope.

It should be pointed out that calculation of the transmissivity, T, of a water-bearing formation must be made from Δs based on the first part of the time-drawdown curve. A numerical value that may represent the slope of the second part of the graph is of no significance in analyzing the pumping test data; the new or second Δs value cannot be used because physical characteristics of the aquifer are somewhat different farther from the well. For example, the aquifer may thicken in one or more directions away from the well, causing reduction in the slope of the time-drawdown curve. The hydraulic conductivity of the aquifer has not changed, only the volume of water available for withdrawal. Thus, the second Δs does not represent the true hydraulic conductivity of the primary aquifer sediments. It represents a geologic anomaly within the radius of influence. Similarly, if a zone of higher hydraulic conductivity is encountered within the cone of depression, the time-drawdown curve will flatten somewhat. Considerable variation in aquifer thickness and hydraulic conductivity violates two of the Theis assumptions. The time-drawdown curve will reflect this departure from idealized conditions.

Recharge from a River

Equilibrium conditions that stabilize the cone of depression around a pumping well may develop in several general situations. One of these is when an aquifer is recharged from a river or lake. Figure 9.17 illustrates this situation after equilibrium has been reached.

During the early part of the pumping period, the cone of depression does not extend to the river and no recharge is evident. The pumping level in the well goes down as pumping continues, which is shown by the first part of the drawdown curve in Figure 9.15. When the cone of depression intersects a river channel, a hydraulic gradient develops between the groundwater in the aquifer and the water in the river. If the streambed is hydraulically connected with the aquifer, river water will percolate downward through the pervious streambed under the influence of the hydraulic gradient. Thus, the river recharges the aquifer at an increasing rate as the cone of depression enlarges. When the rate of recharge to the aquifer equals the rate of discharge from the well, the cone of depression and the pumping level become stable. This condition corresponds to the horizontal part of the drawdown curve in Figure 9.15 and the situation shown in Figure 9.17.

Extension of the drawdown curve in Figure 9.15 shows that the predicted drawdown after 5,000 minutes of continuous pumping would be 61.2 ft (18.7 m). Recharge to the aquifer in this case reduces the drawdown 12.1 ft (3.7 m) from the values shown in Figure 9.14 after the same period of continuous pumping.

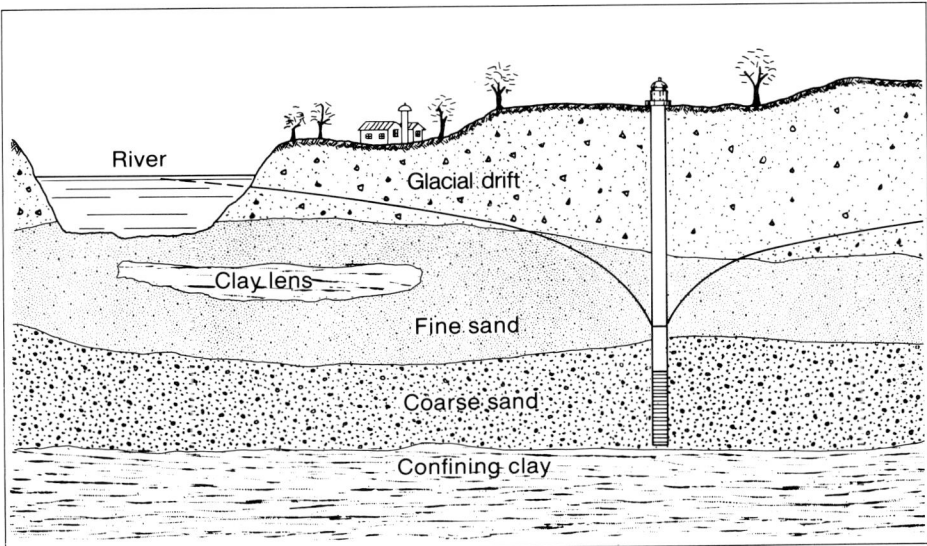

Figure 9.17. Cone of depression expanding beneath a riverbed creates a hydraulic gradient between the aquifer and river. This can result in induced recharge to the aquifer from the river.

Recharge from Vertical Infiltration

Another situation leading to equilibrium conditions is when vertical recharge occurs throughout the area of influence around a pumping well. An example is a well completed in an unconfined aquifer where all the material in the vadose zone from the ground surface to the water table is permeable sand. Assume that infiltration is from rain falling within the cone of depression of the well. When the quantity of water percolating to the water table within that circle equals the discharge from the well, the cone of depression stops spreading and the pumping levels within the well and the aquifer become stable.

The more common type of recharge producing equilibrium conditions is vertical leakage from saturated layers above an aquifer tapped by a production well. The upper strata of saturated material may have a considerably lower hydraulic conductivity than the deeper material in which the well terminates. The difference in hydraulic conductivity between the upper and lower strata may be so great that the upper material is not considered as part of the aquifer. When the area of the cone of depression covers tens of thousands of square feet, the total vertical seepage from the upper material, even though the material is of low hydraulic conductivity, can equal the well discharge and thus bring about equilibrium in the pumping levels. This situation may also develop in lenticular formations in which only the lower part of the formation is screened.

Effect of Recharge on Drawdown in Observation Wells

Discussion of the recharge effect on the shape of the time-drawdown curve has been confined to the graph for a pumped well (Figure 9.15). It should be understood that measurements in an observation well would produce a similar time-drawdown curve, except that the time at which the change in slope occurs would differ according to the relative positions of the pumped well, the observation well, and the recharge source.

For an observation well between the pumped well and the river in Figure 9.17, the flattening of the drawdown curve would occur sometime before 240 minutes. If the observation well were on the opposite side of the pumped well, away from the river, the change in slope would occur sometime after 240 minutes of pumping.

Data from observation wells are usually more reliable and accurate than data from pumped wells, so the time-drawdown plots from observation wells are most often relied upon to reveal the performance of an aquifer. Data from observation wells are not affected by minor changes in well discharge caused by variations in pump speed, or by uncertain measurements of the true water level because of turbulence in the well bore. When transmissivity is calculated, the value of Δs corresponding to the first part of the time-drawdown curve must be used. Theoretically, the drawdown curves of the pumped well and observation well should produce the same value of Δs.

It should be evident from this discussion that accurate and frequent measurements of drawdown at the start of a constant-rate pumping test are extremely important. Sufficient dependable data must be obtained during the early part of the test to reflect clearly both the well and aquifer performances before recharge effects or other outside influences, which invalidate the basic nonequilibrium theory, begin to distort the data.

Sloping Water Table or Potentiometric Surface

Groundwater flow toward and past a well produces another situation that can bring about virtual stabilization of the pumping level. As explained in Chapter 5, groundwater moves because hydraulic gradients develop between areas of recharge and areas of discharge. The gradient is represented by the slope of the water table or potentiometric surface. Most natural slopes of these surfaces are relatively flat and do not affect appreciably the well performance curves. A relatively steep slope, however, causes distortion of the cone of depression around the well. Rather than being circular, the area of influence of the well becomes elliptical. More of the pumped water comes from the up-gradient direction than from all directions equally.

Slow Drainage

In unconfined conditions, slow drainage within the cone of depression can affect the validity of early time-drawdown data. Figure 9.18 shows the effect on the drawdown curve. Typically, the effect of slow drainage lasts only a matter of hours before the slope of the drawdown line reflects true aquifer characteristics.

The causes of this phenomenon are the great difference between the horizontal and vertical hydraulic conductivity in some sediments, and the severely limited hydraulic conductivity of sediments overlying many aquifers. In glacial drift, for example, layers of rather coarse sand or gravel commonly lie between thin layers of silt or even clay.

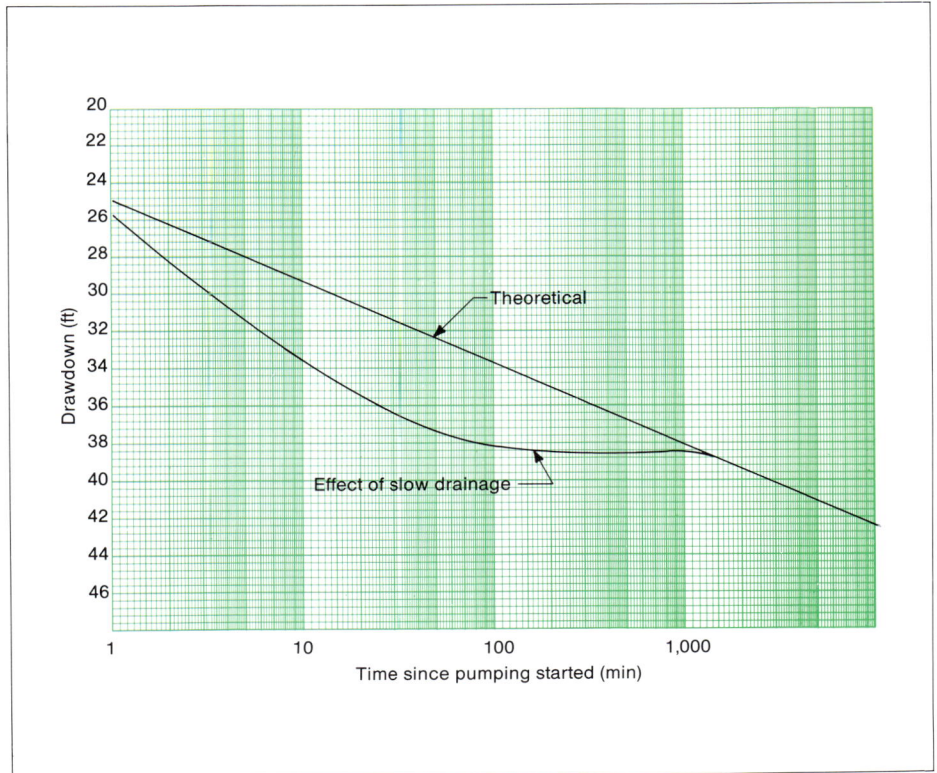

Figure 9.18. Time-drawdown curve showing the effect of slow drainage on the early part of the curve.

Water can flow freely in a horizontal direction, but vertical flow is greatly retarded. When pumping begins, the amount of vertical water movement toward the screen is relatively small, but as time passes and the cone of depression widens, a larger percentage of the water moves toward the well vertically, thereby reducing the slope of the time-drawdown curve.

It is important to recognize the effects of slow drainage because the curve might otherwise suggest that a recharge boundary has been encountered after a few minutes of pumping. Actually, the true transmissivity can be calculated only after several hours of pumping. Earlier transmissivity values may seem appropriate for a particular aquifer, but the storage values will be much lower than expected. Often the storage coefficients are similar to those encountered in a confined aquifer. Abnormally low values are clear indications of slow drainage. Thus, early data in slow drainage situations do not indicate true aquifer conditions.

Vertical Leakage

Like slow drainage, vertical leakage can distort the time-drawdown curve, but at a later time. Vertical leakage occurs when an aquifer is confined geologically by two aquitards. Commonly these aquitards are limestone layers that have numerous vertical joints or faults. Under ordinary conditions, vertical leakage into the confined aquifer is minimal. During pumping, however, reduction in head within the cone of depression may cause leakage from both above and below the aquifer.

Detailed analyses of leakage effects on aquifers have been developed by Jacob (1946a), Walton (1960), and others. Figure 9.19 shows distortion of the Jacob curve by vertical leakage. In a practical sense, it is difficult to account generally for leakage ef-

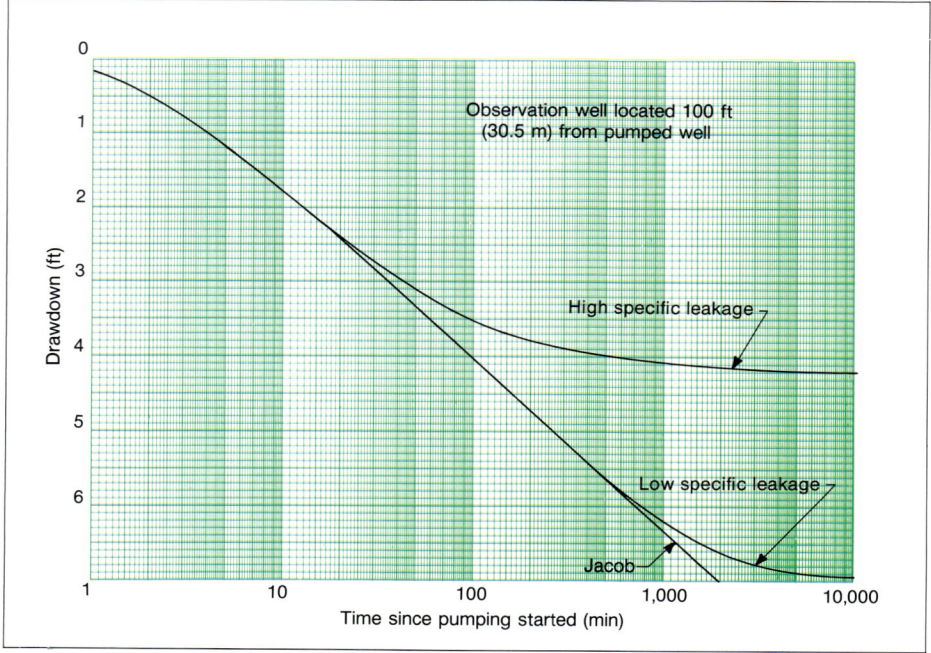

Figure 9.19. Drawdown graphs showing both low and high specific leakage.

fects mathematically because they are often masked by other physical irregularities of the aquifer.

Impervious Boundaries

Relatively few aquifers conform to the basic assumption of infinite extent in all directions from the pumped well. In many localities, definite geologic and hydraulic boundaries limit aquifers to areas ranging from a few to many square miles. This is especially true in glaciated regions.

The effect of an impervious boundary on the time-drawdown graph is opposite to the effect of aquifer recharge. The boundary causes the slope of the drawdown plot to steepen instead of flatten. This is easily understood if we consider how water stored within the aquifer is supplied to the well.

Under the basic assumptions mentioned above, water flows equally from all directions toward a well. When the expanding cone of depression encounters an impervious boundary on one side of a pumped well, it can expand no farther in that direction and no additional water can be supplied from that locality. The cone must expand and deepen more rapidly in all other directions to maintain the yield of the well. The effect on the semilog time-drawdown curve is to steepen its slope, as shown in Figure 9.20.

Care must be used in analyzing aquifer test data when either nearby boundaries or recharge sources are suspected. Aquifer coefficients must be calculated from the early test data if boundary effects are apparent. In no case should calculations be made from the slope of any part of the time-drawdown plot that reflects a boundary. The latest established slope of the time-drawdown plot should be extended to predict water levels after longer periods of continuous pumping.

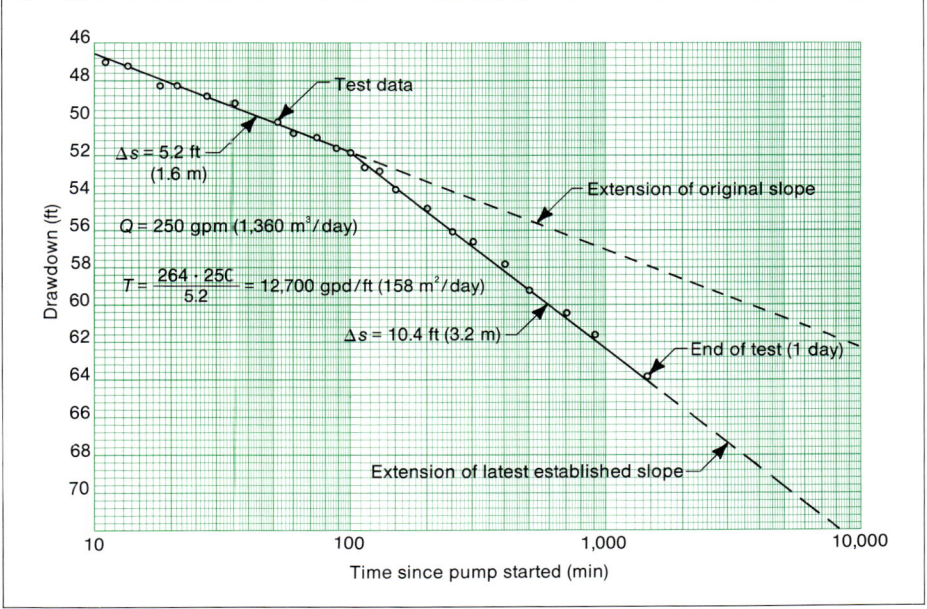

Figure 9.20. Steepening of the time-drawdown, semilog curve indicates a limited aquifer. The expanding cone of depression encountered one or more impervious boundaries at the time shown by the change in slope.

232 GROUNDWATER AND WELLS

It is also pertinent to note the reason for the earlier suggestion that pumping tests last at least 1 day for wells in confined aquifers and at least 3 days for wells in unconfined aquifers. Figure 9.20 shows evidence of a boundary effect 100 minutes after the pump was started. If the test had been conducted for only 100 minutes, the boundary would not have been revealed. Moreover, extension of the initial slope indicates a drawdown of 62 ft (18.9 m) after 7 days (10,000 minutes) of pumping at 250 gpm (1,360 m³/day), whereas the correct estimate is 73 ft (22.3 m) as determined from the extension of the second leg of the curve.

In a confined aquifer, enlargement of the cone of depression during a 24-hour pumping period is usually extensive enough to encounter boundaries that could appreciably affect drawdown predictions from the semilog diagram. In an unconfined aquifer, the cone of depression expands more slowly, and so a longer pumping period is required to detect boundaries that may exist.

Casing Storage

Schafer (1978) suggests that in many instances early pumping test data may not fit Jacob's modification of the nonequilibrium theory, and that calculations based on this early Δs value will be erroneous. These early data reflect the removal of water stored in the casing. When pumping begins, water in the casing is removed first. As the water level in the casing falls, water begins to enter the well from the surrounding formation. Gradually, a greater percentage of the well's yield will be from the aquifer. The Δs value will be higher during the time required to exhaust the casing storage, giving an erroneously low transmissivity value in the early stages of the pumping test. Figure 9.21 shows data from a typical pumping test in which casing storage has

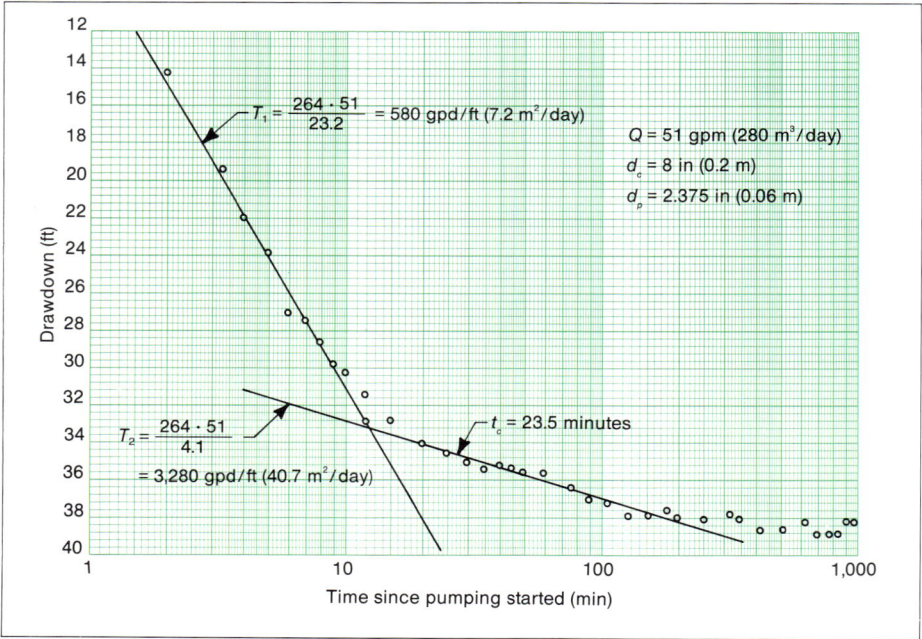

Figure 9.21. Pumping test data in which casing storage has altered the early part of the time-drawdown plot.

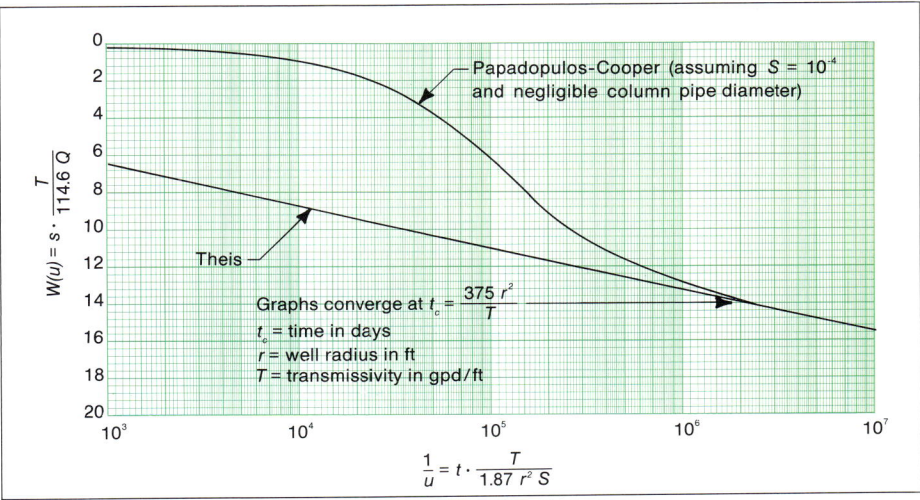

Figure 9.22. Graphic representation of the Papadopulos-Cooper equation which takes into account casing storage.

distorted the early part of the time-drawdown curve.

Before the effect of casing storage on pumping test data was recognized, an interpreter might have mistaken the flattened or second part of the drawdown curve as an indication of aquifer recharge. The duration of the casing storage effect varies greatly from well to well depending on the casing diameter and specific capacity. In general, the storage effect will last longer for wells with large diameters and low specific capacities.

Papadopulos and Cooper (1967) and Ramey et al. (1973) present equations that modify the early part of the Jacob and Theis curves by taking into account casing storage (Papadopulos-Cooper equation for determining t_c is shown in Figure 9.22). These equations indicate the critical time after which casing storage no longer contributes to the yield of a well. Presumably, drawdown data collected after this time will represent the true physical conditions within an aquifer. Unfortunately, these equations can be used only if the transmissivity and well efficiency are known in advance.

Schafer suggests that the critical time can be calculated by the equation:

$$t_c = \frac{0.5\,(d_c^2 - d_p^2)}{Q/s} \qquad\qquad t_c = \frac{0.017\,(d_c^2 - d_p^2)}{Q/s} \qquad (9.9)$$

where

- t_c = time, in minutes, when casing storage effect becomes negligible
- d_c = inside diameter of well casing, in inches
- d_p = outside diameter of pump column pipe, in inches
- Q/s = specific capacity of the well in gpm/ft of drawdown at time t_c

where

- t_c = time, in minutes, when casing storage effect becomes negligible
- d_c = inside diameter of well casing, in mm
- d_p = outside diameter of pump column pipe, in mm
- Q/s = specific capacity of the well in m³/day/m of drawdown at time t_c

Determination of the true transmissivity value depends on being able to identify whether a casing storage effect has occurred or a recharge boundary has been encountered early in the pumping test. As an example, let us assume that it is unknown whether the pumping test data presented in Figure 9.21 represent boundary conditions or casing storage. If an iteration using Equation 9.9 is done, it is possible to delineate whether the change in the slope of the curve is caused by a casing storage effect or actually represents a recharge boundary. If the change in slope is caused by a casing storage effect, the time of the change can be predicted by Equation 9.9.

Equation 9.9 requires that the drawdown at time t_c be known; thus, there appear to be two unknowns, s and t_c. Initially, however, any drawdown value s_1 can be chosen and a trial t_c value can be calculated. Using the trial value of t_c and the time-drawdown graph, a new drawdown value, s_2 is obtained which can be used in Equation 9.9 to calculate a second trial value of t_c. This procedure is repeated two or three times until the calculated value for t_c does not change. The use of this equation does not require knowledge of either the transmissivity or well efficiency.

To use Equation 9.9, an initial drawdown value is selected and a t_c value is calculated. For illustration, assume the initial s is 20 ft; the estimated t_c is then calculated:

$$t_c = \frac{0.6 \ (8^2 - 2.375^2)}{51/20} = 13.7 \text{ minutes}$$

At 13.7 minutes, the drawdown is 33 ft (read from time-drawdown graph). Another calculation is made using 33 ft instead of 20 ft, resulting in a t_c value of approximately 22.7 minutes. The drawdown at 22.7 minutes is 34.2 ft. Substituting this value in Equation 9.9 leads to a t_c value of 23.5 minutes. Note that the change in the t_c value between the last two iterations has been only 0.8 minutes, a small value. Thus, the three iterations using Equation 9.9 suggest that the casing storage effect would have become negligible at approximately 23 minutes. Thus, the initial slope provides an erroneous T value and any predictions of the well's performance should be based on the T value calculated on the basis of the latter part of the curve. If the change in slope were caused by a boundary effect, the time of the change in slope could not be predicted by applying Equation 9.9 and the initial T value must be used to predict well performance.

Analyses of pumping tests in which casing storage is a factor indicate that T_1 and T_2 can be related by the equation:

$$T_2 = \frac{4 \ T_1}{E} \qquad (9.10)*$$

where E is the efficiency of the well, T_2 is the transmissivity reflecting the true aquifer characteristics, and T_1 is the apparent transmissivity calculated from the portion of the graph affected by casing storage. This equation can be used to check calculated transmissivity values and well efficiencies derived from pumping tests, especially when data from the pumped well are the only data available. The numerical value of 4 on the right side of the equation is based on the value of the exponent of the storage coefficient, that is, $S = 10^{-4}$. It will change as the exponent varies.

Careful collection of early time-drawdown and recovery values can enhance the data

---

*This equation can be used with either SI or English terms.

base used to evaluate wells and aquifers. The effect, however, of casing storage on the early measurements cannot be ignored and must be incorporated into the overall data analysis. Estimation of t_c by Equation 9.9 aids in the interpretation by determining which data are influenced by casing storage and are therefore not subject to conventional analysis. Equation 9.10 then provides a useful check on obtained values of transmissivity and efficiency.

CALCULATING DRAWDOWN FOR INTERMITTENT PUMPING SITUATIONS

For intermittent pumping situations, direct calculation of the cumulative drawdown is quite cumbersome after a large number of pumping cycles. If boundary conditions or recharge are encountered, calculation is impossible. Graphical methods as discussed above are not applicable in cases of intermittent pumping. However, an abbreviated method can be used to estimate the drawdown resulting from pump cycling.

To obtain this estimate, the cycling well is assumed to be replaced by two imaginary wells which pump continuously without cycling. One well is assumed to pump continuously throughout the entire operating period at a rate that would produce a volume equal to the volume produced by the cycled well. The other imaginary well is assumed to pump for one pumping cycle only and at a rate equal to the *difference* between the actual pumping rate and that of the first imaginary well. The drawdowns in each imaginary well are determined and added together to give a reliable estimate of the drawdown in the real well.

As an illustration, assume that a 1,000-gpm (5,450-m^3/day) well pumps 75 percent of the time, 18 hours on and 6 hours off each day for 90 days. The drawdown at the end of the 18-hour pumping cycle on the 90th day can be estimated by thinking of the single well as two imaginary wells, one pumping at three-fourths of the actual discharge for the entire duration of pumping and the other pumping at one-fourth the actual discharge for just the last cycle. In other words, one imaginary well pumps 750 gpm (4,090 m^3/day) for 89 days, 18 hours (89.75 days), and the other imaginary well pumps 250 gpm (1,360 m^3/day) for 18 hours (0.75 days). The drawdowns from the imaginary wells are then added together to give the desired estimate. Although this procedure is not mathematically precise, it is intuitively acceptable and produces drawdown values within 1 or 2 percent of those obtained mathematically.

DISTANCE-DRAWDOWN GRAPHS

Applying the second relationship outlined on page 219, a semilog distance-drawdown graph can be constructed. Simultaneous drawdown measurements in at least three observation wells, each at a different distance from the pumped well, are needed to construct an accurate distance-drawdown graph. If these values are plotted on ordinary graph paper, as in Figure 9.23, the familiar trace of the cone of depression near a pumped well is obtained. This is similar to the curves in Figures 9.4, 9.8, 9.9, and 9.17.

If the drawdowns in the same three wells are plotted on a semilog diagram, however, the drawdown curve becomes a straight line, as in Figure 9.24. Points representing drawdowns in other observation wells farther away from the pumped well would fall a little below the straight line in Figure 9.24, because at some distance from the pumped well u is larger than 0.05. If u is greater than 0.05, the straightline relationship of s to log r no longer holds.

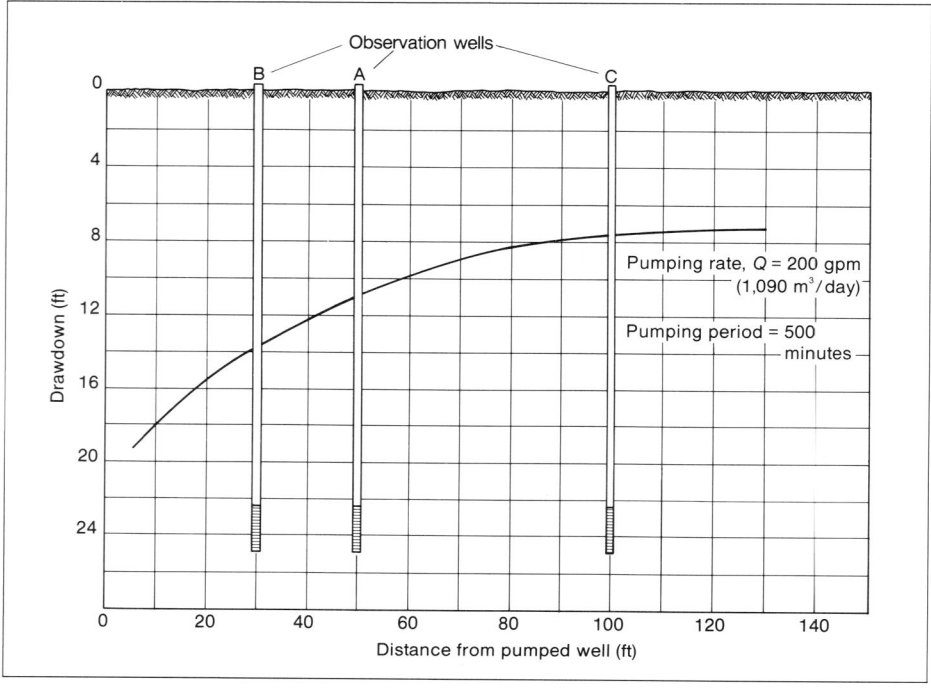

Figure 9.23. Plotting drawdowns for three observation wells defines part of the cone of depression.

The semilog plot of the cone of depression (distance-drawdown diagram) simplifies application of the distance-drawdown relationship. The straight line may be extended to the right to determine the effect of pumping at any distance from the pumped well. For example, Figure 9.24 shows that when the test well was pumped at 200 gpm (1,090 m³/day) for 500 minutes, a drawdown of 2.6 ft (0.8 m) would have been produced in another well 300 ft (91.5 m) away.

Transmissivity

A simple transformation of Equation 9.6 permits calculation of transmissivity from the distance-drawdown diagram. The slope of the straight line is used in a manner similar to the procedure already explained for use of the time-drawdown diagram. The equation for this method is:

$$T = \frac{528\, Q}{\Delta s} \qquad\qquad T = \frac{0.366\, Q}{\Delta s} \qquad (9.11)$$

where
T = coefficient of transmissivity, in gpd/ft
Q = pumping rate, in gpm
Δs = slope of the distance drawdown graph expressed as the change in drawdown, in ft, between any two

where
T = coefficient of transmissivity, in m²/day
Q = pumping rate, in m³/day
Δs = slope of the distance drawdown graph expressed as the change in drawdown, in m, between any two

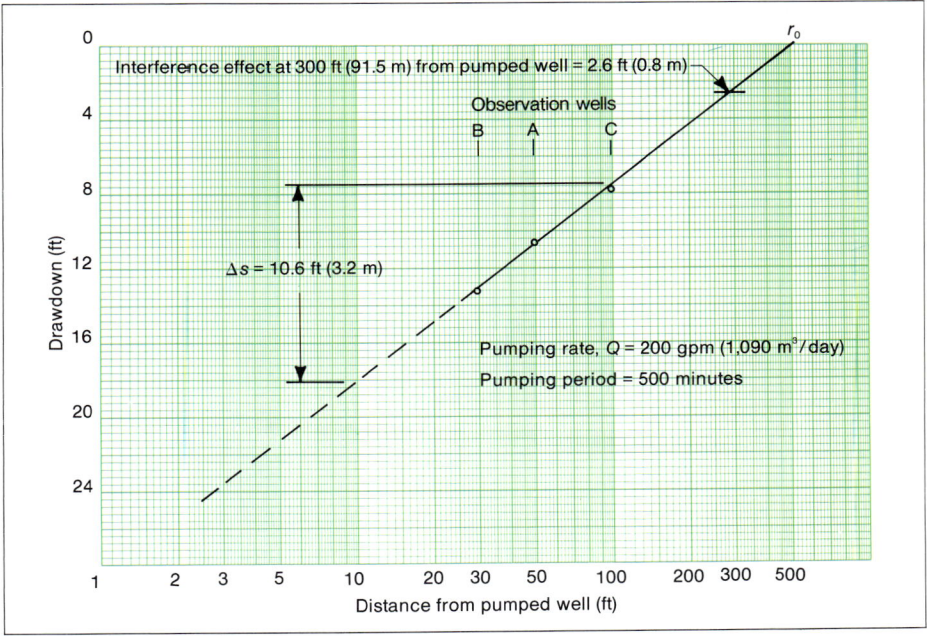

Figure 9.24. Trace of the cone of depression plotted on semilogarithmic graph paper becomes a straight line. Drawdown in each observation well was measured 500 minutes after start of the pumping test.

values of distance on the log scale whose ratio is 10

values of distance on the log scale whose ratio is 10

For the example shown in Figure 9.24:

$$T = \frac{528 \cdot 200}{10.6}$$

$$= 9{,}960 \text{ gpd/ft}$$

$$T = \frac{0.366 \cdot 1{,}090}{3.2}$$

$$= 125 \text{ m}^2/\text{day}$$

Coefficient of Storage

The coefficient of storage can be obtained from the distance-drawdown diagram by using the following equation derived from Equation 9.6:

$$S = \frac{0.3\,Tt}{r_0^2} \qquad\qquad S = \frac{2.25\,Tt}{r_0^2} \qquad (9.12)$$

where
S = coefficient of storage
T = coefficient of transmissivity, in gpd/ft
t = time since pumping started, in days
r_0 = intercept of extended straight line at zero drawdown, in ft

where
S = coefficient of storage
T = coefficient of transmissivity, in m²/day
t = time since pumping started, in days
r_0 = intercept of extended straight line at zero drawdown, in m

From Figure 9.24, the value of r_0 is 500 ft (152 m), T is 9,960 gpd/ft (124 m²/day), and t is 500 minutes (0.347 days). Therefore:

$$S = \frac{0.3 \cdot 9,960 \cdot 0.347}{(500)^2}$$

$$= 4.1 \times 10^{-3}$$

$$S = \frac{2.25 \cdot 124 \cdot 0.347}{(152)^2}$$

$$= 4.2 \times 10^{-3}$$

It is seen, then, that aquifer storage coefficients may be calculated from the following two relationships, revealed by an aquifer test:

1. Rate of lowering of the water level at any place within the cone of depression on the time-drawdown diagram.
2. Shape and position of the cone of depression at any given time on the distance-drawdown diagram.

These calculations are independent of each other, so the result from one may be used to check the other.

Other Uses of Distance-Drawdown Graphs

Note that the numerical constant in Equation 9.11 is twice that in Equation 9.7. This occurs because t appears to the first power in Equation 9.6, whereas r appears to the second power. Since log r^2 is the same as 2 log r, it follows that the value of Δs for the distance-drawdown graph is twice the Δs for the time-drawdown graph. For a given aquifer and a given pumping rate, this ratio for the slopes of the two straight lines is a fixed relationship. Therefore, when Δs is determined from a time-drawdown graph, the slope of the curve on the distance-drawdown graph should be twice as great if the

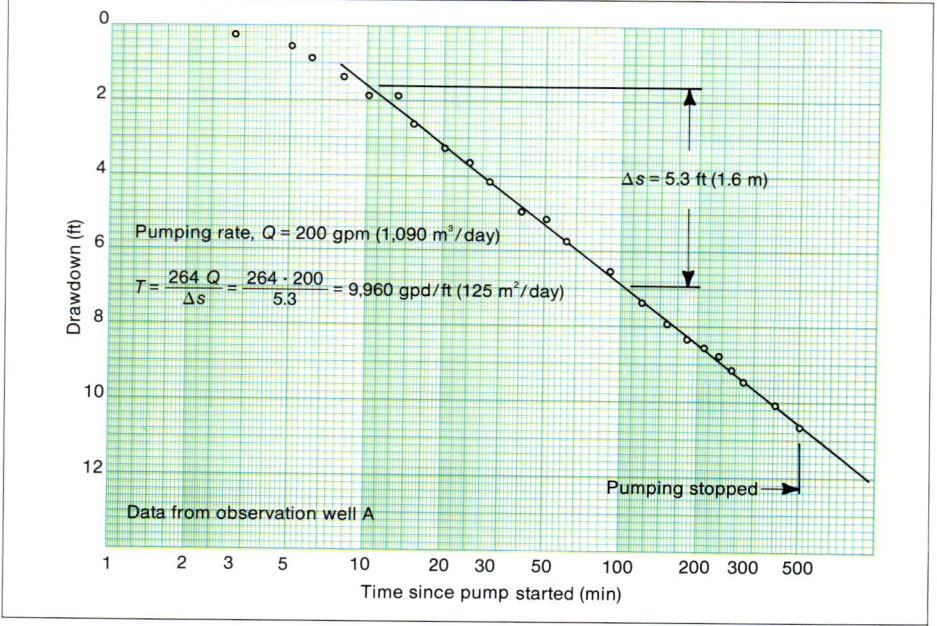

Figure 9.25. Time-drawdown curve for observation well A which is 50 ft (15.2 m) from a pumping well. Aquifer test was conducted at a constant pumping rate of 200 gpm (1,090 m³/day).

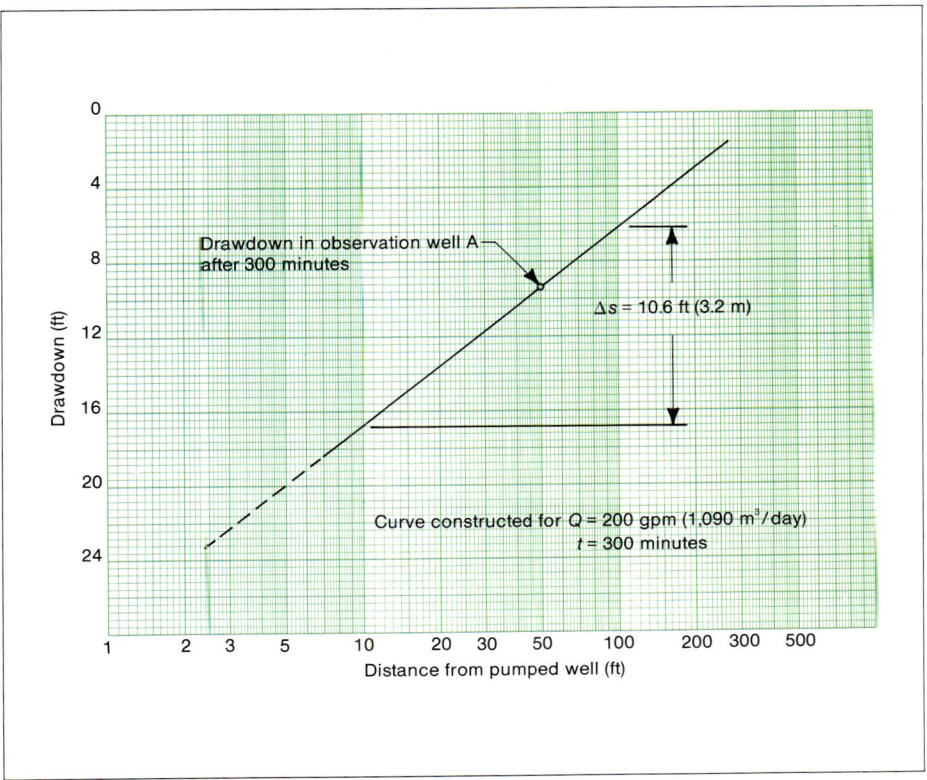

Figure 9.26. Distance-drawdown plot constructed from time-drawdown data, Figure 9.25, represents the position of the cone of depression after 300 minutes of pumping at 200 gpm (1,090 m³/day).

well is pumped at the same rate. This permits construction of the distance-drawdown graph from a time-drawdown graph derived from measurements in a single observation well. However, when drawdown is observed in only one well (for example, the production well), independent calculations of aquifer performance are not possible.

The following example explains how distance-drawdown graphs can be constructed from time-drawdown data. Figure 9.25 shows a semilog time-drawdown graph of data from observation well A, which is 50 ft (15.2 m) from the pumped well as shown in the two previous illustrations. The Δs from the time-drawdown plot is 5.3 ft (1.6 m), exactly one-half the value of Δs from the distance-drawdown curve in Figure 9.24.

If a distance-drawdown plot is constructed after 300 minutes of pumping, the drawdown in well A at 300 minutes is 9.4 ft (2.9 m) as shown in Figure 9.25. This measurement is plotted at the 50-ft distance on a new diagram, as shown in Figure 9.26. A straight line with $\Delta s = 2 \times 5.3 = 10.6$ ft (3.2 m) is then drawn through this point. Thus, a distance-drawdown plot is derived which represents the cone of depression after 300 minutes of pumping at 200 gpm.

In many cases, the drawdown must be calculated at pumping rates different from the one used in the pumping test. These data are needed to calculate the interference effect of one well upon another at various pumping rates. For example, it may be necessary to calculate the interference effect in a well 300 ft (91.5 m) away from a permanent well pumping 400 gpm (2,180 m³/day) that will be constructed at the site of the test well.

Equation 9.11 shows that Δs varies directly with the pumping rate, Q. For an assumed pumping rate of 400 gpm, therefore, the Δs of the distance-drawdown plot would be 2 × 10.6 = 21.2 ft (6.5 m).

From the data shown in Figures 9.25 and 9.26, a series of straight-line graphs can be developed, each one representing the relationship between distance and drawdown at a different pumping rate. To do this, two relationships are applied from the equations already discussed. The first of these relations is that the drawdown at any location outside the pumped well, such as at observation well A, varies directly with the pumping rate. The second is that Δs, the slope of the distance-drawdown graph, also varies directly with the pumping rate.

To illustrate this procedure, distance-drawdown graphs are constructed for pumping rates of 100 gpm (545 m³/day), 250 gpm (1,360 m³/day), and 400 gpm (2,180 m³/day) at 300 minutes. The first step is to determine one definite point through which each of the desired curves must pass. This is done by multiplying the measured drawdown in observation well A, taken from either Figure 9.25 or Figure 9.26, by the ratio of the pumping rates. The drawdown at observation well A, for a pumping rate of 100 gpm, would be one-half of the drawdown as measured at the pumping test rate of 200 gpm. The measured drawdown at 200 gpm was 9.4 ft (2.9 m), so the calculated drawdown for 100 gpm is 4.7 ft (1.4 m) for the same period of pumping. This value plotted on a semilogarithmic graph (Figure 9.27) is a definite point representing conditions when pumping 100 gpm, 50 ft (15.2 m) from the pumped well.

The next step is to calculate the slope of the distance-drawdown plot for 100 gpm.

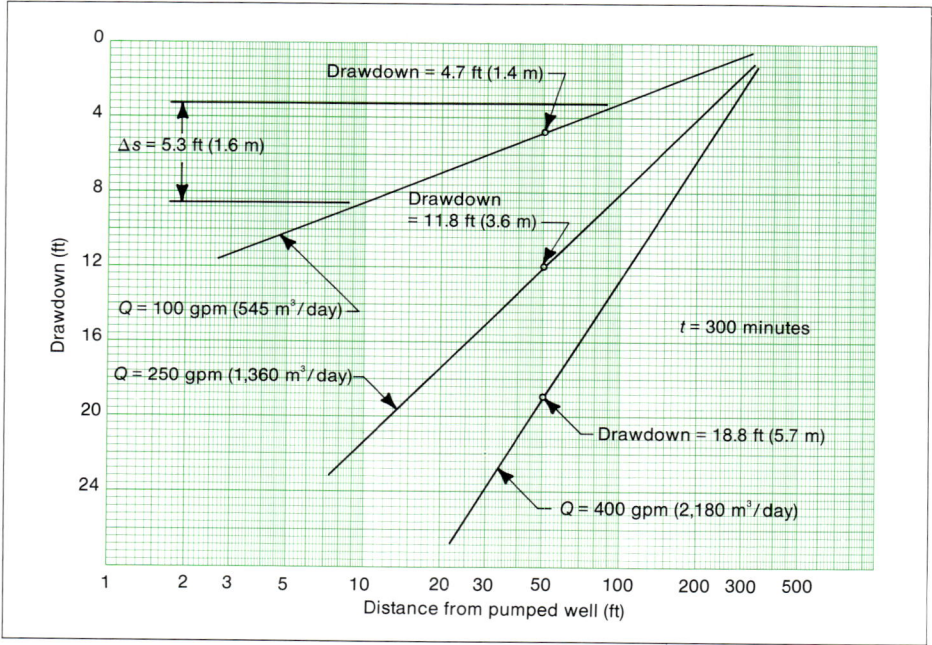

Figure 9.27. Distance-drawdown diagram for various rates of pumping, using data from Figure 9.25. Each plot represents conditions after 300 minutes of pumping at the rate indicated.

This is done by multiplying the slope shown in Figure 9.26 by the ratios of the pumping rates. The value of Δs is 10.6 ft (3.2 m) when the pumping rate is 200 gpm. The value of Δs for a rate of 100 gpm is one-half as much, or 5.3 ft (1.6 m).

With the calculated drawdown of 4.7 ft at 50 ft from the pumped well and the calculated value of Δs, the distance-drawdown curve can now be drawn for a pumping rate of 100 gpm. This is done by constructing a straight line with a slope of 5.3 ft per log cycle through the point representing 4.7 ft of drawdown at 50 ft from the pumped well, as shown by the upper curve in Figure 9.27. Data for plotting the other curves in Figure 9.27 are given in Table 9.2. These were calculated in the manner described above.

Table 9.2. Data for Distance-Drawdown Graphs

Pumping Rate		Drawdown at 50 ft		Values of Δs	
gpm	m³/day	ft	m	ft	m
100	545	4.7	1.4	5.3	1.6
200	1,090	9.4	2.9	10.6	3.2
250	1,360	11.8	3.6	13.3	4.1
400	2,180	18.8	5.7	21.2	6.5

Each graph is constructed by plotting a point representing the drawdown for a certain time at a given distance from the pumped well (50 ft in this case) and drawing

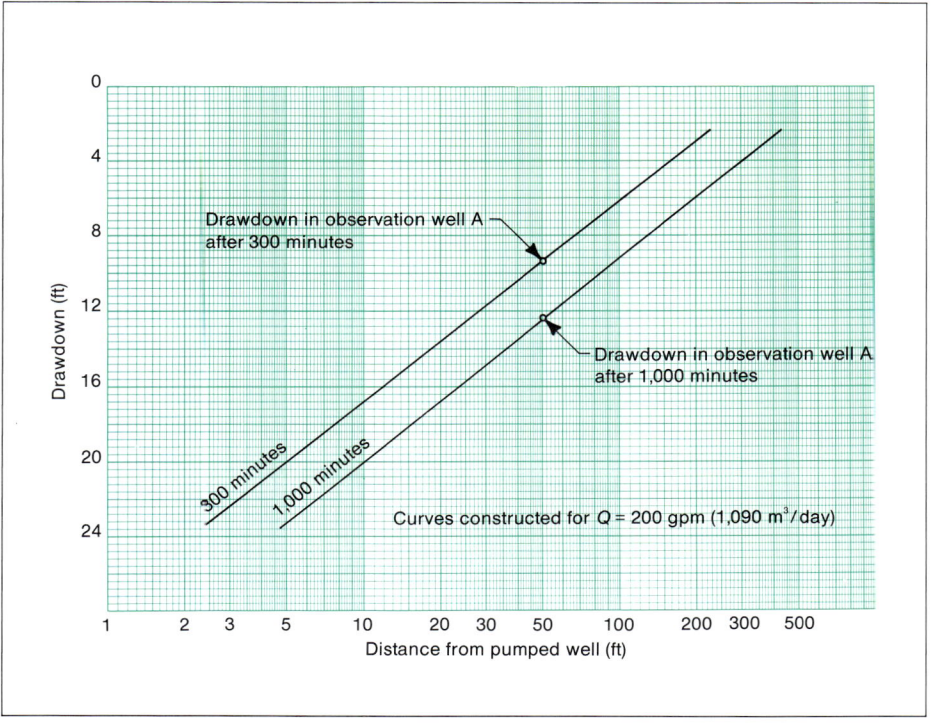

Figure 9.28. Distance-drawdown graph showing the positions of the cone of depression after pumping for 300 minutes and 1,000 minutes at a constant rate of 200 gpm (1,090 m³/day).

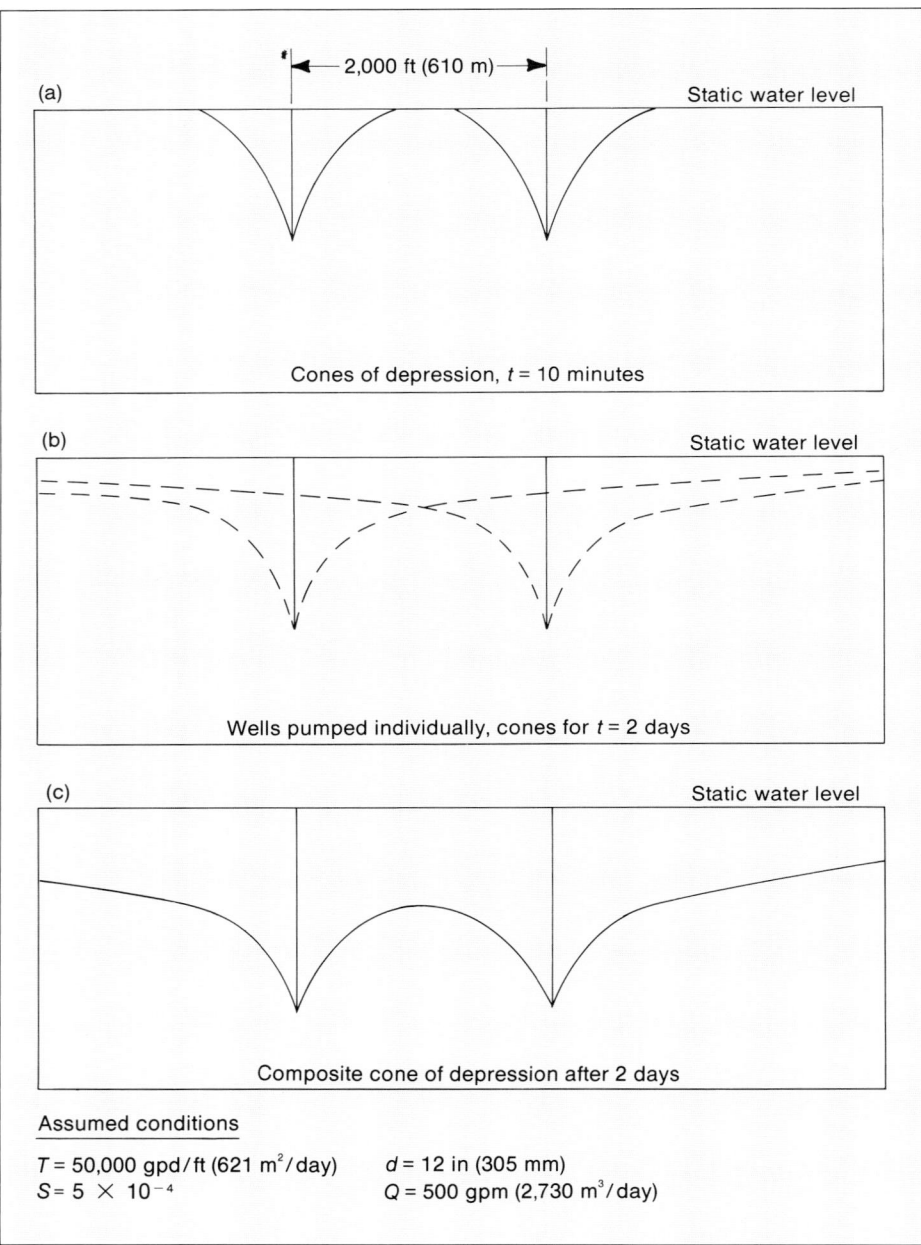

Figure 9.29. Interference between adjacent wells tapping the same confined aquifer. Composite cone is for both wells pumping simultaneously under the assumed conditions.

through this point a straight line having a Δs, or slope, of the calculated value.

WELL INTERFERENCE

The interference or drawdown in another well 300 ft (91.5 m) from the pumped well

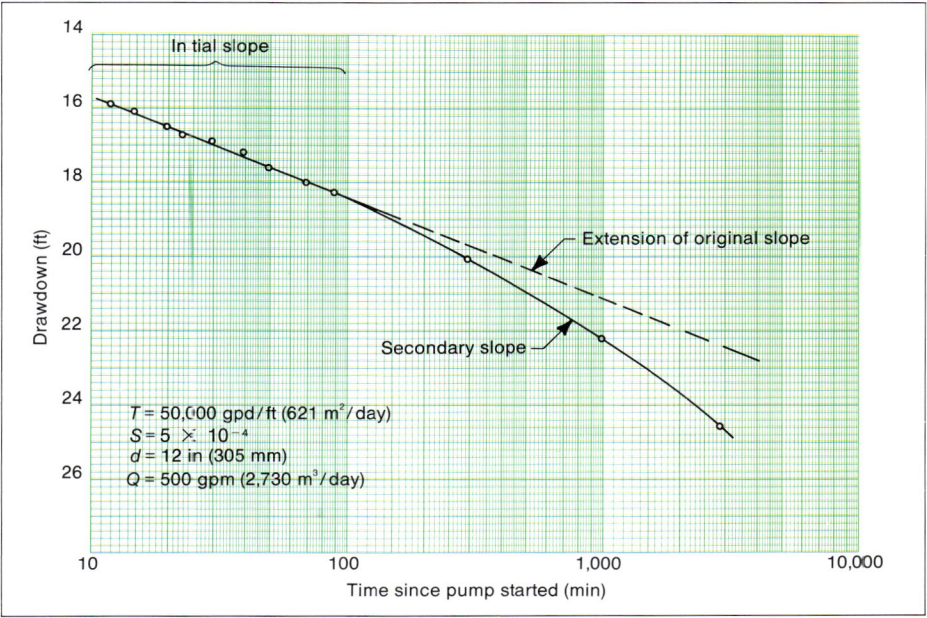

Figure 9.30. Time-drawdown graph for either well in Figure 9.29 shows effect of well interference when the cones of depression intersect.

used in the previous example is obtained directly from Figure 9.27. If the test well were pumped at 400 gpm (2,180 m³/day), the distance-drawdown plot shows that the water level 300 ft away would be drawn down 2.2 ft (0.7 m); the total drawdown would then be that caused by its own pumping plus the 2.2 ft of drawdown interference from the first well. Operation of the second well would also produce the same degree of interference on the first well if both were pumped at 400 gpm. The interference effects of several wells spaced at various distances can be evaluated in the same way. The total effect in any one well is the sum of the drawdown interferences produced by all the others in the group.

In this discussion, the calculations of well interference are for aquifer conditions after pumping continuously for 300 minutes. Figure 9.26 was constructed on this basis. The first point plotted was taken from the time-drawdown graph for observation well A at 300 minutes after pumping began. Because Figure 9.27 was constructed from Figure 9.26, all the distance-drawdown curves in Figure 9.27 are also valid only for 300 minutes after pumping began.

If interference data for a different period of continuous pumping are desired, one must start with another point taken from the original test data. The drawdown in observation well A after 1,000 minutes of pumping, shown in Figure 9.25 as 12.3 ft (3.8 m), might be used. The plot showing conditions after 1,000 minutes of pumping at 200 gpm (1,090 m³/day) is parallel to the plot for a pumping period of 300 minutes as shown in Figure 9.28. Another set of plots similar to those in Figure 9.27 can then be prepared, starting with the data from the 1,000-minute plot in Figure 9.28.

Figure 9.29 should be studied carefully for a clear picture of interference and just what it means in a well field where several cones of influence overlap. In part (a), the profiles of two wells that are 2,000 ft (610 m) apart are shown, each pumping continuously at a rate of 500 gpm (2,730 m³/day) for 10 minutes. Note that at the end of 10 minutes the cones of depression have not reached each other. In part (b), the extent of the cones of depression after 2 days of continuous pumping is shown. If either of the two wells were pumped alone, the cone of depression would be represented by the dashed line for the well being pumped. When both wells are pumped, however, the net result will be as in part (c), which is the sum of the drawdowns at any point within the zone of influence of both wells.

A time-drawdown curve for each of the two wells in Figure 9.29 would appear as shown in Figure 9.30. The correct transmissivity of the aquifer can be obtained only from the initial slope of this curve. The secondary slope, which is about twice as steep, shows the effect of the other well. The net result of the second well, when pumping at the same rate, is mathematically equivalent to an impervious boundary.

WELL EFFICIENCY

The efficiency of a pumped well in some cases can be estimated from the distance-drawdown graph. This can be done by extending the straight line representing the profile of the cone of depression to show the drawdown in the aquifer just outside the well. Such an extension is shown in Figure 9.31.

The intersection of the extended line with the radius of the pumped well shows the theoretical drawdown for a 100-percent efficient well. (In a filter-packed well, the radius is taken from the center of the well to the outside of the filter pack.) The result is valid for a confined aquifer only when the full thickness of the aquifer is screened. The theoretical drawdown for the example shown is 26.3 ft (8 m); the actual drawdown in

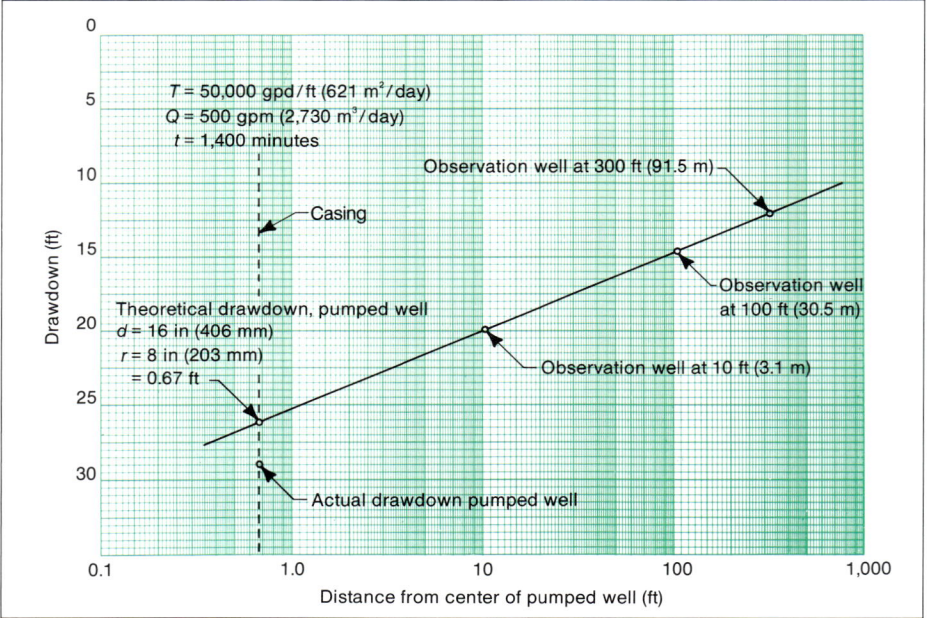

Figure 9.31. Theoretical drawdown of a pumped well can be compared with the actual drawdown by extending the straight line on the distance-drawdown diagram to a point where the radius of the well (outer face of the well) is indicated on the horizontal scale.

the well is 29 ft (8.8 m). The well efficiency is calculated as 26.3/29 = 91 percent.

Extra drawdown occurs when well operation causes considerable reduction in the saturated thickness of the aquifer. The extra drawdown, found primarily in wells in unconfined aquifers, should not be considered as representing inefficiency in the pumped well. An equation is given in Appendix 9.C to correct the theoretical drawdown so that well efficiency can be calculated more accurately when partial dewatering of the aquifer occurs.

The factors contributing to excess drawdown in wells (inefficiency) can be grouped into two classes. One class comprises those factors related primarily to choices made in the design of wells; the other class includes factors related to construction. The following is a summary of the factors in the two classes.

Design Factors

1. Choice of a well screen with insufficient open area makes entrance velocities too high, resulting in greater-than-normal entrance (head) losses.

2. Poor distribution of screen openings causes excessive convergence of flow near the individual openings, and may produce twice as much drawdown as necessary. The accompanying diagrams of flow nets around well screens illustrate this point (Figure 9.32).

3. Insufficient length of well screen, resulting in partial penetration of the aquifer, distorts the flow pattern for some distance around the well (Figure 9.33). Flow to the well screen then includes major vertical components as well as the main horizontal component. Vertical hydraulic conductivity is generally lower than horizontal hydraulic conductivity, so considerable head losses result from the vertical flow. Even though extra drawdown may result, a shorter well screen may be used in many instances because of other design considerations. The effect of partial penetration of the aquifer is discussed later in this chapter; screen length is discussed in Chapter 13.

4. Improperly sized filter packs or those made from angular or platelike materials can restrict flow into a well screen. Particle shape, size, and grain size distribution affect the hydraulic conductivity of the pack.

Construction Factors

1. Inadequate development of a well may leave so much drilling fluid and small particles in the formation around the screen that the original permeability is reduced. Drilling contractors have great difficulty developing wells when they install screens that have insufficient open area or poor distribution of slot openings.

2. Improper placement of the well screen may put it at a depth that does not correspond to the best water-bearing stratum.

RADIUS OF INFLUENCE

The radius of influence of a well can be determined from the distance-drawdown graph in most instances. Many owners of high-capacity wells want this information because it indicates how far the cone of depression extends from the well. For all practical purposes, it is the distance indicated by extending the straight line on the distance-drawdown graph to the point of zero drawdown on the drawdown scale. Note that this distance is the same as r_0 in Equation 9.12.

Some believe that the radius of influence should be the determining factor in well spacing to avoid any interference between wells. This practice is often impractical, es-

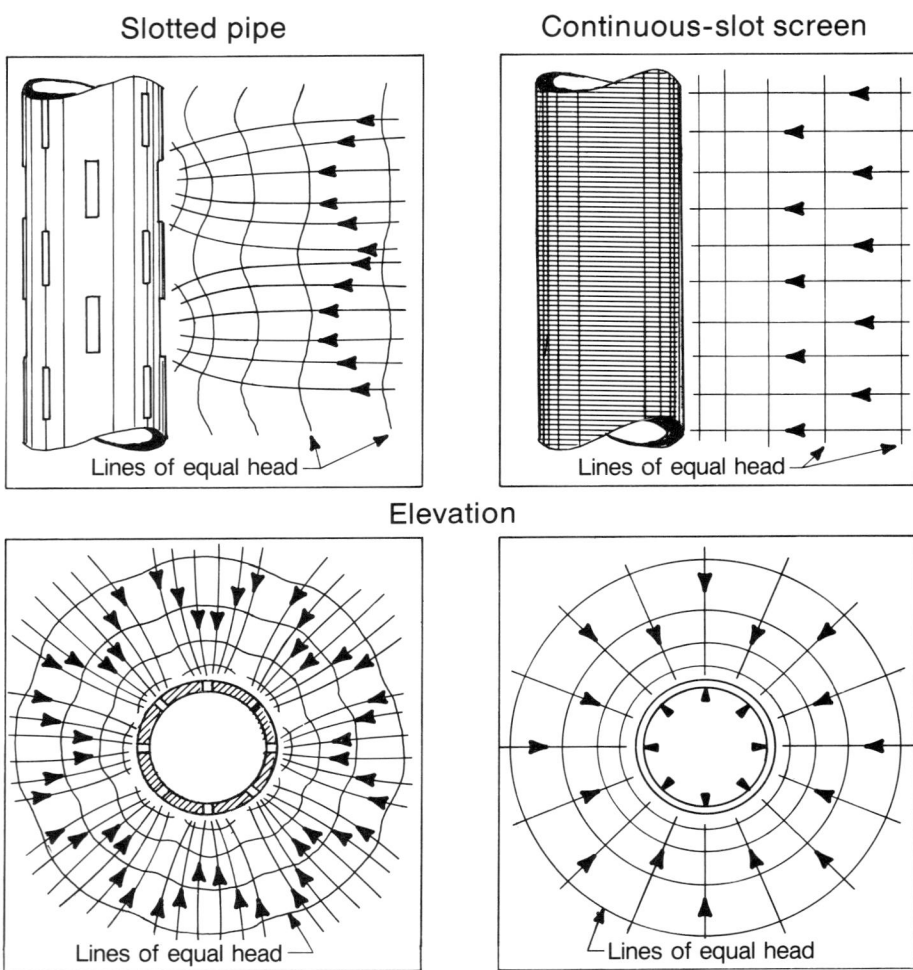

Figure 9.32. Flow nets around screen devices. Water approaches openings along lines indicated by arrows. Flow lines for slotted pipe converge to individual slots; flow lines for continuous-slot well screens are less distorted.

pecially in confined aquifers where cones of depression may extend 1 to 2 mi (1.6 to 3.2 km) or more for high-capacity wells.

As the cone of depression begins to stabilize during a pumping test, further increases in the radius of influence are usually slight. The cone of depression may stabilize because of recharge from precipitation, leakage from saturated strata above or below the aquifer, or natural groundwater flow toward and past the well. Equilibrium may also result from induced recharge from a lake or stream (considered to be a line source) when a hydraulic gradient develops between the line source and the pumped well.

RECHARGE AND BOUNDARY EFFECTS

In the case of induced recharge, the cone of depression is steeper toward the line source than in other directions away from the well. If the lake or stream is nearby, measurements in observation wells between the pumped well and the recharge source produce a distance-drawdown curve that is steeper than normal. This occurs because the recharge effect results in less drawdown in the observation wells nearest the line source as compared with those observation wells on the opposite side of the pumped well. But the slope of the distance-drawdown curve derived from the pumping test will be almost the same in all directions if the recharge source is relatively distant and the observation wells are relatively close to the pumped well. Except in cases of nearby line sources of recharge, the slope of the distance-drawdown curve provides a reliable basis for calculating aquifer transmissivity regardless of recharge effect.

The same cannot be said for calculation of the storage coefficient when recharge occurs. Recharge affects the vertical position of the distance-drawdown curve, which makes the value of r_0 less than it would be in the absence of recharge. This, in turn, makes the calculated storage coefficient greater than its real value. In some cases, the calculated storage coefficient is greater than one — an impossible value — proving that recharge is occurring.

The effect of impervious boundaries on the distance-drawdown graph is essentially opposite to the effect of recharge. The slope of the curve is affected only slightly if the distance to any boundary is much greater than the distances to the observation wells. Observation wells close to a boundary, however, show drawdowns greater than normal, and the resulting distance-drawdown curve is a little flatter than it would be otherwise. This leads to a calculated value of transmissivity that is higher than the true value. The calculated value of the storage coefficient is lower than the true value because the distance-drawdown curve is displaced downward on the semilog diagram because of the influence of the aquifer boundary. An unreasonably low value for the storage coefficient usually indicates an impervious boundary within the zone of influence of the pumping well.

COMBINED USE OF SEMILOG GRAPHS

As shown above, calculations from time-drawdown graphs can be used to check cal-

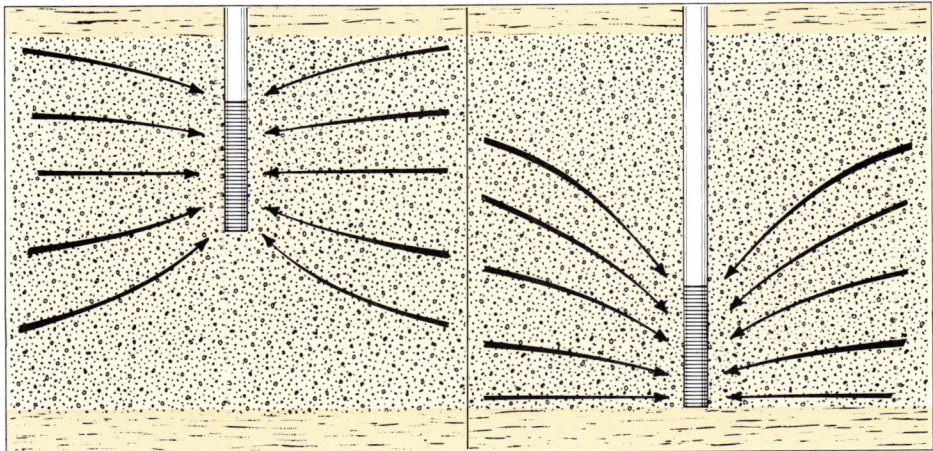

Figure 9.33. Partial penetration occurs when the intake portion of the well is less than the full thickness of the aquifer. This causes distortion of the flow lines and greater head losses.

Table 9.3. Comparisons of Recharge and Boundary Effects on Semilog Diagrams

Recharge Effect During Pumping Test

Time-drawdown graph	Distance-drawdown graph
1. Slope of the curve becomes flatter. Transmissivity calculated from the flatter slope will be higher than the true value.	1. Slope of the straight line remains almost unchanged if the recharge source is relatively distant and the observation wells are relatively close to the pumped well. Transmissivity calculated from the graph usually will be close to the true value.
2. Extending straight line of the flatter slope gives a value for t_0 that is too low. Storage coefficient calculated from this figure will be lower than the correct value.	2. Straight line is displaced upward. Extension to zero drawdown gives a value for r_0 that is too low. Storage coefficient calculated from this figure will be higher than the correct value.

Boundary Effect During Pumping Test

Time-drawdown graph	Distance-drawdown graph
1. Slope of the curve becomes steeper. Transmissivity calculated from the steeper slope will be lower than the true value.	1. Slope of the straight line remains almost unchanged if the distance to a boundary is relatively great compared with the distance to the observation wells. Transmissivity calculated from the graph will be close to the true value.
2. Extending line of the steeper slope gives a value for t_0 that is too high. Storage coefficient calculated from this figure will be higher than the correct value.	2. Straight line is displaced downward. Extension to zero drawdown gives a value for r_0 that is too high. Storage coefficient calculated from this figure will be lower than the correct value.

culations from distance-drawdown graphs, and vice versa. If an extensive homogeneous aquifer supplies a well from storage only, both graphs should give identical results. Calculations from the one diagram are independent of those from the other, assuming that the distance-drawdown graph is drawn from measurements in three or more observation wells.

If a recharge effect appears during the pumping test, or if the expanding cone of depression encounters an impervious aquifer boundary, the effects on the two graphs are quite different. Table 9.3 summarizes these contrasts.

Knowledge of the differing effects is useful in aquifer-test interpretation. Suppose, for example, that a test has been performed and drawdown measurements from three observation wells are available. Suppose, also, that the time-drawdown graphs show only one clearly defined straight line. The transmissivity as calculated from the slope of these plots is about one-half of the transmissivity value from the distance-drawdown graph. This result leads to the suspicion that a boundary effect influenced the drawdown measurements so early in the pumping period that a break in the slope of the time-drawdown plots is not apparent. In the example shown in Figure 9.13, any abnormal deviation during the first 10 minutes of pumping would not be evident on the semilog diagram.

Comparison of calculations of the storage coefficient in our present example, however, supports the suspected boundary condition if the value obtained from the time-drawdown graph is higher than the value from the distance-drawdown graph. In this case, neither value is the true coefficient of storage; it would lie between the limiting values of the two calculated figures. The correct coefficient of transmissivity can be obtained from the distance-drawdown graph if the observation wells are close enough to the pumping well so that the boundary effect does not occur within the first 10 minutes of pumping.

Up to this point, most of our discussion of well performance has assumed that flow to a well bore is entirely radial. Figure 9.2 illustrates this condition, where all flow lines are horizontal and have no vertical component. The flow is radial here because it occurs in (1) a confined aquifer, (2) a screen whose length is the same as the aquifer thickness, and (3) a pumping condition in which the formation near the well is not dewatered.

EFFECT OF PARTIAL PENETRATION

When flow is not strictly radial, the flow pattern and drawdowns differ somewhat from those calculated from the equations presented thus far. Figure 9.34 shows a confined aquifer with only the upper 50 percent screened. Here the arrows represent some typical flow lines or paths of water particles as they move through the formation to the intake portion of the well. The departure from radial flow is evident.

Water in the lower part of the aquifer must move upward along the curved lines to reach the well screen. In doing so, this water must take paths that are somewhat longer than radial flow lines. Also, the flow must converge through a smaller cross-sectional area while approaching the short screen. The result of the longer flow paths and smaller cross-sectional area is an increase in head loss. For a given yield, therefore, the drawdown in a pumping well is greater if the aquifer thickness is only partially screened. For a given drawdown, the yield from a well partially penetrating the aquifer is less than the yield from one completely penetrating the aquifer.

Figure 9.34 shows that drawdowns near the pumped well vary with depth in the

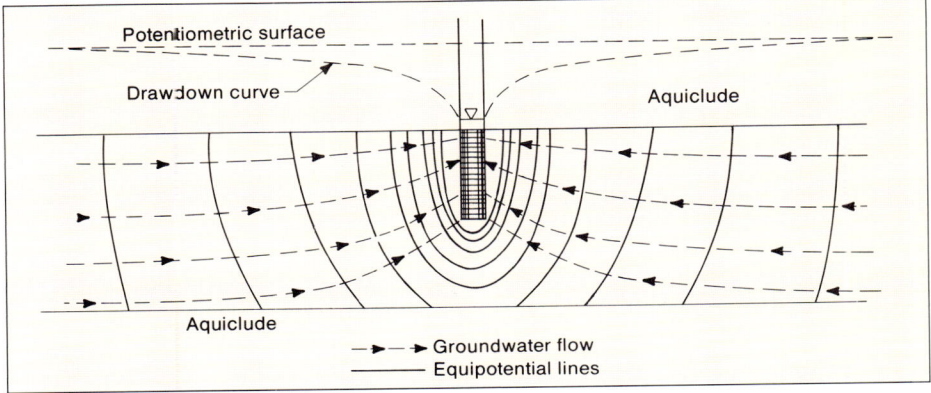

Figure 9.34. When the intake section of a well partially penetrates a confined aquifer, flow lines deviate somewhat from the radial flow pattern associated with a fully penetrating well. *(Water and Power Resources Service, 1981).*

confined aquifer; the potentiometric drawdowns shown here also indicate that pressure varies with depth. This variation in drawdown with depth is most pronounced adjacent to the well and decreases with distance from the well. At a distance equal to about twice the aquifer thickness, the line representing drawdown in the aquifer is vertical, showing that the drawdown is the same at all depths. Radial flow prevails at this distance and beyond.

The mathematical analysis of partial penetration problems is difficult for homogeneous aquifers, but the development of equations for stratified aquifers is almost impossible. The curves in Figure 9.35 provide a simple method for estimating results from partially penetrating wells in confined aquifers that are reasonably homogeneous. This diagram was developed from the Kozeny (1933) equation:

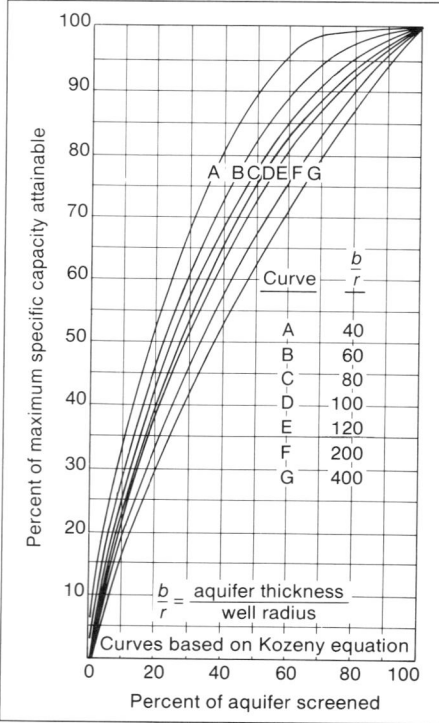

Figure 9.35. Relationship of partial penetration and attainable specific capacity for wells in homogeneous confined aquifers.

$$\frac{Q/s_p}{Q/s} = L\left(1 + 7\sqrt{\frac{r}{2bL}} \cos\frac{\pi L}{2}\right)$$

(9.13)*

where

Q/s_p = specific capacity of a partially penetrating well, in gpm/ft (m³/day/m)

Q/s = maximum possible specific capacity of a fully penetrating well, in gpm/ft (m³/day/m). (Left-hand side of the equation is the ratio of the production of a partially penetrating well to that of a fully penetrating well.)

r = well radius, in ft (m)

b = aquifer thickness, in ft (m)

L = well screen length as a fraction of the aquifer thickness

This equation is not valid when the aquifer thickness is small, the percent of penetration is large, and the well radius is large. Calculated results for some combinations of these conditions would be higher than those for a completely penetrating well, which is theoretically impossible. The curves in Figure 9.35 represent conditions within the valid range of the equation.

To use the curves, the screen length must be expressed as a percentage of the aquifer thickness. After locating this value on the horizontal scale, move upward along a vertical line to the curve that represents the ratio of aquifer thickness to well radius for the case in question. Then move horizontally and read the percentage value on the vertical scale. This represents the specific capacity of the partially penetrating

---

*This equation can be used with either SI or English terms.

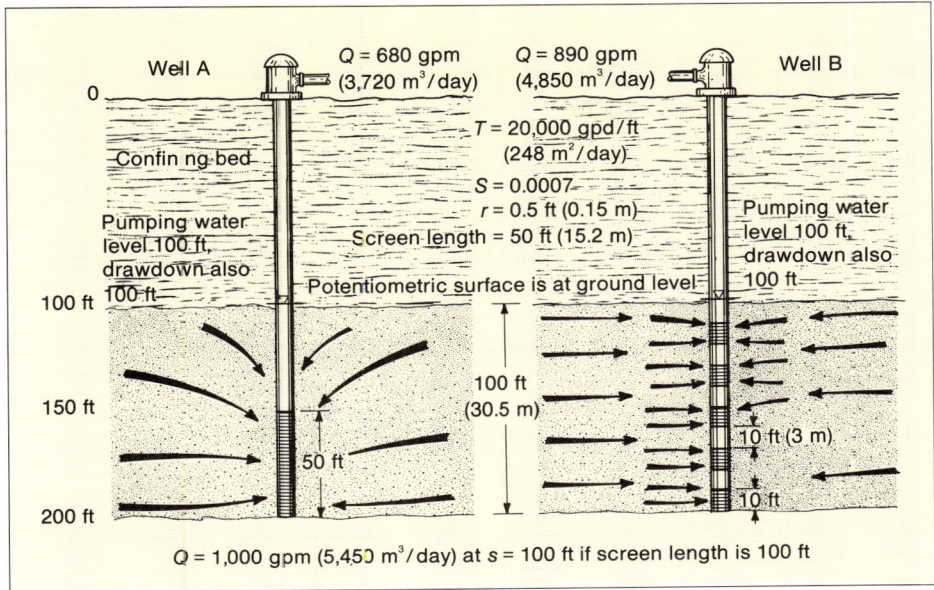

Figure 9.36. Well performance can be improved by using multiple sections of well screen in a thick aquifer to reduce the effect of partial penetration. Total screen length is the same in both wells.

well, expressed as a percentage of the specific capacity that could be obtained with a fully penetrating well. The specific capacity of a fully penetrating well is referred to here as its maximum possible specific capacity.

Analysis of the alternate well designs shown in Figure 9.36 provides a practical example of the use of the curves. Both wells have been constructed with 50 ft (15.2 m) of 12-in (305-mm) diameter well screen. A homogeneous confined aquifer is assumed. In well A, the screen is positioned in the lower half of the aquifer, making $b/r = 200$. Figure 9.35 shows that 68 percent of the maximum possible specific capacity could be expected from this well. Using Equation 9.5, the specific capacity of a fully penetrating well with a 6-in (152-mm) radius is calculated as 10 gpm/ft (179 m³/day/m) of drawdown after 24 hours of pumping. The expected specific capacity for well A, therefore, is 68 percent of this figure, or 6.8 gpm/ft (122 m³/day/m) of drawdown. The available drawdown to the top of the aquifer is 100 ft (30.5 m), so the yield would be $6.8 \times 100 = 680$ gpm ($122 \times 30.5 = 3,720$ m³/day).

In well B, the 50-ft length of screen has been split into 10-ft (3-m) sections. These are separated by four 10-ft sections of blank pipe. With this arrangement, the 100-ft aquifer can now be treated as five aquifers, each 20 ft (6.1 m) thick with each section 50 percent screened; this changes the b/r ratio. The advantage of this screen arrangement is apparent from the curves. If 50 percent of an aquifer that is 20 ft thick ($b/r = 40$) is screened, 89 percent of the maximum specific capacity can be obtained. From this observation, the estimated yield of well B is 890 gpm (4,850 m³/day), a 31-percent improvement over well A. (To simplify this comparison, well entrance losses have been ignored.)

In a well in an unconfined aquifer, the problem of partial penetration must always be considered because the pumping well dewaters the upper part of the water-bearing formation. This reduces the saturated thickness and necessarily shortens the intake portion of the well. This type of well is often pumped so that its drawdown is a large proportion of the aquifer thickness. This results in a greatly distorted flow pattern as compared with strictly radial flow. Figures 9.11 and 9.12 can provide a reasonable estimate of the effects of partial penetration and dewatering for wells in unconfined aquifers.

WATER-LEVEL RECOVERY DATA

When pumping is stopped, well and aquifer water levels rise toward their pre-pumping levels. The rate of recovery provides a means for calculating the coefficients of transmissivity and storage. The time-recovery record, therefore, is an important part of an aquifer test. The time-drawdown measurements taken during the pumping period and the time-recovery measurements taken during the recovery period provide two different sets of information from a single aquifer test. Values obtained from analysis of the recovery record serve to check calculations based on the pumping record.

The water-level recovery data from an observation well will indicate the hydraulic characteristics of the aquifer if the well is located close enough to the pumped well so that the drawdown changes significantly (easily measured) during the pumping test. If no observation well is available, the water-level recovery data from the pumped well can be used for limited calculations of aquifer capability. One observation well should be used if at all possible; two or more are desirable. In all cases, water levels should be measured in the pumped well and in each observation well.

Recovery data can be analyzed as described below only when the test pumping is done at a constant rate. Recovery measurements following a variable-rate test, such as a step-drawdown test, cannot be used (see Chapter 16). The exact time of starting and stopping the pump must be recorded, along with any changes in pumping rate and the time each occurs. The recovery curves reflect the change in aquifer water

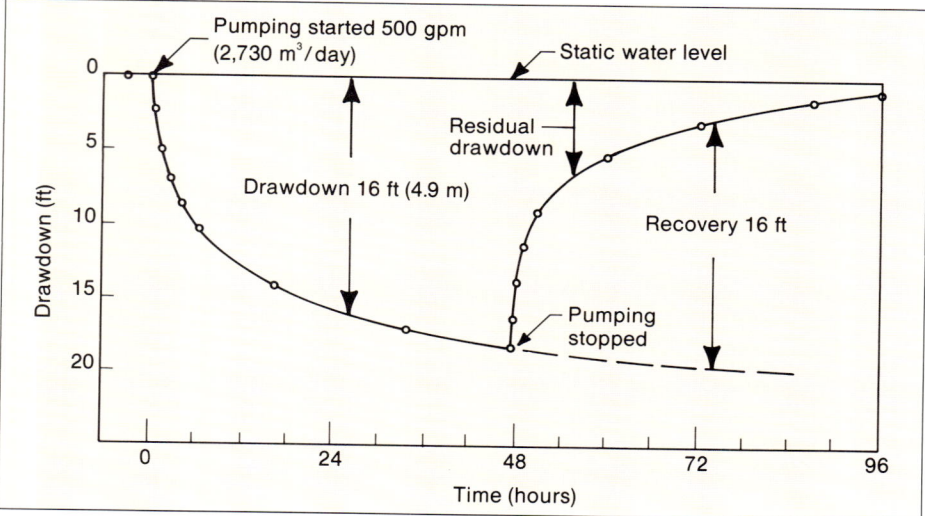

Figure 9.37. Typical drawdown and recovery plots for a well pumped for 48 hours at a constant rate of 500 gpm (2,730 m³/day) followed by a 2-day period for water-level recovery.

level with time. An essential part of every recovery record, therefore, is the exact time at which each measurement is made. All wells do not have to be measured simultaneously, but the time intervals between readings should be about the same for each.

The elevation of the measuring point (often the top of the well casing) should be determined for each well. An electrical or acoustical sounding device or a weighted steel tape are needed to measure the depth to water accurately.

Figure 9.37 shows how water levels in a well change with time. The left half of the diagram corresponds to the pumping period, the right half to the recovery period. The recovery curve is practically an inverted image of the drawdown curve. The shape of each curve is determined by the physical characteristics of the aquifer.

The points plotted for the recovery curve (right half of Figure 9.37) represent the residual drawdown in the well during the recovery period. Each point represents the difference between the original static water level and the depth to water at a given instant during the recovery period.

The recovery measurements of the water level must be based on the pumping water level. The hydraulic theory of well and aquifer performance assumes that water-level changes during the recovery period are the result of input from an imaginary recharge well. If such a well injects water into the aquifer at the rate the real well pumps it out, with both wells working simultaneously after a given instant, the recovery curve will be like the one in Figure 9.37. The rise in water level, because of the imaginary recharge well, is the vertical distance between an extension of the time-drawdown curve and the actual time-recovery curve. Recovery, then, means the difference between the measured water level in an observation well at a given time after pumping stops and the level to which the water would have dropped if pumping had continued until that instant.

When defined in this way, the degree of water-level recovery at any time after the end of the pumping period is theoretically identical with the drawdown for the same time during the pumping period. Putting it another way, the recovery 24 hours after pumping stops should equal the drawdown that was measured 24 hours after pumping started. Complete recovery, however, generally requires a period considerably longer than the previous pumping period, except in cases where recharge to the aquifer occurs during the pumping and recovery periods.

Assume that a 6-in (152-mm) test well and an observation well 50 ft (15.2 m) away are available for an aquifer test, as shown in Figure 9.38. After pumping the

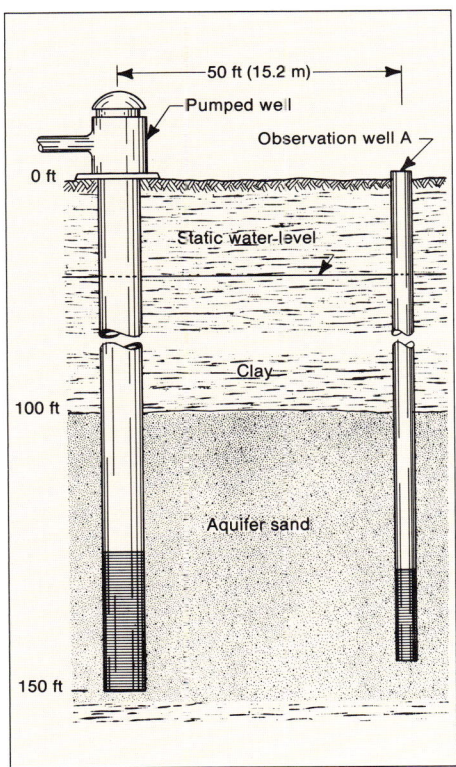

Figure 9.38. Production well and observation well as used in a confined aquifer test.

well at 200 gpm (1,090 m³/day) for 500 minutes, the pump is stopped and water-level measurements are made during the first 400 minutes of the recovery period.

Table 9.4 shows the depth-to-water measurements in the observation well and the residual drawdown for time intervals measured from both the beginning of the pumping test and the beginning of the recovery period. These intervals are designated t and t', respectively. The ratios of the two time periods, t/t', are also shown in the table.

Figure 9.39 shows the recovery curve plotted for the observation well. Extension of the preceding drawdown curve indicates the drawdown that would have occurred if pumping had been continued beyond the pumping period. The water-level recovery for various time intervals is the vertical difference between the curves in this diagram. Values are shown in Table 9.4.

The recovery curve plotted in Figure 9.39 is difficult to portray by mathematical analysis. It can be simplified for analysis, however, in either of two ways: Theis' (1935) corollary to the nonequilibrium equation (presented below), or Jacob's (1946b) modification of the nonequilibrium equation. It has been shown that the time-drawdown curve for the pumping period becomes a straight line on a semilogarithmic diagram. The same simplification can be used for the time-recovery plot, where the horizontal scale represents the logarithm of time during the recovery period and the vertical scale represents water-level recovery, $(s - s')$.

Data from Table 9.4 plotted in this way are shown in Figure 9.40. The result is similar to a time-drawdown plot for the pumping phase of the same aquifer test. Theo-

Table 9.4 Residual Drawdown and Calculated Recovery in the Observation Well

Time since pump started, t	Time since pump stopped, t'	Ratio, t/t'	Depth to water*		Residual drawdown*, s'		Drawdown, s, from pumping curve†		Calculated recovery $(s - s')$	
min	min		ft	m	ft	m	ft	m	ft	m
500	0	—	18.60	5.67	10.60	3.23	10.60	3.23	0.00	0.00
501	1	501.00	18.55	5.66	10.55	3.22	10.60	3.23	0.05	0.01
502	2	251.00	18.50	5.64	10.50	3.20	10.60	3.23	0.10	0.03
503	3	168.00	18.40	5.61	10.40	3.17	10.61	3.23	0.21	0.06
504	4	126.00	18.09	5.52	10.09	3.08	10.61	3.23	0.52	0.15
506	6	84.00	17.72	5.40	9.72	2.96	10.62	3.24	0.90	0.28
508	8	64.00	17.22	5.25	9.22	2.81	10.63	3.24	1.41	0.43
510	10	51.00	16.64	5.07	8.64	2.63	10.64	3.24	2.00	0.61
520	20	26.00	15.27	4.66	7.27	2.22	10.67	3.25	3.40	1.03
530	30	17.70	14.50	4.42	6.50	1.98	10.70	3.26	4.20	1.28
540	40	13.50	13.63	4.16	5.63	1.72	10.73	3.27	5.10	1.55
560	60	9.35	12.95	3.95	4.95	1.51	10.80	3.29	5.85	1.78
590	90	6.55	12.01	3.66	4.01	1.22	10.96	3.34	6.95	2.12
650	150	4.33	10.80	3.29	2.80	0.85	11.15	3.40	8.35	2.55
710	210	3.38	10.70	3.26	2.70	0.82	11.35	3.46	8.65	2.64
770	270	2.85	10.06	3.07	2.06	0.63	11.56	3.52	9.50	2.89
830	330	2.51	9.96	3.04	1.96	0.60	11.76	3.59	9.80	2.99
890	390	2.28	9.60	2.93	1.60	0.49	11.95	3.64	10.35	3.15

*Static water level, 8 ft (2.44 m)
†Average pumping rate during preceding pumping period was 200 gpm (1,090 m³/day)

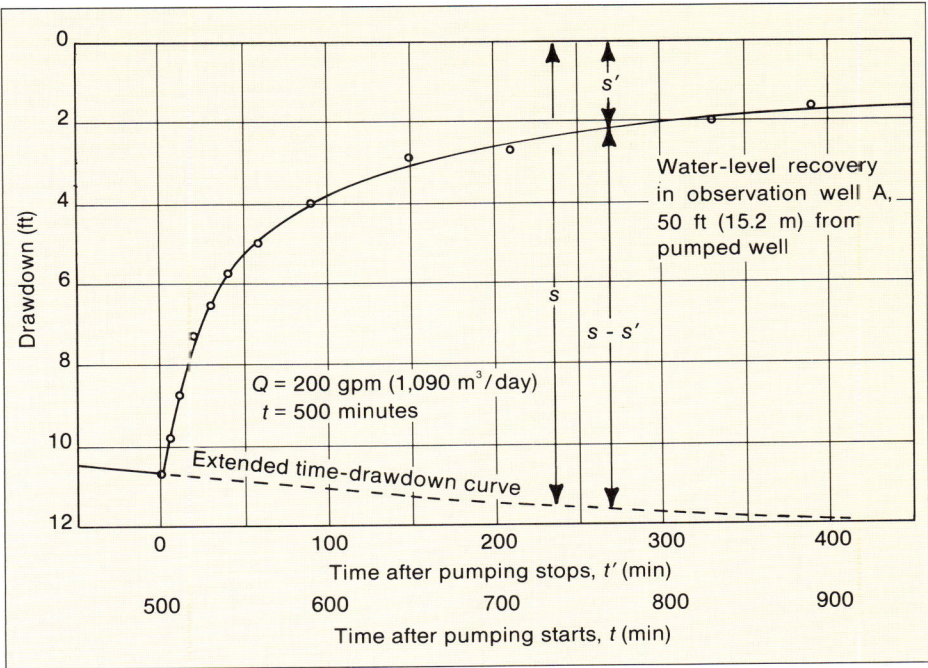

Figure 9.39. Residual-drawdown curve from observation well, with extended time-drawdown curve (on arithmetic scales) showing how calculated recovery is determined at any instant during the recovery period. Producing well pumped 200 gpm (1,090 m³/day) for 500 minutes.

retically, the drawdown and recovery plots should be identical if the aquifer conditions conform to the basic assumptions of the Theis concept.

The time-recovery data from the pumped well can also be plotted by using the method applied to the observation well. The time-recovery plot for the pumped well is more accurate than its time-drawdown plot because the residual-drawdown measurements are more accurate. During the recovery period, water-level measurements can be made without being affected by pump vibrations and momentary variations in the pumping rate.

In analyzing the time-recovery plot, its slope is of primary interest. Two factors determine the slope of the straight line in Figure 9.40. One is the average pumping rate during the preceding pumping period, the other is the aquifer transmissivity.

In Figure 9.40, the slope of the straight line is expressed numerically as the change in the water-level recovery per logarithmic cycle. It is designated by $\Delta(s - s')$. Its value in Figure 9.40 is 5.2 ft (1.6 m), which is the recovery during the period from 10 minutes to 100 minutes after pumping stopped.

The next step is to calculate the transmissivity of the aquifer from the following equation:

$$T = \frac{264\, Q}{\Delta\, (s - s')} \qquad\qquad T = \frac{0.183\, Q}{\Delta\, (s - s')} \qquad (9.14)$$

Note that this equation is similar to Equation 9.7. Figure 9.40 shows the value of T to

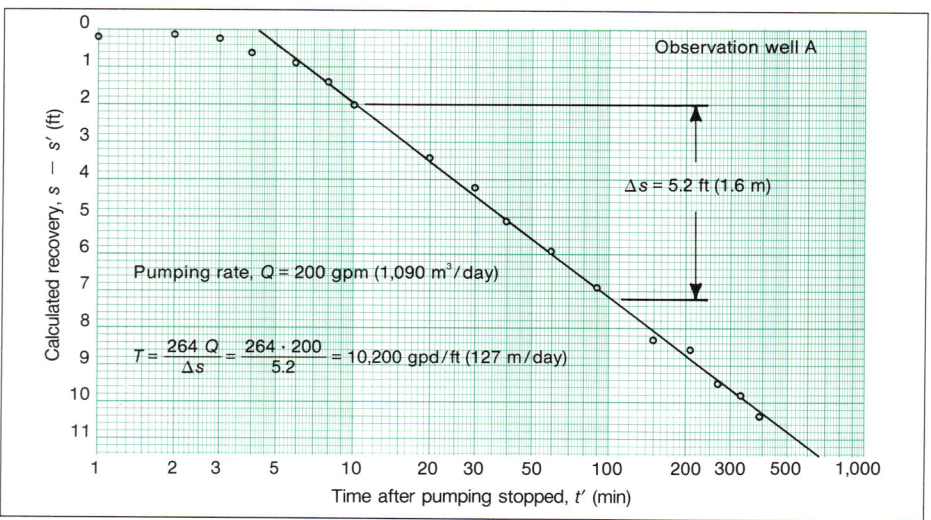

Figure 9.40. Time-recovery plot for observation well becomes a straight line when plotted on a semilog diagram, similar to the time-drawdown diagram for the preceding pumping period.

be about 10,200 gpd/ft (127 m²/day), which may be compared with T as calculated from the time-drawdown data plotted in Figure 9.25. If test conditions meet the required standards and measurements are taken carefully, the two results should agree reasonably well.

A second method of plotting the data permits direct use of the residual drawdown without calculating the recovery from an extension of the time-drawdown plot. It can be shown that the residual drawdown is related to the logarithm of the ratio t/t' as follows:

$$s' = \frac{264 \, Q}{T} \log t/t' \qquad\qquad s' = \frac{0.183 \, Q}{T} \log t/t' \qquad (9.15)$$

Mathematical development of this relationship is given in Appendix 9.D.

This equation shows that when values of s' are plotted against corresponding values of t/t' on semilogarithmic graph paper, a straight line can be drawn through the plotted points. Figure 9.41 shows the data from Table 9.4 plotted on a semilog diagram, with s' indicated on the vertical arithmetic scale and t/t' on the horizontal logarithmic scale. The transmissivity is then calculated from the following equation:

$$T = \frac{264 \, Q}{\Delta s'} \qquad\qquad T = \frac{0.183 \, Q}{\Delta s'} \qquad (9.16)$$

Note from Figure 9.41 that time during the recovery period increases toward the left in this method of plotting, whereas on the time-drawdown and time-recovery plots time increases toward the right.

The residual-drawdown plot as shown in Figure 9.41 is preferred over the recovery plot, Figure 9.40, for calculating transmissivity. The method shown in Figure 9.41 provides a more independent check on the results calculated from the pumping period.

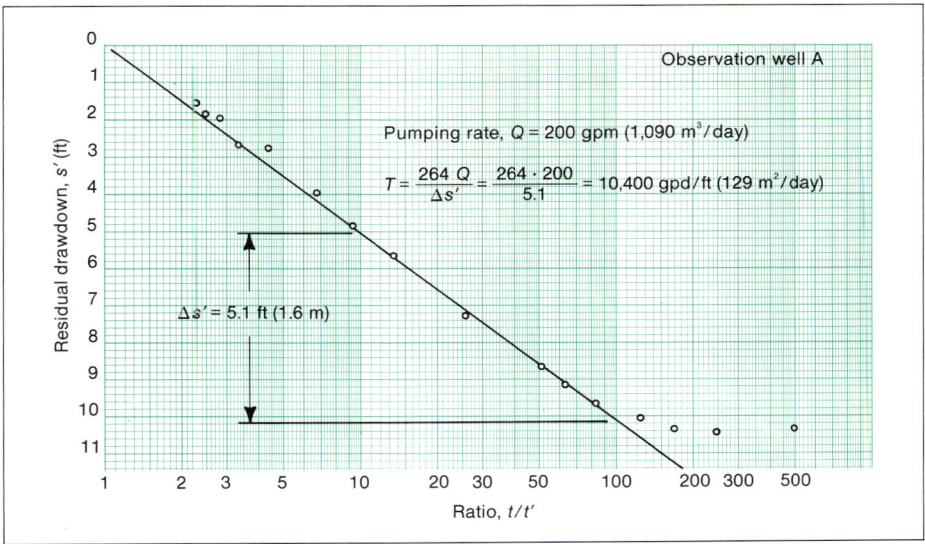

Figure 9.41. Residual drawdown plotted against the ratio t/t' becomes a straight line on semilog graph and permits calculation of transmissivity as shown. Time during recovery period increases toward the left in this diagram.

The method used in Figure 9.40 depends upon extension of the time-drawdown plot through the recovery period; thus, the drawdown plot itself determines the values used in the recovery plot, and any inaccuracies in the drawdown plot are projected into the recovery plot.

If no observation well is available, the recovery data from the pumped well usually provide the best basis for calculating the transmissivity of the aquifer. The residual-drawdown plot, as shown in Figure 9.41, should always be used in such a case.

Determining Storage Coefficient Using Recovery Data

If measurements are made in at least one observation well during the recovery period, the storage coefficient can be calculated from portions of these data. The data must be plotted as shown in Figure 9.40. The residual-drawdown plot cannot be used for determining the storage coefficient, even though that plot is valid for calculating the transmissivity.

Figures 9.42 and 9.43 show the similarity in calculations of the storage coefficient from time-drawdown and time-recovery diagrams. Using Equations 9.7 and 9.8, the time-drawdown data for an observation well, shown in Figure 9.42, give values of $T = 13,000$ gpd/ft (161 m²/day) and $S = 5.7 \times 10^{-4}$, respectively. Parallel calculations from Figure 9.43 using $\Delta(s - s')$ in place of Δs and t'_0 in place of t_0, give values of $T = 13,700$ gpd/ft (170 m²/day) and $S = 4.4 \times 10^{-4}$, respectively. These two sets of results are considered to be in reasonable agreement.

It is apparent from the residual-drawdown curve in Figure 9.41 that t'_0 cannot be obtained from that diagram. The horizontal scale represents a ratio without units. The intercept of this curve at zero drawdown has an entirely different significance on this graph. It is necessary to review the basic assumptions listed on page 218 that were used in developing the equations for both the pumping period and the recovery period

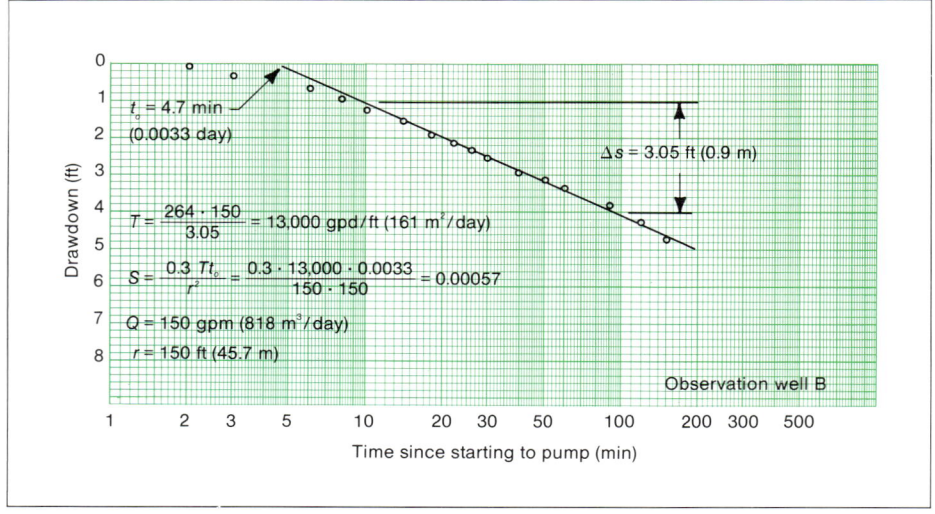

Figure 9.42. Time-drawdown curve plotted from measurements in an observation well 150 ft (45.7 m) from a pumped well. Value of t_0 for calculating the coefficient of storage, S, is obtained by extending straight line to the left to zero drawdown.

of an aquifer test. When an aquifer conforms to these assumptions, the residual-drawdown curve, when extended to the left as in Figure 9.41, should pass through the zero-drawdown point where the ratio t/t' reaches 1. The ratio t/t' approaches 1 as the recovery period is extended.

After a long period of recovery, the water level throughout the aquifer tends to return to the original static level, with the residual drawdown approaching zero as t/t' approaches 1. Theoretically, therefore, the residual-drawdown curve should pass through the upper left corner of the diagram as in Figure 9.41.

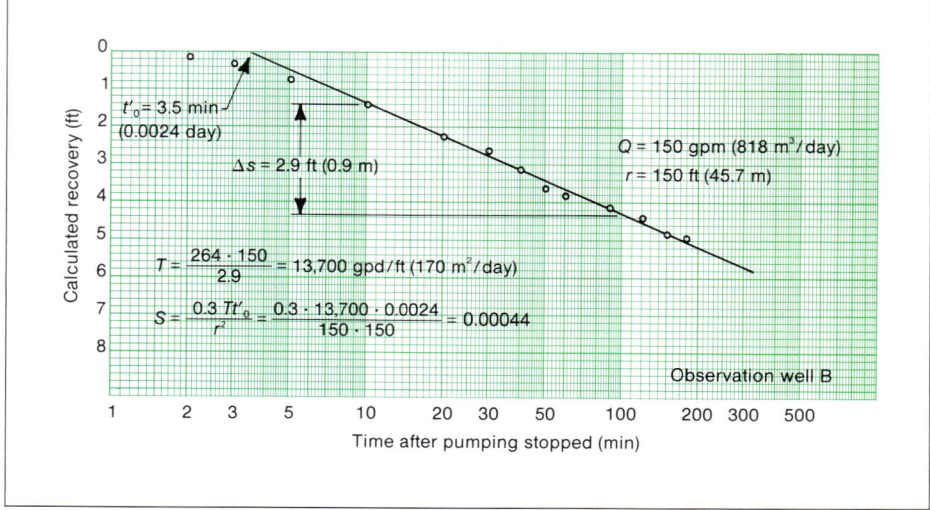

Figure 9.43. Time-recovery curve for an observation well, extended to the left to get t'_0 for computing the storage coefficient, S.

A study of residual-drawdown curves from actual aquifer tests reveals that the curve does not always pass through this point, called the origin of the diagram. When the curve fails to pass through the origin, it is concluded that the aquifer conditions do not conform to the assumed idealized conditions.

Three ways in which the conditions differ from the theoretical aquifer may be indicated by the residual-drawdown plot. If the graph indicates zero drawdown at a t/t' value of 2 or more, it is concluded that some recharge water reached the aquifer during the pumping period. The result of the recharge is to bring about full recovery to the original static level during a relatively short recovery period, long before t/t' approaches 1. The upper plot in Figure 9.44 might be obtained for such a situation.

A different condition is indicated when the plot extended to the left shows a residual drawdown of several inches or more as t/t' approaches 1. This situation would occur in an aquifer of limited extent with no recharge, when pumping permanently lowers the static water level. The lowest plot in Figure 9.44 illustrates this type of result.

The third condition that can account for minor displacement of the residual drawdown plot results from a variation in the storage coefficient, S. In theory, the storage coefficient is assumed to be constant during both the pumping period and the recovery period of the test. In practice, however, S probably varies and is apt to be greater during the pumping period than during the subsequent recovery (Jacob, 1963).

The value of S for a confined aquifer depends upon the elastic properties of the formation. If the aquifer is not perfectly elastic, it does not rebound vertically during recovery of water levels (recovery of pressure) at the same rate that it is compressed as a result of the drawdown during the preceding pumping.

During pumping from an unconfined aquifer, air occupies the voids in the sands within the cone of depression, because that part of the formation is actually dewatered. The volume of water drained per cubic foot of the formation is the value of S. When pumping is stopped, the rising water table may trap some of the air as bubbles in the

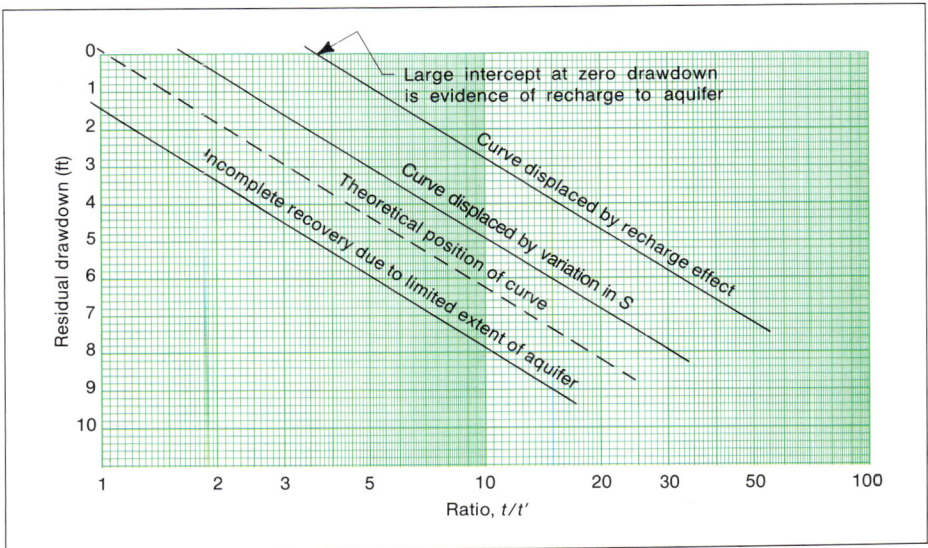

Figure 9.44. When real aquifer conditions differ from theoretical conditions, the residual-drawdown plot may be displaced in any of the three ways shown in this diagram.

voids of the sand. Thus, a slightly smaller volume of water will refill the dewatered portion of the formation, resulting in a correspondingly lower value of S during recovery. The effect of the variation in S on the residual-drawdown graph is shown in Figure 9.44. Note that this plot crosses the zero-drawdown line at a value of t/t' between 1 and 2.

Although analysis of residual-drawdown data is useful in interpreting constant-rate pumping tests, these data cannot be used to obtain a distance-drawdown plot. Readings in two or more observation wells taken at any instant during pumping can be plotted, as has been described, to illustrate the extent of the cone of depression. The nature of water-level recovery in the cone of depression causes the residual drawdown in observation wells away from the pumped well to equalize soon after pumping stops.

It may seem that much emphasis has been placed on variations in the value of S and the fact that S cannot always be calculated from the type of data available from a given pumping test. This does not imply, however, that these limitations invalidate the basic methods of aquifer analysis. Even a sustantial error in S does not affect greatly the calculations of well performance and interference between wells.

If the storage coefficient cannot be calculated from the test data, other information from the test is still useful. An assumed storage coefficient can be used, based on whether a confined aquifer or unconfined aquifer is being considered; geologic information would indicate this in most cases. For a confined aquifer, an assumed value of $S = 5 \times 10^{-4}$ can be used; for an unconfined aquifer, an assumed value of $S = 0.1$ can be used. The calculations will be less accurate than in cases where the real values are known, but close enough for most situations.

Although most aquifers do not conform to all the theoretical conditions assumed by Theis and Jacob, results from the application of these equations and their graphic relationships have been quite satisfactory. In nonuniform aquifers, uneven hydraulic interconnections throughout the formations result in continuous adjustment of flow between local regions of differing hydraulic conductivity. The cone of depression then tends to deepen and spread in a way that reflects the overall transmissivity and storage characteristics of the aquifer. Because T and S are average values for large areas, some variations in yield at individual well sites should be anticipated when predicting future performance from aquifer tests.

THEIS NONEQUILIBRIUM WELL EQUATION

Considerable space has been devoted to applications of Jacob's modification of the more general Theis equation. In many cases it is not necessary to use the more laborious Theis method because the additional accuracy it provides is not significant. But in some cases the pumping time may be limited or the drawdown cannot be measured relatively near the well. Data collected when t is small or r is large do not plot on a straight line in Jacob's method. Therefore, in certain cases, the Theis method is necessary to obtain T and S.

Recall that the simplest form of the Theis equation (9.5) is:

$$s = \frac{114.6 \, Q \, W(u)}{T} \qquad s = \frac{1}{4\pi} \frac{Q \, W(u)}{T}$$

$W(u)$, the well function of u, is an abbreviation for the exponential integral:

$$\int_u^\infty \frac{e^{-x}}{x} dx = W(u) = -0.5772 - \log_e u + u - \frac{u^2}{2 \cdot 2!} + \frac{u^3}{3 \cdot 3!} - \frac{u^4}{4 \cdot 4!} + \cdots \quad (9.17)$$

Equation 9.5a gave the value of u as:

$$u = \frac{1.87 r^2 S}{Tt} \qquad\qquad u = \frac{r^2 S}{4 Tt}$$

where
r = distance, in ft, from the center of a pumped well to a point where the drawdown is measured

where
r = distance, in m, from the center of a pumped well to a point where the drawdown is measured

If the transmissivity and storage coefficients are known, values for these and other terms can be substituted in the equation to obtain an unknown.

For example, suppose the transmissivity of a confined aquifer is 50,000 gpd/ft (620 m²/day) and the storage coefficient is 5×10^{-4}, a typical value for confined conditions. What would be the specific capacity, Q/s, of a 12-in (305-mm) well after 1 day of continuous pumping? In this case, $r = 0.5$ ft (0.15 m) and $t = 1$ day. First, calculate u:

$$u = \frac{1.87 \cdot (0.5)^2 \cdot 5 \times 10^{-4}}{50{,}000 \cdot 1} \qquad\qquad u = \frac{(0.15)^2 \cdot 5 \times 10^{-4}}{4 \cdot 620 \cdot 1}$$

$$= 4.7 \times 10^{-9} \qquad\qquad = 4.5 \times 10^{-9}$$

Next, the value of $W(u)$ corresponding to this value of u is read from Appendix 9.E. $W(u)$ in this case is 18.60 (18.64). Rearranging Equation 9.5 gives:

$$\frac{Q}{s} = \frac{T}{114.6\, W(u)} \qquad\qquad \frac{Q}{s} = \frac{4\pi T}{W(u)}$$

$$= \frac{50{,}000}{114.6 \cdot 18.60} \qquad\qquad = \frac{4 \cdot 3.14 \cdot 620}{18.64}$$

$$= 23.5 \text{ gpm/ft of drawdown} \qquad\qquad = 418 \text{ m}^3/\text{day/m of drawdown}$$

Having calculated the specific capacity, the drawdown in the pumped well for any pumping rate is easily determined. If the pumping rate is 200 gpm (1,090 m³/day), then:

$$s = \frac{Q}{Q/s} = \frac{200}{23.5} = 8.5 \text{ ft} \qquad\qquad s = \frac{Q}{Q/s} = \frac{1090}{418} = 2.6 \text{ m}$$

These calculations assume that the pumped well is 100-percent efficient. If the pumped well is discharging 200 gpm, what would be the drawdown 1,000 ft (305 m) from the pumped well after 1 day of pumping?

$$u = \frac{1.87\, r^2 s}{Tt} \qquad\qquad u = \frac{r^2 s}{4Tt}$$

$$= \frac{1.87 \cdot (1{,}000)^2 \cdot 5 \times 10^{-4}}{50{,}000 \cdot 1} \qquad\qquad = \frac{(305)^2 \cdot 5 \times 10^{-4}}{4 \cdot 620 \cdot 1}$$

$$= 0.0187 \qquad\qquad = 0.0188$$

For this value of u, Appendix 9.E shows that $W(u) = 3.46$. Substituting this value in Equation 9.5 gives:

$$s = \frac{114.6\, Q\, W(u)}{T} \qquad\qquad s = \frac{1}{4\pi}\frac{Q}{T} W(u)$$

$$= \frac{114.6 \cdot 200 \cdot 3.46}{50{,}000} \qquad\qquad = \frac{1{,}090 \cdot 3.46}{4 \cdot 3.14 \cdot 620}$$

$$= 1.59 \text{ ft} \qquad\qquad = 0.48 \text{ m}$$

The calculated drawdown is independent of the efficiency of the pumped well.

Use of the Theis equation to determine T and S from a field pumping test requires measurements of drawdown in at least one observation well. These should be made at proper intervals after the pumping starts; measurements in more than one observation well also can be used. A direct calculation from Equations 9.5 and 9.17 is not possible.

Theis worked out a graphical solution which yields T and S if values of other terms are known. The method involves matching a curve plotted from specific pumping test data with a type curve. The type curve is prepared by plotting values of $W(u)$ against $1/u$ on graph paper with logarithmic scales, as shown in Figure 9.45. Values from Appendix 9.E are used for this. Data from a pumping test are plotted on similar graph paper with logarithmic scales and cycles identical to those of the type curve.

Table 9.1 (page 221) shows data from a pumping test. Drawdown was measured at frequent intervals in an observation well 400 ft (122 m) from a well that was pumped at a constant rate of 500 gpm (2,730 m³/day).

Figure 9.46 shows these data plotted on regular graph paper. The curve shows how drawdown increases rapidly at first, with the rate of increase diminishing as time progresses so that the water level would appear to stabilize with continued pumping. Because the measurements show that the water level was still dropping after 240 minutes of pumping, the equilibrium equation (9.2) cannot be applied. The test data in this case can be analyzed using the Theis nonequilibrium equation.

The test data may be plotted conveniently in any of the following ways:

Method	Vertical log scale	Horizontal log scale
(a)	Drawdown, s	t
(b)	Drawdown, s	t/r^2
(c)	Drawdown, s	$1/r^2$

Method (c) is restricted to cases where data are obtained from three or more obser-

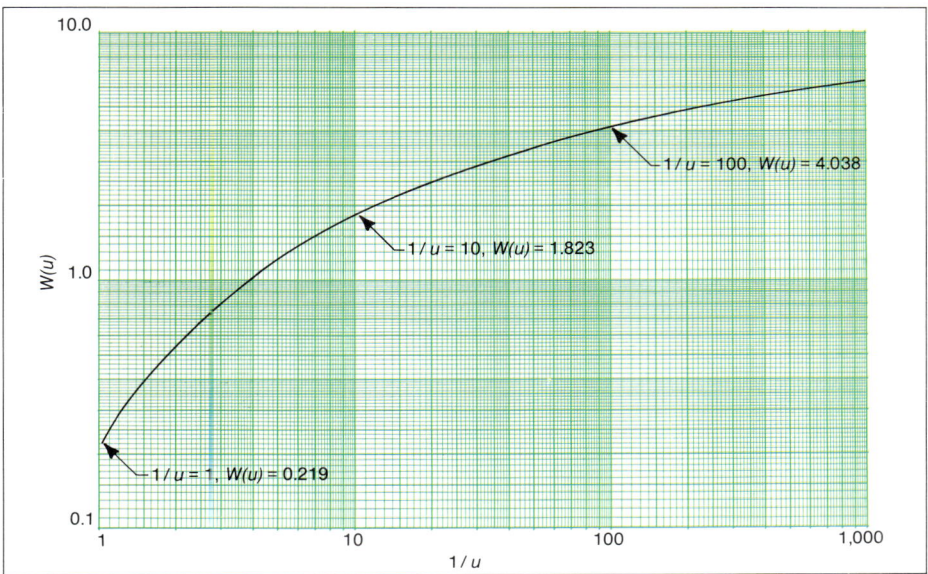

Figure 9.45. Type curve for graphic solution of Theis nonequilibrium equation shows values of $W(u)$, well function of u, corresponding to values of $1/u$. Curve is plotted on logarithmic graph paper.

vation wells during the test.

Use of the nonequilibrium equation requires that the data be plotted on logarithmic graph paper as shown in Figure 9.47. Plotting by method (a) is used here; drawdown

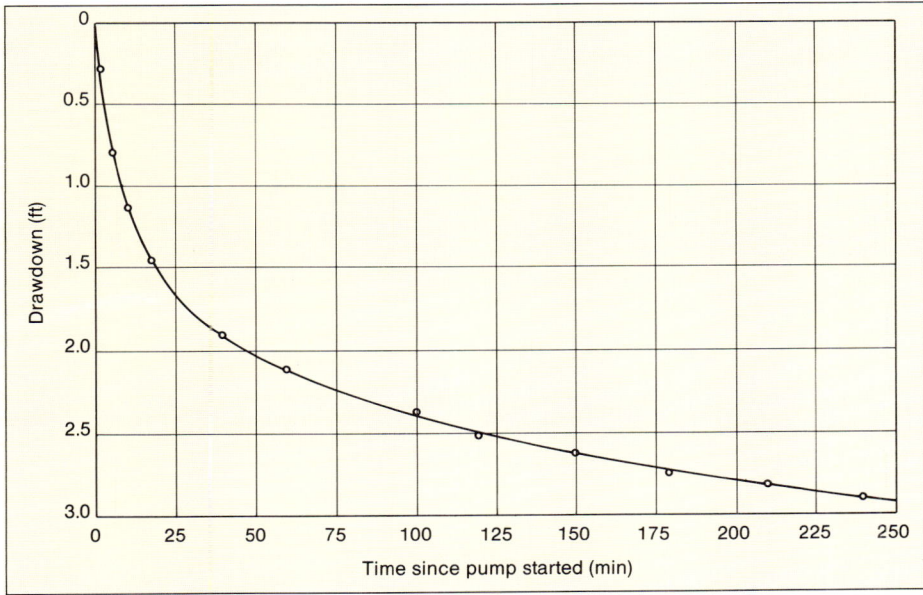

Figure 9.46. Data given in Table 9.1, plotted on ordinary graph paper, show how drawdown varies with duration of pumping. Water level drops rapidly at first, but rate of lowering decreases as pumping continues at 500 gpm (2,730 m³/day). Proper analysis of the test data requires use of the Theis nonequilibrium concept.

is on the vertical axis and the time since pumping began is on the horizontal axis. This graph is then superimposed on the type-curve sheet so the plotted points fall on or fit some portion of the type curve. In finding the position of best fit, the axes of both graphs must be kept parallel.

Once a good matching position is found, a match point is selected. The match point can be any convenient point on the graphs (that is, where u equals 1, 10, or 100, or s equals a whole number), but often the point is selected in the center of the area of best overlap as shown in Figure 9.48. The match point in Figure 9.48 is shown on the type curve where $1/u$ equals 100, and $W(u)$ equals 4.038. At the corresponding point on the time-drawdown diagram, s equals 2.3 ft (0.7 m) and t equals 83 minutes (0.058 days). Substituting in Equation 9.5, we have:

$$T = \frac{114.6 \, Q \, W(u)}{s} \qquad\qquad T = \frac{1}{4\pi} \frac{Q}{s} W(u)$$

$$= \frac{114.6 \cdot 500 \cdot 4.038}{2.3} \qquad\qquad = \frac{1}{4 \cdot 3.14} \frac{2{,}730}{0.7} \cdot 4.038$$

$$= 101{,}000 \text{ gpd/ft} \qquad\qquad = 1{,}250 \text{ m}^2/\text{day}$$

After determining T, we can calculate S from the following relationship:

$$S = \frac{u \, T t}{1.87 \, r^2} \qquad\qquad S = \frac{4 u T t}{r^2}$$

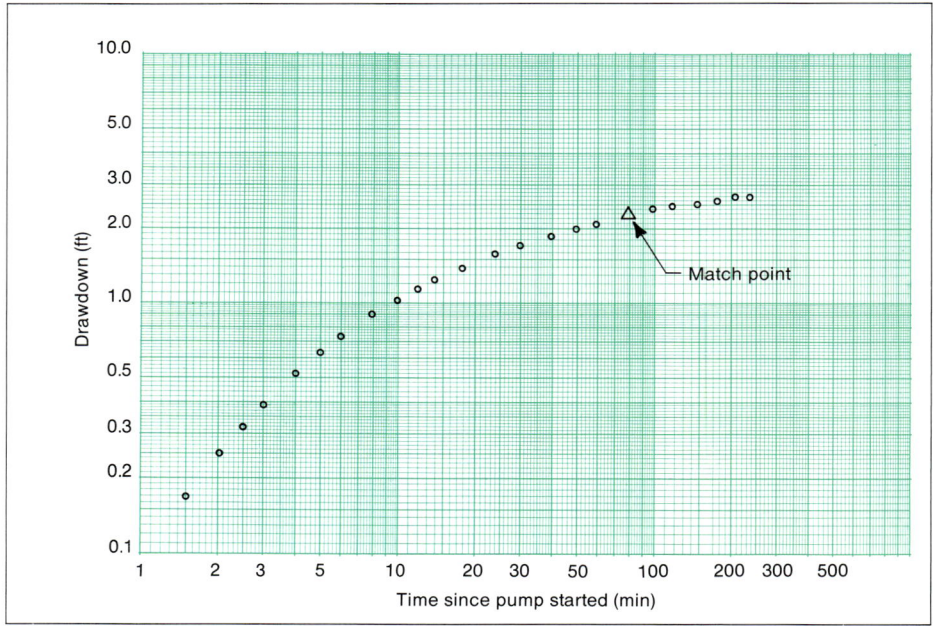

Figure 9.47. Data given in Table 9.1, plotted on logarithmic graph paper, define a curve similar in shape to the type curve in Figure 9.45.

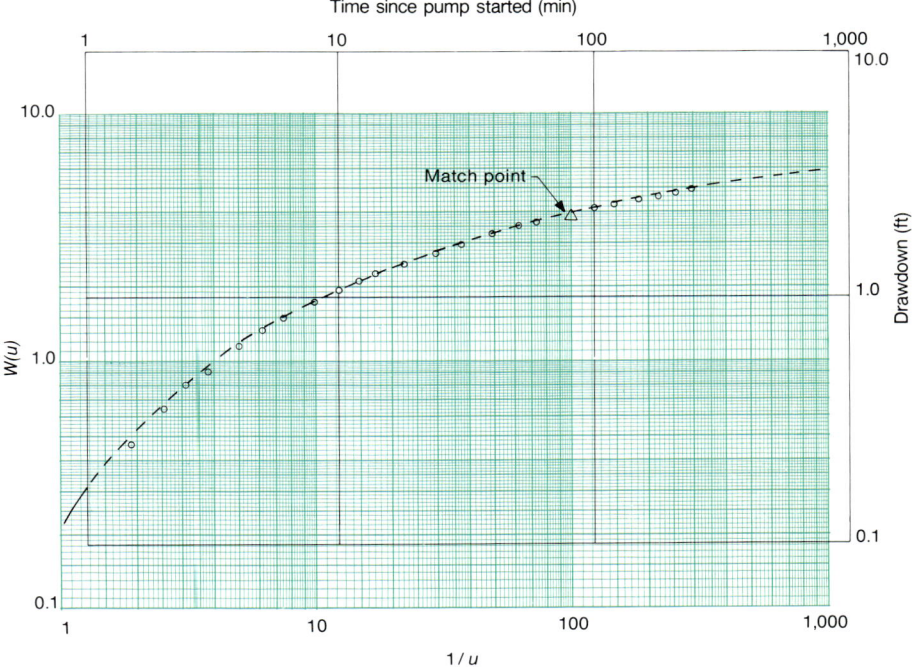

Figure 9.48. Diagram of plotted points representing pumping test data superimposed on the type curve. Match point chosen for $1/u = 100$.

The value of r in this example is 400 ft (122 m), the distance from the pumped well to the observation well. Thus:

$$S = \frac{1 \cdot 101{,}000 \cdot 0.058}{100 \cdot 1.87 \cdot (400)^2}$$
$$= 1.9 \times 10^{-4}$$

$$S = \frac{4 \cdot 1 \cdot 1{,}250 \cdot 0.058}{100 \cdot (122)^2}$$
$$= 1.9 \times 10^{-4}$$

Brown (1953) presents detailed explanations of other ways to solve the nonequilibrium equation by the curve-matching technique.

Under certain hydrogeologic conditions, pumping test data may deviate from the type curve. For example, Figure 9.49 shows that recharge and negative boundary effects can alter the pumping test data sufficiently to cause a distinct deviation from the type curve. Any variance from the type curve indicates that one or more of the assumptions upon which the Theis method is based have been violated. Leakage, casing storage, and mechanical difficulties encountered in the pumping test also can lead to data that do not follow the type curve exactly.

OTHER METHODS OF AQUIFER ANALYSIS

In recent years, many other analytical models (in addition to the Theis and Jacob methods) have been developed to aid in aquifer analysis. Analytical models, as presented in this chapter, are used to solve problems relating to small parts of a total aquifer system. Numerical computer models, on the other hand, are utilized to describe what is happening in the entire aquifer system. Although extremely useful to

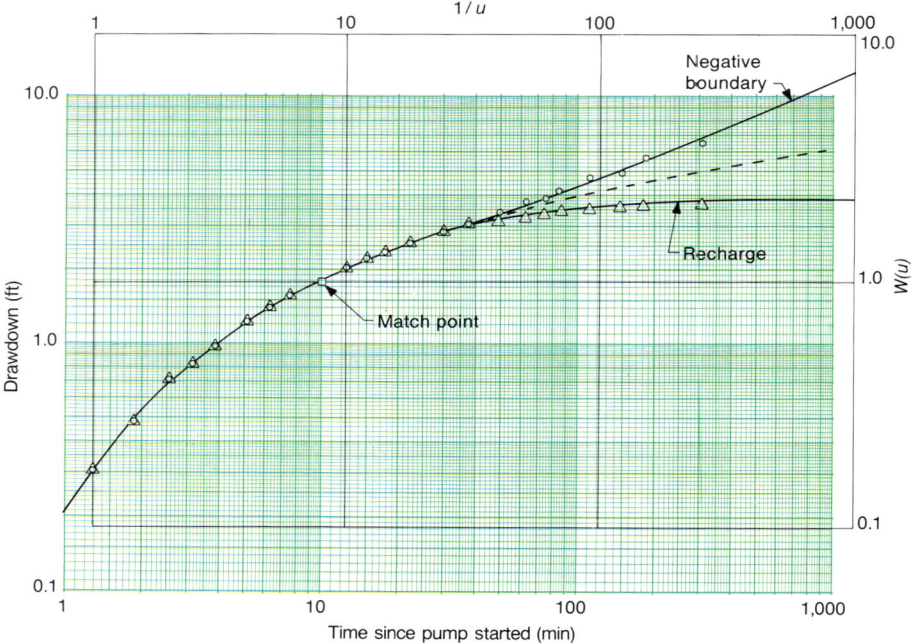

Figure 9.49. Effect of recharge and negative boundaries on pumping test data when using the Theis curve-matching method. In this instance, both boundaries are the same distance from the pumping well.

environmental planners, computer models are rarely of value to engineers in solving site-specific problems, as in the case of a single well. If computer models are based on a large and reliable data base, however, they can predict the response of an aquifer to large-scale (regional) pumping demands.

Analytical models will continue to be one of the most versatile methods for typical design engineers or well drilling contractors to analyze what is happening at specific sites. Generally, the models describe flow conditions in aquifers under more or less standard or ideal conditions. That is, the aquifer is (1) homogeneous, (2) infinite in areal extent, or has easily described geometric limits in the form of barriers or recharge boundaries, (3) uniform in thickness, (4) either isotropic or anisotropic, and (5) partially or fully penetrated by wells.

The models listed in Appendix 9.F describe nonequilibrium, time-drawdown, and distance-drawdown relationships in nonleaky, leaky, and unconfined aquifer systems. Use of these models where applicable may yield more accurate information than can be obtained by either the Theis or Jacob methods. The equations, however, may be somewhat more difficult to use. Long experience shows that judicious use of the Jacob method generally yields data of sufficient accuracy for engineering purposes, although greater accuracy may be obtained with one or more of the sophisticated models noted in Appendix 9.F.

SUMMARY

Three mathematical methods commonly used to calculate the principal aquifer characteristics, T and S, have been discussed in this chapter. Knowledge of these two aquifer parameters permits calculation of specific capacity, potential yield, drawdown at any point away from a well, and well efficiency. In some cases, however, drawdown data from pumping tests are difficult to analyze, and there may be a high degree of uncertainty regarding the actual values of T and S, and hence the real hydraulic potential of the aquifer. The extreme heterogeneity of many geologic formations leads to data that seem to defy rational explanation. Earlier chapters described in detail the formation of various types of aquifers. Recollection of this information is vital in analyzing pumping test data, for it allows the engineer to understand anomalies in the data which occur so often. Chapter 16 presents numerous examples of pumping tests conducted in various geologic media. Combining geologic information with a thorough understanding of well hydraulic theory is essential when analyzing drawdown data for T and S.

CHAPTER 10
Well Drilling Methods

Various well drilling methods have developed because geologic conditions range from hard rock such as granite and dolomite to completely unconsolidated sediments such as alluvial sand and gravel. Particular drilling methods have become dominant in certain areas because they are most effective in penetrating the local aquifers and thus offer cost advantages. In many cases, however, the drilling contractor may vary the usual drilling procedure depending on the depth and diameter of the well, type of formation to be penetrated, sanitation requirements, and principal use of the well. It is obvious, then, that no single drilling method is best for all geologic conditions and well installations. In most cases, the drilling contractor is best qualified to select the particular drilling procedure for a given set of construction parameters. Successful drilling is both an art developed from long experience and the application of good engineering practices.

Well construction usually comprises four or five distinct operations: drilling, installing the casing, placing a well screen and filter pack, if required, grouting to provide sanitary protection, and developing the well to insure sand-free operation at maximum yield. Two or more of these operations may be carried out simultaneously, depending on the drilling method used. For example, when drilling into an unconsolidated formation by the cable tool or drill-through casing driver methods, the casing is installed as drilling proceeds. When a well point (screen) is driven, three operations are performed simultaneously: the borehole is opened, the casing installed, and the well screen set.

Well drilling and installation methods are so numerous that only the basic principles and some of their applications can be described in this chapter. The practical limits of major drilling methods are presented for various geologic conditions. Methods for installing well screens and procedures for well development are explained in Chapters 14 and 15, respectively.

CABLE TOOL METHOD

Developed by the Chinese, the cable tool percussion method was the earliest drilling method and has been in continuous use for about 4,000 years. Using tools constructed of bamboo, the early Chinese could drill wells to a depth of 3,000 ft (915 m), although construction sometimes took two to three generations. Cable tool drilling machines, also called percussion or "spudder" rigs, operate by repeatedly lifting and dropping a heavy string of drilling tools into the borehole (Figure 10.1). The drill bit breaks or crushes consolidated rock into small fragments, whereas the bit primarily loosens the material when drilling in unconsolidated formations. In both instances, the reciprocating action of the tools mixes the crushed or loosened particles with water to form

Figure 10.1. This small cable tool rig operating in Australia is equipped with a bailer to remove cuttings periodically from the borehole. Cable tool machines also are called percussion or "spudder" rigs.

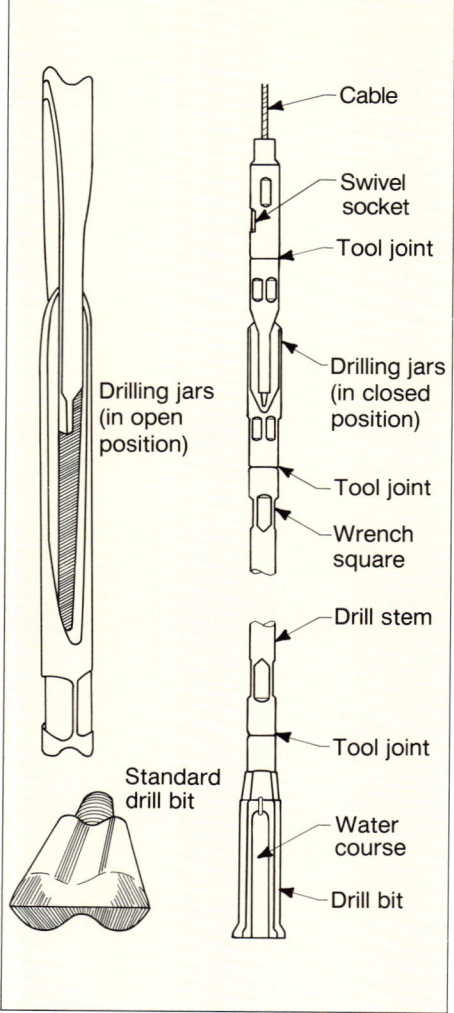

Figure 10.2. A full string of cable tools consists of five components that are necessary for drilling.

a slurry or sludge at the bottom of the borehole. If little or no water is present in the penetrated formation, water is added to form a slurry. Slurry accumulation increases as drilling proceeds and eventually it reduces the impact of the tools. When the penetration rate becomes unacceptable, slurry is removed at intervals from the borehole by a sand pump or bailer.

A full string of cable tool drilling equipment consists of five components: drill bit, drill stem, drilling jars, swivel socket, and cable (Figure 10.2). Each component has an important function in the drilling process. The cable tool bit is usually massive and heavy so as to crush and mix all types of earth materials. The drill stem gives additional weight to the bit, and its length helps to maintain a straight hole when drilling in hard rock.

Drilling jars consist of a pair of linked, heat-treated steel bars. When the bit is stuck, it can be freed most of the time by upward blows of the free-sliding jars. This is the primary function of the drilling jars; except in unusual circumstances, they serve no purpose in the drilling operation itself. The stroke of the drilling jars is 9 to 18 in (229 to 457 mm) and distinguishes them from fishing jars which have a stroke of 18 to 36 in (457 to 914 mm) or longer.

The swivel socket connects the string of tools to the cable; in addition, the weight of the socket supplies part of the upward energy to the jars when their use becomes necessary. The socket transmits the rotation of the cable to the tool string and bit so that new rock is cut on each downstroke, thereby assuring that a round, straight hole will be cut. The elements of the tool string are screwed together with right-hand threaded tool joints of standard API (American Petroleum Institute) design and dimension.

The wire cable that carries and rotates the drilling tool is called the drill line. It is a ⅝- to 1-in (16- to 25-mm) left-hand lay cable that twists the tool joint on each upstroke to prevent it from unscrewing. The drill line is reeved over a crown sheave at the top of the mast, down to the spudding sheave on the walking beam, to the heel sheave, and then to the working-line side of the bull reel (Figure 10.3). Bull reels are generally set up with a separator on the drum to provide a working-line side and a

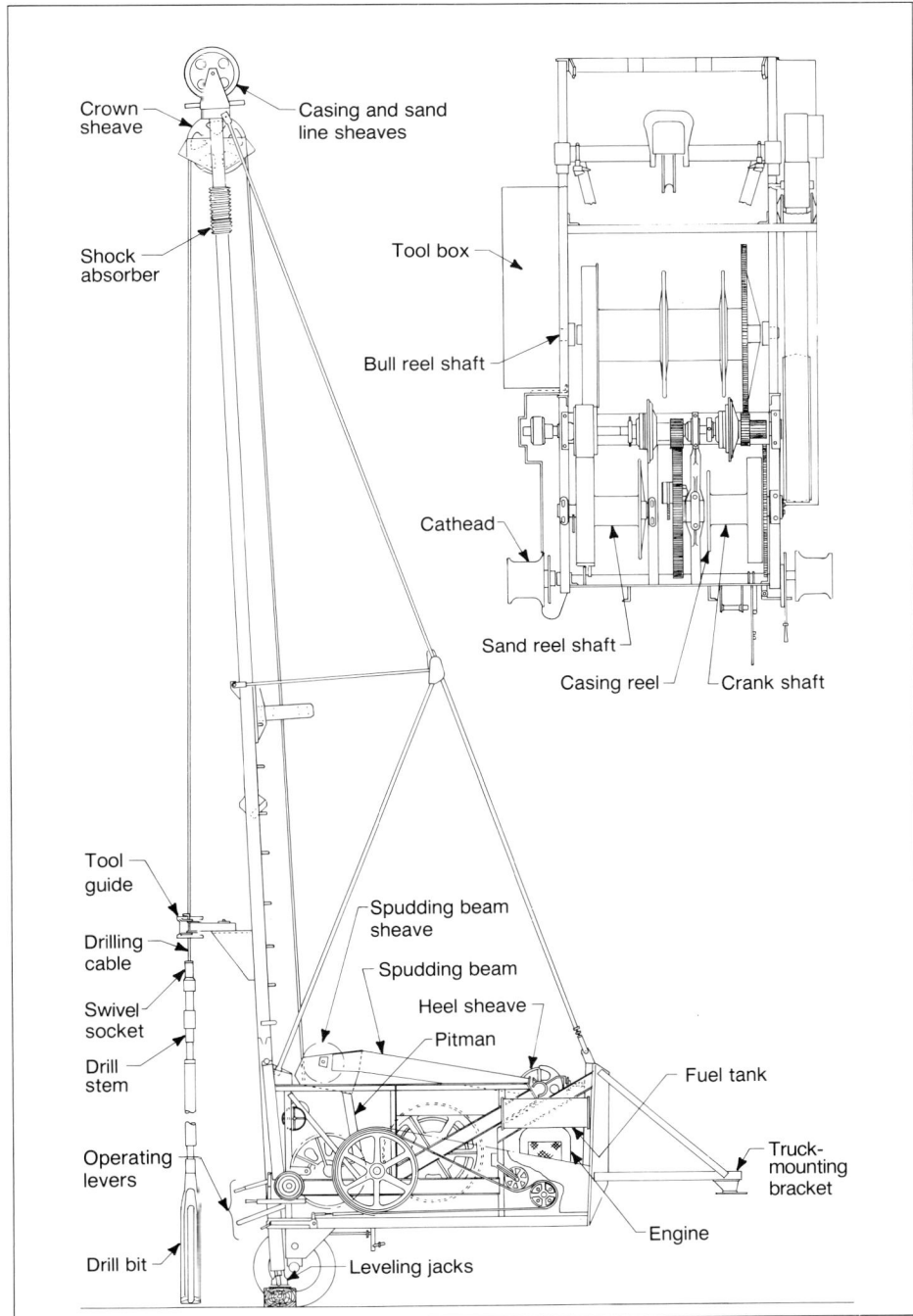

Figure 10.3. Engineering drawing of a Bucyrus-Erie Model 22-W shows how the drill line is reeved in a typical cable tool rig. The spudding action is imparted to the drill line by the vertical motion of the spudding beam. The shock absorber mounted beneath the crown block helps control the impact of the bit on the rock. *(Bucyrus-Erie Company)*

storage-line side.

Bailers used to remove the mud or rock slurry consist of a pipe with a check valve at the bottom. The valve may be either a flat pattern or a ball-and-tongue pattern called a dart valve (Figure 10.4). A bail handle at the top of this tool attaches to a cable called the sand line. The sand line is threaded over a separate sheave at the top of the mast and down to the sand-line reel. The diameter of the sand line can vary according to the anticipated loads.

Another type of bailer is called the sand pump or suction bailer. This bailer is fitted with a plunger so that an upward pull on the plunger tends to produce a vacuum that opens the valve and sucks sand or slurried cuttings into the tubing. The sand pump can have a bit bottom, but more often in water well drilling it has a flat bottom with a flap-type valve (Figure 10.5). Some sand pump bailers have a latch bottom for slurry release. Most sand pumps are either 10 or 20 ft (3 or 6.1 m) long.

Figure 10.4. Dart valve bailers are used periodically to remove the slurry from the borehole. *(Bergerson-Caswell Company)*

The characteristic up and down drilling action of a cable-tool machine is imparted to the drill line and drilling tools by the walking beam. The walking beam pivots at one end while its outer end, which carries a sheave for the drill line, is moved up and down by a single or double pitman connected to a crank shaft. The vertical stroke of the walking beam, and thus the drill tools, can be varied by adjusting the position of the pitman pin on the bull gear and the pitman connection to the walking beam. The number of strokes per minute can be varied by changing the speed of the drive shaft. The bull gear is driven by a pinion mounted on a clutch. This clutch, the friction drive for the sand line (on smaller cable tool rigs only), and the drive pinion for the drill-line reel are all mounted on the same drive shaft assembly.

Figure 10.5. Sand pumps are another common type of bailer that are fitted with a flap-type bottom valve. As the sand pump is alternately lowered and raised, slurry is sucked into the pump barrel. Slurry is released at the surface into a chute that directs the slurry into a container if a sample is desired or onto the ground away from the rig. *(E. H. Renner & Sons, Inc.)*

Another drum, called a casing reel, is frequently added to the basic machine assembly. The casing reel is capable of exerting a powerful pull on a third cable, the casing line. This cable is used for handling pipe, tools, and pumps, or other heavy hoisting. It may be used to pull a string of casing when the cable is reeved with blocks to make two-, three-, or four-part lines. Reinforcement of the derrick by means of a "stiff leg" may be required to utilize the maximum pull that can be applied.

Another commonly used auxiliary hoisting device on a cable tool machine is called a cathead. Use of this drum requires that a heavy line of manila rope be carried on a separate sheave at the top of the derrick. This line may be used for handling light loads and alternately lifting and dropping tools such as a drive block or bumper which are used to drive or lift casing. Synthetic ropes made with nylon or dacron are considerably stronger than manila rope, but they are not resistant to abrasion or heat and therefore cannot be used with a cathead. Two or three loose turns of the free end of the rope are wrapped on the cathead. When the cathead is rotating, the driller pulls on the free end of the rope, causing the coils to tighten and grip the cathead. This raises the load at the other end of the rope. When the driller reduces the pulling pressure on the rope, friction between the rope and the revolving cathead is reduced and the load descends at a controlled rate. The cathead is a live drum; that is, there are no clutches to engage or disengage during use.

Every cable tool machine has certain interdependent limits on borehole depth and diameter. For example, if a hole is relatively small in diameter, it may be drilled to relatively great depth. In large-diameter holes, the weight of the drill string and cable may become so excessive that the machine cannot function, thereby limiting well depth at the initial diameter. Collapsing formations may further limit the effective depth for large-diameter casing, because considerable friction develops between the casing and borehole wall while the casing is being driven. In many cases, the casing size is progressively decreased as the hole is deepened, thereby reducing friction and also the weight of the drilling tools. Friction between the borehole wall and casing can be reduced by the addition of a drilling fluid slurry around the outside of the casing during driving. This small amount of slurry will also decrease the energy required for pulling back casing to expose screens set within the casing. In water well drilling, the depth capability for cable tool rigs ranges from 300 to 5,000 ft (91.5 to 1,520 m).

The drilling motion of the cable tool machine must be synchronized with the gravity fall of the tools for effective penetration. Several factors (thickness of the slurry in the borehole, whip in the cable, hole alignment, and rocks protruding in the borehole) may interfere with the free gravity fall, and the driller must adjust the motion and speed of the machine to the vertical movement of the tools. Effective drilling action is obtained when the engine speed is synchronized with the fall of the tools and the stretch of the cable, while paying out the correct amount of cable to maintain proper feed of the bit. The bit should strike the bottom of the hole at the extreme (elastic) limit of the cable and immediately snap upward so that a sharp blow is given to the earth material by the bit. This requires some resilience and elasticity in the cable and certain parts of the rig mechanism. An elastic snubber or shock absorber is usually installed in the mounting of the drill-line crown sheave to provide part of the resilience in the system. The shock absorber compresses as the walking beam completes its upstroke and starts its pull on the cable. Cable tension then reaches its maximum, because the tools are still moving downward. The shock absorber's rebound helps to lift the tools sharply after they strike bottom. The objective is to give the tools that peculiar whip at the end of the stroke which is essential to rapid drilling. At the surface, the cable will appear to be constantly in tension. When properly done, this technique conserves power and increases drilling speed. The shock absorber also dampens the vibration that occurs when the drill bit strikes the bottom of the hole; it protects the derrick and the rest of the machine from severe shock stresses.

Drilling Consolidated Formations

Most boreholes completed in consolidated formations by the cable tool method are drilled "open hole," that is, no casing is used during part or all of the drilling operation. When drilling in consolidated rock, the cable tool bit is essentially a crusher. Its performance depends on the energy it can deliver to the bottom of the hole when the proper drilling motion is maintained.

Figure 10.6. Several types of bits are used in cable tool drilling depending on the nature of the geologic materials. Standard and star types are the most common bit configurations. *(Keys Well Drilling; E. H. Renner & Sons, Inc.)*

Factors that affect drilling rate or efficiency are: resistance of the rock; dip of the rock structure; weight of drill tools; length of stroke; strokes per minute; diameter, sharpness, and shape of bit; clearance between the tool string and the hole; and density and depth of the accumulated slurry. Each driller relies on the drilling machine manufacturer for guidance on these factors, and adds to this basic knowledge from personal experience. Two common bits used in cable tool drilling are shown in Figure 10.6. A partial listing of bit types and dimensions is given in Appendix 10.A.

Drilling Unconsolidated Formations

Drilling in unconsolidated formations differs from hard-rock drilling in two ways. First, pipe or casing must follow the drill bit closely as the well is deepened to prevent caving and keep the borehole open. Usually the casing has to be driven by an operation similar to pile driving. Second, the drilling action of the bit is largely a loosening and mixing process. Actual crushing is of little importance except when a large stone or boulder is encountered.

Figure 10.7. To drive casing, drive clamps are mounted on the upper wrench square of the drilling stem. The tools provide driving weight. *(NWWA)*

A drive shoe made of hardened and tempered steel is attached to the lower end of the casing string. This shoe prevents damage to the bottom of the casing when it is being driven. For the pipe-driving operation, a drive head is fitted to the top of the casing to serve as an anvil and protect the top of the casing. Drive clamps — constructed of heavy steel forgings made in halves — are attached to the square near the top of the drill stem. Drive clamps act as the hammer face, and the tools provide the weight for driving the pipe (Figure 10.7). The tools are lifted and dropped by the spudding action of the drilling machine.

The usual procedure is to drive the casing initially for 3 to 10 ft (0.9 to 3 m). Material in the casing is then mixed with water by the drill bit to form a slurry. Most of the slurry is bailed out and the pipe is driven again. Each time that the casing is cleaned out, more water must be added if none is encountered in the formation being drilled. In some cases, the hole is drilled 3 to 6 ft (0.9 to 1.8 m) below the casing; the casing is then driven down to the undisturbed material and drilling is resumed. Driving, drilling, and bailing operations are repeated until the casing is at the desired depth.

When friction on the outside of the casing increases to the point where the casing cannot be driven any deeper or further driving might damage it, a string of smaller casing is inserted inside the first one. Drilling is then continued inside the smaller casing. The diameter of the well is thus reduced; two or three reductions may be required in certain cases before reaching the desired depth. If friction problems are anticipated, casing in the upper part of the borehole should be one or two sizes larger than the diameter specified for completion of the well.

Figure 10.8. An alternative type of drive clamp (sometimes called a drive block) is shown being attached to the drill string.

When penetrating most unconsolidated formations, driving the casing occupies as much time as the actual drilling and bailing. The physical nature of clay, silt, sand, gravel, and marl profoundly affects the rate at which casing can be driven. The best driving weight and setting of the spudding motion is determined from experience in a given locality.

Sometimes a drive block is used in place of drive clamps on the drilling tools for driving casing. A drive block assembly, similar to that shown in Figure 10.8, is set on top of the casing each time the casing is driven. The block is lifted and dropped by using the cathead. The cathead and block can also be used to bump back the casing.

Figure 10.9. Hydraulic jacks can be used to pull the casing into the ground while drilling and bailing proceed. *(A. M. Bisley and Company)*

In some cable tool operations, the casing is not driven at all, but is pushed into the ground by hydraulic jacks as drilling and bailing proceeds (Figure 10.9). Casing can also be pulled back by using the jacks. Several advantages of this method are immediately apparent. Drilling proceeds rapidly because it is not necessary to stop the drilling and bailing to drive pipe. Because downward pressure is maintained constantly on the casing during drilling and bailing, caving and overexcavation are minimized. Perhaps most important, the jarring action of ordinary driving procedures, which compacts sand and gravel formations near the casing and causes excessive friction, is avoided. Sixteen-inch (406-mm) casing has been hydraulically jacked to depths of 1,000 ft (305 m) in unconsolidated formations, and subsequently pulled back 150 ft (45.7 m) to expose the screen. In this case, the jacks provided more than 500,000 lbs (227,000 kg) of lifting force.

When drilling in shallow sands, the casing may follow the bit down without being driven. In these areas, the casing may have to be held to prevent it from sinking too rapidly and to maintain plumbness. Also, these formations may be drilled more rapidly by using the sand pump bailer (suction bailer) to remove the sand without a bit. In Argentina, near Buenos Aires, temporary or surface casing has been installed to depths of 200 ft (61 m) using this method.

Another drilling technique, called the open-hole or reverse cable tool method, has been used for many years in Japan and has been used recently in the Western United States. With the borehole full of water or drilling fluid, heavy sand pumps or bailers are operated inside the casing to cut the borehole. Holes to 24 inches (610 mm) in diameter with cobbles to 12 in (305 mm) can be drilled in this manner. Large-diameter irrigation wells have been drilled to 100 ft (30.5 m) and screened within one day using this method. In isolated cases, the hole can be drilled open-hole even in completely unconsolidated formations, because the hydrostatic water pressure prevents caving of the borehole walls.

The cable tool method has survived for thousands of years because it is reliable for a wide variety of geologic conditions. It may be the best, and in some cases the only, method to use in coarse glacial till, boulder deposits, or rock strata that are highly disturbed, broken, fissured, or cavernous. In situations where the aquifers are thin and yields are low, the cable tool operation permits identification of zones that might be overlooked in other drilling methods. The cable tool method offers the following advantages:
1. Rigs are relatively inexpensive.
2. Rigs are simple in design and require little sophisticated maintenance.
3. Machines have low energy requirements.
4. Borehole is stablized during the entire drilling operation.
5. Recovery of reliable samples is possible from every depth unless heaving conditions occur.
6. Wells can be drilled in areas where little make-up water exists.
7. Wells can be constructed with little chance of contamination.
8. The driller maintains intimate contact with the drilling process and the materials encountered by keeping a hand on the drilling cable.
9. Generally, only one person is needed to operate the drilling rig, although a helper is usually available to assist.

10. Because of size, machines can be operated in more rugged, inaccessible terrain or in other areas where space is limited.
11. Rigs can be operated in all temperature regimes.
12. Wells can be drilled in formations where lost circulation is a problem.
13. Wells can be bailed at any time to determine the approximate yield at that depth.

Some disadvantages of the cable tool method include the following:
1. Penetration rates are relatively slow.
2. Casing costs are usually higher because heavier wall or larger diameter casing may be required.
3. It may be difficult to pull back long strings of casing in some geologic conditions, unless special equipment is available.

Other drilling techniques have been devised because of some inherent disadvantages of the cable tool method. Because the method is often slow, each cable tool driller can complete only a limited number of holes per year despite high operating efficiency. In times of high customer demand, the driller may not be able to take on much new business without adding new machines, an expense the long-term economics of the business may not allow. In addition, drillers experienced with cable tool rigs may not be available.

CALIFORNIA STOVEPIPE METHOD

The California stovepipe method of drilling applies the same principles as the usual cable tool method, but differs in three ways: a heavy bailer called a mud scow is used as both drill bit and bailer, laminated steel casing in short lengths is used in place of standard steel pipe, and hydraulic jacks are used to force the casing downward as opposed to driving the casing by impact of the tools.

When the casing reaches the desired depth, a casing perforator is used to puncture holes in the pipe opposite the water-bearing formation (Figure 10.10). The size of the openings produced by the perforator is relatively uncontrolled and wells completed this way often pump large quantities of sand. The telescope method of installing well screens (described in Chapter 14) cannot be utilized because the spot-welded casing joints are too weak to withstand the pull-back forces required to expose the well screen. If line pipe is used for casing, however, well screens that will control sand may be

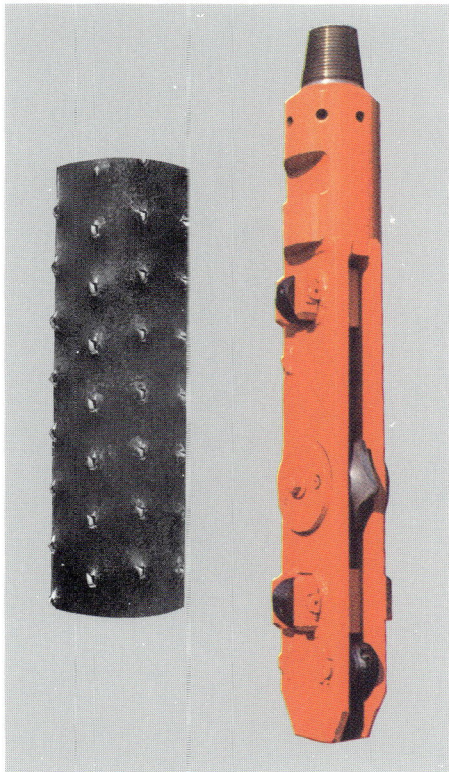

Figure 10.10. Casing perforators offer a crude method to gain access to the aquifer when well screens cannot be used. This perforator is activated by compressed air. *(Driltech, Inc.)*

DIRECT ROTARY DRILLING

The direct rotary drilling method was developed to increase drilling speeds and to reach greater depths in most formations (Figure 10.11). The borehole is drilled by rotating a bit, and cuttings are removed by continuous circulation of a drilling fluid as the bit penetrates the formation. The bit is attached to the lower end of a string

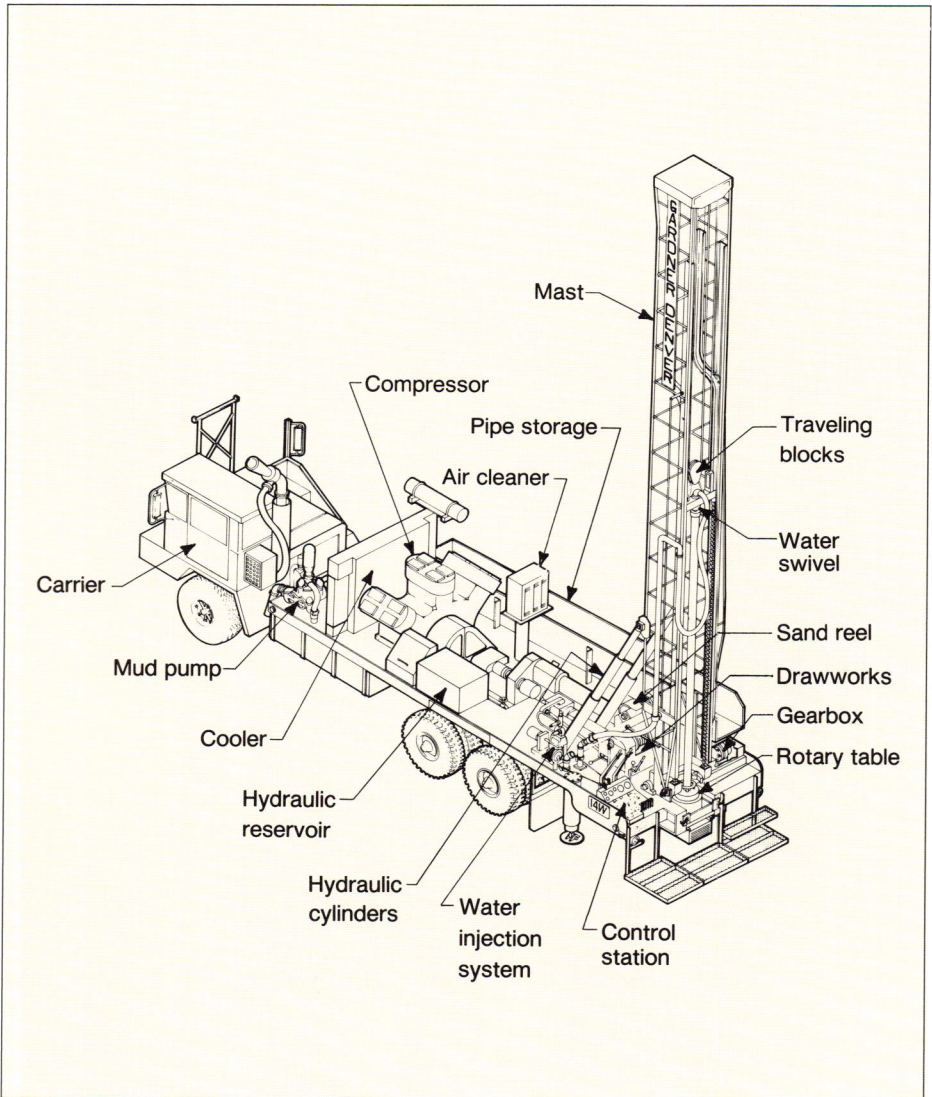

Figure 10.11. Schematic diagram of a direct rotary rig illustrates the important operational components of this truck-mounted drilling machine. This machine, operating with either an air-based or water-based drilling fluid, can drill more rapidly than a cable tool rig. *(Gardner-Denver Company)*

of drill pipe, which transmits the rotating action from the rig to the bit. In the direct rotary system, drilling fluid is pumped down through the drill pipe and out through the ports or jets in the bit; the fluid then flows upward in the annular space between the hole and drill pipe, carrying the cuttings in suspension to the surface. At the surface, the fluid is channeled into a settling pit or pits where most of the cuttings drop out. Clean fluid is then picked up by the pump at the far end of the pit or from the second pit and is recirculated down the hole (Figure 10.12). For relatively shallow wells, 150- to 500-gal (0.6- to 1.9-m^3) portable pits may be used; much larger portable pits, 10,000 to 12,000 gal (37.9 to 45.4 m^3), are used for deeper wells. Mud pits may also be excavated for temporary use during drilling and then backfilled after completion of the well (see Chapter 11 for various mud pit configurations).

Before 1920, the type of rotary drill used in water well drilling was commonly called a whirler. This equipment used the well casing itself as the drill pipe. The lower end of the pipe was fitted with a serrated cutting shoe with an outside diameter a little larger than the drill pipe couplings. The sawteeth of the shoe cut and loosened the materials as the pipe was rotated. Water was pumped under pressure through the pipe to lift the cuttings to the surface. Native clays and silt were depended upon to seal the borehole wall to maintain circulation; prepared drilling fluids were not used. The method was suitable for drilling only relatively small-diameter, shallow wells in unconsolidated formations that did not contain cobbles or boulders.

In the 1930's, shot-hole rotary drills, used for seismograph work in oil exploration, were successfully adapted for drilling small-diameter water wells. Shot-hole machines, however, could not drill the large-diameter holes necessary for water well work because the mud pump and drill pipe were generally too small to circulate enough drilling

Figure 10.12. Drilling fluid from the borehole flows into the larger pit where the cuttings settle out. The fluid then flows into the second pit through a constricted opening. The mud pump on the rig withdraws drilling fluid from this pit to inject down the drill rods to the bit. This Italian driller has lined the drilling fluid pits with polyethylene film to reduce fluid loss into the ground. Note the homemade hole cleaner or scratcher the driller uses to keep the borehole open during drilling.

Figure 10.13. Drag bits are used in rotary drilling for fast penetration in unconsolidated or semiconsolidated sediments.

fluid to efficiently drill even an 8-in (203-mm) well. In time, truck-mounted portable rigs for drilling large-diameter water wells were developed from oil field exploration technology.

The components of the rotary drilling machine are designed to serve two functions simultaneously: operation of the bit and continuous circulation of the drilling fluid. Both are indispensable in cutting and maintaining the borehole. For economic and efficient operation, rotary drillers must acquire considerable knowledge concerning these factors and how they relate to various formation conditions.

In direct circulation rotary drilling for water wells, two general types of bits are used — the drag bit (fishtail and three- and six-way designs) and the roller cone bit, usually called a rock bit. Drag bits have short blades, each forged to a cutting edge and faced with durable metal (Figure 10.13). Short nozzles direct jets of drilling fluid down the faces of the blades to clean and cool them. Drag bits have a shearing action and cut rapidly in sands, clays, and some soft rock formations, but they do not work well in coarse gravel or hard-rock formations.

Roller (cone) bits exert a crushing and chipping action, making it possible to cut hard formations (Figure 10.14). The rollers, or cutters, are made with either hardened steel teeth or tungsten carbide inserts of varied shape, length, and spacing, designed so that each tooth applies pressure at a different point on the bottom of the hole as the cones rotate. The teeth of adjacent cones intermesh so that self-cleaning occurs. Long, widely spaced teeth are used in bits designed to cut soft clay formations, whereas shorter, closer spaced teeth are used for denser formations. Some roller bits are made with carbide buttons for particularly dense and abrasive formations such as dolomite, granite, chert, basalt, and quartzite.

The tricone bit, used as an all-purpose bit in every type of formation, has conically shaped rollers on spindles and bearings set at an angle to the axis of the bit. Another design has four rollers; two are set at an angle and two are normal to the vertical axis of the bit. The cutting surfaces of all roller bits are flushed by jets of drilling fluid directed from the inside (center) of the bit. The jets can be sized so as to maximize the cutting action of the bit. The jets are also effective in breaking up or washing away soft formation materials.

When hole enlargement becomes necessary, two other types of bits are used — reamers and underreamers (Figure 10.15). A reamer is used to straighten, clean, or enlarge a borehole. This tool sometimes consists of a 10- to 20-ft (3- to 6.1-m) section of drill pipe with specially hardened surfaces on vertical ribs. Other types of reamers are constructed of flanges welded on short sections of drill pipe and mounted between the bit and the stabilizer. In the underreaming process, the borehole diameter is enlarged beneath the permanent casing. Underreamers are particularly useful when a filter pack must be placed around a screen, but the cost of drilling the entire borehole at the larger diameter required for the filter pack would be prohibitive.

The bit is attached to the lower end of the drill pipe, which resembles a long tubular shaft. The drill string usually consists of four parts: the bit, one or more drill collars or stabilizers, one or more lengths of drill pipe, and, in table-drive machines, the kelly (Figure 10.16). Selection of the bottom-hole assembly will depend on the physical conditions of the geologic materials; these include dip of the formation, presence of faults or fractures, and drillability of the formation.

Each drill collar is a heavy-walled length of drill pipe; one or more drill collars are used to add weight to the lower part of the drill-stem assembly (Figure 10.17). The concentration of weight just above the bit helps to keep the hole straight, and provides sufficient weight for the bit to maintain the proper penetration rate. Drill collars fitted with stabilizer bars or rollers are even more effective in drilling straight boreholes. Table 10.1 presents representative data on recommended sizes of drill collars.

Stabilizers are an important component of the bottom-hole tools (Figure 10.18). To be effective in maintaining straight holes in soft formations, the stabilizer must have large wall contact. Increased contact can be achieved by using stabilizers with longer and wider blades, or by using longer stabilizers. The flow of drilling fluid upward around the stabilizer must not be restricted too much, however, because cuttings may pack around the stabilizer. This leads to sticking and a possible loss of circulation if back pressure builds up. Weakening of the formation structure can also result from the pressure increase. Accumulation of cuttings around the stabilizer may also cause local zones of erosion in the bore-

Figure 10.14. Roller or cone-type bits are preferred when drilling consolidated rock. The number of teeth on each roller cone depends on the drilling difficulty. As the rock becomes harder and more difficult to drill, the bit should have more teeth on each cone. For particularly dense or abrasive formations, carbide buttons are used instead of teeth on the roller cones. Roller cone bits are often constructed in configurations that will enlarge the borehole in stages as the bit penetrates the formation. For the bit shown, the primary bit is 17½ in (445 mm) and the reamer is 22 in (559 mm).

hole wall. In relatively hard formations, the stabilizer can perform satisfactorily with less wall contact.

Drill pipe is seamless tubing manufactured in joints that are usually 20 ft (6.1 m) long, although other lengths are available. Each joint is equipped with a tool-joint pin on one end and a tool-joint box on the other (Figure 10.19). Outside diameters of drill pipe used for direct rotary drilling generally range from 2⅜ to 6 in (60 to 152 mm). High circulation rates for drilling fluids in water well drilling require that the drill pipe diameter be adequate to hold friction loss in the pipe to an acceptable level so as to reduce the power required for the pump. For efficient operation, the outside diameter of the tool joint should be about two-thirds the borehole diameter; this ratio may be impractical, however, for holes larger than 10 in (254 mm).

In table-drive machines, the kelly constitutes the uppermost section of the drill string column. It passes through and engages in the opening in the rotary table, which is driven by hydraulic or mechanical means (Figure 10.20). The outer shape of the

Table 10.1. Ideal Size Range for Drill Collars

Hole size, in	Casing size to be run, in OD	Calculated ideal drill collar range, in		API drill collar sizes which fall in the ideal range, in
		Min.	Max.	
6⅛	4½	3.875	4.750	4⅛, 4¾
6¼	4½	3.750	4.875	4⅛, 4¾
6¾	4½	3.250	5.125	3½, 4⅛, 4¾, 5
7⅞	4½	2.125	6.125	3⅛, 3½, 4⅛, 4¾, 5, 6
	5½	4.225	6.125	4¾, 5, 6
8⅜	5½	3.725	6.500	4⅛, 4¾, 5, 6, 6¼, 6½
	6⅝	6.405	6.500	6½
8½	6⅝	6.280	6.750	6½, 6¾
	7	6.812*	6.750	6¾
8¾	6⅝	6.030	7.125	6¼, 6½, 6¾, 7
	7	6.562	7.125	6¾, 7
9½	7	6.812	7.625	6, 6¼, 6½, 7, 7¼
	7⅝	7.500	7.625	7⅝†
9⅞	7	5.437	8.000	6, 6¼, 6½, 6¾, 7, 7¼, 7¾, 8
	7⅝	7.125	8.000	7¼, 7¾, 8
10⅝	7⅝	6.375	8.500	6½, 6¾, 7, 7¼, 7¾, 8, 8¼
	8⅝	8.625*	8.500	8¼
11	8⅝	8.250	9.625	8¼, 9, 9½
12¼	9⅝	9.000	10.125	9, 9½, 9¾, 10
	10¾	11.250*	10.125	10
13¾	10¾	9.750	11.250	9¾, 10, 11
14¾	11¾	8.750	12.000	9, 9½, 9¾, 10, 11, 12†
17½	13⅜	11.250	13.375	12†
20	16	14.000	14.750	14†
24	18⅝	15.500	16.750	16†
26	20	16.000	19.500	16†

*In these instances, the equation used to calculate the ideal minimum drill collar size produces an anomalously high value. See Woods and Lubinski (1954) for a complete discussion on how to determine the best collar size for a specific diameter borehole.

†Not API standard size drill collar.

(Drilco, 1979)

Figure 10.15. On the left, a three-tiered 48-in (1,220-mm) reamer bit has just been removed from the borehole. *(Snider Drilling Ltd.)* For soft sediments, underreamers (right) are constructed of blades that extend outward from the bit.

kelly may be square or hexagonal, or round with lengthwise grooves or flutes cut into the outside wall. Made about 3 ft (0.9 m) longer than one joint of drill pipe, the kelly has an inside bore that is usually smaller than that of the drill pipe because of the heavy wall thickness required. The square, hexagonal, or grooved circular section of the kelly works up and down through drive bushings in the rotary table. With the bushings properly in place around the kelly, the entire drill stem and bit are forced to turn with the rotary table. While rotating, the kelly slips down through the drive bushings to feed the bit downward as the hole is drilled. The lower end of the kelly is provided with a replaceable substitute joint (sub), called a "kelly saver," that connects to the drill pipe. The sub saves the tool joint on the kelly from excessive wear resulting from the screwing and unscrewing of innumerable sections of drill pipe. The upper end of the kelly connects to a swivel (by a left-hand threaded joint) that is suspended from a traveling block in the derrick (Figure 10.21). A heavy thrust bearing between the two parts of the swivel carries the entire weight of the drill string while allowing the drill pipe to rotate freely.

Some rotary drilling machines use a top-head drive to rotate the drill string (Figure 10.22). In this system, the rotational unit moves up and down the mast; energy is obtained from a hydraulic transmission unit powered by a motor-driven pump.

In both the rotary table and top-head drive mechanisms, the driller can determine the rotation speed depending on the resistance of the formation and the rate of penetration. For shallow boreholes of 200 to 400 ft (61 to 122 m), pull-down pressure may be applied to the bit. Down-hole pressures on the bit can be increased beyond

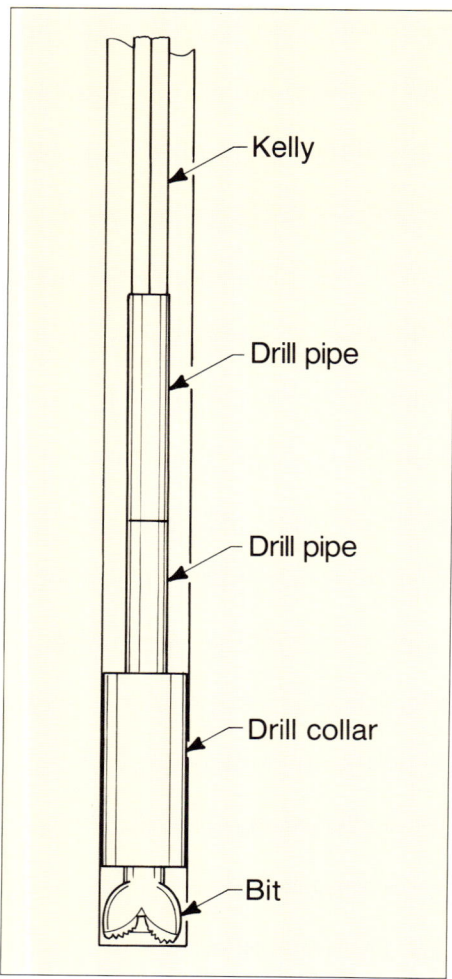

Figure 10.16. The drill string for a direct rotary rig consists of a bit, drill collar or stabilizer, drill pipe, and kelly for table drive units.

the weight of the drill string by exerting a pull-down force derived from the weight of the drilling rig. The chain assemblies (or cables) on the mast are used to transfer part of the weight of the drilling rig to the drill string. Caution should be used to avoid excessive pull-down pressure (weight) because hole deflection (crooked holes) may result. To avoid crooked holes, many drillers will use drill collars that concentrate additional weight on the bit rather than exert pull-down pressure. Rotation speed is adjusted to the pull-down or existing pressures on the bit. In general, the higher the pressure on the bit, the slower the rotation should be. In most deep direct rotary boreholes, the driller must hold back (suspend) part of the drill string weight from the swivel so that the weight on the bit does not become excessive. In general, the driller may start holding back when the weight of the drill string exceeds 10,000 lb (4,540 kg), although the exact figure depends on the bit being used. Bit manufacturers usually indicate the optimum pressure that an individual bit should exert against the formation for maximum cutting rates.

Adding drill rods (pipe) to the drill string or removing rods to change bits or take split-spoon or core samples is a major part of every rotary drilling operation. "Tripping in" and "tripping out" are the terms used to describe the process of running the bit into or pulling the bit from the hole. Most newer drilling rigs have been designed to make this process as fast and automated as possible. With some new machines, it is possible to pull back a 20-ft (6.1-m) rod and remove it from the drill string in approximately 30 seconds. In general, top-head drive machines, especially those equipped with carousels (drill rod storage racks mounted on the mast), offer an advantage in rod handling speed, although recent modifications in table-drive machines have enabled this type of rig to match the speed of the top-head drive rotaries.

When a rod is to be added, the swivel is just above the rotary table (in a table-drive machine). Usually the driller will circulate the drilling fluid for a few minutes to make sure that most of the cuttings are out of the hole to prevent the bit and drill string from sand-locking when the circulation is stopped to add a drill rod. The kelly is raised until the joint between the kelly sub and the uppermost drill rod is just above the drive table. Slips are placed in the table to hold the drill string (Figure 10.23). The kelly is then disconnected and placed out of the way momentarily. A sand line (cable) is joined to another rod section using a quick-release elevator (clamp). The rod is hoisted into place above the rod held in the table and the two are threaded together, usually with the aid of automatic pipe clamps. The slips are removed and the string is lowered by the sand line until the top (tool-joint box) of the just-added drill rod is just above the table. The slips are reinserted, the elevator is removed, and the kelly is rethreaded to the drill string. After lowering the kelly into the drive table, drilling can continue.

In top-head drive machines, no kelly is required and therefore the bottom sub of the hydraulic drive motor is connected directly to the drill rod. Additional rods can be taken directly from a carousel by the top-head drive unit. If the machine is equipped with side storage racks, a sand line must be used to raise the drill rod into position.

Internal pressure created by the drilling fluid can cause a momentary but forceful surge of drilling fluid out of the drill string at the point where the kelly is disconnected from the upper drill rod. Drillers usually break this joint slowly to allow the pressure to dissipate so that drilling fluid is not expelled violently. Occasionally during the addition of a drill rod, drilling fluid may continue to overflow from the top of the rods. Confining pressures within permeable material in the borehole may be causing this flow, but it is more likely that clay "collars" packed around the drill rods are falling deeper into the borehole, thereby pushing drilling fluid back up the center of the rods.

Direct rotary drilling, the most common method, offers the following advantages:
1. Penetration rates are relatively high in all types of materials.
2. Minimal casing is required during the drilling operation.

Figure 10.17. Heavy collars added to the drill string above the bit help keep the borehole straight.

3. Rig mobilization and demobilization are rapid.
4. Well screens can be set easily as part of the casing installation.

Major disadvantages include the following:
1. Drilling rigs are costly.
2. Drilling rigs require a high level of maintenance.
3. Mobility of the rigs may be limited depending on the slope and condition (wetness) of the land surface.
4. Most rigs must be handled by a crew of at least two persons.
5. Collection of accurate samples requires special procedures.
6. Use of drilling fluids may cause plugging of certain formations.
7. Rigs cannot be operated economically in extremely cold temperatures.
8. Drilling fluid management requires additional knowledge and experience.

Figure 10.18. Stabilizers mounted just above the bit in the drill string are important in maintaining a straight borehole. Flat-bar steel plates welded to the stabilizer help maintain borehole diameter and provide channels for the passage of drilling fluid. *(Hydro Drillers)*

DRILLING FLUIDS

Drilling fluid control is essential to efficient rotary drilling. There must be proper coordination of the hole size, drill pipe size, bit type, pump capabilities, and drilling fluid characteristics based on the geologic conditions at the site if drilling is to proceed efficiently. Drilling fluids include air, clean water, and scientifically prepared mixtures of special-purpose materials*. The essential functions of a drilling fluid are to:

1. Lift the cuttings from the bottom of the hole and carry them to a settling pit.
2. Support and stabilize the borehole wall to prevent caving.
3. Seal the borehole wall to reduce fluid loss.
4. Cool and clean the drill bit.
5. Allow cuttings to drop out in the settling pit.
6. Lubricate the bit, bearings, mud pump, and drill pipe.

The viscosity (the degree to which a fluid resists flow under an applied force) of the drilling fluid and the uphole velocity required to remove cuttings will depend on a number of factors that are discussed in Chapter 11. An uphole velocity

---

*Because the majority of rotary drilled holes are completed using a water-based drilling fluid, air will not be discussed here, but is thoroughly covered in Chapter 11.

of 100 to 150 ft/min (30.5 to 45.7 m/min) is used by many drillers (Table 10.2). The ability of the fluid to lift cuttings increases rapidly as viscosity and velocity are increased. After cuttings are brought to the surface, however, it is essential that they drop out as the fluid flows through the settling pit. The desired results are obtained by selecting the appropriate drilling fluid additive, properly designing the mud pits, controlling the viscosity and weight of the drilling fluid, and adjusting the pump speed.

When circulation of the drilling fluid is interrupted for some reason, to add drill pipe for example, the cuttings being carried by the mud column tend to drop back toward the bottom of the hole. Cuttings can bridge on tool joints and build up on top of the bit if they settle rapidly. Excessive pump pressures may then be required to move these cuttings and resume circulation; if the cuttings cannot be removed, the drill pipe and bit become stuck in the hole (sanded in). Many drilling fluids develop gel strength, that is, the ability to suspend cuttings when flow slows or stops. It may be advisable before adding drill pipe to circulate the fluid

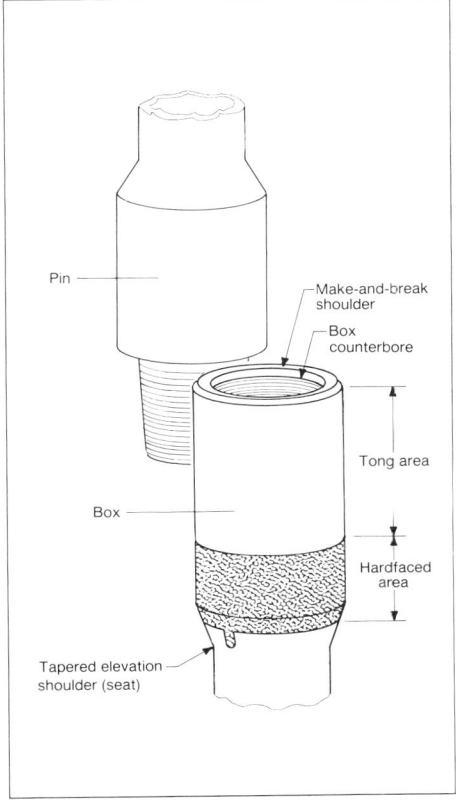

Figure 10.19. Drill pipe is heavy-walled seamless tubing with tool-joint pin and box-end fittings.

for a few minutes without applying bit pressure to clear the hole of most cuttings. This is particularly important for deep holes.

The drilling fluid prevents caving of the borehole because it exerts pressure against the wall. As long as the hydrostatic pressure of the fluid exceeds the earth pressures and any confining pressure in the aquifer, the hole will remain open. The pressure at any depth is equal to the weight of the drilling fluid column above that point.

The weight of the drilling fluid required for a given situation cannot be predicted precisely without test borings. Most water well drillers rely on past experience in making up drilling fluid. If caving occurs while drilling, weighting material may be added to increase the drilling fluid weight or special additives may be added to isolate any swelling clays. To prevent excessive intrusion of fine drilling fluid par-

Figure 10.20. A rotary table rotates the kelly, which is connected to the top of the drill string. The kelly can be square, hexagonal, or round with flutes cut into the outer wall. *(Huron Drilling)*

Table 10.2. Pump Output and Annular Velocities for Different Sizes of Holes

Size of drilling bit		Pump output normally used		Annular velocity of drilling fluid*	
in	mm	gal/min	ℓ/min	ft/min	m/min
6	152	150 to 200	568 to 757	169 to 223	52 to 68
6¾	171	200 to 250	757 to 946	173 to 220	53 to 67
7⅞	200	300 to 400	1,140 to 1,510	169 to 226	52 to 69
8½	216	300 to 400	1,140 to 1,510	164 to 220	50 to 67
9⅞	251	400 to 500	1,510 to 1,890	130 to 170	40 to 52
12¼	311	600 to 700	2,270 to 2,650	115 to 130	35 to 40
15	381	750 to 900	2,840 to 3,410	92 to 115	28 to 35
17½	445	800 to 1,000	3,030 to 3,790	69 to 92	21 to 28

*Annular velocity of drilling fluid may be slightly different depending on the size of the drill pipe.
(Ingersoll-Rand)

ticles into the formation, the drilling fluid weight should be just heavy enough to maintain hole stability. Numerous additives are available for imparting specific properties to drilling fluids. Chapter 11 discusses the various kinds of drilling fluids, with particular reference to their advantages and disadvantages in certain geologic formations.

As drilling progresses, a film of small particles builds up on the wall of the borehole. This flexible lining, which may consist of clay, silt, or colloids, forms when the pressure of the drilling fluid forces small volumes of water into the formation, leaving the fine, suspended material on the borehole wall. In time, the lining completely covers the wall and holds loose particles or crumbly materials in place. It protects the wall from being eroded by the upward-flowing stream of drilling fluid, and acts to seal the wall and reduce the loss of fluids into surrounding permeable formations. Although the flexible lining effectively controls fluid losses in the borehole, it cannot prevent the hole from collapsing if the hydrostatic pressure created by the drilling fluid is not

Figure 10.21. The upper end of the kelly connects to a swivel suspended from a traveling block in the mast. *(E. H. Renner & Sons, Inc.)*

greater than the pressure exerted by the water in the formation.

The drill bit is cooled and cleaned by the jets of fluid that are directed at relatively high velocity over the cutting faces and body section of the bit. A properly prepared drilling fluid is an excellent lubricant, but the viscosity must be controlled so that the concentration of cuttings does not become excessive.

In direct rotary drilling, water and special viscosity-building additives are usually mixed to produce a drilling fluid. Drilling fluids can be mixed in either a portable pit carried from site to site (Figure 10.24) or in a pit excavated next to the drilling rig. Cuttings collecting on the bottom of the pit must be removed periodically to maintain the efficiency of the pit (Figure 10.25). When enough drilling fluid has been mixed and sufficient time has elapsed to insure complete hydration, it is circulated into the hole using a mud pump. The size of the mud pump must be chosen carefully so that the correct uphole velocity can be maintained. Centrifugal and piston mud pumps are discussed in Chapter 17.

In clay-rich formations, the driller may begin drilling with clean water which quickly mixes with the natural clays in the borehole to form a thin clay slurry.

Figure 10.22. On some direct rotary rigs, a top-head drive is used to rotate the drill string. The amount of torque delivered to the bit by a top-head drive is usually somewhat less than that produced by a table drive, but the rod-handling speed is exceptionally good. *(Olson Brothers Well Drilling)*

This drilling fluid is used in the upper portion of the borehole, commonly the first 100 to 300 ft (30.5 to 91.5 m). Thereafter, most drillers will mix fluids with additives of either high-quality clays or natural or synthetic polymers so that proper viscosity and hydrostatic pressure can be maintained in the borehole.

REVERSE CIRCULATION ROTARY DRILLING

In direct rotary drilling, the viscosity and uphole velocity of the drilling fluid are the controlling factors in removing cuttings effectively. Unless cuttings can be removed, drilling cannot continue. Because of limitations in pump capacity and therefore effective cuttings removal, most direct rotary machines used to drill water wells are limited to boreholes with a maximum diameter of 22 to 24 in (559 to 610 mm). This size may not be sufficient for high-capacity wells, especially those that are to be filter packed. Also, as hole diameters increase past 24 in, the rate of penetration by direct rotary machines becomes less satisfactory. To overcome the limitation on hole diameter and drilling rate, reverse circulation machines were designed; originally they were used only in unconsolidated formations. Recently, reverse circulation drilling has been used in soft consolidated rocks such as sandstone and even in hard rocks using both water and air as the drilling fluid.

The design of a reverse circulation rig is essentially the same as that of the direct rotary rig except most pieces of equipment are larger. For example, larger compressors and mud pumps are required because of the larger diameter boreholes. Only table drives are used in reverse circulation drilling because of the large borehole diameter and the torque required to turn the bit. The components of many reverse rotary rigs are mounted on long-bed trailers rather than being mounted on trucks.

In reverse circulation rotary drilling, flow of the drilling fluid is reversed when compared with the direct rotary method. The suction end of the centrifugal pump, rather than the discharge end, is connected through the swivel to the kelly and drill pipe. The drilling fluid and its load of cuttings move upward inside the drill pipe and are discharged by the pump into the settling pit (Figure 10.26). Centrifugal pumps with large passageways are often used to pump the drilling fluid because they can handle cuttings without excessive wear on the pump. In operation, however, most of the cuttings do not actually enter the pump but bypass it by means of an eductor system. An uphole velocity of at least 150 ft/min (45.7 m/min) is recommended. The fluid returns to the borehole by gravity flow. It moves down the annular space between the drill pipe and borehole wall to the bottom of the hole, picks up the cuttings, and reenters the drill pipe through ports in the drill bit.

In the reverse circulation rotary method, the drilling fluid can best be described as muddy water rather than drilling fluid; drilling fluid additives are seldom mixed with the water to make a viscous fluid. Suspended clay and silt that recirculate with the fluid are mostly fine materials picked up from the formations as drilling proceeds. Occasionally, low concentrations of a polymeric drilling fluid additive are used to reduce friction, swelling of water-sensitive clays, and wa-

Figure 10.23. Various types of slips are inserted into the kelly just above the drive table to hold the drill string when adding or removing a drill pipe (joint). In the upper photo, the drill rod is being held by a plain-end wrench. The wrench is also used to set up or break the connections between rods when tripping in or out. The drill pipe in the lower photo has flanged connections and an external air line mounted on the outside of the drill pipe. This type of drill pipe is generally used in reverse rotary drilling. *(Huron Drilling)*

ter loss.

To prevent caving of the hole, the fluid level must be kept at ground level at all times, even when drilling is suspended temporarily, to prevent a loss of hydrostatic pressure in the borehole. The hydrostatic pressure of the water column plus the velocity head (inertia of the water moving downward) outside the drill pipe support the borehole wall. Erosion of the wall is usually not a problem because velocity in the annular space is low.

Water infiltrates the permeable formations surrounding the borehole. Some of the fine particles suspended in the fluid are filtered out on the wall of the hole, resulting in a thin mud deposit that partially clogs the pores and reduces the water loss. A considerable quantity of make-up water is usually required and must be immediately available at all times when drilling in permeable sand and gravel. Under these conditions, water loss can increase suddenly, and if this causes the fluid level in the hole to drop significantly below the ground surface, caving usually results. Water loss can be reduced by mixing clay additives with the fluid, but this is usually avoided unless absolutely necessary. As little as 20 gpm (109 m^3/day) of make-up water is enough in some cases, whereas as much as 1,000 gpm (5,450 m^3/day) may be needed when drilling through a highly permeable aquifer such as coarse, dry gravel.

The settling pit and water supply pit should hold at least three times the volume of the material to be removed during the drilling operation (Figure 10.27). The circulation rate for the water used in drilling is commonly 500 gpm (2,730 m^3/day) or more.

Many reverse rotary drilling rigs are equipped with air compressors to aid in circulating the drilling fluid. When drilling has reached a depth sufficient for proper operation of an air lift within the drill pipe, the mud pump is bypassed. Compressed air is introduced through a 1¼- or 1½-in (32- or 38-mm) plastic or metal air line

Figure 10.24. A portable mud pit can be transported from site to site. *(Oelke Drilling Company)*

suspended inside the drill pipe, or through an external air line attached to the outside of the drill pipe (Figure 10.28). The external air line system may consist of two pipes welded on opposite sides of the drill pipe. The air is injected by means of a manifold into the drill string at the proper depth. In these processes, water is lifted to the surface from the borehole. Air-lift pumping procedures are discussed in more detail in Chapter 15.

Any cobbles or boulders larger than the drill pipe or the openings in the drill bit cannot be brought out in the drilling operation, because most reverse rotary bits cannot break cobbles. Thus, further penetration is impossible when a few large cobbles or boulders collect in the bottom of the hole. One solution is to drive the boulders into the borehole wall as they are encountered by a Zublin-type bit (Figure 10.29). If the boulders are relatively stable

Figure 10.25. Cuttings that have settled to the bottom of a small portable pit are removed periodically to maintain the efficiency of the pit.

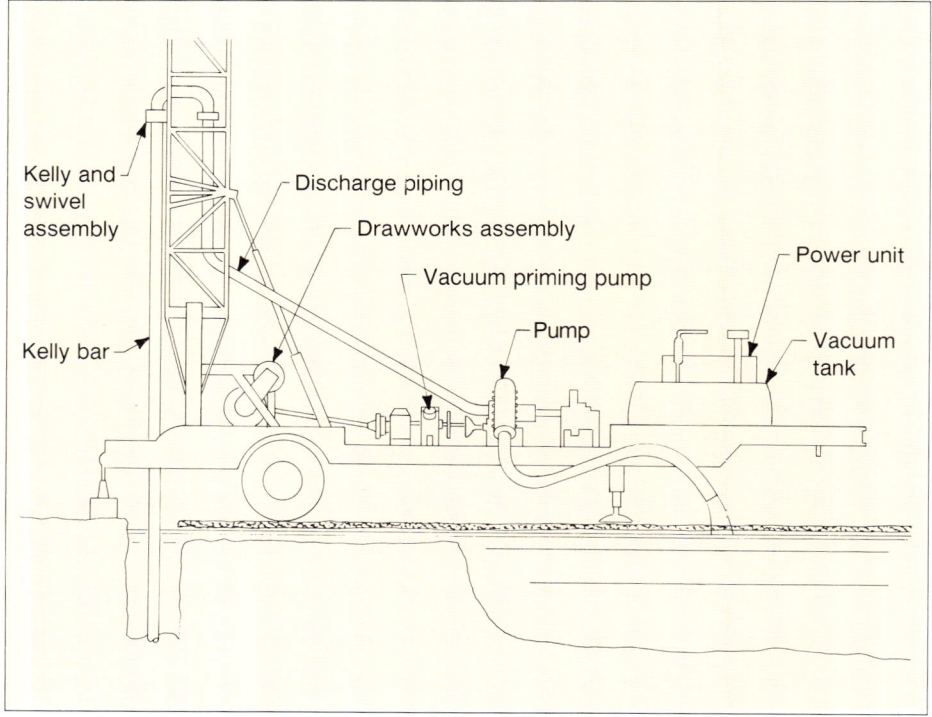

Figure 10.26. In a reverse rotary circulation system, the drilling fluid flows from the mud pit down the borehole outside the drill rods, then passes upward through the bit into the drill rods after entraining the cuttings. After flowing through the swivel and mud pump, it passes into the mud pit where the cuttings settle out.

in the hole, a roller cone bit can be used to grind them into small fragments; cement may be used to stabilize the boulders prior to grinding. Orange-peel buckets or boulder catchers are often used to fish out large boulders or cobbles (Figure 10.30). Common types of drag and roller bits for reverse drilling are shown in Figure 10.31. Note the coarse tooth structure on the roller reamer bit that is suitable for semiconsolidated formations.

Most new drill pipe used in reverse circulation rotary drilling is threaded and coupled pipe that can be as much as 8 inches (203 mm) in diameter and operated at depths of 2,000 ft (610 m) or more. Formerly, most drill pipe used for reverse circulation drilling had flanged joints that were joined by four to eight bolts per flange.

Figure 10.27. The settling pit for a reverse rotary rig is considerably larger than for a direct rotary machine because fluid losses in the borehole are always much higher. The pit should contain at least three times the volume of the material to be removed during drilling. *(Portadrill, Inc.)*

Flanged pipe is suitable for relatively shallow boreholes, but the labor of attaching sections for deep holes led to the adoption of threaded and coupled pipe.

When flanged pipe is used, the smallest practical borehole that can be drilled is about 18 in (457 mm) because the diameter of the flanges is about 11 in (279 mm). The diameter of the hole must be large in relation to the drill pipe so that the velocity of the descending water is about 1 ft/sec (0.3 m/sec) or less to prevent erosion of the borehole wall. If air-conductor pipes are attached to the flanges, they must be aligned properly to allow for air movement.

Reverse circulation drilling is the least expensive method for drilling large-diameter holes in unconsolidated formations. When geologic conditions are favorable, increasing the diameter of the borehole does not appreciably increase the cost of the well. Therefore, most water wells drilled by this method are 24 inches (610 mm) in diameter or larger [to 60 in (1,520 mm)]. Filter packs are installed almost universally in wells drilled by reverse circulation drilling because of the relatively large diameter of the borehole.

Reverse circulation drilling is most successful in soft sedimentary rocks and unconsolidated sand and gravel where the static water level is 10 ft (3 m) or more below ground level. In cases of high static water level, ramps are built above grade to support the drilling rig, or the weight of the drilling fluid is increased to obtain the necessary hydrostatic pressure. The reverse circulation drilling method may not be satisfactory when the static water level is too high and adequate water supplies are not available. Advantages of the reverse circulation method include the following:

1. The porosity and permeability of the formation near the borehole is relatively undisturbed compared to other methods.
2. Large-diameter holes can be drilled quickly and economically.
3. No casing is required during the drilling operation.
4. Well screens can be set easily as part of the casing installation.

Figure 10.28. Compressed air can be added to the drill rods to enhance upward flow of the drilling fluid. The drill pipe shown on the truck has external air lines mounted on the outside of the drill rods to conduct the air down to the injection manifold.

5. Most geologic formations can be drilled, with the exception of igneous and metamorphic rocks.
6. Little opportunity exists for washouts in the borehole because of the low velocity of the drilling fluid.

Disadvantages include the following:
1. Large water supply is generally needed.
2. Reverse-rotary rigs and components are usually larger and thus more expensive.
3. Large mud pits are required.
4. Some drill sites are inaccessible because of the rig size.
5. For efficient operation, more personnel are generally required than for other drilling methods.

Figure 10.29. Boulders collecting at the bottom of the borehole present a significant problem in reverse rotary drilling. Some contractors use a Zublin-type bit to drive the boulders into the borehole wall when they are encountered.

AIR DRILLING SYSTEMS

Two different drilling methods use air as the primary drilling fluid — direct rotary air and down-the-hole air hammer. In conventional reverse circulation methods, air is used as an assist but not as the primary drilling fluid. In the air rotary method, air alone lifts the cuttings from the borehole. A large compressor provides air that is piped to the swivel hose connected to the top of the kelly or drill pipe. The air, forced down the drill pipe, escapes through small ports at the bottom of the drill bit, thereby lifting the cuttings and cooling the bit. The cuttings are blown out the top of the hole and collect at the surface around the borehole (Figure 10.32). Injecting a small volume of water or surfactant and water (foam) into the air system controls dust and lowers the temperature of the air so that the swivel is cooled.

Figure 10.30. Orange-peel buckets are used to remove boulders from the bottom of the borehole in reverse rotary drilling. Two sizes are shown in this photograph. *(Layne Northwest)*

Air drilling can be done only in semi-consolidated or consolidated materials. Therefore, to achieve the capability to operate in completely unconsolidated as well as consolidated formations, air rotary drilling machines are often equipped with a mud pump in addition to a high-capacity air compressor (Figure 10.33). Conventional water-based drilling fluids are then used when drilling through the overlying, caving formations above the bedrock (or more consolidated formations), whereas air is used once bedrock has been reached. Thus drillers are uti-

Figure 10.31. Large-diameter drag and roller cone bits provide satisfactory penetration rates for reverse drilling. Drag bits are used in unconsolidated formations only. Note the air-assist lines mounted in the drag bit and the coarse-tooth structure of the reamer bit. *(Moab Bit and Tool Company)*

lizing various options of drilling technology to adjust to the different physical characteristics of the formation. In many instances, casing may have to be installed through the overburden to avoid caving or excessive erosion of the borehole wall after changing to air circulation.

Cuttings are removed by grinding the material finely enough so that the uphole velocity of the air is sufficient to lift them to the surface. The lifting capacity of the air can be enhanced by adding a small amount of surfactant and water solution to the air. Larger cuttings can then be removed, thereby increasing the drilling rate.

Figure 10.32. In dry-air drilling, cuttings are blown up the borehole and collect at the surface. *(Ingersoll-Rand)*

Foam also reduces loss of air to the formation. Suggestions for appropriate uphole velocities and the use of various drilling fluid additives are presented in Chapter 11.

Roller-type rock bits, similar to those designed for drilling with water-based fluids, can be used when drilling with air. Tricone rock bits up to about 12-in (305-mm) diameter are commonly used. Larger sizes are available. Button bits, made with sintered tungsten-carbide inserts set into the perimeters of steel rollers, are used successfully in many areas. Figure 10.34 lists the formations drilled effectively by carbide and steel-tooth bits in rotary air drilling.

Field tests with various sizes of bits have shown that the penetration rate is often faster and the bit life longer when using air as compared with water-based drilling fluids. Better bottom-hole cleaning is partly responsible for this difference in performance. If too much water comes into the hole during drilling, however, the penetration rate is no better than when drilling with water-based drilling fluids. Air also keeps the bit bearings cool and clean and causes some oxidation of the bearings; the oxidized material then becomes a lubricant. On the other hand, water-based drilling fluids are often abrasive and cause wear on the bearings.

A second direct rotary method using air is called the "down-the-hole" drilling system. A pneumatic drill operated at the end of the drill pipe rapidly strikes the rock while the drill pipe is slowly rotated (Figure 10.35). The percussion effect is similar to the blows delivered by a cable tool bit. The hammer is constructed from alloy steel with heavy tungsten-carbide inserts that provide the cutting or chipping surfaces. Tungsten-carbide is extremely resistant to abrasion, but drill bits do become dull with continued use. The inserts are sharpened by grinding when operating conditions indicate that the bit is not cutting properly. Alternatively, the bits can be provided with carbide buttons that can be periodically replaced when worn.

Rotation of the bit helps to assure even penetration and, therefore, straighter holes even in extremely abrasive or resistant rock types. The rates of penetration in several rock types are higher than those obtained by other drilling methods or other types of tools. Six-in (152 mm) and 6½-in (165 mm) hammer bits are most commonly used, although sizes range up to 17½ in (445 mm). Cuttings are removed continuously by the air used to drive the hammer. Unlike the conventional cable tool bit that is constantly striking previously broken rock fragments, the bit (or buttons) on the air hammer always strike a clean surface. Thus, the air hammer is highly efficient.

Figure 10.33. Many air rotary drilling rigs are equipped with conventional mud pumps so they can be used to drill in unconsolidated overburden. *(Schramm, Inc.)*

298   GROUNDWATER AND WELLS

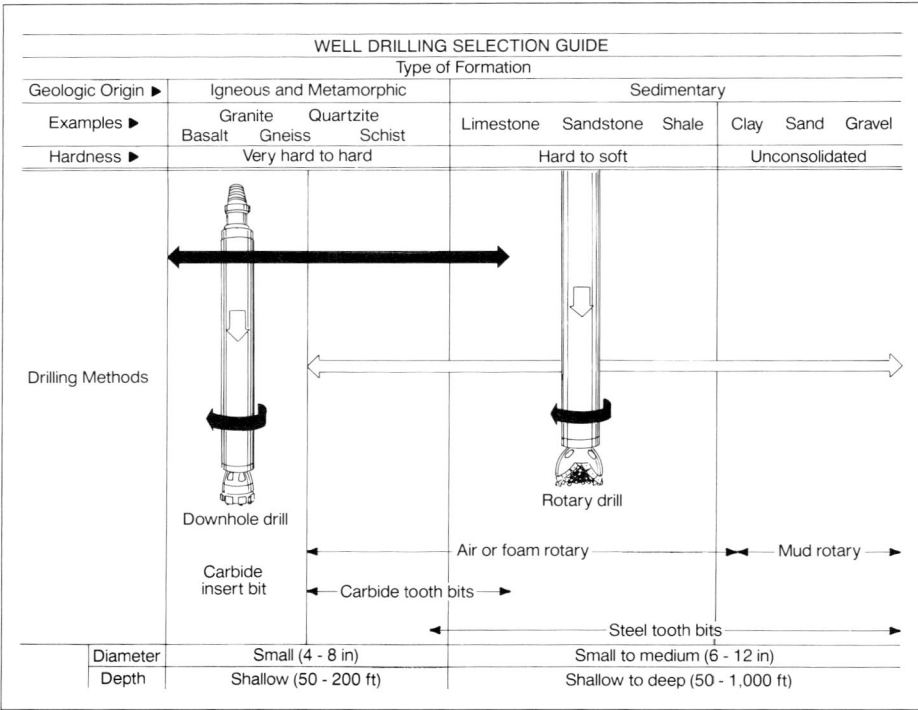

Figure 10.34. Guide for the use of bit types in air-drilling systems. *(Ingersoll-Rand)*

Compressed air must be supplied to the hammer at a pressure of 100 to 110 psi (690 to 758 kPa). Some tools require as much as 200 psi (1,380 kPa). To remove cuttings effectively, the upward velocity in the space outside the drill pipe should be about 3,000 ft/min (915 m/min) or more. For drilling 4-in (102-mm) holes, the air supply must be at least 100 cfm (0.047 m^3/sec) [assuming a 2⅞-in (73-mm) drill rod]; for 6-in (152-mm) holes, at least 330 cfm (0.156 m^3/sec) is needed*. Proper rotation

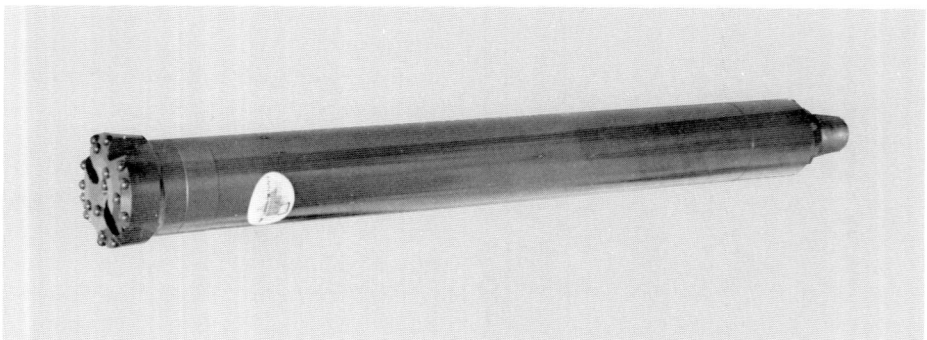

Figure 10.35. A down-the-hole hammer is extremely effective in penetrating dense, resistant formations such as basalt, quartzite, and granite. *(Ingersoll-Rand)*

---

*260 cfm (0.123 m^3/s) is needed when using a 4½-in drill rod. Other recommendations are presented in Chapter 11.

speed is from 10 to 30 rpm; reduced speed is best in harder and more abrasive rock. Advantages of using air drilling methods include the following:

1. Cuttings removal is extremely rapid.
2. Aquifer is not plugged with drilling fluids.
3. No maintenance costs for mud pumps (mud pumps are not used during air drilling).
4. Bit life is extended.
5. Drilling operations are not hampered by extremely cold weather.
6. Penetration rates are high, especially with down-the-hole hammers, in highly resistant rocks such as dolomite or basalt.
7. An estimate can be made during drilling of the yield from a particular formation.

Disadvantages include the following:

1. Restricted to semiconsolidated and well-consolidated materials.
2. Initial cost and maintenance costs of large air compressors are high.

IN-VERSE DRILLING

A recent innovation for the top-head drive, direct rotary machine involves the addition of an air assist by using a special 6-in (152-mm) inside diameter, side discharge swivel assembly and 5⅞-in (149-mm) drill pipe with built-in air channels. This equipment permits compressed air to be injected through an injection stem into air channels mounted outside the drill pipe and then into the drilling fluid as it moves up inside the drill pipe (Figure 10.36). Thus, the drilling fluid and cuttings are assisted to the surface by an airlift inside the 6-in diameter conductor (drill) pipe. This method is known as the In-Verse system and converts a direct rotary, top-head drive machine into a reverse circulation rig.

Use of the In-Verse system can increase the capacity of a direct rotary rig to drill large-diameter wells. Depending on the rig, boreholes from 20 to 30 in (508 to 762 mm) can be drilled routinely. If the pulling capabilities of the rig are suffi-

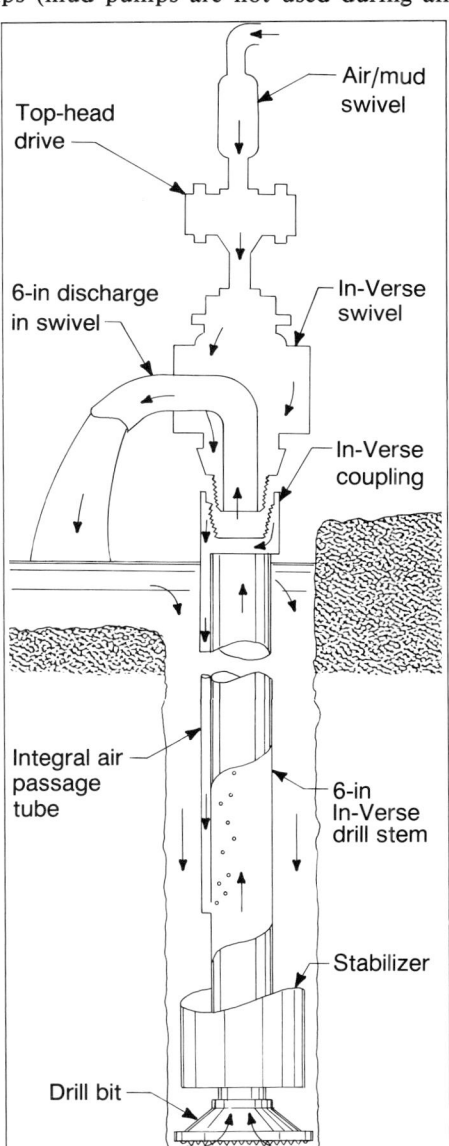

Figure 10.36. In In-Verse drilling, air is injected into a special double-walled drill stem to increase the efficiency of cuttings removal. Use of this system permits top-head drive, direct rotary rigs to drill large-diameter boreholes by reverse circulation methods. *(In-Verse Tool Corporation)*

cient, enough torque is available, and larger bits can be accommodated under the centralizer, boreholes of 30 to 60 in (762 to 1,520 mm) are possible in unconsolidated formations. Boreholes smaller than 12 in (305 mm) are not recommended because the drill pipe has an outside diameter of approximately 9 in (229 mm) at the tool joint and significant erosion of the borehole wall may occur depending on the degree of formation consolidation.

It is recommended that at least a 300 cfm (0.1 m^3/sec) compressor operating at 125 psi (862 kPa) be used for the In-Verse system. At this pressure, the maximum stem submergence is approximately 250 ft (76.2 m). If the borehole must extend past 250 ft (76.2 m), 50 to 60 ft (15.2 to 18.3 m) of drill pipe are pulled and another injector stem is installed in the drill string. Thus, if the drilling rig is equipped with only 250 ft of air channel pipe and the hole will be 500 ft (152 m) deep, the drill pipe with the air channels must be mounted above the conventional drill pipe for any depth over 250 ft. This requirement increases drill pipe handling time somewhat.

The In-Verse equipped rig operates most satisfactorily with a centrifugal pump or a 3 x 4 or 5 x 6 piston pump.* The latter pump will operate at approximately 300 psi (2,070 kPa). This size pump would be required to drill test holes or wells smaller than 12 inches (305 mm) in diameter using direct rotary drilling.

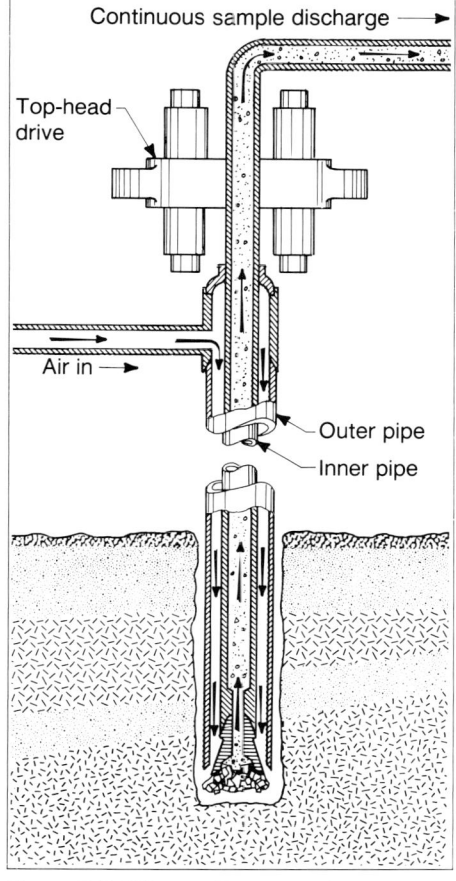

Figure 10.37. Water wells can be drilled to depths of 1,000 ft (305 m) or more using the dual-wall reverse circulation method. The drilling fluid (air or water) is injected down the outer annulus of dual-walled pipe; cuttings are lifted to the surface through the inner pipe. The pipe is connected directly to either a down-the-hole air hammer or a tricone bit. Main use is in test drilling.

Advantages of the In-Verse system include the following:
1. Large-diameter boreholes can be drilled.
2. Penetration rates are high in unconsolidated sediments.
3. Less drilling fluid additives are required to lift the cuttings.
4. Development time is reduced.

Disadvantages include the following:
1. Extra costs for drill pipe, special swivel, and air compressor (if the rig is not equipped with one).
2. Drill pipe handling time may increase for deep holes.

---

*Piston pumps are designated by two dimensions — the diameter of the piston and stroke length.

DUAL-WALL REVERSE CIRCULATION ROTARY METHOD

In mining exploration, a drilling system called the dual-wall method has been used for many years to obtain accurate geologic samples from known depths. The dual-wall method uses flush-jointed, double-wall pipe in which the drilling fluid (air or liquid) moves by reverse circulation (Figure 10.37). Unlike conventional reverse circulation, however, the drilling fluid does not run down the outside of the drill pipe. Instead, the flow is contained between the two walls of the dual-wall pipe and only contacts the walls of the borehole near the bit. Recently this method has been applied to water well exploration and construction in all types of geologic formations, although its principal use is still test drilling.

Available drill pipe diameters for the dual-wall method are:

3½-in OD x 1¾-in ID (89 mm x 44 mm)
4½-in OD x 2½-in ID (114 mm x 64 mm)
5½-in OD x 3¼-in ID (140 mm x 83 mm)
6⅝-in OD x 4¼-in ID (168 mm x 108 mm)
9⅝-in OD x 6¼-in ID (244 mm x 159 mm)

The 4½-in OD size is the most common. Male and female tool joints are used to connect the outer pipes; a connector sleeve with an "O" ring seals the joint between the inner pipes.

Dual-wall pipe can be driven into place in loosely consolidated materials by a steam-, gasoline-, or diesel-operated pile hammer as the formation is being cut by a drive bit. Air or water is forced down the annulus to lift the cuttings to the surface through the inner pipe. If bedrock is reached, drilling may be continued by direct rotary methods using the dual-wall pipe as temporary casing. The pile-driving method is generally not used in the water well industry because the hammering compacts unconsolidated formation materials. In addition, the method may not penetrate deeply enough for most water well applications.

More frequently, dual-wall pipe is set by standard reverse circulation methods using a top-head drive unit. The top-head drive should deliver about 4,500 to 5,000 ft-lb (6,100 to 6,780 J) of torque to be

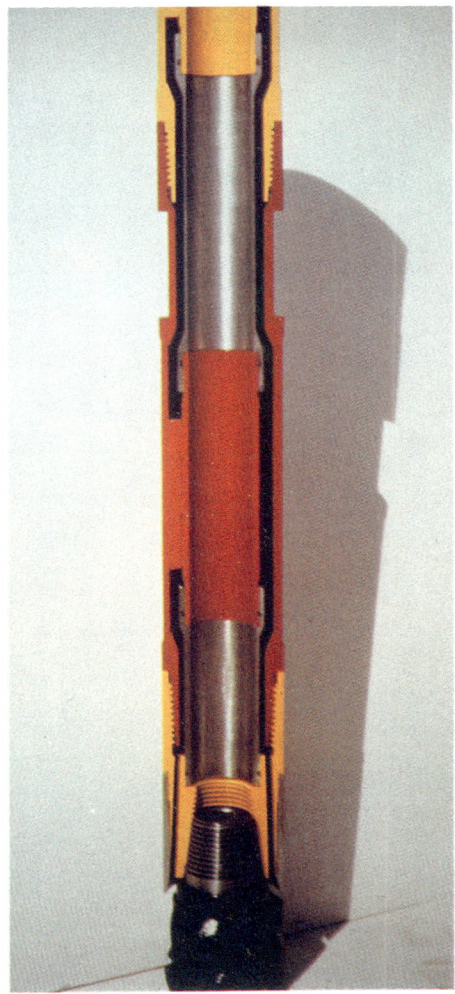

Figure 10.38. When a tricone bit is used in the dual-wall method, the cuttings enter the pipe just above the bit. Thus, samples taken at the surface come from a small vertical section of the formation and are unadulterated by overlying material. The dual-wall system is a popular method for geologic exploration because the samples are highly representative of the formations. *(Drilling Services Company)*

effective. Down-the-hole air hammers and tricone bits can be used to cut the formation. As in the pile-driving method, air or water lifts the cuttings. Surface casing is not needed when the dual-wall system is used.

The outer pipe of the dual-wall system must be able to operate within the normal tensile, column, and collapse pressures associated with rotary drilling. The inner pipe is under little physical stress, but the abrasion caused by earth materials moving up the pipe from the bit causes wear. In practice this abrasion will generally cause the inner pipe to wear out more rapidly than the outer pipe. The inner pipe can be replaced if necessary.

If dual-wall casing is being set by a top-head drive, several different types of bits can be used, but the bit size is normally one nominal size larger than the drill pipe. Thus, the space between the outer pipe and the borehole wall is small and the pipe partially (or totally) supports the wall like a conventional stabilizer. The bit is mounted into a permanent sub that has ports for passage of the drilling fluid. If a tricone bit (either a chisel-tooth or button-tip type) is used, the drilling fluid passes upward through the inner part of the bit. A bit-wear sleeve is attached as close as possible to the cutting face and serves as a wear ring. The drilling fluid passes from the annular space between the two pipes, through a predrilled bit sub, and is discharged toward the cutting surface along the periphery of the bit sleeve; after entraining the cuttings, the fluid passes upward through the inner pipe.

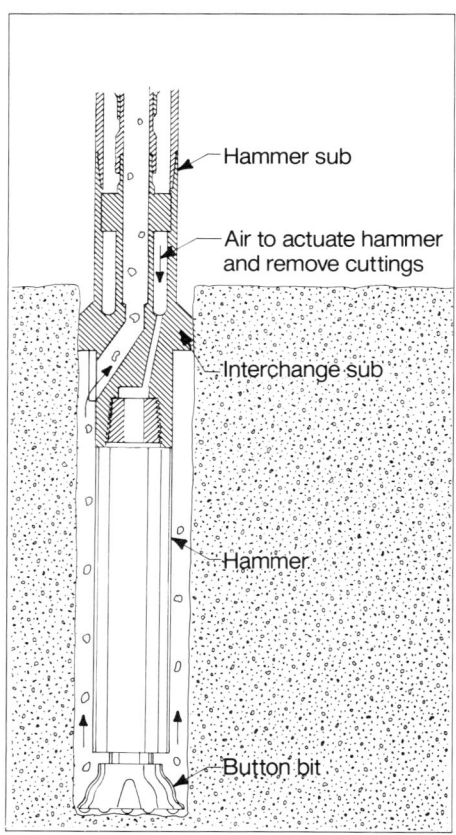

Figure 10.39. The cross-over channel in the interchange sub mounted on top of the hammer permits the cuttings to enter the inner casing. *(Drilling Services Company)*

When a tricone bit is used, the formation sample passing upward through the inner casing originates from a small vertical section of the formation (Figure 10.38). In the use of a down-the-hole hammer, however, the bit extends 4 to 5 ft (1.2 to 1.5 m) out from the bottom of the dual-wall pipe. Air is forced down inside the hammer, out the ports, and then passes up around the outside of the hammer shaft and into a special type of cross-over channel (interchange) sub and then into the inner casing (Figure 10.39). Thus, the formation sample or water sample passing up the pipe can originate over a longer vertical section (3 to 4 ft) of the formation. It must be remembered, however, that this distance is still small when compared with intervals sampled by other types of rotary air drilling.

At the surface, drilling fluid enters the annular space between the inner and outer pipes by a special side inlet swivel.

Drilling fluids can consist of dry air, air and water, air and water with surfactants, or water with clay or polymers. When air is used, velocities in the dual-wall system average 4,500 to 6,000 ft/min (1,370 to 1,830 m/min). After passing down the annular space and up inside the inner pipe, air passes with the formation sample into a cyclone that can be equipped with an automatic splitter (a three-tier sampler is often used) (Figure 10.40). A minicyclone can also be used. The minicyclone is approximately 1:10 scale and is mounted at 180 degrees away from the entrance of the 4-in (102-mm) hose conveying the cuttings from the hole. The sample is collected in a sausage-skin type of sample bag. Under ordinary drilling conditions, 5 ft (1.5 m) of sample bag will be filled for every 20 ft (6.1 m) of hole drilled.

In the past, most boreholes drilled using the dual-wall method rarely exceeded 500 ft (152 m). Recently, however, depths of 800 to 1,400 ft (244 to 427 m) have been reached by using booster compressors.

Screens and conventional casing can be installed when using the dual-wall drilling method. Screen and casing can be washed in over the dual pipes; small-diameter screens [1 to 2 in (25 to 51 mm)] can be installed through the bit; or the dual pipe can be pulled from the hole before a screen and casing are set.

Advantages of the dual-wall system include the following:

1. Continuous representative formation and water samples can be obtained.
2. Estimates of aquifer yield can be made easily at many depths in the formation.
3. Fast penetration rates are possible in coarse alluvial deposits or broken or fissured rock.
4. Problems of lost circulation are either eliminated or reduced drastically.
5. Washout zones are reduced or eliminated.

Disadvantages of the dual-wall system include the following:

1. Initial cost of drilling rig and equipment is high.
2. System is limited to rather slim holes [less than 9 to 10 in (229 to 254 mm)].
3. System is limited to depths of approximately 1,200 to 1,400 ft (366 to 427 m) in alluvial deposits [works best to 600 ft (183 m)] and generally up to 2,000 ft (610 m) in hard rocks.

Figure 10.40. A cyclone segregates the cuttings from the drilling fluid (air or water) and deposits them into a sample splitter. A three-tier sampler is shown here. *(Drilling Services Company)*

4. Well-trained drilling crews are needed.

DRILL-THROUGH CASING DRIVER

Drilling rig manufacturers have long sought to build drilling machines that could combine the hole stability of the cable tool rig and the speed of an air rotary rig. Some manufacturers are now providing casing drivers that can be fitted to top-head drive, direct air rotary rigs (Figure 10.41). The driver can be suspended in the mast independent of the rotary drive unit because of its rather short length. Use of a casing driver permits the casing to be advanced during drilling, but both drilling and driving can be adjusted independently depending on the nature of the formation. Drivers are usually equipped so that they can be used to drive upward to remove casing or expose a screen.

In the casing driver system, the drill pipe and casing are usually preassembled as a unit [must be the same length, usually 20 ft (6.1 m)] and raised into position on the mast. The bottom of the casing is fitted with a forged or cast alloy steel drive shoe as in cable tool operations. A bit that fits inside the casing is attached to the bottom of the drill pipe (Figure 10.42). The top of the casing fits in the bottom of the casing driver by means of an anvil. The casing is driven by a piston that is activated by air pressure (Figure 10.43). Table 10.3 shows the relationship between air pressure, air consumption, and blows per minute for two sizes of drivers.

Three drilling procedures can be followed when using the casing driver: (1) the drill bit and casing advance as a unit, (2) the casing is driven first (in unconsolidated materials only) and then the plug in the casing is drilled out, and (3) the drill bit advances beyond the casing a few feet, is withdrawn into the casing, and then the casing is driven.

As drilling commences using the first procedure, the cone-type bit protrudes out the bottom of the casing, but rarely more than 12 in (305 mm). Cuttings are blown up the short open hole into the casing, and pass out the top through a horizontal tube. During drilling, the casing is simultaneously driven into the ground; that is, the casing advances at the same rate as the drill bit (Figure 10.44). The driller adjusts the pull-down and distance

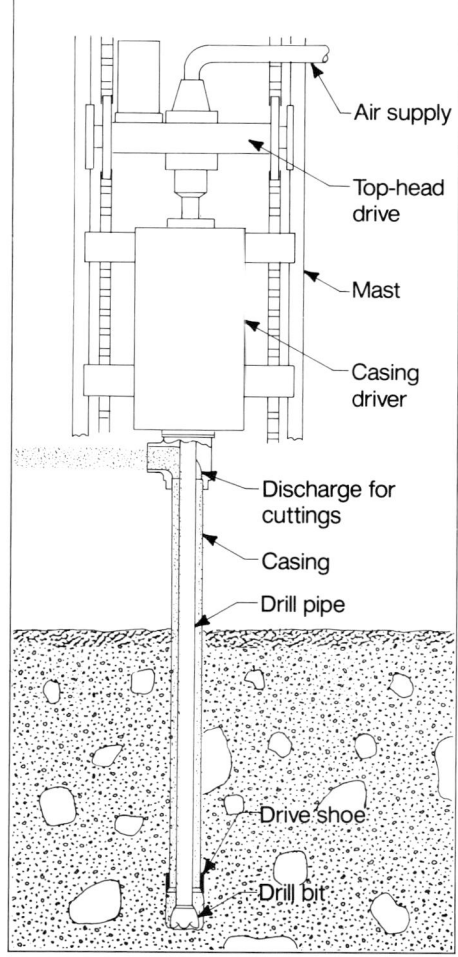

Figure 10.41. Casing drivers can be fitted to top-head drive rotary rigs to simultaneously drill and drive casing. *(Wellen Drill Tools, Inc.)*

the bit is outside the casing according to the rate of advance and speed of cuttings removal. Occasionally the bit may be pulled up within the casing for a few moments to allow the air pressure to blow out the cuttings. Cuttings removal is facilitated by periodically adding small volumes of water if the borehole has not encountered water. This method is particularly suitable for drilling in stratified deposits that have large differences in particle size, for example, sand and silt to boulders.

In the second procedure, the casing is driven into the ground approximately 0.5 to 1.5 ft (0.2 to 0.5 m) and the plug in the casing is then drilled out. The casing is usually driven only short distances so that each formation can be identified and sampled. During the casing-driving procedure, the drill bit is withdrawn inside the casing and rotation is continued. Air is constantly circulated down the drill pipe to prevent clogging of the casing.

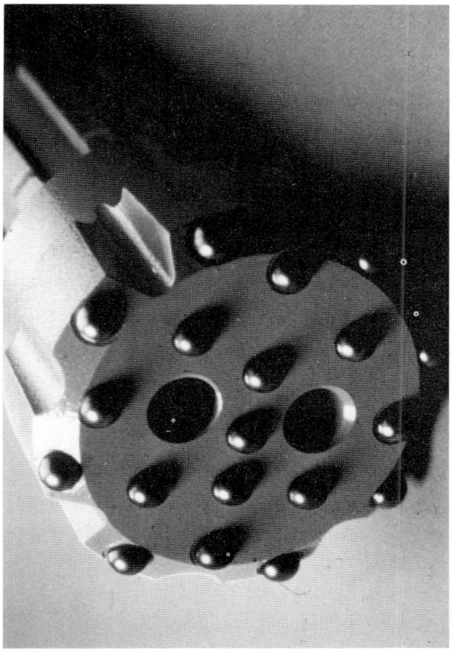

Figure 10.42. Either a down-the-hole hammer equipped with a button bit (shown here) or a roller bit is used in the casing-driver method. *(Wellen Drill Tools, Inc.)*

In the third procedure, the drill bit advances out the end of the casing a few feet. When the hole begins to become unstable, the bit is retracted into the casing and the casing is driven with the air pressure still applied to the borehole. This method is particularly successful in semiconsolidated sands, but also functions well in loose alluvium.

The drill-through casing driver arrangement achieves high drilling rates in most unconsolidated formations, even in bouldery till (Figure 10.45). In fact, welding two joints of casing together often requires more time than drilling and driving a 20-ft (6.1-m) section of casing. When welding casing, some drillers weld straps across the welded joint for added strength. If rock underlies an unconsolidated formation, a down-the-hole hammer can be substituted for the cone bit once the casing is seated in the rock.

Table 10.3. Relationship Between Air Pressure, Air Consumption, and Blows per Minute for Two Sizes of Casing Drivers

Banger Model		Slammer Model	
Air pressure at:		Air pressure at:	
50 blows per minute	40 psi	50- 60 blows per minute	60 psi
120 blows per minute	80 psi	90-100 blows per minute	90 psi
Air consumption	245 cfm	Air consumption	450 cfm
Driving energy	2,100 ft/lb	Driving energy	6,300 ft/lb

(Tigre Tierra, Inc.)

If a screen is to be set, the casing can be pulled back by the top-head drive line, casing line, or, if some simple adjustments are made, by the casing driver (the driver can be adjusted to drive upward). It is wise to add a short piece of riser pipe to the top of the screen to prevent its loss if the casing is pulled back too far.

For some borehole diameters, it is possible to eliminate the casing driver but still drill and install casing at the same time. In loose overburden, an eccentric (off-centered) bit unit can be attached to a down-the-hole hammer (Figure 10.46). In this arrangement, the bit can cut a borehole slightly larger than the casing, allowing the casing to drop into place under its own weight. It may be necessary to drive the casing occasionally if it does not fall into place. This can be done in shallow holes by bringing the down-the-hole hammer out of the hole and driving on a driving cap placed on top of the casing. When consolidated rock is reached and the casing is seated, the rotation of the drill string is reversed for one revo-

Figure 10.43. A casing driver is mounted in the mast of this top-head drive rotary rig. As the top-head drive turns the drill string, the casing driver, activated by air pressure, simultaneously drives the casing. *(Kramer Well Drilling)*

lution, causing the eccentric bit to center itself in the casing. It can then be withdrawn from the borehole and a conventional bit attached to the drill pipe. The new bit will cut a hole slightly smaller than the casing diameter.

The ability to drill and drive casing simultaneously is a major technological advance. It reduces costs and minimizes operational difficulties for the drilling contractor, especially during extremely cold weather. The drill-through casing method is particularly successful in bouldery tills or coarse, highly stratified alluvial deposits where rotary methods are ineffective or cable tool methods too time consuming.

Advantages of using the drill-through casing driver include the following:

1. Wells can be drilled in unconsoli-

Figure 10.44. During drilling, the bit advances just ahead of the casing. If cuttings begin to collect in the casing, the bit can be withdrawn into the casing for a few moments to clear it. Small volumes of water can be added periodically to increase the rate of cuttings removal.

dated geologic materials that may be difficult to drill with cable tool or direct rotary methods.
2. Unlike other rotary methods, the borehole is fully stabilized during the entire drilling operation.
3. Penetration rates are rapid even under difficult drilling conditions.
4. Lost-circulation problems are eliminated.
5. Accurate formation and water samples can be obtained.
6. Casing drivers can be used in all weather conditions.
7. No water-based drilling fluid is required in unconsolidated materials.

Disadvantages include the following:
1. Additional cost of the casing driver.
2. Noise of operation (driving casing).

When air drilling techniques are used, the driller can easily see how much water is being blown out with the cuttings as the hole is deepened. From this observation, the driller can estimate when the borehole is deep enough to produce the desired yield. When static water levels are low, however, the air pressure in the hole may prevent water from entering the borehole.

The cost per foot of drilling with the air rotary system in consolidated formations is sensitive to the life and cost of the bits as well as to penetration rates. Experience in a given locality for specific types of consolidated rock must be depended upon for choosing the bit type that produces the best results economically.

JET DRILLING

There are two methods for installing wells in which a high-velocity stream of water is used in the drilling procedure. One of these, the jet-percussion system, is generally limited to drilling 3- to 4-in (76- to 102-mm) diameter wells to depths of about 200 ft (61 m). Large-diameter wells more than 1,000 ft (305 m) deep have been sunk by

Figure 10.45. When drilling in bouldery till deposits, the time required to weld the casing joints may exceed the time required to drive each 20-ft (6.1-m) length of casing.

jet drilling, but other drilling procedures have displaced this method for deep, large-diameter holes.

Drilling tools for the jet-percussion method consist of a chisel-shaped bit attached to the lower end of a pipe string. Holes on each side of the bit serve as nozzles for water jets that keep the bit clean and help loosen the material being drilled. Water is pumped under moderate to high pressure through the drill pipe and out the drill bit. The drilling water then flows upward in the annular space around the drill pipe, carrying the cuttings in suspension. It overflows at the ground surface and is led into one or more pits where the cuttings settle to the bottom. The water is then picked up by the suction of the pump and recirculated through the drill pipe.

The discharge from the pump is delivered through a pressure hose and water swivel attached to the top of the drill pipe. The fluid-circulation system is similar to that of direct rotary drilling. With water circulation maintained, the drill rods and bit are lifted and dropped in a manner similar to cable tool drilling but with shorter strokes. The chopping action of the bit in combination with the washing action of the water jets opens the borehole. The drill rods are rotated by hand so that the bit cuts a round hole. Casing, fitted with a drive shoe, is normally sunk as drilling proceeds. The pipe is driven by using a drive-block assembly that can be attached to the upper end of the casing string.

Open holes can be drilled to limited depths in unconsolidated materials by the jet method if drilling fluid additives are mixed with the water to form a low-viscosity drilling fluid. The viscosity is useful in lifting cuttings, but cannot be so great that it impedes the force of the jetting action at the bit. Casing must be installed, however, and must follow the bit rather closely whenever the uncased hole tends to cave or is in zones of high fluid loss. Jet-percussion drilling is commonly used for drilling small-diameter wells in water-bearing sand, but can also be used to penetrate some semiconsolidated and consolidated formations that are not too hard.

A second drilling procedure uses small-diameter pipe and well points with open bottoms that can be sunk in sandy formations by the washing action of a water jet without any type of drilling tools. This procedure is described in Chapter 14.

Figure 10.46. For some casing diameters, the use of an eccentric (off-centered) bit, attached to a hammer, can eliminate the continuous need for the casing driver. As the bit underreams the borehole, the casing will ordinarily drop by gravity, although some driving may be necessary. *(Stang Hydronics, Inc.; Atlas-Copco Bit)*

HYDRAULIC-PERCUSSION METHOD

This drilling procedure, often called the hollow-rod method, utilizes a string of drill pipe or drill rod similar to that used in the jet-percussion method. The bit is also similar, except that a ball check valve is provided between the bit and the lower end of the drill pipe. Water is introduced at the surface into the annular space between the drill rods and well casing to keep the hole full of water.

Drilling is done by lifting and dropping the drill rods and bit with quick, short strokes. As the bit drops and strikes bottom, water with cuttings in suspension enters

Figure 10.47. Bucket auger rigs can be used to construct water wells in weakly consolidated, stable formations. *(Gus Pech Manufacturing Co., Inc.)*

the ports of the bit. When the bit is lifted, the check valve closes and traps the fluid inside the drill pipe. Continuous reciprocating motion produces a pumping action to lift the fluid to the top of the string of drill pipe where it discharges into a settling tank. Water is returned to the hole from the settling tank. Casing is driven as drilling proceeds. A driving weight is usually clamped to the drill rods; with this weight attached, the drill rods are lifted and dropped so the weight strikes the top of the casing. The cable tool machine is well adapted to this system because no pressure pump is required.

Figure 10.48. As drilling proceeds, material is excavated by a cylindrical bucket fitted with auger-type cutting blades on the bottom. When the bucket is full, it is raised above the drive table and swung off to the side by the dumping arm.

Advantages of this method include minimum equipment requirements and the ability to obtain accurate samples. Its use is limited, however, to drilling only small-diameter wells through clay and sand formations that are relatively free of cobbles and boulders.

BORING WITH EARTH AUGERS

Earth augers of various sizes and designs are used in certain areas for drilling water wells. Three principal types are used commonly: (1) large-diameter bucket auger, (2) solid-stem auger, and (3) hollow-stem auger.

Bucket Auger

The first method utilizes a large-diameter bucket auger to excavate earth materials. This method is referred to as rotary bucket drilling (Figure 10.47). The excavated material is collected in a cylindrical bucket that has auger-type cutting blades on the bottom (Figure 10.48). The bucket is attached to the lower end of a kelly bar that passes through and is rotated by a large ring gear that serves as a rotary table.

The kelly is square in cross section and consists of two or more lengths of square tubing, one length telescoped inside the other (Figure 10.49). This design permits boring to a depth several times the collapsed length of the kelly bar before having to add a length of drill rod between the kelly and bucket. In drilling with only the telescoping kelly serving as the drill stem, the bucket is lifted from the hole and dumped without disconnecting. If

Figure 10.49. The massive kelly consists of two or more lengths of square tubing, each length telescoped inside the next larger size.

one or more drill rods are used for deeper boring, the drill rods must be removed each time the bucket is brought to the surface.

Wells more than 250 ft (76.2 m) deep have been drilled by this method, although depths of 50 to 150 ft (15.2 to 45.7 m) are more common. Water wells drilled with the bucket auger are from 18 to 48 inches (457 to 1,220 mm) in diameter, but few wells are larger than 36 in (914 mm). Special hardened teeth or tungsten carbide inserts are fixed to the cutting blades on the bottom of the bucket when augering in dense formations.

Rotary bucket drilling of water wells has found primary application in areas of clay formations that stand without caving

Figure 10.50. Large-diameter solid-stem augers are used to drill holes in stable formations. *(Getty Refining and Marketing Company)*

while the borehole is drilled and pipe is installed to serve as well casing. Drilling in sand below the water table is difficult, but not impossible if the hole is kept full of water or drilling fluid. A considerable supply of water may be needed if the sand formation is quite permeable. Thus, many drillers will use drilling fluid additives such as bentonite or polymers to control fluid loss. The maximum demand for water may last for only a few hours, however, because drilling will proceed rapidly under favorable conditions.

Cobbles and boulders can cause much difficulty in the bucket-auger procedure because they must be picked out of the bottom of the hole individually by using an orange-peel bucket, stone tongs, or ram's-horn tool. The hole diameter must be large enough to permit the use of these tools when necessary.

In operation, an auger bit will remove a cylinder of material 24 to 48 in (610 to 1,220 mm) deep in a contiguous mass. Therefore, samples obtained by the bucket-auger method are representative of the formation being drilled, unless sloughing or caving of the borehole walls has occurred.

Solid-Stem Auger

A second boring method uses a solid-stem auger with either a single flight (one section) or continuous flighting (multiple sections)*. Augers having a single section of flighting are commonly called earth augers, construction augers, or large-diameter augers (Figure 10.50). Earth augers with diameters as large as 54 in (1,370 mm) have been used in shallow holes, but 14- to 24-in (356- to 610-mm) single-flight augers are more common. Borehole depths of 60 ft (18.3 m) are not unusual in stable ground using the smaller diameter augers.

Drilling rigs equipped with large-diameter earth augers are similar in most respects to bucket auger rigs. They usually employ a kelly bar drive system. As with bucket

---

*In a strict sense, flighting is the spiral flanges welded to pipe. However, each auger section is commonly referred to as a flight. A single flight consists of one length of auger with flighting; continuous flighting refers to multiple auger sections with continuous spiral flanges.

augers, special hardened teeth or cutters are used when augering through hard ground, cobbles, or soft rock. This method is ineffective in loose ground or when drilling below the water table. It is sometimes used to bore a large-diameter hole to the water table; thereafter, casing is set and other drilling methods are used to complete the well. Shallow water wells are often constructed by augering to the top of a sand aquifer, lowering small-diameter pipe to that depth, and then advancing the pipe into the saturated formation by a bail-down or jetting operation.

Solid-stem augers with continuous flighting are used to advance holes in stable formations. Solid-stem augers are not truly solid, because the continuous flight design is welded onto small-diameter pipe; but the hexagonal pin placed at both ends of the flight (section) makes this type of auger nonhollow. Drill rigs turn the auger sections using a rotary drive head mounted on a hydraulic-feed mechanism that pushes the auger section down or pulls it back. Single auger lengths are generally 5 ft (1.5 m); diameters range from 4 to 24 in (102 to 610 mm), with diameters of 6 to 14 in (152 to 356 mm) used in well drilling. Although depths of 400 ft (122 m) have been recorded with the 6-in auger, auger depths of 40 to 120 ft (12.2 to 36.6 m) are more usual for the common diameters.

For drilling, a special auger bit or cutter head is attached to the leading auger flight section and cuts a hole for the flights to follow. The cutter head is usually 2 in (51 mm) larger in diameter than the flights, providing about 1 in (25 mm) clearance. The cuttings are brought to the top of the hole by the flights which act as a screw conveyor. As the auger drills into the earth, more auger sections are added until the desired depth is reached or penetration is halted by obstructions, hard ground, or caving conditions.

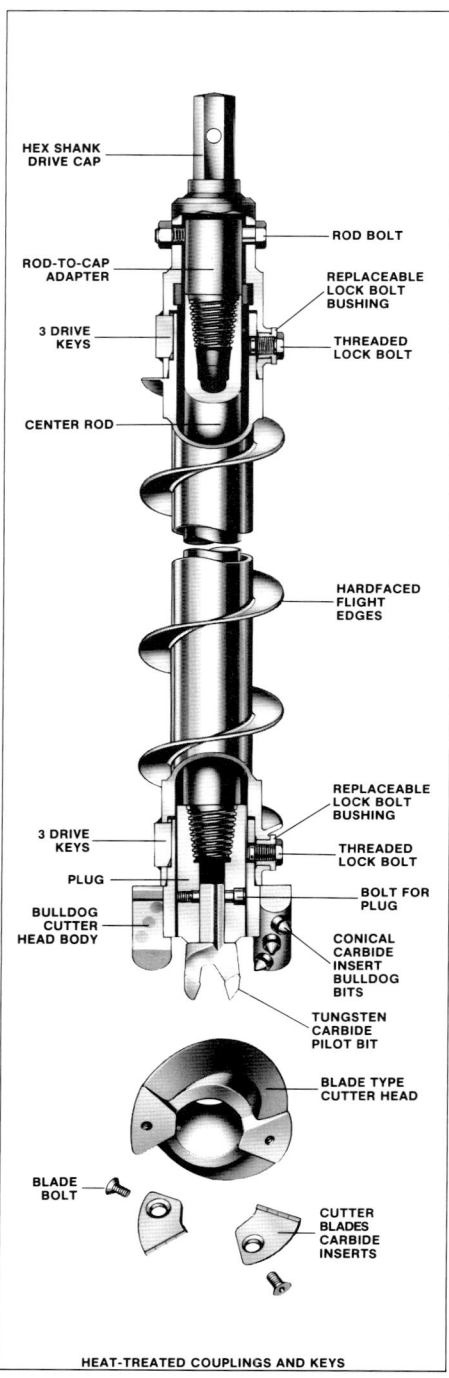

Figure 10.51. The lowest flight in a hollow-stem auger drill string is equipped with a cutter head and a pilot bit. *(Mobile Drilling Company, Inc.)*

Hollow-Stem Auger

The third augering method is the hollow-stem continuous-flight augering method. Although geotechnical and exploration drillers have been using the hollow-stem auger since the early 1950's, its use by the water well drilling industry has been quite limited until recently. The flights for the hollow-stem auger are welded onto larger diameter pipe with a cutter head mounted at the bottom (Figure 10.51). Unlike the solid-stem method, drill rods (drill stems) can pass through the center of the auger sections. A plug is inserted into the hollow center of the cutter head to prevent soil from coming up inside the auger. This center plug has an attached bit that helps advance the auger. The drill rod and plug connect through the auger flights to the top-head drive unit by small-diameter drill rods to insure that the drill rods and plug rotate with the flights. The center plug is omitted in stiff or dense formations because only 2 to 4 in (51 to 102 mm) of earth will rise up inside the hollow stem.

Hollow-stem augers with outside diameters ranging from 6¼ to 22 in (159 to 559 mm) [2½ to 13 in ID (64 to 330 mm ID)] have been used to drill water wells, although the common outside diameters are 6¼ to 13 in (159 to 330 mm) [2½ to 6 in ID (164 to 152 mm ID)]. Auger lengths are usually 5 ft (1.5 m), but on larger hollow-stem rigs, especially those equipped with carousel racks, the auger flights are 10 ft (3 m) long and are stored in 20-ft (6.1-m) sections. Holes as deep as 300 ft (91.5 m) have been drilled with 6¼-in diameter hollow-stem augers; more common depths in stable formations are 120 ft (36.6 m) with 6¼-in diameter hollow-stem augers and 40 ft (12.2 m) with 12-in (305-mm) diameter hollow-stem augers.

Hollow-stem augers are more effective than solid-stem augers because they can be used as temporary casing to prevent caving and sloughing of the borehole wall. The hollow-stem method is a fast and efficient means of drilling and completing small-diameter wells to moderate depths. Screens can be installed and filter packed without using casing or drilling fluids. Use of the hollow-stem auger method is also particularly advantageous in obtaining accurate samples (see Chapter 8). A major disadvantage of this method is the relatively high cost of hollow-stem flight augers.

DRIVEN WELLS

Driven wells can be installed only in unconsolidated formations that are relatively free of cobbles or boulders. Well points can be driven by hand methods to depths of about 30 ft (9.1 m), depending upon the tightness of the soil. Well points driven by hammers weighing 250 to 1,000 lb (113 to 454 kg) reach depths of 50 ft (15.2 m) and even more in favorable situations. In some cases, points are driven out the bottom of larger diameter casing when the aquifer has been reached. Points may be set to greater depths if the screen is protected by casing during driving. At a predetermined depth, the casing is pulled back to expose the screen (Figure 10.52).

Whether driving is to be done by hand or machine, the first step is to bore a hole that is slightly larger than the well point with a hand auger or post-hole digger. The bored hole must be vertical and should be as deep as possible. The drive point and one or more 5-ft (1.5-m) lengths of riser pipe are assembled. Drive couplings should have recessed ends and tapered threads to provide stronger connections than ordinary plumbing couplings. Pipe-thread compound is applied to the threads to make the joint airtight. When ready, the well-point and riser-pipe assembly is set in the bored hole and the hole backfilled. Before driving begins, a malleable iron drive cap is usually screwed to the top of the pipe. Occasionally, a drive coupling may be used when the driving hammer has a centering guide.

Hand driving is best done with a weighted pipe, similar to the type used for driving steel fence posts. It may be operated by one or two persons, depending on its weight. Driving can also be done with a heavy maul, but this is not recommended because it is difficult to deliver square, solid blows with the maul, and glancing blows may break or bend the pipe.

Driving tools can be suspended from a tripod or derrick. The driver must be hung directly over the center of the well so that it will strike square blows. The weight of these tools varies from 75 to 100 lbs (34 to 45 kg) or more. The heavier driving weights are more efficient and are usually operated with a light-duty drilling machine. For example, the spudding action of a cable tool machine is ideally suited for rapid driving with these tools. Other mobile, power-driven drivers are available and are ideal if a large number of points are to be set (Figure 10.53).

To insure that the threaded joints remain tight, the riser pipe is turned slightly with a wrench periodically. Rotation of the riser pipe, however, does not make the well point drive more easily, and in fact the screen can be damaged by excessive twisting. The wrench should only be used to take up slack in the threaded joints (two wrenches are usually used to tighten joints to keep the point from turning).

Driven wells are usually pumped by suction lift; therefore, the static water level must be within about 15 ft (4.6 m) of the surface. If 2-in (51-mm) or larger pipe is used, certain jet or cylinder pumps can be installed to lift water from greater depths.

DRILLING PROCEDURES WHEN BOULDERS ARE ENCOUNTERED

In many formations, boulders or large cobbles can slow or even stop drilling progress regardless of the drilling method being used. If the casing or borehole is being deflected, the driller must do something about the boulders before drilling can be continued. Boulders occur commonly in glacial tills, extremely coarse outwash deposits, former beach zones

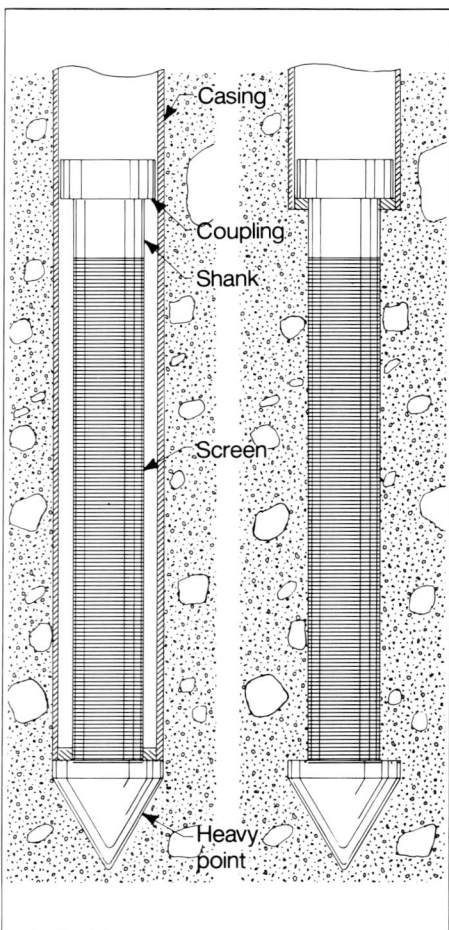

Figure 10.52. In rocky terrains, a drive point can be set at greater depths if it is protected by casing. For this procedure, the driller builds up or purchases a special heavy point that rests just beneath the bottom of the casing. During driving, the casing pushes the point downward, while the body of the casing protects the screened portion of the point.

now buried, conglomerate deposits that formed near the base of steep slopes, alluvial fans, and alluvial deposits in mountainous regions. Drilling costs can rise significantly when boulders are encountered in the hole.

In general, do not drill below a protruding boulder because it may fall partly into the hole causing the bit to become lodged. Whether boulders are removed or destroyed will depend on the drilling method being used. Alternative procedures include the following:

Cable Tool:
1. Change the bit.
2. Increase drill-string weight to break the rocks.
3. Bail out material below the boulder so that it drops into the hole. Fluid levels may have to be increased to keep the borehole open.
4. Blast the boulder.

Direct Rotary:
1. Increase the weight on the bit to grind through or crush the rock or force it to the side of the borehole.
2. Install a new or different bit.
3. Fish out the boulder if it is completely within the borehole.
4. Switch to air and use an air hammer.
5. Cement the boulder if it is sufficiently far above the aquifer, then continue drilling.
6. Blast the boulder.

Reverse Rotary:
1. Install a new bit to either push the boulder into the borehole or grind up the rock; cement can be used to stabilize the rock (if boulders are not near the aquifer).
2. Increase the weight on the bit.
3. Fish out the boulder.

Air Rotary with Casing Driver:
1. Keep the bit close to the bottom of the casing so boulders cannot become lodged between the bit and casing.
2. Drill and drive only short distances.
3. Increase the weight on the bit.
4. Pull back slightly to allow the boulder to fall into the borehole or be pushed into the borehole wall.
5. Change the bit, preferably to a down-the-hole hammer.
6. If boulders are sufficiently far above the aquifer, cement them into position so they can be drilled.

Figure 10.53. A mechanical driving device facilitates placing a large number of well points. *(Crystal Water Products)*

Several general points can be made concerning drilling through boulders:
1. The driller should proceed cautiously to prevent damaging the drive shoe or deflecting the casing.
2. It may be best to case through boulders.
3. Drill at least 5 to 10 ft (1.5 to 3 m) into the rock to make sure that bedrock has been reached.
4. If the casing has been dented by a boulder during driving, the casing diameter should be restored by using a casing swedge (the swedge is also useful in lining up broken casing so it can be lined with a sleeve).

Although blasting is recommended as a method for destroying boulders, it is the least acceptable solution because of possible casing and borehole damage. Therefore, any blasting procedure should be used with caution and only as a last resort. State, municipal, federal, and insurance requirements or codes relating to blasting should be thoroughly understood by the drilling contractor before using explosives. For drillers working in areas with relatively shallow boulder concentrations, conversion to air rotary rigs and the use of drill-through casing drivers are recommended.

If blasting is seen as the only solution, several important aspects should be considered. First, accurate measurements are vitally important in successful blasting procedures. The driller must know the precise depth of the hole, the length of casing in the hole, and the depth at which the charge is placed. Negligence in determining these depths can cause serious economic and safety problems.

Second, it is difficult to set specific rules for the size of the explosive charge. The amount required depends on several factors: general characteristics of the formation, size of the boulders, composition of boulders, depth of the hole, and diameter of the well to be drilled. Many drillers use an experimental approach. A charge of moderate size is used initially; if it proves to be too small, the operation is repeated with a larger one. Drillers with experience in a given local area can often judge the proper amount to use. It is nearly impossible to get into serious trouble when using a charge that is too small, although ineffective charges will slow down the overall drilling operation. Use of too much explosive in the beginning, however, may damage the casing and shatter the formation more than necessary. Never blast closer than 50 to 60 ft (15.2 to 18.3 m) from the surface.

Third, the explosive should always be set as far below the casing (if used) as possible. If necessary, the casing should be pulled back. Maintaining a high fluid level in a cable tool hole will keep the unconsolidated material from heaving into the hole when the casing is pulled back. Rupturing or bending of the casing caused by the explosion may necessitate abandonment of the borehole.

FISHING TOOLS

In most drilling methods, tools can be broken off or dropped into the borehole. The object or tool that is lost in the hole is called the "fish," which the driller retrieves by "fishing." Fishing jars are used in the cable tool method to retrieve tools from the hole. They are placed between the fishing stem [usually 10 ft (3 m) long] and a fishing tool such as a horn socket or center spear. In this position, the stem increases the impact of the jars on the fishing tool during the upstroke. The greater stroke of the fishing jars prevents accidental downstroke hitting during retrieval of the lost tool. Hitting both up and down will usually free the "fish" to be removed from the hole.

In the rotary drilling method, the shear stresses placed on the drill string are often excessive, unlike the cable tool method where only the force of gravity is utilized for drilling. These shearing stresses are magnified because the weight of the entire drill column is augmented by the hydraulic-driven pull-down weight that may be applied by the driller. These pull-down weights may reach 30,000 lb (13,600 kg) or more. Because the torque applied to the drill string can occasionally exceed the breaking strength of the equipment, special fishing tools have been developed to extract pieces of the sheared drill string from the hole.

Six fishing tools are used most commonly in rotary drilling operations: tapered tap, die collar, releasing spear, junk mill, circulating overshot, and magnet (Figure 10.54). Many drillers construct fishing tools that may be particularly suitable for their own equipment. After determining the depth at which the string or tool has been lost, the driller attempts to enter (tapered tap) or overshoot (die collar) the top of the lost drill rod and then rotate the fishing tool until it is firmly attached. Releasing spears can be used in place of a taper tap. They offer the advantage of quick release from the fish and provide easy re-engagement if necessary. If greater force is required to pull the fish, another type of tool called a releasing and circulating overshot is used. It consists of three main components — a top sub, a bowl that houses the engaging and packing-off element, and a guide to center the tool over the fish. A junk mill is used to grind up smaller objects lost in the borehole. Powerful magnets are useful in removing relatively small tools or other parts from the hole. To be successful, circulation must be established or maintained during most fishing operations.

One particularly common fishing operation in large-diameter holes involves retrieval of roller cones that have become detached from the bit. Failure of the bearings on which the cones rotate is the principal cause of cones falling to the bottom of the borehole. Bearing failure is usually attributable to excess weight on the bit, high operating temperatures, or excessive use. The most common techniques for retrieval of lost cones includes the use of a junk basket, a strong magnet, or a button or diamond bit to grind up the cone. Lost cones can sometimes cause abandonment of the well. To avoid this problem, the driller should immediately replace any bit on which a cone has become damaged or locked in place.

GROUTING AND SEALING WELL CASING

In engineering practice, grouting is the act of injecting certain substances into the void space of earth materials to reduce or eliminate their permeability, consolidate them, or increase their strength (Bowen, 1981). Thus, grouting is widely used in constructing tunnels, dams, bridges, and foundations for buildings. Low-viscosity grouting materials are used in soils having low hydraulic conductivity, whereas high-viscosity grouts are used in coarse-grained, highly permeable soils. Although several basic types of grouting materials exist, multiphase (suspension) systems are common in the water well industry.

Grouting (cementing) well casing involves filling the annular space between the casing and the drilled hole with a suitable slurry of cement or clay*. The term "grouting" is used by drillers to describe the process of mixing and placing grout. The length

---

*The terms "grouting" and "cementing" are often used interchangeably, but grouting is the preferred term because it refers to the filling of void spaces. Cement is the most common grouting material and thus cementing has become synonymous with grouting.

of the borehole section to be grouted will vary according to water well codes, aquifer structure, and water quality. Typically, all public water supply wells must be grouted from the surface to a depth of at least 50 ft (15.2 m) to prevent leakage of contaminants from the surface. Water wells constructed in rock that is overlain by relatively thin, loosely consolidated sediment will usually be grouted from the surface to the rock. In some formations where poor-quality aquifers are interspersed with high-quality water zones, the poor-quality aquifers are cemented off. Grouting is also standard practice in monitoring well construction.

The grouting methods described below focus primarily on the use of cement and water (neat cement), although the slurry may contain sand, bentonite, or hydrated lime in certain situations. A clay slurry made with a high-grade bentonite can also serve for grouting, provided it is used at a depth where drying and shrinking of the grout will not occur, and where water movement will not wash away the clay particles. Synthetic materials, especially polymers, are also used as grouting materials, but their extremely low solids content and great shrinkage if dried make them less suitable for sealing wells.

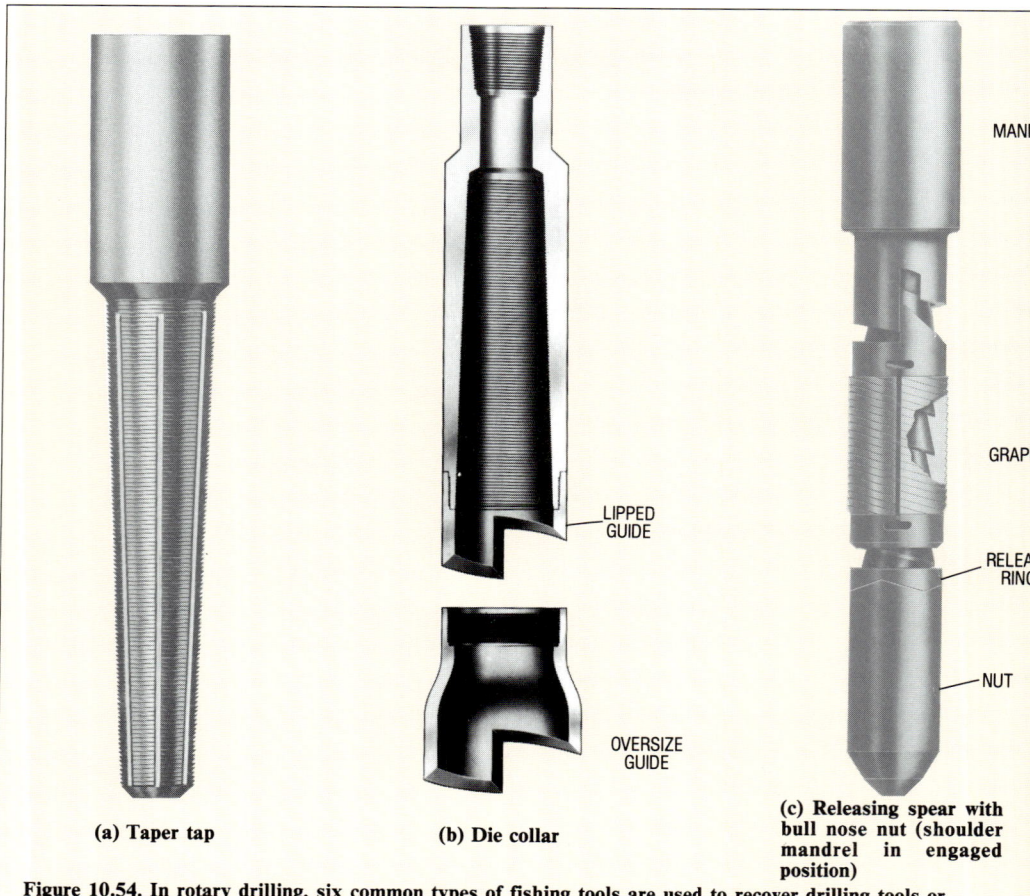

(a) Taper tap (b) Die collar (c) Releasing spear with bull nose nut (shoulder mandrel in engaged position)

Figure 10.54. In rotary drilling, six common types of fishing tools are used to recover drilling tools or casing lost in the borehole. *(Bowen Tools, Inc.)*

Various types of cement are manufactured to accommodate different chemical and physical conditions found in the subsurface environment. Five types are given in ASTM specifications and are used generally at the ground surface. The high pressures and temperatures encountered in deep wells, especially oil wells, has led to the development of eight classes of cement under API specifications. Table 10.4 lists five API cement classes used in water well construction, although Classes A, B, and C are more commonly used. Cement classifications used outside the United States are given in Appendix 10.B. The constituents of these cements are given in API Standard 10A.

The compressive strengths of portland (types A and B) and high-early cements (type C) are shown in Table 10.5 for setting times of 24 and 72 hours at various temperatures. Various compositions of cement have different compressive and tensile strengths after curing; compressive strengths are usually about 10 times greater than tensile strengths. Compressive strengths are determined by crushing small cubes of cement under laboratory conditions. For most drilling operations, the cement should reach a compressive strength of 500 psi (3,450 kPa) before drilling is resumed. The temperature in the borehole, chemistry of the formation water, dilution of the cement,

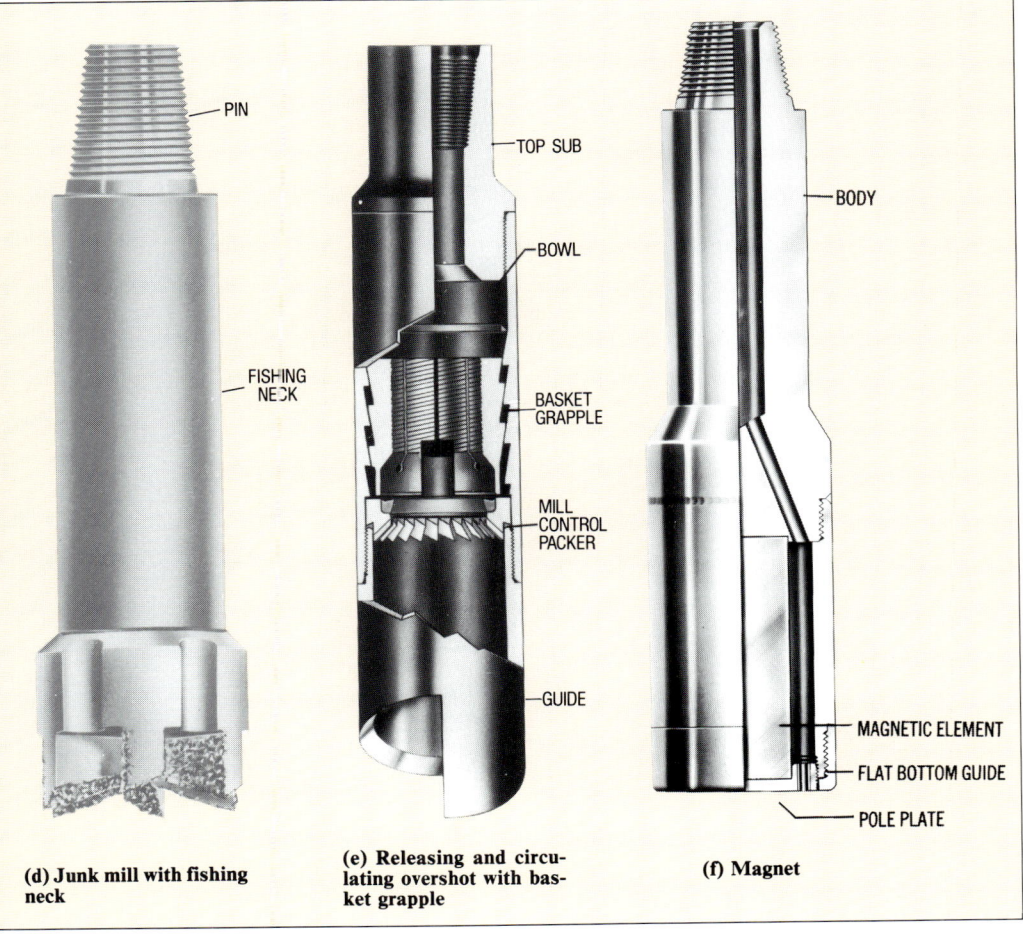

(d) Junk mill with fishing neck

(e) Releasing and circulating overshot with basket grapple

(f) Magnet

Table 10.4. Classifications of Cements Used in Water Wells

API Classification	Special Properties	Recommended Range for Well Depth ft.	m
A (similar to ASTM C150, Type I)	None	0-6,000	0-1,830
B (similar to ASTM C150, Type II)	Moderate to high sulfate resistance	0-6,000	0-1,830
C (similar to ASTM C150, Type III)	High early strength	0-6,000	0-1,830
G	Can be used with accelerators and retarders	0-8,000	0-2,440
H	Can be used with accelerators and retarders	0-8,000	0-2,440

and downhole pressure affect the rate at which the cement cures. Generally, the 500-psi compressive strength is reached between 12 and 24 hours after placement.

Equipment for mixing and placing cement grout need not be elaborate for most water-well work. However, the chemical reaction that causes grout to set and harden begins as soon as cement and water are mixed, and the equipment used to mix and place the grout must be adequate to complete the installation while the grout is still fluid (Figure 10.55).

The size of the annular space required for grouting depends on the method of grouting. Thus, planning the size of the borehole is important. The annular space to be grouted should have a diameter that is 4 to 8 in (102 to 203 mm) larger than the casing. The ideal result is a uniform sheath of cement around the casing for the entire vertical distance to be grouted. Tight places and "dead spots" result where casing not properly centered touches the wall of the hole, causing channeling of the slurry. Some

Table 10.5. Compressive Strengths of Portland and High Early Cement

Temperature		Borehole Pressure		Typical Compressive Strength							
				24 Hours*				72 Hours*			
				Portland		High Early		Portland		High Early	
°F	°C	psi	kPa	psi	kPa	psi	kPa	psi	kPa	psi	kPa
60	15.6	0	0	615	4,240	780	5,380	2,870	19,790	2,535	17,480
80	26.7	0	0	1,470	10,140	1,870	12,890	4,130	28,480	3,935	27,130
95	35.0	800	5,520	2,085	14,380	2,015	13,890	4,670	32,200	4,105	28,300
110	43.3	1,600	11,030	2,925	20,170	2,705	18,650	5,840	40,270	4,780	32,960

*Strengths based on the following criteria:

	Portland		High Early	
Water	5.19 gal/sack	19.6 ℓ/sack	6.32 gal/sack	23.9 ℓ/sack
Slurry weight	15.6 lb/gal	1,870 kg/m³	14.8 lb/gal	1,770 kg/m³
Slurry volume	1.18 ft³/sac	0.03 m³/sack	1.33 ft³/sack	0.04 m³/sack

(Halliburton, 1968; Courtesy of SPE Monograph, Cementing, 1976)

of the design criteria applying to grouting of well casing for sanitary protection are given in Chapter 18.

State or federal laws may dictate the minimum length of grout required for various casing diameters for certain types of wells. The drilling contractor should become familiar with specific regulations for the type of wells drilled.

It is important to recognize that cement grouts exert greater collapse pressure on casing than do either water or drilling fluid. Water alone exerts 0.433 psi (3 kPa) for every 1 ft (0.3 m) of depth in the borehole. Because some solids are added to water making up a drilling fluid, the weight is greater; a typical drilling fluid made with clay additives weighs approximately 9.5 lb/gal (1,140 kg/m^3), versus 8.33 lb/gal (998 kg/m^3) for water. Thus, the required collapse pressures for water well casing are usually calculated on the basis of 0.5 psi per ft (3.4 kPa per 0.3 m) of depth. The specific gravity of cement grouts is about twice that of water; the cement and water slurry will weigh approximately 123 lb/ft^3 or 16.4 lb/gal (1,970 kg/m^3). To calculate the potential pressure at the bottom of the casing (where it is greatest), the pressure increase with depth is assumed to be 0.8 psi per ft (5.5 kPa per 0.3 m) at a minimum. For example, the maximum pressure that could be exerted by the cement grout at the bottom of a 500-ft (152-m) casing is 0.8 times 500, which equals 400 psi (2,760 kPa). A safety factor of two is recommended when selecting the wall thickness for the casing. Any fluid inside the casing will reduce or balance the pressure exerted by the cement column, depending on the relative heights of the fluid and cement columns. But if the inside of the casing contains no water or drilling fluid, the casing must be

Figure 10.55. Three cement trucks are required to provide enough cement to grout the casing in this 500 ft (152 m) borehole. The cement is pumped from the stock tank into the borehole by a positive displacement duplex pump, shown on the left. *(Test Drilling Services)*

strong enough to support the entire grout column.

Proportioning Cement Grout

Laboratory tests indicate that 5.2 gal (19.7 ℓ) of water are needed to hydrolyze one 94-lb (42.6-kg) sack of portland cement. This mixture produces a slurry weight of 15.6 lb/gal (1,870 kg/m^3). An advantage of using the proper water-cement ratio is more effective bridging of cement particles in the pores of permeable formations, which prevents excessive penetration of the grout into these formations. Although thinner mixtures with more than 6 gal (23 ℓ) per sack are used for grouting foundation materials, this ratio is less suitable for water-well work. Shrinkage increases with greater water content, because water is squeezed out of the thinner mixtures by pressure against fine sand or other permeable formation materials. Cement will settle out of the slurry if the ratio is greater than 10 gal (38 ℓ) per sack of cement. Water used for grout should be free of oil and other organic material. Dissolved minerals should be less than 2,000 mg/ℓ; high sulfate content is particularly undesirable.

Bentonite clay can be added to the cement to hold cement particles in suspension,

Table 10.6. Effects of Additives on the Physical Properties of Cement

		Bentonite	Diatomaceous Earth	Pozzolan	Sand	Heavy Minerals	Accelerator	Sodium Chloride	Retarder
Density	Decrease	⊗	⊗	⊗					
	Increase				⊗	⊗		x	
Water Required	Less								
	More	⊗	⊗	x	x				
Viscosity	Decreased						x	x	⊗
	Increased	x	x	x	x	x			
Thickening Time	Accelerated						⊗	⊗	
	Retarded	x	x						⊗
Early Strength	Decreased	x	x	x					⊗
	Increased						⊗	⊗	
Final Strength	Decreased	⊗	⊗	x			x		
	Increased								x
Durabilty	Decreased	x	x						
	Increased					⊗			
Water Loss	Decreased	⊗							x
	Increased			x					

x Denotes minor effects.
⊗ Denotes major effects and/or principal purpose for which used.

(API, 1959; Smith, 1976)

Figure 10.56. Portable grouting machines are capable of performing both the mixing and pumping operations. *(Acker Drill Company, Inc.)*

reduce shrinkage, and improve fluidity of the mixture. Approximately 3 to 5 lb (1.4 to 2.3 kg) of bentonite should be mixed with 6.5 gal (25 ℓ) of water per sack of cement. If the amount of bentonite exceeds 6 percent, excessive shrinkage of the cement will occur. It is best to mix the bentonite and water first, then add cement to the clay-water suspension.

Potential fluid-loss conditions may call for the addition of sand or other bulky material to permit the grout to bridge larger openings without excessive fluid loss. These coarse materials add to the difficulty of handling and placing grout, but they may be necessary to reduce cost of material where large openings are to be filled. The physical effect of additives commonly added to cement are given in Table 10.6.

Mixing the Grout

It is important that grout be mixed thoroughly and be free of lumps. If the mixture is purchased from a ready-mix concrete plant, the correct proportions must be assured. To avoid stones and lumps of concrete, the driller should insist that the delivery trucks be thoroughly cleaned before the grout is transported. It is best to provide a protective strainer on the tank from which the grout is pumped into the well.

Some drillers use small portable grouting machines that combine both the mixing and pumping operations (Figure 10.56). Many of these machines are equipped with a positive displacement pump because this type of pump can work efficiently against much greater head pressures with little loss in emplacement volume. The effective operation of a centrifugal pump is much more limited under high head conditions. Most drillers avoid using the mud pump on their rotary rigs because of the abrasive qualities of the cement and the difficulty in removing all traces of the cement from the pump after completing the cementing operation.

The volume of grout required cannot always be determined accurately. Irregularities in the size of the borehole and losses into fractured rock occur in many wells. Therefore, the driller must be prepared to augment initial estimates on short notice. Table 13.13 (page 445), which gives the volume of filter pack required, can also be used to estimate the minimum amount of grout required between different casing diameters or be-

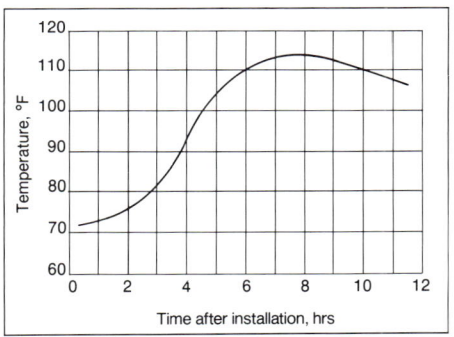

Figure 10.57. Heat is produced when cement is hydrated. The amount of heat released depends on the volume of cement used to grout the casing, the ambient temperature of the formation, and the pressure in the borehole. This graph shows the temperature change over time at a depth of 550 ft (168 m) where the formation temperature was 65° F (18.3°C), the mixing water temperature was 74° F (23.3°C), and the slurry weighed 15.4 lb/gal (1,850 kg/m^3). *(Canadian Institute of Mining and Metallurgy, 1965)*

tween the casing and borehole wall.

When water is mixed with cement and hydration occurs, heat is released (Figure 10.57). The amount of heat released is a function of the volume of cement — the more cement, the more heat. If the formation temperature is high, the hydration process is accelerated and heat is released more quickly. If cement fills a 2-in (51-mm) annulus, the heat produced during hydration creates a maximum temperature rise of 35° to 45°F (19.5° to 25°C) (Smith, 1976).

Slurry Placement Methods

Successful placement of the cement will depend on the temperature and pressure in the borehole, how well the casing is centered in the hole, and the emplacement method. Temperature has a significant effect on how fast the cement slurry hydrates and thus how fast the cement develops strength. Pressures caused by the weight of the drilling fluid can reduce the rate at which the cement can be pumped. At high pressures (only a problem in deep water wells), the hardening time for the cement can be substantially reduced. The use of centralizers is important to assure a uniform thickness of cement around the casing. Centralizers should be placed every 40 ft (12.2 m) on the casing. Several placement methods are described below. Each method is satisfactory but care should be taken to assure that channeling does not occur, thus avoiding gaps in the cement.

Figure 10.58. A cement basket is mounted on the casing to support the column of cement grout to be placed above it. Use of a basket prevents grout from entering weak underlying formations or infiltrating the filter pack. *(Corner S. A.)*

To assure that grout will provide a satisfactory seal, it is necessary to place it in one continuous operation, before setting begins. Regardless of the grouting method used, the grout should be introduced first at the bottom of the space to be grouted. This procedure minimizes both contamination or dilution of the slurry and bridging of the mixture. Suitable pumps with sufficient air or water pressure should be used to force grout into the space to be filled. If the cement is pumped under turbulent flow conditions, drilling fluid removal is enhanced and voids are filled more completely.

Moyno™, diaphragm, and piston pumps are most often used to pump cement grout. The Moyno pump is a positive displacement pump with an effective output pressure of 225 to 250 psi (1,550 to 1,720 kPa); it cannot be permitted to pump sand, however. Diaphragm pumps, although having lower output pressures of 100 to 110 psi (690 to 758 kPa), can handle particles up to $\frac{1}{4}$ to $\frac{3}{8}$ inches (6.4 to

9.5 mm) in diameter. They are not as efficient as the Moyno pump because of higher friction losses. Both types are used for batch mixing.

For larger grouting jobs, either piston pumps or, less frequently, centrifugal pumps are favored. Piston pumps of various sizes (2 x 3, 3 x 4, or 5 x 6) can build pressures to 120 psi (827 kPa), and have been used successfully to place grout to 3,000 ft (915 m) or more with a 2-in (51-mm) tremie pipe. Because they develop less pressure, centrifugal pumps can be arranged so that a hopper feeds the pump under pressure, thereby increasing pump output. The grout is usually placed in one continuous operation because so much water would be required to clean the pumps and pipe after batch mixing.

In cases where an open borehole has been drilled below the depth to which the casing is to be grouted, the lower part of the hole must be backfilled, or a bridge (cementing basket or formation packer shoe) must be set in the hole, to retain the slurry at the desired depth. Backfilling the hole to the proper level with sand is a common procedure. The sand must be fine enough so that cement will not penetrate downward more than a few inches. Controlled experiments indicate that there is no significant penetration by cement into uniform sand with grain size finer than 0.025 in (0.6 mm), or into nonuniform sand with hydraulic conductivity less than 3,000 gpd/ft^2 (122 m/day). Material sold ordinarily as plaster sand or mortar sand is usually satisfactory.

When the borehole cannot be backfilled, external packers combined with a float shoe or cement baskets are used to support the cement column. Cement baskets are installed on the outside of the casing by clamps (Figure 10.58). External packers must be installed in the casing string as the casing is run into the borehole (Figure 10.59); the packers are expanded before cementing begins.

Cement should be allowed to harden for 24 hours before drilling resumes, although some types of cement may require longer curing times. It is false economy to risk damaging a good grouting job by drilling the plug out too soon. If an attempt is made to drill out the plug prematurely, the drilling contractor ordinarily can determine if the cement is still soft.

Tremie Pipe Outside Casing

Grout can be placed through a string of small-diameter pipe (tremie or grout pipe) placed outside the casing. The casing is lowered into the hole with centering

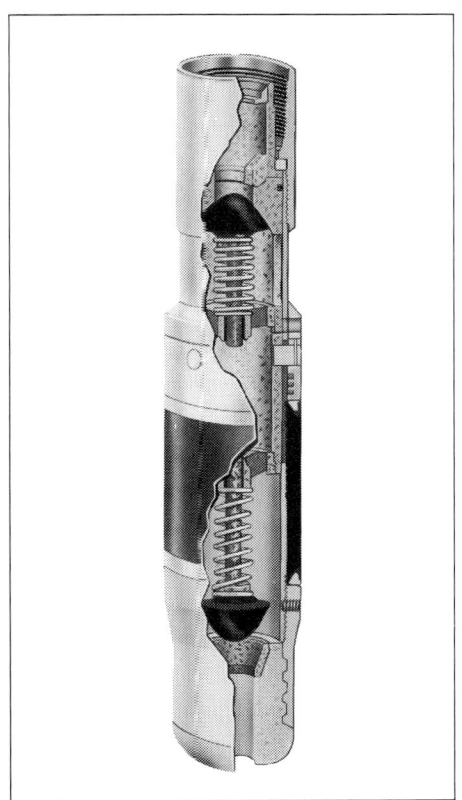

Figure 10.59. An external packer equipped with a float shoe can be installed in the casing string to facilitate placing cement grout. *(Halliburton)*

guides attached. Care must be taken to align the centering guides along the entire length of casing to be grouted so that the tremie pipe can pass by them. The lower end of the casing should be closed with a drillable plug or driven into clay so the grout cannot enter. To overcome the buoyant effect of the slurry, the casing may be filled with water or be held down by the weight of the drill rig. This may be dangerous if the drill rig is used and excessive pressures are built up in the borehole.

Grout can be placed by gravity through a tremie pipe, but pumping is preferred because the required volume of grout can be introduced rapidly and with little chance of leaving voids in the grout. Pump pressure must equal the hydrostatic pressure of the grout plus the fluid friction in the grout pipe and annular space.

For shallow holes where the grout is placed by a positive displacement pump, the cementing operation may be completed in a single step; that is, the position of the tremie pipe is not changed as the annulus is filled. If a centrifugal pump is used or if the hole is deep, the tremie must be raised periodically so the hydraulic head created by the cement does not exceed the working pressure of the pump. Usually the tremie is withdrawn one or more joints at a time, but the bottom of the tremie should always remain beneath the surface of the cement. The rate of tremie withdrawal will depend on the pumping rate and the volume of the annulus. The depth to the top of the grout can be

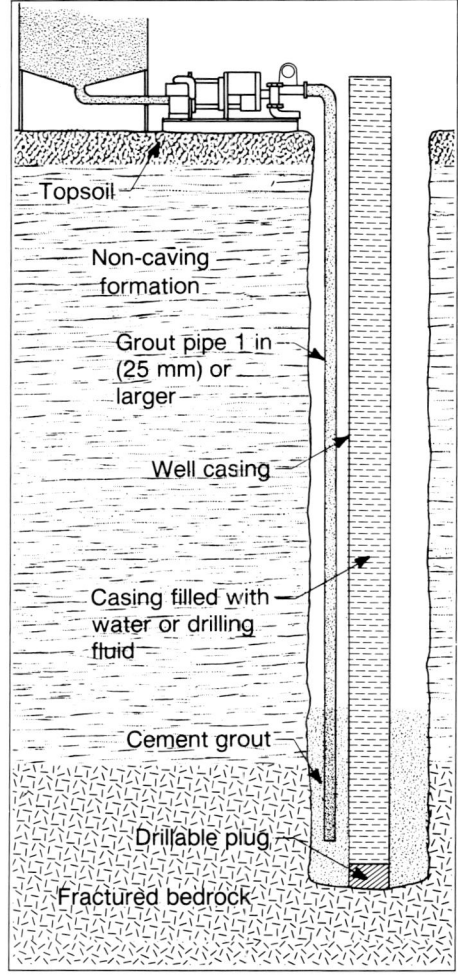

Figure 10.60. Grouting can be accomplished by means of a tremie pipe suspended in the annulus outside the casing. During grouting, the bottom of the tremie should always be submerged a few feet beneath the grout level. As the grout level rises, the tremie should be withdrawn at approximately the same rate.

detected by using a weighted line or a weight indicator. The volume (and therefore the height) of the grout can also be estimated by knowing the volume of material in the hopper before grouting begins.

The grout pipe must be large enough so that all the grout can be placed before hardening begins. A ¾- or 1-in (19- or 25-mm) grout pipe may be used, although 2-in (51-mm) pipe is used for deeper holes. The borehole should be 4 to 8 in (102 to 203 mm) larger than the casing to accommodate the grout pipe. Initially, the pipe should extend to the bottom of the annular space and should remain submerged in the slurry while the grout is being placed (Figure 10.60). Should the tremie become plugged, the output pressure can be increased, the tremie can be raised to reduce the pressure at the bottom of the line, or it can be vibrated or struck to dislodge the stuck material. If operations are interrupted for any reason, the pipe should be raised above the grout level and not be lowered into the slurry again until all air and water in the pipe have been displaced by grout.

When multiple filter-pack screens are separated by grouted casing sections, a larger diameter hole is drilled and placement of the slurry is usually done with a tremie outside the casing. The top of the filter pack is at least several feet above the top of the screen. A low-permeability sand is placed on top of the filter pack to contain the grout until it hardens (Figure 10.61). The grout is run up slightly higher than the formation to be isolated. Once the lower grout has set, filter pack material is again introduced for the next screen, followed by another layer of low-permeability sand, and then another grouted section if necessary.

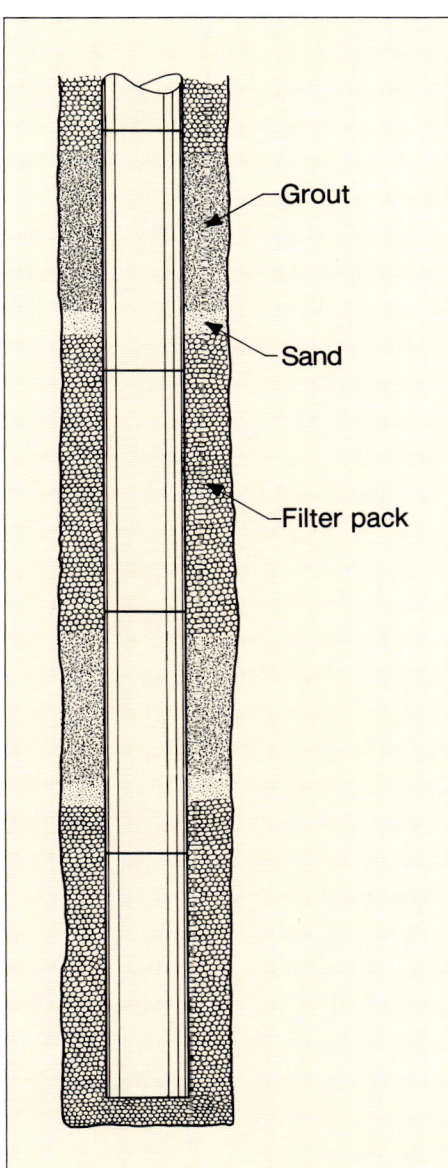

Figure 10.61. Sand is sometimes placed on top of the filter pack to prevent grout from penetrating into the pack.

Tremie Pipe Inside Casing (Inner String Method)

When the use of a grout pipe outside the casing is impractical, grouting may be done by using a grout pipe installed temporarily within the casing (Figure 10.62). In the oil-well industry, this is referred to as the inner-string method of cementing. A cementing plug (float shoe) is attached to the bottom of the casing, which permits the grout to pass into the annular space but prevents it from leaking back into the casing while grouting or after removing the grout pipe. Figure 10.63 shows a cementing plug with a ball-type check valve that prevents reverse flow of the grout. All the internal parts can be drilled out easily upon completion of the cementing.

In the grouting process, the casing is filled with water and suspended just above the bottom of the borehole. Grout is pumped through the grout pipe and float shoe and forced upward around the casing. When cement appears at the surface, displacing all other fluid in the annular space, the grout pipe is disconnected from the float shoe. Cement is washed out of the pipe by pumping water through it before removing it from the well. Because calcium residues may have a deleterious effect on the viscosity-building characteristics of some drilling fluid additives, the casing should be completely flushed with clean water after completing the cementing operation.

Casing Method of Grouting

The casing method of grouting, in which the slurry is forced down the casing and into the annular space (originally called the Halliburton method), has been adopted from the oil-well industry. In one method, two spacer plugs are used. One plug, introduced first, separates the cement slurry above from the drilling fluid in the casing; the other separates the slurry from water pumped in above it to wash the slurry from the casing (Figure 10.64).

After pumping water or drilling fluid through the casing to circulate fluid in the annular space and clear any obstructions from the hole, the first plug is inserted and the casing capped. A measured volume of grout is then pumped in, the casing is opened, a second plug is inserted, and the casing recapped. A measured volume of water is then added and pushed to the bottom of the casing, forcing most of the cement slurry from the casing and into the annular space. The water in the casing is held under pressure to prevent backflow of the slurry until it has set and hardened. When the cement has hardened sufficiently, the second plug and any cement remaining in the casing are drilled out; drilling is continued below the grouted section, through the first plug and into the formation. To protect the physical characteristics of some drilling fluids, it may be necessary to remove residual cement scale from casing walls with brushes or other descaling devices.

A modification of the double-plug procedure is favored by many drillers. After pumping a predetermined quantity of grout into the casing, a plug is installed on top of the grout and enough water is added to force most of the grout from the casing. The usual practice is to leave 10 to 15 ft (3 to 4.6 m) of grout in the casing. If only a single plug is used, that part of the slurry diluted by the drilling fluid

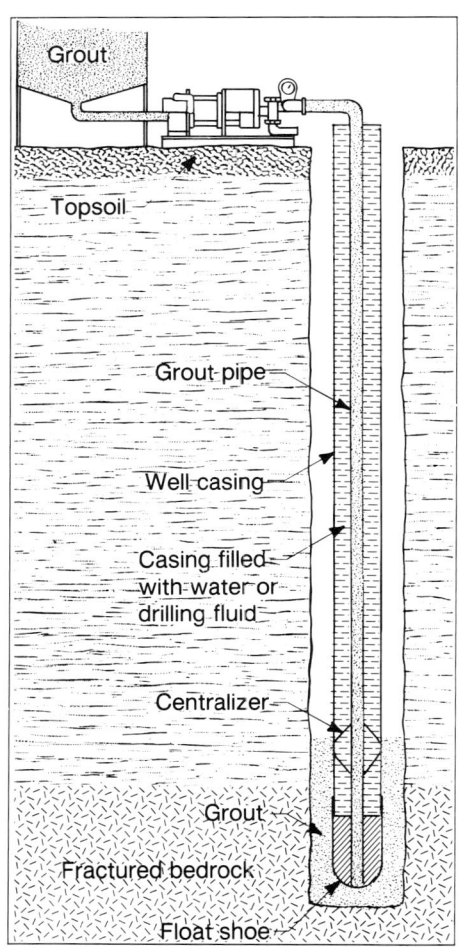

Figure 10.62. In the inner-string method of placing grout, the tremie is suspended in the casing. A cementing (float) shoe is attached to the bottom of the casing before the casing is placed in the borehole. A tremie pipe is lowered until it engages the shoe.

must be expelled to waste at the surface so that a sound, uncontaminated grout seal is achieved at the upper end of the casing. The use of a plug insures slurry and water separation, resulting in a proper grout seal at the lower end of the casing. To eliminate over or under displacement of the cement, a landing collar is set 10 to 20 ft (3 to 6.1 m) above the bottom of the casing to stop the drillable plug at the appropriate depth.

Spacer plugs should be made of materials that can be drilled easily (wood and cement are often used). When a plug settles on sand or clay, the cushioning effect of the soft formation permits the plug to sink into the formation before it is drilled out. Wood and some rubber-fiber combinations have been known to push down through 5 to 20 ft (1.5 to 6.1 m) of clay before being destroyed by the drill bit. Shredded fibers of wood are quite voluminous and can obstruct flow into the well if they are simply pushed aside while the water-bearing formation is drilled. To avoid damaging the bottom of the casing by exerting excessive pump pressure after the plugs have come together, a wire line is sometimes attached to the upper plug so that plug depths can be measured accurately.

In some cases it may be necessary to grout the borehole after the entire hole has been drilled. For example, if a screen

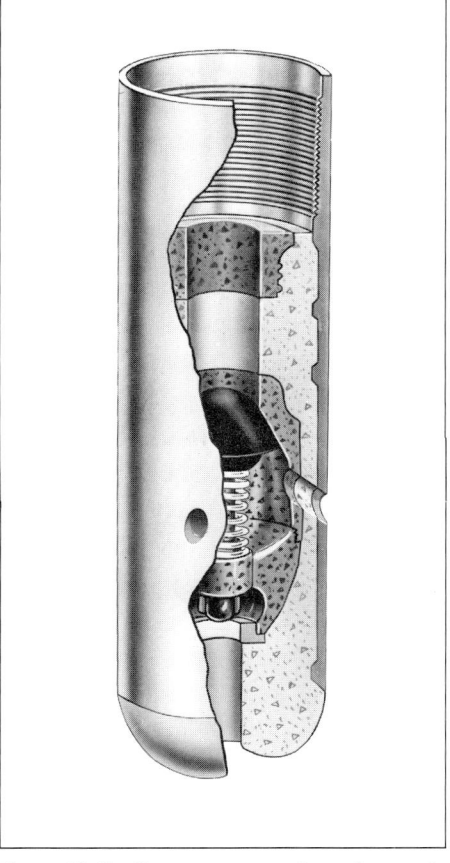

Figure 10.63. Cement can pass from the tremie through the bottom or sides of the shoe into the annulus. The cementing shoe has a ball-type check valve that prevents the grout from reentering the casing when the tremie pipe is withdrawn. The internal parts can be drilled out easily once the grout has hardened. *(Halliburton)*

has been installed in the lower portion of the casing string, and the annular space above the screen must be grouted, a cement basket (or baskets) is used to isolate the screen from the annulus to be filled with cement. Before the casing and screen string is installed, at least one, but preferably two, cement baskets are attached above the screen and a drillable bridge plug is placed in the casing above the screen. Holes are cut into the casing above this plug with a cutting torch, mills knife, or other type of perforator. A cement slurry is introduced into the casing and forced into the annular space above the cement basket (Figure 10.65). Grout is usually extended 5 ft (1.5 m) above and below the formation to be sealed. The bridge plug is drilled out after the cement has hardened. This method of placing grout should not be used in formations where low-quality water must be sealed off in multiple-screen installations or where screens are filter packed. The presence of the baskets interferes with the grouting and filter packing procedures.

Grouting Failures

Several factors may contribute to grouting failures. Some common problems are premature setting, partial setting, insufficient grout column length, voids or gaps in the grout, excessive shrinkage, and casing collapse. Premature setting of the cement can be a serious problem and is usually caused by incorrect assumptions concerning borehole temperature, or by hot mixing water, improper water-to-cement ratios, contaminants in the mixing water, mechanical failures, and interruptions of the pumping operation. Voids within the grouted annulus, another major grouting problem, are usually caused by contact of the casing with the borehole wall or by the presence of washouts.

Testing the Grout Seal

Before drilling out the grout plug, the effectiveness of the grout seal can be checked by three methods: measuring water-level change in the casing over time, pressure testing, and analysis of an acoustic (sonic) cement-bond log. In wells with a low static water level, the casing can be filled with water or drilling fluid and later checked for any water loss. If the static water level is high, the casing can be nearly emptied and any influx of water into the casing can be measured. This procedure should not be used with thin-walled casing. When pressure testing, the grout must be able to contain pressures of 7 to 10 psi (48.3 to 69 kPa) after curing for at least one hour. If the acoustic log is used, it must show that no voids or gaps exist in the grouted annular space (see Chapter 8).

Two other methods of checking the continuity of the grout are available, but they are not used often in the water well industry. The first consists of a temperature

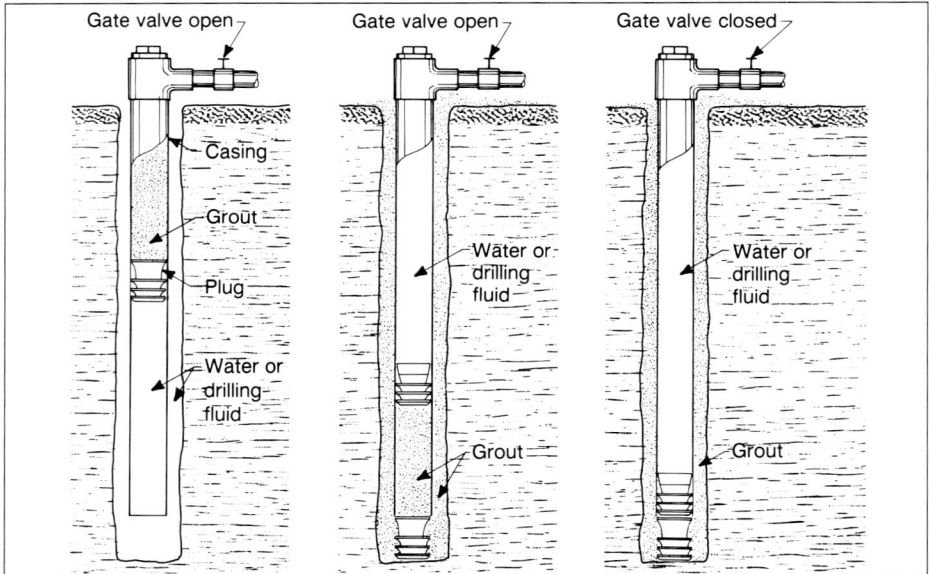

Figure 10.64. Grout can be placed in the casing and then forced out the bottom and up the annulus. This is called the casing method of placing grout. Plugs are used to separate the grout from the drilling fluid and the water used to drive the grout into place. The plugs and float shoe are drilled out after the grout hardens. The casing method of grouting was orginally used in the oil-well industry.

survey that measures the heat produced during the setting of the cement. The temperature survey should be conducted within the first 12 to 24 hours for good results (Smith, 1976). The second method involves mixing a short-lived radioactive tracer into the cement. The radioactivity is then checked to verify the position of the cement. Disadvantages of this method include its high cost, special requirements for handling radioactive materials, and interference with other types of geophysical logs relying on the natural radioactivity of the formations.

Abandoned and improperly constructed wells provide vertical openings or channels through which contaminated water may gain entry into usable fresh-water aquifers. Grouting of abandoned wells is discussed in Chapter 18 in connection with sanitary protection of groundwater resources.

Installation of Bentonite Grout

Bentonite (essentially montmorillonite) is widely used as a grouting material, especially for monitoring wells and water wells where surface contamination may occur, because of its low cost and ease of placement. Commercial bentonite used for grouting is available in either pelletized or granular form. When either of these forms are mixed with water, they begin to hydrate within seconds. Thus, it is impossible to place the granular form by dropping the particles into the annulus. Even pellets dumped down the annulus will begin to stick together and to the walls of the annulus within a few feet of the surface, and therefore may bridge high above the intended depth. It is possible to freeze the pellets first and then carry them to the drilling site in a cooler containing dry ice. In this condition, the pellets will settle a greater distance before sticking. The pellets can also be cooled with liquid nitrogen; in this case, an icy outer layer forms which further protects the pellets so that they may fall 40 ft (12.2 m) or more before hydration begins. In general, the pellets should always be tamped into place to eliminate any bridging that may have occurred.

A much better practice is to pump a prepared bentonite slurry by means of a tremie pipe, using a Moyno pump [40 to 60 gpm (218 to 327 m³/day)] or diaphragm pump [60 to 100 gpm (327 to 545 m³/day)]. If the mixture of bentonite

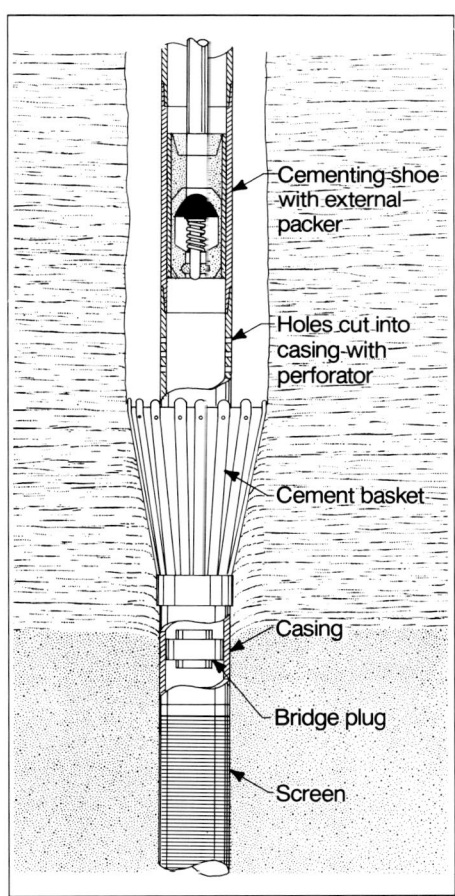

Figure 10.65. A cementing shoe can direct the grout out into the annulus above one or more cement baskets mounted at any position in the casing string. The grout passes through holes cut into the casing by a mills knife or other kind of perforator.

(usually granules) and water is used, only 1 lb (0.5 kg) of bentonite can be mixed per gal (3.8 ℓ) of water because the resulting viscosity will be at the limit of pumping capacity. After being placed, grout with this concentration of bentonite may eventually shrink 25 percent, even though the ground around the grout usually remains somewhat moist. This is a highly unsatisfactory shrinkage rate. Virtually no shrinkage will occur in grout mixed at concentrations of 1.5 lb (0.7 kg) bentonite per gal (3.8 ℓ) of water. This concentration can be pumped only if the water has been pretreated with 1 qt (0.9 ℓ) of polymer per 100 gal (380 ℓ). The polymer prevents the clays from hydrating immediately, and once the particles are evenly distributed in the water the viscosity remains low enough so the slurry can be pumped for about 20 minutes. The granular bentonite should be mixed gently into the water with a paddle, not a mixer or pump; these latter devices will break up the particles and cause the viscosity of the slurry to increase prematurely.

Bentonite grouts should be mixed in batches so they can be pumped before the slurry becomes too viscous. Ideally, the diameter of the suction hose should be as large as possible. In most cases, the slurry reservoir is above the pump intake so that hydrostatic pressure created by the reservoir makes the pump operate more efficiently. The pump and all piping should be flushed with clean water after each batch of grout is pumped into place. The volumes of bentonite, polymer, and water for various annulus sizes, per 100 ft (30.5 m) of depth, are given in Table 10.7.

Bentonite grout has several advantages over cement grout. It has a faster setting time, no heat of hydration, a lower hydrostatic pressure (specific gravity is 9.2 for the grout given in Table 10.7), and the cost is one-third that of cement. Also, bentonite will adhere to both walls of the annulus, whereas cement will adhere firmly only to the soil.

There are several limitations on the use of bentonite grout. Bentonite grouts cannot be used when the borehole is underreamed, because the "set" taken by the grout is not sufficient to withstand the vertical hydrostatic pressures. Thus, the grout may eventually flow into the underreamed section. Another limitation is that bentonite grout should not extend so close to the ground surface that it can dry out and shrink because of low soil moisture. Cement is always used at or near the top of the borehole. The presence of salt water will cause bentonite grout to flocculate and thereby lose viscosity. Organic acids can also destroy the impervious character of the grout seal.

Table 10.7. Amounts of Bentonite, Water, and Polymer Required to Grout 100 ft (30.5 m) of Three Common Annuli

	Bentonite		Water		Polymer*	
	lbs	kg	gal	ℓ	qts	ℓ
2-in (51-mm) pipe in 4-in (102-mm) hole	75	34	50	189	0.5	0.5
4-in (102-mm) pipe in 6-in (152-mm) hole	112	51	75	284	0.75	0.7
5-in (127-mm) pipe in 8-in (203-mm) hole	225	102	150	568	1.5	1.4

*Concentration of polymer recommended by Baroid for their product EZ-Mud®.

PLUMBNESS AND ALIGNMENT

A water well should be both straight and plumb, although in practice any borehole of substantial depth may not be perfectly straight or perfectly plumb. A straight well is one in which each casing section is joined to adjacent sections in a manner that maintains perfect alignment. A borehole that is plumb is one whose center does not deviate from an imaginary vertical line running from the ground surface to the center of the Earth (Figure 10.66). A well bore may be straight, but not plumb; if the borehole is plumb, however, it will be straight. Some tolerance or deviation in straightness (alignment) and plumbness is normally allowed in practice. By custom, a deviation from plumbness of two-thirds the well's inside diameter per 100 ft (30.5 m) of anticipated pump setting is allowed and thought to be reasonable, considering the inherent difficulties of drilling in earth materials (American Water Works Association, 1984). The U.S. Environmental Protection Agency (1975c) has suggested that wells should be constructed so that the borehole deviation from plumbness is 1 degree or less per 50 ft (15.2 m) when using drift indicators. Table 10.8 shows the allowable limits of deviation for various depths.

Of the two factors, straightness of the well bore is the most important, because it determines whether or not a properly sized turbine pump can be installed in the well to the desired depth. If the well is out of alignment beyond a certain limit, the pump cannot be set. A pump can be installed without difficulty in a well that is straight but out of plumb. Too much deviation from the vertical may affect the operation and life of some pumps, however, so plumbness does need to be controlled within reasonable limits. In general, turbine pumps require reasonably straight well bores, whereas submersible pumps can be set in well bores that are more out of alignment.

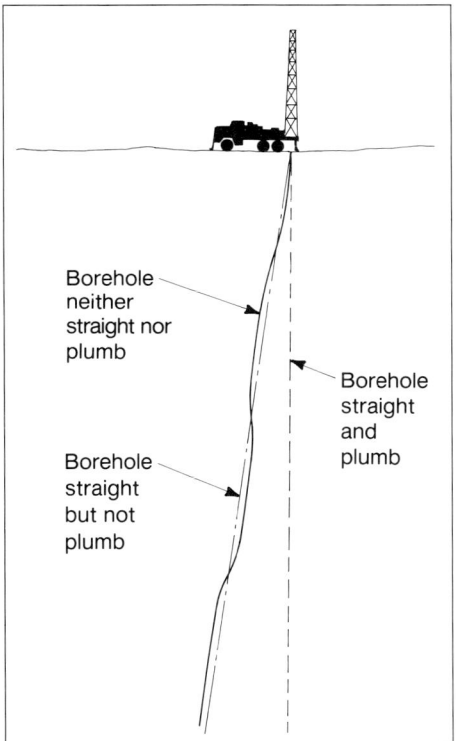

Figure 10.66. A plumb borehole is one that follows a vertical line from the ground surface to the earth's center. A straight borehole is one in which each succeeding casing joint (or length) is aligned with the preceding joint.

Some conditions that cause wells to become misaligned and out of plumb are (1) character of the subsurface material (faults, boulders in the borehole, inclined strata), (2) too much or too little weight on the drill bit, (3) trueness of the casing and drill pipe, and (4) the pull-down force applied to the drill pipe in rotary drilling. While the force of gravity tends to make the drill bit cut a vertical hole, the varying hardness of different materials being penetrated deflects the bit from a truly vertical course. In glacial drift, the edge of a boulder can deflect a cable tool or rotary bit. In cable tool drilling, a boulder may deflect the well casing, causing the hole to drift increasingly as the well is deepened.

When drilling by the rotary method, too much force applied at the top of the drill stem will bend the slender column of drill pipe. This tends to cause the bit to cut off-center. Heavy drill collars in the lower part of the drill stem help to put weight just above the bit, which overcomes the tendency to drift off a true vertical course. They are also more rigid than ordinary drill pipe, and thus help keep the lower part of the drill string straight. Large stabilizers are also used by many drillers to keep holes straight.

Obviously, any variation in the straightness of casing results in a corresponding misalignment of the well. Sections of pipe may be slightly bowed, or the center line of the threaded or bevelled ends may not exactly coincide with the center line of the

Table 10.8. Well Deflection Limits for Drift Indicator Survey

Depth		Allowable Deviation	
(ft)	(m)	(ft)	(m)
50	15.2	0.4	0.1
100	30.5	0.9	0.3
150	45.7	1.3	0.4
200	61.0	1.7	0.5
250	76.2	2.2	0.7
300	91.5	2.6	0.8
350	107	3.1	0.9
400	122	3.5	1.1
450	137	3.9	1.2
500	152	4.4	1.3
600	183	5.2	1.6
700	213	6.1	1.9
800	244	7.0	2.1
900	274	7.8	2.4
1000	305	8.7	2.7
1100	335	9.6	2.9
1200	366	10.5	3.2
1300	396	11.3	3.4
1400	427	12.2	3.7
1500	457	13.1	4.0
1600	488	14.0	4.3
1700	518	14.8	4.5
1800	549	15.7	4.8
1900	579	16.6	5.1
2000	610	17.4	5.3
2100	640	18.3	5.6
2200	671	19.2	5.9
2300	701	20.1	6.1
2400	732	20.9	6.4
2500	762	21.8	6.6

pipe. Commercial tolerances permit certain deviations in the straightness of pipe and the accuracy of threads. These must be considered in specifying the allowable deviation of a completed well.

Most careful drillers check the hole alignment several times when drilling a deep well. This is especially common in cable tool drilling. Time and money can be saved by taking steps to correct the misalignment just as soon as a deviation is discovered. In rotary drilling, the alignment is checked at preselected intervals during drilling [every 100, 500, or 1,000 ft (30.5, 152, or 305 m)]. In many wells, however, the alignment may be checked only after the well has been completed.

In recent years, special deviation instruments have been developed to measure the mislignment that occurs during drilling. A deviation survey is conducted along with the standard suite of logs after the maximum hole depth has been reached. Two typical battery-powered deviation instruments are shown in Figure 10.67. Special centralizers are fitted to the inclinometer to keep it centered within specially grooved plastic or aluminum casing placed in the well. Readings from the downhole inclinometer are transmitted to the surface indicator console where the readouts are given directly as displacement. Other drift indicators are entirely mechanical; a timer is set at the surface, the indicator is lowered at a certain rate into the casing or into drill pipe, and at a predetermined time the timer actuates a mechanism that punches a hole in a paper target inside the indicator. The location of the hole in the target indicates the inclination of the borehole. This type of drift indicator is run inside the drill pipe in rotary drilling, or inside the bailer in cable tool drilling. Drift angles of 1.5 to 90 degrees can be determined with this instrument to an accuracy of 0.1 to 0.5 in per 100 ft (2.5 to 12.7 mm per 30.5 m), depending on the deviation from the vertical. The deviation device is especially useful in deep holes where other methods of determining plumbness or straightness are not accurate or are too time consuming.

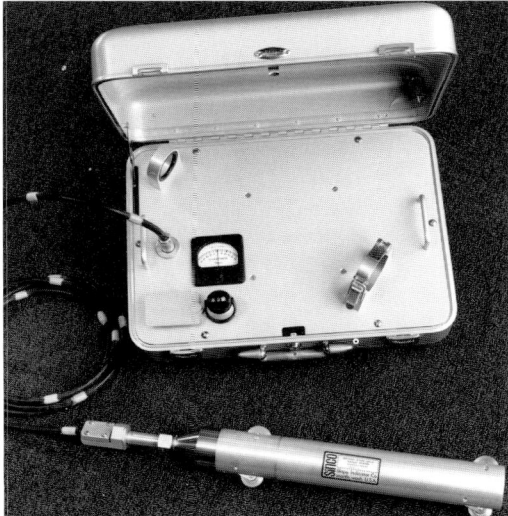
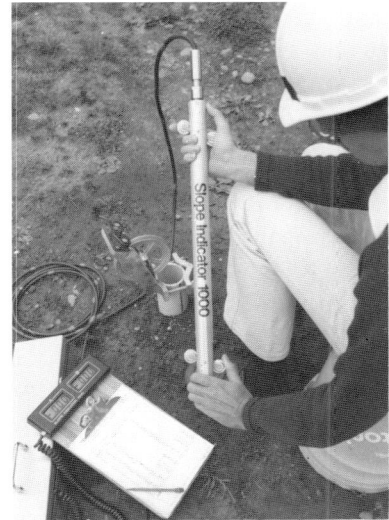

Figure 10.67. The alignment of well casing can be determined by lowering an inclinometer down the well casing inside a specially grooved plastic or aluminum casing. The grooves keep the instrument centered and aligned in the casing. This device can sense a deviation from vertical up to 12 degrees. The deviation instruments shown here are battery-powered. *(Slope Indicator Company)*

Several other more advanced instruments used in oil field work may be applicable to water well drilling, but generally only experienced surveying personnel are equipped to use them effectively. Some instruments can be installed in the drill string and provide continuous information on hole deviation and direction as drilling proceeds. One wireless unit transmits data on hole deviations to a surface recorder by means of pressure surges transmitted through the drilling fluid stream. Downhole instruments equipped with a gyrocompass are widely used to check inclination of the borehole. If computer assisted, these instruments can yield data on dogleg severity and borehole direction at the well head. Magnetic multishot instruments record data on film that can be used for complete directional surveys. To record data the lights in the instrument are turned on, thus exposing the film (a shot is taken on the multiple-shot discs), while the operator records the time and depth of each survey station. Over 1,000 records may be obtained from one run into the hole. The instruments can be hung on a wireline in the borehole or be installed inside a drill collar.

Plumbness of shallow wells can be checked with a special plumb bob, and straightness can be tested with a 40-ft (12.2-m) cylindrical dummy* that is slightly smaller than the inside of the well casing. However, the deviation from plumbness and straightness may be measured by a plumb-bob test alone, as suggested below.

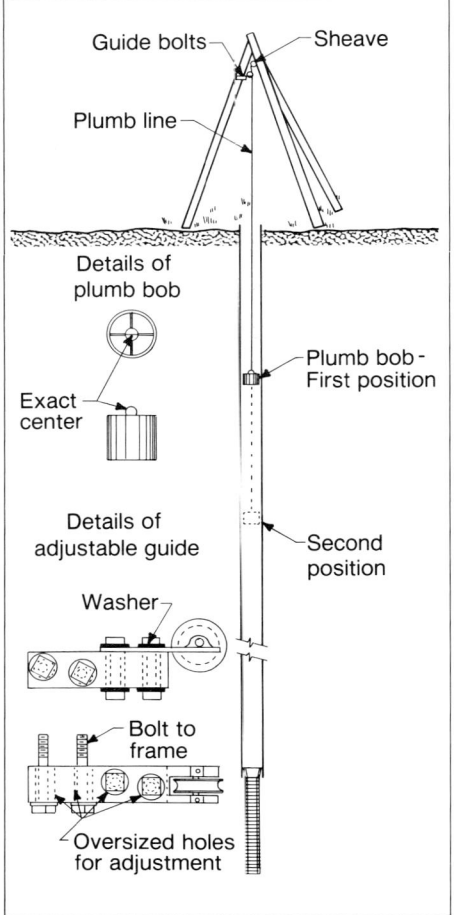

Figure 10.68. The straightness and plumbness of a well can be determined by using a small tripod mounted over the top of a well bore and a plumb bob suspended in the casing.

The device used to check the straightness and plumbness of a well is shown in Figure 10.68. The plumb bob, suspended on a wire line, is a short cylinder with an outside diameter about ¼ in (6 mm) smaller than the inside diameter of the casing. It must be suspended from the exact center of the device. The plumb bob should be heavy enough to stretch the line taut. A ⅛-in (3-mm) stainless steel wire cable makes a suitable line because of its flexibility and resistance to corrosion.

The line may be suspended from the derrick of a drilling machine or from a tripod so that the plumb bob will hang in the exact center of the well casing. The guide block

---

*The typical dummy is 40 ft long because standard casing joints are 20 ft. Other lengths of dummies may be used if the casing lengths are increased. The dummy must extend two casing lengths to be functional.

is mounted so that the vertical distance from the center of the small sheave to the top of the casing is exactly 10 ft (3 m). The guide is then adjusted horizontally so that the plumb bob hangs in the center of the well casing.

The test is started by lowering the plumb bob 10 ft. If the wire line moves away from the center of the well in any direction, the distance it has moved off center is measured. The plumb bob is then lowered another 10 ft, and the distance the wire line has now moved off center is measured. This procedure is re-

Figure 10.69. A plastic template placed on top of the well casing can be used to measure the displacement of the wire line connected to the plumb bob. *(Water and Power Resources Service, 1981)*

peated until the well has been checked to the desired depth. If the casing is exactly round, the measurement to the wire line can be made from the edge of the casing. If the pipe is not exactly round, a plastic template like that shown in Figure 10.69 may be used to measure the displacement of the wire line.

The well is plumb to the depth of the suspended bob as long as the plumb line passes through the center of the template positioned on the top of the casing. Any drift of the well causes the wire line to move off center. Drift at any depth is the measured displacement of the plumb-bob line multiplied by the total length of the line and divided by the fixed distance between the overhead pulley and the top of the pipe. Suppose, for example, that the line is suspended 10 ft above the top of the pipe and it moves ¼ in off center when the plumb is lowered 10 ft into the well. The casing drift, in this case, is ¼ in × 20 ft/10 ft = ½ in. If the rate of drift is the same for each 10-ft interval between any two depths, it means that the well is straight between these points, but out of plumb. A crooked section is revealed by different values of drift for successive 10-ft intervals.

The calculated values of drift may be plotted against depth to obtain a graph of

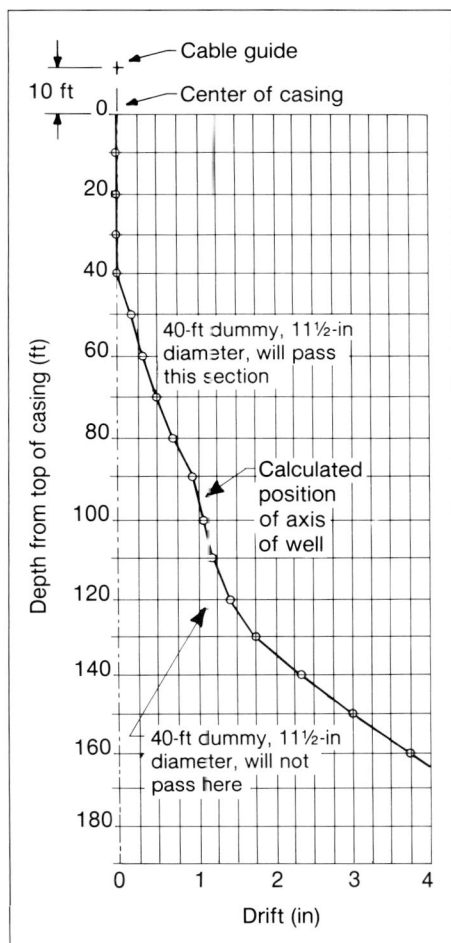

Figure 10.70. A graph of this well shows that it is out of plumb and not straight for much of its length.

Table 10.9. Relative Performance of Different Drilling Methods in Various Types of Geologic Formations

Type of Formation	Cable Tool	Direct Rotary (with fluids)	Direct Rotary (with air)	Direct Rotary (Down-the-hole air hammer)	Direct Rotary (Drill-through casing hammer)	Reverse Rotary (with fluids)	Reverse Rotary (Dual Wall)	Hydraulic Percussion	Jetting	Driven	Auger
Dune sand	2	5	Not recommended →	Not recommended →	6	5*	6	5	5	3	1
Loose sand and gravel	2	5			6	5*	6	5	5	3	1
Quicksand	2	5			6	5*	6	5	5		1
Loose boulders in alluvial fans or glacial drift	3-2	2-1				2-1	4	1	1		1
Clay and silt	3	5	5		5	5	5	3	3	3	3
Firm shale	5	5	3		5	5	5	3			2
Sticky shale	3	5	3		5	3	5	3			2
Brittle shale	5	5	5		5	5	5	3			2
Sandstone—poorly cemented	3	4	5		Not applicable →	4	5	4	Not recommended →	Not recommended →	Not applicable →
Sandstone—well cemented	3	3	3			3	5	3			
Chert nodules	5	3	5			3	3	5			
Limestone	5	5	5	6		5	5	5			
Limestone with chert nodules	5	3	5	6		3	3	5			
Limestone with small cracks or fractures	5	3		6		2	5	5			
Limestone, cavernous	5	3-1	2	5		1	5	1			
Dolomite	5	5	5	6		5	5	5			
Basalts, thin layers in sedimentary rocks	5	3	5	6		3	5	5			
Basalts—thick layers	3	3	4	5		3	4	3			
Basalts—highly fractured (lost circulation zones)	3	1	3	3		1	4	1			
Metamorphic rocks	3	3	4	5		3	4	3			
Granite	3	3	5	5		3	4	3			

*Assuming sufficient hydrostatic pressure is available to contain active sand (under high confining pressures)

Rate of Penetration:
1 Impossible
2 Difficult
3 Slow
4 Medium
5 Rapid
6 Very rapid

the position of the axis or center line of the well bore. Figure 10.70 is a graph for a well that is both out of plumb and crooked. The graph indicates that the casing is straight and plumb to a depth of 40 ft (12.2 m). The deflection at the 40-ft level is caused by a dogleg in the casing. From this point to a depth of about 90 ft (27.4 m), the casing is straight but out of plumb. Below 90 ft, the rate of drift gradually increases and the casing is neither straight nor plumb.

CONCLUSIONS

Selection of the best drilling method for a particular job requires an understanding of the geologic conditions and the physical limitations of the drilling rig. In addition, the value of experience cannot be overestimated, for many drilling difficulties occur because either the driller is unprepared to handle the wide range of subsurface conditions or has pushed the rig beyond safe operating limits. Good record keeping, patience, and a willingness to learn are some important characteristics of good drillers; the age of the machine or the particular drilling method used are of secondary importance in drilling successful wells. Table 10.9 gives the drilling performance of different drilling methods in various geologic formations. The relative performance differences between drilling methods, however, will also depend on the experience of the driller, the presence of geologic anomalies at the site, and the pressure conditions affecting the groundwater.

CHAPTER 11
Drilling Fluids

The technology of drilling fluids has advanced as rapidly and extensively as the rotary drilling machine. In the late 19th century, water alone was the principal fluid used in rotary drilling, although some entrainment of natural clay particles into the fluid must have occurred much of the time. The general term "mud" originated when certain kinds of clays were added to water to form drilling fluid. Recent advances, however, have made the term "mud" somewhat obsolete. Modern mud systems are now referred to as drilling fluids because of the large number of additives that can be used to impart special properties to drilling fluids. Much of the progress in drilling fluid development has occurred in the oil industry and has been applied thereafter in the water well industry. Today, the drilling fluid system can represent a major cost for deeper rotary-drilled holes; therefore, the economic success of the drilling operation may be determined by the contractor's ability to control the physical characteristics of the drilling fluid.

TYPES OF DRILLING FLUIDS

Drilling fluids used in the water well industry include water-based and air-based systems (Table 11.1). Oil-based fluids, commonly used in drilling for oil and gas, cannot be used in the water well industry and are not discussed here. Water-based drilling fluids consist of (1) a liquid phase, (2) a suspended-particle (colloidal*) phase, and (3) cuttings entrained during drilling. The colloidal phase may range from less than 1 percent to as much as 50 percent by volume. Air-based drilling fluids may consist of only a dry air phase, but more often they contain some water to which a surfactant (soap) is added to produce a foam. Occasionally a small amount of clay or polymer may be added to stiffen the foam. Thus, the primary drilling fluids, water and dry air, may be used alone, but a great variety of additives are available to modify their physical and chemical properties so they will perform more satisfactorily.

In this chapter, three major, but vastly different, types of drilling fluid additives are discussed and contrasted — clays, polymers, and surfactants. Clays and polymers are commonly added to water-based systems, and surfactants and occasionally clays

---

*Suspended particles that are approximately 0.0005 to 0.5 microns in size, do not settle out of the liquid rapidly, and are not readily filtered. (There are 25,400 microns per inch.)

or polymers are added to dry air systems. Water with clay additives produces a high-solids drilling fluid, whereas a combination of polymeric additives and water produces a low-solids drilling fluid. Many other special additives, such as flocculants, thinning agents (dispersants), weighting materials, corrosion inhibitors, filtrate reducers, lubricants, preservatives, bactericides, and lost-circulation materials, are used to further adjust the properties of drilling fluids. Thus, the term "drilling fluid" in the water well industry refers variously to clean water, dry air, a suspension of solids or a mixture of liquid additives in water, and droplets of water dispersed in air, or a mixture of water and surfactant or water, surfactant, and colloids dispersed in air.

The exact drilling fluid system selected will depend principally on the rock formation or stratigraphy expected and the equipment available. Remoteness of the drilling site, availability of drilling equipment and water supplies, environmental regulations, and the experience of the drilling crew also play an important part in selecting the fluid system. Drilling in hard rock, for example, requires procedures different from drilling in sedimentary rock or unconsolidated overburden. Water-based drilling fluid systems with clay or polymeric additives are typically used in unconsolidated formations; air is used in well-consolidated or semiconsolidated rocks and sediment; and clean water is used with reverse rotary drilling equipment for large-diameter wells in unconsolidated, semiconsolidated, and nonsensitive (nonswelling) sediments. The success of any drilling fluid system depends mainly on the chemistry of the mix water, the particular additives selected, and the physical and chemical characteristics of both the cuttings and the water in the formation being drilled.

FUNCTIONS OF A DRILLING FLUID

Drilling fluids can perform many functions, depending on the physical and chemical conditions found in the borehole. The primary functions are:

1. Remove cuttings. The primary purpose of the fluid system is to remove cuttings from the borehole during drilling. The rate at which cuttings can be removed depends on the viscosity, density, and uphole velocity of the drilling fluid, and the size, shape, and density of the cuttings. Ideally, the fluid should entrain the cuttings at the bit, carry them to the surface, and allow them to drop into a settling pit or tank before the fluid is recirculated. Inefficient removal of cuttings can reduce the penetration

Table 11.1. Major Types of Drilling Fluids Used in the Water Well Industry

Water Based	Air Based
1. Clean, fresh water	1. Dry air
2. Water with clay additives	2. Mist: Droplets of water entrained in the airstream
3. Water with polymeric additives	3. Foam: Air bubbles surrounded by a film of water containing a foam-stabilizing surfactant
4. Water with clay and polymeric additives	4. Stiff foam: Foam containing film-strengthening materials such as polymers and bentonite

rate of the drill bit, adversely affect the physical properties of the drilling fluid, and increase the energy required to recirculate the drilling fluid.

2. Stabilize the borehole. To maintain an open borehole, the drilling fluid stabilizes the borehole walls and prevents expansion of swelling clays. When using water-based systems, the drilling fluid must provide a pressure greater than that existing in the formations penetrated. The pressure exerted against the borehole wall depends on the height of the fluid column and the weight of the drilling fluid. If water is permitted to flow into the well bore from the penetrated formations, sloughing of the hole may occur, resulting in lost time and increased drilling costs. Occasionally, a portion of the drill string may become buried by a caving formation, requiring an extensive fishing operation; it may even be permanently lost, causing abandonment of the borehole.

Drilling fluids should prevent formation clays from expanding into the borehole during drilling. Some hydrating clays can absorb large volumes of water, thereby increasing the physical dimensions of the clay. To control this problem, the drilling fluid must isolate formation clays from the water in the drilling fluid. This is usually achieved by adding certain chemicals such as potassium chloride to water-based drilling fluids that contain clay additives, or by using polymeric drilling fluid additives which coat the formation clays and minimize swelling caused by hydration (Figure 11.1).

3. Cool and lubricate the drill bit. Fluids circulating through the drill string cool and lubricate the bit, thereby avoiding unnecessary bit wear and reducing maintenance.

Figure 11.1. Polymeric drilling fluid additives effectively coat the borehole walls, thereby preventing the hydration of swelling clays that ordinarily expand into the borehole during well construction. The coating effect can be seen easily by examining clay cuttings brought to the surface in the drilling fluid. Rather than disperse into the drilling fluid, the clay cuttings are carried up the borehole intact.

4. Control fluid loss. All water-based drilling fluid systems must control drilling fluid loss in highly permeable formations by creating a nearly impermeable clay filter cake or polymeric film on the borehole wall. Insufficient filter cake or polymeric film deposition may allow excessive fluid loss or even complete loss of circulation. In an attempt to control fluid loss, some contractors may mix a high-solids drilling fluid that may exert so much pressure on the borehole walls that large volumes of the fluid are forced into permeable zones in the formation. This is called self-induced fluid loss.

5. Drop cuttings into a settling pit. As the drilling fluid is circulated through the settling pit, cuttings should drop out so they are not recirculated. The gel strength of the drilling fluid is the primary factor controlling the rate of settlement. Gel strength is a measure of the fluid's ability to suspend cuttings when the fluid is at rest. The flow rate in the settling pit is also important and is controlled by the shape and size of the settling pit.

6. Facilitate acquisition of information about the well bore. Drilling fluid systems should facilitate the recovery of representative cuttings and permit accurate geophysical logging of the well.

7. Suspend cuttings in the borehole when the drilling fluid is not being circulated. During the time the drilling fluid is not in motion, cuttings tend to settle in the borehole. If the rate of settlement is excessive, cuttings may settle around the drill bit or stabilizer and jam the rotation of the drill string when drilling is resumed. The rate of particle settlement is controlled by the gel strength of the drilling fluid.

No single drilling fluid can fulfill all of these functions perfectly. Usually, a specific drilling fluid and additive system is mixed to create the optimum physical and chemical characteristics essential for control of downhole conditions. During drilling, the principal objective is to maintain the drilling fluid in a suitable condition in spite of changing downhole or surface conditions and the continuous addition of suspended drill cuttings. In most cases, continuous monitoring of the drilling fluid is necessary to achieve the best results.

PROPERTIES OF WATER-BASED DRILLING FLUIDS

Although most drilling fluid systems used in water wells are relatively simple, the drilling fluid properties listed in Table 11.2 should be understood thoroughly by the drilling contractor or project engineer. Regardless of which drilling fluid system is used, its effectiveness will depend upon the contractor's or engineer's ability to anticipate the chemical and physical changes taking place during drilling and to make modifications as required. Attempts to use highly technical drilling fluid systems by inexperienced crews may lead to serious economic losses. At a minimum, all rotary drilling crews should be able to measure drilling fluid density and viscosity, and understand the relationship of these properties to hole stability, cuttings removal, and fluid-loss control.

Table 11.2. Principal Properties of Water-Based Drilling Fluids

1. Density (weight)	4. Gel strength
2. Viscosity	5. Fluid-loss-control effectiveness
3. Yield point	6. Lubricity (lubrication capacity)

The physical and chemical behavior of bentonite and polymers differs significantly. These differences are examined separately as each drilling fluid property is discussed below.

Density

Control of drilling fluid density is a fundamental factor in successful water well drilling. Density is defined as the weight per unit volume of fluid. Thus, the terms "density" and "weight" can be used interchangeably. In the English system, density is expressed in pounds per gallon (lb/gal) or pounds per cubic foot

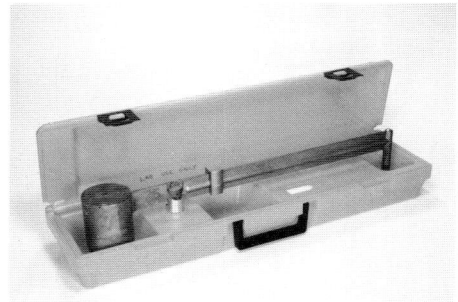

Figure 11.2. The drilling fluid density can be measured easily in the field with a drilling fluid balance. Readings are given in lb/gal and psi/1,000 ft of borehole. The specific gravity and direct weight of the fluid can also be read from the balance arm.

(lb/ft^3); in the metric system, density is expressed in grams per cubic centimeter (g/cm^3) or kilograms per cubic meter (kg/m^3). The densities of several common substances used in drilling fluids are given in Appendix 11.A. The actual pressure exerted at any point in a borehole by a static drilling fluid depends on the fluid density and the height of the fluid column above that point. Many drilling contractors automatically associate high-density drilling fluids with extra downhole pressure required to contain pressurized formations. Specific gravity is another way to express the density of a drilling fluid. It is the ratio of the weight of a given volume of drilling fluid compared with the weight of an equal volume of water.

Drilling fluid density is measured easily with a balance scale (Figure 11.2). The procedure for measuring the weight of a drilling fluid sample is given in Appendix 11.B, and a drilling fluid weight conversion table is given in Appendix 11.C.

Selection and maintenance of proper drilling fluid density prevents collapse of the hole and flow of water into the borehole. To maintain an open borehole, the pressure exerted by the drilling fluid column must exceed the pore pressure (water and gas) in the aquifer. Typically, a minimum excess pressure of 5 psi (34.5 kPa) is desirable, although this pressure requirement may be higher when pressures from confined formations are encountered.

Hydraulic pressure occurs in the pores of both unconfined and confined aquifers. In an unconfined aquifer, the pressure at any point in the aquifer is represented by the head of water above that point. Ordinarily, the water pressure within a freshwater aquifer is 0.433 psi/ft (9.8 kPa/m) of depth, unless the total dissolved solids are abnormally high. Thus, at a depth of 10 ft (3.1 m), the pore pressure is 4.33 psi (29.9 kPa).

Under confined conditions, the potentiometric surface is above the top of the aquifer. Thus, pore pressure at any thickness always exceeds the normal hydraulic pressure of 0.433 psi/ft (9.8 kPa/m) of aquifer thickness. Before the bit penetrates a confined aquifer, the excess pressure partially supports the overlying formations. Typically, the weight of this material exerts a pressure of 1 psi/ft (22.6 kPa/m) of depth. If the overburden were supported entirely by pore pressure, the pressure in the aquifer would have to equal the head of water in the aquifer (0.433 psi/ft) plus the weight of overlying materials (1 psi/ft). Usually, only part of the overburden is supported by confined pressure, and a figure of 0.465 psi/ft (10.5 kPa/m) is often used to estimate pore pressure in confined aquifers.

The drilling contractor should be able to calculate the downhole pressures exerted by the drilling fluid at rest to determine whether the hydrostatic pressure is sufficient to control the pore pressure in the formation. A simple equation for determining the hydrostatic pressure exerted by the drilling fluid in a borehole is given by:

$$\text{Hydrostatic pressure} = \text{fluid density} \cdot \text{height of fluid column} \cdot 0.052 \quad (11.1)$$

where hydrostatic pressure is in psi, density in lb/gal, and height in ft. This value is the hydrostatic pressure when the drilling fluid is at rest. The pump exerts additional pressure in the borehole when the drilling fluid is being circulated; this is called dynamic pressure. The dynamic pressure is relatively small in relation to the static pressure for most water wells and therefore will be ignored.

Under most drilling conditions, the hydrostatic pressure exerted by the weight of the drilling fluid column above the static water level in the borehole is sufficient to create positive pressures in the borehole; that is, the hydrostatic pressure created by the drilling fluid is great enough to keep the borehole open (Figure 11.3). When static water levels are high, however, the weight of the drilling fluid column above the static

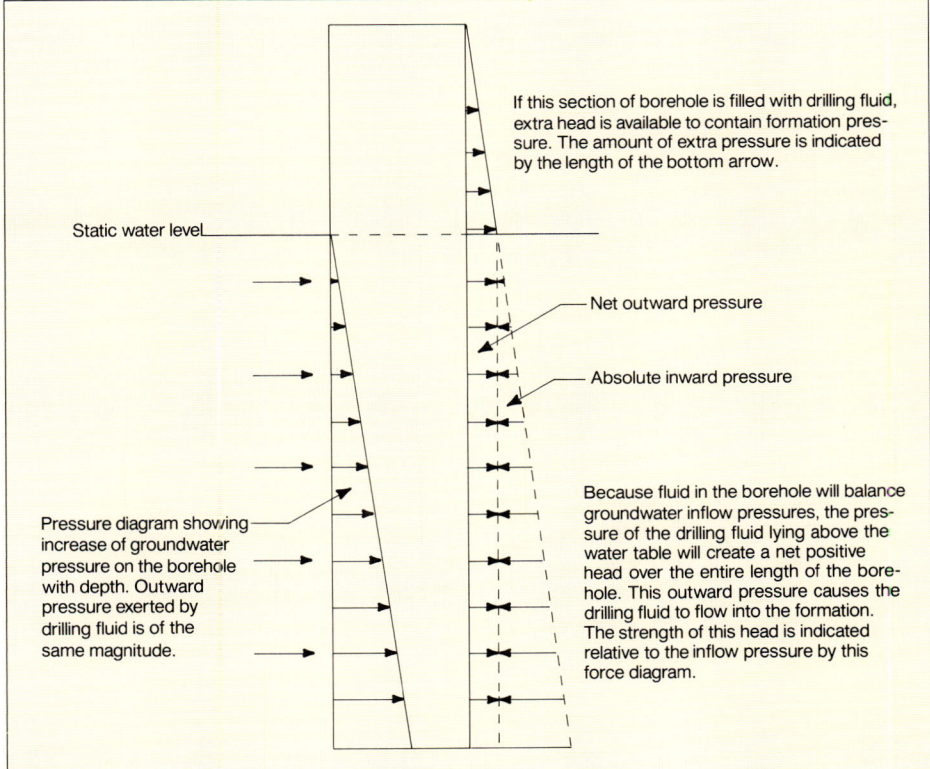

Figure 11.3. Diagram showing fluid forces in a borehole during drilling. The deeper the borehole becomes, the smaller the increase in the inflow pressure (such as those caused by confining pressures) must be to overcome the positive outward pressure. Unstable conditions in the borehole can be prevented by keeping the water level as high as possible and by increasing the density of the drilling fluid.

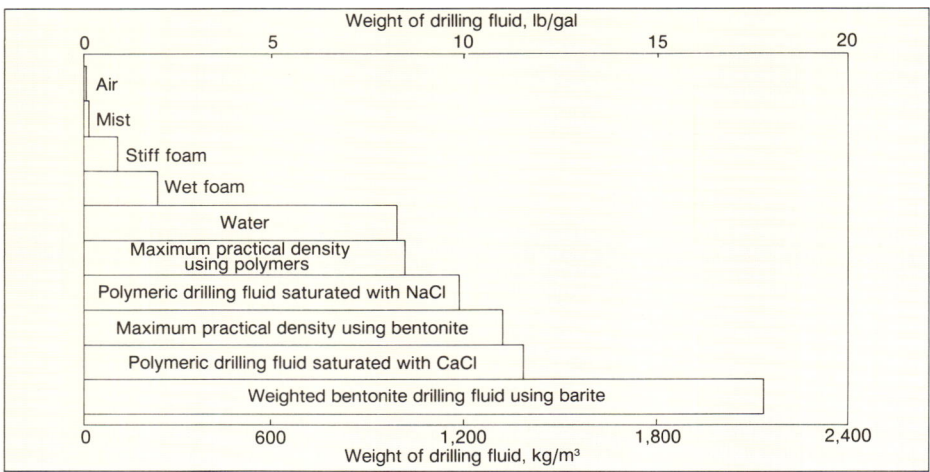

Figure 11.4. Practical drilling fluid densities range from virtually zero for air to well over 15 for bentonite with barite additives. In general, the density of the drilling fluid must be high enough to balance any confined pressure conditions in the borehole. Excessive drilling fluid densities, on the other hand, may cause high fluid losses, plugging of the aquifer, unsatisfactory cuttings removal in the mud pit, and higher-than-necessary pumping costs. *(After Hutchinson and Anderson, 1974; Tschirley, 1978))*

water level may not be sufficient to keep the borehole open.

Under ordinary conditions, the maximum density that can occur in a clay system as a result of the entrainment of solids during drilling is about 11 lb/gal (1,320 kg/m^3). Further increases in density while maintaining a proper solids/fluid ratio requires the introduction of higher density material so that less solids by volume are needed for a specific density. Barite, with a specific gravity of 4.2 to 4.35, is a standard weighting material and is much heavier than clay additives and most formation materials, which have specific gravities of 2.6 to 2.7. Barite particles are sized so as to remain suspended in the drilling fluid, but are not small enough to affect the flow characteristics of the fluid. Another heavy mineral, ilmenite (specific gravity 4.7), is also used for weighting drilling fluids. Soluble salts such as sodium chloride (NaCl) and calcium chloride (CaCl$_2$) are useful for weighting some drilling fluids made with polymeric additives, because these drilling fluids lack gel strength and thus cannot keep heavy minerals in suspension at low uphole velocities or at rest. The density of a fluid made with polymeric additives thus depends less on the amount of suspended solids than on the total dissolved solids. The contractor should contact manufacturers to make sure the polymer they are using is salt-tolerant. The range of drilling fluid densities is given in Figure 11.4.

To control the flow of water into the borehole, the contractor should increase the density of the drilling fluid before reaching the confined formation. The additional drilling fluid density required to equalize the confined pressure is determined by:

$$\textit{Drilling fluid density} = \textit{weight of water} \cdot \frac{\textit{height of water above ground level}}{\textit{depth to top of confined aquifer}} \quad (11.2)$$

The calculated drilling fluid density will only balance the confined pressure, however, and thus a safety factor of 0.3 lb/gal (36 kg/m^3) is usually recommended. The total added density should be enough to control any potential collapse of the formation during circulation of the drilling fluid and withdrawal of the drill pipe.

During the drilling process, solids generally begin to accumulate in the drilling fluid, causing the density to increase. If silt, clay, or weakly consolidated shale are present, the density increase may be significant and water must be added or solids removed to reduce the solids/fluid ratio. Too great an increase in density can affect the drilling and well completion processes in the following ways:

- Large volumes of drilling fluid and cuttings can be forced into the aquifer during drilling. Removal of the drilling fluid and cuttings during development can be extremely difficult, especially if clay additives are used.
- Material costs increase because of high fluid losses, particularly in areas where mix water is expensive or must be hauled long distances.
- Rate of penetration is reduced.
- Sample collection is more difficult and less reliable because cuttings do not drop out of the drilling fluid at the surface.
- Wear on a mud pump is increased because it must keep recirculating the high volume of unnecessary solids.
- Pumping costs increase because solids are continually recirculated.

Rheological Properties of Drilling Fluids

Rheology is the study of the deformation and flow of matter. Understanding rheological concepts is vital because successful completion of the borehole depends on creating the correct physical and chemical qualities for drilling fluids. Drilling fluids behave in predictable ways because of the chemical and physical characteristics produced by specific additives when combined with the base system.

The flow characteristics (rheology) of a drilling fluid — viscosity, gel strength, and yield point — depend primarily on the size, shape, and molecular structure of the particles in the fluid. Clay particles are less than 4 microns in size, silt and barite are 4 to 63 microns, and fine to medium sand is 63 to 500 microns. The silt, and barite if present, provide mainly density, whereas the clay particles enhance the viscosity and filtration characteristics as described below. Polymeric particles are usually much smaller than clay. For example, finely ground polymeric particles made from guar seeds are about 0.0001 micron in size. The addition of even small volumes of polymers to a drilling fluid can have a significant effect on viscosity.

Particle shape is important in determining how a fluid flows. Flat, tabular particles have large surface areas for their sizes and can "tie up" relatively large volumes of water. Some small particles, such as clay colloids, possess powerful electrical charges that affect the fluid both while it is in motion and at rest. In contrast, polymeric particles have a nonionic, long-chained molecular structure that causes distinctive changes in the flow characteristics of a drilling fluid, depending on the amount of stress applied at various points in the circulation system.

Viscosity

Viscosity is the resistance offered by a fluid to flow, or, in this case, to being pumped. It has no relationship to density and is measured in different units. The viscosity and uphole velocity are the primary factors determining the ability of a drilling fluid to remove cuttings from around the bit and move them up the borehole. The viscosity of any drilling fluid depends on many factors: (1) viscosity of the base fluid used, (2) number of particles (solids) per unit volume of drilling fluid, (3) density, size, and shape of particles, and (4) the attracting or repelling forces between the solid particles and between the solids and the base fluid (hydration potential). In general, high-viscosity drilling fluids are required to lift coarse sand or gravel, whereas lower viscosity drilling fluids are adequate to lift fine sand and silt. Separation or settlement of solids at the surface, and effective cleaning of the bit face during drilling, are facilitated by drilling fluids with low viscosity and low gel strength.

Viscosity is defined more precisely as the resistance to a shearing stress (Figure 11.5). Shear stresses occur at various points in the drilling fluid circulation system where flow is accelerated through restricted openings, such as in the pump and at the bit face. The viscosity of a drilling fluid will be highest where the stress on the fluid is minimized. Thus, drilling fluids pumped continuously at a high velocity through numerous small openings will not build maximum viscosity. Some drilling fluids can recover their viscous qualities almost immediately after passing a shear (stress) point. Other fluids may require a longer time or, in some cases, may not be able to recover the original viscosity once they have passed a shear point.

Viscosity of a fluid can be measured by a viscometer or a Marsh funnel. Viscometer measurements are usually done in the laboratory and involve measuring the resistance to shear stress developed between a stationary and a rotating cylinder (Figure 11.6a). The annulus between the cylinders is filled with drilling fluid, the outer cylinder rotated at a specified rpm, and the shear stress or drag that develops between the two surfaces is measured. The drilling fluid between the two cylinders is deformed as indicated in Figure 11.6b. The amount of shear stress developed in a drilling fluid is a function of the number of solid particles in suspension and the flow resistance of the base fluid. These frictional effects are called the plastic viscosity of the drilling fluid. It should be noted that traditional methods of measuring shear rate are not applicable to drilling fluids containing polymers.

A fluid that deforms proportionately to an applied stress, such as water, is called a Newtonian fluid (Figure 11.7). Drilling fluids with clay additives, however, act more like plastic materials; that is, they do not begin to deform until a significant amount of stress has been applied. The measure of this stress, called the yield point, is the

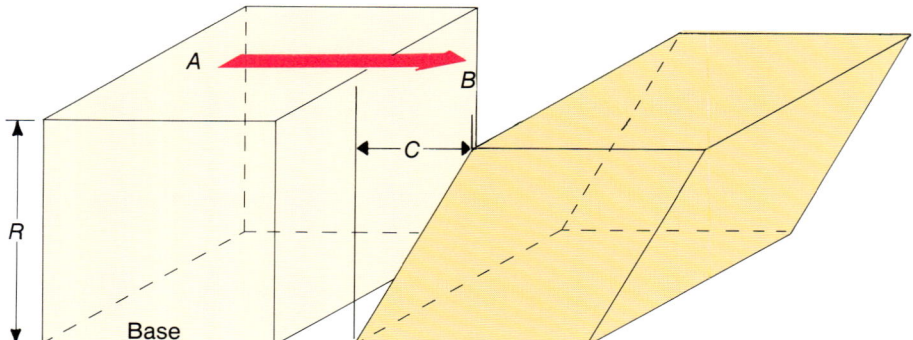

Figure 11.5. Stress and strain can be explained with this illustration. Stress (force) exerted along the A plane in the direction of B tends to distort the fluid to the right. Friction reduces the movement to zero at the base of the block. The amount of horizontal movement R-distance up from the base is given by C; C is the distortion (movement) along the A plane and is called the strain. For most substances, a given stress produces a resulting strain. The resistance of a drilling fluid to a stress is called its viscosity.

pressure at which the pump begins to move the drilling fluid. The yield point is controlled by the strength of the attractive forces between individual particles in the drilling fluid. Drilling fluids become Newtonian when the yield point has been reached; that is, the relationship between stress and strain is more or less constant, indicating that the viscosity does not change significantly with increasing stress. Before the yield point is reached, the viscosity changes continuously in response to increasing stress.

Viscosity is measured in the field with a Marsh funnel (Figure 11.8). A certain volume of drilling fluid is allowed to drain from a special funnel into a cup; the flow time is recorded and calibrated against the time required for an equal volume of water to drain from the funnel [about 26 seconds at 70°F (21.1°C)]. These values, called apparent viscosities, are approximate and are good only in a relative sense. See Appendix 11.D for measurement procedures for the Marsh funnel. More accurate measurements of viscosity can be obtained by using a viscometer, but only for drilling fluids made with clay additives. For most water well applications, however, the Marsh funnel is the principal device used by drillers to determine viscosity; although it is not as accurate as a viscometer, it is acceptable for decisions made in the field. It is especially valuable for judging the relative change in apparent viscosity over time, thereby detecting changes in downhole conditions.

Some of the viscosity testing procedures that have become standard for bentonitic fluids are not applicable for polymeric materials, and new standards will have to be designed. The hydration of high-grade bentonite is rather rapid, but complete hydration of polymeric colloids continues for several hours. Thus, care must be taken when a certain Marsh fun-

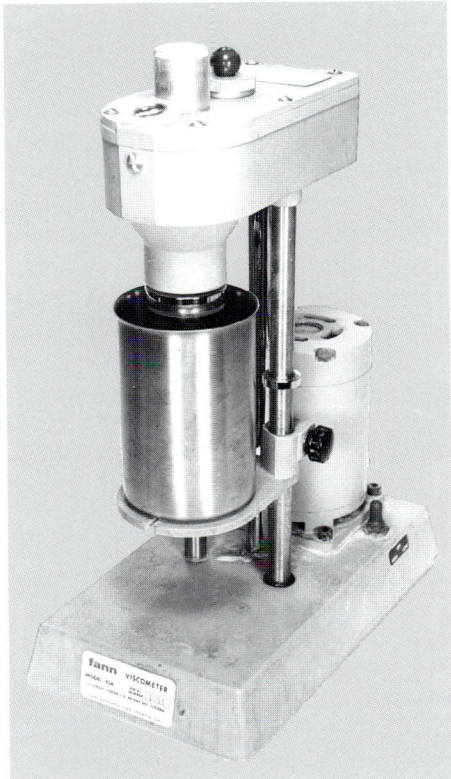

Figure 11.6a. The viscosity of a fluid can be measured under laboratory conditions by a viscometer.

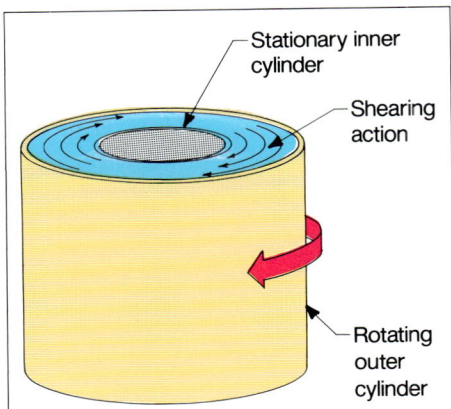

Figure 11.6b. A viscometer measures the amount of shear stress (drag) that develops between a stationary and rotating cylinder when the volume between cylinders is filled with drilling fluid. As the outer cylinder rotates, the amount of shear stress produced is directly proportional to the viscosity of the drilling fluid.

nel viscosity is specified initially for bentonite and later a polymeric drilling fluid is added or substituted. Any viscosity reading taken before 1 to 2 hours may not be indicative of subsequent conditions. Furthermore, excessively high or low pH or temperature may change the hydration rates significantly.

Viscosities should be no higher than necessary to efficiently lift cuttings to the surface and control fluid losses. Although drilling conditions can vary greatly, a Marsh funnel viscosity of 35 to 40 seconds will usually be satisfactory in fine sand formations. If coarse sediment (gravel) is encountered, viscosities must be substantially increased so that coarser particles do not have to be finely ground to be lifted by the drilling fluid (Table 11.3).

Viscosity of Drilling Fluids Made with Clay Additives

The viscous nature of drilling fluids made with clay additives originates from the small size of clay particles (less than 4 microns) and their relatively large surface areas. Most clay particles have a platelike structure; groups of these platelets are common. The edges of clay platelets are positively charged, whereas the flat surfaces are negatively charged. Because clay particles are so small, electrostatic charges govern their

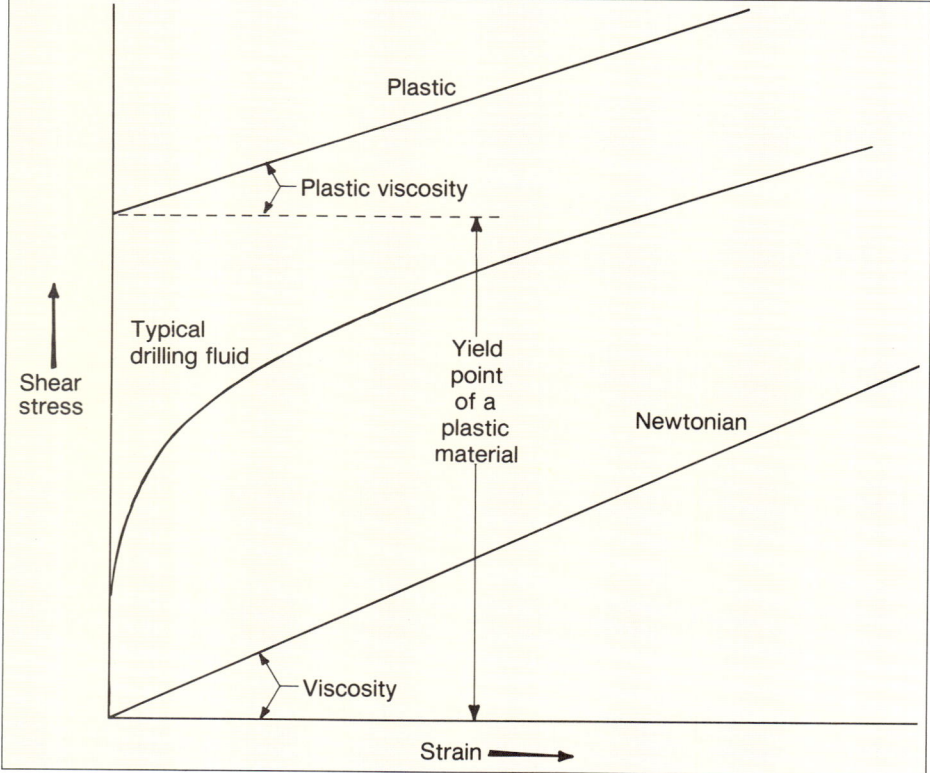

Figure 11.7. Substances deform differently under stress, depending on their physical characteristics. The deformation (flow) characteristics for several different types of substances are shown in this graph. A Newtonian substance, such as water, will deform proportionately to the stress applied, no matter how small the stress. A plastic substance begins to deform only after a certain amount of stress is applied. A drilling fluid acts like a weak plastic — a certain stress must be applied before flow begins. The force (stress) required to induce flow is a measure of the viscosity of the drilling fluid.

Figure 11.8. A Marsh funnel is used at the drilling site to measure the viscosity of the drilling fluid. Marsh funnel viscosity values are not exact, but are extremely useful for comparative purposes.

activity and they are strongly attracted to or repulsed by each other and various other substances. Clay particles generally swell when exposed to water because the electrically unbalanced water molecules are strongly attracted to the plate surfaces and thereby force the plates apart. This results in the clay particles occupying a larger space, which leads to a more viscous fluid. The viscosity of a drilling fluid with clay additives is also a function of the rate the fluid is being pumped. At lower velocities, the viscosity is higher because it is governed by the charges on the plates. At higher velocities, the charges on the plates have less effect. The gelling characteristics of a drilling fluid when at rest also result from these charges.

Different types of clay have a wide range of hydration potential. Clays that hydrate effectively are preferred because they produce a low-solids drilling fluid with high viscosity. Clays such as montmorillonite, kaolinite, and illite are the primary clays used for fresh-water drilling fluids, although montmorillonite is the only clay of these three that is available commercially. The viscosity-building characteristics of sodium montmorillonite are the greatest of any clays, because the sheets of atoms making up the flat clay particles are much thinner and come apart more easily in water than those of other clays. Sodium montmorillonite can swell to approximately ten times its original volume when exposed to water. Water molecules are adsorbed on the platelet surfaces, resulting in a much enlarged clay particle (group of platelets) (Figure 11.9). Kaolinite, illite, and the calcium form of montmorillonite do not swell nearly as much and therefore are not as desirable. Another clay, attapulgite, can be used in both fresh-water and salt-water conditions, unlike montmorillonite which hydrates only in fresh water.

If a clay has a poor hydration potential, many more solids are required to build a

Table 11.3. Approximate Marsh Funnel Viscosities Required for Drilling in Typical Types of Unconsolidated Materials

Material Drilled	Appropriate Marsh Funnel Viscosity (seconds)
Fine sand	35 - 45
Medium sand	45 - 55
Coarse sand	55 - 65
Gravel	65 - 75
Coarse gravel	75 - 85

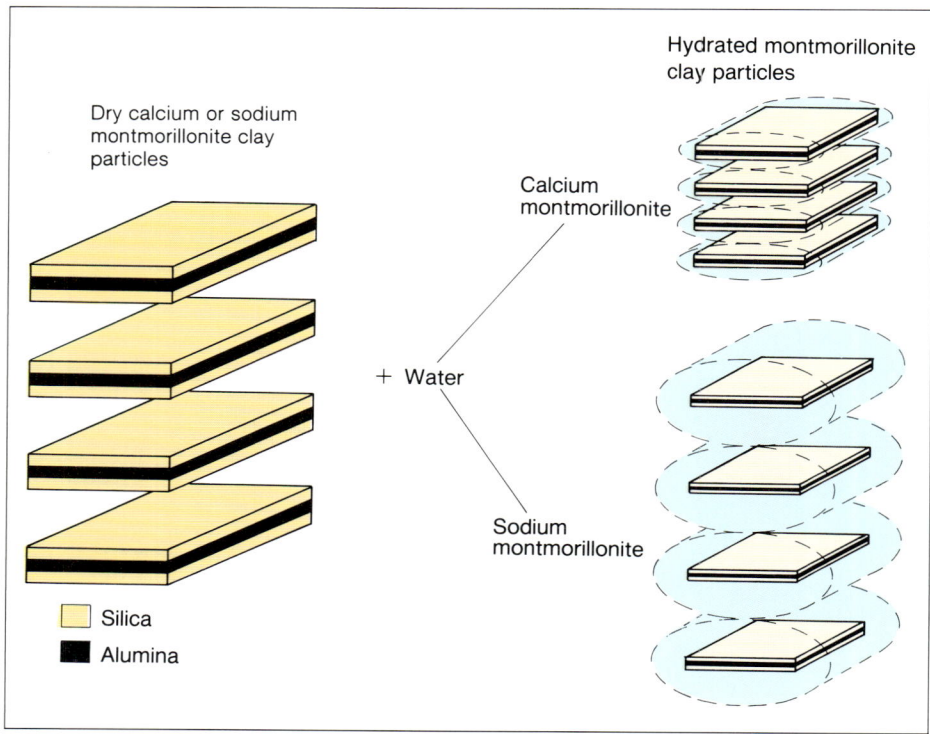

Figure 11.9. Hydration of calcium and sodium montmorillonite. Note that the sodium montmorillonite expands much more than the calcium montmorillonite. Thus, it is much more satisfactory for building viscosity in drilling fluids while keeping the fluid density at a minimum.

given viscosity. Clays used for drilling fluids are rated by their yield, which is defined as the number of 42-gallon (0.16-m^3) barrels of drilling fluid with an apparent viscosity of 15 centipoise produced by 2,000 lb (907 kg) of clay*. Water at 68°F (20°C) has a viscosity of 1.005 centipoise; Figure 11.10 gives centipoise and approximate equivalent Marsh funnel viscosities. Figure 11.11 shows how much more effective Wyoming bentonite (85 percent montmorillonite) is in building viscosity than are other types of clays. The term "bentonite" is used as a commercial name for clays that are predominantly sodium montmorillonite. Wyoming bentonite is the most common drilling fluid additive used in the water well industry.

Viscosity of Drilling Fluids Made with Polymeric Additives

In recent years, the use of natural and synthetic polymeric colloids in drilling fluids has increased. A polymer is a long-chained chemical compound consisting of many small molecular units (monomers) combined together. Polymers can be either natural or synthetic, usually have a high molecular weight, and form chains of monomers several thousand units long. When the chains become tangled, they tend to make a strong film. Polymers may be used as the primary additive or to beneficiate bentonitic drilling fluids. They are described as low-solids or clay-free drilling fluid additives.

---

*This definition of yield originated in the petroleum industry and is in U.S. gallons. No similar expression for yield exists for the water well industry.

Polymers are used to increase drilling rates and drilling fluid yields, thereby decreasing operational costs.

The unusual physical and chemical properties of polymers offer several specific advantages: (1) holes can be drilled with reduced bottom-hole pressures; (2) fluid loss can be controlled without the buildup of a thick filter cake; (3) torque and friction losses are reduced; (4) cores and other samples are not masked by the drilling fluid additive; (5) some polymers are compatible with brackish water or even brine; and (6) cuttings settle rapidly at the surface so it is possible to circulate clean, lightweight, nonabrasive fluids. Polymers also increase the effectiveness of some well-logging methods because of their high resistivity.

Natural Polymeric Gum

Natural polymeric colloids consist of long-chained molecules (polymers) referred to as polysaccharides (Figure 11.12); when hydrated in water, they form a viscous, colloidal dispersion. One type of natural organic colloid can be obtained from the finely ground seed of the guar plant, which is grown in Pakistan, India, Mexico, and certain areas of Arizona, Oklahoma, and Texas. The endosperm portion of the seed is ground into colloid-sized particles (0.0001 micron) and used as food-grade guar gum for building viscosity in many food products. When used as a drilling fluid additive, these nonionic colloids form a low-solids, biodegradable polymeric drilling fluid that does not introduce non-native clay particles into water-bearing strata.

Guar gums possess physical qualities that result in unusual hydration, gelling, and

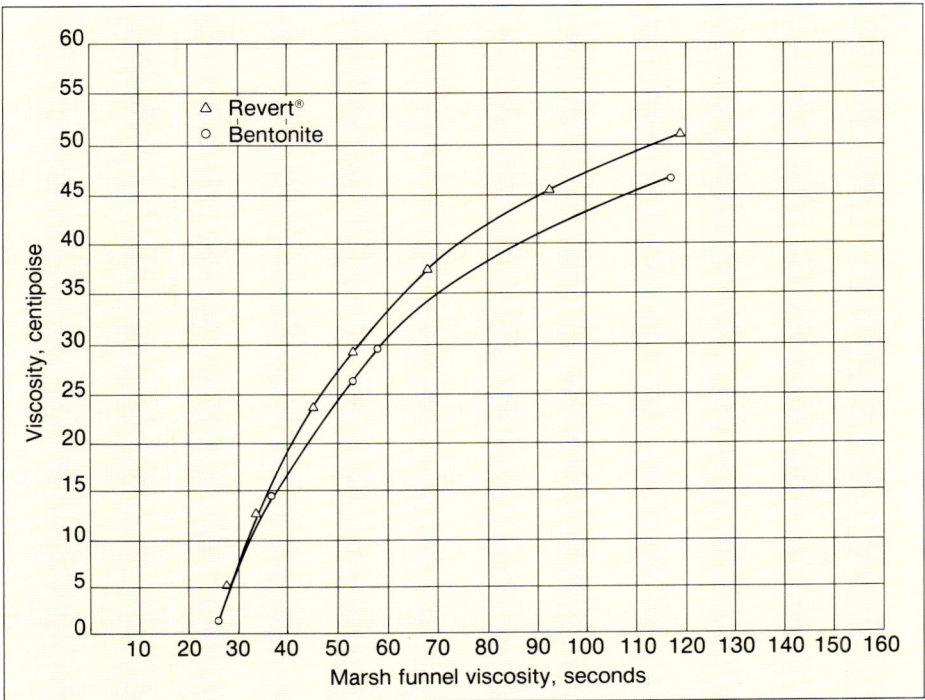

Figure 11.10. These curves show the relationship of viscosity values to Marsh funnel measurements for a polymer and Wyoming bentonite.

viscosity characteristics. Specific recommendations for working with guar additives are provided by the suppliers and should be consulted before using guar as a drilling fluid additive. Specific properties can be controlled by processing and chemical techniques. Viscosity, however, is influenced not only by processing techniques (particle size) but also by purity of the gum product (Goldstein et al., 1973). The large number of evenly distributed galactose side chains in the guar macromolecule prevent coiling that would lead to a tight intermolecular fit and a highly organized molecular structure. This loose molecular configuration gives guar outstanding solubility, even in cold water. Finely ground guar can hydrate significantly more water than even high-grade bentonite. To obtain a Marsh funnel viscosity of 60 requires approximately 10 times as much bentonite by weight as one of the organic drilling fluid additives made from guar gum (Figure 11.13).

Many other factors can affect the viscosity produced by polymeric colloids, including the presence of metallic ions, temperature of mix water, rate of shear, size and concentration of particles, and pH. Figure 11.14 shows how the mix-water temperature affects the Marsh funnel viscosities of a drilling fluid made with a guar additive.

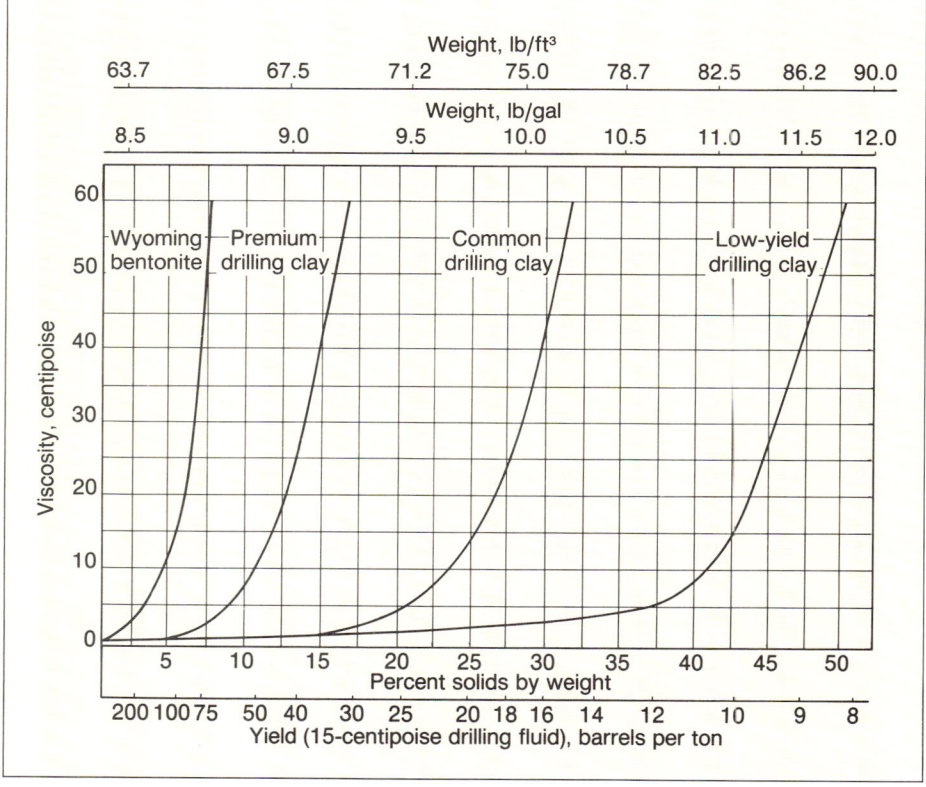

Figure 11.11. The viscosity-building characteristics of the clays on the left are superior because less of these clays is required to build a given viscosity in a drilling fluid. Until a viscosity of 15 centipoise (yield point) is reached, the addition of large amounts of clay has little effect on viscosity. Once the yield point is reached, small additions produce significant increases in viscosity. *(From "Principles of Drilling Fluid Control," 1980. Courtesy of Petroleum Extension Service, University of Texas, Austin.)*

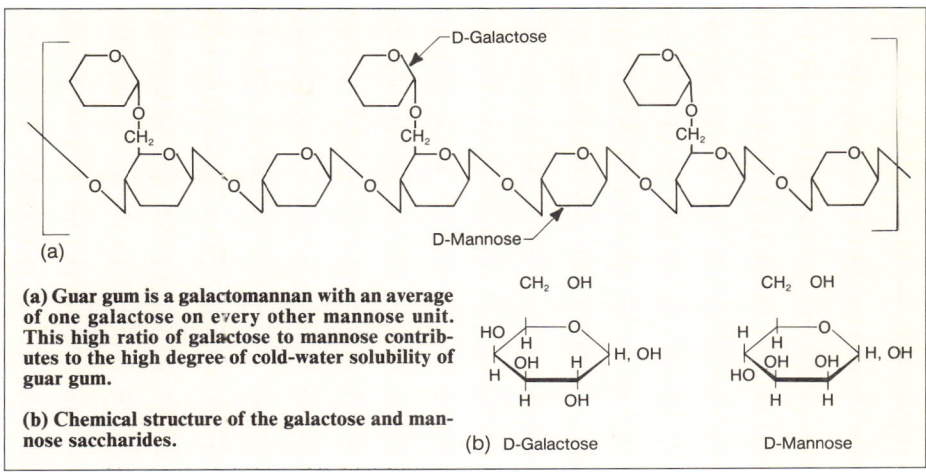

Figure 11.12. General chemical structure of the guar gum molecule.

The viscosity of a drilling fluid with guar additives can be altered physically and chemically. Most hydrated water-soluble polymers roll up in random coils when at rest. These coils become entangled, greatly increasing resistance to flow. Shearing of the drilling fluid as it passes turbulently through the pump, drill stem, bit, and annulus stretches the entangled structures. At high shear rates, the hydrated polymers are stretched out to their full length and slip past one another easily, thereby reducing the viscosity. When the shear stress is reduced, the guar regains its original viscosity.

One of the most remarkable properties of natural polymers is their capacity to lose viscosity through biological activity or by treatment with certain chemicals. The breakdown time for a natural polymer is shown on Figure 11.15 for certain drilling fluid temperatures. The viscosity of organic drilling fluids can be reduced rapidly

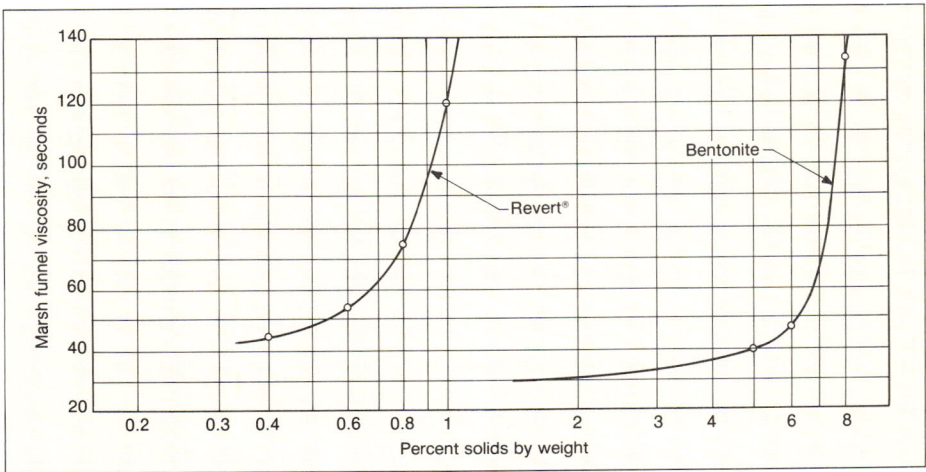

Figure 11.13. The viscosity-building properties of the polymeric additive, Revert®, are about 10 times those of bentonite for the range of Marsh funnel viscosities used for water well drilling. Revert® is the trade name of a guar-gum polymer used as a drilling fluid additive.

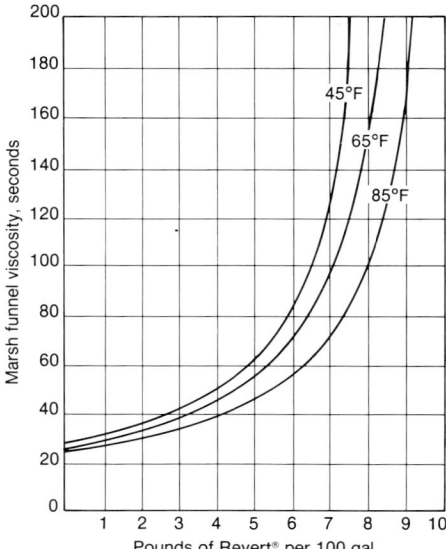

Figure 11.14. Viscosity of drilling fluid containing a given quantity of Revert® will vary with water temperature. The graph above is after 4 hours of hydration. Some drilling fluid additives made with guar contain a lower percentage of guar and will react differently.

to that of water by adding acids, oxidizing agents, or enzymes. These additives depolymerize (break down) the polymer by hydrolysis, thereby destroying the viscous nature of the fluid.

Several other natural polymers and starches are used as drilling fluid additives. Carboxymethylcellulose (CMC), hydroxyethylcellulose (HEC), and xanthan gums are well-known examples. These products are used in oil and mineral exploration, but, because they cost more than guar gum and most synthetic polymers, their use in water well drilling has been limited. CMC is used to control filtration, increase viscosity, maintain hole stability, and enhance cuttings removal. Unlike CMC, HEC is a viscosifier for salt solutions and is even stable in brine solutions; it is also effective in reducing filtration loss. The primary use for xanthan gum is as a viscosifier and suspension agent. Its ability to suspend particles exceeds the capacity of any other polymer currently used as a drilling fluid additive. Usually, CMC, HEC, and xanthan gum are added to the drilling fluid in concentrations of 0.5 to 5 lb per 100 gal (0.2 to 2.3 kg per 0.4 m^3).

Starch has been used in drilling fluids for filtration control for 40 years, and was probably the first organic polymer deliberately added to enhance a drilling fluid. The principal starch source is corn, but wheat, rice, and potato are also used. Since the introduction of the other polymers, the use of these starches has declined. Starch is subject to yeast mold and bacterial degradation, and, unless maintained at a pH of 12 or higher or used in saturated-salt drilling fluids, a preservative must be used. Concentrations of 5 to 24 lbs per 100 gal (5.8 to 27.3 kg per m^3) are effective. Although the material concentration is high, starch is the least expensive polymer available for filtration control.

Synthetic Polymers

Synthetic polymers are of two basic types — natural materials that are produced synthetically, and totally synthetic materials that contain hydroxyl and carboxyl groups similar to natural gums and resins. The latter materials are distinguished by being impervious to the biological decomposition affecting both natural and synthetically produced natural polymers. Some synthetic polymers developed for the oil and exploration drilling industries are designed for high-alkaline fluid systems not commonly used by water well contractors. Many of these polymers are not biodegradable. Some synthetics, however, can be broken down, but observance of manufacturer's instructions is mandatory because complexing (formation of solids) can clog water-bearing formations, filter packs, and well screens. In addition, many of the chemicals used to compound synthetic polymers may not be desirable for use in potable water wells, and the driller or engineer should determine acceptability or suitability beforehand.

The most notable field problem with synthetic polymers has been a lack of shear stability, leading to reduced viscosity as drilling continues. Thus, the initial cost of synthetic polymer additives does not represent the total cost, because repeated additions of new or fresh polymer are required to maintain the desired viscosity. "Freshening up" of a synthetic system can only be done a limited number of times before having to discard the entire system and start over.

Gel Strength of Drilling Fluids Made with Clay Additives

Gel strength is a measure of a drilling fluid's ability to support suspended particles when the fluid is at rest. The gel structure of a drilling fluid made with clay additives is produced when the clay platelets align themselves to join together positive and negative charges. The positively charged edge of a plate aligns itself with the negatively charged flat surface of an adjacent plate. This structure gives the liquid a plastic (quasi-solid) form with strength properties called gel strength. If enough stress (agitation) is applied to the drilling fluid by the pump, the gel will break down.

The actual strength created by the gelling process is a function of how well the clay particles have been mixed into the water and the extent of their hydration. Dispersion of the platelets occurs when they have been thoroughly separated by pumping or some other form of agitation. If mixing is incomplete, the clay platelets may still be loosely

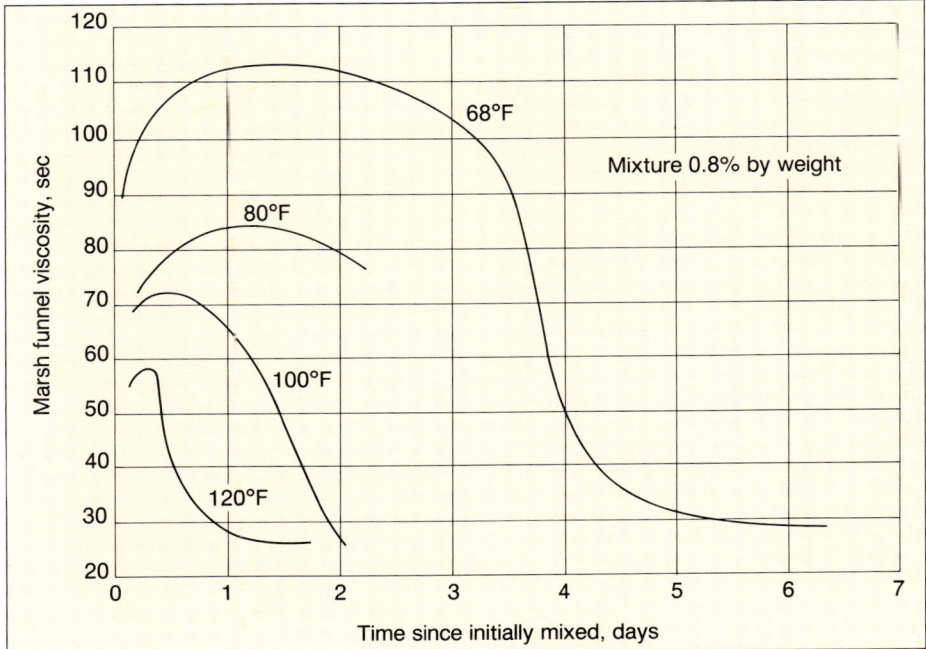

Figure 11.15. The buildup and reduction in viscosity for the natural polymer shown in this figure suggests that formation temperatures have a major effect on the length of time maximum viscosity can be maintained.

assembled as stacked groups in an aggregated state (Figure 11.16). In both the aggregated and the dispersed condition, groups of platelets or individual platelets tend to repel one another when the fluid is in motion. When a drilling fluid is at rest, however, some of the clay plates will orient themselves to balance the electrical charges on the edges and flat surfaces of the plates. This process is called flocculation and is the main cause of gel strength. Occasionally, the gel strength may become excessive even though the fluid is kept in motion. The addition of an appropriate thinner will deflocculate the fluid by reducing the attractive forces and separating the platelets.

A drilling fluid generally exhibits more than one physical condition. The four common drilling fluid states are aggregated-flocculated, aggregated-deflocculated, dispersed-flocculated, and dispersed-deflocculated. The greatest gel strength occurs when the drilling fluid is in a dispersed-flocculated state. For example, if the driller has done a thorough job of mixing the clay additives so the platelets are dispersed, and the drilling fluid is then allowed to remain at rest, the drilling fluid will assume a dispersed-flocculated state leading to a high gel strength and a uniform solids content.

If a drilling fluid with clay additives is left standing in a borehole or mud pit for some time, it gains in gel strength as increasing numbers of clay plates align themselves. This quality is called thixotropy and is a characteristic of many paints and

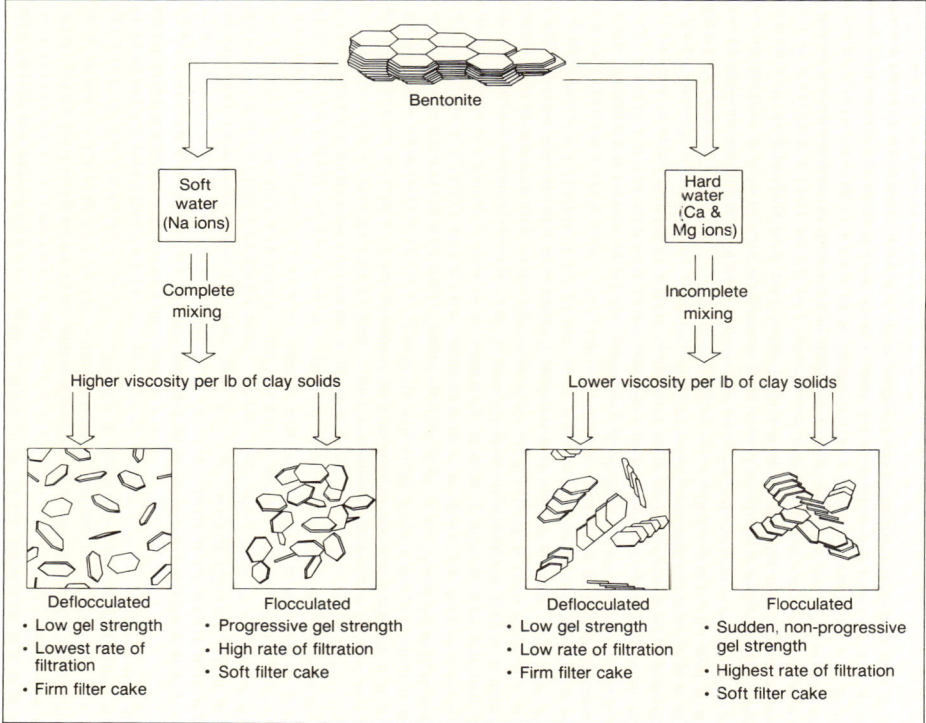

Figure 11.16. When clay particles are mixed into water, they either become dispersed evenly throughout the liquid or exist in an aggregated condition. The individual particles or groups of particles can then become either flocculated or deflocculated in each of these conditions. The hardness of the water and flow rate determine which conditions exist. The highest viscosity occurs when the clays become dispersed and flocculated in the fluid. If the water is hard, the clay platelets remain in an aggregated state, thereby limiting the viscosity.

varnishes. After the drilling fluid has been allowed to remain at rest for some time, excessively high gel strengths may demand so much pump pressure to resume circulation that the drilling fluid may be forced into fractured or weak formations. Gel strength does not instantly break down after the mud pump is started, so continued agitation is needed to break up the gel structure. Eventually, pump agitation will restore the drilling fluid to its original viscosity.

Adding bentonite will increase gel strength, but care must be taken not to add so much that settlement of cuttings at the surface is retarded. Just enough bentonite should be used to lift the cuttings and support any weighting material at the desired pumping rate.

Water chemistry also affects the gel strength of a drilling fluid made with clay additives. The use of soft water helps clay additives attain a well flocculated condition, whereas in hard water groups of clay platelets tend to remain together and gel strengths are somewhat less.

Gel Strength of Drilling Fluids Made with Polymeric Additives

Natural and synthetic polymeric drilling fluids have virtually no gel strength. This lack of gel strength assures that cuttings removal is exceptionally good at the surface, wear on the mud pump by abrasive material is minimized, and pumping pressures are minimized during normal circulation and resumption of circulation. The drilling contractor should, however, clear the borehole of cuttings before circulation is stopped, to prevent them from accumulating around the drill bit.

The reason polymeric additives have little gel strength is that the colloids are non-

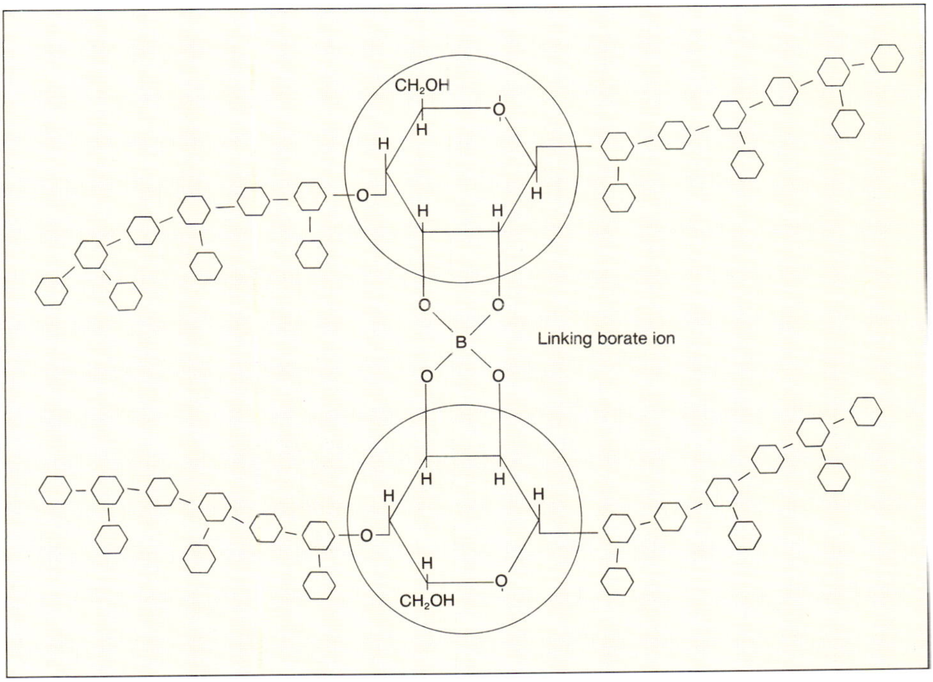

Figure 11.17. When borax is introduced into a natural polymeric drilling fluid made from guar, it crosslinks the molecules. In this process, guaran molecules become linked together to form a strong gel.

ionic; that is, they are not held together by electrical charges. Therefore, a polymeric additive does not produce a thixotropic condition in a drilling fluid. When calculating the yield point of a drilling fluid, it is important to realize that the calculated yield point for bentonite (a true thixotropic drilling fluid) is a direct indication of its gel strength. No similar comparison can be made for a drilling fluid made with polymeric additives; the calculated yield point is a meaningless figure because the drilling fluid does not have any gel strength.

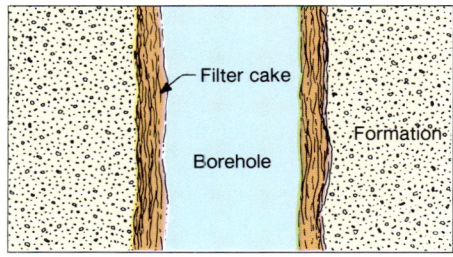

Figure 11.18. Hydrostatic pressure in the borehole forces the drilling fluid into the formation as drilling proceeds. Clay particles in the drilling fluid form a filter (mud) cake on the borehole wall during this fluid loss. As the filter cake becomes thicker, the fluid loss is reduced significantly. The thickness of the filter cake is measured in multiples of 32nds of an inch.

Natural polymers such as Revert® can be strongly gelled chemically when a sudden plugging effect is needed in a borehole (for example, in zones of lost circulation). When borate is added to Revert® and the pH of the water is above 7.5, the borate ion acts as a crosslinking agent with hydrated guar gum to form cohesive three-dimensional gels (Figure 11.17). The gel strength is determined by pH, temperature, and the concentration of reactants (boron and guar colloids). If temperature and concentration are held constant, the gel strength will increase if the pH of the water is raised from 7.5 to 9.2, the optimum pH (Henkel Corporation, Minneapolis, MN). At optimum conditions, the drilling fluid is similar to a firm, food-grade gelatin. The Revert® gel can be liquified by dropping the pH slightly below 7.0, which changes the hydrated borate ion $[B(OH)_4]$ to boric acid $[B(OH)_3]$. Lowering the pH significantly below 7.0 may destroy the viscous characteristics of the drilling fluid.

Filtration

Another of the principal requirements for a drilling fluid is to prevent fluid loss by forming a filter cake or low-permeability film on the porous face of the borehole. The sealing property depends on the amount and nature of the colloidal materials in the drilling fluid. The filter cake produced by clays and the thin film created by polymeric colloids are physically dissimilar, because the size and shape of the particles differ and their ability to hydrate is significantly different. Colloidal particles and suspended cuttings entrained during drilling are important components of the total solids that create a filter cake or film. Thus, the filtration properties of all drilling fluids are, in part, supplied by materials derived from the borehole.

When drilling begins, hydrostatic pressure in the borehole causes the drilling fluid to flow into porous formations. For drilling fluids made with clay additives, the fluid and some clay particles initally enter the formation unhindered; but as the suspended solids and cuttings continue to close off the pores, clay particles filter out and form a cake on the borehole wall. As the remaining pores around the borehole become clogged with particles, progressively smaller volumes of water can pass into the formation (Figure 11.18). In time, a filter cake effectively limits water flow through the borehole wall except in highly permeable zones where lost circulation is apt to occur.

Permeability of the filter cake depends on the type of clay used in the drilling fluid; generally, the higher the number of colloidal particles, the lower the permeability.

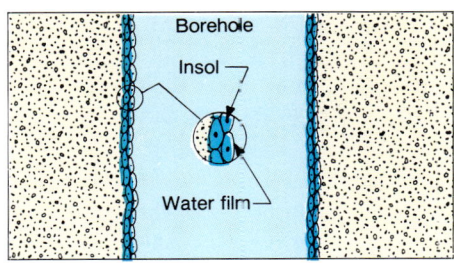

Figure 11.19. When polymers are used as drilling fluid additives, thin polymeric films form on the borehole wall. The plugging is caused by the plastering of the insol (insoluble) particles onto the wall of the formation. Each inscl is surrounded by a viscous film of fluid, and as insols build up on a borehole wall, fluid is prevented from leaving the borehole. In general, polymeric films are much thinner than the filter cakes created by clay particles even though the water loss characteristics are the same.

For example, Gray (1972) points out that, for the same duration of filtration and at the same viscosity for both drilling fluids, the thickness of filter cake formed on a sandstone from a native-clay drilling fluid with a density of 10.3 lb/gal (1,230 kg/m^3) is 36 times greater than from a premium bentonite drilling fluid with a density of 8.6 lb/gal (1,030 kg/m^3).

During drilling, the thickness of the clay filter cake will vary according to the rate of erosion caused by the rotation of the tools and by the uphole velocity of the drilling fluid. In addition, thicker filter cakes will form in formations that have higher hydraulic conductivity. When circulation stops, the filter cake will continue to build up on the wall of the borehole.

The nature of this filter cake or film and the way it forms are quite different when drilling fluids are prepared with polymers. Guar gum is a polysaccharide that provides natural fluid-loss control. This property is derived from both the soluble guar gum particles (sols) dispersed in the drilling fluid and the insoluble cell-wall residue (insols) of the gum. Each of the insol particles is covered with a relatively thick, viscous film of fluid produced by the dissolved sol particles. Thus, fluid loss can be controlled with a relatively thin coating on the wall of the borehole (Figure 11.19). A much thicker layer of clay particles would be required to achieve the same fluid-loss properties.

The filtration rate of a drilling fluid can be determined by a filtration test device (Figure 11.20). It can measure the volume of fluid that may be lost to the formation and the resulting thickness of the filter cake deposited on the permeable walls of the borehole. Bentonite filtration testing procedures, using API standards, are not completely valid for polymers, because the rates of hydration vary for different polymeric materials. Systems that combine bentonite gels and viscosifiers such as polyanionic cellulose powder, carboxymethylcellulose (CMC), and polyacrylamides, however, can be tested for water-

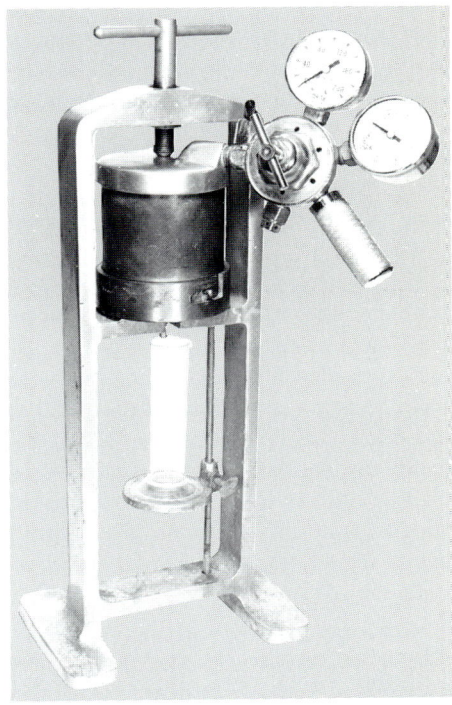

Figure 11.20. A filtration test can be done with this apparatus to measure the flow rate of water through a specific thickness of filter cake.

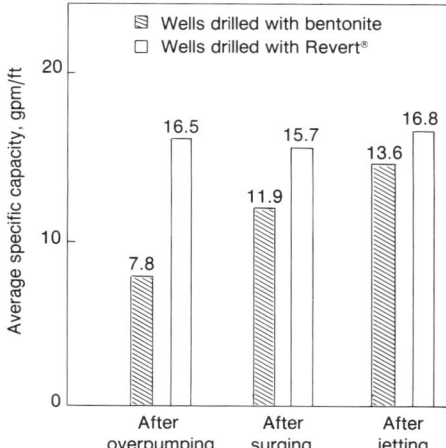

Figure 11.21. Comparison of the average specific capacities after three development episodes for 10 shallow wells constructed in glacial outwash and drilled with either bentonite (7 wells) or Revert® (3 wells). The Revert® was allowed to break down naturally before development procedures were initiated. More powerful development methods are required to remove bentonite drilling fluids. *(Driscoll et al., 1980a)*

loss characteristics with this equipment.

Although a filter cake is required during drilling, any residual clays on the borehole wall and in the aquifer after the well has been completed are highly detrimental to the well's productivity. Well development procedures should be conducted as soon as possible after a well has been drilled. Otherwise, complete drilling fluid removal may become impossible, especially if clay additives are used. Filter cake removal during development can be accomplished primarily by mechanical means, even though the addition of polyphosphates can be helpful.

Polymeric films should also be removed or broken down because some clays will be entrained in the film during normal drilling operations. Many polymeric drilling fluids and films can be broken down chemically within minutes by chlorine or other strong oxidizing materials. Although these chemical processes are not well understood, Figure 11.21 shows that polymeric drilling fluids can be removed in the initial development episode with the least powerful development method because of the presence of naturally occurring enzymes, whereas clay filter cakes require powerful mechanical methods and more time.

TREATMENT OF MIX WATER FOR DRILLING FLUIDS

For maximum drilling efficiency, the drilling contractor should adjust the chemical characteristics of the mix water, if necessary, before mixing begins. If the water has a pH of less than 6.5, for example, the performance of a drilling fluid with clay additives may not be satisfactory. In general, the lower the pH, the more likely the water is to have ions in solution and thus be hard water. In areas where the water has a low pH, it may contain high concentrations of calcium and magnesium. Clay yields can be increased and water losses decreased if hard water used for mixing the drilling fluid is softened beforehand. Soda ash [sodium carbonate (Na_2CO_3)] can be added to the mix water to precipitate the calcium, as calcium carbonate ($CaCO_3$), before the drilling fluid is mixed. The following equation can be used to estimate the soda ash required:

$$Lbs\ Na_2CO_3/100\ gal\ mix\ water = 0.0095 \cdot hardness\ as\ CaCO_3\ in\ mg/\ell \quad (11.3)$$

Under ordinary conditions, 0.5 to 3 lbs (0.2 to 1.4 kg) of soda ash per 100 gal (0.4 m³) is sufficient to soften the mix water. The pH of the drilling fluid should be checked after the addition of the soda ash to verify that the pH is 7.5 or higher.

The hydration potentials of colloidal additives are significantly reduced if metallic ions occur at high concentrations. For example, if iron concentrations are 3 mg/ℓ or greater, the viscosity buildup in a natural organic system may be retarded. Polyvalent metallic ions may be removed from water by chlorination at a 50-mg/ℓ concentration.

It is good practice to chlorinate all water used to mix the drilling fluid, as well as any make-up water added to the system. The mix water should be chlorinated to a concentration of 50 to 100 mg/ℓ, depending on the particular additive selected. An easily measurable free-chlorine residual of approximately 10 mg/ℓ should be maintained during drilling. This concentration can be checked with chlorine paper. Chlorine treatment kills bacteria introduced into the borehole or aquifer during drilling and minimizes the amount of chlorine required upon completion of the well. Mix water should never be taken from wetlands, swamps, or small lakes near the well site. Not only are pathogenic bacteria likely to be in these water supplies, but iron bacteria are also commonly found and their subsequent growth in the well can cause severe problems. Because chlorine is unstable, the chlorine residual should be checked periodically during drilling.

Design of Mud Pits

Drilling fluid is usually mixed adjacent to the drilling rig in either portable or excavated pits. The capacity of portable pits for direct rotary rigs ranges from 200 to 10,000 gal (0.8 to 37.9 m³). Large pits, 20,000 to 80,000 gal (75.7 to 303 m³) are suitable for reverse circulation drilling. Although a mud pit is excavated prior to

Table 11.4. Mud Pit Capacities and Dimensions

A. Rectangular mud pit

   Volume (gal) = length (ft) × width (ft) × depth (ft) × 7.5

B. Pit with sloping sides

   Volume (gal) = length (ft) × average width (ft) × depth (ft) × 7.5

$$\text{Average width} = \frac{\text{width at top} + \text{width at bottom}}{2}$$

C. Ideal dimensions for two basic pits

   In general, the pit should be three times the volume of the finished borehole. Each mud pit should have a settling section and a suction section. The dimensions of the settling pit can be determined by using a basic equation to establish the width. Once the width is known, the length and depth can be calculated.

$$\text{Width} = \sqrt[3]{\frac{\text{hole volume (gal)} \times 2}{2.125 \times 7.5}}$$

Length = 2.5 × width
Depth = 0.85 × width

For the suction pit, the length is 1.25 × width and the depth is 0.85 × width.

drilling with reverse circulation, a premixed drilling fluid is usually not prepared because only clean water is generally used. The size of the mud pit is dictated by the volume of drilling fluid contained in the finished borehole and the need for a reserve volume, which varies according to the particular rotary system used. Usually the volume of the pit is one and a half to three times the volume of the finished hole. For reverse rotary drilling, where drilling fluid losses are usually high, the volume of the pit is generally three times the volume of the finished borehole. Mud pit capacities can be calculated from the information in Table 11.4.

The design of the mud pit should take several factors into consideration (Figure 11.22). The principal objectives of the pit are to store an adequate volume of drilling fluid and to act as an effective settling basin for suspended cuttings. For efficient removal of the suspended cuttings, the pit should be constructed in two sections — the settlement part and the suction part (Figure 11.23). Many drillers, however, use single-reservoir pits that serve both of these functions. The velocity of the drilling fluid as it moves through the mud pit during the storage interval must be as low as possible. This can be achieved by changing the direction of flow as the drilling fluid moves through the pits, as well as by deepening part of the pit or by using baffles and overflows. Deeper, rather than wider, trenches are more satisfactory in reducing drilling fluid velocity, although for large pits, the excavation equipment available may limit the configuration somewhat. When a single pit is used, some drillers will slope the bottom of the pit downward toward the pump suction point to slow the velocity and create a place for the cuttings to settle. The suction hose must be mounted above the bottom of the pit. The bottom of an excavated mud pit should be sealed with a plastic film or a compacted layer of clay.

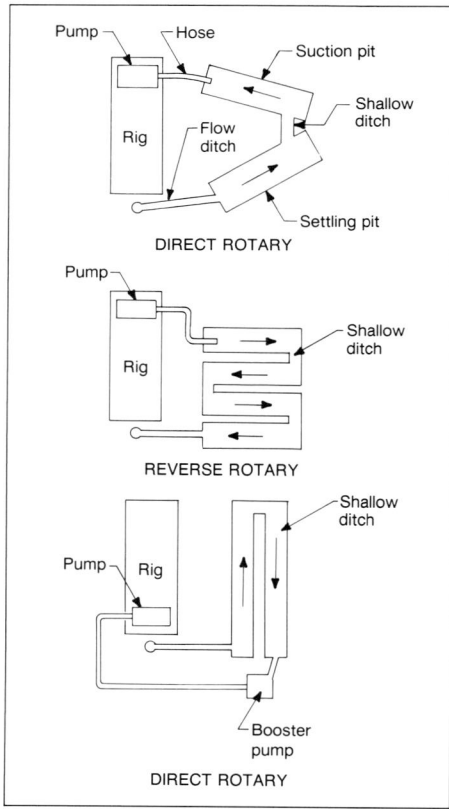

Figure 11.22. Common mud pit designs for rotary systems. The design selected will depend on the type of additive used in the drilling fluid, the depth of the borehole, and the type of drilling rig.

MIXING ADDITIVES INTO WATER-BASED SYSTEMS

Dispersion of the colloids into the drilling fluid and the subsequent rate of hydration depend on the mixing method. Additives can be either poured directly into the mud pit (tank) from the bag or placed in a mud mixer (hopper) connected to a high-velocity water source. Advantages of a mixer include higher mixing speed, greater hydration of the additive particles, and elimination of gum balls (lumps of clay or polymer). Under ordinary conditions, 300 to 500 lb (136 to 227 kg) of clay or polymeric additives can be mixed per hour with a good mixer.

Mechanical-type mixers proportion the amount of clay or polymer to the proper amount of water (Figure 11.24). The air/water mixer is operated by connecting an air line which disperses the fine particles into a high-velocity water stream. In operation, the air supply is turned on first; then the water flow is slowly increased so that the flow rate equals about one-third the air pressure in psi [for example, 30 gpm (164 m^3/day) at 90 psi (621 kPa)]. The recommended operating range is 30 to 40 gpm (164 to 218 m^3/day) with 90 to 120 psi (621 to 827 kPa) of air. The air volume should be about 20 cfm (0.01 m^3/sec). After the proper flow rates are established, the dry polymer or clay is poured into the funnel or cone. Sometimes a suction hose is used to vacuum the additive directly from the bag. Plugging of the mixer after use can be avoided if the water is shut off first, then the air. It is not possible to mix liquid polymers with this type of mixer.

In a water mixer, colloids are entrained from a reservoir into a high-velocity stream that is directed into the pit. Agitation of the particles in the water stream helps assure rapid hydration without the formation of gum balls. Design specifications for two types of water mixers are given in Appendix 11.E. Liquid polymers are mixed with water mixers or added directly to the pit near the hose. Other liquid additives such as chlorine should also be added to the pit near the suction hose to disperse them as quickly as possible.

When additives are mixed in the mud pit, the additive is sprinkled on the surface of the nearly full pit or tank and a powerful water source is directed at the surface to create turbulence and cause hydration (Figure 11.25). If the clay or polymer additive

Figure 11.23. The mud pit is often constructed of two reservoirs to be more effective in settling cuttings. *(Layne Western Company)*

is poured in too quickly, gum balls form which must be broken up by hand. The intake screen on the suction hose can be easily plugged by these gum balls. Furthermore, the yield will be less for a given amount of additive. For these reasons, direct mixing is not recommended.

Various components of an ideal fluid circulating system for direct rotary are presented in Figure 11.26. This particular design is for large-diameter, deep holes, but, except for smaller components, most of the same requirements exist for any fluid system.

AIR DRILLING

Many water wells are now drilled with air because of the relative simplicity and effectiveness of air systems and the increasing number of rotary rigs equipped with air compressors. The earliest attempts to use air as a circulating medium during the 1950's showed that significant increases in penetration rates and bit life could be obtained (Cooper et al., 1977). Air drilling is now recognized as a primary method

Figure 11.24. Mud mixers are extremely effective in hydrating the drilling fluid additive. They prevent the formation of gum balls that impede the viscosity buildup of the drilling fluid.

Figure 11.25. This driller uses both a mechanical mixer and high-velocity jetting hose to hydrate the drilling fluid additive. *(Currie Well Drilling)*

to reduce drilling time and therefore the cost of a well. The mechanics of air drilling are more difficult to understand than typical water-based systems because the drilling fluid is compressible and often contains water, an incompressible fluid, and other special additives.

To drill with air, the drilling rig must be equipped with an adequately sized compressor and a water pump that can inject up to 10 gal (37.9 ℓ) of water and chemicals into the airstream. On some rigs, a special chemical pump is used to inject surfactants and chemicals into water before the water is injected into the drill rods. Use of a chemical pump, although not necessary, allows the driller to add chemicals more accurately and thus to have greater control of the air system. The compressor is usually the key to successful air drilling because insufficient air volume and pressure are the principal problems in air drilling.

Compressors used on water well rigs are either the piston (reciprocating) type or the helical-screw type. Piston-type compressors are efficient at compressions up to 30:1 and possess high pressure capacity. On the other hand, screw-type compressors typically have somewhat lower pressure capacity than piston compressors, but are positive displacement and consequently produce a constant air volume. In general, screw-type compressors operate at about 220°F (104°C), 40 to 50°F (22.2 to 27.9°C) less than piston-type compressors. In screw-type compressors, the rotors operate in an oil bath and do not come into contact with one another or the casing (Figure 11.27).

Compressors are rated to deliver a given air volume at a certain operating pressure (Table 11.5). In practice, air delivery at a rated pressure means that a specific number of cubic feet of air at atmospheric pressure can be compressed and delivered to the rig at that pressure

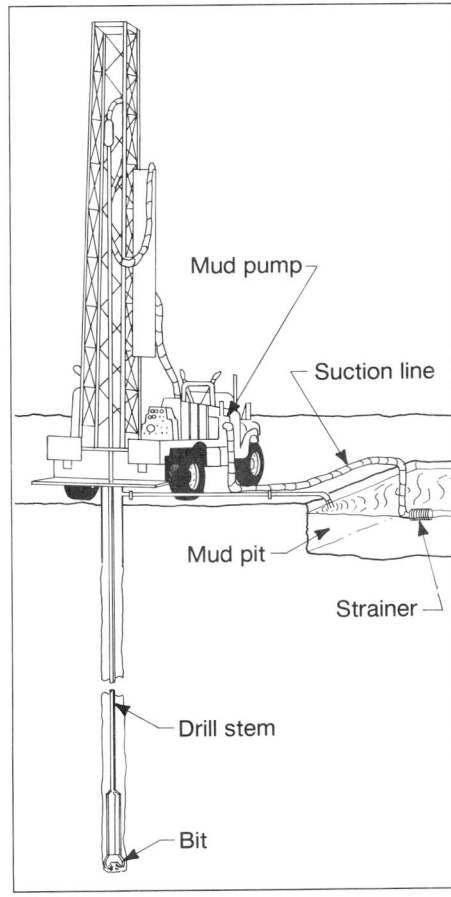

Figure 11.26. Components of a complete drilling fluid circulating system for a direct rotary rig.

Table 11.5. Typical Pressure and Volume Capacities for Compressors Mounted on Water Well Rigs

Number of Stages	Pressure		Volume	
	psi	kPa	cfm	m³/sec
1	125	862	600 - 900	0.28 - 0.42
1	150	1,030	750 - 1,150	0.35 - 0.54
2	275	1,900	750 - 900	0.35 - 0.42
2	350	2,410	900 - 1,050	0.42 - 0.5
2	500	3,450	1,050	0.5

every minute. Once the air has been compressed to the delivery pressure, it no longer has the original free-air volume. For example, 750 ft³ (21.2 m³) of air compressed to 250 psi (1,720 kPa) has a compressed volume of only 42 ft³ (1.2 m³). The compressor rating is valid at sea level (maximum air density) at an air temperature of 60°F (15.6°C). The rating is usually designated at a compressor speed of 1,800 to 2,100 rpm. Correction factors must be used if the air temperature or elevation of the rig varies significantly from standard conditions (Table 11.6). For instance, the rated volume for an air compressor is reduced about 17 percent if it is operated at 60°F in Denver (5,000-ft elevation). Similarly, if the temperature rises significantly over 60°F, air density falls, and although the same volume of air is taken into the compressor, the mass of this air is less. Thus, as atmospheric pressure decreases or temperature increases above 60°F, less standard air volume is compressed at a given rpm. Some temporary factors also affect the ability of a compressor to produce its rated air volume. These include significant changes in atmospheric pressure caused by storm systems, large diurnal changes in air temperature, and the placement of the air intake to avoid pulling in hot air from the engine.

Both piston compressors and screw-type compressors can have one or more stages (compression units). Air volume or pressure demands often require that more than one compressor be used. Compressors connected in series will increase the pressure in an air system, whereas a parallel arrangement will increase the volume. Although

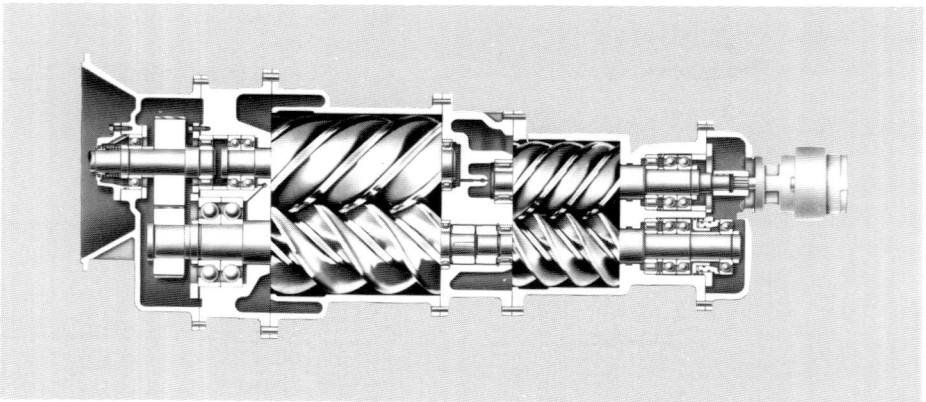

Figure 11.27. This two-stage screw-type compressor can deliver 600 to 750 cfm (0.3 to 0.4 m³/sec) at 250 psi (1,720 kPa). The rotors are oil-flooded and do not touch each other or the housing. *(Ingersoll-Rand Company)*

the output pressure can be regulated on a screw compressor, it should never be reduced when a down-the-hole air hammer is being used in air drilling. The pressure may be adjusted downward for well development work, however, so the screen is not damaged. This is especially important if the pressure is greater than 300 psi (2,070 kPa).

The standard compressors used on water well rigs have increased in air-volume capacity and pressure capability in order to drill deep, large-diameter holes, to achieve high penetration rates, and to maximize drilling rates with down-the-hole hammers. In general, few drillers have too much air available. A good method to verify that enough air is available to remove the cuttings efficiently in dry-air drilling is to check the time needed for the air to clean up after drilling ceases. The time required to clean the cuttings should not greatly exceed 6 to 7 seconds per 100 ft (30.5 m) of borehole (Cooper et al., 1977). More air is needed as boreholes deepen; in addition, 30 to 40 percent more air is required when drilling with air-mist systems (Magcobar, 1979). For down-the-hole air hammers, higher pressures translate into increased penetration rates when drilling with dry air because the hammer action is more rapid. Higher available pressure can also be used to overcome static heads following a temporary cessation of drilling.

Several options exist when air is used as the drilling fluid:
1. Air alone (dry air)
2. Air mist
   a. Air plus a small volume of water
   b. Air, small volume of water, plus a small amount of surfactant
3. Air-foam
   a. Stable foam — air plus surfactant
   b. Stiff foam — air, surfactant, plus high-molecular-weight polymer or bentonite
4. Aerated mud — water-based drilling fluid plus air

Decisions on which of these systems to use depend on the volume of water entering the borehole, the penetration rate, the volume of air available, the nature of the formations being drilled, and environmental conditions affecting the drilling process.

Table 11.6. Temperature and Pressure Correction Factors for Elevations above Sea Level

Elevation above sea level		Atmospheric pressure		Temperature									
ft	m	psi	kPa	0 / -32	20 / -6.7	40 / 4.4	60 / 15.5	70 / 21.1	80 / 26.7	90 / 32.2	100 / 37.8	120 / 48.9	140 F / 60 C
0	0	14.696	101.3	1.130	1.083	1.040	1.000	0.981	0.963	0.945	0.929	0.897	0.867
1,000	305	14.175	97.7	1.089	1.044	1.003	0.964	0.946	0.928	0.911	0.896	0.865	0.835
2,000	610	13.664	94.2	1.051	1.007	0.967	0.930	0.912	0.896	0.879	0.864	0.834	0.806
3,000	914	13.168	90.8	1.012	0.970	0.932	0.896	0.879	0.863	0.847	0.832	0.804	0.777
4,000	1,219	12.692	87.5	0.976	0.936	0.899	0.864	0.848	0.832	0.816	0.803	0.775	0.749
5,000	1,524	12.225	84.3	0.940	0.901	0.865	0.832	0.816	0.801	0.786	0.773	0.746	0.721
6,000	1,829	11.778	81.2	0.905	0.867	0.833	0.801	0.786	0.771	0.757	0.744	0.718	0.694
7,000	2,133	11.341	78.2	0.872	0.836	0.803	0.772	0.757	0.743	0.730	0.717	0.692	0.669
8,000	2,438	10.914	75.3	0.840	0.805	0.773	0.743	0.729	0.716	0.702	0.690	0.666	0.644
9,000	2,743	10.501	72.4	0.808	0.774	0.744	0.715	0.701	0.689	0.676	0.664	0.641	0.620
10,000	3,048	10.108	69.7	0.777	0.745	0.715	0.688	0.675	0.662	0.650	0.639	0.617	0.596

(Mason and Woolley, 1981)

Dry-Air Systems

The simplest air drilling system involves using only dry air as the drilling fluid. Optimum drilling rates are achieved by using dry air because the column of air puts minimum pressure on the bottom of the borehole. As removal of the rock overburden proceeds, the cuttings become progressively easier to chip off because the overlying weight of the rock has been removed. Figure 11.28 shows how the penetration rate is affected adversely by downhole pressure. Water well drillers will ordinarily begin all boreholes with only dry air, but dust problems and influx water will usually create the need to alter the dry-air system.

Compared with water-based drilling fluids, dry-air systems offer the following advantages:

1. Higher penetration rates in dense, consolidated rock
2. Reduced bit wear
3. High solids-carrying capacity
4. Reduced formation damage and self-induced fluid loss
5. Low water requirements
6. Minimized swelling problems associated with water-sensitive clays

Air drilling is extremely effective in drilling hard, stable formations such as igneous and metamorphic rocks, and tough, dense sedimentary rocks such as dolomite where penetration rates are often exceptional. Because air has the lowest density of any drilling fluid and therefore places minimum downward pressure on the formation being drilled, cuttings chip off readily with either a roller or a down-the-hole air hammer bit. Boreholes are kept clean by the high annular fluid velocity, which ranges from 3,000 to 5,000 ft/min (915 to 1,520 m/min) for dry air. Uphole velocities to 7,000 ft/min (2,130 m/min) may be desirable for deep holes drilled at high penetration rates (Kuetzing, 1981). In the range of 3,000 to 5,000 ft/min, cuttings lift easily in spite of air's low density and viscosity. It should be noted, however, that cuttings will not rise up the borehole at the same velocity as the drilling fluid. For example, if the dry air has a velocity of 5,000 ft/min (1,520 m/min), fine (dust) cuttings move up the borehole at about 3,000 ft/min (915 m/min), and the large cuttings may rise at only 1,000 to 2,000 ft/min (305 to 610 m/min). In general, the lifting capacity of air is proportional to its density and to the square of its annular velocity. Thus, the driller adjusts air volume and pressure at the surface to compensate for the weight of the cuttings and for increases in density with depth in order to maintain the required annular velocity. See Appendix 11.F for uphole velocities based on hole diameter, drill

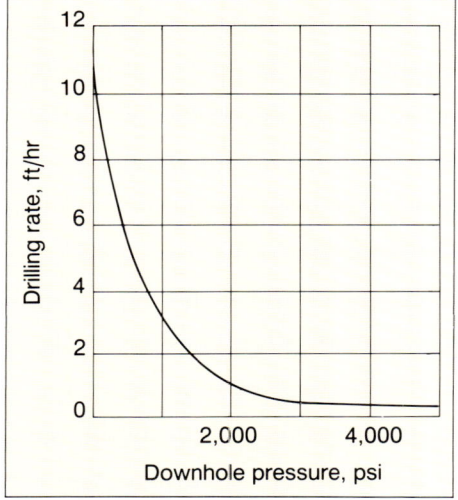

Figure 11.28. Effect of downhole pressure on drilling rate. This test was conducted in Indiana limestone with a two-cone rock bit operating at 50 rpm and having a weight of 1,000 lb (454 kg). *(Cunningham and Eenink, 1958)*

pipe size, and air volume.

Air drilling is particularly advantageous where drill and drive techniques are used for unconsolidated formations such as glacial outwash and alluvial deposits or for semiconsolidated formations such as bouldery tills. In this type of drilling, hole stability is not a major problem and the many benefits of drilling with air make this technique attractive.

Air is helpful in overcoming lost-circulation problems in highly fractured igneous and metamorphic rock and in highly porous formations. Many drillers who use water-based drilling fluid systems will change immediately to an air or air-foam system if large crevices or cavities are encountered. But lost-circulation problems can occur when air systems are used to drill permeable sandstone. In this type of formation, so much air is lost that uphole velocities may be insufficient to lift the cuttings. To overcome this problem in sandstone, polymers and water are sometimes added to the air. The thin polymeric film seals the formation pores so air loss is minimized.

Problems with air drilling, especially dry-air drilling, usually involve an insufficient air supply, resulting in an annular velocity that is not high enough to carry the cuttings to the surface. For a given diameter, hole depth is a primary factor affecting cuttings removal because air-volume requirements are directly related to depth. Erosion of the borehole walls can also create an increase in the demand for air that may exceed the capacity of the compressor. On the other hand, too much air can be a problem because soft spots in a borehole wall can be eroded, resulting in blowouts (borehole enlargement) which, in turn, can lead to even higher air-volume demands. Table 11.7 shows the minimum dry-air volume requirements for various hole and drill pipe sizes when the drilling rate is approximately 60 ft (18.3 m) per hour.

Another drilling problem occurs when small amounts of water begin to enter the borehole during dry-air drilling. Water mixes with the smallest rock cuttings to produce muds that can plug the formation and limit the potential yield of the well. If enough mud is present to form rings, or collars, on the drill string or borehole wall, air flow is restricted and the drill rods may stick in the borehole. When mud collars form, restriction of the annulus causes excessive pressure build up below the collar, and fracturing of the formation materials can occur. Fracturing and blowouts can be minimized if the rig is equipped with compressors that can be controlled by a relief valve at the operator's station. The driller can then take immediate action when the pressure rises suddenly to reduce the chance for blowouts and fracturing in loose formations.

Air-Mist Systems

Adding small amounts of water to air creates an air-mist system. Many drillers add water to the air system to control dust and to help break down any mud collars forming on the drill rods. To help increase the wetting action of the injected water, small amounts of surfactant may also be added to the airstream. Air-volume requirements usually increase substantially when switching from dry air to air misting, because greater downhole pressure prevents immediate expansion of the air as it leaves the bit. Air-mist techniques can be used satisfactorily as long as only small volumes of influx water, 15 to 25 gpm (81.8 to 136 m^3/day), enter the borehole from the aquifer. When the volume of water entering the borehole increases, an air-foam system must replace the mist system.

Air-Foam Systems

Ordinarily, foam is defined as a dispersion of air in water. In an air-foam drilling system, however, air is the continuous phase and water is the dispersed or discontinuous phase. Thus, drilling foam is created when a small volume of water and surfactant is injected into an airstream. Although some foam forms naturally when water enters an airstream, the amount and stability of the foam is enhanced significantly when a surfactant is added. The term "foam drilling" is associated with the introduction, into air, of a surfactant mixed with water. Surfactants include anionic soaps, alkyl polyoxethylene nonionic compounds, and cationic amine derivatives. All

Table 11.7. Minimum Air-Volume Requirements for Dry-Air Drilling

Hole Size	Drill Pipe	Depth & Volume Required S.C.F.M.*					
		500	1000	1500	2000	3000	4000
4-3/4	2-7/8	253	278	303	328	377	427
	2-3/8	293	315	338	360	405	450
6-1/4	3-1/2	461	493	524	557	620	684
	2-7/8	522	551	579	608	665	723
6-3/4	3-1/2	568	601	634	667	733	799
7-3/8	3-1/2	710	745	780	815	885	955
7-7/8	4-1/2	711	752	794	836	918	1000
	3-1/2	834	870	906	943	1015	1088
8-3/4	5	873	920	966	1013	1106	1199
	4-1/2	946	990	1033	1077	1164	1252
	3-1/2	1070	1109	1147	1186	1263	1341
9	5	945	992	1039	1086	1181	1275
	4-1/2	1019	1063	1107	1152	1240	1329
	3-1/2	1142	1182	1221	1261	1340	1419
9-7/8	5-1/2	1131	1183	1235	1287	1391	1495
	5	1212	1261	1311	1360	1459	1557
	4-1/2	1286	1333	1379	1426	1519	1613
11	6-5/8	1299	1361	1423	1485	1609	1733
	5-1/2	1511	1566	1621	1676	1786	1896
	4-1/2	1666	1717	1767	1818	1919	2020
12-1/4	6-5/8	1765	1830	1895	1960	2090	2220
	5-1/2	1977	2037	2096	2156	2275	2394
	4-1/2	2134	2190	2245	2301	2412	2523
15	6-5/8	2980	3056	3131	3207	3358	3509
	5-1/2	3195	3267	3338	3410	3553	3696
	4-1/2	3353	3422	3490	3559	3696	3833
17-1/2	6-5/8	4297	4386	4474	4563	4740	4917
	5-1/2	4513	4599	4684	4770	4941	5112
	4-1/2	4671	4754	4837	4920	5086	5252

*The volume required is based on an estimated penetration rate of 60 ft/hr (18.3 m/hr) and an annular velocity of 3,000 ft/min (915 m/min). Air volumes are given at standard pressure (sea level) and temperature [60°F (15.6°C)].

(Magcobar, 1979)

of these are available as commercial products.

The addition of a surfactant to an air-based drilling fluid has several advantages over the use of air alone. These include:
1. Higher solids-carrying capacity
2. Ability to lift large volumes of water
3. Reduced air-volume requirements
4. Reduced erosion of poorly consolidated formations
5. Effective dust suppression
6. Increased borehole stability

Air-foam systems are not effective, however, when confined formations are intercepted, because the downhole pressure is so low that the borehole may become unstable or so much water may enter the borehole that the air-foam system cannot remove it.

Foams are used primarily to enhance the rate of cuttings removal by preventing them from aggregating so they can be lifted more easily to the surface. Foaming agents are also added to air when the airstream can no longer lift the water entering the borehole. A surfactant injected into the airstream helps break up the water mass by reducing the surface tension of the water droplets. Once a water droplet is broken up, it can be lifted easily. Upon reaching the surface, the foam breaks down rather quickly, depending on the particular foaming agent used and the prevailing environmental conditions.

Foams reduce the uphole velocity requirements to 50 to 1,000 ft/min (15.2 to 305 m/min), considerably less than the rate for drilling with air alone. The lowest uphole velocities are possible when using high concentrations of surfactant together with clay or polymeric additives. Lower uphole velocities reduce the amount of air required (Table 11.8). Lower air-volume requirements, in turn, may also reduce the pressures needed. As a result, the loss of air into the formation is minimized. The use of foam is recommended for weakly consolidated formations where the higher uphole velocities required for dry air would cause excessive erosion of the borehole.

Surfactants are often mixed with water in a large container adjacent to the rig, and then injected slowly into the airstream at a rate sufficient to lift the cuttings. Another mixing method is direct injection of a surfactant through a metered chemical pump into a water stream. The required volume of surfactant will usually range from 1 qt to 3 gal (1 to 12 ℓ) per hour, depending on the type of surfactant, the volume of water entering the borehole, the diameter and depth of the borehole, and the quantity and size of the cuttings. The surfactant concentration commonly varies from 0.25 to 2 percent of the injected water, but may be increased significantly in deep, large-diameter boreholes with large quantities of influx water.

Table 11.8. Air Requirements for Uphole Velocities of 100 to 300 ft/min

Hole Diameter		Air Volume	
in	mm	cfm	m^3/sec
6	152	30 - 60	0.01 - 0.03
10	254	75 - 150	0.04 - 0.07
15	381	200 - 300	0.09 - 0.14
26	660	400 - 600	0.19 - 0.28

Table 11.9. Types of Foam as Determined by Water Content

Type of Foam	Liquid-Volume Fraction (percent)
Dry foam	< 2
Stable foam	2 - 10
Wet foam	10 - 25

Surfactants used in water wells should be biodegradable and nontoxic, but drillers may encounter complaints when using foam in urban areas because some long-lived foams can build up on the ground around the drilling machine. If a strong wind is blowing, large masses of dry foam may travel hundreds of feet before dissipating. Uninformed onlookers may perceive the foam as dangerous to the environment and may alert local health or law enforcement authorities. Foam can be dissipated by injecting defoamers such as aluminum stearate into the flow at the end of the discharge line by means of a spray nozzle.

Foams are classified into several different types based on their resistance to slumping. The initial resistance to slumping depends on the concentration of the surfactant, the volume of injected water, and whether other additives such as polymers or bentonite* are added to the airstream (Table 11.9). The addition of high-molecular-weight polymers or bentonite increases the viscous qualities of the foam, reduces the uphole velocity requirements for effective penetration and adequate cleaning, and helps to stabilize the borehole walls. Polymers also help reduce friction in the borehole and are biodegradable and therefore do not plug the formation after breakdown. If bentonite is used to stiffen the foam, any filter cake forming on the borehole walls must be removed during development. Typically, 30 to 50 lb (13.6 to 22.7 kg) of bentonite is mixed with 100 gal (0.4 m³) of water. Lesser amounts of polymers are required to create similar viscous qualities in the foam. The surfactant is then added to the mixture in a concentration of 0.25 to 2 percent. Always add the surfactant after any clay, polymer, or other chemicals are mixed. The resulting mixture is then injected into the airstream at an ideal rate of 2 percent of the free-air volume (rated capacity of the compressor). The addition of either polymer or bentonite generally raises foam density to 2 to 6 lb/ft³ (32 to 96.1 kg/m³).

Aerated Drilling Fluids

Air is sometimes injected into water-based drilling fluids to lighten the weight of the drilling fluid column. This proce-

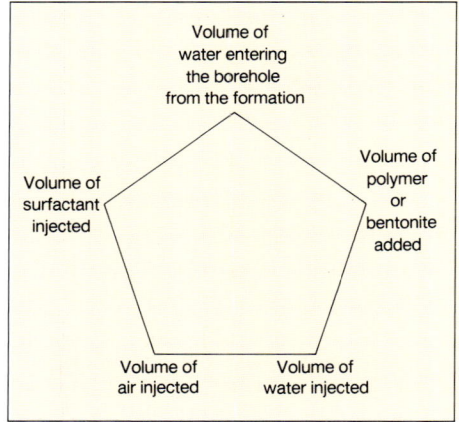

Figure 11.29. Common volumetric variables that must be considered by the driller when using air as the drilling fluid.

---

*Bentonite cannot be added to an air-foam system if a down-the-hole air hammer is being used, because the clay particles will cause the hammer to malfunction within a short time.

dure is common in the Inverse system and for air-assist reverse-circulation rotary drilling. The introduction of air increases the penetration rates 10 to 50 percent over conventional mud drilling (Kuetzing, 1981). In addition, fluid loss to highly permeable zones is often reduced or eliminated. One disadvantage of aerating water-based drilling fluids is that it causes higher rates of drill-pipe corrosion. Because air is not the primary drilling fluid in this method of drilling, it will not be discussed further in this section. See Chapter 10 for a description of the Inverse and air-assist reverse-circulation rotary drilling methods.

Physics of Air Drilling

The design of an air or air-foam drilling fluid system is complicated because the system consists of both gaseous (air) and liquid (water and surfactant) phases. These phases react differently to changes in pressure and temperature. Air, unlike water-based drilling fluids, is subject to pronounced volumetric changes. Because pressure and temperature changes occur as the drilling fluid is circulated, the volume of the fluid at various points in the circulation system also changes. Successful drilling with air requires that the contractor learn how temperature, pressure, and volume change as the fluid circulates in the system, and how the appropriate adjustments are made based on the variables shown in Figure 11.29. In the discussion below, an analysis of air-foam systems is developed; the same physical principles apply to dry-air systems.

The physical response of an air-foam system can be predicted, in part, by using the Ideal Gas law (based on combining Boyle's and Charles' laws):

$$\frac{Pv}{T_a} = \text{a constant} \qquad (11.4)$$

where P is pressure, v is volume, and T_a is the ambient temperature in degrees Kelvin (°C + 273). Computing the changes in one or more of these factors as the air-foam drilling fluid circulates from one point to another can be accomplished by using the general equation:

$$\frac{P_1 v_1}{T_{a_1}} = \frac{P_2 v_2}{T_{a_2}} \qquad (11.5)$$

When pressure and temperature conditions change in the circulation system, the air volume in the drilling fluid will also change. In general, once the foam leaves the drill rods, pressure on the foam is greatest at the bottom of the borehole; therefore, the volume of air at the bit will be the smallest in relation to the liquid component. At the discharge point, the pressure goes to zero (ignoring atmospheric pressure) and the air expands to its maximum volume. Large changes in temperature can also affect air volume, but, because temperature variations affecting the drilling fluid once it is in the borehole are minimal, they can usually be ignored. Examples are presented below to illustrate these points.

The fundamental requirement when drilling with foam is to be able to determine the magnitude of the volumetric changes occurring in a specified volume of foam as it leaves the bit, the most critical point in the circulation system, and rises up the borehole. If the temperature remains constant as the foam rises, and 1 ft³ (0.03 m³) of air is delivered to the bit at 72.5 psi (500 kPa), the air volume at the surface is given by:

$$v_2 = \frac{P_1 v_1}{P_2} = \frac{72.5 \cdot 1}{14.5} = 5 \text{ ft}^3$$

Thus, by decreasing the pressure to 14.5 psi (100 kPa), the volume of air increases to five times its former value. As the air volume increases, the uphole velocity also increases proportionately, neglecting friction and formation losses.

As suggested above, changes in ambient borehole temperature from the bottom of the borehole to the surface are usually insufficient to cause any significant air-volume change. If it is assumed, for example, that the temperature for 1 ft³ (0.03 m³) of air at the bit decreases from 54°F to 45°F (12.2°C to 7.2°C) as it rises from 950 ft (290 m) to ground surface and the pressure is kept constant, the new volume is given by:

$$v_2 = \frac{v_1 T_{a_2}}{T_{a_1}} = \frac{1 \cdot 280.2}{285.2} = 0.98 \text{ ft}^3$$

Thus, by lowering the temperature 9°F (5°C), the volume of air only decreased 2 percent. The general pressure-temperature-volume relationships at various points in an air rotary circulation system are shown in Figure 11.30.

There is one point in the drilling fluid circulation system, however, where temperature changes are quite significant. As air leaves the compressor, it has a temperature of about 220°F (104°C). By the time it reaches the bit, the temperature is approximately 140°F (60°C). When the air leaves the bit, the heat in the air is absorbed almost immediately by the water because the heat capacity of air is so small in relationship to that of water. Thus, the air temperature falls almost instantly to the ambient temperature of the borehole.

These examples illustrate that it is possible to calculate the air volume at any point in the drilling fluid system if the air volume is known at one point for a specific temperature and pressure. It can be assumed that changes in pressure and temperature will have an insignificant effect on water volume in an air-foam system. Changes in pressure will have a significant effect on air volume, however, which is highly compressible and behaves according to Ideal Gas laws. The extent of these changes, and their effect on drilling efficiency, are examined below.

Liquid-Volume Fraction (LVF)

Air-foam systems consist of both air and water:

$$v_f = v_w + v_a \tag{11.6}$$

where v_f is the volume of foam, v_w is the volume of water, and v_a is the volume of free air.

The percentage of the liquid-volume fraction, LVF, is given by:

$$LVF = \frac{v_w}{v_f} \cdot 100 \tag{11.7}$$

And, by substitution from Equation 11.6:

$$LVF = \frac{v_w \cdot 100}{v_w + v_a} \tag{11.8}$$

Because LVF is the most vital factor in controlling the lifting capability of the foam, it is important to develop an idea of how changing pressure conditions in the borehole affect LVF. Experiments have shown that maximum lift is obtained with a 2-percent LVF (Figure 11.31). Thus, once water begins to enter the borehole, the contractor adds enough water and surfactant to the airstream to create a 2-percent LVF based on the free-air volume. While the foam is in the drill rods, the LVF is significantly higher than 2 percent because the air is so compressed. But the moment the air leaves the bit, a large amount of expansion takes place because the pressure in the borehole is much less than the pressure in the drill rods. For dry-air systems when no water is entering the borehole, air expansion will be almost instantaneous as the air leaves the bit. In stiff-foam systems where water is entering the borehole, air expansion occurs much more slowly and may not be complete until the air leaves the borehole. The actual downhole pressure is a function of the weight of the water in the borehole, the weight of the cuttings being carried by the foam, and the weight of the foam. Foam density will ordinarily range from virtually zero to about 16 lb/ft^3 (256 kg/m^3).

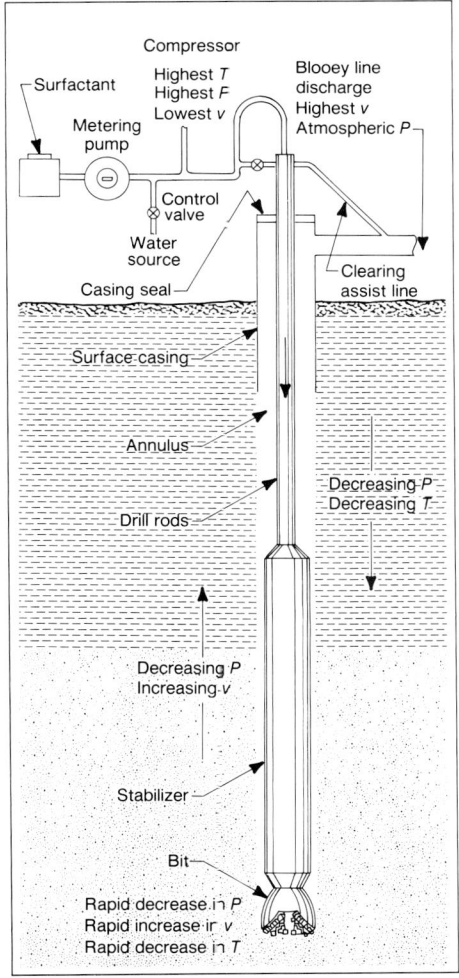

Figure 11.30. Basic components of an operating air rotary circulation system showing the pressure and volume conditions in the drilling fluid at various sites. Greatest pressure and volume changes generally occur at the bit, which is the most critical point in an air drilling-fluid system.

An example will illustrate what changes are occurring in the LVF from the bit to the top of the borehole. Assume that an appropriate volume of surfactant and water will be injected into the airstream to create a 2-percent LVF based on the free-air volume. If an 8-in borehole is 400 ft deep, the average uphole velocity is 200 ft/min, and the volume of water entering the borehole is 100 gpm, the volume of water in the borehole at any instant is 200 gal. Therefore, the downhole pressure attributed to the water is:

$$\frac{200}{7.5} = 26.7 \text{ ft}^3 \cdot 62.4 = 1,666 \text{ lb}$$

The area at the bottom of the borehole is:

$$\pi r^2 = 3.14 \cdot (4)^2 = 50.2 \text{ in}^2$$

Therefore, the pressure exerted at the bottom of the borehole is $1,666/50.2 = 33.2$ psi.

If it is assumed that cuttings in the borehole contribute an additional pressure equal to one-tenth the pressure exerted by the water, the total downhole pressure is equal to $33.2 + 3.3 = 36.5$ psi [gauge pressure (psig) of 22].

The air pressure in the drill rods, which

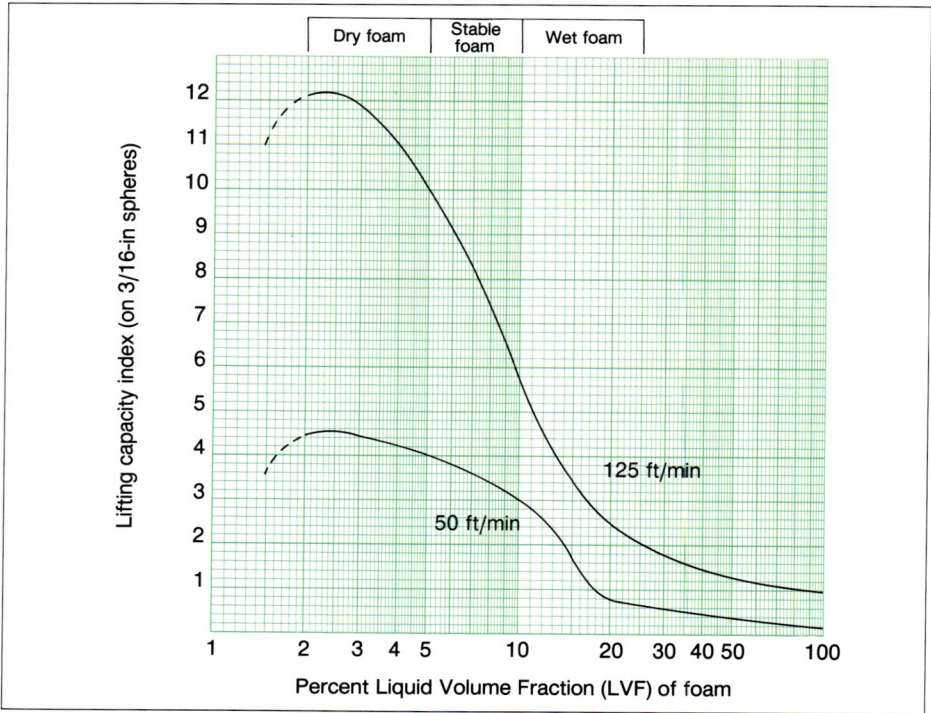

Figure 11.31. The relationship between LVF of a drilling fluid and its lifting capacity on $^3/_{16}$ -in (4.8-mm) spheres is shown on this plot for two uphole velocities. *(DrilChem)*

is about 250 psi (ignoring friction losses), falls almost instantaneously to about 37 psi (22 psig) just outside the bit. The air will then expand to approximately 40 percent of the original volume (Figure 11.32). The actual expansion will be somewhat less than 40 percent because of the accompanying sudden drop in temperature. With expansion to 40 percent, the LVF will be 4.5 percent if no formation water has entered the foam or is being carried by the foam (Figure 11.33). This example shows that, if the injection rate is 2 percent and downhole pressures are reasonable, expansion in the borehole will be sufficient to produce an effective LVF at the bit.

Two changing physical conditions will now affect the LVF of the foam as it rises to the surface. First, the LVF based on the initial volume of water and surfactant added by the driller will decrease to 2 percent by the time the foam reaches the discharge point. This happens because the air expands gradually as it flows up the borehole and regains its original (free-air) volume by the time it reaches the surface. This decrease in LVF is only theoretical, however, because, at the same time, water entering the borehole will increase the LVF by breaking down some of the foam. Therefore, the foam will become less stiff and by the time it leaves the blooey line*

---

*Some drillers install a 6- to 10-in (152- to 254-mm) ID flow line to conduct the cuttings away from the rig. In dry-air drilling, this blooey line reduces dust problems around the rig and, if foam is being used, reduces the accumulation of thick layers of foam in the working area around the well bore. It is important that the blooey line have no bends in dry-air drilling because the dust particles will cut the line. Although blooey lines in oil fields may be 150 to 200 ft (45.7 to 61 m) long, the lines used by water well contractors may be only 20 to 50 ft (6.1 to 18.3 m) long.

it may appear to be quite "runny" in character. This condition indicates that the LVF has risen to 20 to 25 percent because a high percentage of the bubbles produced by the surfactant have been dissipated by the influx water. To avoid an overly runny, inefficient foam, drillers increase the concentration of the surfactant to create the foam stiffness necessary to remove the influx water effectively from the borehole.

Typical Calculations Used in Air-Foam Systems

Pressure and temperature changes affecting the air-foam system are predictable in their impact on the percent LVF, air-volume requirements, and annular velocity. Although most of these relationships can be calculated easily from Equation 11.5, they can also be portrayed by graphs such as those shown in Figures 11.32 and 11.33, which may be somewhat easier to use in the field. Use of these graphs will help the contractor and engineer to visualize what changes are occurring in the drilling fluid system and whether the physical conditions are facilitating or slowing the drilling process.

1. Calculating the injection rate for a 2-percent LVF. To calculate the volume of water needed to produce a specific LVF, multiply the desired LVF by the annular volume of 1 ft of borehole. Assume that an uphole velocity of 100 ft/min will be sufficient to lift the cuttings from the 8-in borehole. The drill rod has a 4-in outside diameter. A 250-psi, 450-cfm compressor is mounted on the rig. Static water level at the drill site is at a depth of 120 ft.

The annular volume between the rods and the borehole wall can be calculated by using the equation for the volume of a cylinder:

$$v = \pi r^2 h$$

Thus:

$$3.14 \left(\frac{4}{12}\right)^2 \cdot 1 - 3.14 \left(\frac{2}{12}\right)^2 \cdot 1 = 0.262 \text{ ft}^3/\text{ft of borehole}$$

The volume of water required per foot of borehole for a 2-percent LVF is $0.262 \times 0.02 = 0.005$ ft^3. Thus, the surfactant and water should be injected initially at

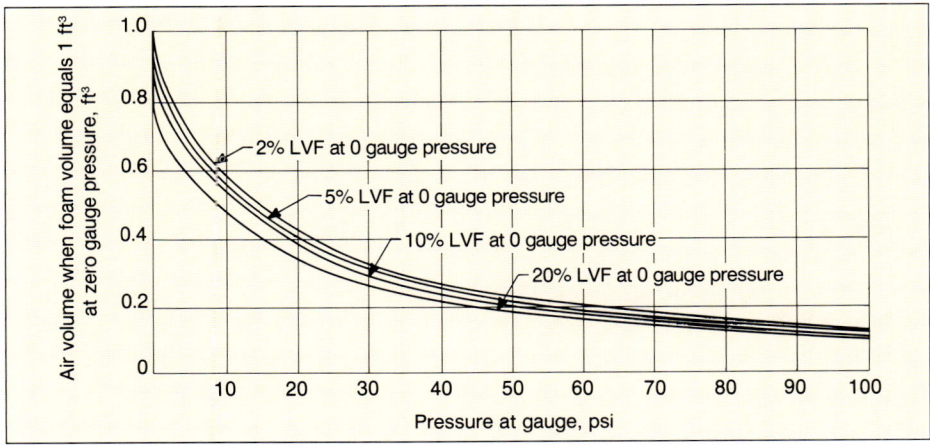

Figure 11.32. This graph shows the relationship between pressure and air volume, where liquid-volume fractions of 2, 5, 10, and 20 percent are given at atmospheric pressure [14.5 psi (100 kPa)]. Note that the higher the initial LVF, the lower will be the relative air volume at a particular pressure.

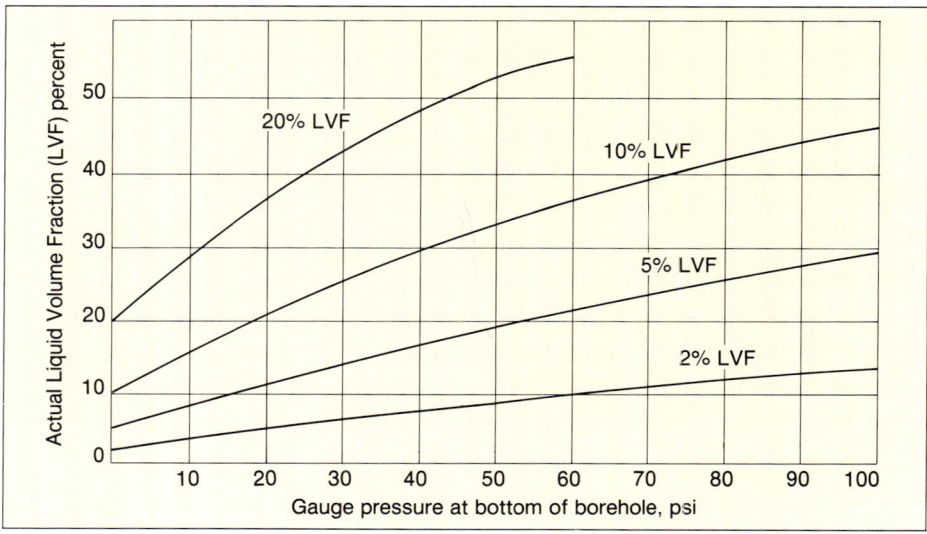

Figure 11.33. To maintain the proper LVF in the drilling fluid, the contractor must be able to estimate how the LVF will change because of pressure fluctuations caused by changes in annular volume, friction losses, pressure in the borehole, and temperature. This figure shows the effect of increasing pressure on LVF. In general, the higher the percent LVF at low-pressure points, the greater will be the chance for the drilling fluid to be inefficient at high-pressure points in the system.

the rate of 0.005 × 100 (annular velocity) = 0.5 cfm = 3.8 gpm. An initial surfactant concentration of 0.5 gal per 100 gal of injection water is selected. As drilling proceeds, the injection rates of both water and surfactant can be adjusted as required.

2. Determining air-volume requirements. For air-foam drilling, most situations require 50 to 500 cfm (0.02 to 0.2 m³/sec) of air at pressures ranging from 100 to 350 psi (690 to 2,410 kPa). The compressor should be able to maintain an uphole velocity of 50 to 1,000 ft/min (15.2 to 305 m/min).

Air-volume requirements can be calculated from the following equation:

$$v_a = \frac{100 \cdot v_w}{LVF} - v_w \qquad (11.9)$$

In the drilling problem discussed above, the water volume required to achieve a 2-percent LVF is 3.8 gpm (0.5 cfm). Thus, using Equation 11.9, the air-volume requirements are:

$$v_a = \frac{100 \cdot 0.5}{2} - 0.5 = 24.5 \text{ cfm}$$

Normally the calculated air-volume requirement should be increased by 25 percent so that losses into the formation will not reduce the annular velocity below the desired rate. The total minimum air volume is then about 30 cfm (0.01 m³/sec), delivered at an assumed pressure of 125 psi* (862 kPa). This volume is much less than the available compressor capacity of 450 cfm (0.21 m³/sec). Therefore, in this example, the air-volume requirements can be met easily.

Air volume also determines annular velocity. Annular velocities will change because of air expansion or, to a lesser extent, compression as the drilling fluid circulates

---

*The driller adjusted the compressor to deliver air at this lower pressure to conserve fuel.

down the inside of the drill pipe and up the annulus. The relationship of air volume, velocity, and cross-sectional area is given by:

$$v_a = VA \qquad (11.10)$$

where v_a is the air volume, V is velocity, and A is the cross-sectional annulus area. Because v_a and V are directly proportional, a change in one caused by pressure variations will cause a corresponding change in the other.

3. Determining pressure requirements. Ordinarily, the pressure requirements for foam systems during drilling are rather low because the fluid column contains so much air. Calculation of the maximum pressure required to operate the drilling fluid system just as the foam leaves the bit will depend on the total weight of the foam, cuttings, water entering the borehole, and any clay or polymeric additives. If drilling ceases for some reason, however, the pressure delivered by the compressor when drilling resumes must be enough to overcome any head of fluid that has accumulated in the borehole. The driller or engineer now must decide whether the original operating pressure and air volume are adequate to restart the drilling process. If drilling ceases in the above example for a few hours after reaching a depth of 480 ft (360 ft beneath the water table), the minimum pressure required to restart the air-foam system is 156 psi (360 divided by 2.31), assuming the borehole is completely filled with water to the static water level and there are no friction losses in the drill pipe. This is higher than the initial operating pressure of 125 psi, but still well within the capability of the compressor, which is 250 psi.

Regulating the Air-Foam Drilling System

In most air-foam drilling operations, the contractor will have an intuitive feeling as to whether the drilling fluid system is functioning properly, mainly on the basis of penetration rate. But just as in water-based drilling fluid systems, adjustments of the physical characteristics of the drilling fluid should be based on more than just intuition if maximum drilling efficiency is to be maintained. Because conditions change as the drilling fluid circulates in the system, the required pressure, air volume, and LVF normally should be established for the most critical point — in the annulus just above the bit. At this point, the LVF should be 2 to 5 percent; in no case should it be more than 15 percent.

Unfortunately, it is rarely possible to actually measure pressure or temperature conditions at many points in the circulation system. Therefore, the driller must observe (usually at the discharge point) various physical characteristics of the air-foam system when possible and learn to relate these characteristics to drilling efficiency. The important visual observations or measurements which indicate when the adjustments should be made are listed below.

1. Water volume, air volume, and pressure.
2. Foam pressure at the standpipe.
3. Percentage of surfactant and other foam stabilizers being injected.
4. Foam consistency at the surface:
    a. Visual assessment for consistency
    b. Density
    c. Percent LVF
    d. Percent and size of solids

e. Volume
5. Regularity of returns at the surface.
6. Drill-string torque requirements

Some of these factors can be measured directly by means of gauges. Other factors such as percent solids, density, and volume, however, must be estimated visually. It is relatively easy for an experienced driller to estimate the LVF by the tendency of the foam to form streams (high LVF) or to billow up around the rig (low LVF) (Figure 11.34).

As a general operating guide for an air-foam system, the driller should attempt to maintain foam consistency once the rates of penetration and water removal are satisfactory. The ability to make the correct adjustments to the system by visual observations and actual measurements will improve as the experience and skill of the contractor increases. When initiating an air-foam system, the driller will generally observe the following (Magcobar, 1979):

1. Before the foam has become stable downhole, a steady rush of air will leave the blooey line. If the foam does not form, fluid (injection) volume should be increased or air volume decreased for a few minutes.

2. When the proper foaming action begins downhole, a gentle puffing of air can be felt at the blooey line.

3. When foam returns begin, the foam should extrude steadily and have good stiffness.

4. Good foam-carrying capacity usually occurs when a gentle surging action is observed at the blooey line. If the foam is too stiff, this surging action will not occur. If the foam surges violently, too much air is being pumped into the borehole.

5. As conditions change, the proper foam concentration, water injection rate, and air volume must be maintained to achieve good penetration (Figure 11.35).

Common problems with foam drilling fluids are indicated by the physical condition of the foam at the surface and by pressure buildup in the borehole. These foam or pressure conditions and suggested adjustments are listed in Table 11.10. Additional

Figure 11.34. The liquid-volume fraction determines how the foam looks as it leaves the borehole. When the foam is relatively dry, it collects around the top of the borehole (left). If the LVF is higher, the foam tends to run away on the surface of the water brought up the borehole. *(Nelson Well Drilling; Test Drilling Service Company)*

practical information on air-drilling problems and procedures is presented in the general guidelines section of this chapter.

In summary, specific suggestions for successful air-foam drilling include the following:

1. The greatest lifting capacity occurs when the LVF is about 2 percent; therefore, the LVF at the bottom of the borehole should be as close to 2 percent as possible. If the LVF exceeds 25 percent, the lifting capacity of the foam will be unsatisfactory. From a practical standpoint, drillers do not generally increase the injected water volume in direct proportion to any increase in air volume requirements. For example, if 3.8 gpm is required to produce an LVF of 2 percent at 30 cfm, the driller will not ordinarily pump 15.2 gpm if the air volume requirements rise to 120 cfm. Limited injection pump capacity and the natural entry of water into the borehole either prevents or makes it unnecessary to increase the injection rate proportionately.

2. The annular velocity at the bottom of the borehole should be at least 50 ft/min (15.2 m/min).

3. To calculate the correct volume of water to be injected for an LVF of 2 percent and an uphole velocity of 100 ft/min, multiply the annular volume for 1 ft of borehole by 2. For example, if the volume of 1 ft of borehole is 0.262 ft³, the injection rate

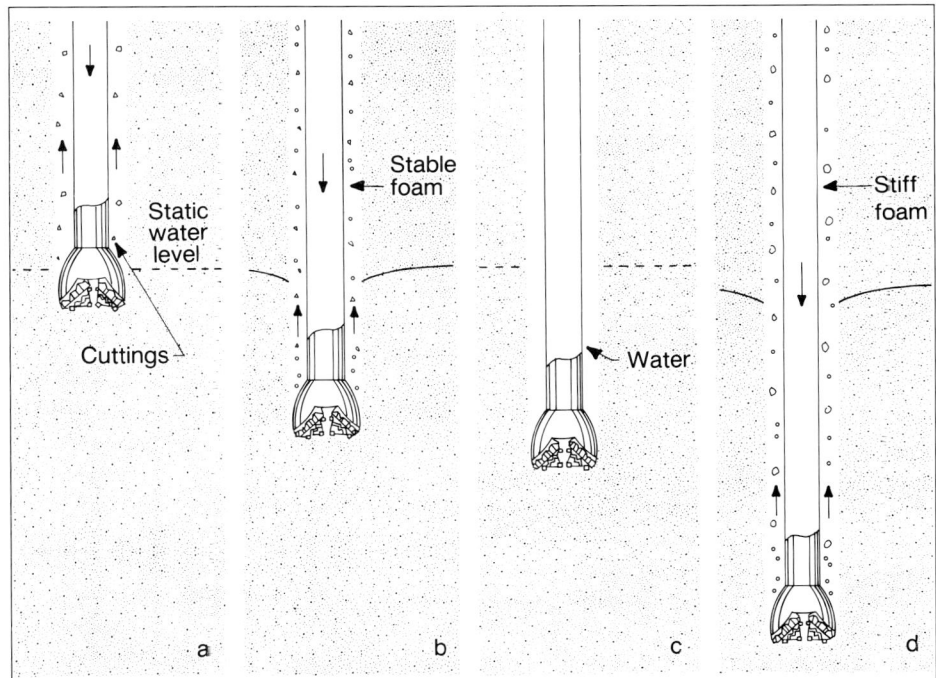

Figure 11.35. Steps in air drilling. (a) Drilling above the water table: start with dry air, and add a small volume of water to the air to control dust. Increase the air volume with depth. (b) Drilling below the water table: add surfactant to injection water to remove water and cuttings. LVF should be maintained between 2 and 10 percent at the bit. (c) Drilling stopped and must be restarted: air pressure must be sufficient to overcome head of water in the borehole. Increase air volume as hole deepens and keep LVF at 2 percent at the bit. (d) Drilling continues: if uphole velocity cannot be maintained by air volume increase alone, the foam can be stiffened by adding clay or polymers and increasing the concentration of surfactant. Increased concentrations of surfactant help maintain the lifting capability.

would be 2 times 0.262 = 0.524 ft³ or 3.9 gpm at an uphole velocity of 100 ft/min. For other annular velocities, multiply the figure for 100 ft/min by the appropriate multiplier. Another good rule of thumb is to inject at a rate of 0.5 to 1 gpm (2.7 to 5.5 m³/day) per 1 in (25.4 mm) of borehole diameter.

4. The air-volume requirements for any borehole can be calculated from Equation 11.9 for a specific LVF. For most shallow water wells, 50 to 500 cfm (0.02 to 0.2 m³/sec) of air are needed at pressures of 100 to 350 psi (690 to 2,410 kPa).

5. A safety factor for calculated air volume is usually 25 percent.

6. A temperature correction for air-volume changes may be necessary, but for most drilling operations the temperature change is not sufficient to necessitate a correction.

Certain chemical and physical conditions in the borehole can adversely affect the efficiency of air-foam systems. Lost circulation, abnormally high borehole pressures, and swelling clays are typical problems. Contaminants such as salt, cement, natural gases, gypsum, and anhydrite reduce the foam's capacity to lift cuttings. Occasionally, chemical treatment of the fluid may be required before injection to correct a contaminant problem. In addition, high borehole or high surface temperatures can cause drastic fluctuations in air volume if the drilling fluid becomes cooled at other points in the circulation system.

Appropriate adjustments to the physical characteristics of an air-foam system require a thorough understanding of what is occurring within the circulation system. Even though measurement of physical conditions in the borehole is normally impossible, a skilled driller will learn to relate visual observations at the surface to what

Table 11.10. Common Problems with Air-Foam Systems

Problem	Reason for Occurrence	Corrective adjustment
Air blowing free at the blooey line with a fine mist of foam	Air has broken through foam mix preventing stable foam formation	Increase liquid injection rate or decrease air injection rate
Foam thin and watery	Formation water entry with possible salts contamination	Increase liquid and air injection rates, and possibly increase percent of foaming agent
Quick pressure drop	Air broken through foam mix preventing formation of stable foam	Increase liquid injection rate or decrease air injection rate
Slow, gradual pressure increase	Increase in amount of cuttings or formation fluid being lifted to surface	Increase air injection rate slightly
Quick pressure increase	Bit plugged or formation packed off around drill pipe	Stop drilling and attempt to regain circulation by moving pipe

(After DrilChem)

DRILLING FLUID ADDITIVES

A proliferation of drilling fluid additives has occurred since 1940. As a brief guide, Appendix 11.G lists the materials potentially required, their commercial names (four companies only), and a brief description of their primary applications. It is beyond the scope of this chapter to discuss the drilling parameters that determine when and how to use these additives. The drilling fluid engineer can contact any of the major drilling fluid companies listed in Appendix 11.H for advice.

For most water well drilling operations, certain standard procedures are followed which depend on the particular type of drilling fluid used. Specific ranges for viscosities, uphole velocities, and additive concentrations are well established and represent the starting point for mixing most drilling fluids (Table 11.11). However, unusual borehole structure or groundwater chemistry may dictate a change from these initial drilling fluid conditions.

Table 11.11. Typical Additive Concentrations, Resulting Viscosities, and Required Uphole Velocities for Major Types of Drilling Fluids Used in Various Aquifer Materials

Base Fluid	Additive/ Concentration	Marsh Funnel Viscosity (seconds)	Annular Uphole Velocity (ft/min)	Observations
Water	None	26 ± 0.5	100 - 120	For normal drilling (sand, silt, and clay).
Water	Clay (High-Grade Bentonite)			Increases viscosity (lifting capacity) of water significantly.
	15-25 lb/100 gal	35 - 55	80 - 120	For normal drilling conditions (sand, silt, and clay).
	25-40 lb/100 gal	55 - 70	80 - 120	For gravel and other coarse-grained, poorly consolidated formations.
	35-45 lb/100 gal	65 - 75	80 - 120	For excessive fluid losses.
Water	Polymer (Natural)			Increases viscosity (lifting capacity) of water significantly.
	4.0 lb/100 gal	35 - 55	80 - 120	For normal drilling conditions (sand, silt, and clay).
	6.1 lb/100 gal	65 - 75	80 - 120	For gravel and other coarse-grained, poorly consolidated formations.
	6.5 lb/100 gal	75 - 85	80 - 120	For excessive fluid losses.
				Cuttings should be removed from the annulus before the pump is shut down, because polymeric drilling fluids have very little gel strength.
Air	None	N/A	3,000 - 5,000	Fast drilling and adequate cleaning of medium to fine cuttings, but may be dust problems at the surface.
			4,500 - 6,000	This range of annular uphole velocities is required for the dual-wall method of drilling.
Air	Water (Air Mist) 0.25-2 gpm	N/A	3,000 - 5,000	Controls dust at the surface and is suitable for formations that have limited entry of water.

Base Fluid	Additive/Concentration	Marsh Funnel Viscosity (seconds)	Annular Uphole Velocity (ft/min)	Observations
Air	Surfactant/Water (Air-Foam)	N/A	50-1,000	Extends the lifting capacity of the compressor.
	1-2 qt/100 gal (0.25-0.5% surfactant)			For light drilling; small water inflow; also for sticky clay, wet sand, fine gravel, hard rock; few drilling problems.
	2-3 qt/100 gal (0.5-0.75% surfactant)			For average drilling conditions; larger diameter, deeper holes; large cuttings; increasing volumes of water inflow; excellent hole cleaning.
	3-4 qt/100 gal (0.75-1% surfactant)			For difficult drilling; deep, large-diameter holes; large, heavy cuttings; sticky and incompetent formations; large water inflows.
				Injection rates of surfactant/water mixture: Unconsolidated formations 3–10 gpm; Fractured rock 3–7 gpm; Solid rock 3–5 gpm
Air	Surfactant/Colloids/Water (Stiff Foam)	N/A	50 - 100	Greatly extends lifting capacity of the compressor.
	3-4 qt/100 gal (0.75-1% surfactant) plus 3-6 lb polymer/100 gal or 30-50 lb bentonite/100 gal			For difficult drilling; deep, large-diameter holes; large, heavy cuttings; sticky and incompetent formations; large water inflows.
	4-8 qt/100 gal (1-2% surfactant) plus 3-6 lb polymer/100 gal or 30-50 lb bentonite/100 gal			For extremely difficult drilling; large, deep holes; lost circulation; incompetent formations; excessive water inflows.

(Compiled partly from information presented in Imco Services, 1975; Magcobar, 1977; and Baroid, 1980.)

GUIDELINES FOR SOLVING SPECIFIC DRILLING FLUID PROBLEMS

The enormous variety of chemical and physical conditions that can exist in boreholes, and the large number of commercial products available to remedy specific problems, preclude a simple prescription for successful use of drilling fluids. Certain problems, however, occur regularly in typical geologic formations when using ordinary additives. Swelling clays, for example, are a major problem when drilling with rotary systems. The Corcoran Clay in the San Joaquin Valley of California, and the Laramie Formation in the Denver Basin, are notable examples. As the clays hydrate and expand, the borehole is partially or completely plugged, and occasionally the drill string may become stuck. The section below contains recommendations for solving clay swelling and other common drilling fluid problems.

PROBLEM: Inadequate cuttings removal from borehole.
RECOMMENDED ACTION:
 1. Clays and polymeric solids in water
   a. Increase uphole velocity of the drilling fluid (Figure 11.36).
   b. Increase viscosity of the drilling fluid by adding more colloidal material.
   c. Increase density of the drilling fluid by adding weighting material (Tables 11.12 and 11.13).
   d. Reduce penetration rate to limit cuttings load.
 2. Air

a. Increase uphole velocity of fluid system by adding air or water.

b. Add surfactant to produce foam or to increase concentration of surfactant.

c. Decrease air injection rate if air is breaking through the foam mix and preventing formation of stable foam.

d. Decrease water content of the foam system.

PROBLEM: The rate at which cuttings will drop out is too low because the inadvertent addition of native clays during drilling has produced excessive viscosity in the drilling fluid.

RECOMMENDED ACTION:

1. Add water to dilute the drilling fluid (Table 11.12).
2. Add commercial thinner to reduce the attractive forces between clay colloids.
3. If using clay additives, convert to a polymeric system.
4. Separate the solids from a clay additive system with shale shakers and desanders connected in series (Figure 11.37), or a shale shaker alone (Figure 11.38). Utilization of a desander or shale shaker may be unnecessary when a polymeric system is being used.
5. Redesign or clean the pit system to increase rate of cuttings settlement.

PROBLEM: Gel strength becomes too great because of strong flocculation, high concentration of solids, or contamination from evaporite deposits or cement. (Excessive gel-strength problems do not occur with polymeric colloids.)

RECOMMENDED ACTION:

1. Add water to dilute the drilling fluid.
2. Add polyphosphate or commercial thinner to reduce electrical charges between clay colloids.
3. Use desander or shale shaker to remove solids from a clay additive system.
4. Lower the pH.

PROBLEM: Excessive fluid loss into the formation, causing thick filter cakes that can produce tight places in the hole, development problems, formation (clay) sloughing, and misinterpretation of electric or gamma-ray logs.

RECOMMENDED ACTION:

1. Increase viscosity by adding bentonite or polymeric colloids to any water-based system.
2. Add commercial viscosifiers such as CMC or HEC.
3. Reduce density of the drilling fluid.
4. Prevent drastic changes in downhole pressures and maintain downhole pressures at a minimum. Suggestions include (Bariod):

a. Raise and lower the drill string slowly.

b. Drill through any tight section;

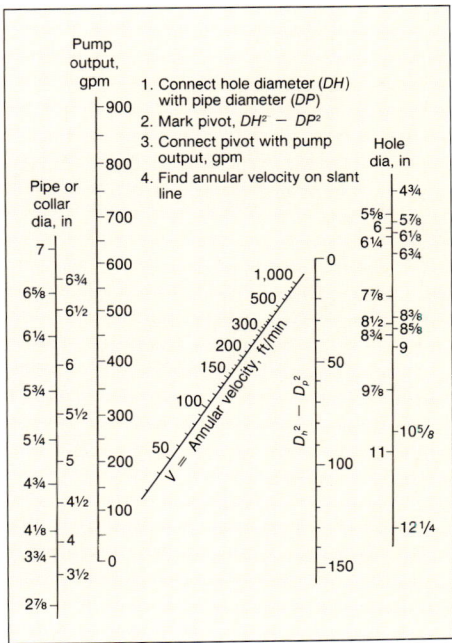

Figure 11.36. This nomogram can be used to calculate uphole velocity in the annulus between the drill rods and the borehole wall.

do not spud.

c. Begin rotation of the drill pipe, and then start the pump at a low rate and gradually increase the rate.

d. Operate the pump at the lowest rate that will assure adequate cooling of the bit and removal of cuttings from the bit face.

e. Prevent balling at the bit; do not drill soft formations so fast that the annulus becomes overloaded and pressure builds up.

PROBLEM: Lost circulation in permeable formations, faulted and jointed rock, solution cavities in dolomite and limestone, or fractures created by excessive borehole pressures in semiconsolidated or well consolidated rock.

RECOMMENDED ACTION:

1. Reduce density of the drilling fluid system.
2. Switch from a clay additive drilling fluid system to an air-foam fluid, or add surfactant to a dry-air system.
3. Gel natural polymeric fluids at the point of fluid loss.
4. Use commercial sealing materials.
5. Drill remainder of the hole with a cable tool rig.
6. Case off, then resume rotary drilling.
7. Fill the borehole with clean sand to the point above lost circulation. Let the material stand in borehole over night. Resume drilling, using low pump pressure.
8. Grout the lost-circulation zone and drill through the plug.

PROBLEM: Confined pressures in the formation.

Figure 11.37. At this drilling site, the drilling fluid passes through two shale shakers mounted in parallel and then connected in series with a desander shown at the right. *(Snider Drilling Co. Ltd.)*

Table 11.12. Drilling Fluid Weight Adjustment with Barite or Water

Initial drilling fluid weight, lb/gal	Desired drilling fluid weight, lb/gal											
	9.5	10.0	10.5	11.0	11.5	12.0	12.5	13.0	13.5	14.0	14.5	15.0
9.0	69	140	214	293	371	457	545	638	733	833	940	1050
9.5		69	143	219	298	381	467	557	650	750	855	964
10.0	43		71	145	221	305	390	479	569	667	769	876
10.5	85	30		74	148	229	312	398	488	583	683	788
11.0	128	60	23		74	152	233	319	407	500	598	700
11.5	171	90	46	19		76	157	240	326	417	512	614
12.0	214	120	69	37	16		79	160	245	333	426	526
12.5	256	150	92	56	32	14		81	162	250	343	438
13.0	299	180	115	75	48	27	12		81	167	257	350
13.5	342	210	138	94	63	41	24	11		83	171	264
14.0	385	240	161	112	78	54	36	21	10		86	176
14.5	427	270	185	131	95	68	48	32	19	9		88
15.0	470	300	208	150	110	82	60	43	29	18	8	

The lower left half of this table shows the number of gallons of water which must be added to 100 gal of drilling fluid to produce desired weight reductions. To use this portion of the table, locate the initial drilling fluid weight in the vertical column at the left, then locate the desired drilling fluid weight in the upper horizontal row. The number of gal of water to be added per 100 gal of drilling fluid is read directly across from the initial weight and directly below the desired weight. For example, to reduce an 11 lb/gal drilling fluid to a 9.5 lb/gal drilling fluid, 128 gal of water must be added for every 100 gal of drilling fluid in the system.

The upper right half of this table shows the number of pounds of barite which must be added to 100 gal of drilling fluid to produce desired weight increases. To use this portion of the table, locate the initial drilling fluid weight in the vertical column to the left, then locate the desired drilling fluid weight in the upper horizontal row. The number of pounds of barite to be added per 100 gal of drilling fluid is read directly across from the initial weight and directly below the desired weight. For example, to raise a 9 lb/gal drilling fluid to 10 lb/gal, 140 lb of barite must be added per 100 gal of drilling fluid in the system.

(After Petroleum Extension Service, 1969)

RECOMMENDED ACTION:

1. Increase density by adding heavy mineral additives such as barite to drilling fluid systems made with clay additives (Table 11.12). To suspend barite, the minimum Marsh funnel viscosity must equal four times the final (desired) drilling fluid weight (in lb/gal). A nomograph is presented in Figure 11.39 for determining the fluid density required to control confined pressures.

2. Increase density by adding a salt solution to polymeric drilling fluid systems (Table 11.13).

PROBLEM: Shale sloughing caused by hydration (swelling and dispersion), pore pressures, and overburden pressure.

RECOMMENDED ACTION:

1. Use polymeric additive to isolate water from shale.
2. Maintain constant fluid pressures in the borehole.
3. Minimize uphole velocities.
4. Avoid pressure surges caused by raising or lowering drill rods rapidly.
5. Add 3 to 4 percent potassium chloride (KCl) to water-based systems.
6. Raise the pH of the drilling fluid to stiffen the clay.

PROBLEM: Presence of contaminants. Contaminants usually consist of cement, soluble salts, and gases (hydrogen sulfide and carbon dioxide). Cement in the hole can cause polymeric drilling fluids to break down, thereby increasing fluid losses. Salts may cause drilling fluids with clay additives to separate into liquid and solid fractions. Gases in water may affect the physical condition of the drilling fluid.

RECOMMENDED ACTION:

1. For cement problems:
   a. Maintain the pH for natural polymeric drilling fluids at 7 or lower.

Figure 11.38. Coarse cuttings collected by the shale shaker are being shoveled away from the shaker outfall. *(Test Drilling Service Inc.)*

b. Add commercial chemicals such as sodium acid pyrophosphate to drilling fluids with clay additives to restore original viscosity.
2. For salt problems:
   a. Change the clay additive from montmorillonite to attapulgite.
   b. Change to a natural polymeric drilling fluid additive.
3. For gas problems:
   a. Add a corrosion inhibitor.

PROBLEM: Drilling at air temperatures significantly below freezing, causing freeze-up of the recirculation system.

RECOMMENDED ACTION:
1. Add sodium chloride (NaCl) or calcium chloride ($CaCl_2$) to a natural polymeric drilling fluid. See Figure 11.40 for the salt concentrations required to prevent freezing at specific temperatures. Salt must not be added to a drilling fluid made with bentonite.

The problems enumerated above are not the only ones that can occur in water well

Table 11.13. Approximate Quantities of Salt Required for Weighting Clean Water or Drilling Fluids Made with Polymeric Additives

Weight		Pressure	NaCl		$CaCl_2$		Multipliers for Final Volumes	
lb/gal	Specific Gravity	psi/100 ft	lb/1,000 gal	kg/m³	lb/1,000 gal	kg/m³	NaCl	$CaCl_2$
8.33	0.966	43.27	—	—	—	—	1.00	1.00
8.5	1.02	44.16	—	—	—	—	—	—
8.7	1.04	45.19	531	63.7	—	—	1.020	—
8.8	1.06	45.71	725	86.9	—	—	1.028	—
9.0	1.08	46.75	1031	123.6	925	110.9	1.041	1.024
9.2	1.10	47.79	1356	162.6	1136	136.2	1.054	1.030
9.4	1.13	48.83	1828	219.2	1587	172.9	1.075	1.040
9.6	1.15	49.87	2106	252.5	1828	219.2	1.090	1.052
9.8	1.18	50.91	2531	303.5	2075	248.8	1.095	1.058
10.0	1.20	51.94	2926	350.8	2363	283.3	1.127	1.070
10.2	1.22	52.99	—	—	2750	329.7	—	1.081
10.4	1.25	54.03	—	—	3104	372.2	—	1.098
10.6	1.27	55.06	—	—	3625	434.6	—	1.109
10.8	1.30	56.10	—	—	3784	453.7	—	1.126
11.0	1.32	57.14	—	—	4174	500	—	1.137
11.2	1.34	58.18	—	—	4486	538	—	1.148
11.4	1.37	59.22	—	—	5156	613	—	1.170
11.65	1.40	60.52	—	—	5554	665	—	1.192

Note: Increasing the fluid density is sometimes required to overcome formation pressures. The addition of a salt will result in solution weighting, as opposed to suspension weighting. Suspension weighting materials are generally used in fluids with high gel-strength properties.

drilling. Many other problems result from inadequate pumping equipment, particle accumulation in the fluid system and at the bottom of the borehole, slumping or expansion of active shales, and caving of resistant shelflike rocks such as limestone and dolomite. Maintenance of borehole stability and careful drilling procedures can help eliminate these problems, but once they occur a solution must be determined rapidly for the specific case.

TYPICAL DRILLING PROBLEM

Drilling efficiency depends largely on the driller's expertise, drilling fluid control, and knowledge of local geology. The example cited below is an actual case history and demonstrates some factors that can influence a drilling operation.

A well contract was awarded to construct a 16-in diameter well to a depth of 350 ft (Figure 11.41). Most wells in the area are small in diameter and yield relatively small quantities of water. The drilling contractor studied two well logs from nearby 6-in wells drilled by the direct rotary method because he was unfamiliar with the particular area. The static water level was at 20 ft.

A direct rotary test hole indicated a confined sand aquifer at a depth of 288 to 325 ft. This aquifer was bounded by dense clay formations at least 30 ft thick.

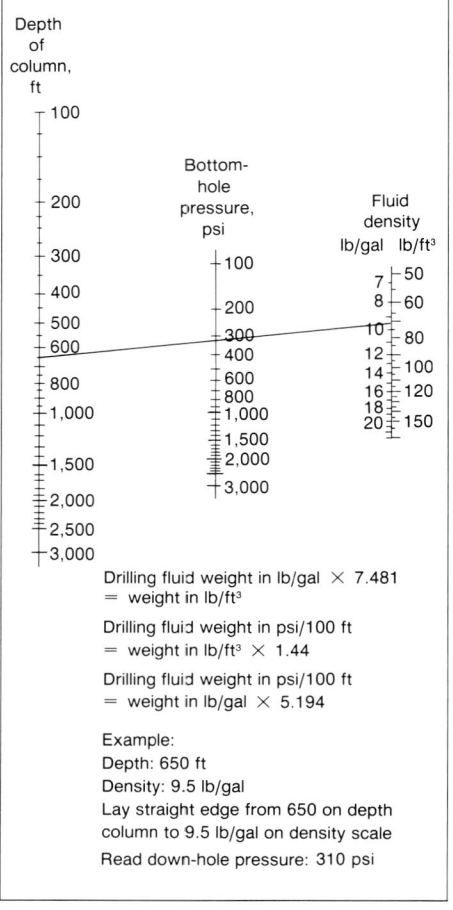

Drilling fluid weight in lb/gal × 7.481
= weight in lb/ft³

Drilling fluid weight in psi/100 ft
= weight in lb/ft³ × 1.44

Drilling fluid weight in psi/100 ft
= weight in lb/gal × 5.194

Example:
Depth: 650 ft
Density: 9.5 lb/gal
Lay straight edge from 650 on depth column to 9.5 lb/gal on density scale
Read down-hole pressure: 310 psi

Figure 11.39. Nomograph for determining the hydrostatic head produced by drilling fluids.

Also, a dense clay formation with interbedded silty to fine-sand layers extended from 14 to 150 ft.

The contractor mobilized his reverse circulation rig at the site in the hilly country. A mud pit 20 by 30 by 8 ft was constructed and drilling began. The presence of the thick clay layers concerned the contractor, who anticipated that they might swell into the borehole (recall that clean water is generally used for a reverse circulation rotary drilling fluid). He planned to use Revert®, a polymeric drilling fluid, to control the clay swelling if it occurred.

The drilling went smoothly until the contractor detected about 1 ft of fill in the bottom of the borehole when he added a drill rod at a depth of 131 ft*. At 141 ft, he had some difficulty cleaning out the borehole so he could add the next drill rod. After a somewhat longer circulation period, however, the driller was able to complete the rod addition and drilled to 151 ft. He was unable to make any rod additions at this

---

*This particular reverse rotary machine was equipped with 10-ft flanged drill stem sections.

depth, and the rods almost became stuck in the hole when, on two occasions, he shut down his pump. During this time, the drilling fluid discharging from the flume into a kitchen-type strainer included only a few lumps of clay, not enough to indicate a major clay-swelling problem. Because this occurred during the night, he decided to trip out and wait until morning to assess the problem. After a few hours, a sounding line indicated that the hole had filled in to 129 ft.

In the morning, the drilling crew determined that no water had been added to the mud pit within the last 12 hours, and the pump used to take water from a nearby river to replenish the pit had run out of fuel. It was now clear that a confined aquifer had been encountered during the night and the drilling fluid density would have to be increased.

To estimate the amount of material needed to weight the drilling fluid, the driller had to determine the shut-in head (confined head). Shutting off the flow pipe between the pit and the borehole caused the fluid to rise in the extended surface pipe to a level 5 ft above grade. The depth of the hole at this time was 122 ft. The fluid weight necessary to control the flow can be calculated by using Equation 11.2:

$$\text{Fluid weight required} = \text{weight of water} \cdot \frac{\text{height of water above top of aquifer}}{\text{depth to top of confined aquifer}}$$

$$= 8.33 \cdot \frac{127}{122}$$

$$= 8.67 \text{ lb/gal}$$

Thus, a drilling fluid density of 8.67 lbs/gal would be required to equal the formation pressures. Some additional weight would have to be added beyond this amount to continue drilling. The driller decided to add salt to increase the fluid weight to 8.8 lbs/gal. Table 11.13 gives the amount of salt needed to raise the drilling fluid weight

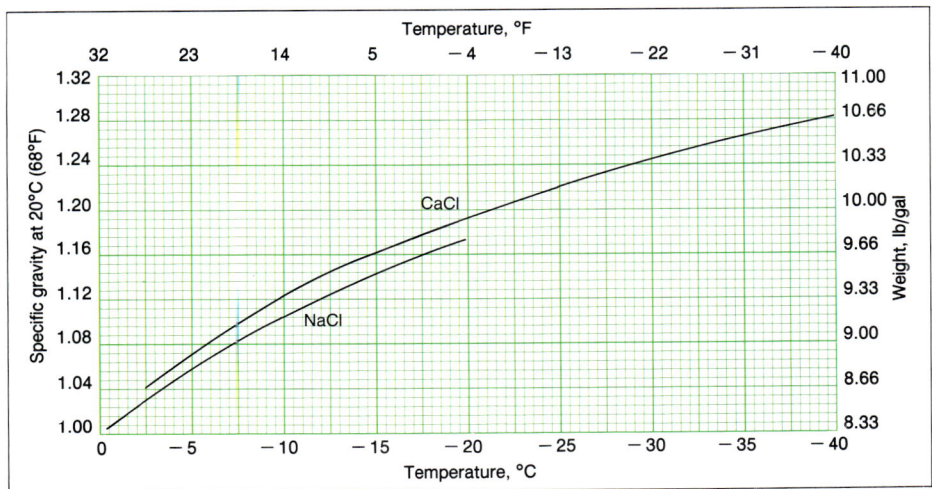

Figure 11.40. "Mud" rotary drilling operations can proceed in temperatures below freezing if either sodium chloride (NaCl) or calcium chloride (CaCl$_2$) are added to the drilling fluid. The specific drilling fluid weights required for subfreezing temperatures are shown. Salts cannot be added to drilling fluids containing bentonite additives without causing a serious reduction in viscosity.

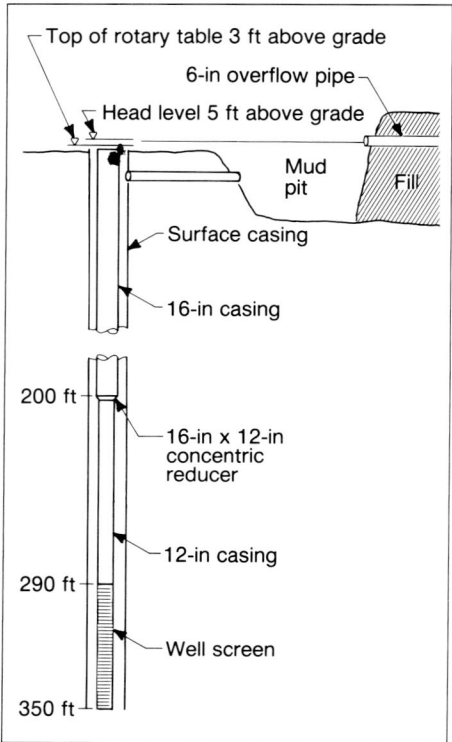

Figure 11.41. Diagram of well to be drilled.

to 8.8 lb/gal. As shown in Table 11.13, 725 lb of salt per 1,000 gal is required to raise the weight of the drilling fluid to 8.8 lb/gal.

Even after increasing the density of the drilling fluid in the mud pit, the borehole continued to flow. Unfortunately, the drilling crew did not realize that the problem was in the borehole, not the mud pit. Salt was then added to the borehole and circulated in a closed system until flow ceased; then the fluid density in the pit was again increased to 8.8 lbs/gal because earlier flow from the borehole had reduced the fluid density. As additional silts and clays were put into suspension after the resumption of drilling, the fluid weight approached 9.0 lb/gal. The hole was completed without further serious complications and produced more than 900 gpm.

In this case, the driller's unfamiliarity with the area and the effect of confined conditions on reverse rotary drilling practices caused delays in hole completion. The driller made the following mistakes:

1. Did not recognize that the hole started to fill when he intercepted the confined conditions (clays overlying sand layers).

2. Was not aware that his pump at the river had stopped pumping.

3. Did not recognize that his pit was taking on water because he was not watching his overflow pipe.

4. Did not take into account that the older, direct rotary, bentonite-drilled wells and test hole were drilled with a higher density fluid that could overcome limited formation pressure, and therefore no flows were noticed.

5. Drilled the reverse rotary hole without giving enough thought to the borehole hydraulics. This was apparent on two occasions: (a) when he had difficulty changing rods, and (b) when he treated the pit before treating the drilling fluid in the borehole, thus allowing his pit to become constantly diluted while trying to increase the pit fluid density.

CONCLUSIONS

This chapter has examined common drilling fluid systems and some typical problems that occur in different geologic materials. Many specific problems associated with drilling can be resolved with the help of the suppliers of drilling fluid products. Many times they can provide in-the-field help, and at least one company sponsors regular training classes in drilling fluid systems. One point should be remembered by both project engineer and drilling contractor — whatever system you use, understand it well.

CHAPTER 12
Well Screens and Methods of Sediment-Size Analysis

A well screen is a filtering device that serves as the intake portion of wells constructed in unconsolidated or semiconsolidated aquifers. A screen permits water to enter the well from the saturated aquifer, prevents sediment from entering the well, and serves structurally to support the unconsolidated aquifer material. The importance of a proper well screen cannot be overemphasized when considering the hydraulic efficiency of a well and the long-term cost to its owner.

Well screens are manufactured from a variety of materials and range from crude hand-made contrivances (Figure 12.1) to highly efficient and long-life models made on machines costing hundreds of thousands of dollars (Figure 12.2). The value of a screen depends on how effectively it contributes to the success of a well. Important screen criteria and functions include:

1. Criteria
   a. Large percentage of open area
   b. Nonclogging slots
   c. Resistant to corrosion
   d. Sufficient column and collapse strength
2. Functions
   a. Easily developed
   b. Minimal incrusting tendency
   c. Low head loss through the screen
   d. Control sand pumping in all types of aquifers

Maximizing each of these criteria in constructing screens is not always possible depending on the actual screen design. For example, the open area of slotted casing cannot exceed 11 to 12 percent or the column strength will be insufficient to support the overlying casing during screen installation. However, open areas of 30 to 50 percent are common for continuous-slot screens with no loss of column strength. In highly corrosive waters, the use of plastic is desirable, but its relatively low strength makes its use impractical for deep wells.

It is clear from earlier chapters in this book that the physical nature of aquifers

 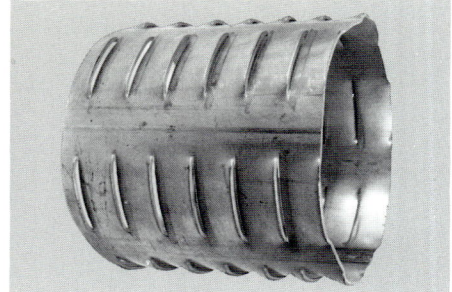

Figure 12.1. Many well screen designs have been developed for use in specific types of aquifers, for use with certain well construction methods, or to meet cost or local fabricating conditions. Because of these differing demands, well screens vary significantly in their hydraulic efficiency, ease of installation, and longevity. Some screen openings are produced by hand cutting and by punching holes or louvers in casing.

and the chemistry of water in the aquifers vary remarkably. The screen design must accommodate these varying physical and chemical characteristics. Extensive experience has shown that screens with the following characteristics provide the best service in most geologic conditions and will satisfy the criteria listed above.

1. Slot openings should be continuous around the circumference of the screen, permitting maximum accessibility to the aquifer so that efficient development is possible.

2. Slot openings should be spaced to provide maximum open area consistent with strength requirements to take advantage of the aquifer hydraulic conductivity.

3. Individual slot openings should be V-shaped and widen inward to reduce clogging of the slots and sized to control sand pumping.

4. Screen construction methods should permit the use of a wide variety of materials that are compatible with differing groundwater environments to minimize corrosion and incrustation.

5. If constructed of metal, screens should be of single-metal construction to minimize galvanic corrosion.

6. Screens must be sufficiently strong to withstand stresses normally encountered during and after installation.

7. A full series of fittings should be available to facilitate screen installation and well completion operations.

Continuous-Slot Screen

The continuous-slot screen is widely used throughout the world for water, oil, and gas wells, and is the dominant screen type used in the water well industry. It is made by winding cold-rolled wire, approximately triangular in cross section, around a circular array of longitudinal rods (Figure 12.2). The wire is attached to the rods by welding, producing rigid

Figure 12.2. Continuous-slot screens are widely used for water, oil, and gas wells. They are constructed by winding cold-rolled, triangular-shaped wire around a circular array of longitudinal rods.

Figure 12.3. Continuous-slot screens are made from stainless steel, low-carbon steel, plastic, and even fiberglass materials. In this photo, a fiberglass screen is being installed.

one-piece units having high strength characteristics at minimum weights. Welded screens are commonly fabricated from Type 304 and Type 316 stainless steel, monel, galvanized or ungalvanized low-carbon steel, and thermoplastic materials, mainly PVC and ABS or alloys of these materials. Other highly specialized materials can be used for waters that are unusually corrosive (Figure 12.3).

Slot openings for continuous-slot screens are manufactured by spacing successive turns of the outer wire to produce the desired slot size. These screens are typically fabricated in slot sizes ranging from 0.006 to 0.250 in (0.15 to 6.4 mm). Width of the openings can be held to close tolerances in the all-welded manufacturing method; allowable variations from the designated (ordered) slot size generally range from 0.001 to 0.002 in (0.03 to 0.05 mm), depending on the screen material and screen size. Most high-quality screen manufacturers are concerned with slot variation because sand pumping problems may occur if too many slots are significantly oversized. Slot control quality is usually checked by comparing the designated size versus the average finished size. The degree of slot variation can be determined by applying a standard deviation analysis (how much the average slot varies from the designated slot). All slots should be clean and free of burrs and cuttings.

Slot openings have been designated by numbers which correspond to the width of the openings in thousandths of an inch. A No. 10 slot, for example, is an opening of 0.010 in. Slot size may also be expressed in metric units; for example, 0.010 in equals 0.25 mm. Both measurement units will be used in this book. For small-diameter screens covered with wire mesh, the number of openings in the mesh per inch are designated by gauze numbers. The size relationship of slot number and gauze number is shown in Figure 12.4.

For continuous-slot screens, individual slot spacing can be varied during fabrication. In fact, a single section of screen may be made with many different slot sizes if geologic conditions require these variations (Figure 12.5). In this way, maximum use of the hydraulic conductivity of each stratum is possible. Principles of slot selection are discussed in Chapter 13.

Each slot opening between adjacent

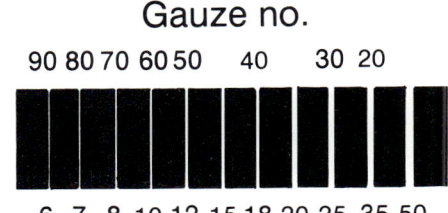

Figure 12.4. A comparison for size of slot openings in inches and gauze numbers is shown for common sizes of screen openings.

wires is V-shaped, resulting from the special shape of the wire used to form the screen surface (Figure 12.6). The V-shaped openings, designed to be nonclogging, are narrowest at the outer face and widen inwardly; they allow only two-point contact by any sand grains with a diameter larger than the slot size. Thus, oversized particles are retained outside the screen and cannot close off the openings. Any sand grain that will pass through the narrow outer part of the V-shaped opening enters the screen without wedging in the slot. In screens with cut slots without V-shaped design, entering particles often turn or twist sideways and, once lodged in the slots, the available intake area of the screen is considerably reduced, causing either lower yield or greater drawdown (Figure 12.7).

Continuous-slot screens provide more intake area per unit area of screen surface than any other type. For any given slot size, this type of screen has maximum open area. Table 12.1 gives representative open areas of various continuous-slot screens with differing outer wire sizes (face widths) and slot sizes. Note that as the wire face width increases for a given slot size, the open area decreases. However, larger wires increase the collapse strength of the screen for any given diameter. Some typical wire profiles are shown in Figure 12.8. Comparisons of

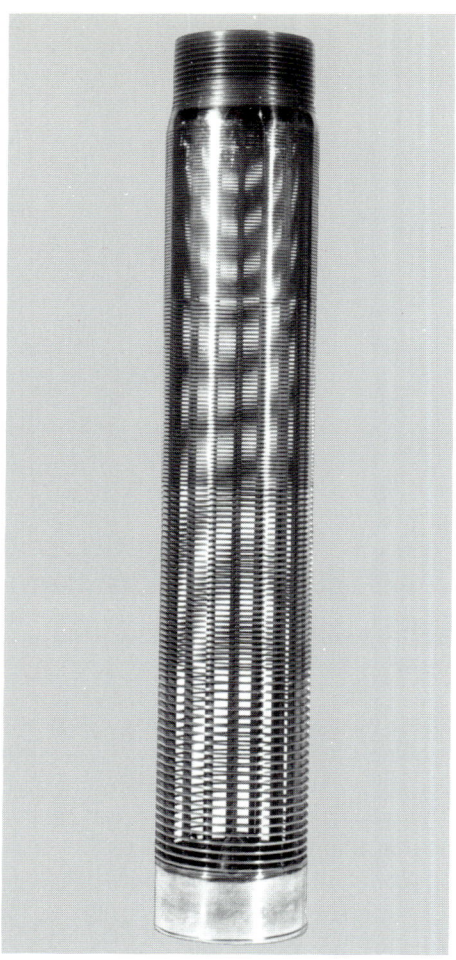

Figure 12.5. Slot openings in continuous-slot screens can be varied to match the size of the aquifer materials.

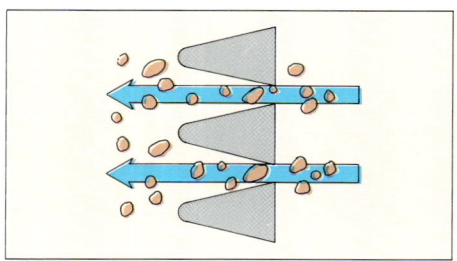

Figure 12.6. Slot openings are V-shaped in continuous-slot screens. The slots are non-clogging because they widen inwardly. Particles passing through the narrow outside opening can enter the screen.

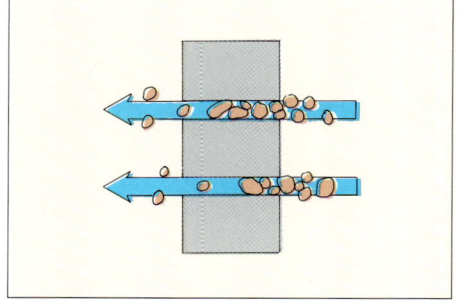

Figure 12.7. Elongate or slightly oversize particles can clog straight-cut, punched, or gauze-type openings.

Table 12.1. Representative Open Areas of Various Continuous-Slot Screens Constructed with Different Wire Shapes

Wire Face Width		Percent Open Area for Indicated Slot Width				
in	mm	0.010 in (0.25 mm)	0.025 in (0.64 mm)	0.050 in (1.27 mm)	0.100 in (2.54 mm)	0.150 in (3.81 mm)
0.047	1.19	17.5	34.7	51.5	68.0	76.1
0.061	1.55	14.1	29.1	45.0	62.1	71.1
0.092	2.34	9.8	21.4	35.2	52.1	62.0
0.110	2.79	8.3	18.5	31.3	47.6	57.7
0.120	3.05	7.7	17.2	29.4	45.5	55.6
0.135	3.43	6.9	15.6	27.0	42.6	52.6
0.156	3.96	6.0	13.8	24.3	39.1	49.0
0.178	4.52	5.3	12.3	21.9	36.0	45.7
0.200	5.08	4.8	11.1	20.0	33.3	42.9
0.215	5.46	4.4	10.4	18.9	31.7	41.1

open areas for other types of screens are given in Appendix 12.A.

For best well efficiency, the percentage of open area in the screen should be the same as, or greater than, the average porosity of the aquifer material. Typical porosities for sandstone and sand and gravel deposits are presented in Table 12.2. Continuous-slot screens often equal or exceed the open area of the natural aquifer material, except where unusually small openings must be used to control fine sand.

Water flows more freely through a screen with a large intake area compared to one with limited open area. The entrance velocity through the larger intake area is low, and therefore the head loss for the screen itself is at a minimum. This, in turn, minimizes drawdown in the well at a given rate of pumping.

Screens with large open areas and low entrance velocities are less subject to incrustation because the pressure drop that occurs in the water as it moves into the screen is minimal. Reduction of the incrusting potential lengthens the useful life of wells. The impact of screen intake area on hydraulic efficiency and screen incrustation is discussed in detail in Chapters 13 and 19, respectively.

Large open area also reduces the ability of a corrosive water to attack the screen openings. Only a small quantity of water will pass through an individual slot as a certain volume of water enters a screen, but the larger the open area, the lower the

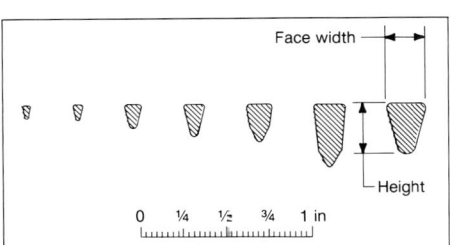

Figure 12.8. Typical wire profiles used in constructing continuous-slot screens. Wire shape is described by its face width and height, although manufacturers usually designate a certain wire profile by its face width only; for example, 156 wire has a face width of 0.156 in (3.96 mm).

Table 12.2. Representative Values of Porosity

Material	Porosity, %
Gravel, coarse	28*
Gravel, medium	32*
Gravel, fine	34*
Sand, coarse	39
Sand, medium	39
Sand, fine	43
Sandstone, fine grained	33
Sandstone, medium grained	37

*These values are for repacked samples, all others are undisturbed.

flow rate through the openings, and thus, the lower the corrosive effect of the water on the walls of the slot. Therefore, the corrosive effect of the water is directly related to the volume of water passing by a metal surface. If the flow rate is minimized, the corrosive attack is also minimized (see Appendix 13.I).

The characteristics of the continuous V-shaped slot openings are vital to successful development and completion of a screened well. Any development method depends on having the smaller sizes of sand and silt pass through the screen openings, which must be nonclogging and closely spaced. Development is most effective when the screen openings are evenly spaced around the circumference of the screen, the open area is as large as possible, and the configuration of the slot openings allows the development energy to reach into the formation. Development methods are described in Chapter 15.

The total cost of well operation also depends in part on the total open area of a screen. In most geologic formations, drawdown is a function of open area — the lower the open area, the greater the drawdown for a certain yield. Lifting water to the surface is usually the largest cost factor in well operation. In general, continuous-slot screens have the largest open areas and thus the lowest drawdown during pumping. Although screens with low open area may be less expensive to purchase, the long-term maintenance and operational costs are usually much higher than are these costs for higher priced continous-slot screens. Therefore, higher initial costs are offset by lower operational costs. Usually the higher cost of continuous-slot, high open area screens can be recouped in one to three years, depending on the volume of water pumped. Thereafter, the more efficient screen can save the high-capacity well owner thousands of dollars in energy costs each year. Over the life of a typical well, the savings may amount to $50,000 or more.

Screen Diameter

Continuous-slot well screens of welded construction are available in two series of

Table 12.3. Telescope-Size Johnson Well Screens

Nominal Casing Size		Screen Outside Diameter		Screen Inside Diameter		Pipe-Size Threaded Fittings		
in	mm	in	mm	in	mm	in		mm
3	76	2¾	70	2	51	2	M or F	51
4	102	3¾	95	3	76	3	M or F	76
5	127	4¾	121	4	102	4	M or F	102
6	152	5⅝	143	4⅞	124	5	M or F	127
8	203	7½	191	6⅝	168	6	M or F	152*
10	254	9½	241	8⅝	219	8	M or F	203*
12	305	11¼	286	10⅜	264	10	M or F	254*
14	356	12½	318	11⅜	289	12	M or F	305*
16	406	14¼	362	13⅛	333			
18	457	16¼	413	15	381	*Special threads are used when connecting multiple screen sections to maintain ID dimensions.		
20	508	18¼	464	17	432			
24	610	22⅝	575	20¾	527			
30	762	27⅞	708	26	660			
36	914	31⅞	810	30	762			

Table 12.4. Pipe-Size Johnson Well Screens

Screen Size		Inside Diameter		Outside Diameter		OD of Female Threaded End	
in	mm	in	mm	in	mm	in	mm
2	51	2	51	2⅝	67	2¾	70
3	76	3	76	3⅝	92	3¾	95
4	102	4	102	4⅝	117	4¾	121
5	127	5	127	5⅝	143	5¾	146
6	152	6	152	6⅝	168	7	178
8	203	8	203	8⅝	219	9¼	235
10	254	10	254	10¾	273	11⅜	289
12	305	12	305	12¾	324	13⅜	340
14	356	13⅛	333	14	356		
16	406	15	381	16	406		
20	508	18¾	476	20	508		
24	610	22¾	578	24	610		
30	762	28¾	730	30	762		

diameters: telescope-size and pipe-size screens. Telescope-size screens are designed to be placed in wells by "telescoping" them through the well casing. The diameters of telescope-size screens allow them to be lowered freely through the corresponding size of standard pipe which serves as well casing. The screens in this series are designated by the nominal* diameter of the pipe into which they will telescope. A 4-in (102-mm) telescope-size screen, for example, is actually 3¾ inches (95.3 mm) in outside diameter, which permits just enough clearance for it to pass through a 4-in standard pipe. Table 12.3 gives the screen dimensions and certain other data for selected telescope-size well screens. Lowering the well screen into place through the well casing is a common method of installation, because it eliminates any possibility of borehole caving. A plate is usually welded or threaded to the bottom of the screen. A bail hook can be mounted on the upper surface of the bottom plate to facilitate lowering the screen into the well. Threaded bottoms facilitate attachment of the hook.

Pipe-size screens have the same inside diameter as the corresponding sizes of standard pipe. Pipe-size screens are used when the well design specifies that the screen be attached directly to the well casing to maintain the same diameter for the full depth of the well. These screens are usually supplied with pipe-size welding rings at each end which can be welded to the corresponding size of pipe. Threaded end fittings are also available, but threaded connections are seldom used on sizes larger than 12 in (305 mm) ID. An end plate is generally threaded or welded to the bottom weld ring of the screen. Table 12.4 shows the dimensions and other details for the pipe-size series of well screens.

Other Types of Well Screens

Several other types of well screens exist. Some of these are manufactured, whereas others are hand perforated from casing or other materials. Under certain conditions, one or more of these screens may be adequate in some geologic formations, but may provide only marginal success under many other hydrogeologic conditions. Limited

*See Chapter 13, page 418.

Figure 12.9. Bridge-slot and louvered screens are generally installed in filter-packed wells because sand particles from the natural formation can clog screen openings.

open area, poor slot configuration, and short-lived screen material contribute to their limited success. Nevertheless, the use of these screens in certain parts of the world requires that a description of each of the major types be presented.

Louvered and Bridge-Slot Screens

Louvered and bridge-slot screens have openings that are arranged in rows (Figure 12.9). The openings, which are oriented either at right angles or parallel to the screen axis, are punched in the wall of a welded tube or are punched in a flat sheet which is then rolled into a tube. A power-punch working against a die that limits the extent to which the metal is forced outward determines the slot size or width of the openings. The number of slot sizes that can be made by each manufacturer depends upon the series of die sets used. A complete series of slot openings is impractical for any one manufacturer.

The shape of the louver or bridge-slot openings does not permit their use in naturally developed wells, because the openings become blocked during the development process if the aquifer material contains an appreciable amount of sand. Therefore, use of these screens is confined almost entirely to filter-packed wells.

The percentage of open area in the louvered and bridge-slot screen is limited because sizable blank spaces must be left between openings. Screens are commonly made in 5- to 20-ft (1.5- to 6.1-m) lengths that can be welded together to make longer sections. Threaded connections are provided only in special cases. Most louvered and bridge-slot screens are made of mild steel, although both types can be made of stainless steel.

Pipe-base Screens

Pipe-base screens are often specified in oil field work because of their strength and also because they can be retrieved from great depths. In some parts of the world, this type of screen is used for water wells. The pipe-base screen is made by using a perforated pipe as a structural core, or base, with a continuous-slot screen mounted on the outside of the pipe base. The outer screen surface is made by winding a trapezoid-shaped wire directly onto the pipe (called the wrapped-on-pipe screen) or winding the wire over a series of longitudinal rods spaced around the circumference of the pipe. This latter construction is more efficient than the wrapped-on-pipe screen because the rods hold the wire away from the pipe surface so that fewer of the openings in the pipe base are blocked. An even better design is a separate unit of welded continuous-slot well screen slipped over the pipe. Such a screen jacket is stronger and less likely to be torn (Figure 12.10).

The pipe-base screen has two sets of openings. The outer openings are the spaces between turns of the wire; the inner openings are the holes drilled or cut in the pipe base. The total area of holes in the pipe is generally less than the area of the slot openings formed by the outer wire. Thus, the hydraulic performance of the screen depends primarily upon the percentage of open area in the pipe base, although the use of a narrow wrapping wire can enhance the overall efficiency. Unfortunately, the efficiency of pipe-base screens is usually low.

Pipe-base screens are commonly fabricated from steel pipe, with an outer screen of stainless steel. A stainless steel screen in contact with ordinary steel pipe always results in some electrolytic (galvanic) action, causing corrosion of the steel pipe. Therefore, pipe-base screens will undergo electrolytic corrosion because of the two-metal construction. One way to prevent this electrolytic action is to use pipe which is the same metal as the screen. The cost may be somewhat higher, but the product is more durable.

Slotted Metal Pipe

Pipe with slots, produced by one of several means, is used in water wells as a screening device (Figure 12.11). Slot openings may be cut with a saw or oxy-acetylene (cutting) torch or punched with a chisel-and-die casing perforator. The casing perforator is a tool used in the field to produce openings in well casing after it is set in the aquifer (Figure 10.10, page 277). Important limitations of perforated pipe are: (1) openings cannot be closely spaced; (2) percentage of open area is low; (3) size of slot openings varies significantly; and (4) openings small enough to control fine or medium sand are difficult or impossible to produce.

Slotted steel pipe is not corrosion resistant and most methods of perforation tend to hasten corrosive attack on the metal when the water is aggressive. Jagged edges and slot surfaces are susceptible to selective corrosion. In general, the use of slotted steel pipe will increase maintenance costs and may significantly reduce the life of the well.

Figure 12.10. A pipe-base screen is constructed by wrapping wire around, or placing a continuous slot screen over, a perforated pipe base. This exceptionally strong construction is often specified for oil wells and occasionally for deep water wells. Both steel and plastic materials are used in this type of construction.

Figure 12.11. Torch-cut or perforated casing can be used for water wells, but the percentage of open area is limited if adequate casing strength is to be maintained.

Slotted Plastic Pipe

Slotted plastic pipe (Figure 12.12) is also used to screen wells in some areas, particularly in clay-rich sediments (for example, clayey tills) where no aquifer zone can be identified. Slotted plastic screens are not affected by corrosive water, are easy to install, and are relatively inexpensive. In cold climates, some plastic materials must be handled with care to avoid breakage. The same limitations that apply to slotted metal pipe also apply to plastic pipe. For example, slotted plastic screens have less than half the open area of continuous-slot plastic screens. In addition, plastic pipe materials are from one-sixth to one-tenth as strong as stainless steel well screens.

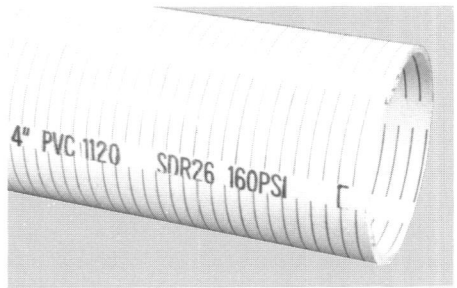

Figure 12.12. The slotted-plastic screen shown above has about half the open area of continuous-slot plastic screens.

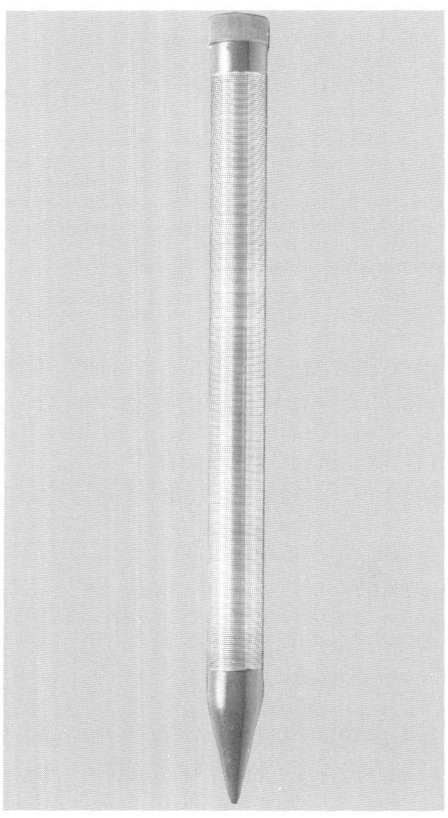

Figure 12.13. Continuous-slot well points are built by attaching a drive point to the lower end of the screen and a threaded pipe shank to the upper end. Driven well points are suitable for shallow domestic wells or for dewatering purposes.

Well Points

Well points are made in a variety of types and sizes. The welded continuous-slot screen is made as a well point by attaching a forged-steel point to the lower end of a screen and a threaded pipe shank to the upper end (Figure 12.13). This type of construction is the most efficient hydraulically. The most common sizes are designed for direct attachment to either 1¼-in (32-mm) or 2-in (51-mm) pipe. Continuous-slot well points are constructed of either low-carbon steel or stainless steel. The size of openings are designated by numbers that correspond to the actual width of openings in thousandths of an inch or the metric equivalent. Although these units can withstand hard driving, they should not be twisted while being driven or used in areas where boulders or large stones are expected unless special installation methods are used.

Another type of well point consists of a perforated brass or stainless steel jacket covering a perforated pipe, with an intervening layer of wire mesh. The perforations in the steel pipe core, less the obstruction of the mesh and outer jacket, constitute the effective intake area for this type of well point. The forged-steel-point bottom has a widened shoulder designed to push gravel or stones aside and reduce the danger of ripping or puncturing the jacket as the well point is being driven into the ground. The relatively low open area of this type of well point may cause high rates of incrustation in hard waters and the bimetal construction often leads to excessive galvanic corrosion in waters with a low pH.

Still another type of well point design is a galvanized pipe with half-moon-shaped perforations (Figure 12.14). A layer of stainless steel mesh is wrapped on a plastic pipe insert which is slipped inside the galvanized pipe. Although the intake area of the screen is not large, the wire mesh is reasonably well protected from rocks and stones during driving. The size of the screen openings is designated by the mesh number, which is the number of openings per linear inch. Common sizes are 40, 50, 60, 70, and 80 mesh.

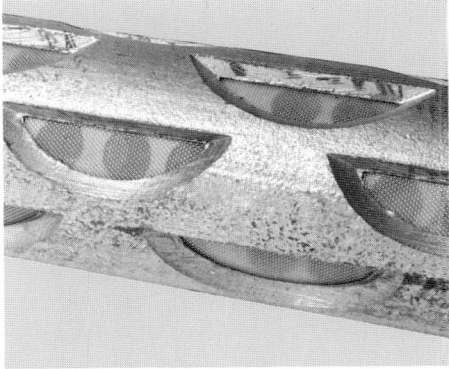

Figure 12.14. Some well points are constructed so that the wire gauze screening material is protected by an outer casing so it will not be damaged during driving. The open area of this type of well point is limited compared to continuous-slot screens.

Optimum Well Screen Open Area

The desirable percentage of open area in a well screen should at least equal the porosity of the water-bearing sand or filter pack. Assume the sand has 30 percent porosity and the well screen installed in the sand has 10 percent open area. The difference causes a constriction of flow as water enters the well. This means more drawdown for a given pumping rate, because additional head loss occurs as water passes through the screen openings. Thus, screened wells perform best when the intake area of the screen is as great as possible for a particular slot opening and strength requirement.

For wells completed in fine-sand formations, most well screens cannot provide a percentage of open area equivalent to the sediment porosity. The size of openings required to control the fine sand is often so small that even the best continuous-slot screen will fall somewhat short of matching the porosity. Porosities for well-sorted sands vary from about 20 to 40 percent. If a No. 10 (0.010 in; 0.25 mm) slot screen is needed, the most efficient screen construction (continuous-slot with narrowest wire) would provide a little more than 18 percent open area. If the slot is increased to No. 20 (0.020 in; 0.51 mm), the open area is 30 percent. However, many inefficient screens have less than 10 percent open area. Slotted pipe, for example, may have as little as 2 percent open area. Thus, the continuous-slot screen is the only type that can approximate the natural porosity of well-sorted aquifer material for the full range of screen slot sizes.

SEDIMENT SIZE ANALYSIS

Selection of slot size is a critical step in assuring maximum well performance. The slot size of the screen is based on a size analysis of the formation samples. By analyzing the component sizes of the grains in the sample, a grain-size distribution curve can

Figure 12.15. The grain-size distribution of the cuttings can be determined by sieving the sample in a special vibration machine.

be drawn. Several methods can be used to obtain information on the grain-size distribution. The most widely used method involves passing the materials through a stacked set of 8-in (203-mm) brass or stainless steel sieves which are shaken in a special vibration machine (Figure 12.15). Smaller diameter sieves [3 in (76 mm)] can be effectively shaken by hand. During the sieving process, each sieve filters out a certain percentage of the entire sample; the finest material collects in the bottom pan. Plotting of these percentages (weights) of the whole sample provides an insight into the physical makeup of the sample (Figure 12.16).

Other methods to determine the grain-size distribution include sedimentation analysis using velocity settling tubes for sediments smaller than 0.003 in (74 microns) and automatic particle size analyzers with computer printouts of grain-size distribution data by x-y plotters. The expense of the automated devices limits their use to only the most sophisticated laboratories. Computer analysis of the data is relatively inexpensive, however, and standard software is available for many micro (personal) computers. Regardless of which method is used, it is important to establish consistent methods for testing sediment samples. Because sieving is the most common method used to determine the grain-size distribution, much of the discussion which follows will focus on the correct procedures to use when sieving.

Sieve analysis not only provides the basis for determining the slot size, but also other factors affecting screened well design. These points are discussed in detail in Chapter 13. In this section, procedures used to characterize the physical makeup of the sediment will be explained.

Conducting a Sieve Analysis

The testing equipment required for the sieving method includes a small hotplate for drying samples, a set of standard testing sieves, and an accurate balance or scale for weighing sample material. Eight-in (203-mm) diameter wire-mesh screen

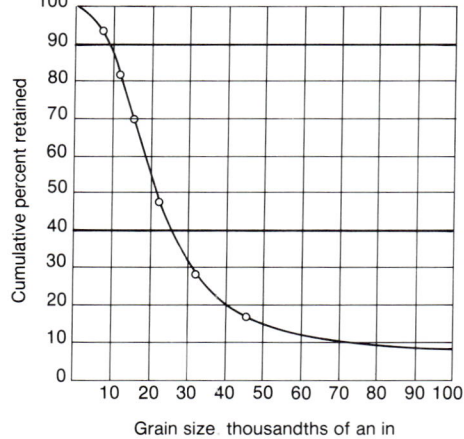

Figure 12.16. Plotting the percent of the sample retained on each sieve provides a graphic illustration of the grain-size distribution.

sieves are normally used, although other sizes are available. The best type of balance weighs in grams and is sensitive to about one gram.

All samples must be dry before the sieve analysis is made. If the sample is wet, dry it over a low heat, stirring frequently. As the material dries, any clay that may be present causes the sand particles to stick together. Any clods that form must be broken down to separate all the particles. If the sample is too large for the sieves [more than 1 cup (224 g)], it is reduced by using a sample splitter or the quartering method. If the material is fine sand, only ½ cup (112 g) should be used so the fine-mesh sieves will not be overloaded. An erroneously coarse analysis will result if the sieves are overloaded. This can lead to a screen design that permits sand pumping.

If a sample splitter is used, the sample is poured into the top of the splitter and automatically divided into two equally representative parts. One of the halves can then be split again, and so on, until the proper size sample is obtained. The reduced sample must contain the same proportions of particle sizes as found in the original sample.

In the quartering method, a well-mixed sample is heaped in a pile on a flat surface and then flattened (Figure 12.17). The flattened pile should be circular; it is then divided into quarters. The sample is reduced to half its original size by removing two opposite quarters and mixing the remaining portions together. If the sample is still too large, the quartering operation may be repeated. Do not attempt to prepare a sample of a certain weight.

Select five to eight sieves with a series of openings that will separate the sample into various grain sizes. The coarsest sieve should not retain more than 20 percent of the sample. Suggested groups of sieves are shown in Figure 12.18. Sieve openings are designated by size in thousandths of an inch, millimeters, and by mesh number of the wire cloth.

Stack the sieves with the finest one resting on the bottom pan and the coarsest at the top. Weigh the dried sample, record this weight, and pour the sediment onto the top sieve. Shake the whole set of sieves with a circular motion and some up-and-down movement, accompanied by jarring to keep the material moving on each sieve and prevent clogging. If possible, samples should be shaken mechanically for at least 5 minutes.

Empty the sample material retained on the top sieve into a pan or onto a large sheet of paper. Then transfer this material to the weighing pan on the balance and weigh it. Record this weight and the size of sieve opening on which the material was retained. Dislodge any particles caught in the sieve, but be careful to avoid damaging the wire mesh.

Add the material retained on the second sieve to that already in the weighing pan on the balance. Record the combined weight. Empty each sieve successively and record the weight of the accumulated sample in each case. Finally, add the fin-

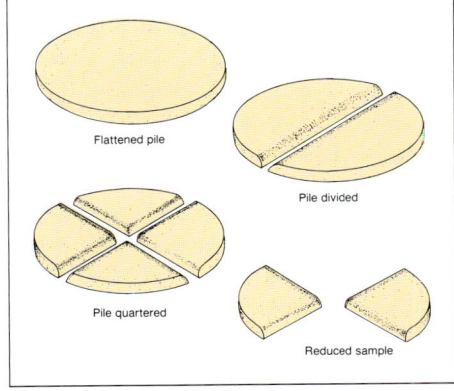

Figure 12.17. The quartering method can be used to obtain a representative fraction from a large sample so that the sieves will not be overloaded.

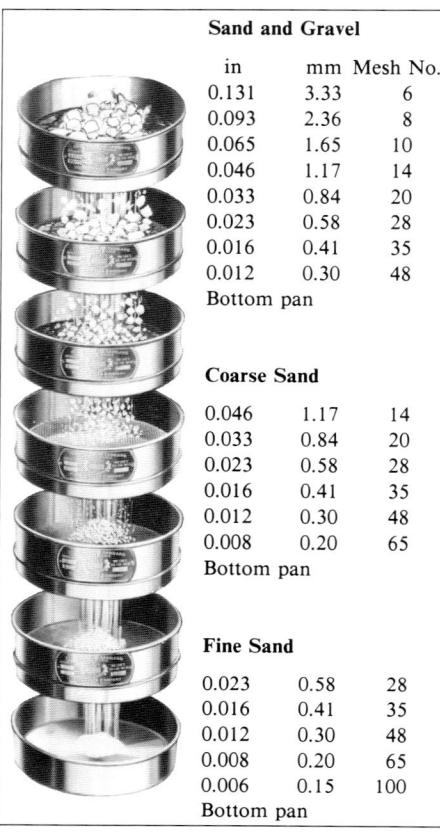

Sand and Gravel		
in	mm	Mesh No.
0.131	3.33	6
0.093	2.36	8
0.065	1.65	10
0.046	1.17	14
0.033	0.84	20
0.023	0.58	28
0.016	0.41	35
0.012	0.30	48
Bottom pan		

Coarse Sand		
0.046	1.17	14
0.033	0.84	20
0.023	0.58	28
0.016	0.41	35
0.012	0.30	48
0.008	0.20	65
Bottom pan		

Fine Sand		
0.023	0.58	28
0.016	0.41	35
0.012	0.30	48
0.008	0.20	65
0.006	0.15	100
Bottom pan		

Figure 12.18. Recommended sieve groups suitable for sieving various classes of unconsolidated sediments.

est material from the bottom pan and weigh. This accumulated weight should equal the weight of the original sample within two or three grams. A sample record of the accumulated weight of material from a series of sieves is shown below. The cumulative percent retained is calculated by dividing the cumulative weight retained by the total weight of the sample (Table 12.5).

These data are then ready for plotting on arithmetic graph paper. The cumulative percent retained on each test sieve is plotted as a point against the sieve opening in thousandths of an inch or mm. Percent retained is given on the vertical scale of the graph and the size of opening, or particle size, is on the horizontal scale. The size of the sieve opening is considered to be the diameter of the smallest particle retained by each sieve. While this is not strictly true because particle shape varies, it is standard practice to plot the size of sieve openings as the grain size on the graph. These plotted points are then connected with a smooth curve, as shown in Figure 12.16.

Under field conditions, all the equipment required to do a sieve analysis may not be available, but a procedure utilizing 3-in (76-mm) sieves, a dried sample, and a graduated 100-mℓ glass cylinder can give good results. A 100-mℓ representative sample is measured and then sieved by hand. The volume of material contained on each sieve is then poured back into the cylinder

Table 12.5. Cumulative Percent Retained

Size of Sieve Opening		Cumulative Weight Retained		Cumulative Percent Retained
in	mm	oz	g	
0.046	1.17	1.0	28.4	17
0.033	0.84	1.6	45.4	28
0.023	0.58	2.6	73.7	47
0.016	0.41	3.9	110.6	70
0.012	0.30	4.6	131.4	82
0.008	0.20	5.3	150.2	94
Bottom pan		5.6	158.8	

Original weight — 5.6 oz (158.8 g)

and the cumulative amount of material measured. These volumes can then be plotted in a manner similar to the procedure discussed above. It is assumed all the material has essentially the same density. This method is not as accurate as a laboratory sieve analysis or sedimentation procedure.

Grain-Size Distribution Curve

The grain-size distribution curve shows at a glance how much of the sample material is smaller or larger than a given particle size. For example, the curve in Figure 12.16 shows that 90 percent of the sample consists of sand grains larger than 0.009 in (0.23 mm) and 10 percent is smaller than this size. Reading the curve in another way, the 40-percent sand size is 0.026 in (0.66 mm); or 40 percent of the sample is coarser than 0.026 in and 60 percent is finer than 0.026 in.

Grain-size distribution curves have many applications other than in the water well industry. They are used to represent the grading of concrete sand, foundry sand, earth materials for embankments and dams, filter sands, and many other types of granular materials. Engineers in these different fields use several variations in plotting the curves. Attention is called to this fact because there may be occasion to use grain-size distribution curves that are plotted differently than those discussed here.

In the most common variation, percent of material passing a given sieve is plotted on the vertical scale instead of percent retained. This has the effect of reversing the curve so it slopes upward from left to right instead of downward. However, plotting percent retained on the vertical scale is the logical procedure because this permits using the cumulative weights in the same manner that they are recorded in the laboratory. To permit plotting percent passing, on the other hand, the percent retained must be subtracted from 100.

A second variation is the use of a logarithmic scale for the particle size or sieve opening. This has the effect of elongating the part of the curve that represents the finer fraction, and squeezing that part of the curve representing the coarser material.

No single term or word can be used to give an overall description of a sand or sand and gravel mixture, because the material consists of the whole range of particle sizes. Between the limits of the smallest and largest particle sizes, the intermediate sizes can be distributed in many different ways, and each distribution changes the shape of the curve.

There are three elements essential to a complete description of a grain-size distribution curve: (1) sediment size (fineness or coarseness); (2) slope of the curve; and (3) shape of the curve. Any of these elements can change independently of the others, and this makes it necessary to use all three for a complete description of the material grading.

Sediment Size

In describing the fineness or coarseness of a granular material, the terms fine sand, coarse sand, fine gravel, and other similar terms are used. Unfortunately, these terms do not apply to specific particle sizes, which results in various scientific and engineering specialties using different terms for sediments of the same size. Therefore, several different grain-size classifications have been developed to define each descriptive term. Each of these systems has been adopted in the special field where it seems to fit the best.

Table 12.6. Grain-Size Classification

Wentworth Classification	Size Range
Boulder	10.08 in & above (256 mm & above)
Cobble	2.52 to 10.08 in (64 to 256 mm)
Pebble*	0.16 to 2.52 in (4 to 64 mm)
Granule (very fine gravel)	0.08 to 0.16 in (2 to 4 mm)
Very coarse sand	0.04 to 0.08 in (1 to 2 mm)
Coarse sand	0.02 to 0.04 in (0.5 to 1 mm)
Medium sand	0.01 to 0.02 in (0.25 to 0.5 mm)
Fine sand	0.005 to 0.01 in (0.125 to 0.25 mm)
Very fine sand	0.002 to 0.005 in (0.063 to 0.125 mm)
Silt	0.0002 to 0.002 in (0.004 to 0.063 mm)
Clay	Below 0.0002 in (Below 0.004 mm)

*The USGS has subdivided this category as follows:

Very coarse gravel	1.26 to 2.52 in (32 to 64 mm)
Coarse gravel	0.63 to 1.26 in (16 to 32 mm)
Medium gravel	0.31 to 0.63 in (8 to 16 mm)
Fine gravel	0.16 to 0.31 in (4 to 8 mm)

(Krumbein & Pettijohn, 1938)

The Wentworth scale, developed in 1922, is still the basic particle size classification used in the groundwater field. The United States Geological Survey (USGS) uses this classification but has taken one size range, 0.16 to 2.5 in (4 to 64 mm), and subdivided it into groups. The Wentworth scale and USGS amendments are shown in Table 12.6.

The curve in Figure 12.16 shows that the sample tested consists of medium and coarse sand, according to the USGS classification. Applying the same system to the four curves in Figures 12.19 to 12.22 gives the following descriptions:

Class A curve — fine sand
Class B curve — fine and very coarse sand
Class C curve — coarse and very coarse sand
Class D curve — coarse sand and very fine gravel

Other Ways to Describe Sediment Size

Often, a specific point on a grain-size distribution curve is used as a general index of fineness. An attempt is then made to correlate this size with the hydraulic conductivity of the sediment. Several selected sizes have been used by various researchers.

The term "effective size" was developed by Allen Hazen in his studies of filter sand in 1893. He defined it as the particle size where 10 percent of the sand is finer and 90 percent is coarser. On all the curves shown here, the effective size is the 90-percent retained size. The effective size of the Class A curve in Figure 12.19 is 0.003 in (0.08 mm). In Figure 12.21, the effective size of the sand is 0.010 in (0.25 mm).

Another curve point often used as an index of fineness is the 50-percent size, which for the curve in Figure 12.16 is 0.022 in (0.56 mm). For the Class A and Class B curves, the 50-percent size is 0.007 inch (0.18 mm) in both cases. The 50-percent size may be referred to as the mean or average particle size for uniform (steep slope) sediments. However, when the general slope of the curve is flatter, such as the Class D curve in Figure 12.22, the 50-percent size is inaccurate as an indicator of fineness or coarseness.

WELL SCREENS AND METHODS OF SEDIMENT-SIZE ANALYSIS

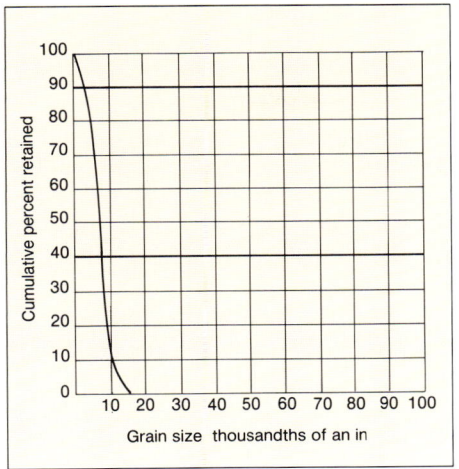

Figure 12.19. Class A curve for fine sand.

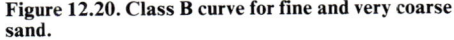

Figure 12.20. Class B curve for fine and very coarse sand.

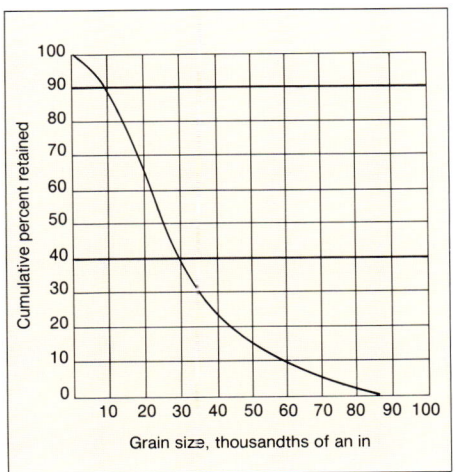

Figure 12.21. Class C curve for coarse and very coarse sand.

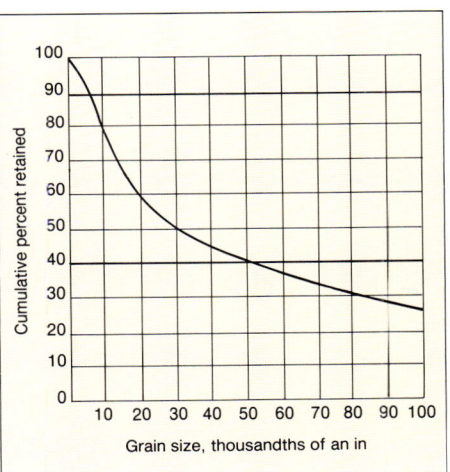

Figure 12.22. Class D curve for coarse sand and very fine gravel.

Slope and Shape of Curve

The slope of the major portion of a grain-size distribution curve can be described in several ways. One term that is used extensively is the uniformity coefficient, which was developed by Hazen at the same time he adopted the idea of effective size. Uniformity coefficient is defined as the 40-percent retained size of the sediment divided by the 90-percent retained size. The lower its value, the more uniform is the grading of the sample between these limits. Larger values represent less uniform grading. The uniformity coefficient is limited in practical application to materials that are rather uniformly graded. It is meaningful only when its value is less than 5. It is well suited for describing the desired uniformity of filter-pack materials. The uniformity coefficient for the sample in Figure 12.16 is 2.9 [0.026 in (0.66 mm) divided by 0.009 in (0.23 mm)]. For the Class B curve, the uniformity coefficient is 2; for the Class C curve, its value is 3.

The grain-size distribution curves for most granular materials deposited by running water and wave action are referred to as S-shaped curves, although this term is properly applied only to the percent-passing curves. The S-shape of the curve becomes distorted when gravel constitutes 15 percent or more of a mixture of sand and gravel. The curve in Figure 12.16 and the Class A and Class C curves are typical S-shaped distributions. The Class D curve has a "tail" of coarse material. Size distributions that result in S-shaped curves usually represent samples having higher porosities than are found in samples with a "tail"-type configuration.

There is yet no accurate way to calculate hydraulic conductivity directly from the grain-size distribution curve. Many tests and research studies have been performed to find a simple relationship between the grading of a sediment and its hydraulic conductivity, but no dependable correlation that may be applied generally has yet been discovered. Nevertheless, with practical experience, it is possible to estimate the relative yields of different sand and sand and gravel mixtures by careful consideration of the three factors described in this section.

CHAPTER 13
Water Well Design

Well design is the process of specifying the physical materials and dimensions for a well. The principal objectives of good design should insure the following:
- The highest yield with minimum drawdown consistent with aquifer capability
- Good quality water with proper protection from contamination
- Water that remains sand free
- A well that has a long life (25 years or more)
- Reasonable short-term and long-term costs

Although well design may appear to be a straightforward procedure, local hydrogeologic conditions and practical considerations complicate many well designs. The design guidelines presented below focus mainly on the design of municipal, industrial, and irrigation wells. These wells must be designed to obtain the highest yield available from the aquifer, achieve high efficiency, and produce water that is free of sediment. These factors influence operating costs directly. Another important cost factor is the economic loss that may result from interruption of service in large water-supply developments. Good design minimizes this danger by building into water wells the features that will assure long and trouble-free life. Special design factors that relate to domestic, farm, and small commercial wells are discussed separately in this chapter. Design of monitoring wells is covered in Chapter 21.

The well-design engineer must use reason in determining requirements for a well. It is obviously not good engineering practice to design a 300-gpm (1,640-m^3/day) well to serve a suburban home when 15 gpm (81.8 m^3/day) will satisfy the owner's needs. It is equally poor engineering to use well casing and well screen of inadequate size, or to choose materials of inferior quality, merely to cut initial costs. This would only burden the owner with higher pumping and maintenance costs, and reduce the useful life of the well. Any additional investments for a properly designed, efficient well will usually reduce operation and maintenance costs and more than pay for themselves during the life of the well.

Important hydrogeologic information required for the design of efficient high-capacity wells includes:
1. Stratigraphic information concerning the aquifer and overlying sediments

2. Transmissivity and storage coefficient values for the aquifer
3. Current and long-term water balance conditions in the aquifer
4. Grain-size analyses of unconsolidated aquifer materials and identification of rock or mineral types if necessary
5. Water quality

Dimensional factors, strength requirements, and costs associated with well construction and maintenance also play a part in establishing the particular design parameters.

Before starting a well design project, the engineer should study the design, construction, and maintenance of other wells in the area. Additional information available to the well-design engineer includes well records maintained by federal and state agencies, local municipalities, agricultural associations, drilling contractors, and some well screen manufacturers.

Every well consists of two main elements, the casing and the intake portion. The casing serves as a housing for the pumping equipment and as a vertical conduit for water flowing upward from the aquifer to the pump intake. Some of the borehole length serving as a conduit may be left uncased when the well is constructed in consolidated rock. The intake portion of wells in unconsolidated and semiconsolidated aquifers is generally screened to prevent sediment from entering with the water and to serve as a structural retainer to support the loose formation material. At the same time, the screen must not obstruct the flow of water into the well. The design of the screen requires careful consideration of the hydraulic factors that influence well performance.

In a consolidated rock aquifer, the intake portion of the well may consist of the open borehole drilled into the aquifer. Some consolidated rock aquifers, however, such as sandstone, may deteriorate over time because high flow rates remove cement that holds sand grains together, thus causing a slow collapse of the borehole wall. In other cases, certain minerals may weather in the borehole. For example, the feldspar crystals in granitic rock disintegrate under aerating conditions. Therefore, screens are often used to protect pumps from loosened formation particles, and to stabilize the aquifer materials in many consolidated formations, especially sandstone, limestone, and some granites.

Standard design procedures involve choosing the casing diameter and material, estimating well depth, selecting the length, diameter, and material for the screen, determining the screen slot size, and choosing the completion method. In addition, the choice of a particular well design hinges on the type of drilling rigs that are available. See Chapter 10 for a description of the major well drilling methods. Design criteria presented below have been developed for typical hydrogeologic conditions. Design practices may vary in different regions, however, because of unusual hydrogeologic conditions; some successful, nonstandard designs are described at the end of this chapter.

CASING DIAMETER

Choosing the proper casing diameter for the well is important because it may significantly affect the cost of the structure, depending on the type of drilling equipment used. The diameter must be chosen to satisfy two requirements: (1) the casing must be large enough to accommodate the pump, with enough clearance for installation and efficient operation, and (2) the diameter of the casing must be sufficient to assure

WATER WELL DESIGN

Table 13.1. Recommended Well Diameters for Various Pumping Rates*

Anticipated Well Yield		Nominal Size of Pump Bowls		Optimum Size of Well Casing†		Smallest Size of Well Casing†	
gpm	m³/day	in	mm	in	mm	in	mm
Less than 100	Less than 545	4	102	6 ID	152 ID	5 ID	127 ID
75 to 175	409 to 954	5	127	8 ID	203 ID	6 ID	152 ID
150 to 350	818 to 1,910	6	152	10 ID	254 ID	8 ID	203 ID
300 to 700	1,640 to 3,820	8	203	12 ID	305 ID	10 ID	254 ID
500 to 1,000	2,730 to 5,450	10	254	14 OD	356 OD	12 ID	305 ID
800 to 1,800	4,360 to 9,810	12	305	16 OD	406 OD	14 OD	356 OD
1,200 to 3,000	6,540 to 16,400	14	356	20 OD	508 OD	16 OD	406 OD
2,000 to 3,800	10,900 to 20,700	16	406	24 OD	610 OD	20 OD	508 OD
3,000 to 6,000	16,400 to 32,700	20	508	30 OD	762 OD	24 OD	610 OD

*For specific pump information, the well-design engineer should contact a pump supplier, providing the anticipated yield, the head conditions, and the required pump efficiency.
†The size of the well casing is based on the outer diameter of the bowls for vertical turbine pumps, and on the diameter of either the pump bowls or the motor for submersible pumps.

that the uphole velocity is 5 ft/sec (1.5 m/sec) or less.

The size of the pump required for the desired yield is the controlling factor in choosing the size of the casing. It is recommended that the casing diameter be two pipe sizes larger than the nominal diameter of the pump. In all cases, however, the casing must be at least one nominal size larger than the pump bowls. Table 13.1 shows casing sizes recommended for various pumping rates. Excessive head losses will occur in the system if the uphole velocity is greater than 5 ft/sec (1.5 m/sec). For the pipe sizes and pumping rates shown in the tables, these head losses will be small.

If the casing size is selected from Table 13.1 and if the well meets typical standards for plumbness, there will be adequate clearance for the pump bowls. For lineshaft pump installations, this clearance allows for proper alignment of the shafting to eliminate binding and excessive wear. If the pump is set below any screened section, there will be sufficient area around the bowls to allow water to pass downward with minimum head loss to the pump intake. However, heat build-up can be a problem for a submersible pump set in a sump beneath the screen, because the intake portion of the pump is located above the motor. The pump manufacturer should be consulted for motor cooling recommendations.

Drilling conditions, drilling methods, or economic factors sometimes make it necessary to complete the lower portion of the well with a smaller diameter casing or screen. When using the cable tool method, drillers must often reduce the size of their casing when the original casing cannot be driven any farther because of side-wall friction (Figure 13.1). A single string of casing can usually be driven 300 to 500 ft (91.5 to 152 m) depending on geologic conditions. The outer casing must be cleaned out completely before smaller diameter casing can be telescoped. Ideally, each casing string should be landed in clay or some other non-heaving sediment. If the casing ends in sand, water should be run continuously into the annulus between the two strings to prevent heaving and potential sand locking. Inner strings are generally set by pull-down jacks, rather than by driving, because too much of the driving energy is lost in the unsupported part of the casing, which is up inside the outer casing.

More than one inner casing can be telescoped depending on well depth. A common diameter sequence is 24 by 20 by 16 in (610 by 508 by 406 mm) for the outer casing and two inner strings. Unless the

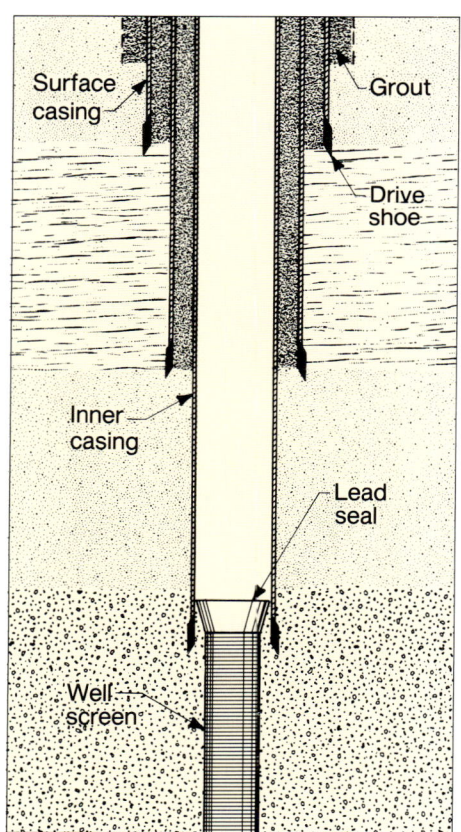

Figure 13.1. A deep well constructed by the cable tool method using successively smaller diameter casing at greater depth. In some installations, the inner and working casings are cut off, after allowing for sufficient overlap.

inner casings must be cut off to accommodate the pump or pump bowls or the value of the casing makes retrieval worthwhile, it is a good idea to leave all casing strings in the hole. Leaving the casing strings in the hole eliminates the need for much of the grouting, the risk of leaks between casing strings, and the chance that inner casings will be damaged during cutoff. If casings are cut off, 20 to 50 ft (6.1 to 15.2 m) of overlap is recommended, depending on the stability of the formation. In stable formations, the overlap may be as short as 20 ft, whereas in unconsolidated sand and gravel formations the overlap may be as much as 50 ft. Grout must then be placed in the annulus between casings if the material will heave or if the two casings are set in materials that have different static water levels. If the casings are landed in clay, no grout may be necessary. Grouting should be done only after the casing strings have been cut off. However, development of the well should always take place before the casings are cut off. Annular space requirements may also be dictated by regulation. For example, grouting regulations usually require at least a 2-in (51-mm) annular space.

In the design process, considerable thought must be given to conditions that will affect the driving of the casing. The design engineer should consult with local cable tool contractors before specifying casing depths or diameters. The casing diameters for individual segments can then be chosen from the bottom of the well upward to accommodate the driving conditions, the overlap requirements, and the annular space necessary for grouting or filter packing.

In deep wells that have both high static and high pumping water levels, the casing diameter can be reduced at a depth below the lowest anticipated pump setting to reduce material costs. This is done in many wells completed in confined aquifers where the pressure is relatively high. Uphole velocity in any smaller diameter casing beneath the pump bowls should be 5 ft/sec (1.5 m/sec) or less. Ordinarily, this requirement is met for the recommended casing sizes shown in Table 13.1. Table 13.2 lists maximum discharge rates for various casing sizes that produce only moderate friction losses.

Table 13.2. Maximum Discharge Rates for Certain Diameters of Standard-Weight Casing, Based on an Uphole Velocity of 5 ft/sec (1.5 m/sec)

| Casing Size | | Maximum Discharge | |
in	mm*	gpm	m³/day
4	102	200	1,090
5	127	310	1,690
6	152	450	2,450
8	203	780	4,250
10	254	1,230	6,700
12	305	1,760	9,590
14	337	2,150	11,700
16	387	2,850	15,500
18	438	3,640	19,800
20	489	4,540	24,700
24	591	6,620	36,100

*Actual inside diameter

CASING MATERIALS

Selection of casing material is based on water quality, well depth, cost, borehole diameter, drilling procedure, and federal, state, and local regulations. The types of casing used in water well construction are steel, thermoplastic, fiberglass, concrete, and asbestos cement (transite). Steel is used most commonly, but thermoplastic materials are gaining a larger share of the water well casing market, especially in areas where groundwater is highly corrosive and wells are less than 1,000 ft (305 m) deep. Table 13.3 lists some common physical characteristics of the different casing types.

Steel Casing

Tubular steel products are variously designated casing, pipe, and tubing. In the water well industry, the terms "pipe" and "casing" are used interchangeably. The number of terms used to describe sizes and other characteristics of these tubular products has grown as the number of applications has increased. In some cases, the terms are difficult to define categorically. Many tubular products are made for specific purposes, and some types are named for their particular application.

Pipe is fabricated either by rolling lengths of skelp (flat sheets) or by hot-working solid steel (welded and seamless). (See Appendix 13.A for construction details.) It can be butt-welded along a hot longitudinal joint by mechanically exerting pressure by means of pass-welding rolls, or electric-resistance welded by joining the two edges using heat and pressure. Seamless pipe is constructed as a tube from hot metal, but may be finished by cold-working to produce the required finished dimensions.

Size Designation for Pipe

The outside dimension for pipe ⅛ in to 12 in (3.2 mm to 305 mm) was originally selected so that, for a standard wall thickness, the inside diameter was approximately equal to a standard size. Depending on the wall thickness, the inside diameter may be less than or greater than the number indicated. Thus, the term "nominal" is used in designating the inside diameter, because the actual size varies somewhat above or below the standard size. For example, 6-in (152-mm) nominal pipe has an outside diameter of 6.625 in (168 mm); for standard wall thickness, the inside diameter is 6.065 in (154 mm).

Pipe larger than 12 in (305 mm) is designated by its outside diameter and is called large OD pipe. The actual inside diameter is controlled by the wall thickness and is significantly less than the indicated outside diameter. For example, the inside diameter of 16-in (406-mm) Schedule 40 pipe is 15 in (381 mm).

To meet certain pressure requirements, pipe is manufactured in three general weight classes: standard, extra heavy, and double extra heavy. The American National Standards Institute (ANSI) has assigned schedule numbers to classify wall thicknesses for different pressure applications. In nominal pipe sizes ⅛ in through 10 in (3.2 mm through 254 mm), ANSI Schedule 40 is the same as standard pipe. Schedule 80 is the same as extra-heavy pipe in nominal sizes ⅛ through 8 in (3.2 mm through 203 mm). In larger sizes, Schedules 40 and 80 may vary greatly from the specifications for standard and extra-heavy pipe. There is no schedule number that correlates with double-extra-heavy pipe.

The American National Standards Institute published criteria for manufacturing various types of casing in 1935. Since then, pipe fabricators have selected certain of

Table 13.3. Comparison of Well Casing Materials

Material	Specific Gravity	Tensile Strength psi	Tensile Modulus 10^5 psi	Impact Strength ft-lb/in	Upper Temperature Limits, °F	Thermal Expansion 10^{-6} in/in °F	Heat Transfer Btu-in/ hr-ft² °F	Water Absorption wt %/24 hrs
ABS	1.04	4,500	3.0	6.0	180	5.5	1.35	0.30
PVC	1.40	8,000	4.1	1.0	150	3.0	1.10	0.05
Styrene Rubber	1.06	3,800	3.2	0.8	140	6.8	0.80	0.15
Fiberglass Epoxy	1.89	16,750	23.0	20.0	300	8.5	2.30	0.20
Asbestos Cement	1.85	3,000	30.0	1.0	250	1.7	0.56	2.0
Low-Carbon Steel	7.85	35,000 (yield)* 60,000 (ultimate)	300.0	†	800–1,000	6.6	333.0	Nil
Type 304 Stainless Steel	8.0	30,000 (yield)* 80,000 (ultimate)	290.0	†	800–1,000	10.1	96.0	Nil

*Yield strength is the tensile stress required to produce a total elongation of 0.5 percent of the gauge length as determined by an extensiometer. Expressed in psi.
†Because testing methods for steel and other materials are not the same and the results are not comparable, the impact strength values for steel are not shown. In any event, the actual impact strength of steel is so high relative to the demands of water well work that it can be ignored in design considerations.

(Purdin, 1980)

these design criteria as a basis for the pipe they manufacture. Ordinarily, a pipe fabricator chooses pipe specifications that match the needs of its individual customers. Thus, no single pipe manufacturer builds all or nearly all of the tubular products listed by ANSI. To complicate matters further, pipe manufacturers have developed their own products with specifications that only approximately follow the design guidelines established by ANSI. Because of these two developments, it is difficult to present a simple analysis of the various types of casing used in the water well industry.

Pipe used for constructing water wells is built to the standards developed for two major pipe markets: (1) plumbing, heating, air conditioning, water, gas lines, and other process systems (ASTM standards), and (2) oil and gas industries (API standards). Table 13.4 gives specific standards for pipe that is manufactured for these two markets and used also in the water well industry. The two major pipe categories are called standard pipe under ASTM designations and line pipe under API specifications.

At the present time, most water well drillers in North America use pipe constructed to ASTM standards rather than to API standards. In oil and gas producing areas, however, pipe used for water wells is generally constructed to API standards. The various ASTM standards for steel pipe cover the following parameters: type of steel used and its chemical composition, tensile and bending requirements for the pipe dimensions, and pipe weights, end finish, galvanizing, and quality control. The latter includes permissible variations in the weight and dimensions of the pipe and in the composition of the steel, and it specifies the performance of the flattening, hydrostatic, longitudinal tension, and transverse weld tests (Handbook of Steel Pipe, 1979).

Appendix 13.B contains data on all sizes of standard and line pipe commonly used for water well casing. Standard wall thicknesses for these pipe classes have been established by the American National Standards Institute (ANSI). Table 13.5 shows, as stated above, that no direct correlation exists between the schedule numbers and the weight classes of standard, extra-heavy, and double-extra-heavy construction. In general, ASTM standard steel pipe is recommended for most drilling situations and typical water quality conditions. Heavier wall casing should be specified for exceptionally corrosive water, for deep wells, and for difficult drilling when using the cable tool method. Difficult drilling is characterized by high-density formation materials and deep boreholes having a large diameter. In every case, the wall thickness must be sufficient to withstand full hydraulic loading if the casing is pumped dry (Table 13.5). The collapse strength must exceed 1 psi (6.9 kPa) for every 2.31 ft (0.7 m) of depth beneath the top of the aquifer.

Standard steel casing used for water wells should conform to ASTM Designation A-53 or A-120, or API Standard Specification 5A or 5L. ASTM specification A-53 for welded and seamless steel pipe covers black and hot-dipped zinc-coated (galvanized) steel pipe in sizes of ⅛ in to 26 in (3.2 mm to 660 mm). Galvanized pipe is rarely used in the water well industry. ASTM A-120 also covers black and hot-dipped zinc-coated (galvanized) welded and seamless steel pipe for ordinary uses in pipe sizes of ⅛ in to 16 in (406 mm).

API Standard Specification 5L includes threaded and coupled line pipe of ⅛-in to 20-in (3.2-mm to 508-mm) OD and plain-end line pipe of ⅛-in to 48-in (3.2-mm to 1,220-mm) OD. API Standard Specification 5A covers threaded and coupled casing of 4.5-in to 20-in (114-mm to 508-mm) OD and tubing of 1.05-in to 4.5-in (26.7-mm to 114-mm) OD. Casing larger than 6 in (152 mm) is rarely threaded and coupled

Table 13.4. Development of Pipe Classification
BASIC PIPE CRITERIA

American National Standards Institute (ANSI)

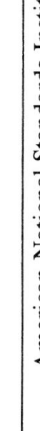

Specific Standards Adopted by the American Society for Testing and Materials (ASTM)

Pipe Designations*

A-53 — Standard specification for steel pipe — covers black and hot-dipped zinc-coated (galvanized) steel pipe in nominal sizes from ⅛ in to 26 in (3.2 mm to 660 mm).

A-120 — Standard specification for black and hot-dipped zinc-coated (galvanized) steel pipe in nominal sizes from ⅛ in to 16 in (3.2 mm to 406 mm).

Use

Standard pipe is used for steam, water, air, and gas lines, and plumbing and heating applications.

Specific Standards Adopted by the American Petroleum Institute (API)

Pipe Specifications

5L — Standard specification includes threaded and coupled line pipe of ⅛ in to 20 in (3.2 mm to 508 mm) OD and plain-end pipe of ⅛ in to 48 in (3.2 mm to 1,220 mm) OD.

5LX — Similar to 5L except built for high-test applications (greater yield and tensile strengths).

5A — Standard specification includes threaded and coupled casing of 4.5-in to 20-in (114-mm to 508-mm) OD and tubing of 1.05-in to 4.5-in (26.7-mm to 114-mm) OD.

Use

Line pipe is used for the conveyance of oil and gas.

Methods of Manufacture†

Continuous Weld — usually lower cost, not suitable for high-pressure applications. Normally limited to pipe sizes ½ to 4 in (13 to 102 mm).

Electric Resistance Weld — uniform wall thickness, limited to a maximum wall of ½ in (13 mm).

Seamless — for high-pressure applications, can be produced with heavy wall thickness and in various alloy grades. Because of the manufacturing method, the walls are not always uniform.

---

*Other designations that are occasionally specified under AWWA and federal guidelines include ANSI/AWWA-C200, ASTM 139, ASTM A-211, ASTM A-714, WWP-406 (comparable to A-120), WWP-404 (comparable to A 53), and ASTM A-409 (stainless steel). Other stainless steel pipe is covered under ASTM A-312.

†All pipe in this table is built by these three manufacturing methods with the exception of API 5LX pipe, which is available in electric resistance welded and seamless only.

because of cost considerations and the strength afforded by welding procedures. Nevertheless, some 7- to 10-in (178- to 254-mm) diameter casing built to API specifications is occasionally threaded and coupled for deep boreholes in oil-producing areas where this type of casing is available. Steel pipe and casing is manufactured to API Standard 5L or 5A in many countries. Western European manufacturers also produce pipe that

Table 13.5. Physical Dimensions of Common Pipe Sizes

Schedule Number	Outside Diameter		Inside Diameter		Wall Thickness		Weight		Collapse Strength		ft of water	m of water
	in	mm	in	mm	in	mm	lbs/ft	kg/m	psi	kPa		
—	6.625	168.3	6.313	160.3	0.156	4.0	10.78	16.04	600	4,137	1,386	423
—	6.625	168.3	6.249	158.7	0.188	4.8	12.92	19.22	1,030	7,102	2,379	725
—	6.625	168.3	6.187	157.1	0.219	5.6	14.98	22.29	1,521	10,487	3,511	1,070
—	6.625	168.3	6.125	155.5	0.250	6.4	17.02	25.33	1,953	13,466	4,510	1,375
40 (STD)	6.625	168.3	6.065	154.1	0.280	7.1	18.97	28.23	2,286	15,762	5,279	1,609
—	8.625	219.1	8.250	209.5	0.188	4.8	16.94	25.21	477	3,289	1,101	336
—	8.625	219.1	8.187	207.9	0.219	5.6	19.66	29.25	750	5,171	1,731	528
20	8.625	219.1	8.125	206.3	0.250	6.4	22.36	33.27	1,092	7,529	2,522	769
30	8.625	219.1	8.071	205.1	0.277	7.0	24.70	36.75	1,422	9,805	3,284	1,001
40 (STD)	8.625	219.1	7.981	202.7	0.322	8.2	28.55	42.48	1,920	13,238	4,433	1,352
—	10.75	273.1	10.374	263.5	0.188	4.8	21.21	31.56	246	1,696	568	173
20	10.75	273.1	10.250	260.3	0.250	6.4	28.04	41.72	579	3,992	1,336	407
30	10.75	273.1	10.136	257.5	0.307	7.8	34.24	50.95	1,048	7,226	2,421	738
40 (STD)	10.75	273.1	10.020	254.5	0.365	9.3	40.48	60.23	1,611	11,108	3,721	1,134
—	12.75	323.9	12.374	314.3	0.188	4.8	25.22	37.53	147	1,014	339	103
20	12.75	323.9	12.250	311.1	0.250	6.4	33.38	49.67	347	2,393	801	244
—	12.75	323.9	12.126	308.1	0.312	7.9	41.45	61.68	673	4,640	1,553	473
30	12.75	323.9	12.090	307.1	0.330	8.4	43.77	65.13	793	5,468	1,830	558
— (STD)	12.75	323.9	12.000	304.9	0.375	9.5	49.56	73.75	1,136	7,833	2,624	800
—	14.0	355.6	13.624	346.0	0.188	4.8	27.73	41.26	111	765	255	78
10	14.0	355.6	13.500	342.8	0.250	6.4	36.71	54.62	262	1,806	604	184
20	14.0	355.6	13.376	339.8	0.312	7.9	45.61	67.87	510	3,516	1,177	359
30 (STD)	14.0	355.6	13.250	336.6	0.375	9.5	54.57	81.20	875	6,033	2,021	616
—	16.0	406.4	15.562	395.2	0.219	5.6	36.91	54.92	117	807	270	82
10	16.0	406.4	15.500	393.6	0.250	6.4	42.05	62.57	175	1,207	404	123
20	16.0	406.4	15.376	390.6	0.312	7.9	52.27	77.78	341	2,351	788	240
30 (STD)	16.0	406.4	15.250	387.4	0.375	9.5	62.58	93.12	592	4,082	1,367	417
40 (XS)	16.0	406.4	15.000	381.0	0.500	12.7	82.77	123.2	1,331	9,177	3,072	937
10	18.0	457.2	17.500	444.4	0.250	6.4	47.39	70.52	122	841	283	86
20	18.0	457.2	17.376	441.4	0.312	7.9	58.94	87.70	239	1,648	552	168
— (STD)	18.0	457.2	17.250	438.2	0.375	9.5	70.59	105.0	417	2,875	962	293
— (XS)	18.0	457.2	17.000	431.8	0.500	12.7	93.45	139.1	970	6,688	2,241	683
10	20.0	508.0	19.500	495.2	0.250	6.4	52.73	78.46	89	614	205	63
—	20.0	508.0	19.376	492.2	0.312	7.9	65.60	97.61	174	1,200	402	123
20 (STD)	20.0	508.0	19.250	489.0	0.375	9.5	78.60	117.0	303	2,089	700	213
30 (XS)	20.0	508.0	19.000	482.6	0.500	12.7	104.1	154.9	716	4,937	1,654	504
10	22.0	558.8	21.500	546.0	0.250	6.4	58.07	86.41	66	455	154	47
—	22.0	558.8	21.376	543.0	0.312	7.9	72.27	107.5	130	896	301	92
20 (STD)	22.0	558.8	21.250	539.8	0.375	9.5	86.61	128.9	227	1,565	525	160
30 (XS)	22.0	558.8	21.000	533.4	0.500	12.7	114.8	170.8	540	3,723	1,248	380
10	24.0	609.6	23.500	596.8	0.250	6.4	63.41	94.35	51	352	118	36
—	24.0	609.6	23.376	593.8	0.312	7.9	78.93	117.4	100	690	231	70
20 (STD)	24.0	609.6	23.250	590.6	0.375	9.5	94.62	140.8	175	1,207	404	123
— (XS)	24.0	609.6	23.000	584.2	0.500	12.7	125.5	186.7	417	2,875	962	293
—	26.0	660.4	25.500	647.6	0.250	6.4	68.75	102.3	40	276	93	28
10	26.0	660.4	25.376	644.6	0.312	7.9	85.60	127.4	79	545	181	55
— (STD)	26.0	660.4	25.250	641.4	0.375	9.5	102.6	152.7	137	945	317	97
20 (XS)	26.0	660.4	25.000	635.0	0.500	12.7	136.2	202.7	327	2,255	756	230
—	28.0	711.2	27.500	698.4	0.250	6.4	74.09	110.2	32	221	74	23
10	28.0	711.2	27.376	695.4	0.312	7.9	92.26	137.3	63	434	145	44
— (STD)	28.0	711.2	27.250	692.2	0.375	9.5	110.6	164.6	110	758	253	77
20 (XS)	28.0	711.2	27.000	685.8	0.500	12.7	146.9	218.6	262	1,806	604	184
—	30.0	762.0	29.500	749.2	0.250	6.4	79.43	118.2	26	179	60	18
10	30.0	762.0	29.376	746.2	0.312	7.9	98.93	147.2	51	352	118	36
— (STD)	30.0	762.0	29.250	743.0	0.375	9.5	118.7	176.6	89	614	205	63
20 (XS)	30.0	762.0	29.000	736.6	0.500	12.7	157.5	234.4	213	1,469	491	150

is similar, but not identical, under other specifications; in a few cases, European pipe is made to metric measurements rather than to dimensions in the English system.

Similar to ASTM A-53, API line pipe is made in Grades A and B; the two designations refer to the tensile strength and yield strength of the material. Grade A pipe must have a tensile strength of at least 48,000 psi (331,000 kPa), whereas Grade B pipe must withstand 60,000 psi (414,000 kPa). Either grade is satisfactory for water well applications, although most of the API line pipe produced for water well use is Grade B. The compositional differences between ASTM and API Grades A and B pipe are presented in Table 13.6.

To provide more specialized products for the water well industry, some pipe fabricators provide other casing varieties. These products were selected to serve specific

Table 13.6. Compositional Differences Between Grade A and Grade B Pipe

Designation	Grade	Maximum Percentage			
		Carbon	Manganese	Phosphorus	Sulfur
ASTM A-53	A	0.25	0.95	0.05	0.06
	B	0.30	1.20	0.05	0.06
API 5L	A	0.22	0.90	0.04	0.05
	B	0.27	1.15	0.04	0.05

segments of the water well market. For example, drive pipe was developed for cable tool operations to insure that the ends of the pipe joints could butt in the coupling. The four specialized products described in Table 13.7 consist primarily of selected pipe classes from both ASTM (standard pipe) and API (line pipe) standards. Some casing sizes, however, are not specified in the basic ANSI criteria and have been developed to complete a special series of casing.

Water well casing is thin walled and has fine- and sharp-threaded recessed couplings. In some cases, lighter casing can be used, as for example the surface casing that will be pulled during grouting or the typical "stovepipe" casing used in the mud scow drilling system in southwestern United States. The use of surface casing is discussed in more detail in the grouting section in Chapter 10.

Reamed and drifted pipe costs more than standard pipe or line pipe because the interior walls of the pipe have been cleaned and straightened. The couplings are much stronger and longer and the recessed ends of the coupling cover the exposed threads on the pipe. This covering reduces the tendency for corrosion to penetrate the threaded portion of the pipe that is not fully engaged in the coupling threads. Reamed and drifted pipe is made from standard, line, or other type of pipe.

Drive pipe has proven unsuitable for cable tool drilling, although it was originally designed for this application. Threads and couplings are designed so that pipe ends butt inside a coupling when the joint is made up. As pipe is driven, the threads loosen slightly. With the pipe ends already butted, a drive-pipe joint cannot be retightened if loosened. On the other hand, line-pipe joints are easily tightened because the pipe ends do not butt and the threads have more taper, but there may be more corrosion at the joints. Threaded and coupled pipe should not be driven, however, because the tapered threads can split the coupling. In general, driven casing larger than 2-in (51-mm) diameter is welded rather than threaded and coupled.

Another minor type of pipe is driven pipe, which ranges from 1 to 2 inches (25 to 51 mm) in diameter. This type of pipe is often used in constructing drive point wells.

Stainless Steel Casing

Engineers occasionally specify stainless steel casing (ASTM A-409) for municipal and industrial wells in highly corrosive environments to increase the life of the well. The addition of chromium (more than 11.5 percent) to regular steel produces a stainless steel capable of resisting corrosion. The presence of nickel and molybdenum also helps prevent certain forms of corrosion. The 300 series of austenitic stainless steels is used widely for well screens and casing, whereas Type 405 stainless steel (ferritic) is suitable in less corrosive environments. Austenitic stainless steels cannot be hardened by heat treatment. They typically contain about 16 to 20 percent chromium and 8 to 14 percent nickel, and are generally the most resistant to corrosion. The initial cost of stainless steel casing is higher than low-carbon steel, but over time the cost of stainless steel may be lower because it lasts longer. The composition of several metals is given in Table 13.8.

Thermoplastic Casing

Plastic casing has been used since the late 1940's in extremely corrosive waters in which steel casing failed within a short time. The use of plastic well casing has grown significantly since the 1960's, when stronger materials were developed and advances

Table 13.7. Special Casing Used for Water Wells

Classification	Description
Water Well Casing	Water well casing is produced by the electric weld or seamless process in sizes of 3½ in OD to 8⅝ in OD (88.9 mm to 219 mm) in walls lighter than standard weight. Because the walls are thinner than standard, the threads are comparatively shallow and short.
Reamed and Drifted Pipe	Reamed and drifted water well pipe is made especially for water well use and may be driven. It is produced in welded and seamless pipe, in diameters, of 1 in through 12 in (25.4 mm to 305 mm), in lengths of 16 ft to 22 ft (4.9 m to 6.7 m), with threaded ends and recessed couplings. The pipe is reamed on the ends to insure inside clearance. Reamed and drifted pipe can be made from standard or line pipe.
Drive Pipe	Drive pipe is electric resistance welded pipe or seamless pipe originally designed for use in cable tool drilling. The ends of the pipe are specially threaded so as to permit them to butt in the coupling when the joint is made up tight. Drive pipe is produced in sizes of 6⅝ in to 16 in OD (168 mm to 406 mm). It is supplied with recessed line couplings, in single-random or half-random lengths.
Driven Well Pipe	Driven well pipe is produced in welded and seamless standard-weight pipe. It is normally threaded and coupled galvanized pipe in sizes of 1 in (25.4 mm), 1¼ in (31.8 mm), 1½ in (38.1 mm), and 2 in (50.8 mm), in lengths of 3 ft to 6 ft (0.9 m to 1.8 m) and 6 ft to 10 ft (1.8 m to 3 m).

(After National Association of Steel Pipe Distributors, Inc., 1979)

in extrusion technology led to a broad range of casing diameters and wall thicknesses. The development of ABS (acrylonitrile butadiene styrene) and PVC (polyvinyl chloride) now provides thermoplastic pipe of much higher strength than earlier plastic materials. Corrosion resistance, light weight, relatively low cost, easy installation, and resistance to acid treatment make plastic casing desirable for many installations where high strength is not required.

Standardization of thermoplastic well casing is covered under ASTM Standard F-480 entitled Thermoplastic Water Well Casing, Pipe and Couplings Made in Standard Dimension Ratios (SDR). Under this standard specification, the minimum physical and chemical characteristics of ABS, PVC, and SR (rubber-modified polystyrene) materials are defined (Appendix 13.C). Unlike steel casing, plastic pipe is built to a physical standard called the standard dimension ratio (SDR). SDR is defined as the outside diameter divided by the minimum wall thickness. As the diameter increases, the wall thickness also increases. In this way, the collapse resistance remains the same

Table 13.8. Basic Composition and Suggested Applications of Various Metals for Water Well Casing and Screens

Metal	Composition		Suggested Applications
Low-Carbon Steel	Carbon	0.8% max	Not corrosion resistant. Will give satisfactory service life where waters are noncorrosive and nonincrusting.
	Iron	Balance	
Type 405 Stainless Steel	Chromium	11.5% min	Moderate corrosion resistance. Applicable for potable waters of low to medium corrosivity. May exhibit slight surface rusting in oxygenated environments.
	Aluminum	0.3% max	
	Iron	Balance	
Type 304 Stainless Steel	Chromium	18% min	Excellent corrosion resistance. Most widely used stainless steel material for water well screens. Occasionally specified for casing in high-capacity wells in corrosive environments.
	Nickel	8% min	
	Manganese	2% max	
	Carbon	0.08% max	
	Iron	Balance	
Type 316L Stainless Steel	Chromium	16% min	For use in groundwaters having a moderate salt content. Molybdenum content helps resist pitting and crevice corrosion in moderate saline solutions. Better resistance to stress-corrosion cracking than Type 304 stainless steel.
	Nickel	10% min	
	Molybdenum	2% min	
	Carbon	0.03% max	
	Iron	Balance	
Carpenter 20Cb-3	Nickel	32% min	Strong resistance to stress-corrosion cracking, pitting, and crevice corrosion. Will provide satisfactory service in very saline waters and at water temperatures over 100°F (38°C).
	Chromium	19% min	
	Copper	3% min	
	Molybdenum	2% min	
	Manganese	2% max	
	Iron	Balance	
Monel 400	Nickel (plus Cobalt)	63% min	Will provide satisfactory service in sea water where a high sodium chloride content is combined with dissolved oxygen. Material is sensitive to hydrogen sulfide.
	Copper	28% min	
	Iron	2.5% max	
	Manganese	2% max	
Incoloy 825	Nickel (plus Cobalt)	38% min	Material has a low corrosion potential in high chloride water and resists cracking, pitting, and crevice corrosion. Satisfactory for use in geothermal, oil and gas, and injection wells.
	Iron	22% min	
	Chromium	19.5% min.	
	Molybdenum	2.5% min	
	Copper	1.5% min	
Inconel 600	Nickel (plus Cobalt)	72% min	For use in geothermal, oil and gas, and injection well applications. Strong resistance to stress-corrosion cracking.
	Chromium	14% min	
	Iron	6-10%	
Hastelloy C	Nickel	51% min	Extremely corrosion-resistant alloy used only in most aggressive environments, such as brine, corrosive gases, and where temperatures exceed 200°F (93°C). Used for geothermal, oil and gas, and injection wells.
	Molybdenum	15% min	
	Chromium	14.5% min	
	Iron	4% min	
	Tungsten	3% min	
	Cobalt	2.5% max	
	Vanadium	0.35% max	

Table 13.9. Typical Physical Properties of Thermoplastic Well Casing Materials at 73.4°F (23°C)

Property	ASTM Test Method	ABS Cell Class, per D-1788		PVC Cell Class, per D-1784		SR Cell Class, per D-1892 4434A
		434	533	12454-B & C	14333-C & D	
Specific gravity	D-792	1.05	1.04	1.40	1.35	1.05
Tensile strength, psi*	D-638	6,000†	5,000†	7,000†	6,000†	3,100†
Tensile modulus of elasticity, (E), psi	D-638	350,000	250,000	400,000†	320,000†	320,000
Compressive strength, psi*	D-695	7,200	4,500	9,000	8,000	5,000
Impact strength (Izod method), ft-lb/in notch	D-256	4.0†	6.0†	0.65	5.0	0.9
Deflection temperature under load (264 psi), °F	D-648	190†	190†	158†	140†	180
Coefficient of linear expansion, in/in·°F	D-696	5.5×10^{-5}	6.0×10^{-5}	3.0×10^{-5}	5.0×10^{-5}	4.8×10^{-5}

*These values have been determined on the basis of short-term tests. Because of the viscoelastic nature of plastic material, the long-term strengths may be only one-third to one-half of the values cited above. For compression strength, a long-term value of one-fifth of the indicated figure should be assumed.

†These are minimum values set by the corresponding ASTM Cell Class designation. All others represent typical values.

(National Water Well Association, 1981)

for all sizes of pipe built to a given strength standard. For example, all sizes of SDR-21 pipe will withstand pressures of 115 psi (793 kPa) (PVC-12454). The exceptions to the standard dimension ratios occur in the regular schedule numbers such as Schedule 40 or Schedule 80, where the wall thickness and diameter have no distinct relationship and are built to the standards for steel. Appendix 13.D gives physical characteristics of commonly used sizes of PVC casing. Because regulations vary concerning the use of plastic casing, drilling contractors should consult the appropriate regulatory agency for current laws.

As with steel casing, the physical characteristics of a plastic material are determined by its chemical composition. For example, in ABS material, acrylonitrile contributes rigidity, strength, hardness, chemical resistance, and heat resistance; butadiene lends toughness; and styrene adds gloss, rigidity, and ease of processing. There are two types of ABS well casing materials: cell class 434 (ASTM F-480) has higher strength and rigidity with good impact resistance, whereas cell class 533 has somewhat lower strength and rigidity, but greater impact resistance. Standard values for important physical properties of the three major plastic types are given in Table 13.9.

A second type of thermoplastic material, polyvinyl chloride (PVC), is commonly used in the water well industry. The standards for PVC are established in ASTM D-1784, Standard Specification for Rigid Poly (Vinyl Chloride) and Chlorinated Poly (Vinyl Chloride) Compounds, under ASTM F-480. In the PVC specification, the various materials are given a cell class designation that describes the minimum physical properties for a particular compound. The few properties contained in the cell classification are (1) type of base resin, (2) impact strength, (3) tensile strength, (4) elastic modulus in tension, and (5) deflection temperature under loading. See Appendix 13.E for the resin types and numerical values for these variables and the methodology for the cell classification scheme.

Some pipe manufacturers may still use the older method of specifying PVC casing, which included the type of polymer, a type number indicating impact strength, a grade number indicating chemical resistance, and the hydrostatic design stress in psi (Appendix 13.F). The new cell classifications along with the older specifications are given in Table 13.10.

SR (styrene-rubber) plastics are also made for water well casing. They are manufactured in cell class 4434 under ASTM D-1892, Standard Specification for Styrene Butadiene Molding and Extrusion Materials. Styrene-rubber plastics are usually used for nonpressure applications (drain pipe and low-head irrigation pipe). Thus, no hydrostatic design stresses are available for this type of material.

Table 13.9 shows that plastic casing will deflect (deform) at relatively low temperatures, is low in compressive strength relative to steel casing, and is flexible. High temperatures can deflect the casing, and care must be used when grouting to minimize

Table 13.10. Cell Classifications for PVC Materials Used for Water Well Construction and Their Previous Designations

Cell Classification	Older Designation
12454-B	PVC 11 (Type I, Grade 1)
12454-C	PVC 12 (Type I, Grade 2)
14333-D	PVC 21 (Type II, Grade 1)

the effects of the heat of hydration by restricting the thickness of grout around the casing (Johnson et al., 1980). Typically the grout should not be more than 2 in (51 mm) thick. PVC has less heat resistance than ABS material and therefore will be deflected more easily in high-temperature environments. At ordinary ground temperatures, however, plastic casing is not subject to deflection.

Plastic material is much more flexible than steel. Thus, plastic casing must be centered in the borehole before backfilling or filter packing is completed. Any voids in the backfill or filter pack material may lead to sudden collapse of formation materials against the casing, causing it to break.

The collapse strength of plastic casing is much less than for steel casing. Short-term hydraulic collapse pressures for thermoplastic casings manufactured under ASTM F-480 are given in Figure 13.2. See Appendix 13.D for collapse strengths of PVC casing. There is some evidence that these collapse pressures may be much too high unless the borehole materials offer a large measure of support (Kurt, 1979). If support is absent, the collapse resistance may be as much as 25 percent lower than the collapse pressure calculated from the equation given in ASTM Standard F-480. Even when the casing is supported by the formation, the actual collapse pressures may be about 10 to 12 percent less than ASTM F-480 (York, 1978; Kurt, 1979). Under ideal conditions where well-compacted materials fully support the casing, the values for collapse resistance given in F-480 are reliable. The actual strength for any situation will depend on the dimension ratio (DR—diameter divided by wall thickness), wall thickness uniformity, roundness of the casing, rate of loading, and temperature of the casing when the loading is applied. Although plastic casing can be set to depths of 1,000 ft (305 m), the drilling contractor should use caution in selecting plastic materials for any well deeper than 300 ft (91.5 m), especially for large-diameter wells.

The short-term strength of plastic casing is usually much higher than its strength over time. For example, pressure class 150 PVC plastic pipe used for water supply distribution may absorb a 1-minute, internal hydraulic overloading of 755 psi (5,210 kPa), more than five times its long-term strength. If the pressure is maintained for only 5 minutes, the pipe could burst (Handbook of PVC Pipe, 1979). In general, short-term test results for strength are not a good indication of long-term strengths.

Other strengths applicable to plastic materials include impact resistance, toughness, and pipe stiffness. When casing protrudes above ground level, for example, it must be protected because it can be severely damaged by moving vehicles or contact with drilling tools. The casing should also be shielded from the sun's ultraviolet rays if exposed above ground for long periods, because the impact strength of the material may be reduced signifi-

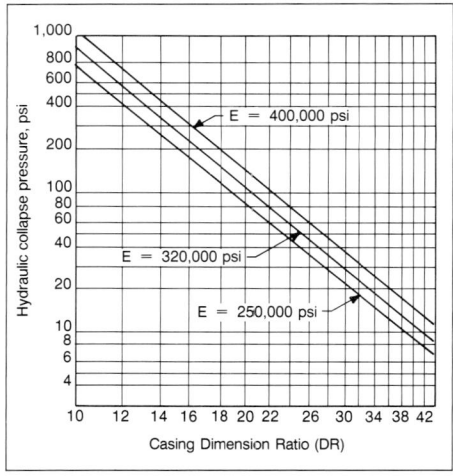

Figure 13.2. Hydraulic collapse pressure at 73°F (22.8°C) vs. casing DR for thermoplastic well casing. DR is the ratio of average casing outside diameter to minimum wall thickness; E is the modulus of elasticity. *(NWWA, 1981)*

cantly over time. Care must be used during cold weather to prevent shattering of the plastic during handling. Because the weight of plastic casing is only one-fifth to one-seventh that of steel, tensile strength is usually less important. Occasionally, it will float in a well during installation, thereby creating special handling problems.

Plastic pipe exhibits one other characteristic that may present a hazard to drinking water quality in areas where groundwater contamination has occurred. If volatile organic chemicals exist in groundwater near a well bore, but above the intake section of the well, some of these chemicals still might move into the discharge by passing through the wall of the casing (Keech, 1984). Although this process is not fully understood, it appears that plastic casing can be permeable in the presence of certain chemicals.

Fiberglass-Reinforced Plastic Casing

Casing for water wells is constructed also from various types of fiberglass-reinforced plastic materials. This type of casing is usually referred to as fiberglass casing. Fiberglass casing is resistant to most forms of corrosion, is not conductive, and for its weight has the strength of steel. It has been used successfully for injection of highly corrosive waters to recharge oil reservoirs. It is also used for water supply wells, especially for irrigation purposes in some areas of the world. Heat, however, can significantly reduce the collapse resistance of fiberglass pipe, but for most water wells any reduction in collapse resistance is minimal. Again, as with stainless steel casing, a higher initial cost is offset by longer well life. Fiberglass casing is somewhat permeable, however. Thus, in formations where poor-quality water is cased off above potable water, some contamination of the water supply may occur. Well fittings such as centralizers, couplings, and surface fittings constructed of fiberglass are available for use with this type of casing.

In summary, casing selection depends on several major factors: strength requirements, corrosion resistance, ease of handling, cost considerations, type of formation, method of drilling, the particular well design, and well construction techniques. Casing must have the column, collapse, and tensile strengths required for a specific borehole. Resistance to collapse is the most critical strength. For boreholes greater than 1,000 ft (305 m), steel casing is usually selected to meet the rigorous strength characteristics demanded of the casing. In general, casing built to ANSI dimensions for standard pipe has sufficient strength for most common applications, but, as the casing diameter increases, the safe depth for setting standard-weight pipe decreases significantly. For corrosive water, PVC or stainless steel casing provides the longest life possible. The use of PVC casing has expanded greatly for relatively shallow, small-diameter wells because of its ease of handling and its low cost. Its use for irrigation wells also has increased. On the other hand, steel casing must always be used when the casing is driven and pulled back by the cable tool method, or when casing is installed in open boreholes that are subject to caving. Because of the many factors involved in selecting the most suitable casing material, the design engineer should consult with the well owner and local drilling contractors before specifying the type of casing.

Methods of Joining Casing

Steel Casing

Steel casing can be joined by threading or welding, although threads are generally not available on casing that exceeds 12 in (305 mm) ID. In some countries, welding is not common practice; therefore, in these countries 16-in (406-mm) casing is threaded. Casing is prepared for welding by beveling the pipe ends at approximately 35 degrees. The welding process is critical to well integrity and should be done by following standardized procedures (American Welding Society, 1981). All welds must be fully penetrating; that is, the entire beveled and flat area must be filled with weld bead in multiple passes around the casing. Care must be taken, however, to avoid burn-through on the first pass so that metal is not deposited on the inside of the pipe. Slag on the inside of the pipe can hinder or prevent tool movement or screen installation.

The proper selection of electrodes is critical in succesfully joining dissimilar metals — for example, when joining low-carbon steel casing to a stainless steel screen. Either E 312-16 (AWS-ASTM classification) or E 309-16 electrodes are recommended for joining low-carbon steel to stainless steel. E 309 electrodes are used most commonly for welding 304 stainless steel to low-carbon steel; these electrodes are readily available and cost less than E 312-16 material. Type E 308-15 or 16 electrodes are used to weld stainless steel to stainless steel. If mild steel electrodes are used to join stainless steel to stainless steel, chromium will precipitate, creating areas of low corrosion resistance and causing eventual structural weakness or failure. See Appendix 13.G for details on field welding procedures and electrode recommendations.

Plastic Casing

Plastic casing can be joined by mechanical means or by solvent welding using a solvent cement to join casing segments. A primer is used to clean and etch the surface before the solvent cement is applied (Figure 13.3). After the solvent has been uniformly applied to the joints and they are joined together, enough time must be allowed for curing. Low temperatures [below 40°F (4.4°C)] and poor fit caused by joint interferences are two major problems that affect the integrity of the joint. Because of the different chemical characteristics of various plastic casing, the drilling contractor must be sure to use the correct solvent cement for a particular casing. When temperatures are below 40°F, plastic components should be joined with a low-temperature solvent cement. A special solvent must be used when joining PVC and ABS casing because of the different chemical composition of these materials (see ASTM D-3138).

Two types of solvent-welded joints are used. In the first type, a coupling (collar) is used to join casing lengths of equal or unequal diameter. The ends of the casing fit tightly into the molded plastic coupling. In the second type, a "bell-end" socket is molded at one end of each casing length to receive the straight (spigot) end of the next casing length. It is important that the bell or coupling be manufactured to close tolerances so that the fit around the pipe is uniformly snug. See Appendix 13.H for solvent-welding procedures.

Plastic casing may also be joined mechanically by threaded couplings, cam-locking lugs, or screws. Threaded connections are used most frequently on small-diameter plastic casing, especially for monitoring wells when joints must be watertight and the use of solvents is not allowed. Special thread lubricants and sealants must be used; avoid using solvent cement on threads, because the cement may set faster than the threads can be tightened.

Some larger diameter plastic casing is joined by cam-locking, using a gasket to make

Figure 13.3. The end of the plastic casing is coated with solvent cement prior to placing it into the socket held in the table.

the seal. There are several advantages to this type of connection: joints can be disassembled and reused if necessary, no solvents are required, and the time needed to join casing is minimized.

Joining casing by use of fully penetrating screws (usually done in conjunction with cementing) is not good practice because their corrosion over time may leave a hole in the casing through which contaminants or bacteria may enter the well. Only stainless steel screws, set partially through the casing, should be used. Depending on the size and spacing of the screws, this method may weaken the casing.

Plastic casing can be joined to both metallic and plastic screens. For screens of 6 in (152 mm) diameter and less, the casing is usually threaded to the screen. For larger diameter screens, special adapters are available. To join a metal screen to plastic casing, a special steel ring or adaptor is attached to the weld ring on the end of the screen. See Chapter 14 for installation procedures.

Fiberglass Casing

Most fiberglass casing is threaded together because of the long setting time required for epoxy or polyester resin cement. In some cases, fiberglass casing is joined by slip joints. Fiberglass casing may also be joined by a flexible key that locks the male and female ends of the casing together. Metallic screens can be adapted to fiberglass casing by using threaded adaptors.

WELL DEPTH

The depth of a well is usually determined from information from the driller's log of a test hole, sample logs of other nearby wells in the same aquifer, geophysical analysis of the formation, and data taken during the drilling of the production well. Generally, a well should be completed to the bottom of the aquifer because:

1. More of the aquifer thickness can be utilized as the intake portion of the well, resulting in higher specific capacity.

2. More drawdown can be made available, permitting greater well yield.

3. Sufficient drawdown is available to maintain well yield even during periods of severe drought or overpumping.

An exception to this rule is made when the well screen is centered between the top and bottom of the aquifer, a practice sometimes followed to make more efficient use of a given length of screen in a uniform, confined aquifer. Furthermore, the extreme upper and lower parts of any aquifer commonly consist of materials that are less uniform than those forming the major part of the aquifer. All pilot holes should penetrate the entire aquifer so that the most productive zones can be identified. Another exception is in extremely thick aquifers, such as the Navajo Sandstone and the sand and gravel deposits along the Rio Grande River in New Mexico, where it may not be economical to drill to the bottom of the aquifer.

A third exception is made when poor-quality water is found in part of an aquifer. In this case, the well should be completed to a depth that will avoid the undesirable water. Any part of the hole drilled into a portion of the aquifer containing poorer quality water should be isolated. Although low-quality water may appear anywhere in an aquifer, it is likely that any gases such as hydrogen sulfide or contaminants having low molecular weight will be concentrated near the top. Heavy ions such as iron and manganese move toward the bottom.

WELL SCREEN LENGTH

The optimum length of well screen is based on the thickness of the aquifer, available drawdown, and nature of the stratification of the aquifer. In virtually every aquifer, certain zones (horizons) will transmit more water than others. Thus, the intake part of the well must be placed in those zones having the highest hydraulic conductivity. Determination of the most productive layers can be made by one or more of the following techniques:

1. Interpretation of the driller's log and comments on drilling characteristics such as fluid loss, penetration rate, and pulldown and chatter.

2. Visual inspection and comparison can be made of samples representing each sediment layer. The relative transmissivity of each layer is estimated from the observed coarseness, lack of silt and clay, and thickness of the layer.

3. Sieve analyses can be made from samples taken from the various layers in the aquifer. Comparison of grain-size curves can indicate the relative hydraulic conductivity of each sample. It is highly recommended that sieve analyses be performed on selected formation samples from any industrial, municipal, or irrigation well. The curves presented in Figure 13.4 indicate the relationship between the grain-size distribution of aquifer materials and the resulting hydraulic conductivity.

4. Laboratory hydraulic conductivity tests can be performed on samples that represent individual layers of the water-bearing formation. In this test, water is caused to flow through a sample of the material. Measurements of the area through which flow occurs, the rate of flow, and the corresponding head loss provide data for cal-

culating the hydraulic conductivity. Aquifer transmissivity can then be determined by adding the individual transmissivity values for all layers of the aquifer (transmissivity equals the hydraulic conductivity times the thickness for each layer).

5. Borehole geophysical logging techniques can help locate zones having the highest hydraulic conductivity. Velocity-meter surveys also are extremely useful. See Chapter 8 for an analysis of the various exploration methods.

Each technique listed above provides useful information on the zones that should be exploited. As many of these techniques should be used as possible. Economic factors governing a well project dictate the cost that can be justified in determining most accurately the productive zones of the aquifer.

Recommended screen lengths for four typical hydrogeological situations are given below.

1. *Homogeneous Unconfined Aquifer.* Theoretical considerations and experience have shown that screening of the bottom one-third to one-half of an aquifer less than 150 ft (45.7 m) thick provides the optimum design for homogeneous unconfined aquifers. In some cases, however, particularly in thick, deep aquifers, as much as 80 percent of the aquifer may be screened to obtain higher specific capacity and greater efficiency, even though the total yield is less.

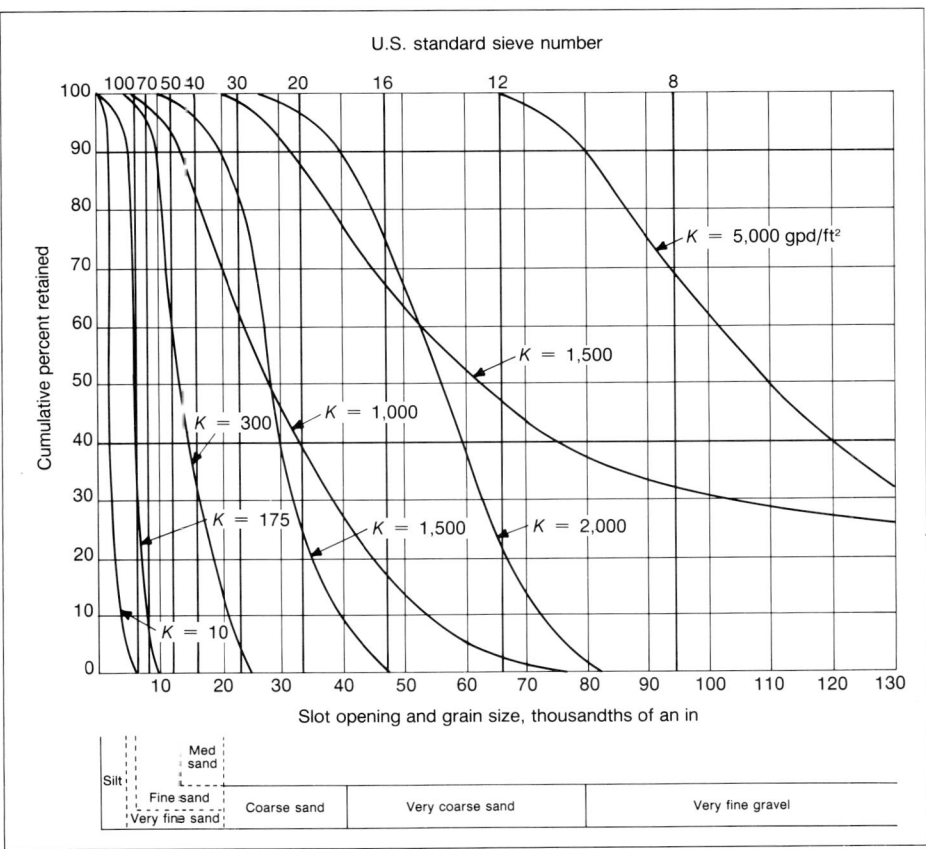

Figure 13.4. Hydraulic conductivity can be estimated on the basis of grain-size-distribution curves.

A well in an unconfined aquifer is usually pumped so that, at maximum capacity, the pumping water level is maintained slightly above the top of the pump intake or screen. The well screen is positioned in the lower portion of the aquifer because the upper part is dewatered during pumping.

For wells in unconfined aquifers, selection of screen length is a compromise between two factors. On the one hand, higher specific capacity is obtained by using the longest screen possible. This reduces convergence of flow and entrance velocity, thereby increasing specific capacity. On the other hand, more available drawdown results from using the shortest screen possible. These two conflicting aims are satisfied, in part, by using an efficient well screen that minimizes the loss in specific capacity as drawdown increases.

As shown in Chapter 9, Figure 9.11, it is impractical from a theoretical (hydraulic) standpoint to pump a well in an unconfined aquifer at a drawdown that exceeds two-thirds the thickness of the water-bearing sediment. Figure 9.11 shows that the well produces 88 percent of maximum yield at 65 percent of the maximum drawdown. If the drawdown were increased to 95 percent of the amount possible, the well yield would increase to 99 percent of its maximum. Thus, 46 percent greater drawdown results in only 12.5 percent greater yield.

2. *Nonhomogeneous Unconfined Aquifer.* The basic principles of well design for homogeneous unconfined aquifers also apply to this type of aquifer. The only variation is that the screen or screen sections are positioned in the most permeable layers of the lower portions of the aquifer so that maximum drawdown is available. If possible, the total screen length should be approximately one-third of the aquifer thickness.

3. *Homogeneous Confined Aquifer.* In this type of aquifer, 80 to 90 percent of the thickness of the water-bearing sediment should be screened, assuming that the pumping water level is not expected to be below the top of the aquifer. Maximum available drawdown for wells in confined conditions should be the distance from the potentiometric surface to the top of the aquifer. If the available drawdown is limited, however, it may be necessary to draw the well down below the bottom of the upper confining layer. When this occurs, the aquifer will respond like an unconfined aquifer during pumping.

Screen lengths chosen according to these rules make it possible to obtain about 90 to 95 percent of the specific capacity that could be obtained by screening the entire aquifer. Best results are obtained by centering the screen section in the aquifer. In the past, screens were often interspaced with blank casing placed in the less permeable zones of the formation. Today, however, higher water demands and lower screen costs have resulted in completely screening most deep wells.

4. *Nonhomogeneous Confined Aquifer.* In this type of aquifer, 80 to 90 percent of the most permeable layers should be screened.

WELL SCREEN SLOT OPENINGS

Screen slot openings for the same formation can differ depending on whether the well is naturally developed or filter packed. Either design is satisfactory and the choice for a particular well will depend primarily on the nature of the grain-size-distribution curve for the aquifer materials. Coarse-grained nonhomogeneous material can be developed naturally, whereas fine-grained homogeneous materials are best developed using a filter pack. Well screen slot openings for either method are selected from a study of sieve-analysis data for samples representing the water-bearing formation.

The design for the slot openings must be based on accurate samples if maximum yields and sand-free water are to be obtained. In rotary drilling, clay added to the drilling fluid can contaminate samples and lead to recommended slot sizes that are smaller than necessary. The log of the well should show if natural clays are present and should be considered in the screen design. In some cases, fine material from formations overlying the aquifer may be kept in suspension in the drilling fluid while the aquifer is being drilled. These fine materials may then be included in the cuttings from the aquifer, although they originate in the overburden. The drilling method may also affect the accuracy of the sample. In air drilling, samples collected at the surface tend to be finer in texture than the materials in the formation. As the bit advances, formation water entering the borehole may pull in finer material differentially, resulting in a higher proportion of fine particles in samples taken at the surface.

On the other hand, highly viscous drilling fluids made with clay additives may entrain fine aquifer materials and prevent them from settling out in the mud pit. Thus, the sample will be coarser than the actual formation materials. The use of a polymeric drilling fluid that has little or no gel strength will minimize these sampling problems. Sampling procedures are discussed in Chapter 8.

Screen Slot Selection for Naturally Developed Wells

In a naturally developed well, the screen slot size is selected so that most of the finer formation materials near the borehole are brought into the screen and pumped from the well during development. This practice results in creating a zone of graded formation materials extending 1 to 2 ft (0.3 to 0.6 m) outward from the screen. The increased porosity and hydraulic conductivity of the graded materials reduces the drawdown near the well during pumping.

To determine the correct slot openings for nonhomogeneous sediments, the typical approach is to select a slot through which 60 percent of the material will pass and 40 percent will be retained. This is usually done when the groundwater is not particularly corrosive and when there is little doubt about the reliability of the sample. On the other hand, the 50-percent-retained size is chosen if the water is extremely corrosive or if there is some doubt about the reliability of the sample. Selecting a smaller slot size is wiser if the water is corrosive or if low-carbon steel screens are used, because enlargement of the openings of only a few thousandths of an inch caused by corrosion could allow the well to pump sand. If the screen is stainless steel, slot enlargement from corrosion is generally not a problem. A conservative slot opening is used in calcareous formations (shell fragments) which dissolve readily if the well is acid treated. Removal of the calcareous materials reduces the amount of bridging material and allows fine clastic material to enter the well.

A more conservative slot selection may be advisable when (1) there is some doubt about the reliability of the samples, (2) the aquifer is thin and overlain by fine-grained loose material, (3) development time is at a premium, and (4) the formation is well sorted. Under these conditions, slot sizes that will retain 40 to 50 percent of the aquifer material are preferred.

When the formation consists of coarse sand and gravel, the designer has greater latitude in selecting the slot openings (Figure 13.5). An increase of a few thousandths of an inch in the slot size allows only a small amount of additional material to pass through the well screen during development. The slot size selected, therefore, may retain from 30 and 50 percent of the aquifer material. If the openings chosen retain only 30 percent, more material is brought through the screen during the development process. This will increase the time required to develop the well.

Offsetting the cost of extra development work, however, are the advantages gained from the larger slot sizes that increase the screen open area. For example, when the water is incrusting, longer service life can be expected before plugging reduces the well yield. Larger slot size and open area permit the development of a thicker, more permeable zone around the screen. This generally increases the specific capacity of the well, making the well more efficient and therefore less expensive to operate.

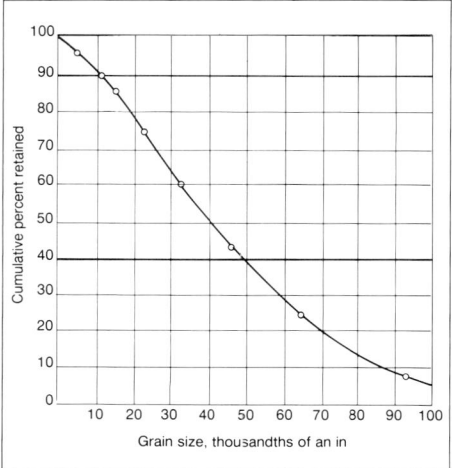

Figure 13.5. A 0.040- to 0.058-in (1.0- to 1.5-mm) slot opening is selected from this grain-size-distribution curve for a naturally developed well. For the larger slot size, the development time will be greater.

Nonhomogeneous formations occur most commonly because the strength of the forces responsible for erosion and deposition of earth materials tend to vary over relatively short periods of time. When designing screens for these formations, slot openings for different sections of the well screen may be chosen according to the gradation of materials in the different layers, if the layers are at least 4 ft (1.2 m) thick and their depths have been determined accurately. Two additional recommendations should be followed, however, when selecting slot openings for a multiple-slot screen:

1. If fine material overlies coarse material, extend at least 3 ft (0.9 m) of the screen designed for the fine material into the coarse layer below.

2. If fine material overlies coarse material, the slot size for the screen section installed in the coarse layer 3 ft (0.9 m) beneath the formation contact should not be more than double the slot size for the overlying finer material. Doubling of the slot size should be done over screen increments of 2 ft (0.6 m) or more.

Application of these two recommendations reduces the possibility that the well may pump sand if the location of each layer has not been determined accurately or if the samples are not reliable. The guide for slot-size selection indicates that about 60 percent of the formation material near the screen will come through the screen during the development process. Removal of this fraction results in some settlement of material around the screen. Thus, the position of the overlying finer layer moves downward slightly as settling occurs.

Figure 13.6 illustrates what can happen if the first of the above recommendations is not followed. Here the section of the screen with openings selected to fit the coarser sand begins at the boundary between the two layers. As the finer fraction of the coarser material is removed during development, slumping of the overlying fine-sand layer may occur. This can easily permit the top section of the screen, which has the larger openings, to come into contact with the fine sand; sand pumping would then occur.

Application of the two recommendations is best illustrated by an example. The curves in Figure 13.7 represent the grain-size distribution for the four layers that make up the lower 65 ft (19.8 m) of an unconfined aquifer 200 ft (61 m) thick. The grain size of the material immediately overlying the aquifer should also be determined. Good design calls for screening of the lower 65 ft of the formation, which means that approximately one-third of the aquifer will be screened. To evaluate the situation, sketch the stratigraphic section and record the information in a design table (Figure 13.8 and Table 13.11). Slot sizes for a screen to be used in a formation of only two layers can be selected readily without a design table. The table is extremely useful, however, for comparing a large number of samples.

The depth and thickness of each layer, for 50-, 40-, and 30-percent-retained sizes of each sample, are recorded. The hydraulic conductivity can then be estimated from the curves in Figure 13.4. A range of slot sizes, above and below the 40-percent size, is then determined.

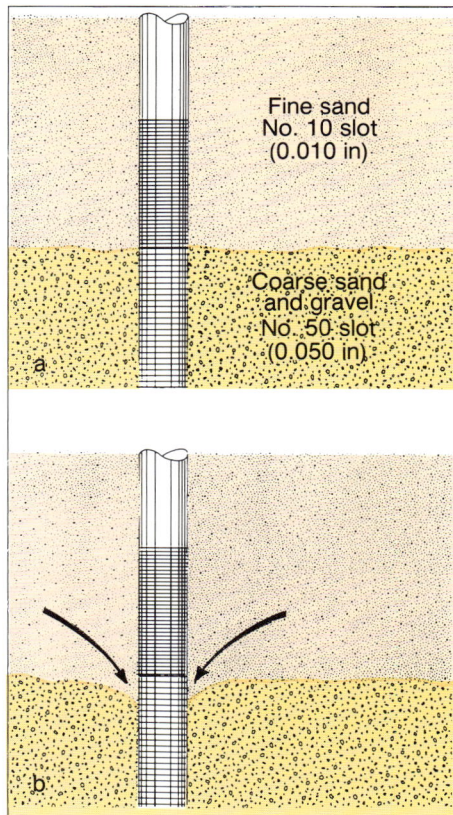

Figure 13.6. Screen in lower part of stratified aquifer (a) should be shorter than the total thickness of the coarser sand, to avoid situation (b) which shows possibility of fine sand entering upper part of the screen after development.

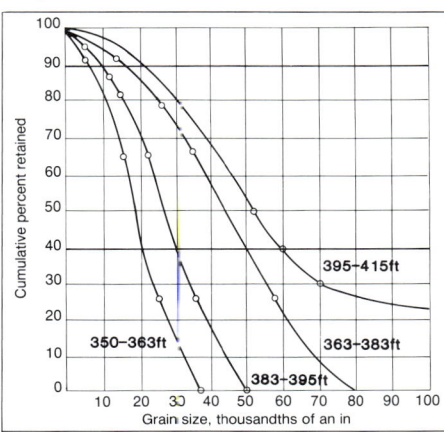

Figure 13.7. Grain-size-distribution curves representing the various layers in a stratified sand aquifer.

These should be values that could be considered for a screen in each isolated layer without considering the other layers.

If a 65-ft screen* is installed in the lower part of the aquifer, the top of the screen is at 350 ft. Average water quality and development conditions are assumed in this example, so the design is based on slot sizes that retain 40 percent of the material. Selection of the slot openings begins at the top of the screen. In this example, the sediments lying just above 350

---

*Most continuous-slot screens are built in lengths of 20 ft (6.1 m) or less that can have many slot sizes. These sections are assembled at the well site by the drilling contractor.

ft can be retained by a screen with 0.020-in slots; that is, the overlying sediments have the same size distribution as those that exist from 350 to 363 ft. Thus, the screen section for the upper layer should have 0.020-in slot openings, the next layer, 0.050 in, the next layer, 0.030 in, and the last layer, 0.060 in.

Applying the first recommendation, the finer openings for the top screen section (350 to 363 ft) must extend at least 3 ft into the underlying coarser material. This puts the lower limit of the section with 0.020-in slot openings at 366 ft. Application of the second recommendation suggests that the slot size can be no more than doubled (0.040 in) from 366 to 368 ft. From 368 to 383 ft, the slot size again follows the 40-percent-retained size. The sediment from 383 to 395 ft is finer and should have a 0.030-in slot size. Even though the grain size becomes larger at 395 ft, 3 ft of 0.030-in-slot screen is dictated for 395 to 398 ft. From 398 to 415 ft, the screen has a 0.060-in slot size. The completed selection of screen openings is shown in Figure 13.8.

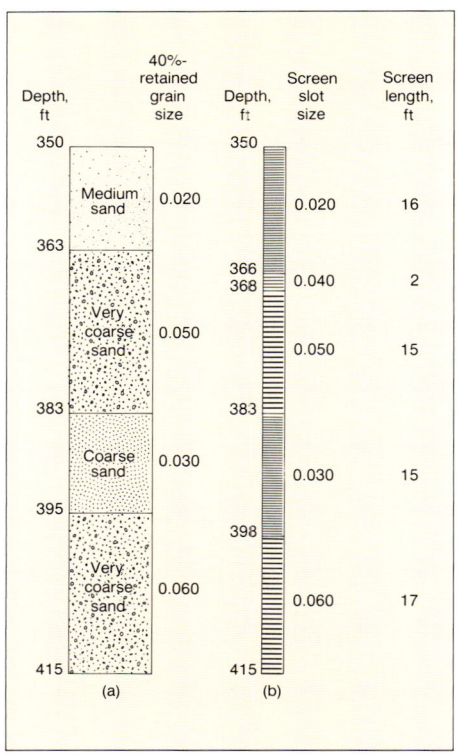

Figure 13.8. (a) Stratigraphic section that will be screened with slot sizes corresponding to various layers. (b) Sketch of screen showing the slot sizes selected based on rules 1 and 2.

If accurate samples from known depths are available, the well designer should custom design the screen to fit the aquifer conditions at the well site, because it costs no more to use a multiple-slot screen. Use of the proper screen openings to fit each sediment layer will help achieve the highest possible specific capacity and will greatly reduce the possibility of pumping sand with the water.

Filter Pack Design

The second primary method for completing wells is by filter packing. In filter-

Table 13.11. Design Table for Screen Slot Size

Depth (ft)	Thickness (ft)	Hydraulic Conductivity (gpd/ft²)	Transmissivity (gpd/ft)	Screen Openings (in)		
				50% Retained	40% Retained	30% Retained
350-363	13	500	6,500	0.019	0.020	0.024
363-383	20	2,000	40,000	0.045	0.050	0.056
383-395	12	1,000	12,000	0.026	0.030	0.034
395-415	20	1,500	30,000	0.052	0.060	0.070
Aquifer Transmissivity			88,500			

packed wells, the zone immediately around the well screen is made more permeable by removing some formation material and replacing it with specially graded material. This relatively thin zone separates the screen from the formation material and increases the effective hydraulic diameter of the well. A filter pack is chosen to retain most of the formation material; a well screen opening is then selected to retain about 90 percent of the filter pack after development. Filter pack materials should be well sorted to assure good porosity and hydraulic conductivity of the materials near the screen. Most commercial filter packs have uniformity coefficients of approximately 2. In certain areas, however, filter packs with uniformity coefficients of 4 to 5 are used occasionally with good results.

Filter packing is especially advantageous when the sediments are highly uniform and fine grained, when the sediments are highly laminated, or when all the materials to be used in the well construction must be on site before drilling begins. A filter pack is also advantageous when the small slot size dictated by natural development limits the transmitting capacity of the screen so that the desired yield cannot be obtained. Moreover, the use of certain drilling rigs may require the installation of a filter pack. For example, reverse rotary rigs will rarely complete a borehole that is less than 14 to 16 inches (356 to 406 mm) in diameter. Thus, the borehole diameter may be much larger than required for the installation of a screen.

Some geologic environments in which filter packs should be considered include:

Fine, uniform sand [glaciofluvial, alluvial, and aeolian (wind blown) aquifers]. In these formations, filter packing should be considered so that larger slot openings can be used to increase the hydraulic efficiency of the well. In general, if a slot opening based on natural development is smaller than 0.010 in (0.25 mm), filter packing may be more desirable because the screen's transmitting capacity may not be great enough to supply the desired yield. If the water is extremely incrusting, a lower limit of 0.015 in (0.38 mm) or 0.020 in (0.51 mm) may be used instead of 0.010. Some deviation from this limit is possible, usually depending on the mineral content of the water. For example, in some areas of the Gulf Coastal Plain of the southern United States, naturally developed wells with screen openings as small as 0.006 in (0.15 mm) are used because experience has shown this to be the best design.

In other situations, filter pack design is dictated by the physical nature of the aquifer. In certain fine-grained, uniformly sorted formations, a naturally developed well may lead either to low yields because screen slot sizes must be reduced, or to high rates of sand pumping. Filter packing of these same wells would generally lead to higher sand-free yields.

Examples of fine-grained formations in which wells are ordinarily filter packed include the Tertiary sands of the Gulf Coastal Plain; the Ogallala Formation in West Texas, Kansas, and Nebraska; the Raritan sand in New Jersey; the Sparta sand in Louisiana; and aquifers of the Indus Plains in West Pakistan.

Semiconsolidated (friable) sandstone. Many productive sandstone aquifers are poorly cemented. The Dakota Sandstone in North and South Dakota, the Jordan Sandstone in some areas of Minnesota, and the Garber and Elk City Sandstones in Oklahoma are examples of this type of formation. If a well is finished as an open hole in these aquifers, some sand particles continually slough from the walls of the hole, resulting in a sand-pumping well. The sloughing may begin immediately after the well is completed or after several months have elapsed, depending on the pumping rate and the amount of cement holding the sand grains together. The higher the pumping rate, the more quickly sloughing will begin. The potential for sand sloughing prompts many well-design engineers to use a well screen. Because most sandstones are fine grained, screen openings of 0.010 in (0.25 mm), or smaller, may be required to screen them correctly as naturally developed wells (based on 50-percent retention). Even screening at the 50-percent retention level may require a long development time, and large amounts of sediment may have to be removed from around the screen. For such formations, therefore, it is good design practice to use specially graded filter packs so that larger screen openings can be used. It is important to recognize, however, that the installation of a filter pack and screen usually reduces the specific capacity of a previously open-hole well. Nevertheless, the reduction in yield is preferable to an unending maintenance problem created by sand pumping.

Another reason for filter packing a sandstone aquifer is that the formation material usually provides little or no lateral support for the screen. Also, the formation does not readily slump against the screen during development, in contrast to unconsolidated sediments. After setting the screen in the open borehole, some void spaces remain between the screen and the borehole wall. It is possible that a section of the formation could break off, fall against the screen, and cause damage. Loose, granular material inserted between the screen and the borehole wall accommodates itself to all the irregularities of the borehole, supporting both the wall and the screen.

Figure 13.9 shows the construction details of a filter-packed well finished in semiconsolidated sandstone. The larger space around the screen is created by an underreamer that increases the borehole diameter only where required. The oversized borehole provides sufficient annular area for a filter pack.

Extensively laminated formations. Some aquifers consist of alternating thin layers of fine, medium, and coarse sediment. Examples include the Magothy Formation of Long Island, New York; the alluvial deposits of the San Joaquin Valley of California; the Santa Fe Formation of central New Mexico; some coastal plain deposits of North Carolina; the Ogallala Formation of the High Plains region; and some highly stratified glacial aquifers. It is often difficult to determine precisely the position and thickness of each individual layer and to design a multiple-slot screen corresponding to the stratification. Therefore, filter packing should be specified for wells in these types of formations to avoid screen placement problems.

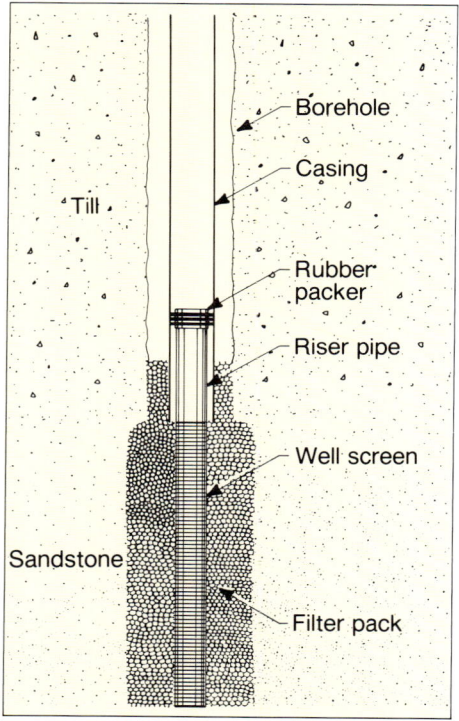

Figure 13.9. Filter pack design is usually preferred for wells completed in semiconsolidated sandstone aquifers such as the Dakota and Lakota Sandstones in South Dakota.

The grading of the filter pack should be based on the grain size of the finest layer to be screened. A filter pack selected in this manner ordinarily does not restrict the flow from the layers of coarsest material. The hydraulic conductivity of the pack is generally several times greater than that of the coarsest layers because the pack is cleaner and more uniform.

Filter pack material should consist of clean, well-rounded grains of a uniform size. These characteristics increase the permeability and porosity of the pack material. Pit-run or crushed materials are usually not satisfactory for filter packs. The chemical nature of the filter pack is as important as its physical characteristics. Filter pack material consisting mostly of siliceous, rather than calcareous, particles is preferred. Up to 5 percent calcareous material is a common allowable limit. This is important because acid treatment of the well might be required later, and most of the acid could be spent in dissolving calcareous particles of the filter pack rather than in removing incrusting deposits of calcium or iron. Similarily, if the groundwater is slightly acidic, partial dissolution of the pack may occur over time. Particles of shale, anhydrite, and gypsum in the filter pack material also are undesirable. Table 13.12 lists the desirable physical and chemical characteristics for a filter pack and the advantages of using these materials.

The steps outlined below are followed in designing a filter pack:

1. Choose the layers to be screened and construct sieve-analysis curves for these formations. Select the grading of the filter pack on the basis of the sieve analysis for the layer of finest material. Figure 13.10 shows the grading of two samples of typical water-bearing material from an aquifer 30 ft (9.1 m) thick. The finest material lies between 75 and 90 ft (22.9 and 27.4 m). The design of the filter pack in this example will be based on this layer. In some instances, it is good practice to ignore unfavorable portions of an aquifer and to use blank pipe between sections of screen positioned in the more permeable sections of the aquifer.

2. Multiply the 70-percent size of the sediment by a factor between 4 and 10. Use 4 to 6 as the multiplier if the formation is uniform and the 40-percent-retained size

Table 13.12. Desirable Filter Pack Characteristics and Derived Advantages

Characteristic	Advantage
Clean	Little loss of material during development Less development time
Well-rounded grains	Higher hydraulic conductivity and porosity Reduced drawdown Higher yield More effective development
90 to 95% quartz grains	No loss of volume caused by dissolution of minerals
Uniformity ccefficient of 2.5 or less	Less separation during installation Lower head loss through filter pack

is 0.010 (0.25 mm) or less. Use a multiplier between 6 and 10 for semiconsolidated or unconsolidated aquifers when formation sediment has highly nonuniform gradation and includes silt or thin clay stringers, as commonly found in arid or semiarid areas. Using multipliers greater than 10 may result in a sand-pumping well. Place the result of this multiplication on the graph as the 70-percent size of the filter material. In Figure 13.10, 0.005 in (0.13 mm) is the 70-percent size of the sand between 75 and 90 ft. Using 5 as the multiplier, the 70-percent size of the filter material is 5 × 0.005 = 0.025 in (5 × 0.13 = 0.65 mm). This is the first point on a curve that represents the grading for the filter pack material.

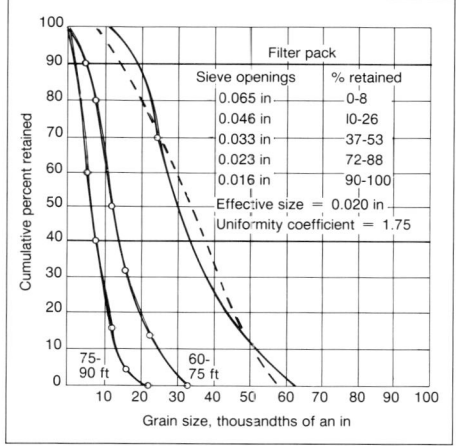

Figure 13.10 Grain-size curves for aquifer sand and corresponding curve for properly selected filter pack material.

3. Through the initial point on the filter pack curve, draw a smooth curve representing material with a uniformity coefficient of approximately 2.5 or less. In Figure 13.10, the curve drawn as a solid line has a uniformity coefficient of about 1.8. It could have been drawn somewhat differently, as shown by the dashed line which has a uniformity coefficient of 2.5. It is good practice to draw the filter pack curve so that the pack is as uniform as practicable. Thus, the material indicated by the solid-line curve is more desirable than the material indicated by the dashed-line curve.

4. Select a commercial filter pack that fulfills the dimensional and chemical requirements listed in Table 13.12. If a proper commercial pack cannot be purchased, but a local source of sand and gravel is available, the following procedure can be used to construct a suitable filter pack.

Prepare specifications for the filter pack material by first selecting four or five sieve sizes that cover the range of values for the curve, and then set down a permissible range for the percentage retained on each of the selected sieves. This range may be eight percentage points below and above the percentage retained at any point on the curve. In the example, the largest sieve would have an opening of 0.065 in (1.7 mm). The curve shows zero percent retained on this sieve, so up to 8 percent of the filter pack may contain 0.065-in material. The next smaller opening in the most commonly used series of sieves is 0.046 in (1.2 mm). The curve, as drawn, shows 18 percent retained on this sieve; 8 percent is added and subtracted to obtain the permissible range. Thus, on the 0.046-in sieve, the range is from 10 to 26 percent. This procedure is repeated until each of the sieves previously selected has been assigned a permissible range. In Figure 13.10, five sizes of sieve openings are shown to cover the desired gradation of the pack material. Giving the filter pack supplier an acceptable range at each of these points makes it possible to produce the desired material at reasonable cost. When designing filter pack material, the designer should keep in mind local sources of filter sand used for rapid sand filters*. Firms that produce these materials

---

*Rapid sand filters consist of sand beds used to filter drinking water supplies in water treatment plants.

have large stocks of clean, uniformly graded sands and gravels that readily fit the requirements for filter packing of water wells. Some firms supply sand materials to oil and gas companies for use as propping materials in hydraulic fracturing of formations. These materials are also suitable for filter packing of water wells. Drilling contractors should obtain grain-size-distribution curves for all local sources of potential filter pack materials. For economic reasons, these packs should be specified if possible.

5. As a final step, select a screen slot size that will retain 90 percent or more of the filter pack material. In our example, the correct slot size is 0.018 in (0.46 mm).

6. Calculate the volume of filter pack required from Table 13.13. The pack should extend well above the screen to compensate for settlement of the pack during development. Use of a caliper log may reveal the presence of washouts in the borehole, necessitating additional filter pack. It is good practice to have extra filter pack on the site, especially if the stability of the borehole is in doubt.

If the well designer and contractor carefully follow the foregoing steps, sand-pumping wells can be avoided. The pack will provide mechanical retention of the formation material and prevent sediment from moving through the filter pack into the well. Occasionally it may be necessary to install more than one size of filter pack in a borehole. For example, thick boulder beds may overlie sand deposits and the yield requirements may dictate that both layers be screened. If the use of more than one filter pack is contemplated, the screen manufacturer should be consulted for specific design recommendations.

Thickness of Filter Pack

The design theory of filter pack gradation is based on the mechanical retention of formation particles; therefore, a pack thickness of only two or three grain diameters is actually needed to retain and control a formation. Laboratory tests made by Johnson Division show that a properly sized pack with a thickness of less than 0.5 in (12.7 mm) successfully retains the formation particles regardless of the velocity of water passing through the filter pack. It is impossible, however, to place a filter pack that is only 0.5 in thick and expect the material to completely surround the well screen. To insure that a continuous layer of filter material will surround the entire screen, the design should specify that the annulus around the screen be at least 3 in (76 mm).

Filter-pack thickness does little to reduce the possibility of sand pumping, because the controlling factor is the ratio of the grain size for the pack material in relation to the formation material. Also, a filter pack that is too thick can make final development of the well more difficult, as explained in Chapter 15. Under most conditions, filter packs should not be more than 8 in (203 mm) thick because the energy created by the development procedure must be able to penetrate the pack to repair the damage done by drilling, break down any residual drilling fluid on the borehole wall, and remove fine particles near the borehole.

It has been suggested that the presence of a filter pack will augment the well yield because water from an overlying aquifer can percolate downward through the filter pack and into the well screen. In practice, however, calculations show this contribution to be insignificant in relation to total yield. For example, assume the conditions shown in Figure 13.11, where 90 percent of a confined aquifer has been screened. The overlying sediments are water bearing and are connected hydraulically to the screened portion of the well by the 6-in (152-mm) filter pack. The theoretical volume of water that can move downward from the upper aquifer to the well screen can be calculated by the Darcy equation:

$$Q = KIA$$

where
Q = vertical flow through the pack material, in gpd
K = hydraulic conductivity of filter pack, in gpd/ft^2
I = hydraulic gradient causing vertical flow in the filter pack
A = cross-sectional area of the filter pack, in ft^2

In this example, the available head is 30 ft — the difference between the pumping level in the well and the static water level in the upper aquifer. The average distance through which the upper water must move is about 28 ft, the distance from the midpoint of the upper aquifer to the top portion of the screen. In this case, $I = 30/28 = 1.1$ and $A = 2.25$ ft^2. The hydraulic conductivity, K, of the filter pack must be estimated. A reasonable upper limit for pack materials is 17,000 gpd/ft^2.

The amount of water transmitted vertically in this example is, therefore:

$$Q = 17,000 \cdot 1.1 \cdot 2.25 = 42,075 \text{ gpd} = 29.2 \text{ gpm}$$

The contribution of 29.2 gpm is a relatively small proportion of the total amount of water that can be pumped from the hypothetical well. If the lower aquifer has a hydraulic conductivity of 1,000 gpd/ft^2, its transmissivity is about 50,000 gpd/ft. An efficient well in this aquifer should develop a specific capacity of about 25 gpm/ft of drawdown. A drawdown of 30 ft would mean that the yield of the lower formation alone is about 750 gpm. The theoretical yield from vertical flow in the filter pack is 29.2 gpm, or about 4 percent of the total for the well.

The actual contribution to the yield through the filter pack will depend on how the pack is placed, how much drilling fluid remains in the borehole, and the physical and chemical changes that take place in the pack over time. When the filter pack material is placed in the well, uneven settlement of the material can create zones of finer particles interspersed with coarser material. This layering effect can reduce significantly the vertical hydraulic conductivity of the pack. The remnants of clay additives left over from the drilling fluid also can decrease the porosity and hydraulic conductivity of the pack. Development methods are effective primarily around the screen, and only

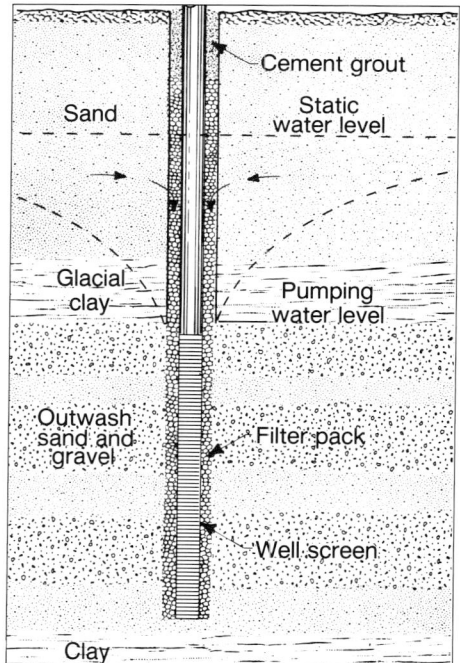

Figure 13.11. In the conditions shown here, the filter pack above the glacial clay will not contribute greatly to the yield of the well.

WATER WELL DESIGN 445

Table 13.13. Volume of Filter Pack Required*

ID of Pipe or Borehole		4 in ft³/ft	102 mm m³/m	6 in ft³/ft	152 mm m³/m	8 in ft³/ft	203 mm m³/m	Outside Diameter of Well Screen											
In	mm							10 in ft³/ft	254 mm m³/m	12 in ft³/ft	305 mm m³/m	16 in ft³/ft	406 mm m³/m	18 in ft³/ft	457 mm m³/m	20 in ft³/ft	508 mm ms³/m
8	203	0.27	0.03	0.15	0.01	—	—	—	—	—	—	—	—	—	—	—	—
10	254	0.47	0.04	0.36	0.03	0.20	0.02	—	—	—	—	—	—	—	—	—	—
12	305	0.70	0.07	0.60	0.06	0.45	0.04	0.24	0.02	—	—	—	—	—	—	—	—
16	406	1.30	0.12	1.20	0.11	1.05	0.10	0.86	0.08	0.62	0.06	—	—	—	—	—	—
20	508	2.10	0.20	2.00	0.19	1.90	0.18	1.65	0.15	1.40	0.13	0.80	0.07	0.42	0.04	—	—
24	610	3.05	0.28	2.95	0.27	2.80	0.26	2.60	0.24	2.35	0.22	1.75	0.16	1.40	0.13	1.00	0.09
30	762	4.85	0.45	4.70	0.44	4.60	0.43	4.40	0.41	4.15	0.39	3.50	0.33	3.15	0.29	2.75	0.26

*Slightly more filter pack is required for telescope-size screens, slightly less for pipe-size screens.

small amounts of fine material can be removed from the filter pack above the screen. Even more important than these two factors are the physical and chemical changes that occur in the upper part of the pack over time. Aeration of the filter pack occurs in the upper aquifer when pumping begins, and minerals in solution can precipitate and plug the filter pack, causing a significant reduction in the yield. Thus, an initial 4-percent contribution to the total yield can become much less after some months or years of pumping. In other situations, the percentage of contribution may vary but is always only a small part of the total yield.

Cost Factors

The cost of filter packing a well depends on the drilling method, the length of development time, the installation procedure, and the availability of filter pack material. The larger hole size required for a filter-packed well generally costs more per foot when drilled with any rig other than a reverse rotary machine. With cable tool drilling equipment, doubling of the well diameter may more than double the drilling cost. In direct rotary drilling, large-diameter holes also cost more, as a rule, because more viscous drilling fluids and higher drilling fluid circulation rates are needed to raise the cuttings to the ground surface. On the other hand, with reverse circulation drilling equipment, an increase in hole diameter is of little concern. Drilling a 36-in (914-mm) hole generally costs only slightly more than a 24-in (610 mm) hole because the only basic items of extra cost are a larger bit, a larger mud pit, and more filter material.

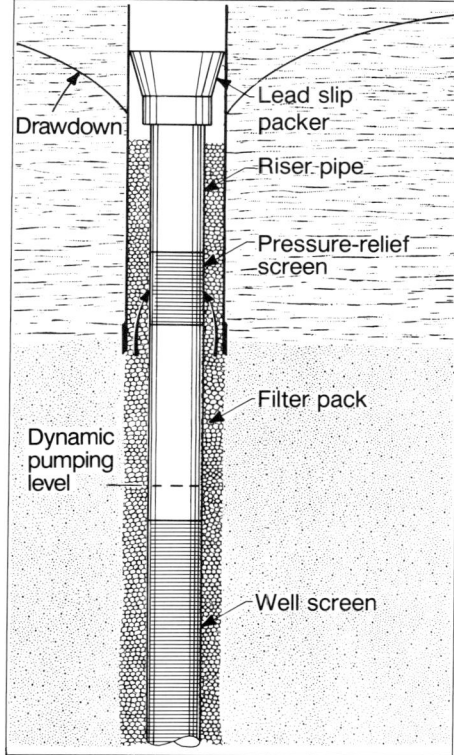

For certain uniform sediments, the cost for natural development of a well may exceed the extra costs for the large borehole and filter pack required for filter-packed wells. In most cases, the drilling contractor will charge approximately 50 percent of the drilling rate for development because the rig is still involved in the development process although there are no bit or casing costs. It is sometimes more economical, therefore, to construct a filter-packed well because the saving in development time may offset the extra initial cost. This is especially true in terrace and alluvial deposits like those found in Oklahoma, Kansas, and Nebraska.

In some areas of the world, filter pack materials are not readily available and must be transported hundreds of miles to the drilling site. Elsewhere, high-grade filter packs are available locally and cost relatively little. Many drilling contractors will stock one or more sizes of filter pack at their place of business because they drill continually in the same aquifers and

Figure 13.12. A pressure-relief screen should be installed just up inside the bottom of the casing to relieve large differential pressures.

can anticipate their needs. In this situation, the cost of the filter material is a small part of the total well cost.

Two design errors are common in filter-packed wells. First, some contractors use the same pack in all their wells regardless of the particular formation, a practice that can lead to low yields or sand-pumping wells. Second, a contractor may use the correct filter pack for a formation, but will use a screen from stock with slot sizes that are too small. Again, low yields are likely because of reduced hydraulic efficiency and a tendency for greater incrustation.

Cost comparisons between wells that are filter packed and those that are naturally developed can be made only for specific wells, depending on a number of factors. No general statement can be made that one method is more costly or will produce more efficient wells than the other. Both methods produce wells that have high specific capacities, are sand free, and will last many years without extensive maintenance.

PRESSURE-RELIEF SCREENS

In some well installations, the drawdown outside the well may be significantly above the pumping level inside the casing. In these circumstances, high pressure differentials may cause a powerful upward flow of water in the filter pack. Occasionally, the difference between the pressure in the casing and the pressure in the borehole is great enough to lift off heavy lead slip packers, or to lift filter pack material up to the pump intake if no packer is used. Tight or highly stratified formations are most likely to lead to high differential pressures. To relieve pressure in the filter pack, a short pressure-relief screen should be installed in the riser pipe so that it is just up inside the bottom of the casing (Figure 13.12). During pumping, pressure differentials are relieved through the screen rather than up through the pack. It is good design practice to install a pressure-relief screen between the top of the production screen and the top of the riser pipe to eliminate the build-up of differential pressures during pumping.

FORMATION STABILIZER

The primary purpose of a formation stabilizer is to keep the borehole open and prevent caving of overlying clays or other fine material into the screened portion of the well. In unstable formations, a stabilizer should be considered if the borehole is more than 2 in (51 mm) larger than the casing or screen. For example, in South Dakota, screened siltstone formations are often packed with stabilizer materials to prevent the pumping of silty water. Stabilizers are also used to prevent premature caving of formation material prior to development. This can be a major problem in oversized boreholes or where confining pressures are present, because caving of highly stratified materials can cause significant reduction in the porosity and permeability of aquifer materials for some distance away from the borehole. A second function of the stabilizer material is to maintain or augment the hydraulic conductivity of the natural formation.

The type of stabilizer used depends on the physical characteristics of the formation materials. For completely unconsolidated formations such as alluvial sands and silts, or glacial sands and gravels, the stabilizer should be chosen with care. Because the well is to be naturally developed, the 40-percent-retained size dictates the screen slot size. The stabilizer is then chosen so that its graded sizes are similar to or slightly larger than the natural formation. In practice, the median grain size (50-percent-retained size) of the formation is used as the basis for selecting the stabilizer. During development, 50 to 60 percent of the stabilizer material will be removed.

In some cases, the physical characteristics of a stabilizer used in an unconsolidated formation are probably more important than the actual grain sizes. The particles should be well rounded, possess a low calcareous content, and be free of contaminating material. The stabilizer should also be well sorted so that its presence will enhance the natural porosity and hydraulic conductivity of the materials immediately outside the well screen. The use of a stabilizer should never reduce the potential hydraulic efficiency of a well.

Local experience plays a large role in determining when the use of a stabilizer is more advantageous than a filter pack. In formations consisting mainly of shell fragments, for example, the installation of a filter pack may reduce significantly the potential yield because of the filter pack's relatively small grain-size distribution. In this instance, some drillers are quite successful when they install a stabilizer material that is much coarser than the recommended filter pack. Even though some calcareous material may be taken into solution, it does not result in freeing a significant number

Table 13.14. Selecting Formation Stabilizers for Unconsolidated and Semiconsolidated Aquifers

Type of Aquifer	Unconsolidated Aquifers: Alluvial and glaciofluvial sands and gravels, and beach deposits.	Semiconsolidated Aquifers: Dirty sandstones and siltstones, and sandy formations containing shells.
Purpose	Provide temporary support for the borehole walls next to the screen.	Permanently hold back the formation without providing any mechanical retention of small particles.
Characteristics of Stabilizer	Grain-size distribution should be equal to or slightly larger than the original formation.	Grain-size distribution is usually greater than 12 to 13 times the 70-percent-retained size.
Development	Approximately 50 to 60 percent of the stabilizer is removed during natural development of the formation.	None of the stabilizer passes through the screen during development.
Result	The part of the formation stabilizer that remains next to the screen has a hydraulic conductivity similar to the natural formation so that flow is unimpeded. The stabilizer also plays a major role in preventing the migration of fine particles into the screen.	The formation cannot slump against the screen even if it becomes weakened over time. The porosity and hydraulic conductivity of the stabilizer are high, reducing drawdown in the immediate vicinity of the screen.

of sand grains. On the other hand, if a coarse stabilizer is used in a formation that consists mainly of sand or silt held together by calcareous cement, a sand-pumping well could result. Guidelines for the selection of a formation stabilizer are given in Table 13.14.

Most stabilizer materials are placed by hand, but placement by pumping through a tremie pipe is also done. To prevent excessive bridging, the use of centralizers is recommended for screens of more than 30 ft (9.1 m). Even though the final depth of the stabilizer in the borehole will vary according to the amount of material removed during development, approximately 30 to 50 ft (9.1 to 15.2 m) of stabilizer should extend above the top of the screen before development begins.

WELL SCREEN DIAMETER

Screen diameter is selected to satisfy a basic principle: enough open area must be provided so that the entrance velocity of the water generally does not exceed the design standard of 0.1 ft/sec (0.03 m/sec). Screen diameter can be adjusted within rather narrow limits after the length of the screen and size of the screen openings have been selected. Screen length depends upon the thickness of the aquifer; screen openings depend upon the gradation of the sediment or the size of the filter pack.

Well yields are affected by screen diameter, although increasing the screen diameter has much less impact on well yield than increasing the screen length. The theoretical increase in yield that results from enlarging the well diameter can be calculated from the relationship developed in Equation 9.1, where:

$$Q = \frac{K(H^2 - h^2)}{1055 \log R/r}$$

This equation can be stated as

$$Q \sim \frac{C}{\log R/r}$$

where C represents all the constant terms.

Table 13.15 shows the figures obtained when $R = 400$ ft (122 m), a typical radius of influence for unconfined conditions. These comparisons indicate that well diameter requires careful consideration because an increase in screen diameter may not enhance specific capacity or well yield significantly. In some cases, however, it may be worthwhile to increase the well diameter to obtain 15 to 25 percent more water, depending on the cost factors involved.

Table 13.15. Well Diameter vs. Yield Ratio, in Gallons (%)

6 in (152 mm)	12 in (305 mm)	18 in (457 mm)	24 in (610 mm)	30 in (762 mm)	36 in (914 mm)	48 in (1219 mm)
100	110	117	122	127	131	137
—	100	106	111	116	119	125
—	—	100	104	108	112	117
—	—	—	100	104	107	112
—	—	—	—	100	103	108
—	—	—	—	—	100	105

Table 13.15 shows that if a 6-in well yields 100 gpm with a certain drawdown, a 12-in well constructed at the same place will yield 110 gpm with the same drawdown; a 48-in well will yield 137 gpm, or 37 percent more water with the same drawdown. Thus, doubling the diameter of the intake section (well screen) can be expected to increase the yield only about 10 percent when other factors are unchanged. For wells in confined aquifers, where R is much larger [5,000 ft (1,520 m) is a typical radius], the increase that results from doubling the well diameter is smaller, generally about 7 percent.

These ratios also apply to specific capacity. For example, if a 12-in well is producing 20 gpm per foot of drawdown, then a 24-in well in the same location would produce 111 percent as much, or 22.2 gpm per foot of drawdown.

OPEN AREA

Most screen manufacturers provide tables that show the open area per foot of screen for each size of screen and for various widths of slot openings. The open area for different slot configurations varies significantly. Table 13.16 shows that continuous-slot screens have much larger open areas than do bridge-slot, louvered, or mill-slotted screens. Limited open area impedes well development, causing increased drawdown and higher pumping costs for a particular yield. The use of continuous-slot screens assures maximum specific capacity.

ENTRANCE VELOCITY

Field experience and laboratory tests show that the average entrance velocity of water moving into the screen should not exceed 0.1 ft/sec (0.03 m/sec)*. At this velocity, the friction losses in the screen openings will be negligible and the rates of

Table 13.16. Open Areas of Screens

Screen Diameter	Slot Size	Continuous Slot		Louvered (Maximum open area)		Bridge Slot		Mill Slotted (Vertical)		Plastic Continuous Slot		Slotted Plastic	
		in²/ft	%	in²/ft	%	in²/ft	%	in²/ft	%	in²/ft	%	in²/ft	%
4" ID	20	44	25	—	—	—	—	—	—	22	13	—	—
	60	90	52	—	—	19	12	8	5	52	30	18	11
8" ID	30	80	25	—	—	—	—	—	—	57	18	26	8
	60	135	41	10	3	17	6	15	5	93	29	47	14
	95	165	51	15	5	—	—	23	7	—	—	—	—
12" ID	30	77	16	—	—	12	3	—	—	—	—	—	—
	60	135	28	20	4	33	7	21	5	—	—	52	11
	95	182	38	30	7	—	—	32	7	—	—	—	—
	125	214	45	39	9	68	14	43	9	—	—	—	—
16" OD	30	97	16	—	—	16	3	—	—	—	—	52	9
	60	169	28	24	4	35	6	27	5	—	—	—	—
	95	228	38	35	6	—	—	41	7	—	—	—	—
	125	268	45	47	8	78	13	55	9	—	—	—	—

See Appendix 12.A for a more complete open-area chart.

---

*The rationale for using a design velocity of 0.1 ft/sec is given in Appendix 13.I.

incrustation and corrosion will be minimal.

The average entrance velocity is calculated by dividing well yield by the total area of the screen openings. If the velocity is greater than 0.1 ft/sec, the screen length and/or diameter should be increased to provide enough open area so that the entrance velocity is 0.1 ft/sec or less. Lengthening a screen in an unconfined aquifer may decrease the available drawdown, thereby reducing the yield. On the other hand, a screen fully penetrating a confined aquifer will enhance yields so long as the aquifer is not dewatered. Occasionally, it may be possible to vary the construction characteristics of a screen to increase the open area. For example, the wire width may be decreased in a continuous-slot screen as long as strength requirements are met.

Example: A 20-ft long, 14-in pipe size, continuous-slot stainless steel screen is to be installed in a well. The width of the outside wrapping wire used to fabricate the screen is 0.156 in, and the recommended slot size is 0.065 in. The anticipated yield is 2,000 gpm.

1. Calculate the surface area for each 1-ft length of screen:

$$\text{Area} = \pi d \cdot 12$$
$$= 3.14 \cdot 14 \cdot 12$$
$$= 528 \text{ in}^2/\text{ft of screen}$$

where d = screen diameter.

2. Calculate total area for 20 ft of screen:

$$20 \cdot 528 = 10,560 \text{ in}^2$$

3. Calculate the percentage of open area of the screen, based on the wire width used to fabricate the screen and the slot size required:

$$\% \text{ Open Area} = \frac{\text{slot size}}{\text{slot size} + \text{wire width}} \cdot 100$$

$$= \frac{0.065}{0.065 + 0.156} \cdot 100$$

$$= 29.4\%$$

Thus, 29.4 percent of the screen's outer surface will be open to the aquifer.

4. Calculate the amount of open area:

$$\text{Open area} = \text{surface area} \cdot \% \text{ open area}$$
$$= 10,560 \cdot 0.294$$
$$= 3,105 \text{ in}^2 \div 144*$$
$$= 21.6 \text{ ft}^2$$

Open area per ft of screen = $21.6 \div 20 = 1.08 \text{ ft}^2$

5. Calculate the average entrance velocity of water moving into the slots, by:

$$Q = VA$$

where
Q = anticipated yield, in ft^3/sec
V = entrance velocity, in ft
A = screen open area, in ft^2

---

*There are 144 square inches in 1 square foot.

Therefore,

$$V = \frac{Q}{A}$$

Convert yield in gpm to ft³/sec:

2,000 gpm ÷ 7.5 gal per ft³ ÷ 60 sec per min = 4.44 ft³/sec

Therefore, the entrance velocity is:

$$V = \frac{4.44}{21.6} = 0.21 \text{ ft/sec}$$

6. Because 0.21 ft/sec is greater than the recommended velocity of 0.1 ft/sec, either the screen length should be increased or the screen diameter enlarged. In this situation, enough drawdown is available to lengthen the screen without limiting the yield.

To calculate the new screen length, determine the amount of open area required at an inlet velocity of 0.1 ft/sec. Therefore:

$$V_1 A_1 = V_2 A_2$$

$$A_2 = \frac{V_1 A_1}{V_2}$$

$$= \frac{0.21 \cdot 21.6}{0.1}$$

$$= 45.4 \text{ ft}^2$$

Open area required is 45.4 ft².

From step 4, the open area is 1.08 ft² per ft of screen. Therefore, the minimum screen length required is:

$$\frac{45.4}{1.08} = 42 \text{ ft}$$

7. As an alternative to adding more screen, the diameter could be increased. If a 20-ft, 36-in pipe size screen is used, the new entrance velocity can be calculated:

$$\text{Area} = \frac{\pi \cdot 36 \cdot 12 \cdot 20}{144} = 188 \text{ ft}^2$$

Open area = 188 · 0.232* = 43.6 ft²

$$\text{Velocity} = \frac{4.44}{43.6} = 0.10 \text{ ft/sec}$$

Thus, a 36-in pipe size, 20-ft screen would have sufficient open area for the anticipated yield. But, this design may not be practical because casing much larger than necessary would be required, construction of the borehole would be more costly, and

---

*On a 36-in-diameter screen, the wire face width would have to be increased to 0.215 in to maintain adequate strength; the open area is then 23.2 percent.

certain drilling methods may be eliminated.

8. Another option is to use a filter pack so that the slot size can be increased.

9. If design criteria for strength permit, the wire width may be decreased, thereby increasing open area.

10. On the other hand, if the calculated entrance velocity is significantly less than 0.1 ft/sec, the screen diameter may be reduced somewhat.

The foregoing statements assume that the pump will be set just above the well screen (the usual case), and thus the head loss associated with the average flow of water upward within the screen section will be small. Unacceptably high head losses can occur when long, small-diameter riser pipes are placed between the screen and the pump. If the upward velocity is less than 5 ft/sec (1.5 m/sec) in both the screen and the riser pipe, the head losses will be reasonable.

SCREEN TRANSMITTING CAPACITY

The transmitting capacity of a well screen, expressed as gpm/ft of screen at the recommended entrance velocity of 0.1 ft/sec, is calculated readily from the open area figures provided by well screen manufacturers. Multiplying the number of square inches of open area, as shown in Table 13.16, by a factor of 0.31 gives the transmitting capacity at a velocity of 0.1 ft/sec. The units conversion factor of 0.31 results from specifying an entrance velocity, V, of 0.1 ft/sec in the equation $Q = VA$. For example, the open area for an 8-in, continuous-slot Johnson well screen (standard construction) with 0.060-in slot openings is 113 in^2/ft of screen; the transmitting capacity is calculated by multiplying 0.31 times the open area of the screen (0.31 × 13 = 35 gpm/ft). Ten feet of this screen would transmit 350 gpm with an entrance velocity of 0.1 ft/sec. It should be recognized that the transmitting capacity of a well screen is a hydraulic characteristic of the screen itself at the assumed entrance velocity, and is not a measure of the yielding ability of the water-bearing formation in which the screen might be installed.

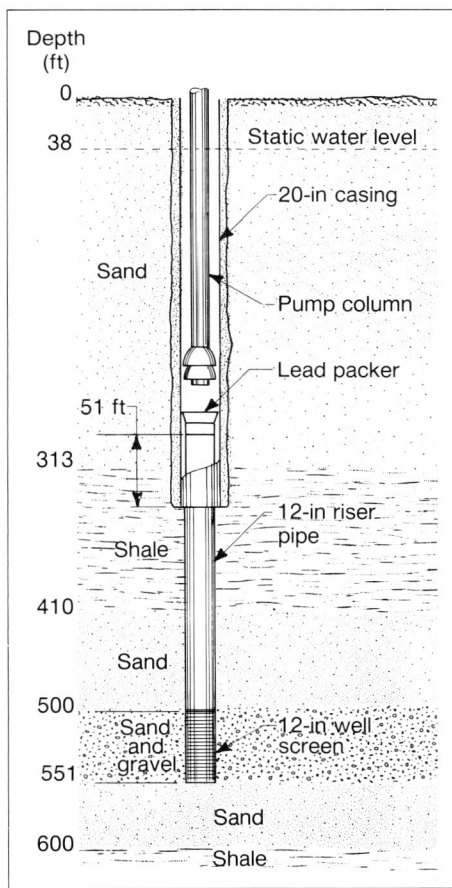

Figure 13.13. Well designed with a screen diameter smaller than the casing diameter.

In nonhomogeneous aquifers, especially those of glaciofluvial and alluvial fan origin, some sediment layers will have much greater hydraulic conductivity than others. Thus, water from these more permeable layers may enter the screen at a higher velocity than the average calculated entrance velocity. Similarly, entrance velocities from the finer grained layers will be lower than average. Experience has shown that these velocity differences along the screen are not serious if the screen is designed to the recommended average entrance velocity of 0.1 ft/sec.

Continuous-slot screens with diameters smaller than the casing are often used in wells. Figure 13.13 shows the installation of a well in the Chicot Formation in southern Louisiana. The expected yield of 1,800 gpm (9,810 m^3/day) or more required the use of 20-in (508-mm) casing to provide adequate space for the pump. Selection of a 12-in (305-mm) diameter pipe-size well screen assured an uphole velocity of less than 5 ft/sec (1.5 m/sec), and provided enough open area so that the entrance velocity would not exceed 0.1 ft/sec (0.03 m/sec).

When the pump is set within pipe that connects two screen sections, the screen diameter for all screen sections set above the lowest anticipated pump setting should be selected from Table 13.1. Any screen section below the pump-setting position, however, can be of smaller diameter if the conditions for entrance and uphole velocities are met.

SELECTION OF MATERIAL

Three factors govern the choice of material used to fabricate well screens. These are (1) water quality, (2) potential presence of iron bacteria, and (3) strength requirements of the screen.

Water Quality

The chemical nature of groundwater is determined from a water quality analysis. The analysis usually shows whether the groundwater is corrosive or incrusting. In some cases, the water may cause both corrosion and incrustation. Corrosion of a low-carbon steel well screen is more likely to be a cause of well failure than is corrosion of casing constructed of similar material. Enlargement of screen openings resulting from removal of only a few thousandths of an inch of metal can permit sediment to enter the well. On the other hand, corrosion could remove 0.125 to 0.250 in (3.2 to 6.4 mm) of the casing wall and still leave enough wall thickness to prevent collapse of the well or entrance of undesirable water and sediment. It is important, therefore, to use a well screen fabricated from corrosion-resistant material (Table 13.9).

The following list of indicators of corrosive water can help the well designer recognize potentially corrosive conditions:

1. Low pH. If the pH value is less than 7, the water is acidic, and corrosive conditions are indicated. Similarly, a Ryznar Stability Index value greater than 7 indicates corrosive conditions.

2. Dissolved oxygen. If dissolved oxygen exceeds 2 mg/ℓ, corrosive water is indicated. Dissolved oxygen may be found in shallow wells in unconfined aquifers.

3. Hydrogen sulfide. Hydrogen sulfide in groundwater can be detected readily by its characteristic rotten-egg odor. Less than 1 mg/ℓ can cause severe corrosion, and this amount can be detected by odor and taste.

4. Total dissolved solids. If total dissolved solids exceed 1,000 mg/ℓ, electrical conductivity of the water is great enough to cause serious electrolytic corrosion. To avoid electrolytic corrosion, metal well screens must be made of a single, corrosion-resistant metal.

5. Carbon dioxide. If the amount of this gas exceeds 50 mg/ℓ, corrosive water is indicated.

6. Chlorides. If the chloride content of the water exceeds 500 mg/ℓ, corrosion can be expected.

The presence of two or more corrosive agents appears to intensify the corrosive attack on metals, compared with the effect caused by individual agents. See Chapter 19 for more information on corrosion and incrustation.

In corrosive waters, metal screens must be constructed of durable materials. In most cases, Type 304 stainless steel will perform satisfactorily for many years. Under extremely corrosive conditions, or for waste disposal wells, other types of material may be required. These include Type 316L stainless steel, Inconel 600, Monel 400, Carpenter 20Cb-3, Incoloy 825, and Hastelloy C. Thermoplastic screens are corrosion resistant, but may not have sufficient strength for some applications. The well-design engineer should contact the well-screen manufacturer for specific recommendations on screen materials for corrosive conditions.

Incrusting water deposits minerals on the screen surface and in the pores of the formation just outside the screen. These deposits plug both the screen openings and the formation. Indicators of incrusting groundwater are:

1. High pH. If the pH value is above 7.5, the water will tend to be incrusting. A Ryznar Stability Index of less than 7 also indicates incrusting conditions.

2. Carbonate hardness. If the carbonate hardness of the groundwater exceeds 300 mg/ℓ, incrustation of calcium carbonate (lime scale) is likely.

3. Iron. If the iron content of the water exceeds 0.5 mg/ℓ, precipitation of iron is likely, although some precipitation may begin at concentrations as low as 0.25 mg/ℓ.

4. Manganese. If the manganese content of the water exceeds 0.2 mg/ℓ and the pH value is high, precipitation of manganese is likely if oxygen is present.

Two well-known analytical methods have been devised to predict the incrusting or corrosive tendencies of a particular water. These methods are called the Langelier and the Ryznar Stability Indices. Langelier was the first to develop a method for predicting the saturation pH (pHs), which he based on the calcium carbonate ($CaCO_3$) equilibrium values taking into account dissociation factors for carbonic acid, bicarbonate, and carbonate, and the theoretical solubility of calcium carbonate (Kemmer, 1979). If the actual pH of a water is below the calculated pHs level, the water has a negative Langelier Index and will dissolve calcium carbonate. Under this condition, the water would probably be corrosive to steel if dissolved oxygen is present. On the other hand, if the measured pH is higher than the pHs, the Langelier Index is positive and, being supersaturated with calcium carbonate, incrustants will probably form. In general, the greater the deviation of the actual pH from the pHs, the more pronounced the chemical instability will be. The Langelier Index is defined as:

$$\text{Langelier Index} = pH - pHs$$

Ryznar adapted the Langelier Index to reflect more accurately the incrusting or corrosive tendencies. This index is used widely for predicting the reaction of metal objects in saturated subsurface environments. A water is corrosive if the index is higher than 7, and incrusting if lower than 7. Determination of the Ryznar Stability Index is explained in Appendix 13.J.

Mineral deposits from incrusting-type groundwater are often removed by putting a strong solution of hydrochloric (muriatic) or sulfamic acid into the well to dissolve the deposits. Metal well screens used in this type of water should be made of corrosion-resistant materials to withstand the corrosive effect of the acid treatment. Type 304 stainless steel well screens have been acid-treated successfully, although common sense dictates limiting the screen's exposure time to the acid. The addition of a corrosion inhibitor to the acid, however, will greatly extend the time a metal screen can be safely exposed to acid (see Chapter 19).

Thermoplastic and fiberglass screens are highly resistant to many forms of corrosion, but are generally as susceptible to incrustation as metal screens. Figure 13.14 shows a small-diameter plastic casing incrusted with iron and manganese oxides. Apparently, the smooth walls of plastic casing do not retard the accumulation of minerals under incrusting conditions. Incrustants can also form in the screen slot openings. Repeated acid treatment of incrustants in plastic or fiberglass screens will usually not harm the screen body.

Figure 13.14. Iron and manganese oxides incrust this small-diameter plastic casing. *(H. Gehrke)*

Bacteria

The most common bacteria affecting the condition of a well are iron bacteria (also see Chapter 19). Iron bacteria are nuisance organisms that cause plugging of pores in water-bearing formations and openings in well screens, but are noninjurious to health. Iron bacteria produce accumulations of slimy material of gel-like consistency, and oxidize and precipitate dissolved iron and manganese. The combined effect of the growing organisms and the precipitating minerals can plug a well almost completely within a short time. Cases have been reported where a 75-percent reduction in well yield has occurred in three months to a year.

Introduction of a strong solution of chlorine is effective in controlling iron bacteria. Other bactericides (oxidants) such as chlorine dioxide also are effective, but may be more expensive to use. Acid is often used following the chlorine treatment to dissolve the precipitated iron and manganese, thus making it possible to remove them from the zone surrounding the well. The chlorine *MUST* be removed from the well, however, before the acid is placed in the well. When iron bacteria are known to exist, a well screen fabricated from a corrosion-resistant material should be selected to withstand the damaging effects of repeated chemical treatments.

Screen Strength

Choice of well screen material may be dictated by strength requirements. The three loads, or forces, imposed on a screen are column load (vertical compression), tensile load (extending forces), and collapse pressure (horizontal forces). While a borehole is open during the installation of the screen and pipe, a screen attached directly to

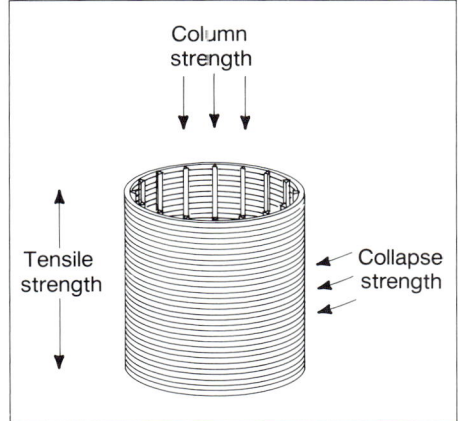

Figure 13.15. All screens have certain tensile, column, and collapse strengths that depend on the material used to construct the screen (plastic, steel, etc.), dimensions of the screen components, and the slot configuration (continuous slot, bridge slot, mill slot, etc.). The constructional components of a continuous-slot screen are selected to withstand the three major stresses that are placed on the screen during construction and use of the well. The column and tensile strength required during well construction are provided by the longitudinal rods. The shape and massiveness of the wrapping wire provide the necessary collapse resistance needed during development and long-term use of the well.

the casing may have to support the entire weight of the pipe. This burden exerts a column load on the screen. A tensile load is exerted on the screen when long sections of screen and casing are installed. The screen must have enough tensile strength to hold temporarily any casing or screen suspended below it. After the borehole materials slough against the screen, earth pressures exert horizontal stresses on the screen, especially during development. The screen must have adequate collapse resistance to withstand both earth and hydraulic pressures (Figure 13.15). Failure to properly analyze these two forces can cause the loss of the well at considerable expense to the well owner, contractor, or engineer. Most collapse failures occur during installation, filter packing, and development, when horizontal forces are maximized (Figure 13.16).

A screen's resistance to both column and tensile loading is directly proportional to the yield strength of the material used in fabricating the screen, whereas the collapse resistance is proportional to the material's modulus of elasticity. For continuous-slot screens, the weight of the pipe column is supported by the cross-sectional area of the vertical rods; for louvered or bridge-slot screens, the casing above is supported by the cross-sectional area of the non-slotted portion of the casing. The screen must remain in alignment with the casing above it, or a severe reduction in column strength will occur. Thus, centering guides are used to keep the screen or slotted casing straight if the screen is more than 20 ft (6.1 m) long. Guides should be installed at least every 40 ft (12.2 m) on long screens. Column strength is important only until formation materials slough around the screen; thereafter, the earth materials hold the casing and screen.

Screen collapse strength is provided by the massiveness of the materials used to build the screen. For continuous-slot screens, the height* of the wrapping wire is the principal factor in determining collapse strength. The amount of open area also has a direct effect on the resistance to collapse pressures. In general, as the

Figure 13.16. The collapse strength of this screen was insufficient to withstand development pressures because too many perforations were made in an attempt to create more open area.

---

*See Figure 12.8, page 399, for the meaning of wire height

458 GROUNDWATER AND WELLS

open area increases, the collapse strength decreases. Note, however, that the actual effect of open area on screen collapse strength is more a function of screen type. For example, a continuous-slot screen with 40-percent open area may have the same collapse strength under similar conditions as a bridge-slot screen with only 12-percent open area. Specific wire shapes and sizes can be selected to accommodate the anticipated (worst case) collapse pressures.

The tensile strength of a screen is important if the screen is interspaced with blank pipe throughout a formation or if the screen is quite long. Long screens built from welded sections must hold together until the entire string of screen and casing is assembled. Large-diameter continuous-slot screens as long as 1,450 ft (442 m) have been installed recently to depths of 3,000 ft (915 m) in Utah (Schafer, 1981). Tensile and column strength can be enhanced by increasing the size and number of vertical rods.

Long-term screen strength is obtained by selecting a screen material on the basis of water-quality conditions. It is poor practice to overdesign screens (providing excess strength), because this reduces the open area of the screen and increases the cost. General information from manufacturers usually focuses on screens built to standard construction specifications only. Therefore, for any large-diameter screen set at depths greater than 300 ft (91.5 m), the screen manufacturer should be contacted for specific design recommendations.

DESIGN OF DOMESTIC WELLS

Many of the design requirements discussed for high-capacity industrial, municipal, and irrigation wells also apply to domestic, farm, and stock wells. The selection of well screen openings, entrance velocity requirements, and recommended screen and pipe material are as important for these wells as for high-capacity wells.

Thousands of wells are drilled every year for homes and farms where the total water requirements may be 5 to 30 gpm (27 to 164 m^3/day). For these requirements, long screens in relatively thick aquifers would be uneconomical. The farmer and the homeowner, however, need a dependable water supply that can be obtained with reasonable drawdown. In these cases, a compromise is necessary between well cost and well efficiency.

The drilling contractor must insure that enough potential drawdown is available to meet present and future yield requirements. The construction represented by well B in Figure 13.17 can yield three or four times as much water as the unscreened well (well A) and has the additional advantage of not pumping sand. The yield from well B can drop off considerably, however, if drought conditions or other wells in the area should cause lowering of the static water level. Well C

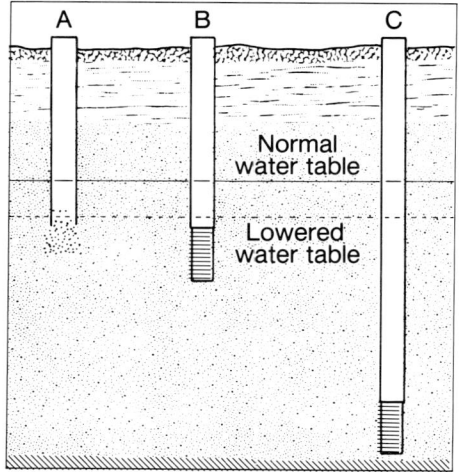

Figure 13.17. Adequate long-term yields are obtained by installing a well screen of adequate length at the proper depth. Enough potential drawdown must be available to meet future yield demands.

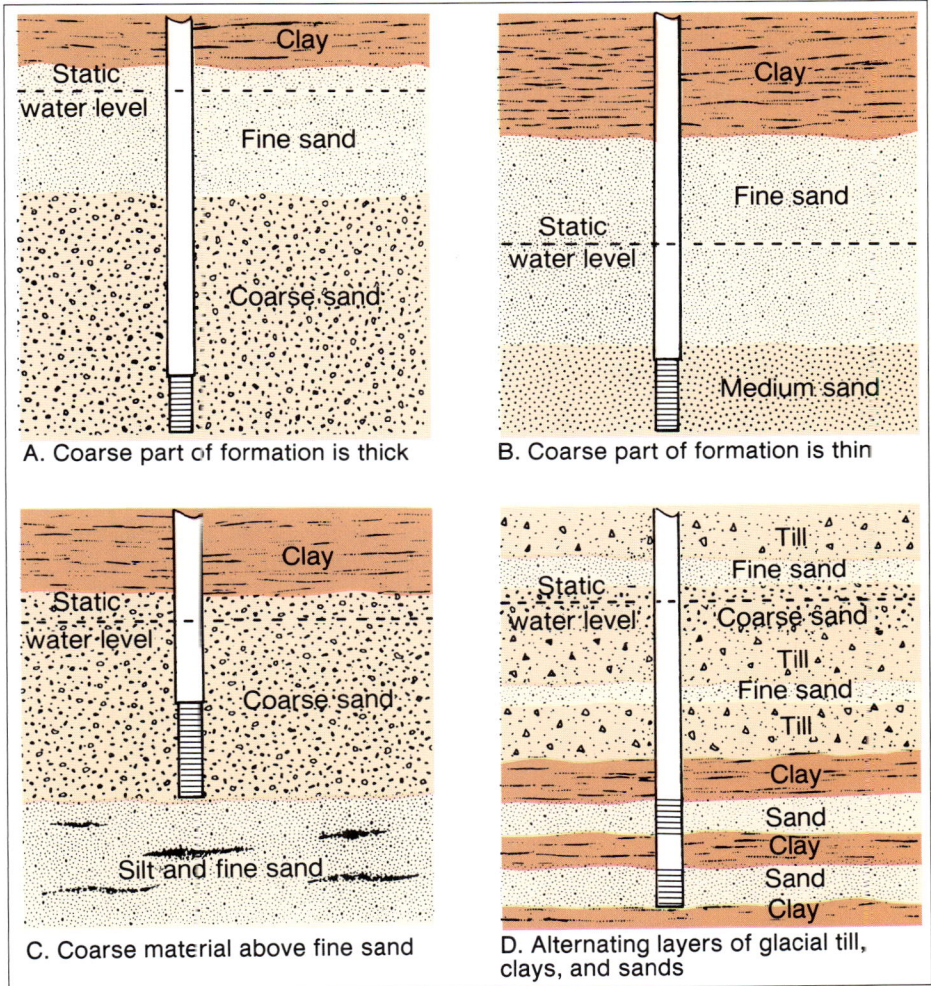

Figure 13.18. Suggested positioning of well screens in various stratified water-bearing formations.

is constructed properly and can supply 20 times as much water as the open-bottom well.

It is difficult to specify exact rules for choosing the screen length for low-capacity wells. For economic reasons, many domestic and farm wells must be constructed in less prolific aquifers and at depths that do not provide maximum hydraulic efficiency. In general, domestic wells should be constructed with screens 4 to 5 ft (1.2 to 1.5 m) long; for farm wells, the screens should be 10 to 15 ft (3 to 4.6 m) long, depending on the hydraulic characteristics of the aquifer and the yield requirements. These recommendations apply to continuous-slot screens only. For other types of slot configurations, much longer screens many be required. The examples cited below demonstrate how short screens are used in typical situations.

For the situation shown in Figure 13.18A, only 4 or 5 ft (1.2 or 1.5 m) of the aquifer needs to be screened for a domestic well because of the relatively high static water level and high hydraulic conductivity of coarse sand. For a farm well, the screen length should be increased to 10 ft (3 m) because the required yield is usually higher. The reduction in available drawdown is not a problem because the transmissivity of the aquifer is adequate.

For the situation in Figure 13.18B, most of the medium sand should be screened. If a screen of this length does not provide sufficient open area for the desired yield, the screen may have to be extended a short distance into the finer sand above, although the contribution to the yield from the added screen footage may not be significant. In this case, sufficient drawdown is available if the screen is lengthened.

The screen for the well in Figure 13.18C should be set at the bottom of the coarse sand layer if adequate drawdown is available (as shown here). The length of the screen should be about one-third the thickness of the coarse sand. Ordinarily, it would not be beneficial to extend the screen deeper into the fine sand, because good yields from highly stratified silt/sand layers are considerably more difficult to obtain.

In Figure 13.18D, the hydrogeologic conditions are not as favorable. Although the static water level is high enough to provide adequate drawdown for a farm well, the thickness of the sand layers is limited. In fact, the two lower sand layers should be partially screened to provide enough open area to the formation. In this case, two 3-ft (0.9-m) sections of screen are placed in the lower portion of the deep sand formations and connected by blank pipe. The top of each screen should be kept 3 ft (0.9 m) below the top of the aquifer to allow for anticipated sloughing of overlying clay during development. This type of installation is relatively common in glaciated terrains, but requires careful logging of the well by the drilling contractor.

Although screens longer than 10 to 15 ft (3 to 4.6 m) may not be necessary in domestic or farm wells to meet present yield requirements, water demands almost invariably increase with time. Contractors should anticipate these greater demands by installing screens of sufficient length to provide for increases in yield, because screen cost is usually a minor part of the total well cost. The yield of a well may be increased almost in proportion to an increase in screen length, provided the well taps a water-bearing formation of reasonable thickness. For example, doubling the screen length can, in most cases, almost double the well yield. Doubling the diameter, however, can be expected to increase yield only about 10 percent, except when the yield is restricted only by pump size and a larger diameter pump would increase yields substantially.

Small wells should also be constructed with screens and casing of sufficient diameter so that effective development methods and tools can be used. Many wells are completed with 2-in (51-mm) drive points driven out of 4-in (102-mm) casing. Although these wells are satisfactory from a design standpoint, they are difficult to develop properly.

DESIGN FOR SANITARY PROTECTION

The design of a water well supplying potable water should include those features that provide continuous sanitary protection. Contaminated water from surface drainage or low-quality water encountered in the well can move downward through the annulus between the casing and borehole wall. Thus, the annulus around the casing must be sealed. Generally, any sealing around the well involves placing a cement grout in the annulus; bentonite is sometimes used in place of cement. Methods for installing grout are given in Chapter 10. Design and installation procedures that will assure sanitary protection are discussed in Chapter 18.

SPECIAL WELL DESIGNS

Various design methods have developed in certain areas because of particular hydrogeologic conditions, the type of drilling equipment used, the availability of filter pack material, and the economic aspects of the wells. These methods and procedures were developed because more water was needed than could be obtained with standard design criteria and because favorable cost/benefit ratios were realized. Although developed locally, these methods can be used successfully in areas where similar hydrogeologic conditions exist. Several examples of alternative well designs are described below.

Case 1

Hydrogeologic conditions: High-quality water found in sinuous, thin alluvial and glaciofluvial deposits in river valleys. Small volumes of water also found in underlying igneous and metamorphic rocks.

Problem: Underlying bedrock aquifers do not offer sufficient volumes of water. Alluvial and glaciofluvial aquifers are small in areal extent and are relatively thin. These aquifers are sustained, however, by high rates of induced filtration from nearby streams or rivers.

Solution: To pump large volumes of water at low incrustation rates, the screen slot size must be as large as possible because the screen length is limited. This is accomplished by placing two filter packs around a large-diameter, but necessarily short, screen segment, thereby increasing the effective diameter of the screen (Figure 13.19). The purpose is to reduce the amount of drawdown required to drive water into the screen by greatly increasing the porosity and hydraulic conductivity of the material adjacent to the screen. The outer pack is selected to control movement of the aquifer materials, whereas the inner pack retains the outer pack material. The inner pack is much larger in particle-size distribution than is the outer pack, so the screen slot size is also much larger. In this design, the intake area of

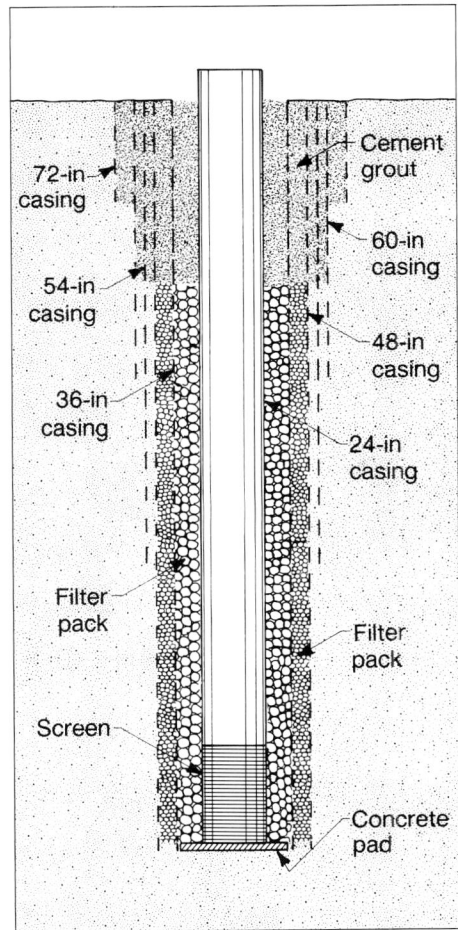

Figure 13.19. Two differently sized filter packs can be installed around a large-diameter screen to enhance the hydraulic conductivity of the formation. This procedure helps maximize the yield from a well installed close to a recharge source such as a river.

the short screen is maximized by increasing the diameter (it cannot be lengthened), and the slot size is considerably larger than could be achieved with a single pack. Thus, high yields are obtainable at low incrustation rates from short screen sections when they are located in highly permeable sediments near sources of recharge.

Case 2

Hydrogeologic conditions: High-quality water exists in near-surface thin sand layers that are underlain by thick clay layers. The underlying bedrock contains only low-quality water that is not suitable for potable supplies.

Problem: Yields from conventional wells are insufficient for even domestic use because the aquifers are thin. Water tables fall enough during fall and winter to cause wells to go dry.

Solution: A 24- to 48-in (610- to 1,220-mm) borehole is drilled through the overlying sand layer into the clay, usually by bucket or earth auger. The borehole is kept open below the water table by keeping it filled with water and by adding drilling fluid additives to reduce fluid losses. A 1- to 2-ft (0.3- to 0.6-m) length of continuous-slot screen is installed in a string of casing so that the screen is placed at the bottom of the aquifer (Figure 13.20). The pump intake is placed in casing (sump) that extends beneath the screen. During periods of nonpumping, water cascades into the sump, which acts as a reservoir. The bottom of the casing string is usually sealed with cement or a steel plate. The screen is filter packed, but the annulus above and below the screen may be filled with gravel, sand, or clay. Yields from this type of installation will vary according to the fineness of the sand layer and the depth of the water table, but sustained yields are usually much greater than can be expected from a typical well installation and are generally adequate for most domestic water demands.

Case 3

Hydrogeologic conditions: Extremely small volumes of water are found throughout thick, dirty, fine-grained sand/siltstone sequences that are buried at depth and may be under confined pressures.

Problem: It is not economical to set large-diameter screens because of the potentially small yields and the difficulty in identifying the most productive zones.

Solution: To obtain a reasonable yield,

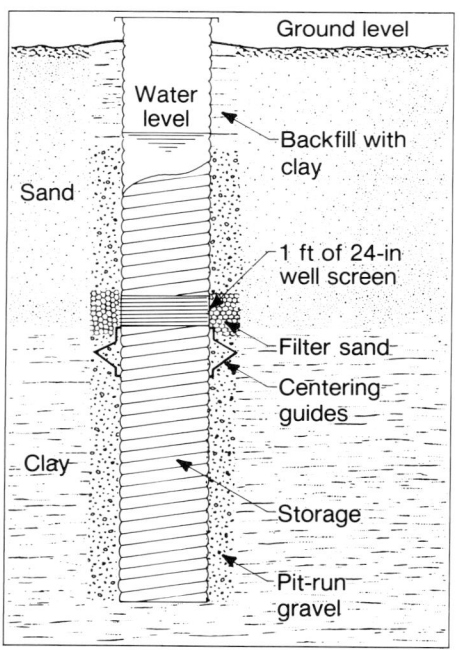

Figure 13.20. To intercept the limited volume of water in a thin, surficial sand aquifer, a short segment of large-diameter continuous-slot screen is installed at the base of the formation. Water entering the well collects in a sump beneath the screen. The pump is ordinarily mounted in the sump to permit removal of the maximum volume. *(Larson, 1979)*

a string of 2-in (51-mm) continuous-slot screen, commonly 40 to 100 ft (12.2 to 30.5 m) long, is installed through the full thickness of the formation. The screen may be filter packed or the annulus filled with a formation stabilizer, depending on how well the aquifer materials are cemented together. Anticipated yields from this type of screen installation are usually 40 to 60 gpm (218 to 327 m^3/day).

Case 4

Hydrogeologic conditions: In coastal areas with high annual rainfall, a thin 2- to 3-ft (0.6- to 0.9-m) veneer of fresh water sometimes overlies saline water. The fresh water is of high quality, but can be easily contaminated with salt water unless pumping rates are kept low.

Problem: Ordinary well designs will cause salt water to cone upward during pumping, leading to contamination of the overlying fresh water.

Solution: To obtain reasonable volumes of high-quality fresh water, it is necessary to eliminate the upconing effect. This is done by installing several widely spaced well points that are then manifolded together and pumped by suction lift. The points are pumped lightly to minimize pressure reductions in the vicinity of the points. The water is stored in a 500-gal (1.9 m^3) tank. A separate pump is used to move the water from the tank to the house system. With conservative use, a single tank may provide enough water for several days for a family.

CHAPTER 14
Installation and Removal of Well Screens

Well screens are required in all unconsolidated and most semiconsolidated formations, and occasionally in consolidated rock. Many different screen installation methods are used, although certain procedures may be more practical or more economical in certain areas or when particular drilling rigs are used. The exact procedures to be followed when installing a well screen depend on the nature of the aquifer materials, the method used to drill the well, the dimensions of the borehole, the hydraulic conditions in the aquifer, and the casing and screen materials. Well completion steps that are done during installation or immediately thereafter include installing the filter pack material, grouting the casing, and developing and sterilizing the well. The most common and successful screen installation methods are described below.

PULL-BACK METHOD

Before the recent increase in the number of direct rotary drilling rigs, the pull-back procedure was used in most wells. It is a safe method of installation that reduces problems resulting from heaving sediment, sloughing of the borehole walls because of swelling clays or insufficient hydrostatic pressures, and setting the screen at the wrong depth. The pull-back method also permits the screen to be removed and replaced, if necessary, without disturbing the sanitary grout seal outside the well casing. The cost of pulling, cleaning, or replacing the screen is usually small in comparison with drilling a new well. The pull-back procedure is particularly suited for wells drilled with a cable tool rig and with air rotary rigs equipped with casing drivers, although some direct rotary drillers use it as their standard method of installation for shallow wells.

The pull-back method involves installing the casing to the full depth of the well, lowering (telescoping) the well screen inside the casing, and then pulling back or lifting the casing far enough to expose the screen to the water-bearing materials (Figure 14.1). The casing must be strong enough to be set the full depth of the well and then be pulled back the length of the screen.

Telescope-size screens are designed for use in the pull-back method. As the term "telescope" implies, the screen is constructed so that it will telescope through standard pipe of the corresponding size, allowing installation of the largest diameter screen possible for a given casing diameter. Occasionally, pipe-size screens that are one or more diameters less than the casing may be telescoped through larger diameter casing.

Some contractors use a riser pipe attached to the top of the screen so the entire screen can be exposed without slipping out the bottom of the casing. The inside diameter of the riser pipe should match as closely as possible the inside diameter of the telescoped screen. Riser pipes can be formed on continuous-slot screens by tight winding the outer wire, thereby eliminating the slots. The use of a 2- to 5-ft (0.6- to 1.5-m) riser pipe is recommended for domestic wells, 10 to 20 ft (3 to 6.1 m) for larger diameter, deeper wells, and 5 to 10 ft (1.5 to 3 m) for small-diameter wells in which the casing can be unintentionally pulled back so quickly that an error in measurement or marker adjustment can result in losing the screen beneath the casing. A riser pipe should always be used in loose or previously drilled sands where the screen may settle during development and drop out the bottom of the casing. Use of a riser pipe assures both extra safety when installing the screen and maximum utilization of the screen intake area.

Packers

A special fitting is required to provide a sand-tight seal between the top of the telescoped screen assembly and the casing. Two types of packers are commonly used:

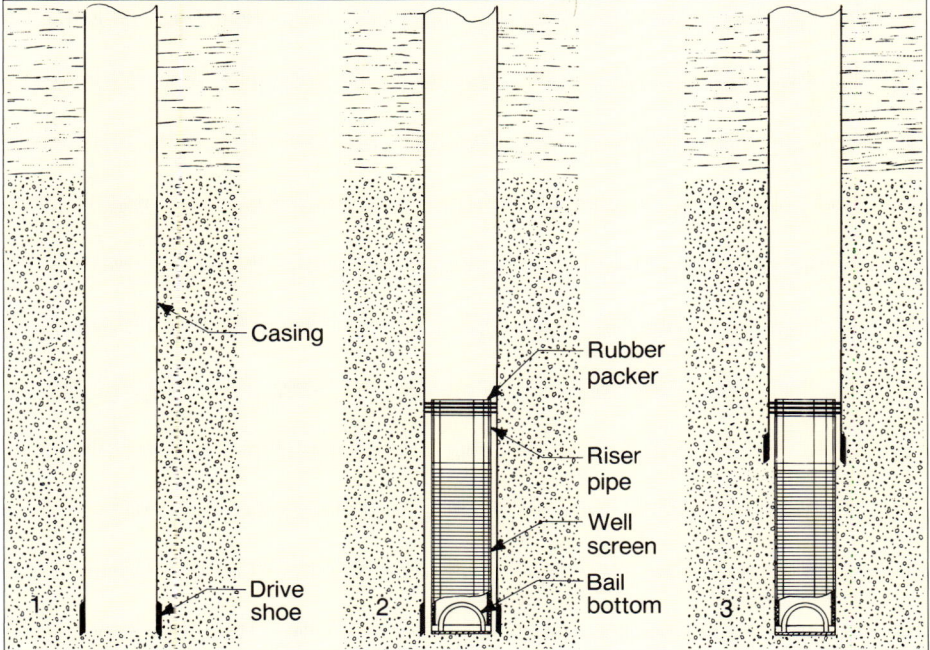

Figure 14.1. Basic operations in setting a well screen by the pull-back method include: (1) driving, bailing, or lowering (rotary method) the casing to the full depth of the well, (2) lowering the screen inside the casing, and (3) pulling the casing back to expose the screen to the aquifer.

Figure 14.2. Self-sealing neoprene rubber packers form an effective seal between the casing and upper end of the screen or riser pipe (left). *(Kramer Well Drilling)* **Petroleum jelly is often applied to the lips of the packer to minimize damage that may occur as the packer and screen assembly is lowered into the casing (center).** *(Burt Well Drilling)* **The screen is lowered into place by the sand line (right).** *(Drillwell Enterprises)*

neoprene rubber (often called K packers) and lead. The packer is attached directly to either the top of the well screen or the top of a riser pipe.

A rubber packer is constructed of flexible neoprene rubber attached to a steel coupling and fits tightly in the casing, sealing the casing to the screen (Figure 14.2). The use of this type of packer has grown rapidly because no expansion of the packer is required. Frequently two or more packers are used in series to eliminate problems caused by small deviations in the dimensions of the casing or packer resulting from improper handling. Furthermore, multiple rubber packers are recommended for screens set at depths exceeding 300 ft (91.5 m) because the rough inner surface of the casing, often caused by weld slag or beads in welded casing, can damage the rubber lips of the packer. Lubricating the lips of the rubber packer with petroleum jelly will reduce damage when the screen is lowered into the casing (Figure 14.2).

A lead packer usually consists of lead molded onto a coupling or weld fitting. When a lead packer is used, it is expanded by a swedge block and bar to make a sand-tight seal between the top of the screen and the casing (Figure 14.3). When using the swedge block to flare the packer, one to three lengths of small-diameter pipe should be attached to the bar that slides

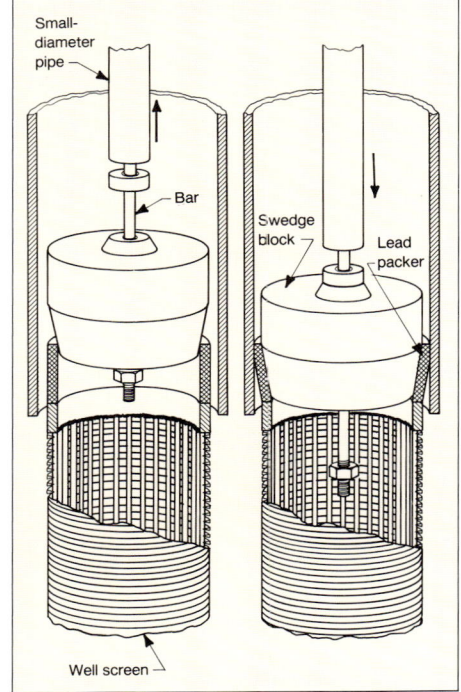

Figure 14.3. A swedge block is used to expand a lead packer attached to the top of a telescoped screen assembly.

through the block. This sliding bar allows a driving stroke of about 1 ft (0.3 m). The assembly should be lowered into the well until the swedge block rests inside the lead packer. The weight provided by the pipe attached to the sliding bar should then be lifted 6 to 8 in (152 to 203 mm) and dropped several times. The swedge block itself should not be lifted out of the packer during the swedging process; it should simply be forced farther down into the packer by successive blows of the bar and the weight above it. To facilitate this process, the swedging pipe should be marked in relation to the casing before swedging begins. Swedging should be stopped after the pipe has moved down approximately 1 to 1.5 in (25 to 38 mm) because continued hammering of the swedge block can cause the lead to break away from the underlying flexible steel matrix and fall into the screen. The extent of the swedging can also be measured by marking the swedge block with driller's chalk. When the swedge block is withdrawn from the borehole, the chalk removal will show the penetration of the block into the lead packer and indicate the extent of swedging. Swedge blocks that have rough surfaces may stick to the packer during swedging. Smoothing of the tapered part of the swedge block with a file or a coating of petroleum jelly will eliminate the problem.

Casing or liner hangers and expandable packers are occasionally used in deep wells, but they are usually more expensive than rubber or lead packers. One type of liner hanger operates by expanding a hollow rubber cylinder and slips placed between one or two threaded devices. The packer is ordinarily attached to a riser pipe and expanded in the casing. The top of the packer is fitted with a left-hand thread so the drill pipe can be disengaged from the packer.

Inflatable packers generally have larger expansion ratios than do casing hangers. The packer is inflated by injecting gas, water, or a solidifying liquid. Thus, the packer can be used for a short time and then retrieved or installed permanently. Some fixed-end packers can be inflated to two times the uninflated diameter, but are generally designed for lower pressure applications than are sliding-end packers (Figure 14.4). Sliding-end packers are used where the differential pressures range from 200 to 2,000 psi (1,380 to 13,800 kPa) or more. Inflatable packers are useful for effecting casing repairs, pumping tests of isolated zones in the borehole, hydrofracturing, and injecting water and gases.

Occasionally a screen much smaller in diameter than the casing is used. For example, when a high yield is anticipated, a relatively large-diameter casing may be

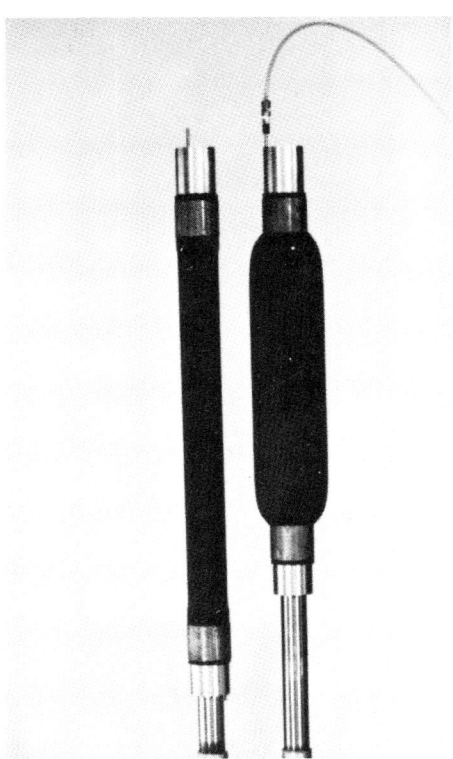

Figure 14.4. Inflatable packers are used as permanent or temporary sealing devices to isolate portions of the well bore. *(Baski Water Instruments, Inc.)*

needed to accommodate the pump bowls. A large-diameter screen may not be needed or practical, however, because the open area of a smaller diameter screen is sufficient to accommodate the expected yield at properly designed entrance and uphole velocities. A special cone adaptor is then used to connect the smaller diameter screen to the larger packer needed for the casing (Figure 14.5). The packer and cone adaptor are attached to the screen before installation.

Setting the Screen in Wells Drilled by the Cable Tool Method

It is important to control sediment movement into the bottom of the casing because the screen should be set as close as possible to the designed depth. This is

Figure 14.5. When the diameter of the screen is two or more sizes smaller than the diameter of the casing, a cone adaptor is mounted on top of the screen. A lead or rubber packer is then attached to the cone adaptor. *(Snider Drilling Co. Ltd.)*

particularly important if the screen has been designed with several slot sizes corresponding to individual sediment layers, or if blank sections have been placed between screen sections. If there is difficulty in keeping sediment from heaving inside the casing, the casing should be filled with water. Fluid losses may be controlled by using prepared drilling fluids. To keep confining pressures in the aquifer under control, the drilling fluid may be weighted with special high-density materials or salt. Sudden vertical movement of tools in the casing also increases the likelihood of heaving problems. Therefore, bailers should be operated slowly to reduce pressure differences at the bottom of the borehole.

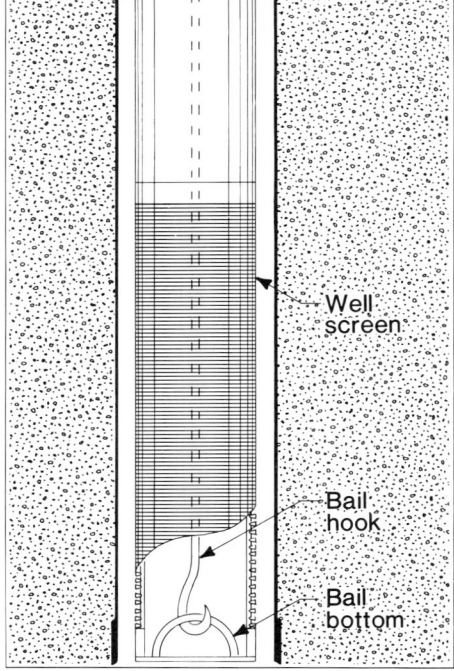

Figure 14.6. When the depth to water is great, the screen can be lowered by attaching the sand line to the bail bottom by a flat hook.

After the casing has been driven to the proper depth, any sediment that has entered or settled within the casing should be removed carefully. Adding water to the casing may aid the settling process. After all sediment has been removed from the casing, the driller must be sure the casing can be withdrawn. If it cannot be withdrawn, the drive shoe is cut off with an inside casing cutter to reduce resistance, and the casing is pulled back a few inches. The screen is then lowered to the bottom of the well. Several devices can be used to

Figure 14.7. Individual screen sections can be hoisted by elevators (left). Once suspended, the screen sections are aligned (center) and then welded or threaded together (right). *(Henkle Drilling Company; Snider Drilling Co. Ltd.)*

lower the screen: bail and hook, eccentric clevis (offset latch), eyelet screws mounted in lead packers, casing lugs (bayonet type), and wash-down bottoms (Figure 14.6).

If the screen is made in two or more sections, the bottom section is lifted by a rope clamp or hitch and suspended inside the casing by a pair of casing clamps or elevators. The next section is then threaded or welded to the top of the first section (Figure 14.7). If the derrick or mast is high enough, several screen lengths may be assembled on the ground, hoisted vertically, and then lowered into the well as a single unit.

If the depth to water is less than 50 ft (15.2 m), short, small-diameter screens may be installed by dropping them inside the casing. If rubber packers are used, screens 4 to 6 inches (102 to 152 mm) in diameter generally will not drop and must be pushed into place, whereas screens 8 in (203 mm) and larger will drop because of their weight. Careful measurements must be kept so the driller will know that the screen is set at the correct depth in the aquifer.

The drill string or a weight attached to the sand line should be placed on the bottom of the screen while the casing is being pulled back. This provides enough weight to keep the screen on the bottom, and the tension in the sand line serves to verify the exact position of the screen during the procedure. If weight is not applied to the screen, any heaving of the formation will force the screen upward at about the same rate that the casing is pulled back.

The casing can be pulled back by one of several methods. Under ideal conditions where the earth materials have not collapsed tightly around the casing, it can be pulled with the casing line on the cable tool drilling machine. Greater lifting force can be obtained by using a block and tackle attached to the casing line. Before pulling is attempted, however, especially for large-diameter casing, guide wires must be in place and tightened to stabilize the mast of the drilling rig. To prevent damage to the drilling rig, the mast should be lowered to its short-masted position. If the casing cannot be withdrawn by the casing line, it may be pulled by jarring with the drilling tools with fishing jars attached, or a bumping block or drive clamps (Figure 14.8). Mechanical

Figure 14.8. In cable tool drilling, bumping blocks or drive clamps are first used to drive the casing and then pull it back to expose the well screen in the aquifer.

or hydraulic jacks may also be required to provide the necessary lifting force (Figure 14.9). If so, a pulling ring or spider with wedges or slips is used to grip the casing. For long casing strings, a vibration hammer is effective in overcoming the skin friction between the casing and formation. Although the cost of this procedure may be high when compared with other methods, it is sometimes the only method powerful enough to withdraw casing (Figure 14.10).

As the casing is being pulled back to expose the screen in the water-bearing formation, depth measurements to the screen are taken. If no riser is attached to the screen, the casing should be pulled back so that the packer is about 12 in (305 mm) above the bottom of the casing. The screen can be fully exposed beneath the casing if a riser pipe is attached to the top of the screen. When a lead packer is used, the packer is flared to make a sand-tight seal between the top of the screen and the inside of the well casing.

Setting the Screen in Wells Drilled by the Rotary Method

When screens are installed in rotary-drilled wells, the pull-back method should be selected if caving conditions exist or lost circulation is a problem. Drill pipe should be withdrawn slowly from the hole after the maximum depth has been reached to eliminate heaving caused by suction. The presence of the drilling fluid column will usually control caving problems, except in relatively shallow, unconsolidated glacial or alluvial sediments. Natural compaction of these sediments is not great, and in confined aquifers the material is close to a condition where any suction or other disturbances in the borehole will cause the sediment to "run" toward the low-pressure (suction) zone.

Setting the casing to the bottom of the hole and then pulling it back may appear to be extra work, but this operation prevents serious problems arising from premature caving which can occur when the drilling fluid viscosity is reduced prior to development. The protection given by the casing is particularly important if a delay is anticipated between drilling and screen installation. During this period, a momentary loss of drilling fluid may cause partial collapse of the borehole.

After the casing is placed in the open borehole, any cuttings that settle inside it

Figure 14.9. A pair of hydraulic jacks working against pulling wedges pulls back 16-in (406-mm) casing after setting the well screen. *(Strasser Drilling Co.)*

are carefully cleaned out. The screen is then lowered to the bottom through the casing and the casing is pulled back to expose the screen. After the casing is pulled back, it must be held in place until the formation has caved around it during development or until the annulus has been backfilled or grouted.

For wells drilled with air rotary machines equipped with casing drivers, two setting procedures can be followed, depending on the stratigraphy. In heaving formations, the casing is usually sunk to the top of a clay layer under the aquifer, the screen set within the casing, and the casing pulled back. If no clay exists and heaving is a problem, it may be necessary to change from air-based to water-based drilling fluids to control fluid pressures in the formation so the screen can be set. Occasionally, stiff foam may be enough to control the formation. If water-based

Figure 14.10. A vibration hammer is sometimes used to drive and pull casing when the skin friction between the casing and formation is excessive. *(J. S. Lee & Sons)*

drilling fluids are used to control the pressures, it is a good idea to have a quick-release valve to add water rapidly through the drill stem so that the drill pipe is not sand locked in the hole during the switchover. In thin aquifers, it is also extremely important that the switchover be accomplished before the borehole is overexcavated. Excessive excavation may cause collapse of weaker overlying sediments, which may destroy the hydraulic characteristics of a thin aquifer in the vicinity of the well. Once heaving has been controlled, the casing can be cleaned by circulating water-based drilling fluids. The screen is then installed and the casing pulled back.

OPEN-HOLE METHODS FOR SCREEN INSTALLATION

Double-String Installation

Use of the pull-back method in rotary drilled wells has declined in favor of open-hole methods. A common procedure for installing screens in high-capacity industrial and municipal wells is described below. Because this is only an example, not all alternative methods for individual steps are included; however, other procedures are presented later in this chapter.

A typical procedure for installing screens in high-capacity wells includes the following steps:

1. Drill a small-diameter test hole at the chosen site, keeping a detailed drilling log and collecting samples at specific intervals. Samples are taken at each formation change and every 5 to 10 ft (1.5 to 3 m), depending on the formation thickness and well depth.

2. Conduct geophysical logs (typically SP, resistivity, natural gamma ray, and hole caliper) in the test hole and determine which aquifer or aquifers are to be screened. Occasionally, side-wall cores are taken to verify the type of formation and its hydraulic characteristics.

3. If more information about the aquifer is needed, such as chemical quality or productivity, a 2- to 5-in (51- to 127-mm) screen is installed in the desired zone and test pumped.

4. After the production zone is chosen from an analysis of the cuttings, driller's logs, geophysical logs, and pumping test data, samples are analyzed and the correct screen-slot openings are determined.

5. If a test well has been drilled, the screen and casing are pulled and the test hole is then used as a pilot hole for the production well. In uniform geologic conditions, the test hole is often left as an observation well and the production well is drilled a suitable distance away. The test hole can also provide make-up water for use in drilling the production well.

6. An open hole is then drilled down to the top of the aquifer to receive the casing (Figure 14.11). The hole diameter should be large enough to allow space in the annulus between the casing and the hole for a minimum grout thickness of 2 in (51 mm).

7. The casing is set in the hole. Generally, a drillable grout shoe (plug) constructed of cast aluminum or cement is installed on the bottom of the casing. The shoe allows the grout to be pumped through the bottom of the casing and should be used on the casing to insure proper centering.

8. The casing is then grouted. At least 24 hours should be allowed for the grout to set.

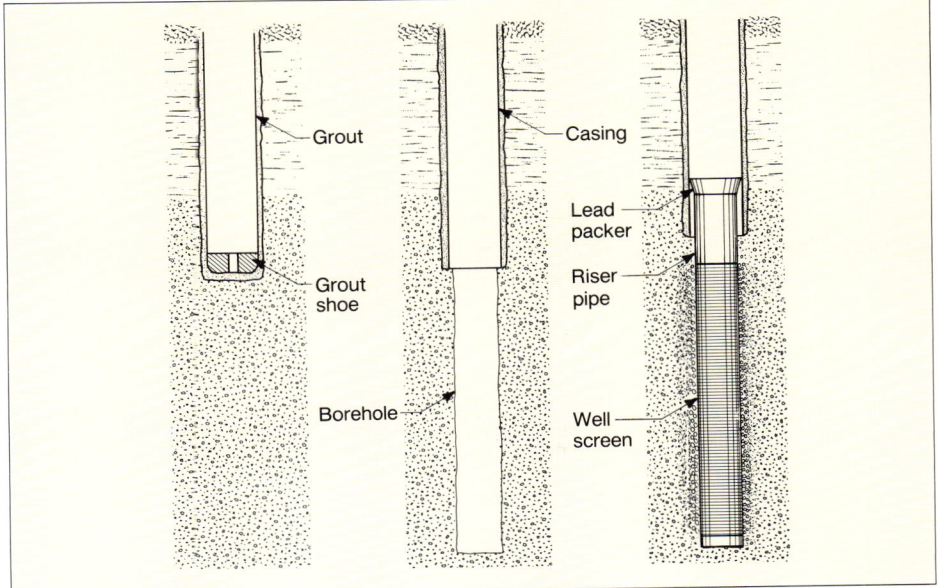

Figure 14.11. A well screen may be set in an open hole drilled below the well casing after the casing has been grouted.

9. Any extra grout in the casing and the drillable shoe is next drilled out, but the aquifer is not penetrated. At this point, all drilling fluid used to drill the grout should be replaced with clean drilling fluid or water if practical. This is an important step; if the grout-contaminated fluid is used to drill the aquifer, it may seal some of the formation and be difficult, if not impossible, to remove from the formation. It may also have an adverse effect on the physical characteristics of the drilling fluid.

10. If the well is to be naturally developed, the aquifer is drilled with a bit that is slightly smaller than the inside diameter of the casing. If the well is to be filter packed and the diameter of the completed borehole is too small to accommodate a proper thickness of filter pack between the screen and formation, the aquifer may be underreamed to obtain a 3- to 8-in (76- to 203-mm) annulus.

11. A suitable length of riser pipe should be used on top of the screen. If there is to be blank pipe between the top of the screen and the bottom of the casing and the well is filter packed, a 5-ft (1.5-m) length of pressure-relief screen should be included in the riser pipe string and set at the bottom of the casing (pressure-relief screens are discussed in the filter-pack section).

12. A screen is then telescoped through the casing into the open hole by an off-centered latch hook (Figure 14.12), bail hook, or left-hand-threaded fittings attached to the plate bottom of the screen or to the top of the riser pipe. The last method is the surest way to position the screen; but in deep holes where the drill pipe exerts high pressure on the threads, the drill string should be held back or a slip section should be used to reduce the weight on the threads while disengaging the drill pipe from the screen. Sometimes a fin (piece of flat metal) is welded to the side of the screen bottom plate to prevent the screen from turning when unthreading the drill pipe. Metal bars or other attachments welded underneath the bottom plate will also

prevent the screen from turning (Figure 14.12). If the well is to be filter packed, it is recommended that one centralizer be attached near the bottom of the screen and another near the top (Figure 14.12). For screens of less than 200 ft (61 m), centralizers should be spaced every 20 ft (6.1 m); for economic reasons, centralizers are often spaced at 40-ft (12.2-m) intervals on screens over 200 ft.

13. If it is anticipated that fill or other borehole material may prevent the screen from reaching the desired depth, a self-closing bottom fitting with internal left-hand threads can be mounted in the screen bottom plate to wash the screen into place. A wash pipe is attached to this fitting and can be used to set the screen. A wash-down bottom provided with a slip-socket wash fitting is sometimes mounted on the bottom of the screen but is not used to set the screen. Instead, these fittings are useful in displacing drilling fluid from the borehole, thereby aiding development. The displacing fluid, often clear water, is directed through the wash-down bottom to remove the drilling fluid from the hole. If an open-bottom screen is used, cement grout can be used to plug the bottom.

14. After the screen is set, the well may be naturally developed or filter packed. If the well is to be naturally developed, the formation is induced to cave around the screen by reducing the hydrostatic pressure in the borehole, by either thinning the drilling fluid or lowering the level of the drilling fluid in the borehole. The setting tool is then removed and the driller begins development. Shale catchers, also called shale traps and formation packers, are sometimes used to prevent sloughing of overlying materials (especially clays) when a well is naturally developed.

If the well is to be filter packed, the setting tool is left in place while filter pack is introduced into the well. A plug can be used to temporarily close the top of the screen during filter packing. After the filter pack is in place, the setting tool or plug is removed and the well is developed. A lead slip packer may be installed once the well is developed to provide a seal between the riser pipe and casing. Normally no packer has to be installed if the riser pipe is lapped 50 ft (15.2 m) or more into the casing and a pressure-relief screen is installed.

Figure 14.12. A screen can be set by attaching it by an off-centered latch hook to the drill rods and then lowering it into the borehole (left). *(Snider Drilling Co. Ltd.)* Metal fins or plates are welded to the bottom of the screen to keep it from turning when the drill rods are disengaged (center). Centralizers are often attached to the screen to keep it centered in the borehole. This step is especially important if the screen is to be filter packed (right). *(LTP Enterprises)*

Single-String Installation

In most small-diameter, rotary-drilled wells completed in unconsolidated sediments, the screens are attached directly to the bottom of the casing. Screens that are smaller in diameter than the casing can be welded or threaded directly to the casing by mounting a cone adaptor or flared weld ring to the top of the screen. Screens that are the same size as the casing can be welded or threaded directly to the bottom of the casing.

The casing and screen are then set in the hole and the drilling fluid is thinned. If the drilling fluid is not thinned, the fluid may not enter the screen as it is lowered into the borehole. High differential pressures can then be created, which may be sufficient to collapse the screen. Under these conditions, it is prudent to fill the screen with water when it is placed in the borehole. For naturally developed wells, the formation is induced to cave in around the screen and casing immediately after the screen is set. When wells are filter packed, the pack material is placed before the formation is induced to cave.

A string of casing and well screen is a slender, flexible column with a high ratio of length to diameter (Figure 14.13). For a 200-ft (61-m) string that is 12 in (305 mm) in diameter, the ratio of length to diameter is 200 to 1 — which is about the same as a 12-in (305-mm) length of $1/16$-in (1.5-mm) wire. A string of casing and well screen of this slenderness has little column strength without lateral support. When the casing and screen are installed in an open hole without side support from the subsurface formations, it is always best to suspend the string from the surface during installation without resting the column on the bottom. As soon as lateral support is attained from the subsurface formations or by placement of the filter pack or formation stabilizer, the full weight of the column may be released safely. Failure to observe this precaution may result in a collapsed or crooked well. Centralizers are extremely effective in maintaining proper alignment, and are

Figure 14.13. Long strings of screens must be handled with care both before and after being placed in the ground.

recommended for all screens that are more than 20 ft (6.1 m) long.

To prevent clay-rich material above the aquifer from sloughing next to the screen during natural development, a formation stabilizer is often installed (Figure 14.14). The formation stabilizer holds the clay in place until the materials cave against the stabilizer. The size gradation of a stabilizer should be similar to the formation material or a little coarser. In most cases, however, a formation stabilizer is not needed if the borehole is only slightly larger than the screen and the top of the screen is placed 3 ft (0.9 m) or more below any clay zone. If a formation stabilizer is not used and there is a large annulus, shale catchers can be mounted on the casing to prevent sloughing.

FILTER PACKED WELLS

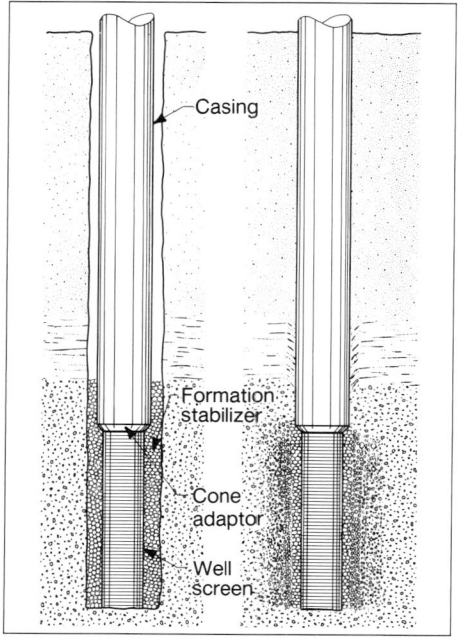

Figure 14.14. A formation stabilizer can be placed next to the well screen to prevent uncontrolled sloughing of the formation against the well screen.

Many wells drilled by cable tool or rotary methods are designed for a filter pack, thereby altering the screen installation process. Filter-packed wells differ from naturally developed wells in that an envelope of specially graded sand or gravel is placed around the well screen to a predetermined thickness. This takes the place of the graded zone of permeable material that is produced by the natural development process. Both types of wells, when properly constructed, are efficient and stable. The geologic conditions, availability of suitable filter pack materials, drilling method, and type of screen determine whether a filter pack should be used.

The thickness of the filter pack is a primary factor in the effectiveness of the development procedures taking place at the interface of the pack and formation. The minimum practical thickness for the pack is 3 in (76 mm). Filter packs thicker than 8 in (203 mm) are not recommended, because the effectiveness of the development procedures may be impaired.

Selection and Placement of Filter Pack

It is important to select a filter pack that will not segregate, because sand pumping can result if fine and coarse particles become separated during placement. It can be demonstrated that a round particle of a given size and density falls through water four times faster than a round particle half as large and with the same density. If filter pack material that is uniformly graded from $1/16$ in (1.6 mm) to $1/8$ in (3.2 mm) is allowed to fall through water as separate grains, the $1/8$-in grains will reach the bottom of the well in one-fourth the time required for the $1/16$-in grains. Thus, well-sorted filter pack material is less apt to segregate than is pack material with a wide range of particle sizes. Sand and gravel mixtures with uniformity coefficients greater than 2.5

are difficult to place without undesirable separation of coarse and fine fractions (see Chapter 12).

All filter pack materials should be treated with a bactericide, usually chlorine, before placement to insure that the well does not become contaminated. All water used in the filter pack operation and any tools or pieces of installation equipment should also be treated with a 50 mg/ℓ free-chlorine solution before use. Whenever possible, the drilling fluid should be thinned before placing the pack material.

Use of a tremie pipe to install the filter pack will minimize the tendency for particle separation and bridging; this is the preferred method for filter pack placement, especially for packs with high uniformity coefficients. A string of 2-in (51-mm) or larger pipe is lowered into the annular space to be filter packed. The filter pack material is fed into a hopper at the well head (Figure 14.15). A liberal supply of water should be introduced with the filter pack to help prevent bridging of the material in the pipe. A typical ratio of water to pack material is 5 to 10 gal (19 to 38 ℓ) for each 1 ft^3 (0.03 m^3) of pack material when a 2-in (51-mm) tremie pipe is used. The tremie system is practical for placing the filter pack in shallow to moderately deep wells [to 2,000 ft (610 m)]. In some cases, filter pack material may be pumped through the tremie pipe along with the water stream instead of being driven by gravity.

During installation of the pack, the tremie pipe is raised periodically as the filter material builds up around the well screen. The tremie pipe or a weighted line inserted through the tremie can be used to feel the top of the filter pack and to measure the depth to the pack as the work progresses.

Another method to determine whether the filter pack has been placed correctly involves using a short piece of telltale (tattletale) screen installed above the production screen. In the filter pack operation, filter pack and water are pumped directly into the annulus, with the return flow passing into the screen and then to the surface through the drill pipe or casing used to suspend the screen assembly. The top of the casing must be sealed tightly to the drill pipe so a pressure gauge can be used to estimate the level of filter pack around the screen. As soon as the level of the material reaches the top of the production screen, the pressure increases somewhat. This is followed by an abrupt pressure rise when the material fills in around the telltale screen. Because filter pack is still falling in the borehole and will add to the filter pack around the screen, the operator stops the filter-pack feed at the moment the filter pack reaches the top of the telltale screen. Knowing the

Figure 14.15. A filter pack can be placed in the annular space between inner and outer casings by using a small hopper attached to the top of a tremie pipe. Water is introduced into the tremie to prevent bridging of the pack material. *(Lang Well Drilling)*

volume of filter pack in suspension and the volume of water in the annular space, the operator can calculate the depth to the top of the filter pack material after all of it has been carried into the space around the riser pipe above the telltale screen.

Filter pack is often placed into large-diameter wells by reverse circulation of the fluid in the well as filter pack material is fed into the annular space by a continuous-feed hopper (Figure 14.16). In this procedure, the borehole is kept full of fluid. Filter pack material is carried down by the water as it flows outside the inner casing and screen. As filter pack fills the space around the well screen, the transporting water passes through the screen

Figure 14.16. Continuous-feed hoppers facilitate the installation of large volumes of pack material. (*W. Hoffstetter*)

openings and is drawn upward. Some drillers use a stinger pipe that forces the water to flow all the way to the bottom of the screen before returning to the surface (Figure 14.17). Use of the stinger pipe is especially advantageous for long screens, where it will reduce bridging and segregation of the filter pack particles.

Contractors can also use direct circulation of clean water to reduce bridging problems. Circulation continues at a slow rate into the screen and up the annulus so as not to restrict gravity flow of pack material. This is a good method to use when a pack is contaminated inadvertently by organic material or contains too much fine material. The uphole velocity will float the organic material or fine sediment to the surface and thereby reduce development time. Care must be taken to

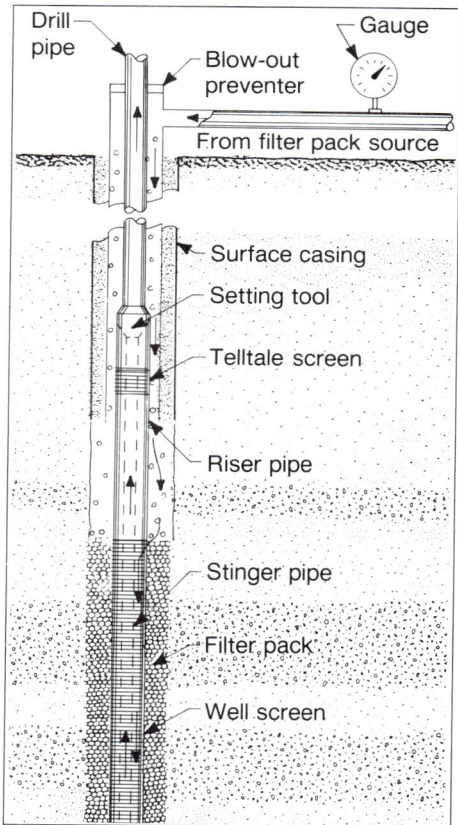

Figure 14.17. When filter packing a long screen, a stinger pipe is installed to force water to flow to the bottom of the screen.

avoid removing the filter cake because the borehole may become unstable.

A more elaborate method of placing filter packs in relatively small-diameter, deep wells has been used in the oil well industry for many years. The filter pack is pumped in with water or thin drilling fluid by an installation device called a cross-over tool. The cross-over tool gets its name from its arrangement of passages and ports. For water wells, the cross-over tool is generally used for deep installations where the annulus is small. Figure 14.18 shows the important features of the tool and the paths of flow when the filter pack is being placed in a typical well. The cross-over tool is connected between the drill pipe and the top of the riser pipe. The screen is suspended from the tool as the assembly is lowered into the well on drill pipe. The top of the riser pipe may be sealed to the casing with a suitable packer after the pack is placed.

During the operation of the cross-over tool, the fluid and filter pack pumped down through the drill pipe discharges below the special rubber packer while the return flow is conducted up through the packer into the annular space around the drill pipe. Pump pressure forces the fluid to move through the screen openings and into the stinger pipe connected to the lower end of the setting tool. The stinger pipe extends to within 3 ft (0.9 m) of the bottom of the screen so that the flow deposits the filter pack initially at the bottom of the hole. Fluid friction and pressure in the circulating system increase slowly as the filter pack material builds up around the well screen. The pressure increases suddenly when the filter pack material reaches the top of the screen. Careful observation is necessary to interpret the pressure-gauge changes correctly and to estimate the volume of material in transit.

Although this description of the operation of a cross-over tool may indicate that the procedure is rather simple, it actually requires elaborate and expensive equipment as well as considerable skill. Every phase of the operation must be co-ordinated properly to assure continuous flow of the filter pack material at a rate that will not plug the drill pipe or bridge in the hole.

Filter Pack Procedures for Wells Drilled by the Cable Tool Method

Several different casing arrangements may be used for filter packing wells drilled by the cable tool method. In one type of installation, the screen is connected to an inner casing, centered in a larger borehole, and surrounded by filter pack material (Figure 14.19). The inner casing will become part of the completed well structure and may accommodate the pump. In deep-well installations, the in-

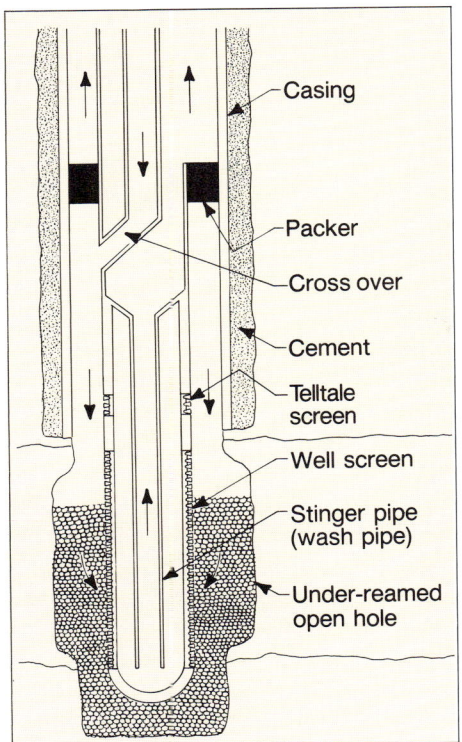

Figure 14.18. Essential features of cross-over tool used for setting well screen and placing filter pack in a deep well. *(Suman et al., 1983)*

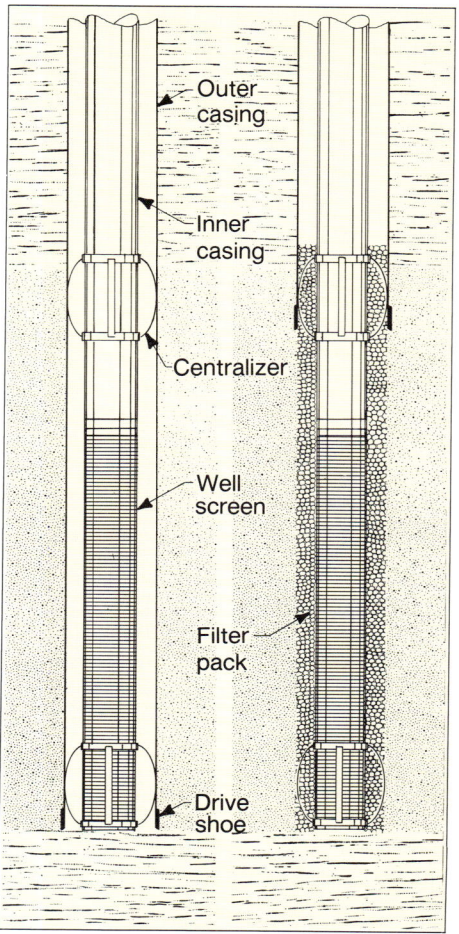

Figure 14.19. In the inner-casing method, the screen is connected to casing centered in a larger, cased borehole. After filter packing, the outer casing is pulled back.

ner casing may not extend all the way to the ground surface. A large-diameter outer casing is first set to the full depth of the well. An inner casing and well screen are then centered in the outer casing, using centering guides. If the screen is not centered, the large slot openings designed to hold back the filter pack may be placed next to the formation, leading to excessive development time or, at worst, uncontrollable long-term sand pumping and caving. The selected filter material is placed in the annular space around the screen and extended high enough above the screen to accommodate settlement of the filter pack after the outer casing has been pulled back. The filter pack material should extend above the top of the screen about one-fourth the screen length.

The pack is usually placed in stages as the outer casing is pulled back. Five feet (1.5 m) of filter pack material should be maintained above the screen as the casing is withdrawn. During withdrawal, depth to the top of the filter pack must be monitored carefully by a sounding line or tremie pipe to insure that the level never drops below the outer casing. Many drillers develop each screen section individually as the casing is withdrawn to prevent bridging of the filter pack. Bridging of the pack can lead to collapse of the formation against the screen and long-term sand pumping. The driller must be careful, however, not to overfill the annulus during withdrawal because a sand lock may develop between the outer casing and screen.

After the filter pack is installed, development work is continued to remove fine sediment from the filter pack and to clean the contact surface between the filter pack and the formation. As development proceeds, some settlement of the filter pack will occur and more filter pack must be added to keep the level above the screen. Permanently installed filter pack (gravel) chutes or conductor pipes for adding material are not necessary if the formation has been adequately developed. After development, the annular space above the filter pack should be sealed by bentonite pellets or cement grout. The outer (surface) casing may be removed or left in place.

Another filter-packing design, although used infrequently because it is laborious, is to place a screen inside an inner casing. The method is particularly applicable to

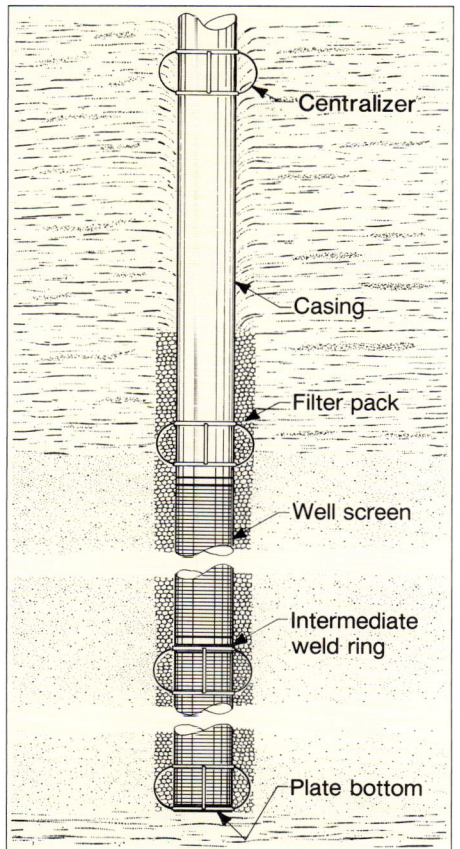

Figure 14.20. The screen and casing are placed into the borehole as a unit in many rotary drilled wells.

the installation of double filter packs in which three or more casing lengths may be used. A surface (outer) casing and working casing are set to appropriate depths and the annulus filled with filter pack material. An inner casing is set inside the working casing to the bottom of the aquifer and the screen telescoped through it. A coarser pack is placed between the inner and working casings. The working casing may be lifted as the second filter pack is placed. Some drillers may place the entire filter pack before the working casing is lifted although this procedure is not recommended because sand locking could occur. If the outer casing is to remain in place permanently after being pulled back, the working and inner casings may be removed entirely if a pipe extension attached to the top of the screen overlaps a minimum of 10 ft (3 m) inside the outer casing.

Drillers can eliminate the inner casing in the installation method described above by using a cone-type cover attached to the setting tool to prevent filter pack material from entering the well screen during installation of the pack. When the setting tool is attached to the bottom of the screen, however, the tool must be removed before filter packing. A plug (or "pig") is then placed temporarily on top of the screen assembly or riser to prevent filter pack material from entering the screen. The filter should be placed with a tremie pipe or by other approved methods. No pack material should extend above the top of the riser pipe or build up around the lowering tool and cover, or a sand lock will develop and prevent the uncoupling of riser pipe and drill rod. The outer (surface) and working casings are pulled back to expose the screen as in a typical pull-back operation. A slip packer is then set on top of the screen assembly.

Filter Pack Procedure for Wells Drilled by the Rotary Method

In most rotary-drilled wells, the screen and casing are placed in the borehole as a unit (Figure 14.20). The screen may be the same diameter as the casing or slightly smaller. Centralizers are generally attached every 20 ft (6.1 m) on the screen body and every 40 ft (12.2 m) on the casing. The casing should be held in tension while the drilling fluid viscosity is reduced as much as possible without allowing collapse of the well bore. Thinning the drilling fluid reduces development time, minimizes flotation effects, and increases settlement rate for pack materials. Filter pack is placed by tremie pipe or other means into the annulus around the screen and usually extends some distance above the top of the screen. Filter pack material should be added as required during development. After development and backfilling of the borehole, the tension on the casing and screen is released.

When telescoping the screen in a rotary-drilled well where the surface casing extends to a limited depth, an inner casing can be set to a depth near the bottom of the aquifer. The casing must be centered in the hole with centering guides. A screen assembly is lowered through the inner casing, and filter pack is placed outside the inner casing to some distance above the screen. The inner casing is pulled back as the filter pack fills the entire annulus (Figure 14.21).

General Guidelines for Installing Filter Packs

For filter pack installations, the pack should extend at least 25 percent of the screen length above the top of the screen. For example, a minimum of 50 ft (15.2 m) of filter pack should be installed at the top of a 200-ft (61-m) screen to provide an adequate reservoir. The top of the riser pipe should be well above the top of the filter pack or outer casing bottom, especially if no packer is used at the top of the riser pipe.

In wells with high pumping rates and large drawdowns, where water-producing formations are present at or near the end of the casing and no packer is used, upward flow in the annular space may lift some filter pack materials into the riser, which then falls into the screen. Sand-pumping wells will result. Use of a pressure-relief screen minimizes this problem. Water that flows upward in the pack material is diverted into the pressure-relief screen because the screen offers lower resistance to flow than does the filter pack (Figure 13.12, page 446). The relief screen equalizes pressure differences created between the formation and the casing by pumping. Under saturated conditions, a relief screen is necessary if any blank riser pipe extends beneath the casing.

In many cases, a slip packer is lowered onto the riser pipe to form a seal to the outer casing, thereby preventing any filter pack from moving into the screen. For wells with large drawdowns or for those in highly stratified deposits where large pressure differentials can occur, slip pack-

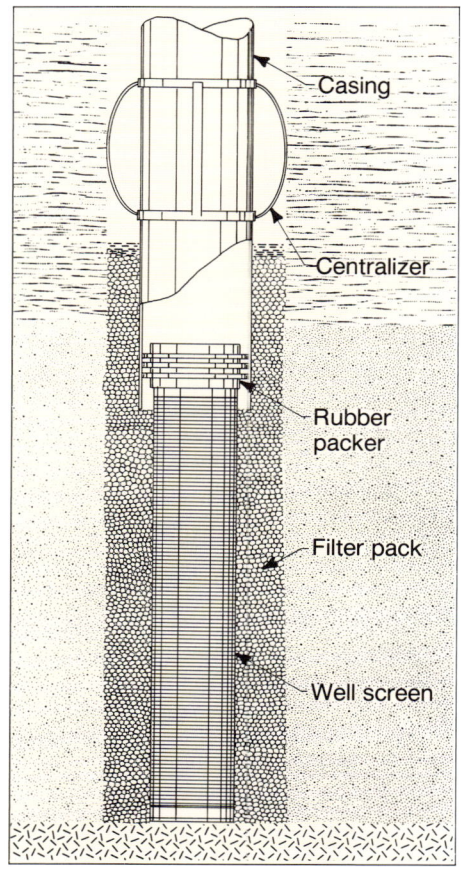

Figure 14.21. A well screen can be easily filter packed when the screen is set by the telescope method in a rotary drilled borehole. The screen is fitted with a rubber packer to seal the top of the screen to the inner casing after the casing has been pulled back.

ers may be lifted off the riser pipe during pumping, thus allowing pack material to enter the well. As suggested above, this problem can usually be prevented by placing a pressure-relief screen in the riser pipe at or near the bottom of the casing.

INSTALLATION OF PLASTIC SCREENS

More plastic materials are now being used for well screens, especially in smaller diameter wells. Plastic materials have predictable physical limitations, but when selected and used properly they provide wells with adequate structural strength, good hydraulic characteristics, and long life. For depth settings exceeding 300 ft (91.5 m), the screen manufacturer should be contacted for recommendations concerning wall thicknesses.

Plastic screens can be set in many of the same ways as steel materials, but plastic does not have the inherent strength of steel and special care must be taken during installation and well completion. In addition, unlike steel, plastic casing may be buoyant when placed in the well, depending on the depth to the water table and the density of the drilling fluid. Thus, setting procedures may have to be modified. It was originally felt that plastic screens would be used only by rotary drillers and be attached directly to plastic casing. Today, however, plastic screens are being installed with steel casing, and steel screens with plastic casing, by both the telescope and the direct-attached methods in rotary-drilled holes (Figure 14.22).

In the past, several plastic materials were used in a variety of different wall thicknesses and schedule numbers. Currently, PVC material constructed to Schedule 40, 80, or SDR-21 specifications are becoming the most common plastic materials used in water wells (see Chapter 13). A wide variety of standard end fittings are available for all types of plastic pipe. Because plastic pipe has standard outside dimensions, end fittings built by different manufacturers are generally interchangeable. There may be extremely small differences in dimensions, however, so for best results and where

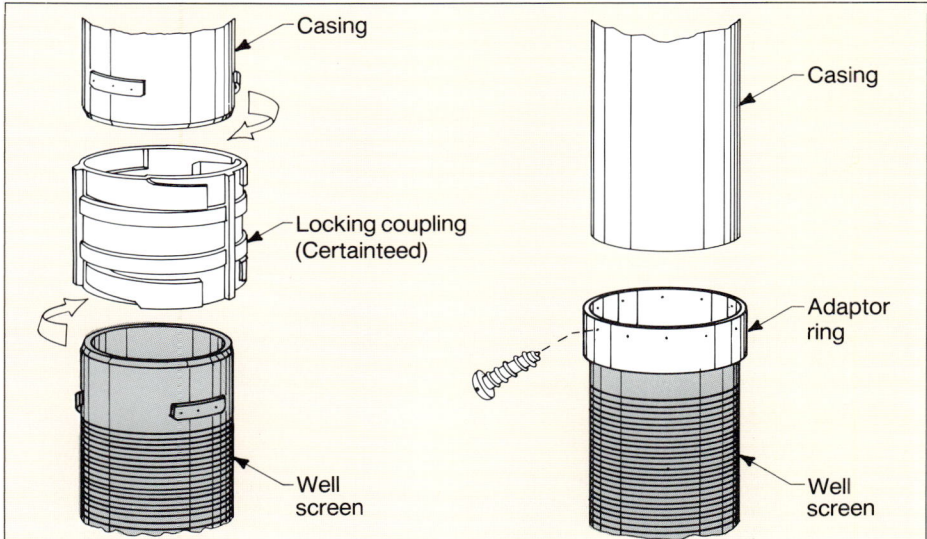

Figure 14.22. Locking couplings and adaptor rings are used to attach metal screens to plastic casing. Couplings can also be used to attach plastic screens to plastic casing.

maximum performance is required, it is advisable to use pipe fittings and pipe from the same manufacturer. PVC fittings are useful in attaching PVC to steel, or vice versa. Fittings such as male adaptors, female adaptors, reducers, and slip couplings are available to fit any threaded fitting on the market.

Telescope Installations

Plastic screens can be telescoped through plastic or steel casing, but the varying inside diameter of plastic casing (depending on schedule or SDR number) may cause some problems in obtaining the best fit between packer and casing. Although lead packers can be swedged out to fit the inside diameter of any suitable plastic casing, they are rarely used. Rubber packers are typically used, but because the outside diameter of the rubber sealing ring is constant the fit may be either tight or loose, depending upon the wall thickness of the plastic casing. For example, in 4-in (102-mm) pipe, there is a 0.046-in (1.2-mm) difference in inside diameters between Schedule 40 and SDR-21. Although the difference in diameter is small, it could cause serious sand pumping in a well completed in fine sand.

Even though plastic-based rubber packers are available, the drilling contractor should make sure that the packer is suitable for use with the particular plastic pipe being installed. Furthermore, if an uncommon size of plastic casing is used, there may be no packer available to properly seal the annular space between screen and casing.

Setting Screens in Open Boreholes (Direct Attached)

Perhaps the simplest method of installing a plastic screen is to attach the screen directly to the casing and then lower the entire assembly into the borehole. Screens can be attached to casing by couplings that use solvent-welding procedures (see Appendix 13.H), adaptor rings that are fastened to the casing, special locking couplings, and threaded connections.

After the casing and screen have been set in the hole, the screen may be filter packed and the annular space above the pack then backfilled as far up the hole as possible. All backfill materials should be clean and may contain large amounts of clay. Placement of backfill material prevents sudden slumping of the borehole walls before or during development. The borehole should be backfilled carefully when plastic casing or screen are used, because plastic casing does not have the collapse resistance of steel casing.

If no special filter pack is used, a rubber shale trap is often fastened to the casing above the top of the aquifer (Figure 14.23). When backfilled with a small amount of material, the shale trap will seal the annular space between the casing and the borehole and prevent loose material from falling into the screened zone. Bentonite can be placed in the annulus

Figure 14.23. A rubber formation packer (shale trap) mounted on the casing above the screen prevents clay from falling into the annulus around the screen.

above the fill material. The casing is then grouted in place with a neat-cement mixture, above the backfill and bentonite materials. The use of too much grout, however, may cause the shale trap to deform or collapse. In addition, the curing temperature of the cement must not be so high that the casing deforms.

The use of plastic casing and screens for completing wells in consolidated rock with overlying unconsolidated materials requires special techniques. When steel casing is used, a drive shoe is normally seated in the rock and open-hole drilling is continued through the consolidated aquifer. Because plastic casing cannot be driven, however, special packers are used to seal the casing to the rock.

When using plastic materials, hydrostatic pressures exerted on both casing and screen must be kept at a minimum during development. Development should begin slowly and gently until all drilling fluid is removed from the annulus around the screen and water is flowing freely into the screen. The casing and screen should never be "blown dry" with an air compressor. If compressed air is used for development, start the development process well above the screen and at first remove only small quantities of water (see Chapter 15).

BAIL-DOWN PROCEDURE

Under some conditions, it may be impossible or undesirable to pull back well casing to expose the screen. For example, side-wall friction on casing by subsurface materials may require too much pulling force, or movement of the casing may disturb the sanitary seal around it. In other situations, the screen cannot be set by direct-attached methods because the static water level may be so high and the aquifer materials so loose that the borehole may not stand open. In these cases, the bail-down method of setting well screens may be used. The objective of the bail-down method is to remove sediment from below the screen so the screen will settle.

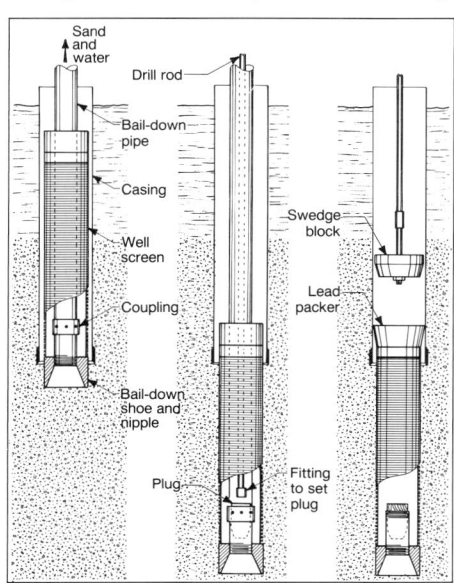

Figure 14.24. In the bail-down method, a screen is fitted with a special shoe that permits material to enter the screen where it is removed by a bailer or air-lift pumping.

In the bail-down method, the casing is generally set before the bailing process can begin. When drilling by the rotary method, the casing must be fixed in a permanent position by grouting or some other sealing method. If cement grout is used, the plug is drilled out of the lower end of the casing before starting the bail-down procedure. When drilling by the cable tool or casing-driver methods, the casing is usually held firmly by side-wall friction. Before setting a screen by the bail-down method, it is often helpful to drill a small-diameter pilot hole through the formation to collect samples and loosen the formation.

The well screen, fitted with a bail-down shoe or an open sleeve at its lower end, is telescoped through the casing (Figure 14.24). A riser pipe may be welded or

Table 14.1. Sizes of Pipe for Bail-Down Operation

Telescope-Size Screens		Pipe-Size Screens		Size of Largest Bailing Pipe	
in	mm	in	mm	in	mm
4	102	3	76	2	51
5	127	4	102	2½	64
6	152	5	127	3½	89
8	203	6	152	5	127
10	254	8	203	6	152
12	305	10	254	8	203
16	406	14	356	10	254
18	457	16	406	12	305

threaded to the top of the screen. If a bail-down shoe is used with special connection fittings, the screen is suspended on a string of pipe called the bailing pipe (Table 14.1). The screen assembly is worked into the formation below the well casing by operating the bailer or drilling tools through the bailing pipe. To make the operation more efficient, the bailer should be as large as possible. Some drillers use an air-lift system to remove materials from below the screen. For this operation, an air line is lowered inside the bailing pipe and the bailing pipe then becomes the discharge or eductor pipe for air-lift pumping. The added weight of the bailing pipe assists in sinking the screen when the weight of the screen alone is insufficient.

When a screen is being bailed down, it is advisable to keep the work progressing as continuously as possible. If the work is stopped for some time, the formation sand may pack tightly around the screen and cause so much friction that the screen can no longer move downward.

Heaving conditions sometimes prevent complete removal of sediment to the bottom of the bailing pipe or well screen after the screen has reached the desired depth. Filling the bailing pipe with water will usually stop the heaving so the bottom can be cleaned out with a small bailer and the plug can be placed. If this is not effective, the bailing pipe and well screen can be filled with a heavier drilling fluid, or a weighted fluid such as salt water, to create greater pressure to counterbalance the tendency of the sand to heave (see Chapter 11).

When the screen has been bailed down to the desired depth, a plug is lowered or dropped through the bailing pipe to seat in the special extra-heavy nipple above the bail-down shoe. Occasionally, cement is used to plug the bottom of the screen. The string of bailing pipe is then disconnected by turning it several turns to the right to unscrew the left-hand joint at the top of the nipple, leaving the plug or cement and extra-heavy nipple to seal the bottom of the screen. In place of a left-hand threaded connection for the bailing pipe, some drillers prefer a lug or bayonet-type connection. After removing the bailing pipe, the lead packer at the top of the screen is expanded with a swedging tool and the well is ready for development.

When bailing down small-diameter screens, the lower end of the screen may be fitted with an open sleeve or short leader pipe instead of the typical bail-down shoe and internal fittings. The screen will generally settle as the bailer removes sediment from below it, although it may be necessary to rest a string of pipe on top of the screen to provide additional weight. The lower end of this string of pipe should be

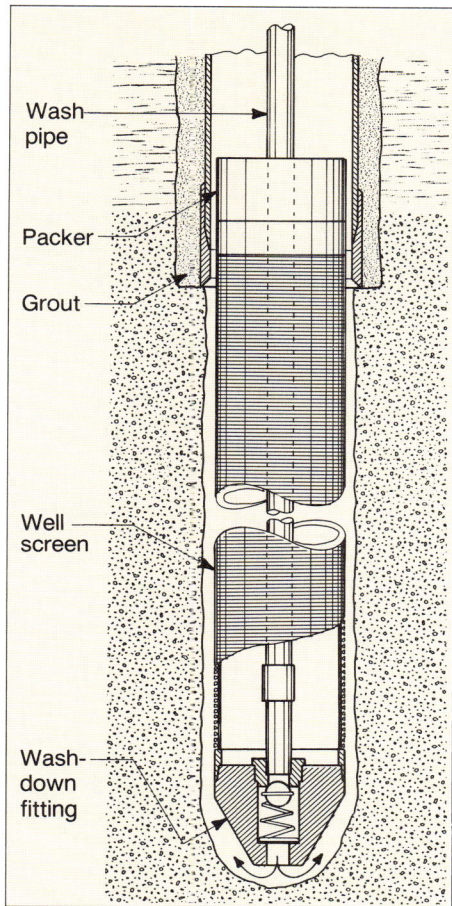

Figure 14.25. A wash-down bottom with spring-loaded valve permits washing a screen into place. Space around the lead packer allows return flow outside the well screen.

fitted with a coupling or flange that will rest squarely on top of the packer without deforming it. A rubber packer may be used so that the pipe can rest on a rigid steel base, rather than on the lead packer. The use of a rubber packer may require additional weight to set the screen, because of friction between the lips of the packer and the inside wall of the casing. The bottom of the screen is plugged with cement.

WASH-DOWN METHOD

Figure 14.25 shows the special fittings required for installing well screens by the wash-down method. The casing is first set to the desired depth and grouted. After the grout has set, the cement plug at the bottom is drilled out. When the casing is in place, a pilot hole can be drilled to obtain formation samples.

A self-closing bottom fitting or back-pressure valve is mounted in the bottom of the screen and connected by a left-hand thread to a string of pipe (usually drill pipe) used as the wash line (Figure 14.26). The screen is lowered to the bottom of the casing and light-weight drilling fluid or water is then pumped through the wash line. A significant fluid loss can occur if water is used as the drilling fluid. Fairly high pump pressure and adequate volume are needed to produce a high-velocity jet of fluid through the self-closing bottom. The jetting action loosens and removes the sediment, and allows the screen to sink. No rotation is applied to the wash line or screen during the jetting operation. The sediment is brought up around the screen and comes up inside the casing with the return flow of the fluid. Some of the larger particles inevitably drop back inside the screen unless a temporary cover plate is mounted on the wash line to cover the top of the screen, because the upward velocity of the fluid decreases suddenly just above the lead packer. Sediment in the screen can be removed by air-lift pumping, bailing, or circulation of drilling fluid after the wash line has been disconnected. If too much sediment is allowed to build up in the screen before disconnecting the wash line, a sand lock may be created and make it difficult to disconnect the wash line from the screen.

When the screen reaches the bottom of the aquifer, clean water should be pumped through the wash line and then circulated at a reduced rate to remove filter cake that

Figure 14.26. The wash-down fitting can be welded or threaded to the bottom of the screen.

may have been deposited on the formation during the jetting operation. It is essential that the formation cave around the screen, holding it so the wash line, when turned to the right, disconnects at the left-hand joint just above the bottom fitting. Metal bars welded to the bottom of the screen help prevent it from turning when being disconnected.

In general, the same wash-down procedure is used if the well casing and screen are attached and set in an open hole. In an open-hole operation, some sloughing can

Figure 14.27. Many large-diameter screens are ordered with wash-down fittings if sloughing formations are anticipated. *(LTP Enterprises)*

occur during the installation of the screen and casing, causing the screen to rest somewhat above the proper depth. The screen and casing are then washed into place. In highly pressurized aquifers when the driller wants to avoid drilling fluid contamination, the casing and screen are placed in an open hole somewhat above the aquifer and then washed into place with no adverse effects from the heaving formation. Many drilling contractors, anticipating these problems, order their screens with a wash-down fitting mounted in the plate bottom (Figure 14.27).

Occasionally, wash-down installation methods are suitable for both the drilling and the screen-setting operations. Under ideal conditions, small-diameter casing and screens that are attached directly can be installed to a depth of 400 ft (122 m) by the wash-down method.

JETTING METHOD

A common method for installing small-diameter screens [usually 2 in (51 mm)] uses a temporary wash pipe that is assembled inside the well screen before the screen is attached to the bottom joint of casing. The wash pipe, commonly ¾ inch (19 mm) in diameter, passes through the screen and extends upward at least 1 to 2 ft (0.3 to 0.6 m) into the casing (Figure 14.28). In some cases, the wash pipe may extend to the surface if the depth of the well is not great. A coupling screwed to the lower end of the wash pipe rests in a conical seat in the self-closing wash-down fitting which is equipped with a plastic-ball closure. A ring seal (doughnut) made of semirigid plastic is slipped over the upper end of the wash pipe and is pushed into the top of the well screen to close the space around the wash pipe and to direct the jetting water into the wash pipe. When the wash pipe extends to the surface, however, the ring seal prevents any return flow of the jetting water in the space between the wash pipe and the screen. All the return flow from the washing or jetting operation takes place outside the screen and casing.

Water pumped into the casing enters the wash pipe and jets out through the wash-down bottom. In sand formations, the jetting action will allow the string of casing and well screen to sink into the water-bearing formation. The screen may thus be set in the desired position without previously drilling a hole into that part

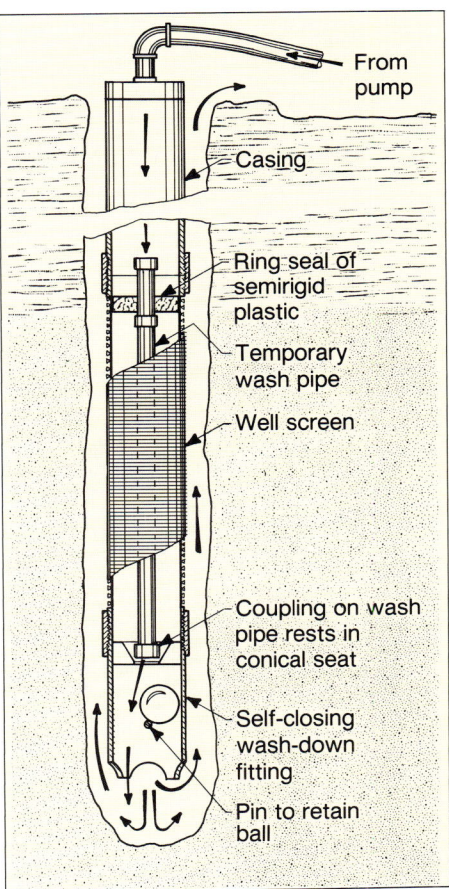

Figure 14.28. Small-diameter screens can be washed into place by jetting through a wash pipe and wash-down bottom with floating-ball valve.

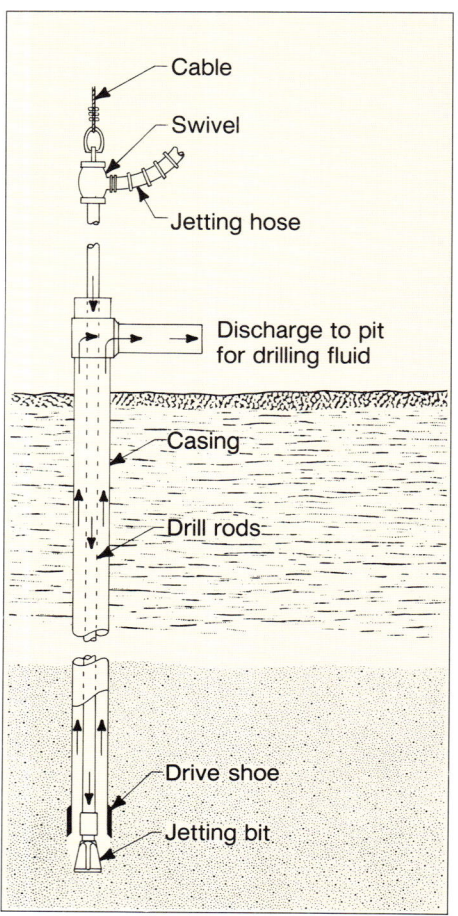

Figure 14.29. A screen can be installed by jetting and simultaneous driving.

of the formation. When water is used for jetting, screens and casing can only be jetted efficiently to depths of 20 to 25 ft (6.1 to 7.6 m), because of fluid loss into the formation and caving of the formation when casing is added. To achieve greater depths, a drilling fluid additive must be mixed with the jetting water to suspend cuttings and stabilize the borehole when circulation is interrupted.

About 5 percent of the jetting water will leak from around the bottom of the wash pipe and flow out through the screen openings. This flow helps prevent fine sand from passing into the screen during the jetting operation and reduces the possibility of sand-locking the wash pipe inside the screen.

When the well screen has been set at the proper depth, the wash pipe is fished out of the well by using a tapered tap or other type of fishing tool. The rigid plastic ball is just light enough to float in water and effectively closes the opening in the wash-down fitting. After the wash pipe has been removed, the well is developed.

Screens may be set in some cases without the wash pipe and seal. Much higher volumes of water are required because some water escapes through the sides of the screen and does not contribute to the jetting action at the bottom of the screen.

A second jetting method, requiring less equipment, utilizes a string of 2-in (51-mm) pipe set adjacent to the well point. Jetting the water through the pipe opens the hole as both the jetting pipe and the well-point assembly settle simultaneously by their own weight. When the final depth is reached, the jetting pipe is withdrawn; but water flow must continue until the jetting pipe is removed from the ground, or caving of the formation may cause it to become lodged in the hole.

A third jetting method involves both jetting and driving 2-in (51-mm) casing by using an internally placed ¾-in (19-mm) drop line with a chisel-point bottom (Figure 14.29). During the jetting operation, the 2-in pipe is driven intermittently; cuttings are washed up the annulus between the ¾-in line and the 2-in pipe. The jetting pipe is rotated and adjusted vertically to increase the effectiveness of the jetting action. It is a good idea to use a regular drilling fluid additive to enhance cuttings removal and sample recovery; heavier cuttings may stay in the hole if only clean water is used. A 1¼-in (31.8-mm) well point (screen) is dropped through the 2-in pipe when the proper depth has been reached. To expose the screen, the casing can be pulled back or the point can be driven out the bottom of the casing. A rubber packer is used to seal the top of the screen to the casing. Depths of 200 ft (61 m) are possible in sand and fine-gravel deposits. This method is especially widespread in Florida, Indiana, and Michigan.

Although jetting is ideally suited to rotary rigs, small cable tool rigs can be equipped with a pump capable of 15 to 25 gpm (81.8 to 136 m^3/day) to provide the necessary water supply. Use of a relatively high-capacity, low-head centrifugal pump is not advisable because they are ineffective beyond a depth of 100 ft (30.5 m).

INSTALLING WELL POINTS

Well points are often installed by some of the same methods already described for larger diameter well screens. For the pull-back method, casing is first set to the full depth. A suitable packer is threaded to the top of the well point or riser pipe. After the well point has been dropped through the casing, the casing is pulled back to expose the screen to the water-bearing sediment (Figure 14.30). The drill tools may have to be placed on the well point to hold it down as the casing is pulled back. Many drilling contractors install 2-in (51-mm) stainless steel well points in 4-in (102-mm) wells by this method.

Occasionally the pull-back method cannot be used because the friction on the pipe is so great that the force required to move the pipe might break it. In this case, a well point can be driven beyond the end of the casing into the sand formation below.

All the sediment in the casing is removed so the well point will not become sand-locked inside the pipe. If the sediment tends to heave, the casing is kept full of water while the screen is being set. The well point, with a self-sealing packer attached, is dropped through the casing. A driving bar, drill stem, or other similar tool is lowered to the top of the packer and alternately raised and dropped to drive the well point out the bottom of the casing. A driving weight of less than 500 lb (227 kg) and operated with a 2-ft (0.6-m) stroke is recommended to minimize potential damage to the screen. Careful measurements must be made so the driller will know when the screen has been driven the correct distance. Use of a riser pipe is advised.

Some well points are manufactured with a drive plate mounted just above the

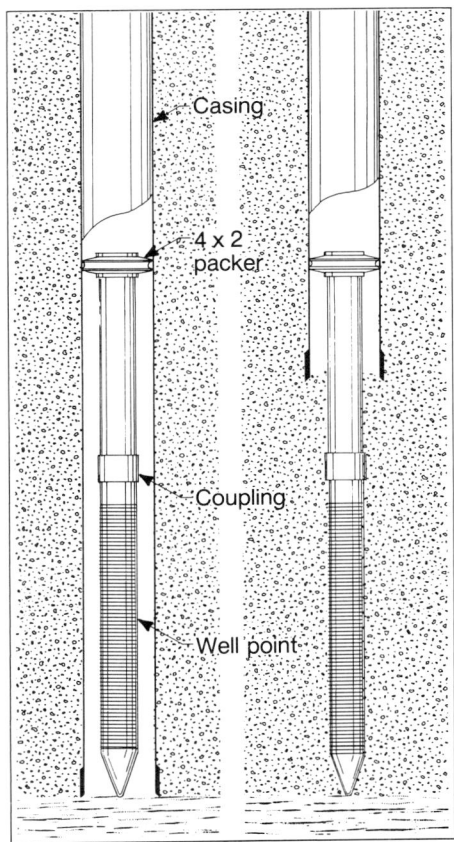

Figure 14.30. Two-inch (51-mm) well points may be set in 4-in (102-mm) diameter wells by the pull-back method. A 4 x 2 rubber packer is mounted on a short section of riser pipe.

point (Figure 14.31). The driving force is directed at the point, and the screen is pulled into place*. This bottom-drive method is preferred when driving relatively long well points through dense materials. A ¾-in (19-mm) solid bar should be used for driving inside 1¼-in (32-mm) well points, and a 1½-in (38-mm) bar inside 2-in (51-mm) well points. The drive rod should be long enough to extend at least 1 or 2 ft (0.3 or 0.6 m) above the packer when resting on the drive plate, to isolate the driving action from the top of the screen. During driving, the drive rod rests continuously on the drive plate as the hammer moves vertically. A string of pipe or the drill stem can be used to provide the required weight. The well point must be equipped with either a lead or a rubber packer. As in the pull-back method, 2-in stainless steel well points can be readily installed in 4-in (102-mm) wells using 4 by 2-in packers. Other methods for installing well points in shallow wells are given in Chapters 10 and 22.

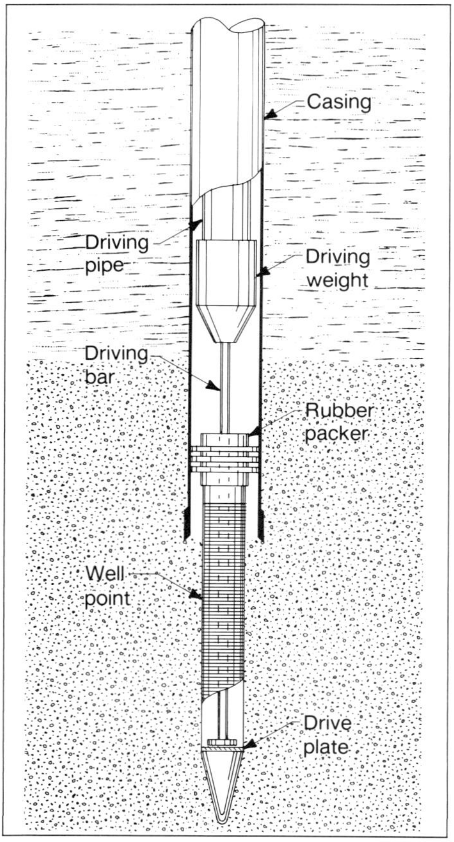

Two-inch (51-mm) well points can be set easily through hollow-stem augers once the auger-flight assembly has reached the proper depth. The screen is attached directly to the casing, and the string is lowered inside the augers to the bottom of the borehole. The auger flights are then pulled back to expose the screen and casing. This method is particularly suitable in shallow, caving formations, and is often used to set monitoring wells. See Chapter 10 for additional information on the auger method.

REMOVING WELL SCREENS

Various circumstances arise that make it necessary to remove a screen assembly from a well. These situations include the following:

1. Inadequate yields at an original screen setting may force reinstallation at another depth.

2. In some areas, declining water tables may require that the well be deepened after some years of use.

3. Incrustation and cementation of the formation around the screen may require that the well be deepened because of the difficulty in chemically treating the screen in situ.

4. The screen must be replaced because corrosion damage is causing the well to pump sand.

Figure 14.31. A driving bar can be used to drive well points from the inside; this procedure is recommended for points longer than 5 ft (1.5 m).

---

*CAUTION: Not all drive points are manufactured to withstand driving from the bottom. If the drive point is built for top driving, inside driving can cause severe damage to the bottom of the screen. All points to be driven from the inside should be identified as such when ordered from the manufacturer.

5. The well is to be abandoned and the screen recovered for use in another well. Under most circumstances, screens that have been telescoped are considerably easier to remove than those placed by other methods.

Almost any screen that has been in place for a long time is certain to be firmly fixed in the formation. To remove the screen safely with no damage, the pulling force should be distributed over a considerable portion of the screen length. This is true regardless of the type of screening device in the well.

Light acid treatment of the well will almost always facilitate removal of the screen by dissolving some of the incrustation on the screen surface. In telescoped installations, incrustants usually develop preferentially around the top of the well screen, inside the lower end of the well casing. Acid treatment loosens these materials and reduces the force needed to pull the screen. See Chapter 19 for the correct procedure for acid treatment of wells. Other fluid additives are also useful in removing screens. For example, some drillers jet a polyphosphate solution into the zone around the screen to loosen clays or other sediments. Polymeric drilling fluid additives are excellent friction reducers, and their use can be helpful in minimizing the force required to pull the screen.

Certain mechanical procedures are successful in freeing the screen from the formation. In underreamed installations, a washover pipe may be used to loosen the filter pack around the screen if the necessary clearances are available. In conventional filter-packed wells, the screen may first be loosened from the formation by milling the plate bottom and removing as much of the pack material as possible. Once loosened, the screen can be removed by one of the methods described below. Overshot tools and various types of spears are also used to remove screens, but may cause damage to the screen. However, they can be used to lift screen assemblies with riser pipes with little chance for damage. Never attempt to remove a screen by pulling on a bail bottom unless the screen has just been set and is held loosely by the sediment.

Screens that are 4 in (102 mm) and larger are best pulled by using a pipe of smaller diameter sand-locked inside the screen. Angular sand particles, placed in the space between the pulling pipe and the inside of the well screen, form the sand joint which serves as a structural

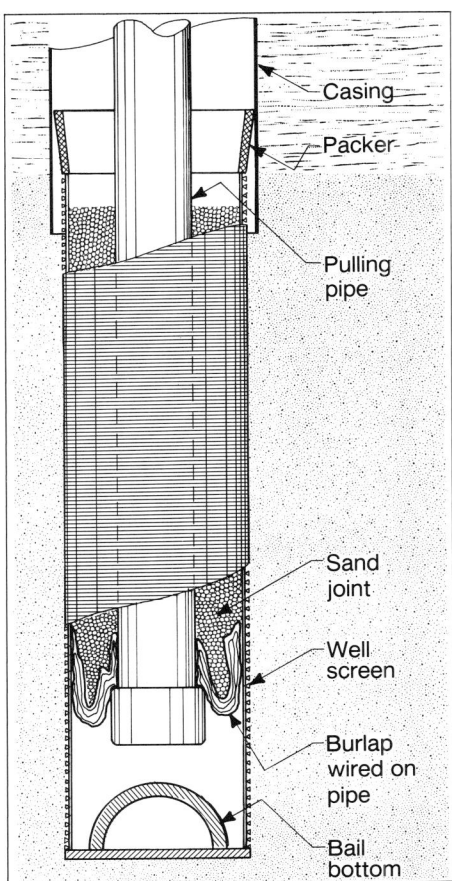

Figure 14.32. To remove telescoped screens from a well, a method called sand locking has been developed. Angular sand placed between the pulling pipe and screen becomes locked when lift is applied to the pulling pipe.

Table 14.2. Sizes of Pulling Pipe

Size of Telescope Screen		ID of Screen		Size of Pulling Pipe		Quantity of Sand	
in	mm	in	mm	in	mm	ft³/ft	m³/m
4	102	3	76	1½	38	0.031	0.003
5	127	4	102	2	51	0.056	0.005
6	152	4⅞	124	3	76	0.070	0.006
8	203	6⅝	168	4	102	0.134	0.012
10	254	8⅝	219	5	127	0.254	0.024
10	254	8⅝	219	6	152	0.187	0.017
12	305	10⅜	264	6	152	0.361	0.034
16	406	13⅛	333	8	203	0.535	0.050
16	406	13⅛	333	10	254	0.308	0.029
18	457	15	381	10	254	0.602	0.056
20	508	17	432	12	305	0.695	0.065
24	610	21	533	12	305	1.52	0.142

connection between the pipe and the screen. The sand joint is the best and most reliable scheme for transmitting the pulling force to the well screen.

Figure 14.32 shows the elements of the sand joint and how it is formed. The following steps should be taken to pull the screen:

1. Obtain high-quality, proper-size pulling pipe (Table 14.2).

Figure 14.33. Strips of burlap are wired to the lower end of the pulling pipe.

2. Attach strips of burlap, 2 to 4 in (51 to 102 mm) wide and about 3 ft (0.9 m) long, by wire to the pulling pipe just above the lower coupling (Figure 14.33). For screens of 16 in (406 mm) diameter and larger, a steel disk slightly smaller than the inside diameter of the screen is attached by four bolts to a plate welded to the pulling pipe.

3. Temporarily tie the upper ends of the burlap against the pipe.

4. Lower the pulling pipe to the top of the strips, cut the strips loose at the upper end, and arrange them evenly around the top of the well casing (Figure 14.34).

5. Lower the pipe so that it penetrates about 70 percent of the screen length.

6. Slowly pour enough sand into the screen to fill two-thirds of its length. Do not fill above the top of the screen. To avoid bridging, wash in the sand or move the pulling pipe horizontally while adding sand. Sand must be clean, angular material with no silt or clay. Generally, use medium sand for pulling small-diameter well screens and coarser material [greater than ¼ in (6.4 mm)] for larger diameter screens.

7. Slowly lift the pulling pipe to create the sand lock. A reasonable load should be applied and held constant for a short time to give the load applied to the pulling pipe a chance to exert a steady pressure on the screen. Jarring the pipe while maintaining the pulling force sometimes helps. The pulling force may then be increased gradually until the screen begins to move. Occasionally, the inside of the well casing is badly corroded and incrusted with rust or tubercles, making it necessary to use lifting jacks against pipe clamps or a pulling ring with slips.

8. After the well screen has been hoisted to the surface, the sand joint should be washed loose with a stream of water or compressed air to allow the pulling pipe to be removed from inside the screen. If a disk is used, the screen bottom is unscrewed or cut off, the four bolts holding the disk to the bottom of the pulling pipe are removed,

Figure 14.34. The upper ends of the strips of burlap are arranged uniformly around the top of the well casing as the pulling pipe is lowered in the well.

and the disk is taken out, thereby facilitating removal of the pulling pipe from the screen.

Some well-drilling contractors have developed innovations in the use of the sand joint that are practical under certain conditions. For safety reasons, two or three slots may be cut in the pulling pipe just above the burlap so that the sand joint can be loosened by backwashing and bailing from inside the pulling pipe if the connection below ground must be broken. Slots can also be cut in the pulling pipe at a level corresponding to the upper part of the screen so that any excess sand will run into the pipe and prevent overfilling of the screen, which could sand lock the drill string. The driller should install right- and left-hand couplings between the bottom of the drill pipe and the pulling pipe to disengage the drill string if sand locking occurs. The screen and part of the pulling pipe would then remain in the ground.

Some drillers weld rings at two or three different levels on the pulling pipe so that friction developed in the sand lock can be more evenly distributed. A series of pipe nipples joined by pipe couplings may also be used. This scheme is especially practical for pulling 6-in (152-mm) diameter and smaller screens.

Small screens, especially 2-in (51-mm) well points and 4- to 6-in (102- to 152-mm) telescoped screens, are pulled most frequently for maintenance or replacement. Several latch-type tools have been designed to facilitate removal of small screens. As in the case of larger diameter screens, it is best to spread the pulling force over as large an area of the screen body as possible. Alternately, the pulling force can be concentrated on the sump-pipe section if the screen is equipped with one. One lifting device is an elliptical plate cut in half and then hinged together. The plate will fold up as it is lowered into the well. But when lifted, the plate unfolds and locks underneath the screen or sump pipe. The folding plate should not be so large that it causes bulging of the sump pipe that might lead to interlocking of the sump pipe and casing.

CONCLUSIONS

In general, the screen-setting procedure is based on the geologic characteristics of the aquifer, the drilling equipment, and whether the screen will be naturally developed or filter packed. Well-design engineers should always consult local drilling contractors before specifying acceptable drilling methods for any high-capacity well. If this is done, unrealistic bid specifications concerning drilling procedures and screen-installation methods will be avoided.

CHAPTER 15
Development of Water Wells

Procedures designed to maximize well yield are included in the term "well development." Development has two broad objectives: (1) repair damage done to the formation by the drilling operation so that the natural hydraulic properties are restored, and (2) alter the basic physical characteristics of the aquifer near the borehole so that water will flow more freely to a well. These objectives are accomplished by applying some form of energy to the screen and formation. Well development is confined mainly to a zone immediately adjacent to the well, where the formation materials have been disturbed by well construction procedures or adversely affected by the drilling fluid. In addition, the undisturbed part of the aquifer just outside the damaged zone may be reworked physically during development to improve its natural hydraulic properties.

All new wells should be developed before being put into production to achieve sand-free water at the highest possible specific capacity. In addition, older wells often require periodic redevelopment to maintain or even improve the original yield and drawdown conditions. Maintaining a high specific capacity assures that the well will be energy efficient.

Another type of development, called aquifer development or stimulation, is done when the aquifer will not yield enough water even after well development procedures have been applied. This form of development is usually limited to semiconsolidated or completely consolidated formations. Well development techniques will be discussed before aquifer development procedures because they apply to every well, regardless of the geologic materials.

WELL DEVELOPMENT

Every type of drilling operation alters the hydraulic characteristics of formation materials in the vicinity of the borehole. These alterations often result in a severe reduction of the hydraulic conductivity close to the well bore. For example, many alluvial and most glacial sand and gravel deposits are relatively young (less than 100,000 years) and have not been thoroughly compacted or cemented by geologic processes. As indicated in Chapter 3, these sediments usually are highly stratified; that is, the deposit consists of many layers, and in each layer the grains are all nearly the same size. These grains are loosely packed, and therefore the deposit has a large percentage of void space — commonly 25 to 35 percent.

If a well is drilled into a sand and gravel deposit with a cable tool rig or a rotary rig equipped with a casing driver, the repeated blows on the casing will change the loosely consolidated and naturally stratified condition of the deposit. Equal-sized grains will pack more closely together and smaller grains from above will move into the underlying void spaces (Figure 15.1). The result is a significant decrease in the porosity of the sediment and a drastic reduction in its natural hydraulic conductivity. The loss of stratification and the resulting mixing of the grain sizes is generally confined to the zone immediately around the borehole. To reduce this type of formation damage, some cable tool drillers sink the casing through the aquifer by bailing methods rather than by drilling and driving the casing.

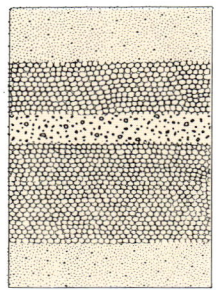 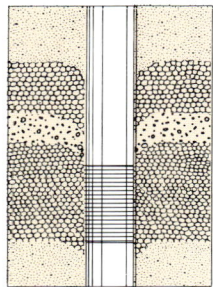

Before drilling begins   After drilling

Figure 15.1. A significant reduction in hydraulic conductivity and porosity of sediments surrounding a borehole can occur during cable tool drilling.

The presence of clay in the formations being drilled or the addition of bentonite to the slurry to suspend the cuttings creates an additional need for development in holes drilled by the cable tool method. In many areas, thin clay lenses are irregularly interspersed with productive sand formations. During drilling, water is added to the cuttings periodically to create the slurry required for bailing. After bailing, some slurry will remain in the casing. When drilling is resumed and the bit protrudes out the end of the casing, some of the residual clay-rich slurry may enter underlying sand formations. In clay-poor formations, drillers add bentonite to form a slurry to suspend the cuttings so that the bailing operation is more efficient. The addition of bentonite may also be necessary to build the required hydrostatic head in the casing to contain heaving formations. As in direct rotary drilling, the bentonite prevents water from moving into the sand aquifers. One other problem occurs when casing is driven through sticky clays. The casing entrains some of the clay and carries it down the borehole, coating the face of potential aquifers. This clay must be removed to restore the original hydraulic conductivity of the aquifer.

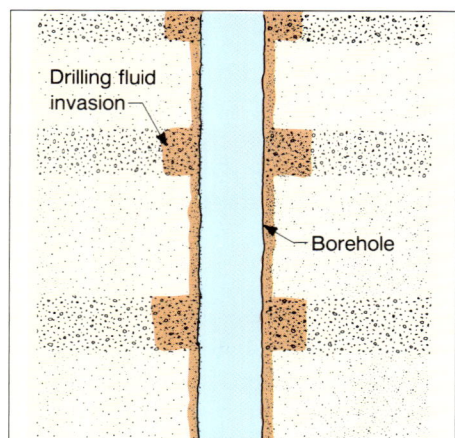

Figure 15.2. When drilling with water-based drilling fluids, some of the fluid will flow into the most pervious parts of the formation, well away from the borehole.

Both types of rotary drilling also cause damage to aquifers. In direct and reverse-circulation rotary drilling, the action of the bit will cause some intermixing of sediments near the borehole, but this disturbed zone is generally not damaged as seriously as the disturbed zone created by the cable tool method. The most serious problem in rotary drilling occurs when drilling fluids containing clay enter the aquifer, often flowing many feet out from the borehole (Figure 15.2). In direct rotary drilling, drilling fluids usually consist of high-grade clay mixed with water. Once in the aquifer, the clay particles become fully hydrated and produce a powerful plugging effect. Even if clear water is used, naturally occurring clays exposed in the borehole can mix with the drilling fluid and plug the pore space of permeable formations. In most reverse rotary holes, fine sand and silt particles will also move into the aquifer because of the lack of a filter cake and the subsequent high fluid losses. Unfortunately, permeable sediments are more susceptible to drilling fluid penetration, and are thereby subject to the greatest potential loss in hydraulic conductivity.

Sloughing of weakly consolidated formations frequently occurs in rotary drilling operations when a sudden loss of drilling fluid momentarily reduces the hydraulic pressure in the borehole. Although the extra volume of material removed from the borehole may not seem significant, the loss of stratification and subsequent mixing of the particle sizes can cause serious damage to the formation's original hydraulic characteristics.

The examples discussed above show that formation damage is unavoidable, regardless of which drilling method is used, and steps must be taken to restore the original hydraulic conductivity of the aquifer. All wells in both consolidated and unconsolidated formations should be developed until they are sand free when pumped at the desired rate. Techniques described below are applicable for all types of common aquifer materials, but the benefits are generally more substantial for unconsolidated sediment. Thus, the emphasis in this chapter will be on development of wells in formations where well screens are required. Development is an essential operation in the proper completion of any water well, however, because maximum specific capacity and well efficiency will rarely be reached without it.

Development procedures have the following beneficial purposes:

1. Reduce the compaction and intermixing of grain sizes produced during drilling by removing fine material from the pore space.

2. Increase the natural porosity and permeability of the previously undisturbed formation near the well bore by selectively removing the finer fraction of aquifer material.

3. Remove the filter cake or drilling fluid film that coats the borehole, and remove much or all of the drilling fluid and natural formation solids that have invaded the formation.

4. Create a graded zone of sediment around the screen in a naturally developed well, thereby stabilizing the formation so that the well will yield sand-free water. Some stabilization of the formation can also be achieved in a filter pack well as long as the filter pack thickness is 8 in (203 mm) or less.

The ultimate result of proper well development is to provide sand-free water at maximum specific capacity.

FACTORS THAT AFFECT DEVELOPMENT

Well Completion Method

There are two major completion methods — natural development and filter packing. The particular completion method is selected on the basis of the geologic character of the aquifer, the type of drilling rig, and the type of screen. The completion method

500 GROUNDWATER AND WELLS

determines to some degree the effectiveness of specific development methods.

In natural development, a highly permeable zone is created around the screen from materials existing in the formation. Creation of this zone is best understood by visualizing what happens throughout a series of concentric cylindrical zones in a sand aquifer surrounding the screen. In the zone just outside the well screen, development removes most particles smaller than the screen openings, leaving only the coarsest material in place. A little farther out, some medium-sized grains remain mixed with the coarse sediment. Beyond that zone, the material gradually grades back to the original character of the water-bearing formation (Figure 15.3). Finer particles brought into the screen in this process are removed by bailing or pumping. Development work is continued until the movement of fines from the formation becomes negligible.

By creating this succession of graded zones around the screen, development stabilizes the formation and prevents further movement of sediment. After development, water moving toward the screen encounters sediment with increasing hydraulic conductivity and porosity. Improving the hydraulic conditions around the well will increase the specific capacity and efficiency. Thus, more water can be obtained from the well, and for any yield the cost of lifting the water to the surface will be minimized.

In filter packing, a specially graded sand or gravel having high porosity and permeability is placed in the annulus between the screen and the natural formation. It

Figure 15.3. Natural development removes most particles near the well screen that are smaller than the slot openings, thereby increasing porosity and hydraulic conductivity in a zone surrounding the screen.

should be emphasized that development of the disturbed formation outside the pack is still mandatory to achieve maximum specific capacity.

Open Area and Slot Configuration

All development methods work best in wells equipped with screens having both maximum open area and the type of slot configuration that permit hydraulic forces exerted inside the well screen to be directed efficiently into the formation. Both factors are equally important in successful development. Screen open areas vary typically from a low of 1 percent for perforated pipe to more than 40 percent for continuous-slot, wire-wound screens. Screens with high open area can be developed more effectively because more of the development energy can reach the formation. (See Chapter 12 for descriptions of various screen types.) Slot configuration also controls how much development energy reaches the formation, and the percentage of the formation that this energy can affect. Thus, more fine material can be removed more quickly if all the available energy can be directed at most or all of the surrounding formation.

Slot Size

Selection of the correct slot size for well screens is essential for successful well development. Slot openings are chosen to permit removal of the fine material from the formation. For naturally developed wells, it is common practice to select a slot width that retains about 40 percent of the sediment in the formation adjacent to the screen. For filter-packed wells, the slot opening is selected to retain about 90 percent of the filter pack material. These points are discussed in detail in Chapter 13.

Slot size may govern the effectiveness of the development procedures. Removal of too much sediment may cause settlement of the overlying surface materials, which can have undesirable effects on the well and produce dangerous conditions for the drilling rig. On the other hand, when well screen openings are smaller than necessary, full development may not be possible and the well yield will be below the potential of the formation. Incomplete development can also lead to cementation or incrustation caused by abnormally high flow velocities and the corresponding pressure drop near the well bore (see Chapter 19).

Drilling Fluid Type

Clay and polymers are the two major drilling fluid additives used in rotary drilling. After a well is drilled, all drilling fluid must be removed from both the borehole walls and the formation by physical or chemical means. Although some polymeric drilling fluid additives will break down naturally over time, it is recommended that they be broken down chemically and removed from the well at the time the well is completed.

The rate and effectiveness of drilling fluid removal depends not only on the type of additive used but also on the physical character of the aquifer, the depth of the well, and the weight and viscosity of the drilling fluid. Results from an experimental well field show that large amounts of development energy are required to remove drilling fluid containing clay additives (Driscoll et al., 1980a). Much less development energy is required to achieve maximum specific capacities for wells drilled with a polymeric drilling fluid additive that has broken down.

Filter Pack Thickness

The thickness of the filter pack has considerable effect on development efficiency. This happens for two reasons. First, the filter pack reduces the amount of energy reaching the borehole wall. The thinner the filter pack, the easier it is to remove all the undesirable fine sand, silt, and clay when developing the well. Second, a filter pack is so permeable that water may flow vertically in the filter pack envelope at places where the formation may be partially clogged, rather than move into or out of the natural formation. To permit the transfer of development energy to the borehole wall, filter packs normally should be no more than 8 in (203 mm) thick and should be properly sized and graded according to the design criteria given in Chapter 13.

Type of Formation

Different types of formations are developed more effectively by using certain development methods. For example, highly stratified, coarse-grained deposits are most effectively developed by methods that concentrate energy on small parts of the formation. In uniform deposits, development methods that apply powerful surging forces over the entire well bore produce highly satisfactory results. Other development methods that withdraw or inject large volumes of water quickly can actually reduce the natural hydraulic conductivity of formations containing a significant amount of silt and clay.

WELL DEVELOPMENT METHODS

Different well development procedures have evolved in different regions because of the physical characteristics of aquifers and the type of drilling rig used to drill the well. Unfortunately, some development techniques are still used in situations where other, more recently developed procedures would produce better results. New development techniques, especially those using compressed air, should be considered by contractors when they buy and equip a new rig. Any development procedure should be able to clean the well so that sand concentration in the water is below the maximum allowable limit set for the particular water use.

Overpumping

The simplest method of removing fines from water-bearing formations is by overpumping, that is, pumping at a higher rate than the well will be pumped when put into service. This procedure has some merit, because any well that can be pumped sand free at a high rate can be pumped sand free at a lower rate.

Overpumping, by itself, seldom produces an efficient well or full stabilization of the aquifer, particularly in unconsolidated sediments, because most of the development action takes place in the most permeable zones closest to the top of the screen. For a given pumping rate, the longer the screen, the less development will take place in the lower part of the screen. After fine material has been removed from the permeable zones near the top of the screen, water entering the screen moves preferentially through these developed zones, leaving the rest of the well poorly developed and contributing only small volumes of water to the total yield. In some cases, overpumping may compact finer sediments around the borehole and thereby restrict flow into the screen. If more powerful agitation is not performed, an inefficient well may result. On the other hand, overpumping may be effective in filter-packed wells in competent, relatively non-stratified sandstone formations because flow toward the well bore is more or less uniform.

There is another objection to overpumping that is commonly overlooked. Water flows in only one direction, toward the screen, and some sand grains may be left in a bridged condition, resulting in a formation that is only partially stabilized (Figure 15.4). If this condition exists and the formation is agitated during normal pump cycles after the well has been completed, sediment may enter the well if the sand bridges become unstable and collapse.

Drillers ordinarily use a test pump for overpumping operations, but when a large quantity of water must be pumped, it may be difficult to obtain equipment of sufficient capacity at reasonable cost.

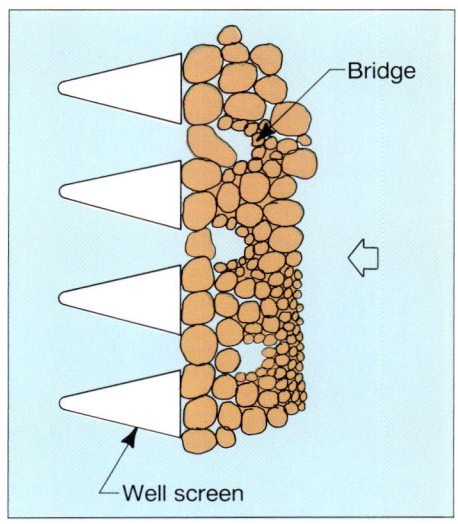

Figure 15.4. During development by overpumping, sand grains can bridge openings because flow occurs in only one direction. Once the well is placed into service, agitation by normal pump cycling can break down the bridges, causing sand pumping.

Therefore, the pumping equipment intended for regular well use is sometimes used for overpumping. Depending on the type of pump, this may be done either by operating the pump at a higher speed or by allowing the pump to discharge at the surface at a lower-than-normal operating pressure. There is one serious objection to performing this work with the permanent pump. Sand pumping will subject the pump to excessive wear, which over time can reduce its operating efficiency. Under severe conditions, the pump may become sand locked, either during pumping or after shut off. Should sand locking occur, the pump must be pulled, disassembled, cleaned, and repaired if necessary before being placed back into service.

Backwashing

Effective development procedures should cause reversals of flow through the screen openings that will agitate the sediment, remove the finer fraction, and then rearrange the remaining formation particles (Figure 15.5). Reversing the direction of flow breaks down the bridging between large particles and across screen openings that results when the water flows in only one direction. The backflow portion of a backwashing cycle breaks down bridging, and the inflow then moves the fine material toward the screen and into the well.

A surging action consists of alternately lifting a column of water a significant distance above the pumping water level and letting the water fall back into the well. This process is called rawhiding. Before beginning the surging operation, the pump should be started at reduced capacity and gradually increased to full capacity to minimize the danger of sand-locking the pump. In the rawhiding procedure, the pump is started, and as soon as water is lifted to the surface the pump is shut off; the water in the pump column pipe then falls back into the well. The pump is started and

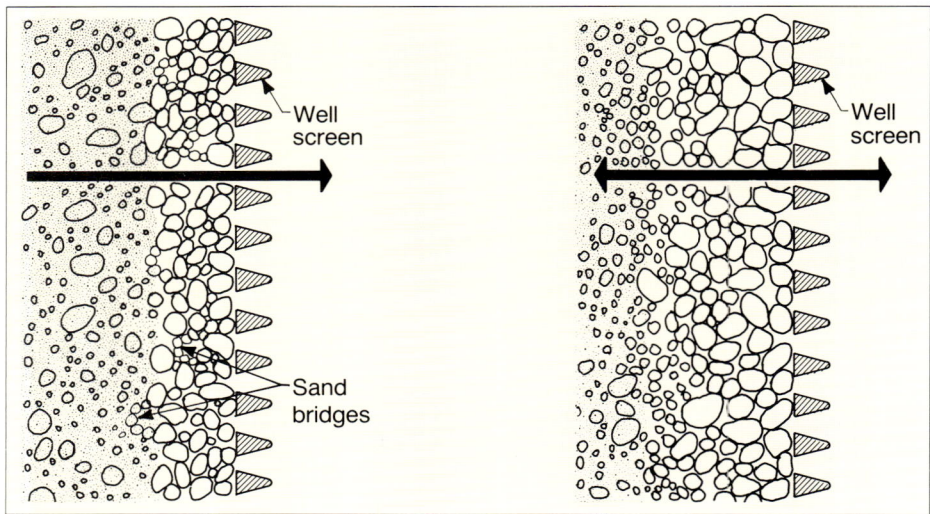

Figure 15.5. Effective development action requires movement of water in both directions through screen openings. Reversing flow helps break down bridging of particles. Movement in only one direction, as when pumping from the well, does not produce the proper development effect.

stopped as rapidly as the power unit and starting equipment will permit. To avoid damaging the pump, the control box should be equipped with a starter lockout so that the pump cannot be started when it is back spinning. During the procedure, the well should be pumped to waste occasionally to remove the sand that has been brought in by the surging action.

Some wells respond satisfactorily to rawhiding, but in many cases the surging effect is not vigorous enough to obtain maximum results. As in the case of overpumping, the surging effects may be concentrated only near the top of the screen or in the most permeable zones. Thus, the lower part of a long screen may remain relatively undeveloped.

Although overpumping and backwashing techniques are used widely, and in certain situations may produce reasonable results, their overall effectiveness in high-capacity wells is relatively limited when compared with other development methods. Other methods, as described below, are capable of removing more fine materials in less time and generally can produce higher specific capacities.

Mechanical Surging

Another method of development is to force water to flow into and out of a screen by operating a plunger up and down in the casing, similar to a piston in a cylinder. The tool normally used is called a surge block, surge plunger, or swab (Figure 15.6). A heavy bailer may be used to produce the surging action, but it is not as effective as the close-fitting surge block. Although some drillers depend on surge blocks for developing screened wells, others feel that this device is not effective and that it may, in some cases, even be detrimental because it forces fine material back into the formation before the fines can be removed from the well. To minimize this problem, fine material should be removed from the borehole as often as possible.

Before starting to surge, the well should be bailed to make sure that water will flow into it. Lower the surge block into the well until it is 10 to 15 ft (3 to 4.6 m) beneath the static water level, but above the screen or packer (Figure 15.7). The water column will effectively transmit the action of the block to the screen section. The initial surging motion should be relatively gentle, allowing any material blocking the screen to break up, go into suspension, and then move into the well. The surge block (or bailer) should be operated with particular care if the formation above the screen consists mainly of fine sand, silt, or soft clay which may

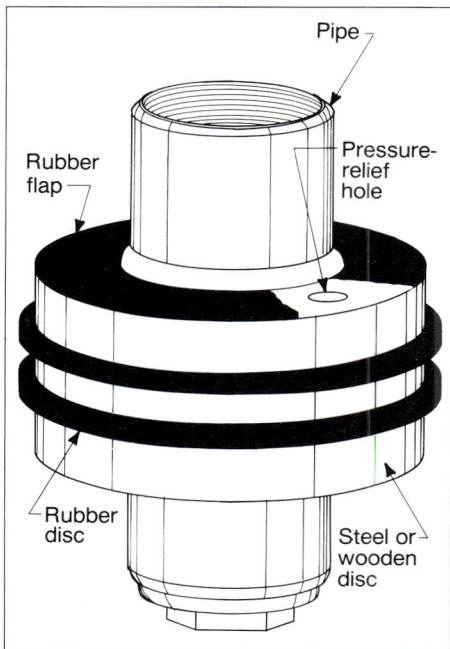

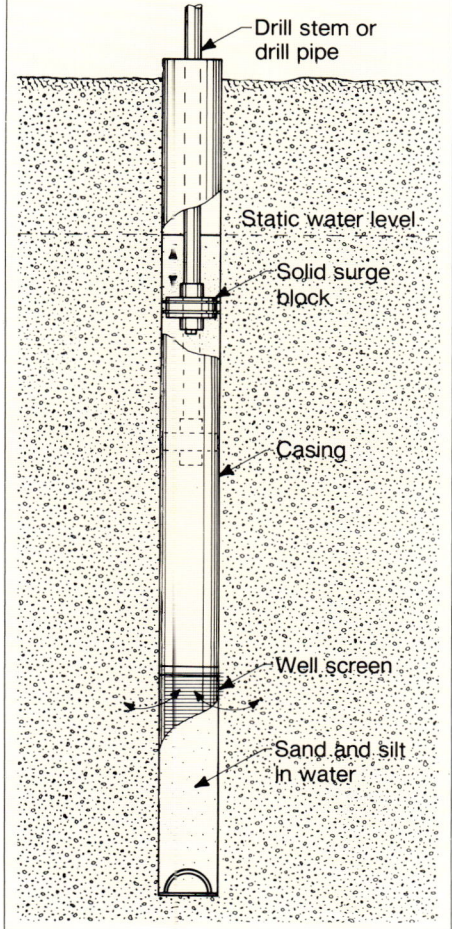

Figure 15.7. For certain types of formations, a surge block is an effective tool for well development. On the downstroke, water is forced outward into the formation; water, silt, and fine sand are then pulled into the well screen during the upstroke.

Figure 15.6. Typical surge block consisting of two leather or rubber discs sandwiched between three steel or wooden discs. The blocks are constructed so that the outside diameter of the rubber lips is equal to the inside diameter of the screen. The solid part of the block is 1 in (25.4 mm) smaller in diameter than the screen.

slump into the screen. As water begins to move easily both into and out of the screen, the surging tool is usually lowered in steps to just above the screen. As the block is lowered, the force of the surging movement is increased. In a well equipped with a long screen, it may prove more effective to operate the surge block in the screen to concentrate its action at various levels. Development should begin above the screen and move progressively downward to prevent the tool from becoming sand locked.

The force exerted on the formation depends on the length of the stroke and the vertical velocity of the surge block. For a cable tool rig, length of the stroke is determined by the spudding motion; the vertical velocity depends on the weight exerted on the block and the retraction speed. A block must be weighted so that it will fall at the desired rate when used with a cable tool rig. During retraction of the block, continue the spudding motion to avoid sand locking the block in the casing. If a rotary rig is being used, the weight on the block is provided by the drill pipe. The speed of retraction and length of pull are governed by the physical characteristics of the rig.

Continue surging for several minutes, then pull the block from the well. Air may be used to blow the sediment out of the well if development is done with a rotary rig or if an air compressor is available. Sediment can be removed by a bailer or sand pump when a cable tool rig is used. The surging action is concentrated at the top of the screen, and this effect is accentuated if the lower part of the screen is continually blocked off by the sand brought in by the development process. In general, development can be accelerated if the amount of sediment in the screen is kept to a minimum. A sump or length of casing installed beneath the screen is helpful in keeping the screen free of sediment. Continue surging and cleaning until little or no sand can be pulled into the well. Total development time may range from about 2 hours for small wells to many days for large wells with long screens.

Occasionally, surging may cause upward movement of water outside the well casing if the washing action disrupts the seal around the casing formed by the overlying sediments. When this occurs, use of the surge block must be discontinued or sediment from the overlying materials may invade the screened zone.

Surge blocks sometimes produce unsatisfactory results in certain formations, especially when the aquifer contains many clay streaks, because the action of the block can cause clay to plug the formation. When this happens a reduction in yield occurs, rather than an increase. Surge blocks are also less useful when the particles making up the formation are angular, because angular particles do not sort themselves as readily as rounded grains. In addition, if large amounts of mica are present in the aquifer, the flat or tabular mica flakes can clog the outer surface of the screen and the zone around the screen by aligning themselves perpendicular to the direction of flow. Clogging by mica can be minimized if the surging procedures are applied rather gently to the well. It is good practice to avoid overdevelopment when mica is present in the aquifer.

One other type of surging tool is called a swab. The simplest type of swab, a rub-

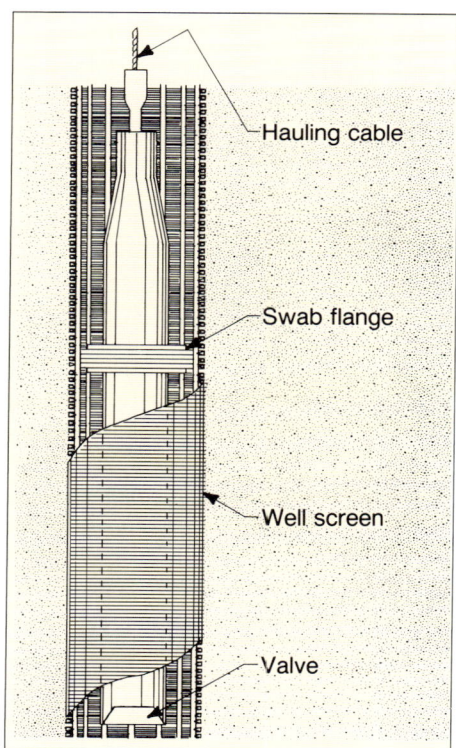

Figure 15.8. Line swabbing is used primarily in consolidated aquifers. As the swab is pulled upward at about 3 ft/sec (0.9 m/sec), high-pressure conditions at the top of the swab force water into the formation. Low-pressure conditions at the base cause flow of sand, silt, and water back into the borehole.

ber-flanged mud scow or bailer, is lowered into the casing to any selected point below the water level and then pulled upward at about 3 ft/sec (0.9 m/sec), with no attempt to reverse the flow and cause a surging effect (Figure 15.8). The length of the swabbing stroke is usually much longer than in surging. As the scow is raised, high pressure is created near the top of the scow, which drives water into the formation. Water is drawn back into the well beneath the swab because the pressure is lower. The scow usually has a valve at the bottom which opens to increase the fall rate in the borehole. This method of swabbing, called line swabbing, is often used to clean fine material from deep wells drilled in consolidated rock aquifers. Swabbing screened wells requires special precautions, however. In

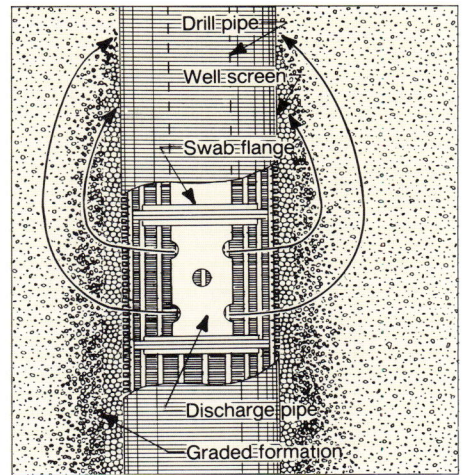

Figure 15.9. When a double-flanged swab is used, water is pumped into the formation between the flanges. Flow reenters the borehole above or below the swab. During pumping, the swab is raised and lowered over short distances.

tight (low-permeability) formations, for example, swabbing can result in collapsed screens, and great care must be taken to insure that the hydraulic conductivity of the formation is capable of yielding sufficient water to keep pressure differentials within reasonable limits. Avoid swabbing wells that have plastic casing or screens. Silt and silty sand formations in which screen-slot sizes are about 0.010 in (0.25 mm) or smaller are particularly troublesome, and use of a swab in this case should be avoided.

A more effective swabbing device is shown in Figure 15.9. With this tool, water is pumped into the formation between two flanges and returns to the well bore either above or below the flanges. During pumping, the swab is raised and lowered in the borehole over short distances. Sometimes a bypass tube is installed in the double-flanged swab to facilitate the movement of water up the borehole from below the tool. The advantage of a double-flanged swab is that the energy of the water being pumped into the tool can be directed at selected parts of the formation.

In summary, surge blocks are inexpensive tools that are convenient to use and, within their limitations, do an effective job. They can be adapted for use on many types of rigs and used in combination with other development methods. In addition, surge blocks can be used for wells of any diameter or depth. Surging procedures produce good results for screen installations in zones having good porosity and hydraulic conductivity.

Air Developing by Surging and Pumping

Many drillers use compressed air to develop wells in consolidated and unconsolidated formations. The practice of alternately surging and pumping with air has grown with the great increase in the number of rotary drilling rigs equipped with large air compressors. In air surging, air is injected into the well to lift the water to the surface. As it reaches the top of the casing, the air supply is shut off, allowing the aerated water column to fall. Air-lift pumping is used to pump the well periodically to remove sediment from the screen or borehole, and is accomplished by installing an air line inside an eductor pipe in the well. Eductor systems are generally required for large-diameter wells, when limited volumes of air are available, or when the static water level is low in relation to the well depth. Most rotary rigs, however, have sufficient air capacity to use the casing as the eductor for 6- to 12-in (152- to 305-mm) diameter wells. Figure 15.10a shows the basic layout of an air-lift system and the appropriate terms.

The uphole velocities required to remove cuttings and water in air drilling were discussed in Chapter 11. Uphole velocities of 3,000 to 5,000 ft/min (915 to 1,520 m/min) are needed for dry-air drilling where little or no water is entering the borehole. For removing large volumes of water and cuttings, a surfactant is mixed into a small volume of water and then added to the airstream. The surfactant breaks up the water masses so they can be lifted to the surface at a rather low velocity [50 to 200 ft/min (15.2 to 61 m/min)], thereby reducing air-volume requirements. During air development, however, surfactants are used only when compressor capacity is insufficient to lift water to the surface. Therefore, the contractor must maintain uphole velocities in the range of 1,000 to 2,500 ft/min (305 to 762 m/min) to achieve a reasonable discharge.

Generally, it is not possible to predict what uphole velocity is actually needed because of submergence factors (discussed later), total pumping lift requirements, and

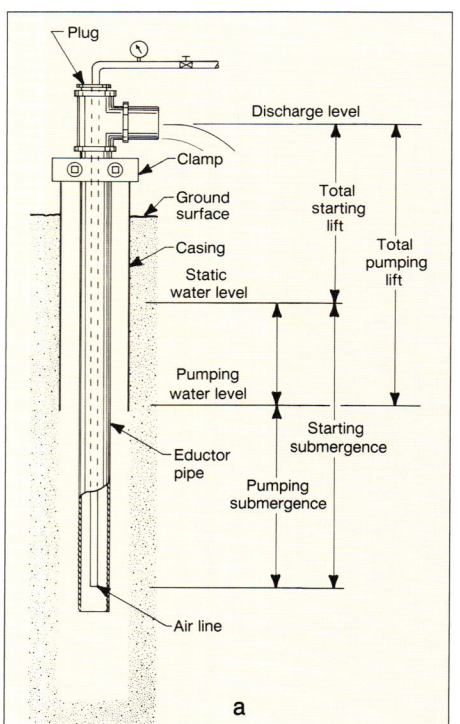
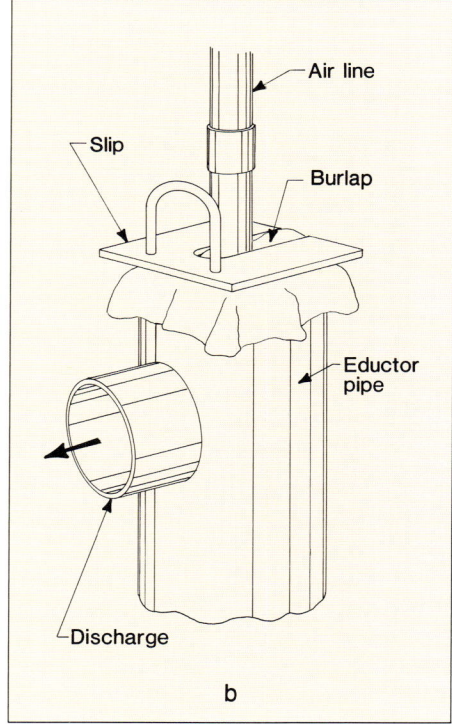

Figure 15.10a. This diagram shows common terms applied to air-lift pumping. *(Ingersoll-Rand, 1971)* **15.10b.** In this installation, the top of the well is plugged during the air-lift with burlap and slips to direct the water to the outlet pipe.

the non-predictable way water will enter the borehole. For example, if water enters as a high-volume, concentrated flow at a discrete point (coarse gravel layer or fracture), the uphole velocity required at that point will be quite large so the water mass can be broken up efficiently. On the other hand, if water is seeping evenly into the borehole over its entire length, the uphole velocity requirement is less because the force (velocity) needed to lift the fine water droplets is less. Thus, it is virtually impossible to predict beforehand the uphole velocities required for air development procedures. In practice, the contractor ignores uphole velocity considerations and concentrates on the air volume needed to lift the water adequately. Fortunately, research on determining the air volume required to lift a certain volume at a specified submergence and total pumping lift has been done. This information, presented later in this chapter, gives the contractor the ability to predict the volume of air that must be available to produce an adequate discharge for certain downhole conditions.

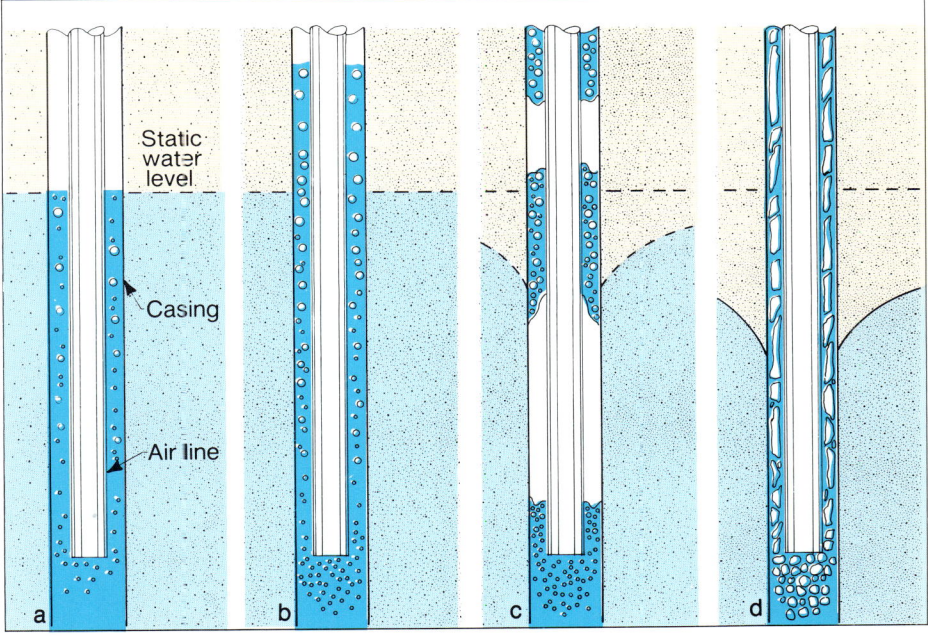

Figure 15.11. The type of discharge produced from a well during air development depends on the air volume available, total lift, submergence, and annular area. In practice, two different flow conditions can be recognized when air is used for water well development, although other flow regimes may exist at much lower or higher velocities in smaller diameter pipes [identified by Taitel and Dukler (1980) and Hestroni (1982)]. This diagram provides a qualitative illustration of how multiphase flow (water and air) occurs in the casing during air development. The percent submergence, total lift, and capacity of the compressor will control the relative proportion of air and water for a particular well. Griffith (1984) points out some of the extreme difficulties in making a rigorous analysis of multiphase systems. (a) Introduction of a small volume of air under a high head causes little change in the water level in the well. In this case, the air pressure available is just sufficient to overcome the initial head exerted by the water column. (b) As air volume increases, the water column becomes partly aerated. Displacement of the water by the air causes the water column to rise in the casing. Drawdown does not change because no pumping is occurring. (c) Further increases in air volume cause aerated slugs of water to be lifted irregularly out the top of the casing. Between surges, the water level in the casing falls to near the static level. (d) If enough air is available, aerated water will continually flow out the top of the well. With average submergence and total lift, the volume of air versus water is about 10 to 1. Higher air volumes may increase the pumping rate somewhat, but still higher rates may actually reduce the flow rate because flow into the well is impeded by excessive air volume.

510　　　　　　　　　　　GROUNDWATER AND WELLS

Both air pressure and air volume are important in initiating and maintaining an air surging or air-lift pumping operation. For typical head conditions found in boreholes 300 to 400 ft (91.5 to 122 m) deep, the compressor used for the air supply should be capable of developing a minimum pressure of 125 psi (862 kPa). This is enough pressure to overcome the initial head created by the submergence of the air line. This head is called the starting submergence. Once the pressure initiates flow, the air capacity (volume) becomes the most important factor in successful air-lift

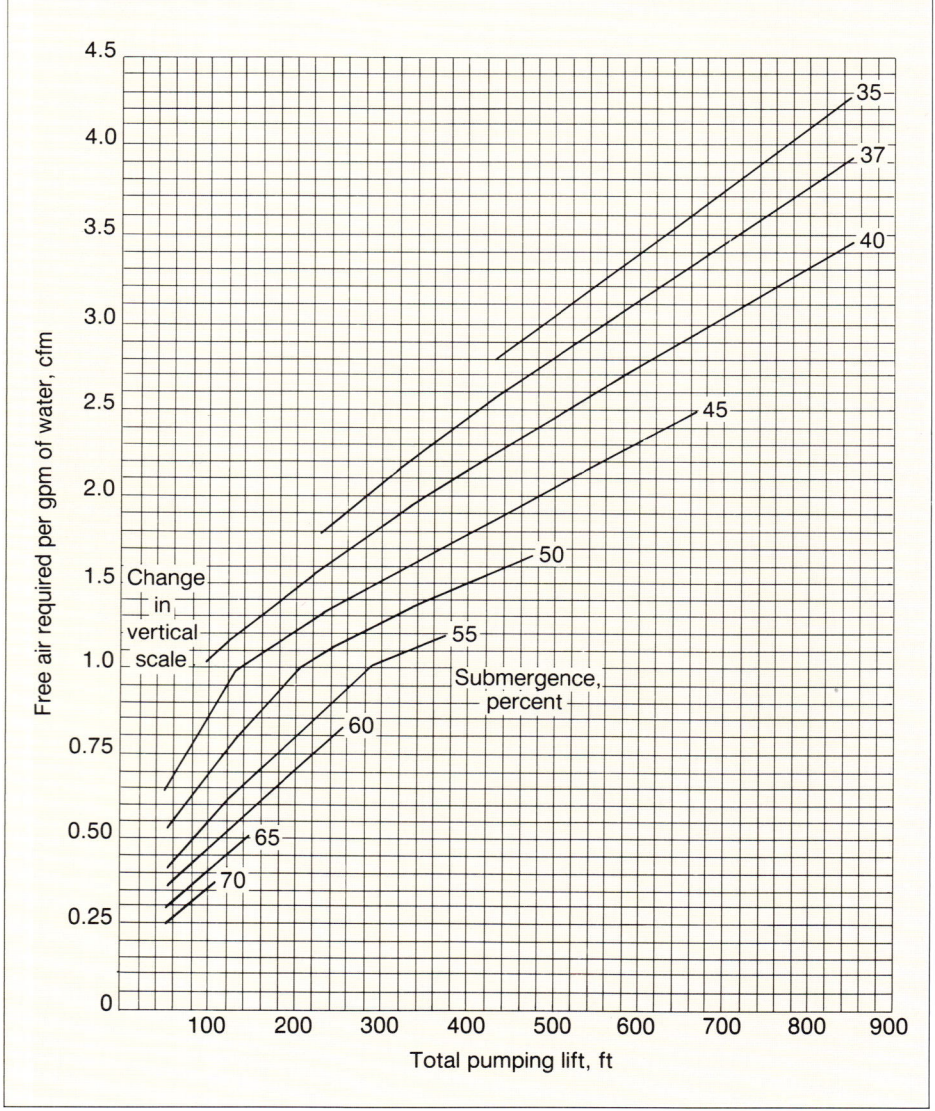

Figure 15.12. Cubic feet of air required to pump 1 gpm (5.5 m³/day) for various submergences and pumping lifts. The ratios shown here between air volume and water pumped are predicated on using the proper size eductor and casing (see Table 15.1). Field experience suggests that these air requirements may be optimistic; in practice, somewhat more air will be required than indicated. *(Ingersoll-Rand, 1971)*

pumping. Figure 15.11 illustrates how the addition of different volumes of air affects the water in the borehole. A useful rule of thumb for determining the proper compressor capacity for air-lift pumping is to provide about ¾ cfm (0.0004 m³/sec) of air for each 1 gpm (5.5 m³/day) of water at the anticipated pumping rate. In practice, a 375-cfm (0.2-m³/sec) compressor can usually pump 400 to 500 gpm (2,180 to 2,730 m³/day) with proper pumping submergence of the air line.

The volume of air required to operate an air-lift efficiently depends on the total pumping lift, the pumping submergence, and the area of the annulus between eductor and casing (Figure 15.12). To calculate pumping submergence, the length of air line below the pumping water level is divided by the total length of air line suspended in the well. For wells with about 100 to 200 ft (30.5 to 61 m) of total pumping lift, air-lift pumping is quite efficient when the air line is submerged about 60 percent of its total length during pumping (60-percent pumping submergence) (Figure 15.13). When the total pumping lift exceeds 200 ft, the pumping submergence may have to be decreased so that the start-up pressure at the bottom of the air line does not exceed the pressure capacity of the compressor. Good results can be obtained by a skillful operator while pumping with a pumping submergence as low as 30 percent. In some cases, acceptable results have been obtained with a pumping submergence as low as 10 percent. The air line should not be placed all the way to the bottom of the well when pumping begins, unless required for proper submergence, because the air must then overcome an unnecessarily high pressure head.

The volume of air needed to lift water also depends on whether intermittent or steady flow is required. For development work, it is not necessary to maintain a steady discharge, and, in fact, some surging of the air lift is beneficial. If steady flow must be maintained (pumping tests, for example), the air-volume requirements will usually be greater than those given in Figure 15.12. For deep boreholes with low static water levels, the actual volume of air required may be two to three times the volume shown in Figure 15.12 to maintain steady flow.

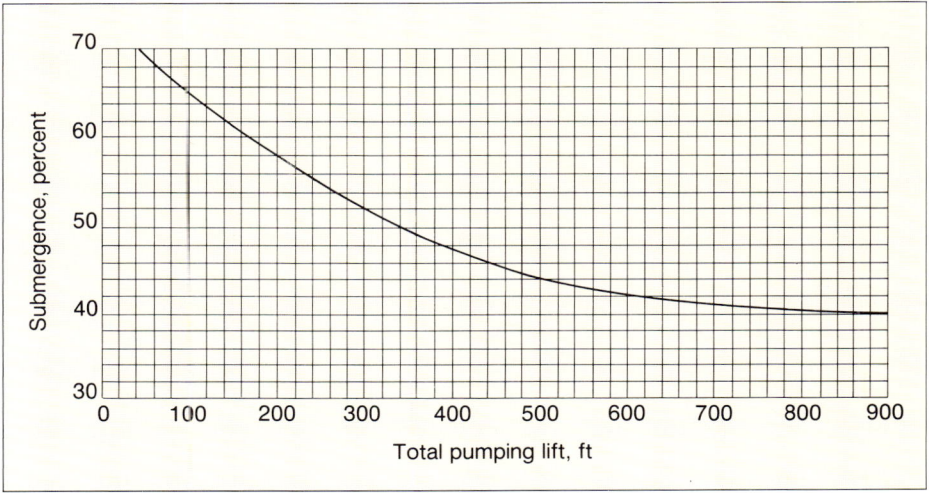

Figure 15.13. Approximate percent pumping submergence for optimum air-lift efficiency. In general, development proceeds most efficiently when the discharge is maximized. Therefore, the submergence should always be as great as possible within practical limits. *(Ingersoll-Rand, 1971)*

Figure 15.10a shows the proper method of placing the eductor pipe and air line in the well. A tee at the top of the eductor pipe is fitted with a discharge pipe at the side outlet. A bushing with the inside opening large enough to clear the couplings of the air line is connected to the top of the tee. Burlap or similar material wrapped around the air line just above the tee, and held by slips, reduces spraying around the top of the well and enables more accurate yield measurements during pumping (Figure 15.10b).

Table 15.1 lists the recommended sizes of eductor pipe and air line for air-lift pumping. Some variation from these sizes may be necessary for practical reasons, but the combinations shown generally give good results.

Designing Air-Lift Pumping Operation

Some drillers and well construction engineers do not take the time to analyze the operational characteristics of an air lift. This practice can lead to inefficient pumping, lost time, and, in some cases, failure to pump any water. The example cited below illustrates the factors that should be considered for designing a successful air-lift pumping operation.

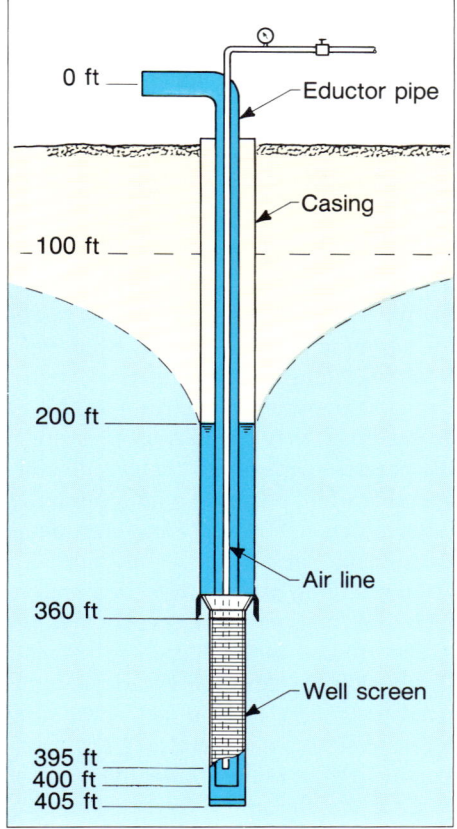

Figure 15.14. Determining the design of an air-lift system for a specific well.

A borehole is completed to a depth of 405 ft and screened from 360 to 405 ft (Figure 15.14). A 10-in diameter casing is selected to minimize friction losses and provide adequate clearance for the pump bowls. Static water level is at 100 ft. Examination of the cuttings suggests that a specific capacity of 6 gpm/ft of drawdown can be obtained. It is decided to air-lift pump at a rate of 600 gpm, 20 percent above the

Table 15.1. Recommended Pipe Sizes for Air-Lift Pumping

Pumping Rate*		Size of Well Casing if Eductor Pipe is Used		Size of Eductor Pipe (or casing if no eductor pipe is used)		Minimum Size of Air Line	
gpm	m³/day	in	mm	in	mm	in	mm
30 to 60	164 to 327	4	102, or larger	2	51	½	13
60 to 80	327 to 436	5	127, or larger	3	76	1	25
80 to 100	436 to 545	6	152, or larger	3½	89	1	25
100 to 150	545 to 818	6	152, or larger	4	102	1¼	32
150 to 250	818 to 1,360	8	203, or larger	5	127	1½	38
250 to 400	1,360 to 2,180	8	203, or larger	6	152	2	51
400 to 700	2,180 to 3,820	10	254, or larger	8	203	3	64
700 to 1,000	3,820 to 5,450	12	305, or larger	10	254	3	64
1,000 to 1,500	5,450 to 8,180	16	406, or larger	12	305	4	102

Actual pumping rate is dependent on percent submergence.

design rate of 500 gpm. Thus, the pumping water level would be at 200 ft. From this information, it is possible to select the proper size equipment for the air lift and to estimate the potential efficiency of the system. Several steps should be followed in the analysis:

1. Determine the diameter of the eductor pipe (if required) and the air line. Table 15.1 shows that an 8-in eductor and a 3-in air line should be used for the 10-in casing. A smaller diameter eductor might be chosen, but the air-pressure requirements would rise significantly because of the additional friction effects.

2. Determine the lengths of the eductor pipe and the air line. In this instance, the air line will always stay inside the eductor pipe. The bottom of the eductor pipe is set at 400 ft, with the air line set at 395 ft. The 400-ft submergence depth is selected so that the lower end of the eductor pipe is set at a reasonable distance above the bottom of the screen.

3. Determine submergence:

$$\% \text{ pumping submergence} = \frac{\text{length of air line below pumping water level}}{\text{total length of air line}} \cdot 100$$

$$= \frac{195}{395}$$

$$= 50\%$$

Experience suggests that the air lift will be reasonably efficient at this pumping submergence; that is, the volume of water pumped per cubic foot of air is acceptable (Figure 15.13). Ideally, pumping submergence should be as great as possible, but may be limited by initial air-pressure requirements as shown in step 5.

4. Determine air-volume requirements. For a total pumping lift of 200 ft, Figure 15.12 shows that 1 cfm is required to pump 1 gpm. Thus, 600 cfm are required to pump at 600 gpm.

5. Determine whether the compressor has sufficient pressure to initiate flow in the air line.

$$\text{Minimum psi requirement} = \frac{\text{length of air line} - \text{static water level}}{2.31}$$

$$= \frac{395 - 100}{2.31}$$

$$= 128 \text{ psi}$$

Thus, at least 128 psi will be needed to start the air lift for the starting submergence selected (a safety factor of 25 percent is usually added to this pressure figure). As drawdown develops, the psi requirement drops substantially because the head acting on the air line decreases.

Proper analysis of the factors affecting the air lift provides some assurance that the air-lift development procedures will function as desired. But the design of an air-lift system and the determination of its operational characteristics are not mathematically precise. Therefore, the performance obtained from a particular air-lift design may be more or less than anticipated. Thus, the contractor or engineer must be prepared to make appropriate adjustments. In general, air-lift pumping will be most efficient when the static water level is high, the casing diameter is relatively small, and the well depth is not excessive in relation to the pressure capability of the compressor.

Air Development Procedures

Air development procedures should begin by determining that groundwater can flow freely into the screen. Application of too much air volume in the borehole when the formation is clogged can result in a collapsed screen. To minimize the initial pumping rate, the air line and eductor (if used) can be placed at a rather shallow submergence. At this setting, even the introduction of large air volumes will produce only a moderate pumping rate and, therefore, will place only low collapse pressures on the well screen. Introduction of small air volumes at greater submergence also will produce low yields.

Once uninhibited flow into the screen has been established, the eductor pipe (if used) is lowered to within 5 ft (1.5 m) of the bottom of the screen, assuming that sufficient pressure is available to overcome the static head. Development can also start near the top of the screen, depending on the preference of the driller. The air line is placed so that its lower end is up inside the eductor pipe at the proper submergence level. Before blowing any water or drilling fluid out of the well with a sudden large injection of air, the air lift should be operated to pump fluids at a reduced rate from the well.

Air is released into the line and the well is pumped until the water is virtually sand free. The valve at the air tank outlet is then closed, allowing the pressure in the tank to build. The actual pressure required will depend on the starting submergence; 43 psi (296 kPa) is needed for each 100 ft (30.5 m) of starting submergence. In the meantime, the air line is lowered so that its lower end is 1 ft or so below the eductor pipe. To initiate surging, the valve is opened quickly to allow air from the tank to rush suddenly into the well. This tends to drive the water outward through the well screen openings. Ordinarily, a brief but forceful head of water will also overflow or shoot from the casing and eductor pipe at the ground surface (Figure 15.15). When the air line is pulled up into the eductor pipe after the first charge of air has been released into the well, the air lift will again pump, thus reversing the flow (water flows into the well) and completing the surging cycle.

The well is pumped until the water clears up, and then another "head" of air is released with the air line set below the eductor pipe. To resume pumping, the air line is again lifted. Surging cycles are repeated until the water is relatively free of sand or other fine particles immediately after the screen has received an air blast. This indicates that development is approaching completion in the region near the bottom of the eductor pipe. The air-lift assembly is then raised to a position

Figure 15.15. During air-lift development, brief but powerful spurts of water will be ejected from the top of the casing. *(Test Drilling Services)*

about 5 ft (1.5 m) higher and the same operations are repeated. In this way, the entire screen is developed in 5-ft intervals. From time to time, the air lift should be lowered to its original position near the bottom of the well and operated as a pump to clean out any sand that has accumulated inside the screen.

Several alternative surging procedures are available. In one method, the air line is always retained within the eductor pipe. Surging is accomplished by letting the water column in the eductor pipe fall periodically. The well is then air-lift pumped until the water becomes clear. This cycle is repeated at different levels in the well screen until the well is developed. In another method, the air line is used in the casing and the well casing acts as the eductor pipe. The surging cycle is the same as described above.

Some drillers use an isolated air lift to remove sediment more effectively from the aquifer. Flanged gaskets are mounted on the top and bottom of the isolation tool (Figure 15.16). The gaskets should be snug fitting, but still allow some sediment to move around them so that sand locking does not occur. After the screen has been cleaned out initially, the isolation tool is lowered to the top of the screen and the air line is set at the proper depth. After each zone is developed by surging and air-lift pumping, the tool is lowered to the next section. Thus, the formation is developed in separate stages.

Two development methods — air lift and mechanical surging — are often combined to repair drilling damage and remove fine material from the formation. This combined method is particularly suited to reverse rotary rigs equipped with air-lift assist. In this technique, a double surge-block assembly attached to the drill pipe is raised and dropped rapidly to produce the required turbulence in and near the screen. Simultaneously, water is lifted with air inside the drill pipe from the zone isolated by the two surge blocks. After the water becomes free of sediment, the assembly is lowered to the next section. This method is especially effective in long screens because it can concentrate the development energy on short sections of the aquifer. It is important to start this procedure at the top of the screen to avoid sand locking the assembly.

Under some conditions, the aquifer may become air locked when a large burst of air is injected into the screened area of the well. Certain kinds of formations are more prone to air locking, especially those formations that consist of stratified,

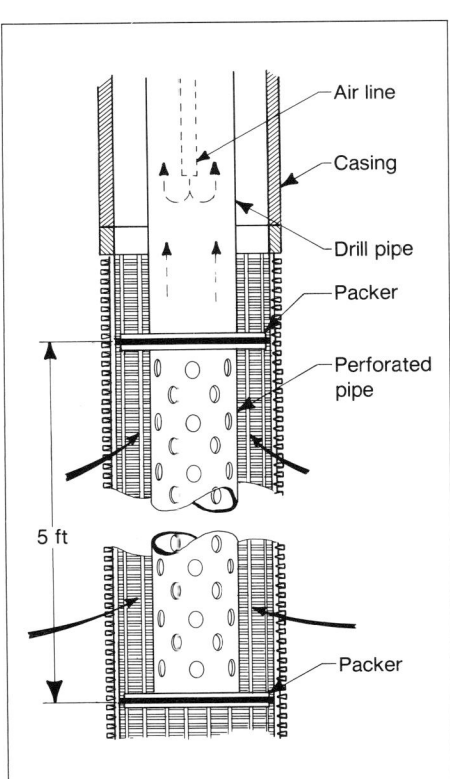

Figure 15.16. Isolation tools are used to focus the energy of air bursts on a specific part of the aquifer and to remove sediment by air lifting.

coarse sand or gravel lenses separated by thin, impermeable clay layers. Aquifers with good vertical hydraulic conductivity are generally not affected. Surging with air usually does not lead to air locking. If some air becomes trapped in the aquifer, however, it may impede the flow of water toward the screen. In formations susceptible to air locking, surging with air should be avoided. Other procedures such as high-velocity jetting with water or air may be more suitable in formations where air trapping is a problem.

High-Velocity Jetting

Development by high-velocity jetting may be done with either water or air. In practice, jetting with water is almost always accompanied by simultaneous air-lift pumping so that clogging of the formation does not occur. This dual process is described later in this section and is one of the most effective methods of well development. The jetting procedure consists of operating a horizontal water jet inside the well screen so that high-velocity streams of water shoot out through the screen openings. Jetting is particularly successful in developing highly stratified, unconsolidated formations.

The equipment required for jet development includes a jetting tool with two or more equally spaced nozzles, high-pressure pump, high-pressure hose and connections, string of pipe, and water tank or other water supply. The high-velocity jets force water through the screen openings, agitating and rearranging the particles of the formation surrounding the screen. The filter cake deposited on the borehole in conventional rotary drilling is effectively broken down and dispersed so the drilling fluid that has penetrated the formation can be pumped out. Jetting will also help correct damage to the formation's porosity and permeabilty resulting from drilling.

Figure 15.17 shows a jetting tool with four nozzles. Nozzles should be spaced equally around the circumference of the jetting tool to hydraulically balance the tool during operation; for example, four nozzles should be spaced 90 degrees apart. Best results are obtained if the nozzles are designed for maximum hydraulic efficiency, but horizontal holes drilled in a plugged pipe or coupling will be reasonably effective. The

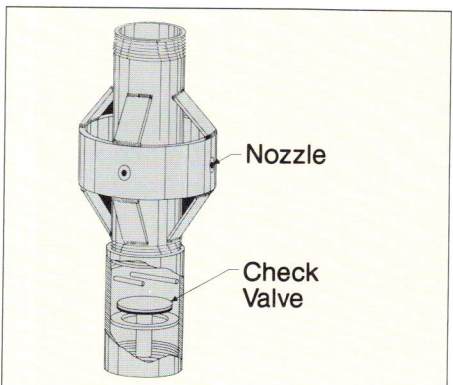

Figure 15.17. Four-nozzle jetting tool designed for jet development of well screens. The check valve allows the jetting tool to be used for intermittent pumping of the well. A plate bottom is often used in place of the check valve if the tool is not to be used for pumping. To avoid breakage, the pipe base of the tool can be made of heavy-wall pipe if it is to be attached to the bottom of the drill pipe. At right, a jetting tool is being lowered into inner casing that has been centered in the surface casing.

jetting tool should be constructed so that the nozzle outlets or holes are as close to the inside diameter of the screen as practical [generally less than 1 in (25.4 mm)].

In many jetting operations, water containing sediment is recirculated through the jetting tool, thereby causing erosion of the nozzle bores and a pronounced pressure reduction at the nozzle face. Erosion of the nozzle bores may be a significant problem if the nozzles are constructed of mild steel. Therefore, all nozzles should be constructed from stainless steel pipe or other abrasion-resistant material. High concentrations of sediment can also damage screens if the jets are directed at one area for long periods. Thus, every effort should be made to limit sediment concentration in water used for jetting.

The lowest nozzle velocity for effective jetting is considered to be about 100 ft/sec (30.5 m/sec). Much better results can be expected when the nozzle velocities are 150 to 300 ft/sec (45.7 to 91.5 m/sec). Velocities higher than 300 ft/sec may not result in sufficient additional benefit to justify the added cost. In fact, velocities obtained by using pressures higher than about 400 psi (2,760 kPa) at the nozzle may cause abrasion, depending on the screen material, the distance between the nozzle and screen body, and the amount of sediment carried in or entrained by the jetting water. In general, 200 psi (1,380 kPa) at the nozzle is the preferred operating pressure for metallic screens. Great care must be exercised in jetting screens constructed of PVC or other less abrasion-resistant materials. All jetting of PVC screens should be done only with clean water to minimize abrasion, and the pressure should not exceed 100 psi (690 kPa). Table 15.2 provides data for nozzles of several sizes at different operating pressures.

The pipe that attaches to the jetting tool should be large enough to minimize friction losses so that velocity at the nozzle is as high as possible. Sizes generally used are given in Table 15.3.

Jetting can be accomplished with virtually any drilling rig, but it is more conveniently done with rotary rigs because the mud pump can supply the required downhole pressure. When using a rotary rig, the jetting tool is attached to the lower end of the drill string and rotation is controlled by the rig. The jetting tool is placed near the bottom of the screen and rotated slowly while being pulled upward at 5 to 15 minutes per foot of screen, depending on the nature of the formation. Material loosened from the formation accumulates at the bottom of the screen as the jetting tool is raised slowly. This material is removed later by air-lift pumping or bailing. By slowly rotating the jetting tool and gradually raising it, the entire surface of the screen is exposed to the vigorous action of the jets. Several passes are made up the screen until the amount of additional material removed from the formation becomes negligible. To avoid erosion of the screen and to expedite development, the jetting tool should never be

Table 15.2. Approximate Jet Velocity and Discharge per Nozzle

Size of Nozzle Orifice		Nozzle Pressure* 100 psi (690 kPa)				Nozzle Pressure* 150 psi (1,030 kPa)				Nozzle Pressure* 200 psi (1,380 kPa)				Nozzle Pressure* 250 psi (1,720 kPa)				Nozzle Pressure* 300 psi (2,070 kPa)			
		Velocity		Discharge		Velocity		Discharge		Velocity		Discharge		Velocity		Discharge		Velocity		Discharge	
in	mm	fps	m/s	gpm	m³/day	fps	m/s	gpm	m³/day	fps	m/s	gpm	m³/day	fps	m/s	gpm	m³/day	fps	m/s	gpm	m³/day
3/16	4.8	100	30.5	8	44	120	36.6	10	55	140	42.7	12	65	155	47.3	13	71	170	51.8	15	82
1/4	6.4	100	30.5	15	82	120	36.6	18	98	140	42.7	21	114	155	47.3	24	131	170	51.8	26	142
3/8	9.5	100	30.5	34	185	120	36.6	41	223	140	42.7	48	262	155	47.3	53	289	170	51.8	58	316
1/2	12.7	100	30.5	60	327	120	36.6	73	398	140	42.7	84	458	155	47.3	94	512	170	51.8	103	561

*To obtain these nozzle pressures, gauge pressure must be somewhat higher to overcome friction losses in the air line.
Note: Laboratory experiments show that the coefficient of discharge through the nozzle is 0.8.

operated in a stationary position.

In general, the effectiveness of the jetting process is controlled by the open area and slot configuration of the screen and the thoroughness of the jetting operation. The screen should have as much open area as possible, and the openings should be evenly distributed around the circumference of the screen. These conditions permit the jetting action to reach a high percentage of the formation around the screen. The configuration of screen openings can either enhance or retard the velocity of the water as it passes through the screen (Figure 15.18). High-velocity jetting works effectively through V-shaped openings in continuous-slot screens, but it is less useful in punched or louvered pipe where slot configuration impedes access to the formation.

Optimal removal of sediment by jetting will depend on the time allotted to the process. Because the jetting energy can focus on only a small part of the formation at a given moment, more time may be necessary than for other methods that affect a larger portion of the formation. Less satisfactory results from jetting almost inevitably occur when not enough time is allowed for a thorough job.

Jetting with Air

Jetting with air is an alternative to water jetting. In areas where water is not readily available, air jetting is an extremely practical procedure that produces good results in both consolidated and unconsolidated formations. Its use requires less set-up time because, in most cases, air has already been used to drill the borehole. In addition, jetting with air initiates air-lift pumping, which helps remove sediment from the well. Casing size and air-line submergence will affect the efficiency of air-lift pumping. The main disadvantage of air jetting is that many rigs are equipped with compressors having limited pressure capabilities. A 125-psi (862-kPa) compressor, for example, can operate only at starting heads that are less than about 285 ft (86.9 m). Furthermore, the pressure near the bottom of the well may be so great that, even if the compressor can pump air, the actual compressed (air) volume is so small that the amount of turbulence generated may not be sufficient to develop the well effectively at that point. In certain localities, however, additional high-pressure, high-volume air com-

Table 15.3. Minimum Size of Pipe Required to Hold Friction Losses to a Total of Approximately 20 ft (6.1 m) of Head

Pumping Rate		Pipe Length							
		100 ft (30.5 m)		200 ft (61 m)		400 ft (122 m)		600 ft (183 m)	
gpm	m³/day	in	mm	in	mm	in	mm	in	mm
35	191	1½	38	1½	38	2	51	2	51
50	273	1½	38	2	51	2	51	2½	64
75	409	2	51	2	51	2½	64	3	76
100	545	2	51	2½	64	3	76	3	76
150	818	2½	64	3	76	4	102	4	102
200	1,090	3	76	3	76	4	102	4	102
250	1,360	3	76	4	102	4	102	5	127
300	1,640	3	76	4	102	4	102	5	127
350	1,910	4	102	4	102	5	127	5	127
400	2,180	4	102	4	102	5	127	5	127

pressors can be rented to help in the development process.

The methodology used in air jetting is the same as that used in water jetting. Because air is a compressible fluid, however, it is not possible to specify anticipated nozzle velocities; the actual (compressed) volume of air emitted from the jetting tool will be controlled by the head of water in the casing (see Chapter 11). Ordinarily, the compressor capacity of typical rotary rigs is sufficient to create enough air to jet effectively in wells 150 to 300 ft (45.7 to 91.5 m) deep. For water well work, the nozzle sizes for an air-jetting tool are approximately the same size as those for a water-jetting tool.

In formations susceptible to air locking, an air-jetting tool with small holes drilled in the bottom should be used to develop the formation. Air coming out the bottom of this special jetting tool enhances the efficiency of the air lift, thereby entraining water, sediment dislodged by the jets, and air that could become trapped in the adjacent sediments. The tool can be pulled up periodically within the casing and used

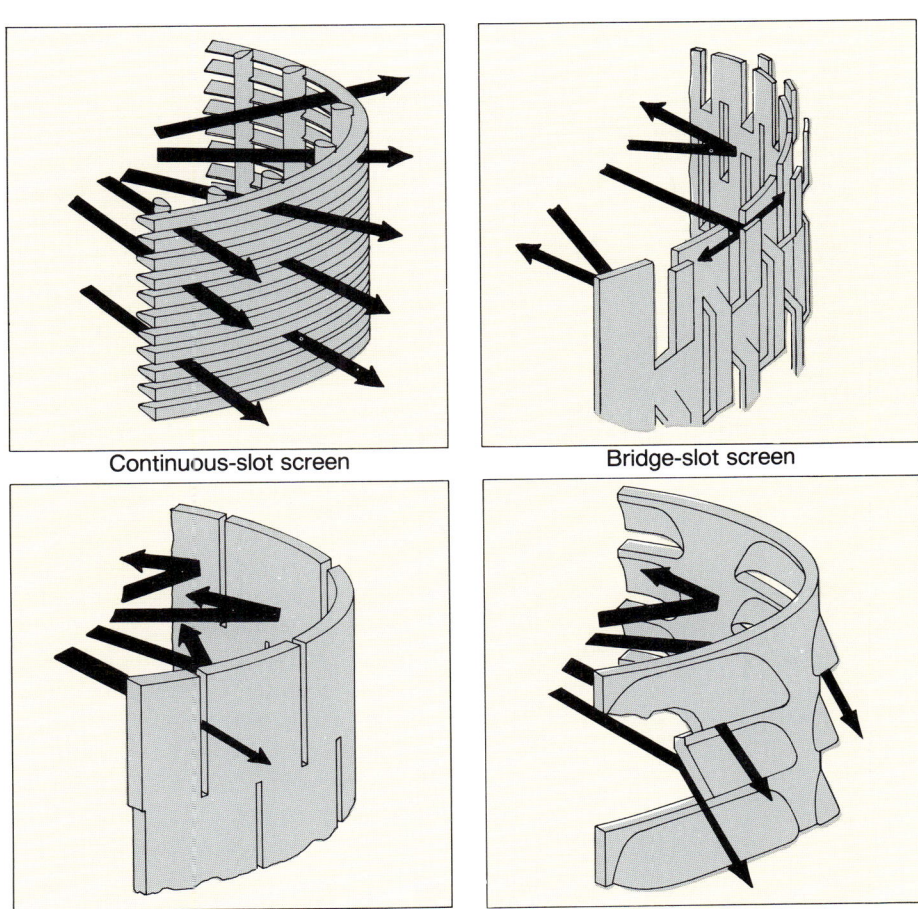

Figure 15.18. The open area of the screen and the configuration of the slot openings are important factors controlling the effectiveness of development procedures using water jetting.

solely as a conventional air-lift device.

In conclusion, development by jetting offers several advantages. Energy is concentrated over a small area with great effectiveness, every part of the screen can be covered selectively, and it is relatively simple to apply and is not likely to cause trouble from overapplication. Although water jetting imparts a more powerful force into the formation compared with air jetting, an air lift usually occurs during air jetting which assures that fines will be drawn out of the formation efficiently. In water jetting, water is being added to the formation so that fine material may not move into the screen quite as readily.

High-Velocity Water Jetting Combined with Simultaneous Pumping

Although water jetting procedures are extremely effective in dislodging material from the formation, maximum development efficiency is achieved when water-jetting procedures are combined with simultaneous air-lift pumping or other pumping methods (Figure 15.19). This combination of development techniques is particularly successful for wells in unconsolidated sands and gravels. In water jetting, water is added to the well at a rate governed by the nozzle size and the pump pressure. The volume of water pumped from the well should always exceed the volume pumped in during jetting, because sediment removal is greatly enhanced with higher discharge. Thus, the water level in the well will be kept below static level and some water will move continuously from the formation into the well screen as the work proceeds. The steady movement of water into the well helps remove some of the suspended material loosened by the jetting operation. The air lift then pumps the sediment from the well before it can settle in the screen.

Table 15.1 lists the air-line diameters recommended to pump certain volumes of water. Occasionally, the size of the air line may have to be decreased somewhat so that it fits into the annulus between the jetting pipe and the casing. Another alternative is to decrease the size of the jetting pipe. The air-lift system operates best when the bottom of the air line is placed just above the jetting tool, because more suspended sediment will be carried up the borehole. When using high-velocity jetting procedures, development should start at the bottom of the screen.

The jetting water is usually clean water hauled to the drill site. In instances where sufficient water supplies are unavailable, the contractor may use the water pumped from the well. To avoid damaging the high-pressure pump, jetting nozzles, and screen, however, the fine sand pumped

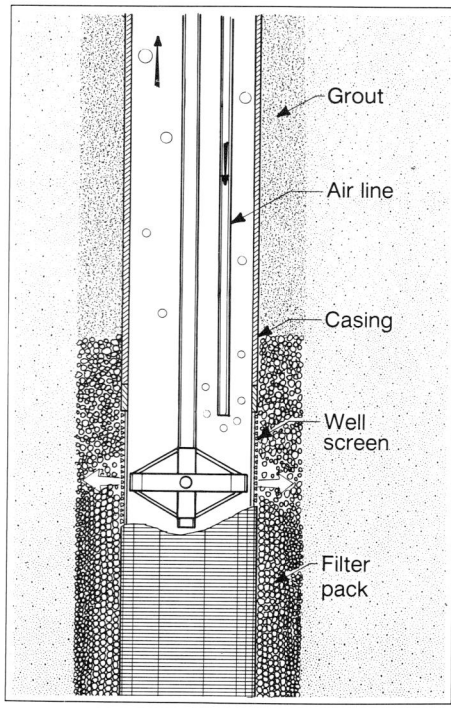

Figure 15.19. The jetting tool and drop pipe are separate from the air line so that jetting and air-lift pumping can be done simultaneously.

from the well should be settled out in a tank or settling pit before this water is recirculated. To enhance the development process, chemicals such as polyphosphates are often added to the jetting water to help break up clays.

When air-lift pumping is impractical, a submersible pump can be used during jetting. Usually the pump must be placed well above the jetting tool so that the amount of sand passing through the pump is minimized. Thus, the pump causes material temporarily placed in suspension by the jetting action to move into the screen, but much of the material falls to the bottom. This sediment must be removed periodically during the jetting and pumping operation so the entire screen is developed.

A COMPARISON OF THREE DEVELOPMENT METHODS

Certain development methods exert more powerful cleaning forces on the formation than do others, and thus are better able to remove the drilling fluid and create a zone of high porosity and hydraulic conductivity around the screen. Results from an experimental well field at the Irrigation Research Center at Staples, Minnesota demonstrate the relative differences in three development methods. Ten irrigation wells were constructed at Staples in surficial glacial outwash, using three types of well screens. This site was chosen because of the highly uniform nature of the aquifer. All wells were drilled by the direct rotary drilling method. Two different drilling fluid additives were used — bentonite and polymer. After well construction, each screen was developed in three stages; first by overpumping, second by mechanical surging, and third by water jetting and simultaneous air-lift pumping. Each development method was continued until the water was essentially sand free. A 24-hour pumping test was conducted after each of the three development episodes; data were acquired, analyzed, and stored with the aid of a computer system. The Staples investigation is one of the few scientific studies conducted in a uniform aquifer that examines many of the well construction and completion variables that affect development.

Results of the study show that the average specific capacities of the wells drilled with bentonite improved 74 percent when overpumping was followed successively by mechanical surging and simultaneous water jetting/air-lift pumping (Figure 15.20). Drawdown data suggest overpumping is the least effective of these three development methods, whereas simultaneous water jetting/air-lift pumping removed sediment and drilling fluid that the other development methods could not dislodge. In general, test data from Staples show that the eventual specific capacity of a well depends to a great extent on what development method is used.

As suggested earlier in this chapter, the improvement in specific capacity

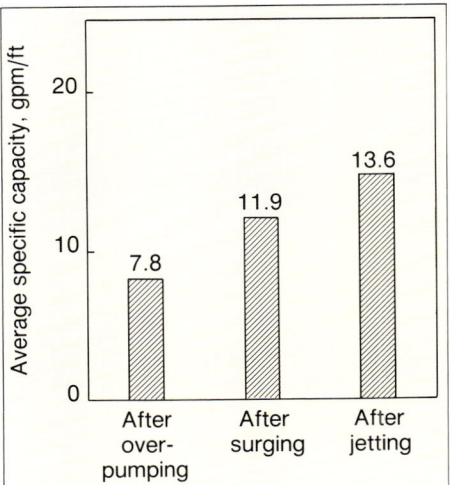

Figure 15.20. Average specific capacities of wells drilled with bentonite, after various development methods. Individual specific capacities were measured when no more sediment could be dislodged from the formation by a particular development method. *(Werner et al., 1980)*

achieved by any development method may be seriously retarded because of limited open area. For the wells drilled with bentonite, those completed with screens having more than 15 percent open area had 52 percent greater specific capacities than those completed with screens having less than 15 percent open area (Figure 15.21). As the open area of the screen is increased, the enhanced effectiveness of the development causes a corresponding increase in specific capacity.

The specific capacity of a well is also a function of which drilling fluid is selected. For each development episode, wells drilled with polymer averaged 56 percent higher specific capacities than did the wells drilled with bentonite (Figure 11.21, page 362) (Werner et al., 1980). This difference occurs because some of the bentonite penetrates farther into the formation during the drilling process than any development method can reach. On the other hand, polymeric drilling fluids beyond the reach of a development method simply break down naturally over time and thus offer no resistance to flow.

USE OF POLYPHOSPHATES IN DEVELOPMENT

Adding a small amount of a polyphosphate before or during development helps considerably in removing clays that occur naturally in the aquifer and those clays introduced into the borehole as part of the drilling fluid. Polyphosphates disperse the clay particles in the formation so they can be removed. Enough time must be allowed between introduction of the polyphosphate and development, usually overnight, so the clay masses become completely disaggregated (see Chapter 11). After the polyphosphate solution is jetted or surged into the screen, water should be added to the well to drive the solution farther into the formation.

Two types of polyphosphates are used in well development — crystalline and glassy. Crystalline polyphosphates that help remove clays from the aquifer are sodium acid pyrophosphate (SAPP), tetrasodium pyrophosphate (TSPP), and sodium tripolyphosphate (STP). Sodium hexametaphosphate (SHMP) is a glassy phosphate that is readily available and therefore often used in developing wells. About 15 lb (6.8 kg) of a polyphosphate should be used for each 100 gal (0.4 m³) of water in the screen. Two pounds (0.9 kg) of sodium hypochlorite (3 to 15 percent chlorine solution) also should be added to every 100 gal of water in the well to control bacterial growth promoted by the presence of polyphosphates.

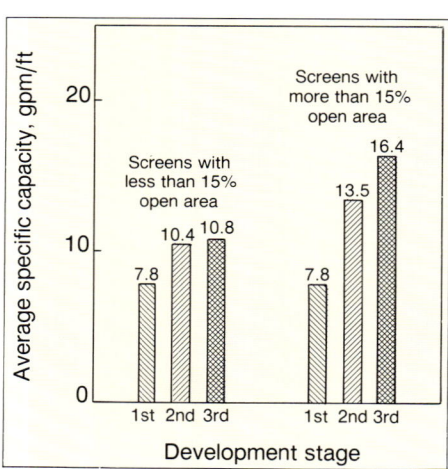

Figure 15.21. Open area of the well screen controls the effectiveness of various development methods for wells drilled with bentonite. In general, development is most effective when the screen open area is the largest. *(Werner et al., 1980; Driscoll et al., 1980a)*

Polyphosphates should be premixed before introduction into the well because they do not mix easily with cold water. Occasionally the mix water is heated to help dissolve the chemical. In general, the lighter the density, the more quickly the polyphosphate will go into solution. Sodium tripolyphosphate, for example, is available in three densities. If the polyphosphate is jetted, some contractors use a solution with twice the recommended concentration. A larger amount of polyphosphate, however, does not give significantly better results when used for drilling fluid dispersion and, under certain conditions, may cause serious problems.

Care must be exercised if SHMP is used because the phosphate can become glassy in the well under certain conditions, causing severe plugging of the formation and screen. For example, an over-rich solution of SHMP can precipitate glassy phosphates on contact with the cold groundwater found typically in northern Europe and southern Canada. These glassy precipitates are gelatinous masses that are extremely difficult to remove because no effective solvents exist. Acids will break down the glassy condition eventually, but the time required is impractical for efficient well development. Therefore, glassy phosphates should never be dumped undissolved into a well. The use of a glassy phosphate also can be undesirable in certain formations. Although little is known about the exact chemical conditions that may cause the phosphate to become semisolid in a borehole, experience has shown that the use of SHMP can be a problem in glauconitic sandstones. SHMP is available in plate form, crushed, or as a powder, which affects the density of the product.

The addition of wetting agents to polyphosphates will increase their effectiveness in disaggregating clays. One pound (0.5 kg) of wetting agent (for example, Pluronic F-68) is added to 100 gal (0.4 m^3) of polyphosphate solution. Wetting agents and polyphosphates should not be used in formations with thinly bedded clays and sands, because these chemicals tend to make the clays near the borehole unstable, causing them to mix with the sand. The hydraulic conductivity of the aquifer near the borehole is then reduced and clay continually passes into the borehole during each pumping cycle.

DEVELOPMENT OF ROCK WELLS

All drilling methods cause some plugging of fractures and crevices in hard-rock formations. In softer formations such as sandstone, the borehole wall may become clogged with finer material. In cable tool drilling, the bit action chips and crushes the rock and mixes it with water and other fine material to form a slurry. The pounding of the bit forces some of this slurry into the openings in the rock outside the borehole. When drilling fluids are used in rock drilling, they also may plug crevices. Even air-drilling methods can blow large quantities of fine material into openings in the rock, causing drastic reductions in yield.

Any material that clogs openings in the rock aquifer must be removed by a development procedure. The full yield of the formation can be realized only if all the fractures and crevices can provide water to the well. Pumping alone sometimes pulls out the remaining sediment because the openings in rock formations are relatively large in comparison with the pores in a sand formation. However, many drillers have found that surging or other means of development for rock wells is needed to obtain maximum capacity.

One of the best methods used to clean rock holes is the water jetting/air-lift pumping method in which inflatable packers are used to isolate the zones that supply water to the well. As in the case of screened wells, the jetting action dislodges loose material from the borehole wall and from the cracks and crevices in the formation. It has been shown that much of the water entering an open borehole enters through these fracture zones. If they are partially plugged with drill cuttings, the water can become turbulent and thus cause higher head losses and significantly reduced specific capacity as yields are increased.

Development of Sandstone Wells

Blasting and bailing techniques have long been used in an attempt to reduce sand pumping and enhance yields from wells constructed in weakly consolidated sandstones. Usually, blasting and bailing techniques are used after other development methods have been applied. To reduce sand pumping, some drillers blast a sandstone aquifer to create a much larger cavern. In theory, the average flow velocity toward the well from the borehole face will be so low that the transport of sand will not occur. Unfortunately, sand pumping does not stop in many cases because the aquifer materials continue to slough off and fill the enlarged borehole. The inability to reduce sand pumping to acceptable levels shows that these procedures may not be cost effective. Recently, an increasing number of friable sandstone wells have been screened rather than blasted to assure control of sand pumping, even though some loss of specific capacity may occur. Screening a well in sandstone may be impractical in some circumstances, because the loss in yield may be unacceptable.

Blasting and bailing are also used to increase the yield by enlarging the effective well diameter. Walton (1962) has shown, for example, that yields increased an average of 29 percent after blasting and bailing techniques had been applied. In many cases, however, the blasting and bailing procedure may take months to perform and add significantly to overall costs. Thus, blasting a sandstone aquifer may not increase the yield enough to justify the additional costs and may not eliminate a sand-pumping problem.

A powerful technique has been developed to clean sandstone wells that combines air-lift pumping, air pressurizing, and rawhiding. This combination of methods overcomes development difficulties caused by the geologic makeup of many sandstone aquifers. These aquifers are often highly stratified, with certain layers cemented by silica, calcium, or iron, or by fine material such as clay. Cementation resulted when minerals were deposited by groundwater flowing along highly permeable lenses of the sandstone after it was uplifted.

After being drilled, a borehole in sandstone with alternating layers of cemented and uncemented material will appear as shown in Figure 15.22. Ledges form from the most resistant, well-cemented portions of the sandstone, whereas the more friable layers erode back away from the borehole. After completion, loosened

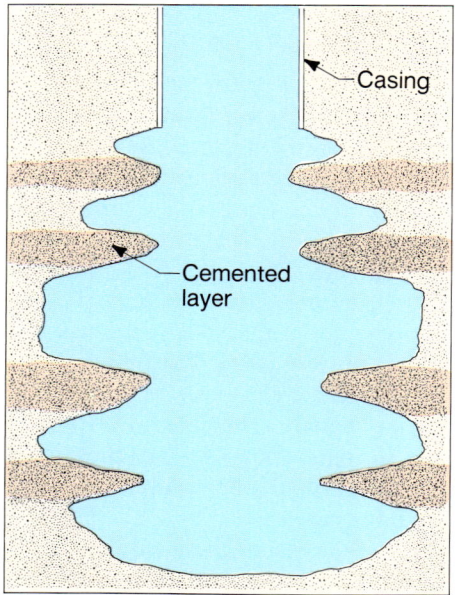

Figure 15.22. Borehole configuration caused by drilling in weakly consolidated sandstone containing well-cemented layers. Loose sand collects on the upper surface of the cemented layers.

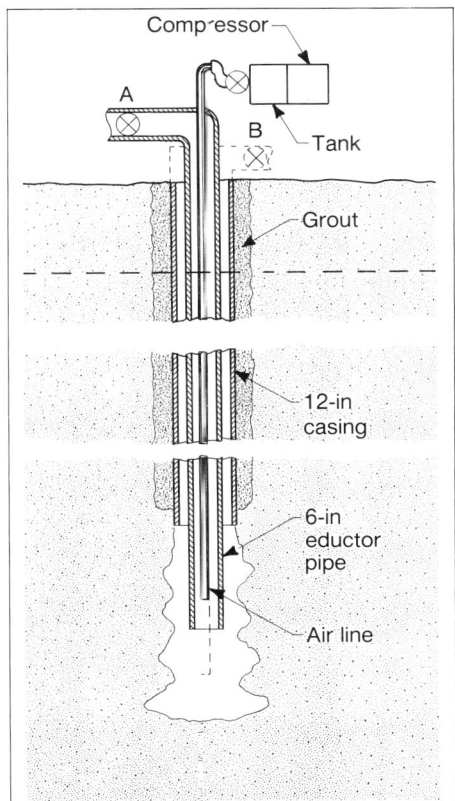

Figure 15.23. With this type of development installation, the development procedures combine air-lift pumping, air pressurizing, and rawhiding. *(Thein Well Company, U.S. Patent No. 4,265,312)*

sand lies on top of the ledges, which can extend away from the borehole for many feet. If the well is put into production after bailing or air development, it will often continue to pump sand. The problem usually stems from the contractor's inability to remove all the loose material in the well, because the development procedures do not extend far enough from the well bore.

A technique that combines air-lift pumping, air pressurizing, and rawhiding has been used successfully to remove virtually all loose sand from sandstone wells to 1,000 ft (305 m) in depth. Use of the equipment illustrated in Figure 15.23 involves three steps that are repeated until the recovery of sediment is negligible. The first step is to air-lift pump with the 6-in (152-mm) eductor pipe and air line. As much material as possible is removed by air-lift pumping. Then valve A at the top of the 6-in pipe is closed, while the air line is raised farther up into the 6-in eductor pipe. Pressurized air building up in the 6-in eductor pipe will eventually force all the water out of the eductor; air then begins to pass up the borehole, around the eductor pipe. Water driven by the air is allowed to escape at the surface through the 12-in (305-mm) well casing. When the water clears, the valve on the eductor pipe is suddenly opened, and a surge of water enters the eductor pipe. The entire water column between the eductor pipe and casing falls down the casing into the borehole with great turbulence. Water is driven into the formation. Because the air lift is still operating, water and sediment are immediately pulled into the eductor pipe and discharged at the ground surface. Uphole velocity in the eductor pipe is kept at approximately 4,000 ft/min (1,220 m/min) so that all sizes of sediment can be removed. The 1,050-cfm (0.5 m³/sec) compressor used in this procedure operates at 425 psi (2,930 kPa); therefore, great care must be taken in working near the well head during operation of the development tool.

Downhole camera (TV) surveys show that loose sand on ledges is removed well away from the borehole. Yields generally increase significantly when air-lift pumping, air pressurization, and rawhiding are used, sometimes producing two to five times more water per foot of drawdown in comparison with predevelopment capacity.

Pressurization is another technique that can produce good results in sandstone boreholes. In this method, the well is pressurized so that the water surface in the borehole falls to just above the bottom of the casing. The contractor should calculate the amount of pressure required so that extra pressure does not force the water below the bottom of the casing, thereby blowing air into the formation. A quick-opening valve mounted on a plate fastened to the top of the casing is then suddenly opened, causing a rush of water into the well. Sand and silt particles move into the borehole and then settle to the bottom. Compression-decompression cycles are repeated until the sediment fills a significant part of the well and must be removed. It can be either bailed or removed by air-lift pumping. The eductor pipe and air line are placed in the well and then the casing cap is attached (Figure 15.24). Pressurization can also be accomplished in the method illustrated in Figure 15.23 by closing valves A and B.

Figure 15.24. In the pressurization method, the eductor and air line are placed in the well and then the casing is capped. *(Bergerson-Caswell, Inc.)*

ALLOWABLE SEDIMENT CONCENTRATION IN WELL WATER

Sediment in water supplies can be destructive to pumps and to water-discharge fittings such as the nozzles on irrigation systems. Although development methods reduce or eliminate high concentrations of sediment in well water, it is impractical to assume that all sediment transport can be eliminated, even by the most powerful development methods. Therefore, some judgment must be used to establish allowable concentrations. The term "sand-free water" as used in this text describes water that contains less than 8 mg/ℓ of sand, silt, or clay.

The concentration of suspended sediment in water is usually estimated by using a large container (Figure 15.25), a centrifugal sand sampler (Figure 15.26), or an Imhoff cone (Figure 15.27). Containers such as the Imhoff cone are less accurate in estimating sediment concentrations because of their small volumes. The sediment concentration is determined by averaging the results of five samples taken at the following times during the final pumping test: (1) 15 minutes after start of test, (2) after 25 percent of the total pumping test time has elapsed, (3)

Figure 15.25. The concentration of suspended sediment can be estimated by collecting water and sediment in a large container. *(Olson Bros. Well Drilling Company, Inc.)*

after 50 percent of the total pumping test time has elapsed, (4) after 75 percent of the total pumping test time has elapsed, and (5) near the end of the pumping test (U.S. EPA, 1975c).

For accurate measurement of sediment concentration, the water sample must be of reasonable volume. The recommended volume of water to be tested for sediment is determined by multiplying the flow rate, in gpm, times 0.05. For example, at 300 gpm, the sample would be 300 × 0.05 = 15 gal. For yields over 1,000 gpm (5,450 m³/day), the test volume should be 50 gal (189 ℓ); for yields under 20 gpm (109 m³/day), 5 gal (18.9 ℓ) should be tested.

The amount of sediment collected in each sample may depend greatly on how the water sample is taken. For flow from a straight, horizontal section of pipe, most of the sediment may be on or near the bottom of the pipe. Therefore, if the sample is collected from the bottom of the flow, the sediment concentration will appear higher than the average of the total stream. Conversely, a sample taken at the top may show a lower concentration than the average. For the most accurate results, the entire flow should be collected over a certain time. But this technique becomes more difficult as the flow increases, and devices such as the centrifugal sand sampler become useful. The sediment should be allowed to settle out, then weighed and compared with the total volume of water collected. When using a centrifugal sand sampler, the sample should be collected through a connection at the midline of the pipe.

Figure 15.26. The concentration of suspended sediment can be estimated by mounting a centrifugal sand sampler on the discharge pipe. *(Henkle Drilling)*

Figure 15.27. The Imhoff cone is used to estimate sediment concentration, but the small volume of the cone limits its accuracy.

Any recommendations for limiting sediment concentration must take into account how the water will be used. The U.S. Environmental Protection Agency and the National Water Well Association (1975) have recommended the following limits:

1. 1 mg/ℓ — water to be used directly in contact with, or in the processing of, food and beverages.
2. 5 mg/ℓ — water for homes, institutions, municipalities, and industries.
3. 10 mg/ℓ — water for sprinkler irrigation systems, industrial evaporative cooling systems, and any other use where

a moderate amount of sediment is not especially harmful.

4. 15 mg/ℓ — water for flood-type irrigation.

Other considerations may also be important when determining allowable sediment concentration. For example, when the maximum concentration is greater than 15 mg/ℓ, so much material is being removed that the aquifer materials and overlying strata may collapse, thus shortening the life of the well.

The limits suggest reasonable goals that can be achieved if good well design, construction, and development practices are followed. In older wells or wells in problem aquifers, a well may pump unacceptable amounts of sediment. If the well cannot be redeveloped by conventional techniques, a special sand separator can be installed as a permanent part of the well system. When desanding systems are installed in the well, material collected in the separator is directed to the bottom of the well by means of a tail pipe (Figure 15.28). Solids can be transported to the surface by an eductor system for periodic removal of material or by an external ejector system for continuous removal. The most common type of separator is installed at the surface near the well head. Although sand separators are efficient, they may not remove all sediment and should not be used as a substitute for good well design and construction practices.

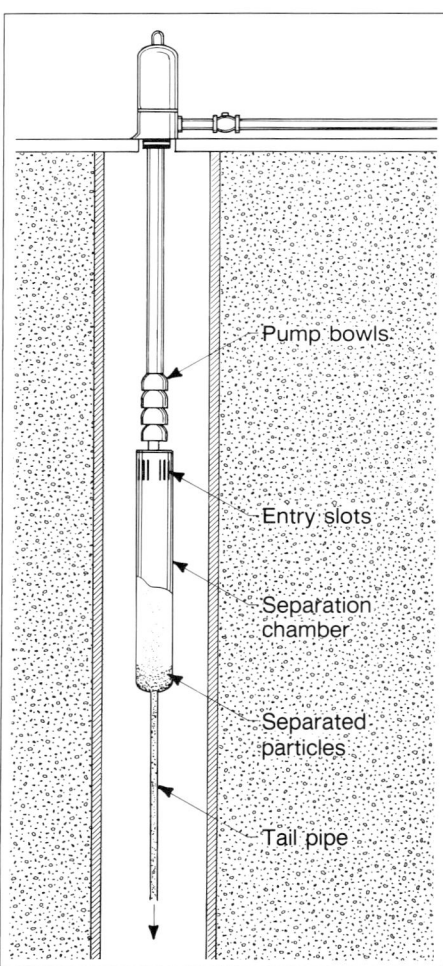

Figure 15.28. A sediment separator can be installed permanently in the well beneath the pump intake. Material collected in the separator falls to the sump at the bottom of the well. *(Claude Laval Corporation)*

AQUIFER DEVELOPMENT TECHNIQUES

In many parts of the world, the only available groundwater comes from bedrock. If the rock is massive, with few joints or faults, the volume of water available is often inadequate. In this type of aquifer, yields can be increased dramatically by applying one or more aquifer development techniques. Aquifer development, also called aquifer stimulation, can be thought of as a second level of development which can increase well yields far beyond those obtained through typical well development. Aquifer development procedures in massive rock are usually cost effective. Some of these methods are described below. Under most circumstances, well development techniques are used before any aquifer development methods are initiated.

Use of Acid

Acid can be used for both well and aquifer development in limestone or dolomite aquifers and in some semiconsolidated aquifers that are cemented by calcium carbonate. Acid dissolves carbonate minerals and opens up the fractures and crevices in the formation around the open borehole, which is the intake portion of this type of well. Some of the acid, however, is forced into cracks and fissures much farther from the well bore. The acid dissolves some material naturally existing in the voids, thereby increasing the overall hydraulic conductivity of the aquifer. Some drillers place the acid in the borehole by jetting, but this procedure can be extremely dangerous and should be performed only by experienced personnel. The types of acids, amounts used, and placement procedures are discussed in Chapter 19.

Use of Explosives*

Explosives are sometimes used to "shoot" rock wells in an attempt to develop greater specific capacity. Good results can be obtained if blasting procedures are appropriate for the rock type and the size and depth of the well. Because of the many unknown factors, however, it is difficult to predict whether the shooting operation will produce beneficial results, especially in sedimentary rocks such as sandstone.

Extreme care must be used when blasting a well. If it is located near a home, the owner should be asked to remove breakable objects from shelves and to otherwise take precautions against a potentially significant seismic shock. People in the general vicinity also should be warned in advance of the blast.

The explosive charge to be used should be carefully planned. Many factors must be considered, including:

1. Depth of water in the hole. Larger charges must be used as the depth of water increases to overcome the hydrostatic pressure.

2. Geologic conditions. Actual or incipient joint systems and faults must be present, or the blasting may be futile.

3. Depth of borehole. For boreholes deeper than 700 ft (213 m), blasting may not be effective because of the weight of the overburden.

4. Desired increase in yield. Larger amounts of explosives will ordinarily produce greater increases in yields when all other factors are equal.

5. Environmental considerations. Is the blast site in an urban area or is it relatively remote?

6. Legal considerations. These include federal, state, and local regulations and spe-

---

Johnson Division makes no guarantee of results and disclaims all liability in connection with the information or the safety suggestions given for the methods described. Water well contractors should contact the manufacturer for information on blasting materials and procedures. In addition, contractors may want to contact other water well companies that have demonstrated expertise in water well and aquifer blasting techniques. It should be understood that not all the acceptable safety procedures are contained herein and that certain circumstances may call for additional precautions. The suggestions given here do not supplement or modify any state, municipal, federal, and insurance requirements or codes relating to blasting.

cial insurance requirements.

In the past, small explosive charges, usually 30 to 100 lbs (14 to 45 kg), were used. More recently, much larger charges of 1,000 to 2,000 lb (454 to 907 kg) have been detonated in igneous rock terrains with excellent results. In northern Minnesota, where igneous and metamorphic rocks outcrop or underlie a thin veneer of glacial drift, one driller has developed the procedure discussed below to extend the natural joint system in the rock (Driscoll, 1978).

Figure 15.29 shows how the explosive agent, sand, and igniting equipment are placed in a well to be blasted. A primer is attached to a primer cord and lowered to the bottom of the well, which is typically 300 to 500 ft (91.5 to 152 m) deep. Four 30-lb (14-kg) bags of jellied blasting agent are then lowered by cable or dropped into the hole. Jellied explosives

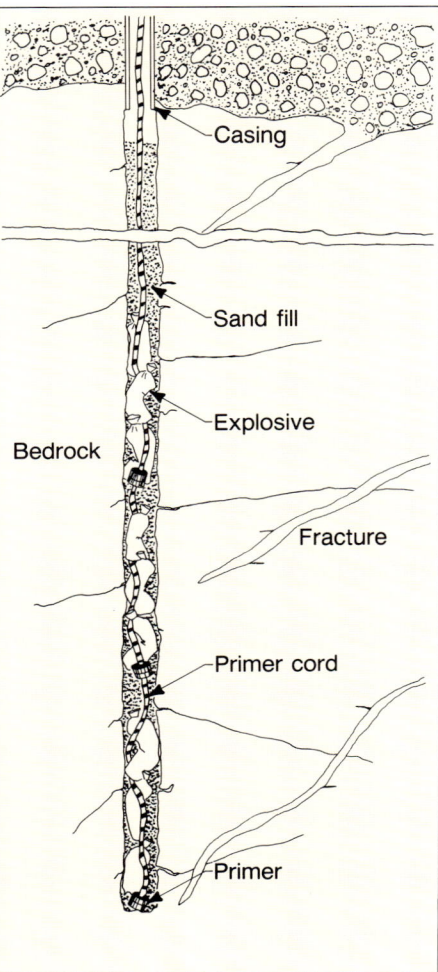

Figure 15.29. After the primers and explosive are in place, 10 to 25 ft (3 to 7.6 m) of sand are added to the well to prevent "rifling" and direct most of the force into the formation.

Figure 15.30. The blasting agent is relatively safe to handle, but should be protected from the sun. *(Fideldy Drilling)*

are packaged in different sizes; a commonly used size is 5 inches (127 mm) in diameter and 36 in (914 mm) long (Figure 15.30). Approximately 1 to 2 ft^3 (0.03 to 0.06 m^3) of medium-grained sand are placed on top of the explosives to act as a propping agent when blasted into the joint systems. This sequence is repeated until 1,000 to 2,000 lbs of blasting agent and sand partly fill the well. Note that a primer is always installed at the bottom of each blasting-agent layer to insure that all of the explosive will ignite. Only one primer cord is used. Individual primers are slipped on the cord and allowed to drop into the borehole.

To maximize the horizontal effects of the blast, a plug of sand 10 to 25 ft (3 to 7.6 m) thick is added to the top of the blasting agent and sand column. The sand layer acts as plugging material to reduce the potential rifling effect that would allow most of the blast force to escape vertically. It is important to concentrate the blast horizontally against the rock and not allow relief out the top of the well.

In practice, 1,000 lb of explosive are placed in a 300-ft (91.5-m) hole and as much as 2,000 lb (907 kg) in a 500-ft (152-m) hole. Each 1,000 lb (454 kg) of blasting agent, plus the layers of sand, will fill approximately 100 ft (30.5 m) of a 6-in (152-mm) hole. Explosives should be placed at a safe distance below the casing to avoid shooting the casing out of the hole. This distance is related to the size of the charge, the casing material, and other factors such as rock type and degree of fracturing.

The blasting procedure described above has produced excellent results in formerly deeply buried igneous and metamorphic rocks. For example, in a 400-ft (122-m) well near Grand Rapids, Minnesota, the yield increased from 3 gpm to 60 gpm (16 to 327 m^3/day) at 60 percent of the maximum drawdown. In another essentially dry hole 500 ft (152 m) deep near Nashwauk, Minnesota, the yield increased to 30 gpm (164 m^3/day) with less than 100 ft (30.5 m) of drawdown. Previously, this well could be pumped dry in less than 3 hours at only 5 gpm (27 m^3/day).

Hydrofracturing

Hydrofracturing has been used successfully since 1947 in oil wells to overcome well bore damage, create reservoir fractures that improve well productivity, aid in secondary recovery techniques, and facilitate injection of brine and industrial wastes (Howard and Fast, 1970). More recently, this method has been used to increase the yields of low-production water wells in rock where joint systems or fracture systems are poorly developed or so tight that little water can move through them (Waltz and Decker, 1981).

In hydrofracturing, high-pressure pumps are used to overcome the pressure of overlying rock and to inject fluids into newly opened fractures (Figure 15.31). For every foot (0.3 m) of depth, the overburden pressure is usually equal to 1 psi (6.9 kPa). Therefore, at 200 ft (61 m) the overburden pressure can be overcome with a pump capable of pumping fluids at pressures greater than 200 psi (1,390 kPa). Oil field hydrofracturing pumps can move fluids at pressures of 20,000 psi (138,000 kPa) and more.

Cleaning the borehole walls before hy-

Figure 15.31. A high-pressure pump provides enough force to overcome the weight of the rock so that fluids carrying sand can be injected into the open joints and fractures. *(Olson Bros. Well Drilling Company, Inc.)*

drofracturing is desirable because it removes drill cuttings, natural clays, or other mineral substances. This is done by mechanically brushing the walls with an oversized wire brush assembly or by jetting the walls with high-velocity water jets. Use of chemical additives such as diluted hydrochloric acid to remove calcium carbonate deposits, or sodium tripolyphosphate to remove clays, can greatly facilitate this process.

The following equipment is needed to apply the hydrofracturing technique: high-pressure water pump; power-takeoff hydraulic pump, quick-coupled to a hydraulic motor; valve assembly directing flow to a packer inflation line and a water-injection line; one or more inflatable packers; and compressed air source to inflate the packers and increase downhole pressure (Figure 15.32).

Hydrofracturing is accomplished by lowering an inflatable packer into a well and inflating it at a depth somewhat above the production zone. Thus, the production zone is isolated from the rest of the well. A fluid, usually water, is then pumped down through the packer into the well at pressures of 500 to 10,000 psi (3,450 to 69,000 kPa). About 800 to 1,000 psi (5,520 to 6,900 kPa) is sufficient to fracture most formations that are already somewhat fractured; much higher pressures may be needed if the rock is massive with few cracks. For example, 2,000 psi (13,800 kPa) was required to fracture massive granite in New Hampshire (Stewart, 1973). In rocks with few fractures, injection rates can be as low as 5 to 10 gpm (27.3 to 54.5 m^3/day), with total water volume injected ranging from 200 to 400 gal (0.8 to 1.5 m^3). In the granitic rocks of New Hampshire, however, 12,600 gal (47.7 m^3) of water were injected at a rate of 300 gpm (1,640 m^3/day) in a 479-ft (146-m) hole.

Pressure in the production zone usually causes small, tight breaks in the rock to open up and spread radially. The newly opened fractures provide effective interconnections between nearby water-bearing fractures and the well bore. However, hydraulically formed fractures tend to heal; that is, they lose their ability to transmit fluids unless propped open. Therefore, sand or small particles of high-strength plastic are introduced into the fractures during pumping to keep the cracks open after the pressure has been released. The optimum concentration of propping particles is related to the well depth and the fracture characteristics of the rock; they must prop open, but not pack, the newly opened fractures.

Sand can be introduced into the well with the high-pressure pump. If the high-pressure pump cannot accommodate sand, the propping agent must be placed in the well before pressurization. Plastic particles can be used in place of sand and may distribute more evenly in the fractures. They are mixed with sand and water and then frozen in 16-oz (454-g) paper cups. In the field, the cups are torn off and the cylindrically shaped packets of frozen sand and plastic beads are dropped into the well. The sand acts as a weighting agent, and, as the ice melts, the plastic beads float slowly upward throughout the

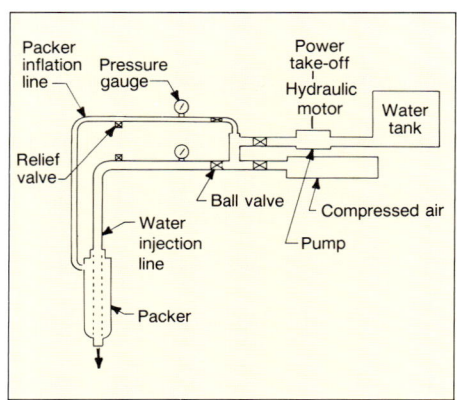

Figure 15.32. A schematic diagram of the equipment used in a small hydrofracturing operation. *(Waltz and Decker, 1981)*

well bore during pressurization.

In Denver, Colorado, yields after hydrofracturing have increased from 50 to 130 percent, although these increases represent only a small volume of water, for example, from 0.5 to 1 gpm (2.7 to 5.5 m^3/day). Improvements in the New Hampshire wells, however, were from 4 to 24 gpm (21.8 to 131 m^3/day) and 4 to 15 gpm (21.8 to 81.8 m^3/day), in 479- and 300-ft (146- and 91.5-m) wells, respectively (Stewart, 1973). These wells are 680 ft (207 m) apart. Before hydrofracturing, pumping one of these wells did not affect the other. After hydrofracturing, the wells became connected hydraulically. This example illustrates how far fracturing can occur away from the borehole.

Hydrofracturing can increase water yield, improve reliability of water yield, reduce suspended sediment in the water, increase water storage in the well, and reduce pumping costs.

CONCLUSIONS

Patience, intelligent observation, and the right tools are required to develop a well correctly. Well development is not expensive, considering the often remarkable results that can be obtained in improving yields and eliminating sand pumping. Similarly, aquifer development is often overlooked as an effective way to increase yields substantially.

CHAPTER 16
Collection and Analysis of Pumping Test Data

Pumping tests may be conducted to determine (1) the performance characteristics of a well and (2) the hydraulic parameters of the aquifer. For a well-performance test, yield and drawdown are recorded so that the specific capacity can be calculated. These data, taken under controlled conditions, give a measure of the productive capacity of the completed well and also provide information needed for the selection of pumping equipment. An accurate test of a well before the pump is purchased pays for itself by assuring selection of a pump that will minimize power and maintenance costs. Many times, high pumping costs and unsatisfactory pump performance are erroneously charged to the well when these conditions really stem from an improperly selected pump.

The second purpose of pumping tests is to provide data from which the principal factors of aquifer performance — transmissivity and storage coefficient — can be calculated. This type of test is called an aquifer test because it is primarily the aquifer characteristics that are being determined, even though the specific capacity of the well can also be calculated. Aquifer tests will predict (1) the effect of new withdrawals on existing wells, (2) the drawdowns in a well at future times and different discharges, and (3) the radius of the cone of influence for individual or multiple wells. Aquifer test data are more valuable today because a better understanding of groundwater hydraulics now exists and new sophisticated methods of data retrieval and analysis have been developed.

An aquifer test consists of pumping a well at a certain rate and recording the drawdown in the pumping well and in nearby observation wells at specific times. There are two primary types of aquifer tests: constant-rate tests and step-drawdown tests. In the constant-rate test, the well is pumped for a significant length of time at one rate, whereas in a step-drawdown test the well is pumped at successively greater discharges for relatively short periods. Data from both types of aquifer pumping tests can be analyzed to determine important hydraulic characteristics of an aquifer and the well. The results from properly conducted tests are the most important tool in groundwater investigations.

Measurements required for both well tests and aquifer tests include the static water levels just before the test is started, time since the pump started, pumping rate, pumping levels or dynamic water levels at various intervals during the pumping period, time of any change in discharge rate, and time the pump stopped. Measurements of water levels after the pump is stopped (recovery) are extremely valuable in verifying the aquifer coefficients calculated during the pumping phase of the test.

For well tests, the yield and drawdown are measured after a certain time has elapsed. Because this procedure has been explained in Chapter 9, it will not be discussed further in this chapter. Although aquifer testing is more involved than well testing, the methods presented below for determining yields and measuring drawdown are used in both well and aquifer pumping tests. These methods and procedures apply primarily to constant-rate and step-drawdown aquifer tests.

The emphasis in this chapter will be on constant-rate and step-drawdown tests, because they are the most suitable methods to analyze highly productive aquifers. Other types of tests can be used when the aquifer has a low hydraulic conductivity which limits the yield from virtually nothing to 1 to 2 gpm (5.5 to 10.9 m^3/day). These latter tests are conducted by monitoring the pressure changes in the formation after they have been disturbed, or by controlled injection. Pressure testing by the drill-stem method is discussed later in this chapter because it is used occasionally for testing low-yield water wells. Testing by injection is mostly limited to sanitary drainage fields, however, and will not be covered in this chapter.

CONDUCTING A PUMPING TEST

Pumping tests will not produce accurate data unless the tests are carried out methodically, carefully recording the time, discharge, and depth measurements. Certain preliminary steps should be taken to assure the reliability of pumping test data recorded during the actual test. For instance, several days before the test is to be conducted, the test well should be pumped for several hours to determine the following:

1. The maximum anticipated drawdown. (For most pumping tests, a major portion of the drawdown will occur in the first few hours of pumping.)
2. The volume of water produced at certain engine (pump) speeds and drawdown.
3. The best method to measure the yield.
4. Whether the discharge from the pump is piped far enough away to avoid recharge.
5. Whether the observation wells are located so that they exhibit sufficient drawdown to produce usable data.

Prior planning and experimentation with the equipment and personnel during preliminary testing can eliminate potential errors that may occur during the actual pumping test. Never begin the actual pumping test, however, until the water level in the aquifer has returned to the normal (pretest) static level following preliminary testing. About 24 to 72 hours should be allowed, depending on the type of aquifer. Beginning a pumping test when the static water level is below normal may eliminate early data that show discharge or recharge boundaries. Without the early drawdown data, it may be impossible to obtain the correct transmissivity and storage parameters for the aquifer.

The accuracy of drawdown data taken during a pumping test depends on the following:

1. Maintaining a constant yield during the test.

2. Measuring the drawdown carefully in the pumping well and in one or two properly placed observation wells.

3. Taking drawdown readings at appropriate time intervals.

4. Determining how changes in barometric pressures, stream levels, and tidal oscillations affect drawdown data.

5. Comparing recovery data with drawdown data taken during the pumping portion of the test.

6. Continuing the test for 24 hours for a confined aquifer and 72 hours for an unconfined aquifer during constant-rate tests. If other wells are being pumped within the potential cone of depression of the well to be test pumped, these wells must be pumped at a constant rate throughout the duration of the test. For step-drawdown tests, 24 hours is usually sufficient for either type of aquifer.

Maintaining a Constant Discharge

Variations in engine (pump) speed are a major cause of erratic drawdown data. If a gasoline or diesel engine is used to drive the pump, the selected yield should be well below the maximum capacity of the engine. Engines running at full throttle tend to vary significantly in rpm, causing variations in the volume of water being pumped. Thus, it is good practice to restrict the engine speed to one-half to two-thirds of the maximum rpm. In this range, the engine will run steadily, producing a more constant yield. Problems with varying rpm and yield can be virtually eliminated if an electric motor is used to drive the pump.

It is vital that a complete set of drawdown data be obtained once the pumping test commences. Therefore, for an aquifer test, the pump and power unit should be capable of operating at a constant pumping rate for at least 48 hours. In cases where the observation wells must be located at considerable distances from the pumped well, the pump must be capable of operating for at least several days. Pump failure during the test is expensive and even if the test is quickly resumed after repairs or refueling, the data are of questionable value. Therefore, the pump should be in good repair and the fuel supply should be adequate for the full term of the pumping test.

The pumping rate should be measured accurately and recorded periodically. Control of the pumping rate during testing requires an accurate device for measuring the discharge of the pump and a convenient means for adjusting the rate to keep it as nearly constant as possible. A valve in the discharge line of the pump provides the best control. The discharge pipe and the valve should be sized so that the valve will be from one-half to three-fourths open when pumping at the desired rate. Unnoticed changes in speed that result from varying line voltage on electric motors, or from variations in air temperature, humidity, or gasoline mixture on gasoline engines, cause less fluctuation of the discharge when the pump is working against the back pressure or head developed in the partially closed valve. Changing the pumping rate by controlling the pump speed is generally unsatisfactory. This is particularly undesirable when the pump is on open discharge and delivering water at low pressure.

Direct Measurement Methods — Containers and Meters

A simple and accurate method for determining the pumping rate is to observe the time required to fill a container of known volume. For example, if it takes 30 seconds to fill a 55-gal (0.2-m^3) barrel, the pump is delivering 110 gpm (600 m^3/day). This method is practical, however, for measuring only relatively low pumping rates.

A commercial water meter is more reliable when measuring large discharges. The dials on the meter show the total volume discharged through the meter up to the time of observation. Subtracting two readings taken exactly one minute apart gives the pumping rate. This is perhaps the easiest apparatus to use. Its only disadvantage is the unavoidable delay in obtaining values at the start of the test, when the pumping rate is being adjusted to the desired level.

Orifice Weir

The circular orifice weir is the device used most often to measure the discharge rate from a high-capacity pump. It will not measure the pulsating flow from a piston pump. Figure 16.1 shows a typical orifice weir.

The orifice is a round hole with clean, square edges in the center of a circular steel plate. The plate must be $1/16$ in (1.6 mm) thick around the circumference of the hole and is fastened against the outer end of a level discharge pipe so that the orifice is centered on the pipe. The end of the pipe must be cut squarely so the plate will be vertical. The bore of the pipe should be smooth and free of any obstruction that might cause abnormal turbulence. The discharge pipe must be straight and level for a distance of at least 6 ft (1.8 m) before the water reaches the orifice plate. This approach channel should be longer if possible. The pipe wall is tapped midway between top and bottom with a $1/8$-in (3.2-mm) or $1/4$-in (6.4-mm) hole exactly 24 in (0.6 m) from the orifice plate. Any burrs inside the pipe resulting from the drilling or tapping of the hole should be filed off.

A device called a piezometer (manometer) tube is fitted to this small hole to measure the water head (pressure) in the discharge pipe. The piezometer consists of a clear

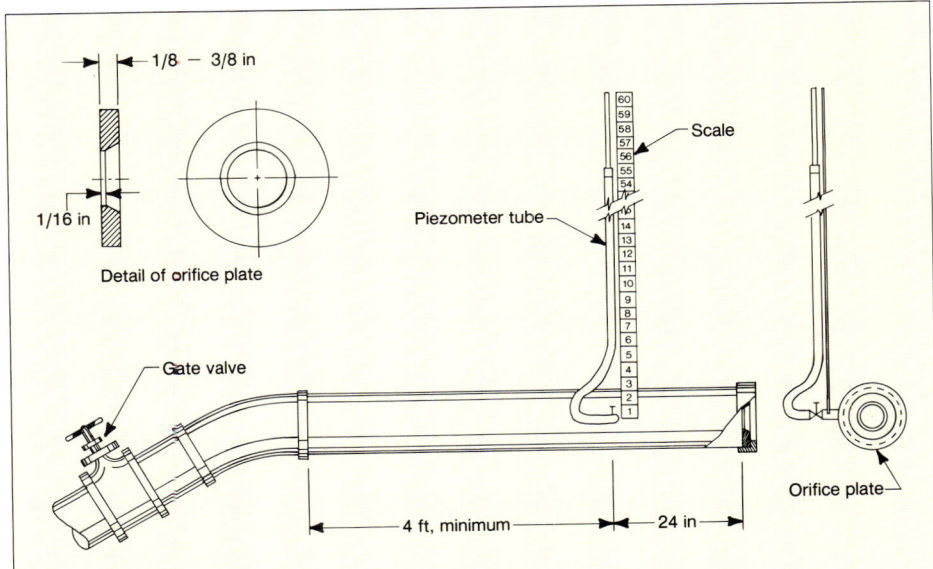

Figure 16.1. Construction diagram of a circular orifice weir commonly used for measuring pumping rates of a high-capacity pump. The discharge pipe must be level and the piezometer (manometer) tube placed exactly 24 in (610 mm) from the end of the pipe.

Table 16.1. Flow Rates Through Circular Orifice Weirs

Head of water in	4-in Pipe		6-in Pipe			8-in Pipe			10-in Pipe			12-in Pipe			16-in Pipe		
	2½-in orifice gpm	3-in orifice gpm	3-in orifice gpm	4-in orifice gpm	4-in orifice gpm	5-in orifice gpm	6-in orifice gpm	6-in orifice gpm	7-in orifice gpm	8-in orifice gpm	6-in orifice gpm	8-in orifice gpm	8-in orifice gpm	10-in orifice gpm	12-in orifice gpm
5	55	89	76	145	131	220	355	310	460	680	300	580	530	880	1,420
6	60	97	82	158	144	240	390	340	500	740	325	640	580	960	1,560
7	65	105	88	171	156	260	420	370	540	830	350	690	620	1,040	1,680
8	69	112	94	182	166	275	450	395	580	880	375	730	670	1,110	1,800
9	73	119	100	193	176	295	475	420	610	940	400	780	710	1,180	1,910
10	77	126	106	204	186	310	500	440	640	990	420	820	750	1,240	2,010
12	85	138	115	223	205	340	550	480	700	1,080	460	900	820	1,360	2,200
14	92	149	125	241	220	365	595	520	760	1,170	500	970	880	1,470	2,380
16	98	159	132	258	235	390	635	555	810	1,250	530	1,040	940	1,570	2,540
18	104	168	140	273	250	415	675	590	860	1,330	560	1,100	1,000	1,670	2,690
20	110	178	150	288	265	440	710	620	910	1,400	590	1,160	1,050	1,760	2,840
22	115	186	158	302	275	460	745	650	950	1,470	620	1,220	1,110	1,840	2,980
25	122	198	168	322	295	490	795	690	1,020	1,560	660	1,300	1,180	1,960	3,180
30	134	217	182	353	325	540	870	760	1,120	1,710	730	1,420	1,290	2,150	3,480
35	145	235	198	380	355	580	940	820	1,210	1,850	790	1,530	1,400	2,320	3,760
40	155	251	210	405	370	620	1,000	880	1,290	1,980	840	1,640	1,490	2,480	4,020
45	164	267	223	430	395	660	1,060	930	1,370	2,030	890	1,740	1,580	2,630	4,260
50	173	280	235	455	415	690	1,120	980	1,440	2,140	940	1,830	1,670	2,780	4,490
60	190	310	260	500	455	760	1,230	1,080	1,580	2,340	1,030	2,010	1,830	3,040	4,920
70	205	350	280	525	490	810	1,280	1,140	1,710	2,530	1,110	2,170	1,970	3,280	5,310

Note: Flow rates indicated below the line are more exact than those above the line because the head developed in the piezometer tube for particular pipe and orifice diameters is large enough to assure the accuracy of results obtained from Equation 16.3.

plastic tube 4 or 5 ft (1.2 or 1.5 m) long. One end is connected to pipe fittings that are tapped into the hole in the discharge pipe. The nipple, which is screwed into the tapped hole, must not protrude inside the pipe. A scale is fastened to a support so that the vertical distance from the center of the discharge pipe up to the water level in the piezometer tube can be measured. The water level in the piezometer tube indicates the pressure head in the approach pipe when water is being pumped through the orifice.

For any given size of orifice discharge pipe, the rate of flow through the orifice varies with the pressure head as measured in this manner. Table 16.1 gives the flow (in gpm) for various combinations of orifice and pipe diameters.

The flow through the orifice is calculated from the basic equation:

$$Q = AVC \qquad (16.1)$$

where Q is the flow per unit of time, A is the area of the orifice, V is the velocity of flow through the orifice, and C is the coefficient of discharge for the orifice. The velocity of the water at the orifice consists of its velocity in the approach channel plus the additional velocity head created by the pressure drop that occurs between the connection for the piezometer tube and the orifice. Because the water discharges at atmospheric pressure, the pressure head indicated by the piezometer tube can be converted to velocity if friction in the pipe is neglected.

The velocity can be related to the head in the piezometer tube by the equation:

$$V = \sqrt{2gh} \qquad (16.2)$$

where V is velocity in ft/sec, g is acceleration of gravity in ft/sec^2, and h is the height of the water in the piezometer tube in ft. To get a value for the actual velocity through the orifice, the value of V from the above equation must be added to the velocity in

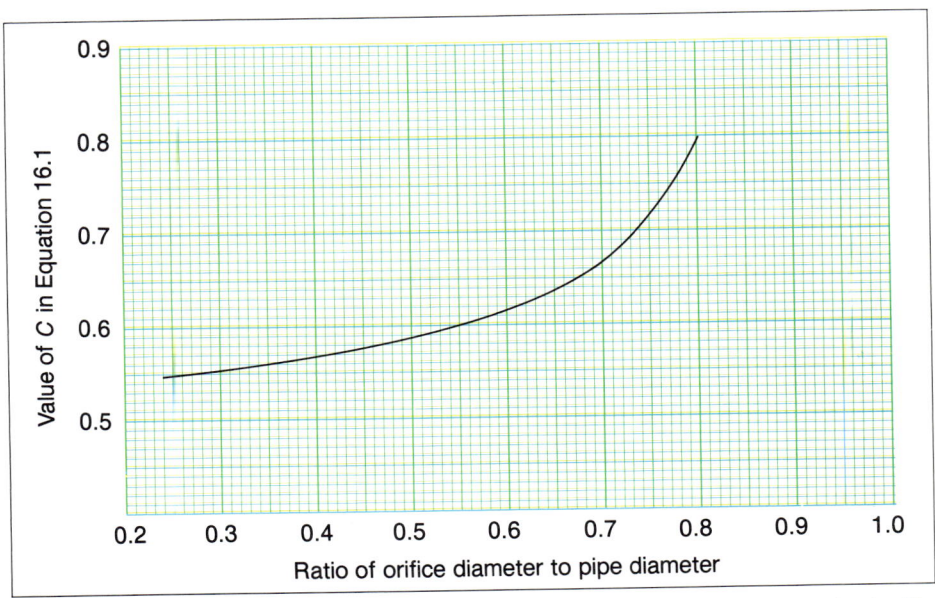

Figure 16.2. The coefficient of discharge, C, in the orifice-weir equation varies with the ratio of orifice diameter to pipe diameter as shown by this curve.

the approach pipe, and the sum of these must be corrected by two factors. One correction is for the contraction of the jet stream just outside the orifice, and the other is for the sudden change in cross-sectional area of flow which is controlled by the size of the orifice relative to the size of the approach channel.

For convenience, the approach velocity and the two correction factors can be combined into a single factor, C, whose value varies with the ratio of the orifice diameter to the approach-pipe diameter as shown by the curve in Figure 16.2.

Combining the foregoing relationships, the equation for the flow through the orifice becomes:

$$Q = CA\sqrt{2gh} = 8.025\ CA\sqrt{h} \qquad (16.3)$$

Values for C may be obtained from Figure 16.2, and Equation 16.3 may be used to calculate the pumping rate for any combination of orifice diameter, approach-pipe diameter, and water height in the piezometer tube. The pumping rate, Q, will be in gallons per minute when the orifice area, A, is in square inches and the water level in the piezometer tube, h, is in inches. The value of C from Figure 16.2 is only valid for use with this combination of units. Extensive calibrations of circular orifice weirs by Purdue University (1949) showed that this device will measure the pumping rate within 2 percent of the true value when constructed and used properly.

Besides making the parts accurately and setting up the device correctly in the field, two precautions must be taken to assure good results. The diameter of the orifice should be less than 80 percent of the inside diameter of the pipe that serves as the approach channel. Figure 16.2 shows that the coefficient, C, changes rapidly at the higher values of the ratio. For this reason, the accuracy of the measurement suffers

Figure 16.3. Circular orifice weir and piezometer tube installed properly and in use during field pumping test.

when this ratio exceeds 0.7.

The piezometer tube must be completely free of air bubbles, obstructions, or constrictions when reading the pressure head. Air bubbles can be eliminated by lowering the tube between readings so that water flows from it. Figure 16.3 shows the correct installation of an orifice weir and piezometer tube.

The gate valve used to control the pump discharge should be installed as shown in Figure 16.1. Turbulence caused by the valve will not interfere with the proper functioning of the orifice weir, provided it is installed some distance ahead of the pipe that serves as the approach channel. It is good practice to place this valve at least 10 pipe diameters from the piezometer connection.

To summarize, certain precautions must be taken in constructing orifice weirs to assure accurate results. These include the following:

1. The pipe on which the orifice plate is mounted must be horizontal and the discharge must be unimpeded.

2. The edges of the orifice opening must be sharp and clean, preferably chamfered to 45 degrees and with the sharp edge upstream.

3. Combinations of pipe and orifice diameters must be assembled so that the anticipated head will be at least three times the diameter of the orifice.

4. The orifice must be vertical and centered in the discharge pipe.

5. The piezometer tube must be free of air bubbles and not protrude beyond the inside surface of the pipe.

Weirs and Flumes

Another method used to measure flow from a well is by means of a constriction placed in a discharge channel originating at the well head. In most cases, the drilling contractor can channelize the flow from a pumping well. A calibrated constriction placed in the channel changes the level of the water in or near the constriction. By knowing the dimensions of the constriction, the rate of flow through or over the constriction will be a function of the water level. A simple depth determination near the constriction provides a discharge measurement.

Two types of constricting structures — weirs and flumes — can be used by the contractor (Figure 16.4). Each has its advantages and can provide discharge measurements to an accuracy of plus or minus 10 percent or better. Selection of which to use depends on a number of criteria, such as cost, silting potential, ease of in-

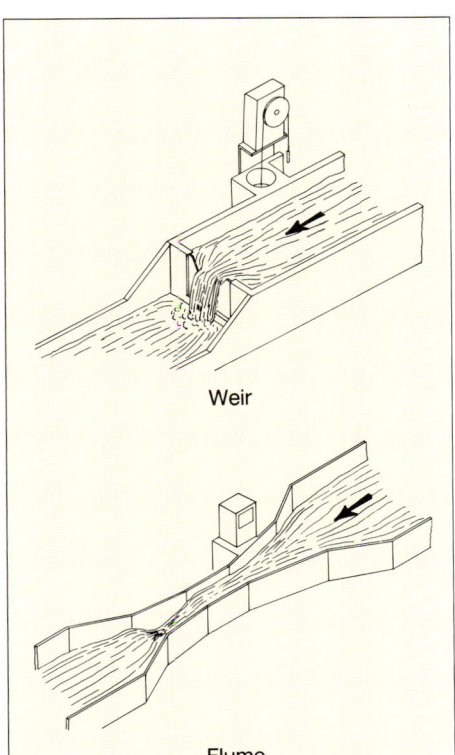

Figure 16.4. Weirs and flumes are the two primary types of constricting structures. *(Instrumentation Specialties Company, 1979)*

stallation, site configuration, flow rates, and accuracy requirements.

A weir is a vertical baffle that restricts the total flow of water in an open or closed channel and is the simplest device used to measure flow. There are two ways of constructing a weir crest — either by using a notch in the weir (Figure 16.5) or by using the weir itself without a notch (Figure 16.6). The crest of a weir is the bottom of the notch or the level to which the water must rise before it spills over.

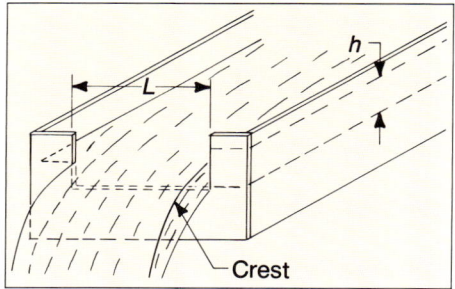

Figure 16.5. Rectangular weir with end contractions.

If a notch is made in the weir, it is then known as a weir with end contractions. When the weir is contracted at the ends and bottom, the ends of the weir should not be closer to the sides of the channel than two times the head on the weir. If the weir itself is used as the crest, then it is known as a weir without end contractions.

Weirs with rectangular or triangular notches are used for measuring flow in open channels. The most commonly used weir is the sharp-crested, rectangular design. The weir is placed flush against the flowing stream, and the notch is made as sharp as possible by using a flat piece of metal with sharp edges forming the weir notch.

The relation between the head and the discharge of a weir varies according to the shape of the weir notch. Different equations are used with different weirs. One of the most common equations used with sharp-crested, rectangular weirs is known as the Francis equation. Francis' equation for a weir without end contractions is:

$$Q = 1495 \cdot L \cdot h \cdot \sqrt{h} \qquad (16.4)$$

where Q is the discharge rate in gpm, L is the length of the crest in ft, and h is the head on the weir in ft. The head on the weir, h, is the difference in elevation between the crest of the weir and the surface of the flowing stream, measured far enough upstream from the weir to eliminate the effect of the increase in velocity as the water spills over the weir. Because small errors in determining h can make a large error in calculating the discharge, it is essential to make the head measurements from 3 to 8 ft (0.9 to 2.1 m) upstream of the crest. The discharge rates for various head conditions based on the Francis equation are presented in Appendix 16.A.

The flow over a rectangular weir with end contractions is less than for an open-ended weir, even though both weirs have the same length and head conditions. Experiments have shown that, for a fully contracted flow (end constriction), the effective length of the crest is reduced one-tenth of the head, $0.1\ h$, at each end of the weir. For example, if a weir is constructed as shown in Figure 16.5 so that

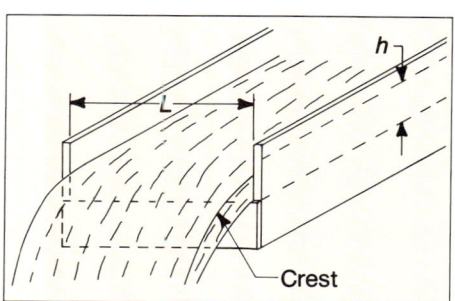

Figure 16.6. Rectangular weir without end contractions.

L is 2 ft, the flow is fully contracted, and h is 1 ft, then the effective length of the outflow is reduced to 1.8 ft. Failure to make this correction when using weirs with end contractions often leads to serious errors in determining discharge.

Advantages of a weir include its low cost, ease of installation, and high accuracy when installed and used properly. To maintain accuracy, weirs must be cleaned to prevent the deposition of sediment upstream of the weir. Variations in the approach velocity can also reduce the accuracy of the flow measurement.

Flumes are the second major device used to measure flow in open channels.

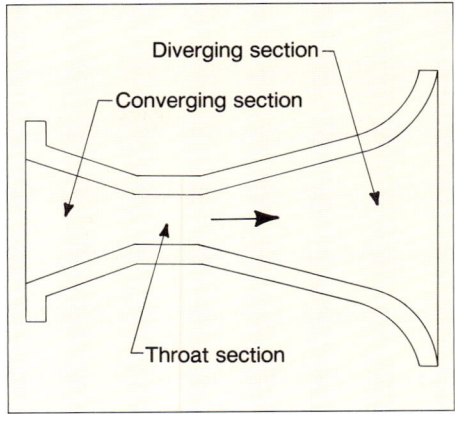

Figure 16.7. General flume configuration. Flumes are suitable for measuring well flow.

A flume is constructed so that a restriction in the channel causes the water to accelerate, producing a corresponding change (drop) in the water level (Figure 16.7). The head can then be related to discharge. Large flumes are used in rivers and streams, but smaller ones are ideal for measuring irrigation flows or discharges from wells. All flumes used for measuring discharges from wells should have the following characteristics (after Grant, 1979):

1. The flume should be located in a straight stretch of the channel (no bends immediately upstream).

2. High approach velocities should be avoided; water should be free of turbulence and waves.

3. Because flow will be restricted in the flume, the channel upstream should have banks high enough to contain the flow.

4. The channel approaching the flume should be regularly shaped so that flow is well distributed in the approach channel.

5. Excessive submergence of the flume throat caused by backwater downstream should be avoided because it reduces the accuracy of discharge measurements.

Several types of flumes have been developed; the most common flume for measuring well discharge is the Parshall flume, originally designed by R. L. Parshall of the U.S. Soil Conservation Service (Parshall, 1950). Construction details for Parshall flumes suitable for measuring water well discharges are presented in Figure 16.8 and Table 16.2. See Appendix 16.B for free-flow discharge for Parshall flumes. They can be constructed of galvanized sheet metal, wood, or fiberglass. Smaller flumes are portable if made of these materials. Concrete is suitable if the flume will be left in place.

Although the Parshall flume operates generally under the same principles as other flume designs, it incorporates a drop in the floor of the flume which produces supercritical flow through the throat of the flume. Under these conditions, the inertial forces in the water are dominant over the gravitational forces, that is, rapid flow is induced through the throat and a hydraulic jump occurs downstream of the throat. The hydraulic jump is caused by the increase in channel width downstream of the restriction and the subsequent slowing of flow. When free-flow conditions occur, the hydraulic jump is not covered by the backwater downstream. Thus, the depth of the backwater is insufficient to reduce the discharge rate. But when the water surface downstream is high enough, and, in effect, "drowns" the hydraulic jump, a backwater effect is created that reduces the discharge through the throat of the flume. This type of flow is called submerged flow. Submerged flow conditions require measuring both the upstream depth (h_a) and the throat depth (h_b) to calculate discharge, whereas, in free flow, only h_a need be measured to determine discharge.

Ideally, the Parshall flume should be designed and constructed to maintain free flow because calculations of flow rate are simplified, compared with submerged flow. Free-flow conditions cease to exist when the h_b/h_a (submergence) ratio exceeds these values (Bureau of Reclamation, 1981):

50 percent for flumes 1 to 3 in (25 to 76 mm) wide

60 percent for flumes 6 to 9 in (152 to 229 mm) wide

70 percent for flumes 1 to 8 ft (0.3 to 2.4 m) wide

A typical set of discharge curves for Parshall flumes operating under free-flow and submerged conditions are presented in Figure 16.9. Note that, beyond the 95-percent submergence mark, Parshall flumes cease to be an effective measuring device because of the difficulty in measuring accurately the small head differential between h_a and h_b (when h_a and h_b are equal, no flow occurs).

Selection of the most suitable size of flume for a particular discharge depends on

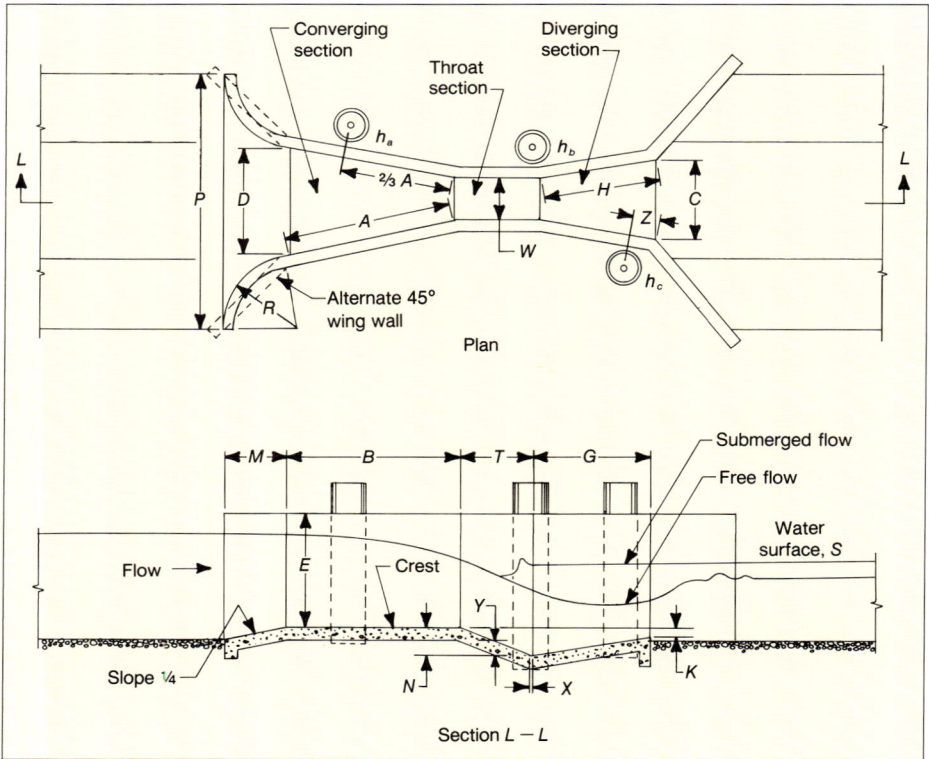

Figure 16.8. Design of a Parshall flume. *(U.S. Soil Conservation Service)*

Table 16.2. Parshall Flume Dimensions

	W		A		2/3A[5]		B		C		D		E		T		G		H		K		M		N		P		R		X		Y		Z		FREE-FLOW CAPACITY MINIMUM	FREE-FLOW CAPACITY MAXIMUM
	FT	IN	FT	IN	FT	IN	FT	IN	FT	IN	FT	IN	FT	IN	FT	IN	FT	IN	FT	IN	FT	IN	FT	IN	FT	IN	FT	IN	FT	IN	FT	IN	FT	IN	FT	IN	CFS	CFS
	0	1[1]	1	2 9/32	0	9 17/32	1	2	0	3 21/32	0	6 19/32	0	6 to 9	0	3	0	8	0	8 1/8	0	3/4	—	—	0	1 1/8		—	—		0	5/16	0	1/2	0	1/8	0.01	0.19
[2]	0	2[1]	1	4 5/16	0	10 7/8	1	4	0	5 5/16	0	8 13/32	0	6 to 10	0	4 1/2	0	10	0	10 1/8	0	7/8		—	0	1 11/16		—	—		0	5/8	0	1	0	1/4	0.02	0.47
	0	3[1]	1	6 3/8	1	1/4	1	6	0	7	0	10 3/16	1 to 1 1/2		0	6	1	0	1	5/32	0	1	—		0	2 1/4		—	—		0	1	0	1 1/2	0	1/2	0.03	1.13
	0	6	2	7/16	1	4 5/16	2	0	1	3 1/2	1	3 5/8	2	0	1	0	2	0	—		0	3	1	0	0	4 1/2	2	11 1/2	1	4	0	2	0	3	—		0.05	3.9
	0	9	2	10 5/8	1	11 1/8	2	10	1	3	1	10 5/8	2	6	1	0	1	6	—		0	3	1	0	0	4 1/2	3	6 1/2	1	4	0	2	0	3	—		0.09	8.9
	1	0	4	6	3	0	4	4 7/8	2	0	2	9 1/4	3	0	2	0	3	0	—		0	3	1	3	0	9	4	10 3/4	1	8	0	2	0	3	—		0.11	16.1
	1	6	4	9	3	2	4	7 7/8	2	6	3	4 3/8	3	0	2	0	3	0	—		0	3	1	3	0	9	5	6	1	8	0	2	0	3	—		0.15	24.6
	2	0	5	0	3	4	4	10 7/8	3	0	3	11 1/2	3	0	2	0	3	0	—		0	3	1	3	0	9	6	1	1	8	0	2	0	3	—		0.42	33.1
[3]	3	0	5	6	3	8	5	4 3/4	4	0	5	1 7/8	3	0	2	0	3	0	—		0	3	1	3	0	9	7	3 1/2	1	8	0	2	0	3	—		0.61	50.4
	4	0	6	0	4	0	5	10 5/8	5	0	6	4 1/4	3	0	2	0	3	0	—		0	3	1	6	0	9	8	10 3/4	2	0	0	2	0	3	—		1.3	67.9
	5	0	6	6	4	4	6	4 1/2	6	0	7	6 5/8	3	0	2	0	3	0	—		0	3	1	6	0	9	10	1 1/8	2	0	0	2	0	3	—		1.6	85.6
	6	0	7	0	4	8	6	10 3/8	7	0	8	9	3	0	2	0	3	0	—		0	3	1	6	0	9	11	3 1/2	2	0	0	2	0	3	—		2.6	103.5
	7	0	7	6	5	0	7	4 1/4	8	0	9	11 3/8	3	0	2	0	3	0	—		0	3	1	6	0	9	12	6	2	0	0	2	0	3	—		3.0	121.4
	8	0	8	0	5	4	7	10 1/8	9	0	11	1 3/4	3	0	2	0	3	0	—		0	3	1	6	0	9	13	8 1/4	2	0	0	2	0	3	—		3.5	139.5
	10	0	—		6	0	14	0	12	0	15	7 1/4	4	0	3	0	6	0	—		0	6	—		1	1 1/2	—		—		1	0	0	9	—		6	200
	12	0	—		6	8	16	0	14	8	18	4 3/4	5	0	3	0	8	0	—		0	6	—		1	1 1/2	—		—		1	0	0	9	—		8	350
	15	0	—		7	8	25	0	18	4	25	0	6	0	4	0	10	0	—		0	9	—		1	6	—		—		1	0	0	9	—		8	600
[4]	20	0	—		9	4	25	0	24	0	30	0	7	0	6	0	12	0	—		1	0	—		2	3	—		—		1	0	0	9	—		10	1000
	25	0	—		11	0	25	0	29	4	35	0	7	0	6	0	13	0	—		1	0	—		2	3	—		—		1	0	0	9	—		15	1200
	30	0	—		12	8	26	0	34	8	40	4 3/4	7	0	6	0	14	0	—		1	0	—		2	3	—		—		1	0	0	9	—		15	1500
	40	0	—		16	0	27	0	45	4	50	9 1/2	7	0	6	0	16	0	—		1	0	—		2	3	—		—		1	0	0	9	—		20	2000
	50	0	—		19	4	27	0	56	8	60	9 1/2	7	0	6	0	20	0	—		1	0	—		2	3	—		—		1	0	0	9	—		25	3000

[1] Tolerance on throat width (w) ± 1/64 inch; tolerance on other dimensions ± 1/32 inch. Sidewalls of throat must be parallel and vertical.
[2] From Colorado State University Technical Bulletin No. 61.
[3] From U.S. Department of Agriculture Soil Conservation Circular No. 843.
[4] From Colorado State University Bulletin No. 426-A.
[5] For flumes 10 to 50 feet, the position of head gage is computed by equation 2/3 $(\frac{W}{2} + 4)$.

(U.S. Soil Conservation Service)

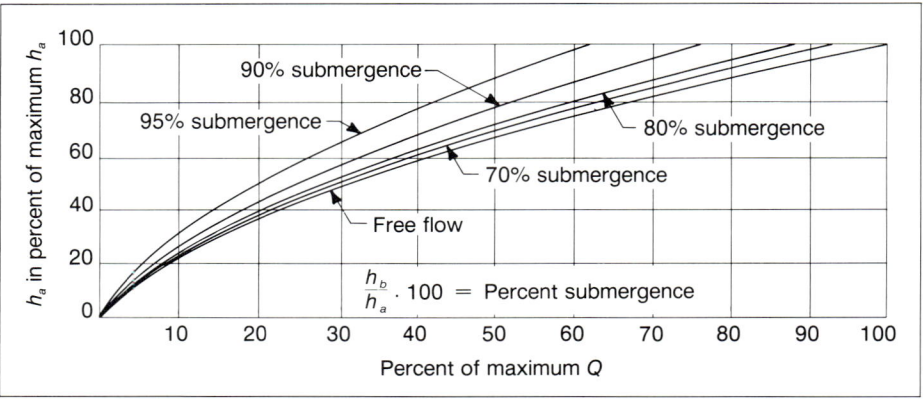

Figure 16.9. Typical discharge curves for Parshall flumes with free-flow and submerged conditions. *(Bureau of Reclamation, 1981)*

range of anticipated discharges, typical upstream channel configurations, and whether the flume is to be portable. See the *Water Measurement Manual*, Bureau of Reclamation (1981), for details on proper selection and installation of Parshall flumes.

A second type of flume has come into use because its construction is simpler than the Parshall flume and it can operate satisfactorily under both free-flow and submerged conditions. This type of flume is called the cutthroat flume because it does not have a throat with parallel walls in its design. It was developed by Skogerboe, Hyatt, Anderson, and Eggleston (1967) at Utah State University. The design of the flume is shown in Figure 16.10. Because the cutthroat flume has a flat floor, it can be placed on a channel bed or inside a concrete-lined channel (Skogerboe et al., 1973). The cutthroat flume is an ideal discharge measuring device for drilling contractors because of its low cost and high accuracy (Figure 16.11). See Appendix 16.C for some typical discharge rates for several sizes of cutthroat flumes that are practical for measuring yields from wells.

Flumes offer several advantages over weirs. The most important of these is the

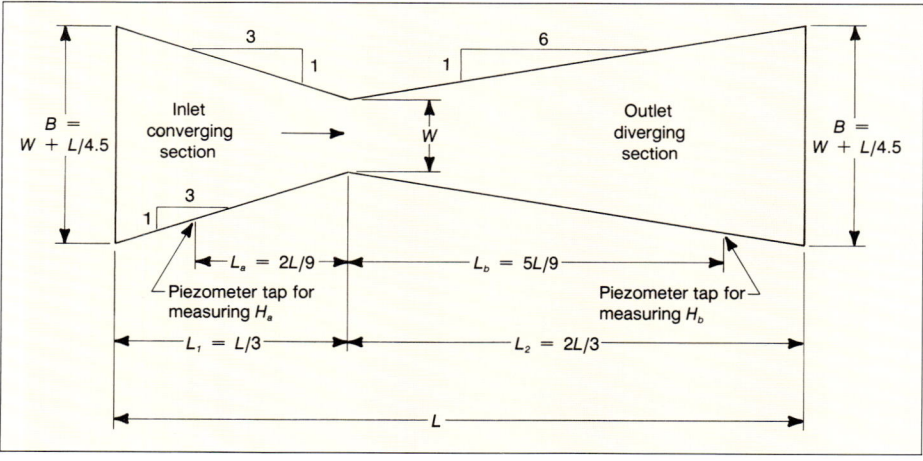

Figure 16.10. Design of a cutthroat flume. *(Skogerboe et al., 1973)*

self-cleaning capacity of flumes compared with sharp-edged weirs. Head losses through a flume are also much less than for a weir, so when the available head is limited, flumes are more desirable. Flumes can function over a wide range of discharges and still require only a single upstream head measurement. On the other hand, flumes require more time to set up and therefore are more expensive to use than the sharp-edged weirs.

Drill-Stem Testing

Drill-stem testing is routinely done in the oil industry to check on the potential yield from a certain formation just after it has been drilled. This type of test is also done by some contractors in the water well industry who have equipped themselves with the necessary tools. A properly run drill-stem test will provide water quality information from the horizon of interest, an estimate of the yield, and an indication of downhole pressures during pumping and periods of no pumping (Earlougher, 1977).

In a drill-stem test, a special tool is attached to the drill string. Packers are installed at one or both ends of the tool so that the intake portion of the tool is isolated from the drilling fluid column in the borehole. The tool is also equipped with pressure sensors. After the tool is lowered to the selected formation, multiple cycles of pumping and non-pumping provide information on pressure, yield, and water quality. Highly sophisticated methods of data analysis are used in the oil industry to evaluate results from drill-stem tests, but these methods are not applicable to drill-stem tests for water wells. Data can simply be used "as is" because boreholes are considerably shallower.

MEASURING DRAWDOWN IN WELLS

Observation Wells

Drawdown data can be taken from both the pumping well and appropriately placed observation wells, but the accuracy of data taken from the pumping well is usually

Figure 16.11. A small cutthroat flume is suitable for measuring discharges from water wells. The throat of this flume can be adjusted from 1 to 8 in (25.4 to 203 mm) to measure flows of 2.5 to 1,000 gpm (13.6 to 5,450 m³/day). (*Baski Water Instruments, Inc.*)

less reliable because of turbulence created by the pump. Thus, at least one observation well should be used when practicable. Furthermore, drawdown data from an observation well are required to calculate the storage coefficient accurately, whereas transmissivity values may be calculated on the basis of drawdown data taken from either a pumping well or observation well.

Observation wells should be just large enough to allow accurate and rapid measurement of the water levels. Small-diameter wells are best, because the volume of water contained in a large-diameter observation well may cause a time lag in drawdown changes. Most observation wells are constructed with screens 3 to 6 ft (0.9 to 1.8 m) long. Longer screens may be desirable, depending on the degree of stratification, but are not absolutely necessary.

When observation wells are too close to the pumped well, the drawdown readings may be affected by the stratification of the aquifer. Stratification distorts the distribution of hydraulic head and drawdown in the vicinity of the pumped well during the aquifer test. Recall that the vertical hydraulic conductivity of a stratified formation is usually much less than its horizontal hydraulic conductivity. This means that changes in head caused by pumping occur more slowly in the vertical direction than in the horizontal direction. At any moment after test pumping is started, the drawdown in a series of observation wells placed at a given distance from the pumped well may vary if the screens are set at different depths within the aquifer. These variations in drawdown become less as time of pumping increases. The distorted pattern of drawdown caused by stratification is eliminated at distances equal to three to five times the aquifer thickness.

For unconfined aquifers, observation wells should be placed no farther than 100 to 300 ft (30.5 to 91 m) from the pumped well. For thick confined aquifers that are considerably stratified, observation wells should be placed within 300 to 700 ft (91 to 214 m) of the pumped well. Locating the wells too far away is not good practice because the pumping test must be continued for a longer time to produce sufficient drawdowns at the most distant points, and small measurement errors may be a significant percentage of the total drawdown in the observation well. In addition, boundary effects may not be noticed.

Screens for observation wells should be installed at about the same depth as the central portion of the screen in the production well. This is especially important if a short screen is used in the pumping well, because the distribution of drawdown is more distorted. If this procedure is followed, the reduction in pressure or water level at the observation well will usually occur within moments of its occurrence in the pumping well (assuming the observation well is spaced at the correct distance from the pumped well). Occasionally, observation wells are terminated in strata above or below the one tapped by the pumped well to see if there is any hydraulic interconnection between the formations. Naturally the response of these observation wells to pumping may be delayed significantly, depending on the degree of hydraulic connection.

The appropriate number of observation wells depends upon the amount of information desired and upon the funds available for the test program. The data obtained by measuring the drawdown at a single location outside the pumped well permit calculation of the average hydraulic conductivity, transmissivity, and storage coefficient of the aquifer. If two or more observation wells are placed at different distances, the test data can be analyzed by studying both the time-drawdown and the distance-drawdown relationships. Using both these analytical methods provides greater assurance that the calculated transmissivity and storage coefficient values are correct. It is usually advantageous to have as many observation wells as conditions allow because the hydraulic conductivity may vary in one or more directions away from the pumping well. Observation wells placed in a circle around the pumping well will reveal this trend.

Before starting the pumping test, a complete program for depth-to-water measurements must be laid out in advance. It is not necessary to make the measurements in all the wells simultaneously. The watches used for timing the measurements, however, should be synchronized so that the time of each reading can be referenced to the exact minute and hour that pumping is started.

Using measurement devices that will give quick and accurate results, drawdown should be measured in the pumping well and all observation wells. Where turbulence is a problem in a pumping well, the best depth indicators are those that provide light or noise signals when probes are immersed in water, although air lines will also provide the necessary accuracy.

When electrical devices are used, a light or ammeter indicates a closed circuit when the probe touches the water (Figure 16.12). Flashlight batteries supply the current. To improve the accuracy of readings, the probe and cable should be left hanging in the well for a series of readings. This eliminates any errors from kinks or bends in the wires which may change the length slightly when the device is pulled up and let down. The change in water level should be measured along the cable with a steel tape, using one of the depth markers as a reference mark. Depth markers are commonly attached to the cable by the manufacturer at about 5-ft (1.5-m) intervals.

Steel tapes will provide accurate results in observation wells for depths to 100 ft (30.5 m), but the number of measurements that can be taken over short pe-

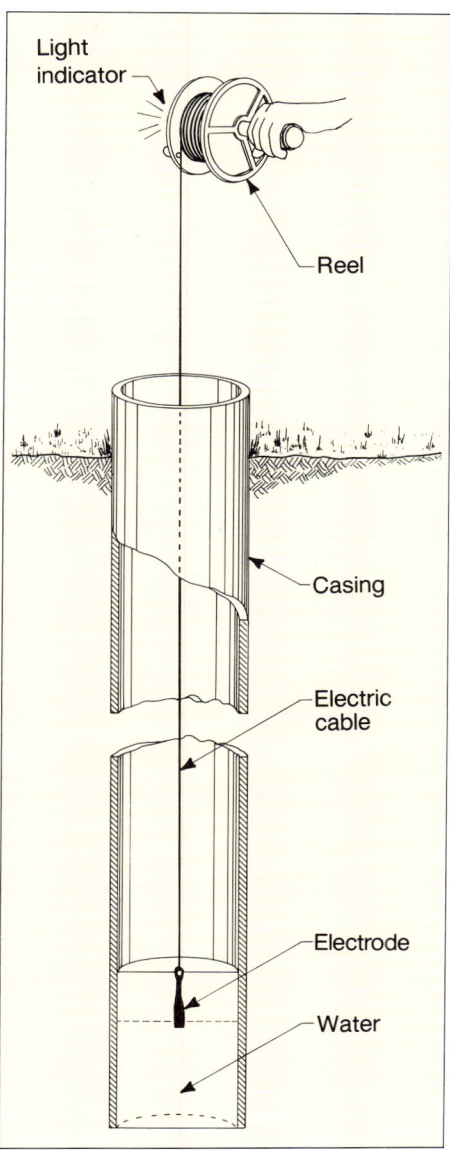

Figure 16.12. Electric sounder for measuring depth to water consists of electrode, two-wire cable, and a light which indicates a closed circuit when electrode touches water.

riods is limited. This can be a problem during the first 10 minutes of a pumping test, when as many measurements as possible are required. When using a steel tape, a lead weight is attached to the bottom. The lower 2 to 3 ft (0.6 to 0.9 m) of the tape is wiped dry and coated with carpenter's chalk before making a measurement. The tape is let down into the well until part of the chalked section is below water and one of the foot marks is held exactly at the top of the casing or at some other measuring point that may have been selected. After withdrawal, the wetted line on the tape can be read to a fraction of an inch on the chalked section. This reading is subtracted from the foot mark held at the measuring point; the difference is the actual depth to the water level.

A disadvantage of this method is that the approximate depth to water must be known so that a portion of the chalked section will be submerged each time to produce a wetted line. The accuracy of this method, however, exceeds other measurement devices, with the exception of some pressure (head) probes as described below.

The gauges ordinarily used to measure the depth of water in a well are:
1. Pressure gauge — reading in pounds per square inch (psi).
2. Altitude gauge — reading in feet and fractions of a foot.
3. Vacuum gauge — reading in inches of vacuum, or difference in pressure between mercury under atmospheric pressure and water being pumped.

Well drillers should become familiar with these gauges and their applications. They are comparatively simple and easy to use, but for important tests they should be calibrated with master gauges. Readings will be more accurate if the range of the gauge only slightly exceeds the range of the anticipated drawdown values. Pressure gauges are most often used by drilling contractors; therefore, only these types of gauges are discussed below.

In English Engineering Units, the depth of a water column is measured in feet or pounds per square inch (psi). For example, when using a pressure gauge, the readings are in psi and must be multiplied by 2.31 to convert to feet of water. To convert a reading in feet to one in psi, the reading is multiplied by 0.433. Table 16.3 contains conversions for the typical types of readings taken during a test.

Figure 16.13 shows the installation of an air line in a well for the purpose of determining the depth to water. The device works on the principle that the air pressure required to push all the water out of the submerged portion of the tube equals the water pressure of a water column of that height. If this pressure is expressed in feet

Table 16.3. Conversions for Typical Readings Taken During a Pumping Test

Unit	psi	ft of water	m of water	inches of mercury	atmospheres
1 psi	1.0	2.31	0.704	2.04	0.0681
1 ft of water	0.433	1.0	0.305	0.882	0.02947
1 m of water	1.421	3.28	1.0	2.89	0.0967
1 in of mercury	0.491	1.134	0.3456	1.0	0.0334
1 atmosphere (sea level)	14.7	33.93	10.34	29.92	1.0

of water, the depth to water can be calculated. The air line consists of a small-diameter plastic (PVC) pipe or tube of sufficient length to extend from the top of the well to a point several feet below the lowest anticipated water level to be reached during the test. Quarter-inch (6.4-mm) copper or brass tubing is occasionally used for the air line. The exact length of the air line must be measured as it is placed in the well. If flexible tubing is used, steps must be taken to assure that the tubing hangs vertically and does not spiral inside the well casing. The air line and connections at the ground surface must be completely air tight. Before starting the pumping test, the line is prepressurized and the gauge pressure recorded. During pumping, the pressure in the line is reduced; this pressure drop can be directly related to feet or meters of water-level fall. The air-line method is generally less accurate than electrical or acoustic sounders.

To make a measurement, the depth from the top of the well casing or from some other reference point to the lower end of the air line must first be determined. After the pressure gauge is connected, air is pumped into the air line. The upper end of the air line is fitted with a valve so that a small compressor or an ordinary tire pump can be used to pump air into the tube. A pressure gauge is connected to a tee in the line to measure the air pressure in the tube. A gauge that measures pressure in feet of water is better than one with a scale reading in psi. During pressurizing, pressure increases until all the water has been forced out of the air line. At this point, the pressure ceases to build and the pressure in the tube balances the water pressure. The stabilized gauge reading shows the pressure necessary to support a column of water equal to the distance from the bottom of the tube to the pumping water level. If the gauge reads in feet of head, it then indicates the submerged length of the air line in feet. Subtracting the submerged length of the air line from the total length gives the depth to water below the measuring point (Figure 16.13). The static water level should be measured before pumping begins. The depth to water is given by the following equation:

$$d = L - h \qquad (16.5)$$

where d is depth to water in ft, L is depth to bottom of air line in ft, and h is pressure head in ft, represented by a column of water equal to the length of the submerged portion of the air line.

For example, suppose the distance from the top of the well casing to the lower end of the air line is 95 ft. As air is

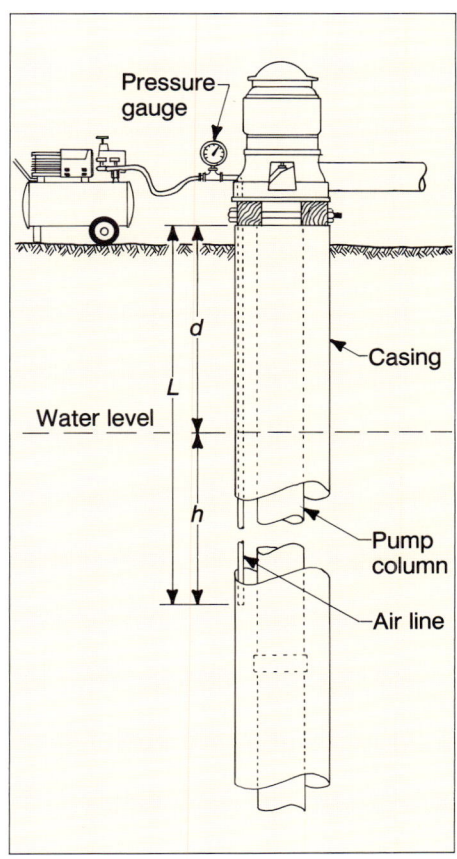

Figure 16.13. Installation of an air-line method for measuring water levels.

pumped into the line, assume that a maximum reading of 46 ft is reached on the gauge. The depth to water is then the difference between 95 and 46, or 49 ft. This is the static water level. After the pump is started, the water level in the well drops, the submerged length of the air line decreases, and the pressure indication on the gauge drops accordingly. A gauge reading of 34 ft, for instance, would mean that the submerged length of the air line has decreased by 12 ft (46 − 34) and the depth to water has changed to 95 minus 34, or 61 ft. This indicates a drawdown of 12 ft below the static water level. If the gauge reads in psi, each reading must be multiplied by 2.31 to convert it to feet of water. For example, a reading of 15 psi corresponds to a pressure head of 34.6 ft of water.

The dependability of measurements made by the air-line device varies with the accuracy of the pressure gauge and the care used in determining the initial pressure reading. Depth to water can usually be determined within 0.2 ft (0.06 m) of the exact value. The air-line method is generally not accurate enough for use in observation wells during an aquifer test, but it is the most practical means for measuring water levels in a pumped well. To avoid disturbances from turbulence near the intake of the pump, the lower end of the air line should be at least 5 ft (1.5 m) above or below the point where water enters the pump.

Because measuring depths during pumping tests is labor intensive, more sophisticated electronic depth indicators have been developed. These instruments record electronic signals from pressure-sensitive transducers placed in the well (Mogg, 1977). Transducers measure feet of water (hydraulic head) above the transducer. The amount of head over the transducer is automatically printed out for any selected time after pumping commences; the time and feet of head can be printed on a continuous tape. This type of electric drawdown recorder has also been coupled with a stand-alone microcomputer (Scherer et al., 1979 and Driscoll et al., 1980b). This system is particularly advantageous for well fields where extensive testing is required.

In automated drawdown measuring systems, transducers are placed in each well and electrically connected to analyzer, computer, and printer-plotter equipment. Data are recorded on paper and also displayed on the cathode-ray tube (CRT). The system can be programmed to do a large number of functions, including automatic data correction for barometric changes during pumping, graphing data in time-drawdown or distance-drawdown plots on the printer-plotter, and calculating transmissivity, storativity, specific capacity, and well efficiency. Thus, the depth analyzer coupled with a computer results in the ability to measure water depths, record data, retrieve data when required, and process the data into useful information. It is the most powerful tool available to measure and analyze pumping test data. It cannot, however, sense anomalies in the data.

Recommended Time Intervals for Measuring Drawdown During a Constant-Rate Pumping Test

All watches of observers should be synchronized before the test begins; times should be recorded to the nearest 10 seconds. Water-level measurements for the pumped well should be recorded at the times suggested in Table 16.4. Of course, drawdown in wells more distant from the pumping well will not occur immediately. Drawdown readings in the observation wells should be taken at the intervals recommended in Table 16.5.

Table 16.4. Recommended Time Intervals for Measuring Drawdown in the Pumped Well During a Pumping Test

Time Since Pumping Started (or Stopped) in minutes	Time Intervals Between Measurement in minutes
0 - 10	0.5 - 1
10 - 15	1
15 - 60	5
60 - 300	30
300 - 1440	60
1440 - termination of test	480 (8 hr)

As suggested in Chapter 9, early test data are extremely important, and as much information as possible must be obtained in the first 10 minutes of pumping for every observation well. The reason for this is that, as the cone of depression moves outward from the well, it may encounter inhomogeneities in the ground which cause either an acceleration or a deceleration of drawdown with increasing time. A form used to record data is in the pocket of this book. Any unusual event (stoppage of pump, onset of weather change, or passage of a train) should be noted on the form, along with the time it occurred.

Ideally, pumping tests should be continued until equilibrium is reached, that is, until the cone of depression stabilizes. In practice this is rarely possible. In confined aquifers, the cone of depression spreads rapidly because no actual dewatering takes place; only a pressure reduction is occurring outward from the well. Thus, 24 hours is usually sufficient to record enough reliable data for confined aquifers. To gain enough information for unconfined aquifers, 72 hours are usually required to dewater the materials within the cone of depression, because of the slow downward percolation of water in many stratified deposits. This time can be reduced if equilibrium conditions are established before 72 hours have elapsed. In no event should pumping

Table 16.5. Recommended Time Intervals for Measuring Drawdown in the Observation Well(s) During a Pumping Test

Time Since Pumping Started (or Stopped) in minutes	Time Intervals Between Measurements in minutes
0 - 60	2
60 - 120	5
120 - 240	10
240 - 360	30
360 - 1440	60
1440 - termination of test	480 (8 hr)

tests be terminated prematurely, however, because the limited data collected may not reveal the true nature of the aquifer.

It is a good idea to plot preliminary drawdown data during the course of the pumping test. Anomalies in the data will become apparent, and the necessary adjustments can be made so that the remaining data will be more useful. Plotting the data will also indicate when equilibrium conditions have been reached for shallow aquifers or for aquifers with high hydraulic conductivities. In this case, the pumping portion of the test can be shortened with no loss of sensitive data.

Barometric or tidal changes can influence drawdown data. For example, a barometric pressure change of 1 in (25.4 mm) of mercury can result in a rise or fall of up to 1 ft (0.3 m) in the potentiometric surface for confined aquifers that have high barometric efficiency. Barometric efficiency refers to the aquifer's ability to transmit changes in atmospheric pressure. Record the nature and time of any weather changes on the drawdown data sheet. Unusually high or low oceanic tides can also affect drawdown data in wells near coastlines.

Recovery Data

Whenever possible, recovery data should be taken to verify the accuracy of pumping data. Often, the recovery data will be more reliable because no pumping is required and any previously inexperienced personnel will have learned proper measurement techniques by the time recovery data can be taken. Recovery measurements should be recorded with the same frequency as those taken during the pumping portion of the aquifer test.

WELL EFFICIENCY

In Chapter 9, the term "well efficiency" was defined and the two major factors influencing efficiency were discussed. For many years prior to the escalation of energy costs in the 1970's, well efficiency was not recognized as a serious well construction objective. Only those contractors who had to deliver the maximum volume of water from limited aquifers were interested.

In recent years, the cost of pumping water has risen precipitately, not only because energy costs have increased, but also because declining water tables in major aquifers have added to lift requirements. At 1984 energy costs, and assuming average efficiency for a pump running 33 percent of the time, the extra fuel expense for lifting 1,000 gpm (5,450 m³/day) an additional 10 ft (3 m) amounts to approximately $1,000 per year. Thus, reducing drawdown should be a major consideration when designing and constructing all high-capacity wells.

The amount of drawdown required to produce a particular yield is determined by the hydraulic nature of the aquifer and the care with which the well was designed, constructed, and developed. Drawdown caused by friction losses in the aquifer as water flows to a well is unavoidable. But substantial head losses sustained as water passes through the disturbed zone around the borehole are avoidable. They are caused by drilling fluid left in the formation, damage to the formation caused by drilling, the presence of a poorly designed filter pack, or use of a well screen with limited open area. Good design practices and enlightened drilling methods can insure that head losses through the zone near the well bore will be minimal. Screens with maximum inlet area, surrounded by a suitable filter pack, and in turn surrounded by a formation developed properly to remove drilling fluid and fine material are necessary for minimizing head losses.

Mogg (1968) defined well efficiency as the ratio of the actual specific capacity at the designed well yield after 24 hours of continuous pumping, to the maximum specific capacity possible, calculated from formation characteristics and well geometry. In this method of defining efficiency, it is possible to identify how much of the total head loss is attributable to natural losses in the formation and those caused by well construction damage to the aquifer and by installation of a screen and filter pack.

The procedure for calculating well efficiency is as follows:
1. Graph the time-drawdown data.
2. Calculate Δs.
3. At a particular time, note the drawdown in an observation well.
4. On a distance-drawdown graph, plot the drawdown for the observation well (for the particular time) at the proper distance from the pumping well.
5. Complete the drawdown curve by using a slope of 2 times Δs (in the Jacob equation, $\log r^2 = 2 \log r$, so the value of Δs in the distance-drawdown graph is twice the value of Δs in the time-drawdown graph).
6. Extend the slope of the data to the radius of the well.
7. In a 100-percent efficient well, drawdown just outside the borehole should equal the drawdown inside the well. It is more likely, however, that the water level inside the well is lower. Therefore, efficiency equals drawdown outside the borehole divided by drawdown inside the casing, times 100.
8. An efficiency of 70 to 80 percent is usually obtainable if good design, construction, and development practices are followed.

An example of the efficiency calculation is shown in Figure 16.14.

STEP-DRAWDOWN TESTS

All conventional well hydraulics theory is based on the assumption that laminar flow conditions exist in the aquifer during pumping. If the flow is laminar, drawdown

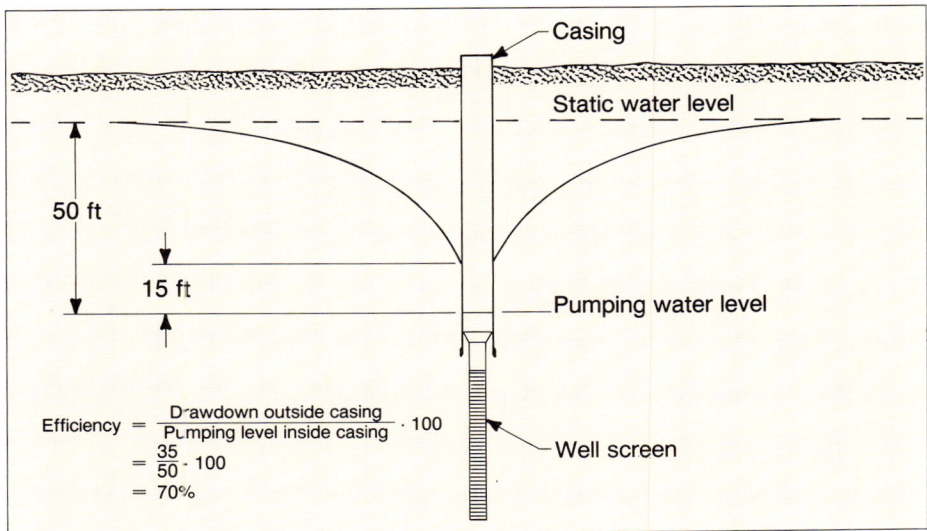

Figure 16.14. Calculating true well efficiency for a high-capacity well.

is directly proportional to the pumping rate. Turbulent flow occurs in some wells, however, when they are pumped at a sufficiently high rate. Under turbulent conditions, the linear relationship between drawdown and pumping rate no longer holds, and part of the drawdown is generally related to the pumping rate raised to some power greater than 1.

When turbulent flow occurs, the specific capacity will decline, often dramatically, as the discharge rate is increased. When this happens, it is useful to have a means of computing the turbulent and laminar drawdown components in order to make proper judgements concerning the optimum pumping rate and pump-setting depth.

The step-drawdown test has been developed to examine the performance of wells having turbulent flow (Jacob, 1946b). In a step-drawdown test, the well is pumped at several successively higher pumping rates and the drawdown for each rate, or step, is recorded. The entire test is usually conducted during one day, and calculations are simplified if all the pumping times are the same for each discharge rate. If time permits, the water level should be allowed to recover to the static level between each step. Usually five to eight pumping steps are used, each lasting 1 to 2 hours. The data from a step test can be used to determine the relative proportion of laminar and turbulent flow occurring at any pumping rate.

Recall from Chapter 9 that, for laminar flow conditions in a perfectly efficient well, drawdown in a confined aquifer can be expressed as follows:

$$s = \frac{264Q}{T} \log \left(\frac{0.3Tt}{r^2 S}\right) \qquad (16.6)$$

This equation is also applicable to unconfined aquifers as long as the drawdown is small in relation to aquifer thickness.

Equation 16.6 can be shortened to:

$$s = BQ \qquad (16.7)$$

where

$$B = \frac{264}{T} \log\left(\frac{0.3Tt}{r^2 S}\right)$$

For a specific well, the value of B is time dependent. However, B changes only slightly after a reasonable pumping duration and can thus be assumed to be a constant. When turbulent flow exists, Jacob suggests that the drawdown in a well can be more accurately expressed as the sum of a first-order (laminar) component and a second-order (turbulent) component:

$$s = BQ + CQ^2 \qquad (16.8)$$

In this equation, Jacob called the laminar term (BQ) the aquifer loss and the turbulent term (CQ^2) the well loss (head loss attributable to inefficiency). Analyses of real wells, however, have shown that this correlation is not correct, because the BQ term almost always includes a major portion of the well losses and the CQ^2 term occasionally includes some aquifer loss. For this reason, computing well efficiency from a step-drawdown test results in an erroneous value. The step test is still useful, however, in evaluating the magnitude of turbulent head loss for the purpose of determining optimum pumping rates.

Using Equation 16.8, Bierschenk (1964) presented a simple graphical method for determining B and C. Dividing Equation 16.8 by Q and rearranging terms yields:

$$\frac{s}{Q} = CQ + B \qquad (16.9)$$

Note that this is a linear equation in s/Q and Q. That is, if s/Q is plotted against Q, the resultant graph is a straight line with slope C and intercept B. Thus, B and C in Equation 16.8 can be calculated from this graph.

Inverting the terms in Equation 16.9 shows how specific capacity declines as discharge increases (only with turbulent flow present):

$$\frac{Q}{s} = \frac{1}{CQ + B} \qquad (16.10)$$

Observing the change in drawdown and specific capacity with increased discharge provides information required to select optimum pumping rates.

A parameter often computed from a step-drawdown test is the ratio of the laminar head loss to the total head loss, expressed as a percentage:

$$L_p = \frac{BQ}{BQ + CQ^2} \cdot 100 \qquad (16.11)$$

Thus, L_p is the percentage of the total head loss that is attributable to laminar flow.

If the assumptions made by Jacob were correct, that is, that aquifer loss equals BQ and well loss equals CQ^2, then L_p would equal the well efficiency. However, testing of hundreds of wells has shown that these assumptions are not correct. Therefore, L_p is not indicative of well efficiency. For example, while turbulent flow may occur near the well and in the well screen, it may also exist in the undisturbed formation around the well. When this happens, a portion of the CQ^2 term actually comes from aquifer loss. Thus, if L_p is used as the well efficiency, a well having turbulent flow may be judged to be inefficient when it may be, in fact, quite efficient. Conversely, in most wells a substantial portion of the well loss is can be attributed to laminar flow rather than turbulent flow. Under these circumstances, part of the BQ term includes well losses rather than only aquifer losses. Thus, when L_p is used as the efficiency value, it appears that a well which has little or no turbulent flow is judged to be efficient, when the true efficiency may be quite low.

An example of a step-drawdown analysis will illustrate these points. Table 16.6 shows discharge and drawdown data from a typical step-drawdown test for a confined aquifer. Using Bierschenk's method of analysis, Figure 16.15 shows s/Q plotted against Q with B and C calculated as 0.0225 and 3.68×10^{-6}, respectively, where C is the slope of the straight-line plot and B is the intercept.

Table 16.6. Discharge and Drawdown Data from Typical Step-Drawdown Test

Yield		Drawdown		s/Q
gpm	m³/day	ft	m	
514	2,801	13.0	4.0	0.0253
1,066	5,810	27.0	8.2	0.0253
1,636	8,916	43.4	13.2	0.0265
1,885	10,273	61.5	18.8	0.0326
2,480	13,516	82.5	25.2	0.0333
3,066	16,710	101.5	30.9	0.0331
3,520	19,184	120.5	36.7	0.0342

Using Equation 16.10, the specific capacity can be computed for any flow rate:
$$\frac{Q}{s} = \frac{1}{3.68 \times 10^{-6}Q + 0.0225}$$

Thus, specific capacity and drawdown can be projected for any discharge rate.

Equation 16.11 can be used to calculate L_p. If a discharge of 2,700 gpm is assumed, then:

$$L_p = \frac{0.0225 \cdot 2700}{0.0225 \cdot 2700 + 3.68 \times 10^{-6} \cdot (2700)^2} \cdot 100$$
$$= 69 \text{ percent}$$

This means that 69 percent of the head loss is attributable to laminar flow. It does not mean that the well efficiency is 69 percent.

A constant-rate pumping test was conducted in order to determine the true well efficiency. Figure 16.16 shows the time-drawdown plot for a discharge of 2,700 gpm. The early slope on the time-drawdown graph shows the effects of casing storage. The secondary slope is used to calculate the transmissivity of 158,400 gpd/ft. Other wells and tests in this same aquifer have verified that the transmissivity value is at least this high. Using the transmissivity value, the theoretical specific capacity can be estimated (see Appendix 16.D):

$$\left(\frac{Q}{s}\right)_{theoretical} = \frac{T}{2000} = 79.2 \text{ gpm/ft of drawdown}$$

The actual 1-day specific capacity is:

$$\left(\frac{Q}{s}\right)_{actual} = \frac{2700}{96} = 28.1 \text{ gpm/ft of drawdown}$$

The true well efficiency is then:

$$E = \frac{28.1}{79.2} = 35\%$$

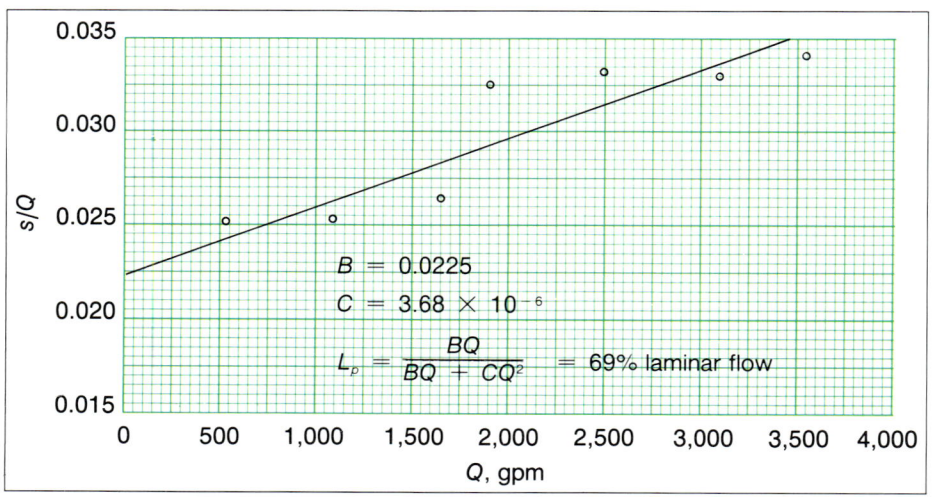

Figure 16.15. Values for *B* and *C* in the step-drawdown equation can be determined from a graph where s/Q is plotted against Q.

This figure is far less than the 69-percent value given for L_p. This means that much of the head loss caused by inefficiency is associated with laminar flow rather than with turbulent flow. Had L_p been used for the well efficiency, the well would have been judged to have a reasonably good efficiency value of 69 percent, whereas the true well efficiency is quite poor.

In summary, the step-drawdown test can be used to determine the following:

1. The specific capacity of the well at various discharge rates. This information can be used to select optimum discharge rates. The length of time the test is run, however, may not be sufficient to encounter boundary effects. Therefore, the actual long-term specific capacity at a certain discharge may be different than originally calculated from the step-drawdown data. Slow drainage can also distort the discharge data in highly stratified aquifers. Depending on the exact nature of the aquifer, the specific capacity may seem to improve with higher discharges and longer pumping time — a highly unlikely situation that will occur rarely, if ever, in natural geologic materials.

2. A ratio, L_p, denoting the percentage of the total head loss attributable to laminar flow, which is not to be confused with the true hydraulic efficiency of the well.

3. Transmissivity and storage coefficient values for the aquifer from time-drawdown and distance-drawdown graphs plotted from data for one of the constant-rate steps of the step test.

PROBLEMS OF PUMPING TEST ANALYSIS

Constant-rate and step-drawdown pumping tests offer the most powerful method for analyzing the hydrogeologic character of an aquifer. Measurements made during

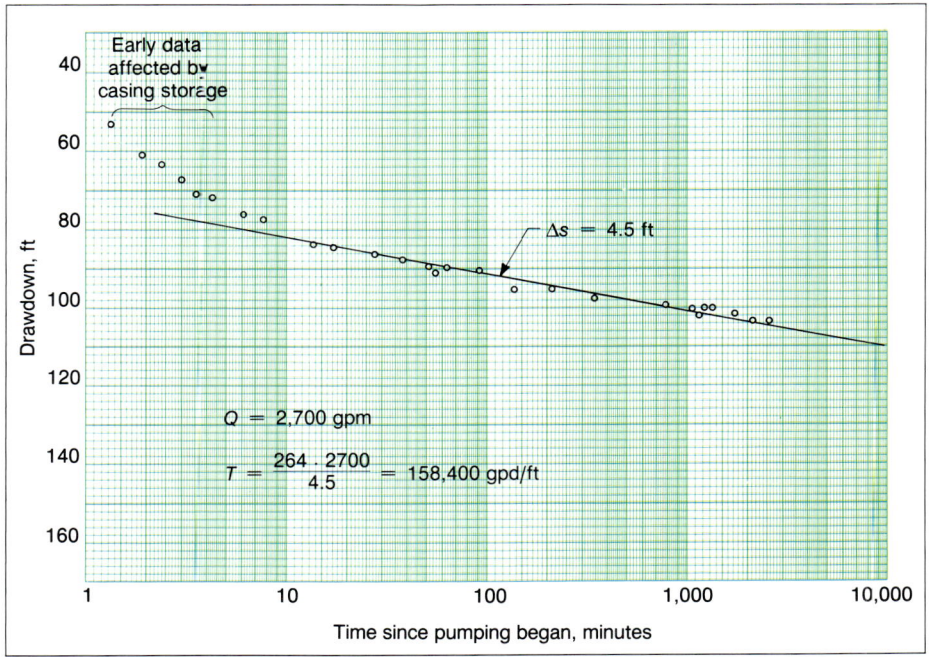

Figure 16.16. Time-drawdown plot for a well discharging 2,700 gpm (14,700 m³/day).

an aquifer test provide values for some of the terms in the Theis equation, permitting calculation of the transmissivity and storage capacity of the aquifer. (See Chapter 9 for a full description of plotting procedures.) The hydraulic conductivity values determined from pumping test data are far more accurate than hydraulic conductivities calculated on the basis of field samples tested in a laboratory, because the samples are rarely representative of the undisturbed formation. The aquifer test measures the performance of the aquifer in its natural state.

In addition to determining hydraulic conductivity, transmissivity, and coefficient of storage, pumping test data can be used to determine any interference between wells at various spacings and rates of pumping other than those employed in the test itself. Under some conditions, pumping test data will also indicate what drawdown may be expected from long-term pumping at different discharge rates, the existence of impervious boundaries that limit the extent of the aquifer, and the existence of recharge sources which may not be apparent otherwise.

In many cases, pumping test data can be interpreted in more than one way. Even in the best controlled test, the data may be confusing unless all hydraulic and geologic factors are taken into account. The analyst must be able to identify unreliable data so that the calculated values for transmissivity and storage coefficient will correctly predict aquifer performance.

The pumping test analyst can be faced with a large number of anomalies in the data, but in the discussions below it is assumed that the following physical conditions for a proper test have been met:

1. The diameter of the pumped well is large enough to accommodate the test pump and provide enough clearance so that the water level can be measured with an air line, electric or acoustic probe, or other device.

2. The test pump can deliver at least 10 percent of the well's maximum yield.

3. In unconsolidated formations, the well screen is at least one-third the thickness of the aquifer, except in extremely thin strata where three-fourths of the formation should be screened.

Although either a constant-rate or a step-drawdown test could have been conducted in the instances discussed below, most of the actual pumping tests are constant-rate tests because the values obtained for transmissivity, storage coefficient, and efficiency are generally more accurate, and different trends in the data are more readily discernible.

The pumping test examples cited below will serve to illustrate some of the major problems that can occur in data interpretation. It is assumed that the reader has thoroughly understood constant-rate pumping test analysis as discussed in Chapter 9. The analyst must also keep in mind several other factors during data interpretation. These include the effect of

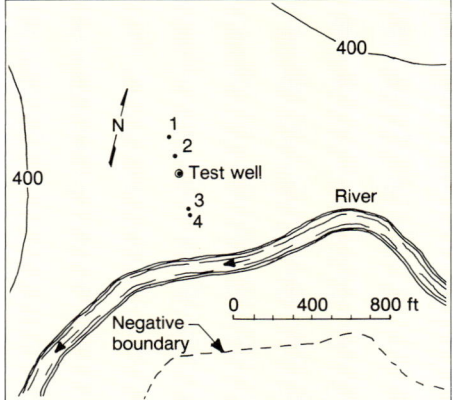

Figure 16.17. Arrangement of 12-in (305-mm) test well and four observation wells for a 72-hour pumping test. This test was made to get sufficient information for designing a well field for a total supply of 2,000 gpm (10,900 m³/day), with wells sized and spaced for best economy and minimum interference.

COLLECTION AND ANALYSIS OF PUMPING TEST DATA   561

the hydrogeologic character of the aquifer on the data, errors in data acquisition, potential omission of important environmental impacts, well design and construction practices, pumping rates, and the theoretical limits of the Jacob theory.

Example 1 — Test Conducted Near a River

In this case, the floodplain of a small river was being considered as the site for an industrial plant (Figure 16.17). A generalized cross section of the two aquifers in this area, and the positions of the various observation wells, are shown in Figure 16.18.

For the aquifer test, a 12-in (305-mm) well was drilled about 400 ft (122 m) from the bank of the river. Both aquifers were penetrated — an unconfined aquifer extending to a depth of about 46 ft (14 m), and a deeper confined aquifer extending from 50 to 105 ft (15.2 to 32 m). A 4-ft (1.2-m) clay layer, from 46 to 50 ft (14 to 15.2 m), separated the aquifers. It was assumed that the 4-ft clay layer effectively isolated the two aquifers. The static water level, however, in both aquifers was 5 ft (1.5 m) below ground level. The arrangement of observation wells for this pumping test illustrates some of the points discussed earlier in this chapter.

Three 2-in (51-mm) observation wells about 90 ft (27.4 m) deep were put down at distances of 100 ft (30.5 m) and 200 ft (61 m) from the 12-in (305-mm) well. Two of these wells were placed away from the river compared with the location of the test well, whereas one well was placed between the river and the test well. A shallow observation well, No. 4, was installed near observation well No. 3 for measuring any water-level changes that might occur in the upper aquifer. This was done primarily to check for any hydraulic connection between the two aquifers, because the continuity of the clay separation was unknown. Any water-level change in the shallow well caused by the pumping of the test well would indicate discontinuity in the clay layer.

The pumping test was conducted for 72 hours. During the early stages of the pumping test, a heavy rainstorm occurred in the upper reaches of the watershed, causing the river level to rise suddenly about 2 ft (0.6 m). No other unusual event took place

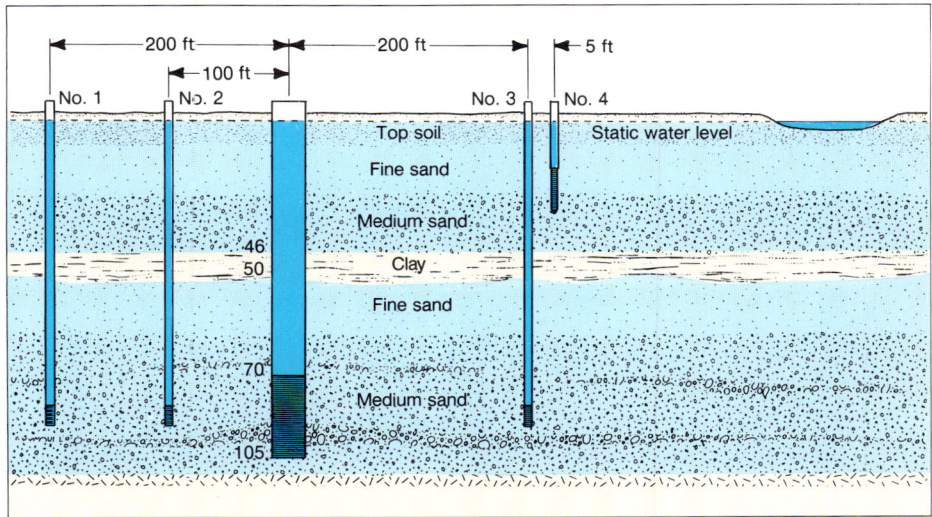

Figure 16.18. Geologic section on a line through test well and observation wells shown in Figure 16.17. The clay layer apparently separates the upper unconfined aquifer from the confined aquifer.

during the remainder of the pumping test.

The consulting engineer was asked to determine transmissivity and storage coefficient values for the aquifer. The consultant decided to use the drawdown data from observation well No. 3 to calculate the transmissivity. Evaluation of the data by the Jacob method indicated that five different slopes were evident in the time-drawdown graph (Figure 16.19). Because the engineer was required, under terms of the contract, to guarantee a certain pumping well efficiency, selection of the correct transmissivity value was important.

Interpretation of Data

After reviewing the data, the consultant reasoned correctly that:

1. Data taken during the first 10 minutes are not valid when using the Jacob technique. (Recall that u in the Theis equation must be less than 0.05 for the Jacob method to be valid.)

2. Slope #1 represents the true aquifer characteristics because the cone of depression had not yet been affected by the subsequent rise in river level or other boundary. If this value of Δs were used in the Jacob equation for transmissivity, the well efficiency could be calculated correctly.

3. Slope #2 is caused when the cone of depression intercepts the river, thereby producing a recharge effect on the drawdown data. Transmissivity values calculated from this Δs value would be too high.

4. Slope #3 represents the sudden rise of the river level, and causes an additional small, but temporary, recharge effect on the data. Transmissivity values calculated on the basis of this part of the curve would also be too high.

5. Slope #4 represents the negative boundary effect as the southern side of the cone

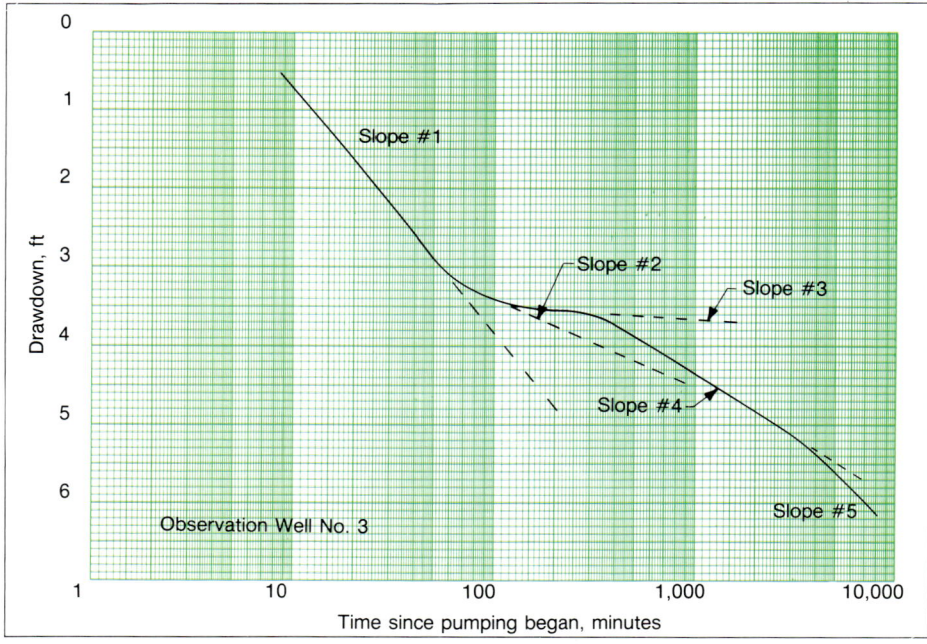

Figure 16.19. Time-drawdown graph for observation well No. 3 showing five different slopes.

of depression intercepts the impermeable rocks of the valley wall.

6. Slope #5 represents the reduction in river level to its pre-rainfall stage. This transmissivity value is still higher than the transmissivity calculated from Slope 1 because the recharge effect of the river is slightly greater than the effect of the negative boundary.

Determination of accurate transmissivity and storage coefficient values from pumping tests will depend not only on how well the test itself has been carried out but also on the analyst's ability to interpret the results. The drawdown data are strongly affected by the hydrogeological character of the aquifer, and the analyst must understand the nature of the aquifer before the data make sense. For example, the pumping test data for the shallow well (4) exhibit the same drawdown characteristics as for the nearby deeper well (3). Thus, the areal extent of the clay layer must be limited in comparison with the extent of the cone of depression, although the clay layer appears to be a major hydraulic factor in the geologic cross section shown in Figure 16.18. In floodplains, clay layers often result from the infilling of old, now-buried oxbow lakes by flooding events; the line of observation wells was, by chance, drilled along this former channel. Because the clay layer is limited (only as wide as the old channel), it does not inhibit recharge effects on the deeper observation and test wells caused by the sudden rise in river level.

The same drawdown trends that occur in well 3 will also occur in wells 1 and 2. But breaks in the drawdown slopes caused by recharge or boundary conditions would reach observation wells at times related to their distances from the river or boundary. For example, the observation well closest to the river would show the recharge effect first.

Example 2 — Pumping Test in Glacial Outwash

Pumping test data taken from wells in aquifers deposited as a result of melting glacier ice often show curious anomalies that at first glance seem to defy rigorous analysis. But experience suggests that each curve indicates important physical characteristics of the aquifer, as this example will demonstrate.

A 6-in (152-mm) test well was installed in glacial outwash (fine glacial gravel) for the city of Brewster, Minnesota. During a 24-hour pumping test, the pumping rate was maintained at 193 gpm (1,050 m^3/day). Drawdown data are presented in Figure 16.20. The part of the graph that best represents the true transmissivity and storage coefficient of the aquifer material is not immediately clear, and the analyst is forced to draw on hydrogeological information to determine which part of the curve should be used.

Interpretation of Data

A quick appraisal of the graph suggests that a negative boundary was encountered at about 100 minutes into the pumping test. The nature of this boundary may not be immediately apparent unless the analyst recalls the way in which glacial outwash is deposited (see Chapter 3). At the glacier terminus, meltwater rushes through loosely consolidated glacial till deposits and entrains a full load of sediment. Upon reaching the land surface beyond the glacier front, which normally has a lower gradient, much of the bedload is deposited in topographically low areas, whereas finer, suspended material moves downriver. The areal extent of a certain lens of outwash is usually

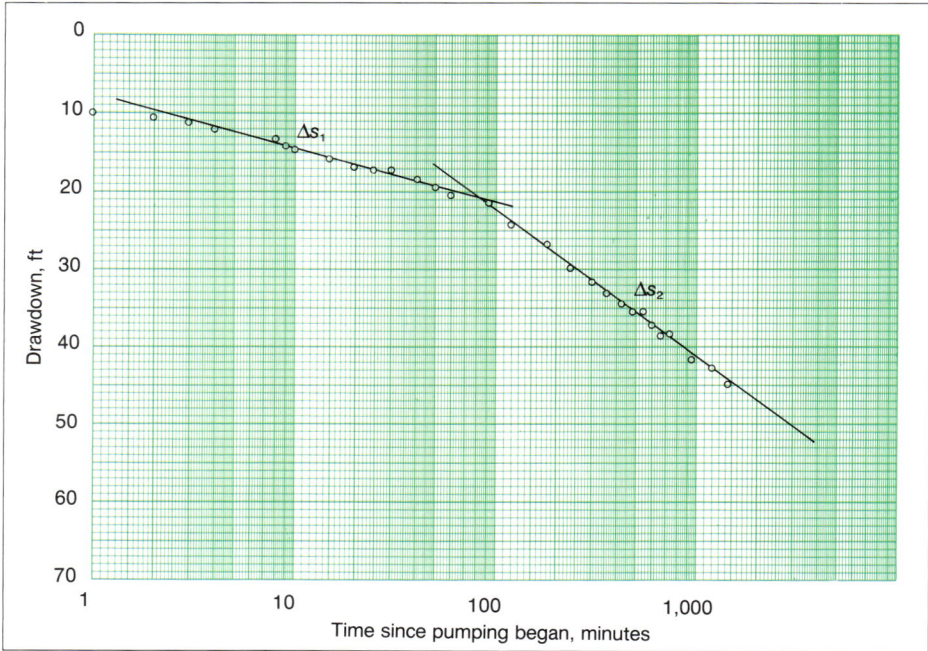

Figure 16.20. Drawdown data for 6-in (152-mm) test well in Brewster, Minnesota.

limited vertically or horizontally because of rapid changes in the stream's direction or ability to carry a certain size bedload. Although the entire outwash deposit may be quite large, the limited extent of a particular sand or gravel lens may have a large effect on drawdown during a pumping test.

During the spread of the cone of depression in the sediment in which the well has been drilled, the original slope (Δs_1) is maintained. At about 100 minutes, however, the cone spreads into finer materials that have a lower transmissivity value (Δs_2). In other words, the gravel has "pinched out" in one or more directions away from the well, and the finer sand of the surrounding material now controls flow through that portion of the aquifer and thus lowers overall transmissivity. A cross section through the well shows a thinning of the lens (Figure 16.21). This type of drawdown curve often occurs in pumping test data from glacial outwash deposits.

Knowledge of how geologic processes form aquifers is vital in analyzing pumping test information. A thorough understanding of the mathematical basis for equations used in pumping test analysis is usually not enough because the analyst must be able to visualize how the physical nature of the aquifer can produce hydraulic anomalies in the pumping test data. Alluvial deposits, for example, are usually highly elongate in character, so any pumping test reveals a negative boundary as the cone of depression encounters the valley walls. Depending on the extent and grain-size distribution of the alluvial material and on the pumping rate, the effect of the valley walls may affect the data only minutes into the test. Glacial outwash deposits, as illustrated above, are almost always deposited erratically, and the topography of the pre-existing land surface plays a major role in defining the thickness and lateral extent of the deposit. Significant variations in thickness can lead to both positive (apparently recharge) and negative boundaries. Identification of the drawdown trend before these boundaries are encountered is important in predicting aquifer performance.

From this example, it may seem to many analysts that slope 2 may yield better results for predicting radius of influence, drawdown after a certain time, and potential yield. But in practice, the hydraulic characteristics of the aquifer will be indicated more accurately by using the Δs_1 value. It is clear, however, that the most realistic Δs value will fall somewhere between the two Δs values. Normally, however, the error induced by using Δs_1 will be less than by using Δs_2. This point should always be kept in mind when the initial data are affected by boundaries and a prediction of aquifer performance must be made.

Example 3 — Pumping Test in Alluvial Deposits

The first two examples show that aberrations in data can be caused by variations in the geologic character of the aquifer materials created in the depositional process. Pumping test data can also be affected by the way the well and pumping equipment are constructed and installed.

In this example, an 8-in (203-mm) well, drilled in deep, fine-to-coarse alluvium, was test pumped at 5.2 gpm (28.3 m³/day). No lakes or rivers occur within the potential radius of influence. The time-drawdown data are plotted in Figure 16.22.

Interpretation of Data

A first hypothesis might suggest that a recharge boundary was encountered after approximately 80 minutes of pumping, but no recharge sources are known to exist. Another hypothesis might be that the pumping rate decreased; it did not.

What is unusual about this pumping test is the relatively low specific capacity. In pumping tests where the specific capacity is low in comparison to the size of the casing, the analyst should anticipate that the data may exhibit casing storage effects (discussed in Chapter 9). Recall that one of the basic assumptions made in deriving Theis's, and subsequently Jacob's, equation is that all of the water pumped from a well during a pumping test comes from the aquifer and none from storage within the well. Papadopulos and Cooper (1967) presented an equation describing the discharge from a pumping well which takes into account the volume of water removed from casing storage when the well is efficient. Drawdown values calculated from this equation differ significantly from the Theis and Jacob equations during the early portion of the pumping test, when a relatively high percentage of the discharge comes from casing storage. During the later stages of the pumping test, when only a negligible

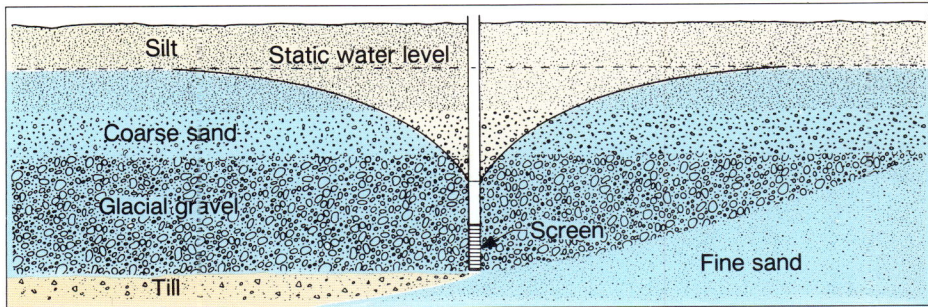

Figure 16.21. Cross section of well showing the thinning gravel deposit.

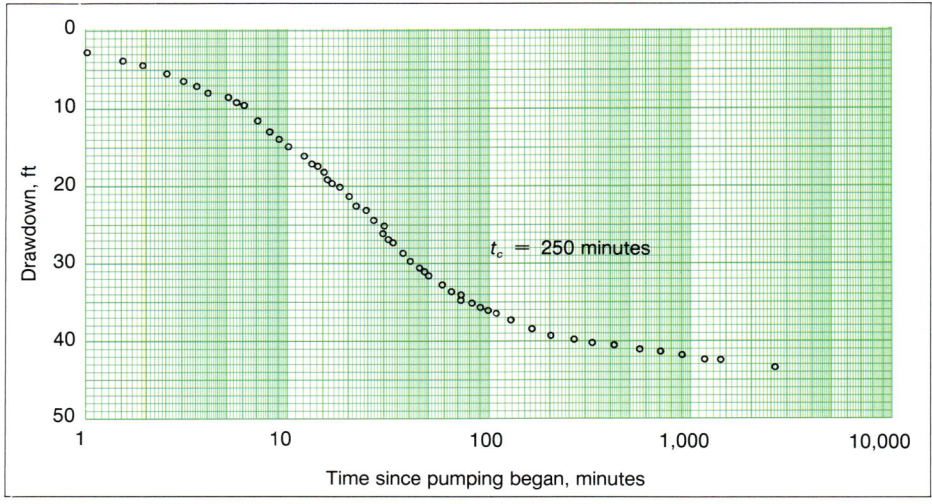

Figure 16.22. In this drawdown curve, an apparent recharge boundary occurs about 80 minutes after pumping began. Yet no obvious recharge sources such as rivers or lakes exist nearby.

quantity of water is obtained from casing storage, the equations produce equal results.

As a result of the casing storage effect on the time-drawdown graph, it is possible to misinterpret the data and to assume that the first slope is the correct one, particularly if the slope persists past the first 10 minutes. The later data (flatter slope) could be interpreted as indicating recharge. Furthermore, it might be possible to have a pumping test of such short duration that only the casing-storage-sensitive data are seen and the "correct" slope never appears. In order to avoid misinterpretation of the data, it is necessary to use the Papadopulus-Cooper equation to calculate t_c, the time at which the casing storage effect becomes negligible on the time-drawdown graph. A slight modification of their equation is:

$$t_c = \frac{375 (r_c^2 - r_p^2)}{T} \qquad (16.12)$$

where
t_c = time, in days, after which the effect of casing storage can be ignored (assuming a 1-percent error in drawdown values)
r_c = radius, in ft, of well casing (inside diameter) over which the water level changes are occurring
r_p = radius, in ft, of pump-column pipe (outside diameter)
T = transmissivity, in gpd/ft

However, this equation is only valid for 100-percent efficient wells and requires prior knowledge of the transmissivity. Schafer (1978) modified Equation 16.12 so that it provides an estimate of t_c which compensates for well efficiency and which does not require prior knowledge of formation and well characteristics.

$$t_c = \frac{0.6 (d_c^2 - d_p^2)}{Q/s} \qquad (16.13)*$$

where

---

*Equation 16.13 was obtained by dividing Equation 16.12 by the well efficiency and making some mathematical substitutions and simplifications. Use of this equation is explained in Chapter 9.

t_c = time, in minutes, when casing storage effect becomes negligible
d_c = inside diameter of well casing, in inches
d_p = outside diameter of pump-column pipe, in inches
Q/s = specific capacity of the well, in gpm/ft of drawdown, at time t_c

In general, t_c will be large when either the well radius is large or the specific capacity is small.

For this example using the Schafer equation, t_c is calculated to be 260 minutes. For the first 80 to 90 minutes of the test, much of the water that is being pumped comes from casing storage. As the water stored in the casing is removed, the contribution from the aquifer steadily increases from nothing to finally providing all the water to the pump. Only after the water has been pulled down sufficiently far in the casing will the aquifer begin to make a major contribution. In this example, the transition took about 260 minutes, which is much longer than in a conventional pumping test, where typically the water in the casing can be drawn down in less than a minute. Thus, the first slope shown in Figure 16.22 is erroneous; the second slope represents true aquifer conditions.

Several conditions contribute to the occurrence of casing storage effects. These are (1) the tightness of the formation, (2) the degree of well and aquifer development, and (3) the relationship of casing diameter, pump column diameter, and pumping rate. Tight formations may occur naturally because of poor grain-size sorting or the existence of stratification. They can also be created during drilling by excessive jarring or vibration of the formation, which tends to compact loosely consolidated sediments. Incomplete development fails to remove drilling clays and fine material from the formation surrounding the screen, thereby preventing water from entering the well except under excessive pumping heads. Low pumping rates from large-diameter casing can also lead to casing storage effects, because the withdrawal rate from the well bore is insufficient to cause a significant volume of water to flow into the well from the aquifer during the early stages of a pumping test. The most serious instances of casing storage occur when all of these factors occur simultaneously.

A question may arise concerning the potential usefulness of early data gathered from certain pumping tests where casing storage may occur. The early data are useful, however, because where well efficiency, E, is greater than about 20 to 30 percent, the following relationship exists between slope 1 and slope 2:

$$T_2 = \frac{4T_1}{E} \qquad (16.14)*$$

Thus, careful collection of early time-drawdown and recovery values can enhance the data base used to evaluate wells and aquifers. The effect of casing storage on the early measurements should be incorporated into the overall data analysis when conditions justify its inclusion. Estimation of t_c aids in the interpretation by determining which data are influenced by casing storage and are therefore not subject to conventional analysis techniques.

Example 4 — Pumping Test from a Bedrock Aquifer

Certain types of aquifers will exhibit marked decreases in specific capacity as drawdown increases. The extent of this decrease depends on the geologic origin of the

---

*For more accuracy, the constant 4 could be replaced by $-\log S$ (negative logarithm of the storage coefficient).

aquifer and on subsequent chemical or physical changes that may produce jointing, cementation, solution cavities, and faulting. Loss of specific capacity is best examined by conducting a step-drawdown test. Examination of Figure 16.23 shows that the specific capacity of a well completed in a deep limestone aquifer is inversely proportional to the discharge rate — the higher the pumping rate, the lower the specific capacity. At a yield of 700 gpm (3,820 m^3/day), the specific capacity is only one-tenth that of the same well pumping at 90 gpm (491 m^3/day). How can the pumping test analyst develop the ability to predict aquifer performance when inhomogeneities in the formation cause drastic changes in the specific yield with increasing drawdown?

Interpretation of Data

As suggested in Chapter 3, rocks that have been exposed at or near the Earth's surface have undergone varying degrees of weathering, with the most intense weathering concentrated in the upper portions of the rock. The only common rock type that would normally exhibit such high transmissivity values is limestone in which large amounts of the rock have been removed by rainwater percolating through the rock along joints or faults; this results in solution cavities that can store and transmit large volumes of water to wells.

The decline of pore space with depth and the resultant loss in hydraulic conductivity combine to cause a change in the flow conditions of the water moving toward the well. Recall that head losses are proportional to the square of the velocity for turbulent flow. Therefore, increasing the flow rate causes an increase in the amount of turbulent flow and, thus, a disproportionate increase in the drawdown. As a result, the specific capacity decreases at increasing drawdowns. The head losses created by this turbulence

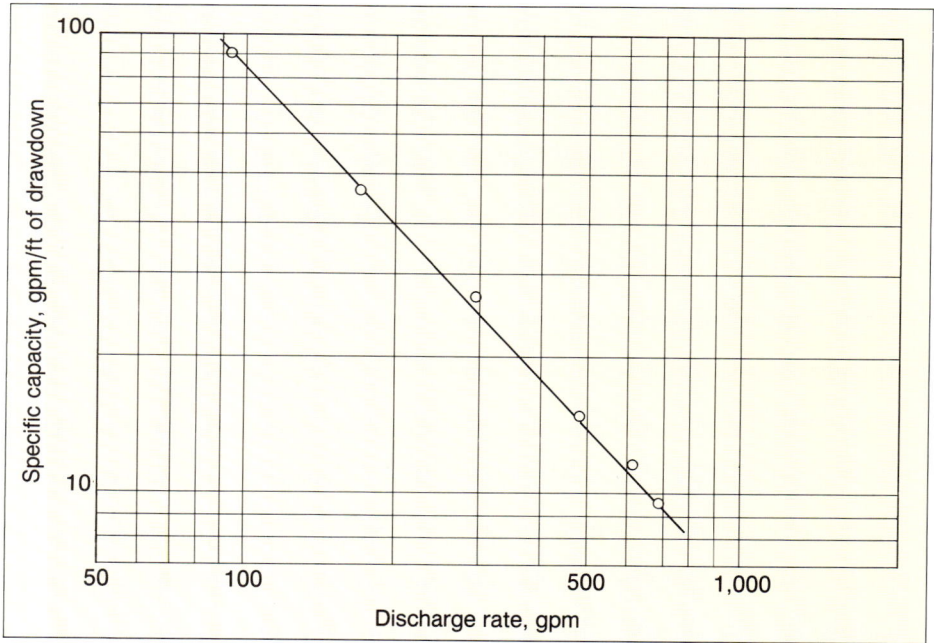

Figure 16.23. Reduction in specific capacity of a well drilled in limestone as the discharge rate increases. *(After Kelly et al., 1980)*

are not a "well loss" as defined earlier in this chapter, but are a formation (head) loss attributable to the physics of the aquifer. Part of the increasing head loss may also be attributable to the natural reduction in porosity (and permeability) with depth.

Whenever possible, details about the drilling conditions that provide information about the aquifer should be available to the person analyzing the pumping test data. This is especially important for weathered limestone, poorly cemented sandstone, igneous rock aquifers, and highly stratified silt, sand, and gravel aquifers. If changing physical conditions are indicated with increasing depth, the rate at which a constant-rate pumping test is conducted should be a higher percentage of the anticipated production rate — perhaps as much as 50 to 60 percent of the production rate — so that a realistic specific capacity can be determined. Step-drawdown tests are especially valuable in defining the falloff in specific capacity as yields increase.

Example 5 — Pumping Tests in Highly Stratified Sediments

Many pumping tests yield data similar to the graph shown in Figure 16.24. During the early stages of the pumping test, the data fall well below those predicted on the basis of the Theis and Jacob equations. But unlike the casing storage phenomenon described above, the persistence of the anomalous plot extends below the ideal curve for a much greater time. In fact, the return to the ideal curve may occur after a typical pumping test is terminated. Thus, the pumping test analyst must be able to recognize from the early portion of the test what is occurring and then extend the test if necessary to collect valid data.

Interpretation of Data

This type of curve, called the slow-drainage curve, is produced when test pumping stratified sediments containing highly permeable coarse sand and gravel layers interspersed with silt deposits. The Theis and Jacob equations assume that there is an instantaneous release of water (specific yield) from aquifers when pumping begins.

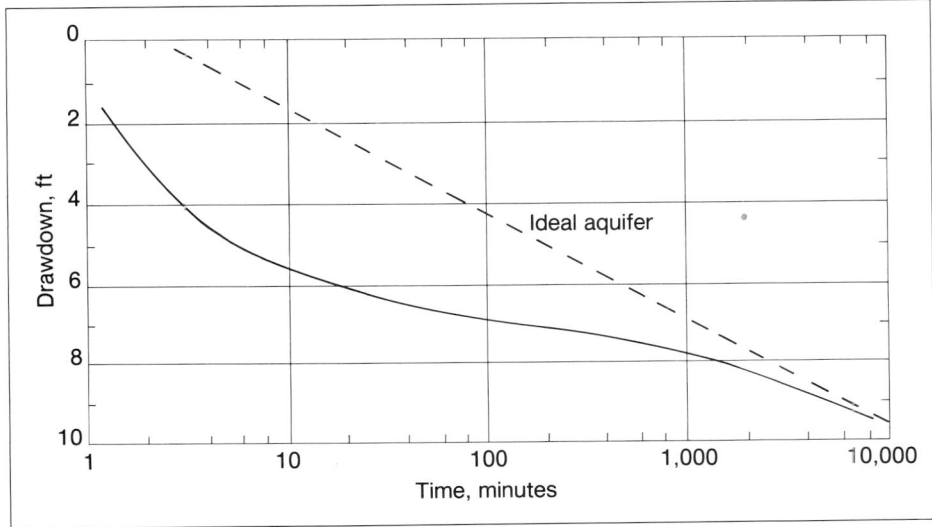

Figure 16.24. The drawdown data for this unconfined aquifer fall below an ideal drawdown curve. $Q = 500$ gpm (2,730 m³/day), $K = 500$ gpd/ft² (20.4 m/day), $H = 100$ ft (30.5 m), $S = 0.1$.

570   GROUNDWATER AND WELLS

In stratified sediments, this is not true and the drawdown data reflect this noncompliance with the theoretical basis for these equations.

The drawdown data presented in Figure 16.24 serve as a prototype for graphs of pumping tests completed in highly stratified, unconfined aquifers. When pumping begins, the aquifer cannot be dewatered as quickly as the cone of depression expands. Therefore, most of the water entering the screen comes from the aquifer at the same depth as the screen, because the water simply cannot percolate vertically at the rate it moves horizontally to the well screen. After a certain amount of time (50 minutes), the downward-percolating water begins to influence the drawdown, causing a significant reduction in the rate of drawdown from 50 to 1,000 minutes. This transient part of the curve appears to act as a recharge boundary. As the hydraulic gradient increases, more water begins to pass through the finer grained layers. After 1,000 minutes, the slow release from storage is virtually over and the drawdown resumes its true slope. Any calculation of transmissivity must be made by using the slope beginning at approximately 2,000 minutes. Use of earlier slopes would yield incorrect values for transmissivity and storage.

Example 6 — Predicting Aquifer Performance for a Well That is Pumped Intermittently

A municipality must supplement the yield from its single well with a second well that will be pumped only intermittently. Because the new well will be placed rather close to the older well in an unconfined aquifer, the city engineer must calculate the potential interference from the new well at the rate it will be pumped. The engineer determines from the pumping test data that, based on an 80-percent efficient well, the drawdown in the new well cannot exceed 150 ft (45.6 m), or the additional drawdown created in the old well will cause the pump to break suction.

Pumping test data are shown in Figure 16.25 for the new municipal well which will be pumped intermittently at 300 gpm (1,640 m³/day), a rate slightly higher than the pumping test rate. If the well is pumped 16 hours per day (16 hours on and 8 hours off) for 180 days, what drawdown will occur at the end of this time? Because recovery of the well for part of each day interferes with the usual long-term drawdown trends, it is not possible to use the drawdown graph directly to calculate the actual drawdown after 180 days and therefore its potential intereference with the older well.

Interpretation of Data

If the well is pumped only two-thirds of the time, the drawdown can be simulated by assuming two imaginary wells; one of these wells is pumped at two-thirds of the yield, 200 gpm (1,090 m³/day), for 179 days and 16 hours (259,000 minutes), and the other is pumped at one-third of the yield, 100 gpm (545 m³/day), for just the last cycle of pumping, 16 hours (960 minutes). See Chapter 9 for a discussion of this procedure. First calculate the drawdown caused by pumping 100 gpm for 960 minutes. Figure 16.25 shows that, at a time of 960 minutes, the drawdown is 62 ft (18.9 m) at 250 gpm (1,360 m³/day). Recall that Q is proportional to s if the aquifer is not dewatered more than 15 to 20 percent of its total depth. Therefore, the drawdown at 100 gpm is determined by multiplying the drawdown at 250 gpm by the ratio of the pumping rates:

$$s \text{ at } 100 \text{ gpm (at 960 min)} = 62 \cdot \frac{100}{250} = 24.8 \text{ ft}$$

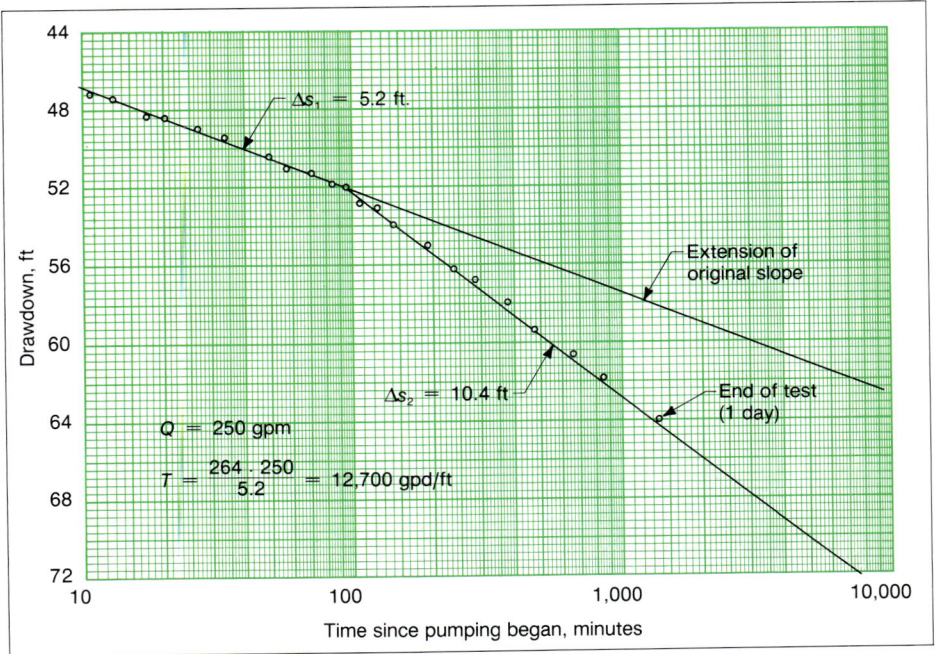

Figure 16.25. These time-drawdown data show a steepening slope, which suggests a limited aquifer. Although the transmissivity value is calculated on the basis of Δs_1, the drawdown at future times is based on Δs_2.

The drawdown for the well pumping 200 gpm (1,090 m³/day) for 259,000 minutes can be found in a similar way by using Figure 16.25 and recalling that Q is proportional to Δs_2. At a time of 2,590 minutes, the drawdown is 66.6 ft (20.3 m); the drawdown at 259,000 minutes can be obtained by adding two Δs_2 values to the drawdown at 2,590 minutes:

$$s \text{ at 250 gpm (at 259,000 min)} = 66.6 + 2(10.4) = 87.4 \text{ ft}$$

The drawdown at 200 gpm is:

$$s \text{ at 200 gpm (at 259,000 min)} = 87.4 \cdot \frac{200}{250} = 69.9 \text{ ft}$$

The final drawdown for the actual well is then the sum of the drawdowns for these two imaginary wells:

$$\text{total drawdown at 180 days} = 24.8 + 69.9 = 94.7 \text{ ft}$$

Thus, the anticipated drawdown in the new well and the resulting cone of depression will not interfere significantly with the operation of the older well.

Example 7 — Pumping Test Data from a Shallow Sand and Gravel Aquifer

Data from some pumping tests will occasionally yield results that are conflicting. Usually the drawdown values are affected by anomalies in the ground that have altered the drawdown data but are not evident to the analyst. Anomalies in drawdown data can sometimes be detected by examining transmissivity and storage values. This example shows this type of situation.

A 500-gpm (2,730-m³/day) pumping test was conducted in a sand and gravel aquifer located at a depth of 140 to 180 ft (42.7 to 54.9 m). The driller's log indicates clay from ground surface to 140 ft and again below 180 ft. The static water level in the well before pumping was 30 ft (9.2 m).

Interpretation of Data

Drawdown measurements during the pumping test at an observation well 150 ft (45.6 m) away are shown in Figure 16.26. Data from the pumped well were unreliable because of excessive turbulence in the well bore. Thus, the time-drawdown graph could not be used to verify data from the observation well. It was determined, however, that the total drawdown in the pumped well was approximately 20 ft (6.1 m).

From the data, the transmissivity was calculated:

$$T = 264 \cdot \frac{500}{4.8} = 27,500 \text{ gpd/ft}$$

An empirical equation can be used in the field to calculate the approximate value for the transmissivity of a confined aquifer on the basis of the pumping rate and drawdown. This equation is:

$$\frac{Q}{s} = \frac{T}{2000} \quad (16.15)$$

From Equation 16.15, the pumping test analyst determines that the specific capacity for this well should be approximately 14 gpm/ft (250 m³/day/m) of drawdown for a 100-percent efficient well. But the actual measured drawdown was 20 ft (6.1 m) at 500 gpm (2,730 m³/day), producing a specific capacity of 25 gpm/ft (447 m³/day/m), nearly twice the approximate value calculated. Thus, the actual value of the trans-

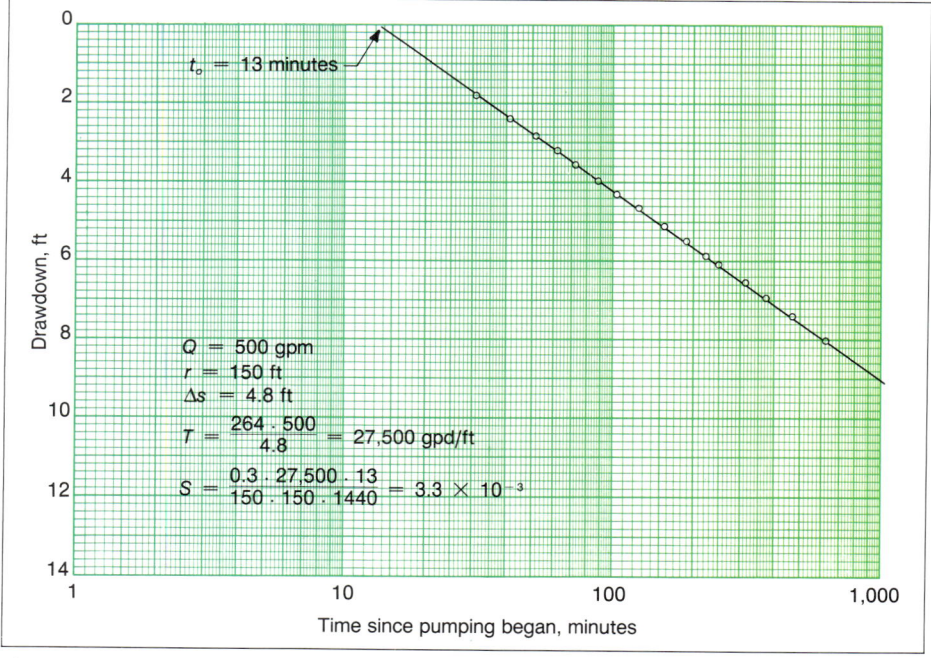

Figure 16.26. Drawdown data taken at an observation well 150 ft (45.6 m) from the production well.

missivity must be much higher than suggested by the pumping test data [roughly 50,000 gpd/ft (621 m²/day)]. This in turn indicates that the time-drawdown graph is too steep and that a negative boundary is affecting the data.

The storage coefficient will also be affected by a negative boundary, and its value can be used to substantiate this hypothesis. The storage coefficient is calculated to be:

$$S = \frac{0.3 \, Tt_o}{r^2} = \frac{0.3 \cdot 27,500 \cdot 13}{(150)^2 \cdot 1440} = 3.3 \times 10^{-3}$$

For confined conditions, the upper limit of storage coefficient values is 10^{-3}, suggesting once again that some sort of undetected negative boundary is present. The slope of the time-drawdown curve is too steep and in itself is a clear indication that something is wrong. The value of t_o may also be indicative of a problem in the data. Usually, t_o will be much smaller in value.

The principal problem in this example is that graphical analysis of the data does not show clearly the effect of the negative boundary. This has occurred because the data taken from the pumping well were too erratic to be useful, and early data were not recorded in the observation well. The presence of the boundary can be detected, however, by comparing the value of transmissivity with the specific capacity, and independently by the value of the storage coefficient. Both values indicate that the data have been affected by an anomaly in the aquifer.

Three conclusions can be reached:

1. Transmissivity is probably 50,000 gpd/ft (621 m²/day) or more, but its actual value cannot be calculated.
2. Storage coefficient is greater than 10^{-5} and less than 10^{-3}, but it cannot be determined from this test.
3. The aquifer has a negative boundary that is encountered by the cone of depression within the first 30 minutes of pumping.

Because the transmissivity and storage values cannot be determined on the basis of this particular pumping test, the analyst should suggest that another test be run.

Example 8 — Use of Pumping Test Data to Solve Interference Problems

For some intensely irrigated regions and for certain municipal well fields, well interference can cause a serious problem in maintaining yields during critical periods of crop growth, drought, or during periods of high domestic and industrial demand. Occasionally lawsuits are brought when new wells interfere with the yields of older wells. This problem occurs frequently when farmers build new wells that adversely impact existing domestic or other low-yield farm wells [30 to 50 gpm (164 to 273 m³/day)]. Substantial interference with other irrigation wells during periods of drought may also be possible. Good pumping test data taken before the construction of a new well, and analyzed using Jacob techniques, will show to what extent the operation of the new well will affect existing wells. If this potential effect is discussed openly with the owners of the older wells, an accommodation can usually be reached. Typically the owner of the proposed well may (1) offer to deepen a well that may go dry, (2) lower the pump setting on the older wells, (3) offer to supply water on a continuous basis from the new well, or (4) work out an arrangement where crops requiring large amounts of water are grown alternately by the farmers so that total withdrawals do not become excessive. In any event, the owners of existing wells should not be made to suffer any economic harm from the operation of the new well.

Pumping test data are plotted in Figure 16.27 for a 6-in (152-mm) test well pumped at 250 gpm (1,360 m³/day) in a confined aquifer.

Interpretation of Data

The procedure for predicting the degree of interference is based on extrapolating the pumping test data of the time-drawdown graph to the distance-drawdown graph (see Chapter 9). Most pumping test data for irrigation wells are obtained from the pumped well because observation wells are rarely available.

If the drawdown in the test well is 59.5 ft (18.1 m) at the end of 2 days of pumping, and the well is assumed to be 70-percent efficient, the actual drawdown in the aquifer just outside the casing is 70 percent of 59.5, or 41.7 ft (12.7 m). Because the slope of the drawdown curve is 2 times Δs from the time-drawdown graph, or 11.5, a distance-drawdown plot can be drawn as shown on Figure 16.28.

The irrigation farmer plans on pumping 1,000 gpm (5,450 m³/day) from the 16-in (406-mm) production well. Using the relationships Q is proportional to s and Q is proportional to Δs, the drawdown plot can be drawn as shown in Figure 16.28. Remember that this is the drawdown plot as it would appear after 2 days of steady pumping. The available drawdown is 200 ft (61 m), so the amount of drawdown is reasonable.

On the basis of the plot in Figure 16.28, a nearby farm well will be affected by the new well. For this relatively deep well, the pump can be lowered so that, aside from some additional lift requirements, the cost of accommodating the irrigation well is minimal. A relatively shallow domestic well belonging to a neighbor presents more

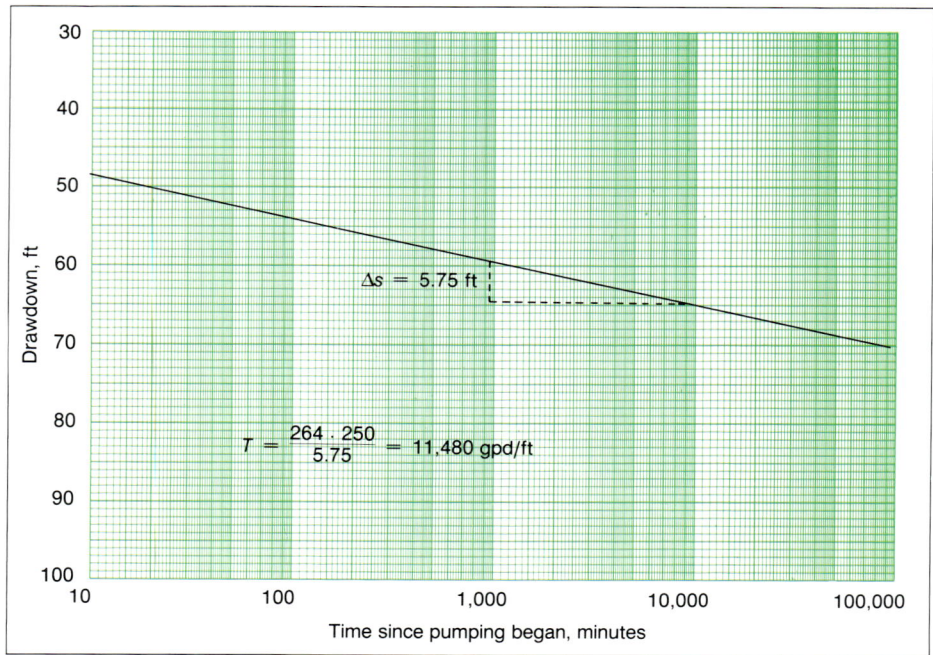

Figure 16.27. Time-drawdown graph for a test well pumped at 250 gpm (1,360 m³/day).

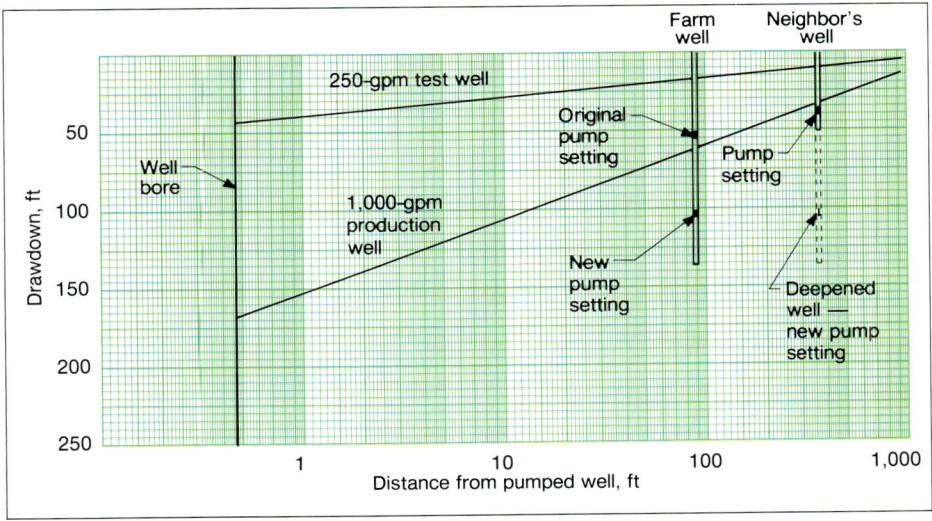

Figure 16.28. Theoretical distance-drawdown graph for a test well, assuming a well efficiency of 70 percent.

of a problem. With normal fluctuations in the water table of 6 to 7 ft (1.8 to 2.1 m) in this area, and the effect of pumping steadily for periods longer than 2 days, the domestic well would certainly go dry from time to time. Therefore, the irrigator elects to deepen the well and to lower the pump at no expense to the neighbor.

The question of possible well interference will be brought up often by people worried about a neighbor who is installing an irrigation well. To help predict well interference, Table 16.7 shows interference in wells located at certain distances from a production well. Typical transmissivities and storage coefficients for various types of sand and gravel formations have been chosen for both confined and unconfined conditions. The time of pumping is 7 days and the calculated well yield is 1,000 gpm (5,450 m³/day). These figures are true when all the assumptions for the Theis and Jacob equations are satisfied. In practice, they offer an approximate guide to anticipated drawdowns.

Table 16.7 Drawdown Produced by Production Well Under Different Hydraulic Conditions

Type of Aquifer	Transmissivity gpd/ft	Drawdown in production well ft	Drawdown at 100 ft ft	Drawdown at 500 ft ft	Drawdown at 1,000 ft ft	Drawdown at 10,000 ft ft
Unconfined	100,000	16.7	6.1	2.4	0.9	0
	50,000	31.8	10.7	3.3	0.1	0
	25,000	60.4	18.2	3.4	0	0
Confined	100,000	22.8	12.2	8.5	6.9	1.6
	50,000	43.9	22.8	15.4	12.3	1.7
	25,000	84.7	42.5	27.7	21.3	0.2

Example 9 — Determining Well Efficiency from Pumping Test Data

Some high-capacity well contracts require that the drilling contractor produce a well with minimum efficiency based on a constant-rate pumping test. In the example

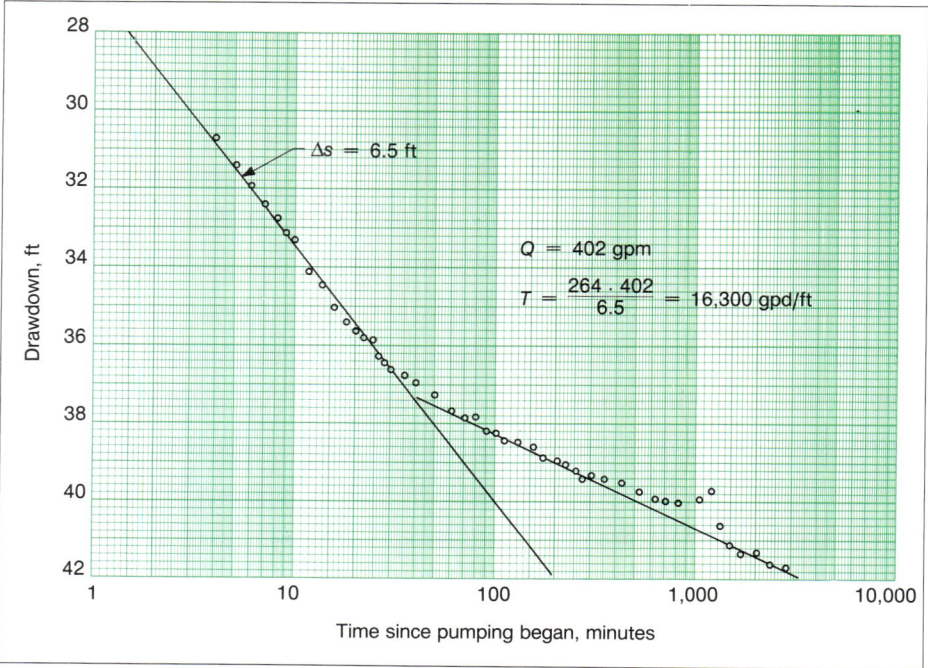

Figure 16.29. Time-drawdown data for a well being pumped at a constant-rate of 402 gpm (2,190 m³/day).

given below, the pumping test data are critical because the contract requires an efficiency of at least 80 percent and calls for penalties for efficiencies below this level.

Time-drawdown data from a constant-rate pumping test for a confined aquifer are plotted for both the pumped well (Figure 16.29) and the observation well (Figure 16.30). The 10-in (254-mm), 126-ft (38.4-m) well was pumped at 402 gpm (2,190 m³/day) during the test. An initial slope is established for each well to calculate transmissivity. One-half hour into the test, however, the slope on each graph flattens, indicating the presence of recharge. As suggested earlier, the transmissivity calculation must be made using the initial slope on each graph. The transmissivity value from the pumped well graph is 16,300 gpd/ft (202 m²/day), whereas that obtained from the observation well is 25,900 gpd/ft (322 m²/day). Each transmissivity value will lead to a different calculated value of well efficiency.

Interpretation of Data

Well efficiency can be estimated by comparing the extrapolated one-day specific capacity with the maximum theoretical specific capacity computed from the transmissivity value. The theoretical specific capacity is calculated without considering the recharge condition. In order to make a fair comparison, the extrapolated one-day specific capacity should also ignore the recharge condition and thus should be extrapolated from the initial slope on the time-drawdown graph.

The extrapolated drawdown after one day of pumping is 47.5 ft (14.5 m), making the one-day specific capacity 8.5 gpm/ft (152 m³/day/m) of drawdown. The maximum theoretical specific capacity can be estimated from the following equation:

$$\frac{Q}{s} = \frac{T}{2000}$$

Using a transmissivity value of 25,900 gpd/ft from the observation well, the theoretical specific capacity would be 25,900/2,000, or 13 gpm/ft (232 m³/day/m) of drawdown. The resulting estimate of well efficiency is then 8.5/13 or 65 percent.

The transmissivity value of 16,300 gpd/ft from the pumped well results in a theoretical specific capacity of 8.2 gpm/ft (147 m³/day/m) of drawdown — somewhat less than the actual capacity. This suggests that for all practical purposes the well is 100-percent efficient. The dilemma facing the contractor, engineer, and well owner is whether or not to invoke the penalty clause for failing to meet the 80-percent efficiency requirement because one calculation falls far short of 80 percent, whereas the other exceeds 80 percent. It is necessary to ascertain which set of data reflect true aquifer conditions.

Recall that the calculations presented above are valid only if Jacob's equation is valid. In turn, Jacob's equation is valid only when the u value is less than 0.05 where:

$$u = \frac{1.87 \, r^2 S}{Tt}$$

Rearranging terms, an expression for the time, t, at which the u value becomes 0.05 can be computed:

$$t = \frac{1.87 \, r^2 S}{0.05 \, T}$$

Because r appears in the numerator, large values of r will require that substantial time must pass before the u-value condition is satisfied. Because of this, the early

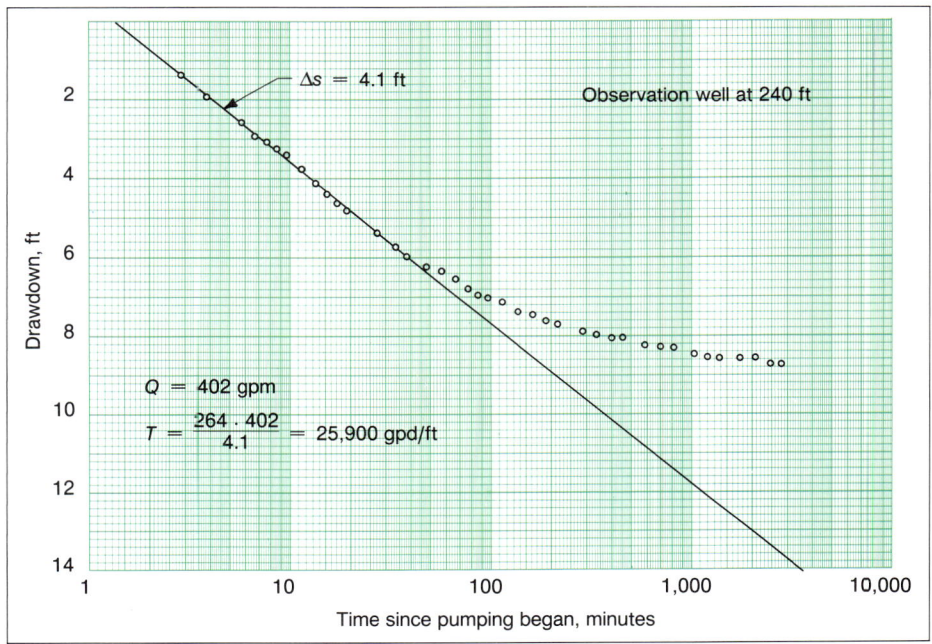

Figure 16.30. Time-drawdown data for the observation well from the constant-rate pumping test.

data collected from observation wells often do not satisfy the u-value condition, and the use of the Jacob analysis is precluded. Thus, the analyst may assume that the transmissivity value of 16,300 is a more believable value than 25,900 because the latter was obtained from the observation well data.

To verify this assumption, it is necessary to calculate the time at which the u value equals 0.05 for the observation well. Using the equation above, and assuming for the moment that the transmissivity value of 16,300 is approximately correct and further assuming a typical storage coefficient value of 5×10^{-4} (based on the site geology), the calculated value for t is:

$$t = \frac{1.87 \cdot (240)^2 \cdot 5 \times 10^{-4}}{0.05 \cdot 16,300} = 0.066 \text{ days or 95 minutes}$$

This means that the Jacob analysis is not valid for data recorded within the first 95 minutes of pumping. If a different storage coefficient were chosen, for example 2×10^{-4}, the result would be a pumping time of 38 minutes. In either case, the calculated transmissivity value of 25,900 has been obtained erroneously. Furthermore, a valid transmissivity cannot be calculated from the later drawdown data because they have been affected by recharge.

When the u value is too large to allow use of the Jacob analysis, it is necessary to use the Theis curve-matching technique (see Chapter 9). Figure 16.31 shows the application of curve matching along with the resultant match point. Note that this procedure uses only the data recorded prior to encountering recharge.

In Figure 16.31, the match point is shown at the point on the type curve where

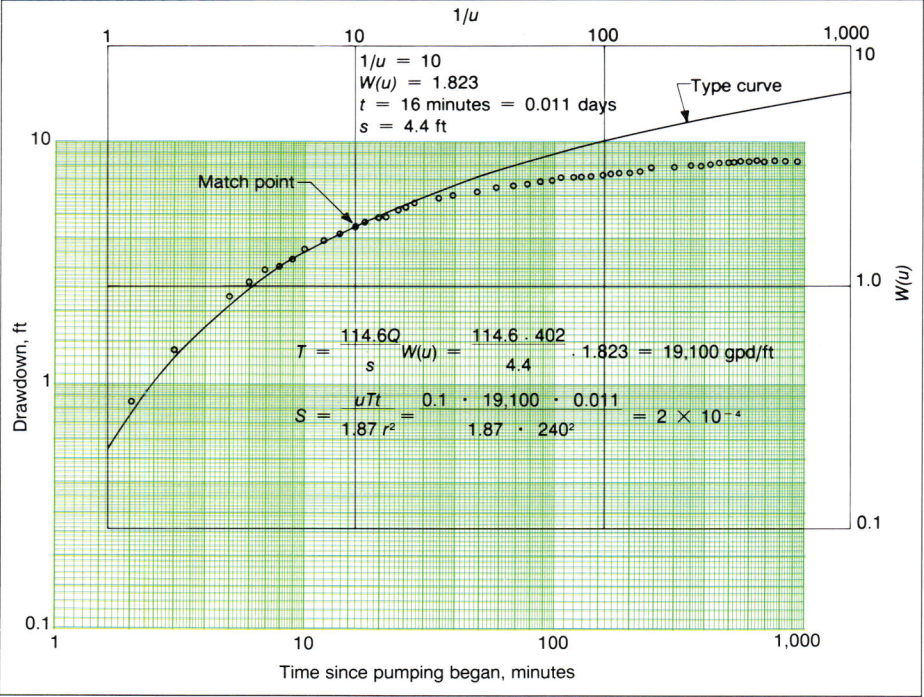

Figure 16.31. Illustration of the Theis curve-matching technique.

$1/u = 10$ and $W(u) = 1.823$. At the corresponding point on the time-drawdown graph, $s = 4.4$ ft and $t = 16$ minutes or 0.011 days. The transmissivity can be computed as follows:

$$T = \frac{114.6Q}{s}W(u) = \frac{114.6 \cdot 402}{4.4} \cdot 1.823 = 19,100 \text{ gpd/ft}$$

This transmissivity value certainly provides better agreement with the value obtained from the time-drawdown graph. The storage coefficient may be calculated as follows:

$$S = \frac{uTt}{1.87 \, r^2} = \frac{0.1 \cdot 19,100 \cdot 0.011}{1.87 \cdot (240)^2} = 2 \times 10^{-4}$$

The best estimate of transmissivity would be obtained by averaging the 16,300 gpd/ft from the pumped well and the 19,100 gpd/ft (237 m²/day) from observation well data using the Theis curve-matching method. This provides an estimated transmissivity value of 17,700 gpd/ft (220 m²/day). The maximum theoretical specific capacity computed from this transmissivity value is 17,700/2,000 or 8.9 gpm/ft (159 m³/day/m) of drawdown. Thus, the efficiency is equal to 8.5/8.9 or 96 percent.

CONCLUSIONS

Analysis of pumping test data requires an appreciation of all the factors that can affect the drawdown data. This appreciation comes chiefly from a thorough understanding of the aquifer, well hydraulic theory, and the practical aspects of conducting the test. The analyst must be able to visualize the physical nature of the aquifer and how it deviates from the basic assumptions on which well hydraulic equations are based. The limitations of hydraulic equations must always be kept in mind while analyzing pumping test data. Most problems in analyzing pumping test data, however, occur because of errors or omissions made in conducting the test. Elimination of time or discharge measurement errors, and the collection of enough data, will generally assure accurate determination of aquifer parameters.

CHAPTER 17
Water Well Pumps

The primary function of a pump is to add hydraulic energy to certain volumes of fluid. This is accomplished when the mechanical energy imparted to the pump from a power source is transferred to the fluid, thereby becoming hydraulic energy. Thus, a pump serves to transfer energy from a power source to a fluid, thereby creating flow or simply creating greater pressures on the fluid. Pumps can serve many different purposes. These include raising a liquid from one level to another, moving a fluid through a pipeline, imparting a high velocity to water, and moving liquids against a resistance.

A pump can impart three types of energy to any fluid: head, pressure, and velocity. The amount of each type of hydraulic energy will vary from place to place in a system. For example, when water is at rest in a storage tank, it possesses head energy but no velocity or pressure energy. Once water starts to flow from the storage tank, it has head, pressure, and velocity energy. As water flows from the end of a pipe or hose, the head and pressure energy are transformed to velocity energy alone.

Pumps are installed in water wells to lift the water to the ground surface and deliver it to the point of use. Many types and sizes of pumps are available, ranging in power from a fraction of one horsepower to several thousand horsepower. In the water well industry, pumps are classified generally into two groups: shallow-well pumps and deep-well pumps. This classification refers to the position of the pump in the well, not to the depth of the well. A shallow-well pump is mounted at ground level and removes water from the well by suction lift. Such a pump can be used as long as the pumping water level is within the suction-lift capability of the pump, regardless of the total depth of the well. A deep-well pump is installed within the well casing, with the pump inlet submerged below the pumping level. Therefore, the inlet is under positive pressure head. The deep-well pump must be used for any well where the pumping level is below the limit of suction lift [approximately 20 to 25 ft (6.1 to 7.6 m)].

A general method of classifying pumps on the basis of engineering design is to divide them into two groups: positive displacement and variable displacement. Although positive displacement pumps are used extensively in groundwater monitoring wells, in hand-pump-equipped wells, and in wind-powered wells, they are used rarely for domestic or large-capacity water wells. Thus, variable displacement pumps will be discussed first in this chapter because of their wide use in the water well industry.

VARIABLE DISPLACEMENT PUMPS

The distinguishing characteristic of variable displacement pumps is the inverse relationship that exists between the rate at which they deliver water and the head against which pumping takes place. For example, as the head increases, the rate of pumping decreases. The greatest input of power for most variable displacement pumps is required at low head because the volume of water increases as the pumping head decreases.

The pumping rate in any variable displacement pump is dependent upon the pressure or number of feet of lift against which the pump is operating. A general term for this pressure is head or static lift, and can be calculated by adding the pumping water level in the well to the lift required above the discharge point (Figure 17.1).

The lift required, however, is only a part of the total head that the pump would be operating against. All friction losses that occur in the pipe during pumping must also be added to the lifting head to determine dynamic head or total dynamic head. These friction losses are sometimes referred to as the friction head. Friction head represents the combined head losses in the pipe, valves, and other fittings caused by flow velocity, viscosity, and specific gravity of the fluid. Tables are published to assist the engineer in calculating the various components of the friction head (see Appendix 17.A).

The major types of variable displacement pumps are:
1. Centrifugal pumps
    a. Suction lift
    b. Deep-well turbine
    c. Submersible turbine
2. Jet pumps

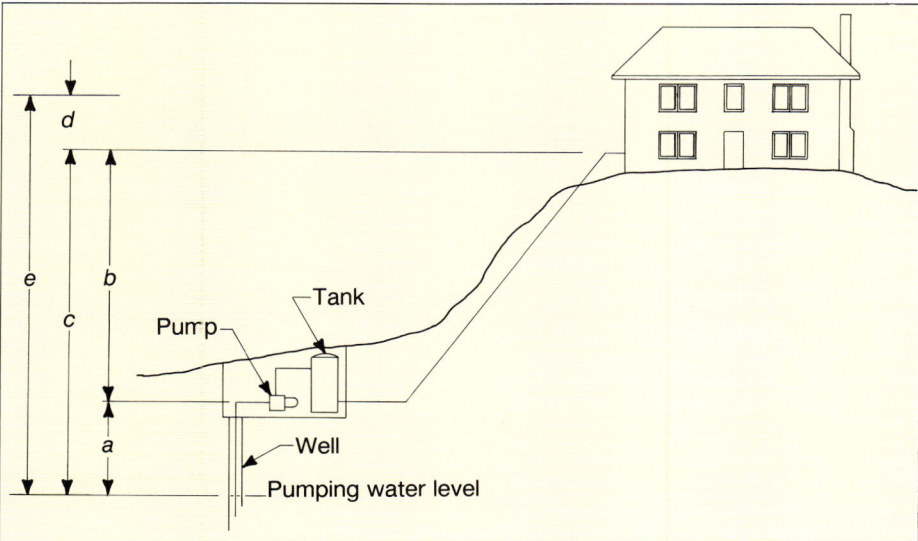

Figure 17.1. The individual head components of a pumping system, where *a* is pumping water level to discharge; *b* is above-ground lift; *c* is total head; *d* is friction head; and *e* is total dynamic head.

3. Air-lift pumps

Centrifugal Pumps

Centrifugal pumps are the most important type of variable displacement pump because of their wide use. The basic principles of the centrifugal pump were recognized and experimentally demonstrated about 300 years ago. Little progress was made in developing practical devices until the latter part of the 19th century, when both the steam turbine and the electric motor were developed as suitable sources of power. The centrifugal pump then became popular as a pumping device. It is capable of delivering large quantities of water, against high as well as low head conditions, with good efficiency. Combining these features with its other attributes — simplicity, compactness, and adaptability to different methods of driving — it was inevitable that its use expanded rapidly.

There are many design variations in centrifugal pumps. Originally designed as a pump to be located at or near ground surface for suction-lift or booster service, it soon was adapted to installation under water in wells, first by long shaft extensions in large caissons (vertical turbine), and later in compact form as the familiar deep-well (submersible) turbine pump.

Various types of centrifugal pumps are used commonly in obtaining water supplies. The basic principle of centrifugal pumping can be illustrated by considering the effect of swinging a pail of water around in a circle at the end of a rope. Centrifugal force causes the water to press against the bottom of the pail, rather than run out at the open end. If a hole were cut in the bottom, water would discharge through the opening at a velocity related to the centrifugal force. Further, if an airtight cover were put on the pail, a partial vacuum would be created inside the pail as water is discharged. This vacuum could draw additional water into the pail through an intake pipe connected to the cover if the lift were not too great.

This simple description fulfills all the basic requirements of pumping by the centrifugal method. The pail and cover represent the casing of the pump; the discharge hole and intake pipe correspond to the pump outlet and intake; the arm that whirls the pail and the rope perform the functions of the energy source and pump impeller*, respectively.

Performance Curves

Most pump manufacturers provide a performance curve for each impeller design. Generally, these curves are drawn for a single stage (pumping unit). They provide information on the discharge rate obtainable against a given total dynamic head, plus the efficiency ratings and required brake horsepower† at that discharge rate. From the performance curves, a specific pump is chosen that can provide the desired yield for the total dynamic head at the highest possible efficiency. The performance curve shown in Figure 17.2 indicates that a single-stage turbine pump with a 10-in (254-mm) impeller operates at a peak efficiency of over 80 percent when pumping at 900 gpm (4,910 m^3/day) against a total head of 92 ft (28 m).

If the total head fluctuates throughout the year, it is difficult to choose a pump that

---

*Impellers are devices which accelerate the water within the pump to build pressure.
†See Page 585 for definition.

will operate efficiently at every head requirement. For example, unit E74 in Figure 17.2 has a peak efficiency of about 80 percent within a total head of 76 to 99 ft (23.2 to 30.2 m) and delivers from 1,170 to 670 gpm (6,380 to 3,650 m^3/day), respectively. If the required head fluctuates within this range, this pump operates efficiently and the power required varies by 7 brake horsepower.

On the other hand, it might be determined that the total head would be as low as 60 ft (18.3 m) during a season of high water level or minimum withdrawal of water; but during another season, the total head might be 100 ft (30.5 m) because the water level in the aquifer has decreased or interference from adjacent wells has increased. Under these conditions, the rate of pumping would range from nearly 1,340 gpm (7,300 m^3/day) down to about 620 gpm (3,380 m^3/day). The efficiency would then vary between about 67 and 77 percent, and the required power would reach a maximum of approximately 30 Bhp and a minimum of about 20 Bhp.

Under these conditions, a pump having operating characteristics different from those shown in Figure 17.2 could be selected so there would be less change in the pumping rate with changes in total head. Thus, the pump selected should have a steeper head characteristic. If this is done, the efficiency and brake horsepower required should be less affected within the head range. Peak efficiency might be lower, but the average pumping conditions would occur at a higher efficiency. The rated horsepower of the motor may remain the same or be slightly lower (compared to a pump having a flatter performance curve) because the maximum brake horsepower requirement would be at the lowest head when the pumping rate rises to a maximum.

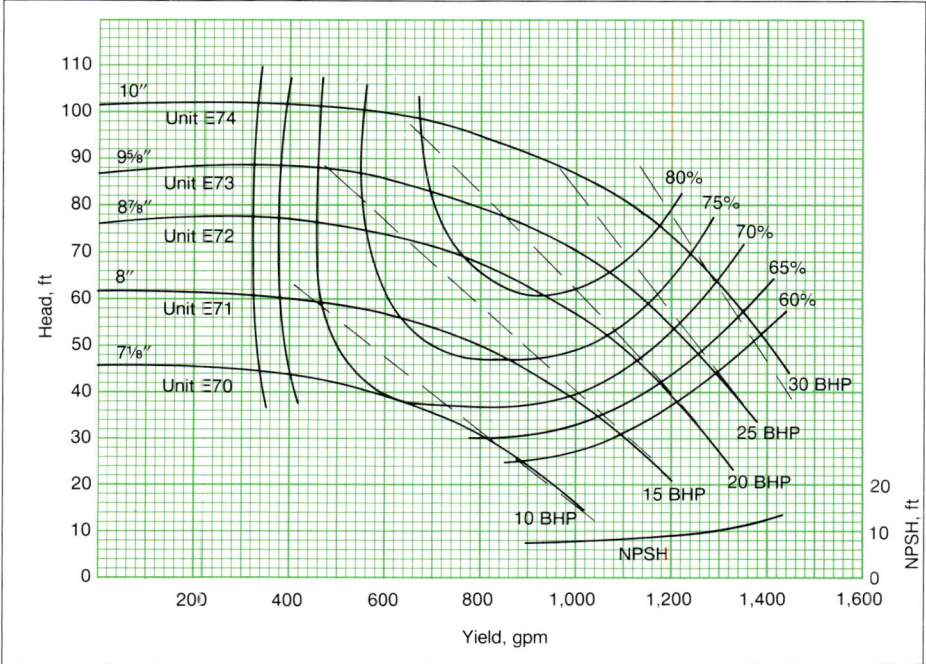

Figure 17.2. Representative performance curve for a vertical turbine pump operating at 1,800 rpm. Head-capacity curve of this pump is relatively flat, which means that small changes in head result in large changes in its pumping rate. *(Deming Division, Crane Company)*

Table 17.1. Sample Data for System-Head Calculations

Pumping Rate gpm	Fixed Head, SWL to Delivery Point, ft	Drawdown in Well, ft		Total Friction Head Loss ft	Total Dynamic Head, ft	
		Q/s = 5 gpm per ft of drawdown	Q/s = 20 gpm per ft of drawdown		Q/s = 5 gpm per ft of drawdown	Q/s = 20 gpm per ft of drawdown
0	80	0	0	0	80	80
200	80	40	10	2	122	92
400	80	80	20	6	166	106
600	80	120	30	12	212	122
800	80	160	40	19	259	139
1,000	80	200	50	29	309	159

Selection of an appropriate pump for a given service or application is made easier by plotting a system-head curve and comparing it with pump curves from several manufacturers. The system-head curve is plotted to show the required dynamic head for various rates of flow through the system. When pumping from a well, major components of the total dynamic head increase with increasing rates of flow; vertical lift increases as more drawdown occurs in the well and friction losses increase with increased discharge.

The shape of the system-head curve varies with the hydraulic characteristics of the well (aquifer) and those of the system served. A relatively flat curve results from the combined effects of high specific capacity for the well, constant elevation of the point to which the water is delivered, and large-diameter piping with low friction losses. A relatively steep curve results from the effects of low specific capacity for the well, added elevation to the point of delivery, and high friction losses in the piping.

Figure 17.3 shows two system-head curves plotted from the data in Table 17.1. With the head-capacity performance curve for a turbine pump plotted on the diagram of the system-head curve, the output of the pump is indicated by the intersection of the curves. Several pump curves may be superimposed to permit selection of the pump that is most efficient for the system conditions.

When selecting a turbine pump, the head-capacity and efficiency curves should be plotted on the system-head diagrams. The efficiencies of two or more pumps at their respective performance points in the system are then easily compared.

It must be remembered that a pump operates only on its own curve. The point of intersection with the system-head curve will change if the total head in the system changes. System head changes may be caused by pump wear, incrustation of the pump column or piping system, or incrustation in the aquifer and well screen. Decreasing pump discharge can be caused by altered conditions in the above-ground piping system, pump wear, or by changing conditions in the well. If the pump is operating off its curve, however, the problem is in the pump. On the other hand, decreased well efficiency caused by plugging or incrustation of the well screen will cause the pump to operate in a less efficient part of its performance curve, but not off the curve.

Pump Efficiency

The most efficient operating point for a centrifugal pump can be selected from the manufacturer's pump performance curves*. Theoretically, the best efficiency point for a centrifugal pump occurs when the energy added to the fluid is divided equally

---

*The efficiency indicated on any performance curve is for the bowl assembly only. Overall pumping plant efficiency can be obtained by multiplying the bowl efficiency times the motor efficiency. A typical pumping plant efficiency is about 70 to 75 percent.

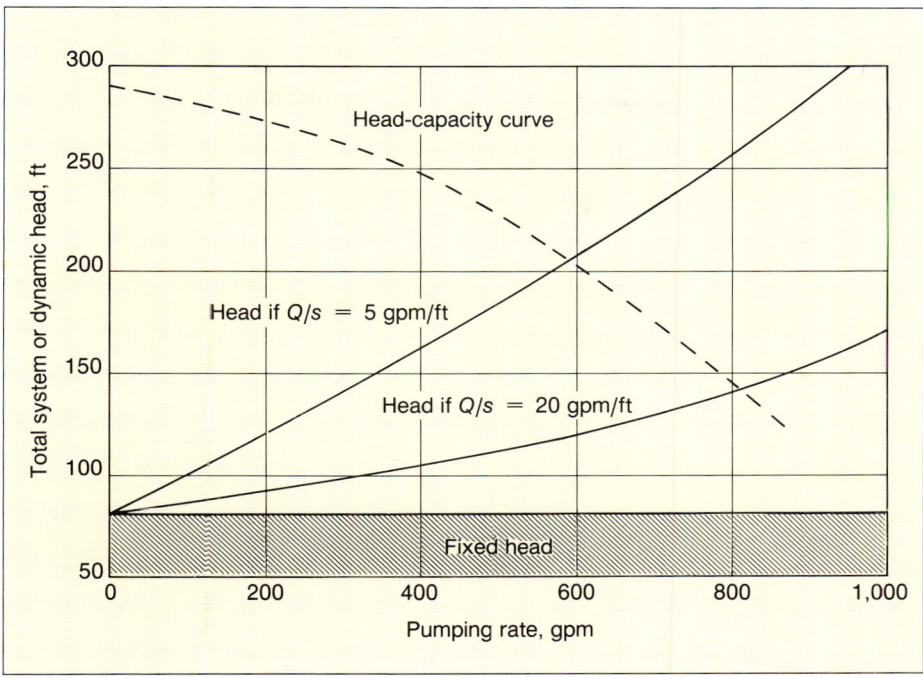

Figure 17.3. System-head curves are obtained by combining the static head, friction head loss in system, and drawdown in a well for each pumping rate. Head-capacity curve for a given turbine pump, superimposed on this diagram, shows that this pump would deliver about 600 gpm (3,270 m³/day) in one case, or about 800 gpm (4,350 m³/day) in another case where the well has a higher specific capacity.

between energy converted to head pressure and energy converted to capacity (Water Systems Council, undated). This situation or point on the performance curve occurs at about 50 percent of the pump's maximum capacity, or 0.7 times the head at which no flow will occur (shut-off head). Operating costs can be held to a minimum if the pump is operated in this range. Depending on the individual pump design, the peak efficiency may vary from these general figures, so it is advisable to consult the manufacturer's performance curves for that pump.

Horsepower

Water horsepower is defined as the energy (horsepower) required to pump water against a given head without consideration of efficiency or friction losses. Water horsepower requirements are calculated as follows:

$$Whp = \frac{\text{gpm} \cdot 8.33 \cdot \text{head} \cdot \text{sp. gr.}}{33,000} = \frac{\text{gpm} \cdot TDH}{3,960} \qquad (17.1)$$

where 8.33 is the weight of 1 gal of water in lb, TDH is total dynamic head in ft, sp. gr. is specific gravity of the fluid (water = 1), and 33,000 is a factor to convert power to horsepower (33,000 ft-lb equals one horsepower).

Because the bowl assembly is less than 100-percent efficient, somewhat greater horsepower than that calculated from Equation 17.1 is required to drive the pump. Brake horsepower takes into account the pump efficiency and is calculated as follows:

$$Bhp = \frac{Whp}{\text{pump efficiency}} = \frac{\text{gpm} \cdot \text{TDH}}{3{,}960 \cdot \text{pump efficiency}} \qquad (17.2)$$

There is another horsepower requirement that must be considered. Any drive unit, whether it is an electric motor, diesel engine, or gas engine, has an efficiency rating. This efficiency has to be considered to determine input horsepower so that the proper size equipment can be selected. For example, a 1,770-rpm, 50-hp vertical-turbine motor may be approximately 90-percent efficient. The brake horsepower then has to be divided by the efficiency of the drive unit to obtain input horsepower:

$$\text{Input horsepower} = \frac{Bhp}{\text{motor efficiency}} \qquad (17.3)$$

Shut-Off Head

Shut-off head is defined as the total head developed by a pump when zero flow occurs. The shut-off head for the pump curve shown in Figure 17.2 for the 9⅝-in (244-mm) pump is approximately 87 ft (26.5 m). It is possible to calculate the shut-off head that a centrifugal pump will develop against a closed valve or calculate the height that the water will reach in a pipe without overflowing. For example, to determine the shut-off head of a single-stage centrifugal pump operating at 1,800 rpm and equipped with a 9⅝-in (244-mm) impeller, first calculate the peripheral velocity, V, by the equation:

$$V = \pi d \cdot \text{rps} \qquad (17.4)$$

where d is the diameter of the impeller in ft, and rps is revolutions per second. The velocity is then:

$$V = 3.14 \cdot 0.8 \cdot 30$$
$$= 75.4 \text{ ft/sec } (23 \text{ m/s})$$

Use the following equation to calculate the shut-off head, H:

$$H = \frac{V^2}{2g} \qquad (17.5)$$

where g is the acceleration of gravity. In this example, the shut-off head is:

$$H = \frac{(75.4)^2}{2 \cdot 32.2}$$
$$= \frac{5685}{64.4}$$
$$= 88.3 \text{ ft } (26.9 \text{ m})$$
$$= 38.2 \text{ psi } (263 \text{ kPa})$$

Most pump manufacturers provide the shut-off head in the pump specification data.

Occasionally, a centrifugal pump may run against a closed valve. This can occur when a gate valve is closed or the pump cannot reach a pressure-head setting that has been inadvertently set too high. The pump will react to the higher total dynamic head and continually follow the performance curve toward the left. It eventually reaches a point where it can no longer pump any water and thus reaches the shut-off head, or where zero production occurs. This suggests that the head requirements are

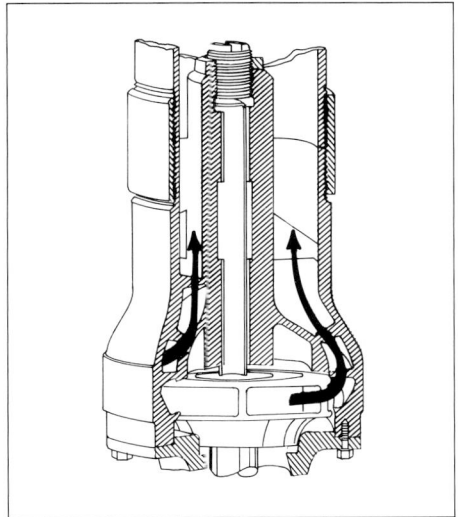

Figure 17.4. In turbine pumps, water leaving the impeller moves out through an impeller that is surrounded by the curved passages between diffuser vanes. As the passages enlarge, the velocity of the water leaving the impeller is reduced, thereby increasing the pressure.

so great that no water can be delivered.

When this happens, the pump will churn and recirculate the water continually within the pump volute. The friction effects of this activity or recirculation will turn mechanical energy into heat energy, and, if the situation persists, enough heat may develop to boil the water; this will cause steam to displace the available water and the impeller will rotate in vapor. The lack of a coolant will cause the bearings, packing, or wear rings to seize, resulting in severe pump and motor damage.

Centrifugal Pump Design

There are five distinct types of centrifugal pumps, each of which can be modified, within limits, by changing the impeller design to provide different operational characteristics. An impeller is the rotary element in a centrifugal pump that imparts a high velocity to the water. The five pump types are:

1. Turbine (diffuser)
2. Volute
3. Mixed flow
4. Axial flow (propeller)
5. Regenerative

Water well contractors generally use only the turbine pump. In this type of pump, the impeller is surrounded by diffuser vanes that provide gradually enlarging passages in which the velocity of the water leaving the impeller is reduced, thereby increasing the pressure (Figure 17.4). Because so many turbine pumps are used in deep-lift installations, the term "turbine pump" is often misapplied to all centrifugal pumps used in these installations. Much of this chapter will focus on the application of turbine pumps to water wells.

The volute pump differs from the turbine pump in that there are no diffuser vanes and the impeller is housed in a spiral-shaped case (Figure 17.5). Similar to a turbine pump, the velocity of the water is reduced upon leaving the impeller, thus transforming velocity head to pressure head. The choice between volute and turbine pumps depends on the conditions of

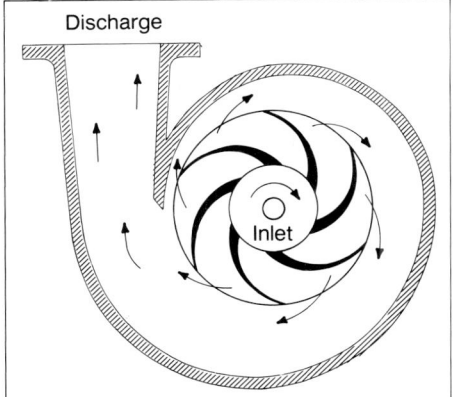

Figure 17.5. Volute-type centrifugal pump has no diffuser vanes or guides.

use. Ordinarily, the volute design is preferred for large-capacity, low-head applications, whereas the turbine design is desirable where high heads are involved.

Mixed-flow centrifugal pumps use both the centrifugal force generated by an impeller and some lifting action produced by a propeller to move water. Mixed-flow pumps are used extensively for large-capacity installations operating against relatively low heads. Axial-flow pumps are often called propeller pumps because they produce most of the flow by the lifting action of propellers. They are used almost exclusively for large-capacity pumping against extremely low heads.

The performance characteristics of a regenerative centrifugal pump lie between those of a conventional centrifugal pump and a rotary (positive displacement) pump. It resembles a rotary pump because of its steep head characteristics. The efficiency of this type of pump is comparatively low, but for many low-capacity, high-head applications, it is superior to other types. However, the clearance fit of the impeller in the housing limits its use to relatively clear fluids.

Impeller Design

The volume of fluid that a centrifugal pump is capable of pumping is dependent upon the design of the impellers and the speed at which they rotate. The motor determines the speed of rotation, whereas the area (opening) of the impeller controls the fluid movement. The impeller area is determined by the width of the vanes between shrouds. The equation relating impeller velocity and area is given as:

$$Q = VA \qquad (17.6)$$

where Q is discharge rate in cfs (m³/sec), V is velocity in ft/sec (m/sec), and A is area in ft² (m²). A change in speed and diameter of an impeller will have the following effects: the capacity will vary directly with the impeller speed and diameter; the head will vary with the square of the impeller speed and the square of the impeller diameter; and the brake horsepower will vary with the cube of the impeller speed and the cube of the impeller diameter.

The specific impeller design determines the velocity of the water as it leaves the impeller surface. This velocity (kinetic energy) is converted into head pressure when the water travels through vanes in the bowl itself. The conversion of this kinetic energy to pressure or head is expressed by Equation 17.5.

To enhance pump efficiencies, another impeller design criteria called specific speed, N_s, must be considered in designing large-capacity pumps. Specific speed is that speed, in revolutions per minute, at which a given impeller would operate if reduced proportionately in size so as to deliver a capacity of 1 gpm (5.5 m³/day) against a total dynamic head of 1 ft (0.3 m) (Colt Industries, 1974). It is calculated from the equation:

$$N_s = \frac{\text{rpm} \cdot \text{gpm}}{H^{0.75}} \qquad (17.7)$$

where $H^{0.75}$ is the head per stage in ft (m).

Specific speed is used to compare one type of impeller or impeller system against another; however, there is no practical value to the number itself. The head and capacity for the pump should be selected at the highest efficiency point of the largest diameter impeller in the pump. Thus, different conditions of head and capacity are accommodated by adjusting the impeller speeds. For example, impellers for high heads usually have low specific speeds, whereas impellers for low heads have high specific speeds.

Specific speed is especially useful in predicting suction-lift capacity for centrifugal pumps. A pump with a low specific speed will avoid cavitation problems by producing a greater suction lift than one with higher specific speed. Slower speeds are used when the suction lift is high [more than 15 ft (4.6 m)]. This requires a larger pump. If the suction lift is low, or a positive head exists on the suction, the specific speed may be increased, thereby reducing the pump size.

Figure 17.6. Semi-open impellers consist of partially unsupported, curved vanes. *(Western Land Roller)*

The impeller performance of centrifugal pumps can be determined by changing the form of the vanes to produce different pumping characteristics, and by selecting (1) the type of enclosure for the vanes; (2) the type of seal at the pump body, and (3) the method of overcoming thrust. The design parameters for impellers are complex, and pump design engineers should always contact pump manufacturers for specific engineering data.

Semi-Open Impellers

Semi-open impellers have a series of curved but partially unsupported vanes that are enclosed at the top only (Figure 17.6). Commonly, this type of impeller is used for pumping liquids carrying some solids. Lack of close clearance or restricted passageways is desirable to prevent plugging. In a deep-well application, this impeller design would be desirable if the well were pumping some sand. Although abrasion would not be as great and the pump would be less likely to sand lock, erosion of the impeller by sand cutting cannot be repaired.

The design of a semi-open impeller allows for some slippage of water from the high-pressure side to the low-pressure side, resulting in a slight reduction in efficiency. The amount of slippage and the efficiency reduction depends on the placement of the impeller in the bowl assembly. If the vanes of the impeller are placed as close to the bottom of the bowl as possible without scraping, the yield and efficiency will be close to the design criteria. In a vertical turbine installation, this placement is accomplished by adjusting the head (adjusting) nut on the head shaft above the motor; this adjustment will be discussed further in the section on vertical turbine pumps.

Closed Impellers

Closed impellers have vanes that are enclosed at the top and bottom (Figure 17.7). This provides a controlled area to channel water through the impeller. Flow enters the eye of the impeller and follows the enclosed vane into the next assembly. The tolerances between the outside skirt of the impeller and the vertical edge of the bowl assembly are quite close and do not allow much leakage past the impeller if placed properly in the bowl. The setting of the impeller within the bowl, however, is not as critical as for the semi-open impeller, so maximum efficiencies and designed yields are much easier to obtain.

Seals

In a vertical turbine pump, there are three basic types of mechanical seals between the impeller and the stationary bowl assembly: side seal, bottom seal, and combined side and bottom seal (Figure 17.8). In the side seal, the seal is created between the vertical skirt of the impeller and the vertical surface of the bowl. The placement of the impeller vertically is not extremely critical because there is some margin for error.

Figure 17.7. Closed impellers have vanes that are enclosed at both the top and bottom. *(Western Land Roller)*

The bottom seal is created between the bottom of the skirt on the impeller and the horizontal ledge of the bowl assembly. Vertical placement of the impeller is critical in this case or leakage will occur and maximum yield and efficiency will not be obtained. This type of seal will not cause the impeller to sand-lock as readily as the side-seal configuration.

The combined side- and bottom-seal arrangement is the most widely used sealing method. If the impeller is set approximately halfway up the side skirt in a new installation, the impeller can be adjusted downward to establish the bottom seal if wearing occurs.

Thrust

Axial thrust (parallel with the shaft) developed by a turbine pump is usually accommodated by the motor bearings. Thrust in vertical turbine pumps originates from two opposing vertical forces — upthrust and downthrust. Under normal operating conditions, the downward component is by far the larger; during starting, however, upthrust forces dominate in some pumps. The forces combining to produce the re-

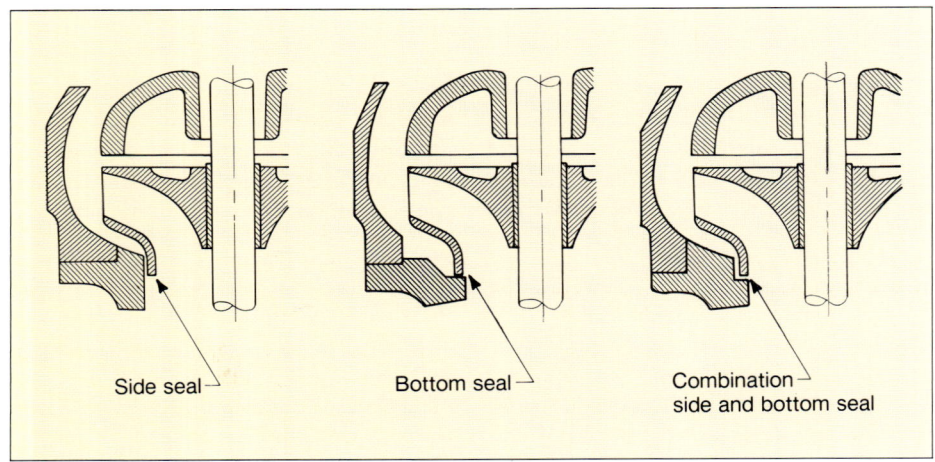

Figure 17.8. The three basic types of mechanical seals between the impeller and bowl assembly are side seal, bottom seal, and combination side and bottom seal.

sultant thrust may be enumerated as follows.

Force I — the weight of the rotating parts (impeller and line shaft). This is a downward force that is constant for a given unit, regardless of the speed, pumping head, or designed capacity.

Force II — the hydraulic force produced by the pressure developed in the pump. This downward force is the result of the pressure differential across the impellers and is proportional to the head developed by the unit.

Force III — the dynamic force produced by the change in the direction of flow from vertical to horizontal as the pumped fluid flows through the impellers. This upward force is proportional to the capacity squared, the angle through which the fluid is deflected from the vertical, and the number of stages in the unit. For a given capacity, this upward force is greatest in a true centrifugal impeller, less in a mixed-flow impeller, and practically negligible in an axial-flow impeller.

Force IV — a second dynamic force produced by the tilt of the vanes and the angle of the waterway through the impeller. This force is also downward and is practically negligible in a true centrifugal impeller and is greatest in an axial-flow impeller.

Whether a pump will operate in an upthrust or downthrust condition depends upon the balance of these four forces. Under normal conditions with a pump operating near the maximum efficiency point, Force I and Force II will exceed Force III even when Force I is at a minimum, as with a close-coupled pump. When a close-coupled pump is started, however, it will operate for a certain period at lower head and correspondingly higher capacities before the normal operating head can be developed. Under these conditions, Force II is reduced and Force III is increased, and the condition of upthrust results. This rarely occurs on units that are set deeper than 20 to 30 ft (6.1 to 9.1 m), because Force I is steadily increased by the weight of the additional shafting to the point where Force III is equalled or exceeded.

The close-coupled pump will continue to operate in upthrust until the head on the unit is increased sufficiently to cause Force II to exceed Force III. This may be an extremely short period when the unit pumps directly through a check valve into a high-pressure main, or may be quite an extended period when the head consists mainly of pipe friction and a long level pipe line must be filled before the head is developed. In rare instances, the pumping cycle of such a unit may include extended periods when it must operate against extremely low heads and will be in upthrust during these periods. Under these conditions, the standard upthrust control provided with the pump may be inadequate, and special provisions must be made in the motor and top-shaft construction to secure satisfactory operation. Because the semi-open impeller has a greater Force II than does the closed impeller, it will be a little less likely to operate in an upthrust condition.

The thrust that is developed must be overcome by hydraulic or mechanical means. In a horizontal centrifugal pump, double-suction impellers are used so that water enters from both sides in equal volumes to balance the thrust hydraulically. In most centrifugal pumps (including vertical turbine), the thrust is taken up by thrust bearings. To assist the well-design engineer, all deep-well turbine manufacturers provide downthrust (K factors) for each impeller design.

Net Positive Suction Head (NPSH)

A centrifugal pump will operate only if a liquid can enter the first-stage impeller under a pressure usually equivalent to atmospheric pressure [14.7 psi (101 kPa)]. The pressure head that is available to the pump consists of two parts, the head of water and the head caused by atmospheric pressure. Pumps using only suction lift can produce flow from atmospheric pressure alone, whereas submerged pumps utilize both pressure sources. The pressure required to operate a pump is referred to as net positive suction head (NPSH), and must be sufficient to prevent vaporization of the water as it enters the pump.

All liquids have a tendency to change from a liquid state into a vapor state. Liquids are said to have high vapor pressures when their tendency to vaporize is great. Many remain as liquids only at pressures greater than atmospheric pressure. Other have extremely low vapor pressure and demonstrate only a slight tendency to vaporize, even under vacuum. Water is somewhere at a midpoint between these extremes. The tendency of water to vaporize increases rapidly with increasing temperature.

Vapor pressure is a significant factor in the behavior of all pumping devices. It is particularly important in variable displacement pumps, such as centrifugal pumps, and less important in positive displacement pumps. In all types of pumps, however, vapor pressure has a limiting effect on suction lift.

As stated above, water must enter the pump at a positive pressure usually equal to atmospheric pressure. If this is not done, the water will tend to vaporize into steam until the vapor pressure in the air surrounding the water reaches a point that prevents further boiling of the water. At higher elevations, water will boil at lower temperatures because the vapor pressure of the liquid can more quickly exceed the lesser atmospheric pressure.

Water circulated in the pump bowls under reduced pressure conditions gains in temperature. As temperatures rise, molecular activity in the water increases, causing more water molecules to escape. At 212°F (100°C), the vapor pressure of water is the same as the normal atmospheric pressure at sea level. Thus, the NPSH available at the surface of the liquid is zero. As long as heat is supplied to the water, it continues to boil in an attempt to create the proper vapor pressure. In summary, the required net positive suction head is the suction head needed to eliminate any boiling of the water under the reduced pressure conditions found near the impellers. Unless this is done, cavitation (vaporization) will occur, causing severe pitting of the impellers and pump housing and shortening of the effective pump life.

Pump manufacturers specify the required NPSH for their various pumps. The engineer computes the available NPSH based on the local altitude, operating temper-

Table 17.2. Atmospheric Pressure at Various Altitudes

Altitude		Barometer Reading		Atmospheric Pressure			
ft	m	in Hg	mm Hg	psia	kPa	ft of water	m of water
-1,000	-305	31.0	787	15.2	105	35.2	10.7
0	0	29.9	759	14.7	101	33.9	10.3
1,000	305	28.9	734	14.2	97.9	32.8	10.0
2,000	610	27.8	706	13.7	94.5	31.5	9.6
3,000	915	26.8	681	13.2	91.0	30.4	9.3
4,000	1,220	25.8	655	12.7	87.6	29.2	8.9
5,000	1,520	24.9	632	12.2	84.1	28.2	8.6
6,000	1,830	24.0	610	11.8	81.4	27.2	8.3
7,000	2,130	23.1	587	11.3	77.9	26.2	8.0
8,000	2,440	22.2	564	10.9	75.2	25.2	7.7
9,000	2,740	21.4	544	10.5	72.4	24.3	7.4
10,000	3,050	20.6	523	10.1	69.6	23.4	7.1

atures, atmospheric pressure changes caused by weather systems, and other minor considerations. The design of the intake system is then adjusted to account for these factors. For operational safety, the available NPSH should exceed the required NPSH by at least 2 to 3 ft (0.6 to 0.9 m), if possible (Walker, 1980). Tables 17.2 and 17.3 show how atmospheric pressure varies with altitude and how vapor pressure changes with temperature.

The equation for calculating available NPSH is:

$$\text{NPSH} = H_a + H_s - H_f - H_{vp} \qquad (17.8)$$

where
H_a = absolute pressure on the liquid surface of the water, in ft (m) of liquid
H_s = elevation of the liquid above or below the impeller eye while pumping, in ft (m). If the level is above the eye, H_s is positive; if below, it is negative.
H_f = friction-head losses in the suction piping, in ft (m)
H_{vp} = absolute vapor pressure of the liquid at the pumping temperature, in ft (m)

Figures 17.9 and 17.10 show the two typical pump settings and the physical relationship of the pump to the groundwater system. The suction-lift pump must be placed within 20 to 25 ft (6.1 to 7.6 m) of the pumping water level, whereas the intake level of the submerged pump is generally well below the pumping water level.

Cavitation

Cavitation is a condition that occurs when the pressure acting on a stream of liquid falls to or below the vapor pressure of that liquid. When water enters the eye of the

Table 17.3. Vapor Pressure of Water

| Temperature | | Absolute Vapor Pressure | | | |
°F	°C	psia	kPa	ft of water	m of water
32	0	0.09	0.62	0.20	0.06
40	4.4	0.12	0.83	0.28	0.09
50	10.0	0.18	1.24	0.41	0.13
60	15.6	0.26	1.79	0.59	0.18
70	21.1	0.36	2.48	0.89	0.27
80	26.7	0.51	3.52	1.2	0.37
90	32.2	0.70	4.83	1.6	0.49
100	37.8	0.95	6.55	2.2	0.67
110	43.3	1.28	8.83	3.0	0.91
120	48.9	1.69	11.7	3.9	1.19
130	54.4	2.22	15.3	5.0	1.52
140	60.0	2.89	19.9	6.8	2.07
150	65.6	3.72	25.6	8.8	2.68
160	71.1	4.74	32.7	11.2	3.41
170	76.7	5.99	41.3	14.2	4.33
180	82.2	7.51	51.8	17.8	5.43
190	87.8	9.34	64.4	22.3	6.80
200	93.3	11.5	79.3	27.6	8.41
210	98.9	14.1	97.2	33.9	10.3

impeller in a turbine pump, the velocity increases, thereby causing a corresponding reduction in pressure. If the pressure falls below the vapor pressure (at the liquid's temperature), the liquid will begin to vaporize, and part of the flow through the pump will consist of vapor pockets. At some point, the liquid will reach an area of higher pressure, causing the pockets to collapse at such a rapid rate that a rumbling noise can be heard. The collapse of these pockets is so violent that it causes pitting on the impeller and bowl surface.

In propeller pumps, water from the larger inlet area entering the throat ahead of the propeller accelerates rapidly. If the head is increased too much, the capacity is reduced to the point where insufficient fluid exists to fill the space between the propeller vanes. Vacuum pockets develop along the vanes momentarily, but almost instantly the space is filled as the liquid crashes against the propeller vanes. The force of the liquid hitting the vane surface can cause severe pitting of the vane.

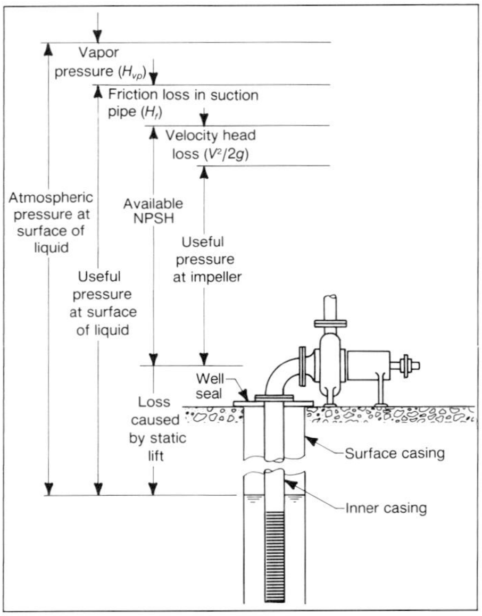

Figure 17.9. Factors that must be considered when determining the NPSH for a centrifugal suction-lift pump (shallow-well installation).

Cavitation is generally indicated by fluctuations or reductions in yield, erratic power consumption (fluctuating amperage readings), and noisy operation. The noise level associated with cavitation suggests how destructive the cavitation may be. Adherance to proper design and operating principles will help avoid most cavitation problems, although total elimination of cavitation may not be possible in some installations. Table 17.4 lists conditions that should be avoided to minimize cavitation.

Suction-Lift Pumps

Suction lift is developed by creating a negative pressure at the pump intake. Atmospheric pressure on the free surface of water in a well forces water up into that part of the pump where the reduced pressure has been developed. The maximum suction lift is limited by four factors: atmospheric pressure, vapor pressure, head losses attributable to friction, and NPSH

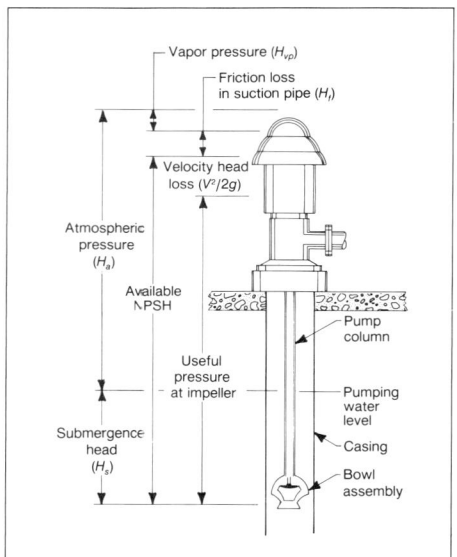

Figure 17.10. Factors that must be considered when determining the NPSH for a vertical-turbine pump (deep-well installation).

Table 17.4. Conditions to Avoid to Minimize Cavitation

Centrifugal Pumps	Propeller Pumps
1. Heads much lower than the head at peak efficiency of pump.	1. Heads much higher than the head at peak efficiency of pump.
2. Capacity much higher than capacity at peak efficiency of pump.	2. Capacity much lower than capacity at peak efficiency of pump.
3. Suction lift higher or positive head lower than recommended by manufacturer.	3. Suction lift higher or positive head lower than recommended by manufacturer.
4. Liquid temperatures higher than that for which the system was originally designed.	4. Liquid temperatures higher than that for which the system was originally designed.
5. Speeds higher than manufacturer's recommendations.	5. Speeds higher than manufacturer's recommendations.

requirements of the pump itself.

Atmospheric pressure varies with atmospheric conditions and elevation. For practical purposes, it is assumed that the Earth's atmosphere normally exerts a pressure of 14.7 psi (101 kPa) at sea level — the equivalent of about 34 ft (10.4 m) of water head. Under normal conditions at sea level, therefore, it might be assumed that a water column would be lifted 34 ft if a perfect vacuum could be produced by a pump. This amount of suction lift cannot occur, however, even if a perfect vacuum could be produced, because of the other limiting factors — vapor pressure and pipe friction. Under field conditions, a centrifugal pump has an average suction-lift capability of 20 to 25 ft (6.1 to 7.6 m).

Deep-Well Turbine Pumps

The centrifugal pump was designed originally as a suction-lift pump, although it was soon adapted for water wells where suction lift was not possible by extending a shaft extension from the above-ground driving unit to the impeller assembly. Today this pump is known as the deep-well turbine pump. It has become one of the most widely used pumps for high-capacity, large-diameter wells. Common types of deep-well centrifugal (turbine) pumps are discussed below.

Vertical Turbine Pumps

The pumping assembly of a vertical turbine pump consists of one or more impellers housed in a single- or multi-stage unit called a bowl assembly. Each stage provides a certain amount of lift; a sufficient number of stages (bowl assemblies) are assembled to meet the head requirements of the system. When designing a pump system, the number of stages needed are proportional to the head and horsepower requirements, whereas the discharge rate and efficiency remain constant. The impellers are suspended on a vertical line shaft (drive shaft) that is housed within the pump column which conducts the water to the surface. The size of the outer column is selected on the basis of the pumping rate. The head losses in the column should not exceed 5 ft per 100 ft (1.6 m per 30.5 m) of column at the designed capacity.

Individual sections of the pump column are generally 10 or 20 ft (3 or 6.1 m) in length and have threaded and coupled or flanged fittings (Figure 17.11). The overall length of column is determined by the pumping water level. For ease of handling during installation and maintenance, a 5-ft (1.6-m) section is normally connected to the pump base and a separate 5-ft section is connected to the bowl assembly. The pump column is attached at the surface to a discharge head which serves several purposes. It provides a base for a driver (electric motor or right-angle gear coupled to a reciprocating engine); creates a packing box around the shaft, that prevents water from entering the motor; acts as an elbow to divert the discharge into the above-ground piping system; and also supports the column pipe in the well.

Pump Intake

The intake of a pump should not be placed within the well screen, because distorted flow patterns will occur in the vicinity of the screen and may result in the following conditions:
1. Increased velocities causing higher incrustation rates.
2. Increased velocities causing corrosion.
3. Increased velocities causing sand pumping.
4. Mixing of air and water from dewatering of the screen, causing incrustation.

The seriousness of these conditions depends upon water quality and well screen design.

Two devices are used to prevent objects that may fall into the well (air lines, water-level indicators, or formation material) from entering the pump. Either a 10-ft (3-m) section of suction (tail) pipe the same diameter as the pump column pipe or a strainer should be attached to the pump intake pipe. Figures 17.12 and 17.13 illustrate the two types of strainers that are generally available, the cone strainer and the basket strainer.

Lubrication

Vertical turbines can be either water lubricated or oil lubricated. In a water-lubricated pump (Figure 17.12), bearings are generally placed every 10 ft (3 m) to permit smooth rotation of the line shaft; these are retained in a centering guide called a spider

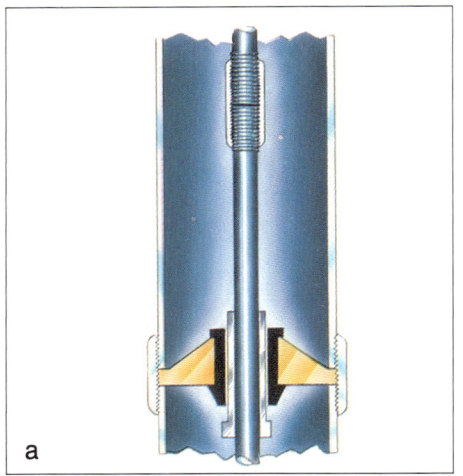
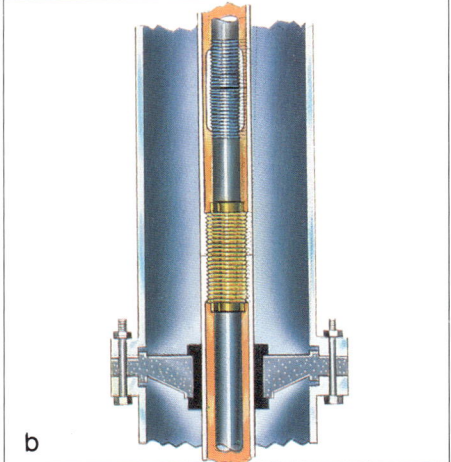

Figure 17.11. Pump column can be (a) threaded and coupled or (b) flanged. As shown here, the spider acts as a seal between the butt ends of the column pipe in addition to keeping the shaft aligned. *(Western Land Roller)*

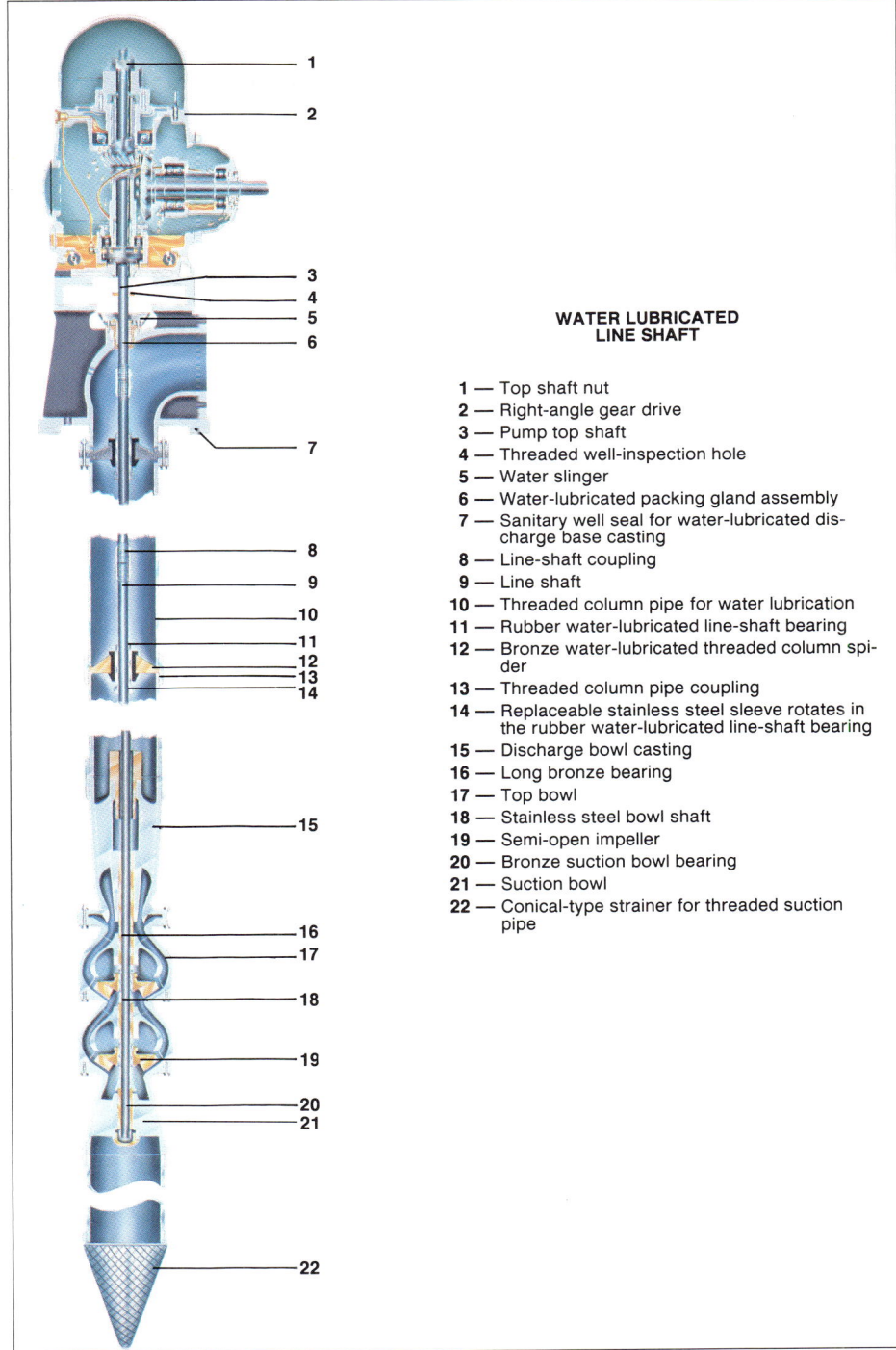

Figure 17.12. In a vertical turbine pump, the bearings can be water lubricated. *(Western Land Roller)*

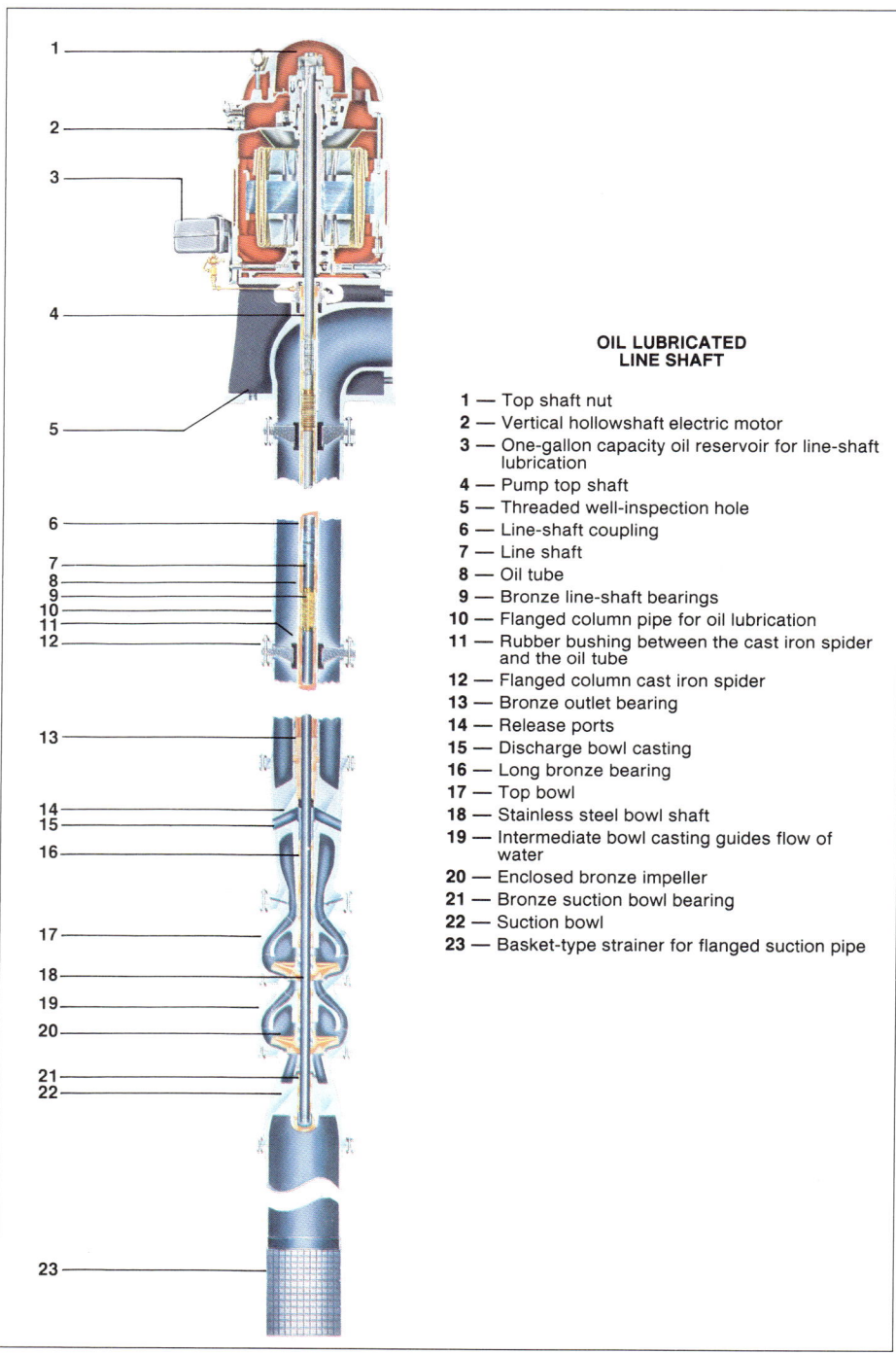

Figure 17.13. The lineshaft in oil-lubricated turbine pumps is placed within a larger diameter pipe called an oil tube. *(Western Land Roller)*

which also acts as a seal between the butt ends of the column pipe (Figure 17.11). The bearing inserts within the spider are generally made of rubber and must be lubricated by the water flowing upward in the pump column. The water prevents the bearings from becoming hot and seizing to the line shaft. Lack of lubrication causes excessive bearing wear and may cause breakage of the spider, producing a whipping action and eventual breakage of the line shaft. If the static water level is lower than 30 ft (9.1 m), these bearings should be lubricated by a prelube line from the surface before the pump is turned on. Generally, this is done from a holding tank at the surface or by means of a small-diameter prelube line leading from the discharge side of a solenoid-controlled check valve on the surface piping. This flow of water provides proper lubrication to the upper bearings until the discharge from the pump reaches them.

In an oil-lubricated pump (Figure 17.13), the line shaft is entirely enclosed within a somewhat larger diameter pipe called an oil tube. There is either a redwood liner or a set of bronze bearings within the oil tube to permit smooth rotation of the line shaft. Bronze bearings are generally recommended where turbine pumps are set at depths of more than 300 ft (91.5 m) because of possible overheating if a redwood liner is used. Oil is fed by gravity from a reservoir at the surface to provide the necessary lubrication. If proper lubrication is not maintained, the bronze bearings will wear, causing premature breakdown. Redwood liners may permit longer operation in this situation, but extended operation is not recommended. In both situations, the oil flows downward and is vented into the well through an oil port in the bowl assembly. The oil will float on the water, and some consideration should be given to the thickness of the oil layer when routine maintenance is performed on the pump. If the pumping water level decreases in a well and comes close to the pump intake or the pump breaks suction, there is the possibility of pumping some oil through the pump and into the distribution system. Therefore, oil lubrication is not recommended for potable water unless the pumping water level is monitored continually.

Figure 17.14. Hollow-shaft electric motors are the most common driving unit for vertical turbine pumps. *(Peerless Midwest)*

Driving Units

The most common driving unit for a vertical turbine pump is a hollow-shaft electric motor that usually operates at 1,800 rpm (Figure 17.14). This power unit is called a hollow-shaft motor because it slips over the head shaft (drive shaft) and is attached by a head nut and keyway. The other kind of electric motor is the solid-shaft type where the motor is attached directly to the drive shaft. This motor also operates at 1,800 rpm. The solid-shaft motor is not used in deep-well installations because it must be turned to thread it to the line shaft during installation.

In many irrigation applications, a diesel, gasoline, or natural-gas engine is used in conjunction with a right-angle gear drive to provide the power to the pump bowls (Figure 17.15). Consideration must be given to the proper rpm of the driver unit so that the required discharge and pressure are attained. The unit should be checked periodically to maintain the efficiency and thus minimize the cost of operation. Another type of driving unit that is not used as commonly is the belt driver from engines or tractors.

Adjustment for Axial Thrust

Thrust can become a major factor in the successful operation of a vertical turbine pump. Downthrust causes a certain amount of stretch in the line shaft, which must be calculated before the start-up of any unit. The head nut is adjusted according to these calculations, and the impeller is adjusted upward within the bowl assembly to compensate for the shaft stretch. This prevents excessive wear caused by the impeller rubbing against the bottom of the bowl assembly. In the intial design of a pump, shaft diameter and length must be considered in relation to the stretch and torque requirements of a given horsepower unit to prevent stretching and twisting. Information can be obtained from the pump manufacturer for these calculations.

Non-Reverse Ratchets

When a pump stops, the weight of the water flowing downward in the pump column causes the pump to run in reverse. The pump shaft normally will not unscrew, because

Figure 17.15. Vertical turbines can be driven by a diesel, gasoline, or natural-gas engine connected to a right-angle gear drive. *(Test Drilling Services)*

it is being driven from the lower end of the shaft by the impellers acting as a water turbine, rather than by the motor. The motor can be damaged or the pump shaft broken, however, if the pump is restarted when the motor is rotating in the reverse direction. This may occur when a water-level or low-pressure sensing device automatically sends a signal to start the motor before the pump has completed its backspin. It can also happen during momentary power losses. Thus, a non-reverse ratchet is generally used in a vertical turbine pump to prevent the motor from spinning backwards after the pump is shut off. There are other devices that can be used with or in place of a non-reverse ratchet. For example, time-delay switches prevent the motor from starting too soon after it has stopped, and rotational switches lock out the starter when the pump is reversing.

A non-reverse ratchet or other device should be used when:
1. Pumping levels for water-lubricated turbine pumps are over 50 ft (15.2 m).
2. Semi-open impellers are used in settings over 100 ft (30.5 m).
3. Closed impellers are used in settings over 200 ft (61 m).
4. The surging effect caused by the water returning to the well causes sediment to enter the well, which can lead to sand pumping.

There are major advantages to using vertical turbine pumps in high-capacity wells:
1. They are highly reliable over long periods of time (if properly designed and maintained).
2. Motor repairs are made easily because of aboveground installation.
3. Motors are not susceptible to failure caused by fluctuations in electric current.

Major disadvantages include:
1. They cannot be used in wells that are out of alignment.
2. They require highly skilled journeymen for proper installation and service.

Submersible Pump

Submersible pumps have bowl assemblies that are the same as those of vertical turbine pumps. The motor, however, is submerged and is directly connected to and located just beneath the bowl assembly. Water enters through an intake screen between the motor and bowl assembly, passes through the stages, and is discharged directly through the pump column to the surface (Figure 17.16).

Submersible motors are extremely compact and generally do not withstand overheating and fluctuations in voltage. They are cooled by water passing by the motor casing and into the intake of the pump, so a free flow of water must be

Figure 17.16. Submersible pumps have the bowl assemblies and motor directly connected. *(Test Drilling Service)*

maintained. Overheating may occur if the well has cascading water or if the pump intake is set into a sump (casing below the screen). This problem can be overcome by placing a shroud over the intake of the pump (Figure 17.17). The water is then forced to pass by the motor and provide the necessary cooling. A shroud is not difficult to make from oversized pipe, and will work quite successfully if the diameter of the casing is large enough.

In most cases, the electric motors used to drive submersible pumps are provided with 2- to 3-ft (0.6- to 0.9-m) cable leads. These lead wires must then be spliced to the length of submersible cable required for each installation. This splice must be watertight to keep the unit operating successfully. Follow manufacturer's recommendations for splicing wires.

Operating voltage is especially important for submersible pumps because they are designed to operate within plus or minus 10 percent of the motor-nameplate voltage. Any variation in the voltage beyond this range, or problems from electrical storms, may cause the motor to fail. Proper starter panels are required to provide overload protection. Manufacturer's recommendations should be followed.

Most small (fractional through 40 hp) submersible pumps operate at 3,500 rpm, although units with other rpm characteristics are available for different design applications. They can be operated successfully to depths of 2,000 ft (610 m) or greater, at pressures of 300 psi (2,070 kPa). Submersible pumps have become the major type of pump used in domestic wells, and increasing numbers of submersible pumps have been installed in large-diameter, high-capacity wells.

Submersible pumps have several advantages:
1. The motor is directly coupled to the impellers.
2. It is easily cooled because of complete submersion.
3. Ground surface noise is eliminated.
4. The pump can be mounted in casings that are not entirely straight.
5. A pump house is not necessary if a pitless adaptor (underground discharge) is used.

Disadvantages of submersible pumps are:
1. Electrical problems caused by submerged cables and splicing of cables to the motor.
2. Overall efficiency is generally lower.
3. They cannot tolerate sand pumping (will sand lock).
4. Motor is less accessible for repairs.
5. They cannot tolerate voltage fluctuations without proper protection.

Jet Pumps

Jet pumps are used in many domestic wells and are a combination of a centrifugal pump and a nozzle-venturi arrange-

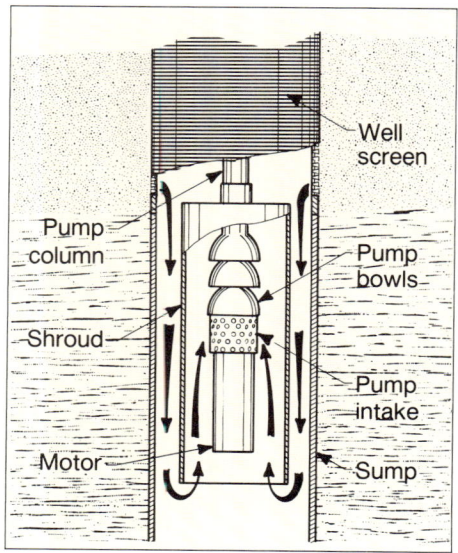

Figure 17.17. A shroud can be placed over the intake portion of a submersible pump to prevent overheating caused by cascading water.

ment (Figure 17.18). Water discharges under pressure through a nozzle inserted in the pipe conveying the water. The nozzle is shaped so that it smoothly, but rather abruptly, reduces the area through which the flow must pass, thus increasing the velocity of flow. In accordance with the Bernoulli law (Chapter 5), the water pressure in a pipe decreases in direct relation to any increase in the flow velocity, and vice versa. That is, if the velocity increases at any point because of a reduction in area, as would occur at point A near the nozzle in Figure 17.18, the pressure decreases proportionately at that point.

If the discharge velocity at the nozzle is great enough, the pressure at point A will be lowered sufficiently to draw water into the venturi assembly through an opening at this point, and this water is added to the total volume of water flowing beyond point A. The gradual enlargement in the venturi tube to the full diameter of the pipe reduces the velocity with a minimum of turbulence, and pressure in the pipe at point B is recovered, minus the head loss caused by friction. The prime mover in a jet pump is a centrifugal pump, which produces the flow to the nozzle and maintains the combined flow through the intake pipe beyond this point. This combined flow is composed of the recirculating water and the water picked up at point A from the well. The additional increment of water obtained from the well continues past the control valve at point C and goes into use or storage, while the volume required for producing the flow is recirculated through the pressure line. The control valve is set (automatically or manually) to maintain the necessary pressure to produce flow at the existing pumping head. No water will be pumped beyond this valve, for use or storage, until enough volume passes through the pressure line to produce the needed pressure at the nozzle.

To increase the discharge head, the work the pump performs must increase. This demand is met by adding stages to the centrifugal pump, which increases the pressure output and horsepower demands in direct proportion to the number of impellers. The volume of flow remains constant, similar to all impeller-type pumps.

A jet assembly can also be added to the intake of a suction-lift installation. It will increase the suction-lift capability considerably beyond that which is practical for

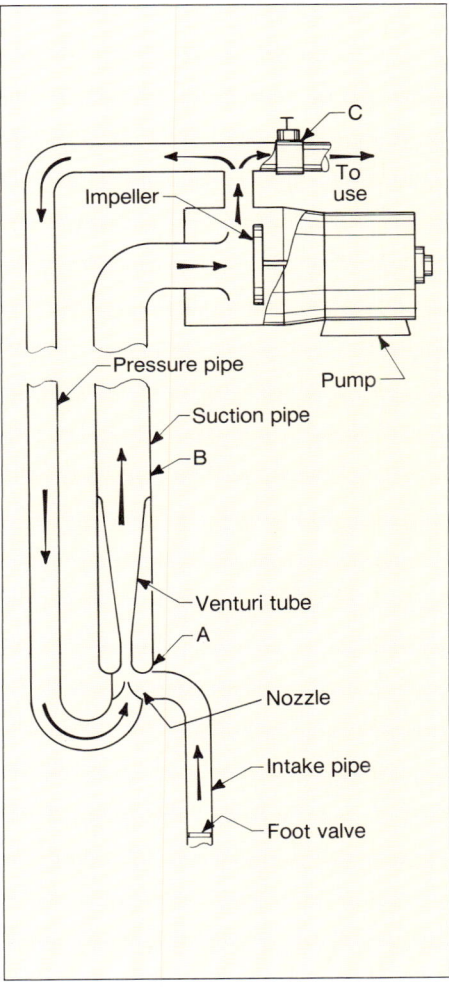

Figure 17.18. A jet pump uses a nozzle-venturi design to pump water.

the small, conventional centrifugal pump itself.

Jet pumps are inefficient when compared with ordinary centrifugal pumps, but this is not necessarily objectionable in domestic installations, because of other favorable features, such as:

1. Adaptable to small wells, down to 2-in (51-mm) inside diameter in deep-lift installations.
2. All moving parts are accessible at the ground surface.
3. Simple design combined with relatively low equipment and maintenance costs.
4. Capable of being installed with the moving parts offset from the well.

In some locations, jet pumps may not be completely satisfactory, such as where water levels are subject to large seasonal variations or where severe corrosion or incrustation causes enlargement or plugging of the nozzle.

Priming Centrifugal Pumps

Priming is necessary to expel air from centrifugal pumps. Many devices and procedures are used to obtain and maintain a primed condition in centrifugal pumps; the literature describing them is fairly extensive. In general, however, all involve one or a combination of the following: (1) a foot valve to hold water in the pump, (2) a means for venting to dispose of entrained air, (3) an auxiliary pumping device to partially fill the centrifugal pump and intake line with water, (4) connection to an outside source of water under pressure for filling the pump, and (5) use of self-priming construction. Self-priming construction retains water for priming in an auxiliary chamber which is integrated into the pump structure in such a way that entrained air is exhausted as the pump circulates the priming water.

POSITIVE DISPLACEMENT PUMPS

Positive displacement pumps discharge the same volume of water regardless of the head against which they operate, although in practice this is not quite true, because some water slips past operating parts. This type of pump must be powered to meet maximum load based on its discharge capacity and the greatest head under which it will operate. When used in a water system, the rate of discharge is essentially the same at both low and high pressure, but the input power varies in direct proportion to the pressure.

There are many designs of positive displacement pumps, but the types used most are:

1. Rotary pumps
2. Peristaltic pumps
3. Piston (reciprocating) pumps

Rotary Pumps

The rotary pump is widely used because there are many design modifications for special applications. Most applications are of the suction-intake type, except when the pump is used for booster purposes in conjunction with another pump to increase the pressure or to pump hot water or other liquids having high vapor pressure; in this case, the pump operates under positive intake pressure. Common designs use cogs or gears, and rigid vanes or flexible vanes; none of these pumps require valves.

The original rotary pump was designed using gears, and is simple in principle and

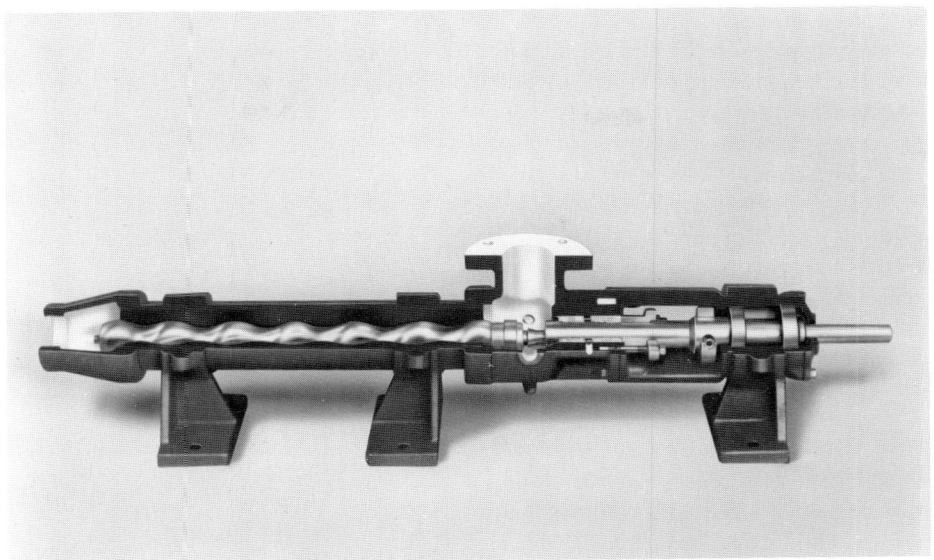

Figure 17.19. Screw or progressing cavity pumps are capable of moving liquids at uniform flow at low-to-high pressures. They are particularly suitable for pumping liquids with high viscosities, such as cement grout. *(Robbins & Myers, Inc., Moyno® Industrial Products)*

construction. It consists of a plain housing with inlet and outlet ports, and openings for shafts which carry the driver gear and a driven or idler gear. The gears are fitted closely in the housing, and mesh with minimum clearance. When rotated, the gears squeeze the water from between the teeth as they mesh together, bringing in a replacement supply of water along the outer surface of the housing at the inlet side of the moving teeth of the gears.

A typical rigid-vane rotary pump has a series of dividers or vanes fitted into a slotted rotor. When rotated, these vanes move radially to conform to the contour of the pump housing, which is eccentric in comparison with the rotor, so that the water is pushed from the pump in a continuous flow ahead of the vanes. Water moves into the housing behind the vanes because a partial vacuum is created. A flexible-vane rotary pump has blades that bend to provide the change in displacement volume which forces the water along its path.

Rotary peristaltic pumps are further modifications of the rotary principle. Typical designs of rotary pumps are shown in Figures 17.19 and 17.20.

Figure 17.20. Rotary peristaltic pumps are often used as chemical feeding devices and are used for taking water samples from monitoring wells. *(Chem-Tech International)*

Piston Pumps

The simplest arrangement for a piston pump is the single-action pump, shown schematically in Figure 17.21. When the piston is drawn upward, the check valve in the piston is closed by gravity and the water pressure above it. A lowering of pressure is therefore produced below the moving piston. Water flows through the intake valve into the pump cylinder as a result of the pressure differential caused by the stroke of the piston.

As the piston moves downward, the intake valve closes when the pressure above it exceeds the pressure below it, and the discharge valve opens when the pressure below it exceeds that above it. Thus, water trapped in the cylinder during the downstroke of the piston is forced upward into the discharge pipe on the next upstroke.

Because water is virtually incompressible, the piston moves the same volume of water at each stroke, regardless of pressure, less any water that slips past the piston and valves. See Appendix 17.B for the volume of water discharged per stroke for various size pumps. Piston pumps must be powered to meet the maximum pressure application, and protection against breakage must be provided by some device, such as a pressure-relief valve, in case the pressure switch or other control mechanism fails.

The basic principles just described apply to all piston pumps; however, there are many design modifications that adapt these pumps to specific uses. Double-action pumps, for example, are constructed with piston and valves arranged so that water is pumped on both inward and outward movement of the piston. These are most commonly suction-lift pumps, but are also available for pressure-intake installations in wells. Duplex and triplex pumps consist of two or three pistons, respectively, and are designed to pump a continuous stream with minimum pulsation, often against high pressure. These pumps are often utilized for pumping drilling fluid and grout.

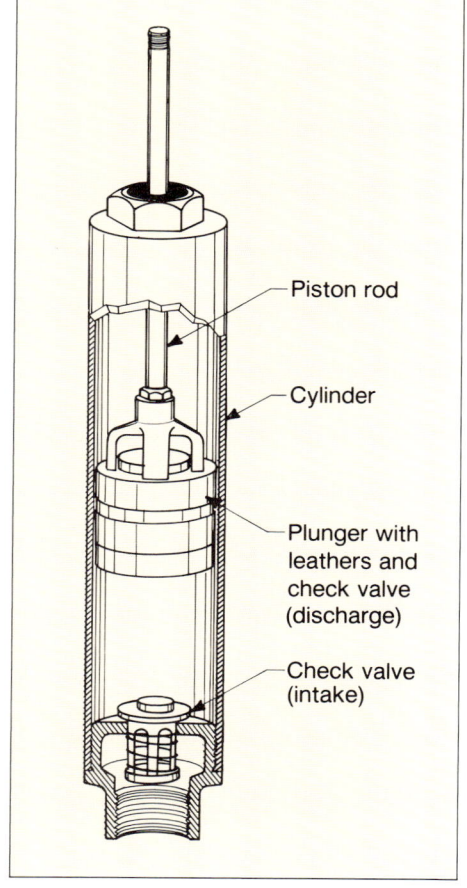

Figure 17.21. A single-acting piston pump, suspended on discharge pipe, can be installed at almost any depth in small-diameter wells.

Priming Positive Displacement Pumps

Positive displacement pumps must be primed only to the extent necessary to stop leakage past pistons, valves, or other working parts. They have the capability to move and compress most fluids, including air, so that water (or another liquid) can be drawn into the pump without priming.

PUMPS USED TO CIRCULATE DRILLING FLUID

Two basic types of pumps are used to circulate drilling fluid — reciprocating and centrifugal. The reciprocating pump can be either single or double acting. Either of these designs can be arranged in parallel by operating the pistons from a common crankshaft. They are then referred to as simplex (1 cylinder), duplex (2 cylinders), and triplex (3 cylinders) pumps (Figure 17.22). Additional pumps can be added if more volume must be moved.

Optimum operating conditions for reciprocating mud pumps are assured if:

1. The suction piping is as short as possible.

2. The diameter of the suction line is large enough so the fluid velocity is less than 3 ft/sec (0.9 m/sec).

3. An intake strainer with two to four times the intake area of the suction hose is installed at the end of the suction line.

4. The diameter of the discharge pipe is large enough so that the fluid velocity is less than 5 ft/sec (1.2 m/sec).

5. The pump operates under proper net positive suction head conditions.

6. A surge suppressor is installed in the suction line as close to the pump as possible.

7. The pump is selected for a higher discharge than the expected demand.

8. Priming time does not exceed 30 seconds in order to minimize friction damage to the plunger.

The operation of a centrifugal pump is described earlier in this chapter. When used for mud pumps, they are designated by the size of the discharge and suction lines and the size and rotation direction of the impeller. Unlike reciprocating designs, the rate of fluid delivery for a centrifugal pump is variable depending on the pressure in the borehole. At maximum discharge, the pressure head is usually about two-thirds the pressure obtainable at zero fluid delivery (Petroleum Extension Service, 1974). The actual drilling fluid discharge is a function of the rotation speed, fluid efficiency, and the available power, as well as the downhole pressure. A centrifugal pump is

Figure 17.22. Reciprocating duplex or triplex pumps can create high pressures and therefore are ideal for pumping drilling fluids. *(Layne Company)*

most suitable for low-pressure, high-volume situations, whereas reciprocating pumps are ideal for high-pressure, low-volume applications. The basic difference in the operational characteristics of the two mud pump types is illustrated in Figure 17.23.

AIR-LIFT PUMPING

Water can be pumped from a well by releasing compressed air into a discharge pipe (air line) lowered into the well. Air bubbles mix with the water and reduce the specific gravity of the water column sufficiently to lift it to the surface. Because air-lift pumping is inefficient in comparison with other pumping methods, and because of the rather cumbersome and expensive equipment required, this method of pumping is rarely used as a permanent pumping system. In those instances where it is used, there is likely to be some special reason, such as the need for aeration to remove an objectionable gas, or the occurrence of a highly corrosive or abrasive water that is destructive to pump parts. Air-lift pumping is used rather extensively in the preliminary testing and development of wells, however, especially in conjunction with the high-velocity jetting method of development. This method is discussed in Chapter 15.

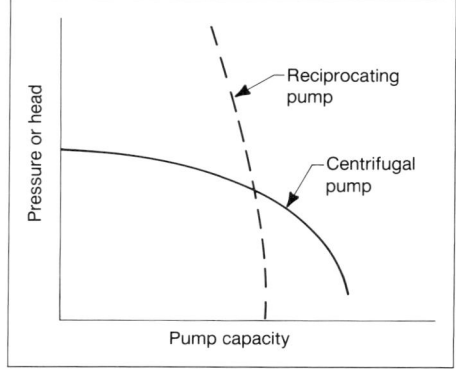

Figure 17.23. Centrifugal versus reciprocating pump pressure and capacity characteristics. *(F. E. Myers Company, 1965)*

For permanent or semipermanent air-lift pumping installations, it is accepted practice to discharge the water at the lowest possible level and use a booster pump for elevation because any additional lift adversely affects the functioning of an air-lift pump. A large tank is required between the air-lift and booster pumps to provide for surges in delivery and deaeration.

Figure 15.12 (page 510) shows the volume of water that can be pumped per cubic foot of air per minute, when total lift and available submergence are known. Table 15.1 (page 512) lists the recommended sizes of eductor and air-line pipes for pumping various quantities of water with an air-lift pump.

PUMP SELECTION

Basic pump design, anticipated pumping conditions, and specific installation procedures are factors that must be considered in choosing a pump for a water well. Pump engineers must outline the general operating conditions for the pump before a specific type is selected. Design parameters include:
1. Well diameter
2. Desired yield
3. Total dynamic head
   a. Pumping water level (shallow- or deep-well conditions)
   b. Above-ground head
   c. All friction losses in column, pipe, fittings, etc.
4. Horsepower requirements
   a. Brake horsepower
   b. Horsepower required to offset shaft losses (vertical turbine)

c. Motor efficiency without thrust load
   d. Losses caused by friction in the thrust bearing
   e. Horsepower curve for varied discharge rates
5. Power source
   a. rpm preferred or required
6. Pumping deviation — system-head-curve parameters
7. Sand-pumping potential
8. NPSH and specific speed, if required
9. Water quality
10. Short- and long-term costs
    a. Initial capital costs
    b. Amortization of investment
    c. Power costs
    d. Supervision and maintenance
    e. Cost of down time and standby equipment

After the most suitable type of pump has been determined from the available data, a specific pump is selected that will best fit pumping requirements at the site. Proper pump selection is particularly important if a well is to serve as a major source of water, and if differences in pumping head are anticipated. Pumping heads often vary because of (1) seasonal variations in static water level, (2) temporary lowering of the dynamic (pumping) water level as a result of long periods of continuous pumping, and (3) interference from other wells in the area. Permanent lowering of the potentiometric surface or water table may also occur if the water is withdrawn from an aquifer at a rate greater than the natural recharge. All of these conditions should be considered before selecting a pump for a well.

Determining the extent of seasonal fluctuations requires collection of regional data from available records, and the integration of these data with those obtained from test pumping the well. The type and thoroughness of the test-pumping procedure varies greatly from one case to another. For domestic purposes, all that is generally needed is a record of drawdown made after test pumping at a rate and for a period of time equal to, or somewhat in excess of, the service requirements. The well driller or pump supplier should be sufficiently acquainted with seasonal fluctuations to make the needed allowance for these factors when selecting the pump size and depth setting. For large industrial, agricultural, or municipal installations, it is necessary to make a more complete study of the situation. Such a study frequently includes pumping tests that are conducted in a manner that will determine the aquifer constants (see Chapters 9 and 16). With adequate data, a pump can then be selected for optimum performance.

Proper selection of a pump for a domestic or low-yield well is not a complicated decision. It can normally be chosen by reference to manufacturer's literature, which provides information on yield versus head conditions for certain stocked units. Proper selection of a high-capacity unit, on the other hand, may become a complex engineering procedure. The volume of water a well will yield efficiently can be enhanced or reduced depending on the decisions made during the design process. As in well construction, the initial cost of the pump and its installation is much less important than pump performance, reliability, and low operating costs over the life of the equipment. It is particularly important during inflationary periods to specify the most

Figure 17.24. Danish water storage tanks are built to accommodate observation platforms and restaurants. *(C. Allen Wortley)*

reliable equipment available because interest, labor, and overall maintenance costs will escalate rapidly over the economic life of the pumping installation.

WATER STORAGE

One of the most critical factors in the life of a pump is how often the pump must cycle to meet water demands placed on the system. Ideally, long pump life is enhanced if maximum volumes of water can be withdrawn from storage before the pump is activated. The design and maintenance of the storage tank determines how often the pump cycles to meet water demands. Several methods are used to provide water from various kinds of storage reservoirs. The gravity-tank method is the oldest means for making water available under pressure in a water system and for providing a stored reserve. Usually this type of tank is used to provide a large storage capacity for municipal systems. Recently, the design of these tanks has been radically changed to make them more useful to communities. For example, some storage tanks are built to accommodate restaurants or provide observation platforms (Figure 17.24). Another means for supplying and maintaining pressure is by pumping from one or more automatically controlled pumps that are directly connected to the system. Controls regulate the pressure and supply by adding to or reducing the number of pumps operating at any time.

Hydropneumatic systems incorporate features of both gravity and direct-pumping systems. A hydropneumatic tank provides storage space for the proper volumes of compressed air and water, so that during intermittent pumping only a small amount of storage is required. Hydropneumatic systems are ordinarily small in size and are used in most recent domestic and small commercial installations; generally they provide 5 to 10 gpm (27.3 to 54.5 m^3/day). Larger systems can produce 100 gpm (545 m^3/day) flow and up to 5,000 gal (18.9 m^3) gross storage.

The basic operation of hydropneumatic systems is simply based on the compression of air in the upper portion of the water tank. The water is pumped into, and withdrawn from, the lower part of the tank. As water is pumped in, air above it is compressed until the selected maximum pressure is reached. A pressure-control switch stops the pump at this point. As water is withdrawn from the tank, the pressure drops, and the pump is automatically started at a predetermined minimum pressure to repeat the cycle.

A minor disadvantage of hydropneumatic systems is the smaller amount of water that can be withdrawn between high- and low-pressure levels, as compared with a conventional compressor-equipped storage tanks. For example, when free air in a 100-gal (0.4-m^3) tank is compressed to 50 psi (345 kPa) by pumping in water, the tank contains about 78 gal (0.3 m^3) of water and about 22 gal (0.08 m^3) of compressed air. If water is withdrawn until the pressure drops to 30 psi (207 kPa), only about 11 gal (0.04 m^3) is delivered. A way of increasing the stored reserve within a selected pressure range is to prepressurize the air in the tank so that a larger volume of water is delivered. The quantity of water delivered from storage increases about 7 percent for each psi of prepressurizing. The air volume can be maintained either by automatic injection during pumping or by periodic addition of air. Controls are available to limit the volume occupied by the air, independent of the pressure.

Many manufacturers now provide tanks equipped with a rubber diaphragm between the water and the air chamber. The advantage of this system is that the air will not become dissolved in the water, thereby eliminating the need to recharge the tank; thus, fewer pump cycles are required for a particular discharge.

CONCLUSIONS

Drilling contractors and pump installers provide pumps based on the head and yield requirements of their customers, which, in turn, are controlled to some extent by local hydrogeology. Selection of a particular pump design should be determined by matching the customer's requirements with the operating characteristics of available pumps. Informed use of the performance curves provided by the manufacturer for different impeller configurations will assure that the most efficient pump and driver combination is selected. Installation of the pump should follow the design guidelines so that the NPSH requirements can be maintained and unusual pumping conditions, such as those produced by cascading water, are avoided. Maintenance schedules should be established and followed closely to assure long pump life (see Chapter 19).

CHAPTER 18
Water-Quality Protection for Wells and Nearby Groundwater Resources

Outbreaks of waterborne disease have increased significantly in the United States since 1970 (Table 18.1). A waterborne disease outbreak is an incident in which (1) two or more persons experience a similar illness after consumption of water or after other use of water intended for drinking, and (2) epidemiological evidence implicates the water as the source of illness. Only outbreaks associated with drinking water are recorded. In 1980, 20,008 cases were reported to the Centers for Disease Control (CDC). Twenty-five states reported at least one outbreak. Even though surveillance by health authorities has improved, many outbreaks still remain unreported.

In the 1971 to 1978 period, most outbreaks occurred in noncommunity systems (58 percent), but the majority of illnesses (69 percent) resulted from outbreaks in municipal or community systems (Craun, 1981). Eighty-one percent of the outbreaks in semipublic and noncommunity systems were attributable to the use of untreated, contaminated groundwater and to treatment deficiencies, principally interruption of treatment and inadequate disinfection (Tables 18.2 and 18.3). Outbreaks of disease

Table 18.1. Waterborne Disease Outbreaks by Year and Type of System, United States — 1971 to 1980

	1971	1972	1973	1974	1975	1976	1977	1978	1979	1980	TOTAL (%)
Community*	5	10	5	11	6	9	12	10	23	23	114 (36)
Noncommunity*	10	18	16	10	16	23	19	18	14	22	166 (53)
Private*	4	2	3	5	2	3	3	4	4	5	35 (11)
TOTAL OUTBREAKS	19	30	24	26	24	35	34	32	41	50	315
TOTAL CASES	5182	1650	1784	8363	10,879	5068	3860	11,435	9720	20,008	77,974

*Community public water systems are public- or investor-owned water systems that have at least 15 service connections or serve 25 year-round residents. Noncommunity public water systems are those in institutions, industries, camps, parks, hotels, or service stations that may be used by the general public. Private systems are those used by single or several residences or by persons traveling outside populated areas.

(U.S. Department of Health and Human Services, 1982)

Table 18.2. Waterborne Disease Outbreaks from Source Contamination of Groundwater Used Without Treatment in the United States — 1971 to 1978

Cause of Outbreak	Number of Outbreaks	Number of Illnesses
Overflow or seepage of sewage	25	3,667
Contaminated springs	9	940
Chemical contamination	6	102
Contamination through limestone or fissured rock	5	880
Contamination by surface runoff	5	231
Flooding	1	88
Data insufficient for classification	23	930
Total	74	6,838

(Craun, 1981. Reprinted from Journal AWWA, Vol. 73, No. 7, by permission. Copyright ©1981, The American Water Works Association.)

caused by contaminated or poorly treated groundwater tend to occur mainly during the period from May to August (Figure 18.1).

In the past, the purity and sanitary quality of groundwater was assumed, and even when groundwater resources were used for drinking water supplies, little or no treatment was thought to be required. But in recent years, it has become obvious that groundwater may not always be safe to drink without adequate treatment. In fact, of the 224 disease outbreaks in the United States from 1971 to 1978, 107 were caused by groundwater that was not chlorinated or treated adequately (Lippy, 1981).

All water that seeps into the ground is contaminated to some degree even before entering the subsurface environment. Rain and melting snow pick up carbon dioxide, minerals, bacteria, and inorganic contaminants, such as oxides of sulfur and nitrogen, from the atmosphere and soil. Once in the ground, any water percolating through landfill sites picks up bacteria, viruses, and toxic substances. In addition, water seeping

Table 18.3. Waterborne Disease Outbreaks from Treatment Deficiencies in Groundwater Systems in the United States — 1971 to 1978

Cause of Outbreak	Number of Outbreaks	Number of Illnesses
Interruption of disinfection		
Chlorine	29	9,669
Iodine	3	71
Inadequate disinfection		
Chlorine	7	1,224
Iodine	1	72
Problems with addition of other chemicals	4	408
Total	44	11,444

(Craun, 1981. Reprinted from Journal AWWA, Vol. 73, No. 7, by permission. Copyright ©1981, The American Water Works Association.)

through soils near certain industrial plants can become heavily contaminated with industrial solvents or chemical residues related to particular manufacturing or processing activities.

Percolation of wastewater from the large number of home septic systems poses a serious threat to the preservation of groundwater quality, especially in suburban and rural areas. In properly designed and constructed systems, the movement of septic discharge through finer grained soils is quite effective in removing pathogenic bacteria over rather short travel distances. However, poor construction practices and the inability of certain soil materials to remove bacteria or inorganic chemicals have led to a proliferation of pollution problems caused by septic systems.

The occurrence and widespread migration of chemicals in the underground environment may ultimately become the most serious threat to groundwater quality. The presence of even tiny amounts of widely used volatile organic chemicals, such as trichloroethylene, is extremely serious. Unfortunately, many of these substances are not adsorbed onto soil particles, and thus can travel great distances in the subsurface environment (see Chapter 21).

Any well, then, can be contaminated biologically and chemically by substances entering the well from the surface or the intake portion of the well. Protection of water quality depends on the well design and the methods and materials used to construct the well. A study by the U.S. Environmental Protection Agency (1975b) indicated that many private water supply wells were constructed improperly. Some of the deficiencies included (1) insufficient or substandard well casing, (2) inadequate seal between the well casing and the borehole, (3) poor welding of casing joints, (4) lack of sanitary protection at the wellhead, and (5) use of well pits. These types of errors in well construction can lead to the introduction of bacteria and other contaminants into the well and subsequently the water supply system.

The water well contractor has a protective responsibility not only to the well owner but also to the community at large; any action the contractor takes must not lead to contamination of the aquifer. Careful site selection, proper well design, good drilling practices, and a thorough disinfection procedure will help assure a sanitary supply of groundwater. In most domestic well situations, the contractor is almost totally responsible for producing a water supply that has less than the maximum contaminant limits for toxic substances and is free of harmful bacteria. For high-capacity wells, however, the contractor may not be in a position to control the exact location of the well and some of the drilling procedures.

CHOOSING A WELL SITE

Groundwater in its natural state is generally sanitary and safe to drink. The contractor must choose a well site so that potential contamination from known nearby pollutant sources is avoided, and construct the well in such a way that surface pollutants cannot reach the aquifer.

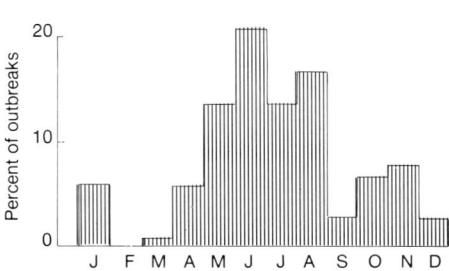

Figure 18.1. Seasonal distribution of waterborne disease outbreaks in systems using untreated groundwater in the United States, 1971 to 1978. *(Craun, 1981. Reprinted from Journal AWWA, Vol. 73, No. 7, by permission. Copyright © 1981, The American Water Works Association.)*

When determining site elevation, well construction methods, and disinfection procedures, the following parameters should be considered (Lehr et al., 1980):

1. Character of local hydrogeology:
   a. Slope of ground surface.
   b. Nature of soil and underlying porous strata.
   c. Thickness of water-bearing formation.
   d. Depth to water table.
   e. Slope of water table.
2. Location, log, and construction details of local wells, whether operating or abandoned.
3. Extent of recharge area likely to contribute water to the supply.
4. Nature of and distance and direction to local sources of pollution.
5. Possibility of surface drainage water or flood water entering the supply; methods of protection.
6. Methods used for protecting the supply against pollution from sewage-treatment and waste-disposal sites.
7. Well construction:
   a. Depth of well.
   b. Casing diameter, wall thickness, material, and length from surface.
   c. Screen diameter, material, construction, location, and length.
   d. Formation seal material (cement, sand, or bentonite), depth of placement, annular thickness, and method of emplacement.
8. Surface protection of well, including presence of sanitary well seal; height that casing projects above ground, pumphouse floor, or flood level; protection of well from erosion and animals; and pumphouse construction.
9. Pump capacity and pumping level.
10. Disinfection: equipment, supervision, and test kits or other types of laboratory control.

In general, a well should be located on the highest ground wherever practicable. It should certainly be on ground higher than nearby sources of pollution such as sewage drainage fields, gasoline stations, farm feed lots, landfill sites, or chemical and industrial plants. The well casing should extend above ground, and the ground surface at the well site should be built up when necessary so that surface water will drain away from the well in all directions.

A well should be located so that it will be accessible for pump repair, cleaning, treatment, testing, and inspection. The top of the well should not be within any building except a pumphouse. When adjacent to a building, the well should be at least 2 ft (0.6 m) away from any projection, such as overhanging eaves.

Minimum distances from a well to possible sources of pollution should be great enough to provide reasonable assurance that subsurface flow or seepage of contaminated water will not reach the well. The following minimum distances are typical of good practice.

A well shall be at least:

1. 150 ft (45.7 m) from a preparation area or storage area of spray materials, commercial fertilizers, or chemicals that may cause contamination of the soil or groundwater.

2. 100 ft (30.5 m) from a below-grade manure storage area*.

3. 75 ft (22.9 m) from cesspools, leaching pits, and dry wells.

4. 50 ft (15.2 m) from a buried sewer, septic tank, subsurface disposal field, grave, animal or poultry yard or building, privy, or other contaminants that may drain into the soil. Recent research has shown, however, that the distance between a septic tank leach field and a down-gradient well should be greater than 100 ft (30.5 m) if the soil is coarser than fine sand and the groundwater flow rate is greater than 0.03 ft/day (0.01 m/day) (AWWA, 1983).

5. 20 ft (6.1 m) from a buried sewer constructed of cast iron pipe or plastic pipe with tested watertight joints, a pit or unfilled space below ground surface, a sump†, or a petroleum storage tank.

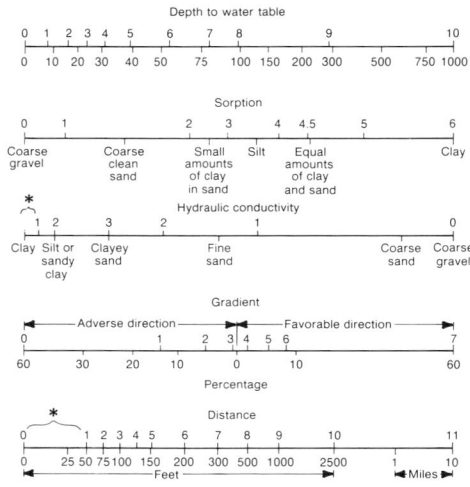

Figure 18.2. Rating chart for pollution potential in unconfined aquifers of unconsolidated alluvial materials. *(After LeGrand, 1964)*

6. Wells with casings less than 50 ft (15.2 m) in depth and encountering less than 10 ft (3 m) of impervious material must be located at least 150 ft (45.7 m) from cesspools, leaching pits, or dry wells and at least 100 ft (30.5 m) from subsurface disposal fields, manure storage piles, or similar sources of contamination.

The distances listed above are minimal, and larger separations between wells and potential contaminating sources are strongly recommended whenever possible. The figures for well separation given above are suitable where the earth materials have the filtering capability of sand. The recommended distances do not apply when highly mobile contaminants such as volatile organic chemicals are present or when the terrain consists of coarse gravel, limestone, or disintegrated rock near the ground surface. Some contaminants can travel great distances in these types of formations with little chance for natural purification.

PREDICTING THE POLLUTION POTENTIAL AT A DRILLING SITE

LeGrand (1964) developed a numerical system that indicates the pollution potential at a site. Although his method is imperfect because of the complex set of variables present at each site, it does suggest the pollution potential based on five parameters for an unconfined alluvial aquifer. The important factors are (1) depth to water table, (2) sorption capability of the soil, (3) hydraulic conductivity, (4) water-table gradient, and (5) distance to a pollution source. Figure 18.2 shows the evaluation graphs for each of the five parameters. A numerical value can be read above the line for each

---

*A below-grade manure storage area may present a special hazard to groundwater quality which may require a greater isolation distance than provided for in this rule depending upon hydrologic and geologic conditions.

†A sump is a watertight tank which receives sewage or liquid waste and which is located below the normal grade of the gravity system and must be emptied by mechanical means.

Table 18.4. Pollution Potential at a Site

Total Points	Possibility of Pollution
0 — 4	Imminent
4 — 8	Probable or possible
8 — 12	Possible but not likely
12 — 25	Very improbable
23 — 35	Virtually impossible

(LeGrand, 1964)

parameter. The sum of these values is then used to estimate the pollution potential at a site (Table 18.4). See Aller et al. (1985) for a more recent rating system.

WELL DESIGN

Construction of a well should be planned and carried out to utilize fully every natural sanitary protection afforded by the geologic and groundwater conditions. Similarly, the well must be designed so that both natural and man-made contamination can be avoided. In selecting the proper well design and construction method, the following principles should be followed:

1. The materials that are to be a part of the permanent well should be durable.
2. The well should be designed so that no unsealed opening will be left around the well that could conduct surface water, contaminated groundwater, or low-quality groundwater vertically to the intake portion of the well.
3. The well should be designed so that formations that are or may be contaminated or formations that have undesirable physical or chemical characteristics are sealed off.

Long-term sanitary protection for a well begins with selection of the casing. When choosing well casing, it is necessary to consider the stress to which it will be subjected during installation, the collapse pressures exerted by the surrounding earth materials, and the corrosiveness of the soil and water. Steel pipe is used most frequently for well casing, although thermoplastic casing has become popular in many parts of the country. When steel casing is installed in corrosive soil or water conditions, longer life can be assured by using casing of greater wall thickness. Under extremely corrosive conditions or when long life is essential, stainless steel casing is used. Although considerably more expensive than low-carbon steel casing, stainless steel casing provides longer service life under most corrosive conditions. In recent years, many municipal and industrial wells have been installed using stainless steel casing. Asbestos cement, concrete, and fiberglass casing are also used in large-diameter wells.

Galvanized sheet-metal pipe is not considered suitable for well casing, because it is extremely thin and its seams and joints are not watertight. If this kind of pipe is used, however, state health departments recommend that it be encased in concrete 6 in (152 mm) thick. The concrete envelope around the casing protects it from corrosive groundwater environments and provides additional collapse strength.

Short-joint concrete pipe, cribbing (galvanized, corrugated steel conduit), or clay tile used in bored wells should also be encased with a 6-in concrete envelope. The envelope should be carried to a depth of 20 ft (6.1 m) or more to exclude surface seepage and shallow contaminated groundwaters. If site conditions are particularly unfavorable, the concrete encasement should extend to a greater depth.

Adequate depth of well casing is an important factor in sanitary protection. It provides vertical protection that augments the horizontal distance between a well site and possible contaminant sources. A complete discussion of the types of casing and typical design recommendations for depths of casing are given in Chapter 13.

Installation of a grout seal around part or all of the casing string is an important factor in assuring that the well will remain free from contamination caused by surface or near-surface sources. Virtually all high-capacity municipal and industrial wells are grouted, as are many domestic wells, especially those constructed in rock. Grouting procedures vary depending on the well-drilling and screen-installation methods.

When drilling by the cable tool method, an outer or surface casing is set so that an annulus can be provided around the permanent casing. The permanent casing is centered in the larger hole and then driven into the formation. As the operation proceeds, the grout space around the permanent casing in the upper borehole should be kept about two-thirds full of clay slurry. Some of this puddled clay is carried downward as the casing is driven, and seals the formation around the pipe below the grout space. Because the clay remains in a viscous state as long as it is in the saturated zone, the clay seal is maintained between the casing and formation, even after the casing is pulled back to expose the screen. Upon completion of this step, the slurry remaining in the annulus between the surface and permanent casing is displaced by grout and the surface casing is removed.

In the direct rotary method, the procedure is somewhat different because the borehole is filled with drilling fluid during the entire drilling operation. Under these conditions, either the telescope or single-string method of well-screen installation can be used advantageously. In the telescope method, the well casing is set to its permanent position and then grouted in place. Using a smaller bit, a hole can then be drilled below the casing to permit setting of the well screen by telescoping it through the casing without disturbing the grout seal. In the single-string operation, the screen and casing are equipped with centralizers and placed in the open hole. Filter pack is installed around the screen. Before grouting, fine sand, or fine sand and then bentonite pellets, are placed on top of the filter pack to prevent grout from seeping downward into the pack. Grout is then installed in the annulus to the surface.

The contractor should select the method most compatible with the equipment available and the local geology. Procedures other than those described here are described in Chapter 10 and can be used to seal the casing to the subsurface materials. State regulatory agencies should be consulted if the contractor is unsure which method will provide the best protection.

Most states require that plans for each public or institutional water-supply project be approved before construction begins. All details of well location and design are reviewed to insure that certain standards of construction are followed. Generally, location requirements for high-capacity wells are more restrictive than those for private wells. Requirements for casing depth and grouting for these larger installations are usually established for each individual well.

DISINFECTION PROCEDURES REQUIRED TO MAINTAIN A SANITARY WELL DURING DRILLING

Before drilling begins, all tools, bits, and pumps should be disinfected thoroughly with a chlorine solution to kill any bacteria remaining from previous drilling operations. Water used in the drilling process and filter pack materials should also be treated with a chlorine solution before introduction into the well. During construction, the well should be disinfected continuously by maintaining a free-chlorine residual of 10 mg/ℓ in the drilling fluid to retard the growth of bacteria introduced into the well while drilling. (No practical way exists to treat a well drilled with dry air.) Because the effectiveness of the chlorine solution is largely dependent on the amount of hypochlorous acid (HOCl) present. In high-pH waters, the sterilizing effect of the solution is less (EPA, 1975c). Thus, stronger chlorine solutions are recommended under these conditions. On the other hand, the addition of chlorine to low-pH waters may lead to excessive corrosion, thereby damaging the drilling equipment used in the hole. The chlorine required to produce a 50 mg/ℓ solution is given in Table 18.5. Chlorine pellets are often added to filter pack material when organic drilling fluid additives are used, because the chlorine helps to break down the viscosity of the drilling fluid residues during well completion. Chlorine technology is discussed later in this chapter.

DISINFECTING WELLS AND PIPING

After the well has been completed, the water well contractor must disinfect the well, pump, and all piping to kill any bacteria that may be present. Materials and tools used in drilling and developing a water well are generally contaminated with earth materials and certain types of bacteria found living in soils. Throughout the well construction operation, some of these contaminants or bacteria may be introduced into the aquifer.

The bacteria and viruses picked up on drilling tools, pipes, and other materials are

Table 18.5. Chlorine Compound Required to Produce a 50-mg/ℓ Solution in 100 ft (30.5 m) of Water-Filled Casing*

Casing Diameter		Volume 100 ft (30.5 m)		65% HTH, Perchloron, etc. (dry weight)†		25% Chloride of Lime (dry weight)†		5.25% Purex, Chlorox, etc. (sodium hypochlorite) (liquid measure)	
in	mm	gal	m³	oz	g	oz	g	oz	ℓ
2	51	16.3	0.06	0.2	5.7	0.5	14.2	2	0.06
4	102	65.3	0.25	0.7	19.8	2	56.7	9	0.3
6	152	147	0.56	2	56.7	4	113	20	0.6
8	203	261	0.99	3	85.1	7	198	34	1.0
10	254	408	1.5	4	113	11	312	56	1.7
12	305	588	2.2	6	170	16	454	80	2.4
16	406	1,045	4.0	11	312	28	794	128	3.8
20	508	1,632	6.2	17	482	43	1,219	214	6.4
24	610	2,350	8.9	24	680	63	1,786	298	8.7

Note: Liquid sodium hypochlorite in a 12-percent solution is often sold for use in water and wastewater treatment plants, and as a commercial bleach or for use in swimming pools. Utilizing a solution of this nature would call for a liquid (chemical) measure equal to one-half the volumes presented in column 5.

*EPA recommends a minimum concentration of 100 mg/ℓ available chlorine. To obtain this concentration, double the amounts indicated.

†Where a dry chemical is used, it should be mixed with water to form a chlorine solution before putting it into the well.

(EPA, 1975c)

commonly those living in the soil at the well site and are usually nonpathogenic. However, the bacteria used as an indicator of possible disease-producing bacteria may be among them. This indicator bacteria is known as coliform bacteria, and is taken as evidence that the water may contain disease-producing organisms that normally live in the intestinal tracts of man and warm-blooded animals. Identification of actual disease-producing micro-organisms is difficult, and the efficiency of any disinfection process is generally not measured by tests for the absence of pathogenic bacteria, but by tests for the number of coliform bacteria (EPA, 1978). The four major types of pathogenic organisms that can affect the safety of drinking water are bacteria, viruses, protozoa, and occasionally worm infections. Typhoid, cholera, and dysentery are caused by bacteria and protozoa. Diseases caused by viruses include infectious hepatitis and polio. Water from a well is considered bacterially safe to drink only when tests show that it contains no more than 1 coliform bacteria per 100 ml.

Coliform bacteria can also be introduced into the water system while installing a pump in the well, connecting the pump to the distribution system, and installing the various elements of the piping system itself. Bacterial contamination can occur any time that the well or piping system is opened for repair or maintenance, because opening any part of the system offers an opportunity for foreign matter to enter it. Therefore, disinfection following construction or repair is necessary.

Disinfection Procedures

Use of a chlorine solution is the simplest and most effective way to disinfect or sterilize wells, pumps, storage tanks, or piping systems*. Chlorine is a powerful oxidizing disinfectant that kills bacteria on contact, although it cannot, at normal dosages, eliminate most viruses (White, 1972). On the other hand, short-wave ultraviolet light is effective in destroying viruses. The effectiveness of the chlorine procedure will depend on (Walker, 1978):

1. Chlorine concentration
2. Free-chlorine residuals
3. pH of the water
4. Retention time
5. Turbidity

The chlorine concentration must be high enough so that a free-chlorine residual remains several hours after treatment; that is, the chlorine demand has been satisfied and some extra chlorine is present after the initial contact period. High-pH waters require higher chlorine dosages than do low-pH waters to obtain the same level of disinfection, because the hypochlorous ions, which have the principal germicidal effect, are increasingly neutralized as the pH rises. Enough retention time must be allowed so that the chlorine can kill the bacteria. High turbidity tends to reduce the effectiveness of the chlorine treatment, but this condition is generally not a problem with groundwater.

Highly chlorinated water may be prepared by dissolving calcium hypochlorite, so-

---

*Chlorine is irritating to skin, eyes, and respiratory tract; wear self-contained breathing apparatus and eye protection when handling chlorine. Dry chlorine does not react with metals, but when moisture is present chlorine becomes strongly corrosive. It is a strong oxidizing agent that will react violently with hydrocarbons (grease and oil) and other organic compounds (turpentine, ethyl alcohol, glycerol, carbon tetrachloride, and charcoal). Do not mix chlorine and acid; large amounts of chlorine gas will be released.

dium hypochlorite, or gaseous chlorine in water. The calcium hypochlorite most commonly used in water well drilling is a white, granular material containing about 65 percent available chlorine by weight. In recent years, this material has also been marketed in tablet form under several trade names, including Pit-Tabs and HTH Tablets*. To distinguish this chemical from chlorinated lime or bleaching powder, it is commonly referred to as high-test calcium hypochlorite.

When dissolved in water, 1 lb (0.5 kg) of calcium hypochlorite with 65 percent available chlorine produces a solution that has the oxidizing power of 0.65 lb (0.3 kg) of chlorine gas dissolved in the same quantity of water. Putting it another way, 1.54 lb (0.7 kg) of calcium hypochlorite is equivalent to 1 lb (0.5 kg) of chlorine gas in a water solution.

The strength of chlorine solutions is usually expressed in milligrams of chlorine per liter of water. A solution of 10 mg/ℓ means a proportion of 10 lb (4.5 kg) of chlorine to one million lb (0.5 million kg) of water. Solution strengths of 50 to 200 mg/ℓ chlorine are used commonly for sterilizing wells and well construction materials. Table 18.6 gives the quantities of chlorine required to make 1,000 gal (3.8 m^3) of sterilizing solution of various concentrations for use in disinfecting wells and pumps.

Table 18.6. Quantities of Chlorine Compounds Required to Produce Chlorine Concentrations of 50, 100, 500, and 1,000 mg/ℓ in 1,000 gal (3.8 m^3) of Water

Strength (mg/ℓ)	Sodium Hypochlorite								Calcium Hypochlorite	
	3%		5%		10%		12½%		65%	
	gal	ℓ	gal	ℓ	gal	ℓ	gal	ℓ	lb	kg
50	1.7	6.4	1.0	3.8	0.5	1.9	0.4	1.5	0.6	0.3
100	3.3	12.5	2.0	7.6	1.0	3.8	0.8	3.0	1.3	0.6
500	16.7	63.2	10.0	37.9	5.0	18.9	4.0	15.1	6.4	2.9
1,000	33.3	126.0	20.0	75.7	10.0	37.9	8.0	30.3	12.8	5.8

If more accurate numbers are required or different volumes or sterilant concentrations are used, the following equations apply:

Calcium Hypochlorite:

$$\text{Wt (lb)} = \text{water volume (gal)} \cdot 8.33 \cdot \frac{\text{required concentration (mg/}\ell\text{)}}{\text{sterilant concentration (\%)}}$$

Sodium Hypochlorite:

$$\text{Volume (gal)} = \frac{\text{water volume (gal)} \cdot 8.33 \cdot \frac{\text{required concentration (mg/}\ell\text{)}}{\text{sterilant concentration (\%)}}}{8.33}$$

Notes:
1. When using the above equations, both required and sterilant concentration should be in decimal form. For example: mg/ℓ of 1,000 = 0.001; percent available chlorine of 5.25% = 0.0525.

2. Mg/ℓ conversion from trade percentages may be determined by using the following equation:

    mg/ℓ = trade percent · 10,000

    Using the above, a 5.25% solution is equivalent to 52,500 mg/ℓ chlorine.

---

*Pit-Tabs are made by Columbia-Southern Chemical Corporation and HTH is made by Olin Mathieson.

Dry calcium hypochlorite is a fairly stable material when stored properly, although it does lose some of its available chlorine over time. When packaged properly and stored in a cool place, it will retain 90 percent of its chlorine content for 12 months after manufacture. If the chemical becomes moist, it is quite corrosive and loses chlorine more rapidly.

Sodium hypochlorite is available only in solution form, because this chemical compound is very unstable. Practically all laundry bleach solutions sold in retail stores are sodium hypochlorite dissolved in water. They are prepared by bubbling chlorine gas through a solution of caustic soda. The solution loses chlorine at a rate such that a 10-percent solution will be reduced to about half strength after 6 months, even though stored under the best conditions. Solutions more than 60 days old should not be counted upon to contain the full amount of available chlorine originally in solution.

Sodium hypochlorite solutions are made in several strengths by many different producers in various localities. The maximum is about 12½-percent available chlorine. Much more common are household laundry bleaches such as Chlorox® and Hilex® that contain 5 to 5.25 percent available chlorine.

Disinfecting solutions can also be prepared by bubbling chlorine gas through water. The chlorine dissolves in the water and forms a mixture of hypochlorous and hydrochloric acid. The pH of the water is reduced and this enhances the disinfecting action of the solution. Use of this procedure is generally confined to water treatment plants because chlorine gas is extremely dangerous to handle.

A solution containing 50 to 200 mg/ℓ available chlorine should be used for disinfecting wells and piping systems. To provide this concentration of chlorine in the well, a stronger solution should be introduced so that, after mixing with the water in the well, 50 to 200 mg/ℓ chlorine will result. All pump parts must be thoroughly cleaned before being placed in the well. Many domestic wells, however, are completed weeks or sometimes months before the pumps are installed and the well is put into service. In the intervening period, a 5- to 10-mg/ℓ free-chlorine residual should be maintained in the well.

Before final disinfection of a well, storage tank, or piping system, the structure should be cleaned thoroughly. Foreign substances such as sediment, soil, grease, joint dope, and scum may harbor and protect bacteria and should be removed.

Chlorine or any other disinfecting agent can destroy only the bacteria it contacts. To simply pour chlorine into a well is not enough to disinfect it completely. The water in the well must be agitated to mix the solution thoroughly. In addition, surfaces of all components above the water level must be flushed or washed down with the sterilizing solution.

Special steps are required to assure chlorination of the entire well bore. One practical scheme is to place dry calcium hypochlorite in a container made from a short length of perforated tubing, capped on both ends and fitted with an eye on one end for suspension on a cable. By lowering and raising the container throughout the full column of water in the well, the chemical will be distributed adequately. The same device may be lowered into a flowing well and moved up and down near the bottom of the well. The natural upward flow will then carry the chlorinated water to the surface. In other cases, a chlorine solution is mixed in a tank at the surface and then circulated in the well by the mud pump. Sometimes chlorine is placed intermittently at high concentrations in the well bore. Agitation by a surge plunger, air pump, or other means mixes the solution at the appropriate strength with the water in the well.

Duration of contact with the chlorine solution is another important factor in effective disinfection. After being agitated in the well, the chemical should be left for at least 4 hours and preferably longer to assure complete disinfection.

After the original well installation and any subsequent repairs, the pumping system, storage tanks, and piping also require disinfection. To do this, the disinfecting solution can be pumped from the well and into the tank and piping system. Steps should be taken to make certain that the chlorinated water is drawn into all tanks and pipes. Faucets, valves, and hydrants should be kept open until the odor of chlorine is detected. These should then be closed and the solution left in the storage and distribution system for 2 hours or more. Care should be taken to wet the entire inner surface of a pneumatic tank with the solution. Finally, it is imperative that all traces of the chlorine residue be thoroughly purged from the well system before placing it into service.

Disinfection Byproducts

The recent improvement in the ability of laboratories to measure small quantities of organic compounds in water has shown that certain byproducts such as trihalomethanes originate from the chlorine treatment process. Some of these byproducts may be hazardous to human health (Rook, 1974; Symons, 1976); however, the extent of this danger is still under study. Fortunately, most water from wells does not require continuous chlorine treatment and, except for the disinfection process described above, no chlorine needs to be added to the water. Exceptions do occur, however. Some treatment is usually advisable for shallow wells or wells located close to known sources of bacterial or chemical contamination. Another exception is when groundwater is obtained from limestone, fractured and jointed igneous and metamorphic rocks, and vesicular basalt; this water is not as safe as water filtered naturally through sand, sandstone, gravel, and clay.

Bacteriological Analysis

The effectiveness of disinfection should be checked after completing the work by testing water samples for the presence of coliform bacteria. The well must be pumped and the piping system flushed out thoroughly to remove all traces of chlorine before collecting water samples for testing. Samples should be collected in containers supplied by the laboratory and in accordance with laboratory instructions*.

At least ½ cup (118 ml) of water is required to check for the presence of coliform bacteria. In many states, sterile sample bottles are provided by health departments for taking samples. This sample bottle should never be rinsed prior to use. Some sample bottles may contain crystals of a chemical that adsorbs chlorine. The sampling procedure is outlined in Table 18.7. State agencies that provide water-sampling services are listed in Appendix 18.A.

Although rarely performed for domestic wells unless contamination is suspected, chemical analysis can be done to assure that the water does not have high concentrations of any toxic substances. This analysis will indicate whether the water must be treated before use. Even though the water may seem to be potable because no

---

*Samples can be taken from the well by pump, air-lift, bailer, or specific type of sampler method (thief, ball) if the plumbing has not been installed.

pathogenic organisms are present, other highly dangerous substances may be in the water. The substances listed below represent many of the constituents that may be dangerous to human health or require water treatment if they occur at certain concentrations:

Common Minerals	Metals and Compounds	Radionuclides	Pesticides/Herbicides
calcium	arsenic	gross alpha	chlordane
chloride	barium	particle emitters	endrin
magnesium	cadmium	radium 226	heptochlor
carbonate	chromium		heptochlor epoxide
bicarbonate	cyanide		lindane
iron	lead		methoxychlor
manganese	mercury		toxaphene
sulfate	nitrate		azodrin
fluoride	selenium		dichlorvos
TDS	zinc		dimethoate
			ethion
			chlorophenoxys

In some cases, other constituents may be important and must be taken into account. The pH of the water may also affect the concentration of individual substances.

SEALING THE WELLHEAD

To prevent the introduction of foreign material into the well after construction, a welded, threaded, or flanged cap or compression seal must be fixed to the top of the well. Regardless of which type of cap or seal is used, the watertight casing should extend at least 12 in (305 mm) above the floor or ground elevation. If occasional floods are anticipated, the top of the casing should extend 1 to 2 ft (0.3 to 0.6 m)

Table 18.7. Checklist for Collecting a Water Sample

1. Select an indoor leak-free cold-water faucet from which to take the sample. Do not take sample from shifter-arm-type faucets.
2. Remove the faucet's aerator or strainer, if one is present.
3. Clean and then flame the inside of the faucet with a propane or homemade torch to disinfect it. Do not wash, wipe or touch after sterilization. Flaming will discolor chrome or gold-finished fixtures.
4. Let water run at full flow for 5 minutes.
5. Reduce faucet flow to a stream of water the size of a pencil and let flow for 1 minute.
6. Fill bottle to three-fourths capacity while holding the bottle's cap in the other hand. Do not let anything touch the inside of the bottle or cap except the water supply.
7. Close bottle immediately after sample is taken.
8. Deliver sample immediately to laboratory or public agency, or store in the manner that they suggest. Samples not delivered within 30 hours must be retaken. Sample should be kept cool, but not frozen.

above the highest flood level recorded for that site. The ground or floor area around the top of the casing should slope away from the wellhead in all directions to prevent ponding of water around the well casing (Figure 18.3). It is recommended that the well be vented with the vent pipe oriented downward and covered with a fine-mesh screen or filled with fiberglass wool. The vent opening must be large enough so that contaminated water is not drawn into the well through electrical conduits or leaks in the seal around the pump base when the pump is turned on. The inlet velocity for air replacing the water being pumped should not exceed 2,000 ft/min (610 m/min) (Keech and Gaber, 1984).

If the casing has been driven or grouted into rock, the effectiveness of the seal can be verified by pressurizing the well. No air loss should occur over a one-hour period under a pressure of 10 psi (6.9 kPa). Another method used to check the seal is the acoustic log which will indicate the integrity of the cement grout for the entire grouted section. The log is usually run 72 hours after emplacement of the grout or when the grout is assumed to have cured.

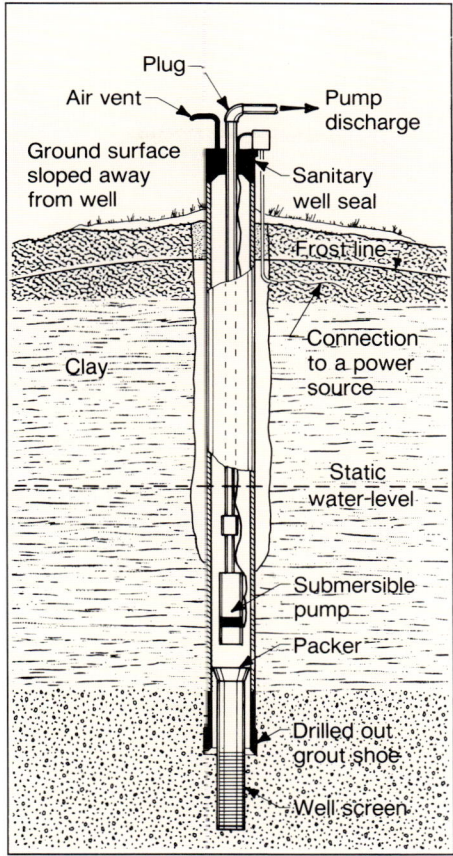

Figure 18.3. This well has been vented and sealed properly. The groundwater surface around the top of the casing has been graded to slope away in all directions. *(After U.S. Environmental Protection Agency, 1973)*

HORIZONTAL SUCTION LINES

Sanitary protection should be provided for horizontal suction lines if the pump is to be installed away from the well rather than inside or directly above the well. In practice, horizontal suction lines are used infrequently because of the popularity of submersible pumps. But if horizontal suction lines are installed, they should be placed below the ground surface inside an outer protective pipe or casing. The space between the outer casing and the suction line should be under the same pressure as the water system if the annulus is filled with water. This protection for suction lines is considered necessary to prevent contaminated surface water from being drawn into the water system if corrosion or shifting of the pipe because of earth movements should cause a leak to develop in the suction line. If the annulus is not filled with water, the protective casing must be arranged for free drainage to the ground surface or into an approved basement. The suction line leading to an offset pump installation may be either the suction pipe of a suction-lift pump or the intake pipe of a jet pump. Under certain operating conditions, the jet-pump intake line may be under negative pressure.

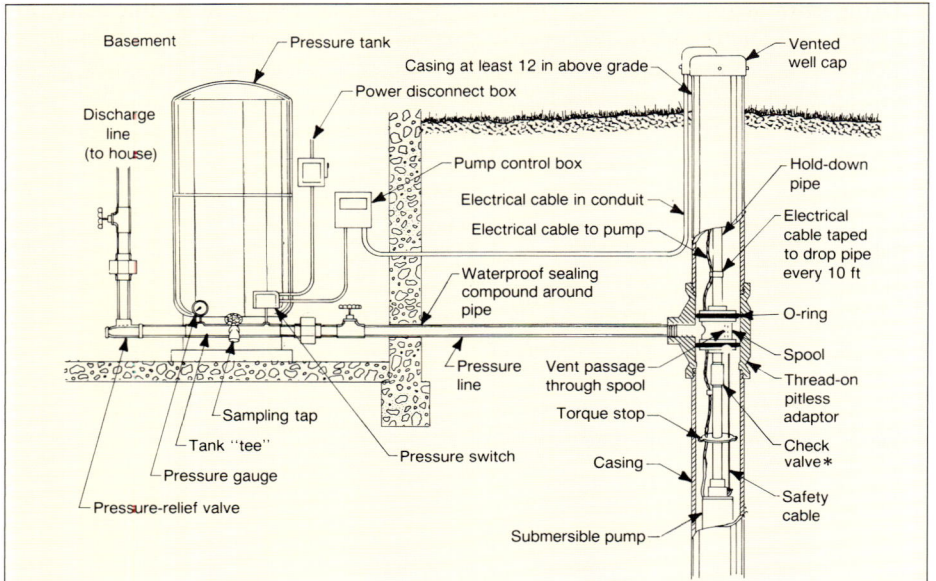

Figure 18.4a. Pitless adaptor and piping to the pressure tank. This well is equipped with a submersible pump. *(Michigan Department of Public Health)*

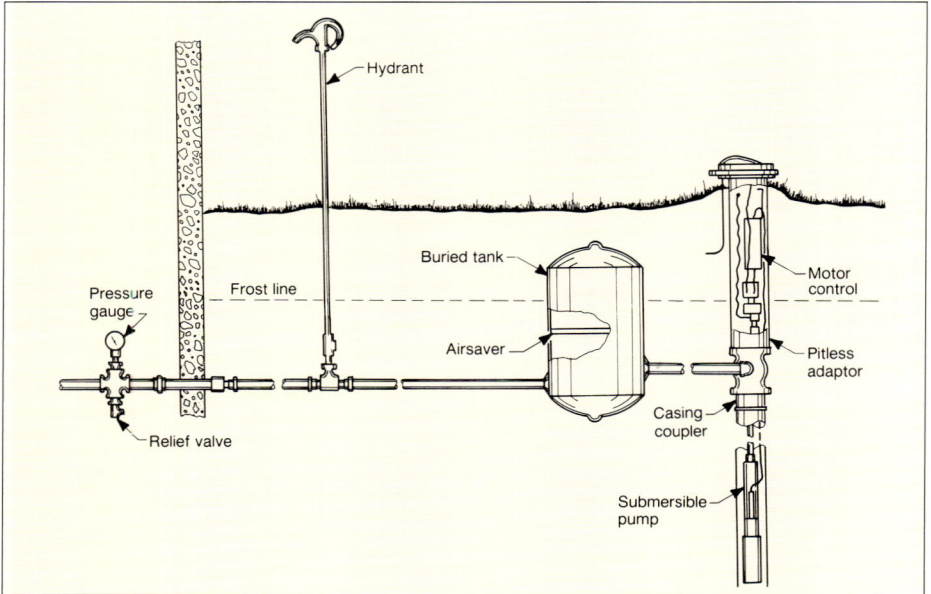

Figure 18.4b. Pitless adaptor connected to underground pressure tank. The entire water system is located in or near the well. The buried pressure tank provides cooler water, eliminates moisture condensation, and does not take up space in the house. *(Baker Manufacturing)*

*If check valve is installed, it must be located below pitless adaptor to maintain positive pressure on line between well and house. Check valve is often installed directly above pump or below spool of adaptor.

Using extra-heavy pipe for the suction line is another way of providing extra protection, especially if used with recessed or sleeve-type couplings. The wall thickness of 4-in (102-mm) extra-heavy pipe (Schedule 80), after threading, is about 50 percent greater than the thickness of 4-in standard-weight pipe (Schedule 40). Recessed couplings provide a short sleeve at each end which covers the threads out to the vanishing point. As rust forms on the threads under these sleeves, the accumulation of the iron oxide slows down the rate of corrosion and thus protects the pipe at these vulnerable points. The life of extra-heavy pipe is increased over that of standard-weight pipe almost in direct proportion to the extra thickness of metal. The cost of 4-in extra-heavy pipe is about 40 percent more per foot than that of standard pipe. Over the long term, this design is probably the most economical way to provide extra protection against corrosion.

PITLESS ADAPTORS

A sanitary underground discharge assembly, called a pitless adaptor, provides the most practical solution to the sanitary completion of the upper part of a well when offset-pump installations are specified. This device, illustrated in Figures 18.4a and 18.4b, attaches directly to the well casing and extends the casing above the ground surface. It provides a watertight subsurface connection for buried pump discharge or suction lines. These pipes must be buried below the frost line to prevent freezing.

Until the development of the pitless adaptor, installing pumps in pits below ground level was common where frost protection for piping was required. Pump pits are always unsanitary, and the pitless adaptor provides a practical means of eliminating them.

Besides their application in offset-pumping installations, pitless adaptors are equally useful where the pump is installed in the well with the power drive mounted either on the well casing or in the well (centrifugal and submersible pumps). The pump may be removed from the well and replaced without disturbing the underground discharge pipe.

A removable device sealed inside the adaptor directs the water into the permanently connected suction or pressure line. This device is suspended from the top of some pitless adaptors and may be lifted out vertically, giving full-diameter access to the well for repair or cleaning.

The installation shown in Figure 18.5 is especially practical for use with jet pumps when state regulations consider the pump-intake line as a suction line and require it to be encased in a pressurized pipe of large diameter. For either the single-pipe or double-pipe jet pump, the intake line can be placed inside the pressurized return line to fully comply with the regulations.

Most sanitation engineers stress the importance of using a second pipe to enclose a suction line where the water or soil may be corrosive. An unprotected line under negative pressure will draw in surface seepage if corrosion causes a leaky joint or a small hole in the pipe.

SEALING ABANDONED WELLS

Abandoned wells need to be sealed carefully to prevent pollution of the groundwater source, eliminate any physical hazard, conserve aquifer yield, maintain confined head conditions, and prevent poor-quality water of one aquifer from entering another.

The principal objective of sealing abandoned wells is to restore, as far as possible, the original hydrogeologic conditions. Before being sealed, the well should be checked to insure that there are no obstructions that may interfere with effective sealing operations. This inspection is especially important in wells that could conduct undesirable water into aquifers yielding potable water. Appendix 18.B contains the recommendations of the American Water Works Association for sealing abandoned wells.

If the casing has not been grouted, it may be removed by hydraulic jacks or by bumping the casing up using a cable tool drilling rig. A vibration hammer can also be used to remove casing. If the casing is in good condition, a trip-type casing spear operated by a fishing string can be used to remove it. Casing cutters are used to separate the drive shoe from the lowest casing to facilitate removal of the casing. Telescope screens can be removed by sand locking as described in Chapter 14.

Removal of liner pipe from some wells may be necessary to assure an effective seal. Liners or casings opposite water-bearing zones should be removed or perforated with a casing ripper beforehand to assure proper sealing throughout these zones. The upper portion of the casing should be removed to assure contact of the grout with the wall of the hole, to form a watertight plug in the upper 15 to 20 ft (4.6 to 6.1 m). Exceptions to this procedure may be permitted where the annular space around the casing was carefully cemented when the well was originally drilled.

To seal an abandoned well properly, the groundwater conditions at the site must be considered. When the groundwater occurs under unconfined conditions, the ob-

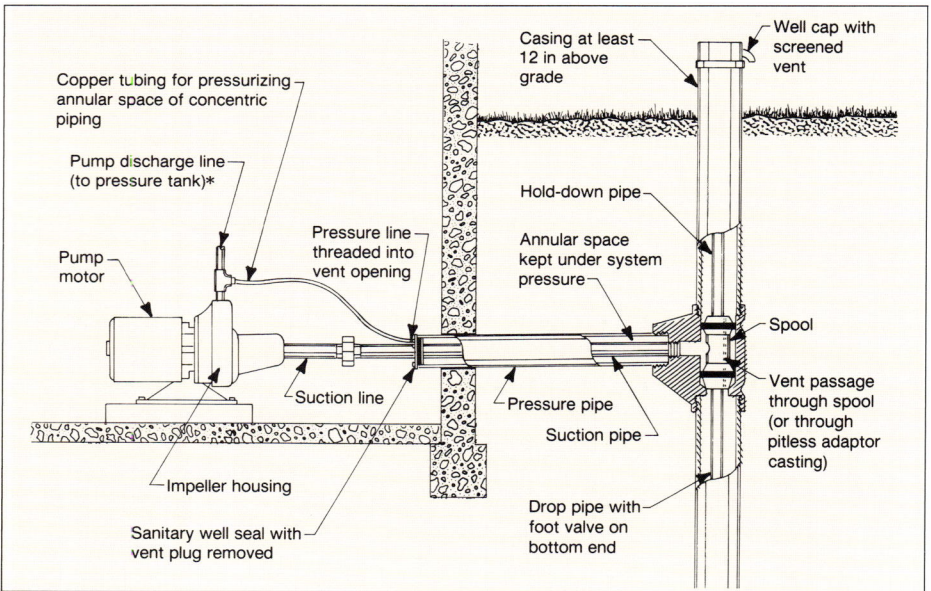

Figure 18.5. Pitless adaptor provides a means for making underground connections to well and pump while, at the same time, extending the well casing above ground surface for proper sanitary protection. Concentric piping for jet pump installation provides protection for horizontal suction line where a pump is offset from the well site. *(Michigan Department of Public Health)*

---

*Sampling tap shall be installed near pump or pressure tank.

jective is to prevent the percolation of surface water through the well bore or along the outside of the casing to the water table. This is accomplished by grouting the entire well bore. When confined conditions exist, the sealing operations must confine the water to the aquifer in which it occurs. This prevents the loss of confining pressure that results from uncontrolled flow from the aquifer.

In flowing wells, the water level must be lowered to control the flow before placing the seal. Flow can be controlled by introducing high-specific-gravity fluids to stop the flow, extending the pipe high enough above the land surface to stop the flow, or by pumping the problem well or nearby wells to create a drawdown in the well to be sealed.

Depending on the flow conditions encountered, three types of seals can be used to properly abandon a flowing well (Figure 18.6). Intermediate seals are placed between confined water-bearing formations having different static heads; thus, no water can pass from one aquifer to another. Bridge seals are cement or weighted wood plugs placed beneath the major aquifers. The lower part of the hole can be filled, if desired, with disinfected fill. Top seals are placed above the aquifers and form the base for the neat cement grout. Concrete or cement grout used to fill the well below the water level should be placed from the bottom up, by methods that will avoid segregation or dilution of material (see Chapter 10).

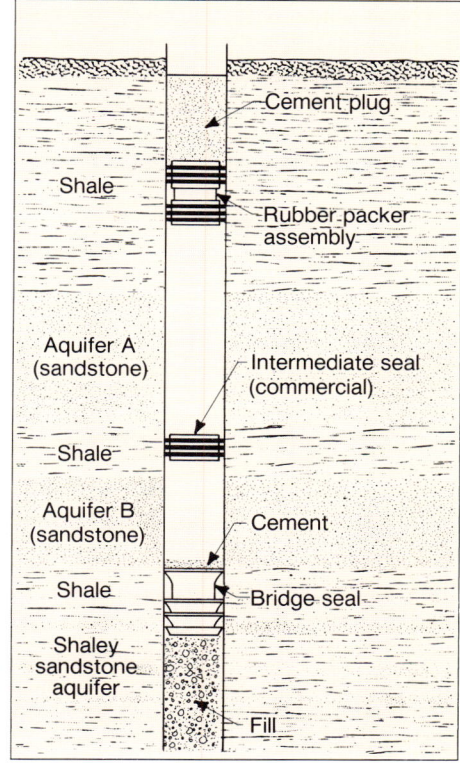

Figure 18.6. Abandoned well in confined aquifer. Aquifer A has higher potentiometric head than Aquifer B.

Employment of a competent well contractor familiar with well-abandonment procedures is advisable. The contractor's knowledge of well construction and the geologic conditions of the region are needed for proper results. The well owner or drilling contractor should report the abandonment of a well, the procedures used to seal the well, and its location to the proper state agency. The site of the abandoned well should be marked if possible.

CONCLUSIONS

Although known cases of groundwater contamination affect only a small percentage of all groundwater resources in the United States, most groundwater that is contaminated lies in highly populated areas. Therefore, the impact of the contamination is greater because of the number of people affected. Water well contractors can take an active role in protecting their customers from groundwater contamination by constructing safe wells and making sure that old wells are abandoned properly.

CHAPTER 19

Well and Pump Maintenance and Rehabilitation

Well rehabilitation is defined as restoring a well to its most efficient condition by various treatments or reconstruction methods. The necessity for well rehabilitation will depend on the effectiveness of the maintenance program and how faithfully it has been followed. In some cases, a major reconstruction of the well may be necessary, such as replacing the screen or lining a portion of the casing. Timely maintenance designed to overcome specific problems can sustain well performance, thereby prolonging well life.

Effective maintenance programs begin with well construction records showing geologic conditions, water quality, and pumping performance, especially specific capacity. A careful study of the operating history of other wells in the local region should suggest logical steps for devising maintenance schedules or rehabilitation procedures. So many variables are involved, however, that a single maintenance program cannot be devised that will work for every hydrogeologic condition and every type of well.

Inspection and routine maintenance schedules must be established on the basis of the individual characteristics of the well and pump. It is important to take note of any changes in the operating characteristics of the well and pump, because both can deteriorate to the point where rehabilitation is difficult, if not impossible. Experience indicates that if the specific capacity of a well declines by 25 percent, it is time to initiate rehabilitation procedures. Further neglect increases costs for maintenance significantly.

To determine any loss in performance, some reference mark will be needed. Performance standards are established by conducting a pumping test as part of the completion of every new well. The data from this test should be given to the owner in the form of a written report. This will allow the well owner to monitor the performance of the well to detect any drop in yield. These data also guide a rehabilitation contractor in devising an appropriate rehabilitation procedure. A form listing the important information to be retained at the well site is in the pocket of this book.

Loss of a major water supply, even temporarily, is intolerable in many cases. Therefore, ongoing performance evaluation of the well is mandatory if well failures are to be avoided. The checklist below can be used to evaluate the performance of a well.

- What is the static water level in the production well?
- What is the pumping rate after a specified period of continuous pumping?
- What is the pumping water level after a specified period of continuous pumping?
- What is the specific capacity after a specified period of continuous pumping?
- What is the sand content in a water sample after a specified period of continuous pumping?
- What is the total depth of the well?
- What is the efficiency of the well?
- What is the normal pumping rate and how many hours per day does it operate?
- What has been the general trend in water levels in wells in the area?
- How much drawdown is created in the production well because of pumping of nearby wells?

A significant change in any of the first 7 conditions listed above indicates that a well or pump is in need of attention. For instance, a decline in the specific capacity might indicate plugging of the screen-slot openings.

Once inspection procedures have been established, they should be followed in every subsequent inspection. The pumping tests, for example, should be run for the same length of time at the same rate, and have the same period of recovery. Local well-drilling or pump-maintenance contractors are helpful in establishing procedures, and sometimes offer maintenance contracts. These individuals retain records of all maintenance they perform and provide written reports to the well owner. A typical pumping (aquifer) test data form that can be used for maintenance evaluation is in the pocket of this book.

After the pumping-test data have been recorded, they can be compared with the original numbers and an evaluation made regarding any decline in the well's performance since the last survey. Storage of well records can be facilitated by the use of computers. For relatively low cost, complete well records can be maintained that can help forecast when maintenance and rehabilitation work should be undertaken.

Table 19.1 lists the most prevalent well problems occurring in various types of aquifers and the typical maintenance frequency required. The maintenance figures in Table 19.1 are based on wells constructed to locally acceptable design and construction standards in the United States that may not be consistent with the best materials or methods available. Therefore, although these maintenance schedules are realistic in light of the materials and construction methods used, they probably indicate greater frequencies than would be anticipated if the best technology were used.

MAJOR CAUSES OF DETERIORATING WELL PERFORMANCE

Five major problems occur with wells over time. The first involves a reduction in the well yield. Well yield may be reduced by chemical incrustation or biofouling of the well screen and the formation materials around the intake portion of the well. Deteriorating screen and formation conditions can be alleviated by the maintenance procedures discussed below. Of course, other environmental factors, either natural or manmade, may lead to reduced yields, but correction of these conditions may be difficult or impossible because of political, engineering, or natural constraints. For example, a general drop in the water table caused mainly by short- or long-term climatic trends will reduce well yield, as will interference from nearby wells. Also, the pumping level may drop over time in wells pumped continuously when the transmissivity of the aquifer limits the amount of water that can reach the wells, even

Table 19.1. Most Prevalent Well Problems Occurring in Various Types of Aquifers and the Typical Maintenance Frequency Required

Aquifer Type	Most Prevalent Well Problems*	Major Maintenance Frequency Requirement (Municipal)
Alluvial	Silt, clay, sand intrusion; iron precipitation; incrustation of screens; biologic fouling; limited recharge; casing failure	2-5 years
Sandstone	Fissure plugging; casing failure; sand production; corrosion	6-10 years
Limestone	Fissure plugging by clay, silt, and carbonate scale	6-12 years
Basaltic lavas	Fissure and vesicle plugging by clay and silt; some scale deposition	6-12 years
Interbedded sandstone and shale	Low initial yields; plugging of aquifer by clay and silt; fissure plugging; limited recharge; casing failure	4-7 years
Metamorphic	Low initial yield; fissure plugging by silt and clay; mineralization of fissures	12-15 years
Consolidated sedimentary	Fissure plugging by iron and other minerals; low to medium initial yield	6-8 years
Semiconsolidated and consolidated sedimentary	Clay, silt, sand intrusion; incrustation of screens in sand and gravel wells; fissure plugging of limestone aquifers in the interbedded sand, gravel, marl, clay, silt formations; biologic fouling; iron precipitation	5-8 years

*Excluding pumps and declining water tables.

Estimates of major maintenance frequencies are based on the following assumptions:
1. Wells are being pumped continuously at the highest sustained rate they are capable of producing.
2. Major maintenance is required when the sustained yield decreases to 75 percent of the initial yield.
3. Major maintenance is considered to represent a cost expenditure of approximately 10 percent of the total current replacement cost. Minor maintenance is excluded.
4. Wells are designed in accordance with current practices, not necessarily in accordance with best available technology.

(After Gass et al.)

though enough water may exist in the aquifer on a regional basis.

Plugging of the formation around the well screen by fine particles is the second factor in deteriorating well performance. Small particles in most unconsolidated formations are disturbed during pump cycling, and while temporarily in suspension they move gradually toward the screen. This same phenomenon apparently occurs in wells constructed in igneous and metamorphic rock, where the original specific capacity is often reduced 10 to 20 percent within a few months of operation. Small particles accumulate in the cracks, fissures, joints, fractures, or cavities that provide most of the water to the well.

The third factor in well failure is the onset of sand pumping. Some wells always pump sand, a condition usually attributable to poor well design or inadequate development. Other wells may begin to pump sand after months or years of service. Localized corrosion of the well screen or casing, or incrustation on only a portion of the screen, can produce higher velocities through either the corroded opening or the nonincrusted areas of the screen. Sand grains moved by these higher velocities may erode and enlarge the screen openings mechanically, allowing larger grains to enter the screen (Figure 19.1). Thus, corrosion and incrustation are major factors in sand pumping problems that develop over time. In some well-cemented sandstones, removal of the cement by water passing into the well can weaken the sandstone to the point where sand particles begin to move into the well. If this situation occurs, sand pumping may increase steadily.

The fourth cause of well failure involves the structural collapse of the well casing or screen. This type of failure is often produced by low-pH (acidic) waters containing high total dissolved solids and carbon dioxide concentrations that combine to cause electrolytic corrosion along the casing below the static water level.

Figure 19.1. Erosion of this well screen resulted from incrustation that caused high flow velocities through the remaining open area.

The last factor affecting well performance, although indirectly, is the condition of the pump. Mistakes in the design and construction of the well can cause severe damage to the pump over time. The impellers, impeller housing, and pump shaft are particularly susceptible to sand pumping. Corrosion of pump parts is also another serious problem in low-pH waters. Either of these conditions can drastically reduce the efficient life of the pump.

WELL FAILURE CAUSED BY INCRUSTATION

Chemical and biological incrustation are major causes of well failure. Water quality chiefly determines the occurrence of incrustation. The surface characteristics of the screen itself may also play a part in regulating the rate at which incrustation occurs. If the screen is constructed of rough-surface metal, for example, incrustants may build up at a faster rate. The kind and amount of dissolved minerals and gases in natural waters determine their tendency to deposit mineral matter as incrustation.

Groundwater normally moves slowly through soil, sand, and gravel, and is in contact with the minerals of these earth materials for hundreds to thousands of years. The time is so long that the water, with its dissolved mineral salts, is in quasi-chemical equilibrium with its environment. Thus, the water may be nearly saturated with the major minerals in the aquifer materials. Any change in the chemical or physical conditions upsets the equilibrium and may cause precipitation of relatively insoluble materials. The chemical equilibrium is upset when the well is pumped; in general, the greater the drawdown, the greater the disequilibrium will be.

Deposition of only a minute fraction of the minerals in the water will cause serious clogging. If material is dropping out of the water entering a screen 20 ft (6.1 m) long, 12 in (305 mm) in diameter, and pumping 500 gpm (2,730 m^3/day) at a rate of 1 mg/ℓ, a deposit of 6 lb (2.7 kg) per 24 hours would result. Assume the material is half calcium carbonate and half magnesium carbonate, with an average specific gravity of 3.0. If the porosity is 20 percent, all of the voids in the sand through a thickness of 6 in (152 mm) outside the screen would be completely filled in 293 days.

The incrustation often forms a hard, brittle, cementlike deposit similar to the scale found in water pipes. Under different conditions, however, it may be a soft, pastelike sludge or a gelatinous material. The major forms of incrustation include: (1) incrustation from precipitation of calcium and magnesium carbonates or their sulfates; (2) incrustation from precipitation of iron and manganese compounds, primarily their hydroxides or hydrated oxides; and (3) plugging caused by slime-producing iron bacteria or other slime-forming organisms (biofouling).

Causes of Carbonate Incrustation

Chemical incrustation usually results from the precipitation of carbonates, principally calcium, from groundwater in the proximity of the well screen. Other substances, such as aluminum silicates and iron compounds, may also be entrapped in the scalelike carbonates that cement sand grains together around the screen. The deposit fills the voids, and the flow of water into the well is reduced proportionately.

The probable explanation for this phenomenon is as follows. Calcium carbonate can be carried in solution in proportion to the amount of dissolved carbon dioxide in the groundwater. The ability of water to hold carbon dioxide in solution varies with pressure — the higher the pressure, the higher the concentration of carbon dioxide. When water is pumped from a well in an unconfined aquifer, the water table is drawn down to produce the necessary gradient or pressure differential in the waterbearing formation to cause water to flow into the well. The hydrostatic pressure in the deeper portions of the water-bearing formation is thus decreased, with the greatest change being at the well. Because of the reduction in pressure, some carbon dioxide is released from the water. When this occurs, the water is often unable to carry its full load of dissolved calcium carbonate and part of this material is then precipitated onto the well screen and in the formation materials adjacent to the well screen. Pumping a well in a confined aquifer produces a similar pressure reduction and resulting precipitation.

Formation of calcium carbonate precipitate from calcium bicarbonate is the classic example:

$$Ca(HCO_3)_2 \xrightarrow{-\Delta P} CaCO_3 \downarrow + CO_2 \uparrow + H_2O$$

where ΔP is a change in pressure. Solubility of calcium bicarbonate on the left side of this equation is about 1,300 mg/ℓ; solubility of calcium carbonate on the right side is about 13 mg/ℓ. Carbon dioxide (CO_2) escapes when the head, or pressure, is reduced.

Magnesium bicarbonate changes to magnesium carbonate in the same manner when the carbon dioxide is released, but magnesium carbonate incrustation occurs only in special instances because it is still soluble at concentrations over 5,000 mg/ℓ (Kemmer, 1979). Precipitation occurs, therefore, only when the carbonate concentration exceeds this level.

Causes of Iron and Manganese Incrustation

Many rocks throughout the world contain iron and manganese, and are the source of iron and manganese ions found in groundwater if the pH is about 5 or less. During pumping, velocity-induced pressure changes can disturb the chemical equilibrium of the groundwater and result in the deposition of insoluble iron and manganese hydroxides. These hydroxides have the consistency of a gel, and may occupy relatively large volumes; over time, they harden into scale deposits. Dissolved iron is affected by pressure reduction as indicated:

$$Fe(HCO_3)_2 \xrightarrow{-\Delta P} Fe(OH)_2 \downarrow + 2CO_2 \uparrow$$

Solubility of ferrous hydroxide on the right side of this equation is less than 20 mg/ℓ. If oxygen is introduced by aeration during pumping, additional precipitation of ferric hydroxide occurs:

$$4Fe(OH)_2 + 2H_2O + O_2 \rightarrow 4Fe(OH)_3 \downarrow$$

Solubility of ferric hydroxide on the right side of this equation is less than 0.01 mg/ℓ.

Soluble manganese becomes insoluble in the same way as iron:

$$2Mn(HCO_3)_2 + O_2 + 2H_2O \rightarrow 2Mn(OH)_4 \downarrow + 4CO_2 \uparrow$$

Further oxidation of the hydroxides of iron and manganese, or an increase in pH, causes the formation of hydrated oxides containing these ions. Ferrous iron in solution, for example, can react with oxygen to form ferric oxide:

$$2Fe^{+2} + 4HCO_3^- + H_2O + \tfrac{1}{2}O_2 = Fe_2O_3 + 4CO_2 + 3H_2O$$

The ferric oxide is a reddish brown deposit similar to rust, whereas the hydrated ferrous oxide is a black sludge. The insoluble manganese oxide is also black or dark brown. Iron and manganese deposits are often found associated with calcium- and magnesium-carbonate scale.

Sometimes the chemical deposits are hardly noticeable. For example, samples of the formation sand adjacent to well screens at a city in Michigan, an industrial plant in northern Indiana, and a plant in southern Illinois revealed no extraneous material in the sand voids, but all the sand particles were coated with hydrated iron oxide. These wells had suffered severe reduction in specific capacity over a period of three or four years. It is also quite possible that ferrous hydroxide, a white and fluffy precipitate, had been lodged in the voids of the formation but was broken up when the samples were taken and was unnoticeable.

In the cone of depression around a well in an unconfined aquifer, air enters the voids and oxidizes iron in the films of water adhering to individual sand grains. If pumping is started and stopped intermittently, a coating of iron oxide can build up, thereby gradually reducing the void space in this part of the formation. This action reduces the formation's storage capacity in the vicinity of the well, and the cone of depression enlarges more rapidly than it would otherwise.

Prevention and Treatment of Incrustation Problems

Thus far, a means of preventing the incrustation of well screens has not been found. One unique method does exist, however, that is designed to reduce the amount of iron incrusting materials reaching the well screen. This method, called the Vyredox™ System, uses a series of injection wells located in a circle around the production well. Oxygenated water is injected into the wells to oxidize iron in solution and promote the growth of iron bacteria so that little iron reaches the production well. See Chapter 23 for a more detailed description of this method.

For most wells where incrusting materials cannot be removed before reaching the well, several actions can be taken to delay incrustation and make it a less serious problem. First, the well screen should be designed to have the maximum possible inlet area to reduce the flow velocity to a minimum through the screen openings. Second, the well should be developed thoroughly. Third, the pumping rate may be reduced and the pumping period increased, thereby decreasing entrance velocities. Fourth, the pumping load may be divided among a larger number of smaller diameter wells instead of obtaining all of the supply from only one or a few larger diameter wells.

Fifth, a more frequent maintenance or cleaning procedure for each well should be practiced wherever local experience shows considerable difficulty from incrustation. In these areas, a qualified water well contractor should be called to perform the necessary maintenance. Corrective measures should not be put off until drastic means must be taken. Contractors generally know the best procedure to use from their past experience in the local area.

In localities where incrustation of wells is prevalent, samples of the incrusting materials and the water should be analyzed. Samples of the incrustants can often be obtained from the outer surfaces of pumps, suction pipes, or well screens. The constituents will normally include calcium carbonate, iron oxide, silica, aluminum silicate, or organic material. The material causing the clogging will usually be a mixture of several things, not a single substance. Recent research has shown, for example, that incrustants on the outside of a well screen may consist of precipitated elements from the groundwater, whereas most of the depositional products on the inside of the screen originate from the screen itself (Figure 19.2a and b). The proportions of the various substances shown by the chemical analysis should indicate the kind of treatment and the type of chemicals that would be most successful in recovering well yield.

Acid Treatment of Wells

Chemical incrustation can best be removed by treating the well with a strong acid solution that chemically dissolves the incrusting materials so they can be pumped from the well. Strong acids are used more often than any other type of chemical for well rehabilitation. Their chief value lies in their ability to dissolve mineral scale as well as some of the iron deposits formed by iron bacteria. The acids most commonly used in well rehabilitation are hydrochloric (HCl), sulfamic (H_3NO_3S), and hydroxyacetic ($C_2H_4O_3$).

Hydrochloric (Muriatic) Acid

Hydrochloric acid (prepared commercially under the name muriatic acid) is one of the most effective acids for removing mineral scale. Commercially prepared hydrochloric acid is a clear to yellowish solution of hydrogen chloride gas dissolved in water. It is available in several strengths that are identified by degrees Baumé*; common strengths are 18 and 20 degrees Baumé which are 28 and 31 percent hydrochloric acid, respectively. Hydrochloric acid is commonly ordered with an inhibitor that minimizes the acid's corrosive effect on metal wells screens, casing, and pump components.

In treating wells, hydrochloric acid is usually introduced into the well screen by conducting it from ground surface through a small-diameter plastic or black iron pipe. It is best to use a quantity of acid equal to the amount of water in the screen plus an additional volume of 25 to 50 percent. To reach farther into the formation, acid volumes of up to twice the screen volume can be used. Table 19.2 shows the proper amount of hydrochloric acid to use in small- and large-diameter wells.

Although it is an extremely effective well cleaner, hydrochloric acid has a number of drawbacks. It is extremely dangerous to handle. Once placed in the well, the acid produces large quantities of toxic fumes that are expelled from the well bore within moments. Inhalation of these toxic fumes will cause death, and contact of the liquid with human tissue can easily result in serious injury.

Figure 19.2a. Incrustants have formed on the inside and outside of this steel pipe-based well screen that is wrapped with a slotted-brass filter. Visual examination of the incrustants suggests that the porous incrustants on the inside of the screen contain different minerals than the dense, well-bonded incrustants on the outside.

Sulfamic Acid

Sulfamic acid† is a dry, white, granular material that produces a strong acid when mixed with water. Its solubility in water increases with temperature, ranging from

---

*Degrees Baumé is a scale referring to the specific gravity of the solution as determined by the acid concentration. As the degrees Baumé increase, the strength of the solution also increases.

†Also known as aminosulfonic, amidosulfonic, and amidosulfuric.

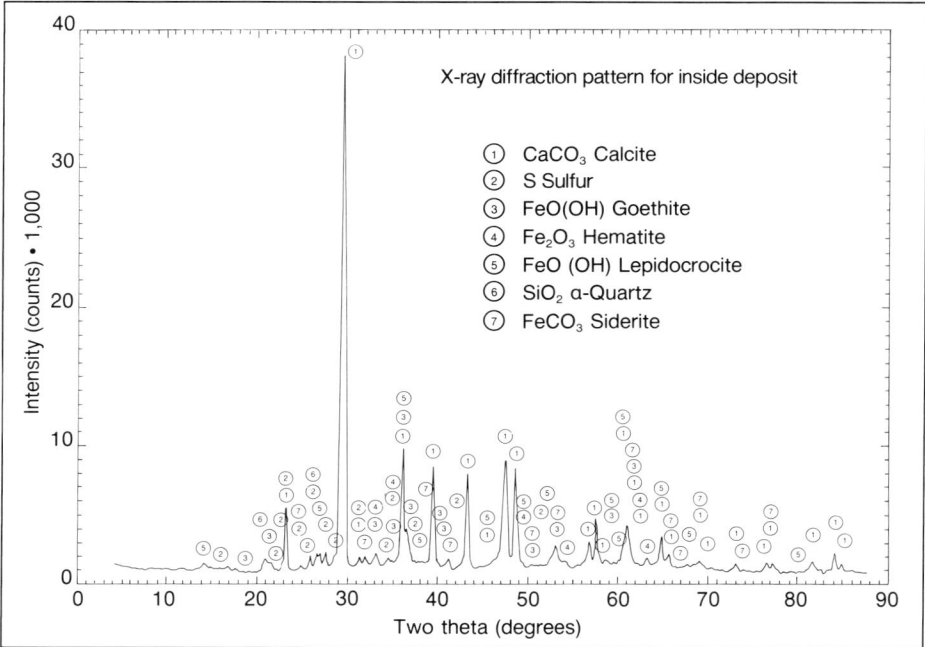

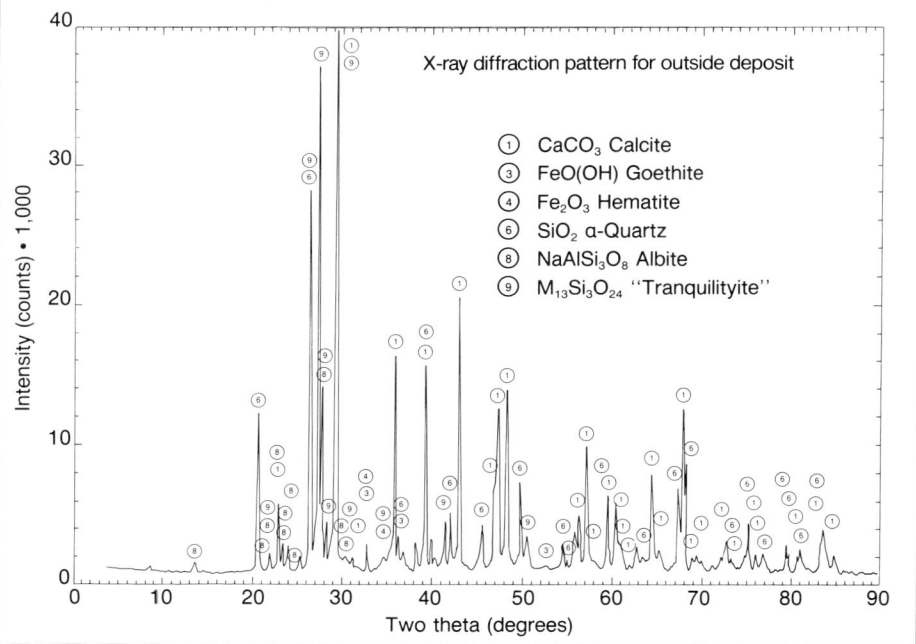

Figure 19.2b. Comparison of the x-ray diffraction patterns illustrates the chemical differences between incrustants. The incrustants on the inside of the screen consist principally of goethite, siderite, and lepidocrocite, which are indicative of iron and steel corrosion. Incrustants on the outside of the screen are derived mainly from the groundwater and include calcite, quartz, hematite, albite, and goethite.

15 to 20 percent by weight at most prevailing groundwater temperatures.

Although it is more expensive than hydrochloric acid and is less aggressive, sulfamic acid offers a number of advantages. In its dry form, it is relatively safe to handle; the dry material does not give off fumes and will not irritate dry skin. If spillage occurs, it may be cleaned up easily and safely, thus providing for safer shipping and handling. If mixed at the surface, however, sulfamic acid should be handled as if it were hydrochloric acid. During treatment, this slowly dissolving acid releases dangerous fumes at a relatively slow rate; nevertheless, proper ventilation should always be provided. Less corrosion of pumps, screens, and casings will occur when an inhibitor is added to the acid. For example, little corrosion results when stainless steel well screens are treated repeatedly with an inhibited sulfamic acid. Sulfamic acid is available in pelletized, granular, and powdered form. The pelletized form is used in wells completed with relatively short screens where the screens are located at the bottom of the well. Because the pellets are heavier than water, they sink through the column of water standing in the casing and then dissolve inside the screen. The pellets should

Table 19.2. Amount of Hydrochloric Acid Required to Treat an Incrusted Screen

Screen Diameter		Amount of HCl Acid (18° to 20° Baumé) per ft (0.3 m) of Screen	
in	mm	Gallons	Liters
1½	38	0.11 - 0.14	0.42 - 0.53
2	51	0.20 - 0.24	0.76 - 0.91
2½	64	0.33 - 0.39	1.25 - 1.48
3	76	0.46 - 0.56	1.74 - 2.12
3½	89	0.63 - 0.75	2.38 - 2.84
4	102	0.81 - 0.98	3.07 - 3.71
4½	114	1.04 - 1.25	3.94 - 4.73
5	127	1.28 - 1.53	4.84 - 5.79
5½	140	1.54 - 1.85	5.83 - 7.00
6	152	1.84 - 2.21	6.96 - 8.36
7	178	2.50 - 3.00	9.5 - 11.4
8	203	3.26 - 3.92	12.3 - 14.8
10	254	5.10 - 6.12	19.3 - 23.2
12	305	7.35 - 8.82	27.8 - 33.4
14	356	10.0 - 12.0	37.9 - 45.4
16	406	13.1 - 15.7	49.4 - 59.4
18	457	16.5 - 19.8	62.6 - 75.1
20	508	20.4 - 24.5	77.2 - 92.7
22	559	24.7 - 29.6	93.5 - 112
24	610	29.4 - 35.3	111 - 133
26	660	34.5 - 41.4	131 - 157
28	711	40.0 - 48.0	151 - 182
30	762	45.9 - 55.1	174 - 208
32	813	52.2 - 62.7	198 - 237
34	864	59.0 - 70.7	223 - 268
36	914	66.1 - 79.3	250 - 300

dissolve in approximately 4 hours if oversaturation does not occur. Agitation of the water in the screen increases the solution rate of the chemical. The proper quantity of pelletized sulfamic acid required to treat the well is generally determined by the length and diameter of the well screen or by the weight of water standing in the screen. Table 19.3 shows the proper quantities of Nu-Well®, a pelletized sulfamic acid, to use for small- and large-diameter screens less than 100 ft (30.5 m) long.

The granular form of sulfamic acid is generally used when acidizing long screens [greater than 100 ft (30.5 m)] or screens separated by casing. It is usually dumped directly into the casing, where it saturates the entire column with acid. The acid goes into solution as the granules descend slowly in the casing. Enough clear water is then added to force the volume of acid standing in the casing above the screen into the formation. For deep wells with high static water levels, granular or powdered acid should be premixed at the surface so it can be piped to the intake portion of the well. A 10-percent solution of granular sulfamic acid is sometimes used for long screens, although a 30-percent solution provides better results.

Sulfamic acid is particularly useful in treating calcium and magnesium incrustants, but is less effective when iron or manganese incrustants are present. The addition of rock salt to sulfamic acid, however, will increase the acid's ability to dissolve iron deposits. Approximately 2 lb (0.9 kg) of rock salt are added to 10 lb (4.5 kg) of Nu-

Table 19.3. Amount of Nu-Well® Required to Treat a Moderately Plugged 1-ft (0.3-m) Section of Screen

Screen Diameter (Pipe Size)		Screen Capacity		Nu-Well Required	
in	mm	gal/ft	ℓ/m	lbs/ft	kg/m
1½	38	0.1	1.2	0.2	0.3
2	51	0.2	2.5	0.4	0.6
3	76	0.4	5.0	0.9	1.3
4	102	0.7	8.7	1.6	2.4
5	127	1.0	12.4	2.6	3.9
6	152	1.5	18.6	3.7	5.5
8	203	2.6	32.3	6.5	9.7
10	254	4.1	50.9	10.2	15.2
12	305	5.9	73.2	14.7	21.9
14	356	8.0	99.3	20.0	29.8
16	406	10.4	129	26.1	38.9
18	457	13.2	164	33.0	49.2
20	508	16.3	202	40.8	60.8
22	559	19.8	246	49.4	73.6
24	610	23.5	292	58.7	87.5
28	711	32.0	397	80.0	119
30	762	36.7	455	91.8	137
32	813	41.8	519	104	156
34	864	47.2	586	118	176
36	914	52.9	657	132	197

The quantities of Nu-Well® are equal to 30 percent of the weight of water in the well screen. This ratio is used for treating relatively short screens that have been affected by moderate incrustation.

Well® (20 percent of the weight of the acid) to create a solution that will treat iron and manganese incrustants.

Sulfamic acid reacts chemically with mineral deposits in the same manner as hydrochloric acid, although at a slower rate. Consequently, longer contact time is usually required to achieve the same results; at least 15 hours is recommended. The effectiveness of the treatment is enhanced considerably if the acid is agitated while and immediately after it dissolves. Forceful agitation is also recommended before the acid is pumped to waste.

Sulfamic acid should not be confused with sulfuric acid. Sulfuric is a strong liquid acid that has been used successfully on rare occasions in well treatment. Its major limitation in well treatment is that when it combines chemically with calcium scale, it forms calcium sulfate which is nearly insoluble in water. Thus, a sulfuric acid treatment may actually reduce the well's performance. In addition, sulfuric acid, even when inhibited, is extremely aggressive in attacking metallic casing and screens.

Hydroxyacetic Acid

Hydroxyacetic acid, also known as glycolic acid, is a liquid organic acid available commercially in 70-percent concentrations. Although not as well known or commonly used as either hydrochloric or sulfamic acid, its use has achieved excellent results in well treatment. It is quite safe to use because it is relatively noncorrosive and produces little or no toxic fumes.

In addition to its ability to dissolve mineral scale, hydroxyacetic acid offers advantages not available with sulfamic or hydrochloric acid. It is an excellent bactericide and therefore may be effective in treating wells with iron bacteria problems. It kills

Table 19.4. Amount of Hydroxyacetic Acid Required per 1 ft (0.3 m) of Screen Length or Open Borehole

Diameter of Well		Amount of 70% Hydroxyacetic Acid per 1 ft (0.3 m) of Screen or Borehole	
in	mm	gal	ℓ
1½	38	0.006 - 0.009	0.02 - 0.03
2	51	0.01 - 0.02	0.04 - 0.08
3	76	0.02 - 0.04	0.08 - 0.15
4	102	0.04 - 0.07	0.15 - 0.27
6	152	0.10 - 0.15	0.38 - 0.57
8	203	0.17 - 0.26	0.64 - 0.98
10	254	0.27 - 0.41	1.02 - 1.55
12	305	0.39 - 0.59	1.48 - 2.23
16	406	0.70 - 1.00	2.65 - 3.79
20	508	1.09 - 1.64	4.13 - 6.21
24	610	1.57 - 2.36	5.94 - 8.93
28	711	2.14 - 3.21	8.10 - 12.1
30	762	2.45 - 3.68	9.27 - 13.9
32	813	2.79 - 4.19	10.6 - 15.9
34	864	3.15 - 4.73	11.9 - 17.9
36	914	3.53 - 5.30	13.4 - 20.1

the bacteria and simultaneously dissolves the bacterial iron deposits as well as other mineral scale.

In addition to its bactericidal properties, hydroxyacetic acid is a chelating or sequestering agent. This means that it has the ability to "surround" metal ions (such as iron, calcium, and magnesium) in solution and keep them from combining chemically with other ions. This insures that all the scale dissolved by the acid remains in solution during the entire treatment period.

Hydroxyacetic acid is placed in the well in the same manner as hydrochloric acid. About 1 gal (3.8 ℓ) of 70-percent hydroxyacetic should be used for every 10 to 15 gal (38 to 56.7 ℓ) of water standing in the well screen. Table 19.4 shows the proper amount of hydroxyacetic acid to use in treating wells of various diameters.

Hydroxyacetic acid is weaker than both hydrochloric and sulfamic acid, and longer contact time is required to achieve the same amount of scale removal. The rate at which an acid removes scale is related to the acid's pH (acid strength). Figure 19.3 shows how pH varies with concentration for the acids described above. Note that hydrochloric acid has the lowest pH and thus will work the fastest, whereas hydroxyacetic has the highest pH and will work more slowly than the other acids.

General Procedure for Acid Treatment

Great care should be taken in placing liquid acid into a well. Only experienced personnel with specialized equipment should attempt to use it in rehabilitating a well. When using any liquid acid, personnel should wear protective rubber clothing and goggles. A breathing respirator should also be used by all personnel handling the acid and by other persons near the well. All mixing tanks, chemical pumps, and piping (tremie pipes) should be constructed of plastic or black iron to minimize reaction to the acid. A large quantity of water, or a water tank with a mixture of sodium bicarbonate, should be available in the event that an accident occurs. Proper ventilation must be maintained because the fumes released from the well during treatment are lethal.

Liquid acid should be introduced into the well through a small-diameter pipe. If the screen is more than 5 ft (1.5 m) long, enough acid should be added to fill the lower 5 ft of screen. Then the pipe should be raised and the next 5 ft of screen filled with acid, continuing in this way until the entire screen is full. Pelletized forms of sulfamic acid dropped into the casing will accumulate in the screen where the pellets dissolve. When the granular forms are

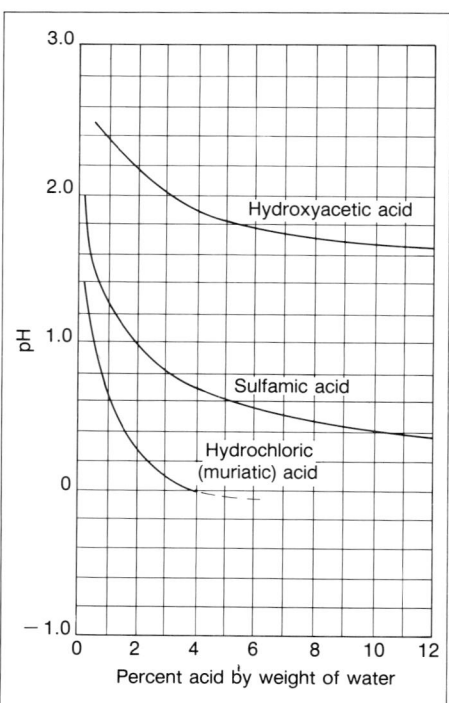

Figure 19.3. Equal concentrations of different acids form solutions with different pH values; pH of an acid-water solution varies with concentration.

poured into the casing, they go into solution throughout the entire column of water in the well.

After the acid is placed in the well (or the pellets dissolve), a volume of water equal to that standing in the well screen is poured into the well to force the acid solution through the screen-slot openings into the formation. Some form of mechanical agitation, such as surging, should be employed while the acid is in the well to help break up the incrustation and improve the overall efficiency of the process. This step is particularly important because it exposes the incrustant to the acid, thereby assuring maximum removal.

The use of surge blocks or jetting tools are effective methods of agitating the well. The agitation time will depend on the amount of incrustant in the well. If a surge block is used, the surging effect drives the acid into the formation and brings loosened material into the screen. In the jetting operation, the acid is first poured into the well. The screen or the face of the well bore can then be jetted with clean water from the surface or acidized water from the well (Figure 19.4). A pump pressure of 100 to 250 psi (690 to 1,720 kPa) is sufficient for this type of operation. Circulation of the acid solution may be corrosive to the jetting pump and other equipment, but the wide use of plastic impellers has eliminated most of this type of corrosion damage. If the job requires recirculating the jetting acid at the surface, it is best to call on a well servicing company that has specialized equipment for this work. Great care should always be exercised whenever acid is being pumped in any well rehabilitation operation.

An extended zone of the formation around the well screen may be wholly or partially clogged. Thus, it must never be assumed that the chemical solution moves uniformly outward into the voids of the water-bearing materials in all directions throughout the full thickness of the formation. The chemical solutions will flow most readily into those areas where the formation or screen is the most open, that is, where resistance to flow is the least. Therefore, it may be extremely difficult and even impossible to diffuse the chemical solution to all points where it can dissolve or otherwise remove the unwanted deposits.

The use of chelating agents is recommended if iron and manganese incrustants are present and the pH of the treatment solution is approximately 3 or less.

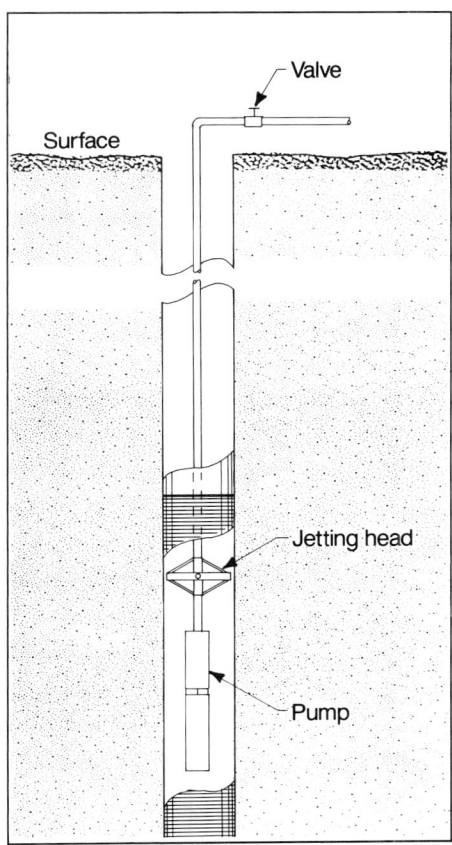

Figure 19.4. To avoid the dangerous practice of pumping acid at the surface, jetting can be accomplished by setting a pump in the well and using the acidified water in the borehole.

At this pH, these cations form insoluble precipitates that settle out and reduce the effectiveness of the acid treatment. Citric, phosphoric, and tartaric acids are three common chelating agents. Four pounds (1.8 kg) of chelating agent are usually added to each gal (3.8 ℓ) of 31-percent (20 degrees Baumé) hydrochloric acid and 1 lb (0.5 kg) of chelating agent to 15 lbs (6.8 kg) of sulfamic (granular) acid.

After mechanical agitation, the solution is left in the well to react with the incrusting materials until the pH is between 6.5 and 7, then agitated again and pumped to waste. The time for the reaction to be completed will vary from a few hours to more than 15 hours, depending on the type of acid used and the amount of incrustants. To minimize disposal problems, the water in the well should be neutralized if necessary before it is removed from the well. In many communities, the water well contractor may be required to haul away the spent acid and dispose of it according to local regulatory agencies. If not, the spent acid should be run onto a sandy section of ground as far away as practicable from the well head. Some contractors neutralize acid wastewater by running it through a limestone-filled container.

Many water well contractors will redevelop the well after it has been acid treated. Solid particles of incrustant can be removed along with any fine sediments that may have entered the zone immediately around the screen after the well was placed in service. In many instances, effective redevelopment of an older, acid-treated well will result in a specific capacity that equals or even exceeds the original specific capacity. The various development procedures are discussed in Chapter 15.

Mechanical Methods to Remove Incrustants

Although removal of most incrustants by acid treatment is extremely effective, several mechanical methods are useful either in preparing for acid treatment or as a primary method of removing incrustants. Wire brushing or other means of mechanical scraping can remove incrustants that have been deposited on the inside of the well screen. The loosened material is then removed from the well by bailing, air-lift pumping, or other means. Removal of these incrustants minimizes the quantity of acid that must be used in any subsequent acid treatment, enhances the effectiveness of this treatment, and reduces the time required for the acidizing process.

Controlled blasting techniques are often useful for temporarily improving well yield by fracturing the incrusting matrix so that water can reach the screen. Incrusting materials are sometimes deposited on formation materials several inches or more away from the screen. The incrustant may become so massive that all voids in the formation become filled and little water can reach the screen. Blasting procedures create cracks in the incrustant, allowing water to enter the well. Some fragments of the incrustant will break away and can be pumped from the well. Unfortunately, the opened cracks eventually will also become incrusted and additional blasting or acidizing treatments will be needed to maintain yield. This technique, when combined with acidizing, is particularly effective. Special service companies have formed to provide this type of blasting service.

Incrustation of Rock Wells

Although this discussion has referred only to screened wells in unconsolidated formations, wells in consolidated rock also suffer from incrustation of the borehole wall or the cracks and fissures leading to the borehole. Many rock wells require treatment from time to time to recover the original yield. Both chemical treatments and blasting have proved to be effective procedures and in some cases both are used.

When blasting incrustant, 5- to 10-lb (2.3- to 4.5-kg) shots of explosive are set at 5-ft (1.5-m) intervals in the production zone of the well. More powerful amounts of explosives are sometimes used at different spacings, depending on the experience of the contractor and the nature of the formation and the incrustant. The explosive charges are set off sequentially, beginning at the bottom of the open hole. Do not set off charges within 50 ft (15.2 m) of a shale layer or the bottom of the casing. After blasting, the loosened material should be removed from the borehole and the well redeveloped completely. Samples of sandstone removed after blasting have shown that most of the incrustation extends only about 0.5 in (12.7 mm) beyond the face of the borehole.

Wells constructed in fissured limestone can be successfully treated with acid. An appropriate quantity of hydrochloric or sulfamic acid is placed in the well and the well head capped. A pressure gauge is installed so that the pressure can be monitored. If the pressure build-up is high, the acid is being contained near the borehole. If the pressure does not build substantially, most of the force is being transmitted away from the well bore by means of cavities or enlarged fissures. The solution effect is still beneficial, nevertheless, even if the pressure build-up is low. When the acid stops working, the gauge will indicate a noticeable pressure drop. Work can then begin on redevelopment by jetting, surging, or other means of agitation. All loosened material should be removed before placing the well back in service.

Johnson Division makes no guarantee of results and disclaims all liability in connection with the information or the safety suggestions given for the methods described. Also, it should be understood that not all the acceptable safety procedures are contained herein and that certain circumstances may call for additional precautions. The suggestions given here do not supplement nor modify any state, municipal, federal, and insurance requirements or codes relating to blasting or acidizing.

Acid Treatment of Municipal Wells in Las Vegas, Nevada

A case history from the Las Vegas (Nevada) Valley Water District demonstrates the effectiveness of acid treatment using sulfamic chemicals. The district wells had become heavily incrusted with calcium and magnesium scale, reducing the yields substantially. Both blasting and dry-ice treatment were used to fracture the incrusted formations. A series of small explosive charges were placed in the well and detonated sequentially. In dry-ice treatment, carbon dioxide gases released by dry ice in the well produce extremely high pressures and cause additional fracturing of the incrustant.

The District then undertook an acid rehabilitation program for five of their most heavily used wells. Each of the five wells had been completed with 16-in casing to an average depth of over 900 ft. Length of perforated areas ranged from 278 to 651 ft. The average yield before treatment was 1,870 gpm per well.

A 10-percent acid solution (by weight of water in the casing) of granular sulfamic acid was determined to be adequate to dissolve the incrustant, which consisted primarily of calcium carbonate. This amounts to 0.75 lb of sulfamic acid per gallon of water in the casing. Six 480-lb loads were placed in each well while the pump was still in place. After each 480-lb load was added, the pump was used briefly to surge the well five times to mix the acid and distribute the solution throughout the casing.

When all the acid had been placed in the wells, the wells were surged ten times every 4 hours for the next 24 hours. The wells were then left for an additional 24 hours to guarantee removal of the most firmly imbedded incrustants and then pumped to waste.

During the treatment, silt and sand that were once cemented together by the incrustants were loosened. In order to obtain optimum yields, it was important to remove these materials so the original permeability of the sediments would be restored. To accomplish this, 300 lb of tetrasodium pyrophosphate were added to each well, surged five times, and allowed to stand in the wells for 24 hours. The addition of phosphates helped break up and disperse silt, clay, and other byproducts of the acid treatment. The wells were again pumped to waste.

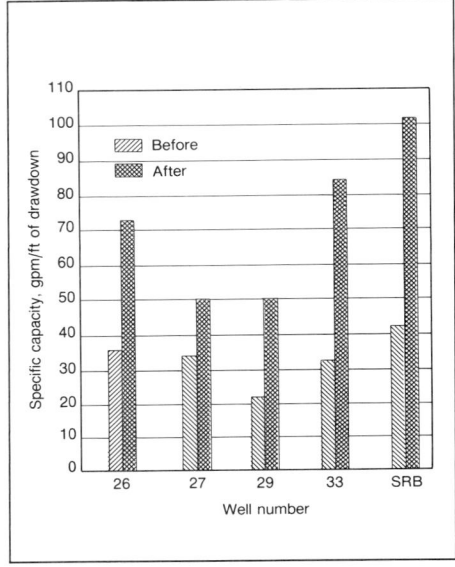

Figure 19.5. Specific capacity of wells before and after acid and polyphosphate treatment.

Before acid treatment, the well characteristics were monitored and recorded as a guide to determine the effectiveness of the treatment. A comparison of specific capacities before and after treatment revealed exceptional results. Figure 19.5 illustrates a range of improvement from 45 to 160 percent of pretreatment specific capacities. After the acid treatment, significant reductions in drawdown resulted in saving the District $16,000 annually in pumping costs alone. The payback time for the investment in materials and labor was estimated to be 1.5 years (Varhol, 1980).

WELL FAILURE CAUSED BY IRON BACTERIA

Iron bacteria occur widely in wells open to the atmosphere when sufficient iron and/or manganese are present in the groundwater in conjunction with dissolved organic material, bicarbonate, or carbon dioxide. Although iron bacteria have been found in wells in all the conterminous United States, the most seriously affected areas include the Southeastern states, the Upper Midwest, and Southern California. In these regions, the principal forms of iron bacteria plug wells by enzymatically catalyzing the oxidation of iron (and manganese), using the energy to promote the growth of threadlike slimes, and accumulating large amounts of ferric hydroxide in the slime (Figure 19.6). In this process, the bacteria obtain their energy by oxidizing ferrous ions to ferric ions, which are then precipitated as hydrated ferric hydroxide on or in their mucilaginous sheaths. Precipitation of the iron and rapid growth of the bacteria create a voluminous material that quickly plugs the screen pores of the sediment surrounding the well bore. Sometimes the explosive growth rates of iron bacteria can render a well virtually useless within a matter of months.

Many other forms of iron bacteria induce the precipitation of iron through nonenzymatic means. Found almost everywhere in both water and soil, these bacteria promote precipitation of iron by four major mechanisms:

1. Raising the pH of the water by (a) metabolizing certain protein or protein-derived materials, resulting in the formation of ammonia, which is alkaline; (b) consuming the salts of organic acids, which can lead to the synthesis of alkaline hydroxyl groups; and (c) assimilation of dissolved carbon dioxide in water by cyanobacteria or algae during photosynthesis.

2. Changing the redox potential of the water by algal photosynthesis. In this process, oxygen given off by plants increases the redox potential, thereby causing the precipitation of iron.

3. Liberating chelated iron by inducing a breakdown in the bond between iron and oxalate, citrate, humic acids, or tannins.

Still other forms of iron bacteria can reduce iron to a ferrous state under anaerobic conditions. Although researchers have not been able to classify many major types of iron bacteria in regard to how they participate in the process of iron deposition, the classification shown in Appendix 19.A provides a tentative guide for enzymatic and nonenzymatic bacteria likely to be found in water wells.

It is unclear whether iron bacteria exist in groundwater before well construction takes place and simply multiply as the amount of iron increases, or whether they are introduced into the aquifer from the subsoil, in mix water during well construction, or by backsiphoning from an affected well to an unaffected well. For example, drilling fluid mix water taken from swamps, marshes, or other stagnant surface-water sources may contain high concentrations of iron bacteria. There is also some evidence to show that iron bacteria can be carried from well to well on drill rods, bits, pumps, and water tanks.

Gallionella, a common enzymatic form of iron bacteria, is usually found in water having certain physical and chemical characteristics. Generally the water:

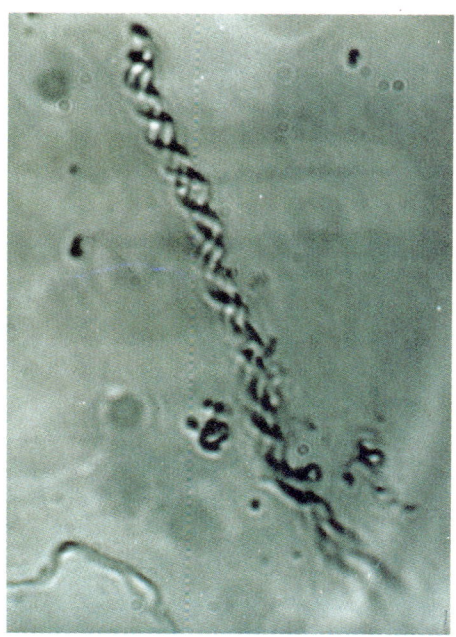

Figure 19.6. Iron bacteria on pump column pipe.

1. Has an iron content of 1 to 25 mg/ℓ and contains only traces of organic matter.

2. Is low in oxygen, typically in the 0.1 to 1 mg/ℓ range.

3. Is usually fresh, although *Gallionella* has been found growing in salt water.

4. Contains over 20 mg/ℓ carbon dioxide.

5. Has a redox potential in the range of 200 to 300 millivolts (mv).

6. Has a pH in the range of 6 to 7.6.

7. Has a temperature from 40 to 60°F (4.4 to 15.6°C).

Presumably, many forms of enzymatic bacteria that could grow in water wells would prefer waters with these same general characteristics. But other forms of iron bacteria, such as *Thiobacillus Ferrooxidans, Sulfolobus Acidocaldarius, Sulfobacillus Thermosulfidooxidans,* and

Leptospirillum Ferrooxidans can grow in waters having extremely low pH (2 to 6) and much higher temperatures [60 to 185°F (15.6 to 85°C)].

A second classification of iron bacteria generally used in the water-well industry is one based on the physical form of these organisms. This method of classification is helpful in identifying which genus of iron-fixing bacteria is contained in a particular water sample. The three general forms recognized are:

1. The capsulated coccoid form, of which only one genus is known, *Siderocapsa*. This organism consists of numerous short rods surrounded by a mucoid capsule. The deposit surrounding the capsule is hydrous ferric oxide, a rust-brown precipitate. This organism probably produces iron precipitates by breaking down the bond between the iron and the chelating agent.

2. The stalked iron-fixing bacteria, composed of twisted bands resembling a ribbon or chain. The genus of this physical form is *Gallionella*, sometimes called *Spirophyllum*, although *Gallionella* is the preferred name. *Gallionella* can be recognized by the twisted stalk and the bean-shaped bacterial cell at the end of the twisted stalk. The only living part of this organism is the bean-shaped cell at the end of the stalk. *Gallionella* is probably the principal enzymatic bacteria occurring in wells.

3. The filamentous group, consisting of four genera: *Crenothrix*, *Sphaerotilus*, *Clonothrix*, and *Leptothrix*. Species of the genus *Crenothrix* have a thin attached end that gradually thickens toward the free end. The separate cells that make up a thread of *Crenothrix* are rod shaped and lie end to end in a sheath. The free end of the filament contains spherical, nonmotile cells called conidia, which are frequently prevented from leaving the sheath. They germinate within the sheath and thrust their filaments through the walls, giving the appearance of numerous branches extending from the parent filament. Members of the genus *Sphaerotilus* exhibit colorless filaments that show false branching. Another iron bacterium that shows false branching is *Clonothrix* (Figure 19.7). This form differs from others in the filamentous group in that its sheath is tapered. The fourth genus in the filamentous group is *Leptothrix*, a simple thread form, usually incrusted with iron along the entire sheath. The sheath of this organism is generally the same width throughout its length and contains colorless cylindrical cells that lie end to end (Figure 19.8). *Leptothrix* and *Sphaerotilus* contain only a relatively small volume of iron in their sheaths and probably do

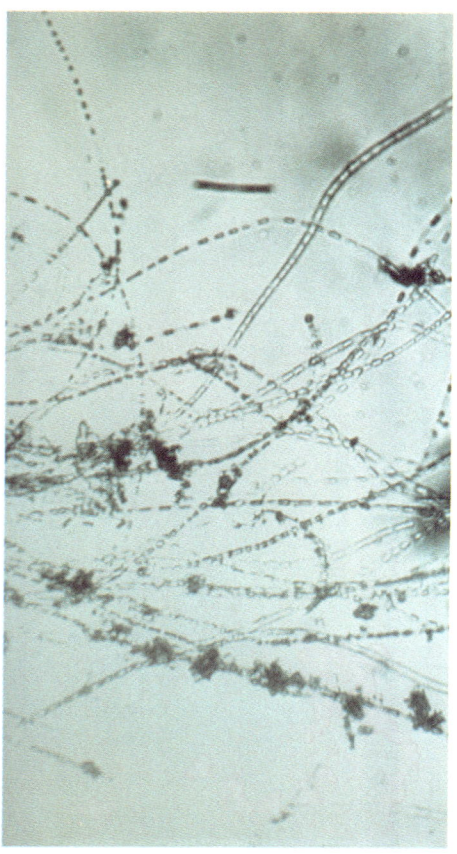

Figure 19.7. Iron bacteria, genera *Clonothrix*.

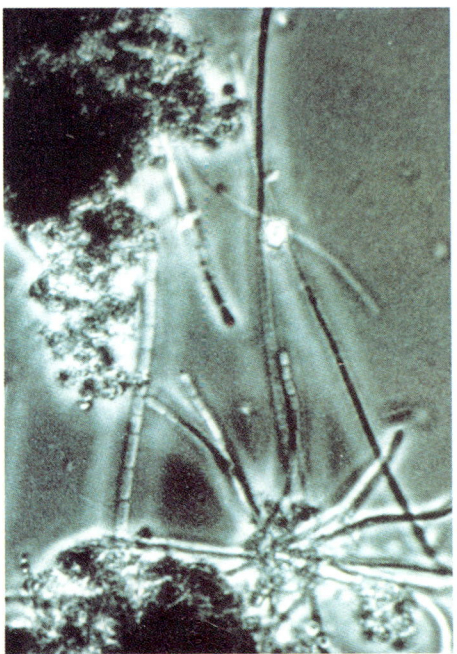

Figure 19.8. Iron bacteria, genera *Leptothrix*.

not derive energy from iron oxidation. This may also be true for *Crenothrix* and *Clonothrix*.

If the presence of iron bacteria is suspected in a well, samples of the organism can be obtained by a filtering device attached to the discharge of the pump for one week. The water passing through the filter during this period leaves a dark brown precipitate on the porcelain cover which can be examined for iron bacteria by a qualified laboratory.

Another method of sample collection is to examine the material scraped from valves or pump discharge lines from suspected wells, pump shaft seals, water closets, or small steel objects suspended temporarily in the well. However, unless a microscope with a magnification of at least 1,000X is available, it is best to send the samples to a state water laboratory or a private firm familiar with iron bacteria identification. Correct identification of iron bacteria is best accomplished by scanning electron or transmission electron microscopy and phase contrast techniques.

Prevention and Treatment of Iron Bacteria

The water well contractor should use great care to avoid introducing iron bacteria into a well during drilling and repair work. All drilling fluid mix water should be chlorinated initially to a 50 mg/ℓ free chlorine concentration, even if secured from a chlorinated municipal water supply. Because chlorine is not stable in a drilling fluid, more must be added periodically to maintain a 10 mg/ℓ free chlorine residual. The drill rods, bits, and tools should be chlorinated thoroughly to eliminate any bacteria remaining from the previous job. Filter-pack material should also be chlorinated before emplacement. This is usually done by adding dry calcium hypochlorite to the pack before it is placed in the well, or chlorinating the water if the pack is pumped into the well. Once the well is completed, it should be sealed immediately to prevent the introduction of airborne bacteria.

Chemical Methods to Control Iron Bacteria

If iron bacteria do grow in a well, they can be controlled by chemical treatments and various types of physical methods (Table 19.5). In general, chemical treatments are more effective and less expensive than physical methods. But for maximum effectiveness, any chemical treatment must be accompanied by physical agitation of the well. Jetting, air surging, air-lift pumping, and valved surge blocks are the principal methods used to agitate the well.

Many effective bactericides are strong oxidizing agents. As this term implies, these

Table 19.5. Methods to Control Iron Bacteria

Chemical	Physical
Oxidizing agents such as chlorine	Heat
pH adjustors such as acids	Vyredox™ technology
Quaternary ammonium compounds	Explosives
	Ultrasonics
	Radiation
	Anoxic blocks

chemicals can oxidize or literally "burn up" organic material. Oxidation is the most common method of killing bacteria, and dissolving and loosening the organic sludge they produce.

Chlorine

Chlorine, a strong oxidizing agent, is used widely to limit the growth of iron bacteria. Chlorine compounds offer significant advantages over other types of bactericides: they are inexpensive, readily available, effective, and generally accepted (actually required in many instances) by health officials as suitable for use in potable water supplies.

The correct chlorine concentration depends on the type of treatment being administered. As little as 50 mg/ℓ free available chlorine is used for routine disinfection of wells and piping following construction, repair, or pump installation, whereas concentrations as high as 500 to 2,000 mg/ℓ are usually desirable for treating wells severely plugged with iron bacteria. A solution strength of 500 mg/ℓ is by definition the strength obtained by dissolving 500 lb (227 kg) of chlorine gas in 1 million lb (454,000 kg) of water. On a smaller scale, this is equivalent to 0.5 lb (0.2 kg) of chlorine gas in 1,000 lb (454 kg) of water [120 gal (0.5 m^3)]. The term "shock chlorination" is reserved for chlorine solutions having a concentration of 1,000 mg/ℓ or more. Table 18.6 (page 621) shows the quantities of chlorine-containing materials necessary to achieve various chlorine solution strengths.

Chlorine gas is the most powerful of the chlorinating agents available commercially. Because it is a gas at normal temperatures and pressures, it must be stored in pressurized cylinders much the same way that propane or acetylene gas is stored. It is extremely corrosive and causes severe damage to human tissues immediately on contact. The use of chlorine gas has generally been restricted to high-capacity municipal and industrial wells because of the skill and equipment required to handle it safely.

During treatment, chlorine gas is usually conducted through a small-diameter plastic tube into the well, where it mixes readily with the water to form the chlorinating solution. A centering device should be used to keep the lower opening of the plastic tube centered in the well screen, because the chlorine gas is so corrosive that holes can form in well screens and casing in a short time, thereby causing sand pumping and ultimately well failure.

After the chlorine solution has been produced in the well, it should be forced through the screen-slot openings into the water-bearing formation by adding water to the well. Then, as with acid treatment, mechanical agitation should be used to enhance the effectiveness of the treatment. As the chlorine disintegrates the organic slime, the mechanical agitation helps dislodge it and move it from the formation into the well, where it can be removed by pumping. Agitation also helps to move fresh chemicals into areas where they may have become expended.

Without some agitation, chlorine may not be effective when treating iron bacteria, because the iron bacteria form a thick, protective slime layer around the cells that is impregnated with oxides and hydroxides of iron and manganese. This layer restricts the movement of chlorine into the cell to the point where the cell may not be inhibited or killed by ordinary lethal doses. In addition, the cells are layered and thus a disinfectant has to penetrate through a series of slime layers, inhibiting and killing the cells as they become exposed. Subsequent disintegration of the dead slime leaves an exposed layer of living iron bacteria beneath which the infestation will continue to grow. Acid treatments are also effective in killing iron bacteria because they generally cannot live at a pH below 2. Figure 19.9 demonstrates that once iron bacteria establish a foothold in a well, they are extremely difficult to eliminate completely by treatment. In this case, the specific capacity of the well is halved in a little over two years.

Agitation can best be achieved by jetting chlorinated water into the formation, because jetting concentrates the greatest amount of energy over the smallest area. Other suitable methods of agitating the chlorine solution include surging by operating a surge plunger in the casing above the screen, or by capping the well and alternately injecting and releasing compressed air, thereby forcing the chlorine solution back and forth through the screen openings. If the pump remains in the well during treatment and there is no foot valve or check valve on the pump, good results may be obtained by pumping and backwashing (alternately starting and stopping the pump). The only requirement is that there not be a net removal of water from the well, because this would

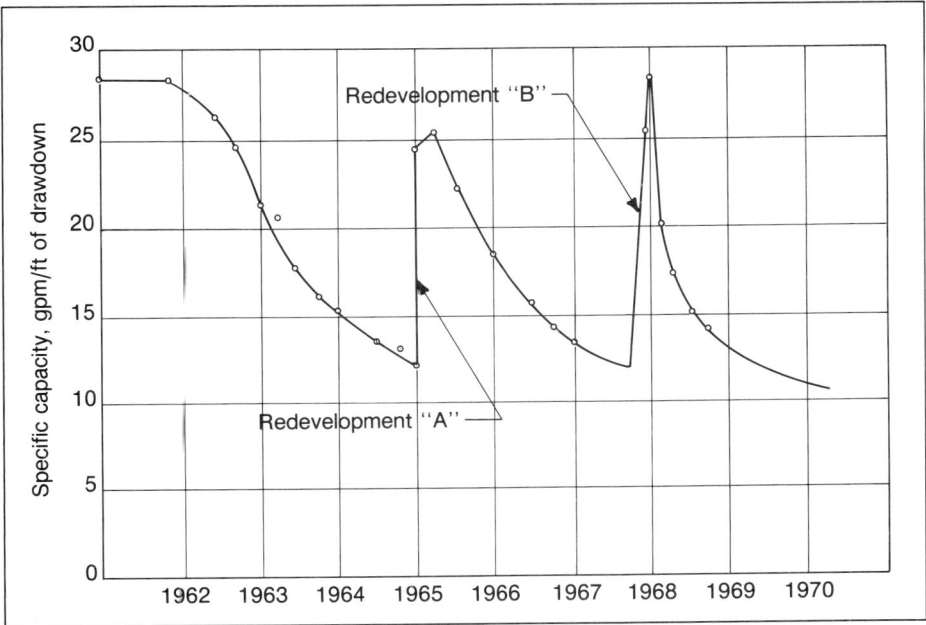

Figure 19.9. Performance record of a well in New Jersey shows declines in specific capacity caused by the growth of iron bacteria, and recovery of specific capacity produced by periodic treatment. Shock treatment with chlorine once a year would help maintain the yield at a satisfactory level. *(A. C. Schultes Company)*

result in removal and waste of the chlorine solution.

Hypochlorites

A relatively safe and convenient alternative to the use of chlorine gas in well treatment is the use of one of several hypochlorite products. In their commercial form, this family of chemicals eliminates some of the dangers inherent in handling chlorine gas and is easily applied to well treatment.

Calcium hypochlorite is a dry mixture containing about 65 percent available chlorine. It can be mixed with water at the surface and poured or piped into the well. Alternatively, the dry material may be poured into the well or suspended in a weighted mesh container, porous sack, or drive point. This latter method is an efficient way to place chlorine at the bottom of an artesian well. If large quantities of dry material are placed directly in the well, some provision should be made for stirring or agitating the water to help dissolve the chemical. Once the chemical has been placed in the well, rehabilitation procedures similar to those used with chlorine gas should be employed. In isolated instances, so much calcium hypochlorite may be introduced that once it combines with the naturally occurring calcium in the water, a precipitate of calcium hydroxide may form that plugs the pores of the formation. For this reason, rehabilitation procedures using calcium hypochlorite may fail to restore the original yield of the well.

Another hypochlorite chemical, sodium hypochlorite, is available in liquid form, typically in solutions of 5 to 15 percent sodium hypochlorite. Pure sodium hypochlorite is highly unstable, actually explosive, and thus cannot be handled safely unless it is dissolved. Even in liquid form, sodium hypochlorite is somewhat unstable and tends to deteriorate with time. During six months storage, a 10-percent solution of sodium hypochlorite loses 20 to 50 percent of its useful chlorine.

Sodium hypochlorite is frequently used in well treatment for routine disinfection of domestic-size wells because it is readily available in the form of household bleach. In addition, it has been used quite successfully to treat iron bacteria problems. The 5.25-percent solution in household bleach contains 5.25 percent available chlorine. Comparing this with calcium hypochlorite which contains 65 percent available chlorine, 1.7 gal (6.4 ℓ) of bleach solution must be used to provide the same disinfecting power as 1 lb (0.5 kg) of calcium hypochlorite.

Chlorine dioxide (ClO_2) is sometimes used for disinfecting drinking water supplies because it produces less trihalomethanes than chlorine except in high-pH waters (Lykins and Griese, 1982). Research has also shown that chlorine dioxide may have stronger oxidative properties than chlorine but its use produces no undesirable organic by-products other than those produced by the use of chlorine. Chlorine dioxide can be used to treat wells or prevent the premature breakdown of drilling fluid made with polymeric additives. In the gaseous form, chlorine dioxide is extremely unstable, and a 10-percent concentration of gas in air is explosive and easily detonated by sunlight. However, in liquid concentrations of 2 to 4 percent, it is relatively stable and can be added to the mix water. The major drawbacks of chlorine dioxide are its relatively high cost and its short lifetime in water (10 minutes).

Table 19.6 contains a list of many of the common chlorinating agents and indicates the amount of each chemical required to provide the same amount of available free chlorine as 1 lb (0.5 kg) of chlorine gas.

Table 19.6. Quantities of Various Chlorine Compounds Required to Provide as Much Available Chlorine as 1 lb of Chlorine Gas

Chemical	% Available Chlorine	Number of lb Equivalent to 1 lb Cl_2
Chlorine Gas	100	1.0
Calcium Hypochlorite	65	1.54
Lithium Hypochlorite	36	2.78
Sodium Hypochlorite	12.5	8.0
Sodium Hypochlorite	5.25	19.05
Trichloroisocyanuric Acid*	90	1.11
Sodium Dichlorcisocyanurate*	63	1.59
Potassium Dichloroisocyanurate*	60	1.67
Chlorine Dioxide	4	25.0
Chlorine Dioxide	2	50.0

*Chlorine compounds that incorporate isocyanuric acid stabilize the chlorine against degradation from sunlight. Except for storage, the advantage offered by the addition of isocyanuric acid is less valuable in water wells.

Potassium permanganate, like chlorine gas, is a strong oxidizing agent that is an excellent bactericide. It has been used successfully to control the growth of iron bacteria in wells. Potassium permanganate is available as a dry, purplish-colored crystal that is both inexpensive and relatively safe to use.

In treating wells infected with iron bacteria, dry potassium permanganate is dissolved in enough water to fill the well screen; the solution is then piped into the screen. A solution strength of 1,000 to 2,000 mg/ℓ has been found to achieve excellent results [1,000 mg/ℓ is equal to 0.83 lb (0.38 kg) in 100 gal (0.4 m^3) of water]. Once the chemical has been placed in the well, vigorous mechanical agitation by surging or jetting should be utilized during treatment to promote loosening and disintegration of the organic plugging material and enhance the overall effectiveness of the procedure.

In treating iron bacteria problems, it must be remembered that the clogging of the well screen and aquifer is caused not only by the organic material produced by the bacteria, but also by the oxides and hydroxides of iron and manganese generally associated with these organisms. In addition, it is usually a matrix of these materials in combination with other mineral scales such as calcium carbonate that causes the problem. Because of the presence of inorganic chemicals, better results are nearly always obtained by treating the well alternately with a bactericide to attack the organic material and a strong acid to dissolve the iron deposits and mineral scale. Between each treatment, the well is pumped to waste. The chlorine and acid must never be in the well at the same time.

Longer time intervals between treatments have been achieved by using a three-step treatment consisting of initial shock chlorination followed by acidizing and then a final shock chlorination of the entire water distribution system. Occasionally, acid is applied first to reduce the thickness of the sheath so that the chlorine is more effective in destroying the tubercles. The added cost of applying three separate treatments is almost always offset by the improved results. A more detailed description of the recommended chlorine-acid treatment process is given at the end of this section.

Physical Methods to Control Iron Bacteria

Pasteurization is a physical method that has been developed to control the growth of iron bacteria. Pasteurization treatments have been shown to be quite effective in maintaining well yield in Saskatchewan in spite of iron concentrations of 1 to 8 mg/ℓ in the well water (Cullimore, 1981). In this treament method, hot water [176°F (80°C)] is circulated continuously in the well until the return water reaches the same temperature. The water is kept at approximately 176°F until temperatures from 113 to 129°F (45 to 54°C) have been reached throughout the layer of iron bacteria. At 113°F the bacterial plugging is dispersed, and at 129°F the bacteria are killed. Tests after pasteurization show a significant drop in the iron bacteria concentration, although bacteria that exist in the formation can quickly reinfest the well.

The cost of treating small-diameter wells by pasteurization is relatively low, because the equipment and procedures are rather simple. However, generating the necessary heat for treating large-diameter wells requires expensive equipment that may make the pasteurization process infeasible economically. Furthermore, depending on the ambient temperatures of the groundwater, the amount of down time required to perform the process may not be tolerable.

Vyredox™ techniques are sometimes used to control the iron content of water and therefore the growth of iron bacteria (see Chapter 23 for a discussion of this technology). By increasing the redox potential of the groundwater around the production well, iron and manganese will precipitate in the aquifer. If the iron concentration can be reduced in the production well to about 0.1 mg/ℓ, iron bacteria probably cannot survive.

The use of explosives and ultrasonic technology to kill iron bacteria have not been effective. Apparently the slime layers can easily absorb the explosive energy or the sound waves with little damage to the bacteria. Although radiation techniques may prove successful in the future in killing bacteria, the use of this technology in wells may not be acceptable to health departments. The effectiveness of creating anoxic blocks in wells to produce anaerobic conditions and thereby kill aerobic iron bacteria has not been ascertained.

Recommended Procedure for Controlling Iron Bacteria

The procedure given below will control the growth of iron bacteria in a large production well. Less complex treatments consisting of only chlorine applications are suitable for most small-diameter wells. It should be noted that virtually no combination of procedures is effective enough to kill all the bacteria in the well. Normally any procedures used will only control the growth of the iron bacteria.

The recommended chlorine-acid procedure is as follows:

1. Inject a mixture of acid, inhibitor, and wetting agent. The addition of a chelating agent such as hydroxyacetic acid may sometimes be beneficial.
2. Agitate the solution with a jetting tool.
3. Pump to waste a volume of solution equal to the volume of the well bore.
4. Determine the pH of the waste. If it is more than 3, repeat steps 1 to 3. (A pH of 3 or less assures that dissolved iron will stay in solution.)
5. Inject a mixture of chlorine and one or more chlorine-stable surfactants (detergents and wetting agents, for example). The concentration of the chlorine should exceed 1 percent.

6. Agitate the solution with a jetting tool.
7. Pump to waste a volume of solution equal to the volume of the well bore.
8. Determine chlorine concentration. If the value is less than 10 percent of the original concentration, repeat steps 5 to 7.
9. Determine the specific capacity of the well. If the specific capacity has improved by more than 5 percent, repeat the entire procedure until the specific capacity does not improve by 5 percent.

WELL FAILURE CAUSED BY PHYSICAL PLUGGING OF SCREEN AND SURROUNDING FORMATION

Over time, almost all screened wells will undergo some loss in specific capacity. Some of this loss is attributable to the slow movement of fine formation particles into the area around the screen. Depending on the type of screen-slot opening, many of these particles may partially plug the screen itself, or even erode the slot openings under certain conditions. Thus, the invasion of small particles reduces the yield, increases the drawdown, and may damage the screen.

Fine-particle movement results from:
1. Improper well design
    a. Poorly designed filter pack
    b. Improper screen placement
    c. Poor slot selection
    d. Inaccurate aquifer sampling techniques
2. Insufficient or improper development when the well was placed in service.
3. Removal of cement holding the sand grains together around the well screen.
4. Corrosion of the screen or casing.
5. Increase in the pumping rate beyond the designed capacity (actually over pumping).
6. Excessive pump cycling.

If the well screen becomes plugged with sediment or incrustants, the entrance velocity of the water passing through the remaining openings increases significantly. As a result, fine sediment is entrained that continually erodes the slot openings. As the slots enlarge, more sediment will pass into the screen. Just how much sand must enter a well to cause failure depends in part on the type of well. Experience indicates that up to 1 mg/ℓ is acceptable in a system with many valves and small orifices, such as a drip-irrigation system. Most industrial and municipal systems can tolerate 2 to 4 mg/ℓ, and some irrigation systems can handle as much as 20 mg/ℓ. At 20 mg/ℓ, a well pumping 700 gpm (3,820 m^3/day) will yield 168 lb (76.2 kg) of sand per day. Over a period of several weeks or months, many tons of sand pass through the pump. To prevent pump damage, the screen may have to be replaced.

Prevention and Treatment of Physical Plugging

Movement of sediment into the formation around the screen can be largely prevented by thorough development of the well during its completion. As suggested in Chapter 15, certain development methods are more suitable for specific types of aquifers. Application of an appropriate development technique for a sufficient length of time will stabilize the formation materials so that subsequent pump cycling and higher discharge rates will not result in sediment movement.

Not all fine-particle problems result from natural formation materials. Occasionally some clay additives used in the drilling fluids may remain in the formation after development. Thus, over time small amounts of these clay residuals enter the well along with other fine material. To completely remove the clay, a chemical treatment may be necessary in the development process.

Polyphosphates and Surfactants

Silt and clay particles tend to adhere strongly to one another in a viscous state, which makes their removal from sand and gravel aquifers quite difficult. Wells that are plugged with silt and clay particles are most effectively restored to efficient conditions by treatment with dispersing and sequestering (chelating) compounds that belong to the polyphosphate family of chemicals. They have the power to separate clay particles. Dispersing agents cause the particles to repel one another, increasing their mobility sufficiently to allow them to move when water is pumped into and out of the well during the development process. Furthermore, the calcium, magnesium, and iron ions adhering to the fine particles can be sequestered (caused to remain in a soluble state) by the use of polyphosphates. Therefore, particles bonded together by these ions can be removed more easily from the aquifer.

Sodium polyphosphates, a family of white, free-flowing dry materials, have been used widely with great success in treating clay-plugging problems. There are two types of sodium polyphosphates, crystalline and glassy. Crystalline polyphosphates that help remove clays from the aquifer are sodium acid pyrophosphate (SAPP), tetrasodium pyrophosphate (TSPP), and sodium tripolyphosphate (STP). Sodium hexametaphosphate (SHMP) is a glassy phosphate that is readily available and therefore often used in rehabilitating wells. Commercial tradenames for sodium hexametaphosphate include Calgon™, Quadrafos™, and Polyphos™. Weltone™ is sodium hexametaphosphate mixed with a chlorinating chemical and wetting agent.

For treating wells, about 15 lb (6.8 kg) of dry polyphosphate should be mixed with 100 gal (0.4 m^3) of water. It is best to mix the material at the surface in warm water in a small container; then dilute with a larger volume of cooler water, chlorinate to 125 mg/ℓ, and put the prepared solution into the well with a tremie pipe, particularly when using the glassy phosphates. If a slug of dry glassy phosphate material is just dumped into the well, it will sink to the bottom and form a large gelatinous mass that could remain undissolved in the well for some time. This mass may plug a significant part of the formation and be extremely difficult to remove. A small amount of hypochlorite should always be used with phosphates because polyphosphates act as a food source for bacteria. This chlorinates the well and kills any bacteria that may be present. About 1.6 lb (0.7 kg) of calcium hypochlorite should be used for each 1,000 gal (3.8 m^3) of water in the well.

Most surfactants are long-chain organic molecules derived from petroleum products. These agents consist of particles that are attracted to oil at one end of the particle and water at the other. Oil can be pulled into a water solution by these particles and removed easily from the porous medium. The presence of a small amount of surfactant speeds penetration of the cleaning chemical by modifying the surface tension of the materials to be cleaned.

The wetting and soil-dispersing properties of surfactants make them ideally suited for use in well cleaning. Those used for wells should be low foaming or used with a defoaming agent to minimize sudsing. Preferably, they should be of the nonionic type — that is, surfactants that do not form ions when dissolved in water. Ionizing surfactants (anionic and cationic types) often react with other chemicals used in the rehabilitation process to form insoluble preciptates that have no cleaning value.

Surfactants are inexpensive to use because only relatively low concentrations of 250 to 500 mg/ℓ are required. They can enhance the dispersing efficiency of the polyphosphates in the removal of silt and clay. Likewise, acidizing is more effective when a surfactant is used with the acid. This is because the surfactant enables the acid solution to soak into all of the pores and cracks of the incrusting deposit, increasing the total contact area between acid and incrustation and thereby speeding the rate of removal of incrustation.

Physical Agitation

Agitation of the phosphate or surfactant solution is important in removing the maximum amount of fine material from the formation. Agitation of the chemical solution during rehabilitation can be done by using a surge plunger, compressed air, well pump, or high-velocity jet. One of the most efficient methods of redeveloping wells with polyphosphates is high-velocity jetting, where the appropriate polyphosphate solution is used as the jetting fluid. If high-velocity jetting is not used, the polyphosphate solution should be placed in the well, forced into the formation adjacent to the screen, and agitated by one of the development techniques described in Chapter 15. Applying these methods in well treatment, however, requires some minor changes in the details of operation. For example, when compressed air is used for surging the chemical solution, the solution must not be discharged from the well before disaggregation of the particles has occurred.

When agitating with a high-velocity jet, it may be desirable to pump the well periodically at a low rate. In operation, jetting adds water to the well at the rate of 25 to 200 gpm (136 to 1,090 m^3/day), depending on the size of the jetting nozzles and the pump pressure. The water pumped from the well can be recirculated to continue the jetting operation. Movement of water through the screen openings into the well carries with it some of the sediment loosened by the jetting process. Thus, material should be settled out in a tank or pit before being recirculated to avoid damaging the screen, pump, or jetting nozzles. Continuous removal of loosened material from the formation will greatly improve the effectiveness of the polyphosphate treatment by allowing the phosphate to reach untreated parts of the formation more quickly. Even though chlorine is used in the phosphate solution, it is good practice to disinfect the well following the polyphosphate treatment to make sure that the well is left in a sanitary condition.

IMPORTANCE OF SCREEN DESIGN ON REHABILITATION

When rehabilitating a well screen, its design will influence considerably the results that can be obtained from various types of chemical treatment and mechanical agitation, particularly horizontal jetting. The force of the jet must be directed through the screen openings. Screens with high open area and uniformly arranged, closely spaced slots that allow direct access to the formation assure the maximum agitation effect from the jetting process. For example, pipe-base and mill-slotted screens offer insufficient open area through the perforations in the pipe. Louver and bridge-slot screens present an almost solid vertical metal surface to the horizontal jet. Continuous-slot screens, on the other hand, have maximum open area and slot configurations that maximize the impact of flow from the jetting tool.

The shape of the screen openings is also important in influencing the effectiveness of the agitation created by the jet. In other words, certain slot configurations will allow the jetting energy to reach deeper into the formation. The best type of opening is a V-shaped slot that widens toward the inside of the screen. When the jet is projected through this V-shaped opening as shown in Figure 15.18 (page 519), the slot opening concentrates the effect of the stream like a second nozzle or venturi. Other slot configurations tend to block or disperse the stream and reduce its force before it reaches the incrusted formation beyond the outer face of the screen.

WELL FAILURE FROM CORROSION

Metals are generally not found in nature in forms that can be used directly by man. They usually exist as ores, that is, stable mineral compounds that are in physical and chemical harmony with the natural environment. These natural minerals must be processed by electrochemical methods to reduce the ores to elemental metals that are suitable for pumps, casing, and well screens. Thus, the chemical and physical properties of ores differ from those of the pure metals. Unfortunately, in the elemental state most metals are not inherently stable. In the environment, elemental metals naturally revert back into more stable mineral compounds. This reaction, called corrosion, is a completely natural process that changes the chemical and physical properties of metals, frequently destroys the usefulness of fabricated metallic articles or structures, and may, over time, reduce or destroy metal products. Corrosion, then, is really the natural reversion of metal to its former state.

Corrosion can severely limit the useful life of water wells in four ways:

1. Enlargement of screen slots or development of holes in the casing, followed by sand pumping.

2. Reduction in strength, followed by failure of well screen or casing.

3. Deposition of corrosion products, thereby blocking screen-slot openings and reducing yield.

4. Inflow of low-quality water caused by corrosion of the casing.

Chemical and Electrochemical Corrosion

Corrosion results from chemical and electrochemical processes. Chemical corrosion occurs when a particular constituent is present in water in sufficient concentration to cause rapid removal of material over broad areas. Commonly, these constituents are carbon dioxide (CO_2), oxygen (O_2), hydrogen sulfide (H_2S), hydrochloric acid (HCl), chloride (Cl), and sulfuric acid (H_2SO_4). Chemical corrosion can cause severe damage in wells, regardless of the amount of total dissolved solids. The number of wells affected by chemical corrosion is small, however, in comparison to wells affected by electrochemical corrosion.

In electrochemical corrosion, flow of an electric current facilitates the corrosive attack on a metal. Two conditions are necessary: a difference in electrical potential on metal surfaces, and water containing enough dissolved solids to be a conductive fluid (electrolyte). A potential (electrical) difference may develop between two different kinds of metals, or between nearby but separate areas on the surface of the same metal. Differences in potential on the same metal surface, such as steel pipe, can occur (1) at heat-affected areas around welded joints, (2) at heated areas around torch-cut slots, (3) at work-hardened areas around machine-cut slots, (4) at cut surfaces of exposed threads at pipe joints, and (5) at breaks in surface coatings such as paint and mill scale. In these cases, both a cathode and an anode develop; metal is removed from the anode.

Bimetallic corrosion results when two different metals are in contact and immersed

Table 19.7. Galvanic Series of Metals and Alloys

Noble or cathodic ↑	Platinum
	Gold
	Graphite
	Titanium
	Silver
	⌈ Chlorimet 3 (62 Ni, 18 Cr, 18 Mo)
	⌊ Hastelloy C (62 Ni, 17 Cr, 15 Mo)
	⌈ 18-8 Mo stainless steel (passive)
	18-8 stainless steel (passive)
	⌊ Chromium stainless steel 11-30% Cr (passive)
	⌈ Inconel (passive) (80 Ni, 13 Cr, 7 Fe)
	⌊ Nickel (passive)
	Silver solder
	⌈ Monel (70 Ni, 30 Cu)
	Cupronickels (60-90 Cu, 40-10 Ni)
	Bronzes (Cu-Sn)
	Copper
	⌊ Brasses (Cu-Zn)
	⌈ Chlorimet 2 (66 Ni, 32 Mo, 1 Fe)
	⌊ Hastelloy B (60 Ni, 30 Mo, 6 Fe, 1 Mn)
	⌈ Inconel (active)
	⌊ Nickel (active)
	Tin
	Lead
	Lead-tin solders
	⌈ 18-8 Mo stainless steel (active)
	⌊ 18-8 stainless steel (active)
	Ni-Resist (high Ni cast iron)
	Chromium stainless steel, 13% Cr (active)
	⌈ Cast iron
	⌊ Steel or iron
	2024 aluminum (4.5 Cu, 1.5 Mg, 0.6 Mn)
Active or anodic ↓	Cadmium
	Commercially pure aluminum (1100)
	Zinc
	Magnesium and magnesium alloys

(Fontana and Greene, 1978)

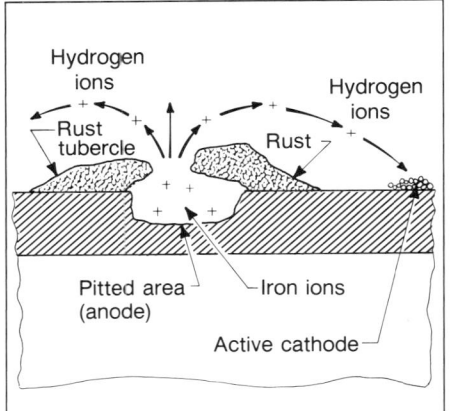

Figure 19.10a. Anodes and cathodes can develop in nearby areas on the same metal surface, resulting in corrosion.

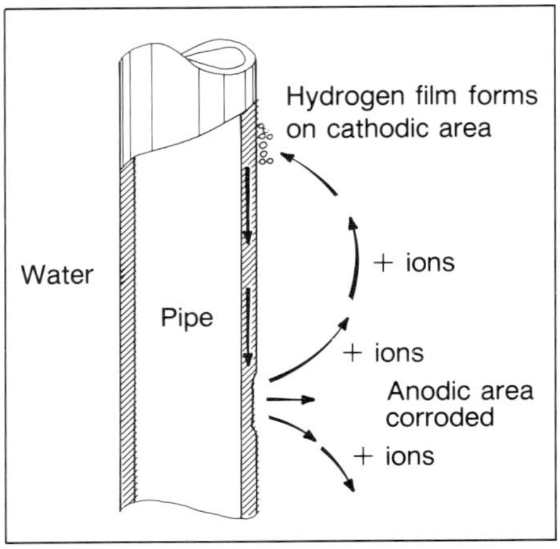

Figure 19.10b. Corrosion of iron at anodic areas leads to the deposition of iron hydroxide and oxide rust at cathodic areas.

in an electrolyte. A galvanic cell is created and corrosion occurs as the electrochemical action proceeds. A well screen made of two different metals, such as low-carbon steel and stainless steel, will be damaged because the mild-steel portion is corroded by the galvanic action.

The relative potential between different metals can be estimated from the galvanic series shown in Table 19.7. The farther apart the position of any two metals in this series, the greater the voltage that will be developed in a galvanic cell. When two dissimilar metals are coupled in a conductive fluid, the metal nearer the bottom of the galvanic series (less noble) becomes the anode and therefore suffers corrosion. The one nearer the top becomes the cathode and usually remains free of attack.

When electrochemical corrosion takes place, corrosion products may be deposited at the cathode (Figure 19.10). These deposits are usually voluminous. If iron or steel is corroded, the corrosion products are iron combined with other elements and are normally ferric hydroxide or ferric oxide. Deposition of corrosion products that results in blocked screen-slot openings and reduced well yields is evidence of electrochemical corrosion.

Prevention and Treatment of Corrosion

The following observations concerning corrosion have been developed from experience and apply to many situations in the groundwater industry:

1. If low-carbon steel is attacked in a particular type of water by chemical corrosion, it will corrode faster if connected to a metal higher in the galvanic series (Table 19.7). This is why stainless steel end fittings should always be used on both ends of stainless steel screens. If they are not, corrosion is concentrated on the relatively small area of mild steel that is welded to the stainless steel. The corrosion rate, however, is considerably lower a short distance away from a connection between two dissimilar metals.

2. Work-hardened material corrodes more rapidly than the same metal that is in an annealed (heat treated) condition.

3. Stressed parts are more likely to corrode than are unstressed parts.

4. Higher temperatures increase corrosion rates. The rate of corrosion accelerates with increasing temperature, generally doubling for each additional 18°F (10°C).

5. High fluid velocities increase corrosion rates in most cases.

6. Dissolved gases such as oxygen, carbon dioxide, hydrogen sulfide, and methane in water increase corrosion rates.

7. Generally, electrochemical corrosion causes loss of material on only parts of well screens and casings, with some of these corrosion products deposited elsewhere. This usually occurs in water that is slightly acidic, with total dissolved solids greater than 1,000 mg/ℓ. A reduction in yield caused by plugging of the screen openings by deposition of corrosion products or by structural failure of the screen can result from corrosion.

In recent years many screen manufacturers have built most of their products out of Type 304 stainless steel. This is because stainless steel is resistant to corrosion caused by underground environments and chemicals added during well rehabilitation, and the price difference between stainless steel and low-carbon steel has narrowed significantly since the mid-1960's. Use of Type 304 stainless steel in most freshwater applications assures the well owner of long-term resistance to corrosion. Certain chemical and physical situations do exist, however, that can lead to severe damage to stainless steel screens; even more resistant materials such as Hastelloy C, Carpenter 20Cb, or Type 316 stainless steel must then be used.

The fundamental resistance of stainless steel to corrosion comes from a protective coating that forms on its surface. This coating is a "passive" film that is resistant to further oxidation or other forms of chemical attack. Although this chemically passive film may be monomolecular in thickness, it is generally protective in oxidizing environments (See Appendix 13.I). The film is produced by the combination of oxygen, water, and constituents of the steel. When the amount of free chromium exceeds 11 percent, steels do not normally form red rust. Thus, chromium is the element that makes stainless steels "stainless."

Stainless steels, however, are subject to corrosion by halogen salts, primarily chlorides, that can easily penetrate the passive film and allow corrosive attack to proceed. Chlorides are generally found only in low concentrations in aquifers used for water supply, but road deicing salts, sewage water, salt-water intrusion, or other forms of subsurface contamination can raise chloride concentrations significantly. Chlorides are soluble, active ions and form electrolytes that can enhance the corrosion rates of stainless steel.

Five basic types of corrosion in stainless steels are recognized:
- General
- Galvanic
- Pitting
- Intergranular
- Stress-corrosion cracking

All corrosion of stainless steel can be related to these five types; the causes of each are discussed below.

General Corrosion

General corrosion is the uniform dissolution of metal over the entire surface exposed to a corrodent and results from the uniform breakdown of the passive film that initially covers the surface. General corrosion may be reduced or even prevented by selecting materials that are resistant in corrosive environments. On the other hand, excessive acid treatment of well screens can cause general corrosion that leads to a widening of the slots. General corrosion can be measured by the rate of metal loss over a surface. This rate is usually given in mils (thousandths of an inch) per year (mpy).

Galvanic Corrosion

Two dissimilar metal alloys connected electrically to a voltmeter and suspended in an electrolyte will show an electrical potential between them, because one alloy will dissolve in the electrolyte more readily than the other. The more active of the two contacting metals will be anodic to the other and will give up metal to the solution, leading to galvanic corrosion. The metal that serves as the cathode will be protected and will show no loss to the corrodent. It would lose metal, however, if it were not in contact with the anodic metal. This is the basis of cathodic protection, where a metal lower in the galvanic series is sacrificed by being electrically connected to the metal to be protected.

Galvanic corrosion is so well known that dissimilar metals of significant difference in voltage (far apart in the galvanic series) are seldom joined. However, when it is necessary to have dissimilar metals in the same corrosive environment, they should be kept apart by nonconductive gaskets (dielectric material) and insulated bolts, or the anodic alloy should have far greater surface area than the cathodic alloy. Most groundwaters have a relatively low concentration of dissolved solids, and thus their ability to conduct electrical current is limited. Consequently, the practice of joining low-carbon steel casing to stainless steel well screens does not cause severe galvanic corrosion to the casing, because the anodic metal has a large surface area. Welding rings fitted to the ends of stainless steel screens should always be made of stainless steel, however. Most corrosion is galvanic in nature, and a thorough understanding of galvanic attack can prevent premature failure of wells.

Cathodic Protection

In groundwater environments where galvanic corrosion poses a potential problem, the well screen, casing, and pump can be protected by means of a cathodic protection system. Cathodic protection for equipment installed underground can be achieved by supplying electrons to the metallic components of the system to be protected. In other words, if an electrical current is caused to flow to a metal casing from the surrounding groundwater, the rate of metal dissolution is greatly retarded. If the current flows the other way — from the metal to the electrolyte — corrosion of the metal casing is accelerated.

The casing, or any metallic structure placed underground, can be protected by initiating current in the proper direction by proper galvanic coupling or by impressing a current from an external power supply. Figure 19.11 shows how a metallic well casing can be protected by the installation of a sacrificial anode. For example, a magnesium anode can be connected to the casing by a coated copper wire. Current flows from the magnesium block through the electrolyte (groundwater), and deposition of the corrosion products occurs on the casing as the anode is slowly taken into solution. The current returns to the anode by way of the copper wire. Some wells are

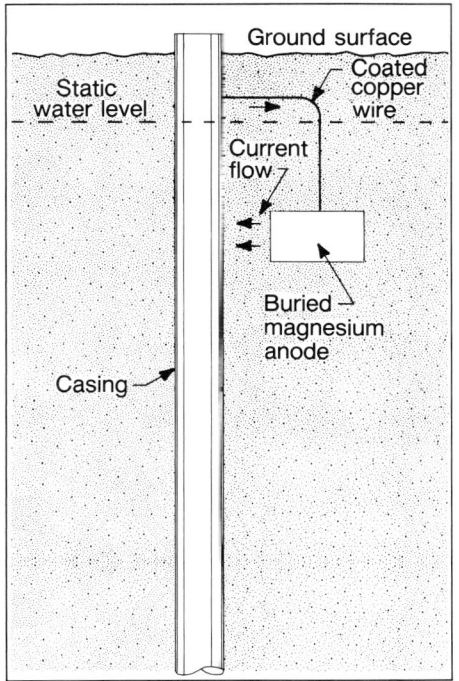

Figure 19.11. A sacrificial anode is connected to a steel casing to prevent corrosion of the casing. Current flows from the anode to the casing through the groundwater. Magnesium ions are thereby transferred to the casing, and eventually the anode is destroyed.

protected by a long cable fastened directly to the casing and suspended in the well below the static water level. This cable, made of a metal less noble than the casing, acts as the anode. Every few years the hanging cable is replaced. Current can also be impressed by an external d.c. power source. The anode is usually placed in a material such as bentonite so that the electrical connection between the anode and the surrounding earth materials is maximized.

The size of the anode used in cathodic protection depends on the current flowing between the anode and cathode. The potential of the cathode can be measured by placing a high-resistance voltmeter in the ground (presumably saturated). This measurement is then used to suggest the size and number of anodes required for complete protection. Spacing of the anodes may be affected by distribution of the currents at various depths adjacent to the well bore. Current flow may be much higher in certain areas because of the presence of clay or other materials that offer less resistance to the applied current.

Pitting

In this form of attack, the rate of corrosion is greater at some areas than others. The depth of pitting is more significant than the number of pits, since one deep pit would limit the well life more than a large number of shallow pits. Pitting is a destructive form of corrosion because only one or two perforations can cause well failure. In many cases, failures caused by pitting often occur over rather short periods, preceeded by years of little or no corrosion.

Pitting of stainless steels occurs where the corrosive environment penetrates the passive film at only a few points rather than over the entire surface area of the metal, as in general corrosion. Pitting is initiated when small areas lose their passivity, become active, and begin to react with surrounding passive areas. Note that "active" stainless steel has a lower place on the galvanic chart that has "passive" stainless steel, with the active being the anodic. Because the passivated surface area, or cathode, is large, the comparatively smaller active area has relatively intense current density and the local metal loss is large. Pitting usually occurs when the corrodent contains one of the halogens (chlorine, bromine, fluorine); these can displace the oxygen in the passive film on the stainless steel and create a local break or a small area where the stainless is unprotected or active. This active area can become a pit if oxygen is excluded by soil accumulations along the surface or is excluded by chemical action

Table 19.8. Resistance of Some Materials to Pitting

↑ Increasing pitting resistance ↓	Type 304 stainless steel Type 316 stainless steel Hastelloy F, Nionel, or Durimet 20 Hastelloy C or Chlorimet 3 Titanium

(Fontana and Greene, 1978)

from surface crevices or irregularities.

The corrosive attack at the active area is exaggerated because of both the current density and the autocatalytic nature of the pitting process in which the pit tends to retain electrolyte and exclude the oxygen that would otherwise repassivate the area. This aspect of pitting corrosion is sometimes referred to as "concentration-cell corrosion." As the depth of the pit increases, conditions for pitting are maintained and the pit will quickly penetrate even thick-walled casing. Each pit acts as a battery that operates by consuming its container.

When halogens are known to be present, pitting may be minimized by using stainless steels alloyed with molybdenum. The higher molybdenum content of certain stainless steels, such as Type 316, is important in preventing pitting. Table 19.8 lists several materials in relation to their resistance to pitting. The various alloying elements that affect pitting resistance are given in Table 19.9.

Intergranular Corrosion

This type of corrosion is restricted, occurring only at grain boundaries. It is often observed when stainless steel is bent, resulting in distinct surface fissures or cracks that follow the grain boundaries.

All austenitic stainless steels contain a small amount of carbon in solution in the austenite. Carbon is precipitated out rapidly at the grain boundaries of the steel within a temperature range of 950°F (510°C) to 1,450°F (788°C); this is called the sensitizing

Table 19.9. Effects of Alloying on Pitting Resistance of Stainless Steel

Element	Effect on Pitting Resistance
Chromium	Increases
Nickel	Increases
Molybdenum	Increases
Silicon	Decreases; increases when present with molybdenum
Titanium and columbium	Decreases resistance in $FeCl_3$; other mediums no effect
Sulfur and selenium	Decreases
Carbon	Decreases, especially in sensitized condition
Nitrogen	Increases

(Greene and Fontana, 1959)

temperature. Carbon combines preferentially with chromium, and precipitates at the grain boundaries in the form of chromium carbide. Thus, the area adjacent to the grain boundary is lower in chromium content and is less resistant to corrosion than the rest of the grain. Under attack, the grain boundaries are dissolved, thus leading to grain disaggregation.

There are three methods available to the corrosion engineer to reduce intergranular corrosion. The first method is to anneal the stainless steel after it has been sensitized. The steel is heated to the proper annealing temperature and quickly cooled through the sensitizing range at such a rate that the carbides are not permitted to precipitate. The second method of avoiding intergranular attack is to minimize the carbon content. If the steel contains less than 0.06 to 0.08 percent carbon, a continuous film of chromium carbide will not precipitate at the grain boundaries during the brief exposure to the sensitizing temperature. Thus, this extra-low-carbon alloy can be welded successfully with no loss of corrosion resistance along the grain boundaries. The third and perhaps surest way to avoid intergranular corrosion is to alloy the metal with titanium or columbium, which are strong carbide formers. The carbon then forms a titanium or columbium carbide rather than a chromium carbide, and the alloy is thus stabilized against chromium carbide formation.

Stress-Corrosion Cracking

All metals are subject to stress-corrosion cracking. If a metal is under tensile stress in a corrosive environment, stress-corrosion cracking may develop. The tensile stress may be residual stress resulting from cold working or quenching after heat treatment, or may originate from externally applied stress, as in the shafting components of a deep-well turbine pump. Solutions containing chloride usually promote stress-corrosion cracking in austenitic stainless steels. Stress-corrosion cracking can be reduced by annealing to relieve the stress developed during fabrication or by selecting materials that are resistant to the potentially corrosive environment.

PUMP MAINTENANCE

Water well pumps are generally built to high engineering standards, utilizing the best materials. Unfortunately, pumps often operate under less than ideal physical and chemical conditions. Thus, they are subject to a wide range of potential maintenance requirements.

The condition of the pumping unit can be evaluated from the following checklist (applying in part to both vertical turbine and submersible pumps):
- Does the pump operate on its original design curve?
- Is there excessive heating of the motor?
- Has there been a change in the bearing-noise level?
- Has there been any change in the pattern of oil consumption of the motor?
- Is there excessive vibration?
- Has the amperage or voltage load to the pump changed?
- Are there cavitation noises or any other unusual sounds?
- Has cracking or uneven settlement of the pad or ground around the pump occurred?

Once a problem is recognized, maintenance must be performed.

A complete analysis of pump maintenance is beyond the scope of this book, but typical malfunction problems and their possible causes in centrifugal and turbine pumps are presented in Appendix 19.B. Pumps can become inoperative for many reasons that may not be directly attributable to the well. The following section, however, discusses only those pump problems that result from deficiencies in well design, construction, and use, and from general long-term deterioration of the well structure. Because the pump can be removed readily from the well, this maintenance problem is less serious than those affecting the screen, casing, or aquifer materials outside the screen.

Figure 19.12. Some of the sand passing through this pump bowl has accreted to the inner surface of the bowl. *(Henkle Drilling)*

Pump maintenance becomes necessary when damage caused by sand pumping, cavitation, high operating temperatures, and low-pH water begins to reduce the yield of the well (Figure 19.12). Sand abrasion is the chief destroyer of pumps and results mainly from:

1. Lack of a screening device in a poorly consolidated formation.
2. Use of a screen with oversized slots.
3. Use of a screen with limited open area (excessive flow rates produced immediately outside the screen cause transport of sediment).
4. Corrosion of the screen.
5. Selection of an oversized filter pack (allows fine grains to pass through it).
6. Inadequate well development.

Correction of these problems can be prohibitively expensive; it may be wiser economically to drill a new well, especially if the well is relatively shallow. If a new or rebuilt pump is to be installed in a sand-pumping well, pump parts in contact with the sand should be built of materials that provide the longest life possible at reasonable cost (Table 19.10). A much better alternative is to replace the screen or place an inner

Table 19.10. The Most Economical Materials Providing Acceptable Service Life for Components of Vertical Turbine and High-Capacity Submersible Pumps

Pump Part	Most Economical Material
Impellers	Bronze (Polycarbonate and R Lexan for submersible pumps that pump less than 20 gpm)
Pump shaft	Stainless steel
Bowl casing	Cast iron
Line shaft	Cold-drawn and polished carbon steel
Wear rings	Hard rubber with interior bronze reinforcement
Diffuser	Cast iron
Column	Steel
Elbow	Cast iron

(After Gass et al.)

screen in the well with a suitable filter pack between the old screen and the new one, although the latter procedure may reduce the well yield somewhat.

Another cause of pump failure is cavitation. Cavitation usually results from not maintaining enough net positive suction head (NPSH) over that head specifically required by the manufacturer for proper operation of the pump. When negative pressure occurs, bubbles of vapor form near the impeller surfaces. Subsequent collapse of the vapor bubbles against the metal surfaces creates extremely high localized stresses that produce pitting of the impeller. Pumps must be run at flow rates close to their maximum efficiency, because flow paths taken by the water in the bowls at other rates lead to destruction of the impeller vanes.

To reduce cavitation effects, centrifugal pumps should be started under reduced or no flow conditions (Walker, 1980). If the pump continues to run under shut-off head conditions, however, the heat build-up may become excessive within a short time and cause the water in the bowls to boil. Any water in the column would be immediately displaced by steam; with no water lubrication, the shaft bearings and any sealing device would be severely damaged. Also, heat expansion of the pump shaft in a deep-well vertical turbine pump may reduce the impeller clearance in the bowls.

Corrosion of pump surfaces by chemically aggressive groundwater accounts for many pump maintenance problems. For example, galvanic cells can be set up between steel shafts and bronze impellers, producing nonuniform corrosion (Figure 19.13). A zinc-free bronze alloy is more resistant to waters with high carbon dioxide concentrations and low pH; use of this alloy will lengthen the serviceable life of the impellers (Clark, 1980).

Changes in flow direction and the compression of gas bubbles in the pump bowls leads to extensive corrosive effects and cavitation. Carbon dioxide contained in the gas bubbles often produces local concentrations of carbonic acid (H_2CO_3) around bubbles. This corrosion effect adds to the metal loss caused by the low pH and differential velocity (erosive) effects when the water changes directions in the pump bowl. Graphitization occurs frequently in cast-iron bowls when iron particles become the anodes in the galvanic process and graphite particles become the cathodes. Under low-pH conditions the iron goes into solution, leading eventually to a greatly weakened structure consisting mostly of graphite and the corrosion products of iron (Figure 19.14). Cast-iron bowls may be destroyed in as little as four years (Clark, 1980). Table 19.11, compiled from the records of many Australian wells, indicates the longevity of pump bowls and impellers, made from similar and mixed metals, in waters with varying pH.

Figure 19.13. Destruction of this pump shaft has been caused by the creation of galvanic cells between the steel shaft and the bronze impeller. *(Bergerson-Caswell Inc.)*

The line shaft in a deep-well turbine is susceptible to corrosion if it is constructed of low-carbon steel. Great stresses are set up in the shaft during pump operation, creating sites for electrolytic attack in low-pH water. The use of Type 304 or 400-series stainless steel will help eliminate this type of electrolytic corrosion. The 400 series is often selected because it can be machined more easily, even though its resistance to corrosion is somewhat less than Type 304 stainless steel.

If the pumping rate declines, it may be because the pump is not operating at its designed efficiency. Increased pumping costs often suggest that there is an efficiency loss in the pump. The wire-to-water efficiency can be recalculated and compared with the original efficiency by using the following equation:

$$\text{Wire-to-Water Efficiency} = \frac{\text{Discharge (gpm)} \cdot \text{Total Head (ft)}}{3{,}960 \cdot \text{Input Horsepower}}$$

Table 19.11. The Effect of pH on the Service Life of Pump Bowls and Impellers

pH	CAST IRON BOWLS BRONZE IMPELLERS	BRONZE BOWLS BRONZE IMPELLERS
8	Safe*	Safe
7	Uncertain	Uncertain
6	Unsafe	
5		Unsafe

*Safe — good chance of pump life exceeding 12 years; uncertain — corrosion erratic; unsafe — definite risk of pump life being less than 12 years.

If a loss in efficiency has occurred, several factors may be responsible:

1. Increased lift (caused by reduction in well efficiency or overall drop in the water table or potentiometric surface).
2. Bowl and impeller damage caused by sand pumping or corrosion.
3. Increased discharge head (rearrangement of piping system or build-up of mineral scale in the piping system).
4. Decrease in engine efficiency.

Greater lift or higher discharge head requirements can usually be identified. If the efficiency loss is not attributable to the factors listed above, the pump or motor must be serviced.

Usually pumps will show some physical signs that maintenance work is needed. Hot motors can be caused by phase imbalance or an overload condition. This overload condition can be caused by poorly adjusted impellers or other types of physical dragging on the bowls. In oil-lubricated pumps, a rapid increase in oil consumption might indicate a hole in the wall of the oil tubing or wear on the packing glands in the tubing. Excessive vibration can be caused by a number of things, including bearing wear, misaligned or crooked shafts, settling of

Figure 19.14. Corrosion of this pump bowl by graphitization has destroyed its usefulness. *(Bergerson-Caswell Inc.)*

the foundation around the well, operating the pump in a crooked well, or improper installation techniques.

Pump maintenance procedures will vary depending on the pump type. In general, the problems are mechanical, that is, those relating to bearings, stuffing boxes, impellers, or bowl assemblies, and are noticeable because of decreased discharge rates or pressure drops.

CONCLUSIONS

Wells are often allowed to deteriorate for such a long time that their specific capacity may be impossible to restore completely, even when using the best chemicals and rehabilitation techniques available. To guard against such a situation, it is essential that the well owner keep good well records so that any decline in performance will not go undetected. The well's specific capacity should be measured at regular intervals, either monthly or bimonthly. The measured specific capacity should be compared with the original specific capacity. As soon as a 10- to 15-percent decrease is observed, steps should be taken to determine the cause and correct the problem. Rehabilitation procedures should be initiated before the specific capacity has declined 25 percent. Declines greater than 25 percent often require much larger expenditures for chemicals and labor without ever regaining the original specific capacity.

The decision whether to rehabilitate an old well or construct a new one can be extremely difficult. The principal items to consider are the cost comparisons between the rehabilitation program and those for the new well, the time required to rehabilitate an old well or to drill and place a new well into service, the projected life of the new well, the economic life of the old well once it has been rehabilitated, and the costs of continuing to use the old well if no maintenance work is performed. With the exception of treatments for iron bacteria and severely corroded screens and casings, systematic well rehabilitation using proper methods and materials can usually restore or may even increase the original specific capacity of the well for a significant length of time. The important point is that a methodical, long-range program of well inspection and monitoring is required to identify problems so that a regular program of preventive maintenance can guarantee a reliable source of water.

CHAPTER 20
Groundwater Law, Water Well Specifications, and Well Contract Problems

Working in the ground is second nature for well drilling contractors, but the inherent difficulties of working under conditions that preclude visual inspection are often extraordinary. The opportunity for drilling equipment to break down is high because of the physical stresses placed on the machine and the people who operate it. Furthermore, mental stress occurs when it is impossible to perform to the letter of a contract because the hydrogeologic conditions encountered in the borehole were not predicted. Under the stress of drilling, the contractor can all too easily omit some common precaution that later can be shown to have caused injury to another. The unpredictability of hydrogeologic conditions can also lead to unintentional damages to the water rights of neighboring landowners. The well contractor and well-design engineer may then be vulnerable to lawsuits based on the principles of negligence and nuisance.

The prudent water well contractor will take precautions to limit legal exposure created by drilling activities. This can be done by:

1. Maintaining drilling rigs in good condition.
2. Thoroughly educating drilling rig operators in the characteristics of both the drilling rig and the local hydrogeology.
3. Taking special care in helping prepare and fully understanding the requirements of a well contract before signing it.
4. Understanding the elements of groundwater law as usually applied in the state, especially those factors that affect the extent to which a person may enjoy a right to water or use of a water resource.
5. Knowing the ways in which the operation of the drilling rig can lead to torts based on negligence and nuisance. (A tort is a private injury* or wrong, not including a breach of contract, for which the injured party is entitled to compensation.)

---

*Under the law, injury is defined as any wrong or damage done to another person or their property, reputation, or rights when caused by the wrongful act of another.

A normal concern for a return on investment usually causes most contractors to follow the first two suggestions on this list. On the other hand, not all design engineers take the time to work effectively with the drilling contractor to develop a well design and construction specification based on all information available. Thus, a large number of change orders commonly occur in the construction of many high-capacity wells. Many of these change orders would be unnecessary if the driller had been consulted during preparation of the well design and construction contract. In other instances, contractors may not be familiar with the basic tenets of groundwater law and how they may affect the drilling operation. Similarly, few contractors take the time to explain to their personnel the legal ramifications of drilling a well. Clear instructions should be given to all personnel on how to avoid causing private or public injury when operating the rig. This is especially important for certain procedures such as blasting, because the courts will view this activity as particularly hazardous and therefore require that the contractor should have taken special precautions to protect public and private interests.

Three subject areas are addressed briefly in this chapter:
1. Important aspects of water law applying to groundwater.
2. Preparation of a typical water well contract for a rotary-drilled well.
3. Common contractual problems.

Each of these subjects is highly complex and the reader is cautioned that considerably more study is necessary to be competent in them. Groundwater law, for example, is a rapidly changing field and the problems affecting contracts are so numerous that only a few can be covered in this chapter. Nevertheless, by studying this chapter, the reader will gain some insight into these subjects and their complexities.

GROUNDWATER LAW

Engineers and water well contractors engaged in development of groundwater resources may become involved in controversies concerning the right to exploit these resources. Litigation can occur if another person's rights to use groundwater are infringed upon, which can result in construction delays and possible professional liability for damages caused. Unfortunately, the evolution of groundwater law has taken a tortuous path that often leads to confusion about which law applies to a particular situation. In addition, groundwater resources are governed by different laws in different states, depending chiefly on the availability of groundwater. To make matters worse, the demands placed on groundwater and the threat of groundwater contamination have led to major changes in the law in almost all parts of the United States. It is clear that the trend is toward increased regulation of groundwater resources. Regulating agencies now evaluate the development of groundwater resources in light of the public good, not just the right of individuals involved. These trends have led to increased conflict between the older concept of private ownership of water and the obligation to protect public rights for the common good of all. It is now evident that public participation and management are vital in establishing an enlightened decision-making system for water (Clark, 1977).

Several factors have combined to hamper the development of a straightforward system of laws governing groundwater. First, the judicial system failed to recognize until recently that groundwater is part of the hydrologic cycle. Second, the original legal doctrines covering groundwater were based on notions that groundwater existed in the same manner as surface streams. A certain body of law then developed from this largely erroneous concept. Much later it was finally realized that, in almost all cases, groundwater percolates through the ground as dispersed flow, not as flow in distinct channels. A different set of laws is generally applied to groundwater that percolates. Furthermore, any legal action taken under these laws will be affected by whether state, federal, or international interests apply.

As noted above, groundwater is treated under two classifications — underground "stream" water and percolating water. Groundwater occurring in streams (a rare occurrence except in karstic terrains or lava flows) is treated under the same general rules as those applying to surface streams — the Riparian Doctrine, the Appropriation Doctrine, and a combination of these two. Under the Riparian concept, two jurisdictional concepts have been applied — the natural flow and the reasonable use concepts. In the first case, the user of water from a groundwater "stream" is entitled to the maintenance of that resource in both quantity and quality. It is clear that almost any use of the groundwater resource may affect the quantity available for other uses, and maintenance of initial water quality conditions over the long term may be impossible. Thus, a more recent version of the Riparian Doctrine developed, called the reasonable use version, in which each owner of land lying above the subterranean stream is entitled to reasonable use of the water. The meaning of "reasonable use" is subjective, however, and often must be determined on a case-by-case basis.

Under the Appropriation Doctrine, the individual who first used the water had the greatest right to it. This is often called the "first in time, first in right" concept. The rights of subsequent users are ranked according to when they first used the water. At some point, all the water will be spoken for and no water will be available for any new users. The Appropriation Doctrine applying to surface water is most strongly developed in the western states where water is scarce. Allocations under the doctrine are done on a priority basis, using permits issued by the courts. A summary and comparison of Appropriation and the Riparian Rights is given in Table 20.1.

In some regions where Riparian rights applied initially, but water availability decreased through time, the Appropriation Doctrine was instituted to establish rights by priority for the remaining water. Extensive use of groundwater in certain parts of Wisconsin, Minnesota, North Carolina, New Jersey, Maryland, Kentucky, Iowa, and Florida has led to the adoption of a dual-doctrine system that can be applied to water in surface streams and subterranean channels.

The laws applying to groundwater that percolates are of much greater interest because virtually all groundwater moves as dispersed flow. Five basic theories are used to allocate rights to withdraw percolated groundwater:

1. The English Rule of Capture, or Absolute Ownership rule. Originating in English common-law doctrine, the owners of land overlying a groundwater resource are allowed to withdraw from their wells all the water they wish for whatever purpose they desire. The water withdrawn can be used for any purpose on or off the owner's land. Under this rule, the landowner could even waste the water, thereby injuring a neighbor, but still have no liability under the law [*Huber* v. *Merkel*, 117 Wis. 355, 94 N.W. 354 (1903)]. States still adhering to this rule no longer go so far as to include malicious actions (Smith, 1980a).

2. The American, or Reasonable Use, Doctrine. Under this doctrine, landowners can withdraw groundwater to the extent that they must exercise their rights reasonably

Table 20.1. Summary and Comparison of the Doctrines of Appropriation and Riparian Rights

Appropriation	Riparian
1. Beneficial use, independent of land ownership, is the basis of the water right.	1. Land ownership is the basis of the water right. Water may be used for any reasonable purpose.
2. Priority of use is the basis of allocation between rival claimants. Rights of the appropriators are not equal.	2. Cosharing equality is the basis of allocation between rival claimants.
3. Rights are to a definite quantity of water.	3. Rights not fixed to a definite quantity of water.
4. Water may be used on nonriparian land.	4. Use of water may be restricted to riparian land.
5. Right may be lost by nonuse or abandonment.	5. Right does not depend on use and is not subject to abandonment.
6. There is no natural flow requirement.	6. There is a qualified right to natural flow in some jurisdictions.

(Tank, 1983)

in relation to the similar rights of others. Furthermore, the owner's use of groundwater for offlying land may be unreasonable and, therefore, unlawful if the withdrawals for the offlying land injure a neighbor.

3. The Restatement of the Law of Torts (2d, Tent. Draft 17, Sec. 858A). This is a version of the reasonable use doctrine that establishes a process to balance competing uses, whether they are on or off the overlying land (White, 1977). In this interpretation, the landowner who withdraws groundwater from the land and uses it for a beneficial purpose is not subject to liability for interference with the use of water by another, unless (Smith, 1980a):

 a. The withdrawal of groundwater causes unreasonable harm through lowering of the water table or reduction of confined pressures.
 b. The groundwater occurs in a distinct underground stream, in which case the rules cited above apply (rules 850A to 857 in Restatement of the Law of Torts).
 c. The withdrawal of water has a substantial effect upon a stream, river, or lake, in which case rules 850A to 857 apply.

4. Correlative Rights. This interpretation derives from the concept that water users will share the resource during droughts, based on the relative areal extent of the land owned by the competing landowners. If no competition for water exists, then correlative rights are the same as reasonable use.

5. Appropriation. In this system, all water is declared to be public and subject to appropriation on the basis of the "first in time, first in right" principle. Control of well use is usually accomplished by permits.

6. Combination. Increased groundwater use among competing interests is leading many states to adopt more than one way of handling the legal aspects of resource

Table 20.2. General Theory of Groundwater Law, by State

Reasonable Use	Correlative Rights	Absolute Ownership	Appropriation
Alabama	California	Connecticut	Alaska
Arizona		Hawaii	Colorado
Arkansas		Indiana	Florida
Delaware		Louisiana	Idaho
Georgia		Maine	Montana
Illinois		Massachusetts	Nevada
Iowa		Mississippi	New Mexico
Kansas		Ohio	North Dakota
Kentucky		Pennsylvania	Oklahoma
Maryland		Rhode Island	Oregon
Michigan		South Carolina	South Dakota
Minnesota		Texas	Utah
Missouri		Vermont	Washington
Nebraska			Wyoming
New Hampshire			
New Jersey			
New York			
North Carolina			
Tennessee			
Virginia			
West Virginia			
Wisconsin			

This table is subject to change through permit legislation and exemptions.
(Smith, 1980a)

allocation and protection. Thus, it is often not possible to say that an individual state operates under a particular groundwater law; there are simply too many exceptions. But Table 20.2 shows the general approach taken by states in regard to groundwater.

In the past, states have taken a negative view of the interstate transport of groundwater. However, in resolving an extremely important case in Nebraska (*Sporhase* v. *Nebraska*), the United States Supreme Court ruled in 1982 that Nebraska, and by implication other states, could not prohibit the transport of groundwater across its borders. In the court's view, groundwater is considered an article of interstate commerce and therefore is subject under the Constitution to regulation by Congress, although Congress has not yet passed such legislation (American Water Works Association, 1982). Prohibiting the transport of groundwater across state boundaries would curtail the affirmative power of Congress to implement its own policies concerning such regulation. Resolution of another similar case, *City of El Paso* v. *Reynolds* (1983), resulted in allowing transport of groundwater from New Mexico for use in the City of El Paso, Texas.

From the short synopsis of groundwater law given above, it is obvious that current laws provide an unsound and confusing base upon which to attempt to resolve groundwater problems of the future. The nation must use all of its water resources more wisely in the future, and water law must adapt to these changing needs. In areas where demand pressures are greatest on water supplies, the following trends can be noted (Tank, 1983):

1. The Riparian Doctrine, with its concept of absolute property rights in water, is becoming obsolete.

2. One principle of the Appropriation Doctrine that suggests that ownership of water rests in the public collectively is becoming the most widely accepted view.

3. The right to appropriate water is being decided on whether the greatest public good is being served.

4. All surface and subsurface water resources are being handled by a single governmental agency with expertise in all aspects of the hydrologic cycle.

WATER WELL SPECIFICATIONS

High-capacity wells costing $250,000 or more are sometimes designed by engineers with only a cursory knowledge of water well technology and the powerful constraints that geology and construction methods place on engineering considerations. To overcome this problem, the U.S. Environmental Protection Agency, with the assistance of the National Water Well Association, published the *Manual of Water Well Construction Practices* in 1975. This manual is an attempt to enumerate the various hydrogeological, constructional, and design parameters that must be considered and how these factors interrelate. This is the most comprehensive analysis of the factors involved in well specifications yet published, and provides an evaluation of the alternative methods used to design, construct, and develop water wells.

Any arrangement by a water-well contractor to perform work for a well owner in return for some sort of payment constitutes a contract. The form of the agreement may vary: a verbal contract; a crude notation on the back of a business card; or a 50-page document signed by both parties in the presence of witnesses. Any agreement between the contractor and the owner is a contract in that there has been mutual agreement on all elements of the project.

The well owner, engineer, and contractor* have well-defined responsibilities in every well construction project. In return for payment, the design engineer and contractor must provide a well that is sand free, produces the desired yield, has the highest possible specific capacity, and provides long service life at low operational and maintenance costs.

A water well and its component parts, such as the pump, motor controls, pumphouse, and discharge piping, may represent a complete construction project under a prime contract. Or, this work may be only one part of a general contract that includes many other components required for more complex projects. In large projects, the specifications for an overall contract are usually written as separate sections, each of which covers a component of the project. Well specifications make up one of these sections. Many specification writers organize each section to cover work customarily performed by a particular trade or specialty contractor. In many cases, it may be best

---

*The owner is defined as a public body (or authority), corporation, association, partnership, or individual for whom the work is to be performed. The engineer (project representative) is the owner's authorized representative assigned to the project. The engineer may be responsible for the specifications and drawings that describe the project and must approve on behalf of the owner the materials and methods used by the contractor. The contractor is defined as the person, firm, or corporation with whom the owner has executed the agreement (contract). For a consideration, the contractor undertakes the risks to construct the project.

to handle test drilling and permanent well construction as a separate prime contract, rather than as a subcontract under a general contract.

Well contracts consist of general conditions, special conditions, and technical standards. General conditions include items such as scheduling, materials and equipment, labor, permits, rights of various parties, tests and inspections, safety, payments, contract bonds, and insurance. The general conditions that will exist between the owner, engineer, and contractor should be established as clearly as possible before work begins. A lawyer should be consulted if the project is large or if it might affect the water use or rights of others. In addition, it is particularly important to agree on how disputes over contract terms will be settled.

Special conditions include the items listed below (U.S. Environmental Protection Agency, 1975c):

a. Scope and general description of work, to include test hole size, well size, depth, and whether it is the intent to obtain either the maximum available production rate or a designated production rate
b. Subsurface information
c. Work schedule
d. Liquidated damages
e. Permits, taxes, legal easements, property boundaries, and other pertinent data
f. Location of existing utilities
g. Availability of construction utilities
h. Specific insurance requirements
i. Bond requirements
j. Submittals
k. Field office
l. Material variations
m. Owner's right to purchase test well
n. Time and notice of all tests
o. Material selection for casing and screens

If conflicts arise between the definition of general conditions and special conditions, the latter shall prevail.

In construction contracts, the technical standards, generally called the specifications, constitute a statement of particulars, that is, a detailed description setting forth the dimensions, materials, acceptable drilling methods, well completion methods, and other details. The plans for the work are graphic specifications.

Many specification details used for other construction work are applicable to well construction. Some important differences, however, require special attention. Each well, or group of wells, is unique even though underground conditions at different locations may seem to be similar. In addition, much of the well structure cannot be inspected visually either during drilling or after completion. Another difference is that the owner is likely to be unfamiliar with the construction methods, particularly the special skills and techniques involved in successful well drilling.

Qualified contractors offer bids on water well projects based on the specifications prepared by the design engineer. When properly prepared, specifications permit the contractor to make intelligent estimates of costs. This encourages fair competition and elicits reasonable bids for the work.

An important federal court decision allows the design engineer to have greater control over the type of material or equipment used by the contractor to construct the well. In the past, the design engineer specified the preferred equipment, but, to avoid antitrust laws, was required to add "or equal" clauses which allowed the substitution of alternative equipment. In many instances, however, contractors were tempted to provide equipment that was inferior to that specified so as to present a lower, and often winning, bid. The design engineer (specification writer) often could not demonstrate clearly enough the unsuitability of the inferior equipment so the bid could be rejected. Thus, the project was constructed using equipment of lower quality than intended. This problem no longer exists because the federal court decision now allows the design engineer to exert much more control over the choice of equipment to be used for the well.

In the *Whitten Corp.* v. *Paddock, Inc.* (1974) decision, four major rulings regarding specifications were established:

1. Proprietary specifications set forth by the specifications writer are not a violation of antitrust laws. Because few makes of equipment are exactly alike, the design engineer has the right to limit the specifications to one or more selected sources and can reject bids that do not conform to the prescribed specifications.

2. If "or equal" is not stated in the specification, other suppliers or manufacturers can qualify as having "or equal" products only when the design engineer chooses to waive specifications or specifically permits these suppliers or manufacturers to bid. Therefore, the suitability of any product proposed as an equal substitution is at the discretion of the design engineer.

3. The design engineer may waive specifications in order to obtain a more desirable product for the client. Thus, any product originally specified can be replaced if the new product would serve the client's interests better.

4. The burden is on the supplier or manufacturer who has not been specified to convince the specifier that their product is equal for the purpose of a particular project.

These changes in the rights of the specifier help assure that well projects will be completed to the quality standards desired. Henceforth, it will be much more difficult to substitute inferior equipment in an attempt to lower the bidding price.

Drafting Specifications for a Well

When drafting the specific design and construction conditions for a high-capacity well project, at least twelve major items should be considered:
1. Test hole logs and samples
2. Well construction procedures
3. Well casing selection and installation procedure
4. Well grouting
5. Well screen selection and installation procedure
6. Filter pack selection and emplacement procedure
7. Well plumbness and alignment
8. Well development
9. Well testing
10. Well disinfection
11. Water sampling and analysis
12. Well abandonment

These items become the technical criteria governing all facets of the well construction. Standards for high-capacity wells vary considerably, however, from one part of the country to another. To a large extent, the hydrogeological conditions and well drilling methods dictate what materials or procedures are best. Thus, no single, standard well specification exists because of the complex array of variables. Nevertheless, a sample set of well specifications is presented below for filter-packed, rotary drilled wells*. This guide includes the major considerations outlined above and some of the many alternative methods that may apply to rotary drilled wells. The standard forms for Notice to Bidders, Proposal, Standard Form of Agreement between Owner and Contractor, and the standard National Water Well Association form for drilling low-capacity wells or for repairs are found in Appendix 20.A.

A contractor who bids for construction work has the right to rely on the plans and specifications for bidding purposes. The rights of the parties are measured by these documents when a contract for the work is concluded. Use of the standard contract form (for domestic and farm wells) is advisable, because its use can eliminate misunderstanding about work performance and can practically guarantee that the contractor will be paid for work completed (McDannald, 1978).

The form of the specification presented below is based on information supplied by water well contractors, design engineers, and the National Water Well Association. All of these specifications are in accordance with the American Water Works Association *Standard for Wells* (AWWA A100-84). Reference is made occasionally in the well specification presented below to various appendices attached to AWWA Standard A100-84. These appendices are not an official part of the standard but do list optional methods and practices that are acceptable under the standards. Consult the appropriate appendices in AWWA A100-84 and the relevant chapters of this book for detailed descriptions of the work listed in the well specification. Blank spaces in the specification form must be filled in based on design criteria for a particular well.

In the sample specification given below, several alternative methods or types of equipment are listed under individual sections. These options constitute some, but not necessarily all, of the ways to accomplish a specific objective. The design engineer should make sure that each option selected does not conflict with other parts of the well design or construction methods. To avoid this problem, the engineer may suggest in the well specification how the individual construction steps are to be accomplished. The engineer should consult with several knowledgeable well contractors before establishing these procedures. In general, this process is more likely to be followed for deep wells or where the well design is more complex than usual. A portion of a tightly written specification for a deep well is presented in Appendix 20.B, in which the engineer not only specifies the exact equipment to be used, but also clearly regulates the type of construction procedures to be followed. If the well construction is relatively uncomplicated, the engineer may omit many construction details.

Specifications for High-Capacity, Filter-Packed Wells Drilled by the Rotary Method

    I. Scope

        These specifications inform all interested parties about the equipment and ma-

---

*The *Manual of Water Well Construction Practices* (EPA, 1975c) contains a matrix diagram listing all of the items to be considered in writing well specifications.

terials needed and the nature of the work required to successfully complete a ____ gpm water well.

II. Personnel and Drilling Equipment
The contractor shall supply capable and experienced personnel and suitable direct-circulation rotary drilling equipment to perform this work. Each bidder shall furnish with the bid additional information which shows:
   A. Experience record of the contractor on wells of this depth and capacity.
   B. Name and experience record of the person or persons likely to serve as construction superintendent.
   C. Manufacturer's name and model number of drilling machine or machines to be used. This includes mud pump and power unit and other pertinent equipment.
Adequate safety equipment such as hard hats, hard-toed shoes, and gloves shall be used by the drilling crew while on the job site.

III. Permits
After the award has been made, the contractor shall apply for and acquire permits from authorized agencies, if required. No field operation shall begin until these approvals have been obtained.

IV. Mobilization and Demobilization
This item in the bid schedule includes moving the materials and equipment for constructing and developing the well to and from the site. It also includes cleaning up the site upon completion of the contract.

Payment for mobilization/demobilization shall be as follows. Fifty percent of the total payment shall be paid on the first pay period after an estimated 10 percent of the total depth of the well has been drilled. The remaining 50 percent will be paid upon completion of the work.

V. Materials
The materials used for the construction of this well shall meet the following requirements:
   A. Drilling Fluid: It is the contractor's responsibility to assure that equipment for measuring fluid properties shall be available at the drilling site. The drilling contractor shall maintain a drilling fluid log showing date, time, depth, Marsh Funnel viscosity, drilling fluid weight, and pH, and shall record any drilling fluid additives used, including time of introduction, as well as any other pertinent comments.
      1. Inorganic drilling fluid made with beneficiated bentonite additives such as Quik Gel™, Kwik-Thik™, IMCO HYB™, or equal shall have the following properties:
         a. Maximum fluid weight: ____ lb/gal.
         b. Maximum Marsh Funnel viscosity: ____ sec.
         c. Maximum filter cake thickness: ____ in.
         d. Maximum 30-minute water loss: ____ cm^3 (Standard API Filter Test).

e. Maximum sand content of drilling fluid entering pump: ____ percent by volume.
f. Any additives to this fluid shall have the engineer's approval.

Modification of any of the above values to meet the requirements of any specific problems encountered shall have the approval of the engineer.

2. (Alternate to 1) Organic drilling fluid made with biodegradable, polymeric drilling fluid such as Revert® or equal shall have the following physical properties:
    a. Maximum fluid weight: ____ lb/gal.
    b. Maximum Marsh Funnel viscosity: ____ sec.
    c. Maximum 30-minute water loss: ____ cm^3 (Standard API Filter Test).
    d. Maximum sand content of drilling fluid entering pump: ____ percent by volume.
    e. Drilling fluid pH range: 6 to 8.
    f. Any additives to this fluid shall have the engineer's approval.

Modification of any of the above values to meet the requirements of any specific problems encountered shall have the approval of the engineer.

The drilling contractor, if unfamiliar with the use of organic drilling fluids, shall contact the manufacturer of these products to determine the proper techniques of drilling fluid control and well development in order to obtain maximum efficiency and safety.

3. (Alternate to No. 1) An organic drilling fluid additive shall be used when drilling swelling clays; other portions of the hole shall be drilled with an inorganic drilling fluid additive.
4. (Alternate to No. 2) An organic drilling fluid additive shall be used when drilling in aquifer materials; the upper portion of the hole shall be drilled with an inorganic drilling fluid additive.
5. (Alternate to No. 1, 2, 3, and 4) The choice of the drilling fluid and its properties and control shall be the complete responsibility of the drilling contractor.

B. Well Casings
1. All well casings shall be new. They shall be manufactured of steel (or other ferrous material) and shall conform to appropriate standards such as those developed by the ASTM or API. The casing shall be joined by either (1) welding in accordance with the standards of the American Welding Society or (2) threaded and coupled joints.
    a. Outer casing. (Also termed surface casing or protective casing.) This casing shall have an external diameter of ____ in and a weight of ____ lbs/ft. It shall be approximately ____ ft in length.
    b. Inner casing. (Also termed working casing, production casing, or riser casing.) This casing shall have an external diameter of ____ in and a weight of ____ lbs/ft. It shall be approximately ____ ft in length.
    c. If required, the contractor shall deliver two copies of the mill certificate to the engineer or owner for approval before delivering casing

to the job site.
: d. (Option) A pitless unit shall be attached to the top of the well casing. This unit shall conform to the standard approved by the local, state, or federal regulatory agency.
 2. (Alternate) The outer (or inner) well casing shall be fabricated of plastic and conform to an appropriate standard such as ANSI-ASTM F480-76. This casing shall have a wall thickness of ___ in and a collapse strength of ___ psi. The connections for joining sections must be approved by the engineer.
C. Grout
 1. Neat cement grout: The grout shall be a mixture of portland cement (ASTM C150, Type I or API-10A, Class A) and not more than 5 gal of water per bag (1 ft^3 or 94 lbs) of cement. The use of bentonite (up to 6 percent by weight of cement) to reduce shrinkage, or of other additives (ASTM C494) to reduce permeability, increase fluidity, and/or control setting time must be approved by the engineer. The water used shall be potable. If the water is questionable, it should be tested in accordance with ASTM C109.
 2. (Alternate to 1) Sand-cement grout: The grout shall be a mixture of portland cement (ASTM C150, Type I or API-10A, Class A), sand, and water in the proportion of one bag of cement and an equal volume of dry sand to not more than 6 gal of clean water.
 3. (Alternate to 1) Clay grout: The grout shall be a mixture of bentonite with the minimum amount of clean water required to facilitate placement.
D. Well Screen
 1. General: The well screen shall be of the continuous slot, wire-wound design to provide maximum inlet area consistent with strength requirements. It shall be fabricated by circumferentially wrapping a triangularly shaped wire around a circular array of rods. The wire configuration must produce inlet slots with sharp outer edges, widening inwardly to minimize clogging. For maximum collapse strength, each juncture between the horizontal wire and the vertical rods shall be fusion welded under water by the electrical resistance method. End fittings shall be welded to the screen body. The well screen shall be manufactured by Johnson Division or an equal approved by the engineer.
   a. Material: The well screen shall be fabricated from Type 304 stainless steel.
   b. End fittings: End fittings provided with the screen shall be fabricated from Type 304 stainless steel and selected on the basis of the well design parameters. The bottom of the screen shall be fitted with a flat plate of the same material as the screen body.
   c. Joining screen sections: Intermediate screen sections shall be joined by welding or by threaded and coupled connections. If welded, the welding rod shall be suitable for joining material so as not to reduce corrosion resistance. The rod selected must be approved by the engineer.

d. Slot Size: The screen slot size shall be selected on the basis of a mechanical size analysis of either the natural water-bearing sediments or the filter pack material.
e. Collapse Strength: The well screen shall have a collapse strength of ____ psi.
f. Tensile Strength: The well screen shall have a tensile strength of ____ psi.
g. Transmitting Capacity: The screen diameter, length, and wire shape shall be chosen so that the maximum average velocity of the water entering the screen is 0.1 foot per second at the desired yield. The screen shall provide a minimum of ____ percent open area with ____ inch slot openings.
h. It shall be the responsibility of the contractor to submit aquifer samples, well logs, and water samples to the well screen manufacturer for analysis. The formation material and water samples shall be analyzed by standard methods, and a report submitted to the contractor and the engineer as a record of the size gradation and water quality of the samples.

2. (Alternate to 1) The well screen shall be of the continuous slot, wire-wound design to provide maximum inlet area consistent with strength requirements. It shall be fabricated by circumferentially wrapping a triangularly shaped wire around a circular array of rods or perforated channels. The wire configuration must produce inlet slots with sharp outer edges, widening inwardly to minimize clogging. For maximum collapse strength, each juncture between the horizontal wire and the vertical rods shall be made by sonic welding. The well screen shall be manufactured by Johnson Division or an equal approved by the engineer.
    a. Material: The well screen shall be fabricated of PVC material.
    b. End fittings: End fittings provided with the screen shall be manufactured from corrosion-resistant materials and selected on the basis of the well design parameters. The end fittings selected must be approved by the engineer.
    c. Joining screen sections: Screen sections shall be joined using methods that are applicable to the type of end fittings used.
    d. Slot size: The screen slot size shall be selected on the basis of a mechanical size analysis of either the natural water-bearing sediments or the filter pack material.
    e. Collapse Strength: The well screen shall have a collapse strength of ____ psi.
    f. Tensile Strength: The well screen shall have a tensile strength of ____ psi.
    g. Transmitting capacity: The screen diameter, length, and wire shape shall be chosen so that the maximum average velocity of the water entering the screen is 0.1 foot per second at the desired yield. The screen shall provide a minimum of ____ percent open area with ____ inch slot openings.
    h. It shall be the responsibility of the contractor to submit aquifer sam-

ples, well logs, and water samples to the well screen manufacturer for analysis. The formation material and water samples shall be analyzed by standard methods, and a report submitted to the contractor and the engineer as a record of the size gradation and water quality of the samples.
3. (Alternate to 1) The well screen shall be of the sawed, punched, or louvered type. The screen shall have a minimum inside diameter of ____ in. Intermediate screen sections shall be joined by welding, using welding rod approved by the engineer, or threaded connections.
4. (Alternate to 1) The well screen shall be a continuous-slot, Type 304 stainless steel type placed over steel casing that has been perforated with uniformly spaced and sized round openings. The outer screen shall conform to the specifications in paragraph 1 above. The casing shall be fabricated to ASTM A-53 or API 5L or equal.
5. (Alternate to b) The bottom of the screen shall be fitted with a self-closing wash-down bottom fitting that permits washing the screen into place.

E. Filter Pack Material
1. The filter pack material shall consist of clean, well-rounded grains that are smooth and uniform. The material should be mostly siliceous with not more than 5 percent calcareous material by weight. The specific gravity of the filter pack material shall be 2.5 or greater.
2. (Alternate to 1) The physical characteristics of the filter pack shall conform with those listed in sections 6.3.1 to 6.3.6 of AWWA *Standard for Water Wells*, AWWA A100-84.
3. The grading of the filter pack shall be determined from sieve analyses of the aquifer materials. The 70-percent-retained size of the filter pack shall be from 4 to 9 times the 70-percent-retained size of the aquifer sample having the finest grain-size distribution. The uniformity coefficient of the filter pack shall not exceed 2.5.
4. (Alternate to 3) The grading of the filter pack material shall be selected by the contractor on the basis of the formation samples and the driller's log.
5. (Alternate to 3) The grading of the filter pack shall be determined by the screen manufacturer.

F. Seal Between Inner Casing and Outer Casing
1. The seal shall be self-sealing rubber (K packer) built on a corrosion-resistant body of the same material as the screen or riser pipe. It shall be attached to the screen or riser pipe by a welded or threaded joint.
2. (Alternate to 1) The seal shall be a lead packer constructed of steel and lead, and shall be attached to the top of the screen or riser pipe.
3. (Alternate to 1) The seal shall consist of a ____ ft concrete plug, formed by placing the cement under water between the inner and outer casing.
4. (Alternate to 1) No seal is required because the inner casing will extend to the land surface or to a point above the screen where the hydraulic conditions within the well will not permit the entrance of filter pack material into the well bore (at least 50 ft).

G. Test Pump and Accessory Equipment for Aquifer Test
1. Pump intake shall be set at ____ ft. The pump and power source shall be capable of discharging ____ gpm (10 to 25 percent of designed well yield) against a total head of approximately ____ ft. All equipment shall be reliable for periods of 24 hours of continuous operation at the design rate.
2. The test pump shall have a foot valve.
3. A ¼- to ½-in diameter metal or plastic air line shall be used to measure water levels. The air line shall terminate at a depth no greater than 2 ft above the pump bowls. The upper end of the air line shall be connected to a direct-reading altitude gauge. There shall also be an air valve for introducing pressurized air. An adequately pressurized air supply must be available.
4. (Alternate to 3) Water levels in the well shall be measured by an electrical sounding device.
5. (Alternate to 3) Water levels in the well shall be measured by an acoustical sounding device.
6. A circular orifice meter shall be installed on the end of the pump discharge line to determine the discharge rate. A control valve shall be installed so that the discharge rate will not vary more than 5 percent from the average rate. The engineer shall approve the equipment and installation.
7. (Alternate to 6) A flume or weir shall be used to determine the discharge rate.

VI. Test Holes and Formation Samples
A. A test hole shall be drilled at the site of or adjacent to the proposed well; the diameter shall be no less than ____ in. It shall be drilled to approximately ____ ft, utilizing a water-based or air-based drilling fluid.
B. An accurate log of the materials penetrated shall be recorded by the driller to determine the depths and thicknesses of the various underlying formations.
C. Formation samples from the test hole shall be collected and handled in a manner selected by the contractor and approved by the engineer.
D. (Alternate to C) Formation samples from the rotary drilled test hole shall be collected and handled in accordance with the procedure given in Chapter 8, *Groundwater and Wells*, 2nd Edition, published by Johnson Division.
E. (Alternate to C) The formation samples shall be obtained by split-spoon methods in accordance with the "Tentative Method for Penetration Test and Split-Barrel Samples of Soils," ASTM D-1586-58T.
F. (Alternate to C) The formation samples shall be obtained by the drive-core or core-barrel method.
G. (Alternate to C) Samples shall be obtained only from water-bearing formations; unsaturated formations shall not be sampled.
H. Water samples shall be collected as outlined in *Standard Methods for the Examination of Water and Waste Water*, 16th edition, 1985, published jointly by the American Public Health Association, American Water Works

Association, and Water Pollution Control Federation.
- I. (Optional) A resistivity log shall be made to the total depth of the test hole. The logging equipment or the service company making the log shall be approved by the engineer.
- J. (Alternate to I) (Optional) A gamma-ray log shall be made to the total depth of the test hole. The logging equipment or the service company making the log shall be approved by the engineer.
- K. It shall be the contractor's responsibility to secure, protect, and deliver all test-hole formation and water samples to the laboratory, engineer, or person designated by the engineer.

VII. Testing of Formation and Water Samples
  A. Aquifer Test
  The test well shall be pumped to estimate the transmissivity, storage coefficient and specific capacity. To obtain this information, at least one and preferably two observation wells shall be drilled at suitable distances from the pumped well. The well shall be test pumped at a constant rate for 24 hours if the aquifer is confined or 72 hours if the aquifer is unconfined.

  The discharge of the test pump will be measured by a circular orifice meter, totalizing meter, flume, or other suitable device, and the water level measured electronically or by air line or tape to the nearest 0.5 in (13 mm). Measurements of the yield and water level will be made every 30 to 60 seconds for the first 10 minutes of the test, every minute for the next 5 minutes, every 5 minutes from 15 to 60 minutes, every 30 minutes from 60 to 300 minutes, every hour from 300 to 1,440 minutes, and every 8 hours from 1,440 minutes until the end of the test. A water sample (approximately 1 qt) shall be obtained within 20 minutes after starting the pump and then about 5 minutes before shutting the pump down.

  After the pump is shut down, recovery measurements of the water level shall be made for a period equal to at least three-fourths of the pumping period or until the water level has reached the original static level. The recovery measurements shall be made at the same time intervals that measurements were made during the pumping portion of the test.
  B. Payment for this item shall be based on the hourly rate beginning when pumping starts and ending when the recovery portion is completed.
  C. All samples shall be sent to an approved laboratory where the following tests will be conducted.
    1. Formation samples:
       a. Grain-size analysis using at least five standard sieves.
       b. Determination of uniformity coefficient.
       c. Determination of hydraulic conductivity.
    2. Water samples:
       a. Chemical analysis including:
          (1) Total alkalinity as calcium carbonate
          (2) Total hardness as calcium carbonate

(3) Calcium hardness as calcium carbonate
(4) Total iron
(5) Total manganese
(6) Sulfate
(7) Chlorides
(8) Trace organics
(9) Total dissolved solids
(10) pH
(11) Ryznar Stability Index
(12) Specific conductance
   b. Temperature of the water when pumped shall be measured by the driller and recorded on the sample container.
   c. (Alternate to a) The chemical analysis shall include the testing of all constituents deemed necessary by the regulatory agencies.
   d. (Alternate to a) The chemical analysis shall include tests for the following substance to determine if the water is suitable for irrigation: irrigation:
      (1) Boron
      (2) Calcium
      (3) Magnesium
      (4) Sodium
      (5) Potassium
      (6) Carbon dioxide
      (7) Bicarbonate
      (8) Sulfate
      (9) Nitrate
      (10) Total dissolved solids
      (11) Sodium adsorption ratio (SAR)
D. (Alternate to B) Some chemical analyses shall be done in the field by the contractor or engineer immediately after the water is discharged. These analyses shall include pH, CO_2, H_2S, and dissolved oxygen. A 1-gal sample of water shall also be sent to an approved laboratory for a more complete analysis. Copies of the field test results shall be sent to the laboratory along with the 1-gal sample.

VIII. Construction of the Well
   A. General:
   The contractor shall drill the well at the exact location designated by the engineer. The contractor shall use direct rotary drilling equipment to drill the hole and shall install the materials previously described so that the finished well conforms to the general design illustrated in Figure 20.1 and to any applicable standards established by local, state, or federal regulatory agencies. The mud pits required shall be positioned at least 10 ft from the proposed pump-foundation pad or pump-house floor.

   The contractor shall dispose of drilling fluid, cuttings, and discharged water in a manner prescribed by the engineer so as not to create damage to public or private property. During the test pumping, the water discharged shall be piped to a point of overland drainage sufficiently far from the well to prevent a recharge effect.

B. Drilling—Outer Casing Hole
The contractor shall drill a ___ in diameter hole to accommodate the outer (surface) casing to a depth of ___ feet. If this depth is adjusted because of underground conditions at the site, payment will be based on the unit price per linear foot of the hole drilled.

C. Setting Outer Casing
The top of the outer casing shall be a minimum of 12 in above the final ground surface or pump-house floor and at least 2 ft above any known flood levels. The casing shall be centered in the hole so a minimum of 3 in exists between the borehole wall and the casing. A casing (cementing) shoe is (is not) required. Centering guides shall be spaced around and along the pipe to insure adequate clearance for the cementing operation. Payment of this item will be based on the unit price per linear foot of outer casing used.

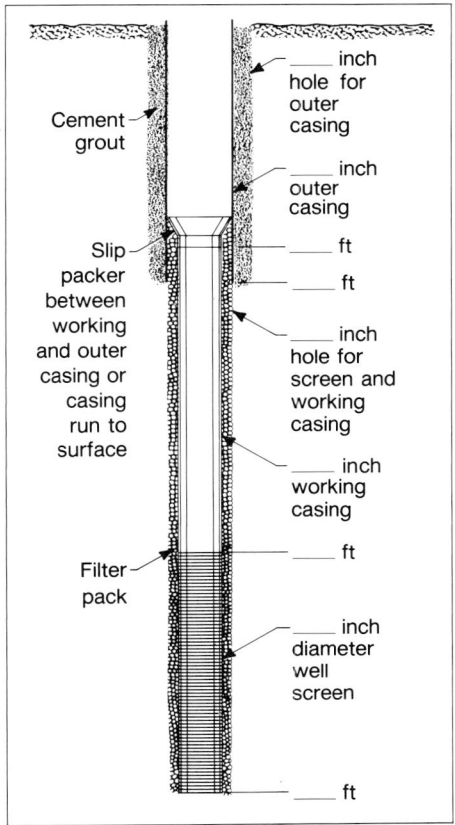

Figure 20.1. Expected completion dimensions of proposed rotary drilled filter-packed water well.

1. After the casing has been lowered into position and before the cementing operation begins, the borehole shall be tested for plumbness and alignment. This test shall be made in accordance with Appendix C, "Testing for Plumbness and Alignment," of AWWA Standard A100-84. Should the casing fail to meet the requirements and interfere with the installation and/or operation of the pumping equipment, the plumbness and/or alignment shall be corrected at the contractor's expense. The AWWA requirement that the plummet or dummy be 0.5 in smaller may be modified to not more than 1 in smaller in large-diameter casings, as approved by the engineer.

2. (Alternate to 1) After the casing has been lowered into position and before the cementing operation begins, the borehole shall be tested for plumbness and alignment by using an inclinometer. The engineer shall judge whether the casing is sufficiently plumb and aligned to permit satisfactory installation and operation of the pumping equipment.

D. Grouting
   1. Prior to grouting, the engineer may require flushing of the annular space with the drilling fluid or with clean water. Grouting of the outer casing shall be completed in the presence of the engineer. It shall be carried out in one continuous operation, filling the annular space between the drilled hole and the casing, from the bottom of the casing to the land surface. The method of placement shall be selected from Appendix B, Sections B.1 to B.6, of AWWA Standard A100-84. The grout returning to the surface shall be wasted until the engineer determines that the grouting is satisfactory.
       a. No further work shall be done on the well until the grout has firmly set (a minimum of 72 hours for neat cement and 24 hours for quick-setting cement).
       b. (Alternate to a) No further work shall be done on the well until the grout has firmly set. The adequate setting time may be determined by placing a sample of the grout, taken at the end of the grouting procedure, in a suitable open container and submerging it in a bucket of water. When the grout has set, drilling can be continued.
   2. Pressure testing of the grout seal shall be performed after the appropriate curing time. A pressure of 7 to 10 psi (48.3 to 69 kPa) of air must be maintained in the well for a period of not less than 1 hour without the addition of more air. Loss of air shall be construed as indicating a defective seal and the owner shall require the contractor to recement and successfully pressure test the well at 15 psi (103 kPa) for 1 hour.
   3. (Option to 2) After the cement has cured, an acoustic (sonic) cement bond log shall be run from the top to the bottom of the borehole to determine the quality of grout placement. An interpretation of the log, given to the owner along with the original log, shall be made by an analyst experienced with this technique.
   4. Payment for grouting will be made at the unit price per bag of cement, and per unit weight of clay or sand used, as shown on the bid schedule.
E. Drilling the Hole for the Screen and Inner Casing
   1. After the grout has set, an ____ in diameter hole below the outer casing will be drilled to a depth of ____ ft. If this depth is adjusted because of underground conditions, payment will be based on the unit price per foot of hole drilled. Measurements of drilling fluid lost to the aquifer, and depths at which these fluid losses occur, shall be recorded by the driller.
   2. (Optional) After the hole that accommodates the screen and inner casing has been drilled, a caliper log shall be run to determine the actual hole size. This is usually required if the hole beneath the outer casing is to be underreamed to a diameter larger than the outer casing. Payment for the caliper log shall be as shown on the bid schedule.
F. Setting the Screen and Inner Casing
   1. The screen and inner casing assembly shall be joined properly and lowered into the well. If welded, procedures conforming to the practices of the American Welding Society shall be followed. The screen may be

lowered by using a temporary string of pipe attached to the top of the inner casing or to the bottom of the screen. The assembly shall be supported so that the screen bottom is slightly above the bottom of the hole to insure that the entire assembly is under tension during placement of the filter pack. Centering guides shall be placed on the screen as recommended by the manufacturer and as required on the inner casing.
   2. Payment for the screen shall be based on the linear feet of screen actually used and the cost of the end fittings. Payment for the inner casing will be based on the number of feet actually used.
G. Filter Packing
   The filter pack shall be introduced uniformly and continuously to minimize or eliminate hydraulic segregation and bridging.
   1. Before emplacement, the filter pack material should be treated with a chlorine solution having a 50 mg/ℓ free-chlorine residual.
   2. The filter pack shall be placed by gravity through a 2-in OD tremie pipe. The pipe shall terminate 5 ft above the bottom of the hole or above the level of the filter pack already placed in the hole. It shall be lifted gradually and sections removed as the level of filter pack rises. The top of the tremie pipe shall be fitted with funnel fittings so the filter pack can be shoveled or dumped into the tremie. Clean water may be mixed with the filter pack to facilitate placement.
   3. The filter pack shall be placed to within 5 ft of the top of the inner casing.
   4. If required, a seal between the inner and outer casing shall be installed and swedged.
   5. (Alternate to 2) The filter pack shall be pumped into place by a closed pressure system.
   6. (Alternate to 2) The filter pack shall be shoveled into place manually. Alternate shovelfuls shall be placed 180 degrees apart at the top of the annular space.
   7. The filter pack shall be placed so that the actual volume used can be calculated to within 10 percent.
   8. Payment for the filter pack shall be made at the unit price per cubic foot for the amount of filter pack actually used to fill the hole.
   9. (Alternate to 8) Payment for filter pack shall be made at the unit price per cubic foot for the total amount of the filter pack delivered to the site.

IX. Developing the Well
   The development of the well shall remove the native silts and clays and drilling fluid residues deposited on the borehole face and in adjacent portions of the aquifer during the drilling process. Development shall also remove a predetermined finer fraction of the filter pack. If organic drilling fluids are used, they must be broken down chemically according to manufacturer's recommendations before or during development. The objective of the development process is to assure maximum specific capacity.
   A. The development process shall be accomplished by high-velocity, horizontal jetting and simultaneous air-lift pumping. The outside diameter of the

jetting tool shall be ½ to 1 in less than the inside diameter of the screen. The minimum exit velocity of the jetting fluid at the jet nozzle shall be 150 ft/sec. The jetting procedure shall proceed from the bottom of the screen to the top. The tool shall be rotated at a speed of 1 rpm. It shall be positioned at one horizon for not less than 2 minutes and then shall be raised to the next horizon which shall be no more than 6 in vertically from the preceding jetting horizon. If the jetting tool is continuously rotated, the withdrawal rate shall be 5 to 15 minutes per ft of screen. The size of the eductor pipe (if used) for air-lift pumping shall be ____ in and the air line ____ in. The air compressor must be capable of delivering ____ cfm at a discharge pressure of ____ psi. The eductor pipe must be placed no more than 5 ft above the top of the jetting tool during development. Pumping from the well shall be at a rate from 10 to 20 percent more than the volume of water introduced through the jetting tool. Water to be used for jetting must be free of turbidity and coliform and iron bacteria.

1. (Option to A) The jetting water shall contain a polyphosphate or other drilling fluid dispersing agent in the proportion of ____ lbs of chemical per 100 gal of clear water.

B. Sediment falling to the bottom of the screen during the jetting procedure shall be removed by either pumping or bailing.

C. Development shall continue until the engineer or contractor decides that further development is unnecessary. Payment for development will be based on the unit prices per hour as shown in the bid schedule. It shall cover only those hours the development tools and equipment are actually being operated.

D. (Alternate to A) The development process shall be carried out by surging and bailing the well. The surging shall be accomplished by a solid or valved surge block. The engineer shall approve the surge block and related apparatus before its use is permitted. When surging in the screen, surging shall be started slowly just below the static water level and continued at a faster rate once free flow into the screen is assured. Thereafter, the surge block can be lowered gradually into the screened area.

E. (Alternate to A) The development process shall be carried out by a combined air-lift pumping and backwashing procedure. The size of the eductor pipe (if required) shall be ____ in and the air line ____ in. Each must have its own suspension apparatus. The air compressor shall be capable of delivering ____ cfm at a discharge presure of ____ psi. Starting from the bottom of the screen, the eductor pipe shall be raised to pumping positions no farther apart than 3 ft.

F. (Alternate to A) The development process shall be accomplished by the use of a test pump capable of pumping 150 percent of the design yield against a total lift of ____ ft. The discharge at particular rates may be held constant for as long as 2 hours. The test pump shall have no check or foot valve, so that backwashing occurs when the power is shut off. Any sand damage to the pump is the responsibility of the contractor.

X. Testing the Production Well
   A. After development, the well shall be pumped to verify that the expected specific capacity has been obtained and the well efficiency is at least ____ percent. The discharge of the test pump will be measured by timing how long it takes to fill a container of known volume, by a circular orifice meter, flume, or by other device, and the pumping water level measured with a suitable measuring instrument to the nearest 0.5 in. The engineer or owner compares this figure with the specific capacity determined from the pumping test and decides whether further development is needed to improve the yield. Sand content shall not exceed ____ mg/ℓ after ____ minutes of pumping at ____ gpm.
   B. Payment for this item shall be based on the hourly rate starting when pumping begins and ending when the test is completed.

XI. Disinfecting the Well
   A. The well shall be disinfected before removing the test pump from the well and before collecting any samples for determining microbiological quality. This shall be done by placing a chlorine solution into the well so that a concentration of at least 50 mg/ℓ of available chlorine exists in all parts of the well at static conditions. All the well surfaces above the static level shall be completely flushed with the solution. The solution shall remain in the well a minimum of 2 hours before pumping the well to waste.
   B. A water sample shall be forwarded to the designated laboratory to verify the potability of the water (1 qt is usually sufficient).
   C. (Alternate to A) The well shall be disinfected according to the procedures established by local, state, or federal regulatory agencies.
   D. (Alternate to A) The well shall be disinfected according to the procedures outlined in Appendix E, Sections E1 to E5, AWWA Standard A100-84.
   E. (Alternate to A) The well shall be disinfected after the permanent pump has been installed and before putting the well into service.

XII. Capping the Well
   During well construction and completion, the contractor shall use all reasonable measures to prevent the entrance of foreign matter into the well. The contractor shall be responsible for any objectionable material that may fall into the well and any effect it may have on water quality or quantity until completion and acceptance of the work by the engineer.
   A. Upon completion of the well, the contractor shall install a suitable screwed, flanged, or welded cap to prevent any pollutants from entering the well. Payment for furnishing and installing the well cap will be based on the lump-sum price shown in the bid schedule.
   B. (Alternate to A) After testing and disinfecting the well, the contractor shall furnish and install a cap on the well. The cap shall consist of ¼-inch plate cut to the outside diameter of the casing and be tack-welded to cover the top of the well. A 1-in diameter threaded nipple and pipe cap shall be attached by threads or welding to a matching hole drilled in the center of the cap.

XIII. Standby Time
   A. Standby time will be credited only for inactive periods resulting from requirements of this contract or for other events over which the contractor has no control. Idle time required for maintenance or failure of equipment shall not be measured for standby time. Standby time shall be based on one work shift per day regardless of the contractor's operating schedule. Standby time will not be paid for Saturdays, Sundays, or national holidays on which work is not customarily performed, unless the contractor has previously agreed to work on such days.
   B. Payment for standby time will be made at the unit price per hour as shown on the bid schedule, and only for that part of a regular ____-hour shift during which the contractor may not continue work because of the requirements of the specifications.

XIV. Guarantees
   A. The contractor shall guarantee that all materials, equipment, structures, and work performed are free from defects in workmanship or materials for a period of one year after completion, and if any part of the work shall fail within this period, it shall be replaced and the unit restored to operation at no cost to the owner. The Surety Bond shall cover this guarantee. Defects to be covered by this guarantee are:
      1. Pumping sand in excess of 2, 5, 10, or 20 mg/ℓ (select one), measured after the pump has been running at the designated rate for ____ minutes.
      2. A reduction in specific capacity in excess of ____ percent over a ____ month period. The specific capacity shall be determined on the basis of a discharge rate and pumping time equivalent to the original pumping conditions.
      3. Collapse of casings or screens from causes other than Acts of God.
   B. (Alternate to A) The contractor shall guarantee a specific capacity of ____ gpm/ft of drawdown at a pumping rate of ____ gpm after ____ hours of continuous pumping. This shall be the condition for acceptance.
   C. (Alternate to A) The contractor shall guarantee a yield of ____ gpm for a period of ____ hours of continuous pumping. This shall be the condition for acceptance.

XV. Abandoned Wells
   If the well fails to conform to these specifications and the contractor is unable to correct the condition or negotiate a mutually acceptable cost reduction for specification deviations, it shall be considered an abandoned hole, and the contractor shall immediately start a new well at a nearby location designated by the engineer.
   A. The abandoned hole shall be treated as follows:
      1. The contractor may salvage as much casing and screen from the initial well as possible and use it in a new well if not damaged. Any casing remaining in the borehole must be perforated with a casing ripper. The upper 5 ft of casing shall be completely removed from the borehole.
      2. Salvaged material shall remain the property of the contractor.

3. The well shall be sealed by concrete, cement grout, or neat cement and shall be placed from the bottom upward by methods that will avoid segregation or dilution of material.
4. The upper 5 feet of borehole shall be filled with native topsoil.
B. (Alternate to A) The well shall be sealed according to the procedures established by local, state, or federal regulatory agencies, or the recommendations given in Appendix I of AWWA Standard A100-84.
C. (Alternate to A) The well shall be filled and sealed in such a manner as to avoid accidents and to prevent it from acting as a vertical conduit for transmitting contaminated surface or subsurface waters into water-bearing formations.

XVI. Records
A. Two copies of the driller's log, signed and dated by the well driller or drilling machine operator will be given to the engineer. The log will record the materials penetrated to the nearest foot.
B. If geophysical logs were made, a copy of each shall be given to the engineer.
C. A record of all static water level measurements, and the times at which they were taken, will be given to the engineer.
D. A complete casing and screen location record will be made by the driller and given to the engineer. This will show the lengths of each casing and screen section and the location of packers, plugs, and seal.
E. Pumping test data shall be supplied to the engineer from all pumping tests conducted on the well and test well. These will show dates, water levels, discharge rates, times of stopping and starting the pump, and other conditions that could affect the test data.

Bid Schedule

Description	Estimated Quantity	Units	Unit Price	Total Price
Mobilization and demobilization				$ _____
Test holes and samples				
a. _____ inch diameter	_____	Lin. ft.	$ _____	$ _____
a. _____ inch diameter	_____	Lin. ft.	_____	_____
b. Geophysical/mechanical logs: (Type) _____	_____	Lin. ft.	_____	_____
c. Stratigraphic logs	_____	Each	_____	_____
d. Formation samples	_____	Each	_____	_____
e. Water (aquifer) samples	_____	Each	_____	_____
f. Services of log analyst	_____	Each	_____	_____
Aquifer test				
a. Installation and removal of testing equipment	_____	Each	$ _____	$ _____
b. Recovery standby time	_____	Per hr.	_____	_____
c. Discharge pipe	_____	Lin. ft.	_____	_____

Description	Estimated Quantity	Units	Unit Price	Total Price
Well construction				
a. _____ inch diameter	_____	Lin. ft.	$ _____	$ _____
a. _____ inch diameter	_____	Lin. ft.	_____	_____
b. Geophysical/mechanical logs: (Type) _____	_____	Lin. ft.	_____	_____
c. Formation samples	_____	Each	_____	_____
d. Services of log analyst	_____	Each	_____	_____
Well casing installation				
a. _____ inch diameter	_____	Lin. ft.	$ _____	$ _____
a. _____ inch diameter	_____	Lin. ft.	_____	_____
b. Pitless unit	_____	Each	_____	_____
c. Pressure testing	_____	Each	_____	_____
Well grouting installation				
a. Grouting installation	_____	Each/bag	$ _____	$ _____
b. Grouting service	_____	Cu. yd.	_____	_____
c. Cement bond log	_____	Lin. ft.	_____	_____
d. Pressure testing	_____	Each	_____	_____
Well Screen				
a. Screen furnished and installed	_____	Lin. ft.	$ _____	$ _____
Filter pack construction				
a. Filter pack furnished and installed	_____	Lin. ft.	$ _____	$ _____
b. Filter pack	_____	Cu. yd.	_____	_____
Well plumbness and alignment				
a. Plumbness and alignment tests	_____	Each	$ _____	$ _____
b. Drift indicator tests	_____	Each	_____	_____
Well development				
a. Well development	_____	Each/hr.	$ _____	$ _____
b. Chemicals	_____	Each/bag	_____	_____
Well test				
a. Installation and removal of testing equipment	_____	Each	$ _____	$ _____
b. Recovery standby time	_____	Per hr.	_____	_____
c. Discharge pipe	_____	Lin. ft.	_____	_____
Well disinfection				
a. Well disinfection	_____	Per hr.	$ _____	$ _____
b. Chemicals	_____	Per bag	_____	_____

Water samples and analyses
   a. Water analyses   _____ Per series $ _____ $ _____

Permanent and test hole well abandonment and temporary capping
   a. Well abandonment   _____ Lin. ft. $ _____ $ _____
   b. Temporary capping   _____ Each   _____   _____

Standby time   _____ Per hr. $ _____ $ _____

Grand Total $ _____

The names and addresses of all persons interested as principals in this proposal are as follows:

Name   Address

_____   _____

_____   _____

_____   _____

_____   _____

Company _____

By _____

Date _____

Lowest Responsible Bidder

Most laws governing the spending of public funds require that the specifications permit free and open competition among responsible bidders. These laws usually provide that specifications should not be unduly restrictive and that contracts be awarded in the best interest of the governmental agency involved.

Interpretation of these laws depends on the judgment of the public officials involved. The law, in most cases, requires that the award be made to the lowest responsible bidder. Water-well contracts are awarded generally to the bidder offering the lowest price on the theory that anyone who can furnish the required performance bond is, by definition, a responsible bidder capable of fulfilling all phases of the contract. Awarding a contract to the lowest bidder relieves public officials of judging the capabilities of the individual bidders and is a practice generally followed by many municipal, state, and federal agencies. However, if the lowest bidder is not capable of fulfilling all phases of the contract or is supplying inferior equipment, experience has shown that the public gets neither the most value for its money nor the lowest cost in the long run.

For guidance in determining the lowest responsible bidder, the National Institute of Municipal Law Officers recommends considering six factors: (l) capacity and skill of the bidder to perform the contract; (2) ability to perform the work within the time specified; (3) sufficient financial resources available to the bidder; (4) ability of the bidder to provide future maintenance and service; (5) number and scope of conditions submitted with the bid; and (6) price. With these things in mind, the specification writer must describe the requirements in terms that will result in a quality job at a fair price.

CONTRACT PROBLEMS

Liability for Faulty Plans

Perfect specifications for construction work of any kind are almost impossible to draft in light of inhomogeneities of the ground and variables of working conditions. The question arises, then, as to where responsibility rests when the desired results are not obtained. In general, if there is a performance problem and the contractor does not fulfill the basic obligations, then payment is not made. Performance problems originate primarily in the sufficiency of either the quantity or the quality of water obtained.

One general rule seems to be that architects and engineers are not liable for construction defects if they have exercised ordinary skill and care in their work. Some states, however, have laws stating that, by implication, architects and engineers warrant that the plans and specifications they prepare will accomplish the intended purpose.

A court decision in South Carolina included the following statement (*Hill* v. *Polar Pantries*): "It seems to be well settled that where a person holds himself out as specially qualified to perform work of a particular character, there is an implied warranty that the work which he undertakes shall be of proper workmanship and reasonable fitness for its intended use. If a party furnishes specifications and plans for a contractor to follow in a construction job he impliedly warrants their sufficiency for the purpose in view. These principles have been applied in building contracts."

What might be called a corollary to this rule is the view that a contractor who has carried out work in accordance with the plans and specifications is not ordinarily liable for defects arising from faulty specifications. Defects that appear as a result of defective plans and specifications are the fault of the party who prepares the plans and specifications.

A decision by the U.S. Supreme Court put it this way (*United States* v. *Spearin*): "... if a contractor is bound to build according to the plans and specifications prepared by the owner the contractor will not be responsible for the consequences of defects in the plans and specifications. This responsibility of the owner is not overcome by the usual clauses requiring builders to visit the site, check the plans, and inform themselves of requirements of the work..."

A Mississippi court defined the seat of responsibility for final results on a construction contract in the following words (*Trustees of the First Baptist Church* v. *McElroy, Miss.*): "If there was an implied warranty of sufficiency, it was made by the party who prepared the plans and specifications because they were his work. In calling for proposals to produce a specific result by following them it may fairly be said to have warranted them to produce that result.

"If I agree to produce a certain result according to my own plan, I impliedly warrant its sufficiency; but, if I agree to produce that result by strictly following the plan prepared by another party, he impliedly warrants its sufficiency. The responsibility rests upon the party who fathers the plan, and presents it to the other with the implied representation that it is adequate for the purpose to be accomplished."

Rulings such as this would seem to apply when determining the extent of a contractor's responsibility regarding a guarantee of quantity or quality of water to be obtained from a well. Whether or not the guarantee is valid would depend upon the wording of the entire contract and upon the owner's action in putting the guarantee requirement in the specifications. Unless a contract contains a provision guaranteeing quantity, the driller does not warrant that a certain quantity will be obtained (*Butler* v. *Davis*, 1903; *Atwood Vacuum Machine Co.* v. *Warner Well and Pump Co.*, 1954; and *Atlas Construction Co., Inc.* v. *Aqua Drilling Co.*, 1977).

A Minnesota court spelled out the rule in one case by saying (*Frederick* v. *County of Redwood*): "Where a contractor makes an absolute and unqualified contract to construct a building or perform a given undertaking, it is the general, and perhaps universal rule, that he assumes the risks attending the performance of the contract and must repair or make good any injury or defect which occurs or develops before the completed work has been delivered to the other party.

"But where he makes a contract to perform a given undertaking in accordance with prescribed plans and specifications, this rule does not apply."

A New York court handed down a more explicit ruling (*MacKnight Flintic Stone Co.* v. *City of New York*) which is referred to in many other cases. The trial resulted from a dispute over a guarantee provision in a construction contract. The City of New York wrote specifications that required the contractor to guarantee a cellar to be absolutely water and damp proof for five years. The contractor agreed to these specifications. Any defect in this respect, then, had to be corrected by him without cost to the city.

The work was inspected as it went along, and upon completion the cellar was dry. Later on, some wetness developed and the city withheld payment on the contract. The contractor sued for his money. He showed the court that he had furnished materials and performed the work in accordance with the specifications. He argued that the guarantee could not be met under the specifications governing his work. The court agreed with the contractor and said:

"The agreement is not simply to do a particular thing, but to do it in a particular way and use specified materials in accordance with the city's design, which is the sole guide. The promise is not to make watertight but to make watertight by following the plans and specifications prepared by the city, from which the contractor had no right to depart . . ."

Limitation on Yield Guarantee

From the cases cited above, it appears that a guarantee of yield in a water well contract has little meaning when the job must be done in accordance with the owner's specifications. Courts do decide occasionally, however, that the addition of a guarantee clause means that the contractor warrants the results and is held liable if the guarantee is not met, even though the plans and specifications are followed to the letter. These decisions against contractors often cause lawyers to advise water well contractors against guaranteeing water in quantity or quality.

Guarantees of yield by water well contractors, nevertheless, continue to be fairly common. Some water well contractors have such a long drilling history in particular areas that they can safely offer yield guarantees. This type of guarantee may give a contractor a competitive advantage over less experienced drillers when competing for large contracts. An example is given below.

"CONTRACTOR GUARANTEES:

(A) The Well: Shall have a minimum producing capacity of not less than ____ U.S. gallons of water per minute when pumped to capacity, as shown by weir or other approved means of measurement and that it remains in good pumping condition for _____. Should the amount of water be unobtainable, Contractor will endeavor to improve or repair said well in an attempt to cause it to produce the stipulated amount of water, or he may sink a new well, or wells, or he may abandon the project entirely, in which event Contractor shall return to the Purchaser all sums paid to him less any sum paid for materials or services that were to have been furnished by Purchaser, and shall have the privilege of removing all machinery, tools, materials and equipment from the premises of the Purchaser, and neither party shall be liable to the other for any sum whatsoever." (Stamm-Scheele)

Unlike the contractor guarantee cited above, many guaranteed-yield contracts for wells really do not provide an absolute guarantee, although they appear to warrant the production of a certain quantity of water. Most such contracts have the effect of setting a price for the guaranteed quantity which then becomes a reference figure for negotiating a settlement at a lower price if the quantity is less than anticipated. There seems to be nothing wrong with this approach, except when the contractor claims that the guarantee offered is absolute, when, in fact, it may not be so.

Despite contract provisions, the contractor cannot provide more water than the formation will yield. If the contractor constructs the well so that it reaches maximum specific capacity, the owner should pay for the water obtained even though it is less than the amount specified in the contract.

In one well contract, the introduction to the agreement is worded as follows: "Whereas the Purchaser requires a supply of well water of such quantity and quality as can be obtained with reasonable efforts from the geologic formations under its premises; and whereas the Contractor is experienced in the development of groundwater and has available the technical personnel, skilled crews and equipment required to construct efficient wells in such water-bearing formations as may exist, it is hereby agreed between these parties as follows . . . ."

This preamble states the desires of the purchaser, acknowledges the expertise of the drilling contractor in fulfilling this desire, and sets the stage for an equitable agreement based on the hydrogeologic conditions at the well site. The specifications and other terms of the contract that can be developed logically within the framework of this philosophy should be fair to both parties, because they recognize the limits natural hydrogeologic conditions place upon the results that can be obtained.

In another example, the following excerpt from well specifications prepared for a midwestern town expresses a fair approach to the guaranteed-yield proposal (Williams, 1955). "In developing a water supply, it is always possible that the water cannot be obtained in the quantities desired at the available well sites. If after test drilling the contractor notifies the Purchaser that suitable water-bearing formations cannot be found and that it is apparently impossible to produce the guaranteed quantity of water, Purchaser shall accept and pay for the capacity actually produced, on a proportionate basis, or cancel the contract without making any payment except for test wells, and with no further liability on either side. The contractor may salvage all his material and equipment, including material already installed."

Responsibility for Site Conditions

An unfair provision in many specifications prepared by public agencies is one which attempts to place all responsibility for estimating subsurface conditions on the contractor. Time and cost usually prohibit a water-well contractor from doing test drilling before bidding. Furthermore, the owner usually cannot make the site accessible to all prospective bidders for test drilling. Yet specifications containing impractical statements about what bidders must do to satisfy themselves regarding drilling conditions at the test site continue to be written. Even information provided by a public agency about a well site may be offered with disclaimers regarding its accuracy, leaving the contractor to verify all conditions independently. Thus, except for occasional data from state or federal water agencies, the contractor must rely mostly on experience.

For example, suppose a water-well contractor agrees to construct a well at a certain site. After setting up, the contractor drills through a buried telephone cable. Although this cable is indicated on the site plan, it is shown to be at least 20 ft from the well site. Is the contractor liable for damage to the cable because its location was not verified? It is obvious that this type of specification requirement places an unreasonable burden on the contractor. In one case (*Morrison-Knudsen* v. *United States*, 1965), the court recognized this fact in stating: ". . . It was not incumbent upon the plaintiff, prior to submitting its bid and entering into the contract to conduct its own investigation in order to ascertain the truth or falsity of the defendant's positive assertions regarding subsurface conditions encountered in drilling holes 260 and 261, even though the contract contained a general condition stating that 'The Contractor further acknowledges that he has satisfied himself as to the character, quality and quantity of surface and subsurface materials to be encountered insofar as this information is reasonably ascertainable from an inspection of the site, including all exploratory work done by the Government,' and also contained in a technical provision stating that 'the Government does not guarantee that materials other than those disclosed by the explorations (i.e., the test borings) will not be encountered.' "

Opinions vary on the question of who should assume the risk for unexpected subsurface conditions on construction contracts. Committees representing engineering and contractor groups have suggested several approaches to the problem. The contract formulated by a joint committee of the American Society of Civil Engineers and the Associated General Contractors suggests the following paragraphs: "Section 7. The owner shall make known to all prospective bidders all information that he may have on subsoil conditions in the vicinity of the work, topographic maps, or other information that might assist the bidder in properly evaluating the amount and character of the work that might be required. Such information is given, however, as being the best factual information available without the assumption of responsibility as to its accuracy or for any conclusions that the Contractor might draw therefrom."

In some cases, it has been established that the owner has a duty to disclose all information regarding the site (*Reamer* v. *City of Swartz Creek*, 1977). Because subsurface conditions may be difficult to specify in the contract before work begins, it may be better to include a statement on how to handle differing subsurface conditions, such as (National Society of Professional Engineers 1910-3): "Section 4.3. Contractor shall promptly notify owner and engineer in writing of any subsurface or latent physical conditions at the site differing materially from those indicated in the Contract Documents. Engineer will promptly investigate those conditions and advise owner in writing if further surveys or subsurface tests are necessary. Promptly thereafter owner shall obtain the necessary additional surveys and tests and furnish copies to engineer and contractor. If the engineer finds that the results of such surveys or tests indicate that there are subsurface or latent physical conditions which differ materially from those intended in the Contract Documents, and which could not reasonably have been anticipated by contractor, a Change Order shall be issued incorporating the necessary revisions."

Note that it is incumbent upon the water well contractor to give prompt notice of changed conditions and for the engineer to make a prompt investigation of these conditions. The owner is then obligated to undertake further investigations and provide the information to both the engineer and the contractor.

The foregoing statements do not satisfy those who feel that the contractor should absorb all additional costs on a particular job. While this may appear to be in the owner's interest, the owner really pays these costs in any case. For every risk taken, the intelligent contractor increases the price for the unknown factors. If unusual conditions do not occur, the contractor's risk estimate provides a profit bonus.

From time to time, extra work may be ordered by the owner during well construction. Typically, the contractor will be paid at the contract rate for extra completed work. When the contractor has fully or substantially performed according to the contract or the owner has accepted the well or otherwise waived full performance, the contractor is entitled to the contract price. If the contractor does not perform, the owner may seek damages for breach of contract (*Kennedy* v. *Reece*, 1964).

Contract Disputes

Occasionally, the completion of a large and complicated well construction contract can lead to disputes between contractor and owner. If these disputes cannot be resolved, the contractor may be forced to file a mechanic's lien to obtain payment for the job. This type of lien is a claim created by law to secure payment of the price or value of work performed and materials furnished in making improvements to a property (Hamilton, 1977). The lien is exercised against real estate, although in isolated cases other types of liens can be created against personal property. Once the lien is in force, the title to the property is not clear because there is a claim against the land. Every state in the United States has mechanic's lien statutes, but great differences exist in these laws.

Two classes of mechanic's lien laws exist — the derivative lien and the direct lien. Under the derivative lien system, a subcontractor cannot recover more than the owner owes the general contractor because the lien is based on the contract between the owner and the general contractor. Under the direct lien system, a subcontractor has a direct claim against the property that is not affected by any agreement between the owner and general contractor. In this case, the owner could be confronted with claims that exceed the original contract price. In some states, however, recent changes in lien laws protect the owner from being subject to liens from parties other than the main contractor.

All contractors should keep abreast of lien statutes in their state so they can comply with all statutory requirements when filing a lien. In all cases, however, a successful lien cannot be brought against a property when the contractor has failed to perform substantially. But if the contractor has substantially performed to the contract, and cannot complete the work because of some uncontrollable factor, a lien can be filed for the value of the work performed under the contract (Hamilton, 1977). The question of what constitutes substantial performance must then be decided by the court.

Warranties

Water wells consist of various components that could fail during initial use of the well. A one-year general warranty applies to most construction work (National Society of Professional Engineers). The warranty commences when well construction is substantially completed and the well placed into service. For equipment placed in the wells, such as a pump, the warranty is provided by the manufacturer. Because sand pumping can quickly ruin a pump and this damage would not be covered by the manufacturer, the owner may add special provisions in the contract to cover this type of problem. Thus, the contractor would assume a greater risk, but the total contract cost would probably be higher to cover this risk. To keep warranties in force, the owner must be willing to take appropriate action when an equipment failure is evident and give prompt notice to the contractor or equipment supplier. In turn, the contractor must respond promptly to the owner's notification that a warranty problem has arisen. If the warrantor (contractor, equipment supplier, or manufacturer) does not act promptly, the owner may have work done at the warrantor's expense. Anticipating this problem, the contractor may want to add special contract provisions to cover damages if such repairs are done improperly.

CONCLUSIONS

It is not easy to write specifications for water well drilling operations and well construction in a clear, unambiguous style. As in all subsurface work, unexpected conditions are encountered much more often than in aboveground construction. In many cases, specifications for underground work may need to be more detailed in regard to methods. This differs with the fundamental view that specifications should detail the end result that is required but not tell the contractor exactly how to carry out the construction. Many engineers like to include in the specifications sufficient details outlining tried and proven methods but allow contractors to use their own methods if they can show the engineer that a satisfactory end result will be obtained.

The great temptation in writing specifications is either to overcontrol or to undercontrol the well construction project. An engineer may restrict the procedures to a rigid and limited set of formula-like conditions, or may describe the requirements in a vague, general way and continually use the expression "as directed by the Engineer." Either extreme will cause excessive stress between contractor and engineer and will inevitably raise costs for both.

CHAPTER 21
Groundwater Monitoring Techniques

Approximately half the population of the United States is dependent upon groundwater for its drinking water supplies. There is growing evidence that this resource, once considered relatively pollution free, is being contaminated locally by municipal and industrial wastes. Groundwater contamination occurs when soluble or insoluble substances are introduced into the hydrogeologic environment as a result of man's activities. Groundwater pollution results when the level of the contaminant concentration restricts the potential use of groundwater. Groundwater contamination is so severe in certain localities that continued use of the water could lead to serious health problems. Even though serious groundwater pollution problems exist over rather small geographic areas, they often occur in areas having high population densities (Clark, 1979). Irresponsible and ignorant waste-disposal practices of the past will continue to affect groundwater quality for many years in spite of major detection and restoration efforts now being pursued.

This chapter focuses on where and why groundwater contamination occurs, the methods used to locate contaminant plumes, the design and construction of wells to monitor groundwater quality, and the procedures used to clean up contaminated aquifers. Many water well contractors have become involved in the installation of groundwater monitoring wells because of their natural interest in aquifer protection and their experience in well design and construction. Other aspects of monitoring such as sampling procedures, equipment used to obtain samples from wells, and procedures to assure quality of the samples are discussed only briefly because most of these activities are performed by environmental consultants, not water well contractors.

Why has groundwater contamination occurred? In the past, people believed that nature provided much better protection for groundwater quality than it actually does. Instances of groundwater pollution such as the Love Canal near Niagara Falls, New York have caused immediate and widespread concern for protection of the underground environment and a realization that chemical contamination of this environment is both serious and widespread (EOS, 1981). In St. Louis Park, Minnesota, for example, creosote contamination of soil and underlying aquifers over a 50-year period has caused the closing of nearby municipal wells. High concentrations of phenols, a potential carcinogen, have been found in these wells. In another example, 100 water wells surrounding a landfill* in Jackson Township, New Jersey were closed because of organic chemical pollution (U.S. Environmental Protection Agency, 1980a). Although the landfill, which was constructed in a porous sand but never sealed properly, was only licensed to accept sewage sludge and septic tank wastes, chemical analyses of the groundwater in the vicinity of the landfill indicate high concentrations of chloroform, benzene, methylene, chloride, trichloroethylene, ethylbenzene, and acetone. Serious health problems have been reported by the well owners. On an even larger scale, 30 mi^2 (77.7 km^2) of the shallow aquifer underlying the Rocky Mountain Arsenal near Denver, Colorado have become contaminated by chemical byproducts resulting from the manufacture of pesticides (Civil Engineering-ASCE, 1981).

In the United States alone, over 250,000 new chemicals are created each year. Of these, some of the most troublesome are the widely used synthetic organic chemicals which are often carcinogenic or toxic to man. Over a million organic chemicals already exist, and several thousand new ones are developed each year. Sources of organic chemicals in groundwater are leaking industrial lagoons, septic tanks, leaking gasoline storage tanks, agricultural chemicals, and residues from paints and solvents.

After certain organic chemicals have entered an aquifer because of inadequate disposal practices, flushing of the aquifer or natural dilution of contaminants is so slow that total cleansing of the aquifer may not occur except over extremely long periods of time — hundreds or even thousands of years. Other organic chemicals have high mobility in the subsurface environment and, once the source is cut off, the water quality returns to normal within 10 to 20 years. The fate of organic compounds in groundwater and their rate of movement through the system depends in part on their sorptive capacity, volatility, dilution, biological activity, and chemical reactions (Pettyjohn and Hounslow, 1982).

Until recently, few people realized the extent of underground contamination or its adverse impact on groundwater quality. Because groundwater contamination is usually difficult to contain or control, governmental policies have been directed at its early detection, treatment, and subsequent elimination. These policies are being expanded to eliminate waste disposal practices that lead to subsurface contamination.

The water well industry must become involved in the successful detection and elimination of threats to groundwater quality. Every drilling contractor should be aware of potential threats to groundwater quality from abandoned wells, leaky sanitary landfills, poorly functioning sewage treatment facilities, and industrial or municipal wastewater ponds.

MAJOR FEDERAL LEGISLATION PERTAINING TO GROUNDWATER QUALITY AND MONITORING PROCEDURES

Several federal laws and much recent state legislation have established groundwater monitoring requirements for various potential contaminant sources. Some states that have assumed the responsibility (primacy) for implementing federal laws may impose more rigorous requirements than the federal law mandates. Thus, contractors installing monitoring wells should ascertain which regulations must be followed in their

---

*Areas in which trash and garbage are buried beneath layers of earth. The bottom of the site is often lined with a layer of bentonite or other impermeable material.

state.

No single federal law deals specifically with the problem of groundwater contamination. Various laws that affect groundwater were drafted to help solve specific environmental problems. The first major federal law that recognized the importance of groundwater was the Safe Drinking Water Act of 1974 (SDWA, PL 93-523), which established standards for insuring the safety of drinking water. Part of this law, the Underground Injection Control (UIC) program, regulates injection wells to prevent contamination of groundwater used for drinking water. Wells injecting wastes into the ground must be monitored to insure that wastes are contained in the prescribed zone. Another aspect of this law protects sole-source aquifers for drinking water. A sole-source aquifer is the dominant or only aquifer in a region.

The Resource Conservation and Recovery Act of 1976 (RCRA, PL 94-480) establishes guidelines for managing solid and hazardous wastes. This is the major federal law relating to groundwater monitoring. The primary objectives of monitoring under this act are to (1) detect whether a facility is discharging hazardous wastes to the highest aquifer, (2) determine whether the concentrations of specific hazardous waste constituents are within prescribed limits, and (3) measure the effectiveness of corrective measures taken at the site. The Toxic Substance Control Act of 1983 (TSCA, PL 94-469) also recognized the significance of groundwater quality protection.

Groundwater monitoring activities are also mandated under the Surface Mining Control and Reclamation Act of 1977 (SMCRA, PL 95-87). This law specifies that pre-mining baseline groundwater data be obtained, as well as data during mining activities and after closure of the facility.

Table 21.1. Major Sources of Groundwater Pollution

Waste Disposal Sources
- Landfills, dumps, and surface impoundments
- Mining wastes
- On-lot wastewater disposal systems
- Radioactive wastes
- Sludge management via land spreading
- Injection wells
- Abandoned sites

Nondisposal Sources
- Abandoned wells
- Accidental spills
- Agricultural chemical practices
- Artificial recharge
- Highway deicing compounds
- Petroleum exploration
- Underground storage tanks and pipelines

Depletion
- Increased salinity
- Salt-water encroachment

(After Canter, 1981)

One other law, the Comprehensive Environmental Response, Compensation, and Liability Act of 1983 (CERCLA, PL 96-510), was created to facilitate clean-up problems at waste sites that resulted from accidents in transporting hazardous wastes and at waste sites where ownership could not be determined. This act set up a trust fund (Superfund) to finance the cleanup of spills and the reclamation of closed sites. Although specific groundwater monitoring requirements are not prescribed in this law, it is likely that requirements developed under this law will eventually follow those given in the Resource Conservation and Recovery Act (Barcelona et al., 1983).

GROUNDWATER CONTAMINATION SOURCES

The major threats to groundwater quality from all contaminant sources are (1) septic tank systems, (2) sanitary landfills, (3) chemical landfills, and (4) wastewater disposal ponds (Canter, 1981). The presence of any of these sources can have a pronounced impact on groundwater quality (Table 21.1). The total number of active hazardous and nonhazardous industrial and municipal waste sites is estimated at 141,000 (U.S. Environmental Protection Agency, 1980b). Furthermore, there may be more than 150,000 inactive sites that may be potential threats to groundwater quality (Hart, 1979). The U.S. Environmental Protection Agency (EPA, 1980b) has indicated that, of the 32,000 to 50,000 disposal sites that may contain hazardous waste, 1,200 to 2,000 could pose serious health or environmental problems. Until recently, about 80 percent of hazardous wastes were being disposed of improperly in landfills or lagoons and they will present a long-term threat to groundwater quality.

Another U.S. EPA report, The Surface Impoundment Assessment, suggests that 181,000 impoundments* exist at 25,800 industrial sites. A study of 8,200 of the industrial sites shows that (U.S. EPA, 1980b):

1. 70 percent of the impoundments are unlined and possibly allow contaminants to enter the ground.

2. 10 percent of the sites that are unlined overlie usable aquifers and are on permeable soils. One-third are within 1 mi (1.6 km) of a water supply well.

3. About 35 percent hold liquid wastes that may contain hazardous constituents.

4. As of 1980, only 5 percent of the sites were known to be monitored.

The degree of the contamination threat to groundwater supplies from landfills and wastewater ponds depends on several factors: toxicity and volume of the contaminant generated at each site, the nature of the geologic medium underlying the site, and the hydrologic conditions dominant in the area.

The recent discovery that many volatile organic chemicals are emanating from landfills and industrial disposal ponds is disturbing because they are known or suspected carcinogens and are not removed easily by natural geochemical processes in the ground. Many of these organic chemicals were found in a high percentage of wells recently tested by the U.S. Environmental Protection Agency. The chemicals listed in Table 21.2 occur in groundwater in many industrial areas and in groundwater adjacent to municipal landfills.

Other less obvious threats to groundwater quality come from a variety of sources. For example, abandoned wells can be a severe problem if poor-quality water enters aquifers having good-quality water via uncemented well bores. This problem is es-

---

*Natural or man-made storage area used for dilution and natural purification of sewage and other municipal and industrial wastewaters.

pecially serious in agricultural areas, where animal wastes, pesticides, and herbicides can easily enter the groundwater system through open well bores. In many coastal communities in Florida and California, salt-water encroachment caused by overpumping of fresh-water supplies is a major problem. In the North, heavy and often indiscriminate applications of road deicing salts, and poorly constructed (uncovered)

Table 21.2. Typical Synthetic Organic Chemicals and Their Major Industrial Uses

Chemical	Uses
Trichloroethylene	A high-volume industrial chemical used extensively as a solvent for degreasing metal and as a septic tank cleaner.
Carbon tetrachloride	Used as a cleaning solvent, pesticide, and intermediate in the production of chlorofluoromethanes.
Tetrachloroethylene	A solvent that is widely used in dry-cleaning and degreasing operations.
1,1,1 Trichloroethane	Also known as methyl chloroform. It is used as an industrial cleaner and degreaser of metals, resin adhesive, and vapor-pressure depressant.
1, 2 Dichloroethane	Used primarily as a raw material for the production of vinyl chloride. Every gallon of leaded gasoline produced in the United States contains dichloroethane as a lead scavenger. This chemical is also used as a paint solvent, cleaning solvent, and grain fumigant.
1, 1-Dichloroethane	Imported for use as a solvent and cleaning agent in specialized processes.
Dichloroethylenes	A group of 3 isomers*. Cis 1,2-dichloroethylene and Trans 1, 2-dichloroethylenes have not had wide industrial usage; 1, 1-dichloroethylene is used as a chemical intermediate in the production of methyl chloroform.
Methylene chloride	Used in the manufacture of paint and varnish removers, insecticides, solvents, pressurized spray products, and Christmas tree bubble lights.
Vinyl chloride	Used for over 40 years in producing polyvinyl chloride, which is the most widely used material for the production of plastics.

*Isomers are two or more chemical compounds, radicals, or ions containing the same numbers of atoms of the same elements in the molecule, radical, or ion, and hence having the same molecular formula but differing in the structural arrangement of the atoms and consequently in one or more properties.

(U.S. Environmental Protection Agency, 1980c; Canter, 1981)

storage areas, can lead to high chloride concentrations in underlying aquifers.

Leaking gasoline storage tanks at automotive service centers, most of which have been installed in the last 35 years, are a serious local problem in an increasing number of communities. This type of pollution is especially detrimental because drinking water becomes unpalatable when it contains extremely low concentrations of petroleum products (Clean Environment Commission, 1976).

Another example is the broad range of pesticides being applied to farmland. This nonpoint source of potential contamination is extremely difficult to control. Any type of injection well can also create a water-quality problem if some of the wastes reach an aquifer containing good-quality water. Especially serious are radioactive wastes. These substances are extremely dangerous to humans for anywhere from 30 to 500,000 years (Winograd, 1974). Yet, while the volume of these wastes has risen dramatically, no safe disposal sites have been identified or built. Early attempts to bury radioactive wastes have been largely unsuccessful because the wastes could not be prevented from contaminating groundwater supplies. Most nuclear-power generators are now storing their wastes aboveground at plant sites until safe sites can be identified.

Efforts to control contamination problems by regulation have been initiated by federal, state, and local governments. Common methods of control include:

1. Reducing the volume of material to be discarded by compaction, incineration, or other pretreatment schemes.

2. Selection of disposal sites that utilize the natural ability of the underground environment to remove contaminants, thereby preserving groundwater quality.

3. Improving engineering aspects of disposal sites, such as the addition of leachate collection systems, installation of double clay liners, or the use of synthetic liners.

EFFECT OF AQUIFER CHARACTERISTICS ON THE SPREAD OF GROUNDWATER CONTAMINATION

In the past, the least expensive and most widely used waste management option for both municipal and industrial wastes has been the sanitary landfill, where wastes are compacted and covered with earth. In any geographic area other than arid zones, the fill is subjected to percolating rainwater or snowmelt which eventually flows out the bottom of the landfill site and moves into the local groundwater system. These percolated waters, known as leachates, can contain large amounts of inorganic and organic contaminants. At some sites, the leachate is collected and treated. But even in the best engineered sites, some leachate escapes into the groundwater system because no permanent engineering solution has been found to isolate the leachate completely from the groundwater.

Common inorganic constituents found in leachates from sanitary landfills are listed in Table 21.3. The concentration of inorganic materials in leachates can be compared with the typical inorganic levels found in groundwater existing in various rock media to determine the leaching effect of water percolating through a waste site (Table 21.4).

It is not known how long leachates continue to contaminate aquifers underlying landfills, but some landfills from the Roman Empire are still producing leachate (Freeze and Cherry, 1979). Contamination plumes can spread thousands of feet downgradient from a source, and once in the ground, they may remain there for many years even if the contaminant source is removed (Schmidt et al., 1981).

The hydrogeologic setting plays a role in determining the degree to which a landfill can alter water quality in local aquifer systems. The type of soils and their ability to adsorb contaminants, how far the landfill is situated above the water table or confined aquifer, and the hydraulic properties of the aquifer contribute to or reduce contaminant concentrations. Two typical geologic settings are presented in Figures 21.1 and 21.2 that demonstrate the interaction between the landfill leachates and the local hydrogeology. Other examples are given in Appendix 21.A. These illustrations are instructive in showing the potential flow paths taken by contaminants.

Drilling contractors and engineers should become familiar with potential contaminant flow paths in any waste project area, because regulations require that most monitoring wells be placed downgradient from the contaminant source. Initially, an estimate of the dimensions of the plume must be made on the basis of assumed hydraulic conductivity, porosity, and dispersion values for the aquifer. Any boundary conditions such as faults or changes in rock type must also be considered. It is important to recognize that some contaminants will become soluble in water, whereas others such as hydrocarbons will float on the surface of the groundwater. The density of a soluble

Table 21.3. Representative Ranges for Various Inorganic Constituents in Leachate from Sanitary Landfills

Parameter	Representative Range (mg/ℓ)
Potassium (K^+)	200 – 1,000
Sodium (Na^+)	200 – 1,200
Calcium (Ca^{+2})	100 – 3,000
Magnesium (Mg^+)	100 – 1,500
Chloride (Cl^-)	300 – 3,000
Sulfate (SO_4^{-2})	10 – 1,000
Alkalinity	500 – 10,000
Iron (Fe) (total)	1 – 1,000
Manganese (Mn)	0.01 – 100
Copper (Cu)	< 10
Nickel (Ni)	0.01 – 1
Zinc (Zn)	0.1 – 100
Lead (Pb)	< 5
Mercury (Hg)	< 0.2
Nitrate (NO_3^-)	0.1 – 10
Ammonia (NH_4^+)	10 – 1,000
Phosphorus (P) as phosphate (PO_4)	1 – 1,000
Organic nitrogen	10 – 1,000
Total dissolved organic carbon	200 – 30,000
COD (chemical oxidation demand)	1,000 – 90,000
Total dissolved solids	5,000 – 40,000
pH	4 – 8

(Leckie et al., 1975; Griffin et al., 1976)

Table 21.4. Typical Chemical Analyses of Groundwater taken from Different Rock Types (mg/ℓ)

	Rhyolite	Granite	Gabbro Basalt	Sandstone	Shale	Limestone	Dolomite	Metamorphic Rocks
SiO_2	49	32	41	23	26	12.8	14.9	23.1
Al	0.62	0.18	0.2	0.1	3.6	0.09	0.13	0.1
Fe	0.32	0.29	0.62	0.74	1.7	0.4	1.1	0.5
Mn	0.0	0.02	0.06	0.06	3.1	0.06	0.07	0.08
Cu	0.0	0.0	0.0	0.0	0.04	0.0	0.0	0.0
Zn	0.07	0.06	0.03	0.0	0.09	0.01	0.0	0.03
Ca	8.4	38.1	25.7	53.2	114.4	71.3	62.0	40.4
Mg	2.2	8.0	26.3	20.8	53.7	19.1	43.7	15.2
Na	20.7	51.2	14.3	51.1	194.3	12.9	27.4	22.4
K	2.3	3.7	9.1	4.3	5.3	2.2	1.8	3.1
HCO_3	77	175	196	252	330	228	272	166
CO_3	0.0	0.0	0.0	2.1	3.0	0.0	0.7	0.0
SO_4	6.9	65.4	17.1	69.0	358.4	60.7	138.2	37.5
Cl	5.1	53.7	22.5	37.3	219.0	19.7	6.9	23.1
F	0.3	1.2	0.2	0.4	0.6	0.3	0.6	0.6
NO_3	2.6	7.6	6.5	4.5	17.2	8.9	6.3	4.4
PO_4	0.1	0.07	0.03	0.02	0.0	0.09	0.0	0.01
TOTAL	176	437	360	519	1330	437	576	337
pH	7.2	7.1	7.5	7.5	7.2	7.5	7.7	7.1

(Synopsis of data from White, Hem & Waring, 1963, Composition of Subsurface Waters, U.S. Geological Survey Prof. Paper No. 440-F)

contaminant relative to that of water will affect its penetration into the aquifer.

In the discussion below, flow of the contaminant through the vadose zone will not be considered, but the reader should be aware that many forms of instrumentation, mainly lysimeters and various types of electrical and nuclear sensors, can be used to monitor this zone. See Morrison (1983) for a thorough discussion of the devices used

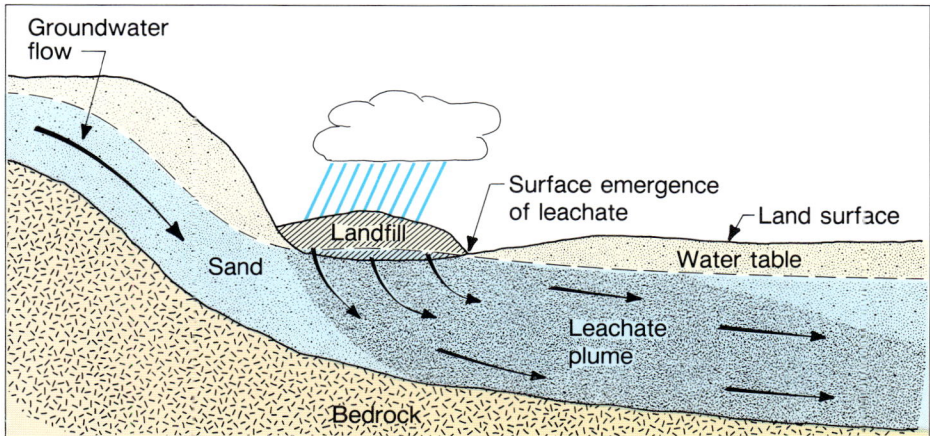

Figure 21.1. Single aquifer intersecting landfill. A steep, shallow groundwater table flows directly into the landfill, generating leachate which flows downward into the aquifer. Surface emergence of leachate may also occur. *(U.S. Environmental Protection Agency, 1977b)*

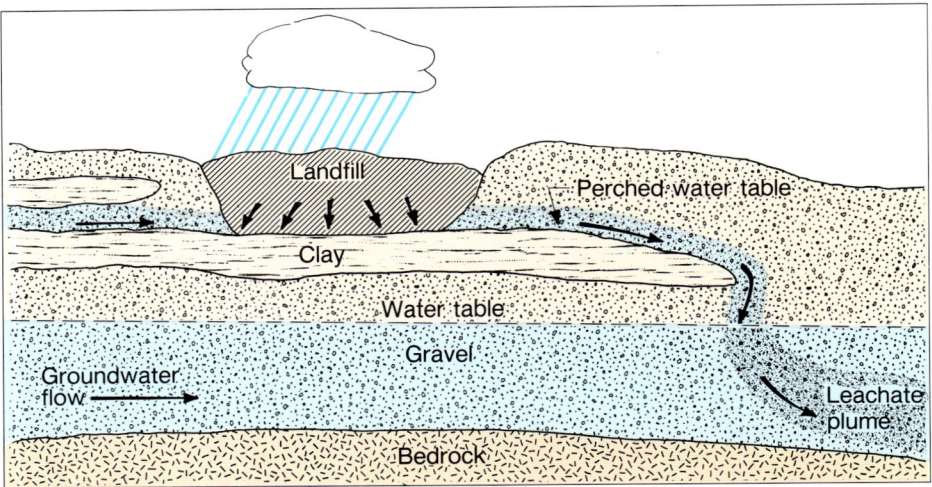

Figure 21.2. Landfill in an abandoned gravel pit with a clay layer at its base. A perched water table containing leachate will build up under the landfill and flow laterally through the ground above the clay until it is free to percolate to the main water table. *(U.S. Environmental Protection Agency, 1977b)*

to monitor pollutant movement in the vadose zone.

Once in the aquifer, the primary driving force for contaminant movement is created by the hydraulic gradient that produces groundwater flow. Contaminants entering the groundwater system are thus carried downgradient, forming a contaminant plume. This type of contaminant movement is termed "advection." Other factors also influence the shape of the plume, including two types of hydrodynamic dispersion — mechanical mixing and molecular diffusion. These two processes cause a spreading (dispersion) of the contaminant over a much larger area than advection alone would produce, and, consequently, a dilution of the contaminant away from the source area. Mechanical mixing processes include velocity differences within the pore openings, velocity differences caused by differences in pore sizes through which the water molecules move, and the degree of tortuosity (length) of the pore channels.

Molecular diffusion (chemical dispersion) can also occur. In the absence of any groundwater movement, a slug of highly concentrated chemical will move outward from its origin toward points of lower concentration. This type of dispersion occurs because of the kinetic activity of the ionic or molecular constituents. The effect of molecular diffusion on contaminant dispersion is usually much less than the effect of mechanical mixing processes, and except in the case of no groundwater movement at all (an improbability), it can probably be ignored in most instances in estimating the spread of contaminant plumes. An exception occurs with light organic chemicals which have moved upgradient in some cases.

It is possible to project how a contaminant plume actually spreads by advection and mechanical mixing. Figure 21.3a shows the theoretical downgradient movement of a plume from a continuous contamination source. Note the marked dispersion of the contaminant as it moves downgradient. Depending on the exact nature of the aquifer, the dispersion may be of even greater magnitude than the longitudinal movement (advection) shown in Figure 21.3a. In Figure 21.3b, a contaminant is injected periodically into the aquifer. Mechanical mixing coupled with advective flow creates

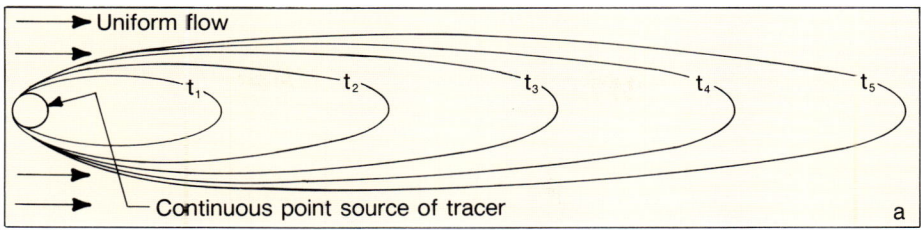
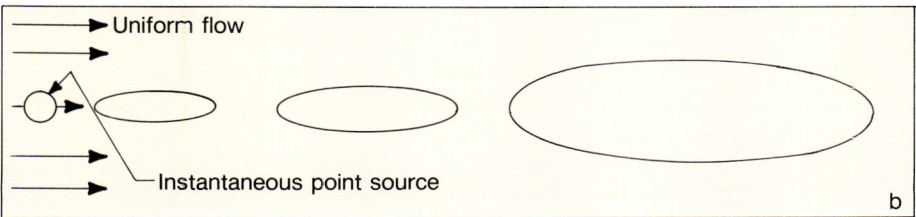

Figure 21.3. Spreading characteristics of a tracer in a two-dimensional uniform flow field in an isotropic sand. (a) Continuous tracer introduced into an aquifer; the plume limits are shown at various times. (b) The distribution of a single slug of contaminant at the instant the material is injected into the aquifer and at three later times. *(Freeze, R. Allan and Cherry, John A., GROUNDWATER, © 1979. Reprinted by permission of Prentice-Hall, Inc., Englewood Cliffs, NJ)*

the ellipsoid-shaped plumes. Clearly, the larger the total area covered by the plume, the more diluted the contaminant becomes.

The density of the contaminant plays a part in determining the vertical dimensions of the plume. If a material entering an aquifer is heavier than water, it sinks slowly as it disperses transversely and longitudinally. The density of the material in relation to water, as well as the hydraulic nature of the aquifer, will govern the vertical penetration of the plume as it moves downgradient (Figure 21.4).

Many chemical and biochemical reactions can take place in the subsurface environment to either augment or, more likely, reduce the concentration of a contaminant. The most important of these are solution-precipitation, oxidation-reduction, adsorption-desorption, acid-base reactions, and microbial cell synthesis. Some of these reactions may take place in the unsaturated zone before the contaminant reaches the aquifer. Once in the aquifer, different contaminants in the same plume may travel at different velocities depending on how they react with the geologic medium. Any investigation of groundwater contamination should include an analysis of the chemical or biological reactions taking place and the effect of these reactions on the strength of the contaminant plume.

In summation, many factors play a part in the spreading and concentration rate of

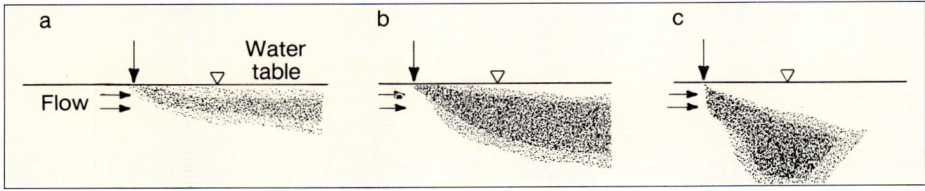

Figure 21.4. Effect of density on the spread of a contaminant solution in a uniform flow field. (a) Slightly more dense than groundwater; (b) and (c) larger density contrasts. *(Freeze, R. Allan and Cherry, John A., GROUNDWATER, © 1979. Reprinted by permission of Prentice-Hall, Inc., Englewood Cliffs, NJ)*

a contaminant: anisotropic/isotropic properties of the rock medium, advection rate, hydrodynamic dispersion processes, and reaction potential with the subsurface materials. Therefore, underground movement of groundwater contaminants is often exceedingly difficult to analyze in a straightforward manner. Many sophisticated methods are now being used to determine contaminant movement, including special mathematical modeling techniques, electrical (surface resistivity) methods, radioactive tracers, various dyes and salts, water temperature, and baker's yeast (Davis et al., 1980). Some of these techniques are discussed later in this chapter.

DELINEATING CONTAMINANT PLUMES

Openings in rocks or unconsolidated materials are not regularly spaced, and the permeability of the aquifer material varies both vertically and horizontally (see Chapter 3). Thus, the flow of contaminants is highly anisotropic (Figure 21.5). In spite of these difficulties, it is necessary to estimate flow direction within the aquifer so that the source of contamination and the direction of plume movement can be determined.

Usually, the general direction of groundwater flow can be established on the basis of the local topography (use of topographic maps or aerial photos) and the presence of streams or rivers which act as groundwater discharge boundaries. Recall that near-surface groundwater flow will generally follow surface drainage patterns. If the flow direction cannot be established, three small-diameter wells are temporarily installed into the aquifer. An analysis of the relative water table or potentiometric surfaces in the wells will reveal the direction of flow (see Chapter 5 for details of this method). For anisotropic aquifers, however, the direction of flow may not be parallel to the hydraulic gradient (Fetter, 1981).

To determine the dimensions of a plume, test borings can be made and water samples taken. The danger of handling many chemicals, even those that are highly diluted, makes this a less desirable procedure than using some form of geophysical method to gain information on the plume. The use of geophysical techniques offers

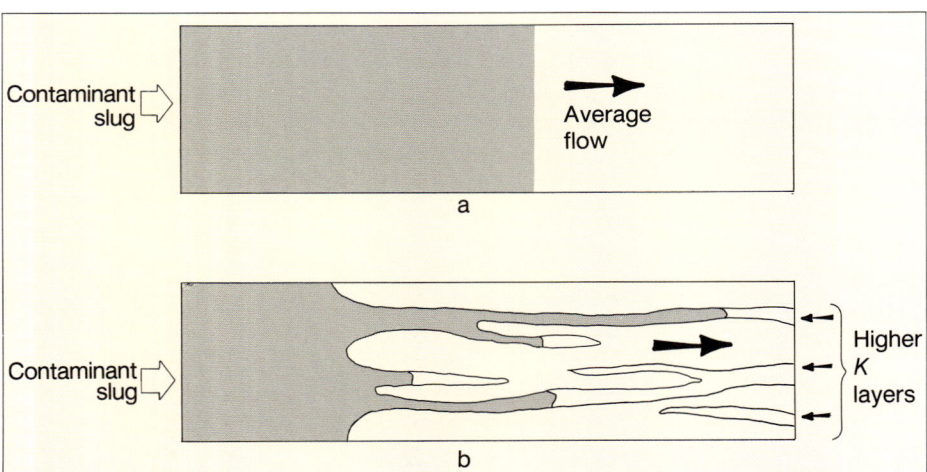

Figure 21.5. Comparison of the advance of contaminant zones influenced by hydrodynamic dispersion. (a) Perfectly homogeneous granular medium; (b) fingering caused by layered beds and lenses. *(Freeze, R. Allan and Cherry, John A., GROUNDWATER, © 1979. Reprinted by permission of Prentice-Hall, Inc., Englewood Cliffs, NJ)*

several advantages: (1) the investigation can proceed rapidly with little danger to health, (2) the near-surface physical characteristics of the aquifer can be determined, and (3) the limits of the plume can be generally defined. At least one and preferably several test borings should be made at the site so the geophysical data can be correlated with data from the borings.

Three of the principal surface geophysical methods used for the reconnaissance of contamination plumes include electromagnetic induction (terrain conductivity), earth resistivity, and ground-penetrating radar. In many investigations, the data from more than one geophysical method are combined to give a clearer picture of both the stratigraphy and the extent of the plume (Taylor and Cherbauer, 1984). Terrain conductivity has proven to be especially useful in tracking hydrocarbon leaks in urban areas, because, unlike resistivity, no stakes need to be placed in the ground (Saunders, 1983). Thus, the use of terrain conductivity is ideal where pavement or structures cover the ground. Electromagnetic (EM) induction methods are often used to track acid mine drainage (Ladwig, 1983). This method is also used to map saltwater intrusion in coastal areas and to circumscribe the extent of hazardous waste sites (Rudy and Caolic, 1984).

Information on plume geometry can be obtained quickly by radar techniques, but interpretation of radar information is difficult for inexperienced personnel and must be correlated with data from strategically placed borings. The cost of a radar investigation is high relative to EM or resistivity methods because of instrument costs and data interpretation (Knowles et al., 1982).

Earth resistivity continues to be a popular device for plume detection, because its depth of detection is significantly greater than with either of the other two methods (Gilkeson and Cartwright, 1982). Furthermore, earth resistivity mesurements are particularly sensitive to the amount of total dissolved solids in the pore fluids. Thus, resistivity readings indicate clearly the areal extent of contaminant plumes that are created by leachate from typical landfill sites. See Chapter 8 for a more detailed explanation of these instruments and others mentioned below.

Other surface methods that are useful in plume detection are magnetic techniques and seismic analysis (both refraction and reflection). Magnetometers are particularly useful in gauging the areal dimensions of a fill site by detecting the location of metal drums. But magnetometers will not indicate the depth of buried metal objects nor provide information on the stratigraphy of the site. In general, the magnetometer is only useful for locating the top of metal objects. Seismic refraction is advantageous in determining the stratigraphy, but only if the density of geologic materials increases with depth. It is especially useful in defining the limits of a hazardous waste site where exploratory drilling would be too dangerous (Knowles et al., 1982). Seismic refraction data also provide information on the location of buried bedrock surfaces, the depths to groundwater, and the lateral lithologic changes within the aquifer (Sendlein and Yazicigil, 1981). This information is helpful in delineating the potential plume dimensions. Data can be gathered much more quickly over a larger area by seismic methods than by test borings.

Several borehole geophysical methods are used in defining the extent of plumes; the most important of these are resistivity, conductivity, neutron, hole caliper, and temperature. Resistivity and conductivity values of the groundwater are affected by the contaminating substance and thus are an indication of the plume's presence at a site. Gamma ray and other nuclear methods provide information on subsurface lithology, particularly zones that have high permeability. It can be expected that contaminant migration will be greatest in these zones. Hole calipers indicate the presence of solution channels penetrated by a borehole in hard terrains (Micham et al., 1984). Temperature logs are useful in tracing the movement of injected water in highly permeable zones and detecting any changes in flow rate over time (Keys and Brown, 1978).

MONITORING CONTAMINANT MOVEMENT (TRANSPORT)

In many instances of groundwater contamination, the ability to predict how the contaminant plume will behave in the future can only be done on the basis of expensive drilling and sampling programs. Many scientists interested in the movement of contaminants in the subsurface believe, however, that it will soon be possible to use mathematical modeling techniques to estimate the spread of a contaminant and its strength at any point in the plume. The steps or processes used to build the model are shown in Figure 21.6. Five basic steps are accomplished in sequence:

1. In the first step, the basic factors affecting contaminant transport are identified — hydraulic characteristics of the aquifer, the physical and chemical properties of the aquifer materials, and the chemical and physical properties of the contaminants entering the groundwater system.

2. The attenuating processes for single chemicals are established. Different chemicals will move at varying rates and therefore occur at different concentrations in the aquifer.

3. In the third step, a mathematical model is set up to account for the attenuation processes, and a method of solving the equation is determined.

4. Predictions are made on the basis of the answers obtained in Step 3 for the occurrence of the various contaminant concentrations in the aquifer at a particular time.

5. In the last step, the validity of the model is assessed by comparing the model's results to any known field data. If the results differ significantly, the various model inputs are adjusted to produce better correlation with the field data.

In this modeling process, the factors identified in Step 1 are the most difficult to determine. This is true because the construction of virtually all aquifers is highly complex, with little uniformity either vertically or horizontally. Thus, it is difficult to predict how fast the contaminants will move through the aquifer, at what depth, and over what area they will be dispersed within a certain time. Furthermore, the geochemical attenuation mechanisms for many chemicals are not thoroughly understood. For these reasons, some transport models have not

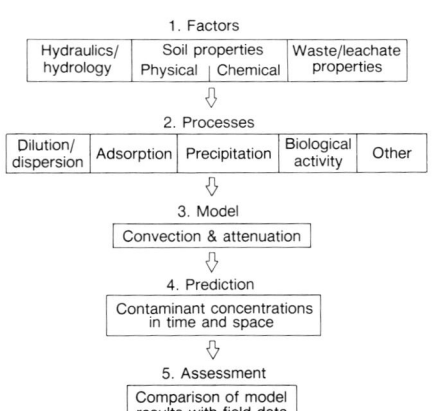

Figure 21.6. Steps in modeling landfill contaminant transport in soil. *(After Farquhar, 1981)*

yielded good results, and field data are much more reliable. Many groundwater scientists are working on ways to improve model accuracy, however, and, because the use of a model can be so much less costly than field work in estimating plume dimensions and contaminant concentrations, the use of contaminant transport models will probably increase significantly in the future.

Even though the mathematical analysis and the complex geochemical relationships that are a fundamental part of any contaminant transport model may be beyond the experience of most drilling contractors, much of this information will probably be available as "canned" models adaptable to a wide range of geologic situations. The contractor or consulting hydrogeologist will then be able to use the models to define a cost-effective field drilling program. Results from the field data can then be used to calibrate the model for the specific site.

LOCATING MONITORING WELLS

Once the areal extent of the plume has been defined, several monitoring wells are installed in or adjacent to the plume. The purpose of a monitoring well is to (Lewis, 1982):

1. Determine the hydrogeologic properties of the formation in which the contaminant exists.
2. Determine the water table or potentiometric surfaces of all aquifers in the system.
3. Permit access for the collection of water-quality samples for detection of contaminants.
4. Monitor the movement of the contaminant plume.

Usually one well is sited near the center of the plume just downgradient from the contaminant source. Another well is installed downgradient of the contaminant source, outside the limits of the plume. For ambient environmental data, one well is placed upgradient of the contaminant source. Other wells may be installed to verify the amount of dispersion taking place. The most difficult decision is rarely where to place the monitoring wells, but at what depths should the samples be taken. Selection of the most appropriate depths will depend on the density of the contaminant, the anisotropic characteristics of the aquifer, and the slope of the water table or potentiometric surface. The design of the monitoring network is extremely important if maximum information concerning the extent of the contamination is to be obtained.

In the past, too few monitoring wells were required for each disposal site. Thus it was not possible to adequately monitor contaminant movement (Clark and Sabel, 1980). In practice, the number of wells required to adequately monitor a specific disposal site will vary greatly, depending on the local hydrogeology. If the disposal site is higher than the surrounding landscape, for example, leachates may flow some distance in all four directions. In this instance, at least four wells would be needed, plus one other to monitor the upgradient chemistry. Ideally, some wells would be installed at more than one depth in the aquifer to verify if vertical flow is occurring or if the spread of the contaminant varies at differernt depths. Proper placement of monitoring wells must be based on accurate information concerning the groundwater flow direction at the waste disposal site and the type of contaminant.

Although monitoring wells can be drilled by virtually any drilling method, some methods may be more suitable in certain situations. Table 21.5 lists the major methods used to install monitoring wells, and their advantages and disadvantages. For details of each method, see Chapter 10.

For monitoring work, many of the objectives of a drilling program are similar to those for a water well, but some of the steps must be done with greater care to insure that the water quality is protected and reliable water samples can be obtained. Specific steps in monitoring well construction include (Luhdorff and Scalmanini, 1982):

1. Ability to penetrate all formation materials at a reasonable rate and to construct a borehole diameter of the proper size, assuring that cross-contamination will not occur.

2. Ability to provide accurate information on all the formations being drilled — either by cuttings, split-spoon samples, or sidewall cores.

3. Containment of cuttings and drilling fluids so they do not contaminate the formation.

4. Collection of water samples at various depths during drilling.

5. Ability to accommodate for lost-circulation problems, confining pressures, and flammable and toxic substances.

6. Construction of the monitoring well either during the drilling process or immediately thereafter.

7. Ability to maintain an open borehole long enough for geophysical exploration (if required) and data analysis.

Table 21.5 Drilling Methods for Monitoring Wells

Type	Advantages	Disadvantages
Hollow-stem auger	• No drilling fluid is used, eliminating contamination by drilling fluid additives • Formation waters can be sampled during drilling by using a screened auger or advancing a well point ahead of the augers • Formation samples taken by split-spoon or core-barrel methods are highly accurate • Natural gamma-ray logging can be done inside the augers • Hole caving can be overcome by setting the screen and casing before the augers are removed • Fast • Rigs are highly mobile and can reach most drilling sites • Usually less expensive than rotary or cable tool drilling	• Can be used only in unconsolidated materials • Limited to depths of 100 to 150 ft (30.5 to 45.7 m) • Possible problems in controlling heaving sands • May not be able to run a complete suite of geophysical logs.
Direct rotary	• Can be used in both unconsolidated and consolidated formations • Capable of drilling to any depth • Core samples can be collected • A complete suite of geophysical logs can be obtained in the open hole • Casing is not required during drilling • Many options for well construction • Fast • Smaller rigs can reach most drilling sites • Relatively inexpensive	• Drilling fluid is required and contaminants are circulated with the fluid • Drilling fluid mixes with the formation water and invades the formation and is sometimes difficult to remove • Bentonitic fluids may absorb metals and may interfere with other parameters • Organic fluids may interfere with bacterial analyses and/or organic-related parameters • During drilling, no information can be obtained on the location of the water table and only limited information on water-producing zones. • Formation samples may not be accurate

Type	Advantages	Disadvantages
Air rotary	• No water-based drilling fluid is used, eliminating contamination by additives • Can be used in both unconsolidated and consolidated formations • Capable of drilling to any depth • Formation sampling is excellent in hard, dry formations • Formation water blown out of the hole makes it possible to determine when the first water-bearing zone is encountered • Field analysis of water blown from the hole can provide information regarding changes for some basic water-quality parameters such as chlorides • Fast	• Casing is required to keep the hole open when drilling in soft, caving formations below the water table • When more than one water-bearing zone is encountered and hydrostatic pressures are different, flow between zones occurs during the time drilling is being completed and before the borehole can be cased and grouted properly • Relatively more expensive than other methods • May not be economical for small jobs
Cable tool	• Only small amounts of drilling fluid are required (generally water with no additives) • Can be used in both unconsolidated and consolidated formations; well suited for extremely permeable formations • Can drill to depths required for most monitoring wells • Highly representative formation samples can be obtained by an experienced driller • Changes in water level can be observed • Relative permeabilities for different zones can be determined by skilled drillers • A good seal between casing and formation is virtually assured if flush-jointed casing is used • Rigs can reach most drilling sites • Relatively inexpensive	• Minimum casing size is 4 in (102 mm) • Steel casing must be used • Cannot run a complete suite of geophysical logs • Usually a screen must be set before a water sample can be taken • Slow

PERSONNEL SAFETY AT MONITORING SITES

Safety should be a primary concern of water well contractors engaged in drilling and constructing monitoring wells. Besides the usual physical hazards of normal drilling activities, chemical, biological, radiological, and explosive hazards are added when drilling monitoring wells. So many toxic chemicals have been placed in the ground, either accidentally or intentionally, that drillers must use extreme caution when drilling in areas of known or suspected waste sites. In the past, many extremely toxic chemicals were mixed indiscriminately into ordinary municipal waste streams. Even the innocent disposal by homeowners of many dangerous organic chemicals has led to their introduction into the groundwater system beneath sanitary landfills. Unfortunately, the exact location or the extent of many former waste disposal sites are not known with precision. Furthermore, many chemicals may appear to be harmless and any injury may be rather insignificant on a short-term basis. Yet long-term effects may be acute, causing premature death, unusual forms of cancer, or generally poor health (Wallace et al., 1982).

Some of the most significant dangers are:

1. Explosions from methane gas produced by the decay of organic materials in sanitary landfills. An explosion potential also exists in monitoring work involving hydrocarbon recovery.

2. Toxic substances used in manufacturing pesticides, herbicides, solvents, paints, and other common products. Sometimes certain nontoxic chemicals placed in a disposal site will react with other chemicals to produce highly toxic chemicals.

3. Biologic wastes from hospitals or medical laboratories at universities that contain bacteria and viruses.

4. Chemical wastes that are corrosive, highly reactive, flammable, or explosive.

5. Vapors from any type of waste.

6. Radioactive wastes from hospitals and industrial and university laboratories.

One vital fact must always be kept in mind — the combination of substances at a waste site may have a more powerfully adverse effect on human health than they would individually. Before attempting to conduct monitoring work at a waste site, the drilling contractor should learn exactly what types of wastes were buried there, provide the necessary protective clothing and training for personnel, and stress that any physical changes in a worker's health may be caused by contact with the waste. Always be prepared for "worst case" conditions.

Any form of drilling is relatively dirty in the sense that it is difficult to avoid contact with cuttings, water encountered in the borehole, and surficial residues at the site. The following practices must be followed at any known or suspected hazardous waste site (Maslansky, 1983):

1. "Personnel should wear properly selected and fitted protective clothing and respirators at all times. Personnel must be given suitable training in the use, limitations, maintenance, cleaning, and storage of protective clothing and equipment." (See Appendix 21.B for a more detailed description of the protective clothing and types of respirators.)

2. "Personnel should not eat, drink, chew gum or tobacco, smoke, take medicines, or perform any other practice that might increase hand-to-mouth transfer of toxic materials from gloves, unwashed hands, or equipment."

3. "Personnel should not have excessive facial hair (heavy mustaches, beards) which can prevent the proper fit of respirators."

4. "Personnel should avoid unnecessary contact with hazardous materials by staying clear of puddles, vapors, mud, discolored surfaces, and containers or site debris."

Carelessness during routine daily activities at the site can lead to serious personal contamination or to contamination of others. Several important habits should be practiced:

1. Always wash hands *before* using restroom.

2. Leave the site for lunch, removing all protective (contaminated) clothing, and wash thoroughly.

3. Wash hands after handling contaminated equipment.

4. Do not take contaminated clothing home to launder.

5. Wear the required protective clothing at all times, even if the need is not apparent. Demand that it be fitted properly. Even a short exposure to a toxic substance can be deadly.

6. Because protective clothing is cumbersome to wear and is often uncomfortable in hot weather, take appropriate rest periods to avoid accidents caused by fatigue or physical irritation.

Even if every safety precaution is taken, an emergency may develop at any time when doing monitoring work. Emergency plans should be well established and understood by everyone involved in the project. First aid equipment should be available, the routes to emergency care centers known, and the necessary personal contacts established at the care centers. All steps of the standard emergency procedures should be practiced so that any team member can take charge.

DESIGN OF MONITORING WELLS

The particular design of a monitoring well will depend on (1) how the well is to be used — whether for taking water samples for measuring the elevation of the water table or potentiometric surface, or for recovering contaminants, (2) the hydrogeologic environment, (3) the chemical nature of the contaminants, and (4) whether the well bore will be used to conduct geophysical investigations. The design consultant should keep in mind that the cost of the best engineered monitoring well constructed of the most suitable materials will be only a fraction of the long-term costs for water-quality analysis. Therefore, the most suitable well materials and construction practices should be selected for monitoring wells.

Many monitoring wells are constructed of 2-in (51.8-mm) casing and screen, although a large number are 4, 6, or 8 inches (102, 152, or 203 mm) in diameter. The most appropriate diameter will depend on numerous site-specific factors. For shallow monitoring wells or those used for measuring water level only, 2-in well screens and casing may be suitable, but for more accurate sampling, better development, deeper wells, and where some form of pumping test or borehole geophysical investigation is necessary, the screen and casing should be at least 4 inches in diameter. Taking representative samples from 4-in wells is more difficult than for 2-in wells because many pumps manufactured for sampling 2-in monitoring wells are technically superior to larger pumps in that they preserve the true chemical character of the sample. They can be pumped at only extremely low rates, however, making their use impractical in wells 4 inches in diameter or larger. If 2-in screens are installed in dirty or tight formations, the driller cannot develop the well properly. Water samples taken from poorly developed wells may not be chemically representative of the water in the formation because recharge to the well is so slow that the person who takes the sample cannot spend the time needed to collect a representative sample.

Screen Criteria for Monitoring Wells

Well screens used for monitoring work should have the following characteristics:
1. Screens should be constructed from a material that is inert in the water being tested.
2. Open area should be maximized to facilitate rapid sample recovery.
3. Slot sizes should retain filter pack or natural formation consistent with the capability to develop the well.
4. Slot openings should be nonplugging in design.
5. Slot openings, slot design, open area, and screen diameter should permit effective development.

Selection of the screen material must be done with care because many common screen or casing materials such as PVC, low-carbon steel, and even stainless steel may react with the groundwater, producing erroneous water-quality data. In general, the following factors should be considered when selecting screen and casing materials (Ramsey and Montgomery, 1982):

1. Contaminants to be sampled
2. Chemical reactiveness/inertness
3. Strength of material
4. Ease of installation
5. Cost of material

Table 21.6 lists the major types of materials used for monitoring wells, along with recommendations for their use. Teflon (Teflon is a registered trademark of E.I. DuPont DeNemours and Co., Inc.) is the most inert material currently being used, but its cost may make its use inappropriate in groundwater environments where less costly materials are satisfactory. PVC materials are suitable for monitoring most landfills unless organic chemicals are present in the groundwater. If they are present,

Table 21.6 Well Casing and Screen Materials

Type	Advantages	Disadvantages
PVC (Polyvinyl-chloride)	• Lightweight • Excellent chemical resistance to weak alkalies, alcohols, aliphatic hydrocarbons, and oils • Good chemical resistance to strong mineral acids, concentrated oxidizing acids, and strong alkalies • Readily available • Low priced compared to stainless steel and Teflon	• Weaker, less rigid, and more temperature sensitive than metallic materials • May adsorb some constituents from groundwater • May react with and leach some constituents from groundwater • Poor chemical resistance to ketones, esters, and aromatic hydrocarbons
Polypropylene	• Lightweight • Excellent chemical resistance to mineral acids • Good to excellent chemical resistance to alkalies, alcohols, ketones, and esters • Good chemical resistance to oils • Fair chemical resistance to concentrated oxidizing acids, aliphatic hydrocarbons, and aromatic hydrocarbons • Low priced compared to stainless steel and Teflon	• Weaker, less rigid, and more temperature sensitive than metallic materials • May react with and leach some constituents into groundwater • Poor machinability — it cannot be slotted because it melts rather than cuts
Teflon	• Lighweight • High impact strength • Outstanding resistance to chemical attack; insoluble in all organics except a few exotic fluorinated solvents	• Tensile strength and wear resistance low compared to other engineering plastics • Expensive relative to other plastics and stainless steel
Kynar	• Greater strength and water resistance than Teflon • Resistant to most chemicals and solvents • Lower priced than Teflon	• Not readily available • Poor chemical resistance to ketones, acetone
Mild steel	• Strong, rigid; temperature sensitivity not a problem • Readily available • Low priced relative to stainless steel and Teflon	• Heavier than plastics • May react with and leach some constituents into groundwater • Not as chemically resistant as stainless steel
Stainless steel	• High strength at a great range of temperatures • Excellent resistance to corrosion and oxidation • Readily available • Moderate price for casing	• Heavier than plastics • May corrode and leach some chromium in highly acidic waters • May act as a catalyst in some organic reactions • Screens are higher priced than plastic screens

stainless steel or Teflon materials must be used. Relatively inert metals such as 304 or 316 stainless steel are not suitable for groundwater in which heavy metals are present, because leaching of chromium or other metallic components may occur. Selection of the screen and casing material generally depends upon the chemical nature of the groundwater, not cost of the screen or casing material. The laboratory doing the analytical chemistry of the water samples should be informed of all materials used in the well.

Screens used for monitoring are almost always placed in materials having extremely low hydraulic conductivity. If possible, the open area of the screen should approximate the natural porosity of the formation, that is, 15 to 20 percent, so that the time required to take a representative sample is minimized. Because most sampling methods require that a water sample be taken only after 3 to 10 well-bore volumes have been removed, the amount of time dedicated to taking a single sample can be excessive if low-open-area screens are installed.

Screen slot sizes must retain a high percentage of the filter pack or natural formation for all 2-in (51.8-mm) wells because effective development of these wells is particularly difficult. For larger diameter monitoring wells, the slot sizes can more nearly follow the recommendations for water wells (see Chapter 13). Development is most effective when the slot openings are distributed uniformly around the circumference of the screen so that as much of the formation and filter pack as possible can be reached by the development action. The configuration of the slot should permit all the development energy to reach the formation.

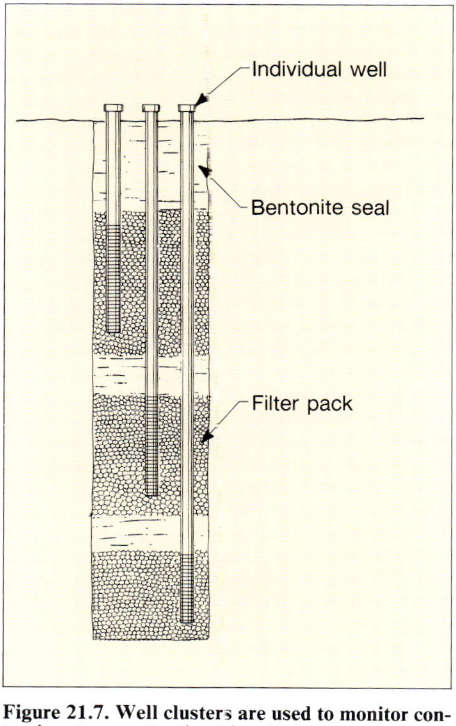

Slot openings should widen inward so that finer formation materials are pulled through the screen during development. Slots that are cut straight through the casing or those of the gauze type will tend to plug with fine material during development, thereby reducing significantly the open area of the screen. This is especially true for 2-in (51.8-mm) screens where any development that is done is relatively inefficient. Plugging of the slots increases the time needed to obtain a representative sample from the formation.

For scientific purposes, water samples from monitoring wells are usually obtained from relatively thin zones in the aquifer. This can be accomplished by using multiple wells with short screen segments. Commonly, a nest of wells will be installed in single or multiple boreholes to gather water samples from several depths in the aquifer (Figure 21.7). Using this method, the vertical dimensions and contaminant strength of the plume can be determined.

Figure 21.7. Well clusters are used to monitor contaminant concentrations in discrete aquifers at a single site or at multiple depths in a single aquifer. *(After Morrison, 1983)*

Screens used for collecting water samples are typically 2 to 5 ft (0.6 to 1.5 m) in length, because samples should come from specific depths and high yields are relatively unimportant. The well yield should be high enough, however, so that a reliable water sample can be collected quickly. Screens that monitor groundwater quality at the top of the water table are usually 10 to 20 ft (3 to 6.1 m) in length, depending on the anticipated long-term changes in groundwater elevation. Some of the screen is always above the water table in the vadose zone. These screens are then used to monitor for the presence of hydrocarbons or other volatile substances that have reached the groundwater table.

Filter Pack Design

Monitoring wells are generally installed in formations having a wide range of particle sizes, which makes it difficult to filter pack effectively. Filter packing procedures recommended for water wells are not suitable for monitoring wells unless the hydraulic characteristics of the formation materials are similar to those of an aquifer. To exclude the entrance of fine silts, sands, and clays into a monitoring well, the grain-size distribution curve for the filter pack is selected by multiplying the 70-percent retained size of the finest formation sample by 3 or 4. This leads to a more conservatively sized filter pack than would be selected for a water well. Selection of too fine a pack will reduce the yield of the well, causing longer sampling times. Uniformity coefficients for filter pack material should range from 1 to 3. All pack material should be purchased from reputable suppliers who have properly cleaned and bagged the material. Some investigators require acid washing of the pack to remove contaminants adhering to the filter pack particles, others use steam cleaning, and still others may use chlorine solutions. At the very least, the pack should be washed with fresh water. A sample of the cleaned filter pack should be collected and chemically analyzed in the event questions are raised regarding possible contamination from the pack (Ramsey and Montgomery, 1982). The design of a typical filter-packed monitoring well is shown in Figure 21.8.

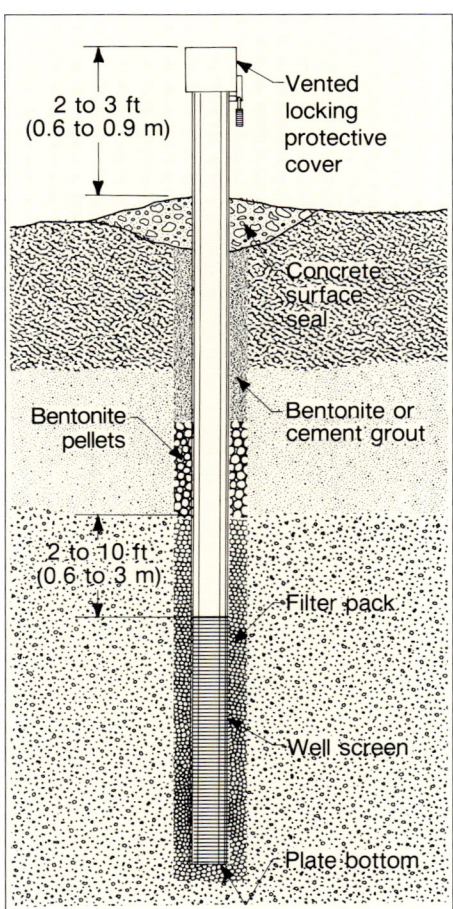

Figure 21.8. This filter-packed monitoring well design will assure sample reliability by eliminating downward vertical movement of contaminants in the annulus around the casing.

Installation Procedures

All screens and casings used for mon-

Table 21.7. Decontamination Solutions

Name of solution	Remarks
Sodium bicarbonate	Effective for acids and bases, amphoteric, 5-15 percent aqueous solution
Sodium carbonate	Effective for inorganic acids, good water softener, 10-20 percent aqueous solution
Trisodium phospate	Good rinsing solution or detergent, 10 percent aqueous solution
Calcium hypochlorite	Excellent disinfectant, bleaching and oxidizing agent, 10 percent aqueous solution

(Richter and Collentine, 1983)

itoring wells should be in a sterile and contaminant-free condition when placed in the ground. Some manufacturers ship their products in this condition, but handling in the field requires a final wash with detergent or other solution. Table 21.7 lists typical decontamination solutions. Some form of steam cleaning, or high-pressure water-spraying technique combined with a low-sudsing soap or detergent, is recommended (Richter and Collentine, 1983). In addition, acetone and hexane are used to clean drilling tools and sampling equipment at hazardous waste sites. Working components of the drilling rig (drill pipe, subs, collars, kelly, and all parts of the rig chassis near the borehole) should also be cleaned. These cleaning operations should be verified by the driller in the well log.

The method of joining screens to casing and of assembling the casing string must also be done so as to prevent contamination of the samples. In general, no solvent

Table 21.8. Fitting Types

Type	Advantages	Disadvantages
Plain square ends (no fittings to weld)	• Readily available in pipe and screen • No need to purchase threads and couplings	• Special equipment and skills needed to field-weld metals • Plastics are welded using solvent cement which causes the following problems: — Cementing procedures are very temperature and moisture sensitive — Cements must be cured after application — Cements may interfere with groundwater quality analysis Time spent welding may cause this type of fitting to actually cost more than threads
Threads and couplings	• No solvents needed • Lengths of pipe and screen joined quickly • Readily available • Reasonably priced	• May be difficult to get filter pack and/or grout past the lip of couplings • May need to wrap threads with Teflon tape to make connections watertight
Flush threads	• No solvents needed • No couplings needed; filter packing and grouting simplified • Lengths of pipe and screen joined quickly • Readily available • Reasonably priced	• May need to wrap threads with Teflon tape to make connections watertight • Threads generally not compatible from manufacturer to manufacturer

welds are recommended; all plastic screens and casing should be joined by threads and couplings or flush threads. The joints are made watertight by wrapping with Teflon tape or by placing a Teflon or Viton (Viton is a registered trademark of E.I. DuPont DeNemours and Co., Inc.) o-ring in the joint (Table 21.8).

A primary objective of monitoring well construction is to make sure that contaminated groundwater does not enter contaminant-free geologic formations. Although some minor amount of cross-contamination may occur during drilling and well installation, the integrity of individual formations must be protected thereafter. This is usually accomplished by placing either bentonite or cement grout in the borehole above the filter pack in both single- and multiple-screen wells. Drill cuttings should not be placed in any open borehole annulus. To prevent downward migration of the bentonite or cement into the screen, the filter pack is extended at least 2 to 10 ft (0.6 to 3 m) above the top of the screen. The filter pack should not extend into an overlying formation, because this would permit downward vertical seepage in the pack and either dilute or add to the contamination of the water being monitored. See Table 21.9 for a comparison of bentonite and cement grouts. Polymeric fluids are not recommended as an alternative to bentonite or cement because they contain so few solids. See Chapter 10 for grout placement procedures.

For monitoring wells drilled by cable tool rigs, contaminant migration in the borehole can be eliminated by the well design shown in Figure 21.9. The 6-in (152-mm) casing is first installed into the clay layer. After flushing the casing and changing the drilling fluid, the borehole is extended using 4-in (102-mm) casing. A 2-in (51.8-mm) monitoring well is then installed and filter packed. Before installing the pack, the well bore should be thoroughly flushed. As the 4-in temporary casing is extracted, cement or bentonite grout is placed as shown in Figure 21.9. A protective surface casing with a locking cap is installed before the cement has hardened. Normally the locking cap will be vented. The well should then be developed as thoroughly as possible.

Table 21.9. Grouting Materials for Monitoring Wells

Type	Advantages	Disadvantages
Bentonite	• Readily available • Inexpensive	• May produce chemical interference with water-quality analysis
		• May not provide a complete seal because:
		— There is a limit (14 percent) to the amount of solids that can be pumped in a slurry. Thus, there are few solids in the seal; should wait for liquid to bleed off so solids will settle
		— During installation, bentonite pellets may hydrate before reaching proper depth, thereby sticking to formation or casing and causing bridging
		— Cannot determine how effectively material has been placed
		— Cannot assure complete bond to casing
Cement	• Readily available • Inexpensive • Can use sand and/or gravel filter • Possible to determine how well the cement has been placed by temperature logs or acoustic bond logs	• May cause chemical interferences with water-quality analysis • Requires mixer, pump, and tremie line; generally more cleanup than with bentonite • Shrinks when it sets; complete bond to formation and casing not assured

If the well is drilled by rotary methods, a 4-in (102-mm) casing can be grouted in an 8-in (203-mm) borehole that is drilled through the contaminated aquifer into an underlying impermeable layer. After the grout has hardened, the borehole is continued inside the 4-in casing to the desired depth. A 2-in (51-mm) casing and screen is filter packed and then grouted in the 4-in casing.

In saline environments, a Dowell seal ring gasket (manufactured by Dow Chemical) may be used in place of bentonite, because bentonite will not hydrate in a highly saline environment (Senger and Perpich, 1983). These gaskets can be made in variable lengths and mounted on the casing just prior to the installation of the screen. They are suitable for use in regularly shaped boreholes and where the organic and inorganic compounds in the gasket do not interfere with the chemical analysis of the water in the well. Cement can be placed above the gaskets to complete the seal.

A cement seal around the top of the well bore is recommended even if the annular seal is carried to the surface. The cement seal is shaped so that surface water flows away from the casing. If plastic casing is used, a short section of metal surface casing should be installed around the top section of the plastic pipe and extended 3 to 5 ft (0.9 to 1.5 m) into the ground. The metal casing prevents accidental damage to the plastic pipe. The top of the casing should be fitted with a locking cap.

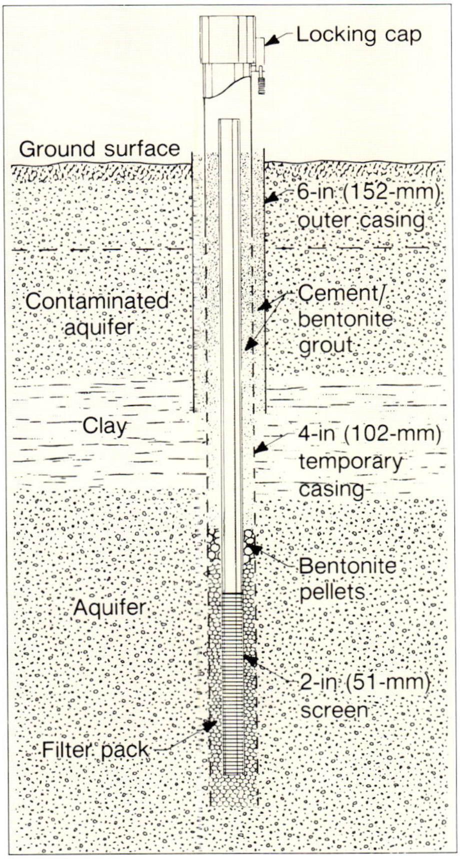

Frost heaving can be a major problem for small-diameter PVC monitoring wells installed in cold climates. As the soil freezes during the winter, it expands upward, occasionally pulling the casing apart. Damage caused by frost heaving can be minimized by placing a metal surface casing to a depth of 5 to 10 ft (1.5 to 3 m). A steeply inclined cement cap should be placed around the surface casing. If frost action exerts pressure on the cement, the surface casing can rise without disturbing the monitoring well casing.

Development is especially important for monitoring wells, because drilling fluid residues remaining in the borehole will affect the chemistry of the water samples (Walker, 1983). Figure 21.10 shows that the presence of bentonite affects the chemical analyses of samples for at least 90 days after completion of the well. More thorough development shortens the time the bentonite will affect water quality. In some cases, the impact of drilling fluid additives on sampling chemistry can last for 1 to 2 years (Walker, 1983).

Residual amounts of polymeric drilling

Figure 21.9. Intraborehole contamination can be prevented by proper well design and construction techniques.

fluid additives can also cause chemical effects; see Appendix 21.C for a list of the chemical constituents of a commonly used polymer. If any of these constituents are found in the first few water samples taken from a monitoring well drilled with the polymeric additive, they are ignored. As in the case of bentonite, the chemical effects of polymeric drilling fluids tend to disappear over time, although they will disappear faster if a breakdown chemical is added to the drilling fluid as the well is developed (Figure 21.11).

It should be stressed that all monitoring wells must be developed as thoroughly as possible, not only to remove all traces of the drilling fluid from the formation, but also to increase the yield so that reliable

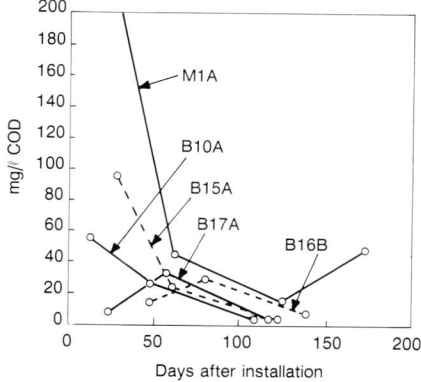

Figure 21.10. Inorganic residues of bentonite remain in the well after development. These residues will affect chemical oxygen demand (COD) of samples for 90 to 100 days after the well has been completed. *(R. Brobst, personal communication)*

samples can be collected in the shortest time. Development is also important to assure that the ambient water quality is maintained in the sample container until the water can be analyzed. Any sediment in the sample container, for example, can react with the water, thereby altering the actual chemical quality.

SAMPLING MONITORING WELLS

Sampling of monitoring wells will usually be done by field personnel from the testing laboratory or by groundwater consultants. Nevertheless, drilling contractors should have an appreciation of the difficulties in sampling protocol. In general, a sample is taken only after the pH, electrical conductivity, and temperature of the water being pumped from the well have stabilized (Wood, 1976). The methodology used in the sampling procedure is critically important if the true chemical nature of the groundwater contamination at that site is to be determined. Samples may not be representative of groundwater conditions for the following reasons:

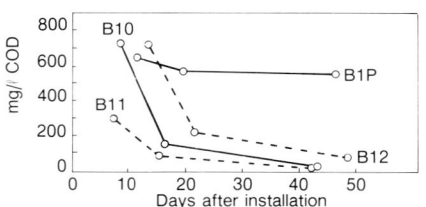

(a) These wells were not developed; the organic residues in the drilling fluid were allowed to break down naturally.

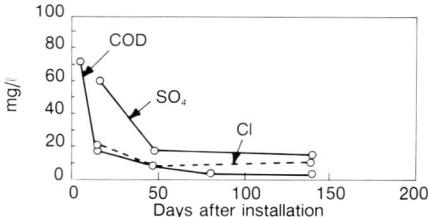

(b) These wells were developed using a breakdown chemical to reduce the viscosity of the drilling fluid.

Figure 21.11. Organic drilling fluid (Revert) residues will affect water-quality data for a certain length of time depending on whether the well is developed. *(R. Brobst, personal communication)*

1. The sample was taken from stagnant water in the well, which is usually different chemically from water in the ground near the well bore. The transmissivity of the aquifer should be determined so that the consultant can estimate the time required to remove enough water to obtain a reliable sample. In most wells, a sample

is taken after 3 to 10 well-bore volumes have been removed (Gibb, 1983). Figure 21.12 shows the length of time required to remove stagnant water (casing storage) from wells of different diameters at a specific pumping rate. Large-diameter wells in formations having low hydraulic conductivities are most susceptible to casing-storage effects. When a sample is taken, enough time should have elapsed so at least 95 percent of the water being pumped comes from the aquifer.

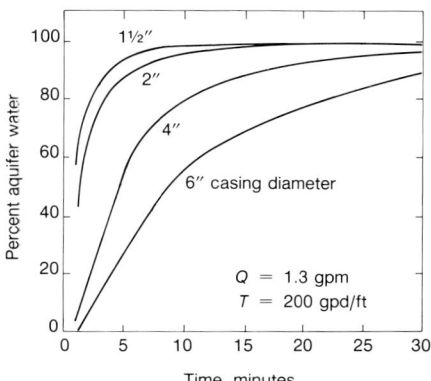

Figure 21.12. Length of time required to remove stagnant water (casing storage) from wells of different diameters at a specific pumping rate. *(Gibb et al., 1981)*

2. Samples were not taken at appropriate intervals. Sampling intervals are usually established based on the hydraulic conductivity of the formation — the faster the rate of contaminant movement, the more often samples are taken.

3. The water sample was contaminated by entrained sediment because the well was not developed properly. When the sample is acidified for preservation, contaminants adhering to the sediment, especially metallic ions, are released into the sample.

4. Sample accuracy was adversely affected by the hydraulic character of the formation. A representative sample may not be obtained when the formation has a high hydraulic conductivity near the screen and the contaminant is concentrated near the static water level in less permeable material. In this case, the proportion of uncontaminated water is so high that the dilution is sufficient to mask the presence of the contaminant.

5. The sample was taken so long after pumping began that it represents water so far from the well site that the groundwater chemistry is not representative.

6. Release of carbon dioxide during pumping caused an increase in pH which in turn caused many metallic ions to come out of solution (iron, manganese, magnesium, cadmium, arsenic, selenium, and boron) (Gibb et al., 1981).

7. Numerous chemical changes took place because the sample was oxidized during recovery. Oxidation can occur in the pump or can be caused by water cascading into a well installed in tight formations (Giddings, 1983). Because many groundwaters are in a reduced state, some of the changes that can be expected include (Stumm and Morgan, 1970):
    a. Oxidation of organics
    b. Oxidation of sulfide to sulfate
    c. Oxidation of ferrous iron and precipitation of ferric hydroxide [$Fe(OH)_3$]
    d. Oxidation of ammonium ion to nitrate
    e. Oxidation of manganese and precipitation of manganese dioxide or similar hydrous oxide

8. The water sample became contaminated by chemical residues in the pump or sampling equipment. If sample recovery equipment is not dedicated to a single well, it must be thoroughly cleaned each time it is used.

9. The sample was not preserved correctly, so chemical changes occurred in the

sample container during storage. Most samples requiring laboratory analysis are preserved by immediately wrapping with foil to prevent light from reaching them and then storing temporarily in a cold environment.

10. The sample was not analyzed quickly enough because either it did not reach the laboratory within a reasonable time or the lab could not perform the analytical tests soon enough because of poor scheduling.

11. The testing procedures at the laboratory were not set up properly and therefore did not yield accurate results. The concentration of a particular contaminant may be in the parts per billion or even parts per trillion range, and the least malfunction in the analytical work can produce erroneous results. Good testing procedures include sending split samples to different labs to compare results and submitting spiked samples periodically to check the reliability of a particular lab's testing procedures. A spiked sample usually contains one or more chemicals that have been purposely added at specific concentrations.

Table 21.10 lists the basic types of equipment used to remove water from monitoring wells. Gibb et al. (1981) have determined that the single most important parameter affecting the chemical quality of groundwater is pH and any disturbance in pH during sampling can cause a distinct change in chemistry. In general, air-lift or nitrogen-lift pumps produce the largest increases in pH during sampling by stripping excess dissolved carbon dioxide from the water. Although Gibb did not test a submersible pump, peristaltic and diaphragm pumps and bailers had less effect on pH. Air-lift pumps also reduce the concentration of volatile organic compounds. It is a good idea to measure oxidation-reduction potential (Eh), pH, and specific conductance in the field in a closed cell to determine their values as accurately as possible (Gibb, 1983).

In conclusion, sampling procedures are highly complex and must be tailored to fit the chemical being monitored, the hydrogeologic situation, and the design of the monitoring well. The presence of stagnant water in the well bore will usually have an adverse effect on the accuracy of the analysis, but time constraints may limit the number of well-bore volumes that can be removed before sampling.

THE TASK OF GROUNDWATER PROTECTION

Contamination of groundwater is more serious than surface-water pollution because it is more difficult to detect in a timely manner, moves more slowly, and requires special expertise to predict the path and rate of contaminant movement. In addition, the complex geochemical reactions taking place in the subsurface between myriad contaminants and earth materials are not well understood, and thus the ability to predict the concentration of a contaminant at any point is limited. The drilling contractor can take an active role in early detection of contaminant problems by being aware of well-known pollution sources and the various chemical and biological indicators that may indicate that contamination is occurring.

Ideally, contamination should be prevented from occurring. Successful prevention means that potential contaminants must be controlled so they cannot react with the groundwater system (Knox et al., 1982). Land-use planning is a major form of prevention in which the producers of hazardous wastes are kept away from areas overlying groundwater resources so that, in the event of an accidental spill, little damage will occur. When potential producers of contaminants are discovered in a community or are allowed to build new facilities, action should be undertaken to develop monitoring networks that will identify ineffective disposal practices that could affect groundwater quality. Vadose-zone sampling equipment should be placed close to waste sites so contaminants can be detected as soon as possible — preferably before they enter the local groundwater system. Monitoring wells should be installed at appropriate places around the waste site to detect any contaminants reaching the groundwater system. Once a contaminant reaches the groundwater, hydrogeologists

Table 21.10. Water-Quality Sampling Devices for Monitoring Wells

Type	Advantages	Disadvantages
Bailer	• Can be constructed in a wide variety of diameters • Can be constructed from a wide variety of materials • No external power source required • Extremely portable • Low surface-area-to-volume ratio, resulting in a very small amount of outgassing of volatile organics while sample is contained in bailer • Easy to clean • Readily available • Inexpensive	• Sampling procedure is time consuming; sometimes impractical to properly evacuate casing before taking samples • Aeration may result when transferring water to the sample bottle
Suction-lift pump	• Relatively portable • Readily available • Inexpensive	• Sampling is limited to situations where water levels are within about 20 ft of the ground surface • Vacuum effect can cause the water to lose some dissolved gas
Air-lift samplers	• Relatively portable • Readily available • Inexpensive • Suitable for well development	• Causes changes in carbon dioxide concentrations; therefore this method is unsuitable for sampling for pH-sensitive parameters • In general, this method is not an appropriate method for acquisition of water samples for detailed chemical analyses because of degassing effect on sample • Oxygenation is impossible to avoid unless elaborate precautions are taken
Gas-operated pump	• Can be constructed in diameters as small as 1 in (25.4 mm) • Can be constructed from a wide variety of materials • Relatively portable • Reasonable range of pumping rates • Driving gas does not contact water sample, eliminating possible contamination or gas stripping	• Gas source required • Large gas volumes and long cycles are necessary when pumping from deep wells • Pumping rates are lower than those of suction or jet pumps • Commercial units are relatively expensive
Submersible pump	• Wide range of diameters • Constructed from various materials • 12-volt pump is highly portable; other units are fairly portable • Depending on size of pump and pumping depths, relatively large pumping rates are possible for wells larger than 2-in (51-mm) diameter • Readily available • 1¾-in (44.5-mm) helical screw pump has rotor and stator construction that permits pumping fine-grained materials without damage to the pump	• With one exception, submersible pumps are too large for 2-in (51-mm) diameter pumps • Conventional units are unable to pump sediment-laden water without incurring damage to the pump • 1¾-in (44.5-mm) pump delivers low pumping rates at high heads • Smallest diameter pump is relatively expensive

should be consulted to determine the direction and rate of plume movement.

After a contaminant or several contaminants are found in groundwater, a decision must be made on whether to rehabilitate the aquifer or find alternative groundwater resources. In some cases, no remedial methods may be undertaken because the areal extent of the contamination is limited or the concentration of the contaminant is below health-effect standards. Occasionally, indirect remedial methods may be most suitable; for example, if a new groundwater supply is available, the contaminated one can be abandoned. In direct remedial methods, the soil and groundwater are treated to eliminate the contamination, or the source is removed and the groundwater allowed to recover naturally through time. In many instances, however, the renovation cost may exceed the community's ability to pay for it.

In the past, the projected costs for restoration were usually sufficient to spur the search for other water resources because new or deeper wells could be constructed at less cost than an aquifer cleanup. When new water resources are not available, however, the costs for restoration become secondary. Fortunately, some techniques used in construction and dewatering practice have been combined with new chemical treatment methods to not only contain the spread of contamination but also to begin the restoration of the aquifer.

AQUIFER RESTORATION

Once contamination of a local groundwater supply has occurred, some action must be taken to (1) find and eliminate the source, (2) contain the contaminants in the area already affected, and (3) restore the water quality of the aquifer. Because groundwater may be the only fresh-water resource in many areas, restoration of the aquifer may be of the highest priority regardless of the costs involved. Protection and restoration of groundwater resources must be a major concern for drilling contractors, and they should become familiar with the options available to handle contamination problems. Not only can drillers advise communities on how to solve their groundwater contamination problems, they may also become involved in the process itself.

All aquifer restoration projects have some general similarities. They are costly to perform, are time consuming to plan and implement, may be only partially effective, and litigation surrounding the contamination may prevent a full disclosure of the facts (Canter, 1981).

Containment of the contaminant source is the first step in aquifer restoration. Recent research has shown that virtually all landfills leak, even if various types of plastic liners or clay layers have been used to retain the leachate. Capping of abandoned landfill sites with bentonite or other low-permeability material prevents rainwater from entering the site, thus eliminating the formation of leachates.

A combined method of containment and abatement is one way to effect aquifer restoration. Containment usually focuses on some hydraulic means of preventing the spread of the contaminant — either through withdrawal of contaminated water or the injection of clean water to create a pressure ridge. Withdrawal of groundwater can reverse the local groundwater gradient, thereby preventing the advance of the contaminant front. The water removed is usually treated before use or is discharged to a surface-water body, where dilution takes place. Contaminant plumes can also be contained by injecting large volumes of water to create a local high in the potentiometric surface.

Slurry walls can also be used to isolate areas of contaminated groundwater. Slurry walls consist of bentonite, water, and backfill material placed in deep trenches. The mixture of soil material and hydrated bentonite can be placed as deep as 100 ft (30.5 m) (Glover, 1982). This method is particularly successful if the slurry walls can be tied into an underlying impermeable formation. Rainwater percolating into the area isolated by the slurry walls is removed by wells to keep the contaminated water from overtopping the walls. This water can be treated and then reinjected downgradient. Steel sheetpiling can also be used to construct cut-off walls to contain groundwater contamination.

Ordinarily, a slurry wall will last 20 to 40 years or even longer. However, the service life of the slurry wall is greatly affected by the type of chemicals in the groundwater. Organic chemicals, for example, can cause a great increase in the hydraulic conductivity of a slurry wall in only a few years.

Some wastes are so dangerous and long lasting that the only effective way to prevent long-term groundwater contamination is to excavate the material, treat it, and replace it or to simply haul it away to a safe disposal site. Where either of these options is not feasible, the wastes are sometimes completely encapsulated by impermeable materials and left permanently at the site. In other cases, chemical alteration of a contaminant in the ground can sometimes be done successfully in relatively small areas. Williams (1982) reports that 4,200 gal (15.9 m^3) of acrylate monomer was immobilized by injecting appropriate catalyst and activator solutions into the contaminated zone. In this process, the chemicals were immobilized by in-situ polymerization.

In many areas where various types of synthetic organic chemicals have contaminated the groundwater, new water treatment methods such as air stripping and activated carbon are used to decontaminate the water, either before use or before it is injected back into the ground. It is fortunate that many organic chemicals have a low affinity for soil particles so that the chemical remains in the water as the plume moves downgradient. Once the contaminant source has been removed and the water cleaned up or normally replaced by recharge, no large-scale contamination remains in the aquifer.

Hydrocarbon contamination results in severe taste and odor problems in wells, as well as infiltration into storm and sanitary sewers and odors in basements. Hydrocarbons can exist as free product at the top of the water table or as dissolved or emulsified product in the aquifer. After the source has been identified and the contamination stopped, removal of hydrocarbons from the surface or near surface of an aquifer involves one or more recovery wells in which a one- or two-pump system is used (Figure 21.13). Additional abatement and cleanup procedures may include the following (Yaniga, 1982):

1. Treatment of contaminated water in an air-stripping tower to remove volatile organics and to induce oxygenation of the contaminated groundwater (Figure 21.14).

2. Recharge of the treated groundwater by infiltration galleries to facilitate flushing and leaching of gasoline adsorbed onto soil particles.

3. Reoxygenate groundwater by means of air compressors and wells to accelerate the growth of aerobic bacteria that metabolize hydrocarbons.

4. Addition of nutrients to wells to stimulate the growth of bacteria.

Standard water well design procedures are used for hydrocarbon recovery wells (Blake and Lewis, 1982). High open area screens are necessary because hydrocarbons

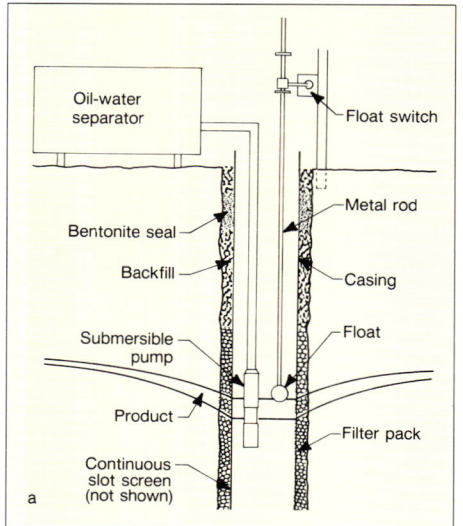
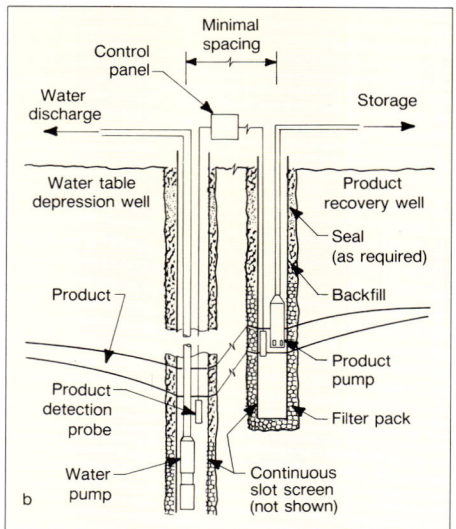

Figure 21.13. (a) One-pump system utilizing a submersible pump and float controls; (b) two-pump system utilizing two small-diameter wells; (c) two-pump system utilizing one recovery well. *(After Blake and Lewis, 1982)*

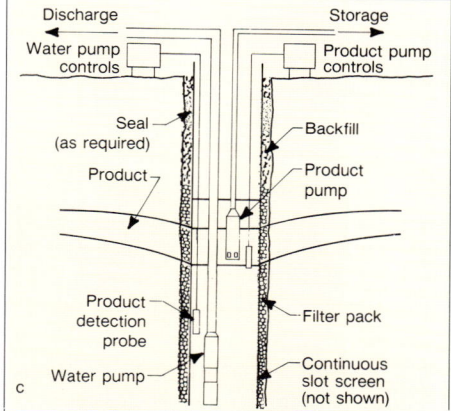

provide an environment in which bacteria can grow and thereby plug the screen. Somewhat longer screens should be used for hydrocarbon recovery because all product (hydrocarbons) floating on the water table must enter the well. Placement of the screen is also critical. Under ordinary circumstances, the cone of depression should be just large enough to recover the product. Steeper drawdown cones cause more of the hydrocarbons to become trapped in the aquifer materials (product retention in the aquifer may range from 8 to 16 percent), reducing the total volume of hydrocarbons that can be recovered. For single-pump systems, the smallest practical casing diameter is 6 in (152 mm), whereas 10 in (254 mm) is a recommended minimum for two-pump systems (Blake and Lewis, 1982).

Drilling contractors and well design engineers should become familiar with these and other effective remedial procedures so they can advise individuals and communities on ways to reduce the spread of contaminants and restore local groundwater quality.

CONCLUSIONS

Cleanup of contaminated aquifers is difficult, time consuming, and occasionally dangerous, depending on the nature of the contaminant. Drilling contractors should be especially careful when asked to drill in areas where various forms of industrial waste have been deposited in the past. Many of these sites may not be posted as old

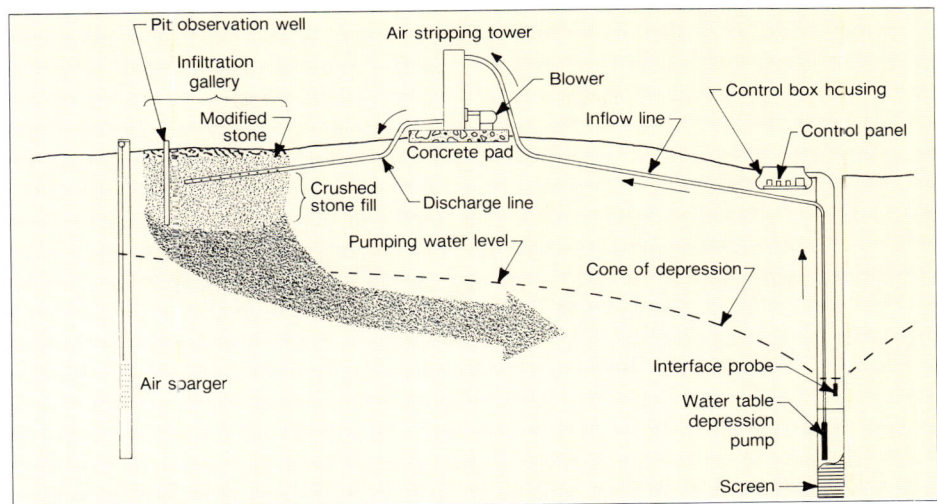

Figure 21.14. Cross section of groundwater recovery operation. *(After Yaniga, 1982)*

dumping sites. Fortunately, the awareness of groundwater contamination problems, coupled with improved scientific technology and increased numbers of well-trained and experienced hydrogeologists, should lead to the solution of many of our present groundwater contamination problems.

CHAPTER 22
Alternative Uses for Wells and Well Screens

To this point, the primary emphasis of this book has been on obtaining potable water supplies from vertical wells. Other well design and construction techniques, however, provide alternative engineering methods for the use or control of groundwater environments. For example, infiltration galleries consisting of one or more horizontally installed screens can take the place of high-capacity vertical wells in some geologic settings where the water-saturated sediments are too thin. In other cases, salt-water intrusion in many coastal communities is being controlled by strategically placed injection wells. Highly engineered injection wells also facilitate the placement of hazardous wastes in deep subsurface environments.

Use of groundwater or aquifer systems for other purposes inevitably leads to environmental conflicts which may have short- or long-range adverse, or beneficial, impact. For example, the recharge or injection of brine wastes resulting from oil-field activities can spoil a near-surface aquifer used for potable water supplies. Dewatering for construction purposes, on the other hand, may have only a short-term adverse impact on nearby shallow wells. Fortunately, recent legislation in many countries, on both national and regional levels, has attempted to recognize the conflicting uses for groundwater and the groundwater environment so that both water quantity and water quality can be protected. The Underground Injection Control program in the United States is an example of this type of legislation. Under this legislation, no wastes can be injected above or into an aquifer being used for water supply. Controls for injecting wastes below an aquifer are also carefully established. Many new laws are being enacted so engineers and contractors must keep abreast of new regulations to avoid litigation.

In this chapter, the principal design criteria are presented for alternative engineering methods to withdraw groundwater, as well as surface water, and to use or control the groundwater environment. Design engineers and many water well contractors will become involved in projects using these methods.

DEWATERING

Dewatering technology has advanced concurrently with improvements in water well design and construction practices, new pump designs, and the development of alternative technologies, such as ground freezing, for controlling groundwater flow. The dewatering industry is large, involving many large and small contracting firms specializing in this business. Dewatering wells are designed primarily to lower the groundwater level to a predetermined depth and to maintain that depth until all below-ground construction has been completed. In the past, almost all temporary or long-term dewatering involved the construction of buildings, subways, tunnels, sewer and water trenches, or earth-sheltered houses. Presently, large quantities of contaminated groundwater are being removed locally and treated as part of groundwater quality restoration programs. Elsewhere, plumes of contaminated groundwater are prevented from spreading by continuous groundwater removal. Dewatering to control the spread of groundwater contaminants is highly effective and has enabled many communities to reduce or eliminate threats to groundwater quality.

The main purposes for construction dewatering include (Cedergren, 1977):

1. Intercepting seepage that would enter an excavation site and interfere with construction activities.

2. Improving the stability of slopes, thus preventing sloughing or slope failures.

3. Preventing the bottoms of excavations from heaving because of excessive hydrostatic pressure.

4. Improving the compaction characteristics of soils in the bottoms of excavations for basements, freeways, and other structures.

5. Drying up borrow pits so that excavated materials can be properly compacted in embankments.

6. Reducing earth pressures on temporary supports and sheeting.

Despite advances in dewatering technology, it is almost impossible to predict subsurface conditions before a project is started. This occurs because near-surface geologic materials are far less uniform both vertically and horizontally than more deeply buried materials. It is therefore difficult to predict how much water will have to be removed and the effect of this removal on the engineering characteristics of the dewatered materials.

It is beyond the scope of this book to investigate the physical effects of dewatering on the dewatered materials, but this aspect of any construction project is vital and anyone involved with a dewatering project must become familiar with the behavior of subsurface materials under stress. Preliminary determination of compressive strengths, Atterberg limits, shear strengths, and grain sizes must be completed for any major dewatering project. Stability of the dewatered materials is of prime concern to the engineer in charge. Unexpected changes in pore-water pressure, caused by dewatering, must never be allowed to produce "quick" conditions in which the dewatered material loses its stability and then flows (for example, quick sand).

In this section, emphasis will be placed on dewatering by both shallow- and deep-well systems. A knowledge of the hydrogeology at a site, basic hydraulic theory, and the practical elements of well design and construction are essential. Experience always plays a major role in successful dewatering engineering. Other dewatering methods are discussed by Powers (1981) and the Water and Power Resources Service (1981).

Determining Aquifer Characteristics

The transmissivity and storage coefficient of an aquifer must be determined accurately, because the volume of groundwater in the area to be dewatered, and the rate at which it can be removed, are the two most important considerations in the design of the well system. These parameters are best calculated on the basis of a constant-rate pumping test.

In order to produce the drawdown necessary to dewater a construction site, a certain volume of water must be removed from the aquifer before construction begins. This volume can be determined from the dimensions of the projected cone of depression and the storage coefficient obtained from pumping test data. The transmissivity, also calculated from pumping test data, is used to estimate the rate at which the wells must be pumped over the long term to maintain the cone of depression.

The most practical method to obtain storage coefficient and transmissivity values from field data is to use the Jacob analysis as described in Chapter 9 and discussed further in Chapter 16. The ideal hydrogeological conditions assumed for aquifers in the basic equations of Thiem, Theis, and Jacob are rarely encountered during a typical dewatering situation. The theoretical assumptions are also violated by several well construction techniques common to the dewatering industry. These violations include the following:

1. Saturated thickness. The saturated thickness of unconfined and occasionally confined aquifers may be seriously reduced during dewatering, causing a drastic reduction in specific capacity. If the saturated thickness is reduced more than 20 percent, a correction factor must be used to predict an accurate transmissivity value or specific capacity.

2. Homogeneity. Anisotropic conditions generally prevail in near-surface materials. That is, the vertical hydraulic conductivity is much less than the horizontal hydraulic conductivity. The drawdown that can be obtained by each well may therefore be much less than for wells installed in aquifers where the vertical hydraulic conductivity is about the same as the horizontal hydraulic conductivity. Therefore, more closely spaced wells may be required in anisotropic aquifers in order to achieve effective dewatering.

3. Partial penetration. Although deep wells may be quite effective in dewatering a certain area, the total volume of water may be too great to be handled practically. Thus, many dewatering designs rely on wells that penetrate only the upper part of the aquifer. Because the head loss is greater as water is forced to flow diagonally across sediment boundaries toward the well, the specific capacity of partially penetrating wells will be less than for fully penetrating wells. As a result, more shallow wells will be required than might be estimated from the hydraulic characteristics of deep wells. Proper use of the Kozeny graph (Figure 9.35, page 250) helps predict what percentage of the specific capacity will be obtained when dewatering a confined aquifer under partially penetrating conditions. It can also be used when dewatering unconfined aquifers in which the drawdown does not exceed 30 percent of the aquifer's original saturated thickness (Powers, 1981).

4. Instantaneous release of water. Release of water from storage in unconfined conditions is usually much slower when dewatering near-surface material than is the instantaneous release assumed for ideal aquifers. If construction starts before drainage is completed, valuable time may be expended because crews must work in unsuitable conditions. Pumping test data will indicate when slow-drainage effects are no longer influencing drawdown.

5. Well efficiency. The primary aim of dewatering is to produce a certain drawdown

at selected points in an often irregular field. This generally involves the installation of multiple wells that are mutually interfering, and therefore the specific capacity is less than optimal. This contrasts with the conventional water well approach, where the number of wells is limited for a certain yield, interference is minimized or avoided altogether, and maximum well efficiency is of paramount importance. Well efficiency is usually of secondary importance in the design of well-point dewatering systems, because most of these systems are installed in highly heterogeneous sediments and are usually pumped for only brief periods. Ease of installation, ruggedness, and flexibility in design are generally more important factors than well efficiency. On the other hand, efficiency is quite important for deep wells which individually are much more expensive to construct and are usually pumped for long periods. High efficiency minimizes the number of deep wells required for a dewatering system.

If pumping test data are not available to provide estimates for transmissivity and storage, the hydraulic conductivity of the surficial material can be estimated from the graphs in Figure 22.1, provided the density of the material is known. Density is obtained from a standard penetration test (ASTM D1586) in which the number of blows per foot are recorded as a split-spoon sampler is driven by a 140-lb (63.5-kg) hammer falling 30 in (762 mm) (see Table 22.1). Transmissivity is then calculated by multiplying the hydraulic conductivity by the estimated thickness of the material to be dewatered.

Storage coefficients can also be estimated if pumping test data are not available. Storage coefficients for unconfined aquifers range from 0.01 to 0.3, and for confined aquifers they range from 10^{-5} to 10^{-3}. For coarse-grained material in an unconfined aquifer, a storage coefficient of 0.2 is generally used. If finer material is present (clays or silts), there may be more water in the pores, but the actual volume of water removed from clays and silts during dewatering may be quite small as compared to coarser grained material. For confined conditions, a storage coefficient of 10^{-5} is assumed for fine-grained sandstone/siltstone formations, whereas 10^{-3} is a good value for clean, coarse-grained sandstone.

Effective dewatering of fine sediments will require much more time and considerably closer well spacings because of significantly lower transmissivity. In practice, only enough water is removed from these fine sediments to increase the density to the point where the material is stable. Some capillary water is needed in the pore spaces to bond or hold the small grains together. Sand drains, electro-osmosis techniques, and well-point systems put under vacuum are methods used to dewater fine-grained materials. Firms familiar with these and other techniques should be consulted when fine-grained materials are to be dewatered.

Table 22.1. Soil Density from Standard Penetration Test (ASTM D1586)*

Granular Soils	Cohesive Soils
0 – 10 Loose	0 – 4 Soft
10 – 30 Medium dense	4 – 8 Medium stiff
30 – 50 Dense	8 – 15 Stiff
Over 50 Very dense	15 – 30 Very stiff

*Blows per foot of a 140-lb hammer falling 30 inches on a standard split-spoon sampler.
(Copyright ASTM, 1916 Race Street, Philadelphia, PA 19103. Reprinted with permission.)

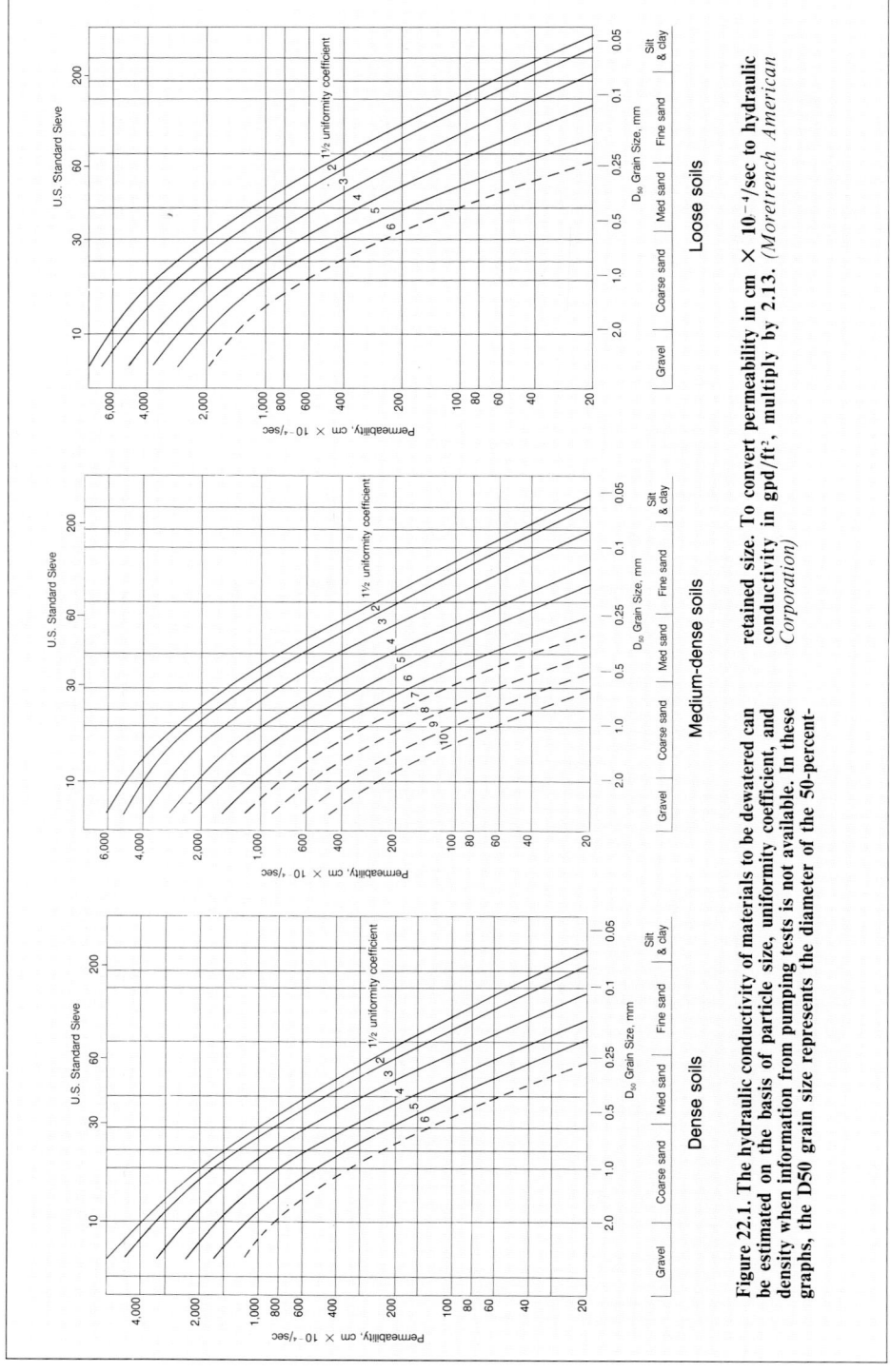

Figure 22.1. The hydraulic conductivity of materials to be dewatered can be estimated on the basis of particle size, uniformity coefficient, and density when information from pumping tests is not available. In these graphs, the D50 grain size represents the diameter of the 50-percent-retained size. To convert permeability in cm × 10^{-4}/sec to hydraulic conductivity in gpd/ft^2, multiply by 2.13. (*Moretrench American Corporation*)

Dewatering Equations

The dewatering contractor is interested primarily in the total volume of water that must be removed to create the required drawdowns throughout the construction area, and the volume of water each well will produce under dewatered conditions. In practice, the volume of water stored in the predicted cone of depression is removed before construction begins; thereafter, the wells must be pumped continuously to remove recharge to the cone. For many dewatering situations, the pumping rate required to establish a predrained condition may be far higher than the rate required to sustain the drawdown.

Under equilibrium conditions, the volume of water that a dewatering system will have to pump from an unconfined aquifer to produce a certain drawdown is given by:

$$Q = \frac{K(H^2 - h^2)}{1055 \log R/r} \qquad Q = \frac{K(H^2 - h^2)}{0.733 \log R/r} \quad (22.1)$$

where
Q = discharge, in gpm
K = hydraulic conductivity, in gpd/ft^2
H = saturated thickness of the aquifer before pumping, in ft
h = depth of water in the well while pumping, in ft
R = radius of the cone of depression, in ft
r = radius of the well, in ft

where
Q = discharge, in m^3/day
K = hydraulic conductivity, in m/day
H = saturated thickness of the aquifer before pumping, in m
h = depth of water in the well while pumping, in m
R = radius of the cone of depression, in m
r = radius of the well, in m

The depth of the water table, h, at any distance, r, from the well (r must be at least 1.5H) is given by:

$$h = \sqrt{H^2 - \frac{1055 Q \log R/r}{K}} \qquad h = \sqrt{H^2 - \frac{0.733 Q \log R/r}{K}} \quad (22.2)$$

Similar equations for confined conditions are:

$$Q = \frac{Kb(H - h)}{528 \log R/r} \qquad Q = \frac{Kb(H - h)}{0.366 \log R/r} \quad (22.3)$$

where
b = thickness of the aquifer, in ft
H = distance from the static water level to the bottom of the aquifer, in ft

where
b = thickness of the aquifer, in m
H = distance from the static water level to the bottom of the aquifer, in m

The drawdown, s, at any point within the cone is given by:

$$s = \frac{528 Q \log R/r}{Kb} \qquad s = \frac{0.366 Q \log R/r}{Kb} \quad (22.4)$$

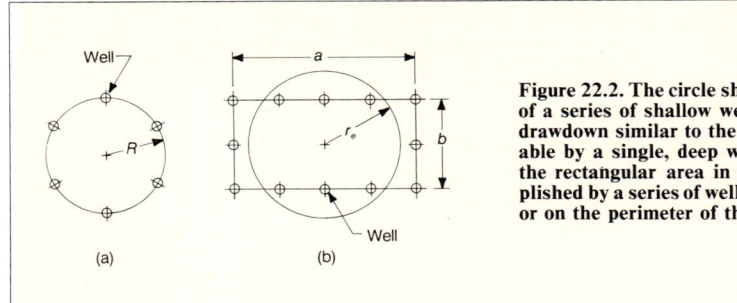

Figure 22.2. The circle shown in (a) consists of a series of shallow wells that produce a drawdown similar to the drawdown obtainable by a single, deep well. Dewatering of the rectangular area in (b) can be accomplished by a series of wells placed on a circle or on the perimeter of the rectangle.

Equations for unconfined conditions are not as accurate as those for confined conditions, because the saturated thickness, H, changes, causing a reduction in transmissivity. Yet, Equations 22.1 and 22.2 are useful in providing approximate values (see Chapter 9).

The four basic equations presented above are applicable to situations where dewatering must be accomplished in an area that can be approximated by a square or circle and where multiple shallow wells are preferred over a single deep well (Figure 22.2). For example, the circle shown in Figure 22.2a consists of a series of wells whose diameter approximates the drawdown obtainable by a single deep well. Dewatering of the rectangular area in Figure 22.2b can be accomplished by a series of wells placed on a circle, or by wells placed on the perimeter of the rectangle where the equivalent radius, r_e, is:

$$r_e = \sqrt{\frac{ab}{\pi}} \qquad (22.5)*$$

Use of Equation 22.5 provides good results when the wells are closely spaced, when the radius of influence, R, is much larger than r_e, and when the ratio a/b is less than 1.5.

Much dewatering is conducted along a line or trench (Figure 22.3). Therefore, flow occurs from a line source and moves toward a drainage trench. For confined aquifers,

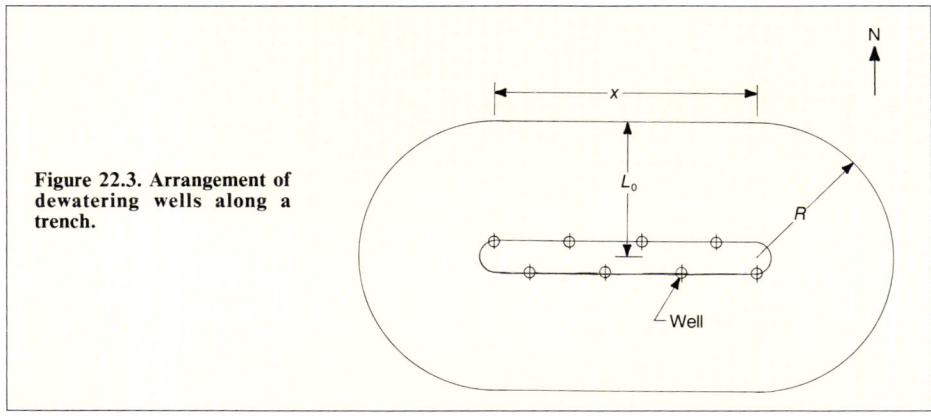

Figure 22.3. Arrangement of dewatering wells along a trench.

*An alternate equation for r_e is $a + b/\pi$, which is based on an equivalent perimeter; Equation 22.5 is based on an equivalent area.

the flow from one side of the trench per unit length is given by:

$$\frac{Q}{x} = \frac{Kb\,(H - h)}{1440\,L_0} \qquad\qquad \frac{Q}{x} = \frac{Kb\,(H - h)}{L_0} \qquad (22.6)$$

where
x = unit length of the trench, in ft
L_0 = distance from point of greatest drawdown to point where there is no drawdown, in ft

where
x = unit length of the trench, in m
L_0 = distance from point of greatest drawdown to point where there is no drawdown, in m

For unconfined aquifers, the flow from one side of the trench per unit length is given by:

$$\frac{Q}{x} = \frac{K\,(H^2 - h^2)}{2880\,L_0} \qquad\qquad \frac{Q}{x} = \frac{K\,(H^2 - h^2)}{2\,L_0} \qquad (22.7)$$

For a trench of finite length, the volume of water that must be pumped can be calculated by combining Equations 22.1 and 22.7:

$$Q = \frac{K\,(H^2 - h^2)}{1055\,\log R/r} + 2\,\frac{xK\,(H^2 - h^2)}{2880\,L_0} \qquad Q = \frac{K\,(H^2 - h^2)}{0.733\,\log R/r} + 2\,\frac{xK\,(H^2 - h^2)}{2\,L_0}$$

In practice, dewatering proceeds along the trench as the excavation is prepared. Thus, each well should be able to pump the larger volume anticipated when the well is at either end of the dewatered area.

Factors in Selecting a Dewatering System

Selection of which major type of dewatering design to use — well points or deep wells — depends on many factors. Some of these factors are hydrogeologic conditions at the site, length of time pumping is required at a site, volume of water to be removed, whether pumping equipment can be installed in the construction area, and availability of drilling and dewatering equipment. Contractor experience is another important factor in selecting the dewatering design.

Characteristics of the water-bearing formation that must be determined before designing a dewatering system are:
1. Whether the aquifer is confined or unconfined.
2. Transmissivity and storage coefficient of the aquifer.
3. Static water level.
4. Depth and thickness of the aquifer.
5. Sources of recharge to the aquifer, and location of these sources.

Information required about the construction site include the following:
1. Dimensions of the area to be dewatered.
2. Depth to which the water levels must be lowered.
3. Plans for disposal of water pumped from the wells.
4. Whether the installation will be permanent or temporary.

The above information, coupled with a firm knowledge of well hydraulics and groundwater-flow theory, allows the contractor to work out a satisfactory design incorporating a minimum number of wells with optimum pumping rates, well depths, and well spacing. In any dewatering design, there is generally a maximum drawdown that can be permitted in each pumped well, and simultaneously there is a minimum drawdown that must be achieved everywhere within the excavation area. For large projects, selecting the proper number of wells, the well spacing, and the well yield, while staying within these two limits, may require thousands of calculations.

Before the use of digital computers in dewatering design, all calculations and comparisons had to be done by hand. Because of time limitations, the design engineer frequently was forced to select the first solution that appeared to work, rather than the optimal design, which would have accomplished the dewatering goals at less cost.

Well-Point Dewatering Systems

Well-point systems are groups of closely spaced wells, usually connected to a header pipe or manifold and pumped by suction lift. Trenches for sewers and water mains and excavations for foundations often have to be carried down into saturated soil below the water table. In most cases, it is more economical to dewater with well points, rather than surround the excavation with a continuous wall of sheetpiling and pump from within the work area. Lowering the water table by pumping a well-point system installed adjacent to the excavation permits working with heavy equipment even in the formerly saturated zones. Proper dewatering also eliminates the hazards of sand boils developing in the bottom of the excavation.

Well-point systems are frequently used because they are relatively easy to install and are adaptable to a wide range of site conditions. They are particularly suited to the following situations:

1. Single-stage dewatering, that is, dewatering jobs where the pumping water level is within suction lift.
2. Formations with low transmissivity (fine sand and silt).
3. Shallow formations that overlie impervious formations.

During operation of a well-point system, a central pump lifts water from each well by producing a partial vacuum in the header and riser pipes. The partial vacuum, or suction lift, that the pump can maintain determines the drawdown that can be obtained in the water-bearing formation. The maximum drawdown is the difference between the suction head and the static water level (Figure 22.4). A pump that has good suction characteristics should be selected to maximize drawdown. In practice, the greatest suction lift

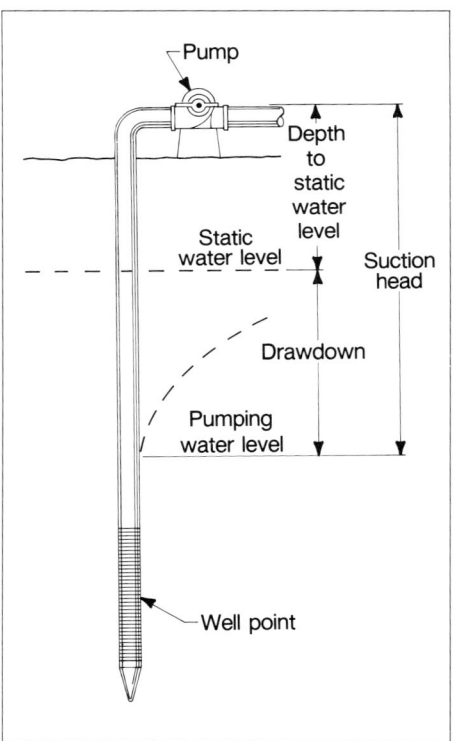

Figure 22.4. When pumping wells by suction lift, the maximum drawdown obtainable is the difference between the suction limit of the pump and the distance from the static water level to the center of the pump. Most well-point systems are engineered on the basis of 15 ft (4.6 m) of suction lift.

that can be developed is about 20 to 22 ft (6.1 to 6.7 m). Some centrifugal pumps, however, will not lift water more than 15 to 16 ft (4.6 to 4.9 m). Thus, most well-point systems are engineered on the basis of 15 ft of suction lift.

The yield from fine soils will always be low — so low, in fact, as to raise the question of whether practical dewatering can be achieved. Experience shows, however, that removal of even a small volume of water will stabilize the material in many cases. With the water moving away from the open slopes of the excavation under the constant driving head created by the well-point system, the surface tension created by the pore water helps stabilize the banks.

The diameter of well points used in dewatering systems is usually either 1½ or 2 in (38 or 51 mm); yield is 10 to 25 gpm (54.5 to 136 m^3/day). Points are typically spaced 3 to 12 ft (0.9 to 3.7 m) apart, depending upon the hydraulic conductivity of the saturated formation, the depth to which the water table must be lowered, and the depth to which the well points can be installed in the water-bearing formation (Powers, 1981). For fine-grained materials, the spacing may be less.

For a given aquifer, the actual spacing of the well points depends on the type of point used, the friction losses in the system, and the volume of water to be removed. For a major dewatering operation, total flow requirements should be estimated from constant-rate pumping tests. The point spacing is then based on the total volume of water to be removed, the average drawdown available, and the anticipated specific capacity. The actual volume of water obtained from each point at this spacing will be reduced somewhat because of head losses in the screen, casing, and header system. Use of smaller diameter points will result in greater friction losses as a rule. In general, pipe-based points that use mesh for slot control have much higher friction losses than do continuous-slot points. If the indicated spacing of the points is closer than 7 ft (2.1 m) when friction losses are taken into account, the contractor may select larger diameter points or consider deep wells.

Most well-point dewatering systems for small projects are installed with little attention to a site-specific design. Length of the screen, and diameter of the screen and pipe, are usually dictated by equipment on hand, which is reused on each project. Spacing of the individual points is generally based on field experience for similar terrains. Because conditions change from one part of a site to another, however, a certain amount of trial and error is required to achieve the necessary drawdown.

Creating the Proper Cone of Depression

Lowering the groundwater level throughout a construction site involves creating a composite cone of depression by pumping from the well-point system (Figure 22.5). The wells must be placed close enough so that they interfere with each other and thus pull the water table down a certain distance at intermediate points between wells. Figure 22.5 shows how the overlapping areas of influence around two small wells produces an enhanced drawdown of the water table. The water level will remain at the levels indicated as long as pumping is continued. Water is drained by gravity from the formation above the lowered water table, and excavation can then take place anywhere within the composite cone of depression.

Complete dewatering of the composite cone of depression will occur well after pumping begins. Although maximum drawdown in the saturated formation around each well point can be obtained in several hours, additional time is required for

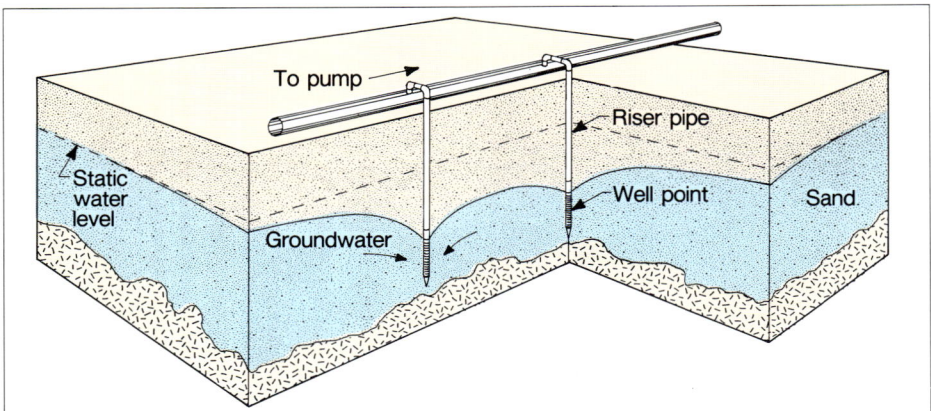

Figure 22.5. Mutual interference between two or more wells lowers the water table for dewatering operations.

vertical drainage of all the water from the saturated zone. In practice, this time lag makes it necessary to start pumping from the well-point system a day or more before excavation begins.

As explained above, the maximum drawdown that can be maintained adjacent to each well is the vacuum or suction head developed by the pumping equipment, minus the distance from the center of the pump to the static water level, and minus the head losses in the piping and well points. The suction head of the pump thus imposes a definite limitation on the depth to which the formation can be dewatered by a single group of well points. An additional reduction in the water table can be achieved by excavating initially almost to the static water level, and then setting the pumps at this lower elevation.

Suction head limitations can also be overcome by using two or more separate systems installed in successive stages. The first-stage system is installed as shown in Figure 22.6. Pumping it lowers the water table enough to permit excavation several feet below the original groundwater level. The second-stage system is then installed at the lower level as illustrated in Figure 22.7. The well points, header pipe, and pumps of the second stage are set as far below the first stage as possible. The second-stage system then lowers the water level so the excavation can be completed. Additional stages may be added if deeper dewatering is needed.

Pumping the second-stage system may drain the soil around the first stage of well points. If this happens, operation of the first stage is stopped. Pumps and other parts of the first stage can then be used on the third- or fourth-stage systems, but they will probably have to be reinstalled after the lower stages have been removed. Under favorable conditions, the water level can be pulled down in steps of about 15 ft (4.6 m) by stage dewatering. Steps of 10 to 12 ft (3 to 3.7 m) per stage are more typical, however, and represent average conditions.

For trench excavations, the contractor must make sure that the current excavation is dry, that the future area of excavation will be dewatered by the time the equipment reaches it, and that the most recently completed part of the construction (trench) is kept dewatered. The length of the point system may be 4 to 8 times the daily progress (Powers, 1981). Three typical dewatering schemes, using well points, are shown in

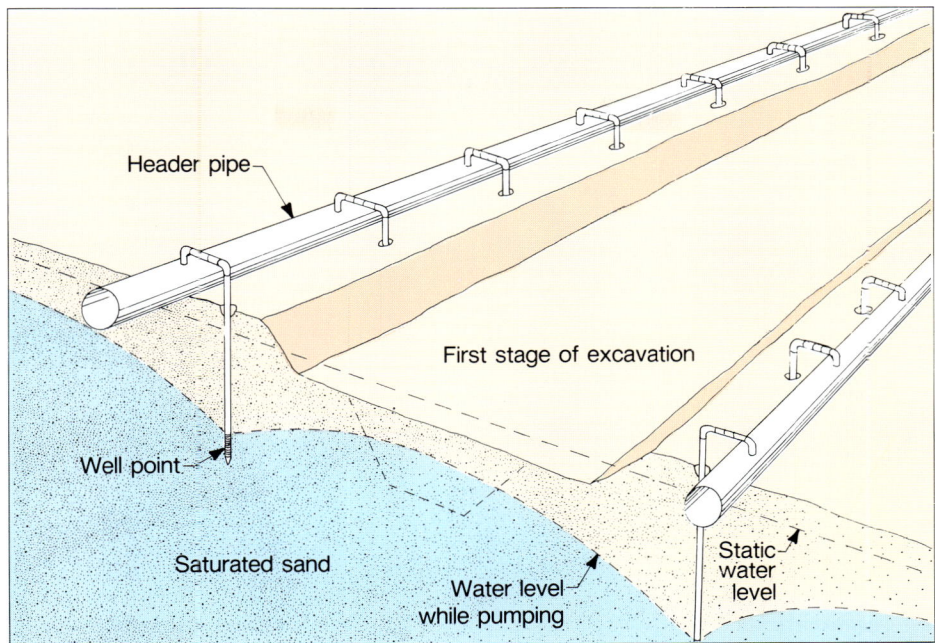

Figure 22.6. One or more stages of a well-point dewatering system may be required, depending upon the depth to which the water table must be lowered. In this figure, the first stage in the dewatering process is illustrated.

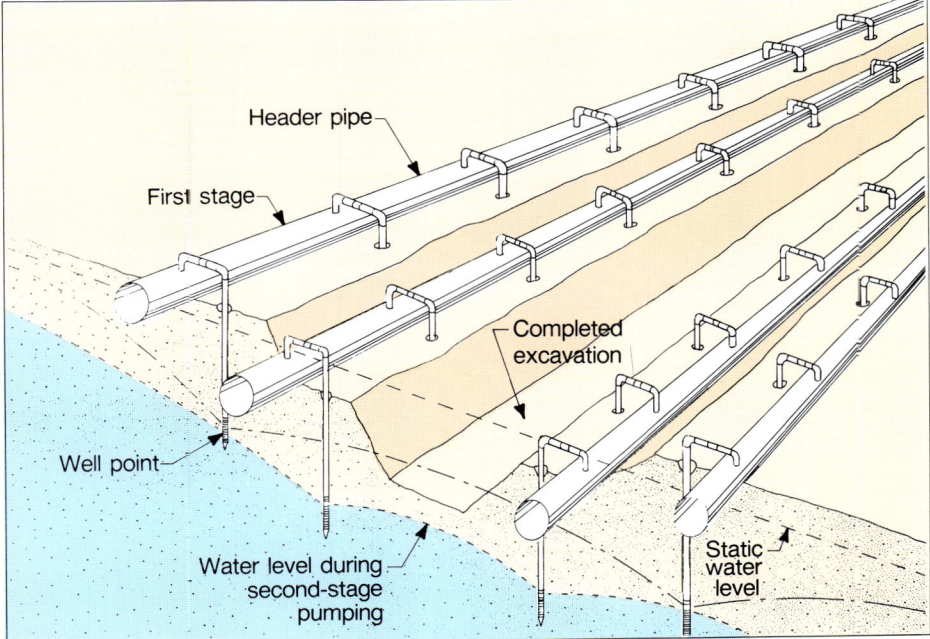

Figure 22.7. A second stage of well points can be installed at a lower level after the initial dewatering and excavation. Pumping the second stage then lowers the water table to the desired depth.

Figure 22.8 for varying subsurface conditions.

The presence of tight layers of silt or clay at various depths in the saturated soil complicates the design of the well-point system, because the layers prevent vertical drainage of the formation lying above them. Even though these layers may be as thin as 1 in (25.4 mm), they may be nearly impermeable. Successful dewatering of a

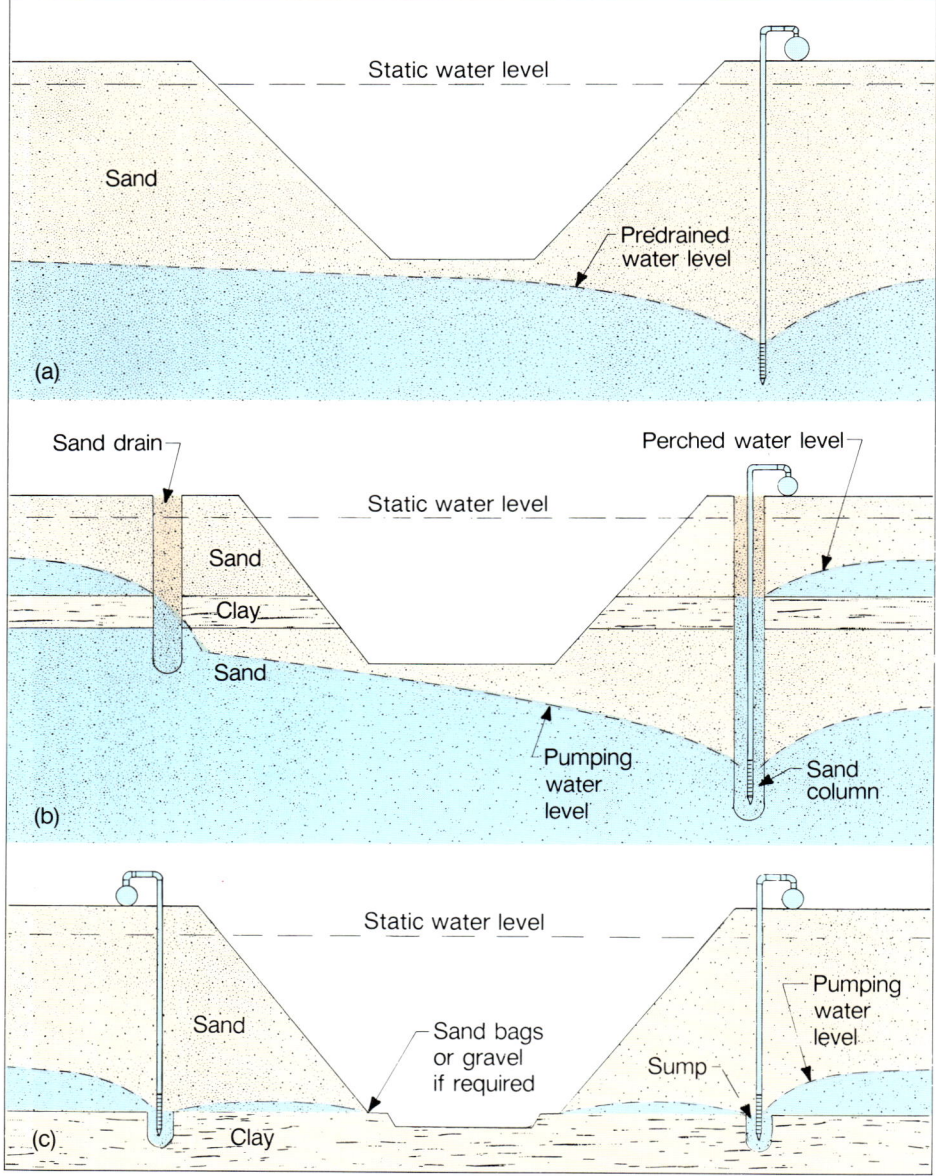

Figure 22.8. Trench dewatering with well points and sand drains. (a) Uniform sediments can be dewatered with well points on one side. (b) A clay layer above subgrade may require sand drains on the opposite side to handle perched water. (c) Clay at and below subgrade may require well points set partially into the clay on both sides of trench. *(Powers, 1981)*

stratified formation requires that each layer of sand be drained separately. Two wellpoint systems may be required where two layers of saturated sand are separated by an impermeable layer of clay. In this case, one set of well points may be sunk to the top of the clay underlying the upper sand and the other set carried down to the desired depth below the impermeable layer in the lower sand.

Where slow drainage is not too troublesome, vertical drainage channels (sand drains) that pierce the impermeable layer can be provided by filter packing each well point and riser pipe in holes 8 to 12 inches (203 to 305 mm) in diameter. A temporary casing is jetted down through all the strata, the well point is centered in it, and the annular space is filled with clean, coarse sand up to the static water level. The outer casing is pulled up as the filter pack is installed. The filter pack will sometimes provide the vertical drainage desired, although this does not always work because a film of silt and clay may form on the wall of the large hole by the "troweling" action of the temporary casing as it is withdrawn. This film cannot be removed above the top of the well point by any development method. In many cases, sand drains are installed without well points to enhance the dewatering process where only a few well points are required to remove the water.

Depth of Setting

The depth at which dewatering well points should be installed is determined by three factors: (1) the maximum depth of the proposed excavation, (2) the existence of clay or rock at this depth, and (3) the existence and depth of impermeable materials above subgrade. If the water-bearing formation extends several feet below the bottom of the excavation, the well points should be sunk deep enough so that their tops are at least 3 ft (0.9 m) below the deepest part of the excavation. If only a single line of well points is to be used for trench excavation, the tops of the well points should be 4 ft (1.2 m) below the trench bottom. It is preferable to set the well points at even greater depths below the bottom of the excavation if conditions permit. Deeper settings insure that the entire length of each screen will always be below the lowest pumping level.

If the site is to be excavated completely through the saturated formation to underlying clay or rock, some special steps need to be taken in setting the well points to insure complete dewatering of the bottom 1 to 2 ft (0.3 to 0.6 m) of the formation. If the well points are set just above the impermeable layer, air may enter the well screen before the water level can be pulled down to the proper depth, causing the pump to break suction. Air enters either at the top of the screen or, if the well points are provided with center drop tubes, at about 1 ft (0.3 m) above the bottom of the saturated formation. Complete dewatering of a formation overlying an impermeable layer is more effective if the well points can be set deeply enough so the pumping level is at the bottom of the aquifer. This is accomplished by filter packing the well points in larger diameter holes cut 2 to 3 ft (0.6 to 0.9 m) into the underlying materials (Figure 22.8c). If the underlying formation is rock or hard shale, it may be necessary to use a well drilling machine to drill the sumps. Although the use of sumps is quite effective, the extra costs imposed on the project should be carefully evaluated.

When highly stratified sediments are anticipated, test wells should be installed to determine the most effective depth settings. The yield of the water-bearing formation should be checked at different depths. A well point can be set to maximum depth first, then pulled up a few feet for a second test. In making the pumping test, a vacuum gauge should be used on the intake of the pump to measure the suction head. The pumping rate is determined by measuring the time required to fill a container of known volume.

Adequate screen length is important in achieving maximum yields. Usually, point systems only partially penetrate the aquifer, resulting in a significant loss in well efficiency. The use of longer screens will increase the efficiency and yield remarkably. Screens should always be set in the most permeable part of the formation if possible.

Screen Configuration and Installation Procedures

Proper selection of a well point can make a large difference in yield and long-term maintenance. Both drive-type and jetting-type well points are used for dewatering work. Because no real development is possible when installing 2-in (51-mm) points, high open area screens are advantageous if the yield is to be maximized. The well points must also be ruggedly constructed to withstand repeated installation and removal. See Chapter 12 for a discussion of well point types.

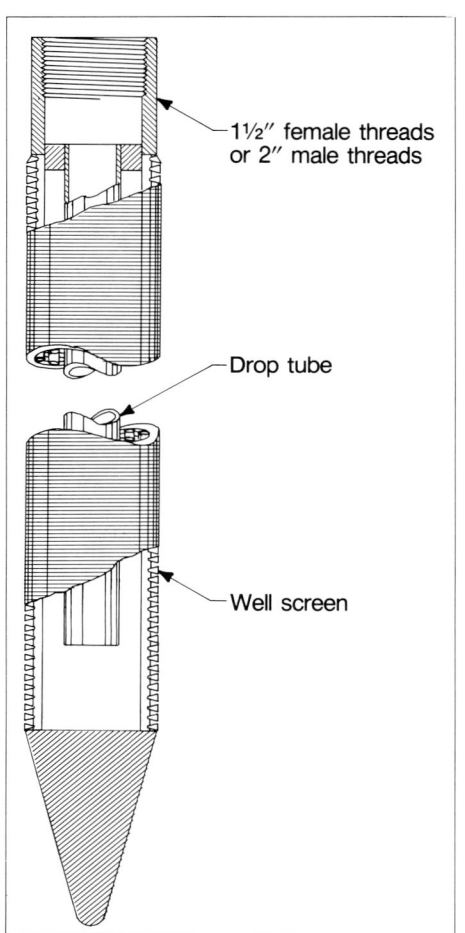

Figure 22.9. A drive-type dewatering well point with center drop tube allows pumping the water level down to near the bottom of the screen without breaking suction.

Both drive-type and jetting-type points can be equipped with a center drop tube if desired (Figure 22.9). The center drop tube serves two useful purposes. First, for a jet-type point, it permits jetting the well point down without running a temporary jetting pipe inside the riser and screen. Second, the drop tube will permit pulling the water down to the bottom of the tube before breaking suction. This is obviously desirable where the well points must be set just above an impermeable layer. One disadvantage of the center drop tube is that it causes additional head loss in the well point. The water entering the screen must flow down outside the tube and then turn 180 degrees to enter and flow up through the inside of the tube. The tube also prevents any effective development of the formation around the well screen.

Well points may be driven to the desired depth by hand or machine, or they may be sunk by one or more jetting methods. Procedures for driving well points are described in Chapter 14. In most dewatering projects, the points are jetted into place.

Well points constructed with self-closing bottoms can be jetted down in sand or other loose formations (Figure 22.10). A temporary jetting pipe is used to deliver the jetting stream at full force through the self-closing bottom. For 1¼-in (31.8-mm) well points, ½-in (13-mm) standard pipe should be used for the jetting line, whereas 1-in (25.4-mm) standard pipe is best for 2-in (51-mm) well points. The jetting pump should have a capacity of about 60 to 100 gpm (327 to 545 m³/day) at 50 to 80 psi (345 to 552 kPa) pressure. Under favorable conditions, a smaller pump may work satisfactorily, but it is always a good idea to have excess pump capacity. Powers (1981) recommends attaching a jetting chain (Figure 22.11) to the bottom of the screen to enlarge the hole so a filter pack can be installed (Figure 22.12).

For wells 20 to 40 ft (6.1 to 12.2 m) deep, the well point and riser pipe are jetted down as a unit. A section of riser pipe

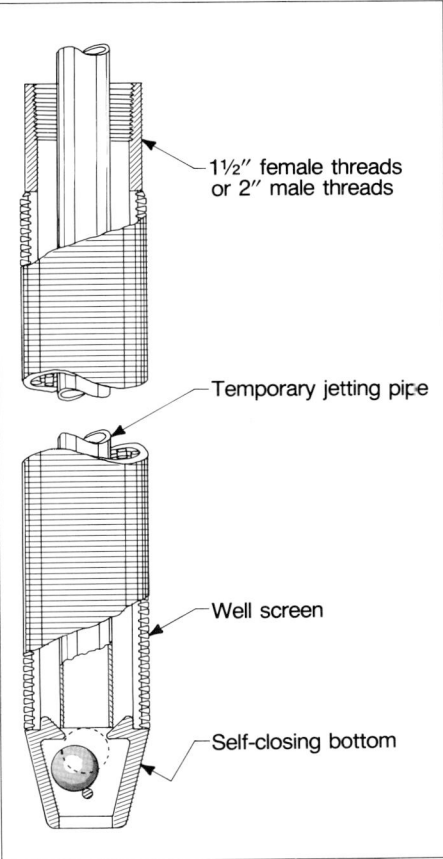

Figure 22.10. Well point with a self-closing bottom of the floating-ball type. Temporary jetting pipe is used to sink the well point and riser pipe to the desired depth.

10 to 20 ft (3 to 6.1 m) long is first coupled to the well point. The jetting pipe is then placed inside the riser pipe and well point so that the bottom of the jetting pipe rests on the self-closing valve. The discharge side of the pump is connected to the jetting pipe by a length of pressure hose at least 1 inch (25.4 mm) in diameter. A valve is usually placed in the line from the pump discharge. All connections are made up tight.

To start the jetting operation, dig a hole about 1 ft (0.3 m) deep, set the well point and riser pipe assembly vertically in the hole, and partially open the valve on the

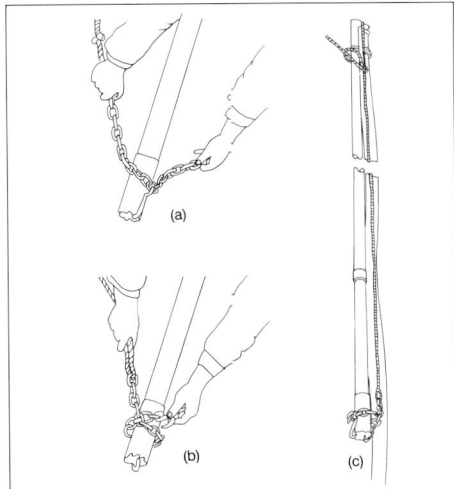

Figure 22.11. A jetting chain attached to the screen and casing helps ream a larger borehole to facilitate filter packing. (a) The middle hook is first attached to the bottom of the point. (b) The end hook is passed over the chain in the form of a half-hitch and placed on the opposite side of the middle hook. (c) A rope and spring-loaded hook secure the chain. The well point with chain in position is now ready for installation. *(Moretrench American Corporation)*

discharge of the pump. As sediment is washed out below it, the well point will sink slowly; up and down movement of the whole assembly will enhance the penetration rate. As jetting continues, increase the water flow by opening the pump valve. If jetting has to be stopped to put on an additional length of riser pipe, the hole around the riser pipe must be kept full of water to prevent caving until jetting is resumed.

To filter pack the well point, the volume of water should be reduced to slow

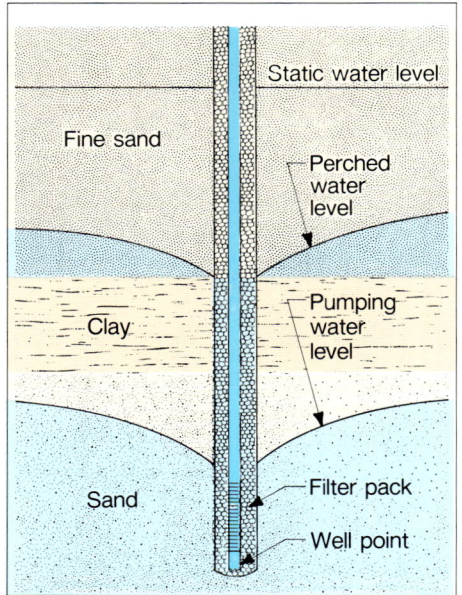

Figure 22.12. The oversized borehole reamed with a jetting chain permits placement of a filter pack to improve vertical drainage from the formation overlying the clay layer and to prevent sand pumping and screen clogging.

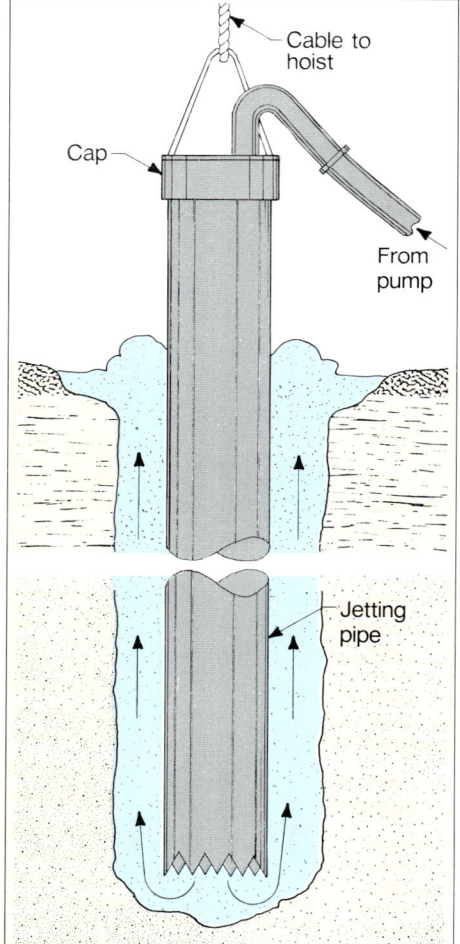

Figure 22.13. Well-point and riser-pipe assembly can be dropped into borehole opened by jetting with separate jetting pipe.

the return flow. Clean, coarse sand is then shoveled into the hole around the riser pipe. The filter sand will fall to the bottom of the jetted hole around the well point while the slow return flow keeps the hole open.

Another jetting method used for well-point installation is illustrated in Figure 22.13. A string of pipe is equipped with a tight cap at the top and roughly cut teeth at the bottom. The pump delivers the jetting stream through a connection in the cap, and a hole is created at the bottom of the pipe. If tight layers of silt and clay are encountered, the pipe is turned back and forth so the teeth break up the harder material. When the hole has been opened to the proper depth, the well point and riser pipe assembly may be dropped down alongside the jetting pipe, or the cap may be removed and the assembly dropped down through the jetting pipe. The jetting pipe is then pulled out of the hole.

Development

When the well point has been installed to the desired depth, the well should be developed to maximize yield. The well can be developed by surging water in and out through the screen openings, to remove silt and fine sand from the formation. A loose plunger worked up and down below the water level inside the riser pipe will surge the well, or a used pump can be connected to the riser pipe and alternately stopped and started to produce a surging action. Pouring water into the well from time to time, to backwash it, also helps to develop and stabilize the formation around the screen. If an air compressor is available, it also can be used to surge the well. Pumping the well periodically will indicate when most of the finer sediment around the screen has been removed.

As soon as the water has cleared, it is important to check for sand inside the well point. Any sand that may have settled in the bottom of the screen must be removed. Sand can be removed by air-lift pumping (see Chapter 15) or, if only a few wells are to be used, by a hand pump attached to a string of ½-in (13-mm) pipe. Pour water into the well bore around the ½-in pipe while operating the hand pump. Continue lowering the pipe as the sand is removed.

Piping, Connections, and Pumps

An important objective in choosing pipe sizes for the riser pipe, well casing, or suction header is to make them generously large so that friction losses in the system are kept to a minimum. This allows more of the total suction head to be used to produce drawdown in the wells. The net effect is to increase the yield of the system. If the drawdown in a well can be increased from 9 ft (2.7 m) to 10 ft (3 m), for example, the yield will go up about 10 percent.

No two wells in the system will pump exactly alike. Hence, it is often necessary to regulate the yield and drawdown in one or more of the wells by adjusting valves installed in the branch connection of each well to balance the system. This regulation process is called "tuning" the system. Ideally, each point should produce its highest yield with minimum air entrainment. The drawdown must be kept above the screen or entry holes of the drop tube. The valve also permits isolating each well for repair or cleaning without shutting down the rest of the wells.

The connecting fittings between the header pipe and each well-point riser pipe are called swing joints. The joints may be made by using 1½- or 2-in (38- or 51-mm) pipe fittings. The connection may be swung horizontally and vertically to meet the top of the riser pipe. Such flexibility is important because the well point and its riser pipe are always a few inches off

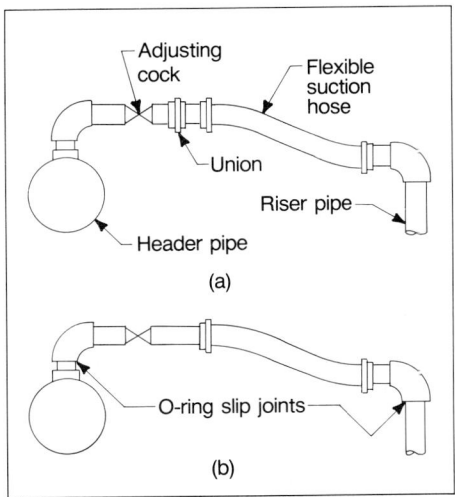

Figure 22.14. Two basic types of well-point connections: (a) standard well-point swing and (b) no-tool swing. The swing configuration is preferred because the well point and header can always be easily connected, even if the header and well point are not spaced properly. *(Powers, 1981)*

the calculated position.

Figures 22.14b and 22.15 show a flexible type of swing joint which avoids the necessity of making up any threaded connection when installing a well-point system. The fittings consist of a smooth, tapered nipple screwed to an elbow at the top of the riser, a length of rubber suction hose, and a second tapered nipple connected to each opening in the header pipe. A shut-off (adjusting) cock is put between the header-pipe tap and the tapered nipple on the header side of the swing joint. When making the connection from the riser to the header, it is only necessary to slip the ends of each rubber suction hose over the two tapered nipples. The vacuum developed during pumping pulls these sleeve joints together so they are airtight. The hookup can be made in a matter of seconds. Installing the points and connecting them to the header pipe are often repeated hundreds of times during a single dewatering project. Reducing the time to connect each unit to the header by only 2 or 3 minutes may become a significant factor in reducing labor costs.

Single or multiple pumps may be used in a single dewatering system; if multiple pumps are used, they can be connected to the system at a central location or be spaced along the headers. To prevent accidental flooding of the excavation during an unintentional pump shutdown, one or more standby pumps (preferably gasoline or diesel powered) are usually installed. They can be started immediately if required.

Figure 22.15. Flexible swing joints are completed by slipping ends of suction hoses over tapered nipples attached to riser and header pipe.

To make the dewatering program more efficient, a centrifugal pump is installed in combination with an auxiliary vacuum pump (Figure 22.16). The vacuum pump is needed to evacuate air from the header system and cause water to rise by suction lift from the well points. The actual lift obtained by the vacuum is controlled by the elevation at the site and by the con-

Figure 22.16. Centrifugal pump installed in combination with an auxiliary vacuum pump.

dition of the pump and well-point system. Friction in the piping system also limits lift.

Header pipe, or aluminum or PVC irrigation pipe, may be used to transport water away from a dewatered site. Ordinarily, the water is conducted to a natural drainage system that will carry it farther away from the work area.

Deep-Well Dewatering Systems

At many construction sites, deep high-capacity wells are used instead of well-point systems. In deep-well installations, each well usually produces many times the quantity of water produced by an individual well point in a well-point system. Design and construction techniques for high-capacity water supply wells have been discussed at length in earlier chapters. These same principles apply to deep wells used for dewatering. Deep-well dewatering schemes work best in the following situations:

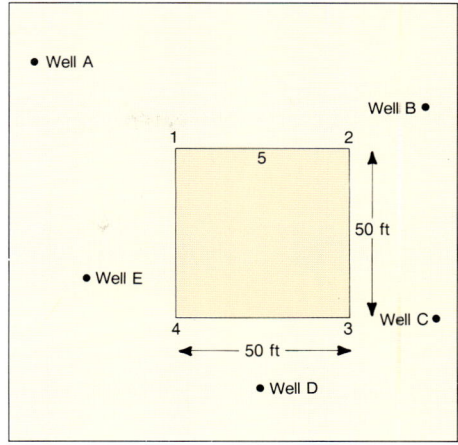

Figure 22.17. Proposed siting of 5 wells to dewater the shaded area.

1. Environments where pumping water levels or dewatered levels in the excavation are beyond the suction-lift capacity of centrifugal pumps, or where multistage well-point dewatering is impractical.

2. Formations in which the transmissivity exceeds 25,000 gpd/ft (310 m^2/day). When the transmissivity is high, the drawdown will be small in relation to the discharge. In tighter aquifer materials, the cone is much steeper and the radius is small.

3. Formations that extend to depths considerably below the depth of the proposed excavation. These formations permit installation of deeper wells, provide more available drawdown, and produce larger yields from fewer wells.

An optimum deep-well dewatering system is one that utilizes a minimum number of efficient wells pumped at proper rates and spaced as strategically as practicable around the job site. In some situations, it is not possible to arrange the wells uniformly enough to simulate the actions of a single well by a series of smaller wells arranged evenly about the hypothetical large well. In these cases, the cumulative-drawdown method provides accurate information on drawdowns within the area of interest if a pumping test has been conducted.

For the example shown in Figure 22.17, the area to be dewatered (shaded area) consists of fine to medium sand deposited in a former floodplain. The depth to the water table is 5 ft (1.5 m) and the saturated thickness is 100 ft (30.5 m). Five deep, 8-in (203-mm) wells are tentatively considered to be enough to dewater the shaded area to the necessary 12 ft (3.7 m). This project area is near the center of a small city and has buildings on three sides. No wells or pumping equipment can be operated in or near the excavation during construction. Therefore, a deep-well installation was selected. The exact siting of the dewatering wells is controlled by existing buildings and roads, and by other environmental considerations.

The transmissivity is calculated to be 10,560 gpd/ft (131 m^2/day) by plotting data

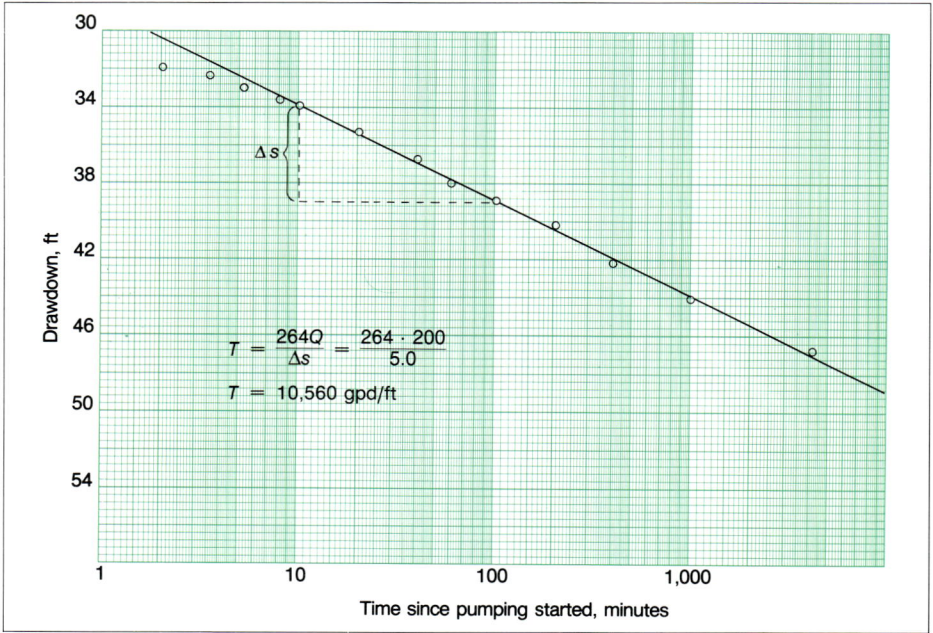

Figure 22.18. Time-drawdown graph for test well.

from a constant-rate pumping test on a time-drawdown diagram (Figure 22.18). Although the transmissivity can be calculated accurately on the basis of a pumping test, it is not possible to predict the specific capacity of a well. The following empirical equation can be used, however, to provide an estimate:

$$\frac{Q}{s} = \frac{T}{1,500} = \frac{10,560}{1,500} = 7 \text{ gpm/ft of drawdown (125 m}^3\text{/day/m)}$$

If each well is pumped for 1 day at 135 gpm (736 m³/day), the drawdown in the pumping well would be approximately 25 ft (7.6 m), assuming no interference from the other wells, a well efficiency of 75 percent, and no loss in specific capacity as the well is drawn down (caused by anisotropic conditions or reduction in saturated thickness). The drawdown just outside the well bore is calculated on the basis of specific capacity and yield of the well (Q/s = 7 gpm; total drawdown = 135/7 = 19.3 ft).

The drawdown data are replotted on a distance-drawdown graph as explained in Chapter 9 (Figure 22.19). Recall that the slope (Δs) calculated from the time-drawdown data is doubled when transferred to the distance-drawdown plot. Furthermore, the slope of the distance-drawdown graph is adjusted for the assumed pumping rate of 135 gpm. The distance-drawdown plot is drawn, using the known drawdown outside the well bore and the slope calculated from the time-drawdown plot. A table is then constructed using this graph to determine the drawdowns at the corners of the dewatered (shaded) area produced by the 5 wells (Table 22.2). The drawdown at any point within the excavation is the sum of the individual drawdowns produced by all the wells in the system at that point. The drawdown required for the construction must be 12 ft (3.7 m) within the shaded area. Thus, it is clear from Table 22.2 that there will be no problem at any of the corners of the area, because the minimum

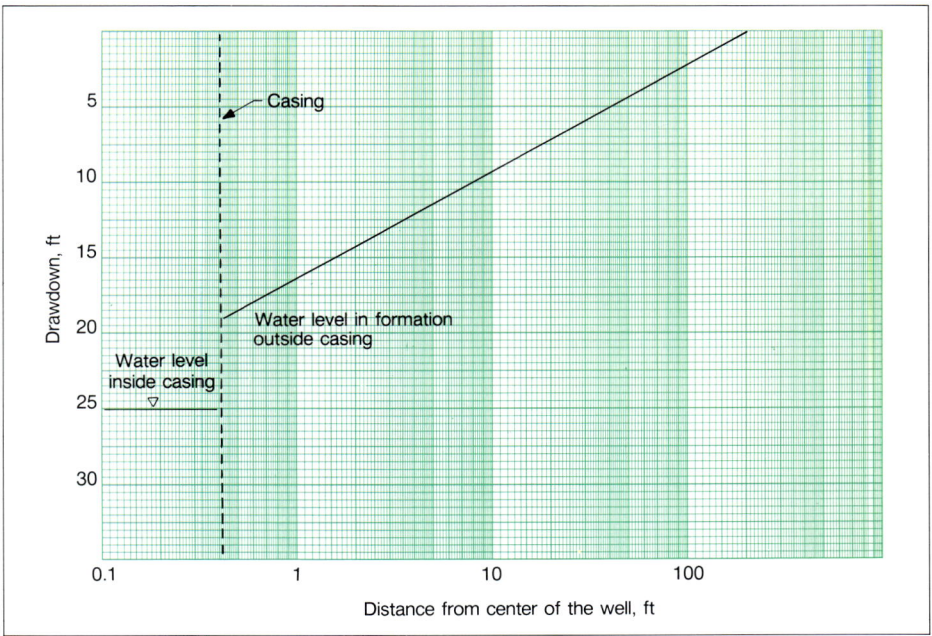

Figure 22.19. Distance-drawdown graph drawn from data obtained from time-drawdown graph.

drawdown is 17.1 ft (5.2 m) at Point 1. Although Point 1 appears to be the critical point, that is, a point where the dewatering may not be sufficient, it is easy to calculate the cumulative drawdown at any other point that appears, from the diagram, to be a problem. For example, assume that Point 5 may be another critical point. Using the same procedure, the total drawdown at Point 5 is calculated to be 17.9 ft (5.5 m).

For illustration purposes, 5 wells were selected for this problem, but it is clear that fewer wells could create the necessary drawdown. Table 22.3 shows that if Well A is eliminated, the drawdown is still enough at the indicated pumping rate to permit construction. Three wells may be sufficient if the pumping rates were increased, although the drawdown may extend into the screen from time to time. Aeration of the screens for extended periods can lead to higher incrustation rates, cascading water, and reduced hydraulic efficiency.

This graphical procedure is simple to use and is an accurate indicator of anticipated drawdown as long as the saturated thickness does not decrease substantially. If the saturated thickness decreases more than 20 percent, the projected drawdowns must

Table 22.2. Drawdown for Various Points in Figure 22.17 when 5 Wells are Pumping

Well	Point #1 Distance from well ft m	Point #1 Drawdown ft m	Point #2 Distance from well ft m	Point #2 Drawdown ft m	Point #3 Distance from well ft m	Point #3 Drawdown ft m	Point #4 Distance from well ft m	Point #4 Drawdown ft m	Point #5 Distance from well ft m	Point #5 Drawdown ft m
A	50 15.2	4.3 1.3	97 29.6	2.2 0.7	120 36.6	1.5 0.5	87 26.5	2.5 0.8	75 22.9	3.2 1.0
B	75 22.9	3.0 0.9	25 7.6	6.4 1.9	65 19.8	3.5 1.1	95 29.0	2.3 0.7	47 14.3	4.4 1.3
C	90 27.4	2.5 0.8	57 17.4	3.8 1.2	25 7.6	6.4 1.9	75 22.9	3.0 0.9	65 19.8	3.5 1.1
D	80 24.4	2.8 0.8	80 24.4	2.8 0.8	35 10.7	5.3 1.6	35 10.7	5.8 1.8	70 21.3	3.2 1.0
E	45 13.7	4.5 1.4	85 25.9	2.5 0.8	77 23.5	2.8 0.8	28 8.5	6.0 1.8	60 18.3	3.6 1.1
TOTAL DRAWDOWN		17.1 5.2		17.7 5.4		19.5 5.9		19.6 6.0		17.9 5.5

Table 22.3. Drawdown for Various Points in Figure 22.17 when 4 Wells are Pumping

Well	Point #1		Point #2		Point #3		Point #4		Point #5	
	Distance from well ft m	Drawdown ft m	Distance from well ft m	Drawdown ft m	Distance from well ft m	Drawdown ft m	Distance from well ft m	Drawdown ft m	Distance from well ft m	Drawdown ft m
B	75 22.9	3.0 0.9	25 7.6	6.4 1.9	65 19.8	3.5 1.1	95 29.0	2.3 0.7	47 14.3	4.4 1.3
C	90 27.4	2.5 0.8	57 17.4	3.8 1.2	25 7.6	6.4 1.9	75 22.9	3.0 0.9	65 19.8	3.5 1.1
D	80 24.4	2.8 0.8	80 24.4	2.8 0.8	35 10.7	5.3 1.6	35 10.7	5.8 1.8	70 21.3	3.2 1.0
E	45 13.7	4.5 1.4	85 25.9	2.5 0.8	77 23.5	2.8 0.8	28 8.5	6.0 1.8	60 18.3	3.6 1.1
TOTAL DRAWDOWN		12.8 3.9			15.5 4.7		18.0 5.4		17.1 5.2		14.7 4.5

be adjusted by the procedure presented in Appendix 9.C. More complex problems should be solved by using a microcomputer so the number of wells can be minimized.

Some dewatering contractors recommend that deep wells not be placed to the bottom of the aquifer, because more water may have to be pumped to achieve the required drawdown. In many alluvial systems, for example, the hydraulic conductivity of lower sediments may be higher than for the more recent overlying sediments. If the screens are not set to the bottom of the aquifer, however, it is impossible to use the predictive methods described above, because they are predicated on the assumption that the wells are fully penetrating. That is, the transmissivity value is established on the total aquifer thickness, with no consideration of partial-penetration effects.

Decisions on the number of wells required, well depths and diameters, and other construction details will depend on numerous site-specific requirements and conditions. The time required to drain the aquifer can be predicted on the basis of the pumping test. Wells must be able to pump the required volumes of water from storage and to maintain drawdowns, even during periods of excess recharge (heavy rainfall events or temporary rises in nearby rivers).

The most critical factor involved in dewatering is the ability to predict drawdown at various pumping rates. Because specific capacity may fall off significantly as drawdown increases and the saturated thickness decreases, the results from a step-drawdown test may be helpful in gauging actual well yield under dewatered conditions. Careful analysis of both constant-rate and step-drawdown tests are important in estimating conditions during dewatering. See Chapters 9 and 16 for the limitations and advantages of both tests.

Computer-Assisted Analysis

For complex dewatering projects involving deep wells, the use of a computer is recommended because this helps assure that the design will produce optimum performance at lowest cost. In view of well construction, completion, and operation costs, the value of the time spent seeking an optimal dewatering design is apparent. Any other design is unnecessarily expensive and therefore less desirable.

To illustrate the use of the computer in a dewatering design, assume that the 400-ft (122-m) square building site shown in Figure 22.20 must be dewatered to a depth of 45 ft (13.7 m) below grade. Also assume that a pumping test and the well logs in the area show the following:

1. Formation is unconfined.
2. Static water level is 10 ft (3 m).
3. Storage coefficient is 0.1.
4. Transmissivity is 90,000 gpd/ft (1,120 m²/day).

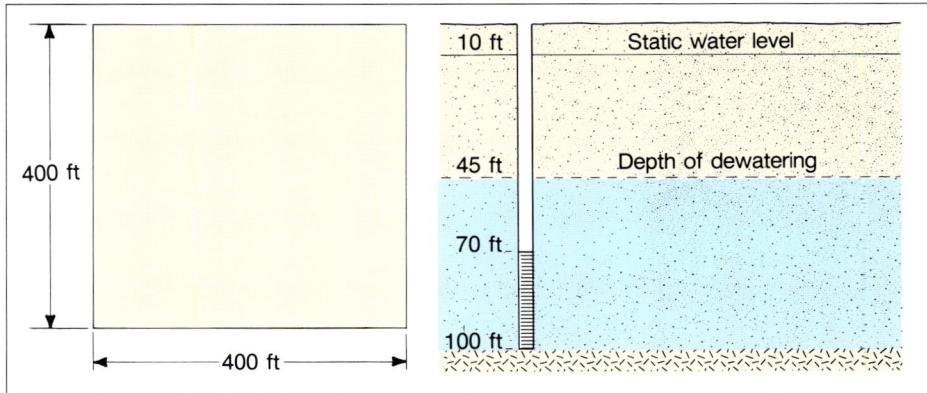

Figure 22.20. Top view and cross section of hypothetical building site that must be dewatered. A digital computer simplifies design of the dewatering scheme.

5. A river more than 800 ft (244 m) away stabilizes the cone of depression after 2½ days of pumping. The radius of influence is 822 ft (251 m).

Following standard well design procedures for unconfined conditions, the dewatering wells will be completed with 30 ft (9.1 m) of screen, allowing a total available drawdown of 60 ft (18.3 m). To satisfy the dewatering requirements, the drawdown everywhere inside the excavation must equal or exceed 35 ft (10.7 m). To insure that the required drawdown is obtained, it is necessary to calculate the drawdown at critical points in the excavation. For rectangular excavations, these critical points occur in the center of the excavation or along an edge. These critical points may be located graphically through experience or by calculations performed by the computer. The reasoning is that the drawdown will exceed 35 ft everywhere within the excavation if it equals or exceeds 35 ft at all critical points.

To arrive at the best design, a preliminary calculation is made to estimate the quantity of water that must be pumped from the excavation. Using the procedure outlined earlier in this chapter, the area to be dewatered [400 ft × 400 ft or 160,000 ft² (122 m × 122 m or 14,900 m²)] is imagined as a circle of equal area with one well located in the center. The equivalent radius, r_c, is calculated to be 226 ft (68.9 m). The hydraulic conductivity of the sediments is 1,000 gpd/ft² (40.7 m/day). The rate the single well must be pumped to produce 35 ft of drawdown at the outermost edge of the circular area is calculated using the given aquifer characteristics and Equation 22.1:

$$Q = \frac{K(H^2 - h^2)}{1055 \log R/r_c} = \frac{1,000 \cdot (90^2 - 55^2)}{1055 \log 822/226} = 8,580 \text{ gpm } (46,800 \text{ m}^3/\text{day})$$

This is the pumping rate that would be required by a system of closely spaced wells. With the more typical larger well spacing, it has been found that the estimated pumping rate should be increased by 10 percent. Thus, the pumping rate required to dewater the proposed construction site is 9,500 gpm (51,800 m³/day).

It is now necessary to estimate the number of wells that could produce the required pumping rate. By using Equation 22.1, and assuming a well radius of 1 ft (0.3 m) and a well efficiency of 80 percent, it can be shown that a single well pumping alone could produce approximately 2,000 gpm (10,900 m³/day) using all 60 ft (18.3 m) of available drawdown. When pumping in conjunction with several other wells spaced around the site, the interference from the other wells would greatly reduce the amount of available drawdown and consequently reduce the possible yield from each production well. In fact, field experience has shown that the actual yield that can be expected from each production well is often less than half the maximum yield that one well pumping alone could produce. Therefore, each production well in the final dewatering system will produce somewhat less than 1,000 gpm (5,450 m^3/day).

A natural first assumption is that 10 wells, each pumping 950 gpm (5,180 m^3/day), would be a reasonable solution to the problem because the yield from each well is less than 1,000 gpm and the total yield would be 9,500 gpm (51,800 m^3/day). Figure 22.21a shows this arrangement of wells. Computer calculations reveal that after 2½ days of pumping, using this number and spacing of wells, the drawdown values in Table 22.4a would be achieved.

These results show that the dewatering design illustrated in Figure 22.21a is unsatisfactory because (1) the drawdown at critical point Y is less than the required 35 ft (10.7 m) and (2) drawdown in some of the pumped wells exceeds the maximum allowable 60 ft (18.3 m). Therefore, it is impossible to obtain a practical scheme with only 10 wells.

Using the computer, the number of wells, the yield, and the spacing are adjusted to produce the proposed design shown in Figure 22.21b. This design utilizes 14 wells spaced around the construction site, each pumping at 680 gpm (3,710 m^3/day). The calculated values for drawdown after 2½ days of pumping are shown in Table 22.4b.

Fourteen production wells provide a solution to the dewatering problem, because the drawdown at critical point Y exceeds 35 ft (10.7 m) and the drawdown in each pumped well remains less than 60 ft (18.3 m). The designer might be tempted to select this design as the optimal one. However, experience suggests that this may be too many wells. According to Table 22.4b, several of the wells in the 14-well scheme have pumping water levels substantially above the top of the screen. Wells A, D, K, and N have more than 10 ft (3 m) of remaining available drawdown, whereas wells B, C, L, and M have nearly 5 ft (1.5 m) of available drawdown. This suggests that it might be possible to pump more water out of each well and thereby reduce the number of wells.

Additional adjustments can be made to the number of wells and the spacing to

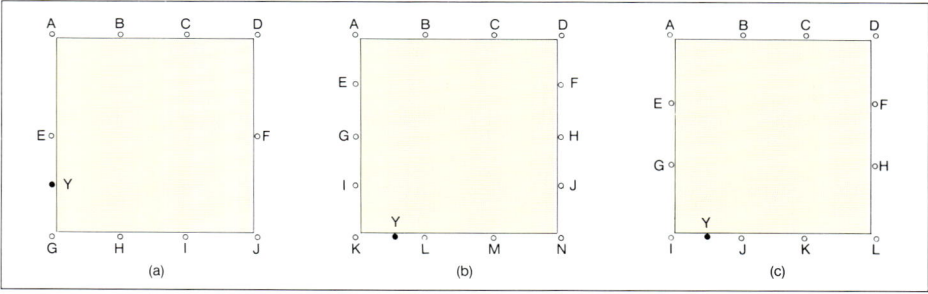

Figure 22.21. Dewatering scheme (a) is unsatisfactory because drawdown at critical point Y has not reached the necessary 35 ft (10.7 m). Critical points usually occur on the perimeter of the dewatering network, midway between wells. Plan (b) will work but requires more wells than necessary. The 12 wells shown in (c) provide the necessary drawdown at all critical points in the excavation while maintaining drawdown in the pumping wells at or slightly above the top of the screen.

Table 22.4. Drawdown for Various Points in Figure 22.21 Using Different Numbers of Wells

Position	Drawdown	
	ft	m
(a) 10-Well Scheme		
Wells A, D, G and J	58.9	18.0
Wells B, C, H and I	72.2	22.0
Wells E and F	67.3	20.5
Critical Point Y	34.0	10.4
(b) 14-Well Scheme		
Wells A, D, K and N	49.3	15.0
Wells B, C, L and M	55.5	16.9
Wells E, F, I and J	57.2	17.4
Wells G and H	59.6	18.2
Critical Point Y	35.1	10.7
(c) 12-Well Scheme		
Wells A, D, I and L	52.0	15.9
Wells B, C, E, F, G, H, J and K	60.0	18.3
Critical Point Y	35.0	10.7

arrive at the proposed design shown in Figure 22.21c. This design requires that 12 wells be placed in a symmetrical pattern around the site. Computer calculations show that the optimal pumping rate for this design is 785 gpm (4,280 m^3/day). The drawdown values shown in Table 22.4c are obtained after 2½ days of pumping. This design produces the required 35 ft (10.7 m) of drawdown at critical point Y, and therefore achieves at least 35 ft of drawdown at all points within the proposed excavation. It also maintains the pumping water level at or above the top of the screen in each well.

Can this design be improved? An earlier analysis shows that 10 wells are not enough to achieve a practical dewatering system, so the only possible improvement on the 12-well solution would be a dewatering system using 11 wells. Drawdown values for the 11-well system do not accomplish the dewatering goals; therefore, the 12-well design would be the one chosen for dewatering the proposed excavation.

Information required for the computer codes usually includes:
1. Data from a constant-rate pumping test utilizing several observation wells.
2. Logs of the subsurface formations at the construction site.
3. Dimensions of the area to be dewatered.
4. Depths to which dewatering is necessary.
5. Distance to potential sources of recharge.

In addition to computer analysis, utilization of specialized crews for well installation and development for large dewatering projects can minimize costs for the contractor and assure a safe working environment. For example, the U.S. Army Corps of Engineers wanted to dewater a 150-acre (60.7-hectare) site in Louisiana for lock-and-dam construction. Computer analysis suggested that 64 wells, pumping at a rate of 100,000 gpm (545,000 m^3/day), would be required to dewater the alluvial deposits.

To minimize costs, the contractors set up four different specialized crews to drill and complete the wells. The first crew drilled each 30-in (762-mm) hole to a depth of approximately 220 ft (67.1 m) in about 3 hours, using an air-assist, reverse-circulation rotary drilling machine. A second crew required about the same amount of time to set the casing and screen. A third crew filter packed the wells. They were followed by a crew that installed an eductor pipe and air line and developed the wells.

Deep wells are ideal for depressurizing confined aquifers and for dewatering thick aquifers that underlie construction sites. Deep wells may be installed outside the limits of excavation, thereby causing little interference with construction activities. The principal limitation of a deep-well dewatering system is its lack of mobility. When excavating a trench for a pipeline, for example, the dewatering system must be moved almost daily to keep ahead of the trenching operation. It is extremely difficult to do this kind of "leap-frog" dewatering with deep wells.

WELL-POINT SYSTEMS USED FOR WATER SUPPLY

For water supply, the layout of the multiple-well system can be adapted to any of the patterns shown in Figure 22.22. Locating the wells on the circumference of a circle gives the system greatest hydraulic efficiency, whereas placing wells in one line gives the lowest efficiency. It is important to space the individual wells for a water-supply system so that their areas of influence overlap only slightly or not at all. Well spacings of 25 to 50 ft (7.6 to 15.2 m) are generally satisfactory. Closer spacings may be used in fine-sand formations, in thin aquifers, or when the maximum drawdown may not exceed 5 ft (1.5 m). Spacings greater than 50 ft may be used where the depth and thickness of the aquifer permit installation of well screens that are 10 ft (3 m) or longer.

Well points with drop tubes should not be used in a water supply system, because the drop tube prevents effective development and causes increased head losses within the well points when the completed wells are being pumped.

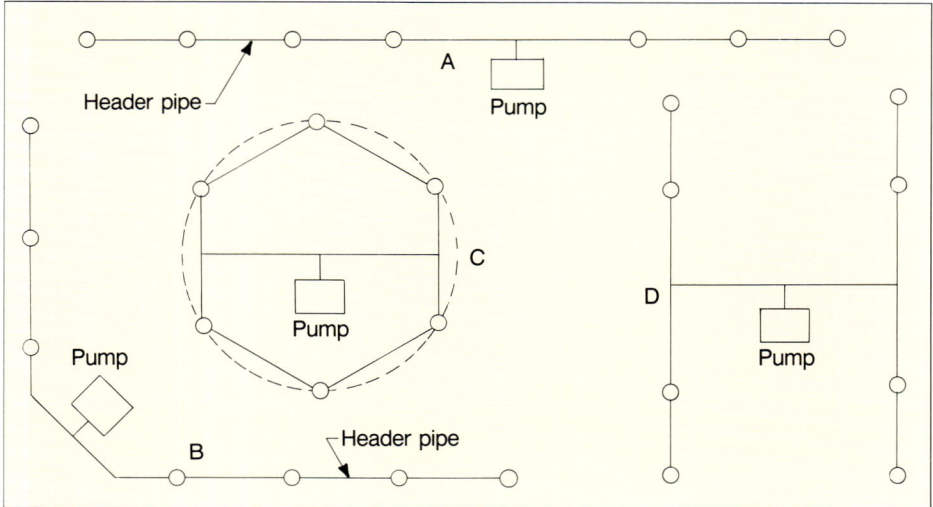

Figure 22.22. Several patterns can be used for the layout of a well-point system to serve as a water supply source. A centrally located pump equalizes suction lift.

The header, or manifold, for a water supply system may be standard galvanized steel pipe, PVC pipe, or cast-iron pipe. It is normally buried below the frost line. Lightweight header pipe, commonly used for temporary dewatering systems, is not suitable for a permanent installation.

INFILTRATION GALLERIES

In some geologic environments, the aquifer thickness may not be sufficient to supply the required volume of water to vertical wells, even though the aquifer is hydraulically connected to a nearby surface-water body. A typical example occurs in a river valley where thin alluvial deposits overlie bedrock. Even though the hydraulic conductivity of the sediment is excellent, the transmissivity is severely limited because the deposits are so thin. In other situations, a thin layer of fresh water may overlie salt water. Deep wells at this site would cause upconing of the salt water, thereby destroying water quality.

Under these hydrogeologic conditions, infiltration galleries, which consist of one or more horizontally laid screens, can be placed in permeable alluvial materials either adjacent to a water body or beneath its bed. A significant quantity of water may be pumped from an infiltration gallery because the hydraulic conductivity of the natural material and the filter pack surrounding the screens is so high that recharge is sufficient to meet the pumping rate. Because the screens are placed in open excavations, the usual practical depth limitation is about 25 ft (7.6 m) (Bennett, 1970). Water entering the screen is often collected in a sump constructed beneath the end of the screens. A large sump can serve as a storage chamber if the infiltration rate is low. The pump intake is placed in the sump (Figure 22.23). In installations where the infiltration rate is high, centrifugal pumps manifolded to the screens negate the necessity for the sump.

This discussion of infiltration galleries will focus on designs that are dependent on the sustained flow provided by a nearby surface-water source. The design and construction parameters are the same, however, for galleries not associated with a direct source of recharge, as for example, those placed in highly permeable sediments that are consistently recharged by rainfall.

The decision of whether to place the gallery adjacent to or under the surface water body depends on several factors:

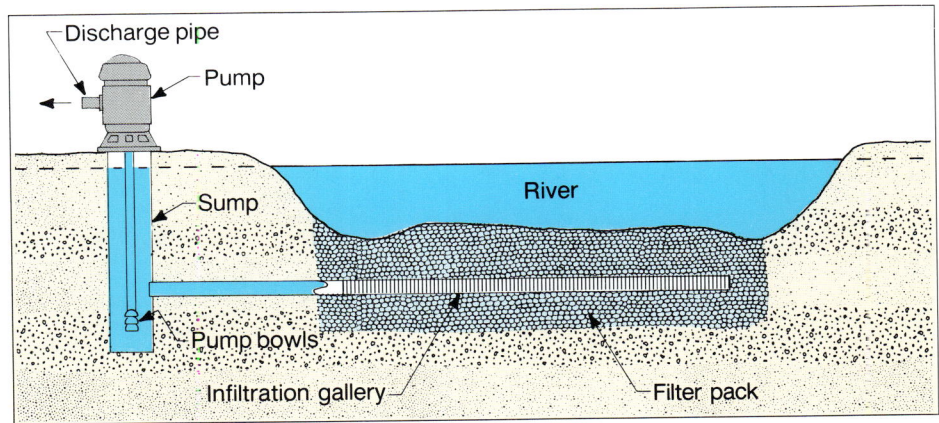

Figure 22.23. Cross section of pump placed in sump of infiltration gallery.

1. Yield requirements. Galleries placed under a water body initially produce twice the yield of galleries placed adjacent to the water body. As the disturbed lake or river bed assumes its normal sedimentation regime, however, the transmissivity values will fall as finer grained particles infiltrate the filter pack material surrounding the screens.

2. Water quality requirements. Galleries located adjacent to a water body usually receive water that has lower turbidity and fewer bacteria than bed-mounted galleries, because the water has been filtered more extensively.

3. Construction difficulties. It is generally more difficult to install a gallery beneath a stream or lake bed.

4. Maintenance considerations. Maintenance and repairs are easier to perform on galleries installed adjacent to a water body. In general, more maintenance is required for bed-mounted galleries because fine material is continually added to the top of the filter pack by stream current.

5. Stability of the river course or lake level. Rivers may meander great distances over relatively short periods, and either carry away a gallery placed on the bank or cover completely a bed-mounted gallery with less permeable material. Changes in the elevation of a water body can also affect where the gallery is placed. For example, the available head may drop considerably in intermittent streams during dry seasons of the year, but flow through the underlying sand and gravel usually continues.

Design Principles

A major design principle for infiltration galleries involves the orientation of the screen relative to the surface water or groundwater flow directions. For bed-mounted galleries, the screen is oriented perpendicular to the stream flow. For bank-mounted galleries, the screen is placed perpendicular to the groundwater flow to minimize the head loss; that is, the screen is placed parallel to the stream or river.

Important design criteria for infiltration galleries include the following:

1. Entrance velocity through the screen slot openings should be 0.1 ft/sec (0.03 m/sec) or less.

2. Axial velocity inside the screen should be 3 ft/sec (0.9 m/sec) or less, so that the head loss, h_L, will be 1 ft (0.3 m) or less. The following equation is used to determine velocity:

$$V = \frac{2.228 \times 10^{-3} Q}{\pi r^2} \qquad\qquad V = \frac{1.16 \times 10^{-5} Q}{\pi r^2} \qquad (22.8)$$

where
V = velocity, in ft/sec
Q = yield, in gpm
r = radius, in ft

where
V = velocity, in m/sec
Q = yield, in m^3/day
r = radius, in m

3. Screen slot size is predicated on the grain-size distribution of the filter pack; always retain 100 percent of the filter pack.

4. Use 304 stainless steel for fresh water, and 316 stainless steel or monel for salt water. Do not use monel if the Ryznar stability index is 9.5 or greater.

5. Filter pack recommendations:

   a. The surface area of the filter pack material is determined on the basis of water entering the pack at a rate of 2 to 5 gpm per ft^2 of surface area (117 to 293 m^3/day per m^2 of surface area). The actual hydraulic conductivity of the pack is usually much higher.

  b. Filter pack design is similar to that for a vertical well, but with a slightly more liberal multiplier of 6 to 7 times the 70-percent-retained size.

  c. Filter pack material should be clean, siliceous, rounded, and uniform.

Bed-Mounted Infiltration Galleries

Typical screen configurations for bed-mounted infiltration galleries are shown in Figure 22.24. Design criteria applying specifically to bed-mounted galleries include the following:

  1. The screen burial depth should be 3 to 5 ft (0.9 to 1.5 m) below the stream bed. There should be 1 ft (0.3 m) of filter pack beneath the screen.

  2. To minimize excessive sedimentation on the gallery surface, the stream selected should have a velocity of at least 1 ft/sec (0.3 m/sec).

  3. Space the screens approximately 10 ft (3 m) apart. Refer to Figure 22.25 for a typical design configuration and suggested dimensions for positioning screens in the infiltration gallery.

  4. If the stream has a large bedload transport, a single screen should be oriented parallel to the bank, but not in the main channel if possible.

  5. Screens should always be placed in the straight reaches of the river or stream, not near the meander bends.

The following equation is used to determine the length of screen required for a trench design installed in a stream or lake bed:

$$L = \frac{528 \, Q \log\left(\frac{1.1 \, d}{r}\right)}{0.25 \, K \, H} \qquad L = \frac{0.366 \, Q \log\left(\frac{1.1 \, d}{r}\right)}{0.25 \, K \, H} \qquad (22.9)$$

where
the width of the trench is approximately equal to 2 times the burial depth, d, that is, the distance between the bottom of the stream and the center of the screen, in ft; and
$L =$ length of the infiltration screen, in ft

where
the width of the trench is approximately equal to 2 times the burial depth, d, that is, the distance between the bottom of the stream and the center of the screen, in m; and
$L =$ length of the infiltration screen, in m

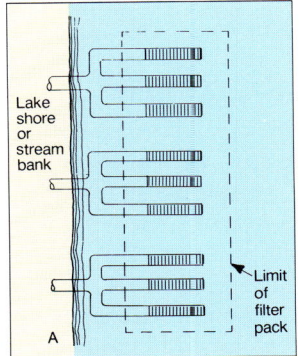
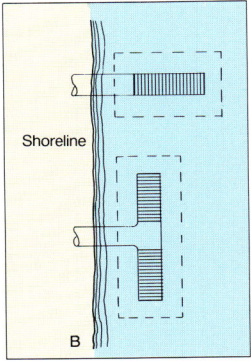
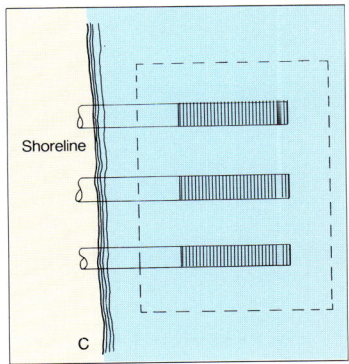

Figure 22.24. Screen arrangements for bed-mounted infiltration galleries.

K = hydraulic conductivity of filter pack material in gpd/ft²

H = submergence of infiltration screen, that is, the distance between the stream surface and the center of the screen (available head), in ft

K = hydraulic conductivity of the filter pack material, in m/day

H = submergence of infiltration screen, that is, the distance between the stream surface and the center of the screen (available head), in m

Equation 22.9 can be used for multiple screens spaced 10 ft (3 m) apart.

Field experience indicates that actual infiltration rates from streams and lakes range from 0.4 to 23 gpd per ft² per ft of head loss (0.05 to 3.07 m³/day per m² per m of head loss) (Walton, 1963; Bennett, 1970). In general, the infiltration rate will be high when the stream gradient is steep and the bedload is coarse. Infiltration rates from lake beds will ordinarily decrease more with time when compared with streams, unless wave activity is particularly vigorous and the bottom is continually disturbed so that fine sediment cannot settle. Wave energy can be transmitted to the bottom if the water depth over the gallery is less than one-half the typical wave length (distance from wave crest to wave crest).

The screens and filter pack material used for infiltration galleries may become partially plugged with sediment over time. Thus, it is good engineering practice to estimate the plugging potential and allow for excess entrance area to maintain the required flow. To maintain yield over time, the actual open area of the screens should be twice the required open area; that is, the screen length should be doubled.

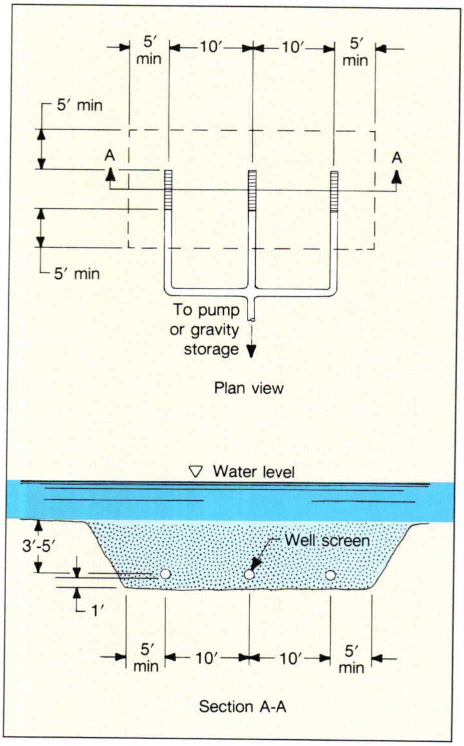

Backwashing capabilities may be specified for some infiltration galleries. The flushing rate is usually twice the pumping rate for the screen configuration. For example, if a series of three infiltration gallery screens were producing 3,000 gpm (16,400 m³/day), each screen should be backwashed at a rate of 2,000 gpm (10,900 m³/day). Backwashing techniques include (1) gravity backwashing, (2) piping and valve systems to pump from several screens while backwashing others, and (3) air back flushing.

On-Land Infiltration Galleries

On-land infiltration galleries are usually installed adjacent to a stream or river (Figure 22.26). A single screen is run parallel to the bank or shore. The screen burial depth should be at least 4 ft (1.2 m) beneath the static water level, but not more than 25 ft (7.6 m). Virtually all of the flow entering the infiltration gallery comes from one side of the screen,

Figure 22.25. Standard spacing and depth setting for infiltration gallery.

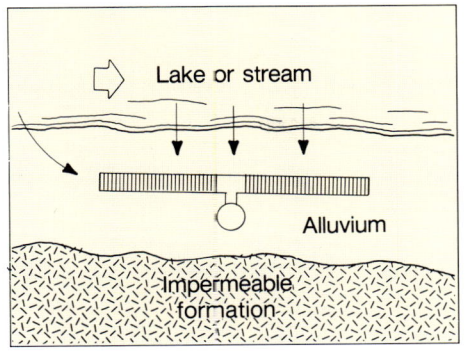

Figure 22.26. On-land infiltration gallery installed adjacent to lake or stream.

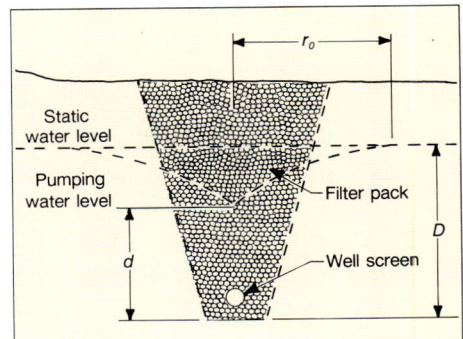

Figure 22.27. Terms used in the equation for determining the flow rate into the screen and the length of the screen.

thus the equation describing the flow rate into the gallery is:

$$Q = \frac{K L (D^2 - d^2)}{2880 \, r_0} \qquad\qquad Q = \frac{K L (D^2 - d^2)}{2 \, r_0} \qquad (22.10)$$

where

K = hydraulic conductivity of the sediments, in gpd/ft²

D = depth of the ditch below static water level, in ft (see Figure 22.27)

d = water above the ditch bottom while operating, in ft (see Figure 22.27)

r_0 = distance to point of no drawdown, in ft

where

K = hydraulic conductivity of the sediments, in m/day

D = depth of the ditch below static water level, in m (see Figure 22.27)

d = water above the ditch bottom while operating, in m (see Figure 22.27)

r_0 = distance to point of no drawdown, in m

The equation used to determine the required length of screen, L, is:

$$L = \frac{2880 \, r_0 \, Q}{K (D^2 - d^2)} \qquad\qquad L = \frac{2 \, r_0 \, Q}{K (D^2 - d^2)} \qquad (22.11)$$

The distance to the point of no drawdown, r_0, is obtained by conducting a pumping test using a series of observation wells laid out in a line from the proposed gallery (screen) location toward the water body, perpendicular to the source of recharge (Kazmann, 1948; Walton, 1962). The distance-drawdown graph from these wells indicates the radius of influence, r_0, in Equation 22.11.

Under some circumstances, a calculation should be made for the volume of water entering one or both ends of the screen. This is particularly important if the screen has a large diameter. The following equation is used to calculate the flow into one end of a screen:

$$Q_{(one\ end)} = \frac{K (D^2 - d^2)}{2111 \log \left(\frac{2r_0}{w}\right)} \qquad\qquad Q_{(one\ end)} = \frac{K (D^2 - d^2)}{1.466 \log \left(\frac{2r_0}{w}\right)} \qquad (22.12)$$

where
w = width of the ditch, in ft

where
w = width of the ditch, in m

Because of the contribution to the yield from one end, the length of the screen can be reduced somewhat for a required yield. The shorter length of screen can be calculated by first estimating how much of the desired discharge can be obtained from the end of the screen and deducting this volume from the required yield. The new screen length is then calculated from Equation 22.11.

The equation for flow into both ends is:

$$Q_{\text{(both ends)}} = \frac{K(D^2 - d^2)}{1055 \log\left(\frac{2r_0}{w}\right)} \qquad Q_{\text{(both ends)}} = \frac{K(D^2 - d^2)}{0.733 \log\left(\frac{2r_0}{w}\right)} \qquad (22.13)$$

Total flow into the screen is the sum of the volumes calculated from Equations 22.10 and 22.13.

Design criteria for infiltration galleries placed in permeable materials not associated with a surface-water body are essentially the same as suggested above; the total volume of flow that could enter a screen placed perpendicular to the flow can be calculated from the Darcy equation:

$$Q = \frac{KA(h_1 - h_2)}{1440 \, L} \qquad Q = \frac{KA(h_1 - h_2)}{L} \qquad (22.14)$$

where
A = cross-sectional area of trench, in ft^2
$\dfrac{h_1 - h_2}{L}$ = hydraulic gradient

where
A = cross-sectional area of trench, in m^2
$\dfrac{h_1 - h_2}{L}$ = hydraulic gradient

Only 60 to 75 percent of this flow can actually be intercepted, however.

If salt water is present, it is important to calculate the maximum drawdown that can be permitted and still exclude salt water from the gallery (Figure 22.28). Using the Ghyben-Herzberg analysis, the drawdown can be determined by the following equation:

$$h_s = \frac{\rho_f}{\rho_s - \rho_f} h_f \qquad (22.15)$$

where
h_s = distance below mean sea level of the salt-water head at the fresh-water/salt-water interface
h_f = distance from top of the water table to mean sea level or the salt-water head
ρ_s = density of salt water (1.025 g/cm^3)
ρ_f = density of fresh water (1 g/cm^3)

The relationship between h_s and h_f is then:

$$h_s = \frac{1.00}{1.025 - 1.00} h_f = 40 \, h_f$$

For a small, low-lying tropical island, this means that the depth to the fresh-

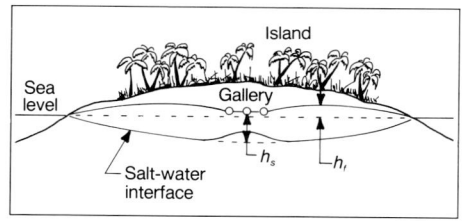

Figure 22.28. Island setting with interface between fresh and salt water.

water/salt-water interface is 40 times the distance between the surface of the water table and the mean sea level. If this distance (h_f) is 4 ft (1.2 m), the interface is 160 ft (48.8 m) below the water table. Reduction of this 4-ft head during pumping causes the interface to rise significantly. For example, if the drawdown is 3 ft (0.9 m), h_f is 1 ft (0.3 m), and the interface would be only about 40 ft below the water table.

Although Equation 22.15 will give an approximate depth to the salt-water interface, two other factors may affect the actual distance. In many situations, the boundary between fresh and salt water is a diffused layer of mixed water that may alter the exact location of the interface. Flow of groundwater toward the sea also affects the position of the interface, as shown in Figure 22.29. When flow occurs (as it normally does), the interface is lower than Equation 22.15 suggests.

Maintenance of Infiltration Galleries

Maintenance of infiltration galleries may be difficult, depending on the configuration of the screen and where it is placed. Therefore, it is important to observe several operational criteria that will reduce or eliminate maintenance problems. These include the following:

1. Never exceed the designed pumping rate. Higher pumping rates may cause fine sediment to enter the filter pack, reducing its permeability. Eventually, sand may enter the screens and block part of the intake openings, causing even more sand pumping.

2. Do not let the screens become aerated, because iron bacteria may cause severe incrustation problems. Near-surface waters are often high in iron, and the rapid growth of iron-ingesting bacteria can clog the gallery. Inorganic deposits of magnesium, calcium, and other ions may also form.

3. Do not let the gallery go unused for long periods. Inactivity tends to lower the hydraulic conductivity of the filter pack and surrounding materials.

If maintenance is necessary, it usually involves backwashing and chemical treatment. For galleries installed beneath a water body, backwashing with 4 to 10 gpm of water per ft^2 (234 to 586 m^3/day of water per m^2) of filter pack provides a reasonable degree of agitation and sufficient flow to remove the fine material from the filter pack. Approximately 5 to 10 minutes of backwashing is usually required to clear the pack, assuming that the flow rate above the infiltration gallery is sufficient to carry away the fine material removed from the pack. The use of compressed air to backwash

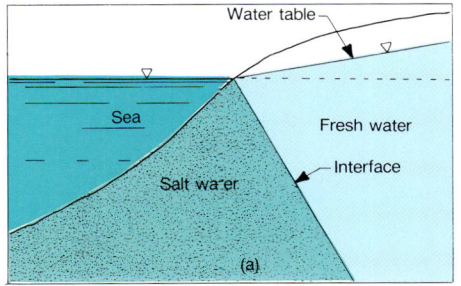 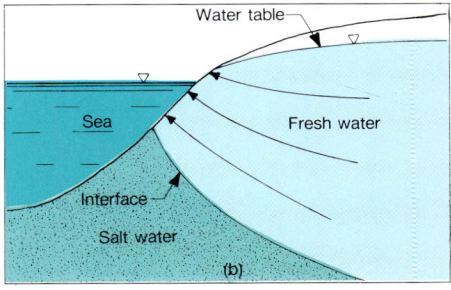

Figure 22.29. The position of the salt-water/fresh-water interface along a coast. In (a), the head created by the groundwater system just equals the head created by the ocean; thus no groundwater flow is occurring and the interface remains stationary. When the groundwater head exceeds that of the ocean, groundwater flows into the ocean forcing the interface to move seaward (b). *(After Hubbert, 1940; Freeze and Cherry, 1979)*

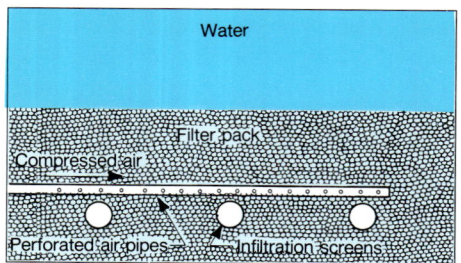

Figure 22.30. Placement of perforated pipes used to backwash infiltration gallery.

the filter pack is also quite effective. During installation of the gallery, perforated air pipes are permanently installed in the filter pack (Figure 22.30). Any degree of agitation necessary to clean the filter pack can be obtained by using appropriately sized truck- or trailer-mounted compressors.

Procedures for removing iron bacteria and inorganic incrustants are the same as for conventional wells (see Chapter 19). Chlorine, hydrochloric acid, and sulfamic acid are the principal chemicals used for treating galleries.

COLLECTOR WELLS

A special adaptation of infiltration galleries is created when the screens extend from a large vertical caisson constructed adjacent to a stream, river, or lake (Figure 22.31). The screens are not installed in trenches, but are jacked out horizontally in sections from the caisson. The steel perforated casing or screen, 16 to 48 inches (406 to 1,220 mm) in diameter, can be jacked out to a distance of 2,000 ft (610 m) beneath the water source, depending on the extent of the alluvial deposit. Plastic screens have been installed successfully by the telescope method. After the screened areas are developed, water infiltrates into the gallery, eventually raising the water level in the caisson to the level of the river or lake. Thus, the caisson can serve as a large storage tank. One or more pumps are installed in the caisson.

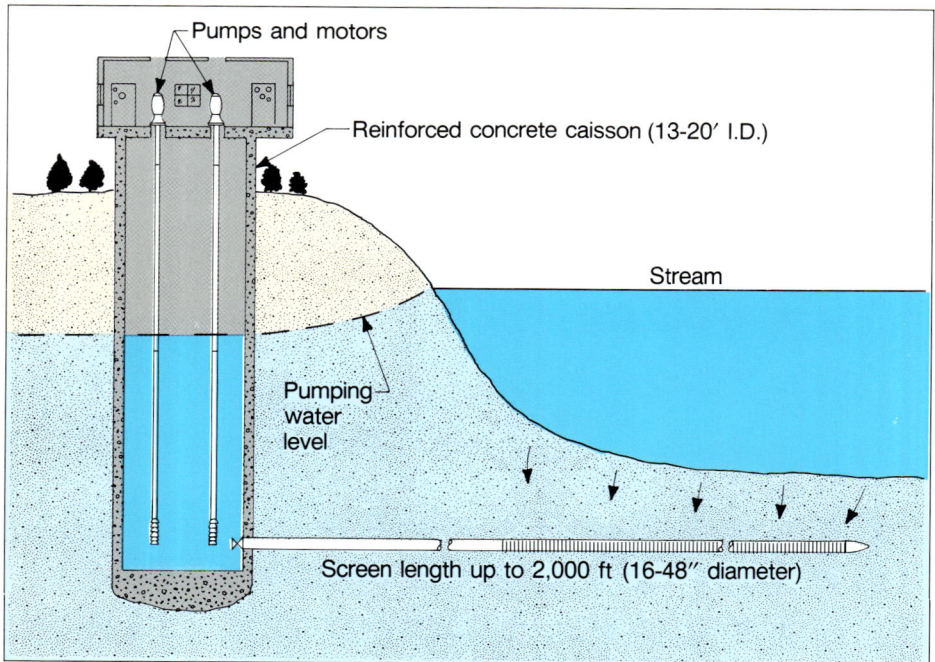

Figure 22.31. Collector well with screen jacked out from a large caisson. *(Hydro Group, Ranney Division)*

A second type of collector system, called a radial collector, uses multiple screens extended horizontally outward from a caisson. This type of system is suitable for floodplain installation and can yield millions of gallons per day with little drawdown. The cost of collector wells, however, is substantial in comparison with vertical wells.

INJECTION WELLS

Injection wells are used for a variety of purposes, encompassing water supply, groundwater control, solution mining, waste disposal, and geothermal energy. Appropriate well design and construction techniques will vary according to the specific purpose of the well, but in all cases the greatest care in design and construction must be taken because injection wells are much more likely to fail than are typical water wells. The consequences of water-chemistry problems, air entrainment, thermal interference, and sand pumping are considerably more serious and common for injection wells. Pumping recharge water with sand at concentrations as low as 1 mg/ℓ can clog injection wells within a short time. A 6-in (152-mm) test well in California, pumping 32 gpm (174 m^3/day) and injecting water containing 3.3 mg/ℓ sand, showed a 30-percent increase in the injection water level within 9 days. A successful recharge well in Nebraska was pumped continuously at 750 gpm (4,090 m^3/day) for 6 months. During this period, the additional head build-up was 6 ft (1.8 m), even though the recharge water contained only 0.004 mg/ℓ of sand (Smith, 1980b). Air entrainment or water-quality changes also could have contributed to the build-up.

With the exception of entrance velocity and screen length, the design criteria given in Chapter 13 apply to injection wells, because most injection wells, except waste-disposal wells, are likely to become pumping wells at some time during their life (Johnson et al., 1966). Clogging of screens is the most serious problem in injection well operation (Olsthoorn, 1982). Thus, screen open area and screen length must be optimal. The average entrance velocity for injection wells should be 0.05 ft/sec (0.015 m/sec). Therefore, screens should be twice as long as for a withdrawal well pumping the same volume of water. The typical injection well should pump sand-free water, be efficient (maximum recharge at minimum pressure build-up), and be cost effective in terms of initial investment and operational costs. In this section, special design features for the major types of injection wells are discussed.

Recharge Wells for Water Supply

Artificial recharge of groundwater increases the rate at which water infiltrates from the land surface to the groundwater system. Water can be injected by means of wells or spread on the land surface to infiltrate. Use of injection wells to recharge groundwater systems has become an accepted method of slowing or stopping overdrafts of groundwater. In water-short areas, surface water may be captured in basins during spring runoff, and allowed to recharge groundwater later in the year by one of several water-spreading methods (Figure 22.32). Elsewhere, treated surface water is stored underground until needed later in the season during peak-demand periods (Olsthoorn, 1982). At Lake Manatee, Florida, for example, a 15-million-gpd (56,800-m^3/day) increase in peak capacity has been achieved by temporarily storing potable water in aquifers containing low-quality water (Water Well Journal, 1983). In this recharge-recovery operation, the treatment facilities can be operated at a steady rate throughout the year. Typical water sources for groundwater recharge are given in Table 22.5.

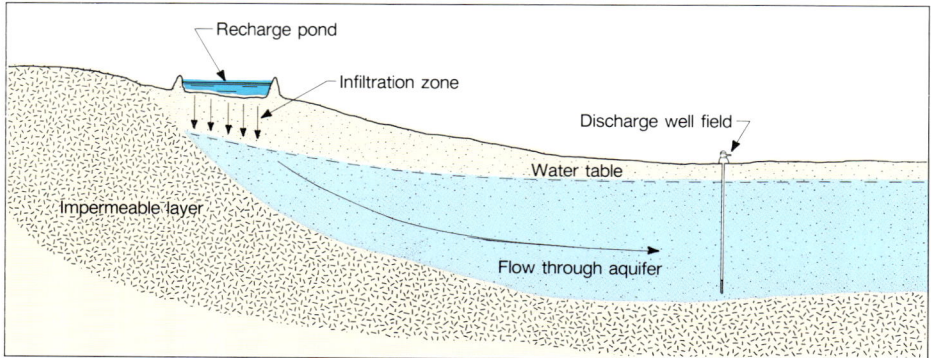

Figure 22.32. Unconfined aquifers can be recharged by infiltration basins.

Injection tubing forms an important part of the design for recharge wells (Figure 22.33). The injection tube must terminate below the static water level in blank casing and must be designed so that positive pressure exists along its entire length. Back-pressure valves should be installed to eliminate negative pressures in the injection tube. Another important criterion is that the injection tube should provide for full flow to eliminate the possibility of air entrainment.

In weakly consolidated, stratified sediments, the injection pressure must be controlled so that the formation is not fractured. If fracturing occurs, there is usually a severe loss in hydraulic conductivity because the bedding planes are disturbed. On the other hand, if injection is into massive consolidated rock, formation fracturing may increase the rate of injection (Howard and Fast, 1970). Pressures that will cause fracturing range from a low of 0.5 psi/ft (11.3 kPa/m) for poorly consolidated coastal plain sediments, to 1.2 psi/ft (27.1 kPa/m) for crystalline rock (Warner and Lehr, 1981). For most recharge wells in unconsolidated sediments, the injection pressure should be controlled carefully so that the positive head does not exceed $0.2 \times h$, where h is the depth from the ground surface to the top of the screen or filter pack (Olsthoorn, 1982).

If water is discharged into an injection well, a cone of recharge will be formed which is similar in shape but the reverse of a cone of depression surrounding a pumping well (Figure 22.34). The equation describing the cone for various discharges can be derived by using the same assumptions applied to a pumping well (page 218). For a confined aquifer with water being recharged into a well completely open to the aquifer at a rate Q_r, the following equation is applicable:

Table 22.5. Comparison of Water Sources Available for Groundwater Recharge

Source	Time of Availability	Predictability of Supply	Proximity to Recharge Site	Volume
Surface storage	Periodic	Good	Peripheral	Large
Natural runoff	Periodic	Poor	Peripheral	Moderate
Imported surface water	Periodic	Good	Distant	Large
Wastewater	Continuous	Good	Close	Small
Flood water	Periodic	Poor	Peripheral	Small
Irrigation deep percolation	Continuous	Good	Close	Large

$$Q_r = \frac{Kb(h_w - H_0)}{523 \log(r_0/r_w)} \qquad Q_r = \frac{Kb(h_w - H_0)}{0.366 \log(r_0/r_w)} \quad (22.16)$$

where

Q_r = rate of injection, in gpm
K = hydraulic conductivity, in gpd/ft^2
b = aquifer thickness, in ft
h_w = head above the bottom of aquifer while recharging, in ft
H_0 = head above the bottom of aquifer when no pumping is taking place, in ft
r_0 = radius of influence, in ft
r_w = radius of injection well, in ft

where

Q_r = rate of injection, in m^3/day
K = hydraulic conductivity, in m/day
b = aquifer thickness, in m
h_w = head above the bottom of aquifer while recharging, in m
H_0 = head above the bottom of aquifer when no pumping is taking place, in m
r_0 = radius of influence, in m
r_w = radius of injection well, in m

For a recharge well penetrating an unconfined aquifer, the following equation is applicable:

$$Q_r = \frac{K(h_w^2 - H_0^2)}{1055 \log(r_0/r_w)} \qquad Q_r = \frac{K(h_w^2 - H_0^2)}{0.733 \log(r_0/r_w)} \quad (22.17)$$

By comparing the discharge equations for pumping and recharge wells, it might be anticipated that the recharge capacity would equal the pumping capacity of a well if the recharge cone were the same size as the cone of depression. Field measurements show, however, that recharge rates seldom equal pumping rates. Theoretically, a properly designed recharge well will recharge as much as the pumping capacity, but problems associated with water quality, turbidity, and high water temperatures reduce the recharge rate over relatively short periods. In an ordinary water well, for example, some fine sediment will be removed continually from the formation, whereas in an injection well, these fines are not removed. In fact, fine material contained in the injection water will continuously collect in the formation or filter pack outside the screen. Over time the formation slowly becomes clogged, reducing the capacity of the aquifer to receive water. Because this phenomenon is inevitable, most injection well designers specify that the screen length be much longer than for a water supply well of equal capacity, assuming the aquifer is thick enough, to lessen maintenance. Even dry sand or gravel zones above the aquifer may be screened if the well is used only for injection purposes.

Plugging of the formation around the screen can be caused not only by sand or incrustants but also by air bubbles entrained in the injected water. When air is entrained with injection water, a serious loss of hydraulic conductivity can be expected because air bubbles can effectively block the outward passage of water by plugging pore spaces within the aquifer.

Plugging by chemical precipitation is

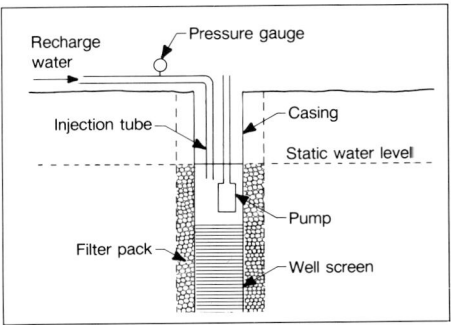

Figure 22.33. Typical design for a recharge well installed in unconsolidated sands.

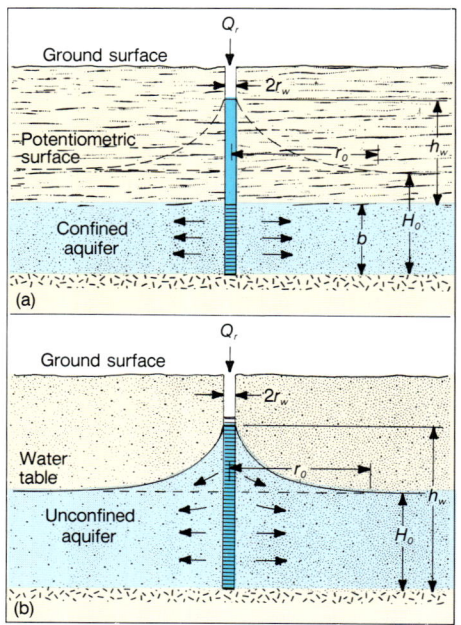

Figure 22.34. Radial flow from recharge wells penetrating (a) confined and (b) unconfined aquifers. *(Espina, 1980)*

another problem common to injection wells. Precipitates can either be carried with the injection water or be formed by mixing waters of different quality. Injection water that has high mineral content can be expected to create incrustation.

Plugging by bacterial action occurs occasionally in recharge wells. Bacterial growth can be promoted by the change in temperature caused by injection, especially when warmer water is added to a cool aquifer. Care should be taken to insure that only treated or bacteria-free water is injected.

If the recharge well is used as a production well after a long period of injection, it is important to pump the injection well to waste for a short time to eliminate turbidity and remove incrustants and other material that may be clogging the well. Some redevelopment of the well prior to use as a production well is recommended to return the well to its original capacity (see Chapter 15).

Environmental Effects of Recharge

The principal benefits of artificial recharge of groundwater can be classified into two categories: relief of overdraft (depleted groundwater supplies) and access to more groundwater by using groundwater basins as cyclic storage and distribution systems (Espina, 1980). By counteracting overdraft of a groundwater basin, artificial recharge may provide the following benefits:

1. Lower operating costs by reducing pumping lifts.
2. Eliminate capital expenditures for deepening wells or lowering pump bowls.
3. Reduce incidence of premature abandonment of wells.
4. Increase farm income and farm value as a result of augmented and more dependable water supplies.
5. Increase the potential for municipal expansion by providing increased groundwater supplies.
6. Prevent sea-water intrusion into coastal aquifers.
7. Prevent intrusion of deep-seated connate brines into the aquifer.
8. Prevent land subsidence by sustaining groundwater levels.

To assure success for a recharge project, field conditions must provide for appropriate storage, movement, and use of recharge water. The California Department of Water Resources has listed the following physical requirements for recharging:

1. Geology — the basin must be suitable from the standpoint of storage capacity and transmissivity of aquifers.
2. Water — adequate recharge water must be available.
3. Infiltration — recharge rates must be maintained at adequate levels.

4. Drainage — where a water table is near ground surface, adequate storage capacity must be provided in the basin.

5. Water quality — recharge water must be chemically compatible with existing groundwater and have a suitable temperature.

6. Recovery efficiency — pumping lifts must not be excessive, pumping capacity must be efficiently used, and quality of water recovered must be satisfactory.

Where conditions allow, recharging the groundwater system can be an effective, efficient means of maintaining adequate water supplies at reasonable costs.

Control of Salt-Water Intrusion

In Florida, southern California, Long Island, New York, and other low-lying coastal areas, the high use of groundwater has allowed the normal fresh-water/salt-water interface to move both inland and nearer the ground surface. Initially, many communities faced with salt-water encroachment problems merely drilled new production wells farther inland. Now, efforts are being made to maintain groundwater levels by ponding surface runoff or river water so it will recharge the groundwater table (Figure 22.35a). Elsewhere, the installation of deep recharge wells has been able to control salt-water intrusion by creating a ridge of groundwater having a potentiometric surface high enough to prevent salt-water intrusion yet allowing pumping below sea level landward of the ridge (Figure 22.35b). Sometimes barrier wells near the shore are used to collect salt water and induce a fresh-water gradient toward the sea (Figure 22.35c). Thus, recharge wells, recharge basins, and barrier wells are extremely useful in maintaining the proper hydraulic gradient between fresh and saline water.

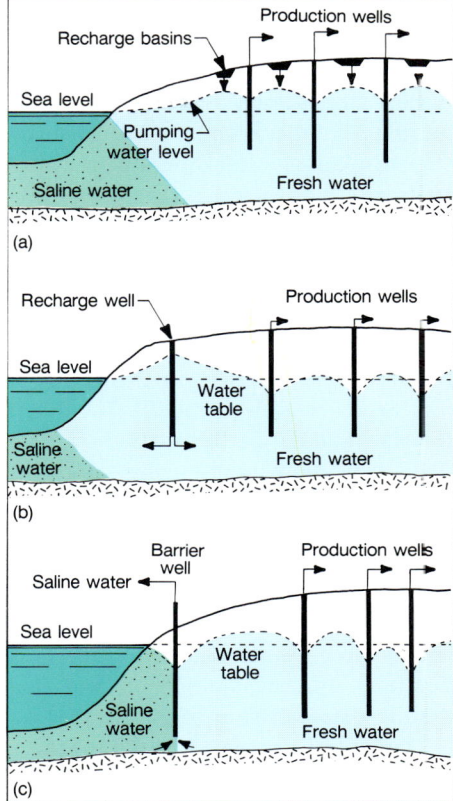

Figure 22.35. (a) Use of artificial recharge in the area of production wells in an unconfined coastal aquifer; the recharged water maintains the water table above sea level to prevent salt-water intrusion. (b) Use of injection wells to form a pressure ridge to prevent salt-water intrusion in an unconfined coastal aquifer. (c) Use of pumping wells at the coastline to form a trench in the water table, which acts as a barrier to further salt-water encroachment. *(From APPLIED HYDROGEOLOGY, by C. W. Fetter, Jr., © 1980 Merrill Publishing Company, Columbus, Ohio)*

Solution Mining

In-situ mining is used to extract deposits of uranium, copper, sulfur, and other minerals from ore bodies located well beneath the Earth's surface without excavation of the overburden. Leaching of the minerals in place is accomplished by injecting chemically treated groundwater into the ore body, allowing the ores to dissolve, and then pumping the groundwater and dissolved minerals to the surface where the minerals are precipitated. Both injection wells and recovery wells are used in this operation. Besides eliminating the need for large excavations, in-situ mining causes significantly less environmental damage, lowers capital and labor costs, increases mining flexibility because wells can be installed relatively quickly compared with traditional mining excavations, and provides greater safety for workers (Tweeton and Connor, 1978).

In-situ mining of uranium, for example, is carried on extensively in Texas by solution methods. There are several steps in the development of in-situ uranium mining fields (Roberts, 1980). The first is to locate and define the ore body. This is done principally by test drilling and core sampling, which determines the areal extent of the ore body, the ore grade, and the ore's chemical equilibrium with its daughter (radioactive decay) products. Core analyses and pumping tests are also conducted to determine hydraulic conductivity, transmissivity, and other hydraulic characteristics of the aquifer.

The second step is to drill and construct production and injection wells. These wells are drilled into the ore body in a pattern best suited for leaching that particular ore. The pattern is determined from the tests run during the exploration phase. Several different patterns of construction have been used in the in-situ mining business, but three of the principal patterns used are the five-spot pattern, the seven-spot pattern, and the line-drive configuration (Figure 22.36). Several injection wells are usually placed around a single recovery well in each pattern. Distances between wells may vary in a spot pattern, depending on the hydraulic characteristics of the ore body, but a distance of 50 ft (15.2 m) is used in a typical production field. In-situ well production varies from 5 to 70 gpm (27.3 to 382 m³/day), depending on the aquifer in which the ore body is contained and its hydraulic characteristics. Well depth

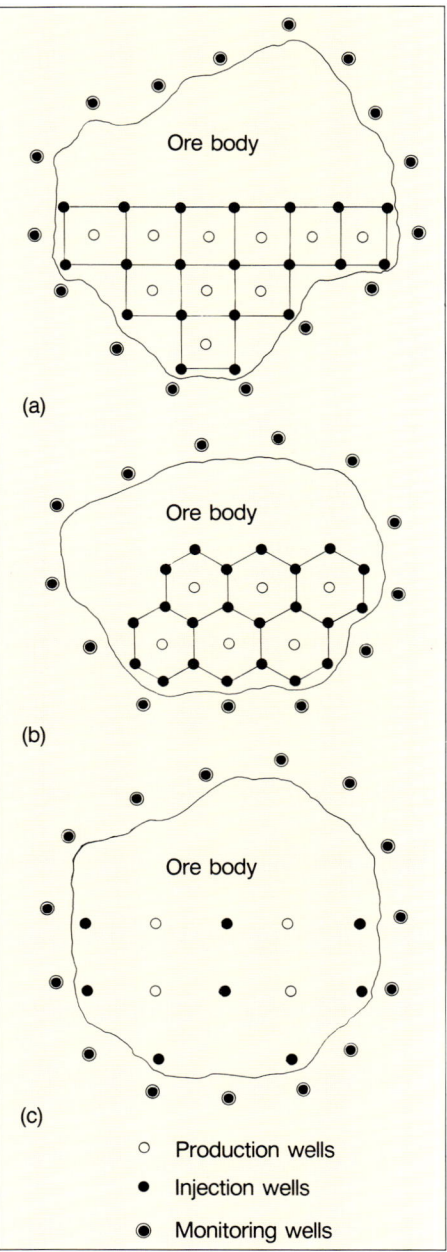

Figure 22.36. There are several construction plans used in in-situ mining: (a) five-spot pattern, (b) seven-spot pattern, and (c) line-drive pattern. Distances vary between each well, although 50 ft (15.2 m) is typical.

ranges from 150 to 600 ft (45.7 to 183 m), although some wells have been drilled to more than 2,500 ft (762 m). The entire ore body can be mined by adding more production and injection wells.

Injection well and recovery well construction are essentially the same; the only difference may be in the size of well casing used. The wells are generally constructed of 4-in (102-mm) plastic well casing; however, 4½-, 5-, and 6-in (114-, 127-, and 152-mm) casing are also used for some applications. The injection well diameter must be large enough to accommodate a pump during development and during restoration of water quality (Tweeton and Connor, 1978). The pump is removed from the well during leaching because of the corrosive chemicals. Recovery wells usually require a larger diameter casing and pump because injection wells outnumber recovery wells.

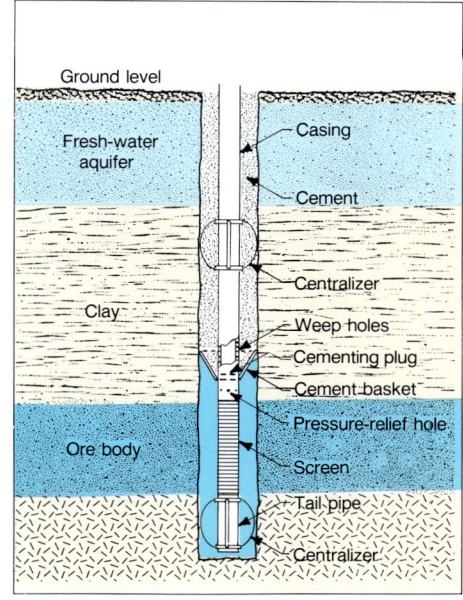

Figure 22.37. Well design used for in-situ uranium mining.

However, some companies interchange injection and production wells, thus requiring that all wells be constructed to the same dimensions.

Wells for in-situ mining consist of a cased portion, an intake portion with tail pipe, a cement basket, a cementing plug, and centralizers (Figure 22.37). The design of the intake portion of the well is vital in assuring a highly efficient well. The configuration of the slot openings must facilitate development by directing water uniformly into the formation. The more open area there is, the faster and more effective the development will be. Close slot tolerances insure that sand pumping will not occur. The tail pipe is attached to the screen so that material entering the screen can fall into the tail pipe and not block any part of the screen. A resistivity log is run to determine the best place to set the cementing basket. A caliper log should be run to detect if any sloughing of the borehole has occurred. The entire casing string is constructed as a single unit that is lowered into the borehole until the intake portion is opposite the ore zone and the cement basket is opposite an impermeable zone above the aquifer containing the ore body. To facilitate installation of the screen and casing, one or two small holes are cut into the casing just below the cement basket so that air can escape from the lower part of the casing as the assembly is being lowered into the borehole. Several 1-in (25.4-mm) weep holes are drilled through the casing above the cementing plug. Centralizers help to center the string in the hole. Upward adjustment of the string during setting is possible only within the narrow range that the basket can slide on the casing [2 to 3 ft (0.6 to 0.9 m)].

Failure to set the cement basket at the correct depth can result in inadequate grouting of the well, partial covering of the well screen by cement moving downward around the basket, or even abandonment of the well. Cement introduced down the inside of the casing passes out the weep holes, and then back up between the casing and the well bore to the surface. Water is used to force the cement out of the weep holes. Some cement should remain in the casing to assure a good seal in the cementing basket area. After curing, the cement plug is drilled out and the well developed.

Because the chemicals used in the injection process are highly corrosive, plastic screens and casing are preferred. Continuous-slot, V-wire screens, both with and without a pipe base, are the two most common types used. Screens built on a pipe base are stronger than rod-based screens, but are less efficient because of limited open area in the pipe base.

The importance of the well screen to successful in-situ solution mining cannot be overemphasized, because of the large number of wells involved in each project. From 500 to 5,000 wells may be drilled into each ore body, using 10,000 to 250,000 ft (3,050 to 76,200 m) of well screen (J. Yelderman, personal communication). Thus, the decisions involving well screens are financially significant.

Every effort is made to avoid plugging the formation during drilling. Polymeric drilling fluid additives are preferred to bentonite additives because bentonite is much more difficult to remove during development. Polymeric additives can be broken down quickly with a minimum amount of physical agitation. Polymers are also more suited to this type of well because they encapsulate clays encountered in the drilling process, thereby limiting their dispersal in the drilling fluid system and eventually into the formation.

After well completion, the next phase is the connection of individual wells into groups. All the injection wells are regulated to insure a specific injection rate and are then tied into an injection solution system. Production wells are metered for flow, connected, and then tied into the production process plant. Submersible pumps are used in the production wells, whereas centrifugal pumps located in the plant control flow in the injection wells.

The solution injected into uranium ore zones is an oxidant, such as gaseous oxygen or hydrogen peroxide along with water, carbon dioxide, and ammonium. As this solution passes through the ore body, it oxidizes the uranium, causing it to go into solution. This uranium-rich solution then travels into the production well and is pumped into the process plant. There, the uranium-rich solution is passed through an ion-exchange process in the plant, where the uranium is removed and the remaining solution is returned to the injection well to start the process over again. The solution injected into the formation is usually extremely incrusting, and both wells and pumping equipment are continually undergoing maintenance.

Government regulations have been enacted to cover the environmental aspects of in-situ mining. In Texas, for example, the principal regulation requires that conditions in an aquifer be determined before mining begins. These conditions must be restored after completion of mining. Water samples from monitoring wells around every mining field are checked periodically to see if injected solutions are being confined within the designated areas. Monitoring wells are also located in the fresh-water aquifer above the formation that is being mined. These monitoring wells are located at various places in the well production field and are monitored according to a prescribed schedule.

Injection Wells for Waste Disposal

Disposal of hazardous and toxic wastes by deep-well injection has become accepted practice in certain regions where the wastes can be isolated from drinking water resources. Typically, the wastes are injected well below aquifers containing fresh water, although, until recently, certain wastes such as brine waters associated with oil or gas wells were injected or allowed to drain into soils above fresh-water aquifers. It is now illegal in the United States to inject hazardous wastes above or into a fresh-water aquifer, and wastes that are injected below an aquifer must be placed at least 1,320 ft (402 m) below the bottom of the aquifer. The geologic formations between the underground injection site and the aquifer should consist, in part, of tight clay or other material that is impervious to the upward migration of the waste (Safe Drinking Water Act, 1974). In most cases, the wastes are injected into aquifers containing salt-water brine or brackish waters because these aquifers are generally considered useless as drinking water resources.

Disposal of wastes by deep-well injection is expensive, and only those wastes that are difficult to render harmless at the surface are injected. Usually these wastes are highly corrosive, and special care must be taken in well design and construction practices to assure adequate well life and reasonable well efficiency. Screens and casing are usually fabricated from the most durable materials to resist corrosive attack. Normally, the casing is stainless steel, whereas screens are constructed of even more resistant, high-strength materials such as Carpenter 20, monel, 316 stainless steel, or Inconel. As a result, the cost of materials for an injection well may be 5 to 10 times higher than for a water well. Greater construction costs are also incurred because of extensive coring requirements, formation testing, geophysical logging investigations, cementing services, and increased supervision.

Injection wells should be designed to be as efficient as typical high-capacity water wells (see Chapter 13). Minimizing pressure build-up during injection is the primary objective of design and construction. Reducing pressure build-up decreases the number of wells required, increases the recharge rate, and decreases well spacings. For example, polymeric drilling fluid additives are often used to prevent plugging of the formation during well construction so that pressure build-up is minimal after the well is placed into service. Screen selection is another critical part of successful injection well design. Some parts of a formation have much higher hydraulic conductivity because of natural characteristics or intentional fracturing. Thus, the open area of the screen should be as great as possible so that wastes can enter these sections of the formation with little head build-up.

Actual pressure build-up can be calculated by using the Darcy equation, the continuity equation, and an equation of state (Matthews and Russell, 1967). Many of the same assumptions used to derive the equations describing pumping wells are also applicable in calculating the pressure build-up over time in an injection well (Warner et al., 1979). It is easy to see that pressure build-up in an injection well is analagous to drawdown in a pumping well. Warner et al. (1979) present three methods for calculating pressure build-up: calculation by solving analytical equations manually, calculation by programmable desk calculator, and calculation by digital computer. Besides determining the radius of pressure influence, the potential for hydraulic fracturing which may permit the escape of wastes can be investigated, as can the flow rate and flow direction of the injected wastes in the reservoir.

PRESSURE-RELIEF WELLS

During the 1940's, it became common practice to relieve hydrostatic pressures resulting from dam or levee construction by installing wells landward of the structures. Dams or levees are often constructed on a thin layer of semipermeable or impermeable materials underlain by a relatively thick layer of highly permeable material, usually alluvium. Driven by hydrostatic pressures in the reservoir, water flows from the reservoir through the underlying permeable sediment to the area landward of the structure (Figure 22.38). Subsurface pressures can become so great that the saturated weight of the surface materials may not be sufficient to contain them. Excess pressure causes localized heaving of the soil; these areas are called sand boils and are characterized by "quick" conditions. The construction of pressure-relief wells prevents the occurrence of excessive seepage and the formation of sand boils at the surface (Middlebrooks and Jervis, 1946). Pressure-relief wells, however, usually have little or no affect on the volume of seepage that passes through a dam or levee (Rutledge, 1947). In a few instances, hydrostatic pressures are not a problem because the dam structure is "keyed" into bedrock. Thus, hydrostatic pressures in the reservoir cannot be transmitted to the area near the toe of the dam.

Prior to the installation of pressure-relief wells, various types of horizontal drains were used to reduce near-surface pressures, but these often proved to be ineffective because much of the pressure build-up occurs well beneath the land surface itself. To be effective, pressure reduction must take place in deep, highly permeable sediments below less permeable surface or near-surface materials. The principal requirements for a relief well system are (Turnbull and Mansur, 1953):

1. Wells must penetrate to the sediment layers having the highest transmissivity values.

2. The wells must be installed close enough so they can intercept a major part of the subsurface seepage, thereby minimizing pressures landward (downgradient) of the well installation.

3. Wells must be efficient so that head losses are minimized.

4. Wells should be designed and constructed so that fine particle movement into the filter pack or screen is prevented.

5. Wells should be resistant to chemical attack.

The design of a relief well system is based on somewhat different principles than for a typical dewatering situation in which pumping capacities can be easily adjusted to create the required drawdown conditions. Wells used to reduce pressure must be designed and installed so that each well in the system can accommodate its share of

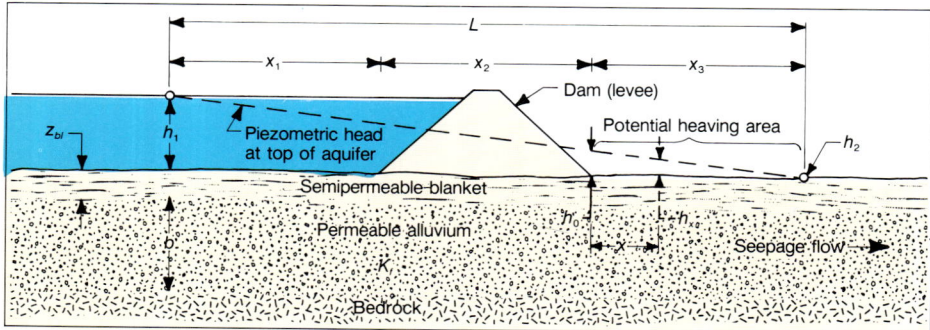

Figure 22.38. Cross section through a dam or levee structure showing the transmission of the static hydraulic head down river or landward of the structure.

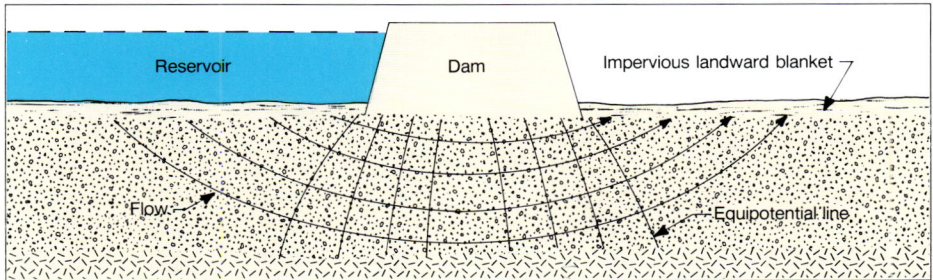

Figure 22.39. Seepage beneath a dam or levee through unconsolidated, homogeneous, isotropic sand.

the total flow in the permeable zone beneath the dam or levee without the use of pumps or the creation of excessive head losses when flowing free. Thus a comprehensive understanding of the flow dynamics in the ground and the effect of this flow on the stability of the soils is required for good relief well design. Because groundwater flow problems associated with relief wells are essentially planar, that is, the groundwater motion is substantially the same in parallel planes away from the structure, it is relatively easy to diagram the problem in two dimensions by the use of streamlines and equipotential lines to construct a flow net (Figure 22.39). The upper and lower boundary conditions controlling the flow are usually easily described. Thus, the unit flow volume through each section along the levee or dam, q, is given by:

$$q = \frac{N_f}{N_c} Kh \qquad (22.18)$$

where
N_f = the number of flow channels in the flow net
N_c = the number of potential drops along each channel
K = hydraulic conductivity
h = total loss in head

Total flow through a section, with a width of w, is given by the equation:

$$Q = \frac{TIw}{1440} \qquad\qquad Q = TIw \qquad (22.19)$$

where
T = transmissivity, in gpd/ft
I = the hydraulic gradient, in ft, and head loss is measured between the seepage entrance and the seepage exit.

where
T = transmissivity, in m²/day
I = hydraulic gradient, in m, and head loss is measured between the seepage entrance and the seepage exit.

As flow occurs, the friction of the water moving through the soil causes an energy transfer from water to soil (head loss). This energy transfer is the seepage force that is responsible for producing "quick" conditions in the soil landward of the dam or levee structure (Figure 22.40). It is vital to be able to predict when surface instability will occur for a certain structure. Terzaghi (1922) was the first engineer to examine this problem and his analysis is still valid today.

The concept of excessive uplift pressures is shown in Figure 22.40. H_1 is the potential head produced by static pressures in the reservoir on the area landward of the dam structure. In this instance, water is being driven up to saturate the soil; excess seepage drains away once it reaches the ground surface. H_2 is the gravity head produced by the saturated soil; z_{bl} is the thickness of the impermeable materials placed landward of the structure.

To evaluate the potential for soil heaving, it is necessary to determine the effect of the reservoir pressure on the surface materials. The upward force on a soil column caused by reservoir pressure, F_w, is given by the following equation:

$$F_w = (H_1 - H_2)\gamma_w \quad (22.20)$$

where
γ_w = the specific gravity of water

The force exerted by the soil is given by:

$$F_s = z_{bl}\gamma_{ss} \quad (22.21)$$

where
γ_{ss} = the submerged specific gravity of the soil

A critical gradient, i_c, (when heaving begins) occurs when:

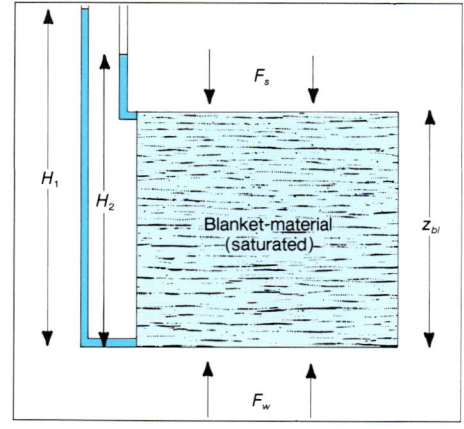

Figure 22.40. Pressure-relief wells are required when the force of the water pushing upward overcomes the submerged weight of the blanket material.

$$i_c = \frac{H_1 - H_2}{z_{bl}} = \frac{\gamma_{ss}}{\gamma_w} = (1 - \eta)(\gamma_s - 1) \approx 1 \quad (22.22)$$

where
η = porosity
γ_s = specific gravity of the soil

In practice, heaving may not occur until the critical gradient is as low as 0.7. The gradient should be at least 1.5 or 2 for minimum safety (Corps of Engineers, 1955).

In the field, the critical gradient can be determined by measuring the hydrostatic head beneath the top (relatively impermeable) layer of soil with a piezometer at the time a sand boil appears. The gradient required to keep a sand boil active is equal to or less than that required to create the sand boil in the first place. In most cases, if the depth of the clay overburden is twice the head at the base of the semipermeable blanket (z_{bl} is equal to or greater than $2 h_x$), few seepage problems will occur. In completely permeable conditions (sand overburden) where the head down river of the dam is equal to or greater than $z_{bl}/2$, seepage will occur and pressure-relief wells must be installed. The actual value of the net head above the ground, h_x, at a distance x from the toe, that may induce heaving can also be determined by:

$$h_x = h_0 e^{-cx} = h_0 e^{-x/x_3} \quad (22.23)$$

where
e = exponential function equal to 2.718
h_0 = head at the toe of the dam or levee
x_3 = distance from the toe of the dam to the point where h_x equals zero

c = a constant equal to: $\quad \dfrac{1}{x_3} = \dfrac{K_b}{K_f z_{bl} b}$

where K_b is the hydraulic conductivity of the blanket materials, K_f is the hydraulic conductivity of the formation, and b is the depth of the permeable formations.

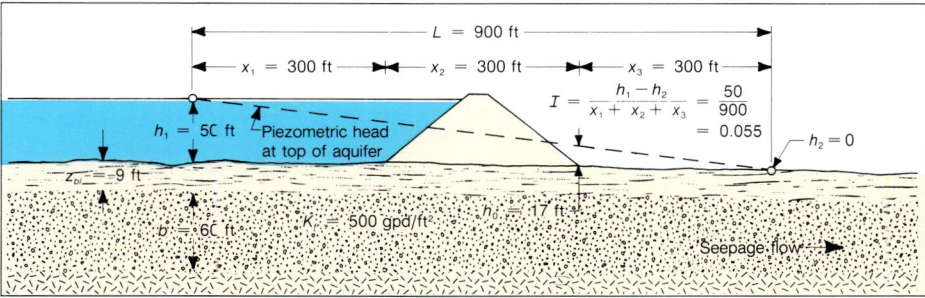

Figure 22.41. Cross section through a dam showing the location of relief wells.

The actual head (h_c) at the toe of the dam or levee depends on several factors:
1. The total head in the reservoir or river.
2. Proximity of the recharge area to the structure.
3. The transmissivity value of the permeable formation.
4. The thickness of the landward surface layer.
5. The hydraulic conductivity of the landward surface layer.

An example will illustrate the major elements in estimating the layout of a pressure-relief system. The reader should be aware that detailed design guidelines have been developed by the U.S. Army Corps of Engineers. Once the need for pressure-relief wells has been established, it is necessary to estimate the volume of flow taking place through a unit length of the structure (Figure 22.41):

$$Q = \frac{KIA}{1440} = \frac{TIw}{1440} = \frac{30{,}000 \cdot 0.055 \cdot 400}{1440} = 458 \text{ gpm } (2{,}500 \text{ m}^3/\text{day})$$

The pressure at the base of the landward blanket at the toe of the dam is calculated to be 17 ft (5.2 m) ($h_x/h_1 = x_3/L$ or $h_x = 50 \times 300/900 = 17$).

Heaving will not occur if the pressure is equal to or less than half the blanket thickness or $9/2 = 4.5$. Thus, the maximum pressure that can be safely tolerated is 4.5 ft (1.4 m) of head. Pressure-relief wells must then reduce the head 12.5 ft (3.8 m) at all points.

If all the flow could be removed by one 12-in (305-mm) diameter well, the specific capacity would be approximately 15 gpm per ft (268 m^3/day per m) of drawdown (Table 22.6). If the actual operating efficiency is 80 percent, the specific capacity is 12 gpm per ft (215 m^3/day per m) of drawdown. At the maximum drawdown required to stabilize the toe [12.5 ft (3.8 m)], the yield would be 150 gpm (818 m^3/day). This figure is about one-third the volume that must be removed from the 400-ft (122-m) wide section.

Although it may appear that a minimum of three wells would be sufficient to accommodate all the flow in the section, experience has shown that it is unrealistic to expect so few non-pumped wells to relieve the pressure build-up efficiently. In this example, approximately 10 wells would be specified. To verify that enough drawdown will be created, a distance-drawdown graph is constructed as shown in Figure 22.42 using the average flow rate expected from each well [$Q = 46$ gpm (251 m^3/day)]. The drawdown in each well would then be $46/12 = 3.8$ ft (1.2 m); Δs is calculated from the Jacob equation using the assumed yield and the given transmissivity value:

Table 22.6. Theoretical Specific Capacity for 100-Percent Efficient Water Wells, in gpm/ft of drawdown

Transmissivity, gpd/ft	Confined Aquifer Storage Coefficient = 5×10^{-4}						Unconfined Aquifer Storage Coefficient = 1×10^{-1}					
	6-in Well			12-in Well			6-in Well			12-in Well		
	12 hrs	1 day	10 days	12 hrs	1 day	10 days	12 hrs	1 day	10 days	12 hrs	1 day	10 days
1,000	0.6	0.5	0.5	0.6	0.6	0.5	0.9	0.8	0.7	1.0	0.9	0.8
2,000	1.1	1.0	0.9	1.2	1.1	1.0	1.6	1.5	1.3	1.9	1.7	1.4
3,000	1.6	1.5	1.3	1.7	1.7	1.5	2.3	2.2	1.9	2.6	2.5	2.1
4,000	2.1	2.0	1.8	2.3	2.2	1.9	3.0	2.9	2.4	3.5	3.3	2.7
5,000	2.6	2.5	2.2	2.8	2.7	2.4	3.7	3.5	3.0	4.2	4.0	3.3
10,000	4.9	4.6	4.2	5.4	5.1	4.5	7.0	6.7	5.7	7.9	7.5	6.2
15,000	7.2	7.0	6.2	7.9	7.5	6.7	10.3	9.7	8.3	11.5	10.8	9.1
20,000	9.5	9.2	8.2	10.3	9.9	8.8	13.3	12.6	10.8	14.9	14.0	11.8
30,000	13.9	13.4	12.0	15.1	14.5	12.8	19.4	18.4	15.9	21.6	20.5	17.4
40,000	18.3	17.6	15.8	19.8	19.0	16.9	25.4	24.0	20.8	28.1	26.7	22.7
50,000	22.6	21.9	19.6	24.3	23.5	20.9	31.1	29.5	25.5	34.6	32.6	27.9
60,000	26.9	26.0	23.3	29.0	27.9	24.8	37.1	35.0	30.4	40.9	38.8	33.1
70,000	31.1	30.9	27.0	33.4	32.3	28.7	42.6	40.6	35.2	47.0	44.2	38.4
80,000	35.4	34.1	30.7	38.0	36.6	32.7	48.2	46.0	39.8	53.5	50.6	44.4
100,000	43.6	42.1	38.0	46.8	45.1	40.3	59.3	56.7	49.4	65.3	62.2	53.4
125,000	53.8	52.2	47.0	57.0	56.0	50.0	73.2	69.8	60.8	80.2	76.5	65.8
150,000	64.4	62.1	56.0	69.0	66.5	59.6	86.7	82.8	72.4	95.3	90.7	78.1
175,000	73.7	71.8	65.0	79.6	76.9	68.9	99.5	95.7	83.5	110	105	90.4
200,000	84.3	81.8	73.7	90.7	87.5	78.5	113	108	95.0	124	118	102

See Appendix 22.A for this table in metric units.

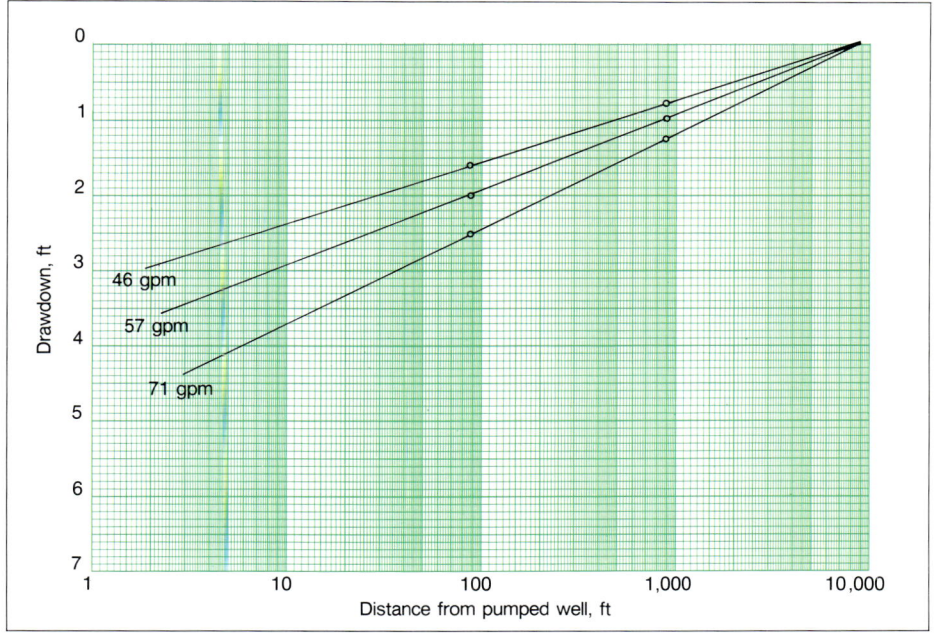

Figure 22.42. Theoretical drawdown plots for various yields from the pressure-relief wells.

$$\Delta s = 528 \cdot 46/30{,}000 = 0.8 \text{ ft } (0.2 \text{ m})$$

To construct the distance-drawdown graph, the zero drawdown point (r_o) is assumed to be equal to L. Once the distance-drawdown plot is drawn, it can be used to estimate the total drawdown in each well caused by its own flow and the interference from other wells. Using the cumulative drawdown method described in the dewatering section, it is determined that the pressure will be relieved adequately if the wells drain at about 60 gpm (327 m³/day). Because the flow is relatively low, 6- or 8-in (152- or 203-mm) diameter wells could be used. If only a few wells are used, the wells should be 12 inches (305 mm) in diameter. Head losses in the riser pipe and screen were not considered in this discussion and all wells were assumed to be installed at the optimum depth.

In the example above, a simple trial-and-error method has been used to arrive at a reasonable design. More rigorous approaches have been taken by Muskat (1937), Middlebrooks and Jervis (1947), Bennett (1945 and 1946), and the Corps of Engineers (1955). Corps of Engineers publication CE1307 (1972) provides a guide to relief-well specifications. In general, pressure-relief wells should be designed and constructed to the same specifications that apply to water-supply wells.

WELLS FOR HEAT PUMPS

A heat pump is a mechanical device that extracts heat energy from a source such as air or water and makes it available for cooling or heating purposes. By definition, a heat pump upgrades the ambient levels of heat in the heat source. That is, it increases or decreases the absolute temperature significantly with respect to that found in the

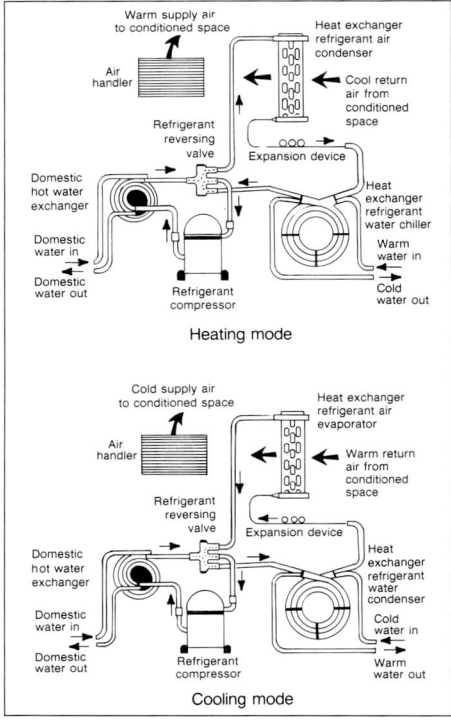

Figure 22.43. Heating and cooling modes of heat pumps. *(Hughes, 1980)*

heat source (Heap, 1979). The function of a heat pump is to concentrate energy and make it available. The amount of energy required to run a heat pump is usually only one-third to one-fifth of the total heating or cooling energy provided by the pump (Kazmann and Whitehead, 1980). The coefficients of performance, that is, the ratio between the number of BTU's moved from the heat source to another location divided by the BTU's used in performing the work, are in the range of 3.5 to 4.5.

Water used in the system serving as the heat source can come from a surface water body (river or lake), groundwater, water contained in a storage tank installed in the ground, or water contained in a closed loop that is circulated in a well or in the ground. Enough heating or cooling can be obtained from these sources to provide all or almost all of the requirements for homes and commerical buildings. This discussion will focus on groundwater as the heating and cooling source.

A description of the various heat-pump designs is beyond the scope of this book, but the basic systems consist of reverse cycle refrigeration units connected to a water circulating loop capable of absorbing or rejecting heat to maintain a comfortable temperature range (Jones, 1980). The basic components of a heat-pump system are compressor, condenser, evaporator, expansion device, and refrigerant (Hughes, 1980).

Heat-pump operation permits the water source to be used as either a sink or a source, depending on whether heating or cooling is required. This can be done because the cold or heat extracted from the water can be stored in, or released from, an intermediate fluid called the refrigerant (a typical refrigerant is Freon™ manufactured by DuPont Corporation). Schematic diagrams of both the heating and the cooling modes for a groundwater heat pump are shown in Figure 22.43.

Groundwater is an excellent source for heating or cooling because the temperature is essentially constant at an approximate depth of 20 ft (6.1 m), mirroring the mean annual temperature at the surface. At depths greater than 100 ft (30.5 m), the temperature of the groundwater rises at a rate of 1°F per 100 ft (0.6°C per 30.5 m). Above 20 ft, the temperature changes slowly throughout the year, depending on weather conditions at the surface.

The volume of water required to run the system will vary according to the size and design of the system. Typical water consumption for a home falls between 1 and 3 gpm (5.5 and 16.4 m³/day) per 12,000 BTU (12.7 million J) output (Gass, 1980). Thus, a 4-ton [48,000 BTU (50.6 million J)] heat pump would require a flow of 4 to

12 gpm (21.8 to 65.4 m³/day). When the water needed to run the groundwater heat pump is obtained from a water supply well, this discharge rate must be added to the peak-demand allowance required for the home. If the well cannot produce this discharge, enough storage must be provided to meet the peak demands; that is, a storage tank must be installed. In bedrock areas where the wells are deep, the volume in storage in the well bore is usually sufficient, even though the specific capacity of the well may be quite low. The severity of the climate is the most critical factor in determining the volume of storage required.

In the majority of groundwater heat-pump installations, the source water can be (a) taken from a well and discharged at the surface to a storm sewer or allowed to infiltrate at the surface or in a drainfield, (b) taken from and recharged to the same well, or (c) taken from one well and recharged to another. For (a) above, the well can be either the well used for potable water supplies or another well specially constructed for the heat pump. The well should be able to deliver 5 to 15 gpm (27.3 to 81.8 m³/day). Because no water will be injected into the well, no modification of the typical water well design and construction standards need be made.

The situation described in (b) is not recommended if the well is being used for potable water, because of the increased potential for contamination and additional maintenance. In addition, if the water is used only for cooling purposes, the temperature of the water near the intake portion of the well may increase.

In the two-well system described in (c), water can be withdrawn from one well and injected into the second, or alternately withdrawn and injected from either well, depending on the need for cold or warm water. If the one-way flow is used, the injection well should be located downgradient from the production well so that the water least desirable for the system is carried downgradient.

The alternating two-well system is ideal for distributing the effect of both heating

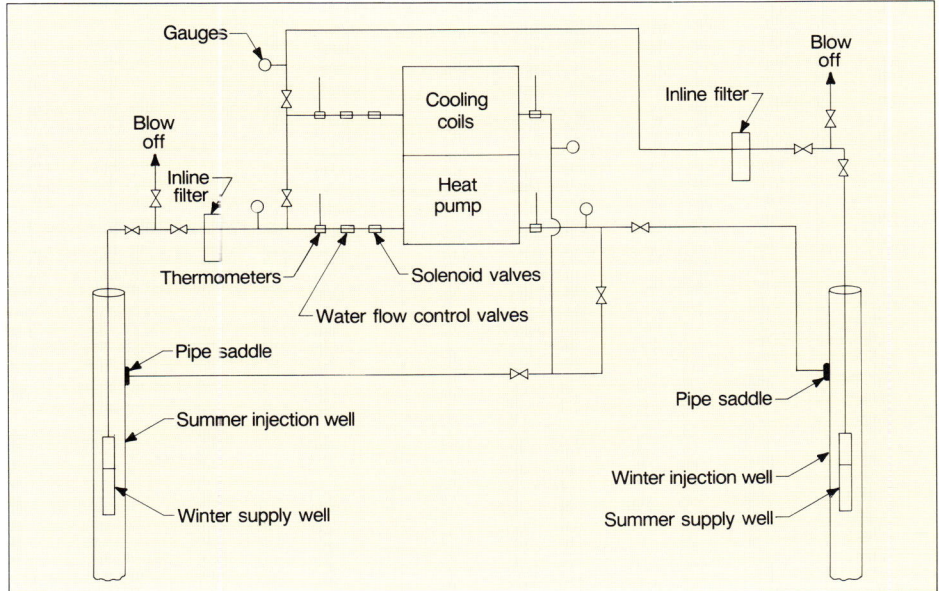

Figure 22.44. Two-well system for heat pumps. *(Schockley, 1980)*

and cooling (Figure 22.44). During the summer, water is removed from the right-hand (summer) well, heat extracted, and the water reinjected into the winter supply well. During the summer, a slug of water with a temperature higher than ambient forms around the summer injection well. During the ensuing winter, water is pumped from the winter supply well. All or nearly all of the slightly heated water is pumped from the ground and reinjected into the winter injection well after passing through the heat pump. Naturally, much heat is removed from this water and, when reinjected, its temperature is less than ambient. If the amount of heating and cooling are about the same, the long-term impact on the temperature of the groundwater at the site is negligible. Energy losses (from either the heated or the cooled water) do occur, however, during the time the slug of water is in the aquifer. These losses are attributable to down-gradient movement of the slug of water, conduction of heat to aquifer materials and groundwater around the slug, and convection.

It is possible to calculate the size of the plume of heated or cooled water, the rate at which it loses or gains heat, and its rate of movement (Schaetzle et al., 1980). The following variables can be used to determine the recommended distance between the supply and the recharge (injection) well: the relative times the water will be heated or cooled, the pumping and injection rate, the porosity of the aquifer and its thickness, and the heat capacity of the aquifer matrix and the groundwater (Table 22.7). It is important to note that the values in this table were calculated on the basis of the standard assumptions concerning aquifer isotropy, homogeneity, constant thickness, and other assumptions noted in Chapter 9. Actual field conditions may produce far different results, as the following example will illustrate.

In Minneapolis, Minnesota, a 6-in diameter well used in a refrigeration/air-conditioning cycle was producing 88 gpm through a screen placed in an aquifer 8 ft thick. The total depth of the well was 115 ft, and the static water level was 45 ft. A well of similar dimensions 500 ft away was used to inject this water, at a temperature of 88°F, back to the aquifer. After 3 days of operation, the temperature of the water from the production well had risen from 48°F to 62°F. The high hydraulic conductivity of the sediments allowed rapid movement of the heated water toward the production well. Under ordinary circumstances, the 500-ft distance between wells would have been appropriate. In this case, however, the wells should have been placed in different formations because of the excessively high hydraulic conductivity.

The injection well in this example also exhibited the common tendency in a two-well system to become much less efficient within a short time. The well overflowed after 6 months of use, indicating that the recharge level had risen about 30 ft. After being treated with polyphosphates and chlorine, the recharge level returned to normal. A chlorinator was installed to treat the water from the production well. Overflow again occurred after 8 months; the well must now be treated every 7 months with acid, polyphosphates, and shock chlorination to maintain the system.

Plugging of injection wells by sand, chemical deposits, or bacterial growths is a major problem that can be alleviated, but usually not completely eliminated, by specifying longer screens for the injection well. As suggested earlier, the injection well screen should be at least twice as long as the production well screen. Where possible, it should be even longer so that maintenance requirements can be minimized.

In summary, the efficiency of two-well heat-pump systems depends on whether the following criteria are followed:

Table 22.7. Required Spacing Between Twin Wells, ft (Porosity = 20%)*

Aquifer Thickness	10 ft			20 ft			30 ft			40 ft			50 ft			80 ft			100 ft		
Pumping Time in Days	100	140	210	100	140	210	100	140	210	100	140	210	100	140	210	100	140	210	100	140	210
Q, gpm																					
10	114	127	145	98	109	127	86	100	117	79	94	108	72	83	99	55	70	83	55	65	78
20	158	177	200	135	153	177	122	140	164	109	126	151	103	117	138	81	98	118	77	90	106
30	195	217	248	171	191	220	149	171	199	135	154	183	128	144	171	103	120	144	94	110	131
40	227	251	289	195	220	253	174	195	231	154	178	213	144	165	197	118	139	168	108	125	150
50	253	281	323	217	245	285	194	219	261	173	201	238	160	185	218	133	155	189	120	141	169
60	280	310	358	237	270	311	208	241	280	189	218	262	176	205	245	147	169	206	132	155	185
70	303	338	389	256	294	338	226	260	304	205	237	283	188	220	264	159	183	222	143	167	200
80	325	360	418	276	315	364	242	278	325	220	254	301	202	235	282	170	196	237	154	178	214
90	345	382	445	293	335	386	256	294	342	234	268	320	213	250	300	179	208	252	162	188	227
100	365	403	472	309	353	408	268	310	362	246	285	338	224	263	308	189	219	266	170	200	240
200							385	445	515	353	411	485	310	369	436	267	310	378	240	283	343
500													510	585	697	420	495	585	382	445	530
1000													712	824	986	590	682	826	540	632	756

*To correct the spacing for a porosity of 10 percent, multiply the tabulated spacing by 1.05. To correct the spacing for a porosity of 30 percent, multiply the tabulated spacing by 0.95. The spacing so calculated is within 1 percent of the spacing calculated more precisely. (Kazmann and Whitehead, 1980)

1. The wells are designed and constructed to the highest water well standards as outlined in Chapters 13 and 10, respectively.
2. The wells are spaced adequately, based on discharge rates and aquifer parameters.
3. In highly stratified, coarse-grained aquifers, the production and injection wells are not placed at the same depth.
4. The screen in the injection well is at least twice as long as the screen in the production well.
5. Maintenance schedules are established and followed where necessary.

Closed-Loop Heat-Pump System

It may not be possible to use any of the three heat-pump systems already discussed because of environmental concerns or local laws. Most municipalities, for example, prohibit the drilling of water wells in areas served by municipal water systems. Groundwater quantity or quality may be a problem in other areas. Furthermore, recharge or surface disposal of water is prohibited in many urban and suburban areas. When these restrictions are present, closed-loop, earth-coupled heat-pump systems provide an environmentally safe way to use geothermal energy.

The two basic types of closed-loop heat pumps consist of a horizontal ground coil buried below the frost zone or a vertical collector loop mounted in a well (Gass, 1983) (Figure 22.45). A closed-loop system consists of three major components: (1) the closed loop, (2) the pump required to move water in the system, and (3) the water-to-water or water-to-air heat pump (Amsterdam, 1983).

If the water in the loop is cooler than the ground or water around it, heat will flow from the ground to the loop. Water in the loop then flows to the heat pump, where the heat is removed. In the cooling mode, the system is reversed; heat is removed

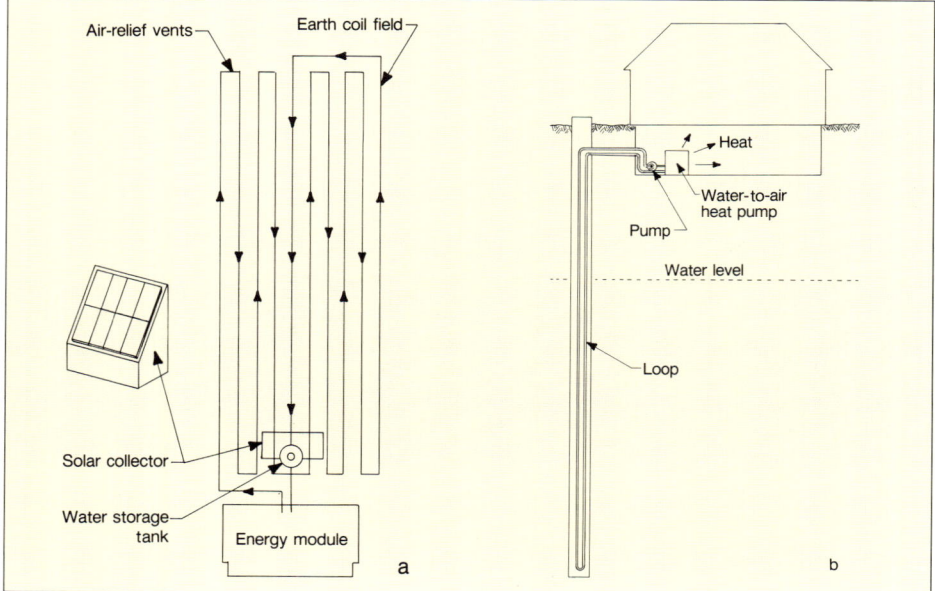

Figure 22.45. (a) Horizontal ground coil and (b) vertical collector loop.

from the building and transferred to the loop. The excess heat is conducted to the ground. The loop itself does not have to be in contact with groundwater to function, but if it is partially immersed in groundwater the transfer of heat is facilitated. Most loops are constructed of plastic materials.

Brine, water with a nontoxic antifreeze additive, or any nontoxic fluid with a low freezing point can be circulated through the coil or loop. The heat pump must be able to operate efficiently at temperatures lower than used for ordinary groundwater heat pumps. If the heating season lasts much longer than the cooling season, ice may form around the well bore or beneath the coil. Solar collectors are sometimes installed and connected to the coil or loop to add heat to the ground, especially during the summer (Gass, 1983). In general, the heating efficiency of an earth-coupled heat pump may be somewhat lower than for a groundwater heat pump.

SURFACE-WATER WITHDRAWAL

The volume of surface water removed for various purposes is far greater than the volume of groundwater. Modern filtering technology developed for groundwater use has been successfully adapted to meet the specific demands of surface-water removal. Drilling contractors located near surface-water bodies may become involved in obtaining water supplies from these resources.

It is clear that the surface-water environment is different from the groundwater environment. The chemical properties of surface water, for example, vary drastically from place to place because of both the natural chemistry and the effect of man-made contaminants. Consideration of biological factors is vital in all surface-water screen installations, whereas biological activity is limited in most groundwater environments. The physical properties of the surface-water environment will differ according to whether it is a river, lake, reservoir, or ocean. Waves, ice formation, salinity, temperature, turbidity, tidal effects, river gradients, and pH are just some of the important characteristics of surface-water bodies. Design of intake systems for specific sites requires careful evaluation of these characteristics.

Fundamental criteria for a water intake system include:
1. A reliable water source.
2. Identification of a withdrawal point from that source.
3. Selection of the most appropriate screening device.

Choice of the water source is usually obvious because it must be close to the processing facility or project.

Three withdrawal-point options exist for most sites: approach channel, onshore, or offshore (Figure 22.46). The withdrawal point should be selected so that sufficient water depth is available, plugging conditions are minimized, the fish habitat will be protected, and costs will be reasonable.

Conventional screening methods include traveling water screens, rotating drums, traditional cylinder screens, and stationary panels. Traveling water screens and rotating drums are used primarily for on-land installations, whereas cylinder screens and stationary panel arrangements are used for both onshore and offshore locations.

Traveling water screens and rotating drum screens are mechanically operated units designed to move the screen surface in a rotating manner out of the water for cleaning. As the screen surface emerges from the water, a jet spray directed outward from inside the screen washes the screen surface. Traveling water screens are used for high-capacity

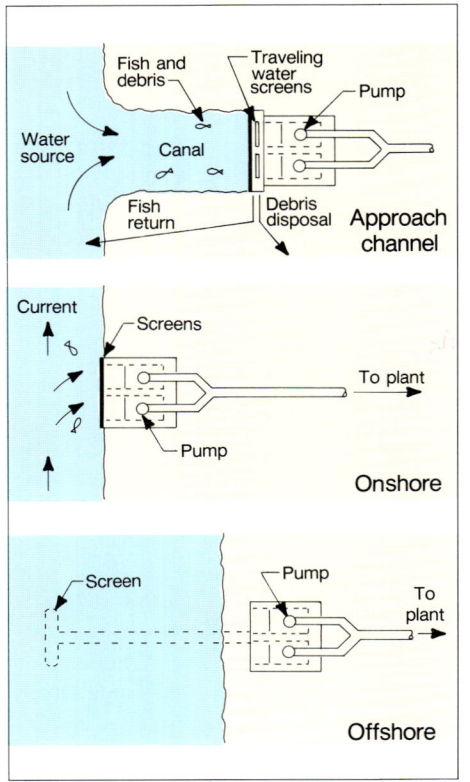

Figure 22.46. Inlet locations with respect to shoreline.

and continuously operating systems, and have significant operating and maintenance costs. Rotating drum screens are expensive and are most frequently used in the fish hatchery industry.

A single-cylinder screen or two cylinders connected in a "T" arrangement are used for offshore bottom installations. These screens can be sized for a wide range of capacities. For the largest systems, multiple screens can be used with a manifolded intake line that can be balanced for even flow. An advantage of the cylinder screen is that it allows flexibility in siting.

Stationary panels are used most frequently for onshore installations at the entrance to intake channels, as shown in Figure 22.47a. Periodic manual cleaning is usually required. Stationary panel screens are also used offshore to screen the entrance of a velocity cap (Figure 22.47b). A velocity cap consists of a vertical intake pipe capped by a large, solid cover to prevent vertical flow from entering the intake pipe. Fish rarely escape from vertical flow but can escape from horizontal flow. Because flow enters horizontally through stationary panels around the velocity cap, the cap prevents fish from being entrained. The screens may be constructed of perforated plate or pipe, slotted plate or pipe, mesh wire, or V-shaped wire. V-shaped-wire screens provide maximum open area and optimum inflow conditions.

Certain types of screens are more useful for specific types of water bodies. Major considerations for screen selection are as follows:

Lake Intakes — Nearshore conditions in lakes usually involve shallow water with large fish populations and debris accumulations. Thus, lake intakes should be offshore in deep water; a cylinder screen or a "T" screen is common because it can be placed in optimum intake locations. In onshore locations, traveling water screens are used to handle the debris and fish accumulations. Dredging is usually required to keep the channel open to the shore facility.

River Intakes — Because stream flow keeps fish and debris from clogging the screens, cylinders or "T" screens oriented parallel to current flow are most effective. End cones on the upstream end of the screen direct current flow past the screen surface. Traveling water screens and stationary panel screens are used for onbank installations.

Reservoir Intakes — Although similar to lakes, these sites undergo wide fluctuations in water levels and thus require offshore intakes. Multilevel withdrawal is possible by using tower or dam-face-mounted screens, as shown in Figure 22.48.

ALTERNATIVE USES FOR WELLS AND WELL SCREENS   791

Sea-Water Intakes — Screens placed in salt water are affected by severe corrosion and biofouling, and thus require access for periodic maintenance and cleaning. Bulkhead-mounted screens on vertical rails provide easy access for maintenance and can be installed either onshore or offshore. Screens constructed of 70-30 percent copper-nickel generally provide the best resistance to biofouling. In applications where bio-

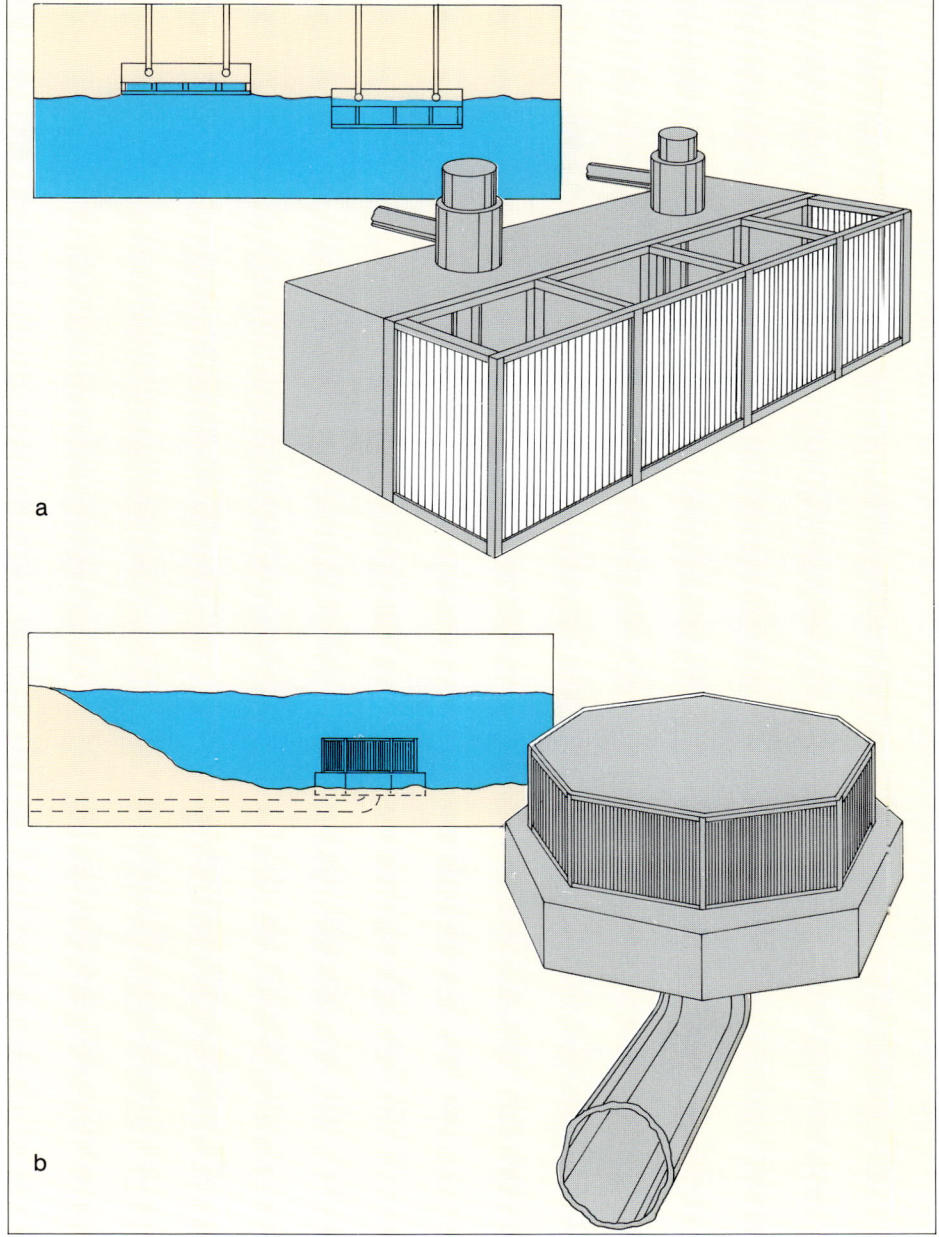

Figure 22.47. (a) Stationary panels in onshore installations; (b) stationary panels in a velocity cap.

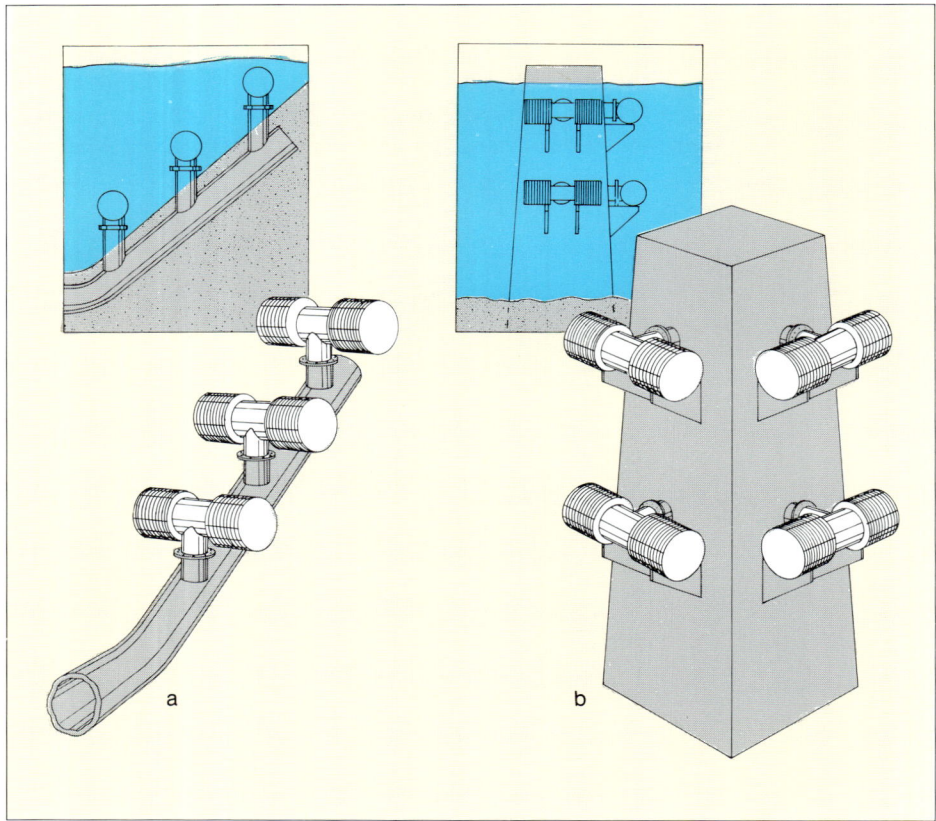

Figure 22.48. (a) Dam-face-mounted screens; (b) tower mounted screens.

fouling is not a problem, 316L stainless steel or an alloy (Carpenter 20) provide adequate resistance to chemical corrosion. Performance of marine intakes is subject to a wide variety of site conditions that are difficult to predict. Thus, a test program to assess specific site conditions is recommended.

Passive Screening

Passive screening is an engineering alternative for removing surface water. Designed to minimize initial costs and to reduce maintenance, passive screening is based on two fundamental concepts: (1) reduced inflow velocity and (2) use of natural currents or force of gravity for cleaning. Limiting the maximum intake velocity to 0.5 ft/sec (0.15 m/sec) minimizes the build-up of solid particles and aquatic organisms on the screen. Materials that do collect on the screen are held so lightly that they are swept away periodically by currents or are permeable to normal inflow. Low inflow velocities and minimum clogging conditions assure that the head loss across the screen surface will be less than 0.1 ft (0.03 m) of water. The screen axis should be oriented parallel to current flow, to provide the most efficient cleaning conditions.

In cases of extremely heavy debris loads or inadequate current conditions, an air-backwash system for auxiliary cleaning may be included as part of the screen assembly. This system consists of an air-distributor pipe within the screen cylinder that will lift materials off the screen surface with a sudden blast of air.

To improve the basic objectives of velocity reduction and natural cleaning, the passive screen is constructed of V-shaped wire to prevent back eddies, lost energy, and excessive head loss. The outer surface of the V wire has a smooth, flat surface that presents no hazard to fish and can easily be kept clean by currents.

The following basic site and pumping parameters must be considered in the design of a passive screen system:

1. Water source: river, lake, reservoir, or sea water.
2. Water depth, including low, high, and mean levels.
3. Approximate frequency of high and low water depths.
4. Velocity and direction of water currents.
5. General water quality.
6. Salinity and pH levels.
7. Icing possibilities.
8. Maximum pumping rate.
9. Pump-screen relationship (direct flow or gravity flow to well pump).
10. Slot opening.
11. Intake line size.
12. Connection type and size.

Inflow velocities can be controlled to a maximum of 0.5 ft/sec (0.15 m/sec), with an average velocity of 0.35 ft/sec (0.11 m/sec), by designing the screen slots and diameter correctly. The ratio of maximum to average velocity is called the uniformity coefficient; for passive screens, this ratio should be 1.25 to 1.6. The basic concept of passive screening is that, by controlling the intake velocity to less than 0.5 ft/sec, the flow velocity away from the screen will be significantly lower. As the distance from the screen surface increases, inflow velocities toward the screen decrease exponentially so that, at a distance equal to one half the diameter of the screen, flow is less than 20 percent of the surface velocity. If a screen has an average inflow velocity of 0.35 ft/sec, flow one-half the screen diameter away will be only 0.06 ft/sec (0.02 m/sec), which is virtually undetectable.

A passive intake system consists of a cylinder screen with a diameter-to-length ratio equal to 1. The cylinder screen has a solid plate on one end and an adaptor plate with the outlet pipe on the opposite end. The final assembly may be a single-cylinder unit or a "T"-screen assembly. A large version of the "T"-screen configuration, called a modular "T", is for deep-water and large-capacity systems.

The drum screen is a large-size adaptation of the single-cylinder screen. It is used for high-capacity withdrawals and, because of size, is normally mounted vertically. Attachment of the screen assembly to the intake line is usually made with either a flange connection or a stud-ring connected to a flange. Other attachment arrangements are welded pipe, slip-on pipe, and victaulic coupling. These screen configurations are shown in Figure 22.49; recommended capacities and dimensions are given in Table 22.8. Screen cylinders can be replaced individually if damaged.

Important environmental conditions that affect the operation of a passive screen include the extent of the open-water areas, water depths, salinity and corrosion levels, fish and debris conditions, and potential for ice formation. The following recommendations should be observed:

1. The main water body must feed the intake evenly and provide good-quality

Table 22.8. Recommended Capacities and Dimensions for Intake Screens

Capacity gpm	Capacity m³/day	Nominal Diameter in	Nominal Diameter mm	Height in	Height mm
Drum Screen Intakes					
100 - 400	545 - 2,180	12	305	14	356
250 - 900	1,360 - 4,910	18	457	20	508
450 - 1,600	2,150 - 8,720	24	610	26	660
700 - 2,300	3,820 - 12,500	30	762	32	813
1,050 - 3,500	5,720 - 19,100	36	914	38	965
1,300 - 4,600	7,090 - 25,100	42	1,070	54	1,370
1,900 - 6,200	10,400 - 33,800	48	1,220	61	1,550
2,250 - 7,800	12,300 - 42,500	54	1,370	68	1,730
3,000 - 9,500	16,400 - 51,800	60	1,520	75	1,910
3,750 - 11,600	20,400 - 63,200	66	1,680	82	2,080
4,600 - 14,000	25,100 - 76,300	72	1,830	89	2,260
5,500 - 16,500	30,000 - 89,900	78	1,980	96	2,440
6,300 - 18,500	34,300 - 101,000	84	2,130	103	2,620
"T" Intakes					
200 - 850	1,090 - 4,630	12	305	42	1,070
300 - 1,400	1,640 - 7,630	16	406	54	1,370
500 - 1,900	2,730 - 10,400	18	457	60	1,520
650 - 2,400	3,540 - 13,100	21	533	69	1,750
900 - 3,250	4,910 - 17,700	24	610	78	1,980
1,200 - 4,000	6,540 - 21,800	27	686	87	2,210
1,450 - 4,700	7,900 - 25,600	30	762	96	2,440
1,800 - 5,800	9,810 - 31,600	33	838	104	2,640
2,100 - 7,000	11,400 - 38,200	36	914	114	2,900
High-Capacity "T" Intakes					
2,500 - 9,250	13,600 - 50,400	42	1,070	142	3,610
3,750 - 12,500	20,400 - 68,100	48	1,220	161	4,090
4,500 - 15,750	24,500 - 85,800	54	1,370	180	4,570
6,000 - 19,250	32,700 - 105,000	60	1,520	199	5,050
Modular "T" Intakes					
2,500 - 9,500	13,600 - 51,800	42	1,070	142	3,610
3,500 - 12,750	19,100 - 69,500	48	1,220	161	4,090
4,500 - 15,700	24,500 - 85,600	54	1,370	180	4,570
6,000 - 19,250	32,700 - 105,000	60	1,520	199	5,050
7,000 - 23,750	38,200 - 129,000	66	1,680	218	5,540
9,000 - 28,000	49,100 - 153,000	72	1,830	237	6,020
11,000 - 33,000	60,000 - 180,000	78	1,980	256	6,500
13,000 - 39,000	70,900 - 213,000	84	2,130	275	6,990

Note: Screen capacities overlap because of the range of slot openings and screen construction.

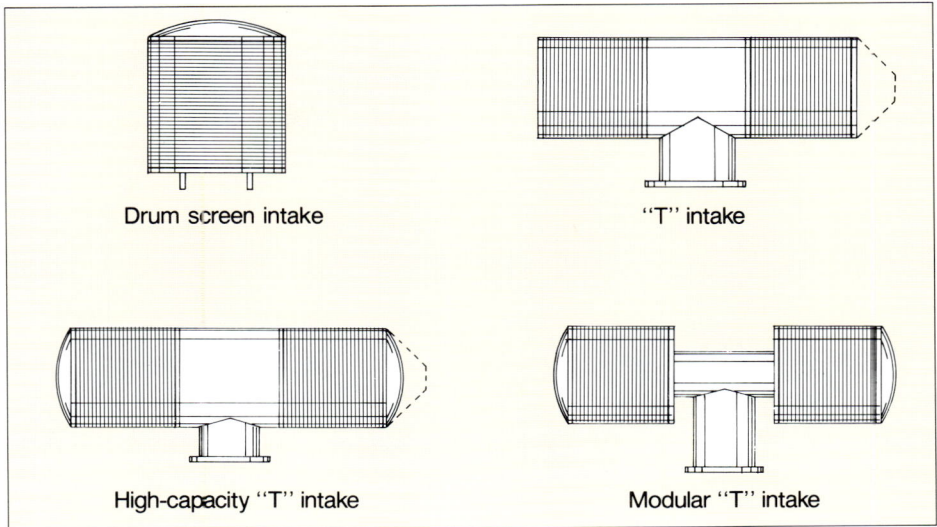

Figure 22.49. Various screen configurations for intake screens.

water. Dead-end approach channels should be avoided because they concentrate flow and collect debris and fish.

2. Adequate water depth is a critical requirement for limiting maximum inflow velocity to 0.5 ft/sec (0.15 m/sec). Screens must be placed a minimum distance of one-half the screen diameter from any boundary, such as a surface, bottom, or solid wall, to achieve this flow. In addition, a distance of one diameter must be provided between adjacent screens.

3. For long and reliable operating life, screens installed in fresh water are commonly made of 304 stainless steel. Marine or brackish waters usually require a higher grade of stainless steel such as 316L; if biofouling is a severe problem, a copper-nickel alloy may be required in order to inhibit rapid build-up of marine organisms.

4. To keep the screen surface free of any foreign substances, the screen axis should be oriented parallel to current flow.

5. Potential icing conditions caused by surface, pack, and frazil (sheet) ice must be considered. Locating screens in deeper water will generally eliminate most icing problems, but, for shallow-water situations, circulation of warm water can prevent build-up of frazil ice. For potential ice-pack problems, appropriately placed trash racks or pilings are practical preventative measures.

CHAPTER 23
Water Treatment

Water well contractors generally do not become involved in the design and construction of municipal water supply or wastewater treatment facilities. Because many water-supply systems depend totally or in part on groundwater, however, contractors should understand the operation of basic treatment processes. Furthermore, contractors should become familiar with techniques designed to reduce or eliminate the contamination threat to groundwater from wastewater plant discharges, and take an active role in promoting their use. In this chapter, the major water treatment processes are described briefly and important terms relating to treatment are defined.

Ordinarily, water well contractors have a much greater opportunity to become involved in treating individual water supply systems to correct water-quality problems. In some parts of the United States, groundwater quality problems exist at virtually all depths, and relatively high amounts of iron, sulfate, manganese, and other inorganic and occasionally organic chemicals can occur in wells. Because the contractor usually knows about these problems before a new well is completed, the customer should be advised about the type of water that can be expected and how to treat the water with various point-of-use treatment systems. If the contractor cannot install the treatment system required, a reputable water treatment specialist should be recommended.

Until the mid-1970's, few people suspected that groundwater could be contaminated. Recent research has shown, however, that groundwater quality is being adversely affected locally by a wide range of volatile organic compounds and inorganic chemicals. Many chemicals contaminating groundwater are man-made, are harmful to human health, and have entered the environment through overuse and careless disposal of waste products. Furthermore, more cases of groundwater contamination caused by pathogenic organisms such as *Giardia lamblia** and *Pseudomona* are now being discovered. Thus, treatment of groundwater is now much more common and new treatment technologies are being adopted.

Many treatment techniques mentioned in this chapter apply to public water supplies and are performed at municipal treatment facilities. Some of these techniques are

---

*This organism is currently responsible for most of the waterborne disease outbreaks in the United States (Craun, 1984).

also used in point-of-use treatment systems for individual wells. These small systems installed in homes have been popular in the past to soften water or to remove a particular inorganic chemical that was staining fixtures or causing taste problems. New types of small-scale treatment systems, especially those using granular activated

Table 23.1. Federal Regulatory Requirements for Establishing Water Quality Standards and Reporting Systems

Regulatory Body

- Federal law is established so that a state or territory can gain primary control over its own drinking water once it shows that it is willing to meet minimum standards set by the U.S. Environmental Protection Agency.

- A total of 57 states or territories are accountable units under the Federal Drinking Water Program:

    49 states have primary enforcement responsibility (primacy)

    8 states or territories have programs that are being administered by EPA regional offices (regional primacy)

Responsibilities

- Establishing Maximum Contaminant Levels (MCL's)

    Maximum contaminant levels for microbiological contaminants, turbidity, and chemical contaminants include over 20 specific chemicals, radiological elements, or organisms which are undesirable in drinking water.

- Developing a Monitoring and Reporting Program

    In order to insure that drinking water is safe, there is required monitoring and reporting for each of these MCL's at a specified frequency according to the characteristics of the public water system.

- Initiating Public Notification when Standards are Exceeded

    Each time an MCL is exceeded, the water supplier is required to provide public notification; however, notification for a monitoring or reporting violation is not required in some primacy states.

- Evaluating Water Quality Data and Storage of Data

    On an annual basis, the states and EPA regional offices are required to forward to EPA's Office of Drinking Water status reports which indicate how successfully regulatory requirements were met. These data are stored in the Federal Reporting Data System.

carbon filtration, are now used extensively to lower the concentration of organic chemicals to acceptable levels and to alleviate taste problems.

Water quality standards have been set in the United States by the U.S. Environmental Protection Agency and are presented in the U.S. Safe Drinking Water Act and the National Interim Primary Drinking Water Regulations. The primary and secondary standards of the drinking water regulations are given in Chapter 6, Tables 6.11 and 6.12, respectively. The general federal requirements are listed in Table 23.1.

Additional standards will be developed as health effects data become available for many volatile organic and synthetic organic chemicals (pesticides) that occur in a large percentage of wells tested. New regulations will be developed by the U.S. Environmental Protection Agency in five phases (Cotruvo and Vogt, 1984):

Phase 1 — Volatile synthetic organic chemicals (VOC's)
Phase 2 — Synthetic organic chemicals (SOC's), inorganic chemicals (IOC's), and microbiological contaminants
Phase 3 — Radionuclides
Phase 4 — Disinfectant by-products, including trihalomethanes (THM's)
Phase 5 — Other SOC's (pesticides) and IOC's not considered previously

COMPONENTS OF WATER TREATMENT AND WASTE TREATMENT SYSTEMS

Water supply systems consist of (1) collection or intake works, (2) purification or treatment works, (3) transmission works, and (4) distribution works (Fair et al., 1971). Water systems can be supplied by surface water, groundwater, or a combination of both. Treatment facilities usually include the ability to:

1. Improve the physical characteristics of the water with regard to odor, taste, color, turbidity, and hardness.

2. Remove or limit the concentration of toxic and nontoxic inorganic contaminants. Toxic contaminants include arsenic (As), barium (Ba), cadmium (Cd), chromium (Cr), fluoride (F), lead (Pb), mercury (Hg), nitrate (NO_3), selenium (Se), and silver (Ag).

3. Remove or limit organic contaminants that are toxic or carcinogenic: endrin, lindane, toxaphene, 2,4-D, 2,4,5-TP (silvex), methoxychlor, trichloroethylene, tetrachloroethylene, carbon tetrachloride, 1,1,1-trichloroethane, vinyl chloride, 1,2-dichloroethane, benzene, 1,1-dichloroethylene, and p-dichlorobenzene.

4. Eliminate disease-producing organisms.

5. Remove radioactive contaminants.

6. Add chemicals, such as fluoride, that contribute to health.

Transmission components of a water supply system carry the water from the source to the treatment facilities and then to the community. In the arid southwestern United States, tranmission lines may be extensive. For example, in Tucson, Arizona, a well field consisting of 24 wells is located 35 mi (56.3 km) from the city.

The last stage of the water supply system is the distribution network in which water is supplied to individual consumers via a piping and reservoir system. A major concern in the distribution system is that sufficient water be available at all times to fight fires and still maintain an adequate flow for normal use.

Another type of community water treatment facility is required to treat wastewater before it is discharged to a surface water body. Wastewater systems serving communities collect, treat, and dispose of the water used for domestic or industrial purposes in the community. In most municipalities, wastewater is collected in a system that is separate from the storm runoff water. Thus, wastewater is routed to a treatment facility while storm runoff is usually discharged to local water bodies (lakes, rivers, or streams) with no treatment. Unfortunately, storm runoff in urban areas contains high amounts of nutrients, often as high as in ordinary wastewater, which can damage aquatic environments (Shapiro and Pfannkuch, 1973). In older cities, such as Boston and New York, storm runoff water and wastewaters are collected in combined systems and all the water is treated before disposal. But in many cases, the volume of storm runoff can be so great for even minor rainfall events that treatment facilities become overwhelmed and the only recourse is to bypass the plant and discharge all water directly into a receiving body.

Treatment facilities for wastewater generally include the ability to remove putrescible material, stabilize degradable substances, remove organics, remove nutrients, and destroy disease-producing organisms. The ability of an individual treatment facility to accomplish these tasks effectively varies enormously, depending on plant design and maintenance, operator training, and the occurrence of natural disasters such as floods and earthquakes. Thus, the degree to which wastewater can be diluted after treatment may be a critical factor in preserving local environmental conditions. In the past, the closer a city was to a large receiving body such as an ocean or one of the Great Lakes, the less likely it would be for its treatment facility to be effective. But the adverse impact of poor treatment and disposal practices on lakes such as Lake Erie and Lake Michigan and on offshore areas near New York City have caused many municipalities to improve their waste management systems.

TREATMENT TECHNOLOGIES APPROPRIATE FOR MEETING DRINKING WATER REGULATIONS

The treatment technologies discussed below are applicable to systems servicing 1,000 or more consumers. Large-scale treatment facilities must achieve the objectives cited above by utilizing one or more of the various treatment technologies described in this section.

Removal of Inorganic Substances

Research by the U.S. Environmental Protection Agency shows that no one treatment method is effective for removing all inorganic chemicals. The most effective treatment methods are presented in Table 23.2. Other metallic ions such as iron, manganese, calcium, and magnesium are not harmful at concentrations found in groundwater, but are often removed or reduced in concentration for taste, color, or other reasons. Table 23.3 lists general treatment processes that remove suspended, colloidal, and dissolved solids. Conventional coagulation and lime softening are the principal methods, although ion exchange and reverse osmosis are sometimes used. Other methods such as distillation may be effective in removing these contaminants in homes, but are too expensive for large-scale systems.

Coagulation

Chemical coagulation occurs when floc-forming chemicals are added to water to destabilize suspended or colloidal solids. Ordinarily, colloidal material will not settle out, because the particles have negative charges on their surfaces and thereby repel

Table 23.2. Most Effective Treatment Methods for Removal of Inorganic Contaminants

Contaminant	Most Effective Methods
Arsenic As^{+3}	Ferric sulfate coagulation, pH 6-8 Alum coagulation, pH 6-7 Excess lime softening Oxidation before treatment required
As^{+5}	Ferric sulfate coagulation, pH 6-8 Alum coagulation, pH 6-7 Excess lime softening
Barium (Ba)	Lime softening, pH 10-11 Ion exchange
Cadmium Cd	Ferric sulfate coagulation, above pH 8 Lime softening Excess lime softening
Chromium Cr^{+3}	Ferric sulfate coagulation, pH 6-9 Alum coagulation, pH 7-9 Excess lime softening
Cr^{+6}	Ferrous sulfate coagulation, pH 7-9.5
Fluoride (F)	Ion exchange with activated-alumina or bone-char media
Lead (Pb)	Ferric sulfate coagulation, pH 6-9 Alum coagulation, pH 6-9 Lime softening Excess lime softening
Mercury (Hg) Inorganic Organic	 Ferric sulfate coagulation, pH 7-8 Granular activated carbon
Nitrate (NO_3)	Ion exchange
Selenium Se^{+4}	Ferric sulfate coagulation, pH 6-7 Ion exchange Reverse osmosis
Se^{+6}	Ion exchange Reverse osmosis
Silver (Ag)	Ferric sulfate coagulation, pH 7-9 Alum coagulation, pH 6-8 Lime softening Excess lime softening

(U.S. EPA, 1978)

Table 23.3. Water and Waste Treatment Processes

Process	General Use or Capability	Advantages	Disadvantages
Coagulation, sedimentation, and filtration	Reduction in suspended solids by 90-98%	Low cost Simplicity	Large area required Does not remove dissolved salts or dissolved organics
Softening	Reduction in hardness level by 95-100%	Relatively low in cost for low hardness waters Simplicity	Requires frequent regeneration Chemical regeneration need increases linearly with hardness Does not remove dissolved organics
Ion exchange	Reduction of dissolved salts by 95-100%	Can reach very low levels of salinity	Does not remove dissolved organics Requires regeneration chemical which adds to waste loading
Biological treatment	Reduction of organic loading by 50-90%	Low in cost	Subject to upsets and variations Cannot remove more than 90% of organic loading which is subject to biological degradation
Carbon columns	Reduction of adsorbable organic loading by 95-100%	Effective way to remove small quantities of organics	Relatively expensive regeneration Removes only specific organics
Evaporation	Removal of dissolved solids and suspended solids	Well demonstrated Can handle wide range of applications	High energy use
Reverse osmosis	Reduction of nearly all contaminants by 90-95%	Simple Low energy utilization	Membrane subject to fouling or deterioration if not properly operated Not effective on very low molecular weight organics Generally requires some form of pretreatment Limited chemical compatibility

(Office of Water Research and Technology, 1979)

other particles. In this stable condition, the particles cannot collide to form larger particles, called flocs, which would settle. By adding coagulants, the negative forces can be neutralized so that the particles become unstable, collide, and then settle. Aluminum, iron salts, and cationic polymers are common coagulants.

The efficiency with which certain inorganic ions are removed depends on the pH of the water, the type and dose of the coagulant, and the initial concentration of the contaminant (U.S. Environmental Protection Agency, 1978). One of the most important variables is pH, because the solubility limits for many substances such as metal hydroxides and carbonates are typically pH-dependent. The valence of the contaminant is also important; for example, it is much easier to remove the oxidized state of arsenic (As^{+5}) than the reduced state (As^{+3}) (Figures 23.1 and 23.2).

Lime Softening

Lime softening reduces hardness by applying hydrated lime [$Ca(OH)_2$] to water to precipitate calcium carbonate ($CaCO_3$), magnesium hydroxide [$Mg(OH)_2$], or both (Kemmer, 1979). Calcium and magnesium hardness and carbon dioxide are precipitated as calcium carbonate, but most waters contain alkalinity in the bicarbonate form. Thus, in the presence of carbon dioxide (CO_2 is almost always present in groundwater), the precipitation of calcium carbonate and magnesium hydroxide necessitates the conversion of carbon dioxide and bicarbonate (HCO_3) to carbonate.

$$Ca(HCO_3)_2 + Ca(OH)_2 \rightarrow 2CaCO_3 \downarrow + 2H_2O \qquad (23.1)$$

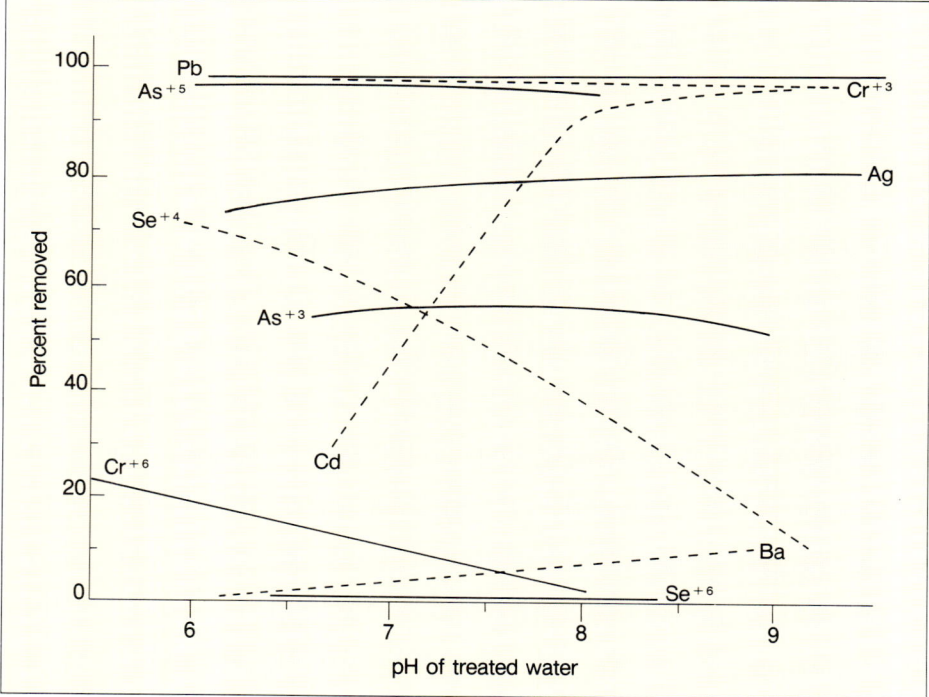

Figure 23.1. Removal of inorganic contaminants by iron coagulation. *(U.S. EPA, 1978)*

$$CO_2 + Ca(OH)_2 \rightarrow CaCO_3 \downarrow + H_2O \qquad (23.2)$$

$$Mg(HCO_3)_2 + 2Ca(OH)_2 \rightarrow Mg(OH)_2 \downarrow + 2CaCO_3 \downarrow + 2H_2O \qquad (23.3)$$

After coagulation or lime softening, inorganic materials are removed from water through sedimentation and filtration. A 1- to 4-hour settling time is usually required followed by filtration. The size of particles and applicable filtration techniques are given in Figure 23.3.

Ion Exchange

Softening can also be accomplished when calcium (Ca), magnesium (Mg), iron (Fe), manganese (Mn), and strontium (Sr) ions in the water are replaced by sodium (Na) ions through ion exchange. Water is passed through a medium, usually a synthetic cation-exchange resin, where sodium ions are exchanged for the contaminant ions. When all the sodium ions are used, a brine solution is passed through the resin to regenerate it — that is, remove the calcium, magnesium, iron, or other ions and replace them with sodium ions. Resins are not affected adversely by regeneration and will usually last for years. Periodic acid treatment may be necessary to remove iron scale. Specific types of resin are discussed in the section on point-of-use treatment systems.

Reverse Osmosis

Osmosis is the spontaneous process in which water with a lower total dissolved

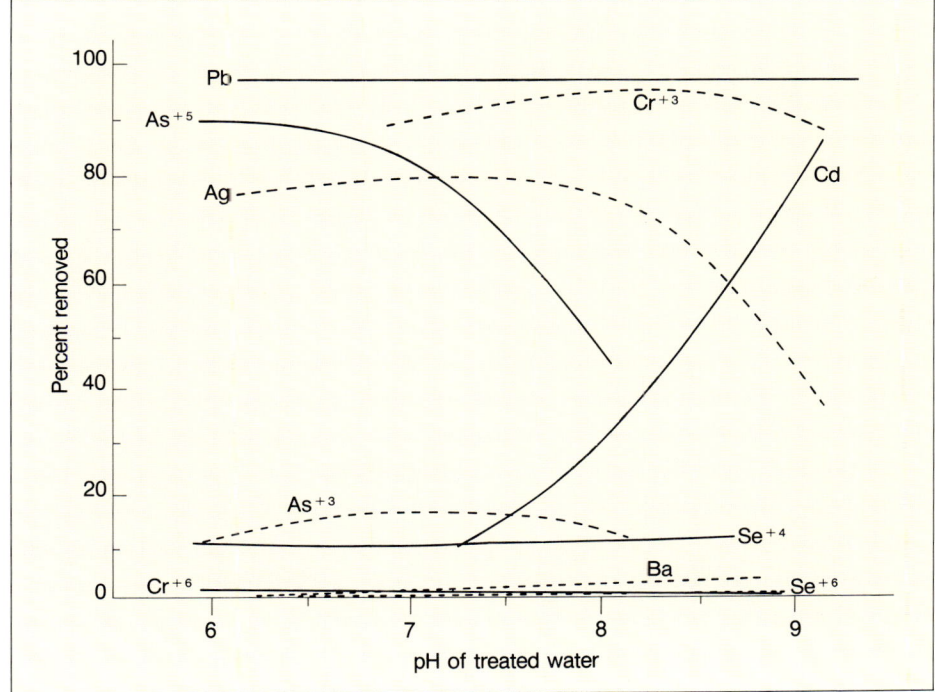

Figure 23.2. Removal of inorganic contaminants by alum coagulation. *(U.S. EPA, 1978)*

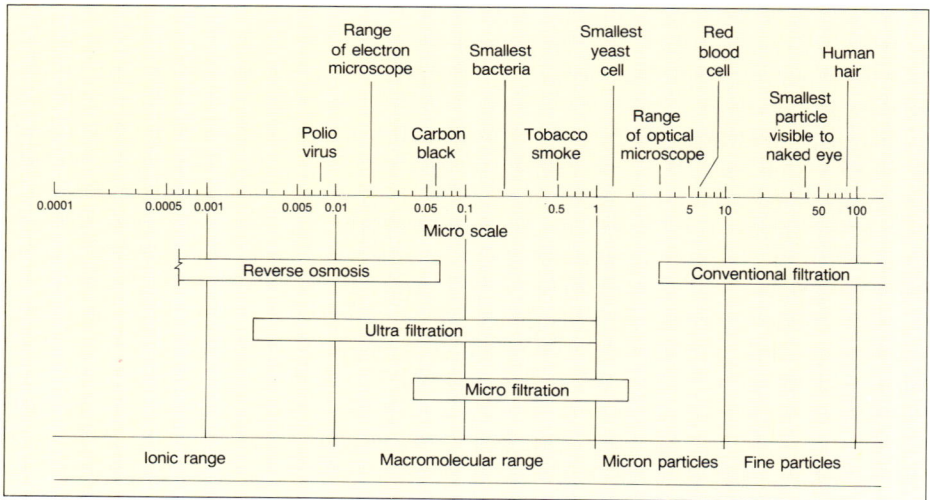

Figure 23.3. Approximate operating ranges of solids/liquids separation devices in treating water. *(Water Equipment Technologies, Inc.)*

solids content will pass through a semipermeable membrane into a solution containing a higher concentration of dissolved solids (Figure 23.4a). Flow continues until the osmotic pressure on the less concentrated side and the liquid head on the concentrated side balance each other. At this point, an equilibrium is reached so that the quantity of water passing through the membrane is equal in both directions. Osmotic pressure can be thought of as a chemical driving force whose strength can be estimated by measuring the increase in head in the solution having the greatest total dissolved solids after equilibrium conditions have been reached. For example, a concentration of 1,000 mg/ℓ of sodium chloride (NaCl) has an osmotic pressure of 11.4 psi (78.6 kPa) and creates a head of 26.3 ft (8 m).

If the pressure equilibrium is changed by applying pump pressure to the concentrated solution, flow is reversed across the permeable membrane; this is called reverse osmosis (Figure 23.4b). A reverse osmosis system consists of semipermeable membrane elements mounted in the pressure tubes, high-pressure water pump, pressure gauges, temperature gauge, and flow meters (Figure 23.5). Pretreatment components

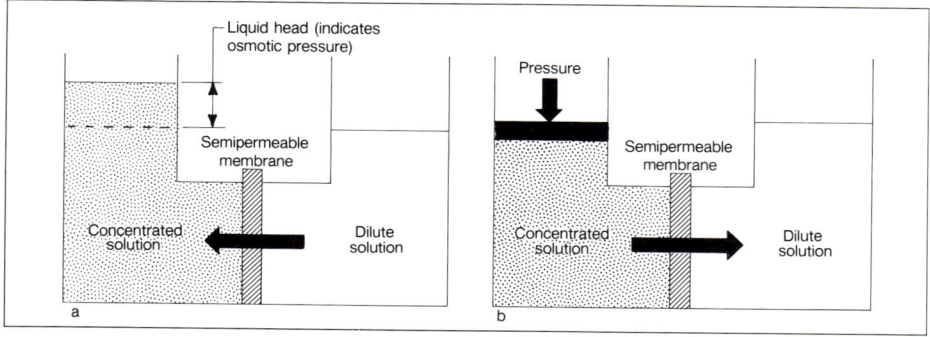

Figure 23.4. (a) Osmosis — normal flow is from low to high concentration. (b) Reverse osmosis — flow is reversed by application of pressure to high-concentration solution.

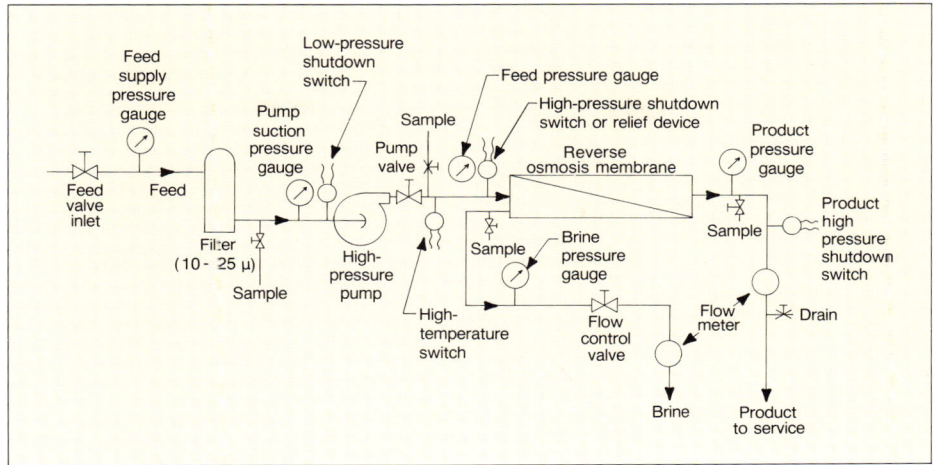

Figure 23.5. Flow schematic of reverse osmosis system.

consisting of filters or pH-adjustment devices are often part of a reverse osmosis installation. Three design configurations for the membranes are available: hollow fiber, spiral wound, and tubular. All designs attempt to maximize surface area relative to the membrane volume. In the hollow-fiber design, the membrane takes the form of thin, hollow fibers which lie parallel to each other and are grounded in an epoxy matrix. Feed water flows around the fibers, and product water seeps into the fibers' hollow centers and flows through them into a header at the end of the unit. Reject water, which is highly concentrated wastewater that results from the process, exits through another header. This method offers a good ratio of surface area to volume, but can easily be fouled by colloidal material.

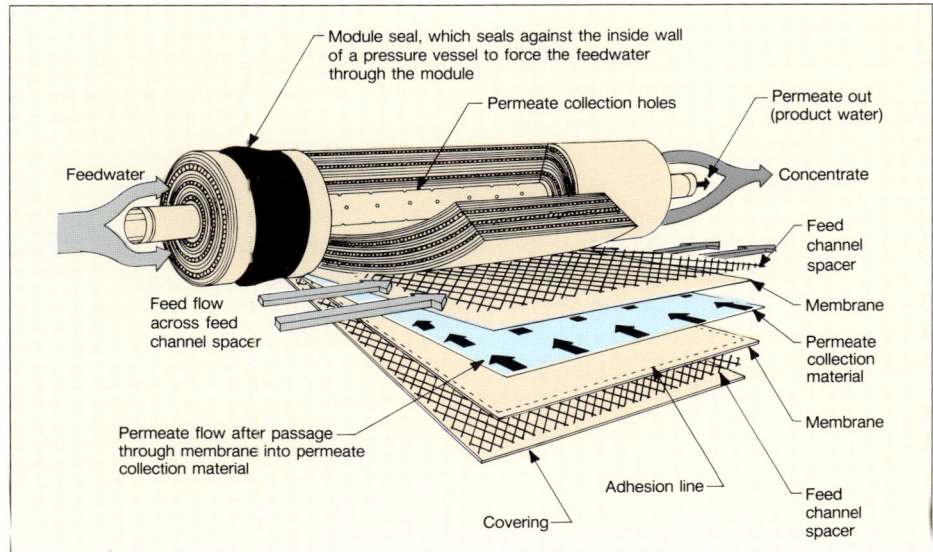

Figure 23.6. Spiral-wound reverse osmosis unit.

The spiral-wound design involves a membrane unit that consists of two layers of membrane formed around a permeate carrier, which in turn is wound around a perforated tube (Figure 23.6). Two of the units are used with a plastic mesh spacer between them. Feed water flows throughout the entire unit, and, after passing through the membrane, product water spirals to the center tube which carries it out of the system. Spiral-wound systems resist fouling and can be operated under turbulent conditions. They also provide a good ratio of membrane area to volume.

The tubular configuration is less frequently used because of the low ratio of surface area to volume. In this process, feed water flows through a hollow, porous tube that is coated either inside or outside with membrane. Under pressure, the permeate flows through the walls of the tube, as the concentrate simply flows away. Although the low ratio reduces the popularity of this method, it is the least susceptible to fouling.

Commonly used membrane materials include cellulose acetate, polyamide, and thin film composites. Polyamide and some thin film composite membranes are susceptible to chemical attack by oxidizing agents, including chlorine. Cellulose acetate is susceptible to bacterial attack and high-pH waters.

The percent salt passage and the percent of water recovery are the two most important design parameters. Percent salt passing is defined as:

$$\% \text{ salt passage} = \frac{\text{salt concentration of product water}}{\text{average salt concentration of feed water}} \cdot 100 \quad (23.4)$$

The percent of feed water recovered is:

$$\% \text{ feed water recovered} = \frac{\text{product water}}{\text{feed water}} \cdot 100 \quad (23.5)$$

The rate of feed flow is based on a particular recovery objective, but the actual rate of recovery is governed by two factors. The first is the water quality of the product water. If the recovery rate is set too high for a specific feed water, the amount of salt passing from the feed water to the product water may be too high. The second factor is the concentration of the feed water. It should not be concentrated so much that mineral compounds begin to precipitate out onto the membrane. Excessive concentration of the feed water can also cause it to lose more salt to the product water than desired. Table 23.4 shows the results of treating brackish well water; an 80-percent recovery is obtained at 373 psi (2,570 kPa).

Large-volume reverse osmosis systems are being installed where supplies of fresh water are limited or the quality of the fresh water must be improved. The data presented in Table 23.5 suggest the effectiveness of reverse osmosis techniques in treating sea water. Drinking water supplies for Jeddah, Saudi Arabia are being augmented by desalting Red Sea water at a rate of 3.2 million gal (12,100 m^3) per day. A much larger installation that treats 73 million gal (276,000 m^3) per day of Colorado River water is located in Yuma, Arizona (Figure 23.7). A reverse osmosis system is shown in Figure 23.8 that can provide enough potable water for a farm or small industrial plant.

Reverse osmosis units are quite efficient, usually removing over 90 percent of the dissolved solids in the feed water and almost all of the colloidal, organic, and other suspended substances. Usually the higher the concentration of dissolved solids in the feed water, the higher will be the total dissolved solids in the product water. Because impurities are constantly removed from the reverse osmosis vessel, no backwashing

Table 23.4. Analysis of Brackish Water Before and After Treatment by Reverse Osmosis

Contaminant	Feedwater	Pretreated Feedwater	Concentrate (Rejected Water)	Permeate (Product Water)
	mg/ℓ	mg/ℓ	mg/ℓ	mg/ℓ
Calcium	61	61	303	0.2
Magnesium	24	24	119	0.1
Sodium	78	78	379	2.4
Potassium	4	4	19	0.1
Ammonia	0	0	0	0
Carbonate	0	0	0	0
Bicarbonate	129	24	115	2.0
Sulfate	201	284	1,414	0.2
Chloride	73	73	348	3.9
Nitrate	0	0	1	0
Fluoride	0	0	1	0
Silica	10	10	47	0.8
Total	581	558	2,747	8.1
Total dissolved solids	516	546	2,689	7.8
pH	7.8	5.7	6.4	4.6
Carbon dioxide	3	80	80	78
pHs		8.8	7.2	

(Fluid Systems Division, UOP Inc.)

Table 23.5. Water Analysis of Seawater Treated by Reverse Osmosis

Contaminant	Feedwater	Pretreated Feedwater	Concentrate (Rejected Water)	Permeate (Product Water)
	mg/ℓ	mg/ℓ	mg/ℓ	mg/ℓ
Calcium	418	418	642	0.4
Magnesium	1,330	1,330	2,044	1.4
Sodium	11,035	11,035	16,885	144.6
Potassium	397	397	607	5.2
Ammonia	0	0	0	0
Carbonate	0	0	0	0
Bicarbonate	146	146	224	0.8
Sulfate	2,769	2,769	4,256	0.3
Chloride	19,841	19,841	30,374	233.6
Nitrate	0	0	0	0
Fluoride	1	1	2	0
Silica	0	0	0	0
Total	35,937	35,937	55,034	386.3
Total dissolved solids	35,864	35,864	54,922	386
pH	8.2	8.2	8.4	5.9
Carbon dioxide	2	2	2	2
pHs		8.2	8.0	

(Fluid Systems Division, UOP Inc.)

808   GROUNDWATER AND WELLS

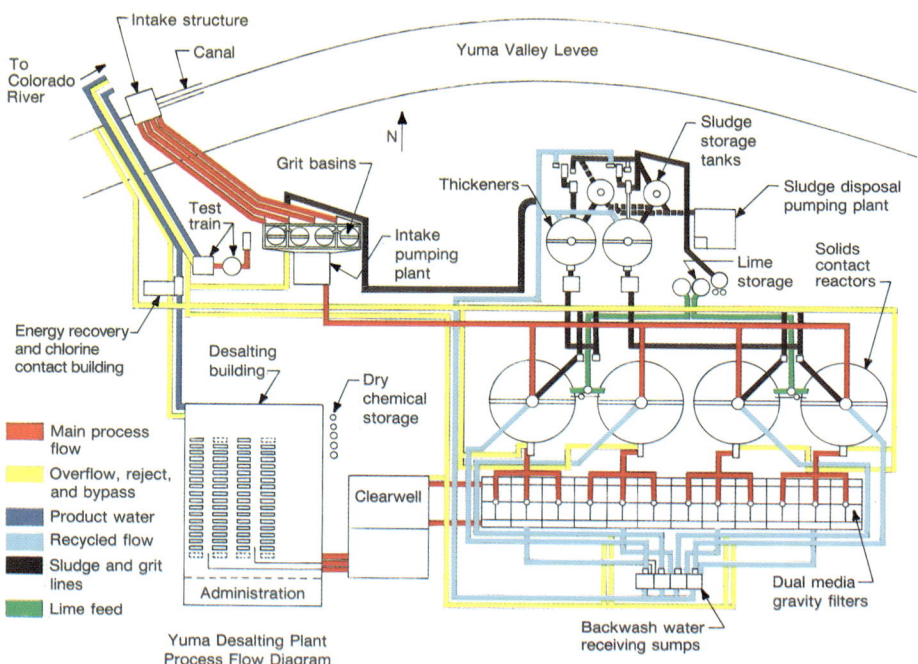

Figure 23.7. A large reverse osmosis treatment plant at Yuma, Arizona reduces the salinity of Colorado River water to meet standards established in treaties with Mexico. *(Fluid Systems Division, UOP Inc.)*

Figure 23.8. This Microfloc reverse osmosis system consists of 10 membrane units that deliver 35 gpm (191 m^3/day) of potable water from a sea-water source. The unit could serve a farm or small industrial plant.

is required.

To make reverse osmosis installations function better, many forms of pretreatment are used, depending on the water chemistry of the feed water. Because reverse osmosis is a concentration process, the efficiency of the system can be seriously impaired by the deposition of suspended material, by iron and manganese precipitates, by slime and algae, by scaling caused by calcium carbonate and calcium sulfate, and by organic matter.

Scale build-ups are particularly worrisome; acidification and sequestration are two important pretreatment methods that help eliminate scale build-up, especially of calcium carbonate. Acidifying the feed water keeps the calcium in solution. Treatment with sodium hexametaphosphate controls scale by inhibiting calcium sulfate precipitation; barium sulfate and strontium sulfate scales can also be prevented by this method.

Lime softening of feed water is used for reducing alkalinity and excessively high levels of hardness. Soluble iron and manganese, which may precipitate within the reverse osmosis modules, can be controlled by oxidation and filtration during pretreatment. One effective method is to pass the feed water through a fine-grained bed of manganese zeolite oxidized with permanganate. Manganese and iron in the reduced states are oxidized to insoluble forms in the bed and are trapped in the media. Tastes, odors, hydrogen sulfide, and carbon dioxide can also be reduced during posttreatment.

Most reverse osmosis modules using fresh surface water or saline water are equipped with filters to retain all particles larger than 5 to 10 microns that have not been removed by other pretreatment methods. Colloidal material such as hydrous metal oxides, organic substances, aluminum silicates, and other suspended matter reacts by coagulation with salts in the influent channels of reverse osmosis modules, causing the colloids to grow much larger and plug the channels.

Usually colloids are not a problem for feed water supplied from wells. If the concentration of colloids is not great, sodium zeolite softening may be satisfactory because removal of multivalent cations in the softening process stabilizes the colloids. If the concentration of colloids is high, a combination of coagulation, clarification, and filtration is a good pretreatment method. Ultrafilters can also be used.

The growth of micro-organisms in reverse osmosis modules must be prevented so the equipment does not become plugged, the treated water is not contaminated, and damage to the cellulose acetate membranes by certain micro-organisms is avoided (Graver Water Division, 1978). Chlorination (1 to 2 mg/ℓ residual), ozonation, and the use of ultraviolet radiation will control the growth of micro-organisms. Care must be taken when using chlorine because continuous exposure of the membrane to higher chlorine residuals reduces their efficiency. Chlorine destroys membranes made of polyamide.

The rate of hydrolysis in the module affects membrane life. For example, cellulose acetate will hydrolyze to cellulose and acetic acid. The rate at which hydrolysis occurs is a function of the feed-water pH and also the temperature. The pH level in the reverse osmosis system is usually held at 5 to 6.5 by an acid-feed system. As the membrane hydrolyzes in use, both the volume of water and the amount of salts that permeate the membrane increase, resulting in lower quality product water (Desai, 1974).

Several posttreatment methods for product water may be necessary. These include pH adjustment, degasification in which excessive carbon dioxide is removed, and chlorination or other disinfection methods to reduce taste and odor problems.

Some reverse osmosis systems suffer rapid performance degradation and permanent membrane fouling in both water and process applications. Reverse osmosis systems depend on compact designs (high membrane densities) which are inherently prone to fouling. This is particularly true with hollow, fine-fiber or spiral-wound units. Where high recovery is needed, foulants often are progressively concentrated in the feed stream, causing performance to decline. Efficiency failure is caused by a variety of factors (Brown, 1983). For example, reverse osmosis systems must operate at high pressure to overcome osmotic pressure; this causes membranes to be more sensitive to fouling mechanisms and therefore may be difficult to clean. Biological degradation in cellulosic membranes, where they are susceptible to bacterial infection, can cause permanent loss of performance. Organic material (humic acid) is often responsible for fouling membranes, although recent research has shown that pretreatment of the feed water by ozone can reduce this problem (Lozier and Sierka, 1985).

From time to time, it may be necessary to restore the materials by cleaning the membranes with citric acid and ammonium hydroxide, detergents, hydrochloric acid, or formaldehyde.

In-Situ Treatment Methods to Remove Iron and Manganese from Groundwater

In some areas, it may be possible to remove iron and manganese from groundwater before it reaches the well. In the Vyredox™ method of treating groundwater, which was developed in Finland, aeration wells and iron-oxidizing bacteria in the ground are utilized to remove iron and manganese some distance from the production well. Iron and manganese are in a soluble form when the groundwater is nearly devoid of oxygen and has a pH less than 6.5. In the Vyredox system, numerous aeration wells are constructed in a ring around the production well. Water is withdrawn from a nearby supply well, degassed (carbon dioxide and methane are removed), aerated with oxygen, and then injected into the groundwater system through the aeration wells. The oxygen-rich water enables the naturally occurring iron and manganese bacteria to increase their metabolic rate, thereby fixing more iron and manganese. In addition, the iron and manganese ions are converted to the insoluble form and precipitation occurs. Periodic treatments are required to maintain the oxygenated state in the groundwater environment near the well. Plugging of the aquifer surrounding the injection (aeration) well by iron deposition does occur, but at a rate that takes much longer to become objectionable than the typical economic life of the well. The layout of a typical Vyredox plant is shown in Figure 23.9.

Removal of Turbidity

Turbidity refers to solids and organic matter that do not settle out of water. Groundwater is rarely turbid, unlike surface water which often contains suspended solids and colloidal or soluble organic matter. Shallow groundwater taken from areas near swamps and marshes, however, may be high in soluble organic matter. Eliminating turbidity not only increases the aesthetic quality of the water, it also helps remove contaminants that cling to suspended solids and inhibits the formation of trihalomethanes, after treatment, by removing organic precursor substances. A maximum contaminant level has been established for turbidity because solids in the water can interfere with disinfection processes and microbiological determinations.

Turbidity is measured by how much light is transmitted or scattered when a beam of light is passed through a water sample. An early type of analysis, called the Jackson Turbidity Unit (JTU), is based on measurements made with a transmitted light beam using a standard candle. This method is not sensitive enough, however, for measuring the turbidity of well water, filtered water, and clarified effluent samples. A light-scattering method is used for these low-turbidity waters. The light is measured in Nephelometric Turbidity Units (NTU), which indicate the light scattered at 90-degree or 270-degree angles to the incident beam. The numerical values produced by this method are not comparable for all types of waters, and therefore instruments cannot be calibrated accurately in terms of absolute turbidity except under special circumstances (ASTM D-1889, 1977).

Turbidity is reduced or eliminated by a combination of coagulation, flocculation, and filtration. These processes will produce a water that meets federal standards for turbidity and organic color. Colloids are so small that they must be coagulated and then flocculated before being passed through a filter media. In passing through the filter media, the colloidal materials are attracted by electrical charges and, to a lesser extent, are filtered. Periodic backwashing, in which the filter media is fluidized, removes the accumulated material.

In the past, slow or rapid sand filters were used for most filtering operations. Improved technology has now led to the adoption of dual- and mixed-media filter beds that are much superior in filtering capacity. Dual-media filters consist of two materials with differing sizes and densities. The coarser materials overlie the finer particles, so the filtration can take place within the media and not merely on the filter surface. Three types of media can also be used to provide filtration within the media bed. Mixed-media filters consist of anthracite (coal) with a density of 1.4, sand with a density of 2.7, and the minerals garnet or ilmenite with a density of about 4. The anthracite particles are about 1 mm in diameter, the sand 0.45 mm, and the minerals 0.3 mm. During backwashing and media agitation, the individual filter materials segment themselves into layers because of differences in density and particle size. Because of this layered structure, filtration occurs within the filter material, head losses are minimized, and high-quality water is produced.

Diatomaceous earth filters are used when the concentration of suspended material is low. In this process, diatomaceous material (precoat) is placed over a thin membrane

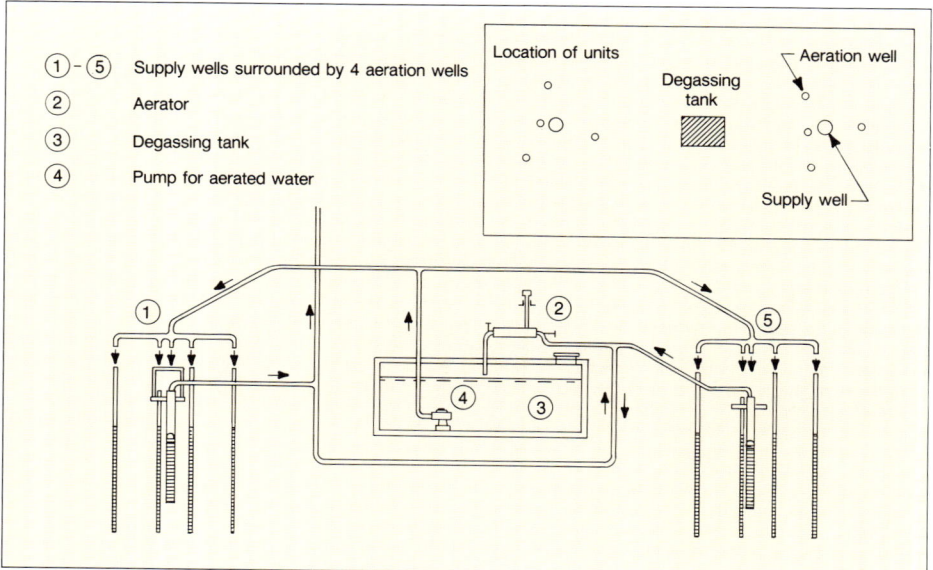

Figure 23.9. Vyredox plant with two supply wells, aeration wells, and oxygenator building. *(Hallberg and Martinell, 1976)*

filter. The diatomaceous earth serves as a filtering surface. Once the diatomaceous earth becomes clogged, it is removed by backwashing and new precoat material is placed on the filter. The filter is also thoroughly backwashed. Usually, some filter material, called body feed, is continually added to the influent water to lengthen the time between precoat renewal. If the addition of body feed is insufficient, the diatomaceous earth filter will plug much more quickly, necessitating more precoat renewals.

Removal of Coliform Organisms

In the past, most groundwater was of such high quality that disinfection, although often required by law, was probably unnecessary. Today, the presence of microbes in groundwater is becoming more widespread, and more rigorous treatment of groundwater is required. For example, *Giardia lamblia* has spread from mountain streams in the Rocky Mountains to wells in Colorado, Montana, and other western states. In large water treatment systems, the *Giardia* cysts generally cannot be removed by disinfection alone, although recent research has shown that ozone is effective in deactivating cysts of some *Giardia* species (Wickramanayake et al., 1985). Diatomaceous earth or granular-media filtration methods must also be used.

The micro-organisms that are most likely to cause problems are listed in Table 23.6. Except for algae and higher life forms, groundwater may be affected by one or more of these microbes.

Identification of specific types of pathogenic bacteria is quite difficult and laborious. Thus, laboratories only attempt to determine the presence or absence of coliform organisms. Coliforms originate primarily in the intestinal tract of warm-blooded animals, including humans. Although the coliform bacteria are normally nonpathogenic, their presence suggests that conditions are appropriate for pathogenic organisms and that fecal contamination is possible. The maximum contaminant level for coliform bacteria in water systems is 1 coliform bacteria per 100 mℓ of water.

Microbes can be controlled by using oxidizing or nonoxidizing biocides. Chlorine is a strong oxidizing agent that reacts with many impurities in water. The chlorine demand of a specific water is the amount of chlorine required to react with these impurities. Chlorine not used immediately is called the chlorine residual. Breakpoint chlorination occurs when the addition of a volume of chlorine satisfies the chlorine demand and produces a free chlorine residual.

Gaseous chlorine is the least expensive form of chlorine for use in water treatment plants where material requirements are large (American Water Works Association, 1973a). For small plants or for field use, materials such as hypochlorites, chlorine dioxide, and chlorinated lime may be safer and more practical. If the water is obtained from a well, hypochlorite can be fed by a solution feed pump when the well is turned on. Figure 23.10 illustrates the effect of pH on the bactericidal ability of chlorine. Chlorine is also helpful in reducing naturally occurring taste and odor problems, and oxidizing nuisance inorganic chemicals such as iron and manganese (Figure 23.11). But the addition of chlorine can cause a taste and odor problem related to the chemical itself.

Although chlorine is the principal chemical used to disinfect water, its use may produce some adverse health effects. When chlorine is added to water containing natural organic materials (humic and fulvic acids and algae), trihalomethanes (THM's) form. Chloroform ($CHCl_3$) is the trihalomethane that occurs in greatest quantity in finished waters (Briley et al., 1984). Because THM's are thought to be carcinogenic, the U.S. Environmental Protection Agency has set a limit of 0.1 mg/ℓ for trihalomethanes in finished waters for communities with populations of 10,000 or more. Treatment methods to reduce trihalomethanes are discussed later in this chapter.

Chlorine dioxide is also effective in disinfecting water and its use produces less trihalomethanes than does chlorine (Lykins and Griese, 1983). Chlorine dioxide (ClO_2) forms from the reaction of aqueous sodium chlorite (Na_2ClO_2) with aqueous chlorine (Cl). Chlorine dioxide is used to control tastes and odors resulting from algal blooms and decaying vegetation in open reservoirs. Thus, its use may be suitable

Table 23.6. Typical Micro-Organisms and Their Associated Problems

Type of Organism	Type of Problem
Bacteria	
Slime-forming bacteria	Form dense, sticky slime with subsequent fouling. Water flows can be impeded and promotion of other organism growth occurs.
Spore-forming bacteria	Become inert when their environment becomes hostile to them. However, growth recurs whenever the environment becomes suitable again. Difficult to control if complete kill is required. However, most processes are not affected by spore formers when the organism is in the spore form.
Iron-depositing bacteria	Cause the oxidation and subsequent deposition of insoluble iron from soluble iron.
Nitrifying bacteria	Generate nitric acid from ammonia contamination. Can cause severe corrosion.
Sulfate-reducing bacteria	Generate sulfides from sulfates and can cause serious localized corrosion.
Anaerobic corrosive bacteria	Create corrosive localized environments by secreting corrosive wastes. They are always found underneath other deposits in oxygen-deficient locations.
Fungi	
Yeasts and molds	Cause the degradation of wood in contact with the water system. Cause spots on paper products.
Algae	Grow in unlit areas in dense fibrous mats. Can cause plugging of distribution holes on cooling-tower decks or dense growths on reservoirs and evaporation ponds.
Protozoa	Grow in almost any water which is contaminated with bacteria; indicate poor disinfection.

(Kemmer, Editor-in-Chief, The Nalco Water Handbook, ©1979, McGraw-Hill Book Co. Reprinted with permission.)

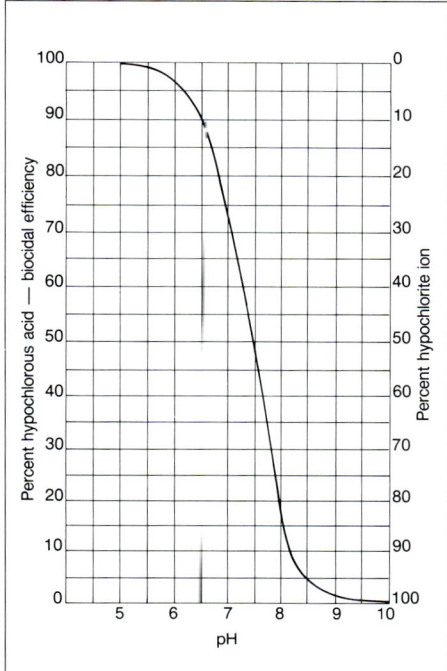

Figure 23.10. The effect of pH on the bactericidal ability of chlorine.

when treating groundwater affected by water from swamps. Chlorine dioxide is more effective than chlorine when the pH of the water rises above 7. Because of its potential explosiveness, chlorine dioxide must be generated on-site.

Ozone is an unstable form of oxygen that can be generated from air or oxygen. Although it has not been used much in the United States, it is an extremely effective oxidizer and disinfectant that is widely used in Canada and Europe. Recently, water treatment specialists in the United States have recognized the usefulness of ozone in not only controlling taste and odors, but in pretreating before biological treatment to aid in flocculation, reducing the need for chemicals such as coagulants, and enhancing filtering rates. Of the 24 water-treatment plants using ozone in the United States in 1985, only 3 plants used it primarily as a disinfectant (Rice, 1985). The other plants utilize it principally as a means to

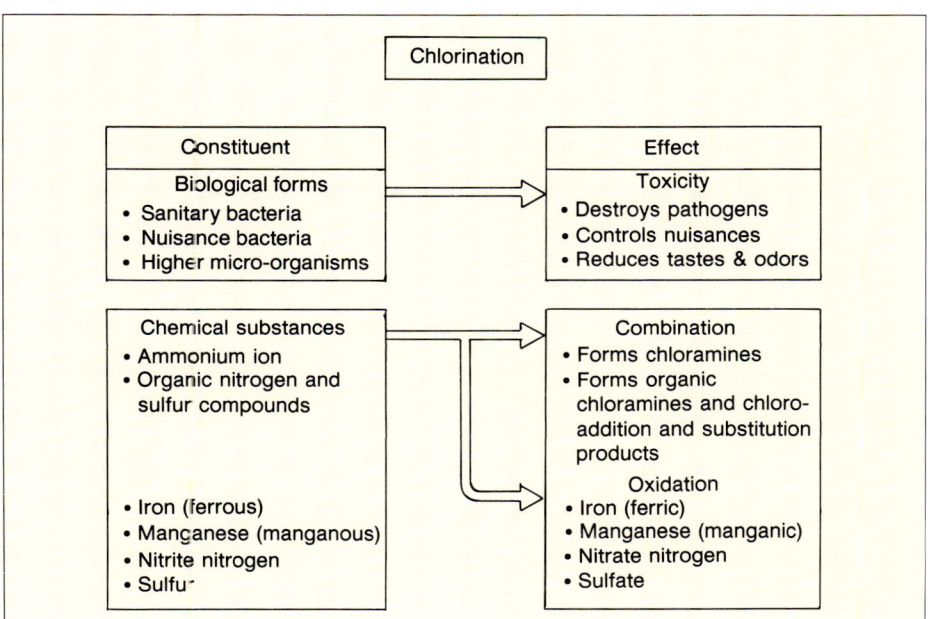

Figure 23.11. Biological and chemical effects of chlorine in water. *(Reprinted from Water Chlorination Principles and Practices, AWWA M20, by permission. Copyright © 1973a, The American Water Works Association)*

enhance other treatment processes. Ozone does not cause the formation of trihalomethanes during disinfection. Because the half life is only about 20 minutes, however, a small amount of chlorine is usually added to water treated with ozone to maintain a chlorine residual. The addition of chlorine to water initially treated with ozone may lead to the formation of trihalomethanes, because ozone does not remove trihalomethane precursors. The layout of a typical treatment system using ozone is shown in Figure 23.12.

Removal of Organic Contaminants

Recent advances in the ability to analyze for extremely low concentrations of contaminants in water have shown that man-made organic chemicals occur in many groundwater supplies. Approximately 113 organic compounds are included in the U.S. Environmental Protection Agency's list of priority pollutants. Because many synthetic organic chemicals are widely used for industrial and domestic purposes, normal use and irresponsible disposal practices have permitted their widespread introduction into groundwater. The first large-scale random testing program by the U.S. Environmental Protection Agency shows that at least one volatile organic chemical occurs in 16.8 percent of water systems serving 10,000 persons or less, and in 28 percent servicing more than 10,000; non-random surveys show higher figures of 22.4 and 37.3 percent, respectively (Westrick et al., 1984). In these studies, the occurrence of 29 volatile organic chemicals was studied along with 5 trihalomethanes (Table 23.7).

The widespread use of organic chemicals for pesticides and herbicides has led to their inadvertent introduction into groundwater. In agricultural areas where the soils are sandy and the groundwater table is relatively high, many organic chemicals can reach the groundwater environment. Overuse of certain chemicals, coupled with extensive irrigation of permeable (sandy) soils, causes them to be driven downward to the groundwater table. In many cases, the ability of the soils to sorb these chemicals is not great enough to prevent their reaching the saturated zone. In some irrigated

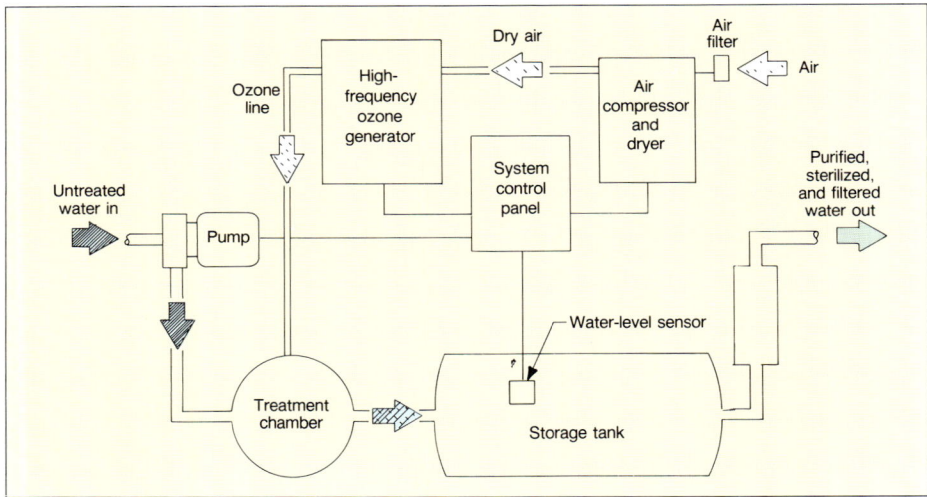

Figure 23.12. Flow diagram of ozone water treatment system. Depending on the flow, several treatment vessels of appropriate size may be used. *(Ground Water Age, 1977)*

Table 23.7. Occurrence of Volatile Organic Chemicals Including Trihalomethanes in Random and Nonrandom Surveys

Chemical	Occurrences from 280 Random Sample Sites for Systems Serving Less than 10,000 Persons		Occurrences from 186 Random Sample Sites for Systems Serving More than 10,000 Persons		Occurrences from 321 Nonrandom Sample Sites for Systems Serving Less than 10,000 Persons		Occurrences from 158 Nonrandom Sample Sites for Systems Serving More than 10,000 Persons	
	Number	Percent	Number	Percent	Number	Percent	Number	Percent
Vinyl chloride	0	0	1	0.5	0	0	6	3.8
1,1-Dichloroethylene	4	1.4	5	2.7	5	1.6	10	6.3
1,1-Dichloroethane	10	3.6	8	4.3	6	1.9	17	10.8
cis- and/or trans-1,2-Dichloroethylene	3	1.1	13	7.0	11	3.4	27	17.1
1,2-Dichloroethane	0	0	3	1.6	3	0.9	4	2.5
1,1,1-Trichloroethane	12	4.3	15	8.1	25	7.8	26	16.5
Carbon tetrachloride	5	1.8	10	5.4	9	2.8	6	3.8
1,2-Dichloropropane	1	0.4	5	2.7	3	0.9	4	2.5
Trichloroethylene	9	3.2	21	11.3	23	7.2	38	24.1
Tetrachloroethylene	13	4.6	21	11.3	27	8.4	18	11.4
Benzene	1	0.4	2	1.1	5	1.6	3	1.9
Toluene	4	1.4	2	1.1	4	1.2	1	0.6
Ethylbenzene	2	0.7	1	0.5	3	0.9	0	0
Bromobenzene	3	1.1	1	0.5	2	0.6	0	0
m-Xylene	6	2.1	2	1.1	8	2.5	0	0
o + p-Xylene	6	2.1	2	1.1	10	3.1	0	0
p-Dichlorobenzene	2	0.7	3	1.6	4	1.2	0	0
1,1,2-Trichloroethane	0	0	0	0	0	0	0	0
1,1,1,2-Tetrachloroethane	0	0	0	0	0	0	0	0
1,1,2,2-Tetrachloroethane	0	0	0	0	0	0	0	0

Table 23.7. (Continued)

Chemical	Occurrences from 280 Random Sample Sites for Systems Serving Less than 10,000 Persons		Occurrences from 186 Random Sample Sites for Systems Serving More than 10,000 Persons		Occurrences from 321 Nonrandom Sample Sites for Systems Serving Less than 10,000 Persons		Occurrences from 158 Nonrandom Sample Sites for Systems Serving More than 10,000 Persons	
	Number	Percent	Number	Percent	Number	Percent	Number	Percent
Chlorobenzene	0	0	0	0	1	0.3	0	0
1,2-Dibromo-3-chloropropane	1	0.4	0	0	0	0	0	0
n-Propylbenzene	0	0	0	0	1	0.3	0	0
o-Chlorotoluene	0	0	0	0	0	0	1	0.6
p-Chlorotoluene	0	0	0	0	0	0	0	0
m-Dichlorobenzene	0	0	0	0	0	0	0	0
o-Dichlorobenzene	0	0	0	0	1	0.3	1	0.6
Styrene	0	0	0	0	0	0	0	0
Isopropylbenzene	0	0	0	0	0	0	0	0
THM's								
Chloroform	104	37.1	106	57.0	155	48.3	100	63.3
Bromodichloromethane	100	35.1	101	54.3	144	44.9	100	63.3
Dibromochloromethane	87	31.1	96	51.6	135	42.1	87	55.1
Dichloroiodomethane	2	0.7	3	1.6	5	1.6	8	5.1
Bromoform	44	15.7	57	30.6	88	27.4	60	38.0

(Westrick et al., 1984)

areas, the presence of an improperly abandoned well can also lead to the direct introduction of flood irrigation waters containing pesticides to the groundwater system. Some organic chemicals also form in water during treatment, or they may enter the water by leaching from certain types of piping systems (Dressman and McFarren, 1978; Larson et al., 1983). Once in the groundwater system, organic chemicals are transported rapidly without retardation or degradation (Roberts et al., 1982).

Three general classes of organic contaminants are now regulated. The first comprises six chemicals used as pesticides that have been identified specifically under the Primary Drinking Water regulations. These are endrin, lindane, methoxychlor, and toxophene, which are chlorinated hydrocarbons, and 2,4-D and 2,4,5-TP (Silvex), which are chlorophenoxys; in 1979, trihalomethanes were added to the regulatory list (U.S. Environmental Protection Agency, 1979 and 1982). Trihalomethanes, however, are produced in the disinfection process when chlorine is added to water containing naturally occurring organics.

Nine volatile synthetic organic chemicals (VOC's) have been recognized by the U.S. Environmental Protection Agency as being dangerous to human health (U.S. Environmental Protection Agency, 1984). These are trichloroethylene, tetrachloroethylene, p-dichlorobenzene, 1,1-dichloroethylene, vinyl chloride, 1,1,1-trichloroethane, 1,2-dichloroethane, benzene, and carbon tetrachloride. These chemicals are not naturally occurring, are extremely stable once in an aqueous environment, and have been shown to be carcinogenic to animals, and in some cases, to humans. Some organics, even in extremely small concentrations, have been shown to be genotoxic, that is, they can affect the DNA molecules in humans.

As more becomes known about other toxic pollutants, additional regulations will be issued. Appendix 23.A lists the U.S. Environmental Protection Agency's priority contaminants. These substances have been identified because of their frequency of occurrence in water, chemical stability and structure, amount of chemical produced, and availability of chemical standards and measurement.

Groundwater affected by volatile organic chemicals usually displays common characteristics. These are (Hess et al., 1983):

1. Affected groundwater supplies contain several organic chemicals.
2. Trichloroethylene is most frequently detected, followed by tetrachloroethylene.
3. In a specific well, one or two organic compounds will dominate at relatively high concentrations of 100 to 500 $\mu g/\ell$, but several others can be detected at less than 50 $\mu g/\ell$.
4. Significant differences in the concentrations of specific contaminants are found in different wells in the same well field.

Conventional treatment techniques such as sand filtration, sedimentation, flocculation, coagulation, and chlorination are not effective in reducing the concentration of volatile organic chemicals in drinking water. Therefore, communities tend to deepen wells or build new wells to avoid contaminated water (Morrison, 1981). Although this may be effective over the short term, it may only delay the time until new treatment facilities are required. Fortunately, relatively inexpensive but highly effective treatment methods are available as described below.

Two treatment technologies are particularly effective in removing or reducing substantially the concentration of organic chemicals in drinking water. These are aeration (air stripping), and adsorption using activated carbon. Fortunately, neither is inordinately expensive nor difficult to install in a water-treatment system. In addition to these technologies, boiling the water is an effective way to reduce or eliminate most organic chemicals, although in practice this method cannot be applied to large volumes of water.

Aeration

In the aeration process, water containing organic chemicals is mixed with air in a special chamber or tower which is filled with packing material that disperses the water to enhance air contact (Figure 23.13). Removal of the chemical depends on the length of contact time, ratio of air to water, temperature, vapor pressure, and solubility of the contaminants (Symons et al., 1981). Aeration is extremely cost effective for removing volatile organics such as trihalomethanes. It is possible to predict, from the solubility of the contaminant and the vapor pressure of the pure material, the minimum air-to-water ratio required to achieve complete removal of the contaminant. In Figure 23.14, the ratios of air and water are shown for four trihalomethanes. The slope of the curve suggests the ease with which the contaminant

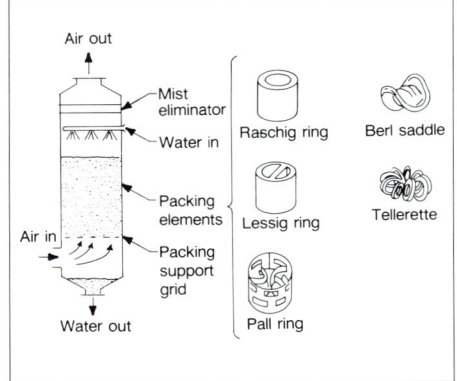

Figure 23.13. Packed tower and types of packing. *(Kemmer, Editor-in-Chief, The Nalco Water Handbook, © 1979, McGraw-Hill Book Company. Reprinted with permission.)*

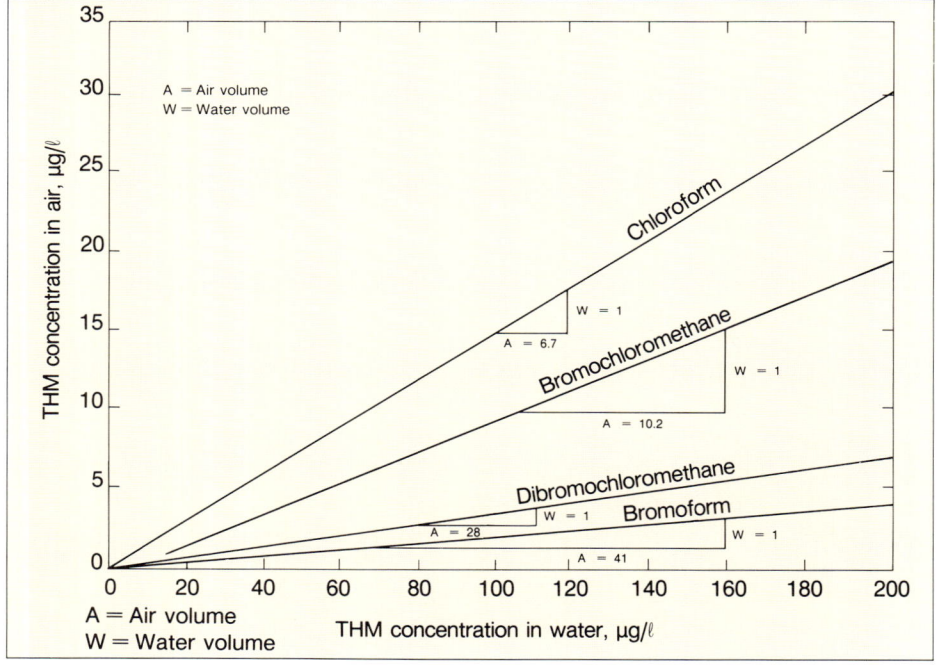

Figure 23.14. Theoretical equilibrium lines for four trihalomethanes (THM). These curves show the minimum air-to-water ratio that will produce complete removal of selected volatile organic contaminants. *(Symons et al., 1981)*

can be removed by aeration in a countercurrent system. The steeper the curve, the less air is required. Theoretically, 6.7 volumes of air must be mixed with 1 volume of water for complete removal of chloroform. In practice, complete removal may not be possible, even with larger volumes of air. Best results can be achieved if the contact time is long (height of tower), the rate of liquid flow and air flow are adjusted properly, and the surface area of the packing material in the tower is optimal.

The aeration tower contains beds of materials such as ¼- to 2-in (6.4- to 51-mm) metal, ceramic, or plastic spheres, or other types of materials such as tellerettes, saddles, or rings (Figure 23.13). Water is introduced across the top of the bed and trickles downward through the bed material. At the same time, air is passed upward through the bed, removing contaminants from the water. In cold climates, the air or water must be heated to prevent freezing.

Adsorption

Two types of materials are used to adsorb organic contaminants — granular or powdered activated carbon and a synthetic carbon-based resin. The high cost of the synthetic resin has limited its use in comparison with activated carbon. Granular activated carbon (GAC) is most often used in treatment systems, and provides huge numbers of sites (pores) where various organic contaminants can become affixed when contaminated water is passed upward through a carbon bed. The water is usually in contact with the GAC for about 10 minutes. In time, the GAC is used up and the contaminant will break through (Figure 23.15). Just before breakthrough, the GAC

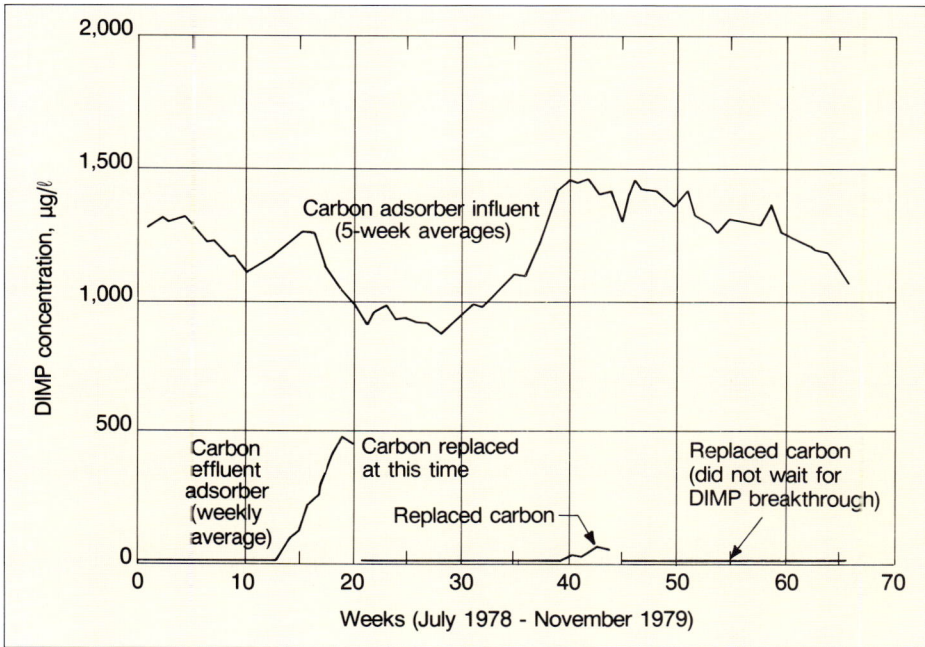

Figure 23.15. Performance of carbon beds in removing the chemical 2,4-di-isomethylphosphonate (DIMP) from groundwater at Rocky Mountain Arsenal, Denver, Colorado. Top curve shows influent concentration of this chemical; bottom curve shows effluent concentration. Di-isomethylphosphonate apparently "breaks through" the carbon beds at intervals of 10 to 20 weeks, so carbon should be replaced with similar frequency. *(Civil Engineering-ASCE, 1981)*

should be replaced or, for large plants, regenerated by heating the material until all the contaminants are driven off. Use of GAC greatly reduces the concentration of organic chemicals in water (note reduction in di-isomethylphosphonate in Figure 23.15). It is also effective in treating nonvolatile organics.

The ability of GAC to remove organic contaminants results from the enormous surface area of the carbon. During activation, select grades of coal are crushed and heated, producing a vast network of pores inside each carbon granule. The internal surface area is from 275,000 to 366,000 ft^2 per oz (900 to 1,200 m^2 per g) for a coal-base adsorber (Symons et al., 1981). Thus, 1 lb (0.5 kg) of carbon granules has an effective surface area equal to that of a 100- to 135-acre (40.5- to 54.6-hectare) farm.

The capacity of the activated carbon can be determined from the appropriate adsorption isotherm. The adsorption isotherm is the relationship between the amount of substance adsorbed and its concentration in the surrounding solution at equilibrium. When the adsorbed material in the GAC is in equilibrium with the influent concentration, the GAC is loaded to capacity and that portion of the bed is exhausted (Symons et al., 1981). For a specific situation, the time it takes to reach exhaustion can be predicted by knowing the influent contaminant concentration, the approach velocity, the adsorber bed depth, the density of the adsorbent, and the equilibrium loading from an adsorption isotherm.

Reactivation of the carbon completely restores its ability to adsorb contaminants, although 5 to 10 percent of the GAC may be lost during regeneration (Lykins and Griese, 1983). The GAC is reactivated in a multiple-hearth furnace at temperatures of approximately 1,800°F (982°C) to burn the organic contaminants from the walls of the pores. The hot carbon is then cooled in a water-filled tank. For larger plants, on-site reactivation may be the most cost-effective method of restoring the adsorptive properties of the spent GAC.

Combined Aeration and GAC Systems

In some communities, both aeration and GAC are used to remove organic chemicals. In Rockaway, New Jersey, for example, a GAC adsorption installation proved unsatisfactory because of the quick breakthrough of di-isopropyl ether (DIPE) and

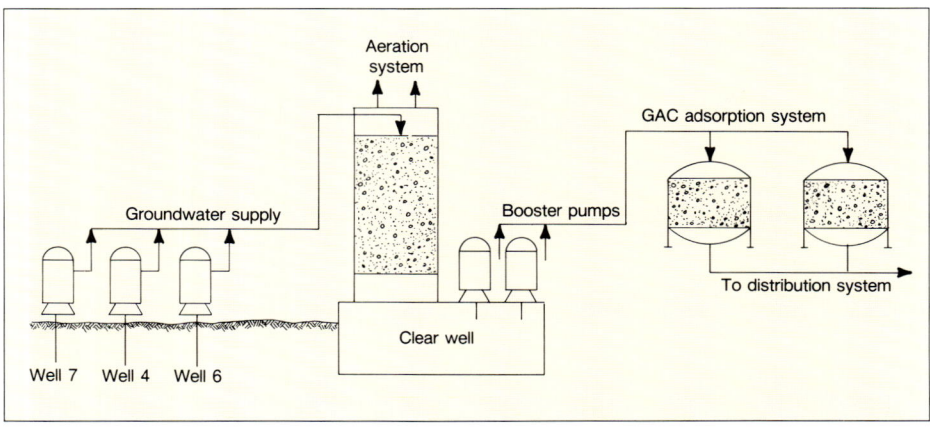

Figure 23.16. Schematic of the Rockaway Township treatment system. *[McKinnon and Dykeson. Reprinted from Journal AWWA, Vol. 76, No. 5 (May 1984), by permission. Copyright © 1984, The American Water Works Association]*

methyltertiary butyl ether (MTBE). Every 4 to 6 weeks the GAC had to be renewed to control the ether compounds, even though trichloroethylene (TCE), another major contaminant present in the water, could not be detected in the effluent water (McKinnon and Dyksen, 1984). Because of the quick breakthrough of the ether compounds and the high cost of GAC renewal, aeration was added to the treatment system (Figure 23.16). The countercurrent system allows water to flow downward by gravity while air is blown upward in a forced draft. The packing material consists of a toroidal-helix design of tellerettes. The aeration system significantly reduces the organic content of the water, so that the GAC system which follows could easily handle the ether compounds (Figure 23.17). The useful life of the GAC is now significantly increased. The use rate for the carbon without aeration was 1 to 2 lb of GAC per 1,000 gal (0.5 to 0.9 kg per 0.4 m^3) of water treated. With aeration, the use rate dropped to less than 1 lb of GAC per 1,000 gal (less than 0.5 kg per 0.4 m^3). Operation of both the aeration and the GAC systems is completely automatic. The GAC system was later deactivated after the strength of the ether compounds in the influent decreased. The aeration system, which costs less to operate, could then remove enough of the ether compounds to make the GAC treatment unnecessary. The GAC system is being kept operable, however, in the event that the strength of the contaminants increases in the future.

Nontreatment Strategies

Nontreatment strategies are sometimes used when organic or inorganic chemicals are discovered in groundwater supplies. Blending uncontaminated water with contaminated water is one solution to lowering the concentration of the contaminant to an acceptable level. Interception is another technique whereby a well is used to intercept the plume of contaminated water before it reaches a water supply well. The contaminated water may be treated and then discharged to a surface water body. In other cases, a contaminated well is simply abandoned and a new groundwater source brought into production.

Drilling contractors should be extremely careful when working with materials containing organic chemicals and when using construction materials that are not resistant to them. Plastic pipe, for example, is permeable to some organic substances when

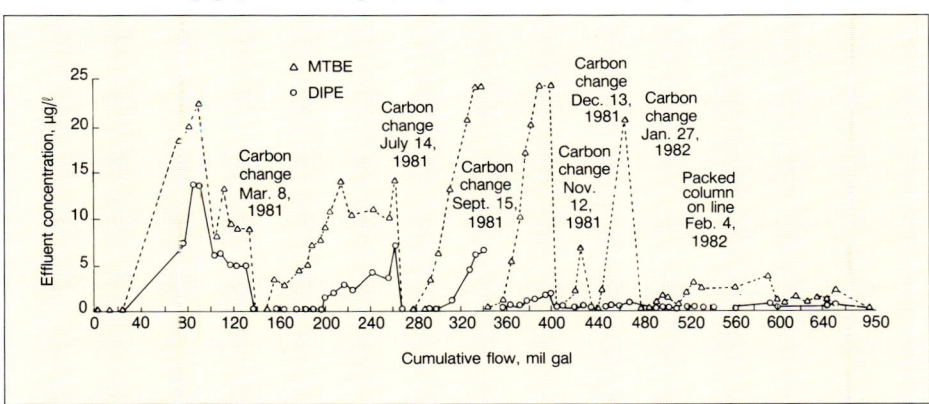

Figure 23.17. Breakthrough curves for di-isopropyl ether (DIPE) and methyltertiary butyl ether (MTBE) in Rockaway Township, New Jersey. The addition of an aeration tower reduced significantly the amount of GAC required to treat the water. *[McKinnon and Dykeson. Reprinted from Journal AWWA, Vol. 76, No. 5 (May 1984), by permission. Copyright © 1984, The American Water Works Association.]*

water in the pipe is under pressure (Keech, 1984). In one Michigan water supply system, organic chemicals from a gasoline spill passed through the walls of polybutylene plastic pipe used for service lines from the water main into homes. Polyvinyl chloride (PVC) may be somewhat more resistant to organic acids than is polybutylene or polyethylene pipe, but organic permeation may still occur (Keech, 1984). In areas where the presence of organic acids is suspected, the use of plastic piping may not be appropriate. Contractors should also be careful not to introduce into the well organic materials such as bitumastic coating compounds, residues from solvents and degreasers, and other materials made from hydrocarbons.

Removal of Radioactive Contaminants

The U.S. Environmental Protection Agency has estimated that 500 public water supplies in the United States may have radium levels exceeding 5 pCi/ℓ (U.S. Environmental Protection Agency, 1975a). Radium is the chief element in groundwater that produces a threat to health from radioactivity. Leaching of radium-rich rocks by percolating groundwater is the principal source of radium in groundwater, especially in Iowa, Illinois, Florida, and Texas. Three treatment methods are used to reduce the radium levels in groundwater supplies: lime or lime-soda softening, ion-exchange softening, and reverse osmosis (U.S. Environmental Protection Agency, 1978). Lime softening can remove up to 90 percent of the radium in waters having concentrations as high as 25 pCi/ℓ if the pH level during treatment is kept above 10. Up to 95-percent radium removal is possible in an ion-exchange process. Reverse osmosis treatment facilities have been shown to remove up to 95 percent of the radium, along with the typically high percentage of total dissolved solids.

Wastes from any of these treatment techniques present a disposal problem, not just because of the radium, but also because of the high salt content. If highly contaminated, disposal of these wastes may be to deep injection wells, evaporation lagoons, or use-recovery systems. Discharge of the wastes may be to a sanitary sewer, stream, or ocean if the waste strength is not excessive.

Removal of man-made radioactivity (strontium 90 and tritium) can also be accomplished by the methods listed above. It is unlikely, however, in light of regulatory controls over the nuclear power industry, that man-made radiation will become a problem in water supplies, especially groundwater.

Emergency Treatment Procedures for Sudden Contamination

From time to time, sudden contamination of a water supply occurs. The public must be notified immediately if the contaminated water is already in the distribution system. When the contaminated water is the only source for the community, certain treatment procedures must be undertaken to remove the contaminant or reduce its toxicity. The emergency treatment measures outlined in Table 23.8 are effective if the contaminant has been positively identified.

POINT-OF-USE WATER TREATMENT SYSTEMS

Drilling contractors have traditionally built bacteriologically safe wells by placing them an appropriate distance from known contaminant sources such as septic systems, and chlorinating the wells before they were placed in service. Most contractors, however, did not take an active interest in treating groundwater that had taste or odor problems, was excessively hard, or contained high concentrations of troublesome ions such as iron or manganese. They left these problems for the homeowner to solve. But the widespread occurrence of volatile organic chemicals in groundwater makes it esssential that contractors become familiar with point-of-use technology, which can remove or reduce the concentration of these chemicals. Fortunately, treatment techniques that are extremely effective in removing organics are also suitable for con-

Table 23.8. Emergency Treatment for Reducing Concentration of Specific Chemicals in Community Water Supplies

Chemical	Treatment
Arsenicals Unknown organic and inorganic arsenicals in groundwater at concentrations of 100 mg/ℓ	Precipitation with ferric sulfate and liming to pH 6.8, followed by sedimentation and filtration.
Cyanides Hydrogen cyanide	Prechlorination to free residual with pH 7, followed by coagulation, sedimentation, and filtration. *Caution*: housed facilities must be adequately ventilated.
	Precipitation with ferrous or ferric salts to form Prussian blue (iron ferric cyanide) followed by coagulation, sedimentation, and clarification. As long as an excess of iron coagulant is applied, the filtered water should be nontoxic even though it is blue.
Acetone cyanohydrin	Same as for hydrogen cyanide.
Cyanogen chloride	Same as for hydrogen cyanide.
Hydrocarbons Kerosene peak concentrations of 140 mg/ℓ	Preapplications of bleaching clay and activated carbon, plus some increase in normal dosages of alum, chlorine dioxide, lime and carbon, to provide treatment enabling continued production of water.
Miscellaneous Organic Chemicals LSD (lysergic acid derivative)	Chlorination in alkaline water, or water made alkaline by addition of lime or soda ash, to provide a free chlorine residual. Two parts free chlorine are required to react with each part LSD.
Nerve Agents (Organophosphorus compounds)	Superchlorination at pH 7 to provide at least 40 mg/ℓ residual after 30-min chlorine contact time, followed by dechlorination and conventional clarification processes.
Pesticides 2,4-DCP (2,4-Dichlorophenol) an impurity in commerical 2,4-D herbicides	Adsorption on activated carbon followed by coagulation, sedimentation, and filtration.
DDT (dichlorodiphenyltrichloroethane), concentrations of 10 g/ℓ	Chemical coagulation, sedimentation and filtration.
Dieldrin, concentration of 10 g/ℓ	Chemical coagulation, sedimentation, and filtration. Supplemental treatment with activated carbon may be necessary.
Endrin, concentrations of 10 g/ℓ	Chemical coagulation, sedimentation, and filtration. Supplemental treatment with activated carbon may be necessary.
Lindane, concentrations of 10 g/ℓ	Application of activated carbon followed by chemical coagulation, sedimentation, and filtration.
Parathion, concentrations of 10 g/ℓ	Chemical coagulation, sedimentation, and filtration. Supplemental treatment with activated carbon may be necessary. Omit prechlorination as chlorine reacts with parathion to form paraoxon, which is more toxic than parathion.

[Reprinted from Opflow, AWWA, Vol. 9, No. 3 (March 1983), by permission. Copyright © 1983, The American Water Works Association.]

trolling other chemical, taste, and odor problems. In this section, various point-of-use systems will be discussed. All of these systems could be installed and maintained by contractors who have acquired the skills and personnel to market and service these treatment systems.

It is well known that certain groundwaters have physical or chemical characteristics that make them aesthetically or chemically objectionable. In the United States, for example, hardness is a significant problem over much of the country (Figure 23.18). Saline water, discolored or turbid water, and water with strong odors also occur. Appendix 23.B lists many of the common water-quality problems encountered in domestic water systems, along with the causes and some suggested treatments.

Although the number of possible contaminants in groundwater is large, only six basic point-of-use treatment systems are used to control them. These are chlorination, ion exchange (softening), filtration, reverse osmosis, distillation, and ultraviolet systems (Table 23.9). Generally, water well contractors are familiar with the groundwater chemistry in their area and are able to suggest which treatment system may be required. The user must recognize, however, that the quality of various treatment devices varies and that poorly designed or functioning treatment devices may not completely remove specific contaminants.

Chlorination, Ozonation, and Ultraviolet Disinfection

If bacteria are known to be in a well, a one-time treatment with chlorine is usually sufficient to eliminate the problem. On the other hand, if the bacteria are in the water, a chemical feed pump must be used to inject chlorine continuously into the well casing, the water pump, or into the line between the pump and the pressure tank. An antisiphon device is used to prevent the chlorine solution from draining into a well

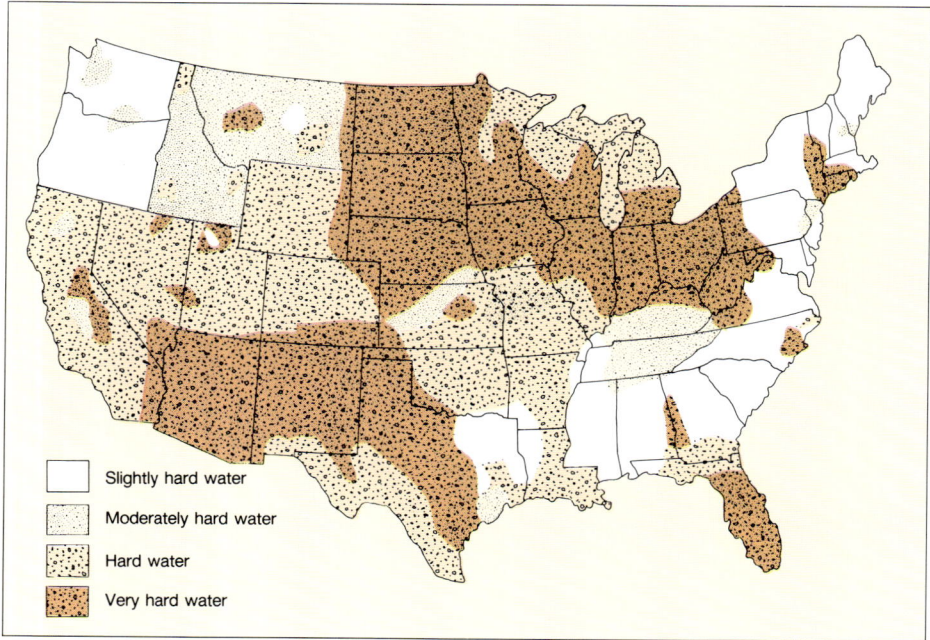

Figure 23.18. Water-hardness map of the United States. *(Brown, 1984)*

when the pump is not operating. Usually the feed pump is wired to the water-well pump-pressure switch so both pumps will operate simultaneously. The free chlorine residual should be maintained between 0.2 and 0.6 mg/ℓ (Mec-O-Matic Co., 1984). The actual contact time required to kill all the bacteria depends on the chlorine residual, temperature, pH, and what organism is present.

The amount of chlorine required to kill bacteria depends on the amount of iron and hydrogen sulfide in the water, because some of the chlorine will be used in oxidizing these substances. For each 1 mg/ℓ of iron, use 1 mg/ℓ chlorine, and for each 1 mg/ℓ of hydrogen sulfide, use 3 mg/ℓ chlorine. If, for example, the water contains 3 mg/ℓ iron and 3 mg/ℓ hydrogen sulfide, the chlorine feed is 12 mg/ℓ ($3 \times 1 = 3$ plus $3 \times 3 = 9$). The iron is oxidized by the chlorine and can then be filtered out. The hydrogen sulfide is oxidized to sulfate. Chlorination by pellets is another way to kill bacteria in the well. A typical installation is shown in Figure 23.19.

The use of ozone and ultraviolet disinfection procedures can also be adapted to individual wells, but the cost makes them prohibitive. Ozone has more disinfecting power than chlorine, yet, unlike chlorine, ozone leaves no residual.

Table 23.9. Processes for Effective Removal of Drinking Water Contaminants by Point-of-Use Systems

General Water Problem	Contaminants or Dissolved Solids	Cation Exchange	Chlorination	Sediment Cartridge Filter	Taste & Odor Cartridge Filter (Carbon)	Reverse Osmosis	Distiller (Vented)	Ultraviolet Disinfection
Particulates	Sand			X			X	
	Silt			X			X	
	Rust particles			X			X	
Inorganics	Arsenic	X				X	X	
	Barium	X				X	X	
	Cadmium	X				X	X	
	Calcium	X				X	X	
	Chromium					X	X	
	Copper	X				X	X	
	Iron	X				X	X	
	Lead	X				X	X	
	Magnesium					X	X	
	Manganese	X				X	X	
	Mercury	X				X	X	
	Radium 226 & 228					X	X	
	Selenium					X	X	
	Silver					X	X	
	Sodium					X	X	
	Strontium 90					X	X	
	Zinc	X				X	X	
	Chlorides					X	X	
	Chlorine				X		X	
	Fluorides						X	
	Nitrates						X	
	Sulfates					X	X	
	Sulfides						X	
Organics	Benzene				X	X	X	
	THM's				X	X	X	
	PCB's				X	X	X	
	Petroleum Solvents				X	X	X	
	Pesticides				X	X	X	
	Herbicides				X	X	X	
	Tannin (Humic Substances)					X	X	
	Odors				X	X	X	
	Swampy taste				X	X	X	
Biological	Algae		X			X	X	X
	Bacteria		X			X	X	X
	Viruses						X	X

If bacteria are present in a domestic water supply, they can be virtually eliminated by passing the water through a device which exposes the water to ultraviolet light. Short-wave (ultraviolet) radiation, with wave lengths of 2,000 to 2,950 angstroms, is extremely effective in destroying bacteria, viruses, and spores. This type of light can be produced by low-pressure mercury vapor lamps. The actual amount of radiation received will be determined by the number of lamps used and the retention time of the water in the purifier. At normal radiation dosages of 635,000 microwatt-seconds per square cm (MWS/cm^2), the retention time should be at least 15 seconds.

The effectiveness of the ultraviolet purifier depends on the clarity of the water. It is often suggested that a filter be installed in the system ahead of the ultraviolet purifier to remove turbidity (Figure 23.20). Iron in the water can also reduce the rate of light transmission, thereby causing a reduction in the kill rate of bacteria and micro-organisms. Although ultraviolet disinfection is used to kill bacteria in water supplies for industrial and institutional purposes, its use for domestic systems has not been widely accepted because no residual exists (Brown, 1984).

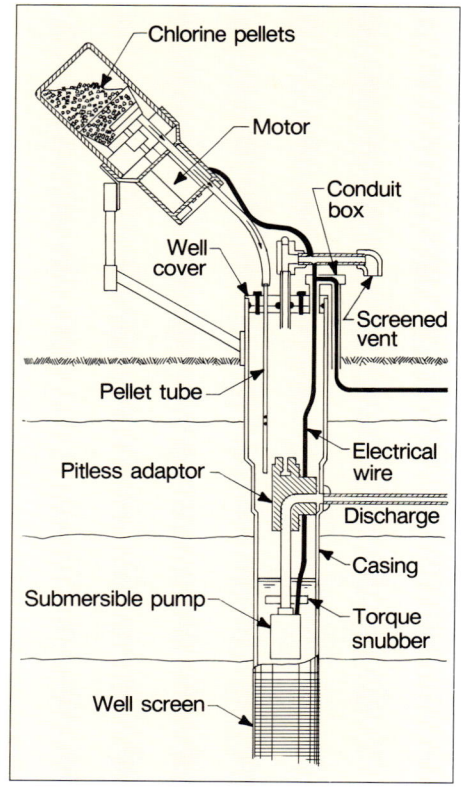

Figure 23.19. Pellet chlorinator. *(Autotrol Corporation)*

Persistent pathogenic contamination of any well should be taken seriously; the source of contamination should be discovered and eliminated if possible. If not, a new water source should be found.

Ion Exchange

The most common type of domestic water treatment is the water softener, which

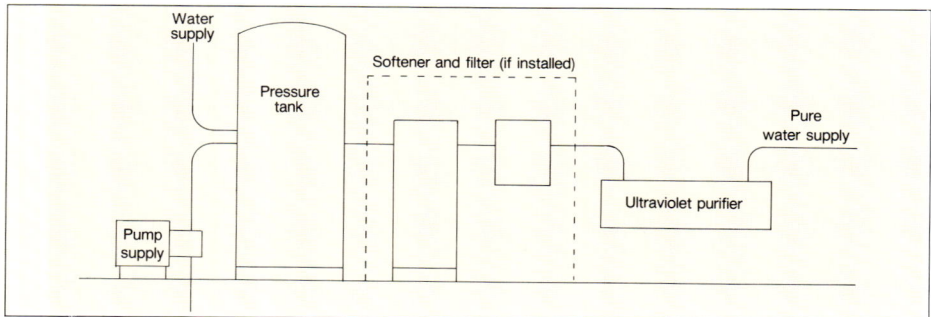

Figure 23.20. Typical residential water system. *(Ion Exchange Products)*

Table 23.10. Hardness Classifications for Water

Hardness		Classification
mg/ℓ	Grains/gal	
0 - 9	0 - ½	Soft
9 - 60	½ - 3½	Slightly hard
60 - 120	3½ - 7	Moderately hard
120 - 180	7 - 10½	Hard
over 180	over 10½	Very hard

is essentially an ion-exchange system used to reduce the hardness of water. For standardization purposes, the Water Conditioning Foundation classifies water hardness as indicated in Table 23.10.

More than 100 years ago, agricultural chemists noticed that certain types of soils containing silica and aluminum compounds had the ability to exchange ions with those in water (Hofheins, 1984). If these soils were converted to a sodium-rich form by treatment with brine, they were capable of softening water by removing calcium and magnesium ions from the percolating water and releasing sodium ions from the exchange sites into the water. Thus, the water would lose magnesium and calcium ions and gain in sodium ions as it passed through the softening media. Even though all these ions have positive charges, the ion exchanger will exhibit a preference for certain ions, depending on their ionic charge, molecular weight, and solution concentration (Table 23.11). Later, special plastic materials, called resins, were developed which could be converted to ion exchangers by chemical processing. From this development came modern demineralizers which can remove a wide range of ions.

Ion-exchange resins are synthetically produced solids in bead form, which are generally about $1/64$ to $1/32$ inch (0.4 to 0.8 mm) in diameter. These porous beads have extremely fine pores, and thus have a large number of exchange sites. Hence, a given volume of resin has an extremely large number of exchange sites that can absorb ions.

Table 23.11. Ion-Exchange Selectivity — Decreasing Preference

Cation Exchanger	Anion Exchanger
Barium	Iodide
Strontium	Nitrate
Calcium	Bisulfate
Copper	Chloride
Zinc	Bicarbonate
Magnesium	Hydroxide
Titanium	Fluoride
Potassium	Bisilicate
Ammonia	
Sodium	
Hydrogen	

(Hofheins, 1984)

The base material of resins is a combination of organic compounds specially formulated to give the resin bead the desirable physical and chemical properties, that is, spherical shape, small size, high porosity, and a large number of sites for the attachment of ions. The exchange properties are created when permanently attached and exchangeable ions are attracted to the resin base. The resin base itself has no exchange properties.

For complete demineralization, two types of resins are required — cation and anion. The cation resin removes cations from water, principally sodium (Na^+), calcium (Ca^{+2}), and magnesium (Mg^{+2}), by exchanging these ions for hydrogen (H^+) and generating acids in the process.

It differs from a water-softening resin in that the exchangeable ion is hydrogen (H^+), rather than sodium (Na^+). The anion resin has hydroxide (OH^+) ions as the exchangeable ion. It removes anions from water, principally chloride (Cl^-), sulfate (SO_4^{-2}), and bicarbonate (HCO_3^-), and releases hydroxide ions (OH^-) into the water in the process. The OH^- and H^+ ions react to form water.

Once all of the exchange sites in the resin have been used, the capacity of the resin is exhausted. The ion-exchange process is reversible, however. With suitable chemical treatment, the resin can be regenerated to its original capacity and is ready for another cycle. Resins can be regenerated many times, but regenerant chemicals for complete demineralization are strong acids or bases and must be handled with care.

Organic matter in water can seriously foul anion resin beds (Hofheins, 1984). Natural organics such as tannic, humic, and fulvic acids come from the decomposition products of animals and plants and are found widely in groundwaters. Organic wastes from industrial and farming activities (lignosalfonic acids, organophosphorous residues, and anionic polymers) also occur in groundwater. In some cases, large organic anions may form in cation resins as a result of degradation of the cation resins by chlorine used in a treatment process. Large organic molecules entering the pores of anion-exchange resin are retained because of their high affinity for the sites. Thus, their diffusion rate through the resin may be slower than for inorganic ions. The organic material can react with metal ions, binding them to the anion bed, but, because of the large size of the organic molecules, many unreacted exchange sites are prevented from participating in the exchange process. In normal regeneration procedures, it is not possible to completely remove the organic substances, thus reducing significantly

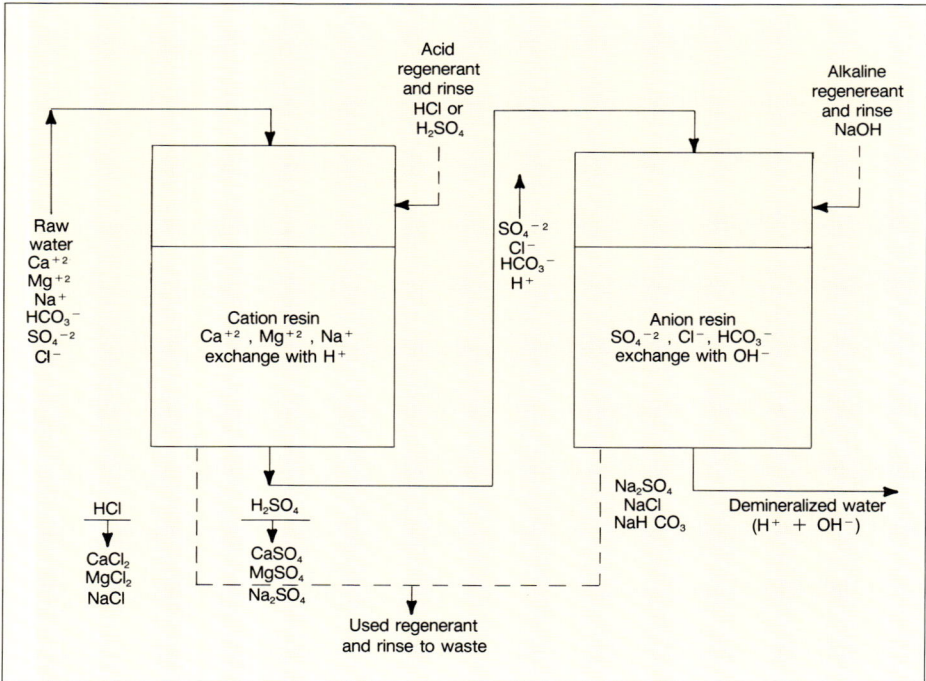

Figure 23.21. Two-bed ion exchange system.

the efficiency of the ion removal process. The resin must then be replaced prematurely even though the resin is not chemically exhausted. It is possible to coagulate the organic material with alum and to filter it out if the system is large enough. Reverse osmosis techniques are also useful in reducing both the organics and the total dissolved solids in the influent.

There are two principal types of systems used in demineralization by ion exchange:

1. A two-bed system has separate beds of cation and anion resins (Figure 23.21). In the service cycle, the raw water flows in series through the cation and anion resins. The ions are exchanged so that the effluent from the anion resin is demineralized. After the service cycle, each bed is independently backwashed, regenerated, and rinsed, and is then ready for the next service cycle.

2. The mixed-bed system has a bed of mixed cation and anion resins in a single tank, with the water flowing through the mixed bed in one pass. In this system, about 60 percent by volume is composed of strongly basic anion resin, and 40 percent by volume is cation resin. Exchange reactions occur sequentially in this type of bed. This system is capable of producing better quality water than the two-bed system. Feed water should be free of suspended solids; therefore, a prefilter is usually included in the system.

Figure 23.22 illustrates the principles of the mixed-bed system operation. Mixed-bed systems are more complex, but usually produce a higher quality water than do two-bed units. During the service cycle, cations and anions are simultaneously removed from the water. To regenerate, the mixed bed is backwashed, separating the cation and anion resins. The beds can be separated easily because of the density difference between the two resins. Regenerant chemicals are fed separately to each bed, followed by rinsing. The beds are then thoroughly mixed with air and settled. After rinsing, the mixed bed is ready for the next service cycle.

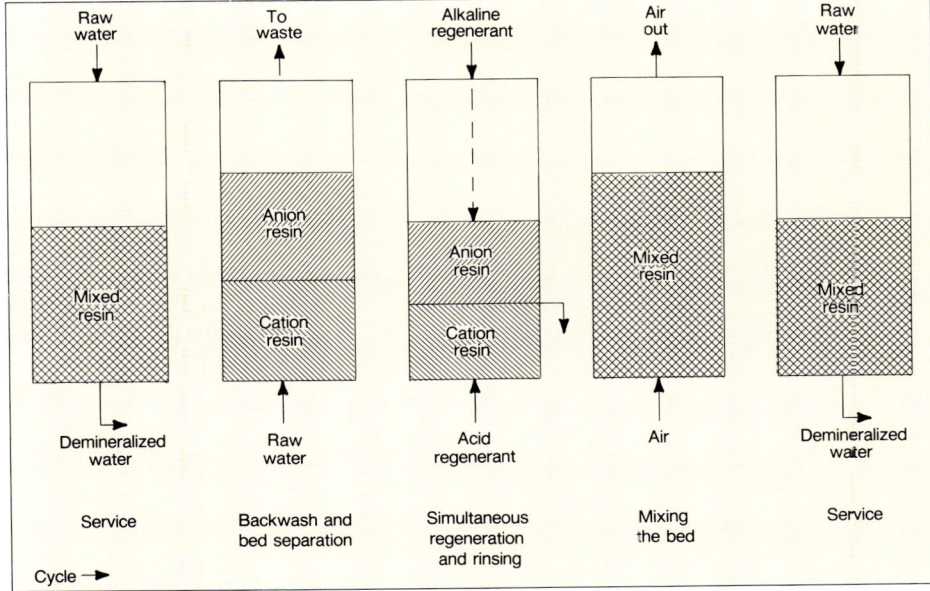

Figure 23.22. Mixed-bed ion-exchange and regeneration system.

Domestic Water Conditioning

Modern home water conditioners are simple ion exchangers and usually consist of two major elements — the resin tank and control unit and the salt storage cabinet. Tank-in-tank units are also common. Ordinarily, the resin beads remove calcium and magnesium ions to soften the water. At some point, all the exchange sites in the softener will be occupied by the calcium and magnesium ions, and the water will no longer be softened as it passes through the media. Conditioners are set so that regeneration takes place automatically, at a time that takes into account the volume of resin, the concentration and type of ions in the water, and the volume of water used per day.

During the recharge cycle, the flow through the conditioner is reversed so that any sediment or turbidity lodged in the resin is flushed out. Backwashing also expands the resin bed somewhat, creating a higher porosity. Normally, evaporated salts in block or briquetted form are most suitable for brine tanks. Evaporated crystals in briquetted form are sold commercially as pellets, beads, gems, or nuggets. Common rock salt is not used because of the presence of calcium sulfate and other insoluble impurities.

The capacity of water softeners is given as grains per gal (gpg) of hardness that can be removed during each cycle of operation. For example, if a household uses 200 gal (0.8 m^3) per day of water containing 7 gpg (120 mg/ℓ) hardness, the softener removes $7 \times 200 = 1,400$ grains of hardness per day. If regeneration occurs every three days, the softener must have a capacity of $1,400 \times 3 = 4,200$ grains. Some adjustment in softener capacity must be made if iron is present, because softener capacity is reduced. In general, the harder the water, the shorter the time between recharging.

Industrial demand has led to development of many specialized types of materials that may be useful in particular domestic situations. For example, if iron, manganese, and hydrogen sulfide are a problem, the use of manganese-treated zeolites may be recommended (Filkins, 1983). Manganese-treated zeolites remove iron and sulfide from water by ion exchange, filtration, oxidation, and catalysis. The three common types of base minerals, synthetic gel zeolite, greensand, and birm, are coated with manganese dioxide which becomes the active substance reacting in several different ways with different ions. This type of water conditioner is regenerated with an oxidizing agent such as potassium permanganate.

Filters

Many types of filters are available to remove one or more types of ions or forms of turbidity. Point-of-use filters are mechanical or chemical screening devices. A physical screen is effective if the suspended material is large enough initially, as with sand or rust, for example. Filters remove coagulated turbidity by passing water through a granular media. Periodic backwashing is essential to remove the filtered material. Small turbidity cartridge filters are also available, but are generally not effective in removing fine turbidity. These can be installed on the main water line or on a single faucet. When plugged, the water pressure will drop, signaling that replacement is necessary.

Some ions must undergo a chemical reaction before they are large enough to be filtered. Oxidizing filters contain a base material coated with manganese dioxide that converts soluble ions into insoluble particles that can be filtered. The filtered products are then removed by backwashing. This type of filter is suitable for medium concentrations of iron, manganese, and hydrogen sulfide (Lindsay Division, 1984). Periodic regeneration by potassium permanganate is required, but a continuous feed of potassium permanganate may be desirable in some systems. Normally, oxidizing filters are placed in the water line ahead of the softener.

Neutralizing filters consist of tanks filled with neutralizing materials that correct for low pH. Low-pH waters can be quite destructive to plumbing systems. Limestone and magnesium oxide are two common materials used to buffer acid waters. Because the neutralizing material is dissolved slowly from the tank, it must be refilled every 6 to 12 months.

Activated carbon filters are extremely effective in removing tastes and odors, and may be useful in mechanical filtration. Usually, the activated carbon is contained in a small cartridge-type filter that can be replaced easily. Some cartridges are so small that they may be effective for only a few hours of use.

Activated carbon filters are installed for temporary or for long-term use in many domestic water systems if organic contamination of the local groundwater is confirmed. In most cases, the contaminant level can be reduced well below the danger level. Great care must be taken, however, to properly design and maintain these units, for if breakthrough occurs, the actual contaminant levels may exceed those in the water because the overloaded filter may release high concentrations of the contaminant. Strict control of installation procedures, and meticulous monitoring of these systems, is essential. Ideally, local health officials or the commercial service installing the filters should be responsible for maintenance.

Distillation

Although distillation is impractical on a large scale, it is economical for drinking water. A distillation unit boils the raw water, collects the steam, and condenses it. Minerals in the water are left in the boiling unit. Small distillers (stills) require about four hours to produce 1 gal (3.8 ℓ) of water. The normal capacity is about 8 to 12 gal (30 to 45.4 ℓ). The system is protected by a float switch which shuts off the unit when the capacity is reached.

Use of distillers can eliminate sodium ions produced by ion-exchange water softeners. For people on a low-sodium diet, this is a significant advantage. The water should be aerated by shaking before use, however, and preferably chilled to improve the taste. Usually distillers are cleaned by acids that are left in the unit overnight and then flushed before the distiller is used.

Reverse Osmosis

Reverse osmosis treatment units have been adapted to point-of-use systems. Like large reverse osmosis systems described earlier in this chapter, small units use semipermeable membrane filters. These membranes can remove a high percentage of dissolved solids (Table 23.12). Many units have a turbidity pre-filter to prevent clogging of the membrane.

The reverse osmosis unit can be attached directly to a faucet so that water is being constantly filtered through the module. A small countertop reverse osmosis unit will produce about 3 gal (11.3 ℓ) per day; slightly larger undersink units produce 5 to 20 gal (18.9 to 75.7 ℓ) per day. When full, the overflow runs into the sink. A water pressure

Table 23.12. Dissolved Solids Removed by a Typical Point-of-Use Reverse Osmosis Method

Dissolved Solid	Removes Up To*
Aluminum	99%
Ammonium	95%
Barium	92%
Bicarbonate	96%
Borate	70%
Bromide	96%
Cadmium	98%
Calcium	98%
Chloride	95%
Chromate	98%
Copper	99%
Detergents	99%
Fluoride	96%
Iron	99%
Lead	90%
Magnesium	98%
Manganese	99%
Nickel	99%
Organic pesticides	99%
Phosphate	99%
Potassium	96%
Selenium	90%
Silicate	97%
Silver	96%
Sodium	96%
Strontium (radioactive)	99%
Sulfate	99%
Zinc	99%

*Removal percentages vary depending on type of water treated, temperature, and process.

(Lindsay Division, 1983)

of 60 psi (414 kPa) is sufficient to operate point-of-use systems. Individual membrane filters can be replaced easily.

Reverse osmosis units will not remove chlorine because it is a gas. If chlorine is present, an activated carbon filter is required.

Recommendations for Point-of-Use Systems

Treatment systems suitable for point-of-use installations vary widely in performance. Unfortunately, many products are successfully marketed that do not improve water quality significantly. Because of the large number of point-of-use treatment systems and their great range in effectiveness, drilling contractors should use care in choosing the most appropriate treatment technique for a specific water-quality problem (Table 23.13). Public health officials are reluctant to rely on a system that is of questionable quality in meeting national drinking water regulations. Once installed, the system must be monitored and maintained so that it operates at peak efficiency. Proper maintenance of point-of-use systems is crucial because malfunction of the system could produce adverse health effects. Thus, all point-of-use treatment systems should meet national standards for performance and be rigorously monitored so that health officials can be confident of their operation. In general, the following recommendations should be followed (Levine and Clark, 1983):

- Monitoring and maintenance should be provided by a water-utility operator employed by a water utility or town, or by a service company or water treatment professional under contract to a utility or town.
- Monitoring should be carried out on approximately one-twelfth of the units each month so that each is checked at least once a year.
- Twice a year, a portion of the samples should be sent to a professional laboratory to be analyzed using approved testing methods. All other samples could be analyzed by local operators. This strategy should be sufficient for compliance, and also for checking the accuracy of on-site testing.
- Testing by a certified laboratory using a nationally recognized protocol is recommended. The laboratory also can be used to develop guidelines on the frequency of

Table 23.13. Contaminants Removed by Point-of-Use Treatment Alternatives

Point-of-Use Treatment	Primary Contaminants	Secondary Contaminants
Reverse Osmosis	Arsenic, barium, cadmium, chromium, lead, mercury, silver	Total dissolved solids, copper, zinc, chloride, sulfate, corrosivity, foaming agents
Cation Exchange	Arsenic, barium, cadmium, chromium, lead, mercury	Copper, zinc, iron,[1] manganese
Anion Exchange	Nitrates, selenium	Chloride, corrosivity, sulfate
Activated Alumina	Fluoride, arsenic	—
Direct Filtration	Turbidity	—
Activated-Carbon Filtration	Organic substances	Color, odor, foaming agents

[1] Low levels.

(Levine and Clark, 1983)

monitoring.

Drilling contractors recommending point-of-use treatment systems should recognize that the cost effectiveness of some point-of-use systems falls substantially as the number of homes on the system increases. Table 23.14 shows the approximate number of homes that can be economically served by point-of-use treatment systems. As the number of homes increases past these recommendations, it is more economical to install public treatment systems (Levine and Clark, 1983).

Table 23.14. Cost-Competitive Point-of-Use Treatment Situations

Point-of-Use Treatment	Cost-Competitive Community Size	Amount of Water Treated
Reverse Osmosis	All sizes < 90 houses	Drinking water only Single tap
Cation Exchange	All sizes < 50 houses < 30 houses	Drinking water only Single tap Whole house
Anion Exchange	< 85 houses	Drinking water only
Direct Filtration	All sizes	All options
Activated Alumina	< 100 houses < 80 houses < 60 houses	Drinking water only Single tap Whole house
Activated-Carbon Filtration	All sizes	All options

(Levine and Clark, 1983)

Table 23.15. Chemical Reduction Requirements*

Substance	Influent Concentration (mg/ℓ)	Maximum Effluent Concentration (mg/ℓ)
THM	0.45 ± 20%	0.10
Lead	0.15 ± 10%	0.025
Fluoride	8 ± 10%	1.4
Nitrate (as N)	30 ± 10%	10.0
Barium	10 ± 10%	1.0
Arsenic		
(added as trivalent)	0.30 ± 10%	0.05
Cadmium	0.03 ± 10%	0.05
Chromium		
(hexavalent)	0.15 ± 10%	0.05
Chromium (trivalent)	0.15 ± 10%	0.05
Selenium	0.10 ± 10%	0.01
Mercury	0.006 ± 10%	0.0002

*Based on National Sanitation Foundation Standard 53
(Mayer, 1984)

The water treatment specialist should be able to tell customers what contaminants will be removed or reduced, how much water can be run through the system before regeneration or replacement is necessary, the recommended flow rate through the system, and the estimated useful life of the treatment device. Any treatment method affecting health must be able to meet or exceed effluent standards for maximum contaminant levels set forth in the Safe Drinking Water Act (Table 23.15). In time, the most efficient point-of-use treatment technologies will probably be recognized by the U.S. Environmental Protection Agency as acceptable technology that is sufficient to meet safe drinking water regulations. Applicable standards for various point-of-use treatment technologies are given in Appendix 23.C.

CHAPTER 24
Wise Use of Groundwater

Wise use of groundwater involves three general principles: (1) development of technologies that will enhance the storage capacity of groundwater reservoirs, (2) protection of groundwater quality, and (3) utilization of groundwater resources for their highest or most valuable use to society.

The first principle of the wise use of groundwater is to maximize the safe yield from aquifers over the long term. Perennial overdrafts of groundwater supplies should not be tolerated based on proposed short-term economic advantages. Excessive groundwater withdrawals that lead to aquifer depletion will increase water supply costs unnecessarily for future generations, because current groundwater management and water treatment practices make it possible to avoid overdrafts. Resource augmentation, for example, is being practiced widely in semiarid regions located close to surface-water sources. Recharge by basins, wells, and irrigation during non-growing seasons are some ways to increase the volume of groundwater in storage. In other water-short areas, conjunctive use and reuse are being utilized to increase the value of limited groundwater supplies. Conservation and reallocation of resources are other ways to make better use of groundwater.

The second principle of wise groundwater use includes efforts made to minimize the adverse impact of man's activities on groundwater quality. In the past, both private and governmental bodies have been slow to recognize how sensitive groundwater quality is to activities carried on at the Earth's surface. Overuse of groundwater near coastlines, for example, has permitted saltwater to contaminate fresh groundwater far inland, requiring the installation of many new wells. Careless handling of brines during oil and gas drilling, and the injection of wastes into poorly designed or poorly maintained wells, have contributed to the contamination of many shallow and deep aquifers. In fact, virtually any waste placed on or in the ground will adversely affect groundwater quality. Many attempts have been made to provide engineering safeguards to isolate wastes from the groundwater environment, but experience shows that complete isolation is nearly impossible to achieve. Thus, alternative methods of waste reduction and disposal and water treatment are being adopted to protect groundwater quality (see Chapters 21 and 23).

Many of the most serious groundwater contaminant problems are caused by synthetic organic chemicals. Fortunately, these chemicals generally move through the subsurface environment without becoming trapped geochemically in the aquifer materials. Once the source of the contaminants is identified and eliminated, a reasonable time is required for the chemicals to be flushed naturally from permeable aquifers. Attempts to accelerate the flushing process by operating high-capacity wells are quite successful because the water can be treated effectively by air stripping or granular activated carbon, or a combination of both methods. This water can then be used for potable water supplies. New federal regulations controlling potentially harmful discharges from industrial or municipal sites stipulate that waste lagoons, spray irrigation, deep underground injection, or other methods be used to protect the subsurface environment. Therefore, the potential for creating future sources for groundwater contamination is being minimized.

The third principle of wise groundwater use involves developing priorities on how groundwater will be utilized in the future. Widespread droughts, shifting populations, and agricultural and industrial expansion have contributed to an increasing stress on groundwater supplies in many areas. Water-resource managers now face the difficult problem of limiting water use for certain purposes, such as agriculture, to provide enough water for drinking or other higher uses. In the coming years, application of the higher use principle may cause severe economic dislocations, as high-volume users of water are forced to curtail their use in favor of other uses that are deemed to be more essential.

Table 24.1 lists the major opportunities and problems affecting individuals responsible for obtaining groundwater supplies, establishing how groundwater is used, and protecting groundwater quality. Of the items cited under opportunities in this table, the recent advances in analytical chemistry have exerted the most powerful effect with regard to understanding groundwater-quality problems. The analytical chemist can now provide the groundwater analyst with extraordinarily detailed information about groundwater quality. Substances that were undetectable just a few years ago are now routinely analyzed in parts per billion or even parts per trillion. Once the presence of contaminants is established, steps can be taken to eliminate their source.

ESTIMATING GROUNDWATER USE, RECHARGE, AND VOLUME IN STORAGE

Groundwater is basically a renewable resource, but the volume of water actually in storage may vary greatly from place to place, depending on the climate, regional hydrogeology, and rate of groundwater used for agricultural, industrial, and domestic purposes. From a practical standpoint, groundwater may not be renewable in terms of a human lifetime if contamination is allowed, recharge areas are eliminated, or extensive mining of groundwater is permitted.

The use of groundwater has escalated significantly worldwide since 1960. During 1984 in the United States, 35 percent of all water withdrawn for municipal supplies came from groundwater (Figure 24.1), whereas 95 percent of drinking water in rural areas comes from groundwater. Twenty-six percent of all water used in industry comes from groundwater. The withdrawals of groundwater versus surface water for the 48 contiguous states are given in Figure 24.2.

Matching long-term withdrawals of groundwater to recharge is the principal objec-

Table 24.1. Opportunities and Problems Affecting Groundwater Development and Water-Quality Protection

Opportunities	Problems
1. A significant increase has occurred in the capability of analytical chemical methods to measure contaminants in extremely small concentrations, that is, in parts per billion and parts per trillion. Recognition that contamination is occurring is the first step in the solution of groundwater-quality problems.	1. Increased demands on groundwater supplies are forcing a shift in how groundwater is used. Prioritizing may cause economic dislocations if not done properly.
2. Improvements in field instrumentation and computer technology now provide more and better information about subsurface conditions.	2. Complexity of most groundwater environments often prevents easy or inexpensive solutions to groundwater quality and supply problems. Basic groundwater equations apply only to the most simple hydrogeologic systems, and rigorous analysis of flow problems may be impossible. Similarly, the rates of many geochemical reactions taking place in the ground are not predictable; thus, predictions of contaminant strength at various points in a plume may not be feasible.
3. Recognition that much current groundwater contamination is temporary because it consists of synthetic organic chemicals which tend to move through the groundwater system almost as fast as the water itself.	3. Toxic effects of many chemicals on human health are not well understood. In the absence of reliable risk-assessment data, the level of a contaminant permitted may be set far too low merely because the contaminant can be measured to extremely small concentrations.
4. Acknowledgement that the price of water should correlate more closely with the actual cost of providing it; that is, water should no longer be considered as almost a free good. A heightened awareness of costs generally leads to the adoption of innovative ways to solve water-supply problems.	4. Decision makers have been slow to respond to new waste-treatment technologies to protect groundwater quality because they appear to be costly and represent a radical departure from long-entrenched waste-management procedures.
5. Recognition that groundwater levels recover more rapidly after severe droughts than earlier thought. This knowledge may permit greater use of the groundwater system in its capacity as a reservoir.	5. Shortage of well-trained and experienced hydrogeologists may result in higher-than-necessary costs for developing water supplies or protecting groundwater quality.

tive of water-resource planners. Three major problems are evident: (1) How can a reasonable long-term use rate be estimated; (2) is it possible to predict when dry periods (drought cycles) will occur and their severity and persistance, so that sufficient groundwater can be supplied when needed; and (3) if the current or projected use rate exceeds this volume, what steps can be taken to develop techniques for conjunctive (multiple) use or to reduce or even eliminate some uses?

The volume of groundwater that can safely be used over either a short or a long period without depleting the resource can usually be estimated from existing records. The volume for both municipal and agricultural purposes can be estimated from records that are usually readily available. Records are kept for both groundwater and surface-water contributions to the total supply. In practice, some part of this water may not actually be delivered to customers because of leaking pipes in the distribution system. In older systems, as much as 25 to 35 percent of the water treated may be lost from leaking pipes before it reaches the customer.

Records for agricultural water use may be more difficult to obtain. If records are not available, however, reasonable estimates can be made based on the crops grown and their water demand. Crops requiring irrigation will usually demand a certain amount of water over the natural rainfall for maximum growth. The actual extra volume depends mainly on soil type, air temperatures, and precipitation patterns. In east-central Minnesota, for example, corn typically requires no more than 2 to 4 in (51 to 102 mm) of extra moisture each year if normal precipitation occurs. In New Mexico, the same crop requires approximately 48 in (1,220 mm) of moisture from an irrigation system. Local agricultural extension services can provide an estimate of the amount of water needed for certain crops, the number of farmers actively irrigating, and the acreage of fields put into specific crops. Some states require the irrigator to record how much water is used. These data are then sent to the appropriate state or county agency.

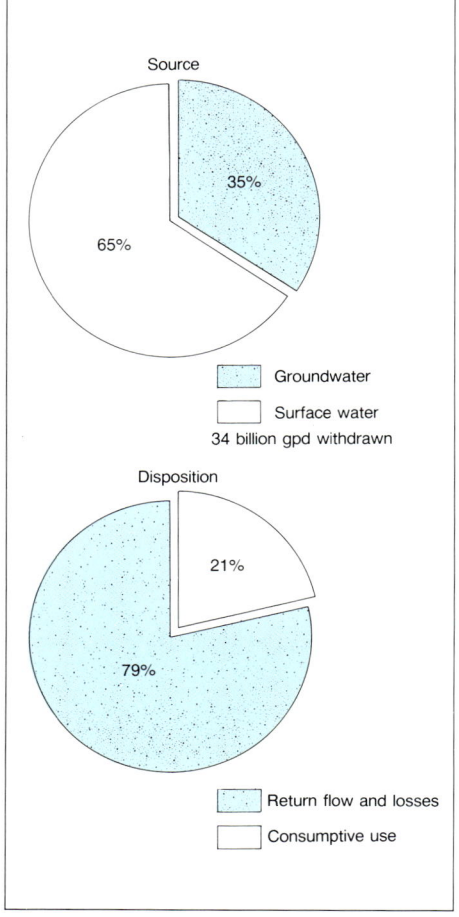

Figure 24.1. Water withdrawn for public supply in 1980, by source and disposition, in percent. *(Solley et al., 1984)*

Once the use rate for groundwater has been determined, the annual recharge and the water volume in the ground must be estimated. Figure 24.3 shows the large precipitation differences over the United States and the time the precipitation occurs. Recharge is generally related to precipitation. The recharge to a groundwater system

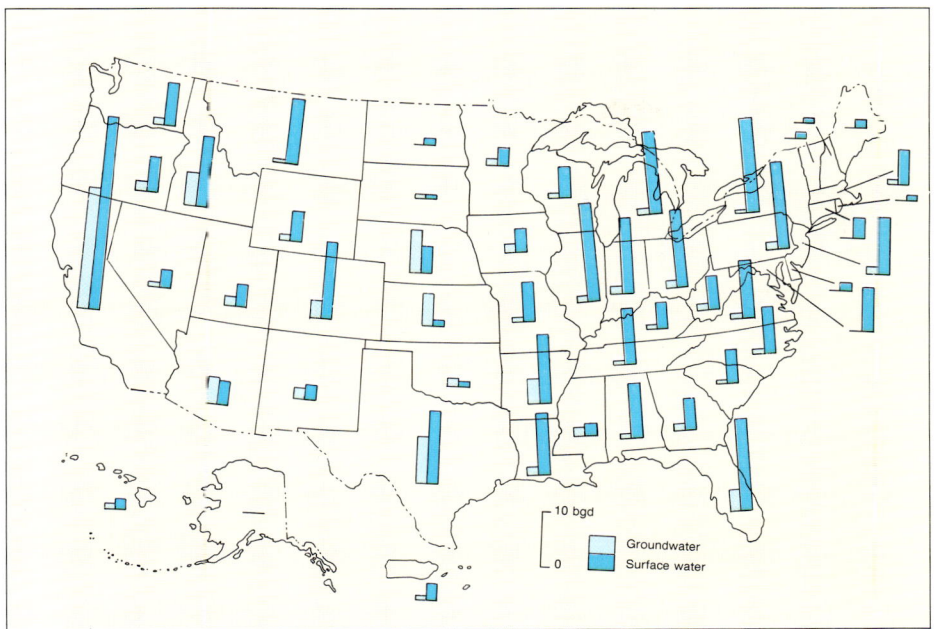

Figure 24.2. Withdrawals for offstream use from groundwater and surface-water sources, by state. (Solley et al., 1983)

varies significantly, but can usually be crudely estimated by measuring the groundwater discharge to local rivers over a period of years. The actual volume flowing to the river depends on the relationship between the river level and the groundwater table. In arid or semiarid regions, for example, the rivers lose water to the groundwater system, and thus they recharge the groundwater table rather than receive water from it. In humid regions, however, rivers recharge the groundwater system only during periodic floods. During the rest of the year when surface runoff is negligible, most of the water flowing in the river is groundwater. In either case, the total volume discharged or recharged by the rivers can be divided by the surface drainage area for these rivers, to arrive at the number of inches of recharge per year. River-discharge records are kept by the U.S. Army Corps of Engineers for all major rivers. State agencies keep records for smaller streams or rivers within their jurisdictions.

The volume of water in the aquifers should also be estimated to determine how much groundwater yields can be increased during periods of drought. The thickness of the aquifer, its areal extent, and its specific yield must be known to determine how much water is available for use. Normally, a specific yield of 0.15 to 0.25 is assumed for most unconsolidated sand and gravel aquifers. A slightly smaller specific yield may be appropriate for sandstones, depending on the degree of cementation. Hence, an alluvial aquifer 5 mi^2 (13 km^2) in area, with a thickness of 250 ft (76.2 m), and having a specific yield of 0.2, can be expected to yield 7 billion ft^3 (200 million m^3) of water. This volume would be sufficient to supply 20 center-pivot systems irrigating 25 percent of the time for 4 months each year at 1,000 gpm (5,450 m^3/day) for 8 years, assuming no recharge. If only 4 in (104 mm) of recharge occurred per year over this 8-year period, pumping could continue for another 3 years.

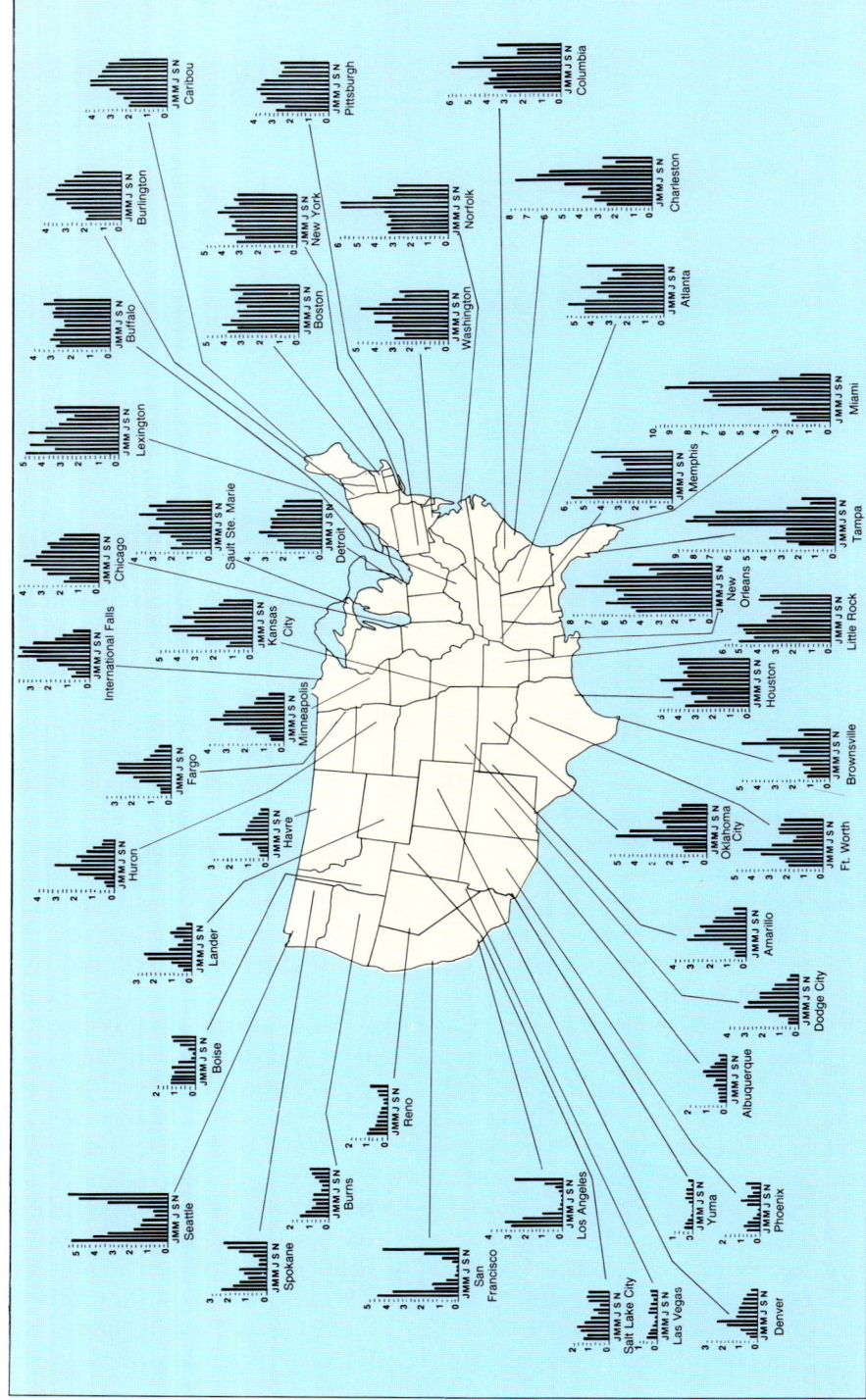

Figure 24.3. Average precipitation by months in different areas of the United States. *(Turner and Anderson, 1980)*

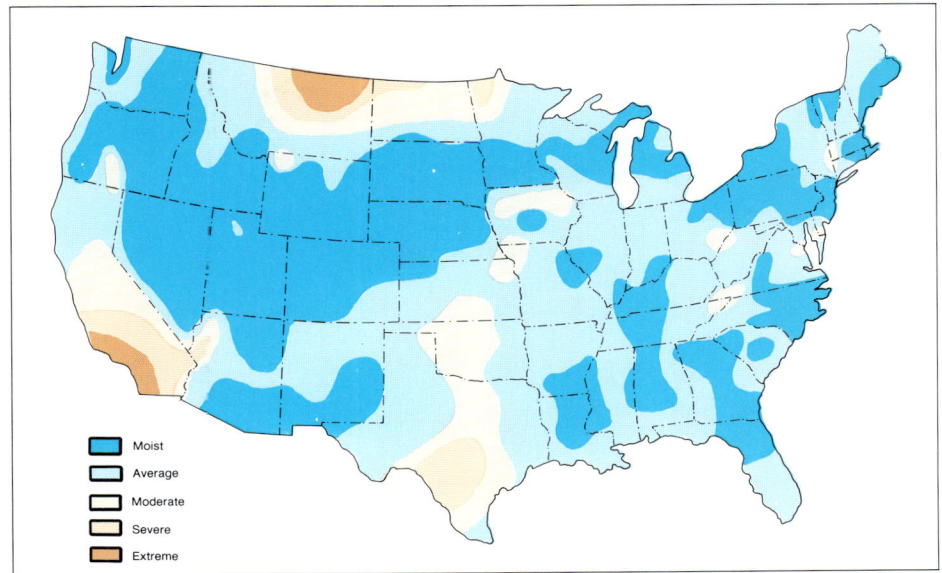

Figure 24.4. Maps prepared on a regular basis portray long-term trends (months, years) of abnormal dryness or moisture. These maps reflect long-term runoff, recharge, and evapotranspiration trends affecting a region, and show general long-term status of water supplies in aquifers, reservoirs, and streams. These maps can be used to estimate drought severity. *(NOAA/USDA Joint Agriculture Weather Facility)*

IMPACT OF DROUGHTS ON GROUNDWATER SUPPLY AND USE

Drought is a recurring climatic condition affecting many areas on Earth. It is usually defined as a period of abnormally dry weather that is sufficiently prolonged to cause serious hydrologic imbalance in the affected area (Huschke, 1959). In the United States, periodic droughts most often occur in the midcontinent regions where the humid East merges into the arid or semiarid Southwest, although serious droughts may occur anywhere in the United States. Areas of drought can range from a few hundred square miles to thousands of square miles. For example, the drought in the United States during 1977 affected 2,200 counties, four-fifths of the nation's total (Crawford, 1978). In the United States, maps are prepared on a regular basis that show the areas most likely to be subjected to drought, based on current climatic conditions (Figure 24.4).

Drought is the third most significant geologic hazard in terms of economic losses produced, ranking only behind floods and frost damage (Haas, 1978). The first and most powerful effect of drought is to reduce agricultural production over a wide area; other adverse economic impacts follow from this loss in production (Rosenberg, 1978). Agencies at various levels of government have taken action to reduce the effect of droughts. Adjustments to drought for agricultural purposes usually involve an attempt to modify the geophysical effects of the drought or to modify some parts of the agricultural system (Table 24.2). Adjustments to urban drought focus on finding new sources of water or on modifying the demand (Table 24.3).

During years of drought, the rate of groundwater recharge is usually insufficient to keep pace with withdrawals or discharges to rivers. Therefore, it is common for the groundwater table to fall several feet or even tens of feet over a period of years,

Table 24.2. Adjustments to Drought for Agriculture

1. Modification of natural event subsystem
    A. Augmentation of water
        Expand use of irrigation
        Increase precipitation by cloud seeding
    B. Conservation of water
        Alter cultivation practices to include strip cropping, minimum tillage, and development of windbreaks
        Protect water supplies by eliminating leakage, evaporation, and transpiration
2. Modification of agricultural subsystem
    A. Alter agricultural characteristics
        Increase operational flexibility by crop and livestock diversification and by selection of drought-resistant crops
        Institute land-use management so water shortages will not develop from uncontrolled municipal and industrial growth
        Develop the ability to predict droughts far enough in the future so that actions can be taken to avoid adverse economic and social impacts
    B. Spread or share losses and costs
        Create financial protection by crop insurance or by setting aside stocks of feed and grain
        Provide relief and rehabilitation programs (usually loans)

(After Haas, 1978)

Table 24.3. Adjustments to Drought for Urban Areas

1. Modification of sources and water systems
    Develop groundwater resources to supplement surface water or existing groundwater supplies
    Integrate wastewater reuse into community water management
    Build desalination facilities so that lower quality water can be used
    Augment precipitation by cloud seeding (rarely attempted)
2. Modification of demand
    Regulate land use so that population growth is controlled consistent with the water supply
    Enact economic incentives and penalties so conservation becomes attractive to consumers
    Institute legal mechanisms that control demand such as spring-loaded faucets, minimum-use toilets and dishwashers, and the prohibition of swimming pools
    Create "voluntary" changes in demand by altering attitudes and values held by the public about water
    Formulate priorities among competing demands so less vital uses can be minimized during drought

(After Haas, 1978)

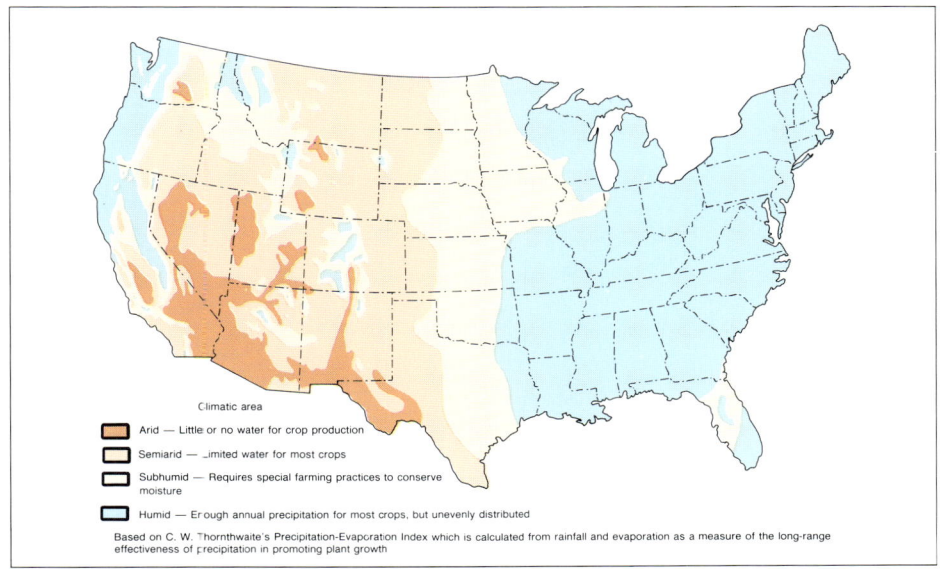

Figure 24.5. Four climatic areas in the United States: arid, semiarid, subhumid, and humid. *(Turner and Anderson, 1980)*

especially in subhumid areas where groundwater use is high (Figure 24.5). But even during severe droughts in these areas, the volume of water remaining in storage is usually enormous, and when rainfall returns to its usual pattern, the groundwater table rapidly returns to its typical level, generally within 6 to 12 months. For example, if the aquifer porosity is 20 percent, the first 5 in (127 mm) of rainfall reaching the saturated zone will cause the groundwater table to rise 25 in (635 mm). As precipitation gradually returns to normal, the groundwater reservoir will be recharged well before subsurface runoff to streams and rivers recovers to long-term levels.

The wise use of groundwater resources can play a significant role in reducing the impact of drought in both urban and rural environments. For maximum benefit, however, it is important to be able to predict when droughts will occur, how long they will last, and how severe they will be, so that the emergency use rate for groundwater during the drought can be estimated to make sure the water will last long enough.

Predicting the occurrence and severity of droughts is difficult because weather data in the United States have only been collected since about 1860 (Bark, 1978). However, records for prehistoric droughts have been established on the basis of tree rings for many parts of the United States. The width of an individual tree ring correlates with variations in climate — the more rainfall, the wider the ring. An example of this kind of data is shown in Table 24.4 for Nebraska. Usually, it is not possible to discern past drought conditions for large regions on the basis of data for only a single state or local area. Nevertheless, Weakly's data (1965) suggest that droughts have been relatively common in his study area over the last 700 to 800 years.

The severity of a drought, or even its occurrence, can be strongly affected by the timing of the annual precipitation (during the crop-raising season, for example), by temperature patterns during the dry period, and by strong winds. It is probable that severe droughts occur less often in areas where humid conditions prevail, such as in

Table 24.4. Drought Periods Identified from Tree-Ring Analyses in Nebraska

Beginning year	Ending year	Duration of period	Years between occurrences
1220	1231	12	
1260	1272	13	29
1276	1313	38	4
1383	1388	6	70
1438	1455	18	50
1493	1498	6	38
1512	1529	18	14
1539	1564	26	10
1587	1605	19	23
1626	1630	5	21
1668	1675	8	38
1688	1707	20	13
1728	1732	5	21
1761	1773	13	29
1798	1803	6	25
1822	1832	11	19
1858	1866	9	26
1884	1895	12	18
1906	1913	8	11
1931	1940	10	18
1952	1957	6	12
	Average	12.8	24.5

(After Weakly, 1965)

eastern United States, but they do occur periodically, as in the serious 1961-to-1966 drought in northeastern United States. Unfortunately, scientists have not been able to establish definite drought cycles or to predict the severity of future droughts (Schneider, 1978). One fact is clear, however. Future droughts will occur, and, when they do, groundwater supplies should be utilized to augment the water supply and therefore reduce the economic and social impact of droughts.

MANAGING GROUNDWATER SUPPLIES

Management Tools

In many areas of the world, virtually all of the local surface water is currently in use. Additional demands must be met by development of groundwater resources. But even groundwater resources are finite and must be managed with care. Several groundwater management procedures are available to either reduce the volume of groundwater required locally or actually increase the groundwater supply by making better use of the aquifer's storage capacity. Advanced water-treatment technologies, recycling techniques, and analytical tools such as groundwater computer models are being applied to assure better use of the groundwater potential. Of these options, groundwater specialists are most likely to become directly involved with computer models. Other types of water-resource specialists are usually better qualified in water recycling and improved water-treatment techniques.

Groundwater Models

For many years, groundwater specialists have used various types of models to simulate how groundwater moves in the subsurface environment. A model is a means of representing a simplified form of reality to help groundwater managers understand and manage the resource. Models are useful (1) for studies that precede field investigations, (2) for the interpretive analysis following the field program, and (3) as predictive tools to estimate how water quality or quantity may change in the aquifer. They provide the geohydrologist with a method of analyzing field data for a better understanding of the aquifer. The wide availability of mini- and micro-computers has facilitated the proliferation and use of special mathematical models describing many types of groundwater situations.

Some models such as laboratory sand-tank models are only valuable in a qualitative sense because they cannot be used to measure flow rates, estimate drawdown, or predict the movement of groundwater. The model is so small that proper scaling of the field situation cannot be achieved. But this type of model does give an overall view of how water moves through the ground toward wells.

Other types of physical models attempt to present analogs for the groundwater situation. Electrical analog models, consisting of electrical networks of resistors and capacitors, have been used extensively in the past but have now been almost completely replaced by high-speed digital computers. The concept of using electricity to model groundwater flow remains valid because Ohm's law for the flow of electricity is analogous to Darcy's law for the flow of groundwater. Individual electrical analog systems, however, were designed specifically for a single groundwater system and therefore could not be adapted to general use. Another type of physical model, called the Hele-Shaw or parallel-plate model, consists of two closely spaced plates containing a viscous fluid, usually oil. As with electrical analog models, the Hele-Shaw apparatus is expensive to set up and models only one situation.

Mathematical models are the most useful type. The development of a mathematical model starts with a conceptual understanding of the physical system to be modeled. The response of an aquifer to being pumped depends on the transmissivity and storage coefficient of the aquifer, the hydrologic and geologic boundary conditions, and the points at which water is being withdrawn or recharged within the system (Bachmat et al., 1980). The term "boundary conditions" refers to the lithologic, geochemical, and hydraulic conditions at the boundaries of an aquifer system. With this information, a set of differential equations can be written to describe groundwater flow.

The conceptual model can then be translated into a mathematical framework (model) that produces governing equations which describe the physical or chemical processes (Figure 24.6). The complicated nature of subsurface conditions can rarely be accommodated completely by mathematical expressions, so simplifying assumptions such as the aquifer being homogeneous, isotropic, and infinite in extent must usually be made to solve flow equations for appropriate boundary and initial hydrologic conditions. If enough simplifying assumptions are made, the descriptive equations can be solved analytically. These simplifying assumptions, however, reduce the accuracy of the model.

More complicated equations that describe the subsurface environment more accurately are often so laborious or so difficult to solve that numerical techniques must be used to solve the equations. Numerical models are especially useful for analyzing aquifers having irregular boundaries, complex structures, and variable pumping or recharge rates (Mercer and Faust, 1981). Use of the computer for solving equations by numerical methods makes it possible to simulate the response of extremely complicated systems to changing conditions.

Two general types of numerical models exist — finite difference and finite element. A pictorial representation of these methods is shown in Figure 24.7. The variables as expressed in the partial differential equations are continuously changing. In the finite-difference or finite-element methods, continuous variables are replaced with discrete variables that are defined at selected points (nodes) within the system domain. These variables are given by a finite number of algebraic equations that define a certain parameter such as hydraulic head. Depending on the magnitude of the investigation, the number of node points may range up to 100,000, although 10,000 is the practical limit (T. Prickett, personal communication). The system of algebraic equations generated at the nodes is solved by designing a program or code run on a digital computer. Thus, numerical solutions involve approximating continuous (defined at every point) partial differential equations with a set of discrete equations in time and space (Mercer and Faust, 1981). Well-known computer models have been developed by Prickett and Lonnquist (1971), Trescott et al. (1976), and McDonald and Harbaugh (1984).

The ability of these models to predict aquifer performance under changing conditions is now well established. Typical model parameters that can be estimated include the effects of (1) varying the rates of withdrawal or injection, (2) changing the well spacing or location, (3) altering the recharge rate, and (4) changing the boundary conditions. These variables can be changed at will to see how the aquifer will respond to new conditions. The ability to perform sensitivity analyses of various groundwater parameters and to test alternative hypotheses about aquifer boundary

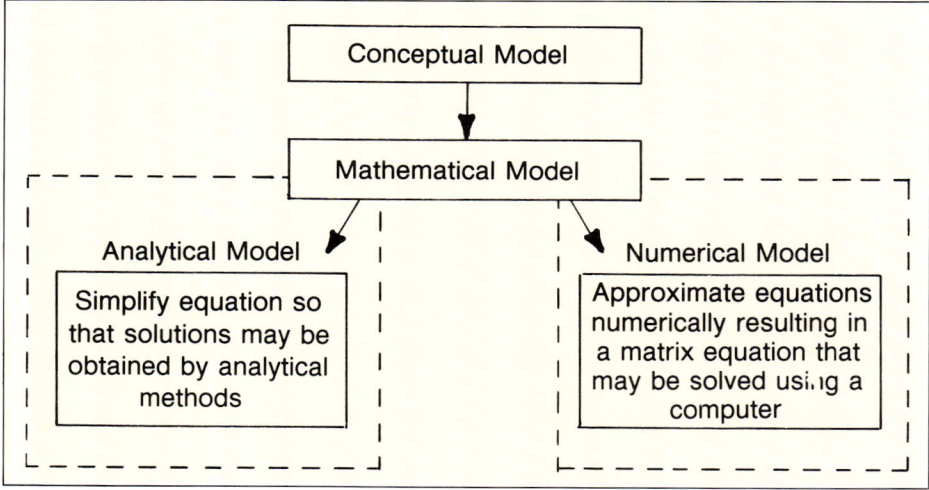

Figure 24.6. A mathematical model is based on a conceptual understanding of the problem as expressed by mathematical equations. These equations can be solved either by analytical means or by numerical methods using a computer. *(Mercer and Faust, 1981)*

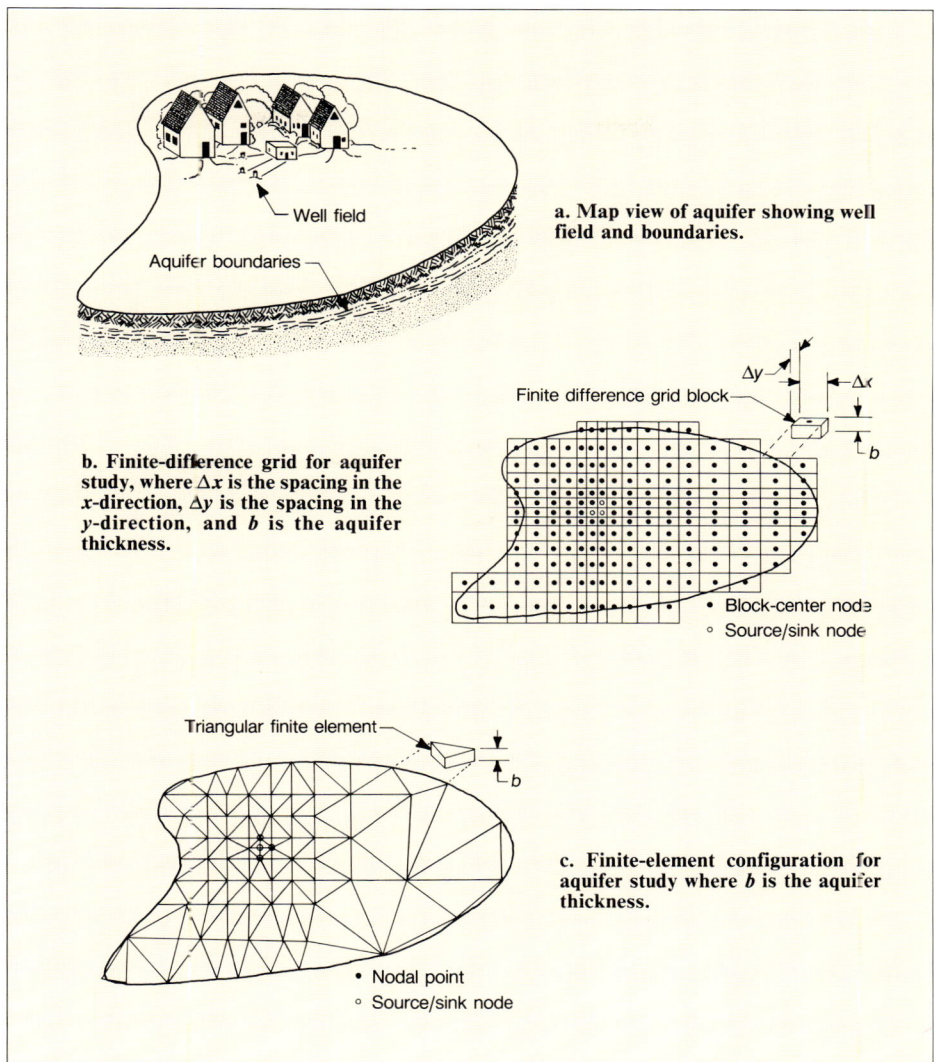

Figure 24.7. Two-dimensional grid systems for finite-difference and finite-element methods. *(Mercer and Faust, 1981)*

computer modeling an important tool in any groundwater investigation (Townley and Wilson, 1980).

Water-resource engineers charged with the analysis of aquifers should become familiar with these methods. In the future, many water well contractors will develop the skills necessary to utilize these techniques, not only for solving groundwater supply problems but also for analyzing dewatering, depressurizing, injection, and contamination problems.

Further description of numerical methods is beyond the scope of this book. Short reviews of the numerical methods can be found in Bachmat et al. (1980) and Mercer and Faust (1981). Wang and Anderson (1982) present a thorough analysis of both finite-difference and finite-element techniques. Javandel et al. (1984) present a listing of numerous finite-element and finite-difference codes and their numerical characteristics. Earlier works by Remson et al. (1971) and Pinder and Gray (1977) are also valuable references for numerical methods. Prickett (1975) presents an overview of modeling techniques for groundwater evaluation.

Augmenting Water Supplies

Water shortages can be relieved if certain measures are taken to fully utilize the storage capacity of groundwater reservoirs, to reuse groundwater, to use the water for multiple purposes, to treat slightly brackish water, to initiate ways to conserve water, and to price water according to its true costs. These procedures are currently being used successfully in areas where groundwater resources are limited. Often these procedures, even those requiring water treatment, do not add prohibitively to the cost of water.

Experience has shown that it is good management practice to draw down the groundwater table during prolonged dry periods in areas where withdrawals normally do not exceed recharge. This reduction in the water table is temporary, and rapid recovery can be expected once precipitation returns to normal. During a prolonged drought, the base flow of groundwater to rivers may be reduced by greater groundwater withdrawals, but use of a part of this base flow may be far more beneficial than maintaining normal stream levels, except in navigable rivers where minimum depths are regulated.

Ambroggi (1978) has estimated that 1,500 mi^3 (6,260 km^3) of water per year are available worldwide from the base flow of rivers in populated areas; this figure includes the flow from lakes and man-made reservoirs. This compares favorably with his estimate of a total of 740 mi^3 (3,090 km^3) now being utilized from all water resources annually. Currently, only about 290 mi^3 (1,200 km^3) of the 740 mi^3 comes from groundwater. In many areas of the world, exceeding the "safe yield" for periods as long as 10 to 15 years is not injurious to the volume of water in storage, because either all or nearly all of the overdraft can be made up in the first year or two of normal rainfall. Ambroggi (1978) cites examples in Tunisia and Morocco where water tables were drawn down 30 to 60 ft (9.1 to 18.3 m), only to be essentially replenished in one year of heavy rains. According to Ambroggi, it may be possible to draw down certain aquifers in these areas for periods of up to 40 years and still have them refilled by natural precipitation within a relatively short time. Up to now, the value of the storage function of aquifers has been somewhat overlooked by planners. For instance, utilization of only a portion of the base flow for irrigation in arid countries could make a large contribution to meeting world food needs.

It is not good management policy, however, to cause severe overdrafts in arid or semiarid areas of the world, where recharge is limited. Pumping from groundwater reservoirs in these areas provides only a short-term solution for a long-term problem. Unfortunately, past and even some current political decisions involving groundwater resources have failed to recognize the long-term effects of overdrafts on groundwater supplies. Too often, short-term economic considerations become the primary factor affecting management decisions concerning groundwater.

Reuse

Reuse of wastewater or use of storm runoff may become a necessity in many parts of the world as increasing populations make one-time use of water unacceptable (Dean and Lund, 1981). Many people view the direct reuse of wastewater with concern, and yet many communities currently use surface water containing a large percentage of sewage effluent as their raw water supply. For example, 100 percent of the water used by the City of London during several past droughts had been in and out of five sewage treatment plants (Blackburn, 1978). Cincinnati, Ohio, and Chanute, Kansas, regularly incorporate wastewater from upstream cities in their water-treatment plants (Dean and Lund, 1981). In fact, many water-treatment specialists now believe that technology is available to treat undiluted sewage so successfully that it can be consumed with the same level of health protection as found in other aspects of life. Naturally, the cost of this high level of treatment exceeds treatment costs for clean water, because more processes are involved and closer attention must be paid to the system. In general, water reuse is an option only for large systems because of increased costs.

Reclaimed wastewater can be used for many purposes. In California's Orange County, reclaimed water is injected into several local aquifers near the coast to contain saltwater intrusion. Reclaimed water can also be used to create lakes when the quality is good enough to meet health standards. In 1977, the Whittier Narrows Reclamation Plant (California) provided over 7,000 acre-ft (8.6 million m³) of tertiary-treated water* for groundwater recharge. This is enough water for almost 9,000 average families for 1 year. Much reclaimed water is also applied to various crops. Treatment processes required will depend on the crop to be irrigated. Industrial reuse, landscape irrigation, and water for cooling towers used for electrical generating plants are other applications in which reclaimed water can be used beneficially.

In rural areas, wastewater is sometimes spread on land, not only to irrigate the crops but also to treat the water. Although mismanagement of many sewage farms has led to their closing, several major cities such as Wroclaw, Poland, Paris, France, Melbourne, Australia, and Braunschweig, Germany, continue to operate their sewage farms (Dean and Lund, 1981). The idea of discharging large quantities of wastewater into spreading basins is gaining favor because the water is

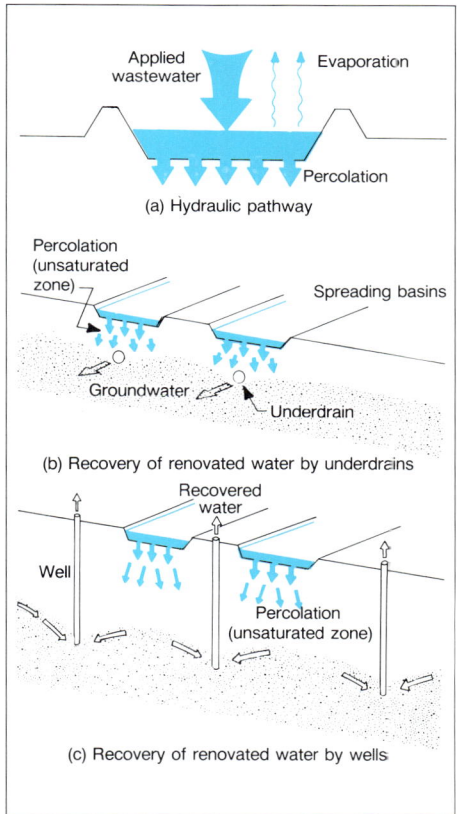

Figure 24.8. Spreading basins for groundwater recharge. *(Crites et al., 1977)*

---

*Tertiary (advanced) water treatment removes all or some of the specific pollutants such as nitrogen, phosphorus, heavy metals, trace organic compounds, viruses, bacteria, dissolved salts, and suspended and dissolved organic solids.

treated naturally in the groundwater system (Crites et al., 1977) (Figure 24.8).

Experimental work at Flushing Meadows, a spreading basin operated by the City of Tucson, Arizona, showed that water collected by wells surrounding the basin was pure enough to be distributed to irrigation ditches (Bouwer et al., 1980). The spreading-basin system will function properly in fine-grained sediments where filtration and adsorption are effective, but periodic maintenance (removal of decomposed organic materials) is vital if the filtration and adsorption processes are to function properly. Some organic substances and volatile organic chemicals are not removed, however, as they pass through the vadose zone.

Conjunctive Use

In many areas of the country, groundwater resources have become so scarce or so fully appropriated that whatever water is available must be used for conjunctive (multiple) purposes. Innovative methods are now required when new demands arise for domestic or industrial water supplies. For example, near Dodge City, Kansas, a community could not expand because both surface and groundwater resources were fully appropriated. To gain additional water, the community decided to trade its treated sewage effluent to an irrigator (rather than discharging it to the Arkansas River) for the right to use the irrigator's groundwater appropriation. Both parties benefited; the community secured the high-quality groundwater from the Ogallala Formation it required to sustain its growth, and the irrigator benefited from the use of nutrient-enriched wastewater.

In this example of conjunctive use, the community treats its wastewater to remove pathogenic organisms; the effluent is then discharged into sewage lagoons. In the lagoon, phosphorus and nitrogen, the key nutrients for plant growth, are taken up during algal production. Some algae, especially blue-green algae, can also fix large amounts of atmospheric nitrogen, thereby adding to the total nitrogen load in the effluent. When the irrigator distributes the wastewater on fields, the algae decay slowly at the surface, releasing phosphorus and nitrogen. Water not transpired by the plants or lost through evaporation percolates through the thick unsaturated zone to the groundwater table. Some or most of this water then returns to the wells that supply the community water system. In this conjunctive-use example, some of the water is lost to the atmosphere by evaporation and transpiration, but the total draft on local groundwater supplies is less than half of what it would have been if both the irrigator and the community had been allowed to withdraw water.

Desalination

Desalination is another way to augment water supplies when local groundwaters have too much total dissolved solids for drinking water. United States government regulations protect, for potable use, groundwaters that contain up to 10,000 mg/ℓ total dissolved solids, because current technology is capable of economically treating water at this solids concentration to produce drinking water. Community water planners may not be aware of the advances made in reverse-osmosis technology that now permit treatment of groundwater with high total dissolved solids. When appropriate, water well contractors and groundwater specialists familiar with reverse-osmosis systems can take a leading role in urging the use of brackish groundwater for a water supply (see Chapter 23). In regions where little groundwater exists, such as along low-lying seacoasts, many communities are using reverse-osmosis systems to treat ordinary seawater (35,000 mg/ℓ). Even at this high concentration of salts, the price consumers pay is reasonable. Indeed, brackish groundwater, or seawater pumped from wells and then treated, may be less expensive than obtaining higher quality surface-water supplies from great distances. In areas where only low-quality groundwater exists, drillers could contribute greatly to solving local water shortages by suggesting appropriate treatment technologies.

Protecting Groundwater Quality by Land-Use Planning

Land-use planning decisions can play a major role in protecting groundwater quality, and therefore supply. In the past, well drilling contractors were often unaware that decisions by governmental planners regarding solid-waste disposal, industrial land use, agricultural zoning, highway maintenance, and sewage treatment could lead to an impairment of groundwater quality over time. This degradation is not intentional, but results from inadequate engineering practices, accidental spills, inaccurate assessment of the ground's capacity to adsorb contaminants, and overuse of many chemicals. Until recently, the susceptibility of near-surface aquifers to the activities of man was not well understood, and it was thought that any degradation of groundwater quality that did occur would not be widespread. Experience now shows that the subsurface environment is highly sensitive to certain activities carried out at the surface, especially those resulting from the improper use or disposal of synthetic organic chemicals.

Recognition that many common human activities can produce groundwater-quality problems has led to regulations concerning the use of agricultural chemicals near wells, the prohibition of certain industrial activities over valuable aquifers, the reduction of salt use on highways and increased care during salt storage, and improvements in how buried municipal wastes are isolated from the groundwater system. These regulations are just being developed, and often involve a high level of public participation because decisions on whether to allow certain types of industrial development or to institute restrictive agricultural zoning practices are vitally important to a community. The significance of these deliberations to the water well contractor cannot be overstated. Growth and prosperity in the community mean a higher level of income for the contractor. On the other hand, growth accompanied by degradation of groundwater quality will limit the potential economic success of the contractor, because alternative sources of uncontaminated (and usually higher priced) surface water will be developed. Few people in a typical community understand the groundwater system as well as the water well contractor. Therefore, the contractor is in a position to exert great influence on decisions relating to the preservation of high-quality groundwater and, indirectly, to the pattern of community development.

Conservation of Groundwater Resources

Drilling contractors can also play a role in water conservation. Understanding how groundwater can be conserved through improvements in irrigation equipment, home plumbing systems, and reuse technology can enable the contractor to sway public opinion in the community. For example, even a small reduction in the groundwater required for irrigation would amount to huge water savings, because 70 percent of the groundwater withdrawn in the United States is used for irrigation (Figure 24.9).

Declining groundwater supplies, especially in the High Plains region of south-central United States, escalating pumping costs, and limited supplies of surface water are forcing greater efficiency in the use of groundwater. Limited tillage, where crop residues are left on the surface after the growing season, increased weed control by chemicals, and elimination of typical plowing techniques reduce soil-moisture losses significantly between growing seasons. Applying irrigation water by means of drop tubes and low-pressure emitters attached to sprinkler frames reduce water losses, because the water is placed directly in the furrow. Furrow diking also helps to distribute the water more uniformly. Specialized irrigation practices coupled with no-till cropping procedures can save large volumes of groundwater without reducing crop yields.

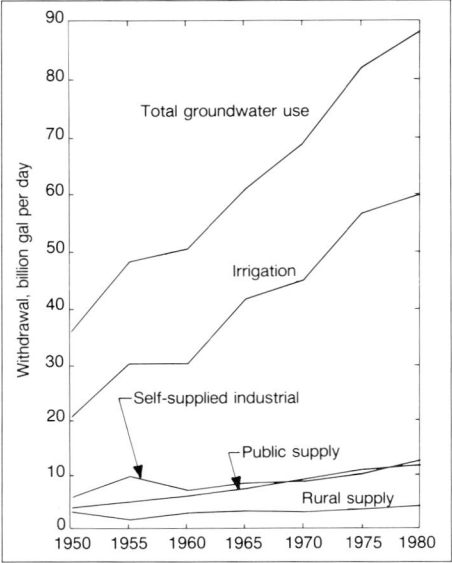

Figure 24.9. Groundwater use in the United States, 1950 to 1980. *(U.S. Geological Survey)*

Ultimately, local groundwater resources may become so diminished or so expensive to pump that only dry-land farming can be practiced. Switching to crops requiring less water, minimum tillage, and furrow diking, however, will still guarantee the farmer an adequate return on investment. Near urban areas, the available groundwater can then be utilized for higher uses, such as municipal water supplies. Some estimates have shown, for example, that groundwater supplies in Arizona are adequate to meet all nonagricultural uses indefinitely.

Practices of typical homeowners also contribute to excessive use of water resources. In Denver, Colorado, the water department estimates that slightly over half of the annual water consumption of a single-family house is used for landscape irrigation (Wall Street Journal, 1985). Many people who move to the desert Southwest attempt to recreate the outdoor environment from which they came. This may include lush lawns, many trees, and large flower gardens. At the outset, the newcomer to Phoenix or Tucson, Arizona, does not see how to use native vegetation (xerophytes*), rocks, and other indigenous materials to create an attractive environment requiring little water. Realistic prices for water and perhaps some restrictions on water use, coupled with education in xeriscape techniques (gardening in dry climates), would help eliminate the excessive use of water for landscape purposes. In another example, the desert City of Palm Springs, California, has a per-capita water use many times the national average. Large volumes are supplied to maintain the city's many golf courses, swimming pools, and lawns because of the almost complete absence of natural precipitation. Where the decision is made to use large volumes of water for special facilities, the cost of providing this water should be fairly assessed against those who benefit the most.

---

*Plants that can survive for extended periods when no soil moisture is available.

Pricing Water Consistent with Its Cost and Highest Use

The most powerful incentive for conservation, price, should reflect the actual cost of the water, treatment plant, and distribution system. In far too many communities, water is almost a free good. Its price bears little relation to its actual cost of delivery. Furthermore, small-volume users pay much more by volume than do large users. In some areas of western United States, surface water is available at pennies per ton, an unreasonably low figure. Even bank interest on construction costs has not been charged until recently, and even then the interest rates charged do not cover the actual costs.

The price of water must reflect its true value for highest use if conservation and the wise use of groundwater are ever to be achieved. Certain users, for example the energy industry in the West, are able to pay many times the rate paid by irrigators because any increased costs to the industry would represent only a tiny fraction of their total costs for power generation. In water-short areas, economic analyses should determine how the available groundwater supply is used, and the price should reflect its true cost so that conservation will be practiced by consumers.

Metering of water service inevitably drives down consumption because users become more conscious of costs. Table 24.5 shows that dramatic reductions in use occur when metering is initiated. In Boulder, Colorado, a 36-percent reduction was attained when metered rates were used (Hanke, 1970). Water used for sprinkling seems particularly sensitive to metering. It is also clear from a study done in Tel Aviv, Israel, that when metering is done on whole buildings, not on individual apartments, the reduction is not as great (Kamen and Dar, 1973). Apparently the more responsible a single family becomes for the water it uses, the less water it consumes.

Table 24.5. Annual Average Per-Capita Domestic (In-House) Water Use for Metered Rate and Nonmetered (Flat) Rate Areas

Location	Metered Rate		Nonmetered Rate	
	gal/day	ℓ/day	gal/day	ℓ/day
Boulder, Colorado	197	748	307	1,160
18 Municipal Agencies, USA				
Leakage and waste	25	94	36	136
Household*	247	937	236	892
Sprinkling	186	703	420	1,588
Total	458	1,731	692	2,616
Maximum day	979	3,700	2,354	8,898
Peak hour	2,481	9,378	5,170	19,580
Toronto, Canada				
Winter	154	581	204	771
Summer	190	719	221	836
Average	172	650	216	815
Tel Aviv, Israel	52	197	63	237

*Household use of water did not decrease when metered because household uses are nondiscretionary (basic) requirements.

(After United Nations, 1980)

Metering by itself is not enough in the long run. To increase the effectiveness of water-conservation measures, metering should be combined with a reasonable base price consistent with the costs of delivering the water. In addition, as more water is used, the price should increase for successively larger volumes.

Governmental Policies and Regulations Affecting Groundwater

In the United States, responsibility for water management is shifting from the federal government to the states. Even though recent federal court decisions have found groundwater to be an article of interstate commerce, management of groundwater can be more logically handled on a regional, state, or local basis. Varying hydrogeological conditions and vastly different legal traditions make it more effective to handle water problems from a local or regional perspective.

Governmental decisions have a profound effect in establishing groundwater management priorities. These include (National Water Alliance, 1985):
1. Correlation between water quantity and water quality.
2. Interdependence of surface water and groundwater.
3. Relationship between water availability and water cost.
4. Coordination among governmental agencies that provide clearly defined policies for public and private activities.
5. Balancing between present and future needs.

In spite of the fact that the federal role in water management is decreasing, it remains the best governmental entity to undertake basic water research. It is particularly appropriate for the federal government to determine the health effects of the numerous contaminants that can appear in groundwater. Using risk assessment and risk management procedures, the federal government can continue to set enforceable maximum contaminant levels for drinking water. Scientifically valid water-quality standards must be established by the federal government to insure consistent health standards across the nation.

One of the most perplexing water-quality problems facing the federal government is how to assess the toxic effects of all the man-made chemicals now determined to be in groundwaters, especially those underlying highly urbanized and industrial areas. Large sums of money are being spent to remove certain chemicals from groundwater used for drinking water, even though they appear in concentrations of only a few parts per billion or even parts per trillion; but the actual health risk of drinking this water if untreated has not been determined. Assessing the risk of drinking a specific water with its array of chemicals is an extremely difficult matter, both scientifically and from the standpoint of what is acceptable in light of economic constraints. Few things in life are absolutely risk-free, and it should not be expected that there be complete safety in drinking all groundwater, just as there is not complete safety in drinking water from all surface sources. Federal drinking-water safety standards must be based on both risk assessment data and risk management principles, that is, what is practical for the community, socially and economically, in avoiding health risks.

Federal and state groundwater policies in the past often ignored or did not balance appropriately many of the conflicting interests affecting groundwater. For example, the interdependence of surface water and groundwater was not recognized. Recently, increased awareness at all levels of government concerning groundwater issues has stimulated a far less fragmented approach to resolving problems affecting groundwater supply and quality.

One of the impediments to changing water policy has been the large number of federal, state, and local agencies with an interest in some aspect of water. In the United States, for example, more than 30 major federal agencies are concerned with water-resources research, education, planning, program information, and overview (Hall, 1979). Inevitably these agencies at times work at cross-purposes, and the existence of so many large agencies produces an inertia in the system that makes it difficult to change the way water is viewed. Recently, however, federal groundwater legislation has mandated that individual states combine those agenices that may have been concerned with groundwater. At the federal level, the Office of Groundwater Protection at the U.S. Environmental Protection Agency has been created to coordinate the federal government's obligations for various aspects of groundwater. Thus, in the future it should be easier to effect changes in groundwater regulations and policies. But changes can only occur when a concerned constituency takes the time to present its views to the appropriate governmental agencies.

Currently, many federal, state, and local policies affect the availability of groundwater. Little uniformity exists in regulations from region to region because of specific hydrogeologic conditions. But, as water demands rise, future legislation must be guided by policies that protect the physical and chemical environment of the groundwater resources. Drilling contractors, individually and through their associations, can exert considerable influence on how local and regional groundwater issues are resolved legislatively. Chief among these issues are (1) establishment of state laws to fund groundwater restoration projects, (2) interbasin transfers of water to move raw materials and to provide additional water resources, (3) new technologies to reduce the adverse impact of municipal and industrial wastes on groundwater quality, and (4) laws that prohibit certain agricultural chemicals in areas where groundwater quality may be affected.

The federal role in groundwater protection is well established by numerous statutory authorities. These include the Resource Conservation and Recovery Act (RCRA); Comprehensive Environmental Response, Compensation, and Liability Act (CERCLA); Safe Drinking Water Act (SDWA); Clean Water Act (CWA); Toxic Substances Control Act (TSCA); Uranium Mill Tailings Radiation Control Act (UMTRCA); Surface Mining Control and Reclamation Act (SMCRA); Atomic Energy Act (AEA); and the Federal Insecticide, Fungicide, and Rodenticide Act (FIFRA) (see Chapter 21). In general, these acts attempt to control sources of groundwater pollution. If the groundwater environment does become contaminated, federal funds can be provided through the federal superfund law to clean up the pollutant source.

The federal government has recognized that control of groundwater supply and quality can best be handled on the state level, with federal agencies acting in an advisory role because of the local and regional complexity of the groundwater environment. Many states are seizing this opportunity to enact legislation that reflects their views on handling groundwater supply and use problems. Arizona and Wisconsin are among the first states to enact comprehensive groundwater management regulations. Some states are also developing superfund laws of their own to help in groundwater quality restoration. Restoration of aquifers contaminated by industrial wastes, leaking landfill sites, and chemical spills should be a top priority of drilling contractors and their state associations. In general, funding for these restoration projects has come from taxes assessed against the chemical industry and the generators of potentially harmful waste.

Many issues now being discussed at a local level relate to waste disposal and are of vital concern to water well contractors because most waste disposal in the ground eventually contaminates groundwater. Recent research has shown, for example, that leachates from waste-disposal sites escape over time from even the best engineered landfills. Wastes in these landfills often contain synthetic organic chemicals from solvents, paints, dry-cleaning fluids, and other common products. These chemicals can destroy within a matter of months or years the holding capacity of clay liners, which control the spread of leachates. Even synthetic liners are not immune from attack. Therefore, these and other chemicals escape into the unsaturated zone above the water table. Because most volatile organic chemicals, and some inorganic chemicals, are not readily adsorbed by soil particles, they continue to percolate down to the groundwater system and then flow downgradient to form a contaminant plume. Many organic chemicals are highly toxic, and if the groundwater is to be consumed by humans, special water-treatment techniques — mainly aeration and activated-carbon filtration — must be used.

In general, wastes should not be placed in the ground where they can lead to groundwater contamination. Other methods of disposal, especially mass-burning and resource-recovery systems, may be far more appropriate. If wastes must be placed in the ground, the contractor should advise the appropriate authorities if the hydrogeologic conditions are not optimal (see Chapter 21).

Highly toxic chemicals can also enter groundwater aquifers by misuse of chemicals. Synthetic organic chemicals are a major constituent of many pesticides. Farmers and others use these chemicals to control insects and rodents. In some hydrogeologic settings, however, overuse of these chemicals can contaminate groundwater. In the San Joaquin Valley of California and the Sand Plains area of central Wisconsin, heavy irrigation and overuse of certain effective chemicals has driven them down through the sandy soils to the relatively shallow groundwater. Reduced rates of irrigation and more moderate use of chemicals could have prevented any contamination. Unfortunately, overuse of chemicals and federal and state policies that (1) provide irrigation water at unreasonably low prices and (2) stipulate that, unless a certain water volume is used, the right to use this volume in the future will be denied, contribute to both resource overuse and groundwater contamination.

Withdrawal of the federal government from direct involvement in water management has reduced the financial resources that are available to states or counties for water and waste management projects. Yet many states and local units of government do not have the capacity to raise the necessary funds for large-scale projects. In the past, the typical method for raising capital for local governmental public works projects has been industrial development bonds, but strict limits are now being placed on how much debt a community may assume using this type of funding. At the same time, the need for these monetary resources is expanding. In older cities, for instance, many parts of the infrastructure* are deteriorating and must be rebuilt. In rapidly growing communities, so many new facilities must be built within such a short time

---

*Infrastructure is the physical facilities necessary to deliver a public service. Water infrastructure includes storage reservoirs, water treatment plants, distribution systems, wastewater collection systems, wastewater treatment plants, and necessary support facilities (Grigg, 1985).

that residents simply cannot afford the additional taxes required for large water projects. Therefore, other sources of capital must be found to extend the financial capability of a community. The answer to the funding dilema may be in the privatization of formerly public facilities.

If private capital is to be attracted to traditionally public projects, however, economic incentives for capital investment must be made available (Kovalic, 1985). Incentives such as investment tax credit and accelerated cost recovery are vital in promoting private investment in public projects. There must be an ongoing effort at the federal level "to remove existing barriers to joint governmental-private ventures, including an examination of tax laws, procurement laws and regulations, and the creation of incentives for state-regional financing mechanisms" (Kovalic, 1985). "Private financing, ownership, and operation of water projects can be a cost effective and efficient approach to meeting present and future water resource goals" (Reilly, 1985). Three factors allow privatization to occur:

1. Localities are willing to privatize because of budget and operational reasons. For example, cost savings of at least 25 percent can typically occur with private ownership and operation.

2. The private sector has the capability and willingness to undertake large water projects because it views privatization as an attractive long-term business opportunity.

3. The investment banking community can arrange low-cost, tax-exempt financing to produce the lowest possible service costs to users.

The principal advantage to the community is that it does not have to create a capital budget or issue local bonds to pay for a large facility; instead, the contractor must arrange the necessary financing (Longest, 1985). In exchange for the business opportunities that are made possible by the appropriate tax incentives, private owner-operators will guarantee that "the plants will perform as intended at controlled, predictable costs" (Reilly, 1985).

CONCLUSIONS

Individual water well contractors and state and national groundwater associations must work together to make sure that groundwater is used wisely, that groundwater quality is protected, and that new technologies are adopted to enhance groundwater supplies. Changing the way water is used is a difficult matter politically, legally, and economically. But as supplies of water shrink or become contaminated in many regions, the appropriate laws must be changed so that water can be used most effectively. Specific actions should assure that:

1. Groundwater used for irrigation is used wisely and efficiently.

2. Urban water-supply systems are well maintained, water loss is reduced, and all users have individual meters.

3. Prices for water are consistent with the cost of supplying the water.

4. Administrative or legislative methods are developed that permit higher beneficial uses of water as economic or social conditions change.

5. Conservation of groundwater resources becomes an ongoing commitment, not one that is practiced only during drought.

6. Reuse and recycling of municipal and industrial wastewaters are expanded to reduce the demand for groundwater resources.

7. Land-use priorities are established so that groundwater quality is protected.

8. Groundwater reserves are developed to offset the effect of droughts.

Drilling contractors and groundwater resource managers in both the private and the public sectors must take the initiative in making sure that today's political decisions will protect groundwater resources for the future. It is not enough to hope that governmental agencies make the correct decisions concerning the allocation of groundwater supplies; groundwater specialists must make their opinions heard during public discussions leading to the formulation of new laws and regulations. Politically active, well-informed water well contractors can make a significant difference in how groundwater is used in their communities and in preserving its quality and quantity.

Groundwater is one of Earth's vital renewable resources, and its value to the economic and social success of society is incalculable. Fortunately, the volume of water stored in the ground is enormous and is the most easily developed source of high-quality water for much of the world's population. Current mitigation efforts are having an immediate and positive effect on restoring groundwater quality in those relatively small areas where groundwater contamination has occurred. The controlling social, economic, political, and legal institutions are taking action to allocate, use, and price groundwater resources more wisely so that overdrafts can be eliminated. Implementation of the principles governing the wise use of groundwater will insure that society will be able to enjoy an undiminished supply of high-quality groundwater now and in the future.

References

Ahrens, T. P., 1957. Well design criteria. *Water Well Journal*, Vol. 11, No. 9 and 11.

Aller, L., Bennett, T., Lehr, J.H., and Petty, R.J., 1985. DRASTIC: A Standardized System for Evaluating Ground Water Pollution Potential Using Hydrogeologic Settings. U.S. Environmental Protection Agency Publication 600/2-85/018. U.S. Government Printing Office, Washington, D.C.

Ambroggi, R. P., 1978. Underground reservoirs to control the water cycle. *Ground Water*, Vol. 16, No. 3, pp 158-166.

American National Standards Institute, 1935. *Pipe Standards*. ANSI, New York, NY.

American Petroleum Institute, 1959. *Oil-Well Cementing Practices in the United States*. American Petroleum Institute, Dallas, TX.

American Petroleum Institute, 1976. *API Specification for Casing, Tubing, and Drill Pipe*. API Spec 5A, American Petroleum Institute, Dallas, TX.

American Petroleum Institute, 1978. *API Specification for Line Pipe*. API Spec 5L, American Petroleum Institute, Dallas, TX.

American Public Health Association, American Water Works Association, and Water Pollution Control Federation, 1985. *Standard Methods for the Examination of Water and Waste Water, 16th Edition*. American Public Health Association, Washington, D.C.

American Society for Testing Materials, 1960. *Manual on Industrial Water, 2nd Edition*. Philadelphia, PA.

American Water Works Association, 1973a. *Water Chlorination Principles and Practices*. AWWA Manual M20, Denver, CO, 84 p.

American Water Works Association, 1973b. *Ground Water*. AWWA Manual M21, Denver, CO.

American Water Works Association, 1982. U.S. Supreme Court rules groundwater is an article of interstate commerce. *Journal American Water Works Association*, Vol. 74, No. 8, Denver, CO.

American Water Works Association, 1983. Handling the threat of contaminated water supplies. *Opflow*, Vol. 9, No. 3, Denver, CO.

American Water Works Association, 1984. *AWWA Standards for Water Wells*. AWWA A100-84, Denver, CO, 75 p.

American Welding Society, 1981. *Structural Welding Code AWS D1.1.* American Welding Society, Miami, FL.

Amsterdam, B., 1983. Closed-loop earth-coupled heat pump systems. *Water Well Journal*, Vol. 37, No. 7, pp 52-53.

Anderson W. L., 1979. Michigan job uses geophysical aids. *Johnson Drillers' Journal*, Nov/Dec, Johnson Division, UOP Inc., St. Paul, MN.

Atlas Construction Co., Inc. v. Aqua Drilling Co., 559 P. 2d 39 (Wyo., 1977).

Atwood Vacuum Machine Co. v. Warner Well and Pump Co., 3 111. App. 2d 571, 122 N.E., 2d 834 (1954).

Ayers, R. S. and Wescot, D. W., 1976. Water quality for agriculture. Food and Agriculture Organization of the United Nations, Irrigation and Drainage Paper 29, Rome, Italy.

Bachmat, Y., Bredehoeft, J., Andrews, B., Holtz, D., and Sebastian, S., 1980. *Groundwater Management: The Use of Numerical Models.* Water Resources Management No. 5, American Geophysical Union, Washington, D.C., 127 p.

Barcelona, M. J., Gibb, J. P., and Miller, R. A., 1983. *A Guide to the Selection of Materials for Monitoring Well Construction and Ground-Water Sampling.* Illinois State Water Survey, ISWS Contract Report 327, Urbana, IL, 78 p.

Bark, L. D., 1978. History of American droughts, *in North American Droughts*, edited by Norman J. Rosenberg, AAAS Selected Symposium 15. Westview Press, Boulder, CO, 177 p.

Baroid, 1980. *Baroid Drilling Fluid Products for Minerals Exploration.* NL Baroid/NL Industries, Houston, TX.

Baver, L. D., 1956. *Soil Physics, 3rd Edition.* John Wiley and Sons, Inc., New York, NY, 489 p.

Bear, J., 1972. *Dynamics of Fluids in Porous Media.* American Elsevier, New York, NY.

Bear, J., 1979. *Hydraulics of Ground Water.* McGraw-Hill Book Company, New York, NY, 569 p.

Beck, A. E., 1981. *Physical Principles of Exploration Methods.* John Wiley and Sons, Inc., New York, NY, 234 p.

Belknap, W. B., Dewan, J. T., Kirkpatrick, C. V., Mott, W. E., Pearson, A. J., and Robson, W. R., 1959. *API Calibration Facility for Nuclear Logs, Drilling, and Production Practice.* American Petroleum Institute, Dallas, TX, pp 289-316.

Bennett, G. D., 1976. *Introduction to Ground-Water Hydraulics. A programmed text for self instruction. Chapter B2, Book 3.* Techniques of Water Resources Investigations of the U.S. Geological Survey, U.S. Department of the Interior, U.S. Government Printing Office, Washington, D.C., 172 p.

Bennett, P. T., 1945. Comments on the design of relief wells. Conference on Control of Underseepage, held at Waterways Experiment Station, Vicksburg, MS.

Bennett, P. T., 1946. The effect of blankets on seepage through pervious foundations. *Transactions, American Society of Civil Engineers*, Vol. 111.

Bennett, T. W., 1970. On the design and construction of infiltration galleries. *Ground Water*, Vol. 8, No. 3.

Bierschenk, W. H., 1964. *Determining Well Efficiency by Multiple Step-Drawdown Tests.* Publication 64, International Association of Scientific Hydrology.

Blackburn, A. M., 1978. Management strategies dealing with drought. *Journal Amer-*

ican Water Works Association, Vol. 70, No. 2, pp 51-59.

Blair, H., 1970. Well screens and gravel packs. *Ground Water*, Vol. 8, No. 1.

Blake, S. B. and Lewis, R. W., 1982. Underground oil recovery, *in Proceedings of the Second National Symposium on Aquifer Restoration and Ground Water Monitoring*, D. M. Nielsen, Editor. National Water Well Association, Worthington, OH.

Boulton, N. S., 1963. Analysis of data from nonequilibrium pumping tests allowing for delayed yield from storage. *Proceedings, Institute of Civil Engineers*, 26 (6693), pp 469-482.

Boulton, N. S. and Streltsova, T. D., 1976. The drawdown near an abstraction well of large diameter under non-steady conditions in an unconfined aquifer. *Journal of Hydrology*, Vol. 30, pp 29-46.

Boulton, N. S. and Streltsova, T. D., 1977a. Unsteady flow to a pumped well in a two-layered water-bearing formation. *Journal of Hydrology*, Vol. 35, pp 245-256.

Boulton, N. S. and Streltsova, T. D., 1977b. Unsteady flow to a pumped well in a fissured water-bearing formation. *Journal of Hydrology*, Vol. 35, pp 257-269.

Boulton, N. S. and Streltsova, T. D., 1978. Unsteady flow to a pumped well in a fissured aquifer with a free surface level maintained constant. *Water Resources Research*, Vol. 14, No. 3, pp 527-532.

Bouwer, H., 1978. *Groundwater Hydrology*. McGraw-Hill Book Company, New York, NY, 479 p.

Bouwer, H., Rice, R. C., Lance, J. C., and Gilbert, R. G., 1980. Rapid-infiltration research at Flushing Meadows Project, Arizona. *Journal of Water Pollution Control Federation*, Vol. 52, October, pp 2457-2470.

Bowen, R., 1981. *Grouting in Engineering Practice, 2nd Edition*. Applied Science Publishers, Halstad Press, New York, NY, 285 p.

Brantley, J. E., 1961. *Rotary Drilling Handbook*. Palmer Publications, Fort Lee, NJ.

Briley, K. F., Williams, R. F., and Sorber, C. K., 1984. Alternative water disinfection schemes for reduced trihalomethane formation Vol. II. Algae as precursors for trihalomethanes in chlorinated drinking water. EPA-600/S2-84-005, Municipal Environmental Research Laboratory, U.S. Environmental Protection Agency, Cincinnati, OH, 5 p.

Brown, N., 1983. The medical applications of reverse osmosis. *Water Technology*, Vol. 6, No. 6, pp 22-32.

Brown, N., 1984. Selling your customers on water treatment. *Ground Water Age*, Vol. 18, No. 9, pp 20-33.

Brown, R. H., 1953. Selected procedures for analyzing aquifer test data. *Journal of American Water Works Association*, Vol. 45, No. 8, pp 844-866.

Brutsaert, W. and Corapcioglu, M. Y., 1976. Pumping of aquifer with visco-elastic properties. *Proceedings, American Society of Civil Engineers*, Journal of Hydraulics Division, 102 (HY11), pp 1663-1675.

Bureau of Reclamation, 1981. *Water Measurement Manual*. U.S. Department of the Interior, U.S. Government Printing Office, Denver, CO, 327 p.

Butler v. *Davis*, 119 Wis 166, 96 N.W. 561 (1903).

California Water Atlas, W. L. Kahrl, Editor, 1979. State of California, William Kaufmann, Los Altos, CA, 118 p.

Campbell, M. D and Lehr, J. H., 1973. *Water Well Technology*. McGraw-Hill Book Company, New York, NY, 681 p.

Canadian Institute of Mining and Metallurgy, 1965. Understanding surface casing waiting-on-cement time, by L. F. Maier. Presented at Canadian Institute of Mining and Metallurgy 16th Annual Technical Meeting, Calgary, Alberta, Canada.

Canter, L. W., 1981. Potential groundwater contamination sources. Presented at Groundwater Quality Protection Institute, University of Wisconsin-Extension, Madison, WI.

Carmichael, R. S., 1976. *Gravity Geophysics for Groundwater Exploration in Glaciated Areas.* Institute of Water Research, Michigan State University, East Lansing, MI, 36 p.

Carr, J. R., 1976. Tacoma's well field might be world's most productive. *Johnson Drillers' Journal,* Sept/Oct, Johnson Division, UOP Inc., St. Paul, MN.

Castle, J. E., 1977. The use of x-ray photoelectron spectroscopy in corrosion science. *Surface Science,* Vol. 68, North/Holland Publishing Company, pp 583-602.

Castle, J. E. and Clayton, C. R., 1977. The use of x-ray photo-electron spectroscopy in the analysis of passive layers. *Corrosion Science,* Vol. 17, Pergamon Press, pp 7-26.

Cedergren, H. R., 1977. *Seepage, Drainage, and Flow Nets, Second Edition.* John Wiley and Sons, Inc., New York, NY, 510 p.

CEMBUREAU, 1967. *Cement Standards of the World — Portland Cement and Its Derivatives.* CEMBUREAU, Paris, France.

Chagnon, R. P., 1981. The 3-D velocity logging system is a valuable tool. *Johnson Drillers' Journal,* Third Quarter, Johnson Division, UOP Inc., St. Paul, MN.

Chebotarev, I. I., 1955. Metamorphism of natural waters on the crust of weathering. *Geochimica et Cosmochimica Acta* 8.

City of El Paso v. Reynolds, Civil No. 80-730-HB, U.S. District Court, New Mexico (1983).

Civil Engineering-ASCE, 1981. Rocky Mountain Arsenal: landmark case of groundwater polluted by organic chemicals and being cleaned up. Vol. 51, No. 9, pp 68-71.

Clark, F. E., 1980. Corrosion and encrustation in water wells; a field guide for assessment, prediction, and control. Food and Agriculture Organization of the United Nations, FAO Irrigation and Drainage Paper, 95 p.

Clark, R. E., 1977. Water law and the public interest: a commentary, *in Water Needs for the Future,* V. P. Nanda, Editor. Westview Press, Boulder, CO, 329 p.

Clark, T. D., 1979. Ground-water pollution — a limited problem. Presented at 4th National Ground Water Quality Symposium, Minneapolis, MN, September, 1978. *Ground Water,* Vol. 17, No. 1.

Clark, T. P. and Sabel, G. V., 1980. Requirements of state regulatory agencies for monitoring ground-water quality at waste disposal sites. *Ground Water,* Vol. 18, No. 2, pp 168-174.

Clean Environment Commission, 1976. Preliminary report on contamination of underground water sources by refined petroleum products. *Ground Water,* Vol. 14, No. 1.

Clemmey, H. and Badham, M., 1982. Oxygen in the Precambrian atmosphere: an evolution of the geological evidence. *Geology,* Vol. 10, No. 3, pp 141-146.

Cloud, P. E., 1976. Beginnings of biospheric evolution and their biogeochemical consequences. *Paleobiology,* Vol. 2, pp 351-387.

Cloud, P. E., 1973. Palaeocological significance of the Banded Iron Formations. *Economic Geology*, Vol. 68, pp 1135-1143.

Cloud, P. E., 1968. Atmospheric and hydrospheric evolution of the primitive earth. *Science*, Vol. 160, pp 729-736.

Collins, W. D., 1923. Graphic representation of water analyses. *Industrial and Engineering Chemistry*, Vol. 15, pp 394.

Colt Industries, 1974. *Hydraulic Handbook*. Fairbanks Morse Pump Division, Kansas City, KS, 246 p.

Cooper, H. H., Jr. and Jacob, C. E., 1946. A generalized graphical method for evaluating formation constants and summarizing well field history. *Transactions, American Geophysical Union*, Vol. 27, No. 4.

Cooper, L. W., Hcok, R. A., and Payne, B. R., 1977. *Air Drilling Techniques*. American Institute of Mining, Metallurgical, and Petroleum Engineers, Inc., Paper No. SPE 6435, Dallas, TX, 16 p.

Corapcioglu, M. Y., 1976. Mathematical modeling of leaky aquifers with rheological properties. Proceedings of Anaheim Symposium on Subsidence, International Association of Hydrology, Scientific Publications No. 121, pp 191-200.

Corps of Engineers, 1955. Relief well design. Civil Works Engineer Bulletin 55-11, June, Department of the Army.

Corps of Engineers, 1972. Relief wells, guide specifications, civil works construction. CE 1307, February, Department of the Army, 30 p.

Costello, R. L., 1980. Identification and description of geophysical techniques. Report DRXTH-TE-CR-80084. U.S. Army Toxic and Hazardous Materials Agency, Aberdeen Proving Grounds, MD.

Cotruvo, J. A. and Vogt, C., 1984. Development of revised primary drinking water regulations. *Journal of American Water Works Association*, Vol. 76, No. 11, pp 34-38.

Cragwall, J. S., Jr., 1979. Remote sensing in hydrology — a challenge to scientists, in *Satellite Hydrology*, M. Deutsch, D. R. Wiesnet, and A. Rango, Editors. Fifth Annual William T. Pecora Memorial Symposium on Remote Sensing, 1981, American Water Resources Association, Minneapolis, MN, 730 p.

Crane Company, 1976. *Flow of Fluids through Valves, Fittings, and Pipe*. Technical Paper No. 410, 16th Edition, Chicago, IL.

Craun, G. F., 1981. Outbreaks of waterborne disease in the United States: 1971-1978. *Journal of American Water Works Association*, Vol. 73, No. 7., pp 360-369.

Craun, G. F., 1984. Waterborne outbreaks of *Giardiosis*: current status in *Giardia and Giardiosis*, S. I. Erlandsen and E. A. Meyer, Editors. Plenum Press, New York, NY.

Crawford, A. B., 1978. State and federal responses to the 1977 drought, in *North American Droughts*, edited by Norman J. Rosenberg, AAAS Selected Symposium 15. Westview Press, Boulder, CO, 177 p.

Crites, R. W., Meyer, E. L., and Smith, R. G., 1977. *Process Design Manual for Land Treatment of Municipal Wastewater*. U.S. Army Corps of Engineers and U.S. Department of Agriculture, EPA 621/1-77-008, U.S. Environmental Protection Agency, Cincinnati, OH.

Culley, R. W., Jagodits, F. L., and Middleton, R. S., 1975. E-phase system for detection of buried granular deposits. Symposium on Modern Innovations in Subsurface

Exploration. 54th Annual Meeting of Transportation Research Board, National Academy of Science.

Cullimore, R., 1981. Controlling iron bacterial plugging by recycling hot water. *Canadian Water Well*. Agri-book Publishing Co., Ltd..

Cunningham, R. A. and Eenink, J. G., 1958. Laboratory study of effect of overburden, formation and mud column pressure on drilling rate. Paper 1094-G, Fall Meeting, AIME, Houston, TX.

Daniels, J. J., 1984. Borehole geophysics for mining and geotechnical application. Short course notes, Short Course on Geotechnical Applications of Borehole Geophysics, Nov. 12-16, Golden, CO.

Darcy, H., 1856. *Les fontaines publiques de la ville de Dijon*. V. Dalmont, Paris, 647 p.

Davis, S. N., 1969. *Flow Through Porous Media*, R. J. M. DeWiest, Editor. Academic Press, New York, NY.

Davis, S. N. and DeWiest, R. J. M., 1966. *Hydrogeology*. John Wiley and Sons, Inc., New York, NY, 463 p.

Davis, S. N., Thompson, G. M., Bentley, H. W., and Stiles, G., 1980. Ground-water tracers — a short review. *Ground Water*, Vol. 18, No. 1.

Davis, S. N. and Turk, L. J., 1964. Optimum depth of wells in crystalline rock. *Ground Water*, Vol. 2, No. 2.

Dean, R. B. and Lund, E., 1981. *Water Reuse: Problems and Solutions*. Academic Press, New York, NY, 264 p.

Deju, R. A., 1971. *Regional Hydrology Fundamentals*. Garden and Breach Science Publishers, New York, NY, 204 p.

Desai, A., 1974. Reverse osmosis technology. W-01-R, Hydranautics, Santa Barbara, CA.

Domenico, P. A., 1972. *Concepts and Models in Groundwater Hydrology*. McGraw-Hill Book Company, New York, NY, 405 p.

Dowdy, D. R. and O'Donnel, T., 1965. Mathematical models of catchment behavior. *ASCE Journal of Hydraulics Division*, Vol. 91, Hy 4.

Dressman, R. C. and McFarren, E. F., 1978. Determination of vinyl chloride migration from polyvinyl chloride pipe into water using improved gas chromatographic methodology. *Journal of American Water Works Association*, Vol. 70, No. 1.

Drilco, Division of Smith International, Inc., 1979. How to select bottom hole drilling assemblies. Publication No. 62, Houston, TX.

Driscoll, F. G., 1978. Blasting — It turns dry holes into wet ones. *Johnson Drillers' Journal*, Nov/Dec, Johnson Division, UOP Inc., St. Paul, MN.

Driscoll, F. G., Hanson, D. T., and Page, L. J., 1980a. Well-efficiency project yields energy-saving data, Part 3. *Johnson Drillers' Journal*, Sept/Oct, Johnson Division, UOP Inc., St. Paul, MN.

Driscoll, F. G., Hanson, D. T., and Page, L. J., 1980b. Well-efficiency project yields energy-saving data, Part 1. *Johnson Drillers' Journal*, March/April, Johnson Division, UOP Inc., St. Paul, MN.

Dunn, T. and Leopold, L. B., 1978. *Water in Environmental Planning*. W. H. Freeman and Company, San Francisco, CA, 818 p.

Earlougher, R. C., Jr., 1977. *Advances in Well Test Analysis*. Society of Petroleum Engineers of AIME, New York, NY, 264 pp.

Eddy, P. A. and Wilbur, J. S., 1980. Radiological status of the ground water beneath the Hanford Project, January - December, 1979. Prepared for the U.S. Department of Energy, Pacific Northwest Laboratory, Richland, WA.

EOS, 1981. The Love Canal: beyond science? *Transactions, American Geophysical Union*, Vol. 62, No. 2, January, p 12.

Espina, J. M., 1980. The case for artificial recharge. *Johnson Drillers' Journal*, Jan/Feb, Johnson Division, UOP Inc., St. Paul, MN.

Fair, G. M., Geyer, C. J., and Okun, D. A., 1971. *Elements of Water Supply and Waste Water Disposal, 2nd Edition*. John Wiley and Sons, Inc., New York, NY, 752 p.

Farquhar, G. J., 1981. Mechanism of leachate movement and attenuation. Paper presented at the Gas and Leachate Management Conference, November, University of Wisconsin-Extension, Madison, WI.

Ferreira, M. G. S. and Dawson, J. L., 1985. Electrochemical studies of the passive film on 316 stainless steel in chloride media. *Journal Electrochemical Society*, Vol. 132, No. 4, Manchester, NH, pp 760-765.

Fetter, C. W., Jr., 1980. *Applied Hydrogeology*. Charles E. Merrill Publishing Company, Columbus, OH, 488 p.

Fetter, C. W., Jr., 1981. Determination of the direction of ground-water flow. *Ground Water Monitoring Review*, Vol. 1, No. 3, pp 28-31.

Filkins, C., 1983. Manganese-treated zeolites: uses and restrictions in conditioning problem water. *Water Technology*, Vol. 6, No. 3, pp 43-47.

Flint, R. F., 1971. *Glacial and Quaternary Geology*. John Wiley and Sons, Inc., New York, NY.

Fontana, M. G. and Greene, N. D., 1978. *Corrosion Engineering*. McGraw-Hill Book Company, New York, NY, 465 p.

Frederick v. *County of Redwood*, 152 Minn. 450, 190, N.W. 801.

Freeze, R. A. and Cherry, J. A., 1979. *Groundwater*. Prentice-Hall, Inc., Englewood Cliffs, NJ, 604 p.

Garrels, R. M. and Mackenzie, F. T., 1971. *Evolution of Sedimentary Rocks*. W. W. Norton and Company, New York, NY.

Garrels, R. M., Mackenzie, F. T., and Hunt, C. A., 1975. *Chemical Cycles and the Global Environment: Assessing Human Influences*. William Kaufmann, Inc., Los Altos, CA, 206 p.

Gass, T. E., 1980. Sizing water well systems for ground-water heat pumps. *Ground Water Heat Pump Journal*, Summer, pp 16-21.

Gass, T. E., 1983. The earth-coupled heat pump. *Water Well Journal*, Vol. 37, No. 7, pp 58-59.

Gass, T. E., Bennett, T. W., Miller, J., Miller, R., and The National Water Well Association, undated. *Manual of Water Well Maintenance and Rehabilitation Technology*. Robert S. Kerr Environmental Research Laboratory, U.S. Environmental Laboratory, Ada, OK, 247 p.

Gearhart Industries, Inc., 1982. *Basic Cement Bond Log Evaluation*. Charleston, WV, 36 p.

Gibb, J. P., 1983. Sampling procedures for monitoring wells. Paper presented at Groundwater Monitoring Technology Conference, Philadelphia, PA, sponsored by University of Wisconsin-Extension, Madison, WI, 32 p.

Gibb, J. P., Schuller, R. M., and Griffin, R. A., 1981. Procedures for the Collection of Representative Water Quality Data from Monitoring Wells. Cooperative Groundwater Report 7, Illinois State Water Survey and Illinois State Geological Survey, Champaign, IL.

Giddings, T., 1983. Bore-volume purging to improve monitoring well performance: an often mandated myth, *in Proceedings of the Third National Symposium on Aquifer Restoration and Ground-Water Monitoring.* D. M. Nielsen, Editor, National Water Well Association, Worthington, OH.

Gilkeson, R. H. and Cartwright, K., 1982. The application of surface geophysical methods in monitoring network design, *in Proceedings of the Second National Symposium on Aquifer Restoration and Ground Water Monitoring.* D. M. Nielsen, Editor, National Water Well Association, Worthington, OH.

Glover, E. W., Jr., 1982. Containment of contaminated ground water: an overview, *in Proceedings of the Second National Symposium on Aquifer Restoration and Ground Water Monitoring.* D. M. Nielsen, Editor, National Water Well Association, Worthington, OH.

Goldstein, A. M., Alter, E. N., and Seaman, J. K., 1973. *Guar Gum in Industrial Gums; Polysaccharides and Their Derivatives.* R. L. Whistler and J. N. BeMille, Editors, Academic Press, New York, NY, 807 p.

Gorham, E., 1961. *Geological Society of America Bulletin*, Vol. 72, pp 795-840.

Grant, D. M., 1979. *Open Channel Flow Measurement Handbook.* Instrumentation Specialities Company, Lincoln, NB, 221 p.

Graver Water Division, 1978. The principle and basics of reverse osmosis. Ecodyne, Technical Bulletin No. D-07-1, Union, NJ, 4 p.

Gray, G. R., 1972. Drilling with mud: simple tests save time and money. *Water Well Journal*, Vol. 26, No. 3, Worthington, OH, pp 33-35.

Greene, N. D. and Fontana, M. G., 1959. Effects of alloying on pitting resistance of stainless steel alloys. *Journal of Corrosion*, Vol. 15, No. 25.

Grichor, C., 1983. The mechanical treatment of solids in drilling fluids. *Water Well Journal*, Vol. 37, No. 11, pp 46-54.

Griffin, R. A., Cartwright, K., Shimp, N. F., Steele, J. D., Buch, R. R., White, W. A., Hughes, G. M., and Gilkeson, R. H., 1976. Alteration of pollutants in municipal landfill leachate by clay minerals. Part 1. Column leaching and field verification. Illinois State Geological Survey Bulletin 78, Urbana, IL.

Griffith, P., 1984. Multiphase flow in pipes. *Journal of Petroleum Technology*, March, pp 361-367.

Grigg, N. S., 1985. The water infrastructure: a survey of staggering statistics and specific cases. *Waterworld News*, March/April.

Ground Water Age, 1977. New ozone technology. Vol. 11, No. 9.

Guyod, H., 1964. An investigation of the factors affecting the SP in soft formations. Transactions of the Society of Professional Well Log Analysts 12th Annual Logging Symposium, 16 p.

Haas, J. E., 1978. Strategies in the event of drought, *in North American Droughts*, edited by Norman J. Rosenberg, AAAS Selected Symposium 15. Westview Press, Boulder, CO, 177 p.

Hall, M. W., 1979. National water policy and the waters of mid-America, *in Waters of Mid-America: Present Demands and Future Development.* Presented at the Public

Responsibilities Symposium, 30th Annual Meeting, American Institute of Biological Sciences, Oklahoma State University, 7 p.

Hallberg, R. O. and Martinell, R., 1976. Vyredox — in situ purification of ground water. *Ground Water*, Vol. 14, No. 2, pp 88-93.

Halliburton, 1968. *Oil Well Cement Manual.* Oklahoma City, OK.

Hamilton, H., 1977. *How to Handle Mechanic's Liens Claims.* Hamilton Press, Inc., Boulder, CO, 105 p.

Handbook of PVC Pipe, 1979. Uni-Bell Plastic Pipe Association, Dallas, TX, 306 p.

Handbook of Steel Pipe, 1979. Committee of Steel Pipe Producers, American Iron and Steel Institute, Washington, D.C., 48 p.

Hanke, S. H., 1970. Demand for water under dynamic conditions. *Water Resources Research*, Vol. 6, No. 5.

Hantush, M. S., 1959. Non-steady flow to flowing wells in leaky aquifers. *Journal of Geophysical Research*, Vol. 64, No. 8, pp 1043-1052.

Hantush, M. S., 1964. Hydraulics of wells, *in Advances in Hydroscience*, V. T. Chow, Editor. Academic Press, New York, NY, pp 281-442.

Hantush, M. S., 1967. Flow to wells in aquifers separated by a semipervious layer. *Journal of Geophysical Research*, Vol. 72, No. 6.

Hantush, M. S. and Jacob, C. E., 1955. Non-steady radial flow in an infinite leaky aquifer and non-steady Green's functions for an infinite strip of leaky aquifer. *Transactions, American Geophysical Union*, Vol. 36, No. 1, pp 95-112.

Hart, Fred C. Associates, 1979. Preliminary assessment of clean-up costs for national hazardous waste problems. New York, NY.

Hazen, A., 1893. Some physical properties of sands and gravels. Massachusetts State Board of Health, 24th Annual Report.

Heap, R. D., 1979. *Heat Pumps.* E. & F. N. Spon Ltd., London, England, 155 p.

Heath, R. C., 1982. Classification of Ground-Water Systems of the United States. *Ground Water*, Vol. 20, No. 4, pp 393-401.

Heath, R. C., 1984. Ground-Water Regions of the United States, *U.S. Geological Survey Water Supply Paper No. 2242.* U.S. Government Printing Office, Washington, D.C.

Heath, R. C. and Trainer, F. W., 1968. *Introduction to Ground-Water Hydrology.* John Wiley and Sons, Inc., New York, NY, 284 p.

Hem, J. D., 1970. Study and interpretation of the chemical characteristics of natural water, Second Edition. *U.S. Geological Survey Water Supply Paper 1473*, U.S. Department of the Interior, Washington, D.C., 363 p.

Henderson-Sellers, A., Benlow, A., and Meadows, A. J., 1980. The early atmospheres of the terrestrial planets. *Royal Astronomical Society Quarterly Journal*, Vol. 21, pp 74-81.

Henry, G., Jr., 1973. Geological investigation of a buried valley in the vicinity of dike A, Caesar Creek Reservoir, Ohio. M.S. Thesis, Wright State University, Dayton, OH.

Hess, A. F., Dyksen, J. E., and Dunn, H. J., 1983. Groundwater contamination's challenge. *Water Technology*, Vol. 6, No. 7, pp 40-46.

Hestroni, G., 1982. *Handbook of Multiphase Systems.* Hemisphere Publishing Corp., Washington, D.C.

Hill v. *Polar Pantries*, 219 S.C. 263, 64 S.E. 2d 885.

Hill, R. A., 1940. Geochemical patterns in Coachella Valley. *Transactions, American Geophysical Union*, Vol. 21, pp 46-53.

Hofheins, W., 1984. Organic fouling of ion exchange resins. *Water Technology*, Vol. 7, No. 1, pp 18-25.

Howard, G. C. and Fast, C. R., 1970. *Hydraulic Fracturing*. American Institute of Mining, Metallurgical, and Petroleum Engineers, Inc. Millet the Printer, Dallas, TX, 203 p.

Howe, C., 1979. Self-recording logging system can save time. *Johnson Drillers' Journal*, Sept/Oct, Johnson Division, UOP Inc., St. Paul, MN.

Huber v. Merkel, 117 Wis. 355, 94 N.W. 354 (1903).

Hubbert, M. K., 1940. The theory of groundwater motion. *Journal of Geology*, Vol. 48, pp 785-944.

Hughes, W. M., 1980. A technical examination of ground water geothermal heat pump components. *Ground Water Heat Pump Journal*, Summer, pp 14-15.

Huisman, L., 1973. *Ground Water Recovery*. University of Technology, Delft, The Netherlands.

Huschke, R. A., 1959. *Glossary of American Meteorology*. American Meteorological Society, Boston, MA.

Hutchinson, S. O. and Anderson, G. W., 1974. What to consider when selecting drilling fluids. *World Oil*, October, pp 83-86.

Hydraulic Institute, 1961. *Pipe Friction Manual, Third Edition*. New York, NY.

Imco Services, 1975. *Applied Mud Technology*. Division of Halliburton Company, Houston, TX.

Ingersoll-Rand, 1970. *Cameron Hydraulic Data, Fourteenth Edition*. Ingersoll-Rand Company, Woodcliff Lake, NJ.

Ingersoll-Rand, 1971. *Compressed Air and Gas Data, Second Edition*, C. W. Gibbs, Editor. Ingersoll-Rand Company, Woodcliff Lake, NJ.

Instrumentation Specialties Company, 1979. *Open Channel Flow Measurement Handbook*. Lincoln, NB, 221 p.

International Atomic Energy Agency, 1971. *Nuclear Well Logging in Hydrology*. Technical Reports Series No. 126, The Working Group on Nuclear Techniques in Hydrology of the International Hydrological Decade, Vienna, 92 p.

Jacob, C. E., 1946a. Radial flow in a leaky artesian aquifer. *Transactions, American Geophysical Union*, Vol. 27, No. 2, pp 198-205.

Jacob, C. E., 1946b. Drawdown test to determine effective radius of artesian well. *Transactions, American Society of Civil Engineers*, Vol. 112, pp 1047-1070.

Jacob, C. E., 1963. Recovery method for determining the coefficient of transmissibility. *U.S. Geological Survey Water Supply Paper 15361*, Washington, D.C.

Jacob, C. E. and Lohman, S. W., 1952. Nonsteady flow to a well of a constant drawdown in an extensive aquifer. *Transactions, American Geophysical Union*, Vol. 33, No. 4, pp 559-569.

Jakosky, J., 1950. *Exploration Geophysics*. Trija Publishing Company, Los Angeles, CA, 1195 p.

Javandel, I., Doughty, C., and Tsang, C., 1984. *Groundwater Transport: Handbook of Mathematical Models*. Water Resources Monograph Series 10, American Geophysical Union, Washington, D.C., 228 p.

Jones, J. W., Sr., 1980. Engineering report: a high-efficiency ground water heat pump system. *Ground Water Heat Pump Journal*, Summer, pp 17-18.

Johnson, A. I., Moston, R. P., and Versaw, S. F., 1966. Laboratory Study of Aquifer Properties and Well Design for an Artificial Recharge Site. *U.S. Geological Survey Water Supply Paper No. 1615-H*, 41 p.

Johnson, R. C., Jr., Kurt, C. E., and Dunham, G. F., Jr., 1980. Well grouting and casing temperature increases. *Ground Water*, Vol. 18, No. 1, Worthington, OH.

Joy Petroleum Equipment, 1978. *What Keeps Your String Together?* Joy Petroleum Equipment, Houston, TX.

Kamen, C. S. and Dar, P., 1973. *Factors Affecting Domestic Water Consumption*. Tel Aviv, Total Water Planning for Israel, Ltd.

Kazmann, R. G., 1948. The induced infiltration of river water to wells. *American Geophysical Union*, Vol. 29, No. 1.

Kazmann, R. G. and Whitehead, W. R., 1980. The spacing of heat pump supply and discharge wells. *Ground Water Heat Pump Journal*, Summer, pp 28-31.

Keech, D. K., 1984. Organic chemicals can be of concern to drillers. *Water Well Journal*, Vol. 38, No. 5, pp 34-35.

Keech, D. K. and Gaber, M. S., 1984. Well vent design. *Water Well Journal*, Vol. 38, No. 8, pp 33-35.

Kelly, J. E., Anderson, K. E., and Burnham, W. L., 1980. Practical problems of confined-aquifer test analyses. American Society of Civil Engineers, New York, 8 p.

Kemmer, F. N., 1977. *Water: The Universal Solvent*. Nalco Chemical Company, Oak Brook, IL, 155 p.

Kemmer, F. N., Editor, 1979. *The NALCO Water Handbook*. Nalco Chemical Company, McGraw-Hill Book Company, New York, NY.

Kennedy v. *Reece*, 225 Col. App. 2d 717, 37 Col Rptr. 708 (1964).

Keys, W. S. and Brown, R. F., 1978. The use of temperature logs to trace the movement of injected water. *Ground Water*, Vol. 16, No. 1, pp 32-48.

Keys, W. S. and MacCary, L. M., 1971. *Application of Borehole Geophysics to Water-Resources Investigations. Chapter E1, Techniques of Water-Resources Investigations of the United States Geological Survey*. U.S. Department of the Interior, Washington, D.C., 126 p.

Knowles, G. D., Lee, G. W., Jr., and Adamowski, S. J., 1982. Hazardous waste site field investigations: geophysical techniques utilized for cost effectiveness, *in Hazardous Waste Management for the 80s*. T. L. Sweeney, H. G. Bhatt, R. M. Sykes, and O. J. Sproul, Editors, Ann Arbor Science, Ann Arbor, MI, pp 183-198.

Knox, R. C., Canter, L. W., and Kincannon, D. F., 1982. Case studies of aquifer restoration. Paper presented at the Spring Convention of the American Society of Civil Engineers, Las Vegas, NV.

Kovalic, J. M., 1985. Alternative financing: who pays?, *in Water Management in Transition 1985*. Freshwater Foundation/Freshwater Society, Navarre, MN, 30 p.

Kozeny, J., 1933. *Theorie and Berechnung der Brunnen*. Wasserkraft u. Wasser Wirtschaft, Vol. 29, 101 p.

Krumbein, W. C. and Pettijohn, F. J., 1938. *Manual of Sedimentary Petrography*. Appleton-Century-Crofts, Inc., New York, NY, 549 p.

Kuetzing, M., 1981. Air mist drilling. *Drilling-DCW*, June.

Kurt, C. E., 1979. Collapse pressure of thermoplastic water well casing. *Ground Water*, Vol. 17, No. 6, pp 550-555.

Ladwig, K. J., 1983. Electromagnetic induction methods for monitoring acid mine drainage. *Ground Water Monitoring Review*, Vol. 3, No. 1, pp 46-51.

Lahee, F. H., 1952. *Field Geology*. McGraw-Hill Book Company, New York, NY, 883 p.

Lai, R. Y. S. and Chen-Wu Su, 1974. Nonsteady flow to a large well in a leaky aquifer. *Journal of Hydrology*, Vol. 22, pp 333-345.

Laney, R. L., Ross, P. P., and Littin, G. R., 1978. Ground water conditions in the eastern part of the Salt River Valley area. *U.S. Geological Survey Water-Resources Investigations 78-61*.

Larson, C. D., Love, O. T., Jr., and Reynolds, G. B., 1983. Tetrachloroethylene leached from lined asbestos cement pipe into drinking water. *Journal American Water Works Association*, Vol. 75, No. 4, pp 184-188.

Larson D. C., 1979. Manitobans find an answer to shallow wells. *Johnson Drillers' Journal*, July/August, Johnson Division, UOP Inc., St. Paul, MN.

Larson, E. E. and Birkeland, P. W., 1982. *Putnam's Geology, Fourth Edition*. Oxford University Press, New York, NY.

Leckie, J. O., Pace, J. G., and Halvadakis, C., 1975. Accelerated refuse stabilization through controlled moisture application. Unpublished report. Department of Environmental Engineering, Stanford University, Stanford, CA.

LeGrand, H. E., 1964. System for evaluation of contamination potential of some waste disposal sites. *Journal American Water Works Association*, Vol. 56, pp 959-974.

Lehr, J. H., Gass, T. E., Pettyjohn, W. A., and DeMarre, J., 1980. *Domestic Water Treatment*. McGraw-Hill Book Company, New York, NY, 264 p.

Levine, M. E. and Clark, J. C., Jr., 1983. Point-of-use treatment to comply with drinking water standards. *Water Technology*, Vol. 6, No. 3, pp 30-31.

Lewis, R. W., 1982. Custom designing of monitoring wells for specific pollutants and hydrogeologic conditions, in *Proceedings of the Second National Symposium on Aquifer Restoration and Ground Water Monitoring*. D. M. Nielsen, Editor, National Water Well Association, Worthington, OH.

Lindsay Division, 1983. *Dissolved Solids Removed by Typical Reverse Osmosis Method*. Ecodyne Corporation, Technical Publication No. 0601979, St. Paul, MN.

Lindsay Division, 1984. *Basic Principles and Treatment of Water*. Ecodyne Corporation, Technical Publication No. 0601881, St. Paul, MN, 35 p.

Lineback, J. A., 1981. Quaternary Deposits of Illinois. Illinois State Geological Survey, Champaign, IL.

Lippy, E. C., 1981. Waterborne disease: occurrence is on the upswing. *Journal American Water Works Association*, Vol. 73, No. 1.

Lohman, S. W., 1979. Ground-water hydraulics. *U.S. Geological Survey Professional Paper 708*, Washington, D.C.

Longest, H. L., 1985. Creative federal participation could pave way for the private sector. *Waterworld News*, March/April.

Lovering, T. S. and Goode, H. D., 1963. Measuring geothermal gradients in drill holes less than 60 feet deep. East Tintic District, Utah. *U.S. Geological Survey Bulletin 1172*. Department of the Interior, Washington, D.C., 49 p.

Lozier, J. C. and Sierka, R. A., 1985. Using ozone and ultrasound to reduce reverse osmosis fouling. *Journal American Water Works Association*, Vol. 77, No. 8, pp

60-65.

Luhdorff, E. E., Jr. and Scalmanini, J. C., 1982. Selection of drilling method, well design and sampling equipment for wells to monitor organic contamination, *in Proceedings of the Second National Symposium on Aquifer Restoration and Ground Water Monitoring.* D. M. Nielsen, Editor, National Water Well Association, Worthington, OH.

Lykins, B. W., Jr., and Griese, M., 1983. Chlorine dioxide disinfection and granular activated carbon adsorption. EPA-600/S2-82-051, U.S. Environmental Protection Agency, Cincinnati, OH, 4 p.

Mabillot, A., 1979. *Le Forage d'Eau, Guide Pratique.* d'Edit Offset a Saint Etienne, Loire, France.

Mackin, J. H., 1948. Concept of the graded river. Geological Society of America Bulletin, Vol. 59, pp 463-512.

MacKnight Flintic Stone Co. v. *City of New York*, 160 N.Y. 72, 54 N.E. 661.

Magcobar, 1977. *Drilling Fluid Engineering Manual.* Magcobar Division, Dresser Industries, Inc., Houston, TX.

Magcobar, 1979. *Air Drilling.* Dresser Industries, Water/Mineral Operations, Aurora, CO, 6 p.

Mairs, D. F., 1967. Surface chloride distribution in Maine lakes. *Water Resources Research*, Vol. 3, pp 1090-1092.

Maslansky, S. P., 1983. Well drilling and hazardous material sites. *Water Well Journal*, Vol. 37, No. 4, pp 46-50.

Mason, K. L. and Woolley, S. T., 1981. How to air drill — from compressor to blooey line: Part I. *Petroleum Engineer International*, March, pp 40-53.

Matthews, C. S. and Russell, D. G., 1967. *Pressure Buildup and Flow Tests in Wells.* Society of Petroleum Engineers of AIME Monograph, Vol. 1, Dallas, TX, 178 p.

Mayer, S., 1984. The industry's growing pains. *Water Technology*, Vol. 7, No. 2, pp 24-36.

McDannald, R. B., 1978. Contracts. *Water Well Journal*, Vol. 32, No. 3, p 35.

McDonald, M. G. and Harbaugh, A. W., 1984. *A Modular Three-Dimensional Finite-Difference Ground-Water Flow Model (MODFLOW).* U.S. Geological Survey, U.S. Department of the Interior, Washington, D.C.

McGowan, W., 1982. Sensitivity: a key water conditioning skill. *Water Technology*, Vol. 5, No. 7, pp 20-23.

McGuinness, C. L., 1963. The Role of Groundwater in the National Water Situation. *U.S. Geological Survey Water Supply Paper 1800.* U.S. Department of Interior, U.S. Government Printing Office, Washington, D.C., 1121 p.

McKinnon, R. J. and Dyksen, J. E., 1984. Removing organics from groundwater through aeration plus GAC. *Journal American Water Works Association*, Vol. 76, No. 5, pp 42-47.

McNeill, J. D., 1980a. Electromagnetic terrain conductivity measurements at low induction numbers. Technical Note TN-6, Geonics Limited, Mississauga, Ontario, Canada, 15 p.

McNeill, J. D., 1980b. Electrical conductivity of soils and rocks. Technical Note TN-5, Geonics Limited, Mississauga, Ontario, Canada, 22 p.

McWorter, D. B. and Sunada, D. K., 1977. *Ground-Water Hydrology and Hydraulics.* Water Resources Publications, Fort Collins, CO, 290 p.

Mec-O-Matic Company, 1984. *Improving Nature's Wonder — Water.* The Marmon Group, St. Paul, MN, 16 p.

Meinzer, O. E., 1923. The Occurrence of Ground Water in the United States, with a Discussion on Principles. *U.S. Geological Survey Water Supply Paper 489*, Government Printing Office, Washington, D.C., 321 p.

Meinzer, O. E., Editor, 1942. *Hydrology.* McGraw-Hill Book Company, New York, NY, 712 p.

Mercer, J. W. and Faust, C. R., 1981. *Ground-Water Modeling.* National Water Well Association, Worthington, OH, 60 p.

Micham, J. T., Levy, B. S., and Lee, G. W., Jr., 1984. Surface and borehole geophysical methods in ground water investigations. *Ground Water Monitoring Review*, Vol. 4, No. 4, pp 167-171.

Middlebrooks, T. A. and Jervis, W. H., 1946. Relief wells for dams and levees. *Proceedings, American Society of Civil Engineers*, June, pp 781-798.

Middlebrooks, T. A. and Jervis, W. H., 1947. Relief wells for dams and levees. *Transactions, American Society of Civil Engineers*, Vol. 112, p 1324.

Milankovich, M., 1969. Canon of insolation and the ice age theory. *Israel Prog. Scientific Translation*, Jerusalem, Israel, 484 p.

Miller, D. W., Geraghty, J. J., and Collins, R. S., 1963. *Water Atlas of the United States, Second Edition.* Water Information Center, Inc., Port Washington, NY, 80 p.

Minning, R. C., 1973. The electrical resistivity method, Part I. Technical Memo Number 3, *Water Well Journal*, Vol. 27, No. 6.

Moench, A. F. and Prickett, T. A., 1972. Radial flow in an infinite aquifer undergoing conversion from artesian to water table conditions. *Water Resources Research*, Vol. 8, No. 2, pp 494-499.

Mogg, J. L., 1959. The effect of aquifer turbulence on well drawdown. *Proceedings, American Society of Civil Engineers*, Hydraulics Division, New York, November, pp 99-112.

Mogg, J. L., 1968. Step-drawdown test needs critical review. *Johnson Drillers' Journal*, July/August, Johnson Division, UOP Inc., St. Paul, MN.

Mogg, J. L., 1977. New instrument expands water well technology. *Johnson Drillers' Journal*, May/June, Johnson Division, UOP Inc., St. Paul, MN.

Moody, L. F., 1944. Friction Factors for Pipe Flow. *Transactions, American Society of Mechanical Engineers*, Vol. 66.

Mooney, H. M., 1981. *Handbook of Engineering Geophysics, Volume 1: Seismic.* Bison Instruments, Inc., Minneapolis, MN.

More, R. J., 1967. Hydrological models and geography, *in Models in Geography.* R. J. Chorley and P. Haggett, Editors. Methuen and Company, Ltd., London, England.

Morrison, A., 1981. If your city's well water has chemical pollutants, then what? *Civil Engineering-ASCE*, Sept., pp 65-67.

Morrison, R. D., 1983. *Ground Water Monitoring Technology — Procedures, Equipment and Applications.* Timco Manufacturing, Inc., Prairie du Sac, WI, 111 p.

Morrison-Knudsen v. *United States*, 170 CT. CL. 712 (1965).

Mossler, J. H., 1972. Paleozoic structure and stratigraphy of the Twin City region, *in Geology of Minnesota: A Centennial Volume.* P. K. Sims and G. B. Morey, Editors. Minnesota Geological Survey, St. Paul, MN, 632 p.

Muskat, M., 1937. *The Flow of Homogeneous Fluids through Porous Media.* McGraw-Hill Book Company, Inc., New York, NY, 763 p.

Myers, F. E. Company, 1965. *Reciprocating Pump Manual: Fundamentals and Application.* Ashland, OH, 40 p.

NL McCullough, 1980. *The Pipe Recovery Manual.* NL McCullough, Houston, TX.

National Academy of Sciences and National Academy of Engineering, 1972. *Water Quality Criteria.* Report of Committee on Water Quality Criteria, U.S. Environmental Protection Agency, Washington, D.C., 594 p.

National Association of Steel Pipe Distributors, Inc., 1979. *Tubular Products Manual.* San Antonio, TX, 135 p.

National Climatic Data Center, 1983. Average Annual Precipitation in the United States. National Oceanic and Atmospheric Administration, Asheville, NC.

National Society of Professional Engineers. Standard Forms of Agreement, Series 1910. Engineering Joint Contract Commission, American Consulting Engineer Council, American Society of Civil Engineers, and Consulting Specifications Institute, Washington, D.C.

National Water Alliance, 1985. Draft consensus statement. Ground Water Task Force, Washington, D.C.

National Water Well Association and Plastics Pipe Institute, 1981. *Manual on the Selection and Installation of Thermoplastic Water Well Casing.* Worthington, OH, 64 p.

Neuman, S. P., 1975. Analysis of pumping test data from anisotropic unconfined aquifers concerning delayed gravity response. *Water Resources Research*, Vol. 11, No. 2, pp 329-342.

Office of Water Research and Technology, 1979. *Reverse Osmosis — Water Research Capsule Report.* U.S. Department of the Interior, U.S. Government Printing Office, Washington, D.C., 20 p.

Okamoto, G. O., 1973. Passive films of 18-8 stainless steel structure and its function. *Corrosion Science*, Vol. 13, Pergamon Press, pp 471-489.

Olson, R. M., 1966. *Essentials of Engineering Fluid Mechanics, Second Edition.* International Textbook Company, Scranton, PA, 448 p.

Olsthoorn, T. N., 1982. *The Clogging of Recharge Wells.* Netherlands Water Works Testing and Research Institute, Communications No. 72, Rijswijk, Netherlands, 131 p.

Papadopulos, I. S., 1965. Nonsteady flow to a well in an infinite anisotropic aquifer. Symposium, International Association of Scientific Hydrology, Dubrovnik.

Papadopulos, I. S., 1967. Drawdown distribution around a large-diameter well. Proceedings, Ground Water Symposium, American Water Resources Association, pp 157-167.

Papadopulos, I. S. and Cooper, H. H., Jr., 1967. Drawdown in a well of large diameter. *Water Resources Research*, Vol. 3, pp 241-244.

Parizek, R. R. and Drew, L. G., 1966. Random drilling for water in carbonate rocks. Institute for Research on Land and Water Resources, Pennsylvania State University, University Park, PA, 22 p.

Parshall, R. L., 1950. *Measuring Water in Irrigation Channels with Parshall Flumes and Small Weirs.* U.S. Soil Conservation Services, Circular 843. U.S. Department of Agriculture, Washington, D.C.

Petroleum Extension Service, 1969. *Principles of Drilling Fluid Control, Twelfth Edition*. The University of Texas at Austin, TX, 215 p.

Petroleum Extension Service, 1974. *Mud Pumps and Conditioning Equipment, Unit 1, Lesson 12*. University of Texas, Austin, TX, 75 p.

Pettyjohn, W. A. and Hounslow, A. W., 1982. Organic compounds and ground-water pollution, *in Proceedings of the Second National Symposium on Aquifer Restoration and Ground Water Monitoring*. D. M. Nielsen, Editor, National Water Well Association, Worthington, OH.

Pinder, G. F. and Gray, W. G., 1977. *Finite Element Stimulation in Surface and Subsurface Hydrology*. Academic Press, New York, NY, 295 p.

Piper, A. M., 1944. A graphic procedure in the geochemical interpretation of water analyses. *Transactions, American Geophysical Union*, Vol. 25, pp 914-923.

Pirson, S. J., 1977. *Geological Well Log Analysis, Second Edition*. Gulf Publishing Company, Houston, TX, 377 p.

Powers, J. P., 1981. *Construction Dewatering: A Guide to Theory and Practice*. John Wiley and Sons, Inc., New York, NY, 484 p.

Prickett, T. A., 1975. Modeling techniques for groundwater evaluation, *in Advances in Hydroscience, Vol. 10*. Academic Press, New York, NY, pp 1-143.

Prickett, T. A. and Lonnquist, C. G., 1971. Selected Digital Computer Techniques for Groundwater Resource Evaluation. Illinois State Water Survey Bulletin No. 55, Champaign, IL, 62 p.

Purdin, W., 1980. Using nonmetallic casing for geothermal wells. *Water Well Journal*, Vol. 34, No. 4, pp 90-91.

Purdue University, 1949. Measurement of water flow through pipe orifice with free discharge. Lafayette, IN.

Rahn, P. H. and Moore, D. G., 1979. Landsat data for locating shallow glacial aquifers in eastern South Dakota, *in Satellite Hydrology*. M. Deutsch, D. R. Wiesnet, and A. Rango, Editors, Fifth Annual William T. Pecora Memorial Symposium on Remote Sensing, 1981, American Water Resources Association, Minneapolis, MN, 730 p.

Ramey, H. J., Jr., Kumar, A., and Gulati, M. S., 1973. *Gas Well Test Analysis Under Water-Drive Conditions*. American Gas Association Monograph, Arlington, VA.

Ramsey, R. H. and Montgomery, J. M., 1982. Monitoring ground-water contamination in Spokane County, Washington, *in Proceedings of the Second National Symposium on Aquifer Restoration and Ground Water Monitoring*. D. M. Nielsen, Editor, National Water Well Association, Worthington, OH.

Reamer, Earl L. v. City of Swartz Creek, 76 Mich. App 227, 256 N.W. 2d 447 (1977).

Reilly, T. A., 1985. Attaining water resource goals: looking to the private sector, *in Water Management in Transition 1985*. Freshwater Foundation/Freshwater Society, Navarre, MN, 80 p.

Remson, I., Hornberger, G. M., and Malz, F. J., 1971. *Numerical Methods in Subsurface Hydrology*. Wiley-Interscience, New York, NY, 389 p.

Restatement (Second) of Torts. American Bar Association, Chicago, IL.

Rhoades, J. D., 1972. Quality of water for irrigation. Soil Science No. 113, pp 277-284.

Rice, R. G., 1985. Trends in ozonation. *Journal American Water Works Association*, Vol. 77, No. 8, p 26.

Richter, H. R. and Collentine, M. G., 1983. Will my monitoring wells survive down there?: design and installation techniques for hazardous waste studies, *in Proceedings of the Third National Symposium on Aquifer Restoration and Ground-Water Monitoring*. D. M. Nielsen, Editor, National Water Well Association, Worthington, OH.

Roberts, P. V., Reinhard, M., and Valocchi, A. J., 1982. Movement of organic contaminants in groundwater: implications for water supply. *Journal American Water Works Association*, Vol. 74, No. 8, pp 408-413.

Roberts, R. B., 1980. In situ mining used in Texas to find uranium. *Johnson Drillers' Journal*, Sept/Oct, Johnson Division, UOP Inc., St. Paul, MN.

Rook, J. J., 1974. Formation of haloforms during chlorination of natural waters. *Water Treatment Examination*, Vol. 23, No. 2, p 234.

Rosenberg, N. J., 1978. Preface, *in North American Droughts*, edited by Norman J. Rosenberg, AAAS Selected Symposium 15. Westview Press, Boulder, CO, 177 p.

Rudy, R. J. and Caolic, J. A., 1984. Utilization of shallow geophysical sensing at two abandoned municipal/industrial waste landfills on the Missouri River floodplain. *Ground Water Monitoring Review*, Vol. 4, No. 4, pp 57-65.

Rutledge, P. C., 1947. Discussion of paper, relief wells for dams and levees by Middlebrooks and Jervis. *Proceedings, American Society of Civil Engineers*, January, pp 65-71.

SPE Monograph, 1976. See Smith, 1976.

Sasman, R. T., Benson, C. R., Ludwigs, R. S., and Williams, T. L., 1982. Water-level trends, pumpage, and chemical quality in the Cambrian-Ordovician aquifer in Illinois, 1971-1980. Illinois State Water Survey Circular 154, Cambridge, IL.

Saunders, W., 1983. Paper presented at Groundwater Monitoring Technology Conference, November, Philadelphia, PA. Sponsored by University of Wisconsin-Extension, Madison, WI.

Sawkins, F. J., Chase, G. C., Darby, D. G., and Rapp, G., Jr., 1974. *The Evolving Earth: A Text in Physical Geology*. Macmillan Publishing Co. Inc., New York, NY, 477 p.

Schaetzle, W. F., Brett, C. E., Grubbs, D. M., and Seppanen, M. S., 1980. *Thermal Energy Storage in Aquifers*. Pergamon Press, Inc., New York, NY, 177 p.

Schafer, D. C., 1978. Casing storage can affect pumping test data. *Johnson Drillers' Journal*, Jan/Feb, Johnson Division, UOP Inc., St. Paul, MN.

Schafer, D. C., 1981. Longest Johnson screen installed in a Utah well. *Johnson Drillers' Journal*, Third Quarter, Johnson Division, UOP Inc., St. Paul, MN.

Scherer, T. F., Werner, H. D., and Bergsrud, F. G., 1979. Aquifer and well characteristics evaluated with an interactive mini-computer. Paper 79-2573 presented at Winter Meeting, American Society of Agricultural Engineers, New Orleans, LA.

Schmidt, K. D., Krancher, J. A. and Bisel, G., Jr., 1981. Brine pollution at Fresno — twenty-six years later. *Ground Water*, Vol. 19, No. 1.

Schneider, S. H., 1978. Forecasting future droughts, *in North American Droughts*, edited by Norman J. Rosenberg, AAAS Selected Symposium 15. Westview Press, Boulder, CO, 177 p.

Schockley, R. L., 1980. Effluent disposal methods. *Ground Water Heat Pump Journal*, Summer, pp 27-28.

Schowengerdt, R., Babcock, E. M., Ethridge, L., and Glass, C. E., 1979. Correlation of geologic structure inferred from computer enhanced Landsat imagery with underground water supplies in Arizona, *in Satellite Hydrology*. M. Deutsch, D. R. Wiesnet, and A. Rango, Editors. Fifth Annual William T. Pecora Memorial Symposium on Remote Sensing, 1981. American Water Resources Association, Minneapolis, MN, 730 p.

Sendlein, L. V. A. and Yazicigil, H., 1981. Surface geophysical methods for ground water monitoring, Part I. *Ground Water Monitoring Review*, Vol. 1, No. 3, pp 42-46.

Senger, J. A. and Perpich, W. M., 1983. An alternative well seal in highly mineralized ground water, *in Proceedings of the Third National Symposium on Aquifer Restoration and Ground-Water Monitoring*. D. M. Nielsen, Editor, National Water Well Association, Worthington, OH.

Shapiro, J. and Pfannkuch, H. O., 1973. The Minneapolis chain of lakes: a study of urban drainage and its effects — 1971-1973. Interim Report No. 9, Limnological Research Center, University of Minnesota, Minneapolis, MN, 250 p.

Silva, S. J., 1981. EPA moving to control industrial toxic pollutants with new NPDES permits. *Civil Engineering-ASCE*, Sept., pp 76-78.

Skogerboe, G. V., Bennett, R. S., and Walker, W. R., 1973. *Selection and Installation of Cutthroat Flumes for Measuring Irrigation and Drainage Water*. Technical Bulletin 120, Colorado State University, Fort Collins, CO., 74 p.

Skogerboe, G. V., Hyatt, M. L., Anderson, R. K., and Eggleston, K. O., 1967. Design and calibration of submerged open channel flow measurement structures: Part 3, cutthroat flumes. Report WG31-4, Utah Water Research Laboratory, Utah State University, Logan, UT.

Slichter, C. S., 1899. Theoretical investigation of the motion of ground water. 19th Annual Report, U.S. Geological Survey, Washington, D.C.

Smith, D. K., 1976. *Cementing*. American Institute of Mining, Metallurgical, and Petroleum Engineers, Inc., Dallas, TX, 184 p.

Smith, R. C., 1963. Relation of screen design to the design of mechanically efficient wells. *Journal American Water Works Association*, Vol. 55, No. 5.

Smith, R. J., 1980a. Current trends in groundwater law. Paper presented at the Water Well and Construction Institute held at the University of Wisconsin-Extension, Nov., Madison, WI.

Smith, A. J., 1980b. Design is the key to a good recharge well. *Johnson Drillers' Journal*, Mar/Apr, Johnson Division, UOP Inc., St. Paul, MN.

Snead, R. E., 1980. *World Atlas of Geomorphic Features*. Robert E. Krieger Publishing Co., Inc., Melbourne, FL, 301 p.

Solley, W. B., Chase, E. B., and Mann, W. B. IV, 1983. Estimated use of water in the United States in 1980. *U.S. Geological Survey Circular 1001*, U.S. Department of the Interior, Alexandria, VA, 56 p.

Solley, W. B., Chase, E. B., and Mann, W. B. IV, 1984. Estimated use of water in the United States in 1981. *U.S. Geological Survey Circular 1001*, U.S. Department of the Interior, Alexandria, VA.

Sporhase et al., Appellants v. *Nebraska*, ex rel. Douglas, Attorney General, 42 CCH S. Ct. Bull. pB4900 (1982).

Stewart, G. W., 1973. Hydraulic fracturing of drilled water wells in crystalline rocks of New Hampshire. Department of Resources and Economic Development, Division of Forests and Lands, New Hampshire.

Stiff, H. A., 1951. The interpretation of chemical water analyses by means of patterns. *Journal Petroleum Technology*, Vol. 3, No. 10, Section 1, 2, 3.

Strahler, A. N., 1969. *Physical Geography, Third Edition*. John Wiley and Sons, Inc., New York, NY, 733 p.

Strahler, A. N., 1975. *Physical Geography, Fourth Edition*. John Wiley and Sons, Inc., New York, NY.

Streltsova, T. D., 1974. Drawdown in compressible unconfined aquifer. *Proceedings, American Society of Civil Engineers, Journal of Hydraulics Division*, Vol. 100 (HY11), pp 1601-1616.

Streltsova, T. D., 1976. Analysis of aquifer-aquitard flow. *Water Resources Research*, Vol. 12, No. 3, pp 415-422.

Stumm, W. and Morgan, J. J., 1970. *Aquatic Chemistry*. Wiley-Interscience, New York, NY.

Suman, G. O., Jr, Ellis, R. C., and Snyder, R. E., 1983. *Sand Control Handbook, Second Edition*. Gulf Publishing Company, Houston, TX.

Summers, W. K., 1972. Factors affecting the validity of chemical analyses of natural water. *Ground Water*, Vol. 10, No. 2, Mar/Apr, pp 12-17.

Symons, J. M., 1976. Interim treatment guide for the control of chloroform and other trihalomethanes. U.S. Environmental Protection Agency Research Laboratory, Water Supply Research Division, Cincinnati, OH.

Symons, J. M., Stevens, A. A., Clark, R. M., Geldreich, E. E., Love, O. T., Jr., and DeMarco, J., 1981. *Treatment Techniques for Controlling Trihalomethanes in Drinking Water*. EPA-600/2-81-156, Drinking Water Research Division, Municipal Environmental Research Laboratory, U.S. Environmental Protection Agency, Cincinnati, OH, 289 p.

Taitel, Y. and Dukler, A. E., 1980. Modeling flow pattern transitions for steady upward gas-liquid flow in vertical tubes. *Journal AL Chemical Engineering*, No. 26, pp 345-352.

Tank, R. W., 1983. *Legal Aspects of Geology*. Plenum Press, New York, NY, 583 p.

Tardy, Y., 1971. Characterization of the principal weathering types by the geochemistry of waters from some European and African crystalline massifs. *Chemical Geology*, No. 7, pp 253-271.

Taylor, R. and Cherbauer, D., 1984. The application of combined seismic and electrical measurements to the determination of the hydraulic conductivity of a lake bed. *Ground Water Monitoring Review*, Vol. 4, No. 4, pp 78-85.

Terzaghi, K. Von, 1922. *Der Grundbruch and Stauwerken and seine Verhutung*, Die Wasserkraft.

Theis, C. V., 1935. The relation between the lowering of the piezometric surface and the rate and duration of discharge of a well using ground water storage. *Transactions, American Geophysical Union*, Washington, D.C., pp 518-524.

Thiem, G., 1906. *Hydrologische methoden*. Leipzig, 56 p.

Thomas, H. E., 1952. Ground water regions of the United States — their storage facilities. United States 83rd Congress, House Interior and Insular Affairs Commission, The Physical and Economic Foundation of Natural Resources, Vol. 3.

Tolman, C. F., 1937. *Ground Water*. McGraw-Hill Book Company, New York, NY.

Townley, L. R. and Wilson, J. L., 1980. *Description of and User's Manual for a Finite*

Element Aquifer Flow Model Aquifem-1. Parsons Laboratory for Water Resources and Hydrodynamics, Report No. 252, 299 p.

Trescott, P. C., Pinder, G. F., and Lercon, S. P., 1976. *Finite Difference Model for Aquifer Simulation in Two Dimensions with Results of Numerical Experiment.* U.S. Geological Survey Techniques of Water Resources Investigations, Book 7, Chapter C1, 116 p.

Trustees of the First Baptist Church v. *McElroy, Miss.*, 78 So. 2d 138.

Tschirley, N. K., 1978. *Manual of Drilling Fluids Technology.* Baroid Petroleum Services Company, Houston, TX, 200 p.

Turnbull, W. J. and Mansur, C. I., 1953. Relief well systems for dams and levees. *Proceedings, American Society of Civil Engineers*, Vol. 79, Separate No. 192, May, 21 p.

Turneaure, F. E. and Russell, H. L., 1901. *Public Water Supplies.* John Wiley and Sons, Inc., New York, NY, 269 p.

Turner, J. H., Editor, and Anderson, C. L., 1980. *Planning for an Irrigation System.* American Association for Vocational Instructional Materials, Athens, GA, 119 p.

Tweenton, D. R. and Connor, K., 1978. Well construction information for in situ uranium leaching. U.S. Department of the Interior, Bureau of Mines, IC 8769, U.S. Government Printing Office, Washington, D.C.

Unesco, 1975. *Legends for Geohydrochemical Maps.* Technical Papers in Hydrology. The Unesco Press, Paris, 62 p.

Unesco, International Association of Scientific Hydrology, International Association of Hydrogeologists, and Institute of Geological Sciences, 1970. *International Legend for Hydrogeological Maps.* 101 p.

United Nations, 1976. *Ground Water in the Western Hemisphere.* Natural Resources/Water Series No. 4, ST/ESA/35, Department of Economic and Social Affairs, New York, NY, 337 p.

United Nations, 1980. *Efficiency and Distribution Equity in the Use and Treatment of Water: Guidelines for Pricing and Regulation.* Natural Resources/Water Series No. 8, United Nations Publications, No. E.80.II.A.11, New York, NY, 176 p.

United States v. *Spearin*, 248 U.S. 132, 398, Ct. 59, 61, 63 L.Ed. 166.

U.S. Department of Agriculture, 1953. Estimated flow from pipes. *Engineering Handbook.* Farmers Home Administration, Washington, D.C.

U.S. Department of Agriculture, 1954. *Diagnosis and Improvement of Saline and Alkali Soils.* Agriculture Handbook 60, L. A. Richards, Editor, Washington, D.C., 160 p.

U.S. Department of Health and Human Services, 1982. Water-related disease outbreaks. *Annual Summary for 1980.* Public Health Service, Centers for Disease Control.

U.S. Environmental Protection Agency, 1973. *Manual of Individual Water Supply Systems.* Office of Water Programs, Water Supply Division, EPA 430/9-74-007, Washington, D.C.

U.S. Environmental Protection Agency, 1975a. National interim primary drinking water regulations, Part IV Water Program. *Federal Register*, Vol. 40, No. 248. Government Printing Office, Washington, D.C.

U.S. Environmental Protection Agency, 1975b. Inventory of public water supplies. Government Printing Office, Washington, D.C.

U.S. Environmental Protection Agency, 1975c. *Manual of Water Well Construction Practices.* EPA-570/9-75-001, Office of Water Supply, Washington, D.C., 156 p.

U.S. Environmental Protection Agency, 1977a. National secondary drinking water regulations. *Federal Register,* 42:17144, March 31. Government Printing Office, Washington, D.C.

U.S. Environmental Protection Agency, 1977b. The report to Congress: waste disposal practices and their effects on ground water. EPA 570/9-77-001, Washington, D.C., pp 294-321.

U.S. Environmental Protection Agency, 1978. *Manual of Treatment Techniques for Meeting the Interim Primary Drinking Water Regulations.* Office of Research and Development, Municipal Environmental Research Laboratory, Water Supply Research Division, EPA-600/8-77-005, Cincinnati, OH, 73 p.

U.S. Environmental Protection Agency, 1979. Amendments to the national interim primary drinking water regulations: control of trihalomethanes in drinking water. *Federal Register,* 44:68624, Nov. 29. Government Printing Office, Washington, D.C.

U.S. Environmental Protection Agency, 1980a. *Planning Workshops to Develop Recommendations for a Ground Water Protection Strategy, Sections I, II, and III.* May, Washington, D.C.

U.S. Environmental Protection Agency, 1980b. Proposed ground water protection strategy. November, Office of Drinking Water, Washington, D.C.

U.S. Environmental Protection Agency, 1980c. *Planning Workshops to Develop Recommendations for a Ground Water Protection Strategy, Sections IV and V.* June, Washington, D.C.

U.S. Environmental Protection Agency, 1982. Proposed amendments to the national primary drinking water regulations: trihalomethanes. *Federal Register,* 47:9796, March 5. Government Printing Office, Washington, D.C.

U.S. Environmental Protection Agency, 1984. National primary drinking water regulations: volatile synthetic organic chemicals: proposed rulemaking. August, Federal Register 49:24330. Government Printing Office, Washington, D.C.

U.S. Geological Survey, 1942. *Water Supply Paper 887.* Government Printing Office, Washington, D.C.

U.S. Geological Survey, 1981. *Water Resources Data for Minnesota, Vol. 1. Great Lakes and Souris-Red Rainy River Basins.* U.S. Geological Survey Water Data Report MN-80-1, St. Paul, MN, 234 p.

Valley, S. L., Editor, 1965. *Handbook of Geophysics and Space Environments.* McGraw-Hill Book Company, New York, NY.

Varhol, B. P., 1980. Restoring wells is not a gamble in Las Vegas! *Johnson Drillers' Journal,* Nov/Dec, Johnson Division, UOP Inc., St. Paul, MN.

Walker, R., 1978. *Water Supply, Treatment and Distribution.* Prentice-Hall, Inc., Englewood Cliffs, NJ, 420 p.

Walker, R., 1980. *Pump Selection: A Consulting Engineer's Manual.* Ann Arbor Science, Ann Arbor, MI, 118 p.

Walker, S. E., 1983. Background ground-water quality monitoring: well installation trauma, *in Proceedings of the Third National Symposium on Aquifer Restoration and Ground-Water Monitoring.* D. M. Nielsen, Editor, National Water Well Association, Worthington, OH.

Wall Street Journal, 1985. Denver turns on to xeriscape, turns off on thirsty grasses. June 18, Dow Jones & Company, New York, NY.

Wallace, D. E., 1970. Some limitations of seismic refraction methods in geohydrological surveys of deep alluvial basins. *Ground Water*, Vol. 8, No. 6, p 8-13.

Wallace, L. P., Martin, W. F., and Whitsell, W., 1982. Drilling monitoring wells may be hazardous to your health. *Water Well Journal*, Vol. 36, No. 9, pp 51-52.

Walton, W. C., 1960. Leaky artesian aquifer conditions in Illinois. Illinois State Water Survey, Report of Investigations, No. 39, Urbana, IL.

Walton, W. C., 1962. Selected Analytical Methods for Well and Aquifer Evaluation. Illinois State Water Survey, Bulletin No. 49, Urbana, IL.

Walton, W. C., 1963. Estimating the infiltration rate of a stream bed by aquifer testing analysis. National Association of Scientific Hydrology, No. 63.

Walton, W. C., 1970. *Groundwater Resource Evaluation*. McGraw-Hill Book Company, New York, NY, 664 p.

Walton, W. C., 1979. Progress in analytical groundwater modeling, *in Contemporary Hydrogeology*, edited by W. Back and D. A. Stephenson. Elsevier Scientific Publishing Company, Amsterdam.

Waltz, J. and Decker, T. L., 1981. Hydro-fracturing offers many benefits. *Johnson Drillers' Journal*, Second Quarter, Johnson Division, UOP Inc., St. Paul, MN.

Wang, H. F. and Anderson, M. P., 1982. *Introduction to Groundwater Modeling — Finite Difference and Finite Element Methods*. W. H. Freeman, San Francisco, CA, 237 p.

Warner, D. L., Koederitz, L. F., Simon, A. D., and Yow, M. G., 1979. *Radius of Pressure Influence of Injection Wells*. Office of Research and Development, U.S. Environmental Protection Agency, EPA-600/2-79-170, Ada, OK, 204 p.

Warner, D. L. and Lehr, J. H., 1981. *Subsurface Wastewater Injection*. Premier Press, Berkeley, CA, 344 p.

Water and Power Resources Service, 1981. *Ground Water Manual*. U.S. Department of the Interior, U.S. Government Printing Office, Washington, D.C., 480 p.

Water Quality Association, 1983. Point-of-use treatment for compliance with drinking water standards. Chicago, IL.

Water Systems Council, undated. Centrifugal pumps. Technical Manual No. 3, Glenview, IL, 23 p.

Water Well Journal, 1981. The water well industry: a study. Vol. 35, No. 1, pp 79-97.

Water Well Journal, 1983. Recharge-recovery introduced in southwest Florida community. Vol. 37, No. 7, p 13.

Weakly, H. E., 1965. Recurrence of drought in the Great Plains during the last 700 years. *Agricultural Engineering*, Vol. 46, p 85.

Wegener, A., 1912. *Die Entstehung der Kontinente*. Petermann's Mitteilungen.

Wentworth, C. K., 1922. A scale of grade and class terms for clastic sediments. *Journal of Geology*, Vol. 30, pp 377-392.

Werner, H. D., Scherer, T. F., and Kajer, T. O., 1980. Effects of irrigation well efficiency on energy requirements. U.S. Department of Energy, EM-78-G-01-5131, Staples, MN.

Westrick, J. J., Mello, J. W., and Thomas, R. F., 1984. The groundwater supply survey. *Journal American Water Works Association*, Vol. 76, No. 5, pp 52-59.

White, D. E., Hem, J. D., and Waring, G. A., 1963. Chemical composition of subsurface waters, *in Data of Geochemistry, 6th Edition*. U.S. Geological Survey Professional Paper 440-F, Washington, D.C, p F1-F67.

White, G. C., 1972. *Handbook of Chlorination*. Van Nastrand, Reinhold Co.. New York, NY.

White, M. D., 1977. Legal restraints and responses to the allocation and distribution of water, *in Water Needs for the Future*. V. P. Nanda, Editor, Westview Press, Boulder, CO, 329 p.

Whitten Corp. v. Paddock, Inc., 376 F. Supp. 125 (1974).

Whitten Corp. v. Paddock, Inc., 508 F.2d 547 (1974).

Wickramanayak, G. B., Rubin, A. J., and Sproul, O. J., 1985. Effects of ozone and storage temperature on *Giardia* cysts. *Journal American Water Works Association*, Vol. 77, No. 8, pp 74-77.

Williams, Clyde E., and Associates, 1955. Well specifications for the town of Milford, Indiana. South Bend, IN.

Williams, E. B., 1982. Contaminant containment by in situ polymerization, *in Proceedings of the Second National Symposium on Aquifer Restoration and Ground Water Monitoring*. D. M. Nielsen, Editor, National Water Well Association, Worthington, OH.

Winograd, I. J., 1974. Radioactive waste storage in the arid zone. *EOS, Transactions, American Geophysical Union 55*, Washington, D.C.

Winograd, I. J. and Robertson, F. N., 1985. Deep oxygenated ground water: anomaly or common occurrence? *Science*, Vol. 216, June, pp 1227-1230.

Witherspoon, P. A., Javandel, I., Newman, S. P., and Freeze, R. A., 1967. Interpretation of aquifer gas storage conditions from water pumping tests. American Gas Association, New York, NY, 273 p.

Wood, W. W., 1976. *Guidelines for Collection and Field Analysis of Ground-water Samples for Selected Unstable Constituents*. Techniques of Water-Resources Investigations of the U.S. Geological Survey, Book 1, Chapter D2.

Woods, H. B. and Lubinski, A., 1954. How to determine best hole and drill collar size. *Oil and Gas Journal*, June.

Yaniga, P. M., 1982. Alternatives in decontamination for hydrocarbon-contaminated aquifers, *in Proceedings of the Second National Symposium on Aquifer Restoration and Ground Water Monitoring*. D. M. Nielsen, Editor, National Water Well Association, Worthington, OH.

York, D. C., 1978. Collapse pressure predictions for thermoplastic water well casings. Unpublished thesis, Auburn University, Auburn, AL.

Zall, L. and Russell, O., 1979. Ground water exploration programs in Africa, *in Satellite Hydrology*. M. Deutsch, D. R. Wiesnet, and A. Rango, Editors. Fifth Annual William T. Pecora Memorial Symposium on Remote Sensing, 1981. American Water Resources Association, Minneapolis, MN, 730 p.

Zohdy, A. A. R., Eaton, G. P., and Mabey, D. R., 1974. Application of surface geophysics to ground water investigations, techniques of water-resources investigations of the U.S. Geological Survey. Washington, D.C.

Glossary

Acid. Any chemical compound containing hydrogen capable of being replaced by positive elements or radicals to form salts. In terms of the dissociation theory, it is a compound which, on dissociation in solution, yields excess hydrogen ions. Acids lower the pH. Examples of acids or acidic substances are hydrochloric acid, tannic acid, and sodium acid pyrophosphate.

Activated carbon. A granular material usually produced by the roasting of cellulose base substances, such as wood or coconut shells, in the absence of air. It has an extremely porous structure and is used in water conditioning as an adsorbent for organic matter and certain dissolved gases. Sometimes called "activated charcoal."

Adsorption. The assimilation of gas, vapor, or dissolved matter by the surface of a solid.

Advection. The process by which solutes are transported by the bulk motion of the flowing groundwater.

Aeration. The process of bringing air into intimate contact with water, usually by bubbling air through the water to remove dissolved gases like carbon dioxide and hydrogen sulfide or to oxidize dissolved materials like iron compounds.

Aggregation. Formation of aggregates. In drilling fluids, aggregation results in the stacking of the clay platelets face to face; as a result, viscosity and gel strength decrease.

Air stripping. A mass transfer process in which a substance in solution in water is transferred to solution in a gas, usually air.

Alkaline. Any of various soluble mineral salts found in natural water and arid soils having a pH greater than 7. In water analysis, it represents the carbonates, bicarbonates, hydroxides, and occasionally the borates, silicates, and phosphates in the water.

Alluvial. Pertaining to or composed of alluvium or deposited by a stream or running water.

Alluvium. A general term for clay, silt, sand, gravel, or similar unconsolidated material deposited during comparatively recent geologic time by a stream or other body of running water as a sorted or semisorted sediment in the bed of the stream or on its floodplain or delta, or as a cone or fan at the base of a mountain slope.

Anion. A negatively charged ion that migrates to an anode, as in electrolysis.

Anion exchange. Ion exchange process in which anions in solution are exchanged for other anions from an ion exchanger.

Anisotropic. Having some physical property that varies with direction.

Annulus. The space between the drill string or casing and the wall of the borehole or outer casing.

Anode. Any positively charged electrode, as in an electrolytic cell, storage battery, or electron tube.

Aquiclude. A saturated, but poorly permeable bed, formation, or group of formations that does not yield water freely to a well or spring. However, an aquiclude may transmit appreciable water to or from adjacent aquifers.

Aquifer. A formation, group of formations, or part of a formation that contains sufficient saturated permeable material to yield economical quantities of water to wells and springs.

Aquifer stimulation. A type of development that is done in semiconsolidated and completely consolidated formations to alter the formation physically to improve its hydraulic properties.

Aquifer test. A test involving the withdrawal of measured quantities of water from or addition of water to, a well and the measurement of resulting changes in head in the aquifer both during and after the period of discharge or addition.

Aquitard. A geologic formation, group of formations, or part of a formation through which virturally no water moves.

Artesian well. A well deriving its water from a confined aquifer in which the water level stands above the ground surface; synonymous with flowing artesian well.

Artificial recharge. Recharge at a rate greater than natural, resulting from deliberate actions of man.

Asthenosphere. The layer or shell of the Earth below the lithosphere which is weak and in which isostatic adjustment takes place, magmas may be generated, and seismic waves are strongly attenuated.

Attapulgite clay. A colloidal, viscosity-building clay consisting of hydrous magnesium aluminum silicates and used principally in salt-water drilling fluids.

Backwash (Water Treatment). The process in which filter beds are subjected to water flow opposite to the service flow direction to loosen the bed and flush solid materials accumulated on the resin bed to waste.

Backwash (Well Development). The surging effect or reversal of water flow in a well. Backwashing removes fine-grained material from the formation surrounding the borehole and, thus, can enhance well yield.

Bactericide. A substance that destroys bacteria.

Barite. Natural finely ground barium sulfate used for increasing the density of drilling fluids.

Basalt. A general term for dark-colored iron- and magnesium-rich igneous rocks, commonly extrusive, but locally intrusive. It is the principal rock type making up the ocean floor.

Base exchange. The displacement of a cation bound to a site on the surface of a solid, as in silica-alumina clay-mineral packets, by a cation in solution.

Bedload. The part of the total stream load that is moved on or immediately above the stream bed, such as the larger or heavier particles (boulders, pebbles, gravel) transported by traction or saltation along the bottom; the part of the load that is not continuously in suspension or solution.

Bedrock. A general term for the rock, usually solid, that underlies soil or other unconsolidated material.

Bentonite. A colloidal clay, largely made up of the mineral sodium montmorillonite, a hydrated aluminum silicate.

Bit. The cutting tool attached to the bottom of the drill stem.

Blowout. An uncontrolled escape of drilling fluid, gas, oil, or water from the well caused by the formation pressure being greater than the hydrostatic head of the fluid in the hole.

Braided stream. A stream that divides into or follows an interlacing or tangled network of several small branching and reuniting shallow channels separated from each other by branch islands or channel bars, resembling in plan the strands of a complex braid.

Bridge. An obstruction in the drill hole or annulus. A bridge is usually formed by caving of the wall of the well bore, by the intrusion of a large boulder, or by filter pack materials during well completion. Bridging can also occur in the formation during well development.

Buried valley. A depression in an ancient land surface or in bedrock now covered by younger deposits, especially a preglacial valley filled with glacial drift.

Capillary fringe. The zone at the bottom of the vadose zone where groundwater is drawn upward by capillary force.

Carbonate. A sediment formed by the organic or inorganic precipitation from aqueous solution of carbonates of calcium, magnesium, or iron.

Carbonate rocks. A rock consisting chiefly of carbonate minerals, such as limestone and dolomite.

Cathode. Any negatively charged electrode, as in an electrolytic cell or storage battery.

Cation. An ion having a positive charge and, in electrolytes, characteristically moving toward a negative electrode.

Cation exchange. Ion exchange process in which cations in solution are exchanged for other cations from an ion exchanger.

Cavitation. A phenomena of cavity formation, or formation and collapse, especially in regard to pumps, when the absolute pressure within the water reaches the vapor pressure causing the formation of vapor pockets.

Cementing. See Grouting.

Chlorine. A gas, Cl_2, widely used in the disinfection of water and as an oxidizing agent.

Clastic. Pertaining to a rock or sediment composed principally of broken fragments that are derived from pre-existing rocks or minerals and that have been transported some distance from their places of origin.

Coefficient of permeability. An obsolete term that has been replaced by the term hydraulic conductivity.

Coefficient of storage. The volume of water an aquifer releases from or takes into storage per unit surface area of the aquifer per unit change in head.

Coefficient of transmissivity. See Transmissivity.

Colloid. Extremely small solid particles, 0.0001 to 1 micron in size, which will not settle out of a solution; intermediate between a true dissolved particle and a suspended solid which will settle out of solution.

Cone of depression. A depression in the groundwater table or potentiometric surface that has the shape of an inverted cone and develops around a well from which water is being withdrawn. It defines the area of influence of a well.

Confined aquifer. A formation in which the groundwater is isolated from the atmosphere at the point of discharge by impermeable geologic formations; confined groundwater is generally subject to pressure greater than atmospheric.

Contamination. The degradation of natural water quality as a result of man's activities. There is no implication of any specific limits, since the degree of permissible contamination depends upon the intended end use, or uses, of the water.

Corrosion. The act or process of dissolving or wearing away metals.

Darcy's law. A derived equation for the flow of fluids on the assumption that the flow is laminar and that inertia can be neglected.

Deflocculation. Breakup of flocs of gel structures by use of a thinner.

Density. Matter measured as mass per unit volume expressed in pounds per gallon (lb/gal), pounds per cubic ft (lb/ft^3), and kilogram per cubic m (kg/m^3).

Desalination. To remove salt and other chemicals from sea water or saline water.

Development. The act of repairing damage to the formation caused by drilling procedures and increasing the porosity and permeability of the materials surrounding the intake portion of the well.

Diatomaceous earth. A light-colored, soft, siliceous earth composed of the shells of diatoms, a form of algae. Some deposits are of lake origin but the largest are marine.

Dispersion. The spreading and mixing of chemical constituents in groundwater caused by diffusion and mixing due to microscopic variations in velocities within and between pores.

Dissociation. A chemical process that causes a molecule to split into simpler groups of atoms, single atoms, or ions. For example, the water molecule (H_2O) breaks down spontaneously into H^+ and OH^- ions.

Drainage basin. The land area from which surface runoff drains into a stream channel or system of channels, or to a lake, reservoir, or other body of water.

Drawdown. The distance between the static water level and the surface of the cone of depression.

Drill collar. A length of extremely heavy steel tube. It is placed in the drill string immediately above the drill bit to minimize bending caused by the weight of the drill pipe.

Drill pipe. Special pipe used to transmit rotation from the rotating mechanism to the bit. The pipe also transmits weight to the bit and conveys air or fluid which removes cuttings from the hole and cools the bit.

Drilling fluid. A water- or air-based fluid used in the water-well drilling operation to remove cuttings from the hole, to clean and cool the bit, to reduce friction between the drill string and the sides of the hole, and to seal the borehole.

Drive point. See Well point.

Effective size. The 90-percent-retained size of a sediment as determined from a grain-size analysis; therefore, 10 percent of the sediment is finer and 90 percent is coarser.

Effluent. A waste liquid discharge from a manufacturing or treatment process, in its natural state or partially or completely treated, that discharges into the environment.

Electrical conductance. A measure of the ease with which a conducting current can be caused to flow through a material under the influence of an applied electric field. It is the reciprocal of resistivity and is measured in mhos per foot (meter).

Electrical resistivity. The property of a material which resists the flow of electrical current measured per unit length through a unit cross-sectional area.

Electrolyte. A chemical which dissociates into positive and negative ions when dissolved in water, increasing the electrical conductivity.

Equipotential line. A contour line on the water table or potentiometric surface; a line along which the pressure head of groundwater in an aquifer is the same. Fluid flow is normal to these lines in the direction of decreasing fluid potential.

Erosion. The general process or group of processes whereby the materials of the Earth's crust are moved from one place to another by running water (including rainfall), waves and currents, glacier ice, or wind.

Evapotranspiration. Loss of water from a land area through transpiration of plants and evaporation from the soil.

Extrusive rocks. Igneous rocks formed from magma that flows out on the Earth's surface. These rocks cool rapidly, producing a fine crystalline structure.

Fault. A fracture or a zone of fractures along which there has been displacement of the sides relative to one another parallel to the fracture.

Filter cake. The suspended solids that are deposited on a porous medium during the process of filtration.

Filter pack. Sand or gravel that is smooth, uniform, clean, well-rounded and siliceous. It is placed in the annulus of the well between the borehole wall and the well screen to prevent formation material from entering the screen.

Filtration. The process of separating suspended solids from their liquid by forcing the latter through a porous medium. Two types of fluid filtration occur in a borehole during the drilling process: dynamic filtration when circulating, and static filtration when at rest.

Fish. Any object lost in the borehole.

Flocculation. The agglomeration of finely divided suspended solids into larger, usually gelatinous, particles; the development of a "floc" after treatment with a coagulant by gentle stirring or mixing.

Floodplain. The surface or strip of relatively smooth land adjacent to a river channel, constructed by the present river and covered with water when the river overflows its banks. It is built of alluvium carried by the river during floods and deposited in the sluggish water beyond the influence of the swiftest current.

Flow lines. Lines indicating the direction followed by groundwater toward points of discharge. Flow lines are perpendicular to equipotential lines.

Foaming agent. See Surfactant.

Formation stabilizer. A sand or gravel placed in the annulus of the well between the borehole wall and the well screen to provide temporary or long-term support for the borehole.

Fouling. The process in which undesirable foreign matter accumulates in a bed of filter media or ion exchanger, clogging pores and coating surfaces and thus inhibiting or retarding the proper operation of the bed.

Gel. A state of a colloidal suspension in which shearing stresses below a certain finite value fail to produce permanent deformation. Gels commonly occur when the dispersed colloidal particles have a great affinity for the base fluid.

Gel strength. The minimum shearing stresses that will produce permanent deformation of a colloidal suspension.

Glacial drift. A general term for unconsolidated sediment transported by glaciers and deposited directly on land or in the sea.

Glaciofluvial. Pertaining to the meltwater streams flowing from wasting glacier ice and especially to the deposits and landforms produced by such streams.

Graded. An engineering term pertaining to a soil or an unconsolidated sediment consisting of particles of several or many sizes or having a uniform or equable distribution of particles from coarse to fine.

Grain per gallon (gpg). A common basis for reporting water analyses in the water-treatment industry in the United States and Canada. One grain per U.S. gallon equals 17.12 milligrams per liter.

Groundwater table. The surface between the zone of saturation and the zone of aeration; the surface of an unconfined aquifer.

Grout. A fluid mixture of cement and water (neat cement) of a consistency that can be forced through a pipe and placed as required. Various additives, such as sand, bentonite, and hydrated lime, may be included in the mixture to meet certain requirements. Bentonite and water are sometimes used for grout.

Grouting. The operation by which grout is placed between the casing and the sides of the well bore to a predetermined height above the bottom of the well. This secures the casing in place and excludes water and other fluids from the well bore.

Hardness. A property of water causing formation of an insoluble residue when the water is used with soap. It is primarily caused by calcium and magnesium ions.

Head. Energy contained in a water mass, produced by elevation, pressure, or velocity.

Head loss. That part of head energy which is lost because of friction as water flows.

Heterogeneous. Nonuniform in structure or composition throughout.

Homogeneous. Uniform in structure or composition throughout.

Hydration. The act by which a substance takes up water by absorption and/or adsorption.

Hydraulic conductivity. The rate of flow of water in gallons per day through a cross section of one square foot under a unit hydraulic gradient, at the prevailing temperature (gpd/ft^2). In the SI System, the units are m^3/day/m^2 or m/day.

Hydraulic gradient. The rate of change in total head per unit of distance of flow in a given direction.

Hydrogeologic. Those factors that deal with subsurface waters and related geologic aspects of surface waters.

Hydrosphere. All waters of the Earth, as distinguished from the rocks (lithosphere), living things (biosphere), and the air (atmosphere).

Igneous rocks. Rocks that solidified from molten or partly molten material, that is, from a magma.

Incrustation. The process by which a crust or coating is formed.

Interference. The condition occurring when the area of influence of a water well comes into contact with or overlaps that of a neighboring well, as when two wells are pumping from the same aquifer or are located near each other.

Intrusive rocks. Those igneous rocks formed from magma injected beneath the Earth's surface. Generally these rocks have large crystals caused by slow cooling.

Ion. An element or compound that has gained or lost an electron, so that it is no longer neutral electrically, but carries a charge.

Isotropic. Said of a medium whose properties are the same in all directions.

Karst topography. A type of topography that is formed on limestone, gypsum, and other rocks by dissolution, and is characterized by sinkholes, caves, and underground drainage.

Kelly. Hollow steel bar that is the main section of drill string to which the power is directly transmitted from the rotary table to rotate the drill pipe and bit.

Laminar flow. Water flow in which the stream lines remain distinct and in which the flow direction at every point remains unchanged with time. It is characteristic of the movement of groundwater.

Landfill. A general term indicating a disposal site of refuse, dirt from excavations, and junk.

Leachate. The liquid that has percolated through solid waste and dissolved soluble components.

Limestone. A sedimentary rock consisting chiefly of calcium carbonate, primarily in the form of the mineral calcite.

Lithosphere. The solid portion of the Earth, as compared with the atmosphere and the hydrosphere. It includes the crust and part of the upper mantle and is about 62 mi (100 km) thick.

Lost circulation. The result of drilling fluid escaping from the borehole into the formation by way of crevices or porous media.

Marsh funnel viscosity. Commonly called the funnel viscosity. The Marsh funnel viscosity is reported as the number of seconds required for 1 qt (946 mℓ) of a given fluid to flow through the Marsh funnel.

Metamorphic rocks. Any rock derived from pre-existing rocks by mineralogical, chemical, and/or structural changes, essentially in the solid state, in response to marked changes in temperature, pressure, shearing stress, and chemical environment, generally at depth in the Earth's crust.

Moho. The boundary surface or sharp seismic-velocity discontinuity that separates the Earth's crust from the underlying mantle.

Molecular diffusion. Dispersion of a chemical caused by the kinetic activity of the ionic or molecular constituents.

Molecule. A stable configuration of atomic nuclei and electrons bound together by electrostatic and electromagnetic forces. It is the simplest structural unit that displays the characteristic physical and chemical properties of a compound.

Moraine. A mound, ridge, or other distinct accumulation of unsorted, unstratified glacial drift, predominantly till, deposited chiefly by direct action of glacier ice.

Naturally developed well. A well in which the screen is placed in direct contact with the aquifer materials; no filter pack is used.

Nominal. Used to describe standard sizes for pipe from $\frac{1}{8}$ in to 12 in (3.2 mm to 305 mm). The nominal size is specified on the basis of the inside diameter. Depending on the wall thickness, the inside diameter may be less than or greater than the number indicated.

Nongraded. An engineering term pertaining to a soil or an unconsolidated sediment consisting of particles of essentially the same size.

Observation well. A well drilled in a selected location for the purpose of observing parameters such as water levels and pressure changes.

Outwash. Stratified sand and gravel removed or washed out from a glacier by meltwater streams and deposited in front of or beyond the end moraine or the margin of an active glacier. The coarser material is deposited nearer to the ice.

Outwash plain. A broad, gently sloping sheet of outwash.

Overburden. The loose soil, silt, sand, gravel, or other unconsolidated material overlying bedrock, either transported or formed in place; regolith.

Oxidation. The combining of an element with oxygen.

Partial penetration. When the intake portion of the well is less than the full thickness of the aquifer.

Pathogenic. Capable of causing disease.

Perched water. Unconfined groundwater separated from an underlying main body of groundwater by an unsaturated zone.

Percolate. The act of water seeping or filtering through the soil without a definite channel.

Permeability. The property or capacity of a porous rock, sediment, or soil for transmitting a fluid; it is a measure of the relative ease of fluid flow under unequal pressure.

pH. A measure of the acidity or alkalinity of a solution, numerically equal to 7 for neutral solutions, increasing with increasing alkalinity and decreasing with increasing acidity. Originally stood for the words potential of hydrogen.

Plate tectonics. A theory of global tectonics in which the lithosphere is divided into a number of plates whose pattern of horizontal movement is that of torsionally rigid bodies that interact with one another at their boundaries, causing seismic and tectonic activity along these boundaries.

Pollution. When the contamination concentration levels restrict the potential use of groundwater.

Polymer. A substance formed by the union of two or more molecules of the same kind linked end to end into another compound having the same elements in the same porportion but a higher molecular weight and different physical properties.

Porosity. The percentage of the bulk volume of a rock or soil that is occupied by interstices, whether isolated or connected.

Potentiometric surface. An imaginary surface representing the total head of groundwater in a confined aquifer that is defined by the level to which water will rise in a well.

Pumping test. A test that is conducted to determine aquifer or well characteristics.

Quick condition. A condition of soil in which an increase in pore-water pressure decreases particle-to-particle attraction and reduces significantly the soil's bearing capacity.

Radius of influence. The radial distance from the center of a well bore to the point where there is no lowering of the water table or potentiometric surface (the edge of its cone of depression).

Recharge. The addition of water to the zone of saturation; also, the amount of water added.

Redox. A chemical reaction in which an atom or molecule loses electrons to another atom or molecule. Also called oxidation-reduction. Oxidation is the loss of electrons; reduction is the gain in electrons.

Regolith. A general term for the layer of fragmental and unconsolidated rock material that nearly everywhere forms the surface of the land and overlies or covers the bedrock.

Residual drawdown. The difference between the original static water level and the depth to water at a given instant during the recovery period.

Runoff. That part of precipitation flowing to surface streams.

Sandstone. A sedimentary rock composed of abundant rounded or angular fragments of sand set in a fine-grained matrix (silt or clay) and more or less firmly united by a cementing material.

Sedimentary rocks. Rocks resulting from the consolidation of loose sediment that has accumulated in layers.

Shale. A fine-grained sedimentary rock, formed by the consolidation of clay, silt, or mud. It is characterized by finely laminated structure and is sufficiently indurated so that it will not fall apart on wetting.

Shear stress. That component of stress which acts tangential to a plane through any given point in a body.

Sieve analysis. Determination of the particle-size distribution of a soil, sediment, or rock by measuring the percentage of the particles that will pass through standard sieves of various sizes.

Slurry. A thin mixture of liquid, especially water, and any of several finely divided substances, such as cement or clay particles.

Specific capacity. The rate of discharge of a water well per unit of drawdown, commonly expressed in gpm/ft or m^3/day/m. It varies with duration of discharge.

Specific gravity. The weight of a particular volume of any substance compared to the weight of an equal volume of water at a reference temperature.

Specific retention. The ratio of the volume of water that a given body of rock or soil will hold against the pull of gravity to the volume of the body itself. It is usually expressed as a percentage.

Specific yield. The ratio of the volume of water that a given mass of saturated rock or soil will yield by gravity to the volume of that mass. This ratio is stated as a percentage.

Static water level. The level of water in a well that is not being affected by withdrawal of groundwater.

Storage coefficient. See Coefficient of storage.

Storativity. See Coefficient of storage.

Stratigraphy. The study of rock strata, especially of their distribution, deposition, and age.

Surfactant. A substance capable of reducing the surface tension of a liquid in which it is dissolved. Used in air-based drilling fluids to produce foam, and during well development to disaggregate clays.

Tensile strength. The resistance of a material to a force tending to tear it apart.

Till. Predominantly unsorted and unstratified drift, generally unconsolidated, deposited directly by and underneath a glacier without subsequent reworking by meltwater, and consisting of a heterogeneous mixture of clay, silt, sand, gravel, and boulders ranging widely in size and shape.

Tortuosity. Sinuosity of the actual flow path in porous medium; it is the ratio of the length of the flow path divided by the length of the sample.

Total dissolved solids, TDS. A term that expresses the quantity of dissolved material in a sample of water, either the residue on evaporation, dried at 356°F (180°C), or, for many waters that contain more than about 1,000 mg/ℓ, the sum of the chemical constituents.

Transmissibility. See Transmissivity.

Transmissivity. The rate at which water is transmitted through a unit width of an aquifer under a unit hydraulic gradient. Transmissivity values are given in gallons per minute through a vertical section of an aquifer one foot wide and extending the full saturated height of an aquifer under a hydraulic gradient of 1 in the English Engineering system; in the International System, transmissivity is given in cubic meters per day through a vertical section of an aquifer one meter wide and extending the full saturated height of an aquifer under a hydraulic gradient of 1.

Transpiration. The process by which water absorbed by plants, usually through the roots, is evaporated into the atmosphere from the plant surface.

Turbulent flow. Water flow in which the flow lines are confused and heterogeneously mixed. It is typical of flow in surface-water bodies.

Unconfined aquifer. An aquifer where the water table is exposed to the atmosphere through openings in the overlying materials.

Uniformity coefficient. A numerical expression of the variety in particle sizes in mixed natural soils, defined as the ratio of the sieve size on which 40 percent (by weight) of the material is retained to the sieve size on which 90 percent of the material is retained.

Vadoze zone. The zone containing water under pressure less than that of the atmosphere, including soil water, intermediate vadose water, and capillary water. This zone is limited above by the land surface and below by the surface of the zone of saturation, that is, the water table.

Viscosity. The property of a substance to offer internal resistance to flow. Specifically, the ratio of the shear stress to the rate of shear strain.

Water table. The surface between the vadose zone and the groundwater; that surface of a body of unconfined groundwater at which the pressure is equal to that of the atmosphere.

Weathering. The in-situ physical disintegration and chemical decomposition of rock materials at or near the Earth's surface.

Well point. A screening device, equipped with a point on one end, that is meant to be driven into the ground.

Well screen. A filtering device used to keep sediment from entering a water well.

Well yield. The volume of water discharged from a well in gallons per minute or cubic meters per day.

Yield strength. The tensile stress required to produce a total elongation of 0.5 percent of the original length as determined by an extensiometer. Expressed in psi.

Appendices

Appendices

3.A	Geologic time scale	898
6.A	Chemical conversion factors	899
6.B	Calculation of adjusted sodium adsorption ratio	900
7.A	Sources of groundwater information in Canada	903
7.B	Sources of groundwater information in Mexico	903
8.A	Publications containing articles on groundwater	904
8.B	Additional references on borehole geophysical methods	905
9.A	Conversion tables	909
9.B-1	Symbols list	914
9.B-2	Abbreviations list	915
9.C	Correction of measured drawdown data for determining true transmissivities for partially dewatered unconfined aquifers	918
9.D	Mathematical development of Equation 9.15	920
9.E	Values of $W(u)$ corresponding to values of u for Theis nonequilibrium equation	921
9.F	Available analytical models simulating flow to wells	924
10.A-1	Dimensions and weights for standard cable tool drill bits	927
10.A-2	Dimensions and weights for standard cable tool drill bits (S.I.)	928
10.A-3	Sizes and weights for tricone roller bits	929
10.B-1	Cement classifications used outside the United States	930
10.B-2	Equivalent cement classifications outside the United States	930
10.C	Recommended rotating speeds for all sizes and types of bits in various formations	931
10.D	Weight on bit and rotary speed	931
10.E	API drill pipe list	932
10.F	Upset ends	933
10.G	Kelly weights in pounds per foot (drive section)	933

APPENDICES

10.H	Ideal size range for drill collars	934
10.I	Drill collar weights in pounds per foot	935
10.J	Correlation between percent of bentonite, volume of water, density, and volume of slurry	936
10.K	Equations used to calculate slurry density	936
11.A	Density and specific gravity of water and three common drilling fluid additives	937
11.B	Procedure for measuring the density of drilling fluid	937
11.C	Drilling fluid weight conversion table	938
11.D	Viscosity measurements using a Marsh funnel	940
11.E	Design specifications for two types of mud mixers	941
11.F	Determining uphole annular velocity using air	942
11.G	Comparable drilling fluid products by tradenames	943
11.H	Major drilling fluid companies	944
11.I	Velocities of water that will cause sand to rise	944
11.J	Ascending velocity of drilling fluid (in ft/min)	945
11.K	Weight, fluid pressure, and buoyancy factors	946
11.L	Volume of water in casing or hole	947
12.A	Representative open areas of screens	948
12.B	Correlation chart of screen openings and sieve sizes	951
13.A	Making pipe and tubing	952
13.B-1	Dimensions, weights, and test pressures for plain-end standard pipe	954
13.B-2	Dimensions, weights, and test pressures for threaded-and-coupled standard pipe	960
13.B-3	Plain-end line pipe dimensions, weights, and test pressures	961
13.B-4	Standard-weight threaded line pipe dimensions, weights, and test pressures	971
13.B-5	ANSI pipe schedules with full metric conversion	972
13.B-6	Pipe data — carbon and alloy steel — stainless steel	976
13.B-7	Short round-thread casing dimensions and weights	983
13.B-8	Water well casing	985
13.C	Chemical and physical standards for thermoplastic material	986
13.D	Dimensions and weights of PVC plastic pipe	987
13.E	Class requirements for rigid poly (vinyl chloride) compounds	990
13.F	Prior designation of PVC pipe material	991
13.G	Recommended procedures for field welding stainless steel well screens	992
13.H	Preparation of pipe surfaces and cement process for plastic pipe	994
13.I	Discussion of appropriate screen entrance velocities	996
13.J	Calculating the Ryznar Stability Index	1007
13.K	Nomogram for determining flow, diameter, head loss, or velocity in black steel pipe	1009
13.L	Head loss characteristics of water flow through rigid plastic pipe	1010
13.M	Thermoplastic water well casing pipe couplings socket dimensions and laying length dimensions	1011

13.N	Tapered sockets for bell-end pipe, in	1012
13.O	Friction loss in feet per 100 feet in asbestos cement class 150-pressure pipe	1013
15.A	Discharge of air through an orifice	1014
16.A-1	Discharge from rectangular weir with end contractions	1015
16.A-2	Discharge from triangular notch weir with end contractions	1017
16.B	Free-flow discharge — Parshall flume, cfs	1018
16.C-1	Free flow discharges in CFS for selected sizes of cutthroat flumes	1019
16.C-2	Submerged flow discharges in CFS for 4 in x 3 ft cutthroat flume	1019
16.C-3	Submerged flow discharges in CFS for 8 in x 3 ft cutthroat flume	1020
16.C-4	Submerged flow discharges in CFS for 12 in x 3 ft cutthroat flume	1020
16.C-5	Submerged flow discharges in CFS for 8 in x 6 ft cutthroat flume	1020
16.D	Empirical equation used to estimate specific capacity and transmissivity	1021
16.E	Estimated flows from pipes	1022
16.F	Flow from vertical pipes	1024
17.A	Friction losses in pipe	1025
17.B	Volume of water discharged per stroke by a reciprocating pump	1037
17.C	Power required for pumping	1038
18.A	State agencies that provide water-sampling services	1040
18.B	Abandonment of test holes, partially completed wells, and completed wells	1044
19.A	Iron oxidizing bacteria	1045
19.B	Malfunctions in high-capacity centrifugal and turbine pumps, possible causes, and solutions	1046
20.A-1	Notice to bidders	1050
20.A-2	Proposal	1052
20.A-3	Standard form of agreement between owner and contractor	1054
20.A-4	Well estimate and/or well drilling contract or repair order agreement	1056
20.B	Portion of tightly written specification	1058
21.A	Typical geologic settings that demonstrate the interaction between landfill leachates and local hydrogeology	1060
21.B	Types of protective clothing and respirators that should be used at hazardous waste sites	1062
21.C	Chemical constituents of a major polymer	1064
22.A	Theoretical specific capacity for 100-percent efficient water wells, m^3/day/m of drawdown	1066

23.A	EPA's priority toxic pollutants	1067
23.B	Common water-quality problems encountered in domestic water systems	1069
23.C	Performance standards for various point-of-use treatment techniques	1072

APPENDIX 3.A.
Geologic Time Scale

Eon	Era	Period		Epoch	Time (Millions of years)
Phanerozoic	Cenozoic	Quaternary		Holocene	0.01
				Pleistocene	1.6
		Tertiary	Neogene	Pliocene	5.3
				Miocene	
					23.7
			Palaeogene	Oligocene	
					36.6
				Eocene	
					57.8
				Paleocene	
					66.4
	Mesozoic	Cretaceous		late Cretaceous	97.5
				early Cretaceous	
					144
		Jurassic		late Jurassic	163
				middle Jurassic	187
				early Jurassic	208
		Triassic			
					245
	Palaeozoic	Permian			
					286
		Carboniferous	Pennsylvanian		320
			Mississippian		360
		Devonian			
					408
		Silurian			
					438
		Ordovician			
					505
		Cambrian			
					570
Proterozoic		Precambrian			
Archaean					4,600

APPENDIX 6.A.
Chemical Conversion Factors

Constituents	mg/ℓ to epm	epm to mg/ℓ	gpg to epm	epm to gpg	mg/ℓ to mg/ℓ CaCo₃
Calcium	0.0499	20.04	0.853	1.172	2.497
Iron	0.0358	27.92	0.612	1.633	1.792
Magnesium	0.0822	12.16	1.406	0.711	4.115
Potassium	0.0256	39.10	0.437	2.286	1.280
Sodium	0.0435	23.00	0.743	1.345	2.176
Bicarbonate	0.0164	61.01	0.280	3.568	0.820
Carbonate	0.0333	30.00	0.570	1.754	1.668
Chloride	0.0282	35.46	0.482	2.074	1.411
Hydroxide	0.0588	17.01	1.005	0.995	2.926
Nitrate	0.0161	62.01	0.276	3.626	0.807
Phosphate	0.0316	31.67	0.540	1.852	1.530
Sulphate	0.0208	48.04	0.356	2.809	1.042
Calcium bicarbonate	0.0123	81.05	0.211	4.740	0.617
Calcium carbonate	0.0200	50.04	0.342	2.926	1.000
Calcium chloride	0.0180	55.50	0.308	3.246	0.902
Calcium hydroxide	0.0270	37.05	0.461	2.167	1.351
Calcium sulphate	0.0147	68.07	0.251	3.981	0.735
Ferrous bicarbonate	0.0112	88.93	0.192	5.201	0.563
Ferrous carbonate	0.0173	57.92	0.295	3.387	0.864
Ferrous sulphate	0.0132	75.96	0.225	4.442	0.659
Magnesium bicarbonate	0.0137	73.17	0.234	4.279	0.684
Magnesium carbonate	0.0237	42.16	0.406	2.465	1.187
Magnesium chloride	0.0210	47.62	0.359	2.785	1.051
Magnesium hydroxide	0.0343	29.17	0.586	1.706	1.715
Magnesium sulphate	0.0166	60.20	0.284	3.520	0.631
Potassium chloride	0.0134	74.56	0.229	4.360	0.671
Sodium bicarbonate	0.0119	84.01	0.203	4.913	0.596
Sodium carbonate	0.0189	53.00	0.323	3.099	0.944
Sodium chloride	0.0171	58.46	0.292	3.419	0.856
Sodium hydroxide	0.0245	40.01	0.427	2.340	1.251
Sodium nitrate	0.0118	85.01	0.201	4.971	0.589
Sodium phosphate	0.0183	54.67	0.313	3.197	0.915
Sodium sulphate	0.0141	71.04	0.241	4.154	0.704

epm = equivalent parts per million
gpg = grains per gallon = 17.1 mg/ℓ
mg/ℓ = milligrams per liter
mg/ℓ CaCo₃ = milligrams per liter of CaCo₃

APPENDIX 6.B.
Calculation of Adjusted Sodium Adsorption Ratio

The adjusted Sodium Adsorption Ratio (adj. SAR) is calculated from the following equation:

$$\text{adj. SAR} = \sqrt{\frac{Na}{\frac{Ca + Mg}{2}}} \, [1 + (8.4 - pHc)] \quad (1)$$

where Na, Ca, and Mg are in meq/ℓ from the water analysis. A nomogram for determining $Na/\sqrt{\frac{Ca + Mg}{2}}$ is presented in Figure 1. The value for pHc* is calculated using the following equation:

$$pHc = (pK'_2 - pK'_c) + p(Ca + Mg) + p(Alk) \quad (2)$$

where
($pK'_2 - pK'_c$) is obtained from Table 1 using the sum of Ca + Mg in meq/ℓ
p(Ca + Mg) is obtained from Table 1 using the sum of Ca + Mg in meq/ℓ
p(Alk) is obtained from Table 1 using the sum of $CO_3 + HCO_3$ in meq/ℓ

An example of pHc calculation follows:

Given: Ca = 2.32 meq/ℓ
Mg = 1.44 meq/ℓ
Na = 7.73 meq/ℓ

Total = 11.49 meq/ℓ

CO_3 = 0.42 meq/ℓ
HCO_3 = 3.66 meq/ℓ

Total = 4.08 meq/ℓ

Using the equation for pHc and substituting the values from Table 1, pHc can be calculated:

$$pHc = 2.2 + 2.7 + 2.4 = 7.3$$

Substituting in Equation 1, the adjusted SAR can be calculated:

$$\text{adj. SAR} = \sqrt{\frac{7.73}{\frac{3.76}{2}}} \, [1 + (8.4 - 7.3)]$$

$$\text{adj. SAR} = 5.64 \cdot 2.1 = 11.8$$

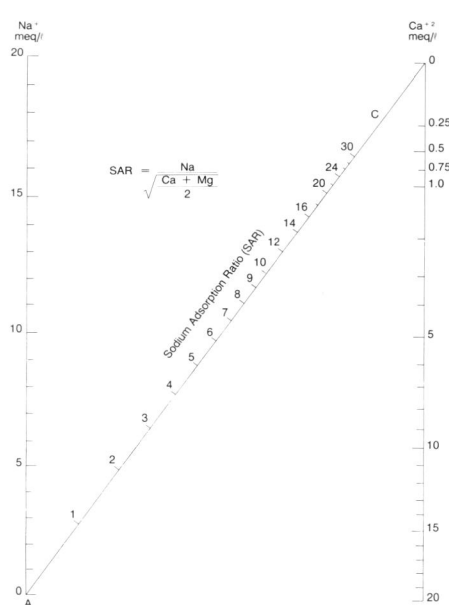

Figure 1. Nomogram for determining the SAR value of irrigation water and for estimating the corresponding ESP value of a soil that is at equilibrium with the water. *(U.S. Department of Agriculture, 1954)*

Figure 2 can be used to classify the soil on the basis of the SAR and the conductivity of the irrigation water. This classification system for evaluating irrigation

---

*pHc is a theoretical, calculated pH of the irrigation water in contact with lime and in equilibrium with soil CO_2.

Appendix 6.B. Continued

Table 1. Values for Three Terms used in Equation 2

Sum of Concentration (meq/ℓ)	$pK'_2 - pK'_c$	$p(Ca + Mg)$	$p(Alk)$
0.05	2.0	4.6	4.3
0.10	2.0	4.3	4.0
0.15	2.0	4.1	3.8
0.20	2.0	4.0	3.7
0.25	2.0	3.9	3.6
0.30	2.0	3.8	3.5
0.40	2.0	3.7	3.4
0.50	2.1	3.6	3.3
0.75	2.1	3.4	3.1
1.0	2.1	3.3	3.0
1.25	2.1	3.2	2.9
1.5	2.1	3.1	2.8
2.0	2.2	3.0	2.7
2.5	2.2	2.9	2.6
3.0	2.2	2.8	2.5
4.0	2.2	2.7	2.4
5.0	2.2	2.6	2.3
6.0	2.2	2.5	2.2
8.0	2.3	2.4	2.1
10.0	2.3	2.3	2.0
12.5	2.3	2.2	1.9
15.0	2.3	2.1	1.8
20.0	2.4	2.0	1.7
30.0	2.4	1.8	1.5
50.0	2.5	1.6	1.3
80.0	2.5	1.4	1.1

potential for particular soils or waters is only a guideline. Each irrigation design requires individual care in adapting the water to soil type and cropping practices.

NOTE: Values of pHc above 8.4 indicate a tendency to dissolve lime from the soil through which the water moves; values below 8.4 indicate a tendency to precipitate lime from the water applied.

(Rhoades, 1972)

Appendix 6.B. Continued

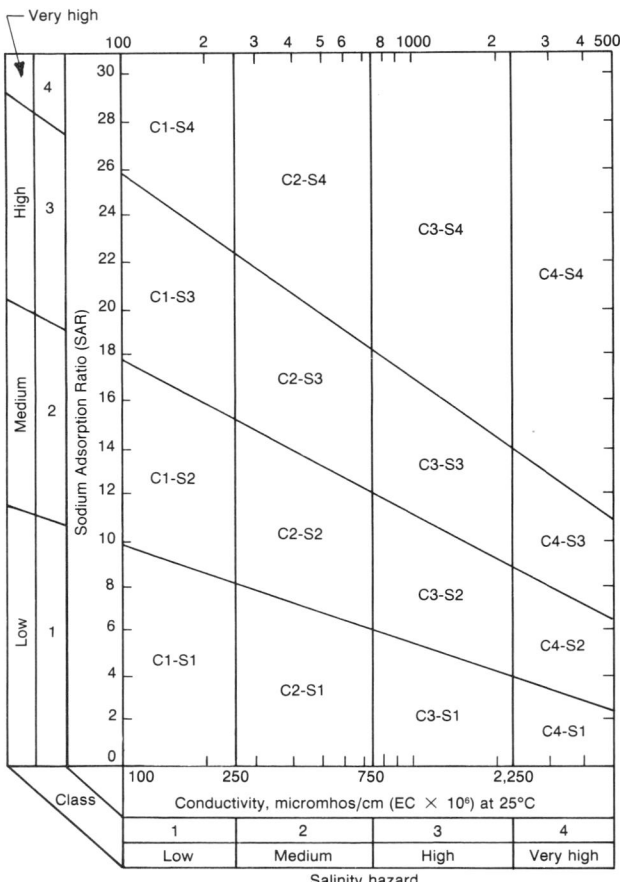

Figure 2. Classification of irrigation waters based on SAR and conductivity. Interpretation of quality-class ratings of water for irrigation purposes is as follows:
Conductivity (Salinity)
Low-salinity water (C1) can be used for most crops and soils with little likelihood that soil salinity will develop. Some leaching is required, but this occurs under normal irrigation on all but the tightest of soils.
Medium-salinity water (C2) can be used where a moderate amount of leaching occurs. Plants with moderate salt tolerance can be grown in most cases without special practices for salinity control.
High-salinity water (C3) cannot be used on soils that have restricted drainage. With adequate drainage, special management for salinity control may be required and plants with good salt tolerance should be selected.
Very-high-salinity water (C4) is not suitable for irrigation under ordinary conditions. If used, the soils must be permeable, drainage must be adequate, considerable excess irrigation water must be applied, and very tolerant crops should be selected.
Sodium Adsorption Ratio
Low-sodium water (S1) can be used with little danger on nearly all soils. Sodium-sensitive crops such as stone-fruit trees and avocados may accumulate injurious concentrations of sodium.
Medium-sodium water (S2) is hazardous for use on fine-textured soils that have high cation-exchange capacity. This water may be used on coarse-textured or organic soils with good permeability.
High-sodium water (S3) may be harmful to most soils and thus requires special soil management: good drainage, high leaching, and addition of organic matter. Chemical amendments may be necessary except for gypsiferous soils.
Very-high-sodium water (S4) is generally unsatisfactory for irrigation purposes, except at low salinity and where calcium from the soil or use of gypsum or other mineral additions may make these waters usable. *(U.S. Department of Agriculture)*

APPENDIX 7.A.
Sources of Groundwater Information in Canada

Alberta
Head of Hydrogeology Branch
Earth Sciences Division
Alberta Environment
14th Floor, Standard Life Centre
10405 Jasper Avenue
Edmonton, Alberta T5J 3N4

British Columbia
Director of Water Management Branch
Planning and Resources Management
  Division
Department of the Environment
Parliament Buildings
Victoria, B.C. V8V 1X5

Manitoba
Water Resources Branch
Department of Natural Resources
1577 Dublin Avenue
Winnipeg, Manitoba R3T 3J5

New Brunswick
Director of Water Resources Branch
Department of the Environment
P.O. Box 6000
Fredericton, N.B. E3B 5H1

Newfoundland
Director of Water Resources Division
Department of the Environment
P.O. Box 4750
St. Johns, Newfoundland A1C 5T7

Nova Scotia
Director of Environmental Assessment
  Division
Department of the Environment
P.O. Box 2107
Halifax, Nova Scota B3J 3B7

Ontario
Director, Water Resources Branch
Environmental Planning Division
Ontario Environment
135 St. Clair Avenue W.
Toronto, Ontario M4V 1P5

Prince Edward Island
Head of Water Resources
Technical Services Division
Department of Community and Cultural
  Affairs
P.O. Box 2000
Charlottetown, P.E.I. C1A 7H8

Quebec
Directeur des Eaux Souterraines
Direction Générale des Inventaires
  et de La Recherche Environement
194 Avenue Saint-Sacrement
Québec, Québec G1N 4J5

Saskatchewan
Manager
Saskatchewan Water Corporation
2121 Saskatchewan Drive
Regina, Saskatchewan S4P 3Y2

APPENDIX 7.B.
Sources of Groundwater Information in Mexico

SEDUE (Secretaria de Desarrollo
  Urbano y Ecologia)
Pizo 4, Vallerta 5
Zona Postal 06030
Mexico D.F., Mexico

Secretaria de Agricultura y Recursos
  Hidraulicos
Paseo Reforma 69
Mexico D.F., Mexico

APPENDIX 8.A.
Publications Containing Articles on Groundwater

The following publications contain articles on groundwater in every issue, several issues a year, or a few issues a year.

Every Issue

Canadian Water Well
Ground Water
Ground Water Age
Ground Water Micro News
Ground Water Monitoring Review
Journal American Water Works Association
Journal of Hydrology
Journal Water Pollution Control Federation
U.S. Water News
Water Resources Bulletin
Water Resources Research
Water Well Journal

Several times per year

Agricultural Engineering
Agua
Drill Bit
Drilling-DWC
Environmental Monitoring and Assessment
Environmental Science & Technology — ES & T
Geothermal Energy
Geothermal Resources Council Bulletin
Irrigation Age
Irrigation Journal
Journal of Soil and Water Conservation
Journal of the Environmental Engineering Division (ASCE)
Journal of the Hydraulics Division (ASCE)
Journal of the Irrigation and Drainage Division (ASCE)
Journal of the National Water Well Association of Australia
Journal of the New England Water Works Association
OPFlow
Soviet Hydrology
Water Engineering and Management
Water Research
Water Technology
Western Water

Few times per year

AAPG Bulletin
Arab Water World
Bottled Water Report
Drilling Contractor
Earthquake Information Bulletin
EPA Journal
EOS
Garber-Wellington Gazette
Geology
Geological Society of America Bulletin
Geotimes
Geophysics
Journal of Petroleum Technology — JPT
Oil and Gas Journal
Pollution Engineering Services
Texas Water Resources
Water & Waste Management
Water Conditioning
Water Engineering & Management
Water World News
World Oil
World Water

APPENDIX 8.B.
Additional References on Borehole Geophysical Methods

GENERAL REFERENCES

Baker, L. E., Campbell, A. B., and Hugen, R. L., 1975. Well-logging technology and geothermal applications: A survey assessment with recommendations. Sandia Laboratories Energy Report SAND 75-0275, Sandia Laboratories, Albuquerque, NM, 76 p.

Dresser Atlas Division, 1975. Log Interpretation Fundamentals. Dresser Industries, Inc., Houston, TX, 250 p.

Hamilton, R. G. and Myung, J. I., (undated). Summary of geophysical well logging. Seismograph Service Corporation Technical Bulletin, Birdwell Division, Tulsa, OK.

Johnson, H. M., 1962. A history of well logging. Geophysics, Vol. 27, pp 507-527.

Pirson, S. J., 1963. Handbook of Well Log Analysis (for Oil and Gas Formation Evaluation). Prentice-Hall, Inc., Englewood Cliffs, NJ, 326 p.

Schlumberger, C., Schlumberger, M., and Leonardon, E. G., 1934. A new contribution to subsurface studies by means of electrical measurements in drill holes. American Institute of Mining and Metalurgical Engineers Transactions, Vol. 110, p 273.

Schlumberger, 1972. Log Interpretation. Schlumberger Ltd., New York.

Sengel, E. W., 1981. Handbook on Well Logging. The Institute for Energy Development, Oklahoma City, OK. IED Press, Inc., Oklahoma City, OK.

Sherriff, R. E., 1970. Glossary of terms used in well logging. Geophysics, Vol. 35, pp 1116-1139.

Van Orstrand, C. E., 1918. Apparatus for the measurement of temperature in deep wells, and temperature determinations in some deep wells in Pennsylvania and West Virginia. West Virginia Geological Survey County Reports of Barbour and Upshur Counties, Wheeling, WV, West Virginia Geological Survey, pp LXVI-CIII.

Welex, (undated). Basic concepts of well log interpretation. Welex, Houston, TX, Publication L-33, 35 p.

RESISTIVITY REFERENCES

Alfano, L. A., 1962. Geoelectrical prospecting with underground electrodes. Geophysical Prospecting, Vol. 10, pp 290-303.

Atkins, E. R., 1961. Techniques of electric log interepretation. Journal of Petroleum Technology, Vol. 13, pp 118-123.

Bowsky, M. C., 1942. The effect of mud resistivities on the intensities of electrical logs. Geophysics, Vol. 7, pp 82-89.

Daniels, J. J., 1977. Three-dimensional resistivity and induced polarization modeling using buried electrodes. Geophysics, Vol. 42, No. 5, pp 1006-1019.

Daniels, J. J., 1977. Extending the range of investigation of borehole electrical measurements. Transactions of the SPWLA 18th Annual Logging Symposium, 17 p.

Daniels, J. J., 1978. Interpretation of buried electrode resistivity data using the layered earth model. Geophysics, Vol. 43, No. 5, pp 988-1001.

Guyod, H., 1952. Electrical Well Logging Fundamentals. Well Inst. Developing Co., Houston, TX.

Guyod, H., 1955. Electric analogue of resistivity logging. Geophysics, Vol. 20, pp 615-629.

Appendix 8.B. Continued

Jones, P. H. and Buford, T. B., 1951. Electric logging applied to ground-water exploration. Geophysics, Vol. 16, No. 1, pp 115-139.

Keller, G. V. and Frishknecht, F. C., 1966. Electrical Methods in Geophysical Prospecting. Pergamon Press, pp 61-89.

Kunz, K. S. and Moran, J. H., 1958. Some effects of formation anisotropy on resistivity measurements in boreholes. Geophysics, Vol. 23, No. 4, pp 770-794.

Lang, W. H., 1972. Porosity-resistivity cross-plotting. Transactions of the 13th Annual Logging Symposium of the SPWLA, May 7-10, Paper F.

Perkins, F. M., Osoba, J. S., and Ribe, K. H., 1956. Resistivity of sandstones related to the geometry of their interstitial water. Geophysics, Vol. 21, No. 4, pp 1071-1086.

Roy, A. and Dhar, R. L., 1971. Radius of investigation in DC resistivity well logging. Geophysics, Vol. 36, No. 4, pp 754-760.

Scott, J. H. and Farstad, A. J., 1977. Electrical resistivity well-logging system with solid state electronic circuitry. U.S. Geological Survey Open-File Report 77-144.

Scott, J. H., 1978. A FORTRAN alogorithm for correcting normal resistivity logs for borehole diameter and mud resistivity. U.S. Geological Survey Open-File Report.

Snyder, D. D. and Merkel, R. M., 1973. Analytical models for the interpretation of electrical surveys using buried current electrodes. Geophysics, Vol. 38, No. 3, pp 513-529.

SP AND RESISTANCE

Columbo, U., Salimbeni, G., Sironi, G., and Veneziani, I., 1959. Differential electric log. Geophysical Prospecting, Vol. 7, pp 91-118.

Dickey, P. A., 1944. Natural potentials in sedimentary rocks. Transactions of the AIME, Vol. 155, p 39.

Doll, H. G., 1949. The SP log: theoretical analysis and principles of interpretation. Transactions of AIME, Vol. 179, p 146.

Frimpter, M. H., 1969. Casing detector and self-potential logger. Ground Water, Vol. 7, No. 6, p 24.

Goudouin, M., Tixier, M. P., and Simard, G. L., 1957. An experimental study on the influence of the chemical composition of electrolytes on the SP curve. Transactions of the AIME, Vol. 210, p 58.

Goudouin, M. and Scala, C., 1958. Streaming potential and the SP log. Transactions of the SIME, Vol. 213, p 170.

Guyod, H., 1964. An investigation of the factors affecting the SP is soft formations. Transactions of the SPWLA Logging Symposium, Paper A.

Hallenburg, J. K., 1971. A resume of spontaneous potential measurements. Transactions of the SPWLA 12th Annual Logging Symposium, 16 p.

Hill, H. J. and Anderson, A. E., 1959. Streaming potential phenomena in SP log interpretation. Petroleum Transactions, Vol. 216, p 203.

Pirson, S. J. and Wong, F. S., 1972. The neglected SP curve. Transactions of the SPWLA 13th Annual Logging Symposium, Paper C.

Schlumberger, C., Schlumberger, M., and Leonardon, E. G., 1934. A new contribution to subsurface studies by means of electrical measurements in drill holes. AIME Transactions, Vol. 110, p 273.

Appendix 8.B. Continued

Vonhof, J . A., 1966. Water quality determination from spontaneous potential log curves. Journal of Hydrology, No. 4, pp 341-347.

GAMMA RAY

Caldwell, R. L., Baldwin, W. F., Bargainer, J. D., Berry, J. E., Salita, G. N., and Sloan, R. W., 1963. Gramma-ray spectroscopy in well logging. Geophysics, Vol. 28, No. 4, pp 617-632.

Conaway, J. G. and Bristow, Q., 1981. Pitfalls in quantitative gamma ray logging. Calibration sleeves and Am-241 temperature stabilization, Transactions of the 22nd Annual Logging Symposium of the SPWLA, Vol. 1, Paper X.

Czubek, J. A., 1976. Recent advances in gamma ray log interpretation. Working paper presented at International AEC Advisory Group Meeting on Evaluation of Uranium Resources, Rome, Italy, Nov. 29-Dec. 3.

Hallenberg, J. K., 1973. Interpretation of gamma ray logs. SPWLA 14th Annual Logging Symposium Transactions, Paper G.

Killeen, P. G. and Bristow, Q., 1976. Uranium exploration by borehole gamma ray spectrometry using off-the-shelf instrumentation. IAEA/NEA International Symposium on Exploration for Uranium Ore Deposits, Paper IAEA-SM-208/4, pp 393-413.

Kokesh, F. P., 1951. Gamma ray logging. Oil & Gas Journal, July 26, 7 p.

Lock, G. A. and Hoyer, W. A., 1971. Natural gamma ray spectral logging. The Log Analyst, Vol. 12, No. 5, pp 3-9.

Lovborg, L., Kirkegaard, P., and Christiansen, E. M., 1976. The design of NaI(Tl) scintillation detectors for use in gamma ray surveys of geological sources. IAEA/NEA International Symposium on Exploration for Uranium Ore Deposits, Paper IAEA-SM-208/21, pp 127-148.

Moxham, R. M., Foote, R. S., and Bunker, C. M., 1965. Gamma ray spectrometer studies of hydrothermally altered rocks. Economic Geology, Vol. 60, No. 4, pp 653-671.

Nargolwalla, S. S., 1973. Nuclear technique for borehole logging of geologic materials. Scintrex Ltd., Applications Brief 73-1, 222 Snidercroft Road, Concord, Ontario, Canada.

Scott, J. H., Dodd, P. H., Droullard, R. F., and Mudra, P. J., 1961. Quantitative interpretation of gamma ray logs. Geophysics, Vol. 26, No. 2, pp 182-191.

Scott, J. H., 1980. Pitfalls in determining the dead time of nuclear well logging probes. SPWLA Transactions of the 21st Annual Logging Symposium.

GAMMA-GAMMA

Alger, R. P., Raymer, L. L., Hoyle, W. R., and Tixier, M. P., 1963. Formation density log applications in liquid-filled holes. Journal of Petroleum Technology of AIME.

Baker, P. E., 1957. Density logging with gamma rays. Transactions of AIME, Vol. 210, p 289.

Pickell, J. J. and Heacock, J. G., 1960. Density logging. Geophysics, Vol. 25, No. 4, p 891.

Snodgrass, J. J., 1976. Calibration models for geophysical borehole logging. Bureau of Mines Report of Investigations, RI 8148, 21 p.

Appendix 8.B. Continued

Scott, J. H., 1979. Borehole compensation algorithms for a small diameter, dual-detector density well logging probe. Transactions of the 20th Annual Logging Symposium of the SPWLA.

Tittman, J. R. and Wahl, J. S., 1965. The physical foundations of formation density logging. Geophysics, Vol. 30, No. 2, pp 284-294.

NEUTRON LOGS

Bivens, H. M., Smith, G. W., Jensen, D. H., Jacobs, E. L., and Rice, L. G., 1976. Pulsed neutron uranium borehole logging with epithermal neutron dieaway. Symposium on Exploration of Uranium Ore Deposits, IAEA, Paper SM/208-48.

Czubek, J. A., 1969. Review paper — neutron methods in geophysics. IAEA Symposium on Nuclear Techniques and Mineral Resources. Proceedings of a Symposium, Buenos Aires, Paper SM-112/5.

Ford, Bacon, and Davies Utah Inc., 1977. Borehole logging with neutron activation — a laboratory assessment. U.S. Bureau of Mines Report No. PB-273-454.

IRT Corporation, 1976. Future research in borehole assaying technology: technology assessment of borehole logging techniques, Vol. 1. Final Report on Contract JO255018, U.S. Bureau of Mines.

Peatross, R. F., 1976. A new lithology compensated capture gamma ray system. Transactions of the SPWLA 17th Annual Logging Symposium, June 9-12.

Senftle, F. E., Moxham, R. M., Tanner, A. B., Boynton, P. W., and Wager, R. E., 1977. Importance of neutron energy distribution in borehole activation analysis in relatively dry, low-porosity rocks. Geoexploration, Vol. 15, pp 121-135.

BOREHOLE EM

Barnett, C. T., Davidson, M. J., McLaughlin, G. H., and Nabighian, M. N., 1978. Exploration with the Newmont EMP system. Geophysical Prospecting, Vol. 26, p 686.

Dyck, A. V., 1975a. Borehole logging (electrical) *in* Report of Activities. Geological Survey of Canada, Paper 75-1a, p 81.

Dyck, A. V., 1975b. Electrical borehole methods applied to mineral prospecting. Geological Survey of Canada, Paper 75-31, pp 13-19.

Macnae, J. C., 1981. Geophysical prospecting with electric fields from an inductive EM source. Ph.D. Thesis, University of Toronto.

McNeill, J. D., 1980. Applications of transient electromagnetic techniques. Geonics Ltd., Technical Note TN-7.

Naokes, J. E., 1951. An electromagnetic method of geophysical prospecting for application to drill holes. Ph.D. Thesis, University of Toronto.

Worthington, M. H., Kuckes, A., and Oristaglio, M., 1981. A borehole induction procedure for investigating electrical conductivity within the broad vicinity of a hole. Geophysics, Vol. 46, pp 65-67.

APPENDIX 9.A.
Conversion Tables

Length

Unit	Equivalent[1,2]					
	millimeters	inches	feet	meters	kilometers	miles
millimeters	1	3.937×10^{-2}	3.281×10^{-3}	1×10^{-3}	1×10^{-6}	6.214×10^{-7}
inches	25.4	1	8.33×10^{-2}	2.54×10^{-2}	2.54×10^{-5}	1.578×10^{-5}
feet	304.8	12	1	0.3048	3.048×10^{-4}	1.894×10^{-4}
meters	1,000	39.37	3.281	1	1×10^{-3}	6.214×10^{-4}
kilometers	1×10^{6}	3.937×10^{4}	3,281	1,000	1	0.6214
miles	1.609×10^{6}	6.336×10^{4}	5,280	1,609	1.609	1

Area

Unit	Equivalent[1,2]						
	square inches	square feet	square meters	acres	hectares	square kilometers	square miles
square inches	1	6.944×10^{-3}	6.452×10^{-4}	1.594×10^{-8}	6.452×10^{-8}	6.452×10^{-10}	2.491×10^{-10}
square feet	144	1	9.29×10^{-2}	2.296×10^{-5}	9.29×10^{-9}	9.29×10^{-8}	3.587×10^{-8}
square meters	1,550	10.76	1	2.471×10^{-4}	1×10^{-4}	1×10^{-6}	3.861×10^{-7}
acres	6.273×10^{6}	4.356×10^{4}	4,047	1	0.4047	4.047×10^{-3}	1.563×10^{-3}
hectares	1.55×10^{7}	1.076×10^{5}	1×10^{4}	2.471	1	0.01	3.861×10^{-3}
square kilometers	1.55×10^{9}	1.076×10^{7}	1×10^{6}	247.1	100	1	0.386
square miles	4.014×10^{9}	2.788×10^{7}	2.59×10^{6}	640	259	2.59	1

Volume

Unit	Equivalent[1,2]						
	cubic inches	liters	gallons	cubic feet	cubic yards	cubic meters	acre-ft
cubic inches	1	1.639×10^{-2}	4.329×10^{-3}	5.787×10^{-4}	2.143×10^{-5}	1.639×10^{-5}	1.329×10^{-8}
liters	61.02	1	0.2642	3.531×10^{-2}	1.308×10^{-3}	0.001	8.106×10^{-7}
gallons	231.0	3.785	1	0.1337	4.951×10^{-3}	3.785×10^{-3}	3.068×10^{-6}
cubic feet	1,728	28.32	7.481	1	3.704×10^{-2}	2.832×10^{-3}	2.296×10^{-5}
cubic yards	4.666×10^{4}	764.6	202.0	27	1	0.7646	6.198×10^{-4}
cubic meters	6.102×10^{4}	1,000	264.2	35.31	1.308	1	8.106×10^{-4}
acre-ft	7.527×10^{7}	1.233×10^{6}	3.259×10^{5}	4.356×10^{4}	1,613	1,233	1

Discharge (flow rate, volume/time)

Unit	Equivalent[1,2]				
	gallons per minute	liters per second	acre-feet per day	cubic feet per second	cubic meters per day
gallons per minute	1	6.309×10^{-2}	4.419×10^{-3}	2.228×10^{-3}	5.45
liters per second	15.85	1	7.005×10^{-2}	3.531×10^{-2}	86.4
acre-feet per day	226.3	14.28	1	0.5042	1,234
cubic feet per second	448.8	28.32	1.983	1	2,447
cubic meters per day	1.369×10^{9}	8.64×10^{7}	6.051×10^{6}	3.051×10^{6}	1

Appendix 9.A. Continued
Velocity

Unit	Equivalent[1,2]				
	feet per day	kilometers per hour	feet per second	miles per hour	meters per second
feet per day	1	1.27×10^{-5}	1.157×10^{-5}	7.891×10^{-6}	3.528×10^{-6}
kilometers per hour	7.874×10^4	1	0.9113	0.6214	0.2778
feet per second	8.64×10^4	1.097	1	0.6818	0.3048
miles per hour	1.267×10^5	1.609	1.467	1	0.447
meters per second	2.835×10^5	3.6	3.281	2.237	1

Mass

Unit	Equivalent[1,2]							
	ounce	pound	kilogram	metric slug	slug	short ton	metric ton	long ton
ounce	1	6.25×10^{-2}	2.835×10^{-2}	2.891×10^{-3}	1.943×10^{-3}	3.125×10^{-5}	2.835×10^{-5}	2.79×10^{-5}
pound	16	1	0.4536	4.625×10^{-2}	3.108×10^{-2}	5×10^{-4}	4.536×10^{-4}	4.464×10^{-4}
kilogram	35.28	2.205	1	0.102	6.852×10^{-2}	1.102×10^{-3}	0.001	9.842×10^{-4}
metric slug	345.9	21.62	9.807	1	0.6721	92.51	9.807×10^{-3}	9.651×10^{-3}
slug	514.7	32.17	14.59	1.49	1	62.17	1.459×10^{-2}	1.436×10^{-2}
short ton	3.2×10^4	2,000	907.2	92.51	62.16	1	0.907	0.8929
metric ton	3.528×10^4	2,205	1,000	102	68.52	1.103	1	0.9842
long ton	3.584×10^4	2,240	1,016	103.7	69.63	1.12	1.016	1

Force

Unit	Equivalent[1,2]			
	dyne	newton	pound$_{force}$	kilogram$_{force}$
dynes	1	1×10^{-5}	2.248×10^{-6}	1.02×10^{-6}
newtons	1×10^5	1	0.2248	0.102
pound$_{force}$	4.448×10^5	4.448	1	0.4536
kilogram$_{force}$	9.807×10^5	9.807	2.205	1

Density

Unit	Equivalent[1,2]				
	pounds per cubic inch	pounds per cubic feet	pounds per gallon	grams per cubic centimeter	grams per liter
pounds per cubic inch	1	1,728	231	27.68	2.768×10^4
pounds per cubic feet	5.787×10^{-4}	1	0.1337	1.6×10^{-2}	16.02
pounds per gallon	4.33×10^{-3}	7.481	1	0.1198	119.8
grams per cubic centimeter	3.61×10^{-2}	62.43	8.345	1	1,000
grams per liter	3.61×10^{-5}	6.24×10^{-2}	8.35×10^{-3}	0.001	1

Appendix 9.A. Continued

Pressure

Unit	pounds per square inch	pounds per square feet	atmospheres	kilograms per square centimeter	kilograms per square meter	inches of water (68°F)[3]	feet of water (68°F)[3]	inches of mercury (32°F)[4]	millimeters of mercury (32°F)[4]	bars	kilo Pascals
					Equivalent[1,2]						
pounds per square inch	1	144	6.805×10^{-2}	7.031×10^{-2}	703.1	27.73	2.311	2.036	51.72	6.895×10^{-2}	6.895
pounds per square feet	6.945×10^{-3}	1	4.73×10^{-4}	4.88×10^{-4}	4.882	0.1926	1.605×10^{-2}	1.414×10^{-2}	0.3591	4.79×10^{-4}	4.79×10^{-2}
atmospheres	14.7	2,116	1	1.033	1.033×10^{4}	407.5	33.96	29.92	760	1.013	101.3
kilograms per square centimeter	14.22	2,048	0.9678	1	1×10^{4}	394.4	32.87	28.96	735.6	0.9807	98.07
kilograms per square meter	1.422×10^{-3}	0.2048	9.678×10^{-5}	0.001	1	3.944×10^{-2}	3.287×10^{-3}	2.896×10^{-3}	7.356×10^{-2}	9.807×10^{-5}	9.807×10^{-3}
inches of water (68°F)[3]	3.609×10^{-2}	5.197	2.454×10^{-3}	2.53×10^{-3}	25.38	1	8.333×10^{-2}	7.343×10^{-2}	1.865	2.49×10^{-3}	0.249
feet of water (68°F)[3]	0.4328	62.32	2.945×10^{-2}	3.043×10^{-2}	304.3	12	1	0.8812	22.38	2.984×10^{-2}	2.984
inches of mercury (32°F)[4]	0.4912	70.73	3.342×10^{-2}	3.453×10^{-2}	345.3	13.62	1.135	1	25.4	3.386×10^{-2}	3.386
millimeters of mercury (32°F)[4]	1.934×10^{-2}	2.785	1.316×10^{-3}	1.36×10^{-3}	13.6	0.5362	4.468×10^{-2}	3.937×10^{-2}	1	1.333×10^{-3}	0.1333
bars	14.5	2,089	0.9869	1.02	1.02×10^{4}	402.2	33.51	29.53	750.1	1	100
kilo Pascals	0.145	20.89	9.869×10^{-3}	1.02×10^{-2}	102	4.022	0.3351	0.2953	7.501	0.01	1

Appendix 9.A. Continued
Energy

Unit	Equivalent[1,2]					
	British thermal unit	foot-pound	horsepower-hour	joules	calorie	kilowatt-hour
British thermal unit	1	777.9	3.929×10^{-4}	1,055	252	2.93×10^{-4}
foot-pound	1.285×10^{-3}	1	5.051×10^{-7}	1.356	0.3239	3.766×10^{-7}
horsepower-hour	2,545	1.98×10^6	1	2.685×10^6	6.414×10^5	0.7457
joules	9.481×10^{-4}	0.7376	3.725×10^{-7}	1	0.2389	2.778×10^{-7}
calorie	3.968×10^{-3}	3.087	1.559×10^{-6}	4.186	1	1.163×10^{-6}
kilowatt-hour	3,413	2.655×10^6	1.341	3.6×10^6	8.601×10^5	1

Temperature conversions

$$F° = 9/5 \ C° + 32° \qquad C° = 5/9 \ (F° - 32°)$$

INSTRUCTIONS: The central figures refer to the temperatures either in degrees Celsius or degrees Fahrenheit which require conversion. The corresponding temperatures in degrees Fahrenheit or degrees Celsius will be found to the right or left respectively.

Example:

C		F
6.67	44	111.2

44° Celsius → 111.2° Fahrenheit
44° Fahrenheit → 6.67° Celsius

°C		°F	°C		°F	°C		°F	°C		°F	°C		°F
−56.7	−70	−94.0	−11.1	12	53.6	11.1	52	125.6	33.3	92	197.6	216	420	788
−53.9	−65	−85.0	−10.0	14	57.2	12.2	54	129.2	34.4	94	201.2	227	440	824
−51.2	−60	−76.0	−8.89	16	60.8	13.3	56	132.8	35.6	96	204.8	238	460	860
−48.4	−55	−67.0	−7.78	18	64.4	14.4	58	136.4	36.7	98	208.4	249	480	896
−45.6	−50	−58.0	−6.67	20	68.0	15.6	60	140.0	37.8	100	212.0	260	500	932
−42.8	−45	−49.0	−5.55	22	71.6	16.7	62	143.6	48.9	120	248	271	520	968
−40.0	−40	−40.0	−4.44	24	75.2	17.8	64	147.2	60.0	140	284	282	540	1004
−37.2	−35	−31.0	−3.33	26	78.8	18.9	66	150.8	71.1	160	320	293	560	1040
−34.4	−30	−22.0	−2.22	28	82.4	20.0	68	154.4	82.2	180	356	304	580	1076
−31.7	−25	−13.0	−1.11	30	86.0	21.1	70	158.0	93.3	200	392	316	600	1112
−28.9	−20	−4.0	0	32	89.6	22.2	72	161.6	104.4	220	428	327	620	1148
−26.1	−15	5.0	1.11	34	93.2	23.3	74	165.2	115.6	240	464	338	640	1184
−23.3	−10	14.0	2.22	36	96.8	24.4	76	168.8	126.7	260	500	349	660	1220
−20.6	−5	23.0	3.33	38	100.4	25.6	78	172.4	137.8	280	536	360	680	1256
−17.8	0	32.0	4.44	40	104.0	26.7	80	176.0	148.9	300	572	371	700	1292
−16.7	2	35.6	5.55	42	107.6	27.8	82	179.6	160	320	608	382	720	1328
−15.6	4	39.2	6.67	44	111.2	28.9	84	183.2	171	340	644	393	740	1364
−14.4	6	42.8	7.78	46	114.8	30.0	86	186.8	182	360	680	404	760	1400
−13.3	8	46.4	8.89	48	118.4	31.1	88	190.4	193	380	716	416	780	1436
−12.2	10	50.0	10.0	50	122.0	32.2	90	194.0	204	400	752	427	800	1472

INTERPOLATION TABLE

°C	0.56	1.11	1.67	2.22	2.78	3.33	3.89	4.44	5	5.56	6.11	6.67	7.22	7.78	8.33	8.89	9.44	10	10.56	11.11
	1	2	3	4	5	6	7	8	9	10	11	12	13	14	15	16	17	18	19	20
°F	1.8	3.6	5.4	7.2	9	10.8	12.6	14.4	16.2	18	19.8	21.6	23.4	25.2	27	28.8	30.6	32.4	34.2	36

Appendix 9.A. Continued
Slot Size

Slot size	inches	millimeters
4	0.004	0.102
6	0.006	0.152
8	0.008	0.203
10	0.010	0.254
12	0.012	0.305
15	0.015	0.381
18	0.018	0.457
20	0.020	0.508
25	0.025	0.635
30	0.030	0.762
35	0.035	0.889
40	0.040	1.016
45	0.045	1.143
50	0.050	1.270
60	0.060	1.524
70	0.070	1.778
80	0.080	2.032
90	0.090	2.286
100	0.100	2.540
125	0.125	3.175
150	0.150	3.810
175	0.175	4.445
200	0.200	5.080
225	0.225	5.715
250	0.250	6.350

Equations for areas and volumes

Circumference of circle = $3.1416 \times$ dia = $6.2832 \times$ radius
Area of circle = $0.7854 \times (dia)^2 = 3.1416 \times (radius)^2$
Area of sphere = $3.1416 \times (dia)^2$
Volume of sphere = $0.5236 \times (dia)^3$
Area of triangle = $0.5 \times$ base $\times$ height
Area of trapezoid = $0.5 \times$ sum of the two parallel sides $\times$ height
Area of square, rectangle, or parallelogram = base $\times$ height
Volume of pyramid = area of base $\times$ 1/3 height
Volume of cone = $0.2618 \times (dia\ of\ base)^2 \times$ height
Volume of cylinder = $0.7854 \times$ height $\times$ dia

[1] Equivalent values are shown to 4 significant figures.
[2] Multiply the value of the given unit by the equivalent value shown to obtain the numerical amount of the equivalent unit.
[3] Water at 68°F (20°C).
[4] Mercury at 32°F (0°C).

APPENDIX 9.B-1.
Symbols List

A	area	Q	discharge
a	array spacing	q	flow through each foot of aquifer width
B	$264/T \log (0.03\ Tt/r^2S)$		
b	aquifer thickness	Q/s	specific capacity
BQ	head loss attributable to laminar flow	R	radius
		R	radius of influence
C	coefficient of discharge	R	resistance
C	constant	r	radius
CQ^2	head loss attributable to turbulent flow	r	distance from center of a pumped well to a point where the drawdown is measured
D	depth		
d	depth to water	S	coefficient of storage
d	diameter	s	drawdown
E	efficiency of well	s'	residual drawdown
e	exponential function	T	temperature
F_s	force exerted by the soil	T	transmissivity
F_w	upward force on a soil column caused by reservoir pressure	t	time
		u	$1.87\ r^2S/Tt$
g	acceleration of gravity	V	velocity
H	total head	V	voltage
h	head	v	volume
h	height	W	width
h_L	head loss	w	width
I	current	$W(u)$	well function of u; represents an exponential integral
I	hydraulic gradient		
i_c	critical gradient	x	unit length
K	hydraulic conductivity	z	elevation above a certain datum
k	intrinsic permeability	z_{bl}	thickness of impermeable materials placed landward of structure
L	length or distance		
N_c	number of potential drops along each channel		
		η	porosity
N_f	number of flow channels in the flow net	γ	specific weight
		μ	dynamic viscosity
N_s	specific speed	π	3.14
P	pressure	ρ	density
p	pressure	ρ	resistivity

APPENDIX 9.B-2.
Abbreviations List

ABS	acrylonitrile butadiene styrene
API	American Petroleum Institute
ANSI	American National Standards Institute
ASTM	American Society for Testing and Materials
AWS	American Welding Society
AWWA	American Water Works Association
Bhp	brake horsepower
Btu	British thermal unit
cfm	cubic feet per minute
cm^3	cubic centimeters
COD	chemical oxygen demand
cpm	counts per minute
°C	degrees Celsius
°F	degrees Fahrenheit
°K	degrees Kelvin
EPA	Environmental Protection Agency
ft	feet
ft^2	square feet
ft^3	cubic feet
ft-lb	foot-pound
ft/min	feet per minute
ft/sec	feet per second
fps	feet per second
g	gram
gpd	gallons per day
gpd/ft	gallons per day per foot
gpd/ft^2	gallons per day per square foot
gpm	gallons per minute
gal	gallons
gpg	grains per gallon
hp	horsepower
hr	hour
Hz	hertz
I.D.	inside diameter
in	inch
in^2	square inches
in^3	cubic inches
J	Joule
kg	kilogram
km	kilometer
km^2	square kilometers
km^3	cubic kilometers
kPa	kilo pascals
ℓ	liter
µg/ℓ	micrograms per liter

Appendix 9.B-2. Continued

lb	pound
LVF	liquid-volume fraction
m	meter
m/day	meters per day — from $m^3/day/m^2$
m^2	square meters
m^2/day	square meters per day — from $m^3/day/m$
m^3	cubic meters
m^3/day	cubic meters per day
m^3/sec	cubic meters per second
m/min	meters per minute
m/sec	meters per second
mb	millibars
mg/ℓ	milligrams per liter
min	minute
mℓ	milliliter
mm	millimeter
mm^2	square millimeters
mm^3	cubic millimeters
MHz	megahertz
mi	mile
mi^2	square mile
mi^3	cubic mile
NPSH	net positive suction head
NWWA	National Water Well Association
O.D.	outside diameter
oz	ounce
Pa	pascal
pHs	saturation pH
pCi/ℓ	picocurie per liter
psi	pounds per square inch
psig	pounds per square inch at gauge
PVC	polyvinyl chloride
PWL	pumping water level
qt	quart
rpm	revolutions per minute
rps	revolutions per second
SAR	sodium adsorption ratio
SDR	standard dimension ratio
sec	second
μsec	microsecond
SOC	synthetic organic chemicals
SP	spontaneous potential
sp.gr.	specific gravity
SR	styrene rubber
SWL	static water level

Appendix 9.B-2. Continued

TDS	total dissolved solids
THM	trihalomethane
U.S. EPA	United States Environmental Protection Agency
USGS	United States Geological Survey
VOC	volatile organic chemicals
Whp	water horsepower
yr	year

APPENDIX 9.C.
Correction of Measured Drawdown Data for Determining True Transmissivities for Partially Dewatered Unconfined Aquifers

If the saturated thickness of an unconfined aquifer decreases by more than 20 percent during pumping, drawdown data must be adjusted if the true transmissivity is to be calculated using the Jacob analysis. Recall that the nonequilibrium equation assumes that the aquifer thickness remains constant during pumping as in a confined aquifer. But in many shallow unconfined aquifers that are pumped at high rates, the cone of depression may become large enough that a significant portion of the aquifer becomes dewatered. As the saturated thickness decreases, the specific capacity of the well also decreases. The resulting extra drawdown reflects the actual reduction in the transmissivity of the aquifer as it becomes partially dewatered. To obtain the true transmissivity of the aquifer (under fully saturated conditions) from the Jacob equation, the measured drawdown is adjusted by the following equation:

$$s_t = s_a - \left(\frac{s_a^2}{2b}\right) \qquad (1)$$

where s_t is the drawdown adjusted to its theoretical value, s_a is the actual or measured drawdown, and b is the saturated thickness of the unconfined aquifer when no pumping is taking place. It is clear that the theoretical (adjusted) drawdown will always be less than the measured drawdown.

An example will illustrate these points. Data from a constant-rate pumping test are

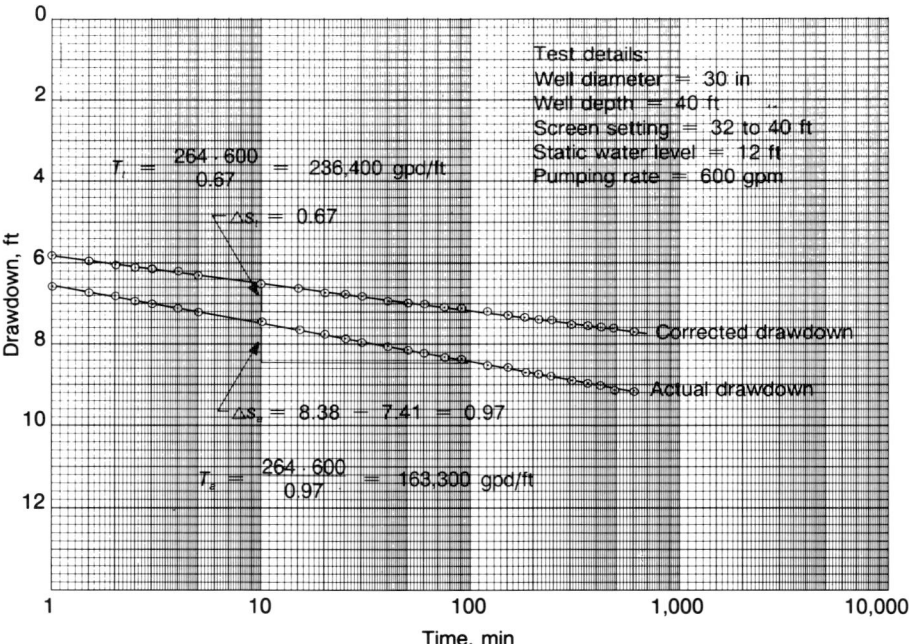

Figure 1. Plot of measured drawdown and adjusted drawdown from a constant-rate pumping test.

Appendix 9.C. Continued

shown in Figure 1. Two plots are shown: measured data and the corresponding adjusted data; the adjusted data are calculated using Equation 1. For example, if $t = 90$ minutes, $s_a = 8.3$ ft, and $b = 28$ ft, then $s_t = 7.05$ ft. The true transmissivity (T_t) is 45 percent larger than the actual transmissivity (T_a) calculated on the basis of drawdown data.

APPENDIX 9.D.
Mathematical Development of Equation 9.15.

A relationship can be established between residual drawdown and the ratio of time since pumping started to time after pumping stopped (t/t'). Drawdown during the pumping period at any distance, r, from the pumped well is given by the modified nonequilibrium equation:

$$s = \frac{264Q}{T} \log \frac{0.3Tt}{r^2S}$$

By simply substituting t' for t in this equation, the recovery of the water level caused by an imaginary recharge well can be calculated. The residual drawdown is the difference between the drawdown caused by the pumped well and the recovery produced by the recharge well. This difference is expressed as:

$$s' = \frac{264Q}{T} \left[\log \frac{0.3Tt}{r^2S} - \log \frac{0.3Tt'}{r^2S} \right]$$

By simplifying this expression, the resulting equation is:

$$s' = \frac{264Q}{T} \log t/t'$$

Because Q and T are constant for a given test, the residual drawdown, s', is directly proportional to $\log t/t'$. Plotting values of residual drawdown against the ratio t/t' on semilog graph paper produces a straight line when conditions are within the limits of validity of the modified nonequilibrium equation.

APPENDIX 9.E.
Values of $W(u)$ Corresponding to Values of u for Theis Nonequilibrium Equation

u \ N	$N \times 10^{-15}$	$N \times 10^{-14}$	$N \times 10^{-13}$	$N \times 10^{-12}$	$N \times 10^{-11}$	$N \times 10^{-10}$	$N \times 10^{-9}$	$N - 10^{-8}$	$N \times 10^{-7}$	$N \times 10^{-6}$	$N \times 10^{-5}$	$N \times 10^{-4}$	$N \times 10^{-3}$	$N \times 10^{-2}$	$N \times 10^{-1}$	N
1.0	33.9616	31.6590	29.3564	27.0538	24.7512	22.4486	20.1460	17.8435	15.5409	13.2383	10.9357	8.6332	6.3315	4.0379	1.8229	0.2194
1.1	33.8662	31.5637	29.2611	26.9585	24.6559	22.3533	20.0507	17.7482	15.4456	13.1430	10.8404	8.5379	6.2363	3.9436	1.7371	.1860
1.2	33.7792	31.4767	29.1741	26.8715	24.5689	22.2663	19.9637	17.6611	15.3586	13.0560	10.7534	8.4509	6.1494	3.8576	1.6595	.1584
1.3	33.6992	31.3966	29.0940	26.7914	24.4889	22.1863	19.8837	17.5811	15.2785	12.9759	10.6734	8.3709	6.0695	3.7785	1.5889	.1355
1.4	33.6251	31.3225	29.0199	26.7173	24.4147	22.1122	19.8096	17.5070	15.2044	12.9018	10.5993	8.2968	5.9955	3.7054	1.5241	.1162
1.5	33.5561	31.2535	28.9509	26.6483	24.3458	22.0432	19.7406	17.4380	15.1354	12.8328	10.5303	8.2278	5.9266	3.6374	1.4645	.1000
1.6	33.4916	31.1890	28.8864	26.5838	24.2812	21.9786	19.6760	17.3735	15.0709	12.7683	10.4657	8.1634	5.8621	3.5739	1.4092	.08631
1.7	33.4309	31.1283	28.8258	26.5232	24.2206	21.9180	19.6154	17.3128	15.0103	12.7077	10.4051	8.1027	5.8016	3.5143	1.3578	.07465
1.8	33.3738	31.0712	28.7686	26.4660	24.1634	21.8608	19.5583	17.2557	14.9531	12.6505	10.3479	8.0455	5.7446	3.4581	1.3098	.06471
1.9	33.3197	31.0171	28.7145	26.4119	24.1094	21.8068	19.5042	17.2016	14.8990	12.5964	10.2939	7.9915	5.6906	3.4050	1.2649	.05620
2.0	33.2684	30.9658	28.6632	26.3607	24.0581	21.7555	19.4529	17.1503	14.8477	12.5451	10.2426	7.9402	5.6394	3.3547	1.2227	.04890
2.1	33.2196	30.9170	28.6145	26.3119	24.0093	21.7067	19.4041	17.1015	14.7989	12.4964	10.1938	7.8914	5.5907	3.3069	1.1829	.04261
2.2	33.1731	30.8705	28.5679	26.2653	23.9628	21.6602	19.3576	17.0550	14.7524	12.4498	10.1473	7.8449	5.5443	3.2614	1.1454	.03719
2.3	33.1286	30.8261	28.5235	26.2209	23.9183	21.6157	19.3131	17.0106	14.7080	12.4054	10.1028	7.8004	5.4999	3.2179	1.1099	.03250
2.4	33.0861	30.7835	28.4809	26.1783	23.8758	21.5732	19.2706	16.9680	14.6654	12.3628	10.0603	7.7579	5.4575	3.1763	1.0762	.02844
2.5	33.0453	30.7427	28.4401	26.1375	23.8349	21.5323	19.2298	16.9272	14.6246	12.3220	10.0194	7.7172	5.4167	3.1365	1.0443	.02491
2.6	33.0060	30.7035	28.4009	26.0983	23.7957	21.4931	19.1905	16.8880	14.5854	12.2828	9.9802	7.6779	5.3776	3.0983	1.0139	.02185
2.7	32.9683	30.6657	28.3631	26.0606	23.7580	21.4554	19.1528	16.8502	14.5476	12.2450	9.9425	7.6401	5.3400	3.0615	.9849	.01918
2.8	32.9319	30.6294	28.3268	26.0242	23.7216	21.4190	19.1164	16.8138	14.5113	12.2087	9.9061	7.6038	5.3037	3.0261	.9573	.01686
2.9	32.8968	30.5943	28.2917	25.9891	23.6865	21.3839	19.0813	16.7788	14.4762	12.1736	9.8710	7.5687	5.2687	2.9920	.9309	.01482
3.0	32.8629	30.5604	28.2578	25.9552	23.6526	21.3500	19.0474	16.7449	14.4423	12.1397	9.8371	7.5348	5.2349	2.9591	.9057	.01305
3.1	32.8302	30.5276	28.2250	25.9224	23.6198	21.3172	19.0146	16.7121	14.4095	12.1069	9.8043	7.5020	5.2022	2.9273	.8815	.01149
3.2	32.7984	30.4958	28.1932	25.8907	23.5881	21.2855	18.9829	16.6803	14.3777	12.0751	9.7726	7.4703	5.1706	2.8965	.8583	.01013
3.3	32.7676	30.4651	28.1625	25.8599	23.5573	21.2547	18.9521	16.6495	14.3470	12.0444	9.7418	7.4395	5.1399	2.8668	.8361	.008939
3.4	32.7378	30.4352	28.1326	25.8300	23.5274	21.2249	18.9223	16.6197	14.3171	12.0145	9.7120	7.4097	5.1102	2.8379	.8147	.007891
3.5	32.7088	30.4062	28.1036	25.8010	23.4985	21.1959	18.8933	16.5907	14.2881	11.9855	9.6830	7.3807	5.0813	2.8099	.7942	.006970
3.6	32.6806	30.3780	28.0755	25.7729	23.4703	21.1677	18.8651	16.5625	14.2599	11.9574	9.6548	7.3526	5.0532	2.7827	.7745	.006160
3.7	32.6532	30.3506	28.0481	25.7455	23.4429	21.1403	18.8377	16.5351	14.2325	11.9300	9.6274	7.3252	5.0259	2.7563	.7554	.005448
3.8	32.6266	30.3240	28.0214	25.7188	23.4162	21.1136	18.8110	16.5085	14.2059	11.9033	9.6007	7.2985	4.9993	2.7306	.7371	.004820
3.9	32.6006	30.2980	27.9954	25.6928	23.3902	21.0877	18.7851	16.4825	14.1799	11.8773	9.5748	7.2725	4.9735	2.7056	.7194	.004267
4.0	32.5753	30.2727	27.9701	25.6675	23.3649	21.0623	18.7598	16.4572	14.1546	11.8520	9.5495	7.2472	4.9482	2.6813	.7024	.003779
4.1	32.5506	30.2480	27.9454	25.6428	23.3402	21.0376	18.7351	16.4325	14.1299	11.8273	9.5248	7.2225	4.9236	2.6576	.6859	.003349
4.2	32.5265	30.2239	27.9213	25.6187	23.3161	21.0136	18.7110	16.4084	14.1058	11.8032	9.5007	7.1985	4.8997	2.6344	.6700	.002969
4.3	32.5029	30.2004	27.8977	25.5952	23.2926	20.9900	18.6874	16.3848	14.0823	11.7797	9.4771	7.1749	4.8762	2.6119	.6546	.002633
4.4	32.4800	30.1774	27.8748	25.5722	23.2696	20.9670	18.6644	16.3619	14.0593	11.7567	9.4541	7.1520	4.8533	2.5899	.6397	.002336
4.5	32.4575	30.1549	27.8523	25.5497	23.2471	20.9446	18.6420	16.3394	14.0368	11.7342	9.4317	7.1295	4.8310	2.5684	.6253	.002073
4.6	32.4355	30.1329	27.8303	25.5277	23.2252	20.9226	18.6200	16.3174	14.0148	11.7122	9.4097	7.1075	4.8091	2.5474	.6114	.001841
4.7	32.4140	30.1114	27.8088	25.5062	23.2037	20.9011	18.5985	16.2959	13.9933	11.6907	9.3882	7.0860	4.7877	2.5268	.5979	.001635
4.8	32.3929	30.0904	27.7878	25.4852	23.1826	20.8800	18.5774	16.2748	13.9723	11.6697	9.3671	7.0650	4.7667	2.5068	.5848	.001453
4.9	32.3723	30.0697	27.7672	25.4646	23.1620	20.8594	18.5568	16.2542	13.9516	11.6491	9.3465	7.0444	4.7462	2.4871	.5721	.001291
5.0	32.3521	30.0495	27.7470	25.4444	23.1418	20.8392	18.5366	16.2340	13.9314	11.6289	9.3263	7.0242	4.7261	2.4679	.5598	.001148
5.1	32.3323	30.0297	27.7271	25.4246	23.1220	20.8194	18.5168	16.2142	13.9116	11.6091	9.3065	7.0044	4.7064	2.4491	.5478	.001021

Appendix 9.E. Continued

u	$N \times 10^{-15}$	$N \times 10^{-14}$	$N \times 10^{-13}$	$N \times 10^{-12}$	$N \times 10^{-11}$	$N \times 10^{-10}$	$N \times 10^{-9}$	$N \times 10^{-8}$	$N \times 10^{-7}$	$N \times 10^{-6}$	$N \times 10^{-5}$	$N \times 10^{-4}$	$N \times 10^{-3}$	$N \times 10^{-2}$	$N \times 10^{-1}$	N
5.2	32.3129	30.0103	27.7077	25.4051	23.1026	20.8000	18.4974	16.1948	13.8922	11.5896	9.2871	6.9850	4.6871	2.4306	.5362	.0009086
5.3	32.2939	29.9913	27.6887	25.3861	23.0835	20.7809	18.4783	16.1758	13.8732	11.5706	9.2681	6.9659	4.6681	2.4126	.5250	.0008086
5.4	32.2752	29.9726	27.6700	25.3674	23.0648	20.7622	18.4596	16.1571	13.8545	11.5519	9.2494	6.9473	4.6495	2.3948	.5140	.0007198
5.5	32.2568	29.9542	27.6516	25.3491	23.0465	20.7439	18.4413	16.1387	13.8361	11.5336	9.2310	6.9289	4.6313	2.3775	.5034	.0006409
5.6	32.2388	29.9362	27.6336	25.3310	23.0285	20.7259	18.4233	16.1207	13.8181	11.5155	9.2130	6.9109	4.6134	2.3604	.4930	.0005708
5.7	32.2211	29.9185	27.6159	25.3133	23.0108	20.7082	18.4056	16.1030	13.8004	11.4978	9.1953	6.8932	4.5958	2.3437	.4830	.0005085
5.8	32.2037	29.9011	27.5985	25.2959	22.9934	20.6908	18.3882	16.0856	13.7830	11.4804	9.1779	6.8758	4.5785	2.3273	.4732	.0004532
5.9	32.1866	29.8840	27.5814	25.2789	22.9763	20.6737	18.3711	16.0685	13.7659	11.4633	9.1608	6.8588	4.5615	2.3111	.4637	.0004039
6.0	32.1698	29.8672	27.5646	25.2620	22.9595	20.6569	18.3543	16.0517	13.7491	11.4465	9.1440	6.8420	4.5448	2.2953	.4544	.0003601
6.1	32.1533	29.8507	27.5481	25.2455	22.9429	20.6403	18.3378	16.0352	13.7326	11.4300	9.1275	6.8254	4.5283	2.2797	.4454	.0003211
6.2	32.1370	29.8344	27.5318	25.2293	22.9267	20.6241	18.3215	16.0189	13.7163	11.4138	9.1112	6.8092	4.5122	2.2645	.4366	.0002864
6.3	32.1210	29.8184	27.5158	25.2133	22.9107	20.6081	18.3055	16.0029	13.7003	11.3978	9.0952	6.7932	4.4963	2.2494	.4280	.0002555
6.4	32.1053	29.8027	27.5001	25.1975	22.8949	20.5923	18.2898	15.9872	13.6846	11.3820	9.0795	6.7775	4.4806	2.2346	.4197	.0002279
6.5	32.0898	29.7872	27.4846	25.1820	22.8794	20.5768	18.2742	15.9717	13.6691	11.3665	9.0640	6.7620	4.4652	2.2201	.4115	.0002034
6.6	32.0745	29.7719	27.4693	25.1667	22.8641	20.5616	18.2590	15.9564	13.6538	11.3512	9.0487	6.7467	4.4501	2.2058	.4036	.0001816
6.7	32.0595	29.7569	27.4543	25.1517	22.8491	20.5465	18.2439	15.9414	13.6388	11.3362	9.0337	6.7317	4.4351	2.1917	.3959	.0001621
6.8	32.0446	29.7421	27.4395	25.1369	22.8343	20.5317	18.2291	15.9265	13.6240	11.3214	9.0189	6.7169	4.4204	2.1779	.3883	.0001448
6.9	32.0300	29.7275	27.4249	25.1223	22.8197	20.5171	18.2145	15.9119	13.6094	11.3068	9.0043	6.7023	4.4059	2.1643	.3810	.0001293
7.0	32.0156	29.7131	27.4105	25.1079	22.8053	20.5027	18.2001	15.8976	13.5950	11.2924	8.9899	6.6879	4.3916	2.1508	.3738	.0001155
7.1	32.0015	29.6989	27.3963	25.0937	22.7911	20.4885	18.1860	15.8834	13.5808	11.2782	8.9757	6.6737	4.3775	2.1376	.3668	.0001032
7.2	31.9875	29.6849	27.3823	25.0797	22.7771	20.4746	18.1720	15.8694	13.5668	11.2642	8.9617	6.6598	4.3636	2.1246	.3599	.00009219
7.3	31.9737	29.6711	27.3685	25.0659	22.7633	20.4608	18.1582	15.8556	13.5530	11.2504	8.9479	6.6460	4.3500	2.1118	.3532	.00008239
7.4	31.9601	29.6575	27.3549	25.0523	22.7497	20.4472	18.1446	15.8420	13.5394	11.2368	8.9343	6.6324	4.3364	2.0991	.3467	.00007364
7.5	31.9467	29.6441	27.3415	25.0389	22.7363	20.4337	18.1311	15.8286	13.5260	11.2234	8.9209	6.6190	4.3231	2.0867	.3403	.00006583
7.6	31.9334	29.6308	27.3282	25.0257	22.7231	20.4205	18.1179	15.8153	13.5127	11.2102	8.9076	6.6057	4.3100	2.0744	.3341	.00005886
7.7	31.9203	29.6178	27.3152	25.0126	22.7100	20.4074	18.1048	15.8022	13.4997	11.1971	8.8946	6.5927	4.2970	2.0623	.3280	.00005263
7.8	31.9074	29.6048	27.3023	24.9997	22.6971	20.3945	18.0919	15.7893	13.4868	11.1842	8.8817	6.5798	4.2842	2.0503	.3221	.00004707
7.9	31.8947	29.5921	27.2895	24.9869	22.6844	20.3818	18.0792	15.7766	13.4740	11.1714	8.8689	6.5671	4.2716	2.0386	.3163	.00004210
8.0	31.8821	29.5795	27.2769	24.9744	22.6718	20.3692	18.0666	15.7640	13.4614	11.1589	8.8563	6.5545	4.2591	2.0269	.3106	.00003767
8.1	31.8697	29.5671	27.2645	24.9619	22.6594	20.3568	18.0542	15.7516	13.4490	11.1464	8.8439	6.5421	4.2468	2.0155	.3050	.00003370
8.2	31.8574	29.5548	27.2523	24.9497	22.6471	20.3445	18.0419	15.7393	13.4367	11.1342	8.8317	6.5298	4.2346	2.0042	.2996	.00003015
8.3	31.8453	29.5427	27.2401	24.9375	22.6350	20.3324	18.0298	15.7272	13.4246	11.1220	8.8195	6.5177	4.2226	1.9930	.2943	.00002699
8.4	31.8333	29.5307	27.2282	24.9256	22.6230	20.3204	18.0178	15.7152	13.4126	11.1101	8.8076	6.5057	4.2107	1.9820	.2891	.00002415
8.5	31.8215	29.5189	27.2163	24.9137	22.6112	20.3086	18.0060	15.7034	13.4008	11.0982	8.7957	6.4939	4.1990	1.9711	.2840	.00002162
8.6	31.8098	29.5072	27.2046	24.9020	22.5995	20.2969	17.9943	15.6917	13.3891	11.0865	8.7840	6.4822	4.1874	1.9604	.2790	.00001936
8.7	31.7982	29.4957	27.1931	24.8905	22.5879	20.2853	17.9827	15.6801	13.3776	11.0750	8.7725	6.4707	4.1759	1.9498	.2742	.00001733
8.8	31.7868	29.4842	27.1816	24.8790	22.5765	20.2739	17.9713	15.6687	13.3661	11.0635	8.7610	6.4592	4.1646	1.9393	.2694	.00001552
8.9	31.7755	29.4729	27.1703	24.8678	22.5652	20.2626	17.9600	15.6574	13.3548	11.0523	8.7497	6.4480	4.1534	1.9290	.2647	.00001390
9.0	31.7643	29.4618	27.1592	24.8566	22.5540	20.2514	17.9488	15.6462	13.3437	11.0411	8.7386	6.4368	4.1423	1.9187	.2602	.00001245
9.1	31.7533	29.4507	27.1481	24.8455	22.5429	20.2404	17.9378	15.6352	13.3326	11.0300	8.7275	6.4258	4.1313	1.9087	.2557	.00001115
9.2	31.7424	29.4398	27.1372	24.8346	22.5320	20.2294	17.9268	15.6243	13.3217	11.0191	8.7166	6.4148	4.1205	1.8987	.2513	.000009988
9.3	31.7315	29.4290	27.1264	24.8238	22.5212	20.2186	17.9160	15.6135	13.3109	11.0083	8.7058	6.4040	4.1098	1.8888	.2470	.000008948
9.4	31.7208	29.4183	27.1157	24.8131	22.5105	20.2079	17.9053	15.6028	13.3002	10.9976	8.6951	6.3934	4.0992	1.8791	.2429	.000008018
9.5	31.7103	29.4077	27.1051	24.8025	22.4999	20.1973	17.8948	15.5922	13.2896	10.9870	8.6845	6.3828	4.0887	1.8695	.2387	.000007185
9.6	31.6998	29.3972	27.0946	24.7920	22.4895	20.1869	17.8843	15.5817	13.2791	10.9765	8.6740	6.3723	4.0784	1.8599	.2347	.000006439

Appendix 9.E. Continued

u	$N \times 10^{-15}$	$N \times 10^{-14}$	$N \times 10^{-13}$	$N \times 10^{-12}$	$N \times 10^{-11}$	$N \times 10^{-10}$	$N \times 10^{-9}$	$N \times 10^{-8}$	$N \times 10^{-7}$	$N \times 10^{-6}$	$N \times 10^{-5}$	$N \times 10^{-4}$	$N \times 10^{-3}$	$N \times 10^{-2}$	$N \times 10^{-1}$	N
9.7	31.6894	29.3868	27.0843	24.7817	22.4791	20.1765	17.8739	15.5713	13.2688	10.9662	8.6637	6.3620	4.0681	1.8505	.2308	.000005771
9.8	31.6792	29.3766	27.0740	24.7714	22.4688	20.1663	17.8637	15.5611	13.2585	10.9559	8.6534	6.3517	4.0579	1.8412	.2269	.000005173
9.9	31.6690	29.3664	27.0639	24.7613	22.4587	20.1561	17.8535	15.5509	13.2483	10.9458	8.6433	6.3416	4.0479	1.8320	.2231	.000004637

NOTE: See page 218 for Theis equation and definitions of terms.

Values of $W(u)$ for u between 1×10^{-15} and 1×10^{-3} computed by R.G. Kazmann assisted by M.M. Evans, U.S. Geological Survey; values for u between 1×10^{-3} and 9.9 adapted from Tables of Exponential and Trigonometric Integrals.

From Water Supply Paper 887, U.S. Geological Survey, 1942.

APPENDIX 9.F.
Available Analytical Models Simulating Flow to Wells

Aquifer system	Isotropic conditions	Well diameter, storage capacity, penetration	Other assumptions	References
Non-leaky confined	isotropic	infinitesimal diameter; no storage capacity; fully penetrating	uniformly porous aquifer	Theis (1935)
Non-leaky confined	isotropic	finite diameter; no storage capacity; fully penetrating	uniformly porous aquifer	Hantush (1964)
Non-leaky confined	isotropic	finite diameter; storage capacity; fully penetrating	uniformly porous aquifer	Papadopulos (1967)
Non-leaky confined	isotropic	infinitesimal diameter; no storage capacity; fully penetrating	uniformly porous aquifer visco-elastic properties	Brutsaert and Corapcioglu (1976)
Non-leaky confined	isotropic	infinitesimal diameter; no storage capacity; partially penetrating	uniformly porous aquifer	Hantush (1964)
Non-leaky confined	isotropic	infinitesimal diameter; no storage capacity; fully penetrating (flowing)	uniformly porous aquifer	Jacob and Lohman (1952)
Non-leaky confined	anisotropic; porous blocks and fractures	infinitesimal diameter; no storage capacity; fully penetrating	fractured-rock aquifer	Boulton and Streltsova (1977a)
Non-leaky confined	anisotropic; porous blocks and fractures	infinitesimal diameter; no storage capacity; production well cased through porous blocks	fractured-rock aquifer	Boulton and Streltsova (1977b)
Non-leaky confined, partial conversion to water table	isotropic	infinitesimal diameter; no storage capacity; fully penetrating	uniformly porous aquifer	Moench and Prickett (1972)
Non-leaky confined	anisotropic permeability varies in two horizontal directions	infinitesimal diameter; no storage capacity; fully penetrating	uniformly porous aquifer	Papadopulos (1965)
Leaky confined	isotropic	infinitesimal diameter; no storage capacity; fully penetrating aquifer wells	negligible aquitard storage and source-bed drawdown changes, uniformly porous aquifer and aquitard	Hantush and Jacob (1955)

Appendix 9.F. Continued

Aquifer system	Isotropic conditions	Well diameter, storage capacity, penetration	Other assumptions	References
Leaky confined	isotropic	infinitesimal diameter; no storage capacity; fully penetrating aquifer wells (flowing)	negligible aquitard storage and source-bed drawdown changes, uniformly porous aquifer and aquitard	Hantush (1959)
Leaky confined	isotropic	finite diameter; storage capacity; fully penetrating aquifer wells	negligible aquitard storage and source-bed drawdown changes, uniformly porous aquifer and aquitard	Lai and Chen-Wu Su (1974)
Leaky confined	isotropic	infinitesimal diameter; no storage capacity; partially penetrating aquitard wells	aquitard storage release; negligible source-bed drawdown changes; uniformly porous aquifer and aquitard	Witherspoon et al. (1967)
Leaky confined	isotropic	infinitesimal diameter; no storage capacity; fully penetrating aquifer wells	negligible aquitard storage and source-bed drawdown changes, uniformly porous aquifer and aquitard visco-elastic properties	Corapcioglu (1976)
Leaky confined	anisotropic	infinitesimal diameter; no storage capacity; partially penetrating aquifer wells	negligible aquitard storage and source-bed drawdown changes, uniformly porous aquifer and aquitard	Hantush (1964)
Leaky confined	isotropic	infinitesimal diameter; no storage capacity; fully penetrating aquifer wells	negligible aquitard storage, uniformly porous aquifer, and aquitard, two mutually leaky confined aquifers	Hantush (1967)
Unconfined	isotropic	infinitesimal diameter; no storage capacity; fully penetrating	uniformly porous aquifer	Boulton (1963)
Unconfined	anisotropic	infinitesimal diameter; no storage capacity; fully penetrating	uniformly porous aquifer	Neuman (1975)
Unconfined	anisotropic	infinitesimal diameter; no storage capacity; partially penetrating	uniformly porous aquifer	Streltsova (1974)

Appendix 9.F. Continued

Aquifer system	Isotropic conditions	Well diameter, storage capacity, penetration	Other assumptions	References
Unconfined	anisotropic	finite diameter; storage capacity; partially penetrating	uniformly porous aquifer	Boulton and Streltsova (1976)
Unconfined	anisotropic; porous blocks and fractures	infinitesimal diameter; no storage capacity; production well cased through porous blocks	fractured rock aquifer	Boulton and Streltsova (1978)
Unconfined in aquitard	anisotropic	infinitesimal diameter; no storage capacity; partially penetrating	uniformly porous aquifer and aquitard	Streltsova (1976)

(After Walton, 1979)

APPENDIX 10.A-1.
Dimensions and Weights for Standard Cable Tool Drill Bits

Hole Size (in)	Approx. Weight (lb)	Approx. Length	Size Pin (in)	Hole Size (in)	Approx. Weight (lb)	Approx. Length	Size Pin (in)
3	50	4'5"	1¾	8	600	6'6"	4¼ or 4½
3½	75	4'6"	1¾	8	650	7'	4¼ or 4½
4	100	4'10"	2¼ or 2⅝	8	700	7'6"	4¼ or 4½
4½	100	4'	2¼ or 2⅝	8	750	8'	4¼ or 4½
4½	125	5'	2¼ or 2⅝	9	550	6'	3¾ or 4¼
4½	150	6'	2¼ or 2⅝	9	600	6'6"	3¾ or 4¼
5	135	4'	2¼ or 2⅝	9	650	7'	3¾ or 4¼
5	165	5'	2¼ or 2⅝	10	400	4'	3 or 3¼
5	190	5'	2⅝ or 3	10	500	5'	3 or 3¼
5	220	6'	2⅝ or 3	10	450	4'	3½ or 3¾
5	250	7'	2⅝ or 3	10	500	4'5"	3½ or 3¾
5⅝	180	4'	2¼ or 2⅝	10	575	5'	3½ or 3¾
5⅝	200	4'6"	3 or 3¼	10	500	4'	4¼ or 4½
5⅝	220	5'	3 or 3¼	10	625	5'	4¼ or 4½
5⅝	260	6'	3 or 3¼	10	750	6'	4¼ or 4½
6	150	3'4"	2¼ or 2⅝	10	875	7'	4½ or 5
6	175	4'	2¼ or 2⅝	10	950	7'6"	4½ or 5
6	200	4'	3 or 3¼	10	1000	8'	4½ or 5
6	250	5'	3 or 3¼	12	500	3'6"	3½ or 3¾
6	300	6'	3 or 3¼	12	700	4'6"	3½ or 3¾
6¼	270	5'	3 or 3¼	12	800	5'	3½ or 3¾
6¼	320	6'	3 or 3¼	12	600	3'6"	4¼ or 4½
6⅝ – 6¼	300	5'	3½ or 3¾	12	800	4'6"	4¼ or 4½
6⅝ – 6¼	350	6'	3½ or 3¾	12	1000	5'6"	4¼ or 4½
6⅝ – 6¼	400	7'	3½ or 3¾	12½	1100	6'	4½ or 5
6⅝ – 6¼	425	7'6"	3½ or 3¾	12½	1200	6'6"	4½ or 5
6⅝ – 6¼	450	8'	3½ or 3¾	12½	1350	7'	4½ or 5
8	250	3'6"	3 or 3¼	12½	1450	7'6"	4½ or 5
8	300	4'	3 or 3¼	14	800	4'6"	4¼ or 5
8	350	4'6"	3 or 3¼	14	1000	5'	4¼ or 5
8	400	5'	3 or 3¼	15½	1550	6'	4½ or 5
8	400	5'	3½ or 3¾	15½	1700	6'6"	4½ or 5
8	450	5'6"	3½ or 3¾	15½	1850	7'	4½ or 5
8	500	6'	3½ or 3¾	15½	2000	7'6"	4½ or 5
8	600	7'	3½ or 3¾	16	1000	3'6"	4¼ or 5
8	400	4'6"	4¼ or 4½	16	1200	4'6"	4¼ or 5
8	450	5'	4¼ or 4½	16	1700	6'	4¼ or 5
8	500	5'6"	4¼ or 4½	16	2000	7'	4¼ or 5
8	550	6'	4¼ or 4½				

(Acme Fishing Tool Co.)

APPENDIX 10.A-2.
Dimensions and Weights for Standard Cable Tool Drill Bits (S.I.)

Hole Size (mm)	Approx. Weight (kg)	Approx. Length (m)	Size Pin (inches)	Hole Size (mm)	Approx. Weight (kg)	Approx. Length (m)	Size Pin (inches)
76.2	22.7	1.35	1¾	203	272	1.98	4¼ or 4½
88.9	34.0	1.37	1¾	203	295	2.14	4¼ or 4½
102	45.4	1.48	2¼ or 2⅝	203	318	2.29	4¼ or 4½
114	45.4	1.22	2¼ or 2⅝	203	340	2.44	4¼ or 4½
114	56.7	1.52	2¼ or 2⅝	229	250	1.83	3¾ or 4¼
114	68.0	1.83	2¼ or 2⅝	229	272	1.98	3¾ or 4¼
127	61.2	1.22	2¼ or 2⅝	229	295	2.58	3¾ or 4¼
127	74.8	1.52	2¼ or 2⅝	254	181	1.22	3 or 3¼
127	86.2	1.52	2⅝ or 3	254	227	1.52	3 or 3¼
127	99.8	1.83	2⅝ or 3	254	204	1.22	3½ or 3¾
127	113	2.14	2⅝ or 3	254	227	1.35	3½ or 3¾
143	81.6	1.22	2¼ or 2⅝	254	261	1.52	3½ or 3¾
143	90.7	1.37	3 or 3¼	254	227	1.22	4¼ or 4½
143	99.8	1.52	3 or 3¼	254	284	1.52	4¼ or 4½
143	118	1.83	3 or 3¼	254	340	1.83	4¼ or 4½
152	68.0	1.01	2¼ or 2⅝	254	397	2.14	4½ or 5
152	79.4	1.22	2¼ or 2⅝	254	431	2.29	4½ or 5
152	90.7	1.22	3 or 3¼	254	454	2.44	4½ or 5
152	113	1.52	3 or 3¼	305	227	1.07	3½ or 3¾
152	136	1.83	3 or 3¼	305	318	1.37	3½ or 3¾
159	113	1.52	3 or 3¼	305	363	1.52	3½ or 3¾
159	145	1.83	3 or 3¼	305	272	1.07	4¼ or 4½
168 – 159	136	1.52	3½ or 3¾	305	363	1.37	4¼ or 4½
168 – 159	159	1.83	3½ or 3¾	305	454	1.68	4¼ or 4½
168 – 159	181	2.14	3½ or 3¾	318	499	1.83	4½ or 5
168 – 159	193	2.29	3½ or 3¾	318	544	1.98	4½ or 5
168 – 159	204	2.44	3½ or 3¾	318	612	2.14	4½ or 5
203	113	1.07	3 or 3¼	318	494	2.29	4½ or 5
203	136	1.22	3 or 3¼	356	363	1.37	4¼ or 5
203	159	1.37	3 or 3¼	356	454	1.52	4¼ or 5
203	181	1.52	3 or 3¼	394	703	1.83	4½ or 5
203	181	1.52	3½ or 3¾	394	771	1.98	4½ or 5
203	204	1.68	3½ or 3¾	394	839	2.14	4½ or 5
203	227	1.83	3½ or 3¾	394	907	2.29	4½ or 5
203	272	2.14	3½ or 3¾	406	454	1.07	4¼ or 5
203	181	1.37	4¼ or 4½	406	544	1.37	4¼ or 5
203	204	1.52	4¼ or 4½	406	771	1.83	4¼ or 5
203	227	1.68	4¼ or 4½	406	907	2.14	4¼ or 5
203	250	1.83	4¼ or 4½				

(Acme Fishing Tool Co.)

APPENDIX 10.A-3.
Sizes and Weights for Tricone Roller Bits

Diameter		Approximate Weight		API Pin Size
in	mm	lb	kg	
2⅞	73.0	4	1.8	N Rod
2¹⁵/₁₆	74.6	4¼	1.9	N Rod
3⅛	79.4	4¼	1.9	N Rod
3½	88.9	7	3.2	2⅜ N Rod
3¾	95.3	7½	3.4	2⅜
3⅞	98.4	7½	3.4	2⅜
4¼	108	10	4.5	2⅜
4½	114	11	5.0	2⅜
4¾	121	13	5.9	2⅞
5	127	17	7.7	2⅞
5⅛	130	19	8.6	2⅞
5⅝	143	22	10.0	3½
6	152	27	12.2	3½
6¼	159	30	13.6	3½
6¾	171	33	15.0	3½
7⅜	187	42	19.1	3½
7⅝ – 7⅞	200	52	23.6	4½
8½	216	75	34.0	4½
8¾	222	75	34.0	4½
9	229	75	34.0	4½
9⅝	244	105	47.6	6⅝
9⅞	251	115	52.2	6⅝
10⅝	270	120	54.4	6⅝
11	279	145	65.8	6⅝
12¼	311	175	79.4	6⅝
13¾	349	219	99.3	6⅝
14¾	375	360	163.3	6⅝
15	381	375	170.1	6⅝ or 7⅝
16	406	450	204.1	6⅝ or 7⅝
17½	445	575	260.8	6⅝ or 7⅝
18½	470	625	283.5	6⅝ or 7⅝
20	508	665	301.6	6⅝ or 7⅝
22	559	810	367.4	6⅝ or 7⅝
24	610	1,014	460.0	6⅝ or 7⅝
26	660	1,217	552.0	6⅝ or 7⅝
30	762	2,320	1,052.35	7⅝ or 8⅝

(Varel)

APPENDIX 10.B-1.
Cement Classifications Used Outside the United States

Abbre-viation	Type of Cement	Similar to ASTM	Similar to API
OC	Ordinary Portland Cement	I	A
RHC	Rapid-Hardening (or High-Early-Strength, or High-Initial-Strength) Portland Cement	III	C
HSC	High-Strength Portland Cement	III	C
LHC	Low Heat (or Slow-Hardening, Low Heat of Hydration) Portland Cement	II	B
SRC	Sulfate-Resisting Portland Cement	V	—
AEC	Air-Entraining Portland Cement	—	—

(CEMBUREAU, 1967; Smith, 1976)

APPENDIX 10.B-2.
Equivalent Cement Classifications Outside the United States

International Designation	Australia	Canada	France	Japan	United Kingdom	West Germany
OC	Type A Ordinary	Normal Portland	CPA-250 CPA-325	Ordinary Portland	Ordinary Portland (B.S. 12:1958)	Z375
RHC	Type B High Early Strength	High Early Strength	CPA-400 CPA-500	Rapid Hardening Portland	Rapid Hardening	—
HSC	—	—	—	—	—	Z450
LHC	Type C Low Heat of Hydration	—	—	Medium Low Heat Portland	Low Heat Portland (B.S. 1370:1958)	—
SRC	—	Sulfate Resisting	—	—	Sulfate Resisting Portland (B.S. 4027:1955)	Z275
AEC	—	—	—	—	—	—
Designation of Standards	AS A2	CSA A5	NF P15-302	JIS R5210	(B.S. 12; 1370; 4027)	DIN 1164
Year Published	1963	1961	1964	1964	1958 and 1966	1969

(CEMBUREAU, 1967; Smith, 1976)

APPENDIX 10.C.
Recommended Rotating Speeds for all Sizes and Types of Bits in Various Formations

(Revolutions per Minute)

Bit Sizes and Types	Sticky Shales or Gumbos	Soft Unconsolidated Shales, Silts, Sandy Shales, etc.	Medium Hard Shales, Sandy Shales, Soft Chalk	Medium Hard Sandstones Hard Very Sandy Shale	Very Hard Sandstones, Quartzite, Angular Limestones, Anhydrite	Hard Brittle Shale and Limestone Conchoidal Fracture
13 to 20 Inch						
Drag	100 – 130	100 – 130				
Zublin	100 – 160	100 – 150	100 – 175	125 – 175		
Rock (Rolling Cutter)		125 – 200	100 – 200	60 – 125	40 – 60	40 – 150
10 to 13 Inch						
Drag	100 – 175	100 – 300				
Disc		110 – 180				
Zublin	125 – 175	125 – 200	125 – 200	125 – 200		
Rock (Rolling Cutter)		150 – 300	100 – 250	80 – 120	40 – 80	60 – 150
6 to 10 Inch						
Drag	125 – 200	100 – 200				
Zublin	150 – 200	100 – 150	150 – 225	150 – 200		
Rock (Rolling Cutter)		150 – 300	100 – 250	80 – 125	40 – 100	60 – 200

The minimum speeds given are for flat lying strata and certain type bits. The maximum speeds are for flat or inclined formations. The maximum allowable weight may be carried in flat beds and the minimum in steeply dipping strata. Slower and faster speeds than these recommended are useful in specific and more or less unusual cases.

(Brantly, 1961)

APPENDIX 10.D.
Weight on Bit and Rotary Speed

Bit Classification	Weight per in (2.54 cm) of Bit Diameter		Rotary Speed rpm
	lb/in	kg/cm	
Soft formation	3,400 to 6,750 4,050 to 7,800	609 to 1,210 725 to 1,400	250 to 100 180 to 60
Medium formation	4,500 to 9,000	806 to 1,610	120 to 40
Hard milled tooth bit	5,600 to 11,250	1,000 to 2,010	70 to 35
Hard insert bit	2,250 to 5,600 4,500 to 9,000	403 to 1,000 806 to 1,610	70 to 35 65 to 35
Hard friction bearing bit	4,500 to 6,750	806 to 1,210	60 to 35

(Ingersoll-Rand)

APPENDIX 10.E.
API Drill Pipe List

1		2	3		4	5		6
Size: Outside Diameter (D)		Nominal Weight, lb/ft	Calculated Plain-end Weight (W_{pe})		Grade	Wall Thickness (t)		Upset Ends, for Weld-on Tool Joints
in.	mm		lb/ft	kg/m		in.	mm	
2⅜	60.3	6.65	6.26	9.32	D, E	0.280	7.11	Ext. Upset
2⅞	73.0	10.40	9.72	14.48	D, E	0.362	9.19	Int. Upset or Ext. Upset
3½	88.9	9.50	8.81	13.12	E	0.254	6.45	Int. Upset or Ext. Upset
3½	88.9	13.30	12.31	18.34	D, E	0.368	9.35	Int. Upset or Ext. Upset
3½	88.9	15.50	14.63	21.79	D, E	0.449	11.40	Int. Upset or Ext. Upset
*4	101.6	11.85	10.46	15.58	E	0.262	6.65	Int. Upset or Ext. Upset
4	101.6	14.00	12.93	19.26	D, E	0.330	8.38	Int. Upset or Ext. Upset
*4½	114.3	13.75	12.24	18.23	E	0.271	6.88	Int. Upset or Ext. Upset
4½	114.3	16.60	14.98	22.31	D, E	0.337	8.56	Ext. Upset or Int.-Ext. Upset
4½	114.3	20.00	18.69	27.84	D, E	0.430	10.92	Ext. Upset or Int.-Ext. Upset
*5	127.0	16.25	14.87	22.15	E	0.296	7.52	Int. Upset
5	127.0	19.50	17.93	26.71	D, E	0.362	9.19	Int.-Ext. Upset
5	127.0	25.60	24.03	35.79	D, E	0.500	12.70	Int.-Ext. Upset
5½	139.7	21.90	19.81	29.51	D, E	0.361	9.17	Int.-Ext. Upset
5½	139.7	24.70	22.54	33.57	D, E	0.415	10.54	Int.-Ext. Upset
6⅝	168.3	25.20	22.19	33.05	D, E	0.330	8.38	†

*These sizes and weights are tentative and marking with API monogram is not permitted.
†Upset requirements for 6⅝ in - 25.20 lb drill pipe not established.
(*API, 1976*)

APPENDIX 10.F.
Upset Ends

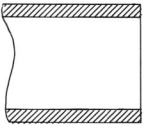
Plain end
(non-upset)

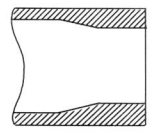
Internal upset

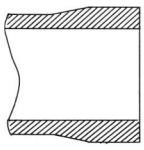
External upset

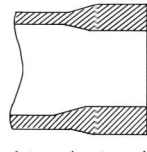
Internal-external upset

APPENDIX 10.G.
Kelly Weights in Pounds per Foot (Drive Section)

Square Kelly

Across Flat	Bore of Square Kelly											
	1¹/₁₆	1¼	1½	1¾	2	2¼	2½	2¾	2⅞	3	3¼	3½
2½	18.3	17.1										
3		25.8	24.0	21.8								
3½			35.6	33.5	31.0	28.2						
4¼						47.9	44.7	41.3	39.3			
5¼								73.5	71.6	69.7	65.5	61.0
6												89.6

(Joy, 1978)

Hexagonal Kelly

Across Flat	Bore of Hexagonal Kelly											
	1¼	1½	1¾	2	2¼	2½	2¾	2⅞	3	3¼	3½	4
3	22.3	20.5										
3½		30.1	27.9	25.4	22.6							
3¾		35.3	33.2	30.7	27.9							
4¼					39.6	36.4	32.9	31.0				
4²⁷/₃₂					56.4	53.3	49.8	47.9	45.9	41.7		
5¼							60.9	59.0	57.1	52.9	48.4	
6										73.2	63.2	

(Joy, 1978)

Round Kelly

Across Flat	Bore of Round Kelly				
	1⅛	1½	1¾	2⅛	2¼
2⅜	10.2				
2⅞		14.2			
3⅜			17.9		
4⅛				28.9	
4⅝					37.5

(Failing)

APPENDIX 10.H.
Ideal Size Range for Drill Collars

Hole size mm	Ideal drill collar range, mm		API drill collar sizes which fall in the ideal range, mm
	Min.	Max.	
156	98.4	121	105, 121
159	95.3	124	105, 121
171	82.6	130	88.9, 105, 121, 127
200	54	156	79.4, 88.9, 105, 121, 127, 152
	107	156	121, 127, 152
213	94.6	165	105, 121, 127, 152, 159, 165
	163	165	165
216	160	171	165, 171
	†173	171	171
222	153	181	159, 165, 171, 178
	167	181	171, 178
241	173	194	152, 159, 165, 178, 184
	191	194	194*
251	138	203	152, 159, 165, 171, 178, 184, 197, 203
	181	203	184, 197, 203
270	162	216	165, 171, 178, 184, 197, 203, 210
	†219	216	210
279	210	244	210, 229, 241
311	229	257	229, 241, 248, 254
	†286	257	254
349	248	286	248, 254, 279
375	222	305	229, 241, 248, 254, 279, 305*
445	286	340	305*
508	356	375	356*
610	394	425	406*
660	406	495	406*

(Drilco, 1979)

*Not API standard size drill collar.

†In these instances, the equation used to calculate the ideal minimum drill collar size produces an anomalously high value. See Woods and Lubinski, 1954, for a complete discussion on how to determine the best collar size for a specific diameter borehole.

APPENDIX 10.I.
Drill Collar Weights in Pounds per Foot

Collar O.D.	BORE OF COLLAR									
	1½	1¾	2	2¼	2½	2¹³/₁₆	3	3¼	3½	3¾	4
3⅜	24.4	22.2								
3½	26.7	24.5								
3¾	31.5	29.3								
3⅞	34.0	31.9	29.4	26.5						
4	36.7	34.5	32.0	29.2						
4⅛	39.4	37.2	34.7	31.9						
4¼	42.2	40.0	37.5	34.7						
4½	48.0	45.8	43.3	40.5						
4¾	54.2	52	49.5	46.7	43.5					
5	60.1	58.5	55.9	53.1	49.9					
5¼	67.5	65.3	62.8	59.9	56.8	53.3				
5½	74.7	72.5	69.9	67.2	63.9	60.5	56.7			
5¾	82.1	79.9	77.5	74.6	71.5	67.9	64.1			
6	89.9	87.8	85.3	82.5	79.3	75.8	71.9	67.8	63.3	
6¼	98.1	95.9	93.5	90.6	87.5	83.9	80.1	75.9	71.5	
6½	106.6	104.5	101.9	99.1	95.9	92.5	88.6	84.5	79.9	
6¾	115.5	113.3	110.8	107.9	104.8	101.3	97.5	93.3	88.8	
7	124.6	122.5	119.9	117.1	113.9	110.5	106.6	102.5	97.9	93.1	87.9
7¼	134.1	131.9	129.5	126.6	123.5	119.9	116.1	111.9	107.5	102.6	97.5
7½	143.9	141.7	139.3	136.5	133.3	129.8	125.9	121.8	117.3	112.5	107.3
7¾	154.1	151.9	149.5	146.6	143.5	139.9	136.1	131.9	127.5	122.6	117.5
8	164.6	162.5	159.9	157.1	153.9	150.5	146.6	142.5	137.9	133.1	127.9
8¼	175.4	173.3	170.8	167.9	164.8	161.3	157.5	153.3	148.8	143.9	138.8
8½	186.6	184.4	181.9	179.1	175.9	168.6	172.5	164.5	159.9	155.1	149.9
8¾	198.1	195.9	193.9	190.6	187.4	183.9	180.1	175.9	171.4	166.9	161.5
9		207.8	205.3	202.4	199.3	195.8	191.9	187.8	183.3	178.5	173.3
9½		232.4	229.9	227.1	223.9	220.4	216.6	212.4	207.9	203.1	197.9
10			255.9	253.1	249.9	246.4	242.6	238.4	233.9	229.1	223.9
10½			283.3	280.4	277.3	273.8	269.9	265.8	261.3	256.4	251.3
11					305.9	302.4	298.6	294.4	289.9	285.1	279.9

(Joy, 1978)

APPENDIX 10.J.
Correlation between Percent of Bentonite, Volume of Water, Density, and Volume of Slurry

Bentonite as percent of cement weight	Density	Mixing water			Volume of slurry	
		gal/94 lb sack	ft³/94 lb sack	ℓ/100 kg cement	ft³/94 lb sack	ℓ/100 kg cement
2	1.73	6.98	0.94	62	1.43	94.5
	1.77	6.53	0.88	58	1.37	90.5
	1.80	6.08	0.82	54	1.31	86.5
	1.84	5.63	0.76	50	1.25	82.5
	1.88	5.18	0.69	46	1.19	78.5
4	1.63	8.79	1.18	78	1.68	111.5
	1.66	8.34	1.12	74	1.62	107.5
	1.68	7.89	1.06	70	1.56	103.5
	1.71	7.44	1.00	66	1.50	99.5
	1.74	6.98	0.94	62	1.44	95.5
6	1.59	9.91	1.33	88	1.84	122
	1.61	9.46	1.27	84	1.78	118
	1.63	9.01	1.21	80	1.72	114
	1.65	8.56	1.15	76	1.66	110
	1.68	8.11	1.09	72	1.60	106
8[1,2]	1.53	11.72	1.57	104	2.10	139
	1.54	11.27	1.51	100	2.04	135
	1.56	10.81	1.45	96	1.98	131
	1.57	10.36	1.39	92	1.92	127
	1.59	9.91	1.33	88	1.86	123

[1]For 8 percent and higher, it is advisable to add a thinner to facilitate placement.
[2]At this percent of bentonite, shrinkage cracks are likely to occur.
Note: All calculations are based on a specific gravity of 3.15 for cement and 2.5 for bentonite.

(Ingersoll-Rand)

APPENDIX 10.K.
Equations used to Calculate Slurry Density

English Engineering Units

$$\text{Slurry density} = \frac{\text{Weight of cement (lb)} + \text{Weight of water (lb)}}{\text{Sacks of cement} \cdot 0.48 \text{ (ft}^3\text{)} + \text{Volume of water (ft}^3\text{)}}$$

$$= \frac{\text{Total weight (lb)}}{\text{Volume of slurry (ft}^3\text{)}} \div 7.5 = \text{lb/gal}$$

Example: If 94 lb of cement is mixed with 5.2 gal of water, the

$$\text{slurry density} = \frac{94 + 43.3}{1 \cdot 0.48 + 0.7} = \frac{137.3}{1.18} = \frac{116.4 \text{ lb/ft}^3}{7.5} = 15.5 \text{ lb/gal}$$

where 0.48 is the actual volume (ft³) of cement in a 94-lb sack

S.I. Units

$$\text{Slurry density} = \frac{\text{Weight of cement (kg)} + \text{Weight of water (kg)}}{\text{Volume of cement (}\ell\text{)} + \text{Volume of water (}\ell\text{)}}$$

$$= \frac{\text{Total weight (kg)}}{\text{Volume of slurry (}\ell\text{)}}$$

Example: If 100 kg of cement is mixed with 46 ℓ of water, the

$$\text{slurry density} = \frac{100 + 46}{\frac{100}{3.15} + 46} = \frac{146}{31.75 + 46} = 1.88 \text{ kg/}\ell$$

where 3.15 is the true specific gravity of powdered cement

APPENDIX 11.A.
Density and Specific Gravity of Water and Three Common Drilling Fluid Additives

	lb/ft³	kg/m³	g/cm³	Specific Gravity
Water	62.4	1,000	1.0	1.0
Barite	268	4,300	4.3	4.3
Clay	156	2,500	2.5	2.5
Salt				
NaCL	135	2,165	2.165	2.165
CaCl₂	134	2,150	2.150	2.150

APPENDIX 11.B.
Procedure for Measuring the Density of Drilling Fluid

1. Set up the instrument base so it is approximately level.
2. Fill the clean, dry cup with the drilling fluid to be weighed.
3. Place the lid on the cup and seat it firmly but slowly with a twisting motion. Be sure some drilling fluid runs out of the hole in the cap.
4. With the hole in the cap covered with a finger, wash or wipe all drilling fluid from the outside of the cup and arm.
5. Set the knife on the fulcrum and move the sliding weight along the graduated arm until the cup and arm are balanced.
6. Read the density of the drilling fluid at the left-hand edge of the sliding weight.
7. Report the result to the nearest scale division in lb/gal, lb/ft³, or specific gravity (or the metric equivalents).
8. Wash the drilling fluid from the cup immediately after each use. It is absolutely essential that all parts of the drilling fluid balance be kept clean if accurate results are to be obtained.

APPENDIX 11.C.
Drilling Fluid Weight Conversion Table

lb/gal	lb/ft³	kg/m³	Gradient psi/100 ft of depth	Gradient kPa/50 m of depth	Specific Gravity*
8.0	59.8	958	41.6	470	0.96
8.1	60.6	971	42.1	476	0.97
8.2	61.3	982	42.6	482	0.98
8.3	62.1	995	43.1	487	0.99
8.4	62.8	1,010	43.6	493	1.01
8.5	63.6	1,020	44.2	500	1.02
8.6	64.3	1,030	44.7	506	1.03
8.7	65.1	1,040	45.2	511	1.04
8.8	65.8	1,050	45.7	517	1.05
8.9	66.6	1,070	46.2	523	1.07
9.0	67.3	1,080	46.8	529	1.08
9.1	68.1	1,090	47.3	535	1.09
9.2	68.8	1,100	47.8	541	1.10
9.3	69.6	1,110	48.3	546	1.11
9.4	70.3	1,130	48.8	552	1.13
9.5	71.1	1,140	49.3	558	1.14
9.6	71.8	1,150	49.9	564	1.15
9.7	72.6	1,160	50.4	570	1.16
9.8	73.3	1,170	50.9	576	1.17
9.9	74.1	1,190	51.4	581	1.19
10.0	74.8	1,200	51.9	587	1.20
10.1	75.6	1,210	52.5	594	1.21
10.2	76.3	1,220	53.0	599	1.22
10.3	77.0	1,230	53.5	605	1.23
10.4	77.8	1,250	54.0	611	1.25
10.5	78.5	1,260	54.5	616	1.26
10.6	79.3	1,270	55.1	623	1.27
10.7	80.0	1,280	55.6	629	1.28
10.8	80.8	1,290	56.1	634	1.29
10.9	81.5	1,310	56.6	640	1.31
11.0	82.3	1,320	57.1	646	1.32
11.1	83.0	1,330	57.7	653	1.33
11.2	83.8	1,340	58.2	658	1.34
11.3	84.5	1,350	58.7	664	1.35
11.4	85.3	1,370	59.2	670	1.37
11.5	86.0	1,380	59.7	675	1.38
11.6	86.8	1,390	60.3	682	1.39
11.7	87.5	1,400	60.8	688	1.40
11.8	88.3	1,410	61.3	693	1.41
11.9	89.0	1,430	61.8	699	1.43

Appendix 11.C. Continued

lb/gal	lb/ft³	kg/m³	Gradient psi/100 ft of depth	Gradient kPa/50 m of depth	Specific Gravity*
12.0	89.8	1,440	62.3	705	1.44
12.1	90.5	1,450	62.9	711	1.45
12.2	91.3	1,460	63.4	717	1.46
12.3	92.0	1,470	63.9	723	1.47
12.4	92.8	1,490	64.4	728	1.49
12.5	93.5	1,500	64.9	734	1.50
12.6	94.3	1,510	65.5	741	1.51
12.7	95.0	1,520	66.0	746	1.52
12.8	95.8	1,530	66.5	752	1.53
12.9	96.5	1,550	67.0	758	1.55
13.0	97.2	1,560	67.5	763	1.56
13.1	98.0	1,570	68.0	769	1.57
13.2	98.7	1,580	68.6	776	1.58
13.3	99.5	1,590	69.1	782	1.59
13.4	100.2	1,610	69.6	787	1.61
13.5	101.0	1,620	70.1	793	1.62
13.6	101.7	1,630	70.6	798	1.63
13.7	102.5	1,640	71.2	805	1.64
13.8	103.2	1,650	71.7	811	1.65
13.9	104.0	1,670	72.2	817	1.67
14.0	104.7	1,680	72.7	822	1.68
14.1	105.5	1,690	73.2	828	1.69
14.2	106.2	1,700	73.8	835	1.70
14.3	107.0	1,710	74.3	840	1.71
14.4	107.7	1,730	74.8	846	1.73
14.5	108.5	1,740	75.3	852	1.74
14.6	109.2	1,750	75.8	857	1.75
14.7	110.0	1,760	76.4	864	1.76
14.8	110.7	1,770	76.9	870	1.77
14.9	111.5	1,790	77.4	875	1.79
15.0	112.2	1,800	77.9	881	1.80
15.1	113.0	1,810	78.4	887	1.81
15.2	113.7	1,820	79.0	893	1.82
15.3	114.4	1,830	79.5	899	1.83
15.4	115.2	1,850	80.0	905	1.85
15.5	116.0	1,860	80.5	910	1.86
15.6	116.7	1,870	81.0	916	1.87
15.7	117.4	1,880	81.6	923	1.88
15.8	118.2	1,890	82.1	929	1.89
15.9	118.9	1,900	82.6	934	1.90

Appendix 11.C. Continued

lb/gal	lb/ft³	kg/m³	Gradient psi/100 ft of depth	Gradient kPa/50 m of depth	Specific Gravity*
16.0	119.7	1,920	83.1	940	1.92
16.1	120.4	1,930	83.6	946	1.93
16.2	121.2	1,940	84.2	952	1.94
16.3	121.9	1,950	84.7	958	1.95
16.4	122.7	1,970	85.2	964	1.97
16.5	123.4	1,980	85.7	969	1.98
16.6	124.2	1,990	86.2	975	1.99
16.7	124.9	2,000	86.7	981	2.00
16.8	125.7	2,010	87.3	987	2.01
16.9	126.4	2,020	87.8	993	2.02
17.0	127.2	2,040	88.3	999	2.04
17.1	127.9	2,050	88.8	1,004	2.05
17.2	128.7	2,060	89.3	1,010	2.06
17.3	129.4	2,070	89.9	1,017	2.07
17.4	130.2	2,090	90.4	1,022	2.09
17.5	130.9	2,100	90.9	1,028	2.10
17.6	131.7	2,110	91.4	1,034	2.11
17.7	132.4	2,120	91.9	1,039	2.12
17.8	133.2	2,130	92.5	1,046	2.13
17.9	133.9	2,150	93.0	1,052	2.15

*In the S.I. system, specific gravity is equivalent to g/cm³.

APPENDIX 11.D.
Viscosity Measurements Using a Marsh Funnel

The procedure for measuring the Marsh funnel viscosity is as follows:
1. Take the sample at the flow line.
2. Cover the end of the tube with a finger and pour the drilling fluid through the screen until the level reaches the bottom of the screen.
3. Remove the finger from the outlet and carefully observe the time required in seconds for 1 qt. of drilling fluid to flow out the funnel*. The number of seconds per quart is recorded as the funnel viscosity.
4. Record the temperature of the sample.

To calibrate the Marsh funnel for the standard API test, fill the funnel to the bottom of the screen with clean water at a temperature between 70 and 80°F (21.1 and 26.7°C) and note the time required for 1 qt (946 cm³) to drain from the funnel. The time for fresh water should be 26 seconds, plus or minus 0.5 second. To insure accurate results, use a clean funnel. If the time is greater, the orifice of the funnel may be partially plugged or damaged.

*In most cases, Marsh funnel viscosities are determined on the basis of the flow time for 1 qt. of liquid, rather than 1 ℓ (1,000 cm³).

APPENDIX 11.E.
Design Specifications for Two Types of Mud Mixers

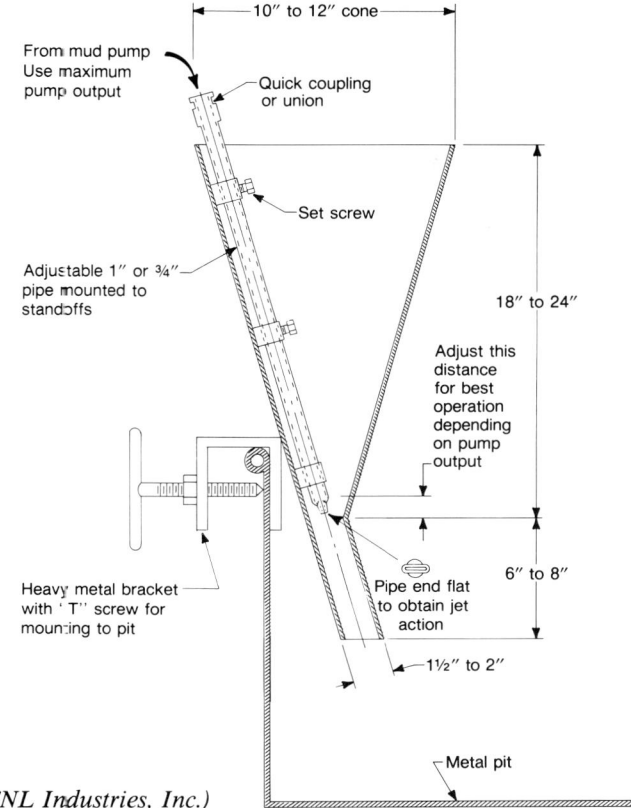

(NL Baroid/NL Industries, Inc.)

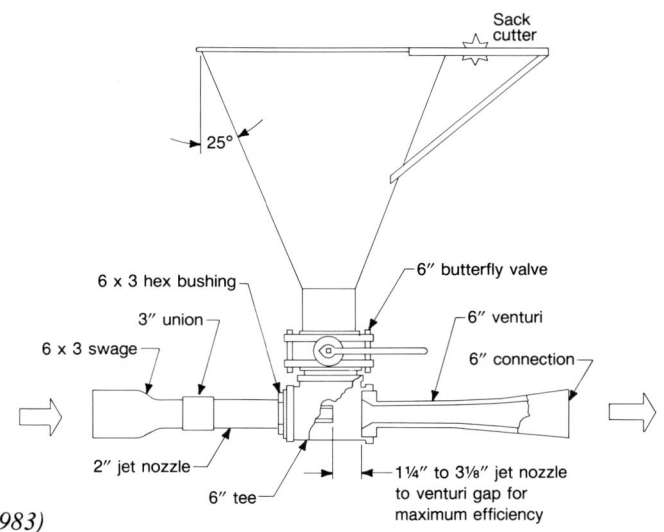

(Grichor, 1983)

APPENDIX 11.F.
Determining Uphole Annular Velocity using Air

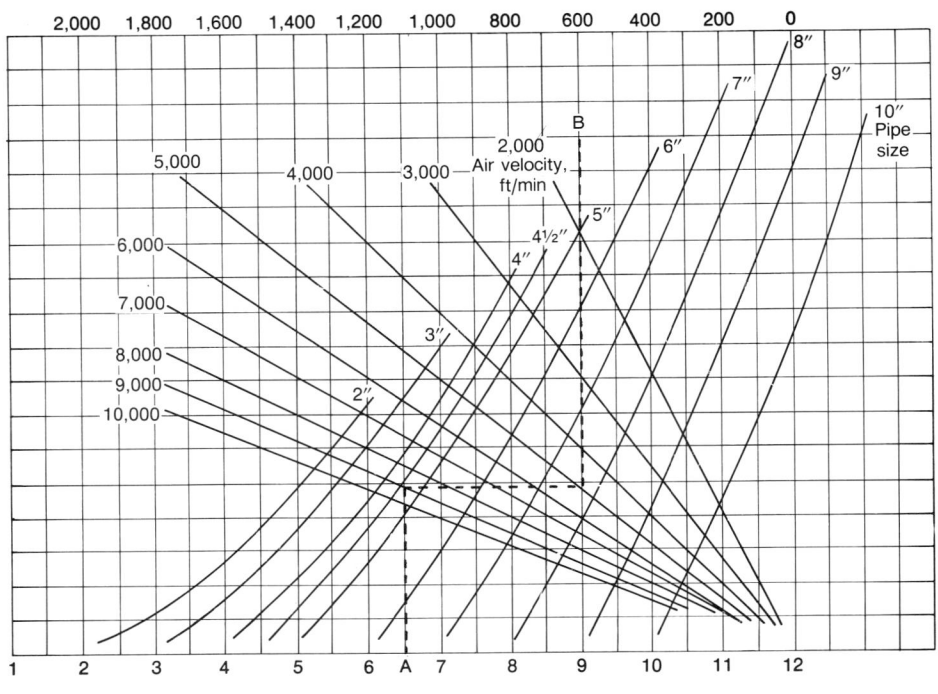

To determine air velocity in the annulus when the pipe size, hole diameter, and air volume are known, follow vertical hole diameter line upward to its intersection with pipe size line. Move horizontally to intersect air volume line. Read air velocity on diagonal air velocity line.

Example: A 6½-in hole is being drilled using 4½-in drill pipe with 600 cfm air volume passing through the annulus. Follow the hole diameter line to Point A, its interesection with the pipe size line. Move horizontally until the line intersects the air volume line, Point B. Read annulus air velocity at Point B (if necessary, interpolate between air volume lines). In this example, the uphole velocity is 5,000 ft/min. An uphole velocity of 3,000 ft/min is the minimum requirement.

Appendix 11.G.
Comparable Drilling Fluid Products by Tradenames

Description	Imco Services	Baroid	Magcobar	Johnson	Primary Application
WEIGHTING AGENTS AND VISCOSIFIERS					
Bentonite	IMCO GEL	Aquagel	Magcogel		Viscosity and filtration control in water-based drilling fluids.
Sub-bentonite	IMCO KLAY	Baroco	High Yield Blended Clay		For use when larger particle size is desired for viscosity and filtration control.
Attapulgite	IMCO BRINEGEL	Zeogel	Salt Gel		Visocosifier in salt-water drilling fluids
Beneficiated Bentonite	IMCO HYB	Quik-Gel	Kwik-Thik		Quick viscosity in fresh water with minimum chemical treatment
POLYMERS					
Natural Organic Polymer				Revert	Light-weight, low-solids drilling fluid for viscosity and filtration control in water-based drilling fluids
Synthetic Organic Polymer		E-Z Mud			Light-weight, low-solids drilling fluid for viscosity and filtration control in water-based drilling fluids made with bentonitic additives
FOAMING AGENTS					
Foaming Agents	IMCO Foamant	Quik-Foam	Magco Rapid Foam	J-Foam	Used in air drilling to increase hole cleaning capability and reduce compressor requirements for a given depth or rate of water inflow
Sodium Tetra-pyrophosphate ($Na_4P_2O_7$)	IMCO PHOS	Barafos	Magco-Phos		Thinner for fresh-water drilling fluids with clay additives
Sodium Acid Pyrophosphate ($Na_2H_2P_2O_7$)	SAPP	SAPP	SAPP		For treating cement contamination in drilling fluids made with clay additives*
DETERGENTS					
Detergent	IMCO MD	Con Det	D-D		Used in water-based drilling fluids made with clay additives to aid in dropping sand. Reduces torque and minimizes bit balling
BACTERICIDES					
Paraformaldehyde	IMCO PRESERVALOID	Aldacide	My-Lo-Jel Preservative		Prevents organic additives from degradation
COMMERCIAL CHEMICALS					
Sodium Hydroxide (NaOH)	Caustic Soda	Caustic Soda	Caustic Soda		For pH control in water-based drilling fluids
Sodium Chloride (NaCl)	Salt	Salt	Salt		For solution weighting of water-based drilling fluids made with polymeric additives
Calcium Chloride ($CaCl_2$)	Salt	Salt	Salt		For solution weighting of water-based drilling fluids made with polymeric additives

*Also used for development of wells drilled with bentonite additives.

APPENDIX 11.H.
Major Drilling Fluid Companies

American Colloid
5100 Suffield Court
Skokie, IL 60077
312/966-5720

IMCO Services
A Division of Halliburton Company
2400 West Loop South
P.O. Box 22605
Houston, TX 77227
713/671-4800

Magcobar Division
Dresser Industries, Inc.
Dresser Tower
P.O. Box 6504
Houston, TX 77265
713/972-6011

NL Baroid
P.O. Box 1675
Houston, TX 77251
713/527-1100

APPENDIX 11.I.
Velocities of Water that will Cause Sand to Rise

Diameter of Grain, in.	Diam. of Grain, mm	Retained on size Screen Number (Tyler Std. Scale)	Velocity of Water, ft/sec.
Up to 0.0097	Up to 0.25	60	0.0 to 0.10
0.0097 to 0.0194	0.25 to 0.50	32	0.12 to 0.22
0.0194 to 0.0368	0.50 to 1.00	16	0.25 to 0.33
0.0368 to 0.0780	1.00 to 2.00	9	0.37 to 0.56
0.0780 to 0.1590	2.00 to 4.00	5	0.60 to 2.60

(W. S. Tyler Company)

APPENDIX 11.J.
Ascending Velocity of Drilling Fluid (in ft/min)

Diameter (in)		Drilling Fluid Circulation (gpm)					
Drill Pipe	Hole	100	200	300	400	500	600
2⅞	4½	209	417				
	5	149	299	448			
	5⅝	107	214	321	428		
	6	90	180	270	361	451	
	6½	74	147	221	294	368	441
3½	6	105	211	316	421		
	6½	83	167	250	333	417	
	7	68	136	204	272	340	408
	7⅝	54	109	163	218	272	327
	7⅞	50	100	151	201	251	301
	8½	42	83	125	167	208	250
	9	36	73	109	145	182	218
	9⅝	31	62	93	124	155	187
4½	7⅝	66	132	198	264	330	396
	7⅞	60	120	180	239	299	359
	8½	48	96	144	192	240	288
	9	41	82	123	165	206	247
	9⅝	35	69	104	138	173	207
	10⅝	27	54	81	108	135	162
	11	25	50	74	99	124	149
	12¼	19	39	58	77	96	116
6⅝	11	32	65	97	130	162	195
	12¼	24	47	71	94	118	141
	13	20	40	60	80	100	120
	14¾	14	29	43	58	72	86
	17	10	20	31	41	51	61
	20	7	14	21	28	35	42
	22	6	11	17	23	28	34
	24	5	9	14	19	23	28
	26	4	8	12	16	20	24

Ascending velocity of drilling fluid can be calculated by using the following equation:

$$V = \frac{25 \cdot Q}{D^2 - d^2}$$

where
V = velocity of drilling fluid, in ft/min
Q = Discharge, gpm
D = diameter of hole, in
d = outside diameter of drill pipe, in

APPENDIX 11.K.
Weight, Fluid Pressure, and Buoyancy Factors

Calculations Based on 1 Cu. Ft. Water = 62.4 lbs.

WEIGHT PER GALLON	WEIGHT PER CU. FOOT	SPECIFIC GRAVITY	FLUID HEAD		*BUOYANCY FACTOR FOR STEEL	WEIGHT PER GALLON	WEIGHT PER CU. FOOT	SPECIFIC GRAVITY	FLUID HEAD		*BUOYANCY FACTOR FOR STEEL
			PRESSURE PER 100 FEET OF HEIGHT	FEET OF FLUID PER POUND OF PRESSURE					PRESSURE PER 100 FEET OF HEIGHT	FEET OF FLUID PER POUND OF PRESSURE	
Pounds	Pounds		Lbs. per Sq. In.	Feet		Pounds	Pounds		Lbs. per Sq. In.	Feet	
8.34	62.4	1.00	43.3	2.31	.873	11.8	88.3	1.41	61.3	1.63	.820
8.4	62.8	1.01	43.6	2.29	.872	12.0	89.8	1.44	62.3	1.60	.817
8.6	64.3	1.03	44.7	2.24	.869	12.2	91.3	1.46	63.4	1.58	.814
8.8	65.8	1.05	45.7	2.19	.866	12.4	92.8	1.49	64.4	1.55	.811
9.0	67.3	1.08	46.8	2.14	.862	12.6	94.3	1.51	65.5	1.53	.807
9.2	68.8	1.10	47.8	2.09	.859	12.8	95.8	1.53	66.5	1.50	.804
9.4	70.3	1.13	48.8	2.05	.856	13.0	97.2	1.56	67.5	1.48	.801
9.6	71.8	1.15	49.9	2.01	.853	13.2	98.7	1.58	68.6	1.46	.798
9.8	73.3	1.17	50.9	1.96	.850	13.4	100.2	1.61	69.6	1.44	.795
10.0	74.8	1.20	51.9	1.93	.847	13.6	101.7	1.63	70.6	1.42	.792
10.2	76.3	1.22	53.0	1.89	.844	13.8	103.2	1.65	71.7	1.39	.789
10.4	77.8	1.25	54.0	1.85	.841	14.0	104.7	1.68	72.7	1.38	.786
10.6	79.3	1.27	55.1	1.82	.838	14.5	108.5	1.74	75.3	1.33	.778
10.8	80.8	1.29	56.1	1.78	.835	15.0	112.2	1.80	77.9	1.28	.771
11.0	82.3	1.32	57.1	1.75	.832	15.5	115.9	1.86	80.5	1.24	.763
11.2	83.8	1.34	58.2	1.72	.829	16.0	119.7	1.92	83.1	1.20	.756
11.4	85.3	1.37	59.2	1.69	.826	16.5	123.4	1.98	85.7	1.17	.748
11.6	86.8	1.39	60.3	1.66	.823	17.0	127.2	2.04	88.3	1.13	.740
						17.5	130.9	2.10	90.9	1.10	.733
						18.0	134.6	2.16	93.5	1.07	.725

*Buoyancy Factor: For converting weight of steel in air to weight in mud, for checking dead weight on hook. Applies only to strings of pipe which are completely filled with mud, while handling.

Actual Load = Length of string × Weight per Foot × Buoyancy Factor.

If the actual weight on the bit must be increased by 25,000 lb by the use of drill collars and the drilling fluid weight is 10 lb/gal, the drill collar weight in air should be:

Weight + 10% safety factor ÷ buoyancy factor

25,000 + 2,500 ÷ 0.847 = 32,500 lb

The 10 percent safety factor is added to offset the lift on the drill collars caused by the ascending drilling fluid.

(NL McCullough, 1980)

APPENDIX 11.L.
Volume of Water in Casing or Hole

Diameter of Casing or Hole (In)	Gallons per foot of Depth	Cubic Feet per Foot of Depth	Liters per Meter of Depth	Cubic Meters per Meter of Depth
1	0.041	0.0055	0.509	0.509×10^{-3}
1½	0.092	0.0123	1.142	1.142×10^{-3}
2	0.163	0.0218	2.024	2.024×10^{-3}
2½	0.255	0.0341	3.167	3.167×10^{-3}
3	0.367	0.0491	4.558	4.558×10^{-3}
3½	0.500	0.0668	6.209	6.209×10^{-3}
4	0.653	0.0873	8.110	8.110×10^{-3}
4½	0.826	0.1104	10.26	10.26×10^{-3}
5	1.020	0.1364	12.67	12.67×10^{-3}
5½	1.234	0.1650	15.33	15.33×10^{-3}
6	1.469	0.1963	18.24	18.24×10^{-3}
7	2.000	0.2673	24.84	24.84×10^{-3}
8	2.611	0.3491	32.43	32.43×10^{-3}
9	3.305	0.4418	41.04	41.04×10^{-3}
10	4.080	0.5454	50.67	50.67×10^{-3}
11	4.937	0.6600	61.31	61.31×10^{-3}
12	5.875	0.7854	72.96	72.96×10^{-3}
14	8.000	1.069	99.35	99.35×10^{-3}
16	10.44	1.396	129.65	129.65×10^{-3}
18	13.22	1.767	164.18	164.18×10^{-3}
20	16.32	2.182	202.68	202.68×10^{-3}
22	19.75	2.640	245.28	245.28×10^{-3}
24	23.50	3.142	291.85	291.85×10^{-3}
26	27.58	3.687	342.52	342.52×10^{-3}
28	32.00	4.276	397.41	397.41×10^{-3}
30	36.72	4.909	456.02	456.02×10^{-3}
32	41.78	5.585	518.87	518.87×10^{-3}
34	47.16	6.305	585.68	585.68×10^{-3}
36	52.88	7.069	656.72	656.72×10^{-3}

1 Gallon = 3.785 Liters
1 Meter = 3.281 Feet
1 Gallon Water Weighs 8.33 lbs. = 3.785 Kilograms
1 Liter Water Weighs 1 Kilogram = 2.205 lbs.
1 Gallon per foot of depth = 12.419 liters per foot of depth
1 Gallon per meter of depth = 12.419×10^{-3} cubic meters per meter of depth

APPENDIX 12.A.
Representative Open Areas of Screens

Screen Diameter	Slot Size	Continuous Slot in²/ft	%	Louvered (Minimum open area) in²/ft	%	Louvered (Maximum open area) in²/ft	%	Bridge Slot in²/ft	%	Bridge Slot in²/ft	%	Mill Slotted* (Vertical) in²/ft	%	Slotted Pipe (Horizontal) slots/ft	in²/ft	%	Plastic Continuous Slot in²/ft	%	Slotted Plastic† in²/ft	%	Concrete in²/ft	%	Fiberglass Reinforced Plastic Continuous Slot‡ in²/ft	%
4" ID	20	44	25	—	—	—	—	—	—	—	—	—	—	—	—	—	22	13	—	—			25	12
	30	58	33	—	—	—	—	—	—	—	—	—	—	—	—	—	31	18	—	—			37	17
	40	72	41	—	—	—	—	13	8	—	—	6	4	—	—	—	40	23	13	8			48	23
	50	78	45	—	—	—	—	19	12	—	—	8	5	—	—	—	47	27	18	11			—	—
	60	90	52	—	—	—	—	—	—	—	—	—	—	—	—	—	52	30	—	—			—	—
	80	102	59	—	—	—	—	—	—	—	—	—	—	—	—	—	—	—	—	—			—	—
	90	105	60	—	—	—	—	29	17	—	—	12	7	—	—	—	—	—	—	—			—	—
	95	106	61	—	—	—	—	—	—	—	—	—	—	—	—	—	—	—	—	—			—	—
	100	112	64	—	—	—	—	—	—	—	—	—	—	—	—	—	—	—	—	—			—	—
	120	99	57	—	—	—	—	41	24	—	—	16	10	90	16	3	—	—	—	—			—	—
	125	100	58	—	—	—	—	—	—	—	—	—	—	130	23	5	—	—	26	16			—	—
6" ID	20	45	18	—	—	—	—	—	—	—	—	—	—	240	43	9	25	10	—	—			—	—
	30	61	25	2	1	7	3	14	6	8	3	9	4	90	22	5	36	14	12	5			14	5
	40	77	31	—	—	—	—	—	—	—	—	11	5	130	31	7	45	18	18	7			—	—
	50	88	35	4	2	11	5	21	8	17	6	17	7	240	58	12	54	22	23	9			29	9
	60	100	40	—	—	—	—	31	12	—	—	—	—	—	—	—	62	25	28	11			—	—
	90	124	50	—	—	15	7	—	—	—	—	23	9	—	—	—	—	—	—	32	13			—	—
	95	127	51	5	2	—	—	43	17	—	—	—	—	—	—	—	—	—	—	—	—			41	14
	100	131	53	—	—	—	—	—	—	—	—	—	—	—	—	—	—	—	46	18			59	20
	120	141	57	—	—	—	—	—	—	—	—	—	—	90	27	6	—	—	51	20			65	22
	125	127	51	—	—	—	—	—	—	—	—	—	—	130	39	8	—	—	—	—			—	—
8" ID	20	58	18	—	—	—	—	—	—	—	—	—	—	240	72	15	41	12	—	—			—	—
	30	80	25	—	—	—	—	—	—	—	—	—	—	—	—	—	57	18	26	8			—	—
	40	98	30	—	—	—	—	—	—	—	—	—	—	—	—	—	72	22	—	—			—	—
	50	114	35	4	1	10	3	—	—	—	—	15	5	—	—	—	86	26	47	14			—	—
	60	135	41	6	2	15	5	—	—	—	—	23	7	—	—	—	93	29	—	—			—	—
	95	165	51	—	—	—	—	—	—	—	—	31	10	—	—	—	—	—	67	21			—	—
	100	169	52	7	2	20	6	—	—	—	—	—	—	—	—	—	—	—	—	—			—	—
	125	166	51	—	—	—	—	—	—	—	—	—	—	—	—	—	—	—	—	—			—	—
10" ID	20	72	18	—	—	—	—	—	—	—	—	—	—	—	—	—	—	—	18	4			—	—
	30	100	25	—	—	—	—	—	—	—	—	—	—	—	—	—	—	—	26	6			—	—
	40	122	30	—	—	—	—	—	—	—	—	15	4	—	—	—	—	—	33	8			—	—
	50	143	35	—	—	—	—	—	—	—	—	—	—	—	—	—	—	—	39	10			—	—

Appendix 12.A. Continued

Screen Diameter	Slot Size	Continuous Slot in²/ft	%	Louvered (Minimum open area) in²/ft	%	Louvered (Maximum open area) in²/ft	%	Bridge Slot in²/ft	%	Bridge Slot in²/ft	%	Mill Slotted* (Vertical) in²/ft	%	Slotted Pipe (Horizontal) slots/ft	in²/ft	%	Plastic Continuous Slot in²/ft	%	Slotted Plastic† in²/ft	%	Concrete in²/ft	%	Fiberglass Reinforced Plastic Continuous Slot‡ in²/ft	%
12" ID	60	135	33	4	1	16	4	—	—	22	5	19	5						45	11				
	90	174	43	7	2	—	6	—	—	—	—	—	7						—	—			32	6
	95	179	44	—	—	24	—	—	—	—	—	28	—						65	16			64	12
	100	186	46	—	—	—	—	—	—	—	—	—	—						72	18			94	18
	120	203	50	—	—	32	8	—	—	—	—	38	9						—	—			138	26
	125	207	51	9	2	—	—	—	—	48	12	—	—						—	—			154	30
	20	69	14	—	—	—	—	—	—	12	3	—	—						—	—				
	30	77	16	—	—	—	—	22	5	—	—	17	4						38	8				
	40	99	21	—	—	20	4	33	7	27	6	21	5						52	11			40	6
	50	117	24	6	1	—	—	49	10	—	—	—	—						—	—			81	12
	60	135	28	—	—	30	7	—	—	—	—	32	7						75	16			119	17
	90	176	37	9	2	—	—	—	—	—	—	—	—						83	18				
	95	182	38	—	—	39	9	68	14	59	13	43	9						—	—				
16" OD	20	68	11	—	—	—	—	—	—	16	3	—	—	104	19	3			35	6			6	
	30	97	16	—	—	—	—	—	—	—	—	—	—	141	29	5			52	9				
	40	124	21	—	—	—	—	—	—	—	—	22	4	192	35	6			69	11			26	26
	50	146	24	—	—	24	4	—	—	35	6	27	5	104	25	4			86	14			29	29
	60	169	28	7	1	—	—	—	—	—	—	—	—	141	34	6			—	—				
														192	46	8								
	80	206	34	—	—	—	—	—	—	—	—	—	—						138	23				
	90	221	37	11	2	35	6	—	—	—	—	41	7	104	31	5			—	—			176	
	95	228	38	—	—	—	—	—	—	—	—	—	—	141	42	7			173	29			196	
	100	238	40	—	—	—	—	—	—	—	—	—	—	192	58	10			—	—				
17" OD (13" ID)	125	268	45	15	2	47	8			78	13	55	9								25	4		
18" OD	190																							
	20	76	11	—	—	—	—	—	—	18	3	—	—	121	22	3								
	30	109	16	—	—	—	—	—	—	—	—	—	—	199	36	5								
	40	137	20	—	—	28	4	—	—	40	6	—	—	336	61	9								
	60	187	28	—	—	—	—	—	—	—	—	—	—											

Appendix 12.A. Continued

Screen Diameter	Slot Size	Continuous Slot in²/ft	%	Louvered (Minimum open area) in²/ft	%	Louvered (Maximum open area) in²/ft	%	Bridge Slot in²/ft	%	Bridge Slot in²/ft	%	Mill Slotted* (Vertical) in²/ft	%	Slotted Pipe (Horizontal) slots/ft	in²/ft	%	Plastic Continuous Slot in²/ft	%	Slotted Plastic† in²/ft	%	Concrete in²/ft	%	Fiberglass Reinforced Plastic Continuous Slot‡ in²/ft	%
	80	228	34	—	—	—	—							121	29	4								
														199	48	7								
														336	81	12								
	95	255	38	11	2	41	6							—	—	—								
	100	263	39	—	—	—	—							121	36	5								
														199	60	9								
														336	101	15								
	125	236	35	15	2	55	8			88	13			—	—	—								
21″ OD (16″ ID)	190																					30	4		
23″ OD (18″ ID)	190																					40	5		

This chart is to give the reader a representative guide to typical open areas for various types of screens. The actual open area may vary somewhat above or below these figures depending on the materials used to build the screen, the collapse and column strength requirements, and the manufacturing techniques. Screen manufacturers should be contacted for specific open area information.

*Diameter of 4″ screen is 4.5″ OD.
†Screen diameters are 4.5″ OD, 6.5″ OD, and 12.5″ OD.
‡Screen diameters are 4.5″ ID, 6.5″ ID, 12″ ID, and 16″ ID.

APPENDIX 12.B.
Correlation Chart of Screen Openings and Sieve Sizes

Geologic Material Grain-size Range	Johnson Slot No.	Gauze No.	Tyler Sieve No.	Tyler Size of Openings Inches	Tyler Size of Openings mm	U.S. Standard Sieve No.	U.S. Standard Size of Openings Inches
clay & silt	—	—	400	0.0015	0.038	400	0.0015
	—	—	325	0.0017	0.043	325	0.0017
	—	—	270	0.0021	0.053	270	0.0021
	—	—	250	0.0024	0.061	230	0.0024
	—	—	200	0.0029	0.074	200	0.0029
fine sand	—	—	170	0.0035	0.088	170	0.0035
	—	—	150	0.0041	0.104	140	0.0041
	—	—	115	0.0049	0.124	120	0.0049
	6	90	100	0.0058	0.147	100	0.0059
	7	80	80	0.0069	0.175	80	0.0070
	8	70	65	0.0082	0.208	70	0.0083
	10	60	60	0.0097	0.246	60	0.0098
medium sand	12	50	48	0.0116	0.295	50	0.0117
	14	—	42	0.0138	0.351	45	0.0138
	16		35	0.0164	0.417	40	0.0165 (1/64)
	18	40	—	0.0180	0.457	—	0.0180
	20		32	0.0195	0.495	35	0.0197
	23		28	0.0232	0.589	30	0.0232
coarse sand	25	30	—	0.0250	0.635	—	0.0250
	28		24	0.0276	0.701	25	0.0280
	31		—	0.0310	0.788	—	0.0310 (1/32)
	33		20	0.0328	0.833	20	0.0331
	35	20	—	0.035	0.889	—	0.0350
	39		16	0.039	0.991	18	0.0394
very coarse sand	47		14	0.046	1.168	16	0.0469
	56		12	0.055	1.397	14	0.0555
	62		—	0.062	1.590	—	0.062 (1/16)
	66		10	0.065	1.651	12	0.0661
	79		9	0.078	1.981	10	0.0787
very fine gravel	93		8	0.093	2.362	8	0.0931
	94		—	0.094	2.390	—	0.094 (3/32)
	111		7	0.110	2.794	7	0.111
	125		—	0.125	3.180	—	0.125 (1/8)
	132		6	0.131	3.327	6	0.132
fine gravel	157		5	0.156	3.962	5	0.157
	187		4	0.185	4.699	4	0.187 (3/16)
	223		3½	0.221	5.613	3½	0.223
	250		—	0.250	6.350	¼	0.250 (¼)
	263		3	0.263	6.680	—	0.263
	312		2½	0.312	7.925	5/16	0.312 (5/16)
	375		0.371	0.371	9.423	3/8	0.375 (3/8)
	438		0.441	0.441	11.20	7/16	0.438 (7/16)
	500		0.525	0.525	13.33	½	0.500 (½)

APPENDIX 13.A.
Making Pipe and Tubing

Steel pipe and tubing are constructed by three major methods. Each process offers certain advantages in producing specific end products. For example, the continuous butt weld pipe mill makes much of the standard pipe used in plumbing. The resistance weld tubing mill is most often utilized in making products of light wall thickness and small diameter, products used widely for mechanical and structural purposes. Seamless products are used as boiler tubing and by the oil and chemical industries.

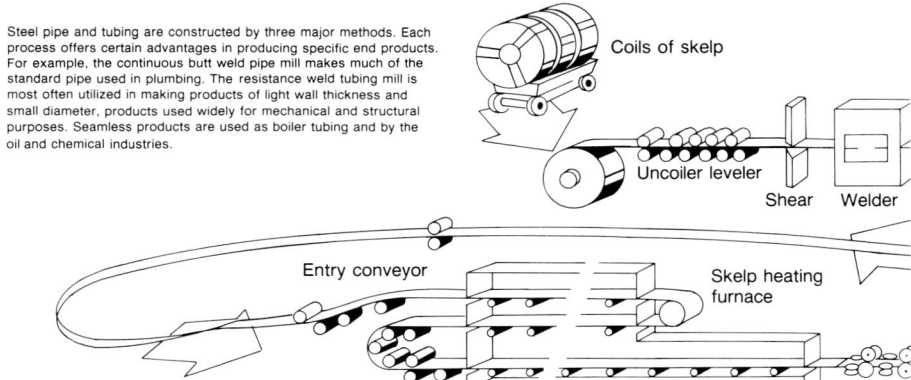

Electric Resistance Welded Tubing Mill
After rotary shears trim the edges of skelp parallel, the contour of the mill-rolls forms tubing, step by step. Carefully guided by rolls, the tube edges are electrically welded together by pressure and the heat caused by steel's resistance to electricity. The welder wheel comprises two copper discs, one positive and the other negative in charge. In continuous operation, provision is made to weld the square-cropped end of the coil to the leading end of the next without stopping the progress of forming and welding.

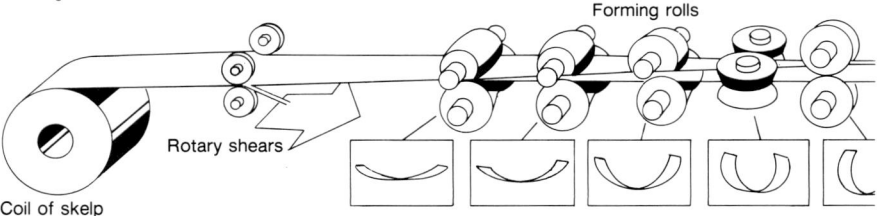

Seamless Tube Mill — Mandrel Type
Round steel billets, often called tube rounds, are first heated to a uniform rolling temperature throughout. A piercing mill rotates the steel under extreme pressure until the metal opens up in the center, permitting the mandrel to enter. Then a cylindrical mandrel bar is inserted into the pieced shell and the desired diameter, wall thickness, and length are obtained as the shell with the bar inside it progress through successive roll stands of the mandrel mill.

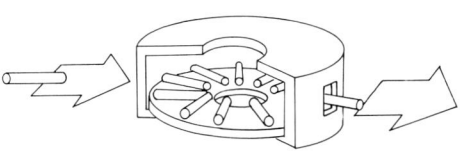
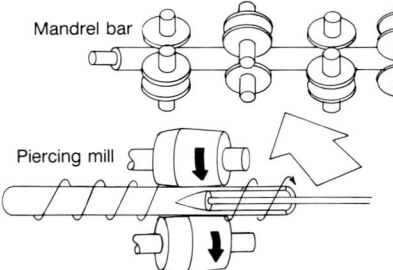

(American Iron and Steel Institute)

Continuous-Weld Pipe Mill

Following the flow of this continuous-weld pipe mill, a coil of flat rolled steel called skelp is drawn through the line by pinch rolls. A long loop of skelp is created as a reserve so that pipemaking can continue while the trailing end of one coil can be welded to the leading end of a new one. The skelp edges are heated to welding temperature as the steel winds through a long furnace. Then the mill forms the tube and squeezes the hot edges together to make a solid weld. Sizing rolls correct pipe diameter before saws cut it to length.

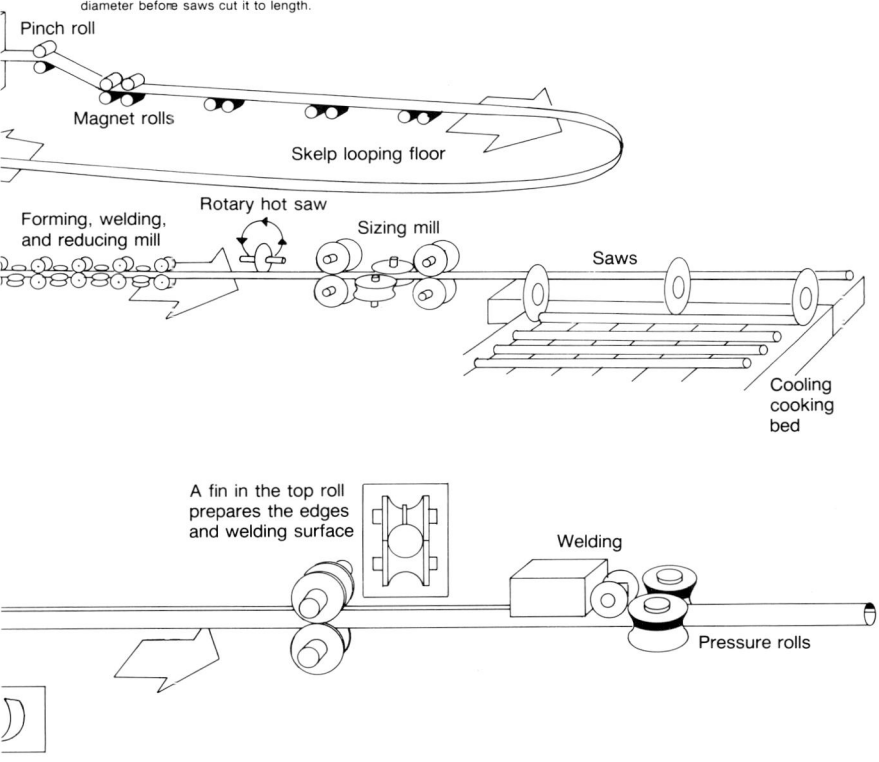

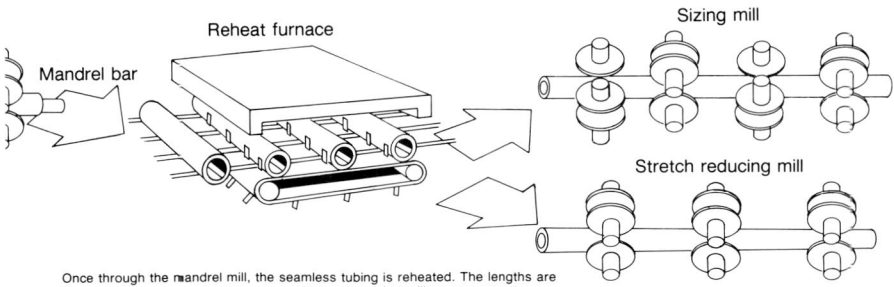

Once through the mandrel mill, the seamless tubing is reheated. The lengths are then rolled to very precise tolerances either on a sizing mill or on a stretch reducing mill. In the latter, each pair of rolls rotates faster than the pair preceding it, thus stretching the steel and reducing the tubes' diameter and wall thickness simultaneously.

APPENDIX 13.B-1
Dimensions, Weights, and Test Pressures for Plain End Standard Pipe

Nominal Size, in.	Outside Diameter, in.[A]	Wall Thickness, in.[A]	Nominal Weight per ft, Plain End, lb[B]	Weight Class	Schedule No.	Test Pressure[C,D], psi		
						Butt-Welded	Grade A	Grade B
1/8	0.405	0.068	0.24	STD	40	700	700	700
		0.095	0.31	XS	80	850	850	850
1/4	0.540	0.088	0.42	STD	40	700	700	700
		0.119	0.54	XS	80	850	850	850
3/8	0.675	0.091	0.57	STD	40	700	700	700
		0.126	0.74	XS	80	850	850	850
1/2	0.840	0.109	0.85	STD	40	700	700	700
		0.147	1.09	XS	80	850	850	850
		0.188	1.31		160	900	900	900
		0.294	1.71	XXS		1000	1000	1000
3/4	1.050	0.113	1.13	STD	40	700	700	700
		0.154	1.47	XS	80	850	850	850
		0.219	1.94		160	950	950	950
		0.308	2.44	XXS		1000	1000	1000
1	1.315	0.133	1.68	STD	40	700	700	700
		0.179	2.17	XS	80	850	850	850
		0.250	2.84		160	950	950	950
		0.358	3.66	XXS		1000	1000	1000
1 1/4	1.660	0.140	2.27	STD	40	1000	1200	1300
		0.191	3.00	XS	80	1300	1800	1900
		0.250	3.76		160	1350	1900	2000
		0.382	5.21	XXS		1400	2200	2300
1 1/2	1.900	0.145	2.72	STD	40	1000	1200	1300
		0.200	3.63	XS	80	1300	1800	1900
		0.281	4.86		160	1350	1950	2050
		0.400	6.41	XXS		1400	2200	2300
2	2.375	0.154	3.65	STD	40	1000	2300	2500
		0.218	5.02	XS	80	1300	2500	2500
		0.344	7.46		160	1400	2500	2500
		0.436	9.03	XXS		1400	2500	2500
2 1/2	2.875	0.203	5.79	STD	40	1000	2500	2500
		0.276	7.66	XS	80	1300	2500	2500
		0.375	10.01		160	1400	2500	2500
		0.552	13.70	XXS		1400	2500	2500
3	3.500	0.125	4.51			800	1290	1500
		0.156	5.57			1000	1600	1870
		0.188	6.65			1000	1930	2260

Appendix 13.B-1. Continued

Nominal Size, in.	Outside Diameter, in.[A]	Wall Thickness, in.[A]	Nominal Weight per ft, Plain End, lb[B]	Weight Class	Schedule No.	Test Pressure[C,D], psi		
						Butt-Welded	Grade A	Grade B
3	3.500	0.216	7.58	STD	40	1000	2220	2500
		0.250	8.68			1300	2500	2500
		0.281	9.66			1300	2500	2500
		0.300	10.25	XS	80	1300	2500	2500
		0.438	14.32		160		2500	2500
		0.600	18.58	XXS			2500	2500
3½	4.000	0.125	5.17			800	1120	1310
		0.156	6.40			1000	1400	1640
		0.188	7.65			1200	1690	1970
		0.226	9.11	STD	40	1200	2030	2370
		0.250	10.01			1300	2250	2500
		0.281	11.16			1500	2500	2500
		0.318	12.51	XS	80	1700	2800	2800
4	4.500	0.125	5.84			800	1000	1170
		0.156	7.24			1000	1250	1460
		0.188	8.66			1200	1500	1750
		0.219	10.01			1200	1750	2040
		0.237	10.79	STD	40	1200	1900	2210
		0.250	11.35			1300	2000	2330
		0.281	12.66			1400	2250	2620
		0.312	13.98			1600	2500	2800
		0.337	14.98	XS	80	1700	2700	2800
		0.438	19.00		120		2800	2800
		0.531	22.51		160		2800	2800
		0.674	27.54	XXS			2800	2800
5	5.563	0.156	9.01				1010	1180
		0.188	10.79				1220	1420
		0.219	12.50				1420	1650
		0.258	14.62	STD	40		1670	1950
		0.281	15.85				1820	2120
		0.312	17.50				2020	2360
		0.344	19.17				2230	2600
		0.375	20.78	XS	80		2430	2800
		0.500	27.04		120		2800	2800
		0.625	32.96		160		2800	2800
		0.750	38.55	XXS			2800	2800
6	6.625	0.188	12.92				1020	1190
		0.219	14.98				1190	1390
		0.250	17.02				1360	1580
		0.280	18.97	STD	40		1520	1780
		0.312	21.04				1700	1980

Appendix 13.B-1. Continued

Nominal Size, in.	Outside Diameter, in.[4]	Wall Thickness, in.[4]	Nominal Weight per ft, Plain End, lb[B]	Weight Class	Schedule No.	Test Pressure[C,D], psi		
						Butt-Welded	Grade A	Grade B
6	6.625	0.344	23.08				1870	2180
		0.375	25.03				2040	2380
		0.432	28.57	XS	80		2350	2740
		0.562	36.39		120		2800	2800
		0.719	45.35		160		2800	2800
		0.864	53.16	XXS			2800	2800
8	8.625	0.188	16.94				780	920
		0.203	18.26				850	1000
		0.219	19.66				910	1070
		0.250	22.36		20		1040	1220
		0.277	24.70		30		1160	1350
		0.312	27.70				1300	1520
		0.322	28.55	STD	40		1340	1570
		0.344	30.42				1440	1680
		0.375	33.04				1570	1830
		0.406	35.64		60		1700	2000
		0.438	38.30				1830	2130
		0.500	43.39	XS	80		2090	2430
		0.594	50.95		100		2500	2800
		0.719	60.71		120		2800	2800
		0.812	67.76		140		2800	2800
		0.875	72.42	XXS			2800	2800
		0.906	74.69		160		2800	2800
10	10.750	0.188	21.21				630	730
		0.203	22.87				680	800
		0.219	24.63				730	860
		0.250	28.04		20		840	980
		0.279	31.20				930	1090
		0.307	34.24		30		1030	1200
		0.344	38.23				1150	1340
		0.365	40.48	STD	40		1220	1430
		0.438	48.19				1470	1710
		0.500	54.74	XS	60		1670	1950
		0.594	64.43		80		1990	2320
		0.719	77.03		100		2410	2800
		0.844	89.29		120		2800	2800
		1.000	104.13	XXS	140		2800	2800
		1.125	115.65		160		2800	2800
12	12.750	0.203	27.20				570	670
		0.219	29.31				620	720
		0.250	33.38		20		710	820

Appendix 13.B-1. Continued

Nominal Size, in.	Outside Diameter, in.[A]	Wall Thickness, in.[A]	Nominal Weight per ft, Plain End, lb[B]	Weight Class	Schedule No.	Test Pressure[C,D], psi		
						Butt-Welded	Grade A	Grade B
12	12.750	0.281	37.42				790	930
		0.312	41.45				880	1030
		0.330	43.77		30		930	1090
		0.344	45.58				970	1130
		0.375	49.56	STD			1060	1240
		0.406	53.52		40		1150	1340
		0.438	57.59				1240	1440
		0.500	65.42	XS			1410	1650
		0.562	73.15		60		1590	1850
		0.688	88.63		80		1940	2270
		0.844	107.32		100		2390	2780
		1.000	125.49	XXS	120		2800	2800
		1.125	139.68		140		2800	2800
		1.312	160.27		160		2800	2800
14	14.000	0.210	30.93				540	630
		0.219	32.23				560	660
		0.250	36.71		10		640	750
		0.281	41.17				720	840
		0.312	45.61		20		800	940
		0.344	50.17				880	1030
		0.375	54.57	STD	30		960	1120
		0.438	63.44		40		1130	1310
		0.469	67.78				1210	1410
		0.500	72.09	XS			1290	1500
		0.594	85.05		60		1530	1790
		0.750	106.13		80		1930	2250
		0.938	130.85		100		2410	2800
		1.094	150.79		120		2800	2800
		1.250	170.22		140		2800	2800
		1.406	189.11		160		2800	2800
		2.000	256.32				2800	2800
		2.125	269.51				2800	2800
		2.200	277.26				2800	2800
		2.500	307.05				2800	2800
16	16.000	0.219	36.91				490	570
		0.250	42.05		10		560	660
		0.281	47.17				630	740
		0.312	52.27		20		700	820
		0.344	57.52				770	900
		0.375	62.58	STD	30		840	980
		0.438	72.80				990	1150

Appendix 13.B-1. Continued

Nominal Size, in.	Outside Diameter, in.[A]	Wall Thickness, in.[A]	Nominal Weight per ft, Plain End, lb[B]	Weight Class	Schedule No.	Test Pressure[C,D], psi		
						Butt-Welded	Grade A	Grade B
16	16.000	0.469	77.79				1060	1230
		0.500	82.77	XS	40		1120	1310
		0.656	107.50		60		1480	1720
		0.844	136.62		80		1900	2220
		1.031	164.82		100		2320	2710
		1.219	192.43		120		2740	2800
		1.438	223.64		140		2800	2800
		1.594	245.25		160		2800	2800
18	18.000	0.250	47.39		10		500	580
		0.281	53.18				560	660
		0.312	58.94		20		620	730
		0.344	64.87				690	800
		0.375	70.59	STD			750	880
		0.406	76.29				810	950
		0.438	82.15		30		880	1020
		0.469	87.81				940	1090
		0.500	93.45	XS			1000	1170
		0.562	104.67		40		1120	1310
		0.750	138.17		60		1500	1750
		0.938	170.92		80		1880	2190
		1.156	207.96		100		2310	2700
		1.375	244.14		120		2750	2800
		1.562	274.22		140		2800	2800
		1.781	308.50		160		2800	2800
20	20.000	0.250	52.73		10		450	520
		0.281	59.18				510	590
		0.312	65.60				560	660
		0.344	72.21				620	720
		0.375	78.60	STD	20		680	790
		0.406	84.96				730	850
		0.438	91.51				790	920
		0.469	97.83				850	950
		0.500	104.13	XS	30		900	1050
		0.594	123.11		40		1170	1250
		0.812	166.40		60		1460	1710
		1.031	208.87		80		1860	2170
		1.281	256.10		100		2310	2690
		1.500	296.37		120		2700	2800
		1.750	341.10		140		2800	2800
		1.969	379.17		160		2800	2800
24	24.000	0.250	63.41		10		380	440
		0.281	71.18				420	490

Appendix 13.B-1. Continued

Nominal Size, in.	Outside Diameter, in.[A]	Wall Thickness, in.[A]	Nominal Weight per ft, Plain End, lb[B]	Weight Class	Schedule No.	Test Pressure[C,D], psi		
						Butt-Welded	Grade A	Grade B
24	24.000	0.312	78.93				470	550
		0.344	86.91				520	600
		0.375	94.62	STD	20		560	660
		0.406	102.31				610	710
		0.438	110.22				660	770
		0.469	117.86				700	820
		0.500	125.49	XS			750	880
		0.562	140.68		30		840	980
		0.688	171.29		40		1030	1200
		0.938	231.03				1410	1640
		0.969	238.85		60		1450	1700
		1.219	296.58		80		1830	2130
		1.531	367.39		100		2300	2680
		1.812	429.39		120		2720	2800
		2.062	483.12		140		2800	2800
		2.344	542.14		160		2800	2800
26	26.000	0.250	68.75				350	400
		0.281	77.18				390	450
		0.312	85.60			10	430	500
		0.344	94.26				480	560
		0.375	102.63	STD			520	610
		0.406	110.98				560	660
		0.438	119.57				610	710
		0.469	127.88				650	760
		0.500	136.17	XS	20		690	810
		0.562	152.68				780	910

[A]1 in. = 25.4 mm.
[B]1 lb/ft = 1.49 kg/m.
[C]1 psi = 6.8948 kPa.
[D]For wall thickness of pipe not listed in Appendix X.XX, the following procedure shall be followed to determine the required test pressure.
 (1) When the wall thickness for a given diameter is between the lightest and heaviest wall thickness listed:
 (a) If the wall thickness is between two wall thicknesses on which the test pressures are identical, use that test pressure as the required test pressure.
 (b) If the test pressure is different for the next lighter and heavier walls listed, interpolate to obtain the required test pressure, using the ratio of the nominal weight per foot ($W = 10.68(D - t)t$) of the desired thickness to the nominal weight per foot of the next heavier thickness (to the nearest 50 lb).
 (2) When the wall thickness is greater than the heaviest wall thickness shown for a given diameter, the test pressure for the heaviest wall listed shall be the required test pressure.
 (3) When the wall thickness is lighter than the lightest shown for a given diameter:
 (a) For Grades A and B in sizes 2 in. and larger, determine the test pressure from the following equation:

$$P = 2St/D$$

where P = minimum hydrostatic test pressure, psi, S = 0.60 times the minimum specified yield point, psi, t = specified wall thickness, in., and D = specified outside diameter, in.
 (b) For Grades A and B in sizes under 2 in. and for all sizes butt welded pipe, use the test pressure given for the lightest wall thickness of the table for the diameter involved.

(Copyright, ASTM, 1916 Race Street, Philadelphia, PA 19103. Reprinted with permission.)

APPENDIX 13.B-2.
Dimensions, Weights, and Test Pressures for Threaded and Coupled Standard Pipe

Nominal Size, in.	Outside Diameter, in.[1]	Wall Thickness, in.[1]	Nominal Weight per ft. Threads and Couplings, lb[B]	Weight Class	Schedule No.	Test Pressure, psi[C] Butt-Welded	Grade A	Grade B
1/8	0.405	0.068	0.24	STD	40	700	700	700
		0.095	0.32	XS	80	850	850	850
1/4	0.540	0.088	0.42	STD	40	700	700	700
		0.119	0.54	XS	80	850	850	850
3/8	0.675	0.091	0.57	STD	40	700	700	700
		0.126	0.74	XS	80	850	850	850
1/2	0.840	0.109	0.85	STD	40	700	700	700
		0.147	1.09	XS	80	850	850	850
		0.294	1.72	XXS		1000	1000	1000
3/4	1.050	0.113	1.13	STD	40	700	700	700
		0.154	1.48	XS	80	850	850	850
		0.308	2.44	XXS		1000	1000	1000
1	1.315	0.133	1.68	STD	40	700	700	700
		0.179	2.18	XS	80	850	850	850
		0.358	3.66	XXS		1000	1000	1000
1 1/4	1.660	0.140	2.28	STD	40	1000	1000	1100
		0.191	3.02	XS	80	1300	1500	1600
		0.382	5.22	XXS		1400	1800	1900
1 1/2	1.900	0.145	2.73	STD	40	1000	1000	1100
		0.200	3.66	XS	80	1300	1500	1600
		0.400	6.41	XXS		1400	1800	1900
2	2.375	0.154	3.68	STD	40	1000	2300	2500
		0.218	5.07	XS	80	1300	2500	2500
		0.436	9.03	XXS		1400	2500	2500
2 1/2	2.875	0.203	5.82	STD	40	1000	2500	2500
		0.276	7.73	XS	80	1300	2500	2500
		0.552	13.70	XXS		1400	2500	2500
3	3.500	0.216	7.62	STD	40	1000	2200	2500
		0.300	10.33	XS	80	1300	2500	2500
		0.600	18.57	XXS			2500	2500
3 1/2	4.000	0.226	9.20	STD	40	1200	2000	2400
		0.318	12.63	XS	80	1700	2800	2800
4	4.500	0.237	10.89	STD	40	1200	1900	2200
		0.337	15.17	XS	80	1700	2700	2800
		0.674	27.58	XXS			2800	2800
5	5.563	0.258	14.81	STD	40		1700	1900
		0.375	21.09	XS	80		2400	2800
		0.750	38.61	XXS			2800	2800
6	6.625	0.280	19.18	STD	40		1500	1800
		0.432	28.89	XS	80		2300	2700
		0.864	53.14	XXS			2800	2800
8	8.625	0.277	25.55		30		1200	1300
		0.322	29.35	STD	40		1300	1600
		0.500	43.90	XS	80		2100	2400
		0.875	72.44	XXS			2800	2800
10	10.750	0.279	32.75				950	1100
		0.307	35.75		30		1000	1200
		0.365	41.85	STD	40		1200	1400
		0.500	55.82	XS	60		1700	2000
12	12.750	0.330	45.45		30		950	1100
		0.375	51.15	STD			1100	1200
		0.500	66.71	XS			1400	1600

[1] 1 in. = 25.4 mm.
[B] 1 lb/ft = 1.49 kg/m.
[C] 1 psi = 6.8948 kPa.

(Copyright, ASTM, 1916 Race Street, Philadelphia, PA 19103. Reprinted with permission.)

APPENDIX 13.B-3.
Plain-End Line Pipe Dimensions, Weights, and Test Pressures

1				2		3		4		5		6		7		8		9	
Size				Plain-End Weight, w_{pe}		Wall Thickness, t		Inside Diameter, d		Test Pressure, min.									
										Grade A				Grade B				Grade A25	
										Std.		Alt.		Std.		Alt.			
Nom.	Desig- nation	Outside Diameter, D																	
in.		in.	mm	lb/ft	kg/m	in.	mm	in.	mm	psi	100 kPa	psi	100 kPa	psi	100 kPa	psi	100 kPa	psi	100 kPa
1/8	Std	0.405	(10.3)	0.24	(0.36)	0.068	(1.73)	0.269	(6.8)	700	(48)	—	(—)	700	(48)	—	(—)	700	(48)
1/8	XS	0.405	(10.3)	0.31	(0.46)	0.095	(2.41)	0.215	(5.5)	850	(59)	—	(—)	850	(59)	—	(—)	850	(59)
1/4	Std	0.540	(13.7)	0.42	(0.63)	0.088	(2.24)	0.364	(9.2)	700	(48)	—	(—)	700	(48)	—	(—)	700	(48)
1/4	XS	0.540	(13.7)	0.54	(0.80)	0.119	(3.02)	0.302	(7.7)	850	(59)	—	(—)	850	(59)	—	(—)	850	(59)
3/8	Std	0.675	(17.1)	0.57	(0.85)	0.091	(2.31)	0.493	(12.5)	700	(48)	—	(—)	700	(48)	—	(—)	700	(48)
3/8	XS	0.675	(17.1)	0.74	(1.10)	0.126	(3.20)	0.423	(10.7)	850	(59)	—	(—)	850	(59)	—	(—)	850	(59)
1/2	Std	0.840	(21.3)	0.85	(1.27)	0.109	(2.77)	0.622	(15.8)	700	(48)	—	(—)	700	(48)	—	(—)	700	(48)
1/2	XS	0.840	(21.3)	1.09	(1.62)	0.147	(3.73)	0.546	(13.8)	850	(59)	—	(—)	850	(59)	—	(—)	850	(59)
1/2	XXS	0.840	(21.3)	1.71	(2.55)	0.294	(7.47)	0.252	(6.4)	1000	(69)	—	(—)	1000	(69)	—	(—)	1000	(69)
3/4	Std	1.050	(26.7)	1.13	(1.68)	0.113	(2.87)	0.824	(21.0)	700	(48)	—	(—)	700	(48)	—	(—)	700	(48)
3/4	XS	1.050	(26.7)	1.47	(2.19)	0.154	(3.91)	0.742	(18.9)	850	(59)	—	(—)	850	(59)	—	(—)	850	(59)
3/4	XXS	1.050	(26.7)	2.44	(3.63)	0.308	(7.82)	0.434	(11.1)	1000	(69)	—	(—)	1000	(69)	—	(—)	1000	(69)
1	Std	1.315	(33.4)	1.68	(2.50)	0.133	(3.38)	1.049	(26.6)	700	(48)	—	(—)	700	(48)	—	(—)	700	(48)
1	XS	1.315	(33.4)	2.17	(3.23)	0.179	(4.55)	0.957	(24.3)	850	(59)	—	(—)	850	(59)	—	(—)	850	(59)
1	XXS	1.315	(33.4)	3.66	(5.45)	0.358	(9.09)	0.599	(15.2)	1000	(69)	—	(—)	1000	(69)	—	(—)	1000	(69)
1 1/4	Std	1.660	(42.2)	2.27	(3.38)	0.140	(3.56)	1.380	(35.1)	1200	(83)	—	(—)	1300	(90)	—	(—)	1000	(69)
1 1/4	XS	1.660	(42.2)	3.00	(4.47)	0.191	(4.85)	1.278	(32.5)	1800	(124)	—	(—)	1900	(131)	—	(—)	1300	(90)
1 1/4	XXS	1.660	(42.2)	5.21	(7.76)	0.382	(9.70)	0.896	(22.8)	2200	(152)	—	(—)	2300	(158)	—	(—)	1400	(96)
1 1/2	Std	1.900	(48.3)	2.72	(4.05)	0.145	(3.68)	1.610	(40.9)	1200	(83)	—	(—)	1300	(90)	—	(—)	1000	(69)
1 1/2	XS	1.900	(48.3)	3.63	(5.41)	0.200	(5.08)	1.500	(38.1)	1800	(124)	—	(—)	1900	(131)	—	(—)	1300	(90)
1 1/2	XXS	1.900	(48.3)	6.41	(9.55)	0.400	(10.16)	1.100	(28.0)	2200	(152)	—	(—)	2300	(158)	—	(—)	1400	(96)

Appendix 13.B-3 Continued

Size				Plain-End Weight, w_{pe}		Wall Thickness, t		Inside Diameter, d		Test Pressure, min.								
										Grade A				Grade B				Grade A25
										Std.		Alt.		Std.		Alt.		
Nom. in.	Desig-nation	Outside Diameter, D																
		in.	mm	lb/ft	kg/m	in.	mm	in.	mm	psi	100 kPa	psi	100 kPa	psi	100 kPa	psi	100 kPa	psi	100 kPa
2	Std	*2⅜	(60.3)	2.03	(3.02)	0.083	(2.11)	2.209	(56.1)	1260	(87)	—	(—)	1470	(101)	—	(—)	600	(41)
		2⅜	(60.3)	2.64	(3.93)	0.109	(2.77)	2.157	(54.8)	—	(—)	—	(—)	—	(—)	—	(—)	800	(55)
		2⅜	(60.3)	3.00	(4.47)	0.125	(3.18)	2.125	(53.9)	—	(—)	—	(—)	—	(—)	—	(—)	1000	(69)
		2⅜	(60.3)	3.36	(5.00)	0.141	(3.58)	2.093	(53.1)	—	(—)	—	(—)	—	(—)	—	(—)	1000	(69)
		2⅜	(60.3)	3.65	(5.44)	0.154	(3.91)	2.067	(52.5)	2330	(158)	—	(—)	2500	(172)	—	(—)	1000	(69)
2	XS	2⅜	(60.3)	4.05	(6.03)	0.172	(4.37)	2.031	(51.6)	2500	(172)	—	(—)	2500	(172)	—	(—)	1100	(76)
		2⅜	(60.3)	4.39	(6.54)	0.188	(4.78)	1.999	(50.8)	2500	(172)	—	(—)	2500	(172)	—	(—)	1200	(83)
		2⅜	(60.3)	5.02	(7.48)	0.218	(5.54)	1.939	(49.2)	2500	(172)	—	(—)	2500	(172)	—	(—)	1300	(90)
		2⅜	(60.3)	5.67	(8.45)	0.250	(6.35)	1.875	(47.6)	2500	(172)	—	(—)	2500	(172)	—	(—)	1400	(96)
		2⅜	(60.3)	6.28	(9.35)	0.281	(7.14)	1.813	(46.1)	2500	(172)	—	(—)	2500	(172)	—	(—)	1400	(96)
2	XXS	2⅜	(60.3)	9.03	(13.45)	0.436	(11.07)	1.503	(38.2)	2500	(172)	—	(—)	2500	(172)	—	(—)	1400	(96)
		*2⅞	(73.0)	2.47	(3.68)	0.083	(2.11)	2.709	(68.8)	1040	(72)	—	(—)	1210	(83)	—	(—)	600	(41)
		2⅞	(73.0)	3.22	(4.80)	0.109	(2.77)	2.657	(67.5)	—	(—)	—	(—)	—	(—)	—	(—)	800	(55)
		2⅞	(73.0)	3.67	(5.46)	0.125	(3.18)	2.625	(66.6)	—	(—)	—	(—)	—	(—)	—	(—)	1000	(69)
		2⅞	(73.0)	4.12	(6.14)	0.141	(3.58)	2.593	(65.8)	—	(—)	—	(—)	—	(—)	—	(—)	1000	(69)
		2⅞	(73.0)	4.53	(6.75)	0.156	(3.96)	2.563	(65.1)	1950	(134)	—	(—)	2280	(157)	—	(—)	1000	(69)
2½	Std	2⅞	(73.0)	4.97	(7.40)	0.172	(4.37)	2.531	(64.3)	2150	(148)	—	(—)	2500	(172)	—	(—)	1000	(69)
		2⅞	(73.0)	5.40	(8.04)	0.188	(4.78)	2.489	(63.5)	2350	(162)	—	(—)	2500	(172)	—	(—)	1000	(69)
		2⅞	(73.0)	5.79	(8.62)	0.203	(5.16)	2.469	(62.7)	2500	(172)	—	(—)	2500	(172)	—	(—)	1000	(69)
		2⅞	(73.0)	6.13	(9.13)	0.216	(5.49)	2.443	(62.1)	2500	(172)	—	(—)	2500	(172)	—	(—)	1100	(76)
2½	XS	2⅞	(73.0)	7.01	(10.44)	0.250	(6.35)	2.375	(60.3)	2500	(172)	—	(—)	2500	(172)	—	(—)	1200	(83)
		2⅞	(73.0)	7.66	(11.41)	0.276	(7.01)	2.323	(59.0)	2500	(172)	—	(—)	2500	(172)	—	(—)	1300	(90)
2½	XXS	2⅞	(73.0)	13.69	(20.41)	0.552	(14.02)	1.771	(45.0)	2500	(172)	—	(—)	2500	(172)	—	(—)	1400	(96)
		*3½	(88.9)	3.03	(4.51)	0.083	(2.11)	3.334	(84.7)	850	(59)	—	(—)	1000	(69)	—	(—)	600	(41)
		3½	(88.9)	3.95	(5.88)	0.109	(2.77)	3.282	(83.4)	—	(—)	—	(—)	—	(—)	—	(—)	800	(55)
		*3½	(88.9)	4.51	(6.72)	0.125	(3.18)	3.250	(82.5)	1290	(89)	—	(—)	1500	(103)	—	(—)	1000	(69)

3	Std	3½	(88.9)	5.06	(7.54)	0.141	(3.58)	3.218	(81.7)	1600	(—)	—	—	1870	(—)	—	—	1000	(69)
		*3½	(88.9)	5.57	(8.30)	0.156	(3.96)	3.188	(81.0)	1770	(110)	—	—	2060	(129)	—	—	1000	(69)
		3½	(88.9)	6.11	(9.10)	0.172	(4.37)	3.156	(80.2)	1930	(122)	—	—	2260	(142)	—	—	1000	(69)
		3½	(88.9)	6.65	(9.91)	0.188	(4.78)	3.124	(79.3)	2220	(133)	—	—	2500	(156)	—	—	1000	(69)
		3½	(88.9)	7.58	(11.29)	0.216	(5.49)	3.068	(77.9)	2500	(153)	—	—	2500	(172)	—	—	1000	(69)
		3½	(88.9)	8.68	(12.93)	0.250	(6.35)	3.000	(76.2)	2500	(172)	—	—	2500	(172)	—	—	—	—
3	XS	3½	(88.9)	9.66	(14.39)	0.281	(7.14)	2.938	(74.6)	2500	(172)	—	—	2500	(172)	—	—	1300	(90)
		3½	(88.9)	10.25	(15.27)	0.300	(7.62)	2.900	(73.7)	2500	(172)	—	—	2500	(172)	—	—	—	—
3	XXS	3½	(88.9)	18.58	(27.67)	0.600	(15.24)	2.300	(58.4)	2500	(172)	—	—	2500	(172)	—	—	—	—
		*4	(101.6)	3.47	(5.17)	0.083	(2.11)	3.834	(97.4)	750	(52)	—	—	870	(60)	—	—	600	(41)
		*4	(101.6)	4.53	(6.75)	0.109	(2.77)	3.782	(96.1)	980	(68)	—	—	1140	(79)	—	—	—	—
		*4	(101.6)	5.17	(7.70)	0.125	(3.18)	3.750	(95.2)	1120	(77)	—	—	1310	(90)	—	—	800	(55)
		*4	(101.6)	5.81	(8.65)	0.141	(3.58)	3.718	(94.4)	1270	(88)	—	—	1480	(102)	—	—	—	—
		*4	(101.6)	6.40	(9.53)	0.156	(3.96)	3.688	(93.7)	1400	(96)	—	—	1640	(113)	—	—	1000	(69)
		*4	(101.6)	7.03	(10.47)	0.172	(4.37)	3.656	(92.9)	1550	(107)	—	—	1810	(125)	—	—	1200	(83)
		*4	(101.6)	7.65	(11.39)	0.188	(4.78)	3.624	(92.0)	1690	(116)	—	—	1970	(136)	—	—	—	—
		*4	(101.6)	9.11	(13.57)	0.226	(5.74)	3.548	(90.1)	2030	(140)	—	—	2370	(163)	—	—	1200	(83)
3½	Std	4	(101.6)	9.11	(13.57)	0.226	(5.74)	3.548	(90.1)	—	—	—	—	—	—	—	—	—	—
		*4	(101.6)	10.01	(14.91)	0.250	(6.35)	3.500	(88.9)	2250	(155)	—	—	2620	(181)	—	—	—	—
		*4	(101.6)	11.16	(16.62)	0.281	(7.14)	3.438	(87.3)	2530	(174)	—	—	2800	(193)	—	—	1700	(117)
3½	XS	4	(101.6)	12.50	(18.63)	0.318	(8.08)	3.364	(85.4)	2800	(193)	—	—	2800	(193)	—	—	—	—
		*4½	(114.3)	3.92	(5.84)	0.083	(2.11)	4.334	(110.1)	660	(45)	—	—	770	(53)	—	—	—	—
		*4½	(114.3)	5.11	(7.61)	0.109	(2.77)	4.282	(108.8)	870	(60)	—	—	1020	(70)	—	—	800	(55)
		*4½	(114.3)	5.84	(8.70)	0.125	(3.18)	4.250	(107.9)	1000	(69)	—	—	1170	(81)	—	—	—	—
		*4½	(114.3)	6.56	(9.77)	0.141	(3.58)	4.218	(107.1)	1130	(78)	—	—	1320	(91)	—	—	1000	(69)
		*4½	(114.3)	7.24	(10.78)	0.156	(3.96)	4.188	(106.4)	1250	(86)	—	—	1460	(101)	—	—	—	—
		4½	(114.3)	7.95	(11.84)	0.172	(4.37)	4.156	(105.6)	1380	(95)	—	—	1610	(111)	—	—	1200	(83)
		4½	(114.3)	8.66	(12.90)	0.188	(4.78)	4.124	(104.7)	1500	(103)	—	—	1750	(121)	—	—	—	—
		4½	(114.3)	9.32	(13.88)	0.203	(5.16)	4.094	(104.0)	1620	(112)	—	—	1890	(130)	—	—	1200	(83)
		4½	(114.3)	10.01	(14.91)	0.219	(5.56)	4.062	(103.2)	1750	(121)	—	—	2040	(141)	—	—	1200	(83)
4	Std	4½	(114.3)	10.79	(16.07)	0.237	(6.02)	4.026	(102.3)	1900	(131)	—	—	2210	(152)	—	—	—	—
		4½	(114.3)	11.35	(16.91)	0.250	(6.35)	4.000	(101.6)	2000	(138)	—	—	2330	(161)	—	—	—	—
		4½	(114.3)	12.66	(18.86)	0.281	(7.14)	3.938	(100.0)	2250	(155)	—	—	2620	(181)	—	—	1700	(117)
		4½	(114.3)	13.96	(20.79)	0.312	(7.92)	3.876	(98.5)	2500	(172)	—	—	2800	(193)	—	—	—	—
4	XS	4½	(114.3)	14.98	(22.31)	0.337	(8.56)	3.826	(97.2)	2700	(186)	—	—	2800	(193)	—	—	—	—
		4½	(114.3)	19.00	(28.30)	0.438	(11.13)	3.624	(92.0)	2800	(193)	—	—	2800	(193)	—	—	—	—
		4½	(114.3)	22.51	(33.53)	0.531	(13.49)	3.438	(87.3)	2800	(193)	—	—	2800	(193)	—	—	—	—
4	XXS	4½	(114.3)	27.54	(41.02)	0.674	(17.12)	3.152	(80.1)	2800	(193)	—	—	2800	(193)	—	—	—	—

Appendix 13.B-3 Continued

1				2		3		4		5		6		7		8		9	
Size										Test Pressure, min.									
Nom.	Desig-nation	Outside Diameter, *D*		Plain-End Weight, w_{pe}		Wall Thickness, *t*		Inside Diameter, *d*		Grade A				Grade B				Grade A25	
										Std.		Alt.		Std.		Alt.			
in.		in.	mm	lb/ft	kg/m	in.	mm	in.	mm	psi	100 kPa	psi	100 kPa	psi	100 kPa	psi	100 kPa	psi	100 kPa
		*5⁹⁄₁₆	(141.3)	4.86	(7.24)	0.083	(2.11)	5.397	(137.1)	540	(37)	—	(—)	630	(43)	—	(—)	—	(—)
		*5⁹⁄₁₆	(141.3)	7.26	(10.81)	0.125	(3.18)	5.312	(134.9)	810	(56)	—	(—)	940	(65)	—	(—)	—	(—)
		*5⁹⁄₁₆	(141.3)	9.01	(13.42)	0.156	(3.96)	5.251	(133.4)	1010	(70)	—	(—)	1180	(81)	—	(—)	—	(—)
		*5⁹⁄₁₆	(141.3)	10.79	(16.07)	0.188	(4.78)	5.187	(131.7)	1220	(84)	—	(—)	1420	(98)	—	(—)	—	(—)
		*5⁹⁄₁₆	(141.3)	12.50	(18.62)	0.219	(5.56)	5.125	(130.2)	1420	(98)	—	(—)	1650	(114)	—	(—)	—	(—)
		*5⁹⁄₁₆	(141.3)	14.62	(21.78)	0.258	(6.55)	5.047	(128.2)	1670	(115)	—	(—)	1950	(134)	—	(—)	—	(—)
		*5⁹⁄₁₆	(141.3)	15.85	(23.61)	0.281	(7.14)	5.001	(127.0)	1820	(125)	—	(—)	2120	(146)	—	(—)	—	(—)
		*5⁹⁄₁₆	(141.3)	17.50	(26.07)	0.312	(7.92)	4.939	(125.5)	2020	(139)	—	(—)	2360	(163)	—	(—)	—	(—)
		*5⁹⁄₁₆	(141.3)	19.17	(28.55)	0.344	(8.74)	4.875	(123.8)	2230	(154)	—	(—)	2600	(179)	—	(—)	—	(—)
5	XS	5⁹⁄₁₆	(141.3)	20.78	(30.95)	0.375	(9.52)	4.813	(122.3)	2430	(167)	—	(—)	2800	(193)	—	(—)	—	(—)
		*5⁹⁄₁₆	(141.3)	27.04	(40.28)	0.500	(12.70)	4.563	(115.9)	2800	(193)	—	(—)	2800	(193)	—	(—)	—	(—)
		*5⁹⁄₁₆	(141.3)	32.96	(49.09)	0.625	(15.88)	4.313	(109.5)	2800	(193)	—	(—)	2800	(193)	—	(—)	—	(—)
5	XXS	5⁹⁄₁₆	(141.3)	38.55	(57.42)	0.750	(19.05)	4.063	(103.2)	2800	(193)	—	(—)	2800	(193)	—	(—)	—	(—)
		*6⅝	(168.3)	5.80	(8.64)	0.083	(2.11)	6.459	(164.1)	450	(31)	560	(39)	530	(37)	660	(45)	—	(—)
		*6⅝	(168.3)	7.59	(11.31)	0.109	(2.77)	6.407	(162.7)	590	(41)	740	(51)	690	(48)	860	(59)	—	(—)
		*6⅝	(168.3)	8.68	(12.93)	0.125	(3.18)	6.375	(161.9)	680	(47)	850	(59)	790	(54)	990	(68)	—	(—)
		*6⅝	(168.3)	9.76	(14.54)	0.141	(3.58)	6.343	(161.1)	770	(53)	960	(66)	890	(61)	1120	(77)	—	(—)
		*6⅝	(168.3)	10.78	(16.06)	0.156	(3.96)	6.313	(160.4)	850	(59)	1060	(73)	990	(68)	1240	(85)	—	(—)
		6⅝	(168.3)	11.85	(17.65)	0.172	(4.37)	6.281	(159.6)	930	(64)	1170	(81)	1090	(75)	1360	(94)	—	(—)
		6⅝	(168.3)	12.92	(19.24)	0.188	(4.78)	6.249	(158.7)	1020	(70)	1280	(88)	1190	(82)	1490	(103)	—	(—)
		6⅝	(168.3)	13.92	(20.73)	0.203	(5.16)	6.219	(158.0)	1100	(76)	1380	(95)	1290	(89)	1610	(111)	—	(—)
		6⅝	(168.3)	14.98	(22.31)	0.219	(5.56)	6.187	(157.2)	1190	(82)	1490	(103)	1390	(96)	1740	(120)	—	(—)
		6⅝	(168.3)	17.02	(25.35)	0.250	(6.35)	6.125	(155.6)	1360	(94)	1700	(117)	1580	(109)	1980	(136)	—	(—)
		6⅝	(168.3)	18.97	(28.26)	0.280	(7.11)	6.065	(154.1)	1520	(105)	1900	(131)	1780	(123)	2220	(153)	—	(—)
		6⅝	(168.3)	21.04	(31.34)	0.312	(7.92)	6.001	(152.5)	1700	(117)	2120	(146)	1980	(136)	2470	(170)	—	(—)
		6⅝	(168.3)	23.08	(34.38)	0.344	(8.74)	5.937	(150.8)	1870	(129)	2340	(161)	2180	(150)	2730	(188)	—	(—)
		6⅝	(168.3)	25.03	(37.28)	0.375	(9.52)	5.875	(149.3)	2040	(141)	2550	(176)	2380	(164)	2800	(193)	—	(—)

6	XS	6⅝	(168.3)	28.57	(42.56)	0.432	(10.97)	5.761	(146.4)	2350 (162)	2800 (193)	2740 (189)	2800 (193)
		6⅝	(168.3)	32.71	(48.72)	0.500	(12.70)	5.625	(142.9)	2720 (187)	2800 (193)	2800 (193)	2800 (193)
		6⅝	(168.3)	36.39	(54.20)	0.562	(14.27)	5.501	(139.7)	2800 (193)	2800 (193)	2800 (193)	2800 (193)
		6⅝	(168.3)	40.05	(59.65)	0.625	(15.88)	5.375	(136.5)	2800 (193)	2800 (193)	2800 (193)	2800 (193)
		6⅝	(168.3)	45.35	(67.55)	0.719	(18.26)	5.187	(131.8)	2800 (193)	2800 (193)	2800 (193)	2800 (193)
6	XXS	6⅝	(168.3)	53.16	(79.18)	0.864	(21.95)	4.897	(124.4)	2800 (193)	2800 (193)	2800 (193)	2800 (193)
		*8⅝	(219.1)	11.35	(16.91)	0.125	(3.18)	8.375	(212.7)	520 (36)	650 (45)	610 (42)	760 (52)
		*8⅝	(219.1)	14.11	(21.02)	0.156	(3.96)	8.313	(211.2)	650 (45)	810 (56)	760 (52)	950 (65)
		8⅝	(219.1)	16.94	(25.23)	0.188	(4.78)	8.249	(209.5)	780 (54)	980 (68)	920 (63)	1140 (79)
		8⅝	(219.1)	19.66	(29.28)	0.219	(5.56)	8.187	(208.0)	910 (63)	1140 (79)	1070 (74)	1330 (92)
		8⅝	(219.1)	22.36	(33.31)	0.250	(6.35)	8.125	(206.4)	1040 (72)	1300 (90)	1220 (84)	1520 (105)
		8⅝	(219.1)	24.70	(36.79)	0.277	(7.04)	8.071	(205.0)	1160 (80)	1450 (100)	1350 (93)	1690 (116)
		8⅝	(219.1)	27.70	(41.26)	0.312	(7.92)	8.001	(203.3)	1300 (90)	1630 (112)	1520 (105)	1900 (131)
		8⅝	(219.1)	28.55	(42.53)	0.322	(8.18)	7.981	(202.7)	1340 (92)	1680 (116)	1570 (108)	1960 (135)
		8⅝	(219.1)	30.42	(45.31)	0.344	(8.74)	7.937	(201.6)	1440 (99)	1790 (123)	1680 (116)	2090 (144)
		8⅝	(219.1)	33.04	(49.21)	0.375	(9.52)	7.875	(200.1)	1570 (108)	1960 (135)	1830 (126)	2280 (157)
		8⅝	(219.1)	38.30	(57.05)	0.438	(11.13)	7.749	(196.8)	1830 (126)	2290 (158)	2130 (147)	2670 (184)
8	XS	8⅝	(219.1)	43.39	(64.63)	0.500	(12.70)	7.625	(193.7)	2090 (144)	2610 (180)	2430 (167)	2800 (193)
		8⅝	(219.1)	48.40	(72.09)	0.562	(14.27)	7.501	(190.6)	2350 (162)	2800 (193)	2740 (189)	2800 (193)
		8⅝	(219.1)	53.40	(79.54)	0.625	(15.88)	7.375	(187.3)	2610 (180)	2800 (193)	2800 (193)	2800 (193)
		8⅝	(219.1)	60.71	(90.43)	0.719	(18.26)	7.187	(182.6)	2800 (193)	2800 (193)	2800 (193)	2800 (193)
8	XXS	8⅝	(219.1)	72.42	(107.87)	0.875	(22.22)	6.875	(174.7)	2800 (193)	2800 (193)	2800 (193)	2800 (193)
		*10¾	(273.0)	17.65	(26.29)	0.156	(3.96)	10.438	(265.1)	520 (36)	650 (45)	610 (42)	760 (52)
		*10¾	(273.0)	21.21	(31.59)	0.188	(4.78)	10.374	(263.4)	630 (43)	790 (54)	730 (50)	920 (63)
		10¾	(273.0)	24.63	(36.69)	0.219	(5.56)	10.312	(261.9)	730 (50)	920 (63)	860 (59)	1070 (74)
		10¾	(273.0)	28.04	(41.77)	0.250	(6.35)	10.250	(260.3)	840 (58)	1050 (72)	980 (68)	1220 (84)
		10¾	(273.0)	31.20	(46.47)	0.279	(7.09)	10.192	(258.8)	930 (64)	1170 (81)	1090 (75)	1360 (94)
		10¾	(273.0)	34.24	(51.00)	0.307	(7.80)	10.136	(257.4)	1030 (71)	1290 (89)	1200 (83)	1500 (103)
		10¾	(273.0)	38.23	(56.94)	0.344	(8.74)	10.062	(255.5)	1150 (79)	1440 (99)	1340 (92)	1680 (116)
		10¾	(273.0)	40.48	(60.29)	0.365	(9.27)	10.020	(254.5)	1220 (84)	1530 (105)	1430 (99)	1780 (123)
		10¾	(273.0)	48.24	(71.85)	0.438	(11.13)	9.874	(250.7)	1470 (101)	1830 (126)	1710 (118)	2140 (147)
10	XS	10¾	(273.0)	54.74	(81.54)	0.500	(12.70)	9.750	(247.6)	1670 (115)	2090 (144)	1950 (134)	2440 (168)
		10¾	(273.0)	61.15	(91.08)	0.562	(14.27)	9.626	(244.5)	1880 (130)	2350 (162)	2200 (152)	2740 (189)
		10¾	(273.0)	67.58	(100.66)	0.625	(15.88)	9.500	(241.2)	2090 (144)	2620 (181)	2440 (168)	2800 (193)
		10¾	(273.0)	77.03	(114.74)	0.719	(18.26)	9.312	(236.5)	2410 (166)	2800 (193)	2800 (193)	2800 (193)
		10¾	(273.0)	86.18	(128.37)	0.812	(20.62)	9.126	(231.8)	2720 (187)	2800 (193)	2800 (193)	2800 (193)

Appendix 13.B-3 Continued

Size									Test Pressure, min.								
1			2		3		4		5		6		7		8		9
Nom.	Desig-nation	Outside Diameter, D	Plain-End Weight, w_{pe}		Wall Thickness, t		Inside Diameter, d		Grade A				Grade B				Grade A25
									Std.		Alt.		Std.		Alt.		
in.		in. mm	lb/ft	kg/m	in.	mm	in.	mm	psi	100 kPa	psi	100 kPa	psi	100 kPa	psi	100 kPa	psi 100 kPa
12		*12¾ (323.8)	23.11	(34.42)	0.172	(4.37)	12.406	(315.1)	490	(34)	610	(42)	570	(39)	710	(49)	
		*12¾ (323.8)	25.22	(37.57)	0.188	(4.78)	12.374	(314.2)	530	(37)	660	(45)	620	(43)	770	(53)	
		*12¾ (323.8)	29.31	(43.66)	0.219	(5.56)	12.312	(312.7)	620	(43)	770	(53)	720	(50)	900	(62)	
		12¾ (323.8)	33.38	(49.72)	0.250	(6.35)	12.250	(311.1)	710	(49)	880	(61)	820	(56)	1030	(71)	
		12¾ (323.8)	37.42	(55.74)	0.281	(7.14)	12.188	(309.5)	790	(54)	990	(68)	930	(64)	1160	(80)	
		12¾ (323.8)	41.45	(61.74)	0.312	(7.92)	12.126	(308.0)	880	(61)	1100	(76)	1030	(71)	1280	(88)	
		12¾ (323.8)	43.77	(65.20)	0.330	(8.38)	12.090	(307.0)	930	(64)	1160	(80)	1090	(75)	1360	(94)	
		12¾ (323.8)	45.58	(67.89)	0.344	(8.74)	12.062	(306.3)	970	(67)	1210	(83)	1130	(78)	1420	(98)	
		12¾ (323.8)	49.56	(73.82)	0.375	(9.52)	12.000	(304.8)	1060	(73)	1320	(91)	1240	(85)	1540	(106)	
	XS	12¾ (323.8)	57.59	(85.78)	0.438	(11.13)	11.874	(301.5)	1240	(85)	1550	(107)	1440	(99)	1800	(124)	
		12¾ (323.8)	65.42	(97.44)	0.500	(12.70)	11.750	(298.4)	1410	(97)	1760	(121)	1650	(114)	2060	(142)	
		12¾ (323.8)	73.15	(108.96)	0.562	(14.27)	11.626	(295.3)	1590	(110)	1980	(136)	1850	(127)	2310	(159)	
		12¾ (323.8)	80.93	(120.55)	0.625	(15.88)	11.500	(292.0)	1760	(121)	2210	(152)	2060	(142)	2570	(177)	
		12¾ (323.8)	88.63	(132.01)	0.688	(17.48)	11.374	(288.8)	1940	(134)	2430	(167)	2270	(156)	2800	(193)	
		12¾ (323.8)	96.12	(143.17)	0.750	(19.05)	11.250	(285.7)	2120	(146)	2650	(183)	2470	(170)	2800	(193)	
		12¾ (323.8)	103.53	(154.21)	0.812	(20.62)	11.126	(282.6)	2290	(158)	2800	(193)	2670	(184)	2800	(193)	
		12¾ (323.8)	110.97	(165.29)	0.875	(22.22)	11.000	(279.4)	2470	(170)	2800	(193)	2800	(193)	2800	(193)	
		*14 (355.6)	27.73	(41.30)	0.188	(4.78)	13.624	(346.0)	480	(33)	600	(41)	560	(39)	700	(48)	
		*14 (355.6)	29.91	(44.55)	0.203	(5.16)	13.594	(345.3)	520	(36)	650	(45)	610	(42)	760	(52)	
		*14 (355.6)	36.71	(54.68)	0.250	(6.35)	13.500	(342.9)	640	(44)	800	(55)	750	(52)	940	(65)	
		*14 (355.6)	41.17	(61.32)	0.281	(7.14)	13.438	(341.3)	720	(50)	900	(62)	840	(58)	1050	(72)	
		14 (355.6)	45.61	(67.94)	0.312	(7.92)	13.376	(339.8)	800	(55)	1000	(69)	940	(65)	1170	(81)	
		14 (355.6)	50.17	(74.73)	0.344	(8.74)	13.312	(338.1)	880	(61)	1110	(76)	1030	(71)	1290	(89)	
		14 (355.6)	54.57	(81.28)	0.375	(9.52)	13.250	(336.6)	960	(66)	1210	(83)	1120	(77)	1410	(97)	
		14 (355.6)	63.44	(94.49)	0.438	(11.13)	13.124	(333.3)	1130	(78)	1410	(97)	1310	(90)	1640	(113)	
		14 (355.6)	72.09	(107.38)	0.500	(12.70)	13.000	(330.2)	1290	(89)	1610	(111)	1500	(103)	1880	(130)	
		14 (355.6)	80.66	(120.14)	0.562	(14.27)	12.876	(327.1)	1450	(100)	1810	(125)	1690	(116)	2110	(145)	

14	(355.6)	89.28	(132.98)	0.625	(15.88)	12.750	(323.8)	1610	(111)	2010	(138)	1880	(130)	2340	(161)
14	(355.6)	97.81	(145.69)	0.688	(17.48)	12.624	(320.6)	1770	(122)	2210	(152)	2060	(142)	2580	(178)
14	(355.6)	106.13	(158.08)	0.750	(19.05)	12.500	(317.5)	1930	(133)	2410	(166)	2250	(155)	2800	(193)
14	(355.6)	114.37	(170.35)	0.812	(20.62)	12.376	(314.4)	2090	(144)	2610	(180)	2440	(168)	2800	(193)
14	(355.6)	122.65	(182.69)	0.875	(22.22)	12.250	(311.2)	2250	(155)	2810	(193)	2620	(181)	2800	(193)
14	(355.6)	130.85	(194.90)	0.938	(23.83)	12.124	(307.9)	2410	(166)	2810	(193)	2800	(193)	2800	(193)
*16	(406.4)	31.75	(47.29)	0.188	(4.78)	15.624	(396.8)	420	(29)	530	(37)	490	(34)	620	(43)
*16	(406.4)	34.25	(51.02)	0.203	(5.16)	15.594	(396.1)	460	(32)	570	(39)	530	(37)	670	(46)
*16	(406.4)	36.91	(54.98)	0.219	(5.56)	15.562	(395.3)	490	(34)	620	(43)	570	(39)	720	(50)
*16	(406.4)	42.05	(62.63)	0.250	(6.35)	15.500	(393.7)	560	(39)	700	(48)	660	(45)	820	(56)
*16	(406.4)	47.17	(70.26)	0.281	(7.14)	15.438	(392.1)	630	(43)	790	(54)	740	(51)	920	(63)
16	(406.4)	52.27	(77.86)	0.312	(7.92)	15.376	(390.6)	700	(48)	880	(61)	820	(56)	1020	(70)
16	(406.4)	57.52	(85.68)	0.344	(8.74)	15.312	(388.9)	770	(53)	970	(67)	900	(62)	1130	(78)
16	(406.4)	62.58	(93.21)	0.375	(9.52)	15.250	(387.4)	840	(58)	1050	(72)	980	(68)	1230	(85)
16	(406.4)	72.80	(108.44)	0.438	(11.13)	15.124	(384.1)	990	(68)	1230	(85)	1150	(79)	1440	(99)
16	(406.4)	82.77	(123.29)	0.500	(12.70)	15.000	(381.0)	1120	(77)	1410	(97)	1310	(90)	1640	(113)
16	(406.4)	92.66	(138.02)	0.562	(14.27)	14.876	(377.9)	1260	(87)	1580	(109)	1480	(102)	1840	(127)
16	(406.4)	102.63	(152.87)	0.625	(15.88)	14.750	(374.6)	1410	(97)	1760	(121)	1640	(113)	2050	(141)
16	(406.4)	112.51	(167.58)	0.688	(17.48)	14.624	(371.4)	1550	(107)	1940	(134)	1810	(125)	2260	(156)
16	(406.4)	122.15	(181.94)	0.750	(19.05)	14.500	(368.3)	1690	(116)	2110	(145)	1970	(136)	2460	(169)
16	(406.4)	131.71	(196.18)	0.812	(20.62)	14.376	(365.2)	1830	(126)	2280	(157)	2130	(147)	2660	(183)
16	(406.4)	141.34	(210.53)	0.875	(22.22)	14.250	(362.0)	1970	(136)	2460	(169)	2300	(158)	2800	(193)
16	(406.4)	150.89	(224.75)	0.938	(23.83)	14.124	(358.7)	2110	(145)	2640	(182)	2460	(169)	2800	(193)
16	(406.4)	160.20	(238.62)	1.000	(25.40)	14.000	(355.6)	2250	(155)	2800	(193)	2620	(181)	2800	(193)
16	(406.4)	169.43	(252.37)	1.062	(26.97)	13.876	(352.5)	2390	(165)	2800	(193)	2790	(192)	2800	(193)
16	(406.4)	178.72	(266.20)	1.125	(28.58)	13.750	(349.2)	2530	(174)	2800	(193)	2800	(193)	2800	(193)
*18	(457.2)	35.76	(53.26)	0.188	(4.78)	17.624	(447.6)	380	(26)	470	(32)	440	(30)	550	(38)
*18	(457.2)	41.59	(61.95)	0.219	(5.56)	17.562	(446.1)	440	(30)	550	(38)	510	(35)	640	(44)
*18	(457.2)	47.39	(70.59)	0.250	(6.35)	17.500	(444.5)	500	(34)	620	(43)	580	(40)	730	(50)
*18	(457.2)	53.18	(79.21)	0.281	(7.14)	17.438	(442.9)	560	(39)	700	(48)	660	(45)	820	(56)
18	(457.2)	58.94	(87.79)	0.312	(7.92)	17.376	(441.4)	620	(43)	780	(54)	730	(50)	910	(63)
18	(457.2)	64.87	(96.62)	0.344	(8.74)	17.312	(439.7)	690	(48)	860	(59)	800	(55)	1000	(69)
18	(457.2)	70.59	(105.14)	0.375	(9.52)	17.250	(438.2)	750	(52)	940	(65)	880	(61)	1090	(75)
18	(457.2)	82.15	(122.36)	0.438	(11.13)	17.124	(434.9)	880	(61)	1100	(76)	1020	(70)	1280	(88)
18	(457.2)	93.45	(139.19)	0.500	(12.70)	17.000	(431.8)	1000	(69)	1250	(86)	1170	(81)	1460	(101)
18	(457.2)	104.67	(155.91)	0.562	(14.27)	16.876	(428.7)	1120	(77)	1400	(96)	1310	(90)	1640	(113)
18	(457.2)	115.98	(172.75)	0.625	(15.88)	16.750	(425.4)	1250	(86)	1560	(107)	1460	(101)	1820	(125)

Appendix 13.B-3 Continued

1			2		3		4		5		6		7		8		9	
Size			Plain-End Weight, w_{pe}		Wall Thickness, t		Inside Diameter, d		Test Pressure, min.									
									Grade A				Grade B				Grade A25	
									Std.		Alt.		Std.		Alt.			
Nom. in.	Desig-nation in.	Outside Diameter, D																
		in.	mm	lb/ft	kg/m	in.	mm	in.	mm	psi	100 kPa	psi	100 kPa	psi	100 kPa	psi	100 kPa	psi	100 kPa
	18	(457.2)	127.21	(189.48)	0.688	(17.48)	16.624	(422.2)	1380	(95)	1720	(119)	1610	(111)	2010	(138)		
	18	(457.2)	138.17	(205.80)	0.750	(19.05)	16.500	(419.1)	1500	(103)	1880	(130)	1750	(121)	2190	(151)		
	18	(457.2)	149.06	(222.02)	0.812	(20.62)	16.376	(416.0)	1620	(112)	2030	(140)	1890	(130)	2370	(163)		
	18	(457.2)	160.03	(238.36)	0.875	(22.22)	16.250	(412.8)	1750	(121)	2190	(151)	2040	(141)	2550	(176)		
	18	(457.2)	170.92	(254.59)	0.938	(23.83)	16.124	(409.5)	1880	(130)	2340	(161)	2190	(151)	2740	(189)		
	18	(457.2)	181.56	(270.43)	1.000	(25.40)	16.000	(406.4)	2000	(138)	2500	(172)	2330	(161)	2800	(193)		
	18	(457.2)	192.11	(286.15)	1.062	(26.97)	15.876	(403.3)	2120	(146)	2660	(183)	2480	(171)	2800	(193)		
	18	(457.2)	202.75	(302.00)	1.125	(28.58)	15.750	(400.0)	2250	(155)	2800	(193)	2620	(181)	2800	(193)		
	18	(457.2)	213.14	(317.47)	1.187	(30.15)	15.626	(396.9)	2370	(163)	2800	(193)	2770	(191)	2800	(193)		
	18	(457.2)	223.61	(333.07)	1.250	(31.75)	15.500	(393.7)	2500	(172)	2800	(193)	2800	(193)	2800	(193)		
	*20	(508.0)	46.27	(68.92)	0.219	(5.56)	19.562	(496.9)	390	(27)	490	(34)	460	(32)	570	(39)		
	*20	(508.0)	52.73	(78.54)	0.250	(6.35)	19.500	(495.3)	450	(31)	560	(39)	520	(36)	660	(45)		
	*20	(508.0)	59.18	(88.15)	0.281	(7.14)	19.438	(493.7)	510	(35)	630	(43)	590	(41)	740	(51)		
	20	(508.0)	65.60	(97.71)	0.312	(7.92)	19.376	(492.2)	560	(39)	700	(48)	660	(45)	820	(56)		
	20	(508.0)	72.21	(107.56)	0.344	(8.74)	19.312	(490.5)	620	(43)	770	(53)	720	(50)	900	(62)		
	20	(508.0)	78.60	(117.07)	0.375	(9.52)	19.250	(489.0)	680	(47)	840	(58)	790	(54)	980	(68)		
	20	(508.0)	91.51	(136.30)	0.438	(11.13)	19.124	(485.7)	790	(54)	990	(68)	920	(63)	1150	(79)		
	20	(508.0)	104.13	(155.10)	0.500	(12.70)	19.000	(482.6)	900	(62)	1120	(77)	1050	(72)	1310	(90)		
	20	(508.0)	116.67	(173.78)	0.562	(14.27)	18.876	(479.5)	1010	(70)	1260	(87)	1180	(81)	1480	(102)		
	20	(508.0)	129.33	(192.64)	0.625	(15.88)	18.750	(476.2)	1120	(77)	1410	(97)	1310	(90)	1640	(113)		
	20	(508.0)	141.90	(211.36)	0.688	(17.48)	18.624	(473.0)	1240	(85)	1550	(107)	1440	(99)	1810	(125)		
	20	(508.0)	154.19	(229.67)	0.750	(19.05)	18.500	(469.9)	1350	(93)	1690	(116)	1580	(109)	1970	(136)		
	20	(508.0)	166.40	(247.85)	0.812	(20.62)	18.376	(466.8)	1460	(101)	1830	(126)	1710	(118)	2130	(147)		
	20	(508.0)	178.72	(266.22)	0.875	(22.22)	18.250	(463.6)	1580	(109)	1970	(136)	1840	(127)	2300	(158)		
	20	(508.0)	190.96	(284.43)	0.938	(23.83)	18.124	(460.3)	1690	(116)	2110	(145)	1970	(136)	2460	(169)		
	20	(508.0)	202.92	(302.25)	1.000	(25.40)	18.000	(457.2)	1800	(124)	2250	(155)	2100	(145)	2620	(181)		
	20	(508.0)	214.80	(319.94)	1.062	(26.97)	17.876	(454.1)	1910	(132)	2390	(165)	2230	(154)	2750	(189)		

20	(508.0)	226.78	(337.79)	1.125	(28.58)	17.750	(450.8)	2020	(139)	2530	(174)	2360	(163)	2750	(189)		
20	(508.0)	238.50	(355.25)	1.187	(30.15)	17.626	(447.7)	2140	(147)	2670	(184)	2490	(172)	2750	(189)		
20	(508.0)	250.31	(372.84)	1.250	(31.75)	17.500	(444.5)	2250	(155)	2750	(189)	2620	(181)	2750	(189)		
20	(508.0)	261.86	(390.04)	1.312	(33.32)	17.376	(441.4)	2360	(163)	2750	(189)	2750	(189)	2750	(189)		
20	(508.0)	273.51	(407.39)	1.375	(34.92)	17.250	(438.2)	2480	(171)	2750	(189)	2750	(189)	2750	(189)		
*22	(558.8)	50.94	(75.88)	0.219	(5.56)	21.562	(547.7)	360	(25)	450	(31)	420	(29)	520	(36)		
*22	(558.8)	58.07	(86.50)	0.250	(6.35)	21.500	(546.1)	410	(28)	510	(35)	480	(33)	600	(41)		
*22	(558.8)	65.18	(97.09)	0.281	(7.14)	21.438	(544.5)	460	(32)	570	(39)	540	(37)	670	(46)		
22	(558.8)	72.27	(107.65)	0.312	(7.92)	21.376	(543.0)	510	(35)	640	(44)	600	(41)	740	(51)		
22	(558.8)	79.56	(118.50)	0.344	(8.74)	21.312	(541.3)	560	(39)	700	(48)	660	(45)	820	(56)		
22	(558.8)	86.61	(129.01)	0.375	(9.52)	21.250	(539.8)	610	(42)	770	(53)	720	(50)	890	(61)		
22	(558.8)	100.86	(150.23)	0.438	(11.13)	21.124	(536.5)	720	(50)	900	(62)	840	(58)	1050	(72)		
22	(558.8)	114.81	(171.01)	0.500	(12.70)	21.000	(533.4)	820	(56)	1020	(70)	950	(65)	1190	(82)		
22	(558.8)	128.67	(191.65)	0.562	(14.27)	20.876	(530.3)	920	(63)	1150	(79)	1070	(74)	1340	(92)		
22	(558.8)	142.68	(212.52)	0.625	(15.88)	20.750	(527.0)	1020	(70)	1280	(88)	1190	(82)	1490	(103)		
22	(558.8)	156.60	(233.26)	0.688	(17.48)	20.624	(523.8)	1130	(78)	1410	(97)	1310	(90)	1640	(113)		
22	(558.8)	170.21	(253.53)	0.750	(19.05)	20.500	(520.7)	1230	(85)	1530	(105)	1430	(99)	1790	(123)		
22	(558.8)	183.75	(273.70)	0.812	(20.62)	20.376	(517.6)	1330	(92)	1660	(114)	1550	(107)	1940	(134)		
22	(558.8)	197.41	(294.04)	0.875	(22.22)	20.250	(514.4)	1430	(99)	1790	(123)	1670	(115)	2090	(144)		
22	(558.8)	211.00	(314.28)	0.938	(23.83)	20.124	(511.1)	1530	(105)	1920	(132)	1790	(123)	2240	(154)		
22	(558.8)	224.28	(334.07)	1.000	(25.40)	20.000	(508.0)	1640	(113)	2050	(141)	1910	(132)	2390	(165)		
22	(558.8)	237.48	(353.73)	1.062	(26.97)	19.876	(504.9)	1740	(120)	2170	(150)	2030	(140)	2500	(172)		
22	(558.8)	250.81	(373.58)	1.125	(28.58)	19.750	(501.6)	1840	(127)	2300	(158)	2150	(148)	2500	(172)		
22	(558.8)	263.85	(393.00)	1.187	(30.15)	19.626	(498.5)	1940	(134)	2430	(167)	2270	(156)	2500	(172)		
22	(558.8)	277.01	(412.61)	1.250	(31.75)	19.500	(495.3)	2050	(141)	2500	(172)	2390	(165)	2500	(172)		
22	(558.8)	289.88	(431.78)	1.312	(33.32)	19.376	(492.2)	2150	(148)	2500	(172)	2500	(172)	2500	(172)		
22	(558.8)	302.88	(451.14)	1.375	(34.92)	19.250	(489.0)	2250	(155)	2500	(172)	2500	(172)	2500	(172)		
22	(558.8)	315.58	(470.06)	1.437	(36.50)	19.126	(485.8)	2350	(162)	2500	(172)	2500	(172)	2500	(172)		
22	(558.8)	328.41	(489.17)	1.500	(38.10)	19.000	(482.6)	2450	(169)	2500	(172)	2500	(172)	2500	(172)		
*24	(609.6)	63.41	(94.45)	0.250	(6.35)	23.500	(596.9)	380	(26)	470	(32)	440	(30)	550	(38)		
*24	(609.6)	71.18	(106.02)	0.281	(7.14)	23.438	(595.3)	420	(29)	530	(37)	490	(34)	610	(42)		
24	(609.6)	78.93	(117.57)	0.312	(7.92)	23.376	(593.8)	470	(32)	580	(40)	550	(38)	680	(47)		
24	(609.6)	86.91	(129.45)	0.344	(8.74)	23.312	(592.1)	520	(36)	640	(44)	600	(41)	750	(52)		
24	(609.6)	94.62	(140.94)	0.375	(9.52)	23.250	(590.6)	560	(39)	700	(48)	660	(45)	820	(56)		
24	(609.6)	110.22	(164.17)	0.438	(11.13)	23.124	(587.3)	660	(45)	820	(56)	770	(53)	960	(66)		
24	(609.6)	125.49	(186.92)	0.500	(12.70)	23.000	(584.1)	750	(52)	940	(65)	880	(61)	1090	(75)		
24	(609.6)	140.68	(209.54)	0.562	(14.27)	22.876	(581.1)	840	(58)	1050	(72)	980	(68)	1230	(85)		
24	(609.6)	156.03	(232.41)	0.625	(15.88)	22.750	(577.8)	940	(65)	1170	(81)	1090	(75)	1370	(94)		

Appendix 13.B-3 Continued

1				2		3		4		5		6		7		8		9	
Size				Plain-End Weight, w_{pe}		Wall Thickness, t		Inside Diameter, d		Test Pressure, min.									
										Grade A				Grade B				Grade A25	
Nom.	Desig-nation	Outside Diameter, D								Std.		Alt.		Std.		Alt.			
in.	in.	in.	mm	lb/ft	kg/m	in.	mm	in.	mm	psi	100 kPa	psi	100 kPa	psi	100 kPa	psi	100 kPa	psi	100 kPa
		24	(609.6)	171.29	(255.14)	0.688	(19.05)	22.624	(574.6)	1030	(71)	1290	(89)	1200	(83)	1500	(103)		
		24	(609.6)	186.23	(277.39)	0.750	(20.62)	22.500	(571.5)	1120	(77)	1410	(97)	1310	(90)	1640	(113)		
		24	(609.6)	201.09	(299.52)	0.812	(22.22)	22.376	(568.4)	1220	(84)	1520	(105)	1420	(98)	1780	(123)		
		24	(609.6)	216.10	(321.88)	0.875	(23.83)	22.250	(565.2)	1310	(90)	1640	(113)	1530	(105)	1910	(132)		
		24	(609.6)	231.03	(344.12)	0.938	(25.40)	22.124	(561.9)	1410	(97)	1760	(121)	1640	(113)	2050	(141)		
		24	(609.6)	245.64	(365.88)	1.000	(26.97)	22.000	(558.8)	1500	(103)	1880	(130)	1750	(121)	2190	(151)		
		24	(609.6)	260.17	(387.52)	1.062	(28.58)	21.876	(555.7)	1590	(110)	1990	(137)	1860	(128)	2300	(158)		
		24	(609.6)	274.84	(409.37)	1.125	(30.15)	21.750	(552.4)	1690	(116)	2110	(145)	1970	(136)	2300	(158)		
		24	(609.6)	289.20	(430.76)	1.187	(31.75)	21.626	(549.3)	1780	(123)	2230	(154)	2080	(143)	2300	(158)		
		24	(609.6)	303.71	(452.38)	1.250	(33.32)	21.500	(546.1)	1880	(130)	2300	(158)	2190	(151)	2300	(158)		
		24	(609.6)	317.91	(473.53)	1.312	(34.92)	21.376	(543.0)	1970	(136)	2300	(158)	2300	(158)	2300	(158)		
		24	(609.6)	332.25	(494.89)	1.375	(36.50)	21.250	(539.8)	2060	(142)	2300	(158)	2300	(158)	2300	(158)		
		24	(609.6)	346.28	(515.78)	1.437	(38.10)	21.126	(536.6)	2160	(149)	2300	(158)	2300	(158)	2300	(158)		
		24	(609.6)	360.45	(536.89)	1.500	(39.67)	21.000	(533.4)	2250	(155)	2300	(158)	2300	(158)	2300	(158)		
		24	(609.6)	374.31	(557.53)	1.562		20.876	(530.3)	2300	(158)	2300	(158)	2300	(158)	2300	(158)		

*These items are special plain-end pipe.

(API, 1978)

APPENDIX 13.B-4.
Standard-Weight Threaded Line Pipe Dimensions, Weights, and Test Pressures

1	2		3	4		5		6		7		8		9		10	
								Calculated Weight				Test Pressure					
Nominal Size	Outside Diameter D		¹Nominal Weight: Threads and Coupling	Wall Thickness t		Inside Diameter d		Plain End w/p_e		²Threads and Coupling ℓ_w		Grade A25		Grade A		Grade B	
in	in	mm	lb/ft	in	mm	in	mm	lb/ft	kg/m	lb/ft	kg/m	psi	100 kPa	psi	100 kPa	psi	100 kPa
⅛	0.405	(10.3)	0.25	0.068	(1.73)	0.269	(6.8)	0.24	(0.36)	0.20	(0.09)	700	(48)	700	(48)	700	(48)
¼	0.540	(13.7)	0.43	0.088	(2.24)	0.364	(9.2)	0.42	(0.63)	0.20	(0.09)	700	(48)	700	(48)	700	(48)
⅜	0.675	(17.1)	0.57	0.091	(2.31)	0.493	(12.5)	0.57	(0.85)	0.20	(0.09)	700	(48)	700	(48)	700	(48)
½	0.840	(21.3)	0.86	0.109	(2.77)	0.622	(15.8)	0.85	(1.27)	0.20	(0.09)	700	(48)	700	(48)	700	(48)
¾	1.050	(26.7)	1.14	0.113	(2.87)	0.824	(21.0)	1.13	(1.68)	0.20	(0.09)	700	(48)	700	(48)	700	(48)
1	1.315	(33.4)	1.70	0.133	(3.38)	1.049	(26.6)	1.68	(2.50)	0.20	(0.09)	700	(48)	700	(48)	700	(48)
1¼	1.660	(42.2)	2.30	0.140	(3.56)	1.380	(35.1)	2.27	(3.38)	0.60	(0.27)	1000	(69)	1000	(69)	1100	(76)
1½	1.900	(48.3)	2.75	0.145	(3.68)	1.610	(40.9)	2.72	(4.05)	0.40	(0.18)	1000	(69)	1000	(69)	1100	(76)
2	2.375	(60.3)	3.75	0.154	(3.91)	2.067	(52.5)	3.65	(5.44)	1.20	(0.54)	1000	(69)	1000	(69)	1100	(76)
2½	2.875	(73.0)	5.90	0.203	(5.16)	2.469	(62.7)	5.79	(8.62)	1.80	(0.82)	1000	(69)	1000	(69)	1100	(76)
3	3.500	(88.9)	7.70	0.216	(5.49)	3.068	(77.9)	7.58	(11.29)	1.80	(0.82)	1000	(69)	1000	(69)	1100	(76)
3½	4.000	(101.6)	9.25	0.226	(5.74)	3.548	(90.1)	9.11	(13.57)	3.20	(1.45)	1200	(83)	1200	(83)	1300	(90)
4	4.500	(114.3)	11.00	0.237	(6.02)	4.026	(102.3)	10.79	(16.07)	4.40	(2.00)	1200	(83)	1200	(83)	1300	(90)
5	5.563	(141.3)	15.00	0.258	(6.55)	5.047	(128.2)	14.62	(21.78)	5.60	(2.54)	—	(—)	1200	(83)	1300	(90)
6	6.625	(168.3)	19.45	0.280	(7.11)	6.065	(154.1)	18.97	(28.26)	7.20	(3.27)	—	(—)	1200	(83)	1300	(90)
8	8.625	(219.1)	25.55	0.277	(7.04)	8.071	(205.0)	24.70	(36.79)	14.80	(6.72)	—	(—)	1160	(80)	1350	(93)
8	8.625	(219.1)	29.35	0.322	(8.18)	7.981	(202.7)	28.55	(42.53)	14.00	(6.36)	—	(—)	1340	(92)	1570	(108)
10	10.750	(273.0)	32.75	0.279	(7.09)	10.192	(258.8)	31.20	(46.47)	20.00	(9.08)	—	(—)	930	(64)	1090	(75)
10	10.750	(273.0)	35.75	0.307	(7.80)	10.136	(257.4)	34.24	(51.00)	19.20	(8.72)	—	(—)	1030	(71)	1200	(83)
10	10.750	(273.0)	41.85	0.365	(9.27)	10.020	(254.5)	40.48	(60.29)	17.40	(7.90)	—	(—)	1220	(84)	1430	(99)
12	12.750	(323.8)	45.45	0.330	(8.38)	12.090	(307.0)	43.77	(65.20)	32.60	(14.80)	—	(—)	960	(64)	1090	(75)
12	12.750	(323.8)	51.15	0.375	(9.52)	12.000	(304.8)	49.56	(73.82)	30.80	(13.98)	—	(—)	1060	(73)	1240	(85)
14D	14.000	(355.6)	57.00	0.375	(9.52)	13.250	(336.6)	54.57	(81.28)	24.60	(11.17)	—	(—)	960	(66)	1120	(77)
16D	16.000	(406.4)	65.30	0.375	(9.52)	15.250	(387.4)	62.58	(93.21)	30.00	(13.62)	—	(—)	840	(58)	980	(68)
18D	18.000	(457.2)	73.00	0.375	(9.52)	17.250	(438.2)	70.59	(105.14)	35.60	(16.16)	—	(—)	750	(52)	880	(61)
20D	20.000	(508.0)	81.00	0.375	(9.52)	19.250	(489.0)	78.60	(117.07)	42.00	(19.07)	—	(—)	680	(47)	790	(54)

¹Nominal weights, threads and coupling (Col. 3) are shown for the purpose of identification in ordering
²Weight gain due to end finishing.

(API, 1978)

APPENDIX 13.B-5.
ANSI Pipe Schedules with Full Metric Conversion

PIPE SIZE	O.D. IN INCHES	5	10	20	30	40	STD.	60	80	XH	100	120	140	160	XXH
⅛ * .317	.405 1.029	.035 .1383 .089 .0628	.049 .1863 .124 .085			.068 .2447 .173 .1111	.068 .2447 .173 .1111		.095 .3145 .241 .1428	.095 .3145 .241 .1428					
¼ † .635	.540 1.372	.049 .2570 .124 .1167	.065 .3297 .165 .1497			.088 .4248 .223 .1929	.088 .4248 .223 .1929		.119 .5351 .302 .2429	.119 .5351 .302 .2429					
⅜ .952	.675 1.714	.049 .3276 .124 .1487	.065 .4235 .165 .1923			.091 .5676 .231 .2577	.091 .5676 .231 .2577		.126 .7388 .320 .3354	.126 .7388 .320 .3354					
½ 1.270	.840 2.134	.065 .5383 .165 .2444	.083 .6710 .211 .3046			.109 .8510 .277 .3863	.109 .8510 .277 .3863		.147 1.088 .373 .4939	.147 1.088 .373 .4939				.187 1.304 .475 .5920	.294 1.714 .747 .7782
¾ 1.905	1.050 2.667	.065 .6838 .165 .3104	.083 .8572 .211 .3892			.113 1.131 .287 .5135	.113 1.131 .287 .5135		.154 1.474 .391 .6692	.154 1.474 .391 .6692				.218 1.937 .554 .8794	.308 2.441 .782 1.108
1 2.540	1.315 3.340	.065 .8678 .165 .3940	.109 1.404 .277 .6374			.133 1.679 .338 .7623	.133 1.679 .338 .7623		.179 2.172 .455 .9861	.179 2.172 .455 .9861				.250 2.844 .635 1.291	.358 3.659 .909 1.661
1¼ 3.175	1.660 4.216	.065 1.107 .165 .5026	.109 1.806 .277 .8199			.140 2.273 .356 1.032	.140 2.273 .356 1.032		.191 2.997 .485 1.361	.191 2.997 .485 1.361				.250 3.765 .635 1.709	.382 5.214 .970 2.367
1½ 3.810	1.900 4.826	.065 1.274 .165 .5784	.109 2.085 .277 .9466			.145 2.718 .368 1.234	.145 2.718 .368 1.234		.200 3.631 .508 1.648	.200 3.631 .508 1.648				.281 4.859 .714 2.206	.400 6.408 1.016 2.909
2 5.080	2.375 6.032	.065 1.604 .165 .7282	.109 2.638 .277 1.197			.154 3.653 .391 1.658	.154 3.653 .391 1.658		.218 5.022 .554 2.280	.218 5.022 .554 2.280				.343 7.444 .871 3.380	.436 9.029 1.107 4.090

Nominal	OD												
2½ 6.350	2.875 7.302	.083 2.475 .211 1.124	.120 3.531 .305 1.603		.203 5.793 .516 2.630	.203 5.793 .516 2.630		.276 7.661 .701 3.478	.276 7.661 .701 3.478			.375 10.01 .952 4.544	.552 13.70 1.402 6.220
3 7.620	3.500 8.890	.083 3.029 .211 1.375	.120 4.332 .305 1.967		.216 7.576 .549 3.439	.216 7.576 .549 3.439		.300 10.25 .762 4.653	.300 10.25 .762 4.653			.437 14.32 1.110 6.501	.600 18.58 1.524 8.435
3½ 8.890	4.0 10.16	.083 3.472 .211 1.576	.120 4.973 .305 2.258		.226 9.109 .574 4.135	.226 9.109 .574 4.135		.318 12.51 .808 5.679	.318 12.51 .808 5.679				.636 22.85 1.615 10.37
4 10.16	4.50 11.43	.083 3.915 .211 1.777	.120 5.613 .305 2.548		.237 10.79 .602 4.899	.237 10.79 .602 4.899	.281 12.66 .714 5.748	.337 14.98 .856 6.801	.337 14.98 .856 6.801	.437 19.01 1.110 8.630		.531 22.51 1.349 10.22	.674 27.54 1.712 12.50
4½ 11.43	5.0 12.70				.247 12.53 .627 5.689			.355 17.61 .902 7.995					.710 32.53 1.803 14.77
5 12.70	5.563 14.13	.109 6.349 .277 2.882	.134 7.770 .340 3.528		.258 14.62 .655 6.637	.258 14.62 .655 6.637		.375 20.78 .952 9.434	.375 20.78 .952 9.434	.500 27.04 1.270 12.28		.625 32.96 1.587 14.96	.750 38.55 1.905 17.50
6 15.24	6.625 16.83	.109 7.585 .277 3.443	.134 9.289 .340 4.217		.280 18.97 .711 8.612	.280 18.97 .711 8.612		.432 28.57 1.097 12.971	.432 28.57 1.097 12.971	.562 36.39 1.427 16.521		.718 45.30 1.824 20.566	.864 53.16 2.195 24.135
7 17.78	7.625 19.37			.250 22.36 .635 10.15	.301 23.57 .764 10.701			.500 38.05 1.270 17.275		.593 50.87 1.506 23.09			.875 63.08 2.222 28.638
8 20.32	8.625 21.91	.109 9.914 .277 4.501	.148 13.40 .376 6.084	.277 24.70 .704 11.21	.322 28.55 .818 12.96	.322 28.55 .818 12.96	.406 35.64 1.031 16.18	.500 43.39 1.270 19.70	.500 43.39 1.270 19.70	.718 60.63 1.824 27.53	.812 67.76 2.062 30.76	.906 74.69 2.301 33.91	.875 72.42 2.222 32.88
9 22.86	9.625 24.45				.342 33.90 .869 15.391			.500 48.72 1.270 22.119					

Appendix 13.B-5. Continued

PIPE SIZE	O.D. IN INCHES	5	10	20	30	40	STD.	60	80	XH	100	120	140	160	XXH
10 25.40	10.75 27.30	.134 15.19 .340 6.896	.165 18.70 .419 8.490	.250 28.04 .635 12.73	.307 34.24 .780 15.54	.365 40.48 .927 18.38	.365 40.48 .927 18.38	.500 54.74 1.270 24.85	.593 64.33 1.506 29.20	.500 54.74 1.270 24.85	.718 76.93 1.824 34.93	.843 89.20 2.141 40.50	1.000 104.1 2.540 47.26	1.125 115.7 2.857 52.23	
11 27.94	11.75 29.84						.375 45.55 .952 20.68			.500 60.07 1.270 27.27					
12 30.48	12.75 32.38	.165 22.18 .419 10.07	.180 24.20 .457 10.99	.250 33.38 .635 15.15	.330 43.77 .838 19.87	.406 53.53 1.031 24.30	.375 49.56 .952 22.50	.562 73.16 1.427 33.21	.687 88.51 1.745 40.18	.500 65.42 1.270 29.70	.843 107.2 2.141 48.67	1.000 125.5 2.540 56.98	1.125 139.7 2.857 63.42	1.312 160.3 3.332 72.78	
14 35.56	14.0		.250 36.71 .635 16.67	.312 45.68 .792 20.74	.375 54.57 .952 24.77	.437 63.37 1.110 28.77	.375 54.57 .952 24.77	.593 84.91 1.506 38.55	.750 106.1 1.905 48.17	.500 72.09 1.270 32.73	.937 130.7 2.380 59.34	1.093 150.7 2.776 68.42	1.250 170.2 3.175 77.27	1.406 189.1 3.571 85.85	
16 40.64	16.0		.250 42.05 .635 19.09	.312 52.36 .792 23.77	.375 62.58 .952 28.41	.500 82.77 1.270 37.58	.375 62.58 .952 28.41	.656 107.5 1.666 48.80	.843 136.5 2.141 61.97	.500 82.77 1.270 37.58	1.031 164.8 2.619 74.82	1.218 192.3 3.094 87.30	1.437 223.5 3.650 101.5	1.593 245.1 4.046 111.3	
18 45.72	18.0		.250 47.39 .635 21.51	.312 59.03 .792 26.80	.437 82.06 1.110 37.25	.562 104.8 1.427 47.58	.375 70.59 .952 32.05	.750 138.2 1.905 62.74	.937 170.8 2.380 77.54	.500 93.45 1.270 42.43	1.156 208.0 2.936 94.43	1.375 244.1 3.492 110.8	1.562 274.2 3.967 124.5	1.781 308.5 4.524 140.0	
20 50.80	20.0		.250 52.73 .635 23.94	.375 78.60 .952 35.68	.500 104.1 1.270 47.26	.593 122.9 1.506 55.80	.375 78.60 .952 35.68	.812 166.4 2.062 75.54	1.031 208.9 2.619 94.84	.500 104.1 1.270 47.26	1.280 256.1 3.251 116.3	1.500 296.4 3.810 134.6	1.750 341.1 4.445 154.8	1.968 379.0 4.999 172.1	
22 55.88	22.0		.250 58.07 .635 26.36	.375 86.61 .952 39.32	.500 114.8 1.270 52.12		.375 86.61 .952 39.32	.875 197.4 2.222 89.62	1.125 250.8 2.857 113.9	.500 114.8 1.270 52.12	1.375 302.9 3.492 137.4	1.625 353.6 4.127 160.5	1.875 403.0 4.762 183.0	2.125 451.1 5.397 204.8	
24 60.96	24.0		.250 63.41 .635 28.79	.375 94.62 .952 42.96	.562 140.8 1.427 63.92	.687 171.2 1.745 77.72	.375 94.62 .952 42.96	.968 238.1 2.459 108.1	1.218 296.4 3.094 134.6	.500 125.5 1.270 56.98	1.531 367.4 3.889 166.80	1.812 429.4 4.602 194.9	2.062 483.1 5.237 219.3	2.343 541.9 5.951 246.0	
26 66.40	26.0		.312 85.60 .792 38.86	.500 136.2 1.270 61.83			.375 102.6 .952 46.58			.500 136.2 1.270 61.83					

28	28.0 71.12		.312 92.26 .792 41.89	.500 146.8 1.270 66.65	.625 182.7 1.587 82.94		.375 110.6 .952 50.21		.500 146.8 1.270 66.65	
30	30.0 76.20		.312 98.93 .792 44.91	.500 157.5 1.270 71.50	.625 196.1 1.587 89.03		.375 118.6 .952 53.84		.500 157.5 1.270 71.50	
32	32.0 81.28		.312 105.6 .792 47.94	.500 168.2 1.270 76.36	.625 209.4 1.587 95.07	.688 230.1 1.747 104.5	.375 126.7 .952 57.52		.500 168.2 1.270 76.36	
34	34.0 86.36		.344 123.7 .874 56.16	.500 178.9 1.270 81.22	.625 222.8 1.587 101.1	.688 244.8 1.747 111.1	.375 134.7 .952 61.15		.500 178.9 1.270 81.22	
36	36.0 91.44		.312 118.9 .792 53.98	.500 189.6 1.270 86.08	.625 236.1 1.587 107.2	.750 282.3 1.905 128.2	.375 142.7 .952 64.78		.500 189.6 1.270 86.08	
42	42.0 106.68						.375 166.7 .952 75.68		.500 22.16 1.270 100.6	
48	48.0 121.92						.375 190.7 .952 86.60		.500 253.6 1.270 115.1	
54	54.0 137.16						.375 214.8 .952 97.51		.500 285.7 1.270 129.7	
60	60.0 152.40						.375 238.8 .952 108.4		.500 317.7 1.270 144.2	

*INCHES — Pipe size in inches
upper figure — wall thickness in inches
lower figure — weight per ft. in lbs.

†METRIC — Pipe size in centimeters
upper figure — wall thickness in centimeters
lower figure — weight per ft. in kilograms

(National Association of Steel Pipe Distributors, Inc.)

APPENDIX 13.B-6.
Pipe Data — Carbon and Alloy Steel — Stainless Steel

Nominal Pipe Size Inches	Outside Diam. Inches	Identification Iron Pipe Size	Identification Steel Sched. No.	Identification Stainless Steel Sched. No.	Wall Thickness (t) Inches	Inside Diameter (d) Inches	Area of Metal Square Inches	Transverse Internal Area (a) Square Inches	Transverse Internal Area (A) Square Feet	Moment of Inertia (I) Inches	Weight Pipe Pounds per foot	Weight Water Pounds per foot of pipe	External Surface Sq. Ft. per foot of Pipe	Section Modulus $\left(2\frac{I}{O.D.}\right)$
1/8	0.405	... STD XS	... 40 80	10S 40S 80S	.049 .068 .095	.307 .269 .215	.0548 .0720 .0925	.0740 .0568 .0364	.00051 .00040 .00025	.00088 .00106 .00122	.19 .24 .31	.032 .025 .016	.106 .106 .106	.00437 .00523 .00602
1/4	0.540	... STD XS	... 40 80	10S 40S 80S	.065 .088 .119	.410 .364 .302	.0970 .1250 .1574	.1320 .1041 .0716	.00091 .00072 .00050	.00279 .00331 .00377	.33 .42 .54	.057 .045 .031	.141 .141 .141	.01032 .01227 .01395
3/8	0.675	... STD XS	... 40 80	10S 40S 80S	.065 .091 .126	.545 .493 .423	.1246 .1670 .2173	.2333 .1910 .1405	.00162 .00133 .00098	.00586 .00729 .00862	.42 .57 .74	.101 .083 .061	.178 .178 .178	.01736 .02160 .02554
1/2	0.840	... ... STD XS ... XXS	... ... 40 80 160 ...	5S 10S 40S 80S	.065 .083 .109 .147 .187 .294	.710 .674 .622 .546 .466 .252	.1583 .1974 .2503 .3200 .3836 .5043	.3959 .3568 .3040 .2340 .1706 .050	.00275 .00248 .00211 .00163 .00118 .00035	.01197 .01431 .01709 .02008 .02212 .02424	.54 .67 .85 1.09 1.31 1.71	.172 .155 .132 .102 .074 .022	.220 .220 .220 .220 .220 .220	.02849 .03407 .04069 .04780 .05267 .05772
3/4	1.050	... ... STD XS ... XXS	... ... 40 80 160 ...	5S 10S 40S 80S	.065 .083 .113 .154 .219 .308	.920 .884 .824 .742 .612 .434	.2011 .2521 .3326 .4335 .5698 .7180	.6648 .6138 .5330 .4330 .2961 .148	.00462 .00426 .00371 .00300 .00206 .00103	.02450 .02969 .03704 .04479 .05269 .05792	.69 .86 1.13 1.47 1.94 2.44	.288 .266 .231 .188 .128 .064	.275 .275 .275 .275 .275 .275	.04667 .05655 .07055 .08531 .10036 .11032
1	1.315	... ... STD XS ... XXS	... ... 40 80 160 ...	5S 10S 40S 80S	.065 .109 .133 .179 .250 .358	1.185 1.097 1.049 .957 .815 .599	.2553 .4130 .4939 .6388 .8365 1.0760	1.1029 .9452 .8640 .7190 .5217 .282	.00766 .00656 .00600 .00499 .00362 .00196	.04999 .07569 .08734 .1056 .1251 .1405	.87 1.40 1.68 2.17 2.84 3.66	.478 .409 .375 .312 .230 .122	.344 .344 .344 .344 .344 .344	.07603 .11512 .1328 .1606 .1903 .2136

Appendix 13.B-6. Continued

1¼	1.660	...	5S	.065	1.530	.3257	1.839	.01277	.1038	1.11	.797	.435	.1250
		...	10S	.109	1.442	.4717	1.633	.01134	.1605	1.81	.708	.435	.1934
		STD	40S	.140	1.380	.6685	1.495	.01040	.1947	2.27	.649	.435	.2346
		XS	80S	.191	1.278	.8815	1.283	.00891	.2418	3.00	.555	.435	.2913
		160	...	.250	1.160	1.1070	1.057	.00734	.2839	3.76	.458	.435	.3421
		XXS	...	.382	.896	1.534	.630	.00438	.3411	5.21	.273	.435	.4110
1½	1.900	...	5S	.065	1.770	.3747	2.461	.01709	.1579	1.28	1.066	.497	.1662
		...	10S	.109	1.682	.6133	2.222	.01543	.2468	2.09	.963	.497	.2598
		STD	40S	.145	1.610	.7995	2.036	.01414	.3099	2.72	.882	.497	.3262
		XS	80S	.200	1.500	1.068	1.767	.01225	.3912	3.63	.765	.497	.4118
		160	...	.281	1.338	1.429	1.406	.00976	.4824	4.86	.608	.497	.5078
		XXS	...	.400	1.100	1.885	.950	.00660	.5678	6.41	.42	.497	.5977
2	2.375	...	5S	.065	2.245	.4717	3.958	.02749	.3149	1.61	1.72	.622	.2652
		...	10S	.109	2.157	.7760	3.654	.02538	.4492	2.64	1.58	.622	.4204
		STD	40S	.154	2.067	1.075	3.355	.02330	.6657	3.65	1.45	.622	.5606
		XS	80S	.218	1.939	1.477	2.953	.02050	.8679	5.02	1.28	.622	.7309
		160	...	.344	1.687	2.190	2.241	.01556	1.162	7.46	.97	.622	.979
		XXS	...	.436	1.503	2.656	1.774	.01232	1.311	9.03	.77	.622	1.104
2½	2.875	...	5S	.083	2.709	.7280	5.764	.04002	.7100	2.48	2.50	.753	.4939
		...	10S	.120	2.635	1.039	5.453	.03787	.9873	3.53	2.36	.753	.6868
		STD	40S	.203	2.469	1.704	4.788	.03322	1.530	5.79	2.07	.753	1.064
		XS	80S	.276	2.323	2.254	4.238	.02942	1.924	7.66	1.87	.753	1.339
		160	...	.375	2.125	2.945	3.546	.02463	2.353	10.01	1.54	.753	1.638
		XXS	...	.552	1.771	4.028	2.464	.01710	2.871	13.69	1.07	.753	1.997
3	3.500	...	5S	.083	3.334	.8910	8.730	.06063	1.301	3.03	3.78	.916	.7435
		...	10S	.120	3.260	1.274	8.347	.05796	1.822	4.33	3.62	.916	1.041
		STD	40S	.216	3.068	2.228	7.393	.05130	3.017	7.58	3.20	.916	1.724
		XS	80S	.300	2.900	3.016	6.605	.04587	3.894	10.25	2.86	.916	2.225
		160	...	.438	2.624	4.205	5.408	.03755	5.032	14.32	2.35	.916	2.876
		XXS	...	.600	2.300	5.466	4.155	.02885	5.993	18.58	1.80	.916	3.424
3½	4.000	...	5S	.083	3.834	1.021	11.545	.08017	1.960	3.48	5.00	1.047	.9799
		...	10S	.120	3.760	1.463	11.104	.07711	2.755	4.97	4.81	1.047	1.378
		STD	40S	.226	3.518	2.680	9.886	.06870	4.788	9.11	4.29	1.047	2.394
		XS	80S	.318	3.364	3.678	8.888	.06170	6.280	12.50	3.84	1.047	3.140
4	4.500	...	5S	.083	4.334	1.152	14.75	.10245	2.810	3.92	6.39	1.178	1.249
		...	10S	.120	4.260	1.651	14.25	.09898	3.963	5.61	6.18	1.178	1.761

Appendix 13.B-6. Continued

Nominal Pipe Size Inches	Outside Diam. Inches	Identification				Wall Thickness (t) Inches	Inside Diameter (d) Inches	Area of Metal Square Inches	Transverse Internal Area		Moment of Inertia (I) Inches	Weight Pipe Pounds per foot	Weight Water Pounds per foot of pipe	External Surface Sq. Ft. per foot of Pipe	Section Modulus $\left(2\dfrac{I}{O.D.}\right)$
		Iron Pipe Size	Steel Sched. No.		Stainless Steel Sched. No.				(a) Square Inches	(A) Square Feet					
4	4.500	STD	40		40S	.237	4.026	3.174	12.73	.08840	7.233	10.79	5.50	1.178	3.214
		XS	80		80S	.337	3.826	4.407	11.50	.07986	9.610	14.98	4.98	1.178	4.271
		...	120		...	.438	3.624	5.595	10.31	.0716	11.65	19.00	4.47	1.178	5.178
		...	160		...	.531	3.438	6.621	9.28	.0645	13.27	22.51	4.02	1.178	5.898
		XXS	...		...	.674	3.152	8.101	7.80	.0542	15.28	27.54	3.38	1.178	6.791
5	5.563	...	...		5S	.109	5.345	1.868	22.44	.1558	6.947	6.36	9.72	1.456	2.498
		...	...		10S	.134	5.295	2.285	22.02	.1529	8.425	7.77	9.54	1.456	3.029
		STD	40		40S	.258	5.047	4.300	20.01	.1390	15.16	14.62	8.67	1.456	5.451
		XS	80		80S	.375	4.813	6.112	18.19	.1263	20.67	20.78	7.88	1.456	7.431
		...	120		...	.500	4.563	7.953	16.35	.1136	25.73	27.04	7.09	1.456	9.250
		...	160		...	.625	4.313	9.696	14.61	.1015	30.03	32.96	6.33	1.456	10.796
		XXS	...		...	.750	4.063	11.340	12.97	.0901	33.63	38.55	5.61	1.456	12.090
6	6.625	...	...		5S	.109	6.407	2.231	32.24	.2239	11.85	7.60	13.97	1.734	3.576
		...	...		10S	.134	6.357	2.733	31.74	.2204	14.40	9.29	13.75	1.734	4.346
		STD	40		40S	.280	6.065	5.581	28.89	.2006	28.14	18.97	12.51	1.734	8.496
		XS	80		80S	.432	5.761	8.405	26.07	.1810	40.49	28.57	11.29	1.734	12.22
		...	120		...	.562	5.501	10.70	23.77	.1650	49.61	36.39	10.30	1.734	14.98
		...	160		...	.719	5.187	13.32	21.15	.1469	58.97	45.35	9.16	1.734	17.81
		XXS	...		...	.864	4.897	15.64	18.84	.1308	66.33	53.16	8.16	1.734	20.02
8	8.625	...	...		5S	.109	8.407	2.916	55.51	.3855	26.44	9.93	24.06	2.258	6.131
		...	...		10S	.148	8.329	3.941	54.48	.3784	35.41	13.40	23.61	2.258	8.212
		...	20		...	.250	8.125	6.57	51.85	.3601	57.72	22.36	22.47	2.258	13.39
		...	30		...	.277	8.071	7.26	51.16	.3553	63.35	24.70	22.17	2.258	14.69
		STD	40		40S	.322	7.981	8.40	50.03	.3474	72.49	28.55	21.70	2.258	16.81
		...	60		...	.406	7.813	10.48	47.94	.3329	88.73	35.64	20.77	2.258	20.58
		XS	80		80S	.500	7.625	12.76	45.66	.3171	105.7	43.39	19.78	2.258	24.51
		...	100		...	.594	7.437	14.96	43.46	.3018	121.3	50.95	18.83	2.258	28.14
		...	120		...	.719	7.187	17.84	40.59	.2819	140.5	60.71	17.59	2.258	32.58
		...	140		...	.812	7.001	19.93	38.50	.2673	153.7	67.76	16.68	2.258	35.65

Appendix 13.B-6. Continued

Nom. Size	OD	Sched.	Sched. (S)	Wall	ID							Weight
		XXS	...	.875	6.875	21.30	37.12	.2578	162.0	72.42	16.10	2.258	37.56
		...	160	.906	6.813	21.97	36.46	.2532	165.9	74.69	15.80	2.258	38.48
10	10.750	...	5S	.134	10.482	4.36	86.29	.5992	63.0	15.19	37.39	2.814	11.71
		...	10S	.165	10.420	5.49	85.28	.5922	76.9	18.65	36.95	2.814	14.30
		20	...	.250	10.250	8.24	82.52	.5731	113.7	28.04	35.76	2.814	21.15
		30	...	.307	10.136	10.07	80.69	.5603	137.4	34.24	34.96	2.814	25.57
		STD	40S	.365	10.020	11.90	78.86	.5475	160.7	40.48	34.20	2.814	29.90
		XS	80S	.500	9.750	16.10	74.66	.5185	212.0	54.74	32.35	2.814	39.43
		80	...	.594	9.562	18.92	71.84	.4989	244.8	64.43	31.13	2.814	45.54
		100	...	.719	9.312	22.63	68.13	.4732	286.1	77.03	29.53	2.814	53.22
		120	...	.844	9.062	26.24	64.53	.4481	324.2	89.29	27.96	2.814	60.32
		140	...	1.000	8.750	30.63	60.13	.4176	367.8	104.13	26.06	2.814	68.43
		XXS	160	1.125	8.500	34.02	56.75	.3941	399.3	115.64	24.59	2.814	74.29
12	12.75	...	5S	.156	12.438	6.17	121.50	.8438	122.4	20.98	52.65	3.338	19.2
		...	10S	.180	12.390	7.11	120.57	.8373	140.4	24.17	52.25	3.338	22.0
		20	...	.250	12.250	9.82	117.86	.8185	191.8	33.38	51.07	3.338	30.2
		30	...	.330	12.090	12.87	114.80	.7972	248.4	43.77	49.74	3.338	39.0
		STD	40S	.375	12.000	14.58	113.10	.7854	279.3	49.56	49.00	3.338	43.8
		...	...	.406	11.938	15.77	111.93	.7773	300.3	53.52	48.50	3.338	47.1
		XS	80S	.500	11.750	19.24	108.43	.7528	361.5	65.42	46.92	3.338	56.7
		60	...	.562	11.626	21.52	106.16	.7372	400.4	73.15	46.00	3.338	62.8
		80	...	.688	11.374	26.03	101.64	.7058	475.1	88.63	44.04	3.338	74.6
		100	...	.844	11.062	31.53	96.14	.6677	561.6	107.32	41.66	3.338	88.1
		XXS	120	1.000	10.750	36.91	90.76	.6303	641.6	125.49	39.33	3.338	100.7
		140	...	1.125	10.500	41.08	86.59	.6013	700.5	139.67	37.52	3.338	109.9
		160	...	1.312	10.126	47.14	80.53	.5592	781.1	160.27	34.89	3.338	122.6
14	14.00	...	5S	.156	13.688	6.78	147.15	1.0219	162.6	23.07	63.77	3.665	23.2
		...	10S	.188	13.624	8.16	145.78	1.0124	194.6	27.73	63.17	3.665	27.8
		10	...	.250	13.500	10.80	143.14	.9940	255.3	36.71	62.03	3.665	36.6
		20	...	.312	13.376	13.42	140.52	.9758	314.4	45.61	60.89	3.665	45.0
		STD	30	.375	13.250	16.05	137.88	.9575	372.8	54.57	59.75	3.665	53.2
		40	...	.438	13.124	18.66	135.28	.9394	429.1	63.44	58.64	3.665	61.3
		XS	...	.500	13.000	21.21	132.73	.9217	483.8	72.09	57.46	3.665	69.1
		60	...	.594	12.812	24.98	128.96	.8956	562.3	85.05	55.86	3.665	80.3
		80	...	.750	12.500	31.22	122.72	.8522	678.3	106.13	53.18	3.665	98.2
		100	...	.938	12.124	38.45	115.49	.8020	824.4	130.85	50.04	3.665	117.8
		120	...	1.094	11.812	44.32	109.62	.7612	929.6	150.79	47.45	3.665	132.8

Appendix 13.B-6. Continued

Nominal Pipe Size Inches	Outside Diam. Inches	Identification				Wall Thickness (t) Inches	Inside Diameter (d) Inches	Area of Metal Square Inches	Transverse Internal Area		Moment of Inertia (I) Inches	Weight Pipe Pounds per foot	Weight Water Pounds per foot of pipe	External Surface Sq. Ft. per foot of Pipe	Section Modulus $\left(2\dfrac{I}{O.D.}\right)$
		Steel		Stainless Steel Sched. No.					(a) Square Inches	(A) Square Feet					
		Iron Pipe Size	Sched. No.												
14	14.00	...	140	...		1.250	11.500	50.07	103.87	.7213	1027.0	170.28	45.01	3.665	146.8
		...	160	...		1.406	11.188	55.63	98.31	.6827	1117.0	189.11	42.60	3.665	159.6
16	16.00	...	...	5S		.165	15.670	8.21	192.85	1.3393	257.3	27.90	83.57	4.189	32.2
		...	...	10S		.188	15.624	9.34	191.72	1.3314	291.9	31.75	83.08	4.189	36.5
		...	10	...		.250	15.500	12.37	188.69	1.3103	383.7	42.05	81.74	4.189	48.0
		...	20	...		.312	15.376	15.38	185.69	1.2895	473.2	52.27	80.50	4.189	59.2
		STD	30	...		.375	15.250	18.41	182.65	1.2684	562.1	62.58	79.12	4.189	70.3
		XS	40	...		.500	15.000	24.35	176.72	1.2272	731.9	82.77	76.58	4.189	91.5
		...	60	...		.656	14.688	31.62	169.44	1.1766	932.4	107.50	73.42	4.189	116.6
		...	80	...		.844	14.312	40.14	160.92	1.1175	1155.8	136.61	69.73	4.189	144.5
		...	100	...		1.031	13.938	48.48	152.58	1.0596	1364.5	164.82	66.12	4.189	170.5
		...	120	...		1.219	13.562	56.56	144.50	1.0035	1555.8	192.43	62.62	4.189	194.5
		...	140	...		1.438	13.124	65.78	135.28	.9394	1760.3	223.64	58.64	4.189	220.0
		...	160	...		1.594	12.812	72.10	128.96	.8956	1893.5	245.25	55.83	4.189	236.7
18	18.00	...	...	5S		.165	17.670	9.25	245.22	1.7029	367.6	31.43	106.26	4.712	40.8
		...	...	10S		.188	17.624	10.52	243.95	1.6941	417.3	35.76	105.71	4.712	46.4
		...	10	...		.250	17.500	13.94	240.53	1.6703	549.1	47.39	104.21	4.712	61.1
		...	20	...		.312	17.376	17.34	237.13	1.6467	678.2	58.94	102.77	4.712	75.5
		STD	...	...		.375	17.250	20.76	233.71	1.6230	806.7	70.59	101.18	4.712	89.6
		...	30	...		.438	17.124	24.17	230.30	1.5990	930.3	82.15	99.84	4.712	103.4
		XS	...	...		.500	17.000	27.49	226.98	1.5763	1053.2	93.45	98.27	4.712	117.0
		...	40	...		.562	16.876	30.79	223.68	1.5533	1171.5	104.67	96.93	4.712	130.1
		...	60	...		.750	16.500	40.64	213.83	1.4849	1514.7	138.17	92.57	4.712	168.3
		...	80	...		.938	16.124	50.23	204.24	1.4183	1833.0	170.92	88.50	4.712	203.8
		...	100	...		1.156	15.688	61.17	193.30	1.3423	2180.0	207.96	83.76	4.712	242.3
		...	120	...		1.375	15.250	71.81	182.66	1.2684	2498.1	244.14	79.07	4.712	277.6
		...	140	...		1.562	14.876	80.66	173.80	1.2070	2749.0	274.22	75.32	4.712	305.5
		...	160	...		1.781	14.438	90.75	163.72	1.1369	3020.0	308.50	70.88	4.712	335.6

Appendix 13.B-6. Continued

20	20.00	...	...	.188	19.624	11.70	302.46	2.1004	574.2	39.78	131.06	5.236	57.4
		...	10S	.218	19.564	13.55	300.61	2.0876	662.8	46.06	130.27	5.236	66.3
		...	10	.250	19.500	15.51	298.65	2.0740	765.4	52.73	129.42	5.236	75.6
		STD	20	.375	19.250	23.12	290.04	2.0142	1113.0	78.60	125.67	5.236	111.3
		XS	30	.500	19.000	30.63	283.53	1.9690	1457.0	104.13	122.87	5.236	145.7
		...	40	.594	18.817	36.15	278.00	1.9305	1703.0	123.11	120.46	5.236	170.4
		...	60	.812	18.376	48.95	265.21	1.8417	2257.0	166.40	114.92	5.236	225.7
		...	80	1.031	17.938	61.44	252.72	1.7550	2772.0	208.87	109.51	5.236	277.1
		...	100	1.281	17.438	75.33	238.83	1.6585	3315.2	256.10	103.39	5.236	331.5
		...	120	1.500	17.000	87.18	226.98	1.5762	3754.0	296.37	98.35	5.236	375.5
		...	140	1.750	16.500	100.33	213.82	1.4849	4216.0	341.09	92.66	5.236	421.7
		...	160	1.969	16.062	111.49	202.67	1.4074	4585.5	379.17	87.74	5.236	458.5
22	22.00	...	5S	.188	21.624	12.88	367.25	2.5503	766.2	43.80	159.14	5.760	69.7
		...	10S	.218	21.564	14.92	365.21	2.5362	884.8	50.71	158.26	5.760	80.4
		...	10	.250	21.500	17.08	363.05	2.5212	1010.3	58.07	157.32	5.760	91.8
		STD	20	.375	21.250	25.48	354.66	2.4629	1489.7	86.61	153.68	5.760	135.4
		XS	30	.500	21.000	33.77	346.36	2.4053	1952.5	114.81	150.09	5.760	117.5
		...	60	.875	20.250	58.07	322.06	2.2365	3244.9	197.41	139.56	5.760	295.0
		...	80	1.125	19.75	73.78	306.35	2.1275	4030.4	250.81	132.76	5.760	366.4
		...	100	1.375	19.25	89.09	291.04	2.0211	4758.5	302.88	126.12	5.760	432.6
		...	120	1.625	18.75	104.02	276.12	1.9175	5432.0	353.61	119.65	5.760	493.8
		...	140	1.875	18.25	118.55	261.59	1.8166	6053.7	403.00	113.36	5.760	550.3
		...	160	2.125	17.75	132.68	247.45	1.7184	6626.4	451.06	107.23	5.760	602.4
24	24.00	...	5S	.218	23.564	16.29	436.10	3.0285	1151.6	55.37	188.98	6.283	96.0
		...	10S	.250	23.500	18.65	433.74	3.0121	1315.4	63.41	187.95	6.283	109.6
		STD	10	.375	23.250	27.83	424.56	2.9483	1942.0	94.62	183.95	6.283	161.9
		XS	20	.500	23.000	36.91	415.48	2.8853	2549.5	125.49	179.87	6.283	212.5
		...	30	.562	22.876	41.39	411.00	2.8542	2843.0	140.68	178.09	6.283	237.0
		...	40	.688	22.624	50.31	402.07	2.7921	3421.3	171.29	174.23	6.283	285.1
		...	60	.969	22.062	70.04	382.35	2.6552	4652.8	238.35	165.52	6.283	387.7
		...	80	1.219	21.562	87.17	365.22	2.5362	5672.0	296.58	158.26	6.283	472.8
		...	100	1.531	20.938	108.07	344.32	2.3911	6849.9	367.39	149.06	6.283	570.8
		...	120	1.812	20.376	126.31	326.08	2.7645	7825.0	429.39	141.17	6.283	652.1
		...	140	2.062	19.876	142.11	310.28	2.1547	8625.0	483.12	134.45	6.283	718.9
		...	160	2.344	19.312	159.41	292.98	2.0346	9455.9	542.13	126.84	6.283	787.9

Appendix 13.B-6. Continued

Nominal Pipe Size	Outside Diam.	Identification				Wall Thickness (t)	Inside Diameter (d)	Area of Metal	Transverse Internal Area		Moment of Inertia (I)	Weight Pipe	Weight Water	External Surface	Section Modulus
		Steel			Stainless Steel Sched. No.				(a)	(A)					$\left(2\dfrac{I}{O.D.}\right)$
		Iron Pipe Size	Sched. No.												
Inches	Inches					Inches	Inches	Square Inches	Square Inches	Square Feet	Inches	Pounds per foot	Pounds per foot of pipe	Sq. Ft. per foot of Pipe	
26	26.00	...	10		...	.312	25.376	25.18	505.75	3.5122	2077.2	85.60	219.16	6.806	159.8
		STD	...		...	.375	25.250	30.19	500.74	3.4774	2478.4	102.63	216.99	6.806	190.6
		XS	20		...	.500	25.000	40.06	490.87	3.4088	3257.0	136.17	212.71	6.806	250.5
30	30.00	...	...		5S	.250	29.500	23.37	683.49	4.7465	2585.2	79.43	296.18	7.854	172.3
		...	10		10S	.312	29.376	29.10	677.76	4.7067	3206.3	98.93	293.70	7.854	213.8
		STD	...		...	.375	29.250	34.90	671.96	4.6664	3829.4	118.65	291.18	7.854	255.3
		XS	20		...	.500	29.000	46.34	660.52	4.5869	5042.2	157.53	286.22	7.854	336.1
		...	30		...	.625	28.750	57.68	649.18	4.5082	6224.0	196.08	281.31	7.854	414.9
36	36.00	...	10		...	.312	35.376	34.98	982.90	6.8257	5569.5	118.92	425.92	9.425	309.4
		STD	...		...	.375	35.250	41.97	975.91	6.7771	6658.9	142.68	422.89	9.425	369.9
		XS	20		...	.500	35.000	55.76	962.11	6.6813	8786.2	189.57	416.91	9.425	488.1
		...	30		...	.625	34.750	69.46	948.42	6.5862	10868.4	236.13	417.22	9.425	603.8
		...	40		...	.750	34.500	83.06	934.82	6.4918	12906.1	282.35	405.09	9.425	717.0

Identification, wall thickness and weights are extracted from ANSI B36.10 and B36.19. The notations STD, XS, and XXS indicate Standard, Extra Strong, and Double Extra Strong pipe respectively.

Transverse internal area values listed in "square feet" also represent volume in cubic feet per foot of pipe length.

(Crane, 1976)

Appendix 13.B-7.
Short Round-Thread Casing Dimensions and Weights

Size: Outside Diameter, in. D	Nominal Weight:[1] Threads and Coupling, lb. per ft.	Wall Thickness, in. t	Inside Diameter, in. d	Calculated Weight	
				Plain End, lb/ft w_{pe}	Threads[2] and Coupling lb e_w
4½	9.50	0.205	4.090	9.40	4.20
4½	10.50	0.224	4.052	10.23	3.80
4½	11.60	0.250	4.000	11.35	3.40
5	11.50	0.220	4.560	11.23	5.40
5	13.00	0.253	4.494	12.83	4.80
5	15.00	0.296	4.408	14.87	4.20
5½	14.00	0.244	5.012	13.70	5.40
5½	15.50	0.275	4.950	15.35	4.80
5½	17.00	0.304	4.892	16.87	4.40
6⅝	20.00	0.288	6.049	19.49	11.00
6⅝	24.00	0.352	5.921	23.58	9.60
7	17.00	0.231	6.538	16.70	10.00
7	20.00	0.272	6.456	19.54	9.40
7	23.00	0.317	6.366	22.63	8.00
7	26.00	0.362	6.276	25.66	7.20
7⅝	24.00	0.300	7.025	23.47	15.30
7⅝	26.40	0.328	6.969	25.56	15.20
8⅝	24.00	0.264	8.097	23.57	23.50
8⅝	28.00	0.304	8.017	27.02	22.20
8⅝	32.00	0.352	7.921	31.10	20.80
8⅝	36.00	0.400	7.825	35.14	19.40
9⅝	32.30	0.312	9.001	31.03	24.40
9⅝	36.00	0.352	8.921	34.86	23.00
9⅝	40.00	0.395	8.835	38.94	21.40
10¾	32.75	0.279	10.192	31.20	29.00
10¾	40.50	0.350	10.050	38.88	26.40
10¾	45.50	0.400	9.950	44.22	24.40
10¾	51.00	0.450	9.850	49.50	22.60
10¾	55.50	0.495	9.760	54.21	20.80
11¾	42.00	0.333	11.084	40.60	29.60
11¾	47.00	0.375	11.000	45.56	27.60
11¾	54.00	0.435	10.880	52.57	25.00
11¾	60.00	0.489	10.772	58.81	22.60
13⅜	48.00	0.330	12.715	45.98	33.20
13⅜	54.50	0.380	12.615	52.74	30.80
13⅜	61.00	0.430	12.515	59.45	28.40

Appendix 13.B-7. Continued

Size: Outside Diameter, in. D	Nominal Weight:[1] Threads and Coupling, lb. per ft.	Wall Thickness, in. t	Inside Diameter, in. d	Calculated Weight	
				Plain End, lb/ft w_{pe}	Threads[2] and Coupling lb e_w
13⅜	68.00	0.480	12.415	66.11	25.80
13⅜	72.00	0.514	12.347	70.60	24.20
16	65.00	0.375	15.250	62.58	42.60
16	75.00	0.438	15.124	72.72	38.20
16	84.00	0.495	15.010	81.97	34.20
18⅝	87.50	0.435	17.755	84.51	73.60
20	94.00	0.438	19.124	91.51	47.00
20	106.50	0.500	19.000	104.13	41.60
20	133.00	0.635	18.730	131.33	30.00

[1] Nominal weights, threads and coupling (Col. 2), are shown for the purpose of identification in ordering.
[2] Weight gain due to end finishing.

(API, 1976)

APPENDIX 13.B-8.
Water Well Casing

		WEIGHTS PER FOOT			DIAMETERS			COUPLINGS		
	Size Inches Nom.	T&C Pounds	Plain Ends, Pounds	Wall Thickness Inches	Outside Inches	Inside Inches	No. of Threads per inch	Length Inches	OD Inches	Calculated Weight Pounds
WATER WELL CASING	3	4.60	4.51	.125	3.500	3.250	14	3⅛	4.000	2.86
	3½	5.65	5.53	.134	4.000	3.732	14	3⅛	4.500	3.24
	4	6.75	6.61	.142	4.500	4.216	14	3⅝	5.000	4.26
	5	9.00	8.79	.154	5.500	5.192	14	4⅛	6.050	6.38
	5½	10.50	10.22	.164	6.000	5.672	14	4⅛	6.625	7.84
	6	13.00	12.72	.185	6.625	6.255	11½	4⅝	7.390	11.88
	8	17.80	16.90	.188	8.625	8.249	11½	5¼	9.625	22.92
REAMED AND DRIFTED PIPE	1	1.70	1.68	.133	1.315	1.049	11½	2¾	1.576	.52
	1¼	2.30	2.27	.140	1.660	1.380	11½	2¾	1.900	.60
	1½	2.75	2.72	.145	1.900	1.610	11½	2¾	2.200	.84
	2	3.75	3.65	.154	2.375	2.067	11½	3⅜	2.750	1.58
	2	4.00	3.94	.167	2.375	2.041	11½	3⅜	2.750	1.58
	2½	5.90	5.79	.203	2.875	2.469	8	3 15/16	3.250	2.32
	3	7.70	7.58	.216	3.500	3.068	8	4 1/16	4.000	3.80
	3½	9.25	9.11	.226	4.000	3.548	8	4 3/16	4.625	5.53
	4	11.00	10.79	.237	4.500	4.026	8	4 5/16	5.200	7.14
	5	15.00	14.62	.258	5.563	5.047	8	4½	6.296	9.57
	6	19.45	18.97	.280	6.625	6.065	8	4 11/16	7.390	12.32
	8	29.35	28.55	.322	8.625	7.981	8	5 1/16	9.625	22.35
	10	41.85	40.48	.365	10.750	10.020	8	5 9/16	11.750	30.61
	12	51.15	49.56	.375	12.750	12.000	8	5 15/16	14.000	47.96
DRIVE PIPE	6	19.45	18.97	.280	6.625	6.065	8	5⅛	7.390	13.35
	8	25.55	24.70	.277	8.625	8.071	8	6⅛	9.625	26.89
	8	29.35	28.55	.322	8.625	7.981	8	6⅛	9.625	26.89
	8	32.40	31.27	.354	8.625	7.917	8	6⅛	9.625	26.89
	10	32.75	31.20	.279	10.750	10.192	8	6⅝	11.750	36.05
	10	35.75	34.24	.307	10.750	10.136	8	6⅝	11.750	36.05
	10	41.85	40.48	.365	10.750	10.020	8	6⅝	11.750	36.05
	12	45.85	43.77	.330	12.750	12.090	8	6⅝	14.000	52.72
	12	51.15	49.56	.375	12.750	12.000	8	6⅝	14.000	52.72
	14D	57.00	54.57	.375	14.000	13.250	8	7⅛	15.000	50.22
	16D	65.30	62.58	.375	16.000	15.250	8	7⅛	17.000	57.17

(National Association of Steel Pipe Distributors, Inc.)

APPENDIX 13.C.
Chemical and Physical Standards for Thermoplastic Material

ASTM Standard F 480-76
Section 4. Materials

4.2. Acrylonitrile-butadiene-styrene (ABS) well casing pipe and couplings plastic shall be virgin plastic produced by the original compounder. It shall contain polymers or blends of polymers, or both, in which the minimum butadiene content is 6%; the minimum acrylonitrile content is 15%; the minimum stryene or substituted styrene content, or both, is 15%; and the maximum content of other monomers is 5% and lubricants, stabilizers, and colorants.

4.3. Poly(vinyl chloride) (PVC) well casing pipe and couplings plastic shall be made of virgin plastic produced by the original compounder. It shall contain poly (vinyl chloride) homopolymer, and such additives — stabilizers, lubricants, processing aids, impact improvers, and colorants — as needed to provide the required processing and toughness characteristics.

4.4. The SR plastics compound shall contain at least 50% styrene plastics, combined with rubbers to a minimum rubber content of 5%, and compounding materials such as antioxidants and lubricants, and may contain up to 15% acrylonitrile combined in the styrene plastics or rubbers, or both. The rubbers shall be of the polybutadiene or butadiene-styrene type, or both, with a maximum styrene content of 25% or nitrile type, or both. The combined styrene plastics and rubber content shall be not less than 90%.

TABLE 1.
Wall Thickness and Tolerances for Thermoplastic Water Well Casing Pipe, in.[A]

Nominal Pipe Size	SDR26		SDR21		SDR17		SDR13.5	
	Minimum	Tolerance	Minimum	Tolerance	Minimum	Tolerance	Minimum	Tolerance
2	...	...	0.113	+0.020	0.140	+0.020	0.176	+0.021
2½	...	...	0.137	+0.020	0.169	+0.020	0.213	+0.026
3	...	...	0.167	+0.020	0.206	+0.025	0.259	+0.031
3½	...	...	0.190	+0.023	0.235	+0.028	0.296	+0.036
4	0.173	+0.021	0.214	+0.026	0.265	+0.032	0.333	+0.040
5	0.214	+0.027	0.265	+0.032	0.327	+0.039	0.412	+0.049
6	0.255	+0.031	0.316	+0.038	0.390	+0.047	0.491	+0.058
8	0.332	+0.040	0.410	+0.049	0.508	+0.061	...	...
10	0.413	+0.050	0.511	+0.061	0.632	+0.076	...	...
12	0.490	+0.059	0.606	+0.073	0.750	+0.090	...	...

[A] The minimum is the lowest wall thickness of the well casing pipe at any cross section. All tolerances are on the plus side of the minimum requirement.
[B] Dimensions below the line meet or exceed Schedule 40 in SDR 13.5, 17, 21 and 26.
[C] Dimensions below the line meet or exceed Schedule 80 in SDR 13.5 and 17.

(Copyright, ASTM, 1916 Race Street, Philadelphia, PA 19103. Reprinted with permission.)

APPENDIX 13.D.
Dimensions and Weights of PVC Plastic Pipe

Pipe Size		Outside Diameter		SDR or Schedule Number	Minimum Wall Thickness		Approximate ID		Weight per ft (0.3 m)		Collapse Strength			
											Cell Class 12454		Cell Class 14333	
in	mm	in	mm		in	mm	in	mm	lbs	kg	psi	kPa	psi	kPa
2	50.8	2.375	60.3	SDR 13.5	0.176	4.47	2.023	51.38	0.78	0.35	436	3,010	349	2,410
				SDR 17	0.140	3.56	2.095	53.21	0.63	0.29	213	1,470	170	1,170
				SDR 21	0.113	2.87	2.149	54.58	0.51	0.23	109	752	87	600
				Sch 40	0.154	3.91	2.067	52.50	0.69	0.31	287	1,980	229	1,580
				Sch 80	0.218	5.54	1.939	49.25	0.94	0.43	862	5,940	689	4,750
2.5	63.5	2.875	73.0	SDR 13.5	0.213	5.41	2.449	62.20	1.14	0.52	436	3,010	349	2,410
				SDR 17	0.169	4.29	2.537	64.44	0.92	0.42	211	1,450	169	1,170
				SDR 21	0.137	3.48	2.601	66.07	0.76	0.34	110	758	88	607
				Sch 40	0.203	5.16	2.469	62.71	1.09	0.49	375	2,590	300	2,070
				Sch 80	0.276	7.01	2.323	59.00	1.44	0.65	995	6,860	796	5,490
3	76.2	3.5	88.9	SDR 13.5	0.259	6.58	2.982	75.74	1.69	0.77	434	2,990	347	2,390
				SDR 17	0.206	5.23	3.088	78.44	1.36	0.62	212	1,460	169	1,170
				SDR 21	0.167	4.24	3.166	80.42	1.12	0.51	110	758	88	607
				Sch 40	0.216	5.49	3.068	77.93	1.43	0.65	245	1,690	196	1,350
				Sch 80	0.300	7.62	2.900	73.66	1.93	0.88	692	4,770	554	3,820
3.5	88.9	4.0	101.6	SDR 13.5	0.296	7.52	3.408	86.56	2.20	1.00	434	2,990	347	2,390
				SDR 17	0.235	5.97	3.530	89.66	1.78	0.81	210	1,450	168	1,160
				SDR 21	0.190	4.83	3.620	91.95	1.46	0.66	109	752	87	600
				Sch 40	0.226	5.74	3.548	90.12	1.72	0.78	186	1,280	149	1,030
				Sch 80	0.318	8.08	3.364	85.15	2.35	1.07	545	3,760	436	3,010

Appendix 13.D. Continued

Pipe Size		Outside Diameter		SDR or Schedule Number	Minimum Wall Thickness		Approximate ID		Weight per ft (0.3 m)		Collapse Strength			
											Cell Class 12454		Cell Class 14333	
in	mm	in	mm		in	mm	in	mm	lbs	kg	psi	kPa	psi	kPa
4	101.6	4.5	114.3	SDR 13.5	0.333	8.46	3.834	97.38	2.79	1.27	434	2,990	347	2,390
				SDR 17	0.265	6.73	3.970	100.84	2.26	1.03	212	1,460	170	1,170
				SDR 21	0.214	5.44	4.072	103.43	1.85	0.84	109	752	87	600
				SDR 26	0.173	4.39	4.154	105.51	1.51	0.68	56	386	45	310
				Sch 40	0.237	6.02	4.026	102.26	2.03	0.92	150	1,030	120	827
				Sch 80	0.337	8.56	3.826	97.18	2.82	1.28	451	3,110	361	2,490
5	127.0	5.563	141.3	SDR 13.5	0.412	10.46	4.739	120.37	4.27	1.94	435	3,000	348	2,400
				SDR 17	0.327	8.31	4.909	124.69	3.45	1.56	211	1,450	169	1,170
				SDR 21	0.265	6.73	5.033	127.84	2.83	1.28	110	758	88	607
				SDR 26	0.214	5.44	5.135	130.43	2.31	1.05	57	393	45	310
				Sch 40	0.258	6.55	5.047	128.19	2.76	1.25	101	696	81	558
				Sch 80	0.375	9.53	4.813	122.25	3.91	1.77	324	2,230	259	1,790
6	152.4	6.625	168.3	SDR 13.5	0.491	12.47	5.643	143.33	6.05	2.74	436	3,010	349	2,410
				SDR 17	0.390	9.91	5.845	148.46	4.89	2.22	212	1,460	169	1,170
				SDR 21	0.316	8.03	5.993	152.22	4.02	1.82	110	758	88	607
				SDR 26	0.255	6.48	6.115	155.32	3.27	1.48	57	393	45	310
				Sch 40	0.280	7.11	6.065	154.05	3.58	1.62	76	524	61	421
				Sch 80	0.432	10.97	5.761	146.33	5.38	2.44	292	2,010	233	1,610
8	203.2	8.625	207.3	SDR 17	0.508	12.90	7.144	181.46	8.30	3.76	212	1,460	170	1,170
				SDR 21	0.410	10.41	7.340	186.44	6.78	3.08	109	752	87	600
				SDR 26	0.332	8.43	7.496	190.40	5.55	2.52	57	393	45	310
				Sch 40	0.322	8.18	7.516	190.91	5.39	2.44	52	359	41	283
				Sch 80	0.500	12.70	7.160	181.86	8.18	3.71	202	1,390	161	1,110

Appendix 13.D. Continued

Pipe Size		Outside Diameter		SDR or Schedule Number	Minimum Wall Thickness		Approximate ID		Weight per ft (0.3 m)		Collapse Strength			
											Cell Class 12454		Cell Class 14333	
in	mm	in	mm		in	mm	in	mm	lbs	kg	psi	kPa	psi	kPa
10	254	10.75	273.1	SDR 17	0.632	16.05	9.486	240.94	12.90	5.85	211	1,450	169	1,170
				SDR 21	0.511	12.98	9.728	247.09	10.50	4.76	109	752	87	600
				SDR 26	0.413	10.49	9.924	252.07	8.60	3.90	56	386	45	310
				Sch 40	0.365	9.27	10.020	254.51	7.64	3.47	39	269	31	214
				Sch 80	0.593	15.06	9.564	242.93	12.10	5.49	173	1,190	138	952
12	304.8	12.75	323.9	SDR 17	0.750	19.05	11.250	285.75	18.10	8.21	211	1,450	169	1,170
				SDR 21	0.606	15.39	11.538	293.07	14.80	6.71	109	752	87	600
				SDR 26	0.490	12.45	11.770	298.96	12.10	5.49	56	386	45	310
				Sch 40	0.406	10.31	11.938	303.23	10.10	4.58	32	221	25	172
				Sch 80	0.687	17.45	11.376	288.95	16.70	7.58	161	1,110	129	889
16	406.4	16.0	406.4	SDR 21	0.762	19.35	14.476	367.69	23.40	10.61	109	752	88	607
				SDR 26	0.616	15.65	14.769	375.13	19.10	8.66	57	393	45	310
				Sch 40	0.500	12.70	15.000	381.00	15.60	7.08	30	207	24	165
				Sch 80	0.843	21.41	14.314	363.58	25.70	11.66	150	1,030	120	827

Collapse strengths were calculated using the equation presented in ASTM F480, Section X2.

APPENDIX 13.E.
Class Requirements for Rigid Poly (Vinyl Chloride) Compounds

Designation Order No.	Property and Unit	Cell Limits								
		0	1	2	3	4	5	6	7	8
1	Base resin	unspecified	poly (vinyl chloride) homopolymer	chlorinated poly (vinyl chloride)	ethylene vinyl chloride copolymer	propylene vinyl chloride copolymer	vinyl acetate-vinyl chloride copolymer	alkyl vinyl ether-vinyl chloride copolymer		
2	Impact strength (Izod) min. J/m of notch ft.-lb./in. of notch	unspecified	<34.7 <0.65	34.7 0.65	80.1 1.5	266.9 5.0	533.8 10.0	800.7 15.0		
3	Tensile strength, min: MPa psi	unspecified	<34.5 <5,000	34.5 5,000	41.4 6,000	48.3 7,000	55.2 8,000			
4	Modulus of elasticity in tension, min: MPa psi	unspecified	<1930 <280,000	1930 280,000	2206 320,000	2482 360,000	2758 400,000	3034 440,000		
5	Deflection temperature under load, min. 1.82 MPa (264 psi): deg C deg F	unspecified	<55 <131	55 131	60 140	70 158	80 176	90 194	100 212	110 230

Note—The minimum property value will determine the cell number although the maximum expected value may fall within a higher cell.

(ASTM D-1784, American Society for Testing and Materials, 1916 Race Street, Philadelphia, PA 19103. Reprinted with permission.)

APPENDIX 13.F.
Prior Designation of PVC Pipe Material

Prior to the development of the cell classification system for PVC compounds defined in ASTM D-1784, PVC pipe compounds were specified by means of a four-digit plastic pipe material code.

As shown in Figure 1, the plastic pipe material code described three properties of a designated PVC compound: (1) impact strength, (2) chemical resistance, and (3) hydrostatic design stress in units of 100 psi. Figure 1 also shows how the material code described the specific properties for a given PVC pipe compound.

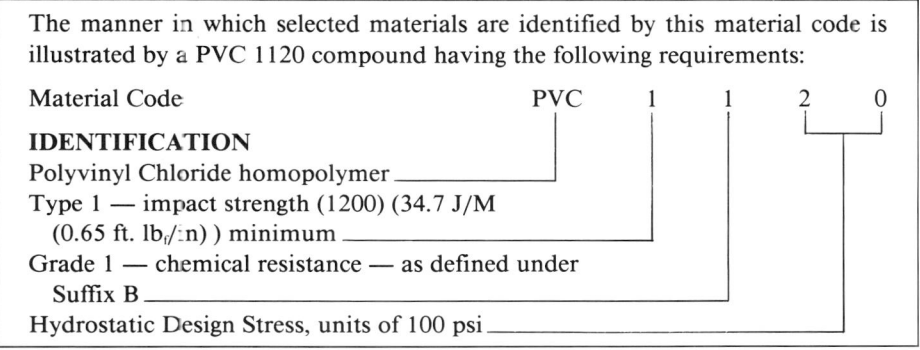

Figure 1. An example of PVC pipe material code. *(American Society for Testing and Materials, 1916 Race Street, Philadelphia, PA 19103. Reprinted by permission.)*

APPENDIX 13.G.
Recommended Procedures for Field Welding Stainless Steel Well Screens

Well screens fabricated of stainless steel are furnished with weld-ring end fittings made of stainless steel. Selection of the proper metal to join screens is based on field experience and knowledge of the local groundwater chemistry. The right choice of electrode reduces corrosion and provides the best joining surface for field welding. Because several combinations of materials are possible, the proper selection of welding electrodes is extremely important.

Welding electrodes are commonly available with two types of coatings. These are designated by number following the standard AWS-ASTM (American Welding Society — American Society for Testing Materials) electrode classification. The 15-type covering has a lime base and must be deposited with a D.C. reverse polarity. Metal transfer occurs as a flutter-drop type and results in a somewhat rough surface deposit. The 16-type covering has a titania base and can be deposited either with A.C. or D.C. reverse polarity. A spray-type metal transfer results in a much smoother deposit. The 16-type electrodes are not recommended for vertical down welding because the fluid slag has a tendency to run ahead of the arc.

Because welding is a skill acquired by experience, personal preference or availability of material or equipment may determine which type of welding rod should be used. In field welding, the importance of proper weld penetration on the first pass cannot be overemphasized. However, care must be taken to avoid burn-through or metal deposit on the inside of the screen or pipe, because this might interfere with the movement of tight-fitting tools or packers.

Welding 304 (AISI*) Stainless Steel to Mild Steel

The use of either E 312 (AWS-ASTM classification) or E 309 electrodes is recommended. Both are designed for dissimilar metal welding, although E 309 is the more commonly used. E 312 is more expensive and less readily available than E 309. Welding technique closely approximates that used for mild steel because either the drag or weave technique may be used. Weaving should not exceed two and a half times the electrode diameter. Warning: Type E 316 electrodes should not be used when welding stainless to mild steel because there is danger of cracking.

Welding 304 Stainless Steel to 304 Stainless Steel

Although either E 308-15 or E 308-16 welding electrodes can be used, the use of E 308-15 is recommended whenever practical. In the majority of applications, the weave technique used in conjunction with D.C. reverse polarity results in the best weld.

Welding 316 (AISI) Stainless Steel

For welding 316 stainless steel to 316 stainless, the use of E 316-16 electrodes is recommended using the weave technique with D.C. reverse polarity. When 316 stainless steel must be welded to mild steel, the use of E 312 or E 309 electrodes is recommended.

These suggestions for welding are only appropriate for the materials noted above. The quality of the welds may not be satisfactory, however, if the joints are prepared improperly.

*American Iron and Steel Institute

Appendix 13.G. Continued

Reference Chart for Coated Stainless Steel Electrodes

AWS Classification	E 308	E 308L	E 312	E 316	E 316L	E 309
Air Products	308	308L	312	316	316L	309
Airco	19-9	19-9ELC	29-9	18-12MO	18-12MO ELC	Airco 25-12
Alloy Rods	Arcaloy 308	Arcaloy 308ELC	Arcaloy 312	Arcaloy 316	Arcaloy 316ELC	Type 309 Arcaloy
Arcos	K	K-LC	29-9	K-MO	KMO-LC	25-12(309)
Hobart	308	308ELC	312	316	316ELC	309
Lincoln	Stainweld 308	Stainweld 308L	—	—	Stainweld 316L	Stainweld B-CB
M&T	308	308L	312	316	316L	Murex 309
NCG	Sureweld 308	Sureweld 308ELC	Sureweld 312	Sureweld 316	Sureweld 316L	Sureweld 309
Reid-Avery	Racolloy 18-8	Racolloy 18-8ELC	Racolloy 29-9	Racolloy 18-12 (2-3MO)	Racolloy 18-12ELC	Racolloy 309 25-12
A.O. Smith	SW308	SW308 ELC	SW312	SW316	SW316 ELC	SW309
Westinghouse	308	308ELC	312	316	316ELC	309
McKay	18-8	18-8ELC	29-9	18-8MO	18-8MO ELC	25-12

Note: E 308L electrode can be substituted for E 308 electrode and E-316L can be used instead of E 316 welding rod.

APPENDIX 13.H.
Preparation of Pipe Surfaces and Cement Process for Plastic Pipe

Joining plastic casing is a relatively simple process, but the integrity of the joint will be a function of the thoroughness of the joining techniques. The materials and tools required include (NWWA, 1981):

1. A fine-toothed saw
2. A mitre box
3. A small knife, file, or beveling tool
4. Fine-grained, abrasive paper
5. Clean, dry cloth or paper towel
6. Cleaner and/or primer
7. The proper solvent cement
8. A natural bristle brush approximately half the diameter of the casing being joined or a specially designed applicator which may be included in the solvent cement container

Preparation of the pipe surfaces and the actual cement process should be accomplished as listed below (NWWA, 1981). Because solvents are highly flammable and their vapors may present a health hazard in enclosed areas, solvent cements should be used with care; avoid contact with eyes and skin and long-term breathing of vapors.

Step 1. Cut casing using a fine-tooth saw and a mitre box to avoid rough or uneven cuts.

Step 2. Smooth the cut end with a knife, file, or sandpaper.

Step 3. Clean the contact surfaces of the pipe end and socket with a clean, dry cotton cloth or paper towel. Grease must be removed with a cleaner recommended by the solvent cement manufacturer. Roughening the contact surfaces with abrasive paper aids in the development of a better bond.

Step 4. Check the fit of the sections to be cemented. A good "dry fit" should show the spigot end entering the socket to about one-half to two-thirds of its depth. Some manufacturers supply casing with marks to designate proper fit. Incorrectly dimensioned pipe, bell, or coupling should not be used.

Step 5. If the casing is made of PVC, apply a primer to the inside surfaces of the socket. The primer may require more time to soften the belled end casing sockets than is necessary to prepare the sockets of a molded coupling.

Step 6. A primer should also be applied to the outside of the PVC casing (spigot) end to prepare it for joining. All surfaces to be cemented should be coated with the primer.

Step 7. Apply a thin coat of solvent cement to the interior surface of the socket. The use of too much solvent could weaken the casing wall.

Step 8. Apply a uniform coat of solvent cement to the spigot end of the casing.

Step 9. Insert the spigot end of the casing section forcefully into the socket to the entire depth of the socket while both the inside socket surface and outside surface of the casing are completely coated with wet cement.

Step 10. Hold the socket and casing sections together for at least 15 to 20 seconds or until an initial set takes place.

Appendix 13.H. Continued

Step 11. Wipe the excess cement from the socket. A properly cemented joint should show a bead of solvent cement around the entire circumference of the casing where it meets the socket.

Step 12. To insure a strong bond, a joint should remain undisturbed until an initial set is reached (Table 1).

Table 1. Solvent Cementing of Belled End Casing — Approximate Initial Set Times*

Temperature Range During Initial Set		Set Time, in Minutes	
°F	°C	2- and 3-in (51- and 76-mm) Casing	Sizes Larger than 3 in (76 mm)
15 to 40	− 9.4 to 4.4	7.5	15
5 to 15	−15.0 to − 9.4	30	60
−20 to 5	−28.9 to −15.0	90	180

*Development of full operating strength requires cure periods about ten times longer than shown by this table.
(NWWA, 1981)

APPENDIX 13.I.
Discussion of Appropriate Screen Entrance Velocities

For many years, authorities on well design have recommended that water should enter the screen at an average velocity of approximately 0.1 ft/sec at the anticipated pumping rate. Screens designed to this specification have shown excellent resistance to chemical incrustation and corrosion in a wide variety of groundwater-quality conditions.

As early as 1937, C. F. Tolman in his classic work *Ground Water* referred to entrance velocities of 0.1 to 0.2 ft/sec in discussing the entrainment of sand into a screen. T. P. Ahrens (1957) stated the following based on his long field experience: "Selection of screen length and diameter is based on average entrance velocities between 0.1 and 0.25 ft/sec."

Robert C. Smith (American Water Works Association, 1963) noted the difficulties in screening sediments with a wide range of particle sizes: "Obviously, it is necessary, if excessive velocities are to be avoided, to establish a standard for safety. By experience, it has been found that the safe entrance velocity in these silty or, as the drillers say, 'dirty' sands and gravels, is 4 ft/min (about 0.066 ft/sec), and this is the entrance velocity the author uses to design wells." He further stated that excessive entrance velocity caused by improper well design has been the cause of more well failures, of more instances of rapid well deterioration, and of more instances of frequent and costly well treatment than any other single design factor.

In its manual *Water Well Construction Practices* (1975c), the U.S. Environmental Protection Agency recommends that screen entrance velocities should be based on the hydraulic conductivity of the aquifer (Table 1). Suggested entrance velocities range from 0.03 to 0.1 ft/sec. This publication was written by 34 nationally known professional groundwater specialists.

Several other authors propose similar velocity recommendations. Hunter Blair (1970) suggests that a maximum entrance velocity of 3 cm/sec (0.1 ft/sec) is appropriate. In his book *Ground Water Recovery*, Huisman (1973) recommends a maximum entrance velocity of 4 cm/sec (0.13 ft/sec). The American Water Works Association Manual M21 (1973) suggests: "It is necessary to assign somewhat more importance to diameter for the purpose of keeping entrance velocities between the desired magnitude of 0.1 and 0.2 ft/sec where a well screen is used." In their book *Water Well Technology* (1973), Campbell and Lehr indicate that: "Desirable entrance velocity is usually considered to be 0.1 to 0.25 ft/sec, based on the open area of the screen." More recently, Mabillot (1979) noted: "Numerous observations indicate the best value of the entrance velocity through the well screen is 3 cm/sec (0.12 ft/sec). At this velocity, the head loss for the flow of water through the openings is negligible."

The collective experience of the Johnson Division's worldwide technical staff suggests that the authorities cited above are correct in recommending low screen entrance velocities. For almost all common hydrogeologic conditions, well screens designed to have an average entrance velocity of 0.1 ft/sec will require less maintenance over the long term, limit sand pumping more effectively, and create minimal head losses as the water enters the screen. To maintain the 0.1 ft/sec entrance velocity, many well design engineers specify that the screen length should be doubled, because they automatically assume that 50 percent of the slots will be blocked. It is clear, however,

Table 1. Recommended Maximum Screen Entrance Velocities

Hydraulic Conductivity of Aquifer		Maximum Screen Entrance Velocity	
(gpd/ft^2)	(m^2/day)	(ft/sec)	(cm/s)
more than 6,000	more than 245	more than 0.10	more than 3.05
6,000	245	0.10	3.05
5,000	204	0.10	3.05
4,000	163	0.10	3.05
3,000	122	0.10	3.05
2,500	102	0.08	2.54
2,000	82	0.08	2.54
1,500	61	0.07	2.03
1,000	41	0.07	2.03
500	20	0.05	1.52
less than 500	less than 20	less than 0.03	less than 1.02

(After U.S. Environmental Protection Agency, 1975c)

that the actual entrance velocity at any point on the screen will vary depending not only on the degree of slot blockage, but also on the hydraulic conductivity of the aquifer materials surrounding the screen. Thus, the entrance velocities during pumping will typically range from much lower than the designed rate to several times higher.

Although a vast amount of practical experience has shown that a designed entrance velocity of 0.1 ft/sec is highly satisfactory, experiments testing the effect of velocity on the rate of screen corrosion and incrustation would be instructive in evaluating the correctness of this figure. To date, much of the corrosion and incrustation research in aqueous media has been limited to industrial applications where the research has been conducted using highly acidic or alkaline process or wastewaters. It is difficult to extract useful information from this type of research that is directly applicable to wells. As a result, Johnson Division initiated research programs to test the effect of entrance velocity on both corrosion and incrustation rates of well screens. The incrustation study is in progress as this book goes to press. The results from the recently completed corrosion study are presented below. In this latter study, a series of experiments tested the effect of entrance velocity on the corrosion rates of screens fabricated of Type 304 stainless steel and low carbon steel.

Corrosion produces two major deleterious effects on screens. Corrosion products can plug screen openings over time causing a reduction in specific capacity. Corrosion is also responsible for slowly increasing slot width so that sand pumping becomes a problem. Corrosion tests are one way to verify the importance of inlet velocity on the efficiency and service life of a well screen.

The corrosion resistance of Type 304 stainless steel screens is excellent in good quality groundwater, but as the chloride content of the water increases, this resistance decreases. In general, screens constructed of low carbon steel are much more susceptible to corrosive attack when compared to stainless steel. Because the chemical reactions producing corrosion are complicated and the results of corrosion tests are difficult to interpret without some understanding of the chemical reactions taking place, a brief explanation of corrosion chemistry is presented below.

Corrosion of metals can be explained using electrochemical theory. When chemical

Appendix 13.I. Continued

reactions occur on a metal surface, an energy change takes place. If no energy is being added to a chemical system, it will tend to assume the lowest energy state. Positive changes in free energy caused by chemical reactions require the addition of energy to the system. The free-energy change produced by an electrochemical reaction is given by the equation:

$$\Delta G = -nFE$$

where ΔG is the free-energy change, n is the number of electrons involved in the reaction, F is the Faraday constant, and E equals the cell potential. Thus, the amount of free-energy change accompanying a specific electrochemical reaction can be calculated from knowledge of the cell potential (in volts) of the reaction (a cell consists of two dissimilar metals in a solution connected by wire in which a voltmeter is connected). In any electrochemical reaction, a definite relationship exists between the free-energy change and the cell potential (Fontana and Greene, 1978).

The electrochemical changes taking place during corrosion include both oxidation and reduction reactions. Some ions are being placed into solution by oxidation processes while other ions are being removed (reduced). But the rate at which the reaction takes place, based on the change in free energy, cannot be determined accurately. Corrosion will not occur unless the electrochemical reaction involves oxidation.

In any electrochemical reaction, the most negative or active half-cell tends to be oxidized, and the most positive or noble half-cell tends to be reduced. The oxidation state (also called the oxidation number) represents the charge that an atom would have if the ion or molecule were to dissociate (Freeze and Cherry, 1979). Oxidation occurs at an anode, whereas reduction occurs at a cathode. Under equilibrium conditions, the rate of oxidation and reduction are equal; that is, the electrons released during oxidation are consumed during reduction. The rate of electron flow is called the current density. Any deviation from an equilibrium condition (occurring at a certain potential) will produce electrochemical polarization. Polarization is defined as the displacement of electrode potential produced from a net current. The magnitude of polarization is often measured as overvoltage. The magnitude of the polarization can be determined by comparing the equilibrium potential of an electrode (normally considered zero) to the displaced potential (in plus or minus millivolts).

There are two types of polarization — activation and concentration. In activation polarization, the rate of the sequential electrochemical reactions is controlled by the rate of the slowest reaction taking place on the surface of the metal. The rate at which concentration polarization takes place depends on the number of reacting ions reaching the anodic or cathodic surface. Thus, the reaction rate is controlled by the diffusion rate of the ions through the solution to the metal surface. The limiting diffusion current density represents the maximum reduction rate that is possible for a certain system and is usually important only during reduction reactions; it can be ignored in most metal-dissolution (oxidation) reactions. In most reactions, both activation and concentration polarization occur at an electrode. Activation polarization usually controls the reaction process at low reaction rates, whereas concentration polarization is the controlling mechanism at higher reaction rates.

During corrosion, the reduction-oxidation (redox) potential of the oxidizing agent and its exchange current density (i_o) on a metal surface are important. As stated above, both oxidation and reduction are involved in corrosion (Figure 1). At low potential overvoltages, reduction takes place ($2H^+ + 2e \rightarrow H_2$ for a hydrogen electrode). At a higher potential, called the equilibrium potential, the rates of oxidation and reduction may become equal. This occurs at a corrosion potential designated E_{corr}. The current density at this point is called the corrosion current density because it represents the equilibrium rate of corrosion. If the metal is isolated electrically (not the case with the vast majority of water wells), the total rate of oxidation at this point must equal the rate of reduction. A current density of 1 microamp/cm^2 (10^{-6} amp/cm^2) usually produces a corrosion rate of approximately 1 mil (thousandth of an inch) per year (mpy) for most metals. Below the potential (voltage) where the rate of oxidation and reduction processes are equal, reduction reactions are dominant (Figure 1). Oxidation reactions take place above this potential.

Predicting the corrosion rate of certain metals becomes difficult when they lose their chemical reactivity under specific environmental conditions. This phenomenon, called passivity, occurs when a thin film of corrosion products form on the surface of the metal. The film, which is typically only 30 to 60 angstroms thick, can reduce the corrosion rate to one thousandth to one millionth of its former rate. The film for many materials is delicate, however, and can be removed or broken down. Figure 2 illustrates the behavior of a metal such as stainless steel which exhibits active and passive corrosion characteristics. Initially, the metal actively corrodes and as the electrode potential increases, the dissolution rate increases exponentially. This is called the active region of the curve. The point at which the corrosion rate changes from active to passive occurs at a certain anodic current density maximum (i_c) at the primary passive potential, E_{pp}. The location of this point is largely controlled by the oxidizing power of the medium.

At slightly more noble (positive) potentials, the dissolution rate decreases rapidly to a small value, and even if the potential continues to increase, the dissolution rate is relatively stable. Over this range of potential values, the curve is in the passive region and the metal is corroding at a much reduced rate. Dissolution (oxidation) may recur at still more noble potentials in the transpassive region where the film has been destroyed (Figure 2).

As temperature and acid concentration increase, the curve shown in Figure 2 will be displaced to the right (Figure 3). These factors greatly increase the critical anodic current density, but not the primary passive potential, and only slightly the passive dissolution rate. For well screens, this effect can be caused by an increase in the chloride concentration of groundwater. Changes in groundwater temperature are usually insignificant and will not affect the corrosion potential.

In many instances, stainless steel will spontaneously passivate as shown in case 3 on Figure 4. As indicated above, passivation occurs when the cathodic reduc-

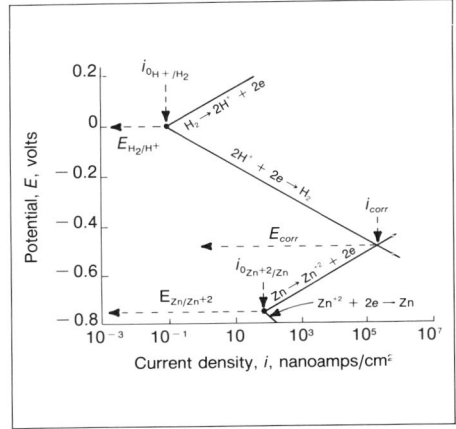

Figure 1. Electrode kinetic behavior of pure zinc in acid solution (e is a designation used for electron). *(Fontana and Greene, 1978)*

Appendix 13.I. Continued

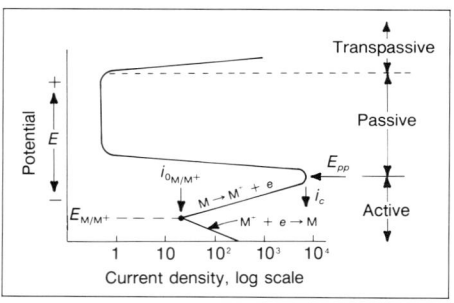

Figure 2. Typical anodic dissolution behavior of an active-passive metal. *(Fontana and Greene, 1978)*

tion rate equals or exceeds the critical anodic dissolution rate. A metal will more likely passivate if it has a small critical anodic current density and an active (more noble) primary passive potential. Case 2 shows how metals can alternate between passive and active states; this situation can lead to rapid localized corrosion. Aerated acid solutions can produce this effect on stainless steel. In case 1, no passive state is reached, but at least the corrosion potential can be predicted.

Practical Application of Electrochemical Theory

Knowledge of the principles of electrochemical theory enables the well design engineer to understand how the corrosion rate of the screen is affected by design. If the assumption is made that groundwater quality remains essentially the same at a site, the screen design engineer must determine what the limits should be on the screen entrance velocities so as to control the corrosion rates within acceptable limits. Velocity effects are illustrated by plotting corrosion rate versus the relative velocity (Figure 5). Initially, the corrosion rate under a diffusion controlled (concentration) process is directly related to the velocity — the higher the velocity of the water entering the screen, the higher the corrosion rate. But as the velocity is increased still more, the corrosion rate stabilizes as the reduction reaction becomes activation controlled. Further increases in velocity past a certain point do not cause an increase in corrosion rate. Fontana and Greene (1978) cite two general rules concerning velocity effects:

a. Solution velocity (velocity of water entering the screen) affects the corrosion rate of a diffusion-controlled system, but has no effect on activation-controlled systems.

b. The corrosion rate of a metal in a diffusion-controlled system becomes independent of the solution velocity at high velocities.

In water well situations, ions in solution are brought to the surface of the screen during pumping (or through ordinary groundwater flow). If the corrosion process is diffusion controlled rather than activation controlled, the corrosion rate will increase to some maximum value. The velocity at which this occurs is critical to calculating corrosion rates for well screens for both low carbon steel and stainless steel.

The key to corrosion protection of stainless steel screens is whether a passive film has formed on the surface and the long term condition of this film. Until the 1970's, it was not possible to make a chemical analysis of the film. Development of x-ray photon electron spectroscopy and Auger Electron Spectroscopy now allows corrosion specialists to ana-

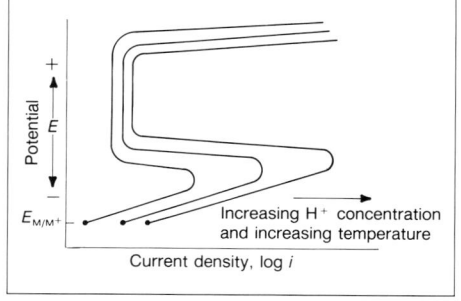

Figure 3. Effect of temperature and acid concentration on anodic dissolution behavior of an active-passive metal. *(Fontana and Greene, 1978)*

lyze the reaction products formed on the surface (Castle and Clayton, 1977). These researchers report that films on stainless steel are about 1.5 to 3 nanometers thick and contain chromium- and iron-rich areas (Cr^{+3}, Fe^{+2}, and Fe^{+5}) that are incorporated in layers having both OH^- and O^{-2} ions. Hydrogen bonded water may also exist in the outer one-third of the film. Apparently the ratio of iron and chromium content do not change in the film compared to the unreacted metal, that is, no solution oxidation occurs. Ferreira and Dawson (1985) have found the passive film may grow in stages; at low potentials the film consists mainly of chromium, whereas at higher potentials iron predominates. Okamoto (1973) determined that surface films on stainless steel are composed primarily of chromium at a potential of 0 to 2 volts. He characterized the film as a hydrated oxide film having a gel-like structure.

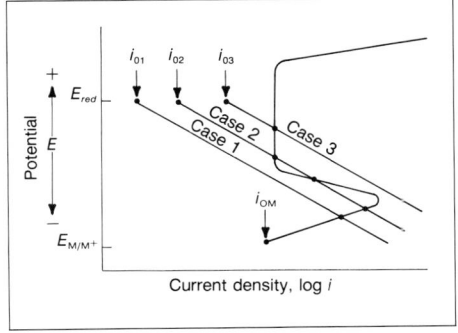

Figure 4. Behavior of an active-passive metal under corrosive conditions. *(After Fontana and Greene, 1978)*

Corrosion resistance of passivated stainless steel is almost completely controlled by the nature of the passive film (Okamoto, 1973). But surface films are quite sensitive to the presence of certain ions such as the chloride ions. In fact, passivity can be lost in the presence of chloride ions (Castle, 1977). In addition, erosion-corrosion results when the protective films are damaged or worn. Sand passing through a stainless steel screen having a passive film can mechanically erode the film, leading to enhanced corrosion rates. In some cases, a passive film may not form at all depending on the water quality. Okamoto (1973) has shown that films do not form in acid solutions. Thus, in low-pH groundwaters the passive film may not exist, or if present, may be only weakly developed.

Low carbon steel screens are much more susceptible to corrosion because only unstable films form on the surface. Therefore, in a diffusion-controlled system, the corrosion rate will continue to accelerate as velocity increases (within the range normally associated with well screens). This point is examined below.

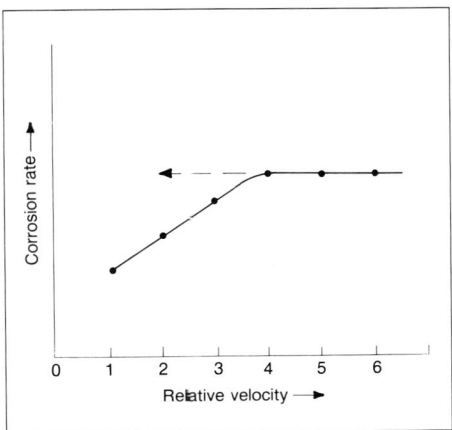

Figure 5. Effect of velocity on the corrosion rate of a normal metal corroding under diffusion control. *(After Fontana and Greene, 1978)*

High velocities may accelerate the corrosion rate by increasing the supply of carbon dioxide, oxygen, or hydrogen sulfide to the metal surface. Higher velocities can also increase the diffusion rate of certain damaging ions such as chloride to the metal surface. Although in certain chemically rich process waters higher velocities may assist in the formation of passive films, high entrance velocities are damaging to well screens for the following major reasons:

1. Sand pumping may occur, causing

Appendix 13.I. Continued

erosion-corrosion.

2. Turbulence is created so that more oxygen, carbon dioxide, and hydrogen sulfide contact the screen surface.

3. Galvanic effects, if present, are enhanced as velocity increases.

To verify the effect of velocity on the rate of corrosion, Johnson Division conducted a series of tests using screen samples made from both 304 stainless steel and low carbon steel. Small [4¼ in (108 mm)] screen coupons (63 wire on 63 rod) having a 0.005 in (0.13 mm) slot were installed in the device shown in Figure 6. Both wire and rod were fabricated from the same coil of wire to eliminate any galvanic effects. Prior to mounting the coupons in the flow cell, they were prepared using the following procedure:

1. Ultrasonic cleaning
2. Brush scrubbing with soap and hot water
3. Rinse in hot water
4. Forced-air dried
5. Agitated in full-strength muriatic acid for 20 seconds
6. Rinse in cold water
7. Acetone rinse
8. Forced-air dried
9. Ethyl or methyl alcohol rinse
10. Forced-air dried

Water used in the experiments had the water-quality characteristics shown in Table 2. To reduce the dissolved oxygen content to that of a typical groundwater, a nitrogen sparge and 10 mℓ hydrazine treatment was used 12 hours before each test. The dis-

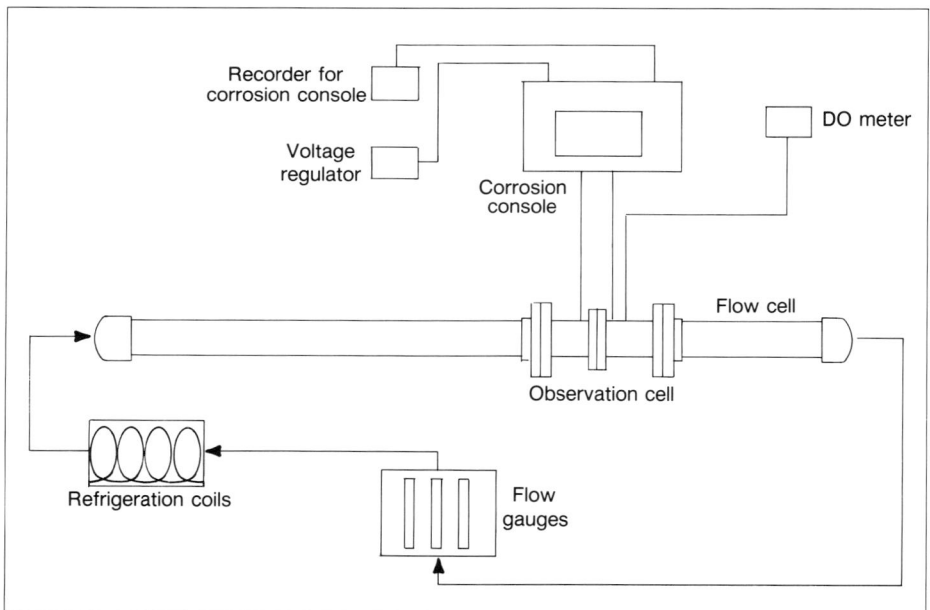

Figure 6. Schematic of corrosion testing apparatus.

Table 2. Analysis of Groundwater used in Corrosion Study

pH value	6.7	
Total alkalinity (as calcium carbonate)	190	mg/ℓ
Total hardness (as calcium carbonate)	186	mg/ℓ
Calcium hardness (as calcium carbonate)	162	mg/ℓ
Chloride	11.5	mg/ℓ
Total iron	0.96	mg/ℓ
Calcium	64.8	mg/ℓ
Sulfate	6.5	mg/ℓ
Specific conductance at 25°C	460	micromhos/cm
Total dissolved solids (calculated)	299	mg/ℓ

solved oxygen ranged from 0.2 to 8 mg/ℓ* depending on the test. The pH of the water ranged from 7.8 to 8.2. Temperature of the water was maintained at 77°F (25°C). Measurements of the corrosion rate in mpy were recorded by a Model 350A Corrosion Measurement System manufactured by EG & G Princeton Applied Research. The device can perform many corrosion measurement techniques, several of which were utilized in the study. In operation, the console is connected to a corrosion cell system in which the screen coupon is mounted (Figure 7).

Because water chemistry varies greatly, the actual corrosion rates determined in the tests are less important than the general trends noted between velocity and corrosion. Figure 8 shows the results of 5 tests on low carbon steel conducted at 0.2 mg/ℓ dissolved oxygen. As flow velocity increased from about 0.1 ft/sec to 2.5 ft/sec, the average corrosion rate accelerated almost linearly from 0.55 to 2.65 mpy. No evidence of a protective film is noted from these data.

Tests conducted with Type 304 stainless steel screen coupons in water containing 2 mg/ℓ (plus or minus 0.2 mg/ℓ) dissolved oxygen showed a trend similar to that apparent in Figure 8 (Figure 9). Up to about 0.4 ft/sec, however, the corrosion rate varies irregularly. Thereafter, it increases so that at 2 ft/sec it is about 3 times the value for velocities ranging from 0 to 0.4 ft/sec. In another series of tests involving

Figure 7. Corrosion cell showing placement of the screen coupon.

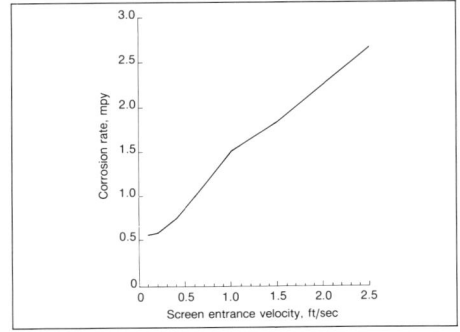

Figure 8. Average corrosion rate in mpy for five screen coupons constructed from low carbon steel at entrance velocities of 0.1 to 2.5 ft/sec. The dissolved oxygen content was 0.2 mg/ℓ.

*Oxygen concentrations in groundwater are ordinarily thought to be quite low although recent research has indicated that oxygen concentrations may range from 2 to 8 mg/ℓ in deep groundwater (Winograd and Robertson, 1985).

Appendix 13.I. Continued

four Type 304 stainless steel coupons in water having 6 mg/ℓ dissolved oxygen, the corrosion rate increased from 0.065 mpy at 0 ft/sec to 0.365 mpy at 2.5 ft/sec (Figure 10).

To check whether any type of protective layer was forming on the low carbon steel screen sample under deaerated conditions, a series of potentiodynamic tests were conducted on seven samples. The data are given in Table 3. It is clear that the corrosion potential increases (the potential becomes less negative) as the velocity increases from 0.1 to 2.5 ft/sec. Therefore, no film or protective layer formed on the low carbon steel sample. The validity of this statement is verified in Figure 11 where no sign of passivation is evident. Table 4 shows the cathodic Tafel constants* compared to flow velocity for samples tested. Based on experience and the value of the potential, these

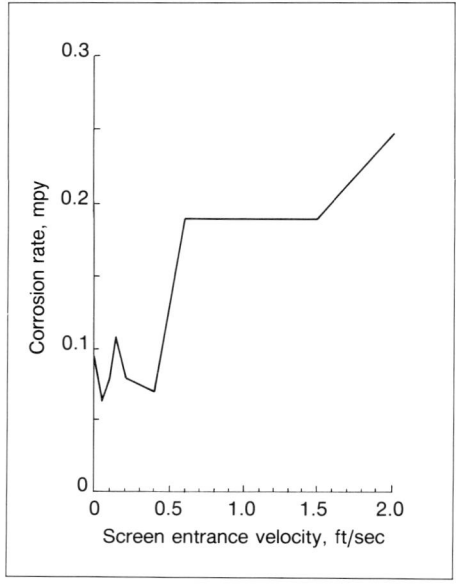

Figure 9. Average corrosion rate in mpy for three screen coupons constructed of Type 304 stainless steel. Entrance velocities ranged from 0 to 2 ft/sec. The dissolved oxygen content was 2 mg/ℓ, plus or minus 0.2 mg/ℓ.

Tafel constants are much higher than the constants expected for hydrogen reduction. Therefore, the corrosion is occurring in the active part of the curve given in Figure 2. The steep Tafel slopes suggest that the reduction reaction is diffusion limited (the

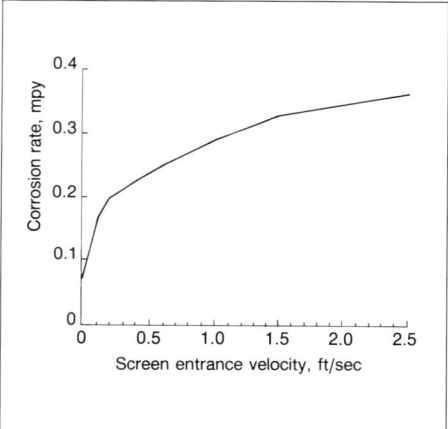

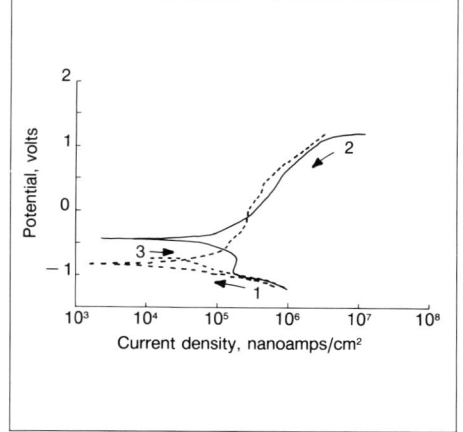

Figure 10. Average corrosion rate in mpy for four screen coupons constructed of Type 304 stainless steel. Entrance velocities ranged from 0 to 2.5 ft/sec. Dissolved oxygen content was 6 mg/ℓ.

Figure 11. Corrosion behavior diagram for low carbon steel screen. The three tests suggest that no protective layer forms on the screen surface. Dissolved oxygen content of the water was 0.5 mg/ℓ.

---

*The Tafel constant indicates the rate of response of the oxidation and reduction reaction to a change in current density and potential.

Table 3. Corrosion Potential Versus Screen Entrance Velocity

Velocity ft/sec	Corrosion Potential in Volts by Coupon Number							
	891	899	905	918	955	925	933	Average
0	−0.552	NT	−0.627	−0.599	NT	NT	NT	−0.593
0.1	−0.340	NT	−0.339	−0.329	(−0.514)	−0.338	−0.375	−0.344
0.2	−0.334	−0.366	−0.326	−0.321	NT	−0.338	−0.353/ −0.354	−0.342
0.4	−0.336	−0.352	−0.331	−0.322	−0.338	−0.331/ −0.338	−0.362	−0.339
0.6	−0.335	−0.339	−0.328/ −0.321	−0.327/ −0.312	−0.331	−0.335	−0.354	−0.331
1.0	NT	−0.326	−0.319	−0.318	−0.323	−0.331	−0.345	−0.327
1.5	−0.318	−0.318	NT	−0.308	−0.317	−0.327	−0.334	−0.320
2.5	NT	−0.290	−0.293	−0.295	−0.304	−0.319	−0.305	−0.301

NT = not tested
( ) = not averaged

cathodic reaction is governed by concentration polarization).

On the basis of the stainless steel data presented above, there is also no indication that a passive film formed during the tests (Figure 12). No passive zone can be seen in the plot. Passivation can be made to occur, however, if sulfuric acid [1 N (normal) concentration] is added to the water used in the aforementioned test (Figure 13). A similar trend is noted for low carbon steel screen samples (Figures 11 and 14).

Table 4. Cathodic Tafel Constants Versus Velocity

Velocity ft/sec	Tafel Constants by Coupon Number							
	933	925	955	918	905	899	891	Average
0	NT	NT	NT	0.040	0.060	NT	NT	0.050
0.1	0.365	0.486	0.371	0.535	0.736	NT	0.813	0.551
0.2	0.300/ 0.235	0.560	NT	0.600	0.765	0.278	1.226	0.456
0.4	0.588	0.907/ 0.410	0.637	0.507	0.654	0.391	2.676	0.583
0.6	0.526	0.745	0.831	0.520/ 0.403	0.620/ 0.589	0.818	V	0.669
1.0	0.416	0.895	0.829	0.503	0.755	0.654	NT	0.675
1.5	0.300	0.790	0.844	0.545	NT	0.601	0.704	0.513
2.5	0.500	0.528	0.616	0.582	0.866	0.357	NT	0.575

NT = not tested
V = vertical

Appendix 13.I. Continued

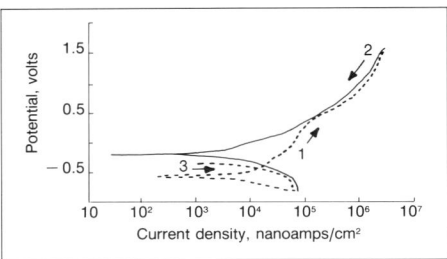

Figure 12. Corrosion behavior diagram for a screen coupon constructed of Type 304 stainless steel. All three tests suggest no passive layer formed. Dissolved oxygen content of the water was 0.5 mg/ℓ.

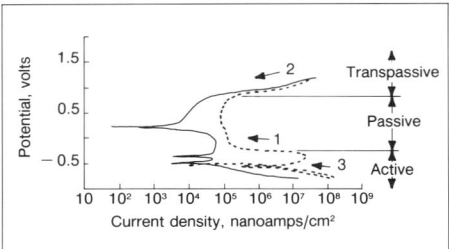

Figure 13. Passivation of this Type 304 stainless steel screen coupon occurred between potential values of −0.2 and 1 when a 1 N (normal) concentration of sulfuric acid was added to the groundwater used in previous tests.

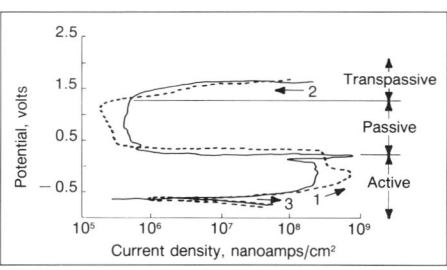

Figure 14. Passivation of this low carbon steel screen sample occurred between potential values of 0.5 and 1.5 in water having a dissolved oxygen content of 0.2 mg/ℓ when a 1 N (normal) concentration of sulfuric acid was added to the groundwater used in the previous tests.

Conclusions

In summary, these tests show that neither Type 304 stainless steel nor low carbon steel screen samples placed in a typical groundwater developed a passive film or oxide layer that retards corrosion in the range of entrance velocities suitable for wells. The screens corrode in an active state. For low carbon steel screens, the corrosion rate increases linearly as velocity increases. At a velocity of 2.5 ft/sec, the corrosion rate is nearly 5 times the rate at 0.1 ft/sec. Thus, for low carbon steel screens, the data indicate that entrance velocities should be kept at about 0.1 ft/sec to minimize corrosion. Higher velocities enhance the corrosion rate significantly.

The corrosion rate for stainless steel screens in deaerated water did not accelerate excessively in one series of tests until a velocity of 0.4 ft/sec was reached. Thereafter, the corrosion rate increases steadily up to a velocity of 1 to 2 ft/sec. As the entrance velocity increases past 1 to 2 ft/sec, however, the rate of increase in corrosion for Type 304 stainless steel diminishes.

In a second series of tests in water having a dissolved oxygen content of 6 mg/ℓ, the average corrosion rate for Type 304 stainless steel screens was somewhat higher than the corrosion rate determined from the tests done in deaerated water. In addition, the corrosion rate accelerated sharply as the screen entrance velocity increased from 0 to 0.25 ft/sec.

Data from these corrosion tests are generally consistent with screen entrance velocities determined to be successful based on field experience. Because actual entrance velocities vary along the screen, prudent well design engineers recognize that maintaining an average entrance velocity of 0.1 ft/sec will provide a safety factor so that the highest velocities will not cause excessive head loss or maintenance problems. Increasing the design velocity much beyond 0.1 ft/sec will increase the operation and maintenance costs for the well unnecessarily.

APPENDIX 13.J.
Calculating the Ryznar Stability Index

The Ryznar Stability Index for a water sample can be calculated from the following equation:

$$I = S - C - pH$$

where I is the Ryznar Stability Index and S and C are factors derived from Figures 1 and 2, based on total dissolved solids, methyl orange alkalinity, and calcium ion concentration (0.4 × calcium hardness).

The steps to determine S and C are:

1. Obtain a value for S using the known total dissolved solids and Figure 1.
2. Obtain a value for C using the methyl orange alkalinity, the calcium ion concentration, and Figure 2.

For example, assume that a water analysis produces the following data:

pH = 7.0
Total dissolved solids = 400 mg/ℓ
Methyl orange alkalinity = 200 mg/ℓ
Calcium hardness = 125 mg/ℓ

Note that the calcium ion concentration for the Ryznar Stability Index equation is 0.4 × 125 = 50 mg/ℓ.

1. The value of S from Figure 1 is 23.12.
2. The value of C from Figure 2 is 8.0.
3. The Ryznar Stability Index, I, is 23.12 − 8.0 − 7.0 = 8.12.

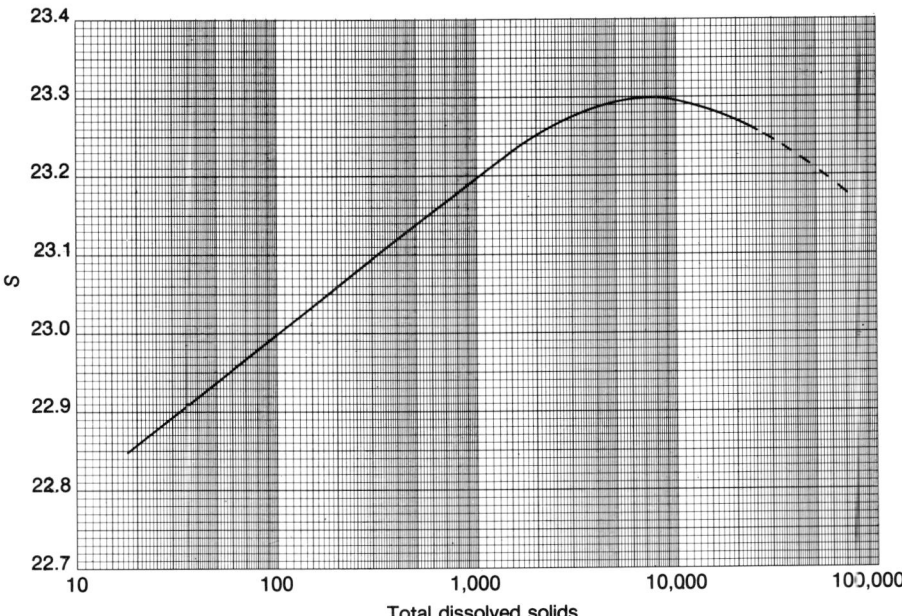

Figure 1. S value as a function of total dissolved solids.

Appendix 13.J. Continued

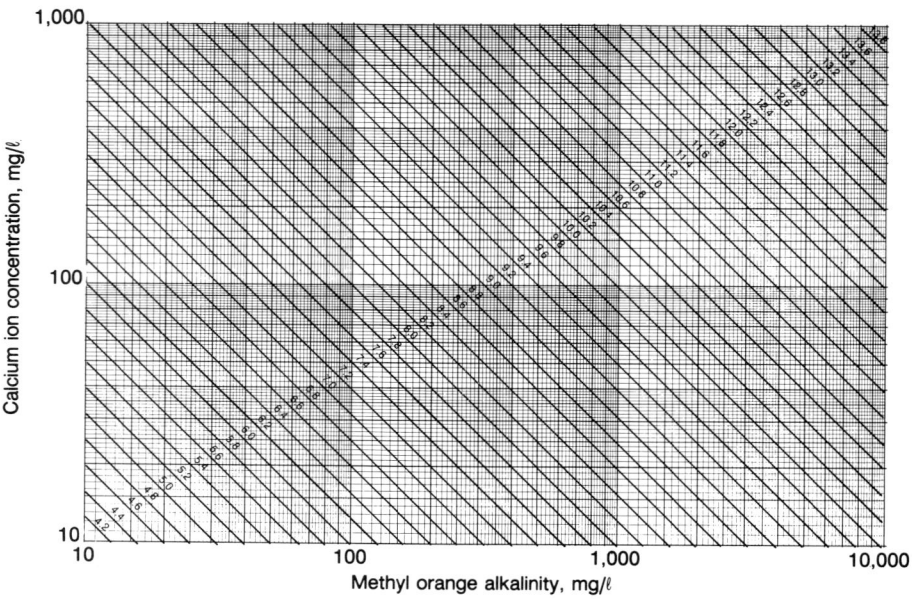

Figure 2. *C* value as a function of calcium ion concentration and methyl orange alkalinity.

APPENDIX 13.K.
Nomogram for Determining Flow, Diameter, Head Loss, or Velocity in Black Steel Pipe

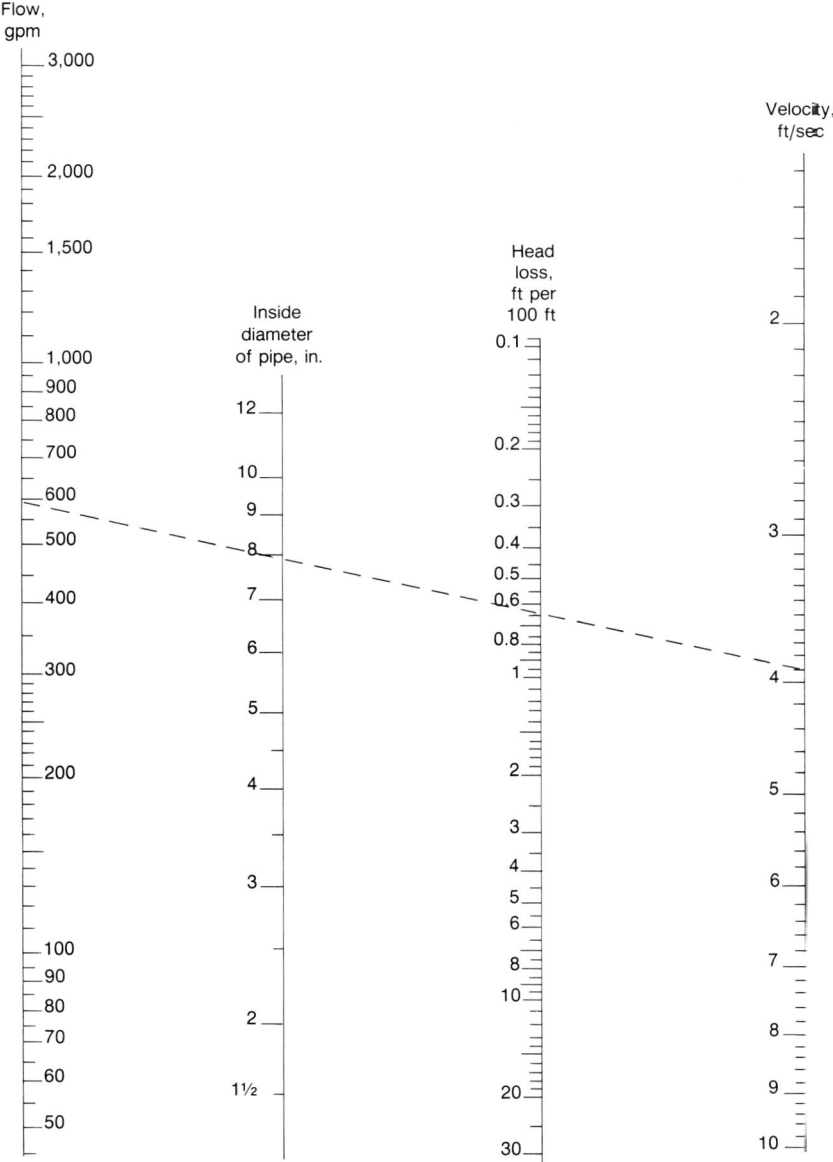

If two of the following values are known — flow, diameter, head loss, and velocity — the other two values can be determined by placing a straight edge on the known factors and reading the others.

APPENDIX 13.L.
Head Loss Characteristics of Water Flow through Rigid Plastic Pipe

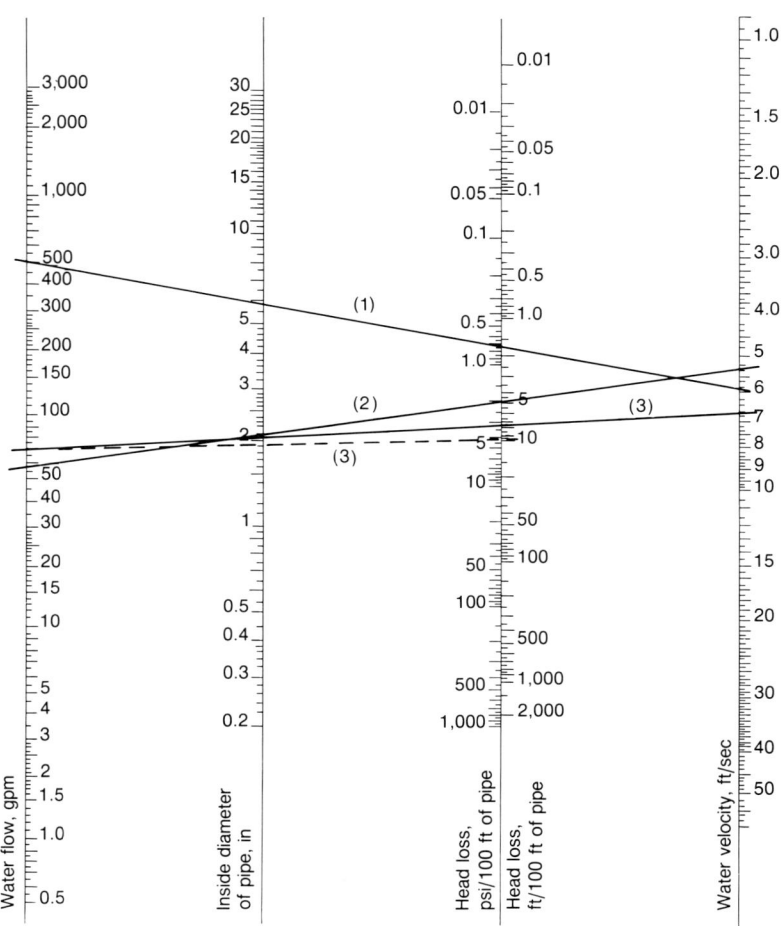

The values of this graph are based on the Hazen-Williams equation.

$$f = 0.2083 \left(\frac{100}{C}\right)^{1.852} \times \frac{g^{1.852}}{d^{4.8655}}$$

where
- f = head loss, in ft of water per 100 ft
- d = inside diameter of pipe, in inches
- g = flow rate, in gpm
- C = constant for inside roughness of the pipe (C = 150 for thermoplastic pipe)

(Plastics Pipe Institute)

APPENDIX 13.M.
Thermoplastic Water Well Casing Pipe Couplings Socket Dimensions and Laying Length Dimensions

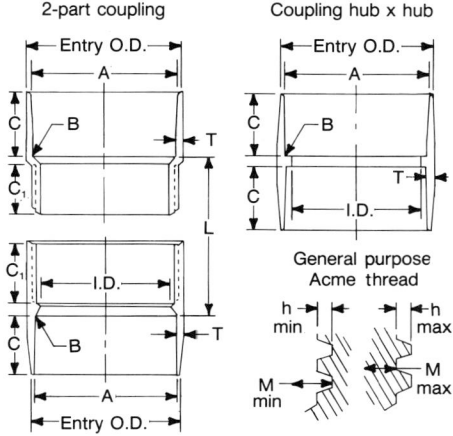

S = nominal pipe size
A = socket entrance diameter
A_1 = tolerance on diameter
B = socket bottom diameter
B_1 = tolerance on diameter
C = socket depth
C_1 = thread length
ID = bore diameter
T = minimum wall thickness
T_1 = tolerance on wall thickness
h = thread height—for ACME 2G screw thread (Note)
M = major diameter of internal thread—for ACME 2G screw thread (Note)
m = minor diameter of external thread—for ACME 2G screw thread (Note)
L = lay length
OD = outside diameter at entry of hub

Inches

S	2	2½	3	3½	4	5	6	8	10	12
A	2.386	2.887	3.514	4.015	4.517	5.584	6.648	8.649	10.796	12.778
A_1	± 0.006	± 0.007	± 0.008	± 0.008	± 0.009	± 0.010	± 0.011	± 0.015	± 0.015	± 0.015
B	2.370	2.869	3.493	3.992	4.491	5.553	6.614	8.613	12.737	12.736
B_1	± 0.006	± 0.007	± 0.008	± 0.008	± 0.009	± 0.010	± 0.011	± 0.015	± 0.015	± 0.015
C	1.500	1.750	2.000	2.250	2.500	3.000	3.500	4.500	5.000	6.000
C_1	1.000	1.250	1.500	1.750	2.000	2.500	3.000	4.000	4.500	5.500
ID	2.149	2.601	3.166	3.620	4.072	5.033	5.993	7.805	9.728	11.538
T	0.113	0.137	0.167	0.190	0.214	0.265	0.316	0.410	0.511	0.606
T_1	+ 0.020	+ 0.020	+ 0.023	+ 0.026	+ 0.032	+ 0.038	+ 0.049	+ 0.061	+ 0.073	
h	0.083	0.100	0.100	0.125	0.125	0.166	0.200	0.250	0.333	0.375
M	2.465	2.982	3.606	4.131	4.630	5.735	6.830	8.878	11.086	13.127
m	2.289	2.772	3.396	3.871	4.370	5.393	6.420	8.368	10.410	12.367
L	1.500	1.750	2.000	2.250	2.500	3.000	3.500	4.500	5.000	6.000
OD	2.612	3.161	3.848	4.395	4.945	6.114	7.280	9.469	11.818	13.990

Millimeters

S	2	2½	3	3½	4	5	6	8	10	12
A	60.60	73.32	89.26	101.98	114.74	141.84	168.86	219.68	274.22	324.56
A_1	± 0.16	± 0.18	± 0.20	± 0.20	± 0.22	± 0.26	± 0.28	± 0.38	± 0.38	± 0.38
B	60.20	72.88	88.72	101.40	114.08	141.04	168.00	218.78	323.52	323.50
B_1	± 0.16	± 0.18	± 0.20	± 0.20	± 0.22	± 0.26	± 0.28	± 0.38	± 0.38	± 0.38
C	38	44	51	57	64	76	89	114	127	152
C_1	25	32	38	44	51	64	76	102	114	140
ID	54.58	66.06	80.42	91.94	103.42	127.84	152.22	198.24	247.10	293.06
T	2.88	3.48	4.24	4.82	5.44	6.74	8.02	10.42	12.98	15.40
T_1	+ 0.50	+ 0.50	+ 0.50	+ 0.58	+ 0.66	+ 0.82	+ 0.96	+ 1.24	+ 1.54	+ 1.86
h	2.10	2.54	2.54	3.18	3.18	4.22	5.08	6.34	8.46	9.52
M	62.62	75.74	91.60	104.92	117.60	145.66	173.48	225.50	281.58	333.42
m	58.14	70.40	86.26	98.32	111.00	136.98	163.06	212.54	264.40	314.12
L	38	44	51	57	64	76	89	114	127	152
OD	66.34	80.28	97.74	111.64	125.60	155.30	184.92	240.52	300.18	355.34

(Copyright, ASTM, 1916 Race Street, Philadelphia, PA 19103. Reprinted with permission.)

APPENDIX 13.N.
Tapered Sockets for Bell-End Pipe, in[1]

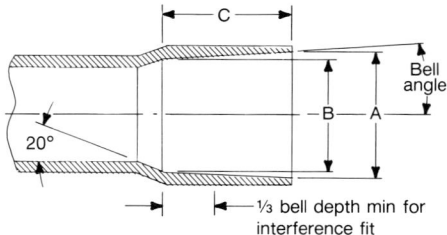

Nominal Pipe Size	A Socket Entrance Diameter			B Socket Bottom Diameter			C Socket Length Minimum
	Diameter	Tolerance on Diameter	Maximum Out-of-Round	Diameter	Tolerance on Diameter	Maximum Out-of-Round	
2	2.386	± 0.006	± 0.012	2.370	± 0.006	± 0.012	3.000
2½	2.887	± 0.007	± 0.015	2.869	± 0.007	± 0.015	3.500
3	3.514	± 0.008	± 0.015	3.493	± 0.008	± 0.015	4.000
3½	4.015	± 0.008	± 0.015	3.992	± 0.008	± 0.015	4.500
4	4.517	± 0.009	± 0.015	4.491	± 0.009	± 0.015	5.000
5	5.584	± 0.010	± 0.030	5.553	± 0.010	± 0.030	6.000
6	6.648	± 0.011	± 0.030	6.614	± 0.011	± 0.030	6.500
8	8.649	± 0.015	± 0.045	8.613	± 0.015	± 0.045	7.000
10	10.776	± 0.015	± 0.050	10.737	± 0.015	± 0.050	7.500
12	12.778	± 0.015	± 0.060	12.736	± 0.015	± 0.060	8.000

[1] Minimum dimensions have zero negative tolerance. The sketches and designs of fittings are illustrative only.

(Copyright, ASTM, 1916 Race Street, Philadelphia, PA 19103. Reprinted with permission.)

APPENDIX 13.O.
Friction Loss in Feet per 100 Feet in Asbestos Cement Class 150-Pressure Pipe

Flow gallons per minute	Nominal pipe diameter in inches				
	4	6	8	10	12
	I.D. = 3.95	I.D. = 5.85	I.D. = 7.85	I.D. = 10.00	I.D. = 12.00
100	.677				
120	.954				
140	1.28				
160	1.55				
180	2.06				
200	2.53	.372			
220	3.03	.447			
240	3.56	.525			
260	4.16	.611			
280	4.77	.705			
300	5.44	.803			
320	6.16	.910			
340	6.91	1.02			
360	7.70	1.14			
380	8.54	1.26			
400	9.40	1.39	.324		
420	10.3	1.52	.355		
440	11.3	1.66	.389		
460	12.3	1.81	.423		
480	13.3	1.96	.458		
500	14.4	2.12	.495		
550	17.2	2.55	.594		
600	20.3	2.99	.701	.214	
650	23.7	3.49	.818	.249	
700	27.3	4.02	.935	.287	
750	31.1	4.57	1.07	.328	
800		5.18	1.21	.370	.152
850		5.81	1.36	.415	.170
900		6.46	1.51	.464	.190
950		7.17	1.68	.511	.210
1000		7.91	1.85	.564	.232
1100		9.45	2.21	.675	.278
1200		11.2	2.62	.800	.328
1300		13.0	3.04	.932	.384
1400		15.0	3.50	1.07	.438
1500		17.1	3.99	1.22	.502
1600		19.3	4.52	1.38	.566
1700			5.06	1.55	.637
1800			5.67	1.73	.710
1900			6.26	1.91	.737
2000			6.90	2.11	.854
2200			8.27	2.53	1.04
2400			9.75	2.98	1.23
2600			11.4	3.47	1.43
2800			13.1	4.00	1.64
3000			14.9	4.56	1.87

From Scobey's formula

$$H_f = K_s \frac{V^{1.9}}{D^{1.1}}$$

where H_f = Head loss in 1000 feet of pipe
K_s = roughness coefficient = 0.32
V = velocity in feet per second
D = inside pipe diameter in feet

(Sprinkler Irrigation Association)

APPENDIX 15.A.

Discharge of Air through an Orifice

In ft³ of free air per minute at standard atmospheric pressure of 14.7 psi absolute and 70°F.

Gauge Pressure before Orifice in psi	Diameter of Orifice									
	1/64"	1/32"	1/16"	1/8"	1/4"	3/8"	1/2"	5/8"	3/4"	7/8"	1"
	Discharge in cubic feet of free air per minute									
1	.028	.112	.450	1.80	7.18	16.2	28.7	45.0	64.7	88.1	115
2	.040	.158	.633	2.53	10.1	22.8	40.5	63.3	91.2	124	162
3	.048	.194	.775	3.10	12.4	27.8	49.5	77.5	111	152	198
4	.056	.223	.892	3.56	14.3	32.1	57.0	89.2	128	175	228
5	.062	.248	.993	3.97	15.9	35.7	63.5	99.3	143	195	254
6	.068	.272	1.09	4.34	17.4	39.1	69.5	109	156	213	278
7	.073	.293	1.17	4.68	18.7	42.2	75.0	117	168	230	300
9	.083	.331	1.32	5.30	21.2	47.7	84.7	132	191	260	339
12	.095	.379	1.52	6.07	24.3	54.6	97.0	152	218	297	388
15	.105	.420	1.68	6.72	26.9	60.5	108	168	242	329	430
20	.123	.491	1.96	7.86	31.4	70.7	126	196	283	385	503
25	.140	.562	2.25	8.98	35.9	80.9	144	225	323	440	575
30	.158	.633	2.53	10.1	40.5	91.1	162	253	365	496	648
35	.176	.703	2.81	11.3	45.0	101	180	281	405	551	720
40	.194	.774	3.10	12.4	49.6	112	198	310	446	607	793
45	.211	.845	3.38	13.5	54.1	122	216	338	487	662	865
50	.229	.916	3.66	14.7	58.6	132	235	366	528	718	938
60	.264	1.06	4.23	16.9	67.6	152	271	423	609	828	1082
70	.300	1.20	4.79	19.2	76.7	173	307	479	690	939	1227
80	.335	1.34	5.36	21.4	85.7	193	343	536	771	1050	1371
90	.370	1.48	5.92	23.7	94.8	213	379	592	853	1161	1516
100	.406	1.62	6.49	26.0	104	234	415	649	934	1272	1661
110	.441	1.76	7.05	28.2	113	254	452	705	1016	1383	1806
120	.476	1.91	7.62	30.5	122	274	488	762	1097	1494	1951
125	.494	1.98	7.90	31.6	126	284	506	790	1138	1549	2023
150	.582	2.37	9.45	37.5	150	338	600	910	1315	1789	2338
200	.761	3.10	12.35	49.0	196	441	784	1225	1764	2401	3136
250	.935	3.80	15.18	60.3	241	542	964	1508	2169	2952	3856
300	.995	4.88	18.08	71.8	287	646	1148	1795	2583	3515	4592
400	1.220	5.98	23.81	94.5	378	851	1512	2360	3402	4630	6048
500	1.519	7.41	29.55	117.3	469	1055	1876	2930	4221	5745	7504
750	2.240	10.98	43.85	174.0	696	1566	2784	4350	6264	8525	11136
1000	2.985	14.60	58.21	231.0	924	2079	3696	5790	8316	11318	14784

Table is based on 100% coefficient of flow. For well rounded entrance multiply values by 0.97. For sharp edged orifices a multiplier of 0.61 may be used for approximate results.

Values for pressures from 1 to 15 lbs gauge calculated by standard adiabatic equation.
Values for pressures above 15 lb gauge calculated by approximate equation proposed by S.A. Moss.

$$W = 0.5303 \frac{ACp_1}{\sqrt{T_1}}$$

Where:
W = discharge in lbs per sec
A = area of orifice in square inches
C = Coefficient of flow
p_1 = Upstream total pressure in psi absolute
T_1 = Upstream temperature in °F abs

Values used in calculating above table were: $C = 1.0$, p_1 = gauge pressure + 14.7 psi, T_1 = 530°F abs.
Weights (W) were converted to volumes using density factor of 0.07494 lbs./cu. ft. This is correct for dry air at 14.7 psi absolute pressure and 70°F.
Equation cannot be used where p_1 is less than two times the downstream pressure.

(Ingersoll-Rand, 1971)

APPENDIX 16.A-1.
Discharge from Rectangular Weir with End Contractions

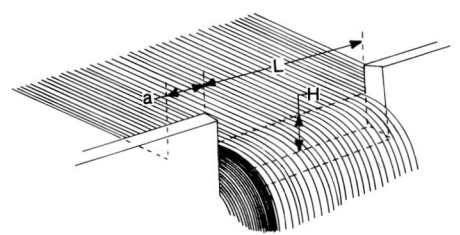

	Figures in Table are in Gallons Per Minute							
	Length (L) of weir in feet					Length (L) of weir in feet		
Head (H) in inches	1	3	5	Additional gpm for each ft over 5 ft	Head (H) in inches	3	5	Additional gpm for each ft over 5 ft
1	35.4	107.5	179.8	36.05	8	2338	3956	814
1¼	49.5	150.4	250.4	50.4	8¼	2442	4140	850
1½	64.9	197	329.5	66.2	8½	2540	4312	890
1¾	81	248	415	83.5	8¾	2656	4511	929
2	98.5	302	506	102	9	2765	4699	970
2¼	117	361	605	122	9¼	2876	4899	1011
2½	136.2	422	706	143	9½	2985	5098	1051
2¾	157	485	815	165	9¾	3101	5288	1091
3	177.8	552	926	187	10	3216	5490	1136
3¼	199.8	624	1047	211	10½	3480	5940	1230
3½	222	695	1167	236	11	3716	6355	1320
3¾	245	769	1292	261	11½	3960	6780	1410
4	269	846	1424	288	12	4185	7165	1495
4¼	293.6	925	1559	316	12½	4430	7595	1575
4½	318	1006	1696	345	13	4660	8010	1660
4¾	344	1091	1835	374	13½	4950	8510	1780
5	370	1175	1985	405	14	5215	8980	1885
5¼	395.5	1262	2130	434	14½	5475	9440	1985
5½	421.6	1352	2282	465	15	5740	9920	2090
5¾	449	1442	2440	495	15½	6015	10400	2165
6	476.5	1535	2600	528	16	6290	10900	2300
6¼		1632	2760	560	16½	6565	11380	2410
6½		1742	2920	596	17	6925	11970	2520
6¾		1826	3094	630	17½	7140	12410	2640
7		1928	3260	668	18	7410	12900	2745
7¼		2029	3436	701.5	18½	7695	13410	2855
7½		2130	3609	736	19	7980	13940	2970
7¾		2238	3785	774	19½	8280	14460	3090

Appendix 16.A-1. Continued

This table is based on Francis equation:

$$Q = 3.33 (L - 0.2H)H^{1.5}$$

where:
Q = flow of water, in ft^3/sec.
L = length of weir opening, in ft (should be 4 to 8 times H).
H = head on weir, in ft (to be measured at least 6 ft back of weir opening).
a = should be at least 3 H.

(Ingersoll-Rand, 1970)

APPENDIX 16.A-2.
Discharge from Triangular Notch Weir with End Contractions

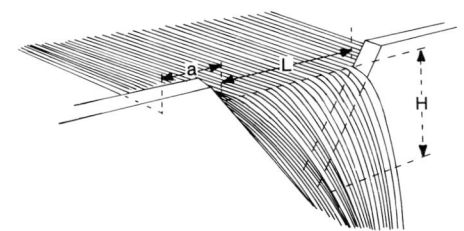

Head (H) in inches	Flow in gallons per min		Head (H) in inches	Flow in gallons per min		Head (H) in inches	Flow in gallons per min	
	90° notch	60° notch		90° notch	60° notch		90° notch	60° notch
1	2.19	1.27	6¾	260	150	15	1912	1104
1¼	3.83	2.21	7	284	164	15½	2073	1197
1½	6.05	3.49	7¼	310	179	16	2246	1297
1¾	8.89	5.13	7½	338	195	16½	2426	1401
2	12.4	7.16	7¾	367	212	17	2614	1509
2¼	16.7	9.62	8	397	229	17½	2810	1623
2½	21.7	12.5	8¼	429	248	18	3016	1741
2¾	27.5	15.9	8½	462	267	18½	3229	1864
3	34.2	19.7	8¾	498	287	19	3452	1993
3¼	41.8	24.1	9	533	308	19½	3684	2127
3½	50.3	29.0	9¼	571	330	20	3924	2266
3¾	59.7	34.5	9½	610	352	20½	4174	2410
4	70.2	40.5	9¾	651	376	21	4433	2560
4¼	81.7	47.2	10	694	401	21½	4702	2715
4½	94.2	54.4	10½	784	452	22	4980	2875
4¾	108	62.3	11	880	508	22½	5268	3041
5	123	70.8	11½	984	568	23	5565	3213
5¼	139	80.0	12	1094	632	23½	5873	3391
5½	156	89.9	12½	1212	700	24	6190	3574
5¾	174	100	13	1337	772	24½	6518	3763
6	193	112	13½	1469	848	25	6855	3958
6¼	214	124	14	1609	929			
6½	236	136	14½	1756	1014			

Based on Thompson equation:

$$Q = (C)(4/15)(L)(H)\sqrt{2gH}$$

where:

Q = flow of water, in ft³/sec
L = width of notch, in ft, at H distance above apex
H = head of water above apex of notch, in ft

Appendix 16.A-2. Continued

C = constant varying with conditions, 0.57 being used for this table
a = should not be less than ¾L.

For 90° notch the equation becomes

$$Q = 2.4381 \, H^{5/2}$$

For 60° notch the equation becomes

$$Q = 1.4076 \, H^{5/2}$$

(Ingersoll-Rand, 1970)

APPENDIX 16.B.
Free Flow Discharge — Parshall Flume, cfs

Head, H_a Feet	Size of Flume, W										
	3"	6"	9"	1'0"	1'6"	2'0"	3'0"	4'0"	5'0"	6'0"	7'0"	8'0"
0.1	0.028	0.05	0.09	—	—	—	—	—	—	—	—	—
0.2	0.082	0.16	0.26	0.35	0.51	0.66	0.97	1.26	—	—	—	—
0.3	0.154	0.31	0.49	0.64	0.94	1.24	1.82	2.39	2.96	3.52	4.08	4.62
0.4	0.241	0.48	0.76	0.99	1.47	1.93	2.86	3.77	4.68	5.57	6.46	7.34
0.5	0.339	0.69	1.06	1.39	2.06	2.73	4.05	5.36	6.66	7.94	9.23	10.51
0.6	0.450	0.92	1.40	1.84	2.73	3.62	5.39	7.15	8.89	10.63	12.36	14.08
0.7	0.571	1.17	1.78	2.33	3.46	4.60	6.86	9.11	11.36	13.59	15.82	18.04
0.8	0.702	1.45	2.18	2.85	4.26	5.66	8.46	11.25	14.04	16.81	19.59	22.36
0.9	0.843	1.74	2.61	3.41	5.10	6.80	10.17	13.55	16.92	20.29	23.66	27.02
1.0	0.992	2.06	3.07	4.00	6.00	8.00	12.00	16.00	20.00	24.00	28.00	32.00
1.1	—	2.40	3.55	4.62	6.95	9.27	13.93	18.60	23.26	27.94	32.62	37.30
1.2	—	2.75	4.06	5.28	7.94	10.61	15.96	21.33	26.71	32.10	37.50	42.89
1.3	—	—	4.59	5.96	8.99	12.01	18.10	24.21	30.33	36.47	42.62	48.78
1.4	—	—	5.14	6.68	10.10	13.48	20.32	27.21	34.11	41.05	47.99	54.95
1.5	—	—	—	7.41	11.20	15.00	22.64	30.34	38.06	45.82	53.59	61.40
1.6	—	—	—	8.18	12.40	16.58	25.05	33.59	42.17	50.79	59.42	68.10
1.7	—	—	—	8.97	13.60	18.21	27.55	36.96	46.43	55.95	65.48	75.08
1.8	—	—	—	9.79	14.80	19.90	30.13	40.45	50.83	61.29	71.75	82.29
1.9	—	—	—	10.62	16.10	21.63	32.79	44.05	55.39	66.81	78.24	89.76
2.0	—	—	—	11.49	17.40	23.43	35.53	47.77	60.08	72.50	84.94	97.48
2.1	—	—	—	12.37	18.80	25.27	38.35	51.59	64.92	78.37	91.84	105.40
2.2	—	—	—	13.28	20.20	27.15	41.25	55.52	69.90	84.41	98.94	113.60
2.3	—	—	—	14.21	21.60	29.09	44.22	59.56	75.01	90.61	106.20	122.00
2.4	—	—	—	15.16	23.00	31.09	47.27	63.69	80.25	96.97	113.70	130.70
2.5	—	—	—	16.13	24.60	33.11	50.39	67.93	85.62	103.50	121.40	139.50

NOTE: Approximate values of flow for heads other than those shown may be found by direct interpolation in the table.

(Colt Industries, 1974)

APPENDIX 16.C-1.
Free Flow Discharges in CFS for Selected Sizes of Cutthroat Flumes

h_a, ft	4 in x 3 ft	8 in x 3 ft	12 in x 3 ft	8 in x 6 ft
0.02	0.00	0.00	0.00	0.00
0.10	0.02	0.04	0.07	0.06
0.20	0.08	0.15	0.23	0.17
0.30	0.16	0.32	0.49	0.34
0.40	0.27	0.55	0.83	0.55
0.50	0.41	0.83	1.26	0.79
0.60	0.57	1.16	1.76	1.07
0.70	0.76	1.54	2.33	1.37
0.80	0.97	1.97	2.98	1.71
0.90	1.20	2.45	3.71	2.08
1.00	1.46	2.97	4.50	2.47
1.10	1.74	3.54	5.36	2.89
1.20	2.04	4.15	6.29	3.33
1.30	2.36	4.81	7.29	3.80
1.40	2.71	5.52	8.36	4.29
1.50	3.07	6.26	9.49	4.81
1.60	3.46	7.05	10.69	5.35
1.70	3.87	7.88	11.95	5.91
1.80	4.30	8.76	13.27	6.49
1.90	4.75	9.68	14.66	7.09
2.00	5.22	10.63	16.11	7.72

(Skogerboe et al., 1973)

APPENDIX 16.C-2.
Submerged Flow Discharges in CFS for 4 in x 3 ft Cutthroat Flume

h_a, ft	$h_a - h_b$, ft							
	0.02	0.10	0.20	0.30	0.40	0.50	0.60	0.70
0.02	0.0	0.0	0.0	0.0	0.0	—	—	—
0.20	0.1	0.1	0.1	0.1	0.1	—	—	—
0.40	0.2	0.3	0.3	0.3	0.3	—	—	—
0.60	0.3	0.5	0.6	0.6	0.6	—	—	—
0.80	0.5	0.8	0.9	1.0	1.0	—	—	—
1.00	0.7	1.2	1.4	1.4	1.5	—	—	—
1.20	0.9	1.5	1.8	2.0	2.0	2.0	2.0	2.0
1.40	1.1	2.0	2.4	2.6	2.7	2.7	2.7	2.7
1.60	1.4	2.4	2.9	3.2	3.4	3.4	3.5	3.5
1.80	1.7	2.9	3.5	3.9	4.1	4.2	4.3	4.3
2.00	1.9	3.4	4.2	4.6	4.9	5.1	5.2	5.2

(Skogerboe et al., 1973)

APPENDIX 16.C-3.
Submerged Flow Discharges in CFS for 8 in x 3 ft Cutthroat Flume

h_a, ft	$h_a - h_b$, ft							
	0.02	0.10	0.20	0.30	0.40	0.50	0.60	0.70
0.02	0.0	0.0	0.0	0.0	0.0	—	—	—
0.20	0.1	0.2	0.2	0.2	0.2	—	—	—
0.40	0.4	0.5	0.6	0.6	0.6	—	—	—
0.60	0.7	1.1	1.2	1.2	1.2	—	—	—
0.80	1.0	1.7	1.9	2.0	2.0	—	—	—
1.00	1.4	2.4	2.8	2.9	3.0	—	—	—
1.20	1.9	3.1	3.8	4.0	4.1	4.2	4.2	4.2
1.40	2.3	4.0	4.8	5.2	5.4	5.5	5.5	5.5
1.60	2.8	4.9	6.0	6.5	6.9	7.0	7.1	7.1
1.80	3.4	5.8	7.2	7.9	8.4	8.6	8.7	8.8
2.00	4.0	6.9	8.5	9.4	10.0	10.3	10.5	10.6

(Skogerboe et al., 1973)

APPENDIX 16.C-4.
Submerged Flow Discharges in CFS for 12 in x 3 ft Cutthroat Flume

h_a, ft	$h_a - h_b$, ft							
	0.02	0.10	0.20	0.30	0.40	0.50	0.60	0.70
0.02	0.0	0.0	0.0	0.0	0.0	—	—	—
0.20	0.2	0.2	0.2	0.2	0.2	—	—	—
0.40	0.5	0.8	0.8	0.8	0.8	—	—	—
0.60	1.0	1.6	1.7	1.8	1.8	—	—	—
0.80	1.5	2.5	2.9	3.0	3.0	—	—	—
1.00	2.1	3.6	4.2	4.4	4.5	—	—	—
1.20	2.8	4.8	5.7	6.1	6.2	6.3	6.3	6.3
1.40	3.5	6.0	7.3	7.9	8.2	8.4	8.4	8.4
1.60	4.3	7.4	9.0	9.9	10.4	10.6	10.7	10.7
1.80	5.1	8.9	10.9	12.0	12.7	13.0	13.2	13.3
2.00	6.0	10.4	12.8	14.2	15.1	15.7	15.9	16.1

(Skogerboe et al., 1973)

APPENDIX 16.C-5.
Submerged Flow Discharges in CFS for 8 in x 6 ft Cutthroat Flume

h_a, ft	$h_a - h_b$, ft						
	0.02	0.10	0.20	0.30	0.40	0.50	0.60
0.02	0.0	0.0	0.0	0.0	0.0	—	—
0.20	0.2	0.2	0.2	0.2	0.2	—	—
0.40	0.4	0.5	0.5	0.5	0.5	—	—
0.60	0.8	1.0	1.1	1.1	1.1	—	—
0.80	1.2	1.6	1.7	1.7	1.7	—	—
1.00	1.6	2.2	2.5	2.5	2.5	—	—
1.20	2.0	2.9	3.3	3.3	3.3	3.3	3.3
1.40	2.5	3.6	4.1	4.3	4.3	4.3	4.3
1.60	3.0	4.4	5.0	5.3	5.3	5.3	5.3
1.80	3.6	5.2	6.0	6.3	6.5	6.5	6.5
2.00	4.2	6.1	7.0	7.4	7.7	7.7	7.7
2.20	4.8	7.0	8.0	8.6	8.9	9.0	9.0
2.40	5.4	7.9	9.1	9.8	10.2	10.4	10.5

(Skogerboe et al., 1973)

APPENDIX 16.D.
Empirical Equations Used to Estimate Specific Capacity and Transmissivity

Two empirical equations have been developed from the modified nonequilibrium (Jacob) equation to estimate the potential specific capacity and transmissivity of a well. These equations are derived by assuming an average well diameter, average duration of pumping, and typical values for the applicable storage coefficient. The equations are useful for quickly checking the accuracy of values obtained for transmissivity and specific capacity during pumping tests.

Recall Jacob's equation (9.6):

$$s = \frac{264Q}{T} \log \frac{0.3Tt}{r^2 S}$$

where
s = drawdown in the well, in ft
Q = yield of the well, in gpm
T = transmissivity of the well, in gpd/ft
t = time of pumping, in days
r = radius of the well, in ft
S = storage coefficient of the aquifer

This equation is based on the simplifying assumptions listed on page 218.

By rearranging terms, the specific capacity is:

$$\frac{Q}{s} = \frac{T}{264 \log \frac{0.3Tt}{r^2 S}} \tag{1}$$

If typical values are assumed for the variables in the log function of the equation such as $t = 1$ day, $r = 0.5$ ft, $T = 30,000$ gpd/ft, and $S = 1 \times 10^{-3}$ for a confined aquifer and $S = 7.5 \times 10^{-2}$ for an unconfined aquifer, the specific capacity of the confined aquifer is given by:

$$\frac{Q}{s} = \frac{T}{2000} \tag{2}$$

The specific capacity for an unconfined aquifer is given by:

$$\frac{Q}{s} = \frac{T}{1500} \tag{3}$$

These empirical equations can be used to check the transmissivity of wells where the specific capacity is known, or the specific capacity where the transmissivity is known.

It may appear to be presumptuous to use an average transmissivity value or even to assume a transmissivity value at all before one is known. However, because it appears in the log term of Equation 1, its affect on the value of the divisor in either derivation is minimal. For example, if a transmissivity of 120,000 gpd/ft is assumed, the divisor increases from 2,000 to 2,133, a difference of less than 7 percent.

Estimates of Q/s using Equation 3 for unconfined aquifers will nearly always be optimistic because part of the aquifer is dewatered during pumping, resulting in a lower transmissivity as the saturated thickness decreases. Therefore, some estimates for unconfined aquifers may be more accurate if Equation 2 is used.

1022   GROUNDWATER AND WELLS

APPENDIX 16.E.
Estimated Flows from Pipes

Pipe must be horizontal for best results
All quantities obtained are approximate

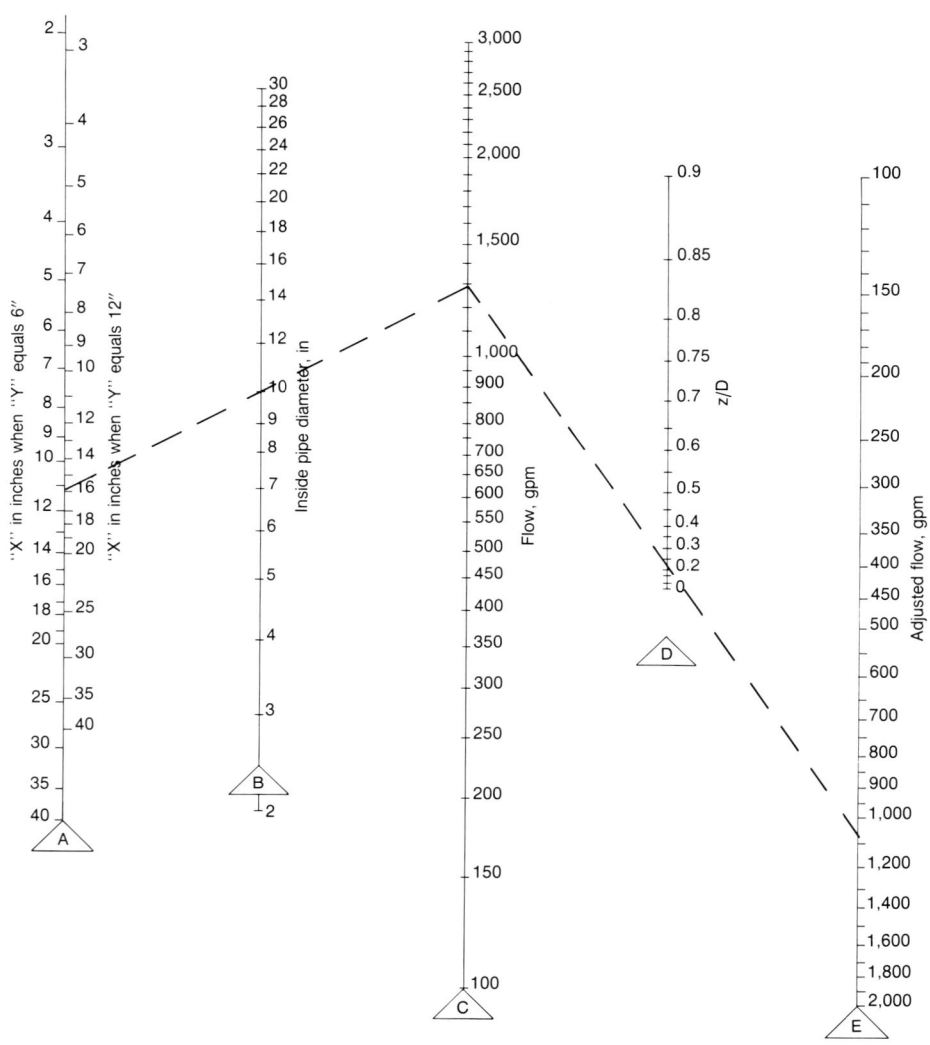

Appendix 16.E. Continued

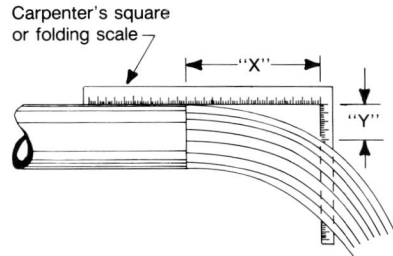

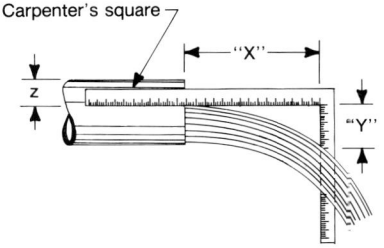

Example 1 (pipe flowing full):

Assumptions:
Pipe Diameter = 10 in
"X" = 11 in
"Y" = 6 in

With a straight edge, connect 11 in on scale A ("Y" = 6 in) with 10 in on scale B and read 1300 gpm on scale C.

Example 2 (pipe flowing partly full)

Assumptions:
Pipe Diameter = 10 in
"X" = 11 in
"Y" = 6 in
"Z" = 2 in; thus Z/D = 2/10 = 0.20

Assume pipe is flowing full and proceed as in Example 1. Then with a straight edge, connect 1300 gpm on scale C with 0.20 on scale E and read 1100 gpm on scale E.

(U.S. Department of Agriculture, 1953)

APPENDIX 16.F.
Flow from Vertical Pipes

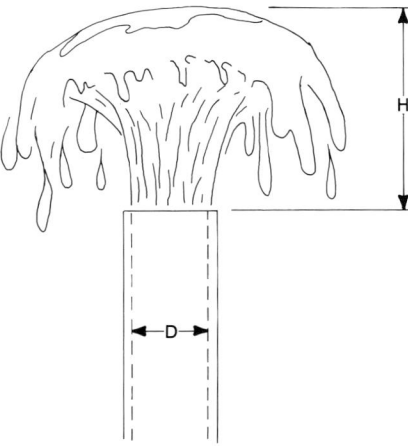

Flow from a vertical pipe can be estimated by measuring the vertical height, H, as shown in Figure 1.

$$Q = 5.68 \, KD^2 \, H^{1/2}$$

where

Q = discharge, gpm
D = inside diameter of pipe, in
H = vertical height of water jet, in
K = a constant, varying from 0.87 to 0.97 for pipes 2 to 6 in diameter and H equal to 6 to 24 in

TABLE 1.
Flow from Vertical Pipes, gpm

Nominal I.D. Pipe, in.	Vertical Height, H, of Water Jet, in										
	3	3.5	4	4.5	5	5.5	6	7	8	10	12
2	38	41	44	47	50	53	56	61	65	74	82
3	81	89	96	103	109	114	120	132	141	160	177
4	137	151	163	174	185	195	205	222	240	269	299
6	318	349	378	405	430	455	480	520	560	635	700
8	567	623	684	730	776	821	868	945	1020	1150	1270
10	950	1055	1115	1200	1280	1350	1415	1530	1640	1840	2010

(Colt Industries, 1974)

APPENDIX 17.A.
Friction Losses in Pipe

Pressure (head) losses in piping systems are caused by various forms of friction. These include the losses caused by (Crane, 1976):

a. roughness of the interior wall of the pipe,
b. the distance (diameter) between the walls of the pipe; the smaller the diameter, the greater the friction losses for a given discharge,
c. velocity, density, and viscosity of the fluid,
d. directional changes in the flow path,
e. obstructions in the flow path, and
f. gradual or sudden changes in the cross-sectional area and slope of the flow path.

In general, roughness of the pipe interior causes the greatest percentage of friction in a piping system.

There are two forms of analysis used to calculate friction or head losses in pipe. The first, called the Darcy-Weisbach equation, was derived from a dimensional analysis of the factors influencing friction losses. The head loss caused by friction, h_f, is given by:

$$h_f = \frac{\Delta p}{\gamma} = f\left(\frac{L}{D}\right)\frac{V}{2g} \tag{1}$$

where
Δp is the change in pressure
γ is the specific weight of water
f is the Fanning friction factor
L is the length of the pipe
D is the diameter of the pipe
V is the average flow velocity
g is the acceleration of gravity

The Fanning friction factor, f, can be determined analytically for laminar flow and semi-empirically for turbulent flow conditions. For turbulent flow, the roughness of the pipe surface, k/D (height of the grain or roughness divided by the diameter of the pipe), is the principal factor in determining f because the variation in the viscosity of water is relatively small and the Reynolds number is high. The friction factor is defined as the ratio of the local wall shear stress to the dynamic pressure of the mean flow (Olson, 1966). Experiments on commercial pipes using an equivalent sand grain roughness have provided good approximations for the friction losses to be expected for pipes with certain roughness (Figure 1).

The friction factor derived from this figure can be used in Equation 1 to determine friction loss. The friction losses for various size pipes and discharges are given in Table 1. Experience has shown that the h_f calculated from Equation 1 may be good only for pipes that are well maintained or carry water low in total dissolved solids. In many cases, however, the friction factor will rise significantly over time as the pipes age (Table 2). Therefore, many engineers use a value for f, called the Darcy-Weisbach friction factor, that is four times the value of the Fanning friction factor. Use of this factor will yield greater values for friction loss.

The second major method used to calculate friction loss is the empirically derived Hazen-Williams equation:

$$V = 1.318C\,(R_h)^{0.63}\,S^{0.54} \tag{2}$$

Appendix 17.A. Continued

and
$$Q = 1.318C \, (R_h)^{0.63} \, S^{0.54} \, A \tag{3}$$
where
Q is the discharge
R_h is the hydraulic radius of the pipe (area divided by perimeter)
S is the slope of the energy grade line (h_f/L)
A is the pipe cross-sectional area
C is the roughness coefficient

Table 3 presents the friction loss of water in feet per 100 ft length of pipe using a roughness coefficient of 100. Other Hazen-Williams roughness coefficients given in Table 4 are used to modify the h_f results given in Table 3. For example, the results given in Table 3 must be multiplied by 100/130 for new steel and cast iron pipe.

The values for h_f produced by the Hazen-Williams equation are usually twice as large as those given by the Darcy-Weisbach equation using the Fanning friction factor. Many engineers use the Hazen-Williams values when designing a water supply system that is likely to incrust over time. In this case, the pipes will be somewhat larger in diameter and higher horsepower pumps will be required. For in-plant water systems or where the water is conditioned and the system maintained periodically, engineers often use the less conservative values given by the Darcy-Weisbach equation to obtain a significantly lower over-all cost.

Valves and other types of fittings in pipe lines add to the friction losses. Table 5 indicates the friction loss for fittings expressed as an equivalent length of straight pipe. Head losses reflect values based on the Darcy-Weisbach analysis.

Appendix 17.A. Continued

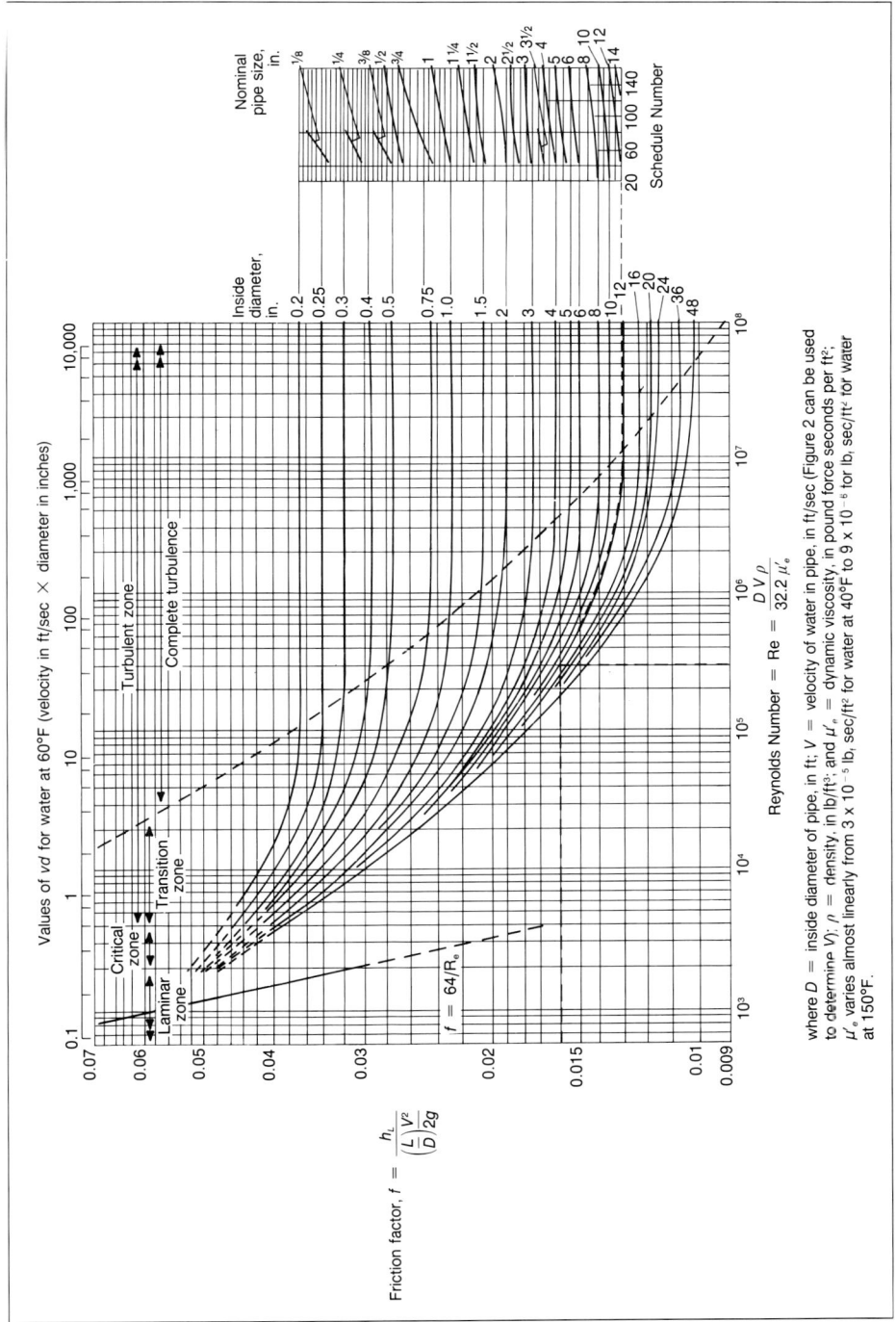

Figure 1. Friction factors for clean commercial steel pipe. *(Moody, 1944; Crane, 1976)*

Appendix 17.A. Continued

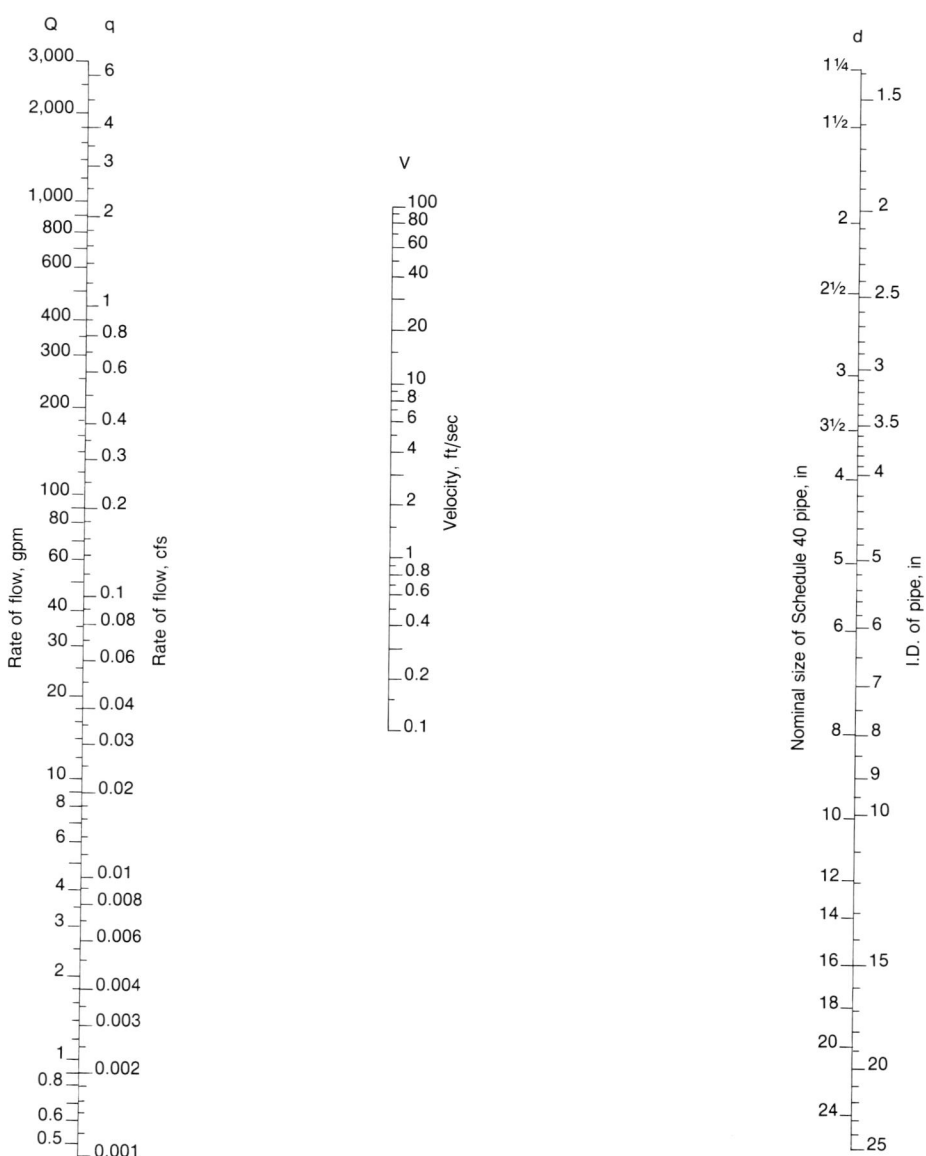

Figure 2. Nomogram for determining the velocity of liquids in pipe. *(Crane, 1976)*

Appendix 17.A. Continued

Table 1. Flow of Water Through Schedule 40 Steel Pipe

Velocity and friction loss in ft per 100 ft in Schedule 40 pipe for water at 60°F

Discharge			1/8"		1/4"		3/8"		1/2"		3/4"		1"		1 1/4"		1 1/2"	
Gallons per Minute	Cubic Ft. per Second		Velocity Feet per Second	Friction loss Feet per 100 ft.	Velocity Feet per Second	Friction loss Feet per 100 ft.	Velocity Feet per Second	Friction loss Feet per 100 ft.	Velocity Feet per Second	Friction loss Feet per 100 ft.	Velocity Feet per Second	Friction loss Feet per 100 ft.	Velocity Feet per Second	Friction loss Feet per 100 ft.	Velocity Feet per Second	Friction loss Feet per 100 ft.	Velocity Feet per Second	Friction loss Feet per 100 ft.
.2	0.000446		1.13	4.30	0.62	0.83												
.3	0.000668		1.69	9.75	0.92	2.09	0.50	0.37										
.4	0.000891		2.26	16.1	1.23	3.72	0.67	0.80										
.5	0.00111		2.82	24.3	1.54	5.52	0.84	1.25										
.6	0.00134		3.39	34.0	1.85	7.60	1.01	1.73										
.8	0.00178		4.52	57.8	2.46	12.6	1.34	2.89										
1	0.00223		5.65	85.9	3.08	19.1	1.68	4.27	1.06	1.39								
2	0.00446		11.29	310	6.16	69.5	3.36	15.2	2.11	4.85	0.60	0.36						
3	0.00668				9.25	148	5.04	32.1	3.17	10.0	1.20	1.22						
4	0.00891				12.33	257	6.72	55.2	4.22	17.1	1.81	2.52						
5	0.01114			2"			8.40	84.8	5.28	25.9	2.41	4.23						
6	0.01337		0.57	0.10		2 1/2"	10.08	120	6.33	36.5	3.01	6.35						
8	0.01782		0.77	0.17	0.67	0.11	13.44	210	8.45	64.0	3.61	8.87						
10	0.02228		0.96	0.25	1.01	0.22			10.56	97.9	4.81	15.2	2.23	2.70				
15	0.03342		1.43	0.52	1.34	0.36	0.87	0.13		3 1/2"	6.02	23.1	2.97	4.60		1 1/4"		
20	0.04456		1.91	0.87			1.09	0.19	0.81	0.09	9.03	49.9	3.71	6.91				
25	0.05570		2.39	1.30	1.68	0.54	1.30	0.26	0.97	0.13	12.03	87.3	5.57	14.7				
30	0.06684		2.87	1.82	2.01	0.76	1.52	0.35	1.14	0.17		4"	7.43	25.2				
35	0.07798		3.35	2.43	2.35	1.01	1.74	0.44	1.30	0.22	0.88	0.09	9.28	38.6	0.44	0.10	0.47	0.10
40	0.08912		3.83	3.12	2.68	1.28	1.95	0.55	1.46	0.27	1.01	0.12	11.14	55.0	0.64	0.21	0.63	0.16
45	0.1003		4.30	3.86	3.02	1.54					1.13	0.15	12.99	74.4	0.86	0.35	0.79	0.24
50	0.1114		4.78	4.69	3.35	1.94	2.17	0.67	1.62	0.33	1.26	0.18	14.85	95.9	1.07	0.52		
60	0.1337		5.74	6.63	4.02	2.73	2.60	0.94	1.95	0.47	1.51	0.25		5"	1.29	0.71	0.95	0.33
70	0.1560		6.70	8.87	4.69	3.67	3.04	1.25	2.27	0.60	1.76	0.33			1.72	1.20	1.26	0.56
80	0.1782		7.65	11.5	5.36	4.69	3.47	1.59	2.60	0.77	2.02	0.42	1.12	0.11	2.15	1.79	1.58	0.83
90	0.2005		8.60	14.3	6.03	5.84	3.91	1.99	2.92	0.96	2.27	0.52	1.28	0.14	3.22	3.77	2.37	1.74
100	0.2228		9.56	17.5	6.70	7.14	4.34	2.43	3.25	1.18	2.52	0.63	1.44	0.17	4.29	6.42	3.16	2.96
125	0.2785		11.97	27.2	8.38	10.9	5.43	3.72	4.06	1.78	3.15	0.96	1.60	0.21			3.94	4.46
150	0.3342		14.36	38.6	10.05	15.5	6.51	5.17	4.87	2.49	3.78	1.34	2.01	0.31	5.37	9.75	4.73	6.28
175	0.3899		16.75	51.5	11.73	20.7	7.60	6.93	5.68	3.33	4.41	1.79	2.41	0.44	6.44	13.7	5.52	8.41
200	0.4456		19.14	66.5	13.42	27.0	8.68	8.94	6.49	4.27	5.04	2.28	2.81	0.58	7.51	18.2	6.30	10.7
225	0.5013				15.09	33.8	9.77	11.2	7.30	5.36	5.67	2.84	3.21	0.75	8.59	23.7	7.09	13.5
250	0.557						10.85	13.7	8.12	6.56	6.30	3.37	3.61	0.93	9.67	29.6		
275	0.6127						11.94	16.5	8.93	7.85	6.93	4.13	4.01	1.14			7.88	16.5
																	9.47	23.6
300	0.6684						13.00	19.3	9.74	9.29	7.56	4.87	4.41	1.35	10.74	36.2	11.05	31.7
325	0.7241						14.12	22.8	10.53	9.45	8.19	5.71	4.81	1.58	12.89	51.3	12.62	40.6
													5.21	1.84		6"	14.20	50.8
															1.11	0.08	15.78	62.1
															1.39	0.13	19.72	95.6
															1.67	0.18		8"
															1.94	0.24	1.44	0.10
															2.22	0.30	1.60	0.12
															2.50	0.37	1.76	0.14
															2.78	0.45	1.92	0.17
															3.05	0.54	2.08	0.19
															3.33	0.64		
															3.61	0.71		

Appendix 17.A. Continued

Velocity and friction loss in ft per 100 ft in Schedule 40 pipe for water at 60°F

Discharge			2½"		3"		3½"		4"		5"		6"		8"	
Gallons per Minute	Cubic Ft. per Second	Velocity Feet per Second	Friction loss Feet per 100 ft.	Velocity Feet per Second	Friction loss Feet per 100 ft.	Velocity Feet per Second	Friction loss Feet per 100 ft.	Velocity Feet per Second	Friction loss Feet per 100 ft.	Velocity Feet per Second	Friction loss Feet per 100 ft.	Velocity Feet per Second	Friction loss Feet per 100 ft.	Velocity Feet per Second	Friction loss Feet per 100 ft.	Velocity Feet per Second	Friction loss Feet per 100 ft.
350	0.7798							11.36	12.5	8.82	6.56	5.62	2.12	3.89	0.85	2.24	0.22
375	0.8355							12.17	14.3	9.45	7.51	6.02	2.43	4.16	0.96	2.40	0.25
400	0.8912							12.98	16.2	10.08	8.50	6.42	2.75	4.44	1.09	2.56	0.28
425	0.9469							13.80	18.2	10.71	9.52	6.82	3.07	4.72	1.22	2.73	0.31
450	1.003							14.61	20.3	11.34	10.6	7.22	3.42	5.00	1.36	2.89	0.35
475	1.059									11.97	11.8	7.62	3.79	5.27	1.51	3.04	0.38
500	1.114									12.60	13.1	8.02	4.18	5.55	1.66	3.21	0.42
550	1.225									13.85	15.7	8.82	5.01	6.11	1.99	3.53	0.51
600	1.337									15.12	18.6	9.63	5.89	6.66	2.36	3.85	0.60
650	1.448											10.43	6.88	7.22	2.73	4.17	0.70
700	1.560			2.01	0.11							11.23	7.92	7.78	3.12	4.49	0.79
750	1.671			2.15	0.12							12.03	9.06	8.33	3.58	4.81	0.91
800	1.782			2.29	0.14							12.83	10.2	8.88	4.04	5.13	1.02
850	1.894			2.44	0.16							13.64	11.6	9.44	4.53	5.45	1.15
900	2.005			2.58	0.17							14.44	12.9	9.99	5.04	5.77	1.28
950	2.117			2.72	0.19	2.02	0.10					15.24	14.3	10.55	5.59	6.09	1.42
1 000	2.228			2.87	0.21	2.13	0.11					16.04	15.8	11.10	6.19	6.41	1.56
1 100	2.451			3.15	0.25	2.25	0.12					17.65	19.0	12.22	7.44	7.05	1.86
1 200	2.674			3.44	0.30	2.37	0.13							13.33	8.80	7.70	2.19
1 300	2.896			3.73	0.35	2.61	0.15							14.43	10.3	8.33	2.56
1 400	3.119			4.01	0.40	2.85	0.18							15.55	11.9	8.98	2.96
1 500	3.342			4.30	0.45	3.08	0.21							16.66	13.5	9.62	3.37
1 600	3.565			4.59	0.51	3.32	0.25	2.54	0.13					17.77	15.3	10.26	3.81
1 800	4.010			5.16	0.64	3.56	0.28	2.72	0.15					19.99	19.3	11.54	4.80
2 000	4.456			5.73	0.78	3.79	0.32	2.90	0.16					22.21	23.8	12.82	5.89
						4.27	0.40	3.27	0.20								
						4.74	0.48	3.63	0.25								
2 500	5.570			7.17	1.19	5.93	0.74	4.54	0.38	2.58	0.12			15.55	11.9	16.03	9.10
3 000	6.684			8.60	1.69	7.11	1.04	5.45	0.54	2.87	0.14			16.66	13.5	19.24	12.9
3 500	7.798			10.03	2.27	8.30	1.40	6.35	0.72	3.59	0.21			17.77	15.3	22.44	17.5
4 000	8.912			11.47	2.93	9.48	1.82	7.26	0.93	4.30	0.30			19.99	19.3	25.65	22.6
4 500	10.03			12.90	3.70	10.67	2.29	8.17	1.16	5.02	0.40			22.21	23.8	28.87	28.2

Discharge		10"		12"		14"		16"		18"		20"		24"			
Gallons per Minute	Cubic Ft. per Second	Velocity Feet per Second	Friction loss Feet per 100 ft.	Velocity Feet per Second	Friction loss Feet per 100 ft.	Velocity Feet per Second	Friction loss Feet per 100 ft.	Velocity Feet per Second	Friction loss Feet per 100 ft.	Velocity Feet per Second	Friction loss Feet per 100 ft.	Velocity Feet per Second	Friction loss Feet per 100 ft.	Velocity Feet per Second	Friction loss Feet per 100 ft.		
500	1.114	1.93	0.12														
550	1.225	2.03	0.14														
600	1.337	2.24	0.16														
650	1.448	2.44	0.19														
		2.64	0.22														
700	1.560	2.85	0.26														
750	1.671	3.05	0.29														
800	1.782	3.25	0.33														
850	1.894	3.46	0.37														
900	2.005	3.66	0.41														
950	2.117	3.86	0.46														
1 000	2.228	4.07	0.50														
1 100	2.451	4.48	0.60														
1 200	2.674	4.88	0.71														
1 300	2.896	5.29	0.82														
1 400	3.119	5.70	0.94														
1 500	3.342	6.10	1.08														
1 600	3.565	6.51	1.22														
1 800	4.010	7.32	1.53														
2 000	4.456	8.14	1.87														
2 500	5.570	10.17	2.86	7.17	1.19												
3 000	6.684	12.20	4.07	8.60	1.69												
3 500	7.798	14.24	5.50	10.03	2.27												
4 000	8.912	16.27	7.11	11.47	2.93												
4 500	10.03	18.31	8.94	12.90	3.70												
5 000	11.14	20.35	10.9	14.33	4.50	11.85	2.80	9.08	1.43	7.17	0.79	5.77	0.46	3.99	0.18		
6 000	13.37	24.41	15.6	17.20	6.40	14.23	3.95	10.89	2.03	8.61	1.12	6.93	0.65	4.79	0.26	3.19	0.12
7 000	15.60	28.49	21.0	20.07	8.64	16.60	5.34	12.71	2.73	10.04	1.51	8.08	0.87	5.59	0.35	3.59	0.15
8 000	17.82			22.93	11.2	18.96	6.91	14.52	3.49	11.47	1.94	9.23	1.13	6.38	0.44		
9 000	20.05			25.79	14.1	21.34	8.69	16.34	4.39	12.91	2.43	10.39	1.40	7.18	0.56		
10 000	22.28			28.66	17.2	23.71	10.6	18.15	5.41	14.34	2.96	11.54	1.71	7.98	0.68		
12 000	26.74			34.40	24.7	28.45	15.2	21.79	7.69	17.21	4.23	13.85	2.45	9.58	0.96		
14 000	31.19					33.19	20.5	25.42	10.4	20.08	5.66	16.16	3.30	11.17	1.30		
16 000	35.65							29.05	13.5	22.05	7.35	18.47	4.27	12.77	1.67		
18 000	40.10							32.68	16.9	25.82	9.31	20.77	5.36	14.36	2.10		
20 000	44.56							36.31	20.9	28.69	11.4	23.08	6.61	15.96	2.59		

Appendix 17.A. Continued

Table 2. Increase in Friction Loss Due to Aging of Pipe
(Multipliers for use with Table 1)

Age of Pipe in Years	Small Pipes 4-10 in (102-254 mm)	Large Pipes 12-60 in (305-1,520 mm)
New	1.00	1.00
5	1.40	1.30
10	2.20	1.60
15	3.60	1.80
20	5.00	2.00
25	6.30	2.10
30	7.25	2.20
35	8.10	2.30
40	8.75	2.40
45	9.25	2.60
50	9.60	2.86
55	9.80	3.26
60	10.00	3.70
65	10.05	4.25
70	10.10	4.70

It is obvious that there is no sudden increase in aging effect between 12-in (305-mm) and 10-in (254-mm) pipe as indicated in Table 2. The values shown are composites of many tests grouped by the experimenter. A reasonable amount of interpretation and logic must be used in selecting and applying a multiplier for each specific problem. It should also be borne in mind that some test data on aging of pipe may vary up to 50 percent from the averages shown in Table 2.

(Colt Industries, 1974)

Appendix 17.A. Continued

Table 3. Friction Loss of Water in Feet per 100 Feet Length of Pipe Based on Hazen–Williams Equation, Using Contact of 100

Pipe Size U.S. Gals. per Minute	1/2" Vel. ft. per Sec.	1/2" Head Loss in Ft.	3/4" Vel. ft. per Sec.	3/4" Head Loss in Ft.	1" Vel. ft. per Sec.	1" Head Loss in Ft.	1 1/4" Vel. ft. per Sec.	1 1/4" Head Loss in Ft.	1 1/2" Vel. ft. per Sec.	1 1/2" Head Loss in Ft.	2" Vel. ft. per Sec.	2" Head Loss in Ft.	2 1/2" Vel. ft. per Sec.	2 1/2" Head Loss in Ft.	3" Vel. ft. per Sec.	3" Head Loss in Ft.	4" Vel. ft. per Sec.	4" Head Loss in Ft.	5" Vel. ft. per Sec.	5" Head Loss in Ft.	6" Vel. ft. per Sec.	6" Head Loss in Ft.	Pipe Size U.S. Gals. per Minute
2	2.10	7.4	1.20	1.9																			2
4	4.21	27.0	2.41	7.0																			4
6	6.31	57.0	3.61	14.7	1.49	2.14	.86	.57															6
8	8.42	98.0	4.81	25.0	2.23	4.55	1.29	1.20	.63	.26													8
10	10.52	147.0	6.02	38.0	2.98	7.8	1.72	2.03	.94	.56	.61	.20											10
12			7.22	53.0	3.72	11.7	2.14	3.05	1.26	.95	.82	.33	.52	.11									12
15			9.02	80.0	4.46	16.4	2.57	4.3	1.57	1.43	1.02	.50	.65	.17	.45	.07							15
18			10.84	108.2	5.60	25.0	3.21	6.5	1.89	2.01	1.23	.79	.78	.23	.54	.10							18
20			12.03	136.0	6.69	35.0	3.86	9.1	2.36	3.00	1.53	1.08	.98	.36	.68	.15							20
25					7.44	42.0	4.29	11.1	2.83	4.24	1.84	1.49	1.18	.50	.82	.21	.51	.06					25
30					9.30	64.0	5.36	16.6	3.15	5.20	2.04	1.82	1.31	.61	.91	.25	.64	.09					30
35					11.15	89.0	6.43	23.0	3.80	7.30	2.55	2.73	1.63	.92	1.13	.38	.77	.13	.49	.04			35
40					13.02	119.0	7.51	31.2	4.72	11.0	3.06	3.84	1.96	1.29	1.36	.54	.89	.17	.57	.06			40
45					14.88	152.0	8.58	40.0	5.51	14.7	3.57	5.10	2.29	1.72	1.59	.71	1.02	.22	.65	.08	.57	.04	45
50							9.65	50.0	6.30	18.8	4.08	6.6	2.61	2.20	1.82	.91	1.15	.28	.73	.09	.62	.05	50
55							10.72	60.0	7.08	23.2	4.60	8.2	2.94	2.80	2.04	1.15	1.28	.34	.82	.11	.68	.06	55
60							11.78	72.0	7.87	28.4	5.11	9.9	3.27	3.32	2.27	1.38	1.41	.41	.90	.14	.68	.06	60
65							12.87	85.0	8.66	34.0	5.62	11.8	3.59	4.01	2.45	1.58	1.53	.47	.98	.16	.74	.076	65
70							13.92	99.7	9.44	39.6	6.13	13.9	3.92	4.65	2.72	1.92	1.66	.53	1.06	.19	.74	.076	70
75									10.23	45.9	6.64	16.1	4.24	5.4	2.89	2.16	1.79	.63	1.14	.21	.79	.08	75
80							15.01	113.0	11.02	53.0	7.15	18.4	4.58	6.2	3.18	2.57	1.91	.73	1.22	.24	.85	.10	80
85							16.06	129.0	11.80	60.0	7.66	20.9	4.91	7.1	3.33	3.00	2.04	.81	1.31	.27	.91	.11	85
90							17.16	145.0	12.59	68.0	8.17	23.7	5.23	7.9	3.63	3.28	2.17	.91	1.39	.31	.96	.12	90
95							18.21	163.8	13.38	75.0	8.68	26.5	5.56	8.1	3.78	3.54	2.30	1.00	1.47	.34	1.02	.14	95
100			8" PIPE				19.30	180.0	14.71	84.0	9.19	29.4	5.88	9.8	4.09	4.08	2.42	1.12	1.55	.38	1.08	.15	100
110									14.95	93.0	9.70	32.6	6.21	10.8	4.22	4.33	2.55	1.22	1.63	.41	1.13	.17	110
120			.90	.08					15.74	102.0	10.21	35.8	6.54	12.0	4.54	4.96	2.81	1.46	1.79	.49	1.25	.21	120
130			.96	.09					17.31	122.0	11.23	42.9	7.18	14.5	5.00	6.0	3.06	1.77	1.96	.58	1.36	.24	130
140									18.89	143.0	12.25	50.0	7.84	16.8	5.45	7.0	3.31	1.97	2.12	.67	1.47	.27	140
150									20.46	166.0	13.28	58.0	8.48	18.7	5.91	8.1	3.57	2.28	2.29	.76	1.59	.32	150
160			1.02	.10					22.04	190.0	14.30	67.0	9.15	22.3	6.35	9.2	3.82	2.62	2.45	.88	1.70	.36	160
170			1.08	.11	10" PIPE						15.32	76.0	9.81	25.5	6.82	10.5	4.08	2.91	2.61	.98	1.82	.40	170
180			1.15	.13							16.34	86.0	10.46	29.0	7.26	11.8	4.33	3.26	2.77	1.08	1.92	.45	180
190			1.21	.14							17.36	96.0	11.11	34.1	7.71	13.3	4.60	3.61	2.94	1.22	2.04	.50	190
200			1.28	.15							18.38	107.0	11.76	35.7	8.17	14.0	4.84	4.01	3.10	1.35	2.16	.55	200
220			1.40	.18							19.40	118.0	12.42	39.6	8.63	15.5	5.11	4.4	3.27	1.48	2.27	.62	220
											20.42	129.0	13.07	43.1	9.08	17.8	5.62	5.2	3.59	1.77	2.50	.73	
240			1.53	.22							22.47	154.0	14.38	52.0	9.99	21.3	6.13	6.2	3.92	2.08	2.72	.87	240
260			1.66	.25							24.52	182.0	15.69	61.0	10.89	25.1	6.64	7.2	4.25	2.41	2.95	1.00	260
280			1.79	.28							26.55	211.0	16.99	70.0	11.80	29.1	7.15	8.2	4.58	2.77	3.18	1.14	280
300			1.91	.32			12" PIPE						18.30	81.0	12.71	33.4	7.66	9.3	4.90	3.14	3.40	1.32	300
320			2.05	.37									19.61	92.0	13.62	38.0	8.17	10.5	5.23	3.54	3.64	1.47	320
340			2.18	.41									20.92	103.0	14.52	42.8	8.68	11.7	5.54	3.97	3.84	1.62	340
													22.22	116.0	15.43	47.9							

Appendix 17.A. Continued

Pipe Size U.S. Gals. per Minute	8" Vel. ft. per Sec.	8" Head Loss in Ft.	10" Vel. ft. per Sec.	10" Head Loss in Ft.	12" Vel. ft. per Sec.	12" Head Loss in Ft.	14" Vel. ft. per Sec.	14" Head Loss in Ft.	1½" Vel. ft. per Sec.	1½" Head Loss in Ft.	2" Vel. ft. per Sec.	2" Head Loss in Ft.	2½" Vel. ft. per Sec.	2½" Head Loss in Ft.	3" Vel. ft. per Sec.	3" Head Loss in Ft.	4" Vel. ft. per Sec.	4" Head Loss in Ft.	5" Vel. ft. per Sec.	5" Head Loss in Ft.	6" Vel. ft. per Sec.	6" Head Loss in Ft.	Pipe Size U.S. Gals. per Minute
360	2.30	.45	1.47	.15		14" PIPE							23.53	128.0	16.34	53.0	9.19	13.1	5.87	4.41	4.08	1.83	360
380	2.43	.50	1.55	.17		.069							24.84	142.0	17.25	59.0	9.69	14.0	6.19	4.86	4.31	2.00	380
400	2.60	.54	1.63	.19	1.08	.075							26.14	156.0	18.16	65.0	10.21	16.0	6.54	5.4	4.55	2.20	400
450	2.92	.68	1.84	.23	1.14	.095									20.40	78.0	11.49	19.8	7.35	6.7	5.11	2.74	450
500	3.19	.82	2.04	.28	1.28	.113	1.04	.06							22.70	98.0	12.77	24.0	8.17	8.1	5.68	2.90	500
550	3.52	.97	2.24	.33	1.42	.135	1.15	.07							24.96	117.0	14.04	28.7	8.99	9.6	6.25	3.96	550
600	3.84	1.14	2.45	.39	1.56	.159	1.25	.08									15.32	33.7	9.80	11.3	6.81	4.65	600
650	4.17	1.34	2.65	.45	1.70	.19	1.37	.09							77.23	137.0	16.59	39.0	10.62	13.2	7.38	5.40	650
700	4.46	1.54	2.86	.52	1.84	.22	1.46	.10									17.87	44.9	11.44	15.1	7.95	6.21	700
750	4.80	1.74	3.06	.59	1.99	.24	1.58	.11		16" PIPE							19.15	51.0	12.26	17.2	8.50	7.12	750
800	5.10	1.90	3.26	.66	2.13	.27	1.67	.13									20.42	57.0	13.07	19.4	9.08	7.96	800
850	5.48	2.20	3.47	.75	2.27	.31	1.79	.14	1.36	.08							21.70	64.0	13.89	21.7	9.65	8.95	850
900	5.75	2.46	3.67	.83	2.41	.34	1.88	.16	1.44	.084							22.98	71.0	14.71	24.0	10.20	10.11	900
950	6.06	2.87	3.88	.91	2.56	.38	2.00	.18	1.52	.095		20" PIPE							15.52	26.7	10.77	11.20	950
1000	6.38	2.97	4.08	1.03	2.70	.41	2.10	.19	1.60	.10	1.02	.04							16.34	29.2	11.34	12.04	1000
1100	7.03	3.52	4.49	1.19	2.84	.49	2.31	.23	1.76	.12	1.12	.04							17.97	34.9	12.48	14.55	1100
1200	7.66	4.17	4.90	1.40	3.13	.58	2.52	.27	1.92	.14	1.23	.05							19.61	40.9	13.61	17.10	1200
1300	8.30	4.85	5.31	1.62	3.41	.67	2.71	.32	2.08	.17	1.33	.06									14.72	18.4	1300
1400	8.95	5.50	5.71	1.87	3.69	.78	2.92	.36	2.24	.19	1.43	.064									15.90	22.60	1400
1500	9.58	6.24	6.12	2.13	3.98	.89	3.15	.41	2.39	.21	1.53	.07		24" PIPE							17.02	25.60	1500
1600	10.21	7.00	6.53	2.39	4.26	.98	3.34	.47	2.56	.24	1.63	.08	1.28	.04		30" PIPE					18.10	26.9	1600
1800	11.50	8.78	7.35	2.95	4.55	1.21	3.75	.58	2.87	.30	1.84	.10	1.42	.05									1800
2000	12.78	10.71	8.16	3.59	5.11	1.49	4.17	.71	3.19	.37	2.04	.12	1.56	.06									2000
2200	14.05	12.78	8.98	4.24	5.68	1.81	4.59	.84	3.51	.44	2.25	.15											2200
2400	15.32	14.2	9.80	5.04	6.25	2.08	5.00	.99	3.83	.52	2.45	.17	1.70	.07	1.09	.02							2400
2600			10.61	5.81	6.81	2.43	5.47	1.17	4.15	.60	2.66	.20	1.84	.08	1.16	.027							2600
2800			11.41	6.70	7.38	2.75	5.84	1.32	4.47	.68	2.86	.23	1.98	.09	1.27	.03							2800
3000			12.24	7.62	7.95	3.15	6.01	1.49	4.79	.78	3.08	.27	2.13	.10	1.37	.037							3000
3200			13.05	7.80	8.52	3.51	6.68	1.67	5.12	.88	3.27	.30	2.26	.12	1.46	.041							3200
3500			14.30	10.08	9.10	4.16	7.30	1.97	5.59	1.04	3.59	.35	2.49	.14	1.56	.047							3500
3800					9.95	4.90	7.98	2.36	6.07	1.20	3.88	.41	2.69	.17	1.73	.05							3800
4200					10.80	5.88	8.76	2.77	6.70	1.44	4.29	.49	2.99	.20	1.91	.07							4200
4500			15.51	13.4	11.92	6.90	9.45	3.22	7.18	1.64	4.60	.56	3.20	.22	2.04	.08							4500
5000					12.78	8.40	10.50	3.92	8.01	2.03	5.13	.68	3.54	.27	2.26	.09							5000
5500					14.20		11.55	4.65	8.78	2.39	5.64	.82	3.90	.33	2.50	.11							5500
6000							12.60	5.50	9.58	2.79	6.13	.94	4.25	.38	2.73	.13							6000
6500							13.65	6.45	10.39	3.32	6.64	1.10	4.61	.45	2.96	.15							6500
7000							14.60	7.08	11.18	3.70	7.15	1.25	4.97	.52	3.18	.17							7000
8000									12.78	4.74	8.17	1.61	5.68	.66	3.64	.23							8000
9000									14.37	5.90	9.20	2.01	6.35	.81	4.08	.28							9000
10000									15.96	7.19	10.20	2.44	7.07	.98	4.54	.33							10000
12000											12.25	3.41	8.50	1.40	5.46	.48							12000

Note: Sizes of standard pipe in inches.

Appendix 17.A. Continued

Table 4. Hazen-Williams Roughness Values

Types of Pipe	Roughness Value
Extremely smooth pipes	140
New steel or cast iron	130
Wood, average concrete	120
New riveted steel, clay	110
Old cast iron, brick	100
Old steel riveted	95
Badly corroded cast iron	80
Very badly corroded iron or steel	60

(Olson, 1966)

Appendix 17.A. Continued

Table 5. Equivalent Length of Straight Pipe for Various Fittings, Turbulent Flow Only

Fittings			1/4	3/8	1/2	3/4	1	1¼	1½	2	2½	3	4	5	6	8	10	12	14	16	18	20	24
Regular 90° Ell	Screwed	Steel	2.3	3.1	3.6	4.4	5.2	6.6	7.4	8.5	9.3	11.0	13.0										
		C.I.										9.0	11.0											
	Flanged	Steel			0.92	1.2	1.6	2.1	2.4	3.1	3.6	4.4	5.9	7.3	8.9	12.0	14.0	17.0	18.0	21.0	23.0	25.0	30.0
		C.I.										3.6	4.8		7.2	9.8	12.0	15.0	17.0	19.0	22.0	24.0	28.0
Long Radius 90° Ell	Screwed	Steel	1.5	2.0	2.2	2.3	2.7	3.2	3.4	3.6	3.6	4.0	4.6										
		C.I.										3.3	3.7											
	Flanged	Steel			1.1	1.3	1.6	2.0	2.3	2.7	2.9	3.4	4.2	5.0	5.7	7.0	8.0	9.0	9.4	10.0	11.0	12.0	14.0
		C.I.										2.8	3.4		4.7	5.7	6.8	7.8	8.6	9.6	11.0	11.0	13.0
Regular 45° Ell	Screwed	Steel	0.34	0.52	0.71	0.92	1.3	1.7	2.1	2.7	3.2	4.0	5.5										
		C.I.										3.3	4.5											
	Flanged	Steel			0.45	0.59	0.81	1.1	1.3	1.7	2.0	2.6	3.5	4.5	5.6	7.7	9.0	11.0	13.0	15.0	16.0	18.0	22.0
		C.I.										2.1	2.9		4.5	6.3	8.1	9.7	12.0	13.0	15.0	17.0	20.0
Tee-Line Flow	Screwed	Steel	0.79	1.2	1.7	2.4	3.2	4.6	5.6	7.7	9.3	12.0	17.0										
		C.I.										9.9	14.0											
	Flanged	Steel			0.69	0.82	1.0	1.3	1.5	1.8	1.9	2.2	2.8	3.3	3.8	4.7	5.2	6.0	6.4	7.2	7.6	8.2	9.6
		C.I.										1.9	2.2		3.1	3.9	4.6	5.2	5.9	6.5	7.2	7.7	8.8
Tee-Branch Flow	Screwed	Steel	2.4	3.5	4.2	5.3	6.6	8.7	9.9	12.0	13.0	17.0	21.0										
		C.I.										14.0	17.0											
	Flanged	Steel			2.0	2.6	3.3	4.4	5.2	6.6	7.5	9.4	12.0	15.0	18.0	24.0	30.0	34.0	37.0	43.0	47.0	52.0	62.0
		C.I.										7.7	10.0		15.0	20.0	25.0	30.0	35.0	39.0	44.0	49.0	57.0
180° Return Bend	Screwed	Steel	2.3	3.1	3.6	4.4	5.2	6.6	7.4	8.5	9.3	11.0	13.0										
		C.I.										9.0	11.0											
	Reg. Flanged	Steel			0.92	1.2	1.6	2.1	2.4	3.1	3.6	4.4	5.9	7.3	8.9	12.0	14.0	17.0	18.0	21.0	23.0	25.0	30.0
		C.I.										3.6	4.8		7.2	9.8	12.0	15.0	17.0	19.0	22.0	24.0	28.0
	Long Rod Flanged	Steel			1.1	1.3	1.6	2.0	2.3	2.7	2.9	3.4	4.2	5.0	5.7	7.0	8.0	9.0	9.4	10.0	11.0	12.0	14.0
		C.I.										2.8	3.4		4.7	5.7	6.8	7.8	8.6	9.6	11.0	11.0	13.0
Globe Valve	Screwed	Steel	21.0	22.0	22.0	24.0	29.0	37.0	42.0	54.0	62.0	79.0	110.0	150.0	190.0	260.0	310.0	390.0					
		C.I.										65.0	86.0											
	Flanged	Steel			38.0	40.0	45.0	54.0	59.0	70.0	77.0	94.0	120.0		150.0	210.0	270.0	330.0						
		C.I.										77.0	99.0											

Appendix 17.A. Continued

TABLE 5. Equivalent Length of Straight Pipe for Various Fittings, Turbulent Flow Only

Fittings			1/4	3/8	1/2	3/4	1	1 1/4	1 1/2	2	2 1/2	3	4	5	6	8	10	12	14	16	18	20	24
Gate Valve	Screwed	Steel	0.32	0.45	0.56	0.67	0.84	1.1	1.2	1.5	1.7	1.9	2.5										
		C.I.										1.6	2.0										
	Flanged	Steel								2.6	2.7	2.8	2.9	3.1	3.2	3.2	3.2	3.2	3.2	3.2	3.2	3.2	3.2
		C.I.										2.3	2.4		2.6	2.7	2.8	2.9	2.9	3.0	3.0	3.0	3.0
Angle Valve	Screwed	Steel	12.8	15.0	15.0	15.0	17.0	18.0	18.0	18.0	18.0	18.0	18.0										
		C.I.										15.0	15.0										
	Flanged	Steel			15.0	15.0	17.0	18.0	18.0	21.0	22.0	28.0	38.0	50.0	63.0	90.0	120.0	140.0	160.0	190.0	210.0	240.0	300.0
		C.I.										23.0	31.0		52.0	74.0	98.0	120.0	150.0	170.0	200.0	230.0	280.0
Swing Check Valve	Screwed	Steel	7.2	7.3	8.0	8.8	11.0	13.0	15.0	19.0	22.0	27.0	38.0										
		C.I.										22.0	31.0										
	Flanged	Steel			3.8	5.3	7.2	10.0	12.0	17.0	21.0	27.0	38.0	50.0	63.0	90.0	120.0	140.0					
		C.I.										22.0	31.0		52.0	74.0	98.0	120.0					
Coupling or Union	Screwed	Steel	0.14	0.18	0.21	0.24	0.29	0.36	0.39	0.45	0.47	0.53	0.65										
		C.I.										0.44	0.52										
Bell Mouth Inlet		Steel	0.04	0.07	0.10	0.13	0.18	0.26	0.31	0.43	0.52	0.67	0.95	1.3	1.6	2.3	2.9	3.5	4.0	4.7	5.3	6.1	7.6
		C.I.										0.55	0.77		1.3	1.9	2.4	3.0	3.6	4.3	5.0	5.7	7.0
Square Mouth Inlet		Steel	0.44	0.68	0.96	1.3	1.8	2.6	3.1	4.3	5.2	6.7	9.5	13.0	16.0	23.0	29.0	35.0	40.0	47.0	53.0	61.0	76.0
		C.I.										5.5	7.7		13.0	19.0	24.0	30.0	36.0	43.0	50.0	57.0	70.0
Re-entrant Pipe		Steel	0.88	1.4	1.9	2.6	3.6	5.1	6.2	8.5	10.0	13.0	19.0	25.0	32.0	45.0	58.0	70.0	80.0	95.0	110.0	120.0	150.0
		C.I.										11.0	15.0		26.0	37.0	49.0	61.0	73.0	86.0	100.0	110.0	140.0
Sudden Enlargement			$h = \dfrac{(V_1-V_2)^2}{2g}$ feet of fluid; if $V_2 = 0$ $h = \dfrac{V_1^2}{2g}$ feet of fluid																				

(Hydraulic Institute, 1961)

APPENDIX 17.B.
Volume of Water Discharged per Stroke by a Reciprocating Pump

Diam. of Pump Cyl., in	Length of stroke in inches with capacity per stroke in U.S. gallons													Area of Circle (Pump Cyl.) in^2
	1	2	3	4	5	6	7	8	9	10	12	18	24	
1	.0034	.0068	.0102	.0136	.0170	.0204	.0238	.0272	.0306	.0340	.0408	.062	.082	.7854
1¼	.0053	.0106	.0159	.0212	.0266	.0319	.0372	.0425	.0477	.0531	.0637	.096	.127	1.2271
1½	.0076	.0153	.0229	.0306	.0382	.0459	.0535	.0612	.0688	.0765	.0918	.138	.184	1.7671
1¾	.0104	.0208	.0312	.0416	.0521	.0625	.0729	.0833	.0937	.1041	.1249	.188	.25	2.4043
1⅞	.0120	.0239	.0359	.0478	.0598	.0718	.0748	.0957	.1080	.1196	.1435	.215	.287	2.7612
2	.0136	.0272	.0408	.0544	.0680	.0816	.0952	.1088	.1224	.1360	.1632	.244	.326	3.1416
2¼	.0172	.0344	.0516	.0688	.0860	.1033	.1205	.1377	.1549	.1721	.2071	.310	.413	3.9760
2½	.0212	.0425	.0637	.0850	.1062	.1275	.1487	.1700	.1912	.2125	.2550	.384	.51	4.9087
2¾	.0257	.0514	.0771	.1028	.1285	.1543	.1800	.2057	.2314	.2571	.3085	.462	.617	5.9395
3	.0306	.0612	.0918	.1224	.1530	.1836	.2142	.2448	.2754	.3060	.3672	.550	.734	7.0686
3¼	.0359	.0719	.1078	.1438	.1795	.2156	.2515	.2875	.3232	.3594	.4313	.648	.862	8.2957
3½	.0416	.0833	.1249	.1666	.2082	.2499	.2915	.3332	.3748	.4165	.4998	.750	1.000	9.6211
3¾	.0479	.0957	.1435	.1914	.2393	.2871	.3350	.3828	.4302	.4785	.5743	.858	1.147	11.044
4	.0544	.1088	.1632	.2176	.2720	.3264	.3808	.4352	.4896	.5440	.6528	.98	1.306	12.566
4¼		.123	.184	.246	.307	.368	.430	.491	.553	.614	.737	1.106	1.473	14.186
4½	.0688	.1377	.2065	.2754	.3442	.4131	.4819	.5508	.6196	.6885	.8262	1.24	1.652	15.904
4¾	.0767	.1534	.2301	.3068	.3835	.4602	.5369	.6137	.6904	.7671	.9205	1.38	1.84	17.720
5	.0850	.1700	.2550	.3400	.4250	.5100	.5950	.6800	.7650	.8500	1.0200	1.530	2.04	19.635
5¼		.187	.281	.375	.467	.562	.656	.75	.842	.937	1.124	1.684	2.248	21.648
5½	.1028	.2057	.3085	.4114	.5142	.6171	.7199	.8228	.9256	1.0285	1.2342	1.852	2.468	23.758
5¾	.1124	.2248	.3372	.4496	.5620	.6744	.7868	.8992	1.0116	1.1240	1.3488	2.024	2.696	25.967
6	.1224	.2448	.3672	.4896	.6120	.7344	.8568	.9792	1.1016	1.2240	1.4688	2.264	2.938	28.274
7	.1666	.3332	.4998	.6664	.8330	.9996	1.1662	1.3328	1.4994	1.6660	1.9992			38.484
8	.2176	.4352	.6528	.8704	1.0880	1.3056	1.5232	1.7408	1.9584	2.1760	2.6112			50.265
9	.2754	.5508	.8262	1.1016	1.3770	1.6524	1.9278	2.2032	2.4786	2.7540	3.3048			63.617
10	.3400	.6800	1.0200	1.3600	1.7000	2.0400	2.3800	2.7200	3.0600	3.4000	4.0800			78.540

(Dempster Industries, Inc.)

To obtain the capacity of a pump with a cylinder diameter given in the table, but with a stroke longer than 24 in, add or multiply the capacity of shorter strokes to represent the required length of stroke. For example, the capacity of a cylinder with a 30-in stroke would be the same as that of a 24-in stroke cylinder of the same diameter added to the capacity of a 6-in stroke cylinder. The same result may be obtained by multiplying the capacity of a cylinder with a 6-in stroke by 5.

To obtain the amount of water discharged per minute, multiply the capacity per stroke by the number of strokes per minute.

To obtain the amount of water discharged per minute for single-cylinder, double-action pumps, multiply gallons per stroke by 2. For duplex, double-action pumps, multiply gallons per stroke by 4.

APPENDIX 17.C.
Power Required for Pumping

Gals. per Min.	Theoretical Horsepower Required to Raise Water (at 60 F) To Different Heights											
	5 feet	10 feet	15 feet	20 feet	25 feet	30 feet	35 feet	40 feet	45 feet	50 feet	60 feet	70 feet
5	0.006	0.013	0.019	0.025	0.032	0.038	0.044	0.051	0.057	0.063	0.076	0.088
10	0.013	0.025	0.038	0.051	0.063	0.076	0.088	0.101	0.114	0.126	0.152	0.177
15	0.019	0.038	0.057	0.076	0.095	0.114	0.133	0.152	0.171	0.190	0.227	0.265
20	0.025	0.051	0.076	0.101	0.126	0.152	0.177	0.202	0.227	0.253	0.303	0.354
25	0.032	0.063	0.095	0.126	0.158	0.190	0.221	0.253	0.284	0.316	0.379	0.442
30	0.038	0.076	0.114	0.152	0.190	0.227	0.265	0.303	0.341	0.379	0.455	0.531
35	0.044	0.088	0.133	0.177	0.221	0.265	0.310	0.354	0.398	0.442	0.531	0.619
40	0.051	0.101	0.152	0.202	0.253	0.303	0.354	0.404	0.455	0.505	0.606	0.707
45	0.057	0.114	0.171	0.227	0.284	0.341	0.398	0.455	0.512	0.568	0.682	0.796
50	0.063	0.126	0.190	0.253	0.316	0.379	0.442	0.505	0.568	0.632	0.758	0.884
60	0.076	0.152	0.227	0.303	0.379	0.455	0.531	0.606	0.682	0.758	0.910	1.061
70	0.088	0.177	0.265	0.354	0.442	0.531	0.619	0.707	0.796	0.884	1.061	1.238
80	0.101	0.202	0.303	0.404	0.505	0.606	0.707	0.808	0.910	1.011	1.213	1.415
90	0.114	0.227	0.341	0.455	0.568	0.682	0.796	0.910	1.023	1.137	1.364	1.592
100	0.126	0.253	0.379	0.505	0.632	0.758	0.884	1.011	1.137	1.263	1.516	1.768
125	0.158	0.316	0.474	0.632	0.790	0.947	1.105	1.263	1.421	1.579	1.895	2.211
150	0.190	0.379	0.568	0.758	0.947	1.137	1.326	1.516	1.705	1.895	2.274	2.653
175	0.221	0.442	0.663	0.884	1.105	1.326	1.547	1.768	1.990	2.211	2.653	3.095
200	0.253	0.505	0.758	1.011	1.263	1.516	1.768	2.021	2.274	2.526	3.032	3.537
250	0.316	0.632	0.947	1.263	1.579	1.895	2.211	2.526	2.842	3.158	3.790	4.421
300	0.379	0.758	1.137	1.516	1.895	2.274	2.653	3.032	3.411	3.790	4.548	5.305
350	0.442	0.884	1.326	1.768	2.211	2.653	3.095	3.537	3.979	4.421	5.305	6.190
400	0.505	1.011	1.516	2.021	2.526	3.032	3.537	4.042	4.548	5.053	6.063	7.074
500	0.632	1.263	1.895	2.526	3.158	3.790	4.421	5.053	5.684	6.316	7.579	8.842

Appendix 17.C. Continued

Gals. per Min.	Theoretical Horsepower Required to Raise Water (at 60 F) To Different Heights										
	80 feet	90 feet	100 feet	125 feet	150 feet	175 feet	200 feet	250 feet	300 feet	350 feet	400 feet
5	0.101	0.114	0.126	0.158	0.190	0.221	0.253	0.316	0.379	0.442	0.505
10	0.202	0.227	0.253	0.316	0.379	0.442	0.505	0.632	0.758	0.884	1.011
15	0.303	0.341	0.379	0.474	0.568	0.663	0.758	0.947	1.137	1.326	1.516
20	0.404	0.455	0.505	0.632	0.758	0.884	1.011	1.263	1.516	1.768	2.021
25	0.505	0.568	0.632	0.790	0.947	1.105	1.263	1.579	1.895	2.211	2.526
30	0.606	0.682	0.758	0.947	1.137	1.326	1.516	1.895	2.274	2.653	3.032
35	0.707	0.796	0.884	1.105	1.326	1.547	1.768	2.211	2.653	3.095	3.537
40	0.808	0.910	1.011	1.263	1.516	1.768	2.021	2.526	3.032	3.537	4.042
45	0.910	1.023	1.137	1.421	1.705	1.990	2.274	2.842	3.411	3.979	4.548
50	1.011	1.137	1.263	1.579	1.895	2.211	2.526	3.158	3.790	4.421	5.053
60	1.213	1.364	1.516	1.895	2.274	2.653	3.032	3.790	4.548	5.305	6.063
70	1.415	1.592	1.768	2.211	2.653	3.095	3.537	4.421	5.305	6.190	7.074
80	1.617	1.819	2.021	2.526	3.032	3.537	4.042	5.053	6.063	7.074	8.084
90	1.819	2.046	2.274	2.842	3.411	3.979	4.548	5.684	6.821	7.958	9.095
100	2.021	2.274	2.526	3.158	3.790	4.421	5.053	6.316	7.579	8.842	10.11
125	2.526	2.842	3.158	3.948	4.737	5.527	6.316	7.895	9.474	11.05	12.63
150	3.032	3.411	3.790	4.737	5.684	6.632	7.579	9.474	11.37	13.26	15.16
175	3.537	3.979	4.421	5.527	6.632	7.737	8.842	11.05	13.26	15.47	17.68
200	4.042	4.548	5.053	6.316	7.579	8.842	10.11	12.63	15.16	17.68	20.21
250	5.053	5.684	6.316	7.895	9.474	11.05	12.63	15.79	18.95	22.11	25.26
300	6.063	6.821	7.579	9.474	11.37	13.26	15.16	18.95	22.74	26.53	30.32
350	7.074	7.958	8.842	11.05	13.26	15.47	17.68	22.11	26.53	30.95	35.37
400	8.084	9.095	10.11	12.63	15.16	17.68	20.21	25.26	30.32	35.37	40.42
500	10.11	11.37	12.63	15.79	18.95	22.11	25.26	31.58	37.90	44.21	50.53

Note: For fluids other than water, multiply table values by specific gravity. In pumping liquids with a viscosity considerably higher than that of water, the pump capacity and head are reduced. To calculate the horsepower for such fluids, pipe friction head must be added to the elevation head to obtain the total head.

(Crane, 1976)

APPENDIX 18.A.
State Agencies that Provide Water-Sampling Services

State	Homeowner should first contact	Name, address, phone number for more information
Alabama	County health department	Water Supply Program Department of Environmental Management 1751 Federal Drive Montgomery, AL 36130 205/832-3170
Alaska	Private laboratory	If an illness is involved that might be waterborne, the Department of Environmental Conservation should be notified. Telephone 907/465-2606.
Arizona	Private laboratory	Department of Health Services Office of Waste and Water Quality Management 2005 North Central Avenue Phoenix, AZ 85004 602/271-5453
Arkansas	County health	State Department of Health 4015 West Markham Little Rock, AR 72201 501/661-2171
California	District office of Sanitary Engineering	State Department of Health Services Sanitary Engineering Branch 2151 Berkeley Way Berkeley, CA 95814 415/540-2173
Colorado	County health	Colorado Department of Health Water Quality Control Division 4210 East 11th Denver, CO 80220 303/320-8333, ext 3453
Connecticut	Private laboratory or local health department	Water Supplies Section Connecticut Department of Health Services 150 Washington Street Hartford, CT 06106 203/566-1253
Delaware	County health	Local county health laboratory
Florida	County health	Local county public health unit or Department of Health and Rehabilitative Services Environmental Health Program 1317 Winewood Boulevard Tallahassee, FL 32301 904/488-4070
Georgia	State	Department of Human Resources Environmental Health Section Attn: W. O. Garrett 41 Trinity Avenue Atlanta, GA 30334 404/656-7045
Hawaii	State	Department of Health Drinking Water Program 645 Halekauwila Street Honolulu, HI 96813 808/548-2235
Idaho	Local district health dept.	Bureau of Water Quality Statehouse Boise, ID 83720 208/334-4255
Illinois	State regional office or local health dept.	Illinois Department of Public Health Laboratories 134 North 9th Street Springfield, IL 62701 217/782-6562, or

Appendix 18.A. Continued

State	Homeowner should first contact	Name, address, phone number for more information
		2121 West Taylor Street Chicago, IL 60612 312/793-4760, or Chautauqua & Oakland Streets P. O. Box 2797 Carbondale, IL 62901 618/457-5131
Indiana	State	Indiana State Board of Health Water and Sewage Laboratory 1330 West Michigan Street Indianapolis, IN 46206 317/633-0232
Iowa	Hygienic laboratory	Hygienic Laboratory University of Iowa, Oakdale Campus Iowa City, IA 52242 319/353-5990
Kansas	County or regional health	Office of Laboratories and Research Kansas Department of Health and Environment Forbes Building 740 Topeka, KS 66620 913/862-9360
Kentucky	County health	Local county health department
Louisiana	Parish health	Louisiana Office of Health Services and Environmental Quality Division of Environmental Health P. O. Box 60630 New Orleans, LA 70160 504/568-5100
Maine	State health	Public Health Laboratories Statehouse, Station 12 221 State Street Augusta, ME 04333 207/289-2727
Maryland	County health	County health departments
Massachusetts	Local health	Local board of health
Michigan	County health department	Sanitary Bacteriology & Chemistry Section Laboratory and Epidemiological Services Bureau Michigan Department of Public Health Box 30035, 3500 North Logan Lansing, MI 48909 517/373-1428
Minnesota	Local community health agencies	Minnesota Department of Health Analytical Laboratory, Room 405 717 Delaware Street SE Minneapolis, MN 55440 612/623-5301
Mississippi	County health	County health departments
Missouri	District health unit	Bureau of Community Sanitation 1407 SW Boulevard Jefferson City, MO 65101 314/751-3696
Montana	State	Water Quality Bureau Cogswell Building, Capitol Station Helena, MT 59601 406/444-2406
Nebraska	County agent	County Cooperative Extension Service

Appendix 18.A. Continued

State	Homeowner should first contact	Name, address, phone number for more information
Nevada	State	Consumer Health Protection Services Kinkead Building, Room 103 Capitol Complex Carson City, NV 89710 702/784-4750
New Hampshire	State	Water Supply and Pollution Control Laboratories P. O. Box 95, Hazen Drive Concord, NH 03301 603/271-3445
New Jersey	County health	County health department
New Mexico	EID (regional office)	Environmental Improvement Division Water Supply Section Crown Building, P. O. Box 968 Santa Fe, NM 87504-0968 505/827-9805
New York	County health or state district office	Division of Environmental Protection Bureau of Community Sanitation and Food Protection Empire State Plaza Albany, NY 12237 518/474-3968
North Carolina	County health	County health departments
North Dakota	District health	State Department of Health Laboratory Services Section P. O. Box 1618 Bismarck, ND 58502 701/224-2384
Ohio	Municipal health or county health dept.	Ohio Environmental Protection Agency Division of Public Water Supply 361 East Broad Street Columbus, OH 43215 614/466-8307
Oklahoma	County health	County health departments or 405/271-5220 (Bacteriological) 405/271-5240 (Chemical)
Oregon	County environmental health	County environmental health
Pennsylvania	Local regional office of Bureau of Community Environmental Control, Department of Environmental Resources	Division of Water Supplies Bureau of Community Environmental Control P. O. Box 2357 Harrisburg, PA 17120 717/783-3795
Rhode Island	State health	Department of Health 50 Orms Street Providence, RI 02908 401/274-1011
South Carolina	County health dept/EQC district office	EQC Analytical Services Laboratory South Carolina Department of Health and Environmental Control 8231 Parklane Road Columbia, SC 29223 803/758-4986

Appendix 18.A. Continued

State	Homeowner should first contact	Name, address, phone number for more information
South Dakota	County health	South Dakota Department of Water & Natural Resources Office of Drinking Water Joe Foss Building Pierre, SD 57501 605/773-3754
Tennessee	County health	Department of Health and Environment Division of Groundwater Protection 150 – 9th Avenue North TERRA Building, 7th Floor Nashville, TN 37219 615/741-0690
Texas	Regional health	Division of Water Hygiene Texas Department of Health 1100 West 49th Street Austin, TX 78759 512/458-7497
Utah	Local health	Local health district offices
Vermont	Local health department	Vermont State Department of Health 60 Main Street, P. O. Box 70 Burlington, VT 05401 802/863-7220
Virginia	Bureau of Water Supply	Virginia Department of Health 109 Governor Street Richmond, VA 23219 804/786-5566
Washington	Private laboratory or county health	Department of Social and Health Services Water Supply and Waste Section M.S. LD-11 Olympia, WA 98504 206/753-3466
West Virginia	County health	Bacteriological: State Hygienic Laboratory 167 – 11th Avenue South Charleston, WV 25363 304/348-3530 Chemical: Environmental Health Services Laboratory Attn: James Rosencrance 151 – 11th Avenue South Charleston, WV 25363 304/348-0197
Wisconsin	Local Dept. of Natural Resources	State Laboratory of Hygiene 465 Henry Mall Madison, WI 53706 608/262-1210
Wyoming	State health	Wyoming Department of Agriculture State Chemistry Laboratory Box 3228 Laramie, WY 82071 307/742-2984

APPENDIX 18.B.
Abandonment of Test Holes, Partially Completed Wells, and Completed Wells

Sec. 13.1 General

Abandoned test holes, including test wells, uncompleted wells, and completed wells shall be sealed.

13.1.1 *Need for sealing of wells*:
1. Eliminate physical hazard
2. Prevent contamination of groundwater
3. Conserve yield and hydrostatic head of aquifers
4. Prevent intermingling of desirable and undesirable waters.

13.1.2 *Restoration of geological conditions.* The guiding principle to be followed by the contractor in the sealing of abandoned wells is the restoration, as far as feasible, of the controlling geological conditions that existed before the well was drilled or constructed.

Sec. 13.2 Sealing Requirements

A well shall be measured for depth before it is sealed to ensure freedom from obstructions that may interfere with effective sealing operations.

13.2.1 *Liner-pipe removal.* Removal of liner pipe from some wells may be necessary to ensure placement of an effective seal.

13.2.2 *Exception to removing liner pipe.* If the liner pipe cannot be readily removed, it shall be perforated to ensure the proper sealing required.

13.2.3 *Sealing materials and placement.* Concrete, cement grout, or neat cement shall be used as primary sealing materials and shall be placed from the bottom upward by methods that will avoid segregation or dilution of material.

Sec. 13.3 Records of Abandonment Procedures

Complete accurate records shall be kept of the entire abandonment procedure to provide detailed records for possible future reference and to demonstrate to the governing state or local agency that the hole was properly sealed.

13.3.1 *Depths sealed.* The depth of each layer of all sealing and backfilling materials shall be recorded.

13.3.2 *Quantity of sealing material used.* The quantity of sealing materials used shall be recorded. Measurements of static water levels and depths shall be recorded.

13.3.3 *Changes recorded.* Any changes in the well made during the plugging, such as perforating casing, shall be recorded in detail.

(Reprinted from AWWA Standard for Water Wells, AWWA A100-84, by permission. Copyright© 1984, The American Water Works Association.)

APPENDIX 19.A.
Iron Oxidizing Bacteria

Enzymatic Iron Bacteria		Nonenzymatic Iron Bacteria
Probable	**Possible**	
Thiobacillus Ferrooxidans	Leptothrix	Gallionella
Sulfolobus Acidocaldarius	Sphaerotilus Natans	Sphaerotilus
Leptospirillum Ferrooxidans	Crenothrix Polyspora	Crenothrix
Sulfobacillus Thermosulfidooxidans	Clonothrix	Clonothrix
Gallionella	Lieskeela Bifida	Leptothrix
	Siderocapsaceae	Siderocapsaceae
	Pedomicrobium	Naumanniella
	Hyphomicrobium	Ochrobium
	Acholeplasma Laidlawii	Siderococcus
		Pedomicrobium
		Herpetosyphon
		Seliberia
		Toxothrix
		Archangium
		Pseudomonas
		Bacillus

APPENDIX 19.B.
Malfunctions in High-Capacity Centrifugal and Turbine Pumps, Possible Causes, and Solutions

Problem	Amp Load	Other Observations	Trouble Source	Probable Cause	Remedy
Insufficient capacity or pressure	Below normal	Pumping water level higher than normal	Gate valve partially closed or closed more than normal		Open gate valve and monitor pumping water level to prevent cavitation
	Below normal	Pumping water level higher than normal	Worn impeller exit vanes, bowl vanes, and shrouds or pitting on bowl casing	Abrasion or corrosion	Install, replace, or redevelop well screen to reduce sand content; replace impeller or bowls
	Below normal	Pumping water level higher than normal	Worn or pitted bowl casing	Cavitation may have caused	Replace pump bowls and insure proper submergence (see Cavitation under Trouble Source)
	Erratic	Pumping water level fluctuates, noisy operation (sounds like rocks passing through pump)	Cavitation	Inadequate submergence	Provide sufficient NPSH for pump by reducing pumping rate, changing pump impellers to low NPSH design, or replacing bowl assembly with model capable of operating within available NPSH. If inadequate submergence results from well incrustation, use appropriate well treatment.
	Below normal or fluctuating	Pumping water level higher than normal	Pitting on entrance vanes of impeller	Cavitation	Replace impeller and insure proper submergence (see Cavitation under Trouble Source)
	Below normal (above normal during wearing)	Pumping water level higher than normal	Wear on impeller skirts and/or bowl seal ring area	Abrasion or excess bearing wear	Install new wear rings; install, replace, or redevelop well screen to reduce sand content
				Impellers set too high	Adjust impeller setting to manufacturer's recommendations
	Below normal	Pumping water level higher than normal	Impeller trimmed incorrectly		Replace impeller
	Below normal	Pumping water level higher than normal	Impeller set too high		Adjust impeller setting to manufacturer's recommendations
	Below normal	Pumping water level higher than normal; gradual capacity or pressure reduction over time; vibration may occur	Impeller, bowl assembly, or discharge piping plugged	Mineral or bacterial incrustation	Use appropriate chemical treatment
	Below normal	Pumping water level higher than normal	Impeller loose on shaft	Repeated shock load by surge in discharge line; foreign material jamming impeller; differential expansion caused by temperature change; improper machining or assembly of parts; torsional loading of submersible pump	Refit impeller and tighten collets
	Generally below normal	Pumping water level higher than normal	Motor operating in reverse	Electrical connections reversed	Reverse motor leads
	Normal or slightly higher	Pumping water level higher than normal	Hole in column pipe or bowl assembly	Corrosion	Repair or replace column or bowls; consider corrosion-resistant material
	Normal or slightly higher	Pumping water level higher than normal	Leaking joints	Insufficient tightening during installation; dirt in joint connections; excessive vibration; misalignment; corrosion	Clean and tighten joint connections; correct vibration or misalignment trouble; or replace column (consider corrosion-resistant material)

Appendix 19.B. Continued

Problem	Amp Load	Other Observations	Trouble Source	Probable Cause	Remedy
	Normal	Pumping water level at or near normal; capacity or pressure only apparently reduced	Faulty capacity meter or pressure gauge		Check and calibrate gauge or meter
	Below normal		Pump operating against total dynamic head greater than design value	Change in or plugging of surface piping; increase in head requirements at surface; decline in local water levels; decrease in well specific capacity	Check head conditions and pumping rate against pump curve; clean surface piping with appropriate chemicals; reduce head requirements; decrease pumping rate; use appropriate well treatment
	Below normal		Air in water	Cascading water in well — pumping water level below top of well screen or upper areas of open bore-hole contributing water to well	Raise pump intake out of screen and reduce pumping rate; use pump shroud for submersible pump
	Below normal		Clogged pump strainer	Mineral or bacterial incrustation; debris in well	Use appropriate chemical treatment; clean strainer and well
	Below normal	No water discharged; noisy operation	Broken shaft	Worn line-shaft bearing; bent shaft	Replace shaft bearing; straighten or replace shaft
				Corrosion	Replace shaft or coupling; consider corrosion-resistant material
				Pipe wrench fatigue on reused coupling	Replace coupling
			Unthreaded shafts	Power applied to shafts not butted in couplings	Check for galling on shaft ends; check for momentary power failure; repair or replace faulty check (foot) valve or starting timers; replace shaft or coupling
				Bearing seized	Check lubrication, bearing temperature limits; replace bearing and shaft or coupling
				Foreign material locking impellers or galling wear rings	Clean well; replace pump strainer; eliminate source of foreign material; install or replace well screen; reline well; replace shaft or coupling
				Metal fatigue caused by vibration	Eliminate cause of vibration; replace shaft or coupling
				Improper impeller adjustment or extended periods of upthrust causing impeller to drag on bowl case	Adjust impeller setting to manufacturer's recommendations; reduce pumping rate to pump at peak efficiency; replace shaft or coupling; check condition of impellers and bowls
Vibration	Below normal	No water discharged	Shaft coupling unscrewed	Pump operated in reverse rotation	Check shafts and other couplings; tighten; correct rotation
	Above normal		Bent shafting	Mishandling in transit or assembly; extended periods of upthrust	Straighten or replace bent section of shaft; reduce pumping rate to pump at peak efficiency
			Excessive shaft wear under rubber bearings	Swelling of rubber bearings in hydrocarbon, hydrogen sulfide, or high temperature conditions	Change bearing material; replace shaft
			Excessive but uniform wear on bearings and shaft	Abrasion	Replace parts
				Shafts not butted in couplings; dirt or grease between shafts	Clean joints and assemble correctly
				Shaft ends not properly faced	Reface shaft ends parallel and concentric or replace shaft

Appendix 19.B. Continued

Problem	Amp Load	Other Observations	Trouble Source	Probable Cause	Remedy
Abnormal noise	Above normal		Uneven wear on bearings, uniform wear on shaft	Nonrotating pump parts misaligned	Replace bearings and repair or replace shaft; check pump foundation, discharge pipe connection, column joints, and bowl connections — correct misalignment; if well is misaligned, avoid misaligned section; replace deep-well turbine with submersible pump, or abandon well.
	Above normal	Noisy operation	Bearing seized, failed or galling on shaft	High temperature; running dry without lubrication	Replace bearing — correct lubrication problems
	Above normal	Noisy operation*	Impellers dragging on bowl case	Improper impeller adjustment	Check condition of impellers and bowls; adjust impeller setting to manufacturer's recommendations
	May be above normal		Electric motor	Motor imbalance	Balance — add shims or washers under drive bolts; have motor shop-tested and balanced
	Above normal		Motor noise	Fatigued or worn motor bearings; pitted bearings from long idle periods; misalignment or eccentricity of rotating parts	Replace bearings; correct lubrication temperature or alignment problems; align or adjust to concentricity
				Unstable or uneven mounting	Secure or correct mounting
				Worn bearings	Replace bearings; correct lubrication, temperature or alignment problems
				Misalignment or eccentricity of rotating parts	Align or adjust to concentricity
				Unstable or uneven mounting	Secure or correct mounting
				Motor rotor rubbing on stator — caused by extended periods of upthrust	Repair motor and correct upthrust problem (reduce pumping rate to pump peak efficiency capacity)
	Above normal		Pump bearings running dry	Lubrication failure	Correct lubrication problem
			Broken column bearing retainers		Replace
	Below normal		Broken shaft or enclosing tube	See Problem: Insufficient capacity or pressure — Broken shaft or coupling	Replace shaft or tube; correct cause of problem
	Above normal	Excessive vibration*	Impellers dragging on bowl case	Improper impeller adjustment	Check condition of impellers and bowls and adjust impeller setting to manufacturer's recommendations
	Erratic	Fluctuating pumping water level (sounds like gravel passing through pump)	Cavitation	Inadequate submergence	Provide sufficient NPSH for pump by reducing pumping rate, changing pump impellers to low NPSH design, or replacing bowl assembly with model capable of operating within available NPSH; if inadequate submergence results from well incrustation, use appropriate well treatment
	Erratic		Foreign material in pump	Debris in well; hole in well screen or casing; unstable open hole	Clean well; replace pump strainer; eliminate source of foreign material — install or replace well screen, reline well; redevelop well; check pump parts for wear
Excessive power consumption	Above normal			Abnormally low voltage	Contact power supplier
	Above normal	Excessive vibration, noisy operation*	Impellers dragging on bowl case	Improper impeller adjustment	Check condition of impellers and bowls and adjust impeller setting to manufacturer's recommendation

Appendix 19.B. Continued

Problem	Amp Load	Other Observations	Trouble Source	Probable Cause	Remedy
	Above normal	Noisy operation, excessive vibration	Bent shafting	Mishandling in transit or assembly; continuous periods of upthrust	Straighten or replace bent section of shaft; reduce pumping rate to peak efficiency capacity
		Noisy operation, excessive vibration	Bearings binding or galling on shaft	High temperature; running dry without lubrication	Replace bearings; consider different material; check condition of shaft; correct lubrication problem
				Nonrotating pump parts misaligned	Replace bearings; check condition of shaft; check pump foundation, discharge pipe connection, column joints and bowl connections — correct misalignment; if well is misaligned, avoid misaligned section; replace deep-well turbine with submersible, or abandon well
			Foreign material in pump	Debris in well; hole in well screen or casing; unstable open hole	Clean well; replace pump strainer; eliminate source of foreign material — install or replace well screen, reline well; check pump parts for wear
			Lubricating oil too viscous		Use lower viscosity lubricant
			Excessively worn or failed thrust bearing in motor	Extended periods of upthrust	Replace bearing and correct upthrust problem (reduce pumping rate to peak efficiency)
	Above normal	Pumping rate higher than normal, pumping water level below normal	Pump operating against total dynamic head lower than design value	Change in or plugging of surface piping; increase in head requirements at surface; decline in water levels; decrease in well specific capacity	Check head conditions and pumping rate against pump curve

*Excessive noise may be heard initially; thereafter, the noise may dissipate as the shaft stretches because of downthrust. If the impellers are set too low, the noise may not become excessive until several seconds after the pump is started.

APPENDIX 20.A-1.
Notice to Bidders

1. Sealed bid, addressed to _____ (Agency) _____ , will be received at the _____ office at _____ (address) _____ (time) until _____ (date) , and then publicly opened and read for furnishing all plant, labor material and equipment and performing all work required for the construction of:

for _____ (Name) , _____ (County) , _____ (State)
hereinafter called Owner.

2. Bids shall be submitted in sealed envelopes upon the blank form of proposal furnished. Sealed envelopes shall be marked in the upper left hand corner as follows:
Sealed Bid
Bid for _____

To be Opened _____ (Time)

_____ (Date)

3. All proposals shall be accompanied by a cashier's or certified check or bid bond upon a national or state bank or surety company in the amount _____ % of the total amount for the base bid payable to _____ (Owner) , as a guarantee that the bidder will enter into a contract and execute performance bond within _____ (No.) (_____) days after notice of award and that his bid will not be withdrawn within _____ (No.) (_____) days after the date of opening of bids without the consent of the Owner. Bids without check will not be considered.

4. All bid securities will be returned to the respective bidders within _____ (No.) (_____) days after bids are opened, except those which the Owner elects to hold until the successful bidder has executed the contract and furnished the required bonds. Thereafter all remaining securities, including security of the successful bidder, will be returned within _____ (No.) (_____) days.

5. The successful bidder must furnish both a performance bond and a payment bond upon approved standard form in the amount of _____ % of the contract price from an approved surety company licensed by the state of _____ , or acceptable, according to the latest list of companies holding certificates of authority from the Secretary of the Treasury of the United States, or other surety or sureties acceptable to the Owner.

6. The right is reserved, as the interest of the Owner may require to reject any and all bids, and to waive any informality in bids received.

7. Prospective bidders and suppliers may obtain specifications and bidding documents from the office of _____ (Agency)
on deposit of: _____ (Price)

Appendix 20.A-1. Continued

per set, which sum so deposited will be refunded to bidders provided all documents are returned in good condition to the office from which they were issued, not later than _____ (_____) days after the time bids are received. Drawings
           (No.)
and specifications may be examined at the _____ .
                                                        (Location)

8. Bidders should carefully examine the plans, specifications and other documents, visit the site of the work, and fully inform themselves as to all conditions and matters which can in any way affect the work or the cost thereof. Should a bidder find discrepancies in, or omission from the plans, specifications, or other documents, or should he be in doubt as to their meaning, he should at once notify the Owner and obtain clarification prior to submitting any bid.

9. Addenda to the specifications and revised drawings issued to bidders prior to the receipt of bids shall be considered part of the contract documents. Bidders shall acknowledge receipt of addenda and revised drawings on the proposal form.

(U.S. Environmental Protection Agency, 1975c)

APPENDIX 20.A-2.
Proposal

For furnishing all Plant, Labor, Equipment and Materials and Performing all Operations necessary for the construction of

_____ in _____
            (Project)   (Location)

_____
(Address of Owner or Owner's Representative)

_____
_____
_____
_____
_____

Gentlemen:

The undersigned, as Bidder, declares that he has carefully examined the Notice to Bidders, Specifications, and the drawings herein referred to for a _____ (Project) _____ for the _____ (Agency) and has carefully examined the locations, conditions, and classes of material of the proposed work, and agrees that he will provide all the necessary machinery, apparatus, tools, and other means of construction, and will do all the work and furnish all materials called for in the contract and specifications in the manner prescribed therein and according to the requirements of the engineer as therein set forth for the amount below.

_____
(Description of Project)

Item No.	Qty.	Unit	Description of item with unit bid price written in words	Total amount bid

It is understood and agreed that the work shall be completed in full within _____ calendar days after the date on which work is to be commenced as established by the Contract Documents.

Accompanying this proposal is a certified check, cashier's check or bid bond, in the amount of $ _____
made payable to _____.
                                       (Owner)

It is understood that the bid security accompanying this proposal shall be returned to the undersigned unless, in case of the acceptance of this proposal, the undersigned should fail to enter into a construction contract and execute bonds as required it is understood and agreed that the bid security shall be forfeited to the Owner and shall be considered as payment for damages due to delay and other inconveniences suffered by the Owner as a result of such failure on the part of the undersigned.

It is understood that the Owner reserves the right to reject any and all bids.

In the event of Award of the Contract to the undersigned, the undersigned agrees to furnish Performance and Payment Bonds, as provided in the Specifications.

The undersigned certifies that the bid prices contained in this proposal have been carefully checked and are submitted as correct and final.

Appendix 20.A-2. Continued

Date _____ Signed _____
                                              (Company)
                          By _____
                                              (Title)
                             _____
                                              (Address)

Witness _____
Seal (If Bidder is a Corporation)
Acknowledge Receipt of Addenda Below:
Addendum No. _____   _____   _____   _____
Date Received _____   _____   _____   _____

(U.S. Environmental Protection Agency, 1975c)

APPENDIX 20.A-3.
Standard Form of Agreement
Between Owner and Contractor

THIS AGREEMENT made as of the _____ day of _____ _____ in the year 19 _____ by and between, _____ (hereinafter called the OWNER) and _____ (hereinafter called the CONTRACTOR)

WITNESSETH THAT the OWNER and CONTRACTOR in consideration of the mutual covenants hereinafter set forth, agree as follows:

Article 1. WORK. The CONTRACTOR will perform all Work as shown in the Contract Documents for the completion of the Project generally described as follows: Construction of _____

Article 2. ENGINEER. The Project has been designed by _____

(Agency)

Article 3. CONTRACT TIME. The Work shall be completed within _____ calendar days after the date which the OWNER shall designate in writing to the CONTRACTOR as the date on which it is expected that the CONTRACTOR will start the Work.

Article 4. CONTRACT PRICE. The OWNER will pay the CONTRACTOR for performance of the Work and completion of the Project in accordance with the Contract Documents subject to adjustment by Modifications as provided therein as follows:

Article 5. PROGRESS AND FINAL PAYMENTS. The OWNER will make progress payments on account of the Contract Price as provided in the General Conditions as follows:

5.1 Progress and final payments will be on the basis of the CONTRACTOR's Applications for Payment as approved by the OWNER.

5.2 On or about the _____ th day of each month during construction: _____ percent of the Work completed and, _____ percent of material and equipment not incorporated in the Work but delivered and suitably stored, less in each case the aggregate of payments previously made.

5.3 Upon Substantial Completion, a sum sufficient to increase the total payments to the CONTRACTOR to _____ percent of the Contract Price less retainages as the OWNER shall determine for all incompleted work and unsettled claims.

5.4 Upon final completion of the Work and Settlement of all claims, the remainder of the Contract Price.

Article 6. CONTRACT DOCUMENTS. The Contract Documents which comprise the contract between the OWNER AND THE CONTRACTOR are attached hereto and made a part hereof and consist of the following:

6.1 This Agreement (Pages _____ to _____ inclusive),
6.2 Exhibits to this Agreement (Pages _____ to _____ inclusive)

Appendix 20.A-3. Continued

    6.3 Specifications consisting of:
        Notice to Bidders   (Pages _____to _____ inclusive)
        General Conditions   (Pages _____to _____ inclusive)
        Special Conditions   (Pages _____to _____ inclusive)
        Technical Provisions   (Pages _____to _____ inclusive)
    6.4 Drawings and Plans as listed in the attached Exhibit _____ .
    6.5 Addenda numbers (_____ to _____ inclusive), and
    6.6 Any Modifications, including Change Orders, duly delivered after execution of this Agreement.

Article 7. MISCELLANEOUS

    7.1 Terms used in this Agreement which are defined in Article 1 of the General Conditions shall have the meanings indicated in the General Conditions.

    7.2 Neither the OWNER nor the CONTRACTOR shall, without the prior written consent of the other, assign or sublet in whole or in part his interest under any of the Contract Documents and, specifically, the CONTRACTOR shall not assign any monies due or to become due without the prior written consent of the OWNER.

    7.3 The OWNER and the CONTRACTOR each binds himself, his partners, successors, assigns and legal representatives to the other party hereto in respect of all covenants, agreements and obligations contained in the Contract Documents.

    7.4 The Contractor Documents constitute the entire agreement between the OWNER and the CONTRACTOR and may only be altered, amended or repealed by a duly executed written instrument.

    IN WITNESS WHEREOF, the parties hereto have executed this Agreement the day and year first above written.

OWNER _____ CONTRACTOR _____
By _____ BY _____
      (CORPORATE SEAL)   (CORPORATE SEAL)
Attest: _____ Attest: _____

(U.S. Environmental Protection Agency, 1975c)

APPENDIX 20.A-4.
National Water Well Association Standard Form
Well Estimate and/or Well Drilling Contract or Repair Order Agreement

Purchaser's Name _____ Date _____
Address _____
City _____ State _____

	WELL PERMIT COST	$ _____
1. WELL AND PRICE _____		
2. PUMP AND PRICE		$ _____
3. SCREEN AND PRICE		$ _____
4. DEVELOPING, SURGING AND PRICE		$ _____
5. CHLORINATING AND PRICE		$ _____
6. REPAIR WORK AND PRICE		$ _____
7. TOTAL CASH PRICE		$ _____
8. TOTAL DOWN PAYMENT		$ _____
* 9. AMOUNT FINANCED		$ _____
10. **FINANCE CHARGE**		$ _____
11. TOTAL OF PAYMENTS		$ _____
12. DEFERRED PAYMENT PRICE (7 + 10)		$ _____

13. **ANNUAL PERCENTAGE RATE _____ % _____

14. In this Contract the term WELL shall consist of the HOLE and CASING ONLY.

15. The depth of the well shall be measured from the top of the ground to the bottom of the well, and it is agreed that if the depth be less than 60 feet, the minimum charge will be for 60 feet.

16. Drilling Contractor reserves the right to reduce the size of the casing and/or abandon drilling. If drilling is abandoned the cash down payment shall be retained by the Drilling Contractor to liquidate operation costs.

17. All electric wiring, installations and connections, and any and all non-drilling work will be furnished by the Purchaser at no charge to the Drilling Contractor.

18. LEGAL DESCRIPTION _____ of Section, _____ Twp. _____ Range or Lot _____ in Block _____ Plat _____
_____ in the City of _____ State _____
and that said property is free and clear of all encumbrances except _____

19. The undersigned agrees that title in and to any and all materials furnished by the Drilling Contractor whether in the ground or attached to the premises shall remain with the Drilling Contractor and the Drilling Contractor has the right to withdraw the casing from the well and materials from the premises. A mechanic's lien is hereby acknowledged to secure the amount of contract or repairs. The cash down payment will be retained by the Drilling Contractor to liquidate damages for the breach of the contract. This clause is binding until full payment is received.

20. Interest at the rate of _____ percent will be charged on all money not paid when due.

21. All agreements and understandings are contained herein and there are no verbal representations or agreements not herein contained.

22. I hereby authorize the above contract and/or repair work together with the necessary materials and will pay $ _____ on signing this contract. Unless deferred payment is provided for in paragraphs 9 through 13 above, I will pay for the drilling and casing when completed and the balance immediately on completion of the WELL SYSTEM.

23. BANK REFERENCES _____

THIS CONTRACT ACCEPTED BY PURCHASER   DRILLING CONTRACTOR
_____   _____
Address _____   Address _____
Telephone No. _____   Telephone No. _____

Appendix 20.A-4. Continued

*Where a finance charge is imposed or where payment is made in more than four installments and where Purchaser is an individual and where the well or repair of well is for personal, family, household or agricultural purposes, the attached Notice of Rescission form must be completed and be given to Purchaser with a signed copy of this contract. The Notice of Rescission need not be completed in cases where emergency repairs are required and where the Purchaser provides a handwritten waiver as follows: "I wish you to come on my premises and make repairs to my water system for the reason that my health, safety and property are in danger by non-operation of my water system. Because of this emergency I waive my rights to cancel or rescind this repair transaction.
Signature _____ "

**Calculation of the Annual Percentage Rate can be extremely complex. You should consult your bookkeeper or accountant to insure calculation of this rate.

NOTICE TO PURCHASER: You are entitled to a copy of the contract you sign. You have the right to pay in advance any unpaid balance of this contract and obtain a partial refund of the finance charge, if any.

This Well Estimate and/or Well Drilling Contract or Repair Order Agreement has been prepared for the use of well contractors by the National Water Well Association, Worthington, Ohio 43085.

NOTICE OF RESCISSION

_____   _____
         Owner's Name   Date

_____
         Owner's Name

_____
         Address
                                          _____
                                               Contractor-Creditor

_____
                    (Identification of Transaction)

Notice to Customer Required By Federal Law:

   You have entered into a transaction on _____
                                                      (Date)
which may result in a lien, mortgage, or other security interest on your home. You have a legal right under federal law to cancel this transaction, if you desire to do so, without any penalty or obligation within three business days from the above date or any later date on which all material disclosures required under the Truth in Lending Act have been given to you. If you so cancel the transaction, any lien, mortgage, or other security interest on your home arising from this transaction is automatically void. You are also entitled to receive a refund of any downpayment or other consideration if you cancel. If you decide to cancel this transaction, you may do so by notifying

_____
                          (Name of Creditor)

at _____
       (Address of Creditor's Place of Business)
by mail or telegram sent not later than midnight of _____
                                                              (Date)

You may also use any other form of written notice identifying the transaction if it is delivered to the above address not later than that time. This notice may be used for that purpose by dating and signing below.

                          I hereby cancel this transaction.

_____   _____
         (Date)   (Customer's Signature)

Note: Each owner of property on which well is to be drilled must receive a copy of this form.
(U.S. Environmental Protection Agency, 1975c)

APPENDIX 20.B.
Portion of Tightly Written Specification

In this partial specification, the engineer has taken great care in describing the equipment to be used and outlining the exact procedures to be followed in placing the screen and casing. The thoroughness of this specification is justified because multiple deep, high-capacity wells were constructed using this specification as a guide.

b. Well Screen.

1. General. The well screen shall be of the continuous slot, wire-wound design in order to provide maximum inlet area consistent with strength requirements. It shall be fabricated by circumferentially wrapping a triangularly shaped wire around a circular array of internal rods. The wire configuration must produce inlet slots with sharp outer edges, widening inwardly so as to be nonclogging. For maximum collapse strength, each juncture between the horizontal wire and the vertical rods will be fusion welded under water by the electrical resistance method. End fittings will be welded to the screen body.

2. Material and Fittings. The well screen and attached end fittings shall be fabricated from corrosion-resistant Type 304 stainless steel. The blank welding pieces on the ends of the screens shall be the minimum length required for connections, welding, satisfactory fabrication, and adequate strength. The ends of the screen and blank sections shall be beveled for welding and shall meet the same standards as the beveled ends of the blank casing.

The contractor is responsible for insuring that the materials utilized will be adequate for the actual conditions encountered. For bidding purposes, however, the 14-inch diameter screen section provided must meet the following minimum requirements:

Above 700 feet — The screen sections installed above 700 feet must have a minimum collapse strength of 160 psi and a minimum weight of 29 pounds per linear foot excluding the end fittings. The total cross-sectional area of the vertical rods must equal or exceed 3 square inches. The allowable open area must equal or exceed 128 square inches per linear foot, calculated on the basis of nominal slot size, wire width, and OD of the screen.

Below 700 feet — Any screens installed below 700 feet must have a minimum collapse strength of 230 psi and must weigh a minimum of 32 pounds per foot, excluding the end fittings. The total cross-sectional area of the vertical rods must equal or exceed 3 square inches. The allowable open area must equal or exceed 117 square inches per linear foot as above.

3. Slot Size. The screen slot size shall be 0.050 inches.

4. Payment. Payment for furnishing the well screen will be at the respective unit price per linear foot bid, which price shall include all necessary end fittings and delivery of the casing to the well site. The Contractor shall submit the bid price of the well screen based on the quoted delivered price to the well site from the screen manufacturer including all required taxes.

The Contractor shall use a screen manufacturer listed herein for supplying the well screens or shall, prior to submitting a bid, submit to the Engineer proof of compliance with the design requirements herein specified for the screen. The pre-qualified screen manufacturer(s) listed below have met the specified design requirements and submitted proof of compliance with them to the Engineer for review and approval prior

Appendix 20.B. Continued

to the specifications being written.

Prequalified screen manufacturer(s):

1) Johnson Division

Any other screen manufacturer's product bid by the Contractor shall have their proof of compliance with the design requirements submitted to the Engineer prior to the bid opening.

12. Installing Casing and Screen. The casing shall be placed in the reamed hole by approved methods in a manner that will insure no damage to the casing.

Positive type centering guides in sets of 3 or 4 shall be installed on the casing and screen at 40-foot maximum intervals equally spaced circumferentially. One set of centering guides shall not be more than 5 feet above the top of the reducer, and another set not more than 5 feet above the bottom of the well. The purpose of the guides is to maintain the casing and screen in the center of the drilled hole during cementing and installation of the filter pack. The type of centering guides and their positions shall be approved by the Engineer prior to assembly and placement of the casing and screen. The centering guides on the screen section shall be mild steel and may be attached by welding or by a bolt-on-split-band or other means approved by the Engineer. The band, if used, shall be placed near the weld joining two lengths of screen and shall not cover any of the screen open area. An approved type of centering guide is shown on accompanying Drawing No. _____ . Payment for furnishing and installing centering guides will be included in the other appropriate bid items.

The casing shall be suspended from the top by means of a clamp which will adequately support the entire weight of the casing. The bottom of the casing shall be a sufficient distance above the bottom of the drilled and reamed hole to insure that none of the casing weight will be supported from below. The Contractor shall provide for slight relaxation of the tension clamp during cementing to insure that the casing is not subjected to excessive (beyond failure) tensile loads in the event part of the cement load is inadvertently transferred to the casing. Special care shall be exercised to insure that the casing is installed straight and true.

Measurement for payment for installing 18-inch casing will be taken as the length from the finished top of the casings to the bottom of the reducer section. Measurement for payment for installing 14-inch stainless steel well screen will be taken as the assembled length of the respective types as inserted into the well from the bottom of the casing to the bottom of the taper section. Payment for installing well casing will be made at the applicable unit price per linear foot given in the Bidding Schedule, which shall include all compensation to be received by the Contractor for all labor and equipment required for installing the casing, taper section, insulated coupling, and screen.

13. Welding. All welding shall be a full, continuous, running weld and shall conform to the American Welding Society or American Petroleum Institute specifications. Welding stainless steel casing to mild steel casing will not be permitted. Welds for connecting stainless steel sections shall be made using AWS Type E308-15 lime-coated or Type E308-16 titanium-coated rods. Reverse polarity d.c. current shall be used with E308-15 rods and a.c. or d.c. reverse polarity with E308-16 rods.

Welders shall be qualified in accordance with "Welding Procedures," ASME Boiler Construction Code or AWS Standard Qualification Procedure.

APPENDIX 21.A.
Typical Geologic Settings that Demonstrate the Interaction between Landfill Leachates and Local Hydrogeology

The examples presented below illustrate how contaminants are transported in common hydrogeologic settings.

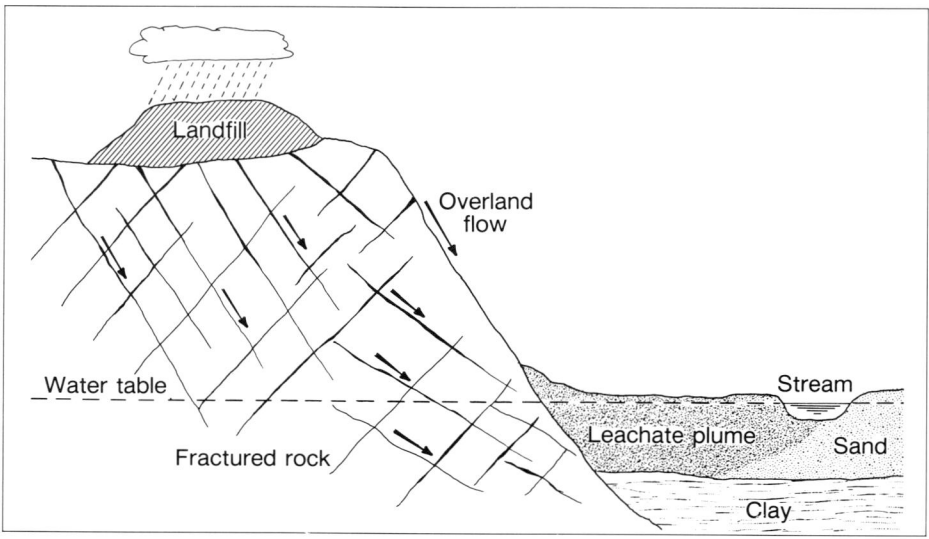

Figure 1. Fractured rock surface with a deep water table. Leachate flows into and through interconnecting fractures and discharges either at the surface or into the subsurface where it moves with the groundwater to some more distant discharge point. *(U.S. Environmental Protection Agency, 1977)*

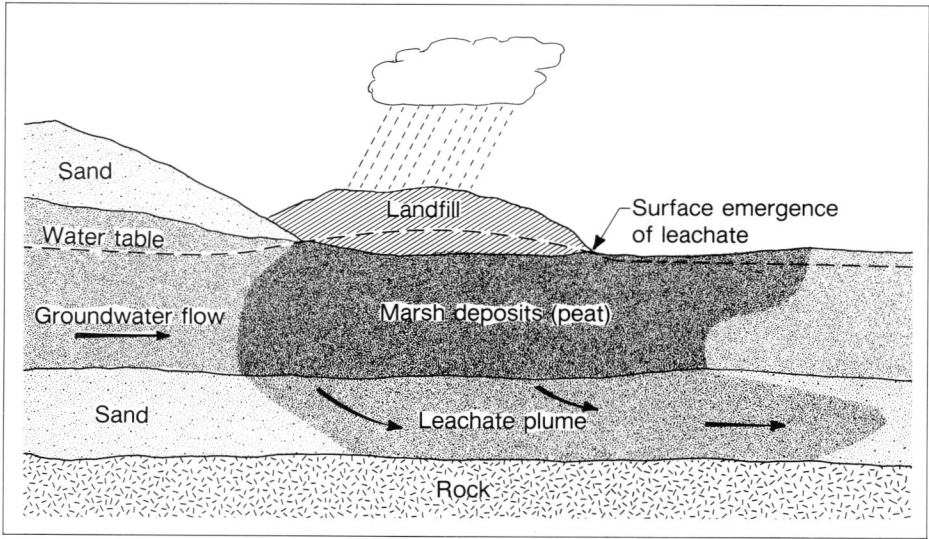

Figure 2. Marsh deposit underlain by an aquifer. The water table is high and a mound is formed at the base of the landfill. Leachate migrates downward through the marsh material to the aquifer. In many cases, surface emergence of leachate will occur at the toe of the slope. Some contaminants may be attenuated within the marsh deposits. The portion reaching the water table moves through the aquifer with the groundwater to some surface discharge point. *(U.S. Environmental Protection Agency, 1977)*

Appendix 21.A. Continued

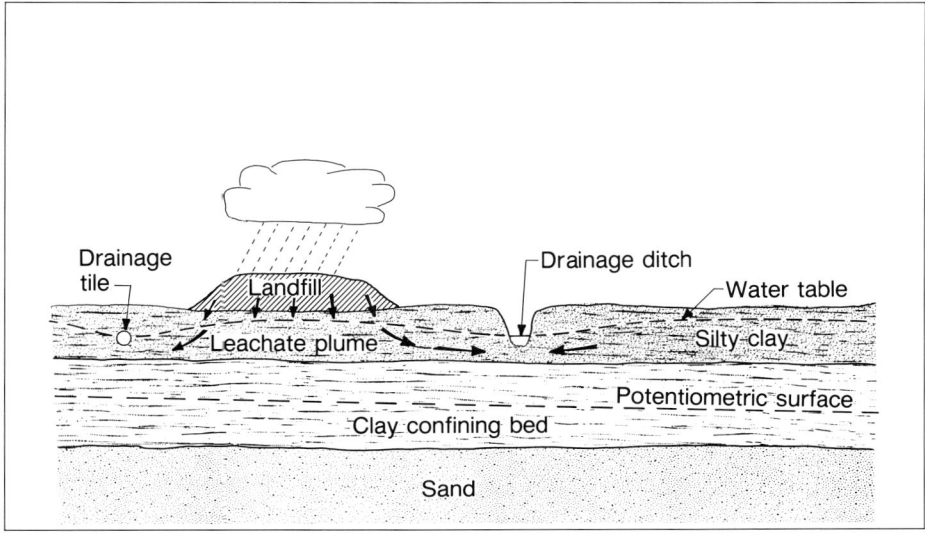

Figure 3. Single aquifer interbedded with clay lenses. The leachate plume is split into two plumes by a clay lens. One plume discharges into a stream near the landfill while the other plume moves deeper into the aquifer and flows to a more distant discharge point. *(U.S. Environmental Protection Agency, 1977)*

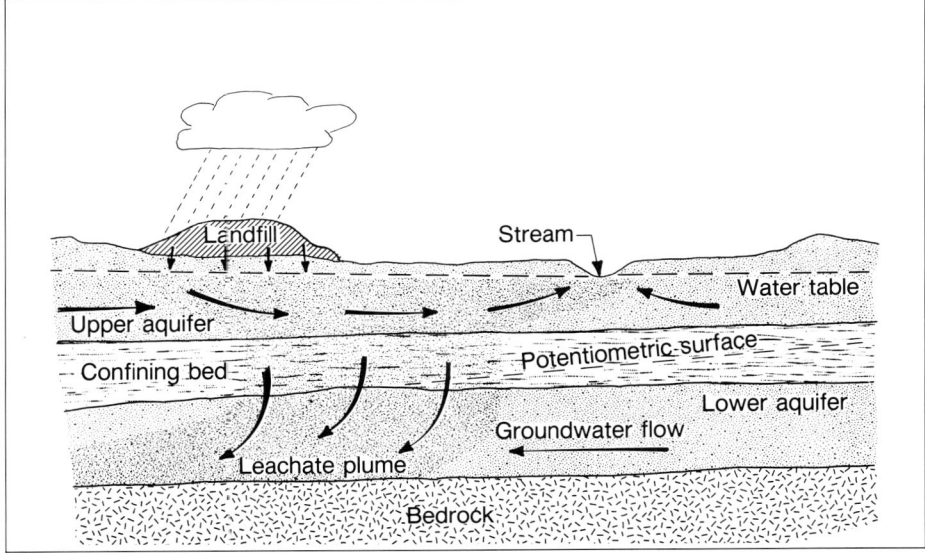

Figure 4. Two-aquifer system with opposite flow directions. Leachate first moves into and flows with the groundwater in the upper aquifer. Some of the leachate eventually moves through the confining bed into the lower aquifer where it flows back beneath the landfill and away in the other direction. *(U.S. Environmental Protection Agency, 1977)*

APPENDIX 21.B.
Types of Protective Clothing and Respirators that Should be Used at Hazardous Waste Sites

The degree of hazard is based on the waste material's physical, chemical, and biological properties and anticipated concentrations of the waste. The level of protective clothing and equipment worn must be sufficient to safeguard the individual. A four-category system is described below (Maslansky, 1983).

LEVEL A consists of a pressure-demand SCBA (air supplying respirator with back-mounted cylinders — see description below), fully encapsulated resistant suit, inner and outer chemical resistant gloves, chemical resistant steel safety boots (toe, shank, and metatarsial protection), and hard hat. Optional equipment might include cooling systems, abrasive resistant gloves, disposable oversuit and boot covers, communication equipment, and safety line. Level A is worn when the highest level of respiratory, skin, and eye protection is required. Most drillers will never wear Level A protection unless engaged in such practices as installation of investigatory or recovery wells within an area contaminated with highly toxic or corrosive materials.

LEVEL B protection is utilized in areas where full respiratory protection is warranted, but a lower level of skin and eye protection is sufficient (only a small area of head and neck is exposed). Level B consists of SCBA, splash suit (one or two piece) or disposable chemical resistant coveralls, inner and outer chemical resistant gloves, chemical resistant safety boots, and hard hat with face shield. Optional items include glove and boot covers and inner chemical resistant fabric coveralls. Many monitoring and recovery wells have been installed in Level B protection, particularly at abandoned hazardous waste sites or at spill sites where the concentration, lack of warning properties, or breakthrough characteristics of the contaminants precluded the use of an air-purifying respirator.

LEVEL C permits the utilization of air-purifying respirators. Level B body, foot, and hand protection is normally maintained. Many organizations will permit only the use of approved full-face masks equipped with a chin- or harness-mounted canister. However, many sites are drilled by personnel wearing a half-mask cartridge respirator. If allowed by the client, it becomes a trade-off of decreased protection versus increased comfort.

LEVEL D protection consists of a standard work uniform of coveralls, gloves, safety shoes or boots, hard hat, and goggles or safety glasses. Some organizations have personnel carry emergency escape masks (supplies 5 to 15 minutes of air) or place these around the site and by work vehicles when personnel are in Level C or D protection. The higher the level of protection required, the greater the chance of accident or heat stress and the greater the amount of training needed.

Respirators

Respirators are of two basic types — air-purifying and air-supplying. Air-purifying respirators are designed to remove specific contaminants by means of filters and/or sorbents. Air-purifying respirators come in various sizes, shapes, and models and can be outfitted with a variety of filters, cartridges, or canisters. Each mask and cartridge or canister is designed for protection against certain contaminant concentrations. Just because a cartridge says it is for use against organic vapors does not mean that it is

Appendix 21.B. Continued

good for all organic vapors.

Air-supplying respirators are utilized in oxygen-deficient atmospheres (less than 19.5 percent) or when an air-purifying device is not sufficient. Air is supplied to a face-mask from an uncontaminated source of air via an air line from stationary tanks (cascade system), from a compressor, or from air cylinders worn on the back (SCBA). Rated capacities of the SCBA's are normally between 30 and 60 minutes. Only positive pressure (pressure demand) respirators should be used in high concentration hazardous environments.

Contact lenses are not permitted for use with any respirator. Contact lenses should not be worn at any site since they tend to concentrate organic materials around the eyes; soft plastic contact lenses can absorb chemicals directly. In addition, rapid removal of contact lenses may be difficult in an emergency. Although eye glasses can prevent a good seal around the temple when wearing goggles or full face masks, spectacle adapters are available for masks and goggles. Respirators often malfunction during cold weather or after continued use. Only NIOSH (National Institute for Occupational Safety and Health)—MSHA (Mine Safety and Health Administration) approved respirators should be used.

APPENDIX 21.C
Chemical Constituents of a Major Polymer

Residues from drilling fluid additives can affect the water quality of samples taken from monitoring wells. If Revert® is used as the drilling fluid additive, traces of certain chemicals may be found in the water sample that should be attributed to the drilling fluid additive, not the groundwater. Table 1 lists the various substances and oxygen demands produced by a 0.8-percent mixture of Revert®. The procedures used to obtain these results are listed in Table 2.

Table 1. Chemical composition of a 0.8% (by weight) mixture of Revert analyzed after six days incubation at 68°F

Parameter	Concentration (mg/ℓ)	Detec. Limit (mg/ℓ)
Butanoic acid	3	0.002
Hexanoic acid	0.04	0.002
Hexadecanoic acid	0.003	0.002
4-methyl-4-hydroxy-2-pentanone	0.003	0.002
1-hexanoe	0.002	0.002
Butanoic acid-3-methylbutylester	0.002	0.002
4-methoxybenzene methanol	0.002	0.002
Nanoic acid	0.002	0.002
Butanoic acid-1-methylpropylester	0.002	0.002
Diester $C_{22}H_{42}O_4$	0.003	0.002
Large unidentified polysaccharide	0.006	0.002
Chemical Oxygen Demand	6000	1
Biochemical Oxygen Demand	2300	1
Suspended Solids	1900	1
Total Organic Carbon	2300	1

Appendix 21.C. Continued

Table 2. Procedures for chemical analyses performed on Revert-spiked water samples

Analysis	Procedure
Biochemical Oxygen Demand	Standard Methods
Chemical Oxygen Demand	Standard Methods
Suspended Solids	Standard Methods
Total Organic Carbon	Oxidation Method (Oceanographic 524B T.O.C. Analyzer)
Base/Neutral Fraction Organic Scan	GC/MS (Direct Injection) Method 625 (Finnegan 4021 GC/MS with Incos data system)
Acid Fraction Organic Scan	GC/MS (Direct Injection) Method 625 (Finnegan 4021 GC/MS with Incos data system)
Volatile Organic Fraction Scan	GC/MS (Purge and Trap) Method 624 (Finnegan 3021 GC/MS with Incos data system)

APPENDIX 22.A.
Theoretical Specific Capacity for 100-Percent Efficient Water Wells, m³/day/m of drawdown

Transmissivity m²/day	Confined Aquifer Storage Coefficient = 5 × 10⁻⁴						Unconfined Aquifer Storage Coefficient = 1 × 10⁻¹					
	152-mm Well			305-mm Well			152-mm Well			305-mm Well		
	12 hrs	1 day	10 days	12 hrs	1 day	10 days	12 hrs	1 day	10 days	12 hrs	1 day	10 days
12.4	10.7	9.0	9.0	10.7	10.7	9.0	16.1	14.3	12.5	17.9	16.1	14.3
24.8	19.7	17.9	16.1	21.5	19.7	17.9	28.6	26.9	23.3	34.0	30.4	25.1
37.2	28.6	26.9	23.3	30.4	30.4	26.9	41.2	39.4	34.0	46.5	44.8	37.6
49.7	37.6	35.8	32.2	41.2	39.4	34.0	53.7	51.9	43.0	62.7	59.1	48.3
62.1	46.5	44.8	39.4	50.1	48.3	43.0	66.2	62.7	53.7	75.2	60.9	59.1
124	87.7	82.3	75.2	96.7	91.3	80.6	125	120	102	141	134	111
186	129	125	111	141	134	120	184	174	149	206	193	163
248	170	165	145	184	177	158	238	226	193	267	251	211
372	249	240	215	269	260	229	347	329	285	387	367	312
497	328	315	283	354	340	303	455	430	372	503	478	406
621	405	392	351	435	421	374	557	528	457	619	584	499
745	482	465	417	519	499	444	664	627	544	732	695	593
869	557	553	483	598	578	514	763	727	630	841	791	687
993	634	610	550	680	655	585	863	823	712	958	906	795
1,242	780	754	680	823	807	721	1,062	1,015	884	1,169	1,113	956
1,552	963	934	841	1,020	1,002	895	1,310	1,249	1,088	1,436	1,369	1,178
1,862	1,153	1,112	1,002	1,235	1,190	1,067	1,552	1,482	1,296	1,706	1,624	1,398
2,173	1,319	1,285	1,164	1,425	1,377	1,233	1,781	1,713	1,495	1,969	1,871	1,618
2,483	1,509	1,450	1,319	1,624	1,566	1,405	2,023	1,933	1,701	2,220	2,112	1,826

APPENDIX 23.A.
EPA's Priority Toxic Pollutants

1. *acenaphthene
2. *acrolein
3. *acrylonitrile
4. *benzene
5. *benzidine
6. *carbon tetrachloride
   (tetrachloromethane)
   ***chlorinated benzenes** (other than dichlorobenzenes)
7. chlorobenzene
8. 1,2,4-trichlorobenzene
9. hexachlorobenzene
   ***chlorinated ethanes**
10. 1,2-dichloroethane
11. 1,1,1-trichloroethane
12. hexachloroethane
13. 1,1-dichloroethane
14. 1,1,2-trichloroethane
15. 1,1,2,2-tetrachloroethane
16. chloroethane
    ***chloroalkyl ethers** (chloromethyl, chloroethyl and mixed ethers)
17. bis (chloromethyl) ether
18. bis (2-chloroethyl) ether
19. 2-chloroethyl vinyl ether (mixed)
    ***chlorinated naphthalene**
20. 2-chloronaphthalene
    ***chlorinated phenols** (other than those listed elsewhere; includes trichlorophenols and chlorinated cresols)
21. 2,4,6-trichlorophenal
22. parachlorometa cresol
23. *chloroform (trichloromethane)
24. *2-chlorophenol
    ***dichlorobenzenes**
25. 1,2-dichlorobenzene
26. 1,3-dichlorobenzene
27. 1,4-dichlorobenzene
    ***dichlorobenzidine**
28. 3,3-dichlorobenzidine
    ***dichloroethylenes**
29. 1,1-dichloroethylene
30. 1,2-trans-dichloroethylene
31. *2,4-dichlorophenol
    ***dichloropropane and dichloropropene**
32. 1,2-dichloropropane
33. 1,3-dichlorpropylene (1,3-dichloropropene)
34. *2,4-dimethylphenol
    ***dinitrotoluene**
35. 2,4-dinitrotoluene
36. 2,6-dinitrotoluene
37. *1,2-diphenylhydrazine
38. *ethylbenzene
39. *fluoranthene
    ***haloethers** (other than those listed elsewhere)
40. 4-chlorophenyl phenyl ether
41. 4-bromophenyl phenyl ether
42. bis (2-chloroisopropyl) ether
43. bis (2-chloroethoxy) methane
    ***halomethanes** (other than those listed elsewhere)
44. methylene chloride (dichloromethane)
45. methyl chloride (chloromethane)
46. methyl bromide (bromomethane)
47. bromoform (tribromomethane)
48. dichlorobromomethane
49. trichlorofluoromethane
50. dichlorodifluoromethane
51. chlorodibromomethane
52. *hexachlorobutadiene
53. *hexachlorocyclopentadiene
54. *isophorone
55. *naphthalene
56. *nitrobenzene
    ***nitrophenols**
57. 2-nitrophenol
58. 4-nitrophenol
59. 2,4-dinitrophenol
60. 4,6-dinitro-o-cresol
    ***nitrosamines**
61. N-nitrosodimethylamine
62. N-nitrosodiphenylamine
63. N-nitrosodi-n-propylamine
64. *pentachlorophenol
65. *phenol
    ***phthalate esters**
66. bis (2-ethylhexyl) phthalate
67. butyl benzyl phthalate
68. di-n-butyl phthalate
69. di-n-octyl phthalate
70. diethyl phthalate

Appendix 23.A. Continued

71.	dimethyl phthalate	98.	endrin
	*polynuclear aromatic hydrocarbons	99.	endrin aldehyde
			*heptachlor and metabolites
72.	benzo (a) anthracene (1,2-benzanthracene)	100.	heptachlor
		101.	heptachlor epoxide
73.	benzo (a) pyrene (3,4-benzopyrene)		***hexachlorocyclohexane (all isomers)**
74.	3,4-benzofluoranthene	102.	Alpha BHC
75.	benzo (k) fluoranthane (11,12-benzofluoranthene)	103.	Beta BHC
		104.	Gamma BHC
76.	chrysene	105.	Delta BHC
77.	acenaphthylene		***polychlorinated biphenyls (PCB's)**
78.	anthracene		
79.	benzo (ghi) perylene (1,12-benzoperylene)	106.	PCB-1016 (Arochlor 1016)
		107.	PCB-1221 (Arochlor 1221)
80.	fluorene	108.	PCB-1232 (Arochlor 1232)
81.	phenanthrene	109.	PCB-1242 (Arochlor 1242)
82.	1,2,5,6-dibenzanthrancene	110.	PCB-1248 (Arochlor 1248)
83.	indeno (1,2,3-cd) pyrene (2,3-ophenylenepyrene)	111.	PCB-1254 (Arochlor 1254)
		112.	PCB-1260 (Arochlor 1260)
84.	pyrene	113.	*toxaphene
85.	*tetrachloroethylene	114.	*antimony (total)
86.	*toluene	115.	*arsenic (total)
87.	*trichloroethylene	116.	*asbestos (fibrous)
88.	*vinyl chloride (chloroethylene)	117.	*beryllium (total)
	pesticides and metabolites	118.	*cadmium (total)
89.	*aldrin	119.	*chromium (total)
90.	*dieldrin	120.	*copper (total)
91.	*chlordane	121.	*cyanide (total)
	*DDT and metabolites	122.	*lead (total)
92.	4,4-DDT	123.	*mercury (total)
93.	4,4-DDE (p,p-DDX)	124.	*nickel (total)
94.	4,4-DDD (p,p-TDE)	125.	*selenium (total)
	*endosulfan and metabolites	126.	*silver (total)
95.	Alpha endosulfan	127.	*thalium (total)
96.	Beta endosulfan	128.	*zinc (total)
97.	endosulfan sulfate	129.	**2,3,7,8-tetrachlorodibenzo-p-dioxin (TCDD)
	*endrin and metabolites		

*Specific compounds and chemical classes listed in the Natural Resources Defense Council (NRDC) consent decree and referenced in the Clean Water Act.

**This compound was specifically listed in the consent decree; however, due to its extreme toxicity EPA recommends that laboratories not acquire an analytical standard for this compound.

(Silva, 1981)

APPENDIX 23.B.
Common Water-Quality Problems Encountered in Domestic Water Systems

Impurity or Contaminant	Symptom	Cause	Treatment
Hard water	Soap curd and scum in wash basins and bathtub. Whitish scale deposits in pipes, water heater, and tea kettle.	Calcium and magnesium salts in raw water measuring 42.8 mg/ℓ (2.5 grains per gal) or higher as calcium carbonate.	All calcium and magnesium salts removed with cation-exchange water softener. General limit 2,740 mg/ℓ (160 grains per gal).
Grittiness	Abrasive texture to water when washing, or residual left in sink.	Excessively fine sand, silt in water passing through well screen or coagulation treatment step.	Sand trap or ultrafiltration. See Turbidity
Odor	Musty, earthy, or wood smell.	Generally, harmless organic matter.	Activated-carbon filter.
	Chlorine smell.	Excessive chlorination.	Dechlorinate with activated-carbon filter.
	Rotten-egg odor.	Dissolved hydrogen sulfide gas in water supply.	Manganese greensand filter up to 6 mg/ℓ H_2S*, with pH not lower than 6.7. Over 6 mg/ℓ H_2S, constant chlorination followed by filtration/dechlorination.
		Presence of sulfate-reducing bacteria in water supply.	Constant chlorination followed by activated-carbon filter.
		Action of magnesium rod in hot water heater and soft water.	Remove magnesium rod from heater.
	Detergent odor, water foams when drawn.	Seepage of septic discharge into underground water supply.	Locate and eliminate source of seepage—then shock chlorinate well. Activated-carbon filter will adsorb limited amount of detergent.
	Gasoline or oil smell.	Leak in fuel oil tank or gasoline tank seeping into water supply.	No residential treatment. Locate and eliminate seepage. Activated carbon will adsorb oil and gasoline for short periods.
	Methane gas (caution required).	Caused by naturally decaying organics found in (a) shallow water wells near swamps, (b) housing areas built above or near former landfills and (c) aquifers overlying well fields.	Residential/commercial aeration system and repump.
	Phenol (chemical) smell.	Industrial waste seeping into aquifers.	Activated-carbon filter will adsorb for short time.
Pesticides, herbicides (DDT, 2-4 D, Silvex, methoxychlor, lindane, endrin, chlordane, etc.)	Sharp chemical odor in water.	Excessive agricultural spraying and surface water run-off to lakes and ponds.	Activated-carbon filter will absorb limited amount. Must continue to monitor treated water.
Taste	Salty or brackish.	High sodium content.	Deionize drinking water only with disposable mixed-bed (anion/cation) resins. Reverse osmosis for drinking water only. Home distillation system for drinking and cooking water. No economical residential treatment for sodium content over 1,800 mg/ℓ.
	Alkali taste.	High total dissolved solids alkalinity.	No economical residential treatment for total compensated hardness over 3,080 mg/ℓ (180 gpg). Reduce to low concentration by reverse osmosis for drinking water.
	Metallic taste.	Very low pH water (3-5.5)	Correct with calcite-type filter (see Acid water).
		Heavy iron concentration in water above 3 mg/ℓ.	See red (iron) water.
Turbidity	Silt, clay or suspended particles in water.	Suspended matter from surface water.	Calcite-type filter up to 50 mg/ℓ.
		Silt or sand from well.	Sand filter or new well screen.
		Organic matter such as algae in water.	Constant chlorination followed by calcite or activated-carbon filter to dechlorinate.

Appendix 23.B. Continued

Impurity or Contaminant	Symptom	Cause	Treatment
Acid water	Green stains on sinks and other porcelain bathroom fixtures. Blue-green cast to water.	Water with high carbon dioxide content (pH below 6.8) reacts with brass and copper pipes and fittings.	Neutralizing calcite filter if pH is above 5.5.
			Calcite/magnesia-oxide mix (5 to 1 ratio) to correct for very low pH water.
			Soda ash chemical feed followed by filtration.
Discolored water Reddish (iron) water	Reddish-brown stains on fixtures, dishes, and laundry. Water turns reddish-brown in cooking or upon heating. Clothing becomes discolored.	More than 0.3 mg/ℓ dissolved iron in water causes staining. Water appears clear when first drawn at cold-water faucet.	Can remove 0.5 mg/ℓ of iron for every 17 mg/ℓ of hardness up to 10 mg/ℓ with water softener and minimum pH of 6.7 (unaerated water).
			Over 10 mg/ℓ iron, chlorination with sufficient retention-tank time for full oxidation followed by filtration/dechlorination.
			In warm climates, residential aerator and filtration will substantially reduce iron content.
		Precipitated iron (water not clear when drawn).	Up to 10 mg/ℓ iron removed by manganese greensand filter if pH is 6.7 or higher.
			Manganese treated, nonhydrous aluminum silicate filter where pH is 6.8 or higher and oxygen is 15 percent of total iron content.
			Downflow water softener with good backwash up to 1 mg/ℓ iron. From 1 to 10 mg/ℓ, use calcite filter followed by downflow water softener.
		Iron dissolved from old pipe with water having a pH below 6.8.	Calcite filter to remove precipitated iron.
	Brownish cast; does not precipitate.	Organic (bacterial) iron.	Treat well with acid to destroy iron bacteria, then constant chlorination followed by activated-carbon filtration to dechlorinate.
			Chemical feed of potassium permanganate followed by filtration.
	Reddish color in water sample after standing 24 hours.	Colloidal iron.	Constant chlorination followed by activated-carbon filter to dechlorinate.
Yellow water	Yellowish cast to water after softening and/or filtering.	Tannins (humic acids) in water are picked up when water passes through peaty soil and decaying vegetation.	Adsorption by anion-exchange resin regenerated with salt (NaCl) up to 3 mg/ℓ.
			Chlorination with full retention time followed by filtration to dechlorinate (over 3 mg/ℓ).
Black cast to water	Blackish staining of fixtures and laundry.	Interaction of carbon dioxide or organic matter with manganese-bearing soils. (Manganese content above 0.05 mg/ℓ causes stains.) Usually found in combination with iron.	Manganese greensand or manganese-treated sodium-aluminosilicate-type filter to limit of 6 mg/ℓ and 15 mg/ℓ, respectively (combined iron and manganese, with pH not lower than 6.7).
			Manganese treated non-hydrous aluminum silicate under proper conditions.
Milky water	Cloudy water when drawn.	Some precipitated sludge created during heating of water.	Blowdown water-heater tank periodically.
		High volume of air in water from poorly functioning pump.	Water will usually clear quickly upon standing.
		Excessive coagulant-feed being carried through filter.	Reduce coagulant quantity being fed, service filters properly.

Appendix 23.B. Continued

Impurity or Contaminant	Symptom	Cause	Treatment
High chloride content in water	Blackening and pitting of stainless steel sinks and stainless ware in commercial dishwashers.	Excessive salt content. High-temperature drying creates chloride concentration, accelerating corrosion.	Use chloride-resistant metals. Reduce total dissolved solids by distillation, ion exchange, or reverse osmosis.
Excess fluorides	Yellowish mottled teeth of children. No visible color, taste, or odor in water.	Fluoride above 1 - 2 mg/ℓ in natural water supply.	Reduce to 0.2 mg/ℓ with activated alumina resins or bone-char filter. Distillation system for drinking and cooking water. Complete water deionization, by disposable mixed-bed for drinking water only.
Nitrates	No visible color, taste, or odor in water. Maximum contaminant level set by U.S. EPA. (Above 10 mg/ℓ as N considered health hazard for infants.)	Nearby human or animal waste disposal sites located near wells. Heavy use of commercial fertilizers with residual NO_3 getting into groundwater. Disposal of corrosion inhibitors containing nitrates (from boiler cleanout), which enter groundwater supplies.	For water with less than 3 mg/ℓ nitrate, remove with strong anion exchanger, regenerated with NaCl. Verify treatment efficiency with water quality analysis. For drinking and cooking water, reverse osmosis should remove up to 65 percent N. Limit nitrate influent to 25 mg/ℓ as N. Home distillation system for drinking and cooking water
Radioactive contaminants	Notices by public health authority. No visible color, taste, or odor. Radium 226 above 5 piC/ℓ and strontium-90 above 10 piC/ℓ considered health risk.	Naturally occurring in deep wells caused by leaching of radium from phosphate rock or radium-bearing rock strata. Atmospheric fallout from man-made explosions producing contamination of surface water supply sources; or stray isotopes getting in water supply from escape of nuclear waste. Radon gas given off by decaying radium dissolved in water.	Can remove most cationic radioactivity with residential cation-exchange water softener. Treat with mixed-bed deionizer for complete removal of both anionic and cationic radioactive contaminants. Reverse osmosis should remove 70 percent. Aeration by good faucet aerator to dispell dissolved radon.
Heavy metals: lead, zinc, copper, cadmium	No visible color, taste, or odor in water. Maximum contaminant level set by U.S. EPA for many metals.	Industrial waste pollution. Corrosion products from piping caused by low-pH waters.	Reverse osmosis for drinking and cooking water. Complete removal by disposable mixed-bed deionizer for drinking water. Water softener will remove copper, cadmium, and zinc under proper conditions.
Arsenic†	No visible color, taste, or odor. Maximum contaminant level set by U.S. EPA. (Above 0.05 mg/ℓ considered health risk.)	Natural groundwater contaminant in local areas. Industrial waste contaminating water supply. Herbicides/pesticides.	Reverse osmosis will remove up to 90 percent. Remove by complete deionization using disposable mixed bed.
Barium	No visible color, taste, or odor. Maximum contaminant level set by U.S. EPA. (Above 1 mg/ℓ considered health risk.)	Naturally occurring in certain geographic regions.	Remove by cation-exchange water softener using a strong brine solution.
Boron	Inhibits normal plant growth. (Above 1 mg/ℓ considered undesirable.)	Naturally occurring in Southwest and other areas.	Remove by selective anion-exchange resin.
Cyanide	No visible color, taste, or odor. (Above 0.2 mg/ℓ considered health risk.)	Industrial waste pollution from electroplating, steel, and coking facilities.	Continuous chlorination and activated-carbon filtration of metals after pH adjustment.
Trichloroethylene (TCE)	Notice from Public Health Department.	Waste degreasing and dry cleaning solutions entering surface or groundwater supply.	Activated-carbon filters in series, with constant monitoring between units for breakthrough.

*Water must be tested at source for accurate determination because H_2S gas escapes quickly.
†Test for arsenic form, either As^{+3} or As^{+5}, to determine treatment step.

(After McGowan, 1982)

APPENDIX 23.C.
Performance Standards for Variouis Point-of-Use Treatment Techniques

Point-of-Use Treatment	Applicable Standard
Filtration	WQA S-200
	NSF 42
	NSF 53
	ASTM
Cation Exchange	WQA S-100
	WQA S-101
Activated Alumina	NSF 53
Anion Exchange	WQA S-100
	WQA S-101
	NSF 53 (nitrate removal)
Reverse Osmosis	WQA under development
	ASTM (membrane only)

Note:
WQA S-100-81: Voluntary Industry Standards for Household, Commercial and Portable Exchange Water Softeners
WQA S-101-81: Voluntary Industry Standards for Salt Efficiency Ratings of Household, Commercial and Portable Exchange Water Softeners
WQA S-200-73: Recommended Industry Standards for Household and Commerical Water Filters
WQA Under Development: Voluntary Industry Standards for Point-of-Use Low Pressure Reverse Osmosis Drinking Water Systems
NSF 24: Plumbing System Components for Mobile Homes and Recreational Vehicles
NSF 42: Filtration Devices Relating to Supplementary Treatment of Potable Water
NSF 53: Drinking Water Treatment Units — Health Effects

(Water Quality Association, 1983; Mayer, 1984)

Index

ABS casing, 425
  methods to join, 430
Abandonment of wells, 627-629
Abbreviations list, 915-917
Ablation till, 36, 38, 40
Absolute Ownership Rule, 672
Acid
  definition, 885
  effect on well screens, 456, 639, 641
  for iron bacteria, 654-655
  general procedure for acid treatment, 642
  hydrochloric, 637
  hydroxyacetic, 641
  in development, 529
  safety precautions, 642
  sulfamic, 637, 639-641
  sulfuric, 641
  treatment of wells for incrustation, 637-646
Acid rain, 18
Acidity, 94
Acoustic log, 195-196, 625
Activated carbon, 731, 821-823, 827, 833, 835, 885
  contaminant breakthrough in, 822-823
Adiabatic temperature change, 50
Adsorption, 885
  in water treatment, 821-823
Advection, 710, 885
Aeration
  definition, 885
  removal of organic chemicals, 731, 819-821, 822-823
  in water treatment, 820-821
  wells for removal of iron and manganese (*see* Vyredox system)
  of well screens, 755
Aeration, zone of (*see* Vadose zone)
Aerial photography, 157-160
Aggregation, 885
  of clays, 358
Air compressors, 367-369

Air development, 507-516
Air line (drawdown measurement), 550-552
Air-lift pumping, 507-513, 608
  air volume required, 510
  pipe sizes for, 512
  use in well development, 507-513
Air rotary drilling, 295-299
Air stripping, 885 (*see also* Aeration)
Air used as drilling fluid, 366-385
Alignment test, 335-339
Alkalinity, 94, 885
Allowable sediment concentration, 526-528
Alluvial, 885
Alluvial deposits, 21-26
  gradation of, 77
  hydraulic characteristics, 24-26
  pumping test in, 565-567
Alluvial fan, 24, 25
  materials of, 24
  structure, geologic and hydrologic, 24, 25
Alluvium, 21, 885
American Doctrine (*see* Reasonable Use Doctrine)
Analog models, 847
Analytical model, 848
Analysis of sand samples (*see* Sediment size analysis)
Anion, 87, 89, 885
Anion exchange, 885 (*see also* Ion exchange)
Anisotropic, 885
Annular velocity, 385-386, 387, 942, 945
Annulus, 885
Anode, 93, 659, 660, 885
Apparent resistivity, 178
Appropriation Doctrine, 672
Aquiclude, 62, 885
Aquifer, 19, 61, 885
  alluvial, 21-26
  confined, 62, 64-66
  formation of, 19-45
  functions of, 66-79

geographical distribution of, 118-149
glacial, 31-42
gradation of, 77
igneous and metamorphic, 42-45
perched, 64
restoration, 730-733
sedimentary, 26-31
stimulation, 497, 528-533, 885
test, 212-267, 885
unconfined, 62, 63-64
Aquifer development, 528-533
  hydrofracturing, 531-533
  use of acids, 529
  use of explosives, 529-531
Aquifer loss, 556, 557
Aquifer stimulation (*see* Aquifer development)
Aquitard, 62, 886
Artesian well, 65, 886
Artificial recharge, 886 (*see also* Recharge)
Asbestos cement pipe, friction loss in, 1013
Asthenosphere, 11, 886
Atmosphere
  composition of, 17
  formation of, 16-18
Attapulgite clay, 351, 886
Auger drilling, 165-166, 310-313
  bucket auger, 310-311
  hollow-stem auger, 166, 313
  solid-stem auger, 166, 311-312
Augmenting water supplies, 850

Backwashing
  definition, 886
  infiltration galleries, 764, 767, 768
  in well development, 503-504
Bacteria, 619-620
  coliform, 109, 620, 813-816
  Giardia lamblia, 796
  iron, 99, 456, 646-655
  Pseudomona, 796
Bactericide, 649-653, 886
Bail bottom, 468, 469
Bail-down method of installation, 485-487
Bail-down shoe, 485
Bailer test, (*see* Slug test)
Bailer, flat and dart valve, 167, 272
  use for sampling, 167
Bajada, 24
Barite, 346, 389, 886
Basal groundwater, 139
Basalt, 8, 886
Base exchange, 886 (*see also* Ion exchange)
Baumé, 637
Bedload, 21, 886
Bedrock, 886
Benioff zone, 11
Bentonite, 352, 886 (*see also* Clay)
Bernoulli equation, 62, 603
Bit, 886
  button, 305
  cable tool, 274
  dimensions and weights, 927, 928, 929

drag, 280, 296
eccentric, 308
hammer, 298
reamer, 283
recommended rotating speeds, 931
roller (tricone), 280, 281, 296, 298
underreamer, 283
weight on, 931
zublin, 292, 295
Blasting
  aquifer stimulation, 529-531
  and bailing, 524
  boulders, 316
  removing incrustation, 644, 645
Blowout, 886
Bored wells, 310-313
Borehole geophysical logs (*see* Geophysical exploration methods)
Borehole permeability tests, 164-165
Borehole resistivity, 181-188
Bottomset beds, 26, 27
Boulder catcher, 293, 295
Boulders, drilling procedures when encountered, 314-316
Bouldery till, 307
Boundary conditions, 246-247, 573
  effects on drawdown, 247-249
Braided stream, 25, 886
Brake horsepower, 585, 586
Breakpoint chlorination, 813
Bridge seal, 629
Bridge-slot screen, 402
Bridging of sand grains, 503, 504, 886
Bucket auger, 310-311
Bumping block, 469-470
Buried valley, 40-41, 886

Cable tool drilling, 268-277
  in consolidated formations, 274
  effect on aquifer, 498
  effect on sanitary protection, 618
  for removing well casing, 628
  sampling, 166
  string of drill tools, 270
  in unconsolidated formations, 274-276
  well screen installation, 468-470
Calcite, 28
Calcium hypochlorite, 652
Calcium-rich deposits, hydraulic properties of, 31
California stovepipe drilling, 277-278
Caliper logs, 200
Calorie, 47
Camera, downhole, 525
Capillarity
  effect of grain size, 60
  effect upon water movement, 60
Capillary fringe, 60, 886
Capillary water, 59, 60
Carbonate, 886
Carbonate rocks, 28, 31, 886
Carbonic acid, 31

Carbon dioxide, 105-106, 455
Casing (*see also* Pipe)
　alignment, 333-339
　diameter, 414-417
　driver, 164, 304-307
　fiberglass reinforced plastic, 429
　hanger, 467
　low carbon steel, 418-424
　materials, 418-429
　materials for monitoring wells, 720
　methods to join, 429-431
　perforator, 277
　size designation, 418-424
　stainless steel, 424
　steel, 418-424
　thermoplastic, 424-429
Casing storage, 232-235, 565-567
Cathode, 93, 659, 660, 886
Cathodic protection, 662-663
Cation, 87, 89, 886
Cation exchange, 886 (*see also* Ion exchange)
Cavitation, 593-594, 595, 667, 886
Cement
　classifications used outside United States, 930
　equivalent classifications outside United States, 930
　neat, 318
　strengths, 319-320
　types, 319-320
Cement basket, 324
Cement-bond log, 195, 196
Centering guides, 480, 481, 482
Centrifugal pump
　axial flow (propeller), 588
　basic principle, 582
　cavitation, 593-594
　impeller design, 588-590
　mixed flow, 588
　for mud pump, 606-608
　net positive suction head (NPSH), 591-593
　non-reverse ratchet, 600-601
　priming, 604
　regenerative, 588
　selection of, 587-588
　thrust, 590-591, 600
　turbine (diffuser), 587
　volute, 587
Centrifugal sand sampler, 526, 527
Check valve, 626
Chelating agent, 642, 643, 644
Chemistry of groundwater, 86-117
Chlorination, shock, 650
Chloride, 455
Chlorine, 613, 619, 620-623, 650-653, 886
　breakpoint chlorination, 813
　calcium hypochlorite, 620, 621, 622, 652
　demand, 620
　dioxide, 652
　gas, 622, 650, 653
　quantities required, 621, 653, 827
　residual, 620, 622, 827

　sodium hypochlorite, 621, 622, 652
Clastic, 887
Clay
　additives for drilling fluid, 350-352, 357-359, 360-362
　attapulgite, 351, 886
　bentonite, 352, 886
　collars, 285
　hydraulic properties of, 30-31
　kaolinite, 351
　illite, 351
　montmorillonite, 351
　size of, 27
Coagulation, 799-802
Coefficient of permeability, 71, 887
Coefficient of storage, 67, 209, 887
　determination of, 221-222, 237, 257-260
　estimating, 737
Coefficient of transmissivity (*see* Transmissivity)
Coliform bacteria, 109, 620, 623, 813-816
Collapse strength, 456-458
Collector wells, 768-769
Colloid, 90, 340, 887
Column strength, 456-458
Combined aeration/GAC systems, 822-823
Competence, 25
Computer modeling (*see* Numerical models)
Cone adaptor, 468
Cone of depression, 211-212, 887
　composite, 242, 243-244, 743
　in confined aquifer, 208
　enlargement of, 211-212
　formation of, 208
　in unconfined aquifer, 208
Confined aquifer, 62, 64-66, 434, 572, 887
Conjunctive use of water supplies, 852
Conservation of groundwater resources, 853-854
Constant-rate pumping test (*see* Pumping test)
Contamination
　definition, 887
　delineating plumes, 712-714
　effect of aquifer characteristics on spread of, 707-712, 1060-1061
　emergency treatment procedures, 824, 825
　from sanitary landfills, 708, 709, 710, 1060-1061
　how to reduce, 707
　monitoring movement, 714-715
　plumes, 707
　reactions of contaminants with subsurface environment, 711
　sources of, 613, 705-707
　types of protective clothing and respirators, 1062-1063
Continental crust, formation of, 10
Continuous-slot screen, 396-401
　open area, 399
Contracts, 696-701
　disputes, 700-701
　liability for faulty plans, 696-697

limitation on yield guarantee, 697-699
responsibility for site conditions, 699-700
warranties, 701
Convection, 51
Conversion factors
  chemical, 899
  drilling fluid weight, 938-940
  mathematical, 909-913
Coral reefs, 27-28
Core samples, 165, 166, 168
Coriolis effect, 49
Correlative Rights, 673
Corrosion, 658-665, 887, 996-1006
  bi-metallic, 659-660
  cathodic protection, 662-663
  chemical, 658-660
  electrochemical, 658-660
  factors affecting corrosion rate, 660-661, 996-1006
  galvanic, 662-663
  galvanic cell, 660
  galvanic series, 659
  general, 661-662
  intergranular, 664-665
  passive film, 661, 663, 999-1001, 1004-1006
  pitting, 663-664
  prevention and treatment, 660-661
  of stainless steel, 661
  stress-corrosion cracking, 665
  structural collapse, 633
  of well screens, 454, 996-1006
Cross-over tool, 479
Crust, 7
Cyclonic wedging, 52

Dams, pressure relief of, (*see* Pressure-relief wells)
Darcy-Weisbach equation, 1025, 1027-1031
Darcy's law, 73, 887
Dart valve bailer, 167, 272
Deep wells for dewatering, 753-760
Decontamination solutions for monitoring work, 723
Deflocculation of clays, 358, 887
Degrees Baumé (*see* Baumé)
Delta, 25, 26, 27
Density, 4, 887
  of drilling fluids, 344-347
Depression storage, 56
Desalination, 852-853, 887
Desander, 387, 388
Development (*see* Well development or Aquifer development)
Deviation instruments (*see* Inclinometer)
Dewatering, 734-760
  computer-assisted analysis, 756-760
  deep-well systems, 753-760
  depth of setting for well points, 747-748
  determining aquifer characteristics, 735-738
  developing well points, 751
  equations, 739-741
  factors in selecting a dewatering system, 741-742
  installing well points, 748-750
  jetting well points, 748-750
  piping, connections and pumps, 751-753
  sand drains, 746, 747
  spacing of wells, 753-756
  stage dewatering, 742-743, 744, 745
  stratified formations, 746-747
  swing joints, 751, 752
  trench dewatering, 740-741, 744
  tuning, 751
  well-point selection, 748
  well-point spacing, 743-747
  well-point systems, 742-753
Diatomaceous earth, 887
Diatomaceous earth filters, 812-813
Diffusion, 710
Direct air rotary method, (*see* Air rotary)
Direct rotary drilling, 278-286, 338
  drilling fluid, 286-289, 340-394
  drilling fluid invasion in aquifer, 287, 521-522
  drilling problems, 315, 334
  effect on sanitary protection, 618
  filter cake, formation of, 288, 360-362
  installing well screens, 470-476
  mud pits, 279
  sampling, 162-164
  table drive, 282
  top-head drive, 283
Discharge, 81
Disinfection
  bacteriological analysis, 623-624
  byproducts, 623
  chlorine solutions, 620-623
  coliform bacteria, indication of pollution, 620
  during drilling operation, 618-619
  procedure for completed wells, 619-623
  of wells, 619-620
Disinfection of water supply
  chlorination, 813-815, 826-827
  ozone, 815-816, 827
  ultraviolet ray, 827-828
Dispersing agents, 522-523
Dispersion, 710, 887
Dissociation, 3, 887
Dissolved gases in water
  carbon dioxide, 105-106, 455
  dissolved oxygen, 104-105, 454
  hydrogen sulfide, 105, 454
Distance-drawdown graph, 236-242, 246-247
  for dewatering predictions, 755
  predicting radius of influence, 238-242
Distillation, 833
Dolomite, 31
Down-the-hole air hammer, 164, 297-299
Downthrust, 590-591
Drainage basin, 887
Drawdown, 26, 206, 887
  correction for partially dewatered unconfined aquifer, 918-919
  curve, 208

INDEX 1077

effect of aquifer boundary, 247-249
effect of recharge, 223-228, 246-249
effect upon system-head curve, 584, 585
interference, 242-244
measurement of, 547-554
predicting from time-drawdown graph, 222-223
residual, 206, 253
variation with time of pumping, 222-232
Drift indicator, 335-336
Drill collar, 281, 282, 285, 887, 934, 935
Drill pipe, 282, 887
Drill-stem test, 547
Drill-through casing driver, 304-307
Driller's log, 160-162
Drilling fluid, 286-289, 340-394, 501, 887
  additives, 385-386, 937
  aerated, 374-375
  air-based, 366-385
    air-foam, 372-374
    air-mist, 371
    calculations used in air-foam systems, 379-381
    dry air, 370-371
    liquid-volume fraction (LVF), 376-381
    physics of air drilling, 375-379
    regulating the air-foam drilling system, 381-385
  companies, 944
  density, 344-347, 937
  filter cake, formation of, 288, 360-362
  filtration, 360-362
  functions of, 286, 341-343
  gel strength, 357-360
  guidelines for solving drilling fluid problems, 386, 392
  invasion into formation, 360-362, 501, 522
  mixing additives into water-based systems, 364-366
  mud mixers, 941
  mud pit design, 363-364
  products, 943
  properties, 285, 343-362
  treatment of mix water, 362-363
  types, 340-341
  typical problem, 392-394
  uphole velocities, 163, 385-386, 508, 509
  viscosity, 347-357
  water-based, 340-366
  well development, effect on, 501, 520
Drilling methods (see Well drilling methods)
Drilling mud (see Drilling fluid)
Drinking water standards, 107-110
Drive block assembly, 469, 470
Drive point (see Well point)
Driven wells, 313-314
Drought
  agricultural adjustments to, 844
  impact on groundwater supply, 843-846
  impact on groundwater use, 843-846
  severity, 845
  urban adjustments to, 844

Dry-ice treatment for rehabilitation, 645
Dual-wall reverse circulation rotary drilling, 301-304
  sampling, 165

Earth augers, 310-313
  sampling, 165-166
Earth resistivity (see Surface resistivity)
Eductor, 508, 512
Effective grain size, 78, 887
Efficiency (see Well efficiency)
Effluent, 887
Electric logging, 181-188, 713-714
Electrical analog models, 847
Electrical conductance of water, 92-94, 887
Electrical potential, 658
Electrical profiling, 178-180
Electrical resistivity, 177-180, 181-183, 713-714, 887
Electrical resistivity log, 181-188, 713-714
Electrical sounder, 549
Electrolyte, 658, 887
Electromagnetic surveys, 174-177, 713
English Rule of Capture (see Absolute Ownership Rule)
Entrance velocity
  appropriate, 996-1006
  how to calculate, 450-453
  recommended maximum, 996-1006
Ephemeral rivers, 24
Equilibrium well equations, 212-214
Equipotential lines, 79, 888
Erosion, 5, 20, 888
Estimating groundwater use, 838-840
Evaporite, 101
Evapotranspiration, 54, 888
Expandable packers, 467
Explosives, use of, 316, 529-531
Extrusive rocks, 11, 888

Fanning friction factor, 1025
Fault, 888
Fiberglass-reinforced plastic casing, 429
  methods to join, 431
Field capacity, 56
Filter cake, 183, 360-362, 888
Filter pack, 888
  cost factors, 446-447
  design, 438-446
  installation of, 476-479, 482-483
  for monitoring wells, 722
  specifications for, 441-443, 476
  thickness, 443-446, 476, 502
  vertical flow in, 443-444, 447
  volume required, 445
Filters
  dual media, 812
  mixed media, 812
Filtration, 888
Finite difference, 848-850
Finite element, 848-850
Fish, 316, 888

Fishing tools, 316-317, 318, 319
Float shoe, 325, 329
Flocculation
 definition, 888
 in drilling fluids, 358
 mechanism of, 358
 removal of turbidity, 812
 in water softening, 799
Floodplain, 23, 24, 25, 888
Flow from pipes, 1022-1023, 1024
Flow of groundwater
 adjacent to streams, 57, 79
 from aquifer to well, 207
 direction of, 81, 82
 rates of (see Hydraulic conductivity)
 to wells, models simulating, 924-926
Flow lines, 79, 888
Flow meters, 201-202
Flow models, 847
Flow net, 79-81
Flow velocity, 73, 75, 81-85, 92, 207
 measuring groundwater flow velocities, 85
Flowing artesian well, 64
Fluid movement log, 201
Flumes, 543-547
 cutthroat, 546-547, 1019-1020
 Parshall, 543-546, 1018
Fluoride, 102-103
Foam (see Drilling fluid)
Foreset beds, 26, 27
Formation packers, 474, 484
Formation sampling
 handling of samples, 168
 uphole velocities, 163
Formation sampling methods, 160-168
 auger, 165-166
 cable tool, 166-168
 core barrel, 166
 direct rotary, 162-164
 drive core, 168
 dual wall, 165
 side-wall coring, 164-165
 split spoon, 166
 wireline, 166
Formation stabilizer, 447-449, 888
Forms for well drilling
 notice to bidders, 1050
 NWWA standard forms — well estimate and/
  or well drilling contract or repair order
  agreement, 1056
 proposal, 1052
 standard form of agreement between owner
  and contractor, 1054
Fouling, 888
Fracture-trace analysis, 158
Friction loss in asbestos cement pipe, 1013
Friction losses in fittings, 1035-1036
Friction losses in pipe, 1025-1036

Galvanic corrosion (see Corrosion)
Galvanic series of metals, 659
Gamma logging method, 191

Gamma-gamma logging method, 193
Gauze numbers, 397
Gel, 360, 888
Gel strength
 definition, 888
 of drilling fluids made with clay additives,
  357-359
 of drilling fluids made with polymeric
  additives, 359-360
Geohydrochemical maps, 156, 157
Geologic maps, 153, 154-155
Geologic time table, 898
Geophysical exploration methods, 168-202
 borehole geophysical methods, 180-202
  acoustic (sonic) log, 195-196
  caliper log, 200
  equipment for measuring resistivity in
   borehole, 185
  fluid movement log, 201
  gamma log, 191
  gamma-gamma log, 193
  interpretation of apparent resistivity
   values, 187
  neutron log, 194
  resistivity log, 181-188, 713-714
  spontaneous potential log, 188-191
  temperature log, 196-200
 references, 905-908
 surface geophysical methods, 170-180
  electrical resistivity method, 177-180
  electromagnetic survey, 174-177, 713
   radar, ground penetrating, 174-176
   terrain conductivity, 176-177
  gravimetric survey, 173-174
  seismic refraction/reflection, 170-173, 713
 use in contaminant transport studies, 712-714
Geothermal gradient, 89, 198
Geothermal heat (see Groundwater heat
 pumps)
Ghyben-Herzberg equation, 766-767
Glacier
 causes of, 32
 deposits, 31-42
 drift, 888
 hydraulic properties of glacial deposits, 38-41
 moraines, 35-37, 38
 outwash, 37-38, 40, 563-565
  gradation of, 77
 quarrying, 34
Glaciofluvial, 888
Glassy phosphates, 522-523, 656
Governmental policies affecting supply of
 groundwater, 856-859
Graded, 888
 river, 21
 sediment, 77
Gradient
 hydraulic, 73, 74, 75
 river, 21
Grains per gallon, 89, 888
Grain-size analysis (see Sediment size analysis)
Grain-size classification, 410

Grain-size distribution curve, 409
Granulated activated carbon (GAC), 821-823, 827, 833, 835
Gravel pack (*see* Filter pack)
Gravity survey (gravimetric), 173-174
Ground penetrating radar, 174-176, 713
Groundwater, 61
  augmenting supply, 850
  changes in ionic content, 96
  chemistry, 86-117
  computer modeling (*see* Groundwater models)
  conjunctive use, 852
  conservation of, 853-854
  energy contained in, 62-66
  estimating recharge, 840-841
  estimating use, 838-840, 841
  estimating volume in storage, 841
  flow velocities, 73, 75, 81-85
  governmental policies affecting, 856-859
  hydraulics (*see* Well hydraulics)
  information for Canada, 903
  information for Mexico, 903
  law, 671-675
  managment, 845-860
  models, 847-850
    conceptual model, 847
    finite difference, 848-850
    finite element, 848-850
    mathematical model, 847-848
    numerical model, 848-850
  pricing, 855-856
  protection, 728-730, 853
  publications, 904
  quality, 107-116
  reuse, 850-852
  rights to, 671-675
  storage (*see* Coefficient of storage)
  table, 60, 888
  temperature, 89
  use, 854
  wise use of, 837-860
Groundwater constituents, 97-107
  acidity, 94
  alkalinity, 94
  calcium, 91
  chloride, 101-102, 455
  dissolved gases, 104-106
    carbon dioxide, 105-106, 455
    dissolved oxygen, 104-105, 454
    hydrogen sulfide, 105, 454
  fluoride, 102-103
  hardness, 90-92, 455
  hydrogen ion concentration (pH), 94-96, 454, 455
  inorganic, 97
  iron, 98-99, 455
  magnesium, 91
  manganese, 99-100, 455
  minor constituents, 107
  nitrate, 103-104
  origin of, 88-89
  radionuclides, 106-107
  silica, 88, 100
  sodium, 101
  sulfate, 104
  total dissolved solids (TDS), 89, 187, 454
  units of measurement, 89-90
Groundwater monitoring techniques, 702-733
  federal legislation pertaining to monitoring and groundwater quality, 703-705
  locating monitoring wells, 715-716
  personnel safety at monitoring sites, 717-719
Groundwater monitoring wells
  design, 719-728
  development of, 725-726
  diameter, 719
  drilling methods, 716-717
  filter pack design, 722
  grouting materials, 724
  installation, 722-726
  location, 715-716
  materials for, 720
  sampling procedures, 726-728
  screen criteria, 719-722
Groundwater regions
  Canada, 140-146
    Appalachian, 141-142
    Canadian Shield, 143
    Cordilleran, 144-145
    Interior Plains, 143-144
    Northern, 145-146
    St. Lawrence Lowlands, 142
  Mexico, 147-149
    Central Highlands, 147-148
    Coastal and Deltaic Areas of the Pacific Ocean, 149
    Coastal Plain of the Gulf of Mexico, 148-149
    Peninsula of Baja California, 149
    Sierra Madre del Sur, 148
    Sierra Madre Occidental, 148
    Sierra Madre Oriental, 148
    Yucatan Peninsula, 149
  United States, 119-140
    Alaska, 138-139
    Alluvial Basins, 123-126
    Alluvial Valleys, 138
    Atlantic and Gulf Coastal Plains, 135-137
    Colorado Plateau and Wyoming Basin, 127-128
    Columbia Lava Plateau, 126-127
    Glaciated Central, 132-133
    Hawaii, 139-140
    High Plains, 128-130
    Nonglaciated Central, 130-132
    Northeast and Superior Uplands, 134-135
    Piedmont and Blue Ridge, 133-134
    Southeast Coastal Plain, 137-138
    Western Mountain Ranges, 121-123
Grout, 888 (*see also* Grouting)
Grout shoe (*see* Float shoe)
Grouting, 317-333, 888
  casing method of grouting, 328-329

curing temperatures, 199
grouting failures, 330
installation of bentonite grout, 331-332,
materials for monitoring wells, 724
mixing the grout, 323-324
  adding bentonite, 936
proportioning cement grout, 322-323
slurry density, equations to calculate, 936
slurry placement methods, 324-329
testing the grout seal, 195, 196, 199, 330-331, 625
tremie pipe inside casing (inner string method), 327-328
tremie pipe outside casing, 326-327
Guar gum, 353-356

Hardness in water, 90-92, 455, 829, 888
Hardpan, 39
Hazen-Williams equation, 1026, 1032-1034
Head
  components of, 62
  definition, 888
  elevation, 62, 207
  energy, 62, 63, 73
  pressure, 62
  velocity, 62
Head capacity curve, 585
Head loss, 63, 208, 209, 568, 889
  friction, 73
  vertical flow in filter pack, 443-444, 447
  well entrance, 450
Heat of vaporization, 2, 50
Hele-Shaw model, 847
Heterogeneous, 889
Hole caliper log, 200-201
Hollow-shaft electric motor, 599
Hollow-stem auger, 313
Homogeneous, 889
Horizontal suction lines, 625-627
Hydration, 351, 352, 889
Hydraulic conductivity, 24, 71-79, 210, 889
  determining, 74, 76-79, 215
  effect of grain size and uniformity, 71, 72
  estimating, 737, 738
  field measurement of, 76-79
  laboratory measurement of, 74, 78-79
  range of values, 74, 75
  variations in, 71
Hydraulic fracturing, 531-533
Hydraulic gradient, 73, 74, 75, 710, 889
Hydraulic head (see Head)
Hydraulic jacks, 470, 471
Hydraulic-percussion method, 309-310
Hydrocarbons, removal of, 731-732, 733
Hydrochloric acid, 637
  amount required to treat screen, 639
Hydrodynamic dispersion, 710
Hydrogen bonding, 4
Hydrogen ion concentration, 94-96, 454, 455, 620
  effect of carbon dioxide on, 95
  effect on chlorine efficiency, 620

indicator of corrosive water, 454
indicator of incrusting water, 455
saturation (pHs), 455
Hydrogen sulfide, 454
Hydrogeologic, 889
Hydrogeologic maps, 157
Hydrogeologic reports, 151-153
Hydrographs, 69, 70, 71
Hydrologic cycle, 53-56
Hydropneumatic, 610, 611, 623
Hydrosphere, 8, 889
  formation of, 14-16
Hydroxyacetic acid, 641-642
  amount required to treat screen, 641
Hypochlorites (see Chlorine)

Igneous rock
  aquifers, 42-45, 530
  definition, 889
  extrusive, 11
  hydraulic properties, 45
  intrusive, 11
Illite, 351
Ilmenite, 346
Imaginary wells, 235-236, 570-571
Imhoff cone, 526, 527
Impeller
  design, 588-590
  enclosed, 589-590
  seals, 590
  semi-open, 589
Impervious boundary, 231-232
Impoundment, 705
Inclinometer, 335-336
Incrustation
  acid treatment, 637-644, 645-646
  causes of, 91, 634-635
  definition, 889
  mechanical methods to remove, 644
  prevention and treatment, 636
  of rock wells, 644-645
  types of, 633-636
  of well screens, 455, 633-636
Infiltration galleries, 761-768
  bed-mounted, 763-764
  design principles, 762-763
  maintenance of, 764, 767-768
  on-land, 764-767
Inflatable packers, 467
Injection wells, 769-777
  control of salt-water intrusion, 773
  environmental effects of recharge, 772-773
  solution mining, 773-776
  waste disposal, 776-777
  water supply, 769-773
In-situ uranium mining, 773-776
Insol, 361
Installation of well screens (see Well screen installation)
Interference, well, 242-244, 573, 575, 889
Intermediate seal, 629
Intermediate vadose zone, 59, 60

Intermittent pumping, 235-236, 570-571
  calculating drawdown, 235-236
Intrinsic permeability, 71
Intrusion of salt water, 766-767, 773
Intrusive rocks, 11, 889
In-Verse drilling method, 299-300
Ion, 3, 86, 889
Ion exchange, 803, 828-832
Iron bacteria, 99
  chemical methods to control, 649-653, 654
  enzymatic and nonenzymatic, 1045
  methods to classify, 647, 648
  physical methods to control, 654
  prevention and treatment, 649-655
  recommended procedure to control, 654-655
  well failure caused by, 456, 646-655
Iron incrustation, 635-636, 637, 638
Iron in water, 98
  ferric compounds, 98
  ferrous compounds, 98
  method of removal, 811, 812, 832
  recommended limit, 98, 455
  relation to iron bacteria, 99, 646-649
Isopach maps, 156
Isostatic equilibrium, 13, 14
Isotropic, 889

Jackson Turbidity Unit, 811
Jacob method of analysis, 219-244, 252-260
Jet drilling, 307-308
  installation of well points, 749-750
Jet-percussion drilling, 307, 308
Jetting, high velocity, in well development, 516-518
Jetting tool, 516, 517, 643
Juvenile water, 15

K-packers, 465, 466
Kaolinite, 351
Karst topography, 31-32, 889
Kelly, 282, 283, 889
  weights, 933
Kozeny equation, 250

Laminar flow, 72, 556-559, 889
Laminar head loss, 556-559
Landfill, 703, 708, 709, 889
Land-use planning, 853
Langlier index, 455
Lapse rate, 49
Law, groundwater, 671-675
Leachate, 707, 708, 709, 889
Leaching, 96
Lead packer, 465-467
Lead slip packer, 474
Leaky aquifer, 66
Liability for faulty plans, 696-697
Liens, 700-701
Lime softening, 802-803
Limestone aquifer, 567
Limestone deposits, 31, 889
Line pipe (*see* Pipe)

Line swabbing, 506-507
Lineaments, 159
Liner hangers, 467
Lithologic log (*see* Driller's log)
Lithosphere, 11, 889
Lodgment till, 34, 35
Loess, 41
Logging, well (*see* Geophysical exploration methods)
Lost circulation, 360, 889
Louver screen, 402
Low-velocity zone, 10

Magma, 11
Magnetometers, 713
Maintenance frequency, 632
Managing groundwater supplies, 846-859
Manganese incrustation, 635-636, 637, 638
Manganese in water, 99, 455
Mantle, 7
Maps, geologic, topographic, hydrogeologic, 153-156
Marsh funnel viscosity, 348, 349, 350, 351, 352, 889
Mathematical models, 847-848
Maximum contaminant level (MCL), 109
Measuring drawdown in wells, 547-554
  observation wells, 547-549
  recovery data, 554
  time intervals, 552-554
Mechanical methods to remove incrustants, 644
Mechanical mixing, 710
Mechanic's lien, 700-701
Membranes for reverse osmosis systems, 805-806
Metamorphic rocks, 42-45, 889
  hydraulic properties of, 45
Methods to join casing, 429-431
  fiberglass casing, 431
  plastic casing, 430-431
  steel casing, 429-430
Minerals in groundwater (*see* Groundwater constituents)
Mix water for drilling fluids, 362-363
Modified nonequilibrium equation, Jacob, 219-244, 252-260
Moho, 14, 889
Molecular diffusion, 710, 889
Molecule, 4, 889
Monitoring wells (*see* Groundwater monitoring wells)
Montmorillonite, 351
Moraine
  definition, 889
  depositional features, 36-37
  formation of, 35-36
  recognition of, 38
Mud (*see* Drilling fluid)
Mud cake (*see* Filter cake)
Mud pit, design of, 363-364
Mud pump, 606-608
Mud scow, 277

Muriatic acid (*see* Hydrochloric acid)

Natural development, 499-500, 889
Nephelometric Turbidity Unit, 811
Net positive suction head (NPSH), 591-593
Neutron logging method, 194
Newtonian fluid, 348
Nitrate in water, 103-104
Nominal diameter, 418, 889
Nonequilibrium well equation, Theis, 218-219, 260-265
  modified, Jacob, 219-220
Nongraded, 77, 890
Non-reverse ratchet, 600-601
Nozzles for jetting tools, 516-517
Numerical models, 848-850
  finite difference, 848-850
  finite element, 848-850

Observation wells, 216, 547-549, 890
Ocean floor, formation of, 10
Open area, 501, 948-950
Orange-peel bucket, 293, 295
Organic contaminants, removal of, 816-824
Orifice, discharge of air through, 1014
Orifice tables, 538
Orifice weir (*see* Weir)
Orographic effects, 52, 53
Osmotic pressure, 804
Outwash
  definition, 890
  formation of, 37-38
  gradation of, 77
  hydraulic properties of, 40
  plain, 37, 38, 890
  recognition of, 38
Overburden, 890 (*see also* Regolith)
Overdrafts (*see* Management of groundwater supplies)
Overpumping, 502-503
Oxidation, 890 (*see also* Redox potential)
Oxidizing agents (*see* Chlorine and Potassium permanganate)
Oxygen, dissolved, 454

Packers, 465-468
Parallel plate model (Hele-Shaw), 847
Parshall flume (*see* Flume)
Partial penetration, 249-252, 890
Passive film, 661, 663, 999-1001, 1004-1006
Pasteurization treatment for iron bacteria, 654
Pathogenic, 620, 624, 890
Perched groundwater, 64, 890
Percolation, 56, 890
Performance curves, 582-584
Peristaltic pumps, 605
Permeability, 26, 890 (*see also* Hydraulic conductivity)
Permeameter, 74, 78-79
Pesticides/herbicides, 624, 707, 816, 819
pH, 890 (*see also* Hydrogen ion concentration)
pHs, 455

Phreatic water, 59, 61
Physical plugging, 655-657
  causes, 655
  prevention and treatment, 655-657
  use of polyphosphates and surfactants, 656-657
Pipe
  API line, 423, 424
  collapse strength, 422
  dimensions and weights
    ANSI pipe, 972-975
    plain-end line, 961-970
    plain-end standard, 954-959
    short round-thread, 983-984
    stainless steel, 976-982
    threaded and coupled standard, 960
    threaded line, 971
    water-well casing, 985
  drill, 282, 932
  drive, 423, 424
  driven, 424
  eductor, 508, 512
  fiberglass-reinforced, 429
  grades, 423
  header, 745, 751-752, 753
  manufacture of, 952
  nomogram for determining flow, diameter, head loss, or velocity in black steel pipe, 1009
  reamed and drifted, 423, 424
  sizes for air lift, 512
  sizes for sand lock, 494
  specifications, 418-429
  stainless steel, 424
  standard, 420-422
  steel, 418-424
  thermoplastic, 424-429
  water well, 423, 424
Pipe-base well screens, 402-403
Pipe-size screens, 401
Piston pumps, 605-606
Pitless adaptors, 626, 627, 628
Plastic pipe (*see* Thermoplastic casing)
Plate tectonics, 8-14, 890
Pleistocene, 32
Plucking (*see* Glacier quarrying)
Plumbness and alignment, 333-339
  causes of misalignment, 333-334
  methods of testing, 335-339
Plumes (*see* Contamination)
Point-of-use water treatment systems, 824-836
  chlorination, 826-827
  distillation, 833
  filtration, 832-833
  ion exchange, 828-832
  ozonation, 827
  performance standards, 1072
  reverse osmosis, 833-834
  ultraviolet, 828
Pollution
  definition, 890
  EPA's priority toxic pollutants, 1067-1068

predicting potential at a drill site, 616-617
sources, 704, 705-707
Polymer
  definition, 890
  chemical constituents of, 1064-1065
  natural, 353-356
  synthetic, 356-357
Polyphosphates
  crystalline, 522-523, 656
  glassy, 522-523, 656
  sodium, 522-523, 646, 656
  use in rehabilitation, 646, 656-657
  use in well development, 522-523
Porosity, 66-68 399, 890
Positive displacement pumps
  duplex, 606, 607
  for mud pumps, 606-608
  peristaltic, 605
  piston (reciprocating), 605-606, 607, 608
  priming, 606
  progressing cavity, 604, 605
  rotary, 604-605
  screw, 604, 605
  simplex, 607
  triplex, 606, 607
Potassium permanganate, 653
Potentiometric surface, 64, 75, 890
  sloping, 229
Precipitation, 49-53
  causes, 51-52
  effect on groundwater, 52-53
Pressure energy, 62
Pressure melting, 33
Pressure-relief screens, 447, 482
Pressure-relief wells for dams and levees, 777-783
Pricing water, 855-856
Primary drinking water regulations, 108
Priming pumps, 604, 606
Privatization of water supplies, 858-859
Procedure for acid treatment, 642-644
Propping agents, 530, 532
Protective clothing and respirators for hazardous waste sites, 1062-1063
Pull-back method of screen installation, 464-472
  cable tool drilling, 468-470
  rotary drilling, 470-472
Pumps
  air-lift, 507-516
  cavitation, 593-594, 667
  centrifugal, 582-604
  corrosion of, 667
  deep-well turbine, 595-604
  driving units 599-600
  efficiency, 584-585
  graphitization, 667
  horsepower, 585-586
  impeller design, 588-589
  intake, 596
  jet, 602-604, 627
  lubrication, 596-599

maintenance, 665-669
malfunctions, possible causes, and solutions, 1046-1049
mixed flow, 588
net positive suction head (NPSH), 591-593
performance, checklist to evaluate, 665
performance curves, 582-584
peristaltic, 605
piston, 605-606, 607
positive displacement, 604-606
power required for pumping, 1038-1039
priming, 604, 606
progressing cavity, 604, 605
regenerative, 588
rotary, 604-605
sand abrasion of, 666
screw, 604, 605
selection, 608-610
setting, 593, 594
shallow-well, 580
shroud, 602
shut-off head, 586-587
submersible, 601-602
suction lift, 594-595
suction lines, 625-627
system-head curve, 584-585
total dynamic head, 581
variable displacement, 581-604
vertical turbine, 595-601
volume of water discharged per stroke by a reciprocating pump, 1037
wire-to-water efficiency, 668
Pumping tests
  analysis of aquifer, 202-204, 205-267
  aquifer test, 212-267
  bailer, 203
  conducting a test, 535-554
  constant-rate test, 203, 212-267, 535-554
  curve matching, 260-265
  definition, 890
  determining transmissivity, 76
  disposal of water, 535
  drill-stem test, 547
  duration, 536, 553, 554
  maintaining a constant discharge, 536
  measuring discharge, 536-547
  measuring drawdown, 547-554
  observation wells, 216, 547-549
  problems of pumping-test analysis, 559-579
  pump-in test, 203
  recommended time intervals for measuring drawdown, 552-554
  recovery data, 554
  step-drawdown test, 203, 204, 555-559
  well-performance test, 534, 535 (see also Constant-rate pumping test)
Pumping test examples, 559-579
Pumping water level, 206
PVC casing (see Thermoplastic casing)

Quality of groundwater (see Water quality)
Quarrying (see Glacier quarrying)

Quick conditions, 778, 779, 890

Radar, ground penetrating, 174-176, 713
Radial flow to a well, 207-208
Radioactive contaminants, removal of, 824
Radioactive contaminants, water-quality standards, 106-107
Radius of influence, 209, 238-242, 245-246, 890
Rawhiding, 503-504
Reamer bit, 283
Reasonable Use Doctrine, 672
Recharge, 81, 890
  effect on drawdown, 223-228, 246-249, 562
  from a river, 57, 227, 562
  infiltration basins, 770
  vertical infiltration, 228
  wastewater spreading, 850-851
  well performance affected by, 223-228
  wells, 769-773
Recovery, 252-260, 554
Redox potential, 890, 998-999
Reduction, 996-1006
Regelation, 34
Regeneration, 830, 831, 832
Regolith, 59, 890
Regressive seas, 28-29
Rehabilitation of wells (*see* Well maintenance)
Removal of well screens, 492-496
Residual drawdown, 206, 890
Residual-drawdown graphs, 252-255
Resins, 829, 830
Resistivity, 177-188
  values, 177
Respirators for hazardous waste sites, 1062-1063
Restatement of the Law of Torts, 673
Reuse of groundwater supplies, 850-852
Reverse circulation rotary drilling, 289-295
Reverse osmosis, 803-811
Reynolds number, 73
Rheology, 347
Rift zone, 10
Riparian Doctrine, 672
Riser pipe, 465
River
  bedload, 21, 25, 26
  competence, 25
  ephemeral, 24
  graded, 21
  point bar, 25-26
  pumping test near, 561-563
  solution load, 21
  suspended load, 21
  valley, development of, 22
Rotary drilling (*see* Direct rotary drilling or Reverse rotary drilling)
Rotary pumps (*see* Pumps)
Rotary table, 282
Runoff, 56, 890
Ryznar stability index, 455
  calculating, 1007-1008
  indicator of corrosive water, 454
  indicator of incrusting water, 455

Salinity, 89, 112
Salt-water intrusion, 773
Salt-water/fresh-water interface, 766-767
Sample splitter, 303
Sampling (*see* Water or Formation sampling)
Sand, hydraulic properties of, 29-30
Sand content, 526-528, 655
Sand drains, 746, 747
Sand joint, for removal of well screens, 493-496, 628
Sand pumping, 633
Sand separators, 303
Sandstone, 30, 524-526, 890
Sanitary protection measures
  grouting and sealing casing, 317-332, 624-625
  horizontal suction lines, 625-627
  pitless adaptors, 627
  well location, 614-616
Satellite imagery, 159-160
Saturation pH, 455
Screens, (*see* Well screens)
Screw pump (*see* Pumps)
Sealing abandoned wells, 627-629, 1044
Sealing the wellhead, 624-625
Seals
  for abandoning wells, 628, 629
  well head, 624-625
Secondary drinking water regulations, 110
Sediment concentration, allowable, 526-528, 655
Sediment-size analysis, 76-78, 405-412
  conducting a sieve analysis, 406-409
  effective size, 78, 410
  estimating hydraulic conductivity, 76-78
  grain-size distribution curve, 409
  interpretation of grain-size distribution curve, 411-412
  other ways to describe sediment size, 410
  preparation of sample, 407
  sediment size, 409-410
  slope and shape of curve, 411-412
  uniformity coefficient, 78, 411
Sedimentary rocks, 26-31, 890
Seismic reflection, 172, 713
Seismic refraction, 170, 713
Septic systems, 705
Settling pit (*see* Mud pit)
Shale, 31, 890
Shale traps (catchers), 474, 485
Shear stress, 348, 890
Sheetwash, 11
Shock chlorination, 650
Shooting (*see* Blasting)
Shroud, 602
Shut-in head, 65
Shut-off head, 586-587
Sieve analysis, 890 (*see also* Sediment-size analysis)
Sieve sizes and screen openings, correlation of, 951

Silica, 88, 100
Silt, size of, 410
Slot openings
  distribution, 501
  gauze equivalents, 397
  shape, 501
  size, 501
Slotted pipe
  metal, 403
  plastic, 404
Slow drainage, 229-230, 569-570
Slug tests, 203
Slurry, 891
  cable tool drilling, 270
Slurry wall, 731
Soda ash, 362
Sodium adsorption ratio, 113, 900-902
Sodium hypochlorite, 652
Softening of water, processes
  ion-exchange process, 803, 828-832
  lime-softening process, 802-803
Soil water, 59, 60
Sol, 361
Solid-shaft electric motors, 599-600
Solid-stem augers, 311-312
Solvent welding, 430-431
Sonic (acoustic) logging method, 195, 625
Specific capacity, 207, 216-217, 558, 646, 891
  calculation of, 217, 572
  effect of partial penetration, 250-251
  empirical equations used to estimate, 1021
  reduction in, 567-569, 651
  theoretical specific capacities, 782, 1066
Specific electrical conductance, 92-94
  relation to dissolved solids, 93
Specific gravity, 891
Specific retention, 67, 891
Specific speed, 588-589
Specific yield, 67, 891
Specifications, 675-696
  bid schedule, 693-695
  drafting well specifications, 677-678
  general conditions, 676
  lowest responsible bidder, 695-696
  portion of tightly written specification, 1058-1059
  sample, 678-693
  special conditions, 676
Split-spoon samplers, 166, 737
Spontaneous potential (SP) logging method, 188-191
Squeeze pumps (*see* Pumps)
Stabilizers (rotary drilling), 281, 286
Stabilizer sand (*see* Formation stabilizer)
Stainless steel
  casing, 424
  corrosion resistance, 661
  well screens, 455-456
Standard dimension ratio (SDR), 425
Static water level, 206, 891
Step-drawdown test, 555-559
Storage of water, 610-611

Storage coefficient (*see* Coefficient of storage)
Storativity (*see* Coefficient of storage)
Stovepipe casing, 277
Strain, 348
Straightline method (*see* Jacob method)
Stratigraphy, 891
Stress, 348
Structure contour map, 158
Styrene rubber (SR) casing, 425
  methods to join, 430
Subduction, 11
Submergence, 508-511
Submersible pump (*see* Pumps)
Suction lift, limits of, 580, 594-595
Sulfamic acid, 637, 645, 646
  amount required to treat screen, 640
Sulfuric acid, 641
Surface casing, 416
Surface geophysical methods, 170-180
Surface resistivity, 177-178, 713
Surface water withdrawal, 789-795
  passive screening, 792-795
Surfactants
  definition, 891
  in drilling fluids, 372-374
  in rehabilitation, 656-657
Surge blocks, 504-506, 643
Surge plungers (*see* Surge blocks)
Surging
  air, 507-512, 514-516
  mechanical, 504-507
Suspended load, 21
Swabbing, 506-507
Swedge block and bar, 466, 467
Swedging, 466, 467
Swing joint, 751, 752
Symbols list, 914
Synthetic organic chemicals, 706, 731
System-head curves, 584, 585

Table drive rotary rig, 282-285, 287
Talus cone, 5
Tattletale screen (*see* Telltale screen)
Teflon, 720
Telescope installation, 464-474
Telescope-size screen, 400
Telltale screen, 477, 478, 479
Temperature logging method, 196-200
Tensile strength, 456-458, 891
Terrace deposits, 24
Terrain conductivity, 176-177, 713
Theis method, type curve, 260-265, 573, 579
Theis nonequilibrium equation, 218-219, 260-265
Thermal logging, 196-200
Thermoplastic casing, 424-429
  acrylonitrile butadiene styrene (ABS), 425
  chemical and physical standards, 986
  class requirements for rigid PVC, 990
  collapse strength, 428-429
  couplings socket dimensions and laying length dimensions, 1011

dimensions and weights of PVC pipe, 987-989
head loss characteristics of water flow through rigid plastic pipe, 1010
joining, 430, 994-995
physical properties, 426
polyvinyl chloride (PVC), 425
prior designation of PVC pipe material, 991
styrene rubber (SR), 425
tapered sockets for bell-end pipe, 1012
Thiem, 212-214
Thixotropy, 358
Thrust, 590-591
Till, 891
 ablation, 36, 38, 40
 lodgement, 34, 35, 38
Time-drawdown graphs, 220-235
 for dewatering predictions, 754
 hydrogeologic conditions that affect, 223-235
 predicting drawdown from, 222-223
Time-recovery graphs, 255-256
Topographic maps, 152-153
Topset beds, 26, 27
Top-head drive drilling rigs, 283-285, 289
Tort, 670
Tortuosity, 85, 710, 891
Total dissolved solids (TDS), 89, 187, 454, 891
Total dynamic head, 581
Toxic pollutants, 1067
Tracers, 84, 85
Transgressive seas, 26-28
Transmissibility (*see* Transmissivity)
Transmissivity, 76, 209, 210, 891
 calculation of, 220-221, 237, 256, 572
 determining, 76
 empirical equations used to estimate, 1021
 estimating, 737
Transmitting capacity, 453-454
Transpiration, 54, 891
Tremie pipe, 477
Trihalomethanes, 623, 652, 816, 818, 819, 820
Troposphere, 49, 50
Turbidity, 620, 811
 removal of, 811-813
Turbulent flow, 72, 556-559, 891
Turbulent head loss, 556-559, 568

Ultraviolet-light water purifiers, 828
Unconfined aquifer, 62, 63-64, 433, 434, 891
Underreamer bits, 283
Uniformity coefficient, 411-412, 891
Units of measure for water quality, 89-90
Uphole velocity, 163, 385-386, 416, 508, 509
Upset ends, 933
Upthrust, 590-591

Vacuum pump, 752, 753
Vadose zone, 59, 891
Valence, 88
Valley train deposits, 41
Vanes, 587-588
Vapor pressure, 593, 594, 595

Variable displacement pump (*see* Pumps)
Velocity energy, 62
Velocity of groundwater, 81-85
Velocity of water that will cause sand to rise, 944
Vertical leakage, 230-231
Vertical turbine pump (*see* Pumps)
Vesicles, 44
Vibration hammer, 470, 471, 628
Viscometers, 348, 349
Viscosity, 286, 347-357, 891
 measurement of, 348-350, 940
 of drilling fluids made with clay additives, 350-352
 of drilling fluids made with polymeric additives, 352-357
Volatile synthetic organic chemicals, 817, 818, 819
Volume of water in casing or hole, 947
Volute pump, 587-588
Vyredox system, 636, 811, 812

Warranties, 701
Wash-down bottom, 487, 488, 489
Wash-down method of installation, 487-489
Wastewater stabilization ponds, 705
Wastewater treatment system components, 798-799
Water
 horsepower, 585
 law, 671-675
 storage, 610-611
  hydropneumatic tanks, 610, 611, 623
 table, 60, 64, 891
 weight, fluid pressure, and buoyancy factors, 946
Water-level measurement, 547-552
Water-level recovery, 252-260
Water quality
 for agricultural use, 112-114
 for domestic use, 107-110
 for industrial use, 110-112
 methods to present, 114-116
 problems in domestic water systems, 1069-1071
 standards, 97-114
 units of measure, 89-90
Water-quality protection, 612-629
 choosing a well site, 614-616
 disinfection procedures, 618-623
 predicting pollution potential at a drilling site, 616-617
 sealing the wellhead, 624-625
 well design, 617-618
Water sampling, 623, 624, 726-728, 729
 devices, 729
 state agencies that provide, 1040-1043
Water treatment
 for removal of coliform organisms, 813-816
 for removal of inorganic contaminants, 799-813
 for removal of iron and manganese, 811, 812

INDEX 1087

for removal of organic contaminants, 816-820
for removal of radioactive contaminants, 824
for removal of turbidity, 811-813
methods of, 799-836
  activated carbon, 731, 821-823, 827, 833, 835
  adsorption, 821-823
  aeration, 731, 819-821, 822-823
  chlorination, ozonation, and ultraviolet disinfection, 826-828
  coagulation, 799-802
  combined aeration and GAC, 822-823
  distillation, 833
  filtration, 832-833
  ion exchange, 803, 828-832
  lime softening, 802-803
  reverse osmosis, 803-811, 833-834
  zeolite process, 832
point-of-use systems, 824-836
system components, 798-799
Waterborne disease, 612-613, 614
Weather patterns, 46-53
Weathering, 19-20, 891
Weirs, 537-543
  discharge from, 1015, 1017
  orifice weir, 537-541
Welding steel casing, 429-430, 992-993
Well
  abandonment, 628, 629, 1044
  casing for, 418-429
  cementing or grouting of, 317-332
  construction, 268-317, 464-496, 497-533
  contamination of, 705-712
  corrosion of, 658-665, 996-1006
  depth, 431-432
  design of domestic, 458-460
  design for sanitary protection, 460-461
  development of (*see* Well development)
  diameter, 414-418
  disinfection of, 619-624
  distance from pollution source, 615-616
  efficiency, 244-245, 554-555, 556, 558, 567, 575-579
  filter packed, 476-483
  for heat pumps, 783-789
    closed-loop system, 788-789
    two-well system, 785-788
  incrustation of, 633-645
  interference, 242-244
  logging procedures, 180-202
  loss, 556, 557
  observation, 547-552
  partially penetrating, 249-252
  performance, checklist to evaluate, 631
  plumbness and alignment, 333-339
  point, 404-405, 491-492, 891
  recharge, 769-777
  records, 630-631
  rehabilitation (*see* Well maintenance)
  sealing of, 624-625
Well design
  diameter of casing, 414-417

diameter of well screen, 400-401, 449-450
domestic wells, 458-460
entrance velocity, 450-453
filter pack, 438-446
formation stabilizer, 447-449
important hydrogeologic information required, 413-414
intake area of well screen, 398-400, 405, 450
length of well screen, 432-434
materials for casing, 418-429
of monitoring wells, 719-728
objectives for good design, 413
pressure-relief, 447
recommended diameters for various pumping rates, 415
for sanitary protection, 460-461, 617-618
special designs, 461-463
strength considerations for well screens, 456-458
total depth, 431-432
uphole velocity, 416
well screen openings, 434-446
Well development
  air development methods, 507-516
  air-lift pumping, 507-513
  air pressurizing, 525
  air surging, 507-512, 514-516
  backwashing, 503-504
  blasting and bailing, 524
  dispersing agents, 522-523
  factors that affect, 499-502
    drilling fluid type, 501
    filter pack thickness, 502
    open area, 501
    slot configuration, 501
    slot size, 501
    type of formation, 502
    well completion method, 499-501
  field demonstrations, 521-522
  filter packed, 500-501, 502
  high-velocity jetting combined with simultaneous air-lift pumping, 520-521
  high-velocity jetting with air, 518-520
  high-velocity jetting with water, 516-518
  mechanical surging, 504-507
  of monitoring wells, 725-726
  natural, 499-500
  objectives, 497
  overpumping, 502-503
  removal of filter cake, 499, 501
  rock wells, 523-526
  sandstone wells, 524-526
  swabbing, 506-507
  use of polyphosphates, 522-523
Well drilling methods
  air rotary, 295-299
  auger, 310-313
  cable tool, 268-277
  California stovepipe, 277-278
  direct rotary, 278-286
  drill-through casing driver, 304-307
  dual-wall, 301-304

hydraulic-percussion, 309-310
In-Verse, 299-300
jet, 307-308
methods for monitoring wells, 716-717
performance of, 338
reverse circulation rotary, 289-295
Well function of u, $W(u)$, 577-578
  tables of values, 921-923
Well hydraulics, 205-267
  coefficient of storage, 221-222, 237-238, 257-260
  combined use of semilog graphs, 247-249
  cone of depression, 211-212
  definition of terms, 206-207
  determining aquifer hydraulic conductivity, 215-216
  distance-drawdown graphs, 236-242
  efficiency, 244-245
  equilibrium well equations, 212-218
  hydrogeologic conditions that affect time-drawdown graphs, 223-235
    casing storage, 232-235
    impervious boundaries, 231-232
    recharge, 223-228
    sloping water table or potentiometric surface, 229
    slow drainage, 229-230
    vertical leakage, 230-231
  interference, 242-244
  intermittent pumping, 235-236, 570-571
  modified nonequilibrium equation, 219-220
  nature of converging flow, 207-211
  nonequilibrium well equation, 218-219, 260-265
  other methods of aquifer analysis, 265-266
  partial penetration, 249-252
  predicting drawdown from time-drawdown graph, 222-223
  radius of influence, 245-246
  recharge, effect of, 223-229, 246-247
  recovery data, 252-260
  relationship of drawdown to yield, 215-218
  Theis nonequilibrium well equation, 218-219, 260-265
  transmissivity, 76-79, 220-221, 237
Well maintenance, 630-665
  acid treatment, 637-646, 654-655
    general procedure for acid treatment, 642-644
    hydrochloric (muriatic) acid, 637, 639
    hydroxyacetic acid, 641-642
    sulfamic acid, 637-641
  causes of deteriorating well performance, 631-633
  chlorine treatment, 650-653
  corrosion, 658-665
    cathodic protection, 662-663
    chemical and electrochemical, 658-660
    galvanic, 662
    general, 661-662
    intergranular, 664-665
    pitting, 663-664
    prevention and treatment of, 660-661
    stress-corrosion cracking, 665
    importance of screen design, 657-658
  incrustation, 633-646
    causes of carbonate, 634-635
    causes of iron and manganese, 635-636
    mechanical methods to remove incrustants, 644
    prevention and treatment of, 636-646
    of rock wells, 644-645
  iron bacteria, 646-655
    chemical methods to control, 649-653
      chlorine, 650-652
      hypochlorites, 652-653
    physical methods to control, 654
    prevention and treatment of, 649
    recommended procedure for controlling, 654-655
  physical agitation of cleaning solutions, 650-651, 657
  physical plugging, 655-657
    prevention and treatment of, 655-656
    polyphosphate treatment, 646, 656-657
    physical agitation, 657
  pump maintenance, 665-669
Well point systems
  area of influence, 740
  for dewatering, 742-753
  layout of system, 760
  piping connections, 751-753
  for water supply, 760-761
Well screen
  advantages, 395
  alternative uses, 734-789
  characteristics, optimum, 396
  construction, 395-405
  corrosion resistance of, 658-661, 996-1006
  definition, 891
  design, 432-460 (*see also* Well design)
    effect on development, 519, 522
    effect on rehabilitation, 657-658
    for monitoring wells, 719-722
  diameter, 400-401, 449-450
  entrance velocity, 450-453, 996-1006
  materials available, 397, 402, 403, 404, 405
  for monitoring wells, 720
  methods of installation, 464-496 (*see also* Well screen installation)
  open or intake area, 398-400, 405, 450, 501, 948-950
  removal, 492-496, 628
  selection of material, 454-458
  slot size, 397, 434-443, 501
  strength, 456-458
  transmitting capacity, 453-454
  types of, 396-405
    bridge slot, 402, 519
    continuous slot, 396-401, 519
    fiberglass-reinforced plastic, 397
    louvered, 402, 519
    pipe base, 402-403
    slotted metal, 403, 519

slotted plastic, 404
Well screen installation, 464-496
  bail-down method, 485-487
  filter packed method, 479-482
    cable tool drilling, 479-481
    rotary drilling, 481-482
  jetting, 489-491
  plastic, 483-485
  pull-back method, 464-472
    cable tool drilling, 468-470
    rotary drilling, 470-472
  open-hole method, 472-476
    double string, 472-474
    single string, 475-476
  wash-down method, 474, 487-489
  well points, 491-492
Well screen length
  homogeneous confined aquifer, 434
  homogeneous unconfined aquifer, 433-434
  monitoring wells, 722
  nonhomogeneous confined aquifer, 434
  nonhomogeneous unconfined aquifer, 434
Well screen open area, 399, 948-950
  optimum ranges, 405
Well screen slot selection, 434-443
  filter packed wells, 438-443
  naturally developed wells, 435-538
Well specifications, 675-696, 1058-1059
Well yield, 207, 891
  for confined aquifer, 214
  for unconfined aquifer, 213
  methods of measuring, 536-547
  methods of testing
    bailer, 203
    constant-rate, 203
    slug, 203
    step-drawdown, 203
Wetting agents, 523

Yield (*see* Well yield)
Yield guarantee, limitation on, 697-699
Yield point, 348-349
Yield strength, 419, 891

Zone of saturation, 61